T0178481

Analysis and Quantum Groups

Lars Tuset

Analysis and Quantum Groups

Lars Tuset
Department of Computer Science
Oslo Metropolitan University
Oslo, Norway

ISBN 978-3-031-07248-2 ISBN 978-3-031-07246-8 (eBook)
https://doi.org/10.1007/978-3-031-07246-8

Mathematics Subject Classification: 47-XX, 46-XX, 20-XX, 43-XX, 22-XX, 28-XX

This Springer imprint is published by the registered company Springer Nature Switzerland AG
The registered company address is: Gewerbestrasse 11, 6330 Cham, Switzerland

To my beloved wife Maria and my wonderful kids Oliver and Evelyn

Preface

Quantum groups from the point of view of analysis have been studied in some form or other for more than a generation. I feel, and have felt for some time now, that this active area of research has reached a certain stage of maturity, which calls for, especially in order to reach out to new generations, a fairly comprehensive and self-contained account of the elaborate theory of locally compact quantum groups.

The hope is that this textbook will render the subject more accessible to a wider community. T. Timmermann's book, *An Invitation to Quantum Groups and Duality*, has quite a different focus, leaving open several of the topics we discuss. However, the main problem with Timmermann's text is that it is far from self-contained. In fact, the part concerning locally compact quantum groups, which is actually rather short, contains few proofs, and this has also been the rule in surveys on the topic. The reason is obvious, the required background in analysis is considerable only to begin to understand the nitty-gritty in the original, scattered articles and preprints, some of which are written only in French. To really get to the bottom of things, you easily find yourself on a very long chase, and in my experience, this is not what conscious students want. I do believe there is some comfort knowing that it is all contained in one volume.

This is why I have included a great deal of functional analysis and operator algebras, even some set theory, topology, and measure theory, rendering the whole text self-contained in the extreme. Only familiarity with real and complex numbers, and with some elementary results from linear algebra, is assumed for the persistent reader. Every statement beyond this point is proved without reference to anything else, the only exceptions being the introduction; the Brouwer fixed point theorem, which is used mildly; and one result in the appendix, namely Green's theorem (from calculus) for sufficiently general boundaries, and finally James' theorem is proved only in the separable case, which is hardly a restriction and has no consequences for the further development in the text.

That said, the reader one has in mind will preferably have taken a course in real analysis, perhaps even an introductory course in operator algebras. Parts of the first half and the appendix of this book could serve as curricula for such courses. A semester on the modular theory of weights, for example, from this book, would be useful, leading up to a full course on locally compact quantum groups. It is nevertheless perhaps unrealistic to expect to cover the whole second half of this book in such a course.

The text is supposed to bring the reader up to date on locally compact quantum groups, and it covers basically everything a student needs to know in analysis, and probably more, to work comfortably in the area. At some point as one includes more and more background material, it becomes natural to present it as a subject on its own terms, and not merely as a prerequisite for something else. This means that we have not always taken the shortest possible route, instead included material to make a rounder presentation serving a larger audience. For this reason it might be a useful reference text for experts.

The reader interested only in the quantum group material in the second half of the book, and who wants to travel light, can manage without reading Sects. 2.8, 2.10, and 2.11; the whole of Chap. 3; Sects. 6.4, 6.5, and 6.6; Sects. 7.5 and 9.4; and Chap. 11. Chapter 14 is also strictly speaking not needed in order to read the material on quantum groups.

To limit the number of pages, especially in the second half, I have kept the style brief and head on, making some of the steps almost like small exercises. Computations within locally compact quantum groups can be lengthy, but are to some extend repetitive, and become standard after a while, and can thus be skipped.

I have not included some of the more extended examples of locally compact quantum groups that draw on other parts of mathematics. This would in any case require a separate volume, similar to the study of representations of specific examples of locally compact groups. Only examples that emerge from our general constructions without much effort are discussed.

As always, research is a collective effort with many smaller results leading to bigger results thanks to a certain joint focus within the area. In writing this book, I am indebted to the quantum group and the operator algebra communities, which have brought the research to a level worth writing a textbook about. These communities have been my scientific and social platforms within academia. I would like to thank everybody for many joyful moments and fruitful discussions. Special thanks go to all my collaborators over the years.

I have adapted the convention that whenever words or phrases are emphasized in the text, a mathematical result is named or a notion is defined at that point.

Oslo, Norway — Lars Tuset
June 2021

Contents

Chapter 1
Introduction

Quantum groups disclosed themselves to us as holders of R-matrices via quantum inverse scattering methods. What soon become an area of intense research experienced tremendous development through the work of V. Drinfeld and others in the 80s and 90s with spectacular links to knot theory, topology of 3-manifolds and conformal field theory, and has since expanded considerably in all sorts of directions.

In the early days, the term quantum groups was almost synonymous with the rich assembly of Hopf algebras of deformed universal enveloping Lie algebras of semisimple Lie groups discovered by Drinfeld and M. Jimbo. The study of their tensor categories of finite dimensional representations culminated in the famous Drinfeld-Khono theorem.

In the axiomatic approach of S. Woronowicz to compact quantum groups as Hopf algebras of 'regular functions', he proved the existence of a 'Haar integral' and developed a generalized Peter-Weyl theory. He also established a generalized Tannaka-Krein theorem characterizing concrete tensor categories as representation categories of compact quantum groups. This theorem allows regarding the 'compact forms' of the deformed universal enveloping Lie algebras of Drinfeld and Jimbo as compact quantum groups. The 'regular functions' are then the matrix elements of certain finite dimensional representations of the Drinfeld-Jimbo algebras.

An abstract harmonic analysis approach in the locally compact case is fraught with analytic difficulties reflected partly by the fact that e.g. the representation theory for non-compact Lie groups is highly non-trivial, invoking infinite dimensional irreducible representations. Nevertheless, great advances in operator algebras in the 70s paved the way for Kac algebras, which offered a category of Hopf-algebra-like-objects that contained, in a precise sense, all locally compact groups, and moreover allowed an extension of Pontryagin duality for locally compact abelian groups to all Kac algebras. Unfortunately, this category fell short of including even the 'compact forms' of Drinfeld and Jimbo.

L. Tuset, *Analysis and Quantum Groups*,
https://doi.org/10.1007/978-3-031-07246-8_1

One sought therefore a sufficiently broad self dual category hosting generalized Pontryagin duality, by relaxing the definition of a Kac algebra. Thanks to the effort of many mathematicians, in the late 90s, the community settled on the beautiful axiomatic definition, and the subsequent theory, of locally compact quantum groups by J. Kustermans and S. Vaes. This fundamental definition has since been consolidated by its ability to handle various intricate examples well, and to accommodate important constructions in the field.

Locally compact quantum groups are set in the language of von Neumann algebras; they are actually von Neumann algebras with some extra structure. The theory of von Neumann algebras, or operator algebras, belongs to functional analysis, and should be thought of as a quantum analogue of measure theory, or quantum probability. F.J. Murray and J. von Neumann introduced these algebras while creating a rigorous foundation for quantum mechanics. At the heart of the matter lies the spectral theorem for unbounded self adjoint operators. Such 'infinite matrices' typically occur in quantum mechanics. One is interested in their eigenvalues, or spectra, which are the possible outcomes of observing the operator, thus called observable. The probability of getting specific real values in an experiment, when the physical system is in a given state, depends on the corresponding eigenspaces, or spectral projections, which are operators belonging to a von Neumann algebra, say the one generated by the possible spectral projections of all possible observables of the system. The state is actually a positive linear functional on this von Neumann algebra, and the probability is gotten simply by evaluating it on the specific spectral projection.

Sloppy approaches to quantum physics have since backfired, for instance with the advent of quantum information theory, where a more subtle understanding of quantum mechanics with its delicate interpretive issues, have led to increased recognition and appreciation for operator algebras and its techniques as the appropriate setting for quantum physics.

Identifying the simplest von Neumann algebras, called factors, as the 'building blocks' for quantum physics, and having shown how to decompose von Neumann algebras into factors, von Neumann and F. Murray undertook a thorough study of factors with the aim of classifying them. They pushed their study amazingly far into this abstract realm of mathematics at a stage when the logical foundations had barely been laid, leaving the area in a state of digestion.

The development of operator algebras picked up during WWII with the monumental work of I. Gelfand and M. Naimark on representations of C*-algebras, which are close relatives to von Neumann algebras. While we can think of the latter as algebras of 'bounded Borel measurable functions' on 'locally compact spaces', the former are algebras of 'continuous functions vanishing at infinity'. When these algebras are commutative, meaning that their elements commute pairwise with each other, then they are indeed algebras of functions on proper spaces, but in general these spaces do not exist as classical measure or topological spaces. In the commutative case the von Neumann factor is just the field of complex numbers, whereas in non-commutative cases the factors might range from the algebra of all complex square matrices of a given size, to something much more complicated, constituting a wild family which is a major challenge to even classify.

With the invasion of Tomita-Takesaki theory, spurred on by quantum statistical mechanics, the subject, in the hands of A. Connes, peaked in the 70s. Excellent treatises were written at the time, with further noteworthy later additions. However, only two of these are advanced enough on relevant material to serve as basis for the theory of locally compact quantum groups, namely the three volume series *Theory of Operator Algebras* by M. Takesaki, and the book by S. Stratila on modular theory for weights, which is a continuation of another he had with L. Zsido on operator algebras. Weights are generalizations of integrals, and since 'Haar weights' play a crucial role in the theory of locally compact quantum groups (their existence are now imposed rather than proved), it is perhaps not surprising that the weight theory of U. Haagerup and Connes with its ensuing Tomita-Takesaki modular theory, is used with full force. Connes used the same theory to push the classification of factors further. His brilliant survey on operator algebras in his book Noncommutative Geometry, while quite comprehensive, does not contain enough detail for our purpose. The books by Stratila and Takesaki also assume a fair bit of functional analysis, complex function theory, measure theory and topology, and in this sense are not self-contained.

In the remaining part of this introduction I will give a non-chronological taste of some of the material treated in this book.

Chapters 2 and 3 deal with functional analysis. This area of mathematics concerns the study of function spaces and the linear maps between them, called operators. Here a function, which maps points in one space to points in another space, is itself seen as a point, or vector, in an infinite dimensional vector space. A century ago pioneers such as D. Hilbert, S. Banach, F. Riesz and von Neumann developed this area as a meta model in order to establish various properties of a whole range of differential- and integral operators by considering general classes of abstract continuous operators acting on abstract infinite dimensional topological vector spaces; one had realized that most function spaces were examples of such spaces and that most operators from applications were continuous in an appropriate topology. One also realized that the collection of all such abstract operators themselves formed topological vector spaces, and in many cases algebras with algebraic structures respecting topologies, and one could in turn consider operators between these spaces. This is the main idea, whatever trivial, behind operator algebras, to study whole algebras of operators.

Many important results in functional analysis center around the notion of convexity; a subset of a vector space is convex if it contains the line segment between any two elements of the subset. An example is the closed triangle in the plane. Its three vertices are not on the segment of any two other points of the triangle, rendering the vertices extreme points, whereas the smallest closed convex subset that contains the vertices is the whole triangle. This kind of thinking carries all the way to infinite dimensions, yielding the powerful Krein-Milman theorem, which holds for the extreme points of any compact convex subset provided the topology is compatible with convexity; one then speaks of a locally convex topological vector space. Recall that compactness is the topologists version of finiteness. To see the usefulness of the theorem, say you want to verify a property, and that the steps involved are continuous and respect convexity. Then the property holds for any point in the convex subset if it holds for the extreme points, which might be easier to check if you have identified these latter points and know their nature. Another type of result in functional analysis with more obvious applications are various fixed point theorems for single maps or whole families of them.

Returning to the plane, if we in addition have an open disc disjoint from the triangle, then these two geometric objects can always be separated by a straight line. This statement has also a passage to infinite dimensions leading to the Hahn-Banach separation theorem, which roughly says that two convex subsets, one of which is open, in a locally convex topological vector space can be separated from each other by a continuous linear functional. Recall that a linear functional is an operator into the real or complex numbers.

Another theme in functional analysis is the interplay between classes of linear functionals and the topologies they induce. This gets particularly exciting when one considers the vector space of linear functionals on the vector space of linear functionals on a vector space, the so called bidual of the vector space. There is a host of results centered around this.

Functional analysis is actually one of few disciplines in mathematics where general topology is applied properly. For instance, in geometry one studies manifolds that are locally Euclidean, while in functional analysis more than ten different topologies might occur naturally in studying say an algebra of operators, and what topological candidate you choose can be decisive.

While general topology, treated in the appendix, investigates the mathematical concept of approaching a point, or converging to it, in the broadest possible sense by declaring certain subsets to be neighborhoods, or open, measure theory, also treated in the appendix, deals with the aspect of measuring the amount of something by systematically attaching a non-negative extended number to it. By something we mean a measurable subset of a given set. We require that when you measure a countable collection of disjoint such subsets, the measured amounts must add up accordingly. If there is a topology around, one usually focuses on measures, so called Borel measures, that can measure the open sets, and in fact, as little as possible beyond that.

Once you have a measure μ and a well enough behaved positive function, a so called measurable function, you may partition the whole set into finitely many measurable subsets, measure them, then rescale each one of them by a non-negative number that is not larger than the function at any point within the subset, then add up, and finally take the least upper bound of such finite sums for all possible such partitions. You are then integrating this function with respect to the measure. The collection of measurable complex functions f with $|f|^p$ having finite integral is a topological vector space, coined $L^p(\mu)$, for every $p \in [1, \infty\rangle$, then with the topology induced, or controlled, by a norm; the norm $\|f\|_p$ of f is gotten by taking the p-th root of the finite integral above. A sequence of such functions f approaches the zero function if their norms eventually creep below any given small positive number. These function spaces are important examples of Banach spaces.

A Banach space is a normed vector space which is complete, meaning that all Cauchy sequences, i.e. sequences where the vectors eventually tend to each other, will converge to vectors in the space. Any normed space can be completed much in the same way as one gets the real numbers $\mathbb{R}$ from the rational ones. The cornerstone results, namely, the open mapping theorem, closed graph theorem, and the principle of uniform boundedness, are frequently used and add depth to the topic although they might look trivial at first glance, namely, they state that continuous operators between Banach spaces have closed graphs, continuous inverses, if these exist, and that there is a bound on the image of the unit ball under a family of continuous operators if this holds for any vector in the ball.

Locally convex topological vector spaces are more general than normed spaces in that their topology is controlled by a family of seminorms. Since a single seminorm can doom non-zero vectors to have zero length, the entire family is required to tell when vectors are zero in order to get a reasonable topology. Some of the deeper results in Banach space theory investigate the relationship between boundedness, convexity and compactness with respect to weak topologies induced by seminorms given by sets of linear functionals. The study of bases in Banach spaces is more recent, and starts from the observation that there exists no infinite dimensional vector space with a countable linear basis, only approximate ones requiring infinite sums. This makes sense in Banach spaces, and as it turns out, is a notion that ties deeply with the structure of the spaces.

The completion of the space of Riemann integrable functions from calculus is realized as $L^1(\mu)$, where μ is the so called Lebesgue measure on $\mathbb{R}$. The determining property of this measure is that intervals can be translated without changing their lengths. This is just saying that μ is the Haar measure, or an invariant Borel measure, of the abelian locally compact additive group on $\mathbb{R}$, which happens to be its own Pontryagin dual. Harmonic analysis here means the study of the Fourier transform.

Compactifying $\mathbb{R}$ simplifies matters and gives the unit circle group $\mathbb{T}$. Decomposing a signal represented by a function f on $\mathbb{T}$ into its harmonics is the Fourier series $\sum_{n=-\infty}^{\infty} c_n v_n$ of f, where $v_n\colon z \mapsto z^n$ and $c_n = (f|v_n)$ is the integral of $f\overline{v_n}$ with respect to the Haar measure μ on $\mathbb{T}$. This series converges to f in $L^2(\mu)$, which is the crux of the Peter-Weyl theorem in this case. The irreducible representations are just the group endomorphisms v_n, which by definition form the Pontryagin dual. As the labeling suggests, this group is the additive group $\mathbb{Z}$ of integers with discrete topology, where convergence eventually means identification. The Pontryagin dual of $\mathbb{Z}$ is $\mathbb{T}$. Taking the dual twice brings you back; this holds even for locally compact quantum groups.

We have $(v_n|v_m) = \delta_{n,m}$ and $\|f\|_2 = \sum |c_n|^2$. So we use the inner product $(f|g)$, which makes sense as the finite integral of $f\overline{g}$ for $f, g \in L^2(\mu)$ and any measure μ, to decomposes f into orthogonal components with respect to the basis $\{v_n\}$, just like we use the scalar product in Euclidean space to decompose vectors.

Chapter 4 deals with operators on Hilbert spaces. Hilbert spaces are Banach spaces with norms $\|v\| = (v|v)^{1/2}$ given by inner products defined axiomatically. The continuous linear functionals are then of the form $(\cdot|v)$. The continuous operators on a Hilbert space V form a unital $*$-algebra $B(V)$ under pointwise linear combinations, composition as product, and with $*$-operation sending an operator T to its adjoint T^* given by $(T^*v|w) = (v|Tw)$ for all $v, w \in V$. We define the norm of T as the radius of the smallest closed ball (with center at the origin) containing the image under T of the unit ball, which is automatically finite. When this image has compact closure, we say that T is a compact operator. They form a norm closed $*$-subalgebra $B_0(V)$ of $B(V)$, and bear the closest resemblance to matrices since compact operators can be approached in norm by matrices, and if they are also normal, that is, commute with their adjoint, then they can be diagonalised with eigenvalues having orthogonal eigenvectors.

In Chap. 5 we discuss spectral theory. One checks that $\|T^*T\| = \|T\|^2$ for $T \in B(V)$, and that the norm of a product is never greater than the product of the norms of each factor. Banach spaces which are $*$-algebras having norms with such properties are called C*-algebras, which form a convenient abstract framework for spectral theory. Instead of merely looking at eigenvalues of T, one is better off considering its spectrum $\mathrm{Sp}(T)$, consisting of those $\lambda \in \mathbb{C}$ with $T - \lambda$ not invertible in $B(V)$; a notion that moreover makes sense for members of (unitisations of) abstract C*-algebras. Thanks to complex function theory the spectrum is always non-empty and compact, and the maximal distance from the origin to the spectrum will be the norm of any self adjoint operator in a C*-algebra. This means that $*$-homomorphisms between such algebras are automatically continuous. A commutative C*-algebra W is $*$-isomorphic to the algebra $C_0(X)$ of continuous functions vanishing at infinity on a locally compact Hausdorff space X. This is due to the Gelfand transform that sends $w \in W$ to the function on X that takes $x \in X$ to $x(w)$, where X consists of the non-zero $*$-homomorphims $W \to \mathbb{C}$ and topologized so that the Gelfand transform becomes continuous. When w is normal, it generates a commutative C*-algebra $C^*(w)$, and then X can be identified with $\mathrm{Sp}(w)$. The inverse Gelfand transform of $f \in C_0(X)$ is then written $f(w) \in W$. This is the continuous functional calculus generalizing the usual one applied to diagonalizable operators, and extends further to Borel measurable functions. It is in turn used to prove the spectral theorem for self-adjoint unbounded operators in Hilbert spaces. We say 'in' because such operators will be bounded if they are defined everywhere; in general their domains are only dense. A self adjoint operator is said to be positive if its spectrum is contained in $[0, \infty]$. We devote a separate chapter, Chap. 12, to the study of unbounded operators.

If one takes for any normed closed unital $*$-algebra of $B(V)$ its closure in the weak operator topology, meaning the one induced by all functionals of the form $(\cdot v|v)$, we get in general a much larger C*-algebra called a von Neumann algebra. These algebras are studied in Chaps. 5, 7 and 11. A von Neumann algebra contains all spectral projections of its members, and by the spectral theorem, is generated by orthogonal projections. The commutant Y' of a subset Y in $B(V)$ consists of the members in $B(V)$ that commute with every element of Y. When $Y^* \subset Y$, then Y' is a von Neumann algebra. The bicommutant theorem says that $W'' = W$ for any von Neumann algebra W.

Hilbert spaces are identified with each other using unitary operators, so an operator u on such a space is unitary if and only if $u^*u = 1 = uu^*$. A unitary representation of a locally compact group G on a Hilbert space V is a group homomorphism $s \mapsto u_s$ from G into the group of unitary operators on V such that $s \mapsto u_s(v)$ is continuous for all $v \in V$. The commutant $R(u)$ of the image of such a representation is a von Neumann algebra. Any orthogonal projection $p \in R(u)$ reduces the representation in that it restricts to a unitary representation on the image of p, and conversely, any reduction produces an orthogonal projection in $R(u)$. Hence we have an irreducible representation exactly when $R(u) = \mathbb{C}$. Orthogonal projections p_i are said to be equivalent if there is $w \in R(u)$ with $w^*w = p_1$ and $ww^* = p_2$, and then the corresponding reduced representations are equivalent. We say they are disjoint if there is an orthogonal projection $p \in R(u) \cap R(u)'$ such that $pp_1 = p_1$ and $(1-p)p_2 = p_2$; otherwise they are called isotypic, which happens if the intersection is $\mathbb{C}$. A von Neumann algebra having trivial intersection with its commutant is by definition a factor.

Finite dimensional C*-algebras decompose into full matrix algebras, so if we have a finite dimensional unitary representation $s \mapsto u_s$ of some group G, then $R(u)$ is a direct sum of full matrix algebras corresponding to the decomposition of the representation into isotypic ones. Each such representation is again a multiple of an irreducible representation with multiplicity equaling the size of the matrices, and corresponds to a minimal projection, like the matrix that is one in the upper left corner and otherwise zero. Think of the decomposition of natural numbers into a product of prime powers, which correspond to the isotypic components with the power being the multiplicity of the specific irreducible representation. In general a factor with a minimal projection is isomorphic to $B(V)$ for some Hilbert space V, and an isotypic representation u has an irreducible subrepresentation u' precisely when $R(u)$ is a factor with a minimal projection, and then u is a multiple of u'.

An important unitary representation of a group G with Haar measure μ is the (left) regular one, denoted by λ, with λ_s acting on $L^2(G) \equiv L^2(\mu)$ by $\lambda_s(f)$ at $t \in G$ given by $f(s^{-1}t)$. When $G = \mathbb{T}$, then $\lambda_z(v_n) = v_{-n}(z)v_n$, so $\lambda = \oplus_{n\in\mathbb{Z}} v_n$. Similarly, the regular representation of any compact group G decomposes into finite dimensional irreducibles, each with multiplicity equaling their dimension, and they exhaust the irreducible ones. This is proved by using the Haar integral to construct a compact member of $R(\lambda)$. A similar proof works for compact quantum groups.

The group von Neumann algebra $W^*(G)$ of a locally compact group G is $R(\lambda)'$. When G is discrete, then $R(\lambda)$ is a factor if and only if the only finite conjugacy class $\{sts^{-1} \,|s \in G\}$ occurs for $t = 1$. Indeed, for the orthonormal basis $\{\delta_s\}$, we have $\lambda_s(\delta_t) = \delta_{st}$, and $R(\lambda)$ is generated as a von Neumann algebra by the operators ρ_s given by $\rho_s(\delta_t) = \delta_{ts^{-1}}$. The vector δ_1 is both cyclic and separating for $R(\lambda)$, meaning respectively that $R(\lambda)\delta_1$ and $R(\lambda)'\delta_1$ are dense in the Hilbert space. For $w \in R(\lambda) \cap R(\lambda)'$, the vector $w\delta_1$ is constant on any conjugacy class, so it must be proportional to δ_1, and w is a scalar. Hence λ is isotypic, but it has no irreducible subrepresentation because it has a (unique and faithful) normalized trace $(\cdot\delta_1|\delta_1)$, meaning a positive unital linear functional which is unaltered on any product subject to cyclic permutations. There exists no normalized trace on $B(V)$ when V is infinite dimensional since elements in the latter algebra are sums of commutators. The conjugacy condition holds for the free group on more than one generator.

The set of equivalence classes of orthogonal projections in a factor on a separable Hilbert space is partially ordered under comparison of representatives, and is a partial semigroup under addition of representatives that are orthogonal to each other, and as such is isomorphic to a subsemigroup of $[0, \infty]$. These are classified, and there is a dimension function d from the set of projections into $[0, \infty]$ preserving the structure. The various images of d results in the type classification of factors. The type I-factors are the countably many $B(V)$, each with d-image $\{1, \ldots, \dim V\}$, whereas there are two uncountable families of type II-factors; the ones with d-images $[0, \infty]$, and the others with d-images $[0, 1]$, like the free group von Neumann algebras. In both these cases the dimensions are continuous, and one is no longer dealing with geometries of lines and planes etc. having integer dimensions. The final type III-factors have d-images $\{0, \infty\}$. They are even more mysterious, and there is a continuum of them. The best understood factors are the hyperfinite ones, i.e. those that are limits of matrix algebras. There is a complete classification for them with only one among each of the two type II-factors.

The formula $\mathrm{Tr}(T) = \sum(Tv_i|v_i)$ for a positive operator T on a Hilbert space V with an orthonormal basis $\{v_i\}$ makes sense as an element of $[0, \infty]$, and this extended real number is the same when we replace T by UTU^* for any unitary U on V. Hence we call it a trace. Letting $B^p(V) = \{T \in B(V) \,|\, \mathrm{Tr}(|T|^p) < \infty\}$, we get the trace class operators $B^1(V)$ and the Hilbert-Schmidt operators $B^2(V)$. They are self adjoint ideals of $B(V)$, and $B_f(V) \subset B^1(V) \subset B^2(V) \subset B_0(V)$, where the first self adjoint ideal consists of the finite rank operators, i.e. those with finite dimensional image. The trace class operators form a Banach algebra under the 1-norm, see above for $p = 1$, and its dual Banach space is $B(V)$ under the pairing given by the trace. It turns out that von Neumann algebras are the C*-algebras that are duals of other Banach spaces called their preduals. The predual consists of the normal bounded functionals on the von Neumann algebra, those that respect infinite sums of orthogonal projections that are pairwise orthogonal, thus reminding of a measure. The formula $(S|T) = \mathrm{Tr}(T^*S)$ makes $B^2(V)$ a Hilbert space, and these operators often occur in applications as integral operators.

One can take these ideas further, namely given any positive linear functional x on a unital C*-algebra W, then $(v'|v) = x(v^*v')$ defines a pre-Hilbert space structure on W with associated Hilbert space V_x and linear map $q\colon W \to V_x$. One uses left multiplication of w on v to define an operator π_x on V_x, and gets what we call a GNS-representation π_x of W on V_x with a cyclic vector $v_x = q(1)$ such that $x(w) = (\pi_x(w)v_x|v_x)$. The relationship between states and representations is studied in Chap. 6. Recall that a representation of W on a Hilbert space V is a $*$-homomorphism $W \to B(V)$. One may extend the linear functional to a normal one on the von Neumann algebra $\pi_x(W)''$. This links topology to measure theory, just like we associate to a positive linear functional x on $C(X)$ for a compact Hausdorff space X, a complex Borel measure μ on X with an integral extending x to the (classes of essentially) bounded measurable functions on X, which as a von Nemann algebra $L^\infty(\mu)$ acts on $L^2(\mu)$ by pointwise multiplication.

Taking the direct sum of GNS-representations for all possible unital positive linear functionals, or states, and using that they separate points in W by the Hahn-Banach theorem, one gets a faithful, hence isometric, representation of W on a large Hilbert space. This shows that any C*-algebra is a closed $*$-subalgebra of $B(V)$ for some Hilbert space V. In studying representations of C*-algebras one investigates the relationship between states, their GNS-representations, and their kernels as left-ideals of W, preferably with some natural topology on these sets. For instance, the pure states, which are the extreme points of the convex set of states of a C*-algebra W, correspond to the irreducible representations of W, and their kernels are prime ideals of the algebra W. When W is commutative, the pure states are the unital homomorphisms $W \to \mathbb{C}$, and the kernels are the maximal ideals.

Modular theory is investigated in Chap. 10. Say W is a von Neumann algebra on a separable Hilbert space V with an orthonormal basis $\{v_n\}$. Then $x = \sum 2^{-n}(\cdot v_n|v_n)$ is a faithful normal state on W with an injective GNS-representation π_x having a cyclic and separating vector v_x. This is the standard representation of W. The Tomita-Takesaki theory consists of polar decomposing $J_x\Delta_x^{1/2}$ the well-defined dense and closable operator on V_x given by $wv_x \mapsto w^*v_x$ into an antiunitary J_x and an (unbounded) positive operator $\Delta_x^{1/2}$. Then $J_x\pi_x(W)J_x = \pi_x(W)'$, and $\sigma_t^x \equiv \Delta_x^{it}(\cdot)\Delta_x^{-it}$ leaves the von Neumann algebra $\pi_x(W)$ invariant for any real scalar t. Moreover, the one-parameter group of automorphisms σ_t^x of $\pi_x(W)$, called the modular group, controls the deviation from tracialness of x through an analytic boundary KMS-condition familiar from quantum statistical mechanics, and this condition uniquely determines the modular group. When $W = B(V)$ the faithful normal state $\mathrm{Tr}(\cdot h)$ with h positive and invertible, has as modular group $\sigma_t = h^{it}(\cdot)h^{-it}$.

This can also be done for unbounded functionals, more precisely, for semifinite normal faithful weights on a von Neumann algebra W, generalizing the theory of non-finite integrals in measure theory. Given such a weight x, the set of $w \in W$ with both $x(w^*w)$ and $x(ww^*)$ finite is a left Hilbert algebra with inner product $(w'|w) = x(w^*w')$. A left Hilbert algebra is a $*$-algebra that is a pre-Hilbert space such that left multiplication gives a non-degenerate representation, and such that the adjoint map is closable, and can thus be polar decomposed as above.

The modular group is unique up to conjugation by a one-parameter family of unitaries u_t satisfying a cocycle condition $u_{s+t} = u_s\sigma_s^x(u_t)$, which means that the modular group is up to inner automorphisms independent of the weight. It is an intrinsic dynamic, or a canonical flow, driven by non-commutativity, thus allowing for natural invariants. One is the Connes spectrum $S(W)$, which is the intersection of the spectra of Δ_x for all semifinite normal faithful weights x on W. It is a closed multiplicative semigroup of $[0, \infty\rangle$, and as such must either be the whole semigroup, or $\{0, 1\}$, or $\{0\} \cup \{\lambda^n \mid n \in \mathbb{Z}\}$ for $\lambda \in \langle 0, 1\rangle$. Accordingly a finer division of type III-factors was obtained with exactly one hyperfinite factor in each class.

We have devoted Chap. 8 to an important notion in dealing with quantum groups, namely tensor products. These are algebraic counterparts of direct products of classical spaces, and are as vector spaces obtained through linearisations of bilinear maps. For instance, if G is a finite group, then the bilinear map $C(G) \times C(G) \to C(G \times G)$ given by $(f, g) \mapsto fg$ with $(fg)(s, t) = f(s)g(t)$ factors uniquely through the tensor product $C(G) \otimes C(G)$ as a linear map $f \otimes g \mapsto fg$. In this case the tensor product is isomorphic to $C(G \times G)$, not only as a vector space, but as a C*-algebra with product and $*$-operation given by $(f \otimes g)(f' \otimes g') = ff' \otimes gg'$ and $(f \otimes g)^* = f^* \otimes g^*$, where by the $*$-operation we mean pointwise complex conjugation.

The algebraic tensor product makes sense for any two C*-algebras, say V and W, producing a $*$-algebra $V \otimes W$ as above, and the latter needs to be completed with respect to some C*-norm. This is a delicate matter. In general there can be a plethora of such norms with a maximal and a minimal one, leading to non-isomorphic completions. When one of them is nuclear, there is by definition only one possibility. The C*-algebra of compact operators on any Hilbert space is nuclear, and so are commutative C*-algebras, and the C*-tensor product $C_0(X) \otimes C_0(Y)$ is isomorphic to $C_0(X \times Y)$ for locally compact Hausdorff spaces X and Y. We can also form tensor products of Hilbert spaces. If W_i are von Neumann algebras on Hilbert spaces V_i, then the von Neumann tensor product $W_1 \bar{\otimes} W_2$ is the bicommutant of the minimal one of W_1 and W_2, which lives in $B(V_1 \otimes V_2)$, and one has then the following fundamental result $(W_1 \bar{\otimes} W_2)' = W_1' \bar{\otimes} W_2'$. Representation theory is indispensable in studying tensor products.

Another and more general construction occurs naturally in the study of dynamical systems. We study such classical systems in Chap. 14. Say one is given a continuous action $\alpha \colon G \to \mathrm{Aut}(W)$ of a locally compact group G on a von Neumann algebra $W \subset B(V)$. Letting $\alpha(w) \colon t \mapsto \alpha_{t^{-1}}(w)$, we may view the action as a coaction $\alpha \colon W \to L^\infty(G, W) = L^\infty(G) \bar{\otimes} W$, where ∞ means essential bounds of pointwise norms of measurable functions. We may then define the crossed product $G \bar{\ltimes}_\alpha W$ as the von Neumann algebra generated by $\alpha(W)$ and $W^*(G) \otimes \mathbb{C}$ in $B(L^2(G) \otimes V)$. When G is abelian, we have a dual action $\hat{\alpha}$ of the Pontryagin dual $\hat{G}$ on $G \bar{\ltimes}_\alpha W$. The crossed product biduality theorem says that $\hat{G} \bar{\ltimes}_{\hat{\alpha}} (G \bar{\ltimes}_\alpha W)$ is isomorphic to the von Neumann algebra $B(L^2(G)) \bar{\otimes} W$, and the bidual action of the former is conjugate by a unitary cocycle to the tensor product action $1 \otimes \alpha$ of the latter. Using the modular group, one can reduce the study of type III-factors to crossed products of type II-factors; these latter factors have tracial weights. One relates properties of the groups in action to those of the C*-algebras obtained as crossed products. These can also be seen as generalizations of Weyl quantization.

Given a finite group G, being discrete, the counting measure is the Haar measure with faithful integral $x \colon C(G) \to \mathbb{C}$ given by $x(f) = \sum_{t \in G} f(t)$. Here the Haar measure integral is both left and right invariant, a property that breaks down for general locally compact groups. Left and right invariant integrals are related by a modular homomorphism from the group to the multiplicative group $\langle 0, \infty \rangle$. This is basically the $*$-operator from Tomita-Takesaki theory, which is hinted at by noting that $u_t^* = u_{t^{-1}}$ for any unitary representation u of the group. We study Haar integrals for locally compact groups in the appendix.

Transposing the multiplication in the finite group G into a $*$-homomorphism $\Delta\colon C(G) \to C(G) \otimes C(G)$ given by $\Delta(f)(s,t) = f(st)$, associativity of the product in G becomes $(\Delta \otimes \iota)\Delta = (\iota \otimes \Delta)\Delta$, whereas left and right invariance of x becomes $(\iota \otimes x)\Delta = x = (x \otimes \iota)\Delta$. The group can be completely recovered from such a coproduct Δ and faithful invariant x. The counit $\varepsilon(f) = f(1)$ and the coinverse $S(f)(t) = f(t^{-1})$ satisfy $(\varepsilon \otimes \iota)\Delta = \iota = (\iota \otimes \varepsilon)\Delta$ and $m(S \otimes \iota)\Delta = \varepsilon(\cdot)1 = m(\iota \otimes S)\Delta$, where m is the multiplication on $C(G)$ extended to a linear map on the tensor product. These identities form axioms of Hopf algebras, which hold automatically here due to the presence of x. In general these Hopf algebra identities are difficult to make sense of analytically.

Hopf algebras are related to quantum groups and both notions are studied at length in Chap. 12. One defines locally compact quantum groups simply by replacing $C(G)$ and Δ by a general von Neumann algebra W with a coassociative normal $*$-homomorphism $\Delta\colon W \to W \bar{\otimes} W$ and left invariant and right invariant semifinite normal faithful weights x and y, respectively. Then x and y are unique up to multiplication with positive scalars, and there exists a coinverse with polar decomposition $S = R\tau_{-i/2}$. Here R is the unitary coinverse satisfying an identity being the transpose of $(st)^{-1} = t^{-1}s^{-1}$ in the group case, namely, the identity $f(R \otimes R)\Delta = \Delta R$ for the flip f given by $f(w \otimes w') = w' \otimes w$. Hence xR is proportional to y. Now $\{\tau_t\}$ is a one-parameter group of automorphisms on W, called the scaling group. Its analytic extension to $-i/2$ is unbounded. When the scaling group is trivial, one gets a Kac algebra, but already for the compact quantum groups of Drinfeld and Jimbo the scaling groups are non-trivial. In general there are nice relations between the scaling group and the modular groups, including the ones of the invariant weights of the dual locally compact quantum group $(\hat{W}, \hat{\Delta})$, which always exists.

A pivotal role is played by the fundamental multiplicative unitary M of a locally compact quantum group (W, Δ). Its existence is one of the first things one proves. When $W = L^\infty(G)$ with Δ given by the transpose of the group multiplication, then $M(g)(s,t) = g(s, s^{-1}t)$ for $g \in L^2(G \times G)$, so M acts like the left regular representation λ in the second variable. In general it satisfies the pentagonal equation $M_{12}M_{13}M_{23} = M_{23}M_{12}$ in $B(V_x \otimes V_x \otimes V_x)$, where $M_{12} = M \otimes 1$ etc., and W is generated by the elements $(\iota \otimes \omega)(M)$ for all normal states ω on $B(V_x)$, and moreover we have $\Delta = M^*(1 \otimes (\cdot))M$. It turns out that $\hat{M} \equiv M_{21}^*$ is the fundamental multiplicative unitary of the dual locally compact quantum group $(\hat{W}, \hat{\Delta})$. In the group case the elements $(\omega \otimes \iota)(M)$ generate $W^*(G)$ and $\hat{\Delta}(\lambda_t) = \lambda_t \otimes \lambda_t$. Then the map $\omega \mapsto (\omega \otimes \iota)(M)$ is an averaging of λ going from the Banach algebra $L^1(G)$, which is the predual of $L^\infty(G)$, to $W^*(G)$. When G is abelian we may view this homomorphism as the Fourier transform $L^1(G) \to L^\infty(\hat{G})$ converting convolution products, being smeared out linearizations of the group multiplication, to pointwise products of functions. We have here used the Plancherel isomorphism $L^2(G) \to L^2(\hat{G})$ to identify $W^*(G)$ with $L^\infty(\hat{G})$. In general we are tacitly using a generalized version of the Plancherel isomorphism when we are saying that $V_x = V_{\hat{x}}$, where $\hat{x}$ is the left invariant weight of the dual quantum group.

Abstract conditions have been put on multiplicative unitaries for them to be the fundamental ones of locally compact quantum groups. Any multiplicative unitary produces coproducts on the von Neumann algebras generated by its slices. Manageable/modular multiplicative unitaries, in the language of Woronowicz, also exhibit coinverses with polar decomposition, and some other nice properties. While one has candidates for the invariant weights, no general proof yields semifiniteness, though this holds for many examples. This would be like proving the existence of a Haar integral for locally compact groups. It should be noted that the fundamental multiplicative unitary of a locally compact quantum group is manageable.

There are also topological, or C*-algebraic, versions of locally compact quantum groups. One takes then norm closures of the $*$-algebras generated by the slices of M using the ω's. The invariant weights restrict to weights on these C*-algebras satisfying certain properties modeling axioms used in the C*-context. This is how locally compact quantum groups were first introduced, but weight theory really belongs to von Neumann algebras, and one has since reverted to that setting. In any case, any locally compact quantum group has a minimal, or reduced, and a maximal, or universal, C*-version, with a whole range between, just like for tensor products. The situation is pretty much like the reduced C*-envelope $C_r^*(G)$ and the universal C*-envelope $C_u^*(G)$ of $L^1(G)$ for a locally compact group G. These are C*-versions of cocommutative locally compact quantum groups, which are all of the form $W^*(G)$, and have by definition the range of the coproduct in the symmetric part of the tensor product. There is also an inverse procedure taking C*-algebraic quantum groups to von Neumann ones, basically the same procedure linking topology to measure theory.

Unitary representations of G correspond to non-degenerate representations of $C_u^*(G)$, and these are weakly contained in λ when $C_u^*(G) \cong C_r^*(G)$, which happens precisely when G is amenable, that is, has a left invariant state on $L^\infty(G)$. For quantum groups there is also a notion of amenabilty and a dual one of coamenability, the latter requiring that there is a bounded counit on the reduced version (W_r, Δ_r), and this then is isomorphic to the universal one (W_u, Δ_u), which always has a bounded counit. The dual correspondence between coamenability and amenability is not yet proven to be perfect. The forward direction is OK, but we do not know if amenability of a locally compact quantum group implies that the dual one is coamenable. This has been shown for discrete and compact quantum groups. A compact quantum group is a locally compact one where W_r is unital; then its locally compact dual is discrete. These matters are discussed in Chap. 13.

In the universal setting there is always a bijective correspondence between non-degenerate representations of a quantum group and corepresentations of its dual, and there exists a universal corepresentation that facilitates this. A corepresentation is an element K of say $C(G) \otimes B(V)$ gotten by viewing a unitary representation $t \mapsto u_t$ of a finite group G on V simply as an element of $C(G, B(V))$. Then $u_{st} = u_s u_t$ translates as $(\Delta_r \otimes \iota)(K) = K_{13}K_{23}$. When G is locally compact, then K is rather an element of $M(C_0(G) \otimes B_0(V))$.

Here $M(A)$ stands for the multiplier algebra of a C*-algebra A. It is a unital C*-algebra that contains A as an essential ideal, so it coincides with A when A is unital. By definition it consists of the adjointable maps on A seen as a Hilbert module, which is an A-module generalization of a Hilbert space with a module compatible 'inner product' taking values in A. While the unitization of a C*-algebra corresponds to the one point compactification of a locally compact Hausdorff space X, the multiplier algebra of the C*-algebra corresponds to the Stone-Chech compactification of X since $M(C_0(X))$ is the C*-algebra of all bounded continuous functions on X. Similarly $M(B_0(V)) = B(V)$. Non-degenerate $*$-homomorphisms between C*-algebras extend uniquely to unital $*$-homomorphisms between the multiplier algebras of the C*-algebras. Multiplier algebras are important when dealing with C*-versions of locally compact quantum groups since for instance $\Delta(f)(t, t^{-1}) = f(e)$ for a continuous function f on X, so in general $\Delta(C_0(X))$ is not contained in $C_0(X \times X)$, but rather in $M(C_0(X \times X))$. The theory of multiplier algebras and Hilbert modules can be found in Chap. 8.

We study coactions in the von Neumann setting in Chap. 15. The theory of crossed products of coactions of locally compact quantum groups is as outlined in the group case, with dual coactions and a crossed product biduality theorem. The theory of dual weights also carries through. This requires perhaps some explanation. Given a locally compact group (W, Δ) with left invariant semifinite faithful normal weight x, then a coaction of (W, Δ) on a von Neumann algebra U is a normal unital $*$-monomorphism $\alpha\colon U \to W\bar{\otimes}U$ such that $(\iota \otimes \alpha)\alpha = (\Delta \otimes \iota)\alpha$. The dual coaction $\hat{\alpha}$ of α is a coaction of $(\hat{W}, \hat{\Delta}^{\mathrm{op}})$ on the crossed product $W \ltimes_\alpha U$. Then $E = (\hat{x} \otimes \iota \otimes \iota)\hat{\alpha}$ is a semifinite faithful normal operator valued weight from the crossed product to the fixed point algebra $\alpha(U)$ of $\hat{\alpha}$, and $\tilde{y} = y\alpha^{-1}E$ is the dual weight of a semifinite faithful normal weight y on U. The unitary $K = J_{\tilde{y}}(J_{\hat{x}} \otimes J_y)$ in $W\bar{\otimes}B(V_y)$ implements the coaction in that $\alpha = K(1 \otimes (\cdot))K^*$, and using Connes relative modular theory one sees that K is a corepresentation of $(W, \Delta^{\mathrm{op}})$ called the unitary implementation of α. This is a generalization of the standard implementation of an action on a von Neumann algebra, and plays a crucial role in the study of coactions.

In Chap. 17 we establish a correspondence between coactions of quantum groups and inclusions of factors. Given an inclusion $U_0 \subset U_1$ of von Neumann algebras and a faithful semifinite normal weight z on U_1, then by representing U_1 on V_z, we get $U_1 \subset U_2 \equiv J_z U_0' J_z$. Repeating this for a weight on U_2 produces the Jones tower $U_0 \subset U_1 \subset U_2 \subset \cdots$ of inclusions. One might wonder to what extend such a tower can be constructed from a coaction α of a locally compact quantum group (W, Δ) on a von Neumann algebra U_1 as repeated crossed products with the fixed point algebra of α as U_0 and with the crossed product as U_2. There is a complete correspondence when the coaction is outer, the inclusion $U_0 \subset U_1$ is irreducible of depth 2, and there exists a regular semifinite faithful normal operator valued weight from U_1 to U_0. The technical proof of this relies on relative modular theory of weights and their spatial derivatives, and connects to a generalization of the celebrated index theory of V. Jones. When irreducibility is skipped a similar result holds, but one needs then quantum groupoids, which are not treated in this book.

The theory of crossed products for coactions can be generalized to cocycle crossed products of cocycle coactions; again we have dual cocycle coactions, a cocycle crossed product biduality theorem, a dual weight and a canonical unitary implementation by a generalized corepresentation of the quantum group with opposite coproduct. This is done in Chap. 16. We also have a generalized Landstad theory, which for abelian groups characterizes actions that are dual actions.

There are analytic versions of standard algebraic constructions involving quantum groups, namely cocycle bicrossed products, which are dealt with in Chap. 16, and doublecrossed products (generalizations of quantum doubles), which can be used to get new locally compact quantum groups from old ones, with various properties inherited. Doublecrossed products are studied in Chap. 19.

There are also topological versions of these constructions. One should then restrict to locally quantum groups with regular fundamental multiplicative unitaries. Regular multiplicative unitaries were studied as an alternative approach to quantum groups, much like the manageable ones, but fundamental multiplicative unitaries of locally compact quantum groups are in general not regular (nor even semiregular), while those of Kac algebras and compact quantum groups are know to be regular, and duals of regular ones are always regular. These issues are discussed in Chap. 16.

Another fundamental and exciting construction with an algebraic origin, is the one related to a Galois object for a locally compact quantum group (W, Δ), say with a left invariant faithful semifinite normal weight x. Such an object is a certain right coaction α of (W, Δ) that is ergodic, i.e. with trivial fixed point algebra, and integrable, meaning that $(\iota \otimes x)\alpha$ is a semifinite weight. One can then reflect $(\hat{W}, \hat{\Delta})$ across α to obtain another locally compact quantum group with surprisingly different features. To illustrate, consider a unitary dual 2-cocycle of (W, Δ), that is, a unitary $X \in \hat{W}\bar{\otimes}\hat{W}$ such that $(1 \otimes X)(\iota \otimes \hat{\Delta})(X) = (X \otimes 1)(\hat{\Delta} \otimes \iota)(X)$. From the theory of cocycle crossed products one can then construct a Galois object α for (W, Δ) such that the reflection of $(\hat{W}, \hat{\Delta})$ across α is a locally compact quantum group with $\hat{W}$ as von Neumann algebra and with coproduct $X\hat{\Delta}(\cdot)X^*$; the so called twisted coproduct $\hat{\Delta}_X$ of $\hat{\Delta}$. In particular, one concludes that $(\hat{W}, \hat{\Delta}_X)$ is indeed a locally compact quantum group. How to construct invariant semifinite weights directly is not obvious. Galois objects are studied in Chap. 18.

Quantization schemes from geometry tend to deform the product on some algebra of bounded observables subject to symmetries. In such situations one often has from the outset a classical theory, say with a commutative locally compact quantum group (W, Δ) and a continuous coaction α on a C*-algebra U. The deformed product can in some cases be extracted as a dual unitary 2-cocycle X of (W, Δ), and while this quantum group does not act anymore on the algebra with deformed product, the dual of $(\hat{W}, \hat{\Delta}_X)$ does. So quantum groups are really needed as 'symmetries' in the quantum world, especially when the underlying configuration spaces are curved.

One may start in the opposite end and construct the deformed algebra from a coaction and a dual unitary 2-cocycle. This theory works well when the fundamental multiplicative unitaries are regular, and then one might even start with a locally compact quantum group. Repeated deformation yields a duality theorem that ties in nicely with those of (cocycle) crossed products. This is all explained in Chap. 18.

Given a locally compact quantum subgroup of a locally compact quantum group, corepresentations of the former can be induced up to corepresentations of the latter; an approach that also relies on modular theory. Such induction is studied in Chap. 20.

Chapter 2
Banach Spaces

This chapter deals with what could be called geometric functional analysis. Results from plane geometry are generalized to infinite dimensional vector spaces, including function spaces, yielding powerful, general results with a wide range of applications from within optimization theory to physics.

While topology plays no role in the finite dimensional case, it is essential in infinite dimensions, where most results hold true only approximately. We include the necessary background from topology in the appendix, urging the reader to look up things there whenever needed. In the appendix we also introduce measure theory, and treat L^p-spaces in detail, which are the most important examples of Banach spaces.

The L^p-spaces are the normed spaces where sequences that converge in some larger normed space, actually converge in the original space. Recall that the norm of a point is the length of the vector, seen as an arrow from the origin of the vector space to the point. The distance, or metric, between two points, is then the length of any of the two vectors between the points. Distance yields neighborhoods, and thus, topology, which is the study of sets of points with more general neighborhoods.

The study of infinite matrices, or operators between normed spaces, begins with the bounded ones. While it makes no sense to say that a (linear) operator is bounded if it has a norm-bounded image, we may require that the image of any ball should be bounded, that is, contained in another ball. This actually means that the operator is continuous, in the sense that the image of any converging sequence, converges to the image of what the original sequence converges to.

After having established the basic properties for normed spaces, we include standard result about bounded operators between Banach spaces, which, as opposed to their counterparts in linear algebra, are non-trivial due to the necessity of topology.

L. Tuset, *Analysis and Quantum Groups*,
https://doi.org/10.1007/978-3-031-07246-8_2

Normed space are special cases of topological vector spaces, that is, vector spaces with a topology making the basic vector space operations continuous. They are moreover locally convex, meaning that they have enough convex neighborhoods to built up the entire topology as unions from them. The topology on a locally convex topological vector space is the weak one induced by some separating family of seminorms, rather than just one norm. Such topologies can be weak enough to get approximation results for a much larger class of function spaces occurring naturally in applications, such as in differential geometry.

In many cases the family of seminorms is given by pointwise absolute values of linear functionals on the normed space. Linear functionals play the role of hyperplanes, and can be used to separate two subsets in a vector space. These separating properties are thus intimately linked to the topology on the space, yielding results like the Hahn-Banach separation theorem. Convexity plays an important role both here, and in results like the Krein-Milman theorem on the extremal points of a convex compact subset of the vector space.

Having established various standard fixed point results for (a collection of) maps between vector spaces, like the ones of Markow-Kakutani, Rull-Nardzewski and Schauder, we study in depth dual spaces of Banach spaces. The dual space of a Banach space consists of the continuous linear functionals on the space, which itself becomes a Banach space under pointwise vector space operations. The evaluation of a continuous linear functional on a vector is a continuous linear functional on the dual space, that is, an element of the bidual space; the dual of the dual space. This gives a canonical embedding from a Banach space into its bidual. In finite dimensions this embedding is surjective, but in infinite dimensions surjectivity holds by definition only for reflexive spaces. The w^*-topology on the dual space is the weak topology induced by the vectors of the space itself, then regarded as a subspace of the bidual. Alaoglu's theorem asserts that the closed unit ball in the dual of a Banach space is compact in this topology, whereas the Krein-Smulian theorem tells us that the w^*-closed subspaces are those having w^*-compact intersection with the closed unit ball.

The Eberlein-Smulian theorem characterizes the subsets of a Banach space with compact closure in the weak topology induced by the dual space in terms of sequences in the subsets having weakly convergent subsequences. We then investigate the relation between reflexivity of a Banach space and continuous linear functionals on the space attaining their norms culminating in James' reflexivity theorem.

Towards the end of the chapter we return to general continuous operators, focusing on aspects of compactness and weak compactness of such operators, leading to a host of results. This includes establishing relations of such operators to complemented and invariant subspaces, and their relation to finite rank operators.

Recommended literature for this and the next chapter is [1, 9, 22, 33, 41].

2.1 Normed Spaces

Analysis enters naturally when dealing with infinite dimensional vector spaces. Topology enters more heavily in the section on weak topologies and onward. In the first few sections we consider vector spaces over ground fields more general than the real numbers $\mathbb{R}$ and the complex numbers $\mathbb{C}$, just to indicate how one would proceed then.

Definition 2.1.1 An *absolute value* on a field F is a function $a \mapsto |a|$ from F to the non-negative real numbers that is only zero at zero and satisfies $|ab| = |a| \cdot |b|$ and $|a + b| \leq |a| + |b|$ for all $a, b \in F$. It is called *non-archimedean* or a *valuation* if it satisfies the stronger inequality $|a + b| \leq \max\{|a|, |b|\}$ for all $a, b \in F$.

Observe that $|0| = 0$ and $|1| = 1$ and $|a| = |-a|$ and $|a^{-1}| = |a|^{-1}$ for any non-zero $a \in F$. The *trivial absolute value* on any field is the non-archimedean absolute value that is one for every non-zero element.

One can obviously define Cauchy sequences and convergence of sequences with elements in F just as one does for the rational numbers, but now with respect to the given absolute value on F. This allows also for the formation of a *completion* of F in the same way as one forms the real numbers from the rational numbers. Such a completion is clearly the unique field which contains F and has the property that any element a of it can be approximated by some sequence $\{a_n\} \subset F$ in the sense that $|a - a_n| \to 0$ as $n \to \infty$. The field F is contained in the completion as the classes having Cauchy sequences that eventually become constant with the constants being the elements in F, and the absolute value was extended to an absolute value on the completion by declaring $|a| = \lim |a_n|$, which exists as $\{|a_n|\}$ is Cauchy in $\mathbb{R}$.

Definition 2.1.2 A *seminorm* on a vector space V over a field F with absolute value $|\cdot|$ is a function $\|\cdot\|: V \to [0, \infty)$ such that $\|av\| = |a| \cdot \|v\|$ and $\|u+v\| \leq \|u\|+\|v\|$ for $u, v \in V$ and $a \in F$. It is a *norm* if it vanishes only for $0 \in V$, and V is then said to be *normed*.

Given a second norm $\| \cdot \|'$ on V, then the two norms are *equivalent* if there are positive numbers r, s such that $r\|v\| \leq \|v\|' \leq s\|v\|$ for all $v \in V$.

The property $\|u+v\| \leq \|u\|+\|v\|$ for all elements $u, v \in V$ is called the *triangle inequality*.

A sequence $\{v_n\}$ in V *converges* to $v \in V$ with respect to a norm $\| \cdot \|$ if to any $\varepsilon > 0$ there is a natural number N such that $\|v - v_n\| < \varepsilon$ for all $n > N$. We then write $v = \lim v_n$ or $v_n \to v$ as $n \to \infty$. Clearly sequences converge to the same vectors with respect to equivalent norms.

A sequence $\{v_n\}$ in a normed space V is *Cauchy* if to any $\varepsilon > 0$ there is a natural number N such that $\|v_n - v_m\| < \varepsilon$ for all $n, m > N$. Obviously every convergent sequence is Cauchy, but the converse is not always true.

Definition 2.1.3 A *Banach space* is a normed vector space where all Cauchy sequences converge.

Obviously non-trivial absolute values are examples of norms on the fields themselves with $\mathbb{R}$ and $\mathbb{C}$ Banach spaces. These norms can also be extended to products of fields, namely, if F is a field with an absolute value $|\cdot|$, then F^n is a normed vector space over F with norms

$$\|(a_1,\ldots,a_n)\|_\infty = \max|a_i| \quad \text{and} \quad \|(a_1,\ldots,a_n)\|_1 = \sum |a_i|$$

for $a_i \in F$.

Proposition 2.1.4 *If F is a complete field, then F^n for any $n \in \mathbb{N}$ is a Banach space with norm $\|\cdot\|_\infty$.*

Proof If $\{v_i\}$ is Cauchy in F^n with respect to $\|\cdot\|_\infty$ with $v_i = (a_1^i, \ldots, a_n^i)$, then for each fixed k, the sequence $\{a_k^i\}$ is Cauchy in F with respect to an absolute value $|\cdot|$ for which F is complete. Thus to $\varepsilon > 0$ there are $a_k \in F$ and natural numbers N_k such that $|a_k - a_k^i| < \varepsilon$ for all $i > N_k$. Let N be the greatest of the finitely many N_k and let $v = (a_1, \ldots, a_n)$. Then $\|v - v_i\|_\infty < \varepsilon$ for all $i > N$. □

A similar proof shows that F^n with F complete is a Banach space with respect to the norm $\|\cdot\|_1$.

Definition 2.1.5 A linear map $A\colon V \to W$ between normed vector spaces over a field is *bounded* if there is a positive number c such that $\|Av\| \leq c\|v\|$ for all $v \in V$. The greatest lower bound of all such positive numbers c is the *(operator) norm of* A, written $\|A\|$.

Note that $\|A\| = \sup_{v\neq 0} \|A(v)\|/\|v\|$. The bounded linear maps $V \to W$ between normed spaces over a field is a vector subspace of $\mathrm{End}(V, W)$ denoted by $B(V, W)$. It is easy to see that the operator norm is indeed a norm on $B(V, W)$ which also satisfies $\|A(v)\| \leq \|A\|\cdot\|v\|$ for $v \in V$. The operator norm is *submultiplicative*, meaning $\|AB\| \leq \|A\| \cdot \|B\|$ for $A \in B(V, W)$ and $B \in B(W, U)$. So $B(V) = B(V, V)$ is a unital algebra over the field with product given by composition of bounded linear maps.

Proposition 2.1.6 *The normed space $B(V, W)$ is a Banach space when W is a Banach space.*

Proof If $\{A_n\} \in B(V, W)$ is a Cauchy sequence with respect to the operator norm, then $\{A_n v\}$ is a convergent Cauchy sequence in W for any $v \in V$. Define a map $A\colon V \to W$ by $A(v) = \lim A_n(v)$, which is evidently linear. It is also bounded and $A_n \to A$ because

$$\|(A - A_n)(v)\| = \lim \|(A_m - A_n)(v)\| \leq \limsup \|A_m - A_n\|\|v\|$$

for $v \in V$ and n. □

Definition 2.1.7 A normed space which is an algebra for which the norm is submultiplicative $\|vw\| \leq \|v\|\|w\|$, is called a *normed algebra*. If the normed space is complete, we call it a *Banach algebra*. A normed algebra is *unital* if it has an identity element 1 of norm one.

Submultiplicativity implies that an identity element in a non-trivial normed algebra has norm not less than one, but in general it can be larger than one, so our assumption is not vacuous.

Note that $B(V)$ is a unital Banach algebra under operator norm when V is a Banach space.

Definition 2.1.8 Given a linear map $A: V \to W$ between normed vector spaces over a field. Then A is *continuous at* $v \in V$ if $A(v_i) \to A(v)$ for every sequence $\{v_i\}$ that converges to v, and if this holds for every v, we say that A is continuous (everywhere).

Clearly all bounded linear maps are continuous, and all linear maps that are continuous at zero are continuous. Compositions of continuous linear maps are evidently continuous.

In a normed algebra the identity

$$\|vw - v'w'\| \leq \|v\|\|w - w'\| + \|v - v'\|\|w'\|$$

shows that the product $(v, w) \mapsto vw$ is (jointly) continuous.

Proposition 2.1.9 *A linear map between normed vector spaces over a field with non-trivial absolute value is bounded if it is continuous at zero.*

Proof Let $A: V \to W$ be such a map. If there was $\varepsilon > 0$ such that for all i we could find $v_i \in V$ with $\|A(v_i)\| > \varepsilon$ but $\|v_i\| < 1/i$, then $v_i \to 0$ while $A(v_i) \not\to 0$. So there is $\delta > 0$ such that $\|A(u)\| \leq 1$ whenever $\|u\| \leq \delta$. Since the absolute value $|\cdot|$ on the field F is non-trivial, there is $a \in F$ with $|a| < 1$. For any $w \in V$ let n be the least integer such that $\|a^n w\| \leq \delta$. Then $\|a^{n-1}w\| > \delta$ and $\|A(a^n w)\| \leq 1$, so $\|A(w)\| \leq \delta^{-1}|a|^{-1}\|w\|$. □

Remark 2.1.10 In the first part of the proof above sequential continuity of A is shown to be equivalent to topological continuity between the metric spaces V and W with metrics given by $d(u, v) = \|u - v\|$. A *metric on a set* X is any map $d: X \times X \to [0, \infty\rangle$ such that $d(x, y) = 0 \Rightarrow x = y$ and $d(x, y) = d(y, x)$ and $d(x, z) \leq d(x, y) + d(y, z)$ hold for all $x, y, z \in X$.

The unions of all *(open) balls* $B_r(x) = \{y \in X \,|\, d(x, y) < r\}$ is then a *topology on* X, and by this is meant a family of so called *open* subsets of X closed under formations of unions and finite intersections, and which contains X and ϕ. A function $f: X \to Y$ between two topological spaces is then (topologically) continuous if $f^{-1}(U)$ is open for every open set U. In the proof we show $B_\delta(0) \subset A^{-1}(B_1(0))$. ◇

The next result shows that for convergence it is irrelevant what norm one uses on finite dimensional vector spaces over complete fields with non-trivial absolute values.

Proposition 2.1.11 *All norms are pairwise equivalent on a finite dimensional vector space over a complete field with a non-trivial absolute value. Such normed vector spaces are Banach spaces.*

Proof Let $\|\cdot\|$ be a norm on a finite dimensional vector space V over a field F with non-trivial absolute value $|\cdot|$. Pick a basis $v_1, \dots v_n$ for V and define a linear isomorphism $A\colon F^n \to V$ by $A(a_1, \dots, a_n) = \sum a_i v_i$.

If $x_i \to 0$ in F^n with respect to the norm $\|\cdot\|_\infty$, then $A(x_i) \to 0$ with respect to $\|\cdot\|$ because $\|A(x_i)\| \leq \sum |a_k^i| \cdot \|v_k\|$, where $x_i = (a_1^i, \dots, a_n^i)$.

We claim that A^{-1} is continuous at zero. We first prove that it preserves Cauchy sequences; proceeding by induction, the case $n = 1$ being trivial. Assuming this holds up till n, but that it fails for n, so that we have a Cauchy sequence $\{A(x_i)\}$ without $\{x_i\}$ being Cauchy. Then there is $\varepsilon > 0$ and natural numbers n_i and m_i for every i such that $\|x_{n_i} - x_{m_i}\|_\infty > \varepsilon$ for all i. Since we have $A(x_{n_i}) - A(x_{m_i}) \to 0$ as $i \to \infty$, to derive a contradiction, we may from the outset assume that $A(x_i) \to 0$ while $x_i \not\to 0$.

So there is $c > 0$ such that, say $|a_1^i| > c$, for arbitrary large i. By going to a subsequence, we may assume that $|a_1^i| > c$ for all large enough i, and for these

$$u_i \equiv \sum_{k=2}^{n} \frac{a_k^i}{a_1^i} v_k = A(x_i)/a_1^i - v_1 \to -v_1.$$

In particular, the sequence $\{u_i\}$ is Cauchy, and by our induction hypothesis, the sequence $\{a_k^i/a_1^i\}$ is therefore Cauchy for each k, and converges to some element in the complete field F. From the equation above we see that this is impossible by linear independence of the vectors v_i, and we have established the induction step.

Hence, if $\|y_i\| \to 0$, then $\{A^{-1}(y_i)\}$ converges by Proposition 2.1.4, and since A is continuous and injective, it must converge to zero, so A^{-1} is continuous at zero.

As A is continuous at zero with respect to any other norm on V as well, we see that $AA^{-1}\colon V \to V$ is continuous at zero with respect to any two norms, and is thus bounded due to Proposition 2.1.9. Interchanging the roles of the norms, we get the required equivalence.

The last assertion follows now from Proposition 2.1.4. □

The following result is clear from the proposition above and Proposition 2.1.6.

Corollary 2.1.12 *For normed vector spaces V and W over a complete field with a non-trivial absolute values, the space $B(V, W)$ of bounded operators is a Banach space when* $\dim W < \infty$.

Example 2.1.13 Let $A \in M_n(F)$ be a square matrix with entries in a field F having absolute value $|\cdot|$, and consider F^n as a normed space under $\|\cdot\|_\infty$. For the associated linear map $A\colon F^n \to F^n$ we have

$$\|Av\|_\infty = \max|\sum_j a_{ij}v_j| \le \max\sum_j |a_{ij}|\cdot|v_j| \le n\max|a_{ij}|\cdot\|v\|_\infty,$$

so A is bounded with operator norm $\|A\| \le n\|A\|_\infty$, where we consider the non-submultiplicative norm $\|\cdot\|_\infty$ on $M_n(F)$.

With respect to the norm $\|\cdot\|_1$ on F^n again A is bounded with the same naive estimate $\|A\| \le n\|A\|_\infty$.

If F is complete with a non-trivial absolute value all these normed spaces will be Banach, and $M_n(F)$ will be a Banach algebra with respect to the two operator norms. ◇

Remark 2.1.14 For normed spaces U and V over the complex (or real) field the operator norm of $A \in B(U, V)$ can be written as $\|A\| = \sup_{\|v\|=1}\|Av\|$. ◇

The following result is immediate.

Proposition 2.1.15 *Let W be a real or complex unital Banach algebra, and let $w \in W$ with $\|w\| < r$. If $\sum_{n=0}^{\infty}|a_nz^n| < \infty$ for $|z| < r$, then $\sum_{n=0}^{m}a_nw^n$ form a Cauchy sequence that converges to an element of W denoted by $\sum_{n=0}^{\infty}a_nw^n$.*

If $f(z) = \sum_{n=0}^{\infty}a_nz^n$, we write $f(w)$ for $\sum_{n=0}^{\infty}a_nw^n$, so we ascribe elements in the Banach algebra to holomorphic functions, see the appendix for more one holomorphic functional calculus.

Since $\sum_{n=0}^{\infty}\|w^n/n!\| \le e^{\|w\|} < \infty$ for w in any real or complex unital Banach algebra W, we have $e^w \in W$. In particular, for a real or complex $n \times n$-matrix A, we can therefore form the matrix e^A.

Lemma 2.1.16 *If $v = \sum v_n$ and $w = \sum w_n$ with $\sum\|w_n\| < \infty$ in a real or complex Banach algebra, then $wv = \sum u_n$, where $u_n = \sum_m w_mv_{n-m}$.*

Proof Let $x_N = -\sum_{N+1}^{\infty}v_n$ and $y_N = \sum^N w_kx_{N-k}$. Since

$$\sum^N u_n = \sum^N w_n\sum v_k + y_N,$$

it suffices to show that $y_N \to 0$. To $\varepsilon > 0$ pick M such that $\|x_N\| < \varepsilon$ for $N > M$. For such N we have

$$\|y_N\| \le \|\sum^M x_kw_{N-k}\| + \varepsilon\sum\|w_n\|,$$

which shows that $y_N \to 0$. □

Corollary 2.1.17 *Let v, w be commuting elements of a real or complex unital Banach algebra W. Then $e^{w+v} = e^w e^v$ and $t \mapsto f(t) \equiv e^{tw}$ takes addition in $\mathbb{R}$ to multiplication in W. The* derivative $f'(t) \equiv \lim_{s\to 0}(f(t+s) - f(t))/s$ *of f makes sense and $f'(t) = we^{tw}$.*

Proof By the binomial theorem and the lemma, we get

$$e^w e^v = \sum_{n=0}^{\infty} \frac{1}{n!} \sum_{k=0}^{n} \frac{n!}{k!(n-k)!} w^k v^{n-k} = e^{w+v}.$$

Thus $f'(t) = f(t) \lim(f(s) - 1)/s = f(t)w$ as $\sum_{n=2}^{\infty} \|s^{n-2} w^n\|/n! < e^{\|w\|}$ when $|s| < \|w\|$. □

Definition 2.1.18 If all elements of a normed linear space are limits of sequences of a subset, then the subset is *dense* in the vector space. A subset is *closed* if it contains all limits of sequences in it that converge in the space. The *closure* $\overline{A}$ of a subset A in a normed space is the intersection of all closed subsets containing A.

Proposition 2.1.19 *Every bounded operator $U \to V$ into a Banach space has a unique extension to an operator $W \to V$ with the same norm whenever U is dense in W.*

Proof Let $A \in B(U, V)$. To $v \in W$ pick a sequence $\{v_i\}$ in U that converges to v. The Cauchy sequence $\{A(v_i)\}$ converges to an element $Av \in V$ which is independent of the sequence, so we have a linear map $v \mapsto Av$ that extends A. If it is bounded the norm cannot be less, and for any $\varepsilon > 0$ and large enough i, we have

$$\|A(v)\| \leq \varepsilon + \|A(v_i)\| \leq \varepsilon + \|A\| \cdot \|v_i\| \leq \varepsilon + \|A\|(\|v\| + \varepsilon),$$

so $\|A(v)\| \leq \|A\| \cdot \|v\|$. Uniqueness is clear. □

Isometries are obviously injective.

Proposition 2.1.20 *Every normed space is dense in a Banach space (called its completion) uniquely determined up to linear isomorphisms that are isometric, that is, preserve norms of vectors.*

Proof For the existence one mimics the construction of the reals from the rationals, considering all Cauchy sequences modulo null sequences.

If V and W are two completions of a normed space U with linear isometries $A\colon U \to V$ and $B\colon U \to W$ having dense images, the isometry $BA^{-1}\colon A(U) \to W$ has a unique extension $C\colon V \to W$ by the proof of the previous proposition, to an isometry such that $CA = B$, and which is surjective since its image is both closed and dense. □

The open mapping theorem, to be established in the next section, says that any bounded bijective operator between Banach spaces (over the real or complex numbers) has a bounded inverse.

Let W be a subspace of a linear space V. Let $q(v) = v + W = \{v + w \mid w \in W\}$ be the *coset of* W *in* V *that contains* $v \in V$. The *quotient space of* V *modulo* W is the family V/W of cosets of W in V regarded as a vector space with operations $q(u) + q(v) = q(u + v)$ and $aq(v) = q(av)$. These are well-defined operations in the sense that if $q(u') = q(u)$ and $q(v') = q(v)$, then $q(u') + q(v') = q(u) + q(v)$ and $aq(v') = aq(v)$ as $v' - v, u' - u \in W$. The zero element in V/W is the coset W, and clearly q is a linear surjective map $q: V \to V/W$ called the *quotient map*.

Proposition 2.1.21 *Let W be subspace of a normed linear space V, and let $q: V \to V/W$ be the quotient map. Then $\|q(v)\| = \inf_{w \in W} \|v - w\|$ is a seminorm on V/W. It is a norm if and only if W is closed in V, and in this case V/W is a Banach space whenever V is Banach.*

Proof For any $v_i \in V$ and $\varepsilon > 0$, there are $w_i \in W$ such that

$$\|q(v_1)\|+\|q(v_2)\|+\varepsilon \geq \|v_1-w_1\|+\|v_2-w_2\| \geq \|v_1+v_2-(w_1+w_2)\| \geq \|q(v_1+v_2)\|,$$

so q is a subadditive seminorm.

If $\|q(v)\| = 0$, there is a sequence $\{w_i\}$ in W such that $\|v - w_i\| \to 0$. Hence the seminorm on V/W is a norm if and only if W is closed in V.

If in addition V is Banach and we have a Cauchy sequence in V/W, then we can find a subsequence $\{q(v_n)\}$ such that $\|q(v_{n+1}) - q(v_n)\| < 1/2^n$, and by induction we can arrange so that $\|v_{n+1} - v_n\| < 1/2^n$. Letting $v = \lim v_n$, which exists since V is Banach, we see that $q(v_n) \to q(v)$ as q decreases norms of vectors. But then the original Cauchy sequence also converges to $q(v)$, since any Cauchy sequence with a convergent subsequence is obviously convergent. □

For any subspace W of a normed space V the quotient map $q: V \to V/W$ takes the unit ball onto the unit ball because if $\|q(v)\| < 1$, there is $w \in W$ such that $\|v - w\| < 1$ and $q(v - w) = q(v)$. So q takes open sets to open sets.

Proposition 2.1.22 *Let $A: V \to U$ be linear. Then $\tilde{A}: V/\ker A \to A(V)$ given by $v + \ker A \mapsto A(v)$ is a well-defined linear isomorphism. If U and V are normed spaces and $A \in B(V, U)$, then* $\ker A$ *is closed in V and $\|\tilde{A}\| = \|A\|$.*

Proof The first statement is easily checked. That $\ker A$ is closed in V is clear since A is continuous, so $V/\ker A$ is a normed space by the previous proposition.

Since

$$\|\tilde{A}(q(v))\| = \|A(v)\| = \|A(v - w)\| \leq \|A\| \cdot \|v - w\|$$

for every $w \in \ker A$, we see that $\|\tilde{A}(q(v))\| \leq \|A\| \cdot \|q(v)\|$. To prove the reverse inequality $\|A\| \leq \|\tilde{A}\|$, note that

$$\|A(v)\| = \|\tilde{A}(q(v))\| \leq \|\tilde{A}\| \cdot \|q(v)\| \leq \|\tilde{A}\| \cdot \|v\|.$$

□

We have the following converse result.

Proposition 2.1.23 *If W is a closed subspace of a normed space V and both V/W and W are Banach spaces, then V is Banach.*

Proof Let $q: V \to V/W$ be the quotient map. If $\{v_n\}$ is a Cauchy sequence in V, then there is $v \in V$ such that $q(v) = \lim q(v_n)$ as V/W is Banach. For each n we can find $w_n \in W$ such that $\|v_n - v - w_n\| < \|q(v_n - v)\| + 1/n$. Hence

$$\|w_n - w_m\| \leq \|q(v_n - v)\| + \frac{1}{n} + \|q(v_m - v)\| + \frac{1}{m} + \|v_n - v_m\|,$$

so $\{w_n\}$ is Cauchy in the Banach space W and converges to some $w \in W$. Then

$$\|v_n - (v + w)\| \leq \|w_n - w\| + \|v_n - v - w_n\| \leq \|w_n - w\| + \|q(v_n - v)\| + 1/n$$

shows that $v_n \to v + w$. □

This result together with an obvious induction argument lead to another proof of Proposition 2.1.11.

Suppose V be a vector space with a seminorm $\|\cdot\|$. Then $W = \{v \in V \mid \|v\| = 0\}$ is clearly a subspace of V. Let $q: V \to V/W$ be the quotient map. If $q(u) = q(v)$, then $|\|u\| - \|v\|| \leq \|u - v\| = 0$, so $\|q(v)\|' = \|v\|$ gives a well-defined norm $\|\cdot\|'$ on V/W called the *normification of* V.

If W is closed ideal of a Banach algebra V, then the norm on the Banach space V/W is clearly submultiplicative for the well-defined product $q(v)q(w) \equiv q(vw)$ rendering V/W a Banach algebra in a natural way.

We may form products of normed spaces.

Proposition 2.1.24 *Let $\{V_i\}_{i \in I}$ be a family of Banach spaces over the same field with an absolute value. Consider the canonical projections $\pi_i: \prod V_i \to V_i$, and let $p \in [0, \infty]$. By $\|v\|_p$ of $v \in \prod V_i$ we mean its p-norm as a function from I with the counting measure, and let W_p consist of those v with finite p-norm. Then W_p is the Banach space completion of $\oplus V_i$ when $p \in [0, \infty\rangle$, whereas for $p = \infty$, the Banach space completion of $\oplus V_i$ is the closed subspace of W_∞ consisting of those v with $\|\pi_i(v)\| < \varepsilon$ for all but finitely many i.*

Proof That all W_p are normed spaces is clear from the Minkowski inequality for sequences of numbers, see the appendix for a proof (this does not require that the normed spaces V_i are Banach). If $\{v_n\}$ is a Cauchy sequence in W_p, then $w_i \equiv \lim \pi_i(v_n) \in V_i$ as π_i is norm decreasing and V_i is Banach.

Let $p < \infty$. Then for any finite subset J of the index set, we have $\sum_J \|w_i\|^p = \lim \sum_J \|\pi_i(v_n)\|^p \leq \lim \|v_n\|_p^p$ which is finite as $\|v_n\|_p < \infty$ and $\{v_n\}$ is Cauchy. Hence we can define $w \in W_p$ such that $\pi_i(w) = w_i$ for all i. Also, as

$$\sum_J \|\pi_i(w) - \pi_i(v_n)\|^p = \lim \sum_J \|\pi_i(v_m - v_n)\|^p \leq \limsup \|v_m - v_n\|_p^p,$$

we see that $\|w - v_n\|_p \to 0$ as $n \to \infty$. So W_p is a Banach space and $\oplus V_i$ is clearly dense in W_p.

For $p = \infty$, the element $w \in \prod V_i$ such that $\pi_i(w) = w_i$ satisfies

$$\|\pi_i(w) - \pi_i(v_n)\| = \lim \|\pi_i(v_m - v_n)\| \leq \limsup \|v_m - v_n\|_\infty,$$

so $\|w - v_n\|_\infty \to 0$ and $w \in W_\infty$. If all $v_n \in \oplus V_i$, then clearly w satisfies $\|\pi_i(w)\| < \varepsilon$ for all but finitely many i. Such elements clearly form a closed subspace of W_∞. □

The Banach space W_∞ is the *normed direct product of the spaces* V_i, whereas the Banach subspace completion of $\oplus V_i$ is the *normed direct sum of the spaces* V_i.

Remark 2.1.25 When all $V_i = \mathbb{C}$ and the index set is countable, we get the usual l^p-sequences, which are L^p-spaces with respect to the counting measure, see the appendix. ◇

Remark 2.1.26 One can always *adjoin* an identity element of norm one to a normed algebra W over a field F. This can be seen as the counterpart of one-point compactification. Namely, consider the direct sum $W \oplus F$ Banach space with norm $\|(w + a)\| = \|w\| + |a|$ and product $(w, a)(v, b) = (wv + av + bw, ab)$, so W can be regarded as an ideal in it and $1 = (0, 1)$. In concrete cases normed algebras are often from the outset embedded in unital ones. ◇

2.2 Operators on Banach Spaces

We need the following version of *Baire's theorem.*

Lemma 2.2.1 *Any countable intersection of open dense subsets of a Banach space is dense.*

Proof Call the subsets A_n, and consider a ball B_0 of radius r in the Banach space. We are done if we can show that B_0 intersects $\cap A_n$, and indeed we will find an element w in this common intersection. Since A_1 is dense and open in the Banach space, the set $A_1 \cap B_0$ is open and non-empty, so we can find a ball B_1 of radius less than $2^{-1}r$ having closure within it. Proceeding this way by induction, we find balls B_n of radii less than $2^{-n}r$ with $\overline{B_n} \subset A_n \cap B_{n-1}$. The centers of the balls form a Cauchy sequence, and by completeness of the Banach space, this sequence has a limit w in the Banach space such that $\{w\} = \cap \overline{B}_n \subset B_0 \cap (\cap A_n)$. □

An *open map* between topological spaces takes open sets to open sets. An open linear map between normed spaces must be surjective since for any w not in its subspace image, the sequence $\{w/n\}$ intersects any ball at the origin.

Here is the *open mapping theorem.*

Theorem 2.2.2 *Any surjective bounded linear map between Banach spaces is open. In particular, every bounded linear bijection between Banach spaces has a bounded inverse.*

Proof Say $A: V \to W$ is a linear bounded surjection between Banach spaces. By linearity it suffices to show that if B is a neighborhood of 0 in V, then $A(B)$ contains a neighborhood of 0 in W. Denote the ball $B(0, 2^{-n}r)$ in V by B_n, where $r > 0$ is picked such that $B_0 \subset B$. As $W = \cup_{m=1}^{\infty} mA(B_{n+1})$, we see from the lemma that the closure of $A(B_{n+1})$ contains an open set, so $\overline{A(B_n)}$ contains a neighborhood of 0 since $B_{n+1} - B_{n+1} \subset B_n$.

To $w_1 \in \overline{A(B_1)}$, pick $v_n \in B_n$ and $w_n \in \overline{A(B_n)}$ such that $w_{n+1} = w_n - A(v_n)$. By what we have just said, this is always possible since for any $w_n \in \overline{A(B_n)}$, the sets $w_n - \overline{A(B_{n+1})}$ and $A(B_n)$ are never disjoint. By completeness of V the Cauchy sequence $\{v_1 + \cdots + v_n\}$ converges to some $v \in B_0$, and

$$A(v) = \lim \sum_{i=1}^{n} A(v_i) = \lim \sum_{i=1}^{n} (w_i - w_{i+1}) = w_1 - \lim w_{n+1} = w_1$$

by boundedness of A. So $\overline{A(B_1)} \subset A(B)$. □

The next two results is the *closed graph theorem* and the *principle of uniform boundedness.*

Corollary 2.2.3 *Any linear map $A: V \to W$ between Banach spaces is bounded if and only if it's* graph $G(A) = \{(v, w) \in V \times W \mid Av = w\}$ *is closed.*

Proof The forward implication is clear. For the opposite direction, note that $G(A)$ is a Banach space. The canonical projection $\pi_1: G(A) \to V$ is bijective and has norm not greater than one, so it's inverse is bounded by the open mapping theorem. The projection $\pi_2: G(A) \to W$ is also bounded, and $A = \pi_2\pi_1^{-1}$. □

Corollary 2.2.4 *Given a family of bounded linear maps $A_i: V \to W$ between Banach spaces such that $\{A_i(v)\}$ is bounded for each $v \in V$, then $\{\|A_i\|\}$ is bounded. In particular, if the family is a sequence and $A_i(v)$ converges for each v, so boundedness of $\{A_i(v)\}$ is automatic, then $A(v) = \lim A_i(v)$ defines a bounded linear map $A: V \to W$.*

Proof By assumption the formula $B(v) = \{A_i(v)\}$ defines a linear map from V to the normed direct product W_∞ of the Banach spaces $W_i = W$. Since $A_i = \pi_i B$ for the canonical projections $\pi_i \colon W_\infty \to W$, we need only show that the graph of B is closed. Given a sequence $\{v_n\}$ in V such that $v_n \to v$ and $B(v_n) \to w$, then $\pi_i B(v) = A_i(v) = \lim A_i(v_n) = \lim \pi_i B(v_n) = \pi_i(w)$, so $B(v) = w$. □

2.3 Linear Functionals

Definition 2.3.1 For a vector space V over a field F, the *dual* vector space V^* of V consists of all linear maps $V \to F$, or *linear functionals on* V. We regarded it as a vector space over F under pointwise operations. If F has an absolute value and V is normed, then the *normed dual* of V is the normed space $V^\star \equiv B(V, F)$.

Note that $V^\star$ is a Banach space when F is complete.

Proposition 2.3.2 *Let $\{v_i\}$ be a basis of a vector space V, and define elements $x_i \in V^*$ by $x_i(v_j) = \delta_{ij}$. Then $\{x_i\}$ is a basis for V^*, called the* dual basis *of $\{v_i\}$, if and only if V is finite dimensional, and then* $\dim V^* = \dim V$.

We have a linear injective map $V \to V^{**}$ given by $v \mapsto f_v$, where $f_v(x) = x(v)$ for $v \in V$ and $x \in V^*$. This map is a linear isomorphism if and only if V is finite dimensional.

Proof If $\dim V < \infty$, then any $x \in V^*$ can be written uniquely as a finite sum $x = \sum x(v_i)x_i$, which is checked by evaluating at each v_j.

If $\dim V = \infty$, then the element $x \in V^*$ which is 1 on every element v_i cannot be written as a finite linear combination of x_i's. So if $\{x_i\}$ is a basis for V^*, then V has to be finite dimensional, and as $\{x_i\}$ and $\{v_i\}$ have the same cardinality, we see that V and V^* have the same dimension.

Clearly f_v is linear on V^* by definition of the vector spaces operations on V^*. It is also evident that $v \mapsto f_v$ is linear by definition of the vector spaces operations on $V^{**} \equiv (V^*)^*$.

To see that $v \mapsto f_v$ is injective, say $x(v) = 0$ for all $x \in V^*$, and observe that any $v \in V$ can be written as $v = \sum x_i(v)v_i$.

If $\dim V < \infty$, then from what we have said, we know that $\dim V^{**} = \dim V$, and $v \mapsto f_v$ is also surjective.

Conversely, if $V \cong V^{**}$, then $\dim V = \dim V^{**}$, which implies that $\dim V = \dim V^*$ since we always have $\dim V \leq \dim V^*$ due to the non-canonical linear map $V \mapsto V^*$; $v_i \mapsto x_i$, which is injective as $\{x_i\}$ is linear independent. Hence this injection is also surjective and $\{x_i\}$ is a basis, so V is finite dimensional. □

For bounded functionals the situation is more delicate. Recall that an *extension of a map* from a set X to a superset $Y \supset X$ is any map from Y that coincides with the original map on X.

The following result is the *Hahn-Banach theorem* for real vector spaces. Further below we also include the more important complex case, then with a useful application that holds in both cases. Both theorems say that a linear functional can always be extended without increasing the norm.

Theorem 2.3.3 *Any bounded functional x on a subspace of a normed real vector space V has an extension $y \in V^\star$ with $\|y\| = \|x\|$.*

Proof Say x is defined on W and that $\|x\| = 1$. The family of all possible extensions (U, z) of x with $W \subset U \subset V$ and $|z(\cdot)| \leq \|\cdot\|$ is clearly a partially ordered set when $(U, z) \leq (U', z')$ means $U \subset U'$ and $z'|U = z$. It is also inductively ordered since to any totally ordered collection $\{(U_i, z_i)\}$ we have a majorant $(\cup U_i, z)$ with a well-defined map z given by $z(v) = z_i(v)$ for $v \in U_i$. By Zorn's lemma there is a maximal extension (V', y). If there is a non-zero $v \in V \backslash V'$, define y' on $V' + \mathbb{R}v$ by $y'(v' + cv) = y(v') + cb$, where b is any real number in the non-empty interval

$$[\sup_{u \in V'}\{y(u) - \|u - v\|\}, \inf_{w \in V'}\{-y(w) + \|w + v\|\}].$$

Then $|y'(\cdot)| \leq \|\cdot\|$, and we have a contradiction. □

If V is a complex vector space and y is the real part of $x \in V^*$, meaning $y(v) = \mathrm{Re}\,(x(v))$ for $v \in V$, then $x(v) = y(v) - i\,y(iv)$ as $c = \mathrm{Re}\,c - i\mathrm{Re}(ic)$ for $c \in \mathbb{C}$. Conversely, if y is any real linear functional on V, then the formula above defines a complex linear functional x on V. Hence if V is also normed, then $x \in V^\star$ if and only if its real part y is bounded on V, now considered as a real vector space. Moreover, every real linear bounded functional on V is the real part of a unique *complexified functional* $x \in V^\star$.

Theorem 2.3.4 *The statement of the previous theorem holds also for complex normed vector spaces. In particular, for every vector v in a real or complex normed vector space V, there is $x \in V^\star$ with norm one such that $x(v) = \|v\|$.*

Proof Say z is a bounded functional on a complex subspace of V. The real part x of z has by the theorem above an extension to a real linear functional y on V with $\|y\| = \|x\|$. By the previous remarks the complexification z' of y is an extension of z, and to every $v \in V$ there is $c \in \mathbb{C}$ with $|c| = 1$ such that

$$|z'(v)| = cz'(v) = y(cv) \leq \|y\|\,\|cv\| = \|x\|\,\|v\| \leq \|z\|\,\|v\|.$$

For the last statement, define z on the subspace spanned by v according to $z(v) = \|v\|$. The extension x to V has evidently the required properties. □

This means that the bounded linear functionals on a real or complex normed space V separate points in V. In particular, we have a canonical vector space inclusion $V \mapsto V^{\star\star}$ into the *bidual* given by $v(x) = x(v)$, which is moreover isometric, so V is a norm closed subspace of $V^{\star\star}$. When $V = V^{\star\star}$, it is called *reflexive*.

Remark 2.3.5 Given a measure μ and conjugate exponents q and $p \in [1, \infty\rangle$, see the appendix. Hölder's inequality $\|fg\|_1 \le \|f\|_p \|g\|_q$ shows that $[f] \mapsto \int f \cdot d\mu$ is a norm decreasing linear map from $L^q(\mu)$ to $L^p(\mu)^\star$. This map is an isometric isomorphism whenever μ is σ-finite, that is, when μ is defined on a countable union of measurable sets having finite measure. So in this case the L^p-spaces are reflexive.

Hilbert spaces (see later) are obviously reflexive, and the same is of course true for finite dimensional normed spaces. However, the normed space $L^\infty(\mu)$ is not reflexive when μ is the Lebesgue measure on $\mathbb{R}^n$. ◇

Corollary 2.3.6 *If W is a closed subspace of a real or complex normed space V and $v \in W^c$, there is $x \in V^\star$ of norm one such that $x(v) = \inf \|v - W\|$ and $W \subset \ker x$.*

Proof Let $y(w + av) = a \inf \|v - W\|$ for every $w \in W$ and scalar a. Then y is a linear functional on the span of W and v such that $y(v) = \inf \|v - W\|$ and $y(W) = \{0\}$. Also $|y(w + av)| \le |a| \|v - (-a^{-1}w)\| = \|w + av\|$ when $a \neq 0$, so $\|y\| \le 1$. As $\|y\| \, \|v - w\| \ge |y(v - w)| = \inf \|v - W\|$ we also get $\|y\| \ge 1$. Extend y by the Hahn-Banach theorem to $x \in V^\star$ without increasing the norm. □

If $A \in B(V, W)$ for normed spaces V and W, then we may define the *adjoint operator* $A^*\colon W^\star \to V^\star$ by $A^*(x)(v) = x(A(v))$ for $x \in W^\star$ and $v \in V$. Then $A \mapsto A^*$ is clearly a norm decreasing linear map $B(V, W) \to B(W^\star, V^\star)$, and if $B \in B(W, U)$, then $(BA)^* = A^* B^*$. Due to the Hahn-Banach theorem the map is isometric when V and W are real or complex vector spaces, and when they are also reflexive, the map is involutive.

Proposition 2.3.7 *Linear maps $A\colon V \to W$ and $B\colon W^\star \to V^\star$ between real or complex Banach spaces such that $x(A(v)) = B(x)(v)$ for all $x \in W^\star$ and $v \in V$ are automatically bounded and $A^* = B$.*

Proof Say $v_n \to v$ in V and $A(v_n) \to w$ in W. Then

$$x(A(v)) = B(x)(v) = \lim B(x)(v_n) = \lim x(A(v_n)) = x(w)$$

for $x \in W^\star$, and as $W^\star$ separates points in W, we get $A(v) = w$. By the closed graph theorem we conclude that A is bounded. The rest is obvious. □

2.4 Weak Topologies

It is not always convenient to treat normed spaces as such. Here is a more general and flexible notion.

Definition 2.4.1 A vector space V over a field F with an absolute value is a *topological vector space* if it is a topological group under addition such that scalar multiplication $F \times V \to V$ is a continuous map. A real or complex vector space is *locally convex* if every open set is a union of open convex sets.

Recall that a subset U of a real or complex vector space is convex whenever $av + (1-a)w \in U$ for $v, w \in U$ and $a \in [0, 1]$. For a real or complex topological vector space to be locally convex, by translation it suffices that every neighborhood of the origin contains a convex neighborhood.

Definition 2.4.2 A family S of seminorms on a vector space V over a field with an absolute value is *separating* if S vanishes identically only on zero. The *weak topology on V induced by S* is the initial topology induced by the functions $v \mapsto s(v-w)$ on V as (s, w) ranges over $S \times V$.

The weak topology on V induced by a separating family S seminorms on V is Hausdorff, and by the triangle inequality for seminorms, every neighborhood of $v \in V$ contains a neighborhood of the form

$$\cap\{w \in V \mid s(v-w) < r\},$$

where $r > 0$ and we intersect over finitely many $s \in S$. Moreover, the maps $(u, v) \mapsto s(u+v-w)$ and $(a, v) \mapsto s(av-w)$ are clearly weakly continuous, so V is a topological vector space over the field F. When F is $\mathbb{R}$ or $\mathbb{C}$, then V is also locally convex because $s(av+(1-a)w) \leq as(v) + (1-a)s(w)$ for $a \in [0, 1]$.

Lemma 2.4.3 *Let C be a convex neighborhood of the origin in a real or complex topological vector space V. Then $m(v) = \inf\{a > 0 \mid a^{-1}v \in C\}$ defines a function $m \colon V \to [0, \infty\rangle$ that is subadditive and satisfies $m(av) = am(v)$ for $v \in V$ and $a \geq 0$, and $C = m^{-1}([0, 1\rangle)$.*

Proof Now $(n^{-1}, v) \to (0, v)$ and $n^{-1}v \to 0v = 0$ as $n \to \infty$, so $n^{-1}v \in C$ eventually. Thus $m(v) < \infty$. For subadditivity, take $v, w \in V$ and $a, b > 0$ such that $a^{-1}v, b^{-1}w \in C$. Then $(a+b)^{-1}(v+w) = a(a+b)^{-1}(a^{-1}v)+b(a+b)^{-1}(b^{-1}w) \in C$ by convexity, so $m(v+w) \leq a+b$ and $m(v+w) \leq m(v)+m(w)$. Clearly also $m(av) = am(v)$. If $v \in C$, then $(1+r)v \in C$ for some $r > 0$ by openness, so $m(v) \leq (1+r)^{-1} < 1$. Conversely, if $m(v) < 1$, then $a^{-1}v \in C$ for $a \in [0, 1\rangle$, so $v = (1-a)0 + a(a^{-1}v) \in C$ by convexity since $0 \in C$. □

We say that a subset A of a vector space over a field F with an absolute value is *balanced* if $aA \subset A$ for all $a \in F$ with $|a| \leq 1$. Recall also that the *interior* X^o of a subset X in a topological space is the union of all open sets contained in X, so it is the largest open set in X.

Theorem 2.4.4 *Every neighborhood of the origin in a locally convex topological vector space contains a balanced convex neighborhood of the origin.*

Proof Suppose A is a convex neighborhood of the origin in a (real or complex) locally convex topological vector space. Since scalar multiplication is continuous, there is $\delta > 0$ and a neighborhood B of the origin such that $aB \subset A$ when $|a| < \delta$. The union C of all such sets aB is clearly a balanced neighborhood of the origin that is contained in A.

Let D be the intersection of all sets aA with $|a| = 1$. Since C is balanced, we have $C = aC$ for such a, so $C \subset aA$ and D^o is a neighborhood of the origin. Since D is an intersection of sets including A, we see that $D^o \subset A$, and since the sets being intersected are convex, so is D. But then D^o is also convex since $bD^o + (1-b)D^o$ is an open subset of D when $b \in \langle 0, 1\rangle$. It remains to show that D^o is balanced, and by the same reasoning we only need to show that D is balanced. Pick $c \in [0, 1]$ and d with $|d| = 1$. Then $cdD = \cap_{|a|=1} caA$ and $caA \subset aA$ as aA is a convex set that contains the origin. Hence $cdD \subset D$. □

Corollary 2.4.5 *The topology on any locally convex topological vector space is the weak one induced by some separating family of seminorms.*

Proof The function m from the lemma above which is associated to a balanced convex neighborhood of the origin is a seminorm due to balancedness. Consider the family S of all such seminorms. By the triangle identity these seminorms are continuous. So the weak topology induced by S is weaker than the original topology, but according to the lemma and the theorem above it cannot be strictly weaker. The family S is automatically separating as the topology is Hausdorff. □

Corollary 2.4.6 *Every topological finite dimensional real or complex vector space is normed.*

Proof Say V is a topological n-dimensional complex vector space. Pick a linear isomorphism $f : \mathbb{C}^n \to V$, so $f(v) = \sum v_i f(e_i)$ for $v = \sum v_i e_i$ in the standard basis $\{e_i\}$. This shows that f is continuous. Thus if S^n is the unit sphere in $\mathbb{C}^n$, then S^n is compact by the Heine-Borel theorem, so $f(S^n)$ is compact. And $0 \notin f(S^n)$ as f is injective. We can therefore find a neighborhood A of the origin disjoint from $f(S^n)$. In the first part of the proof of the theorem above we did not use convexity of A, so we may assume that A is balanced. Then $f^{-1}(A)$ is balanced and disjoint from S by linearity and injectivity of f, so $f^{-1}(A)$ is a connected neighborhood of the origin, and therefore must be contained in the open unit ball. Thus f^{-1} is an n-tuple of linear functionals, each bounded on a neighborhood of the origin, and is therefore continuous by arguments similar to those used for normed spaces. We can then use f to define a norm on V from the norm on $\mathbb{C}^n$ which induces the original topology on V. The case of a real field is proved the same way. □

Any non-trivial continuous functional x on a real or complex topological vector space V is open because x is a composition of the quotient map $V \to V/\ker x$ and a linear isomorphism from $V/\ker x$ to the field, and both these maps are open; the first simply by construction, and the second by the corollary above together with the fact the all norms on a finite dimensional space are equivalent.

When the seminorms are of the form $|x(\cdot)|$ for a family X of functionals x on V that separate points in V, we say the weak topology is *induced by* X rather than from the family of seminorms. On any topological vector space V we denote by $V^\star$ the *dual space*; this is the vector space consisting of all continuous linear functionals on V.

Definition 2.4.7 The *weak topology* on a topological vector space V is the one induced by $V^\star$, whenever $V^\star$ separates the points in V.

When V is normed with a non-trivial absolute value, then $V^\star$ consists of the bounded linear functionals. We will soon see that $V^\star$ is sufficiently rich when V is a locally convex topological vector space.

Lemma 2.4.8 *Consider functionals $x, x_1, \ldots, x_n$ on a vector space V over a field F. Then $x = \sum a_i x_i$ for some $a_i \in F$ if and only if $|x(\cdot)| \leq a \max|x_i(\cdot)|$ for some $a > 0$ if and only if $\cap \ker x_i \subset \ker x$.*

Proof The forward implications are obvious. Assuming the last statement, then $f = (x_1(\cdot), \ldots, x_n(\cdot))$ defines a linear map $f \colon V \to F^n$ with $\ker f \subset \ker x$. So there is a functional g on F^n such that $x = gf$. If $\{e_i\}$ is the standard basis for F^n, then $x(v) = g(\sum x_i(v)e_i) = \sum x_i(v)g(e_i)$, so the first statement holds. □

We interrupt the flow with a result needed later, called *Helly's theorem.*

Proposition 2.4.9 *Let x_i be finitely many bounded functionals on a real or complex normed space, let c_i be scalars and let $r > 0$. If there is M such that $|\sum a_i c_i| \leq M\|\sum a_i x_i\|$ for every linear combination $\sum a_i x_i$, then there is v with $\|v\| \leq M+r$ such that $x_i(v) = c_i$ for all i.*

Proof We can assume that at least one c_i is non-zero, and then some x_i is non-zero. After reordering we may assume that we have a maximal linear independent subcollection $x_1, \ldots, x_m$, so we have scalars a_{ij} such that $x_i = \sum_{j=1}^m a_{ij} x_j$ for all i. Suppose we had proved the proposition only for linear independent x_i's. Then there is v with $\|v\| \leq M + r$ such that $x_i(v) = c_i$ for $i \leq m$. But

$$|x_i(v) - c_i| = |\sum_{j \leq m} a_{ij} c_j - c_i| \leq M\| \sum_{j \leq m} a_{ij} x_j - x_i\| = 0$$

also for $i > m$. So we may assume that all the x_i's are linear independent.

By the lemma above we know that $\cap_{j \neq i} \ker x_j$ is not contained in $\ker x_i$, so there is w_j such that $x_i(w_j) = \delta_{ij}$. Hence there is w such that $x_i(w) = c_i$ for all i. Now w cannot belong to the kernels of all x_i, so by Corollary 2.3.6 there is a norm one functional x on the whole space such that $x(w) = \inf \|w - \cap \ker x_i\|$ and $\cap \ker x_i \subset \ker x$. By the lemma above there are therefore scalars b_i such that $x = \sum b_i x_i$. Hence

$$\inf \|w - \cap \ker x_i\| = x(w) = \sum b_i c_i \leq M\| \sum b_i x_i\| = M,$$

so there is u in $\cap \ker x_i$ such that $\|w - u\| \leq M + r$. Then $v = w - u$ will do. □

Proposition 2.4.10 *If V is a real or complex vector space with weak topology induced by a vector space X of functionals on V that separate points in V, then the weakly continuous functionals on V are exactly those in X, in other words $V^\star = X$.*

Proof The elements in X are certainly weakly continuous. Conversely, if x is a weakly continuous on V, there are finitely many $x_i \in X$ and $r > 0$ such that

$$\cap\{v \in V \mid |x_i(v)| < r\} \subset \{v \in V \mid |x(v)| < 1\}.$$

Hence $|x(v)| \leq r^{-1}\max|x_i(v)|$ for every $v \in V$, so x is a linear combination of the x_i's by the lemma, and thus belongs to the vector space X. □

Proposition 2.4.11 *If W is a subspace of a locally convex topological vector space V, then any continuous linear functional x on W with the relative topology has a continuous linear extension to V.*

Proof There are finitely many continuous seminorms s_i on V such that $|x(w)| \leq s(w) \equiv M \sup s_i(w)$ for all $w \in W$. Let f be the quotient map from V to the vector space $V/s^{-1}(0)$ with norm induced by s. Then $f(w) \mapsto x(w)$ well-defines a norm decreasing linear functional on $f(W)$. By the Hahn-Banach theorem it has a norm decreasing extension y' to $f(V)$, and $y = y'f$ satisfies $|y(v)| \leq s(v)$ for all $v \in V$, and thus is a continuous extension of x. □

The following result is known as the *Hahn-Banach separation theorem.*

Theorem 2.4.12 *Let A and B be non-empty disjoint convex subsets of a real or complex topological vector space V, and assume that A is open. Then there exist $y \in V^\star$ and $t \in \mathbb{R}$ such that* $\operatorname{Re} y(v) < t \leq \operatorname{Re} y(w)$ *for every $v \in A$ and $w \in B$.*

Proof Let us consider the real case first. Let u be the difference of an element in B and one in A, which exists as both sets are non-empty. Let m be the functional from Lemma 2.4.3 associated to the convex neighborhood $C = A - B + u$ of the origin. Note that u does not belong to this neighborhood since $A \cap B = \phi$, so $m(u) \geq 1$. Define a linear functional x on $\mathbb{R}u$ by $x(u) = 1$. Then $x \leq m$ on $\mathbb{R}u$.

It is easy to see that the proof of the Hahn-Banach theorem also works when the norm is replaced by m, so that x can be extended to a linear functional y on V in such a way that it is dominated by m on the whole space V. Now $y(v) \leq m(v) < 1$ for $v \in C$, so $|y| < r$ on the neighborhood $-rC \cap rC$ of the origin. Thus $y \in V^\star$. Also, for $v \in A$ and $w \in B$, we have $y(v)-y(w)+1 = y(v-w+u) \leq m(v-w+u) < 1$, so $y(v) < y(w)$. Now if $y(v) \in y(A)$, then $av \in A$ for some $a > 1$ by openness, so $ay(v) \in y(A)$. Hence $t = \sup y(A)$ will evidently do.

In the complex case, regarding first the spaces as real, we can by the first part of this proof find a real functional y such that $y(A) < t \leq y(B)$. Then the complex functional $z \in V^\star$ defined by $z(v) = y(v) - iy(iv)$ does the trick since $\operatorname{Re} z = y$. □

Since any two distinct points in a locally convex topological vector space V can be separated by convex neighborhoods, the theorem above certainly implies that $V^\star$ separates points in V.

Corollary 2.4.13 *Let A and B be non-empty disjoint convex subsets of a locally convex topological vector space with A compact and B closed. Then there is a continuous linear functional x such that* $\sup \mathrm{Re} x(A) < \inf \mathrm{Re} x(B)$.

Proof First note that $B - A$ is closed because if we have a sequence $\{v_n - w_n\}$ with $v_n \in B$ and $w_n \in A$ converging to u, then by going to a subsequence in the compact set A, we may assume that w_n converges to some $w \in A$. Then $v_n = (v_n - w_n) + w_n \to u + w \in B$ since B is closed, so $u = (u + w) - w \in B - A$. As $0 \notin B - A$ there is a convex neighborhood C of 0 such that $C \cap (B - A) = \phi$. Thus $A + C$ is an open convex set disjoint from the convex set B. The theorem above then gives a continuous linear functional x such that $\mathrm{Re} x(v) < \inf \mathrm{Re} x(B)$ for each $v \in A \subset A + C$. As the continuous function $\mathrm{Re} x$ attains its maximum on the compact set A, we are done. □

Proposition 2.4.14 *The weak closure of a convex subset of a locally convex topological vector space is closed in the original topology.*

Proof Let A be a weak closure of a convex subset of the space in question. For $v \notin A$ there is a convex neighborhood B of v that does not intersect A. By the Hahn-Banach separation theorem there is a bounded functional y such that $\mathrm{Re}\, y(v) < t \leq \mathrm{Re}\, y(A)$. Those w such that $\mathrm{Re}\, y(w) < t$ form a weak neighborhood of v disjoint from A. □

Definition 2.4.15 The *w^*-topology* on the dual $V^\star$ of a normed vector space V is the weak topology induced by the subspace V of the bidual $V^{\star\star}$.

Thus a net $\{x_i\}$ in the dual $V^\star$ converges to x in the w^*-topology if and only if $x_i(v) \to x(v)$ for all $v \in V$.

The following fundamental result is known as *Alaoglu's theorem.*

Theorem 2.4.16 *Let V be a real or complex normed space. Then the closed unit ball of $V^\star$ is w^*-compact.*

Proof We use here the notion of a universal net as defined in the appendix. Take a universal net $\{x_i\}$ in the closed unit ball $\overline{B}$ of $V^\star$, and let $v \in V$. Then $|x_i(v)| \leq \|v\|$ and the Heine-Borel theorem shows that the universal net $\{x_i(v)\}$ belongs to a compact set, and thus converges to a scalar $x(v)$. By continuity this defines an element $x \in \overline{B}$ such that $x_i \to x$ in the w^*-topology, so by the appendix $\overline{B}$ is compact. □

The theory of real or complex compact Banach spaces is somewhat limited, and the property of being locally compact is also fairly restrictive.

Proposition 2.4.17 *The closed unit ball in a real or complex normed space is compact only when the vector space is finite dimensional. In particular, any locally compact real or complex normed space is finite dimensional.*

Proof Let B be the unit ball in the vector space V. By compactness there are finitely many vectors v_i such that the open sets $\{v_i + \frac{1}{2}B\}$ cover $\overline{B}$. The span W of $\{v_i\}$ is finite dimensional, and thus closed. By construction

$$B \subset W + \frac{1}{2}B \subset W + \frac{1}{2}W + \frac{1}{4}B \subset W + \frac{1}{4}B \subset \cdots \subset \cap_{n=1}^{\infty}(W + 2^{-n}B) \subset \overline{W} = W,$$

so $V = \cup nB \subset W$. □

The converse result is of course true by the Heine-Borel theorem.

By definition the w^*-topology is always weaker than the norm topology, and the two results above show that it is strictly weaker for infinite dimensional real and complex normed spaces.

Proposition 2.4.18 *Every real or complex Banach space is isometrically isomorphic to a closed subspace of $C(X)$ for some compact Hausdorff space X.*

Proof By Alaoglu's theorem the closed unit ball X in the dual of a real or complex Banach space V is Hausdorff and w^*-compact. Moreover, the map $f: V \mapsto C(X)$ defined by $f(v)(x) = x(v)$ for $x \in X$ and $v \in V$ is an isometric isomorphism by the Hahn-Banach theorem. □

Recall that a net $\{v_i\}$ converges weakly to v in a normed space V if and only if $x(v_i) \to x(v)$ for all bounded linear functionals x on V. Of course the weak topology is weaker than the norm topology. Later we will see that the weak topology on a Hilbert space in a certain sense is a w^*-topology, so for infinite dimensional Hilbert spaces the weak topology is strictly weaker than the norm topology. Nevertheless the vector space of weakly continuous functionals on a real or complex normed space V coincides with the space of bounded ones.

Proposition 2.4.19 *Let X be a w^*-closed subspace in the dual $V^\star$ of a real or complex normed space V. For every $x \in X$ and $y \in V^\star \backslash X$ there is $v \in V$ such that $x(v) = 0$ and $y(v) \neq 0$.*

Proof There is a convex w^*-neighborhood of y disjoint from X. By the Hahn-Banach separation theorem there is therefore $v \in V$ and real t such that $\text{Re}\, y(v) < t \leq \text{Re}\, X(v)$. Since X is a vector space we get $X(v) = \{0\}$ and $t \leq 0$. □

Corollary 2.4.20 *Given a w^*-closed subspace X of the dual $V^\star$ of a real or complex normed space V. Then there is a norm closed subspace W of V such that X consists of those $x \in V^\star$ that vanish on W.*

Proof Let W consist of those vectors in V killed by the functionals in X. □

Definition 2.4.21 Let V be a normed space. The *annihilator* $X^\perp$ of $X \subset V^\star$ is the closed subspace $\cap_{x \in X} \ker x$ of V. The *annihilator* $Y^\perp$ of $Y \subset V$ is the w^*-closed subspace of $V^\star$ consisting of those $x \in V^\star$ that vanish on Y.

By the last proposition, we see that when V is a real or complex normed space, then $X^{\perp\perp}$ is the w^*-closure of the span of $X \subset V^\star$. Using the Hahn-Banach theorem we similarly see that $Y^{\perp\perp}$ is the closure of the span of $Y \subset V$.

Proposition 2.4.22 *Let V and W be real or complex Banach spaces. Then the adjoint $A^* \in B(W^\star, V^\star)$ of $A \in B(V, W)$ is w^*-to-w^* continuous. Conversely, every w^*-to-w^* continuous linear map $T: W^\star \to V^\star$ is bounded and is the adjoint A^* of some $A \in B(V, W)$.*

Proof Let $v \in V$. Then $vA^* = A(v) \in W$ is w^*-continuous, and so is A^*.

Also vT is a w^*-continuous functional on $W^\star$, so $vT \in W$ and we can define a linear map $A: V \to W$ by $A(v) = vT$. For $x \in W^\star$ we have $x(A(v)) = T(x)(v)$, so $T = A^*$ and both A and T are bounded by Proposition 2.3.7. □

Similarly, as $A^{**}(v)$ is w^*-continuous for any bounded operator $A: V \to W$ between real or complex normed spaces and $v \in V$, we see that the *biadjoint* A^{**} of A restricts to A on V. Note also that $\ker A^* = A(V)^\perp$ and $\ker A = A^*(W^\star)^\perp$.

Proposition 2.4.23 *Let $A: V \to W$ be a bounded operator between real or complex normed spaces. Then A^* is injective if and only if $A(V)$ is dense in W, whereas A is injective if and only if $A^*(W^\star)$ is w^*-dense in $V^\star$.*

Proof Just note that $(\ker A)^\perp = A^*(W^\star)^{\perp\perp}$ is the w^*-closure of $A^*(W^\star)$, and that $(\ker A^*)^\perp = A(V)^{\perp\perp}$ is the closure of $A(V)$. □

Proposition 2.4.24 *Let A be a bounded operator between real or complex Banach spaces. Then A is a an (isometric) bounded isomorphism if and only if A^* is an (isometric) bounded isomorphism.*

Proof The forward implication is clear as $(A^*)^{-1} = (A^{-1})^*$ when A is a bounded isomorphism, and $\|A^*(x)\| = \sup_{\|v\|=1} |xA(v)| = \|x\|$ when A is isometric.

Conversely, if $(A^*)^{-1}$ takes the closed unit ball into the closed ball of radius $r > 0$ with center 0, then $\|A(v)\| = \sup_{\|x\|=1} |A^*(x)(v)| \geq r^{-1}\|v\|$, which shows that A is injective and has as closed image the whole space by the previous proposition. Boundedness of A^{-1} is immediate from the open mapping theorem. When A^* is also isometric, then $r = 1$, so A is also isometric. □

Proposition 2.4.25 *Let W be a closed subspace of a real or complex normed space V. Then the adjoint of the quotient map $V \to V/W$ is an isometric isomorphism from $(V/W)^\star$ onto $W^\perp \subset V^\star$, which moreover is a w^*-to-w^* homeomorphism. While the adjoint of the inclusion map $W \to V$ can be identified with the quotient map $V^\star \to V^\star/W^\perp$.*

Proof Clearly the adjoint f^* of the quotient map $f: V \to V/W$ has image inside $W^\perp$. Since f maps the open unit ball of V onto that of V/W, we have $\|f^*(x)\| = \sup_{\|v\|<1} |xf(v)| = \|x\|$. Any $x \in W^\perp$ defines $y \in (V/W)^\star$ such that $yf = x$, so $\operatorname{im} f^* = W^\perp$. Now f^* is w^*-to-w^* continuous by Proposition 2.4.22, and that its inverse map respects w^*-convergent nets is obvious.

If $g: W \to V$ is the inclusion map, then $\ker g^* = W^\perp$, so there is a norm one map $h: V^\star/W^\perp \to W^\star$ yielding g^* when composed with the quotient map $V^\star \to V^\star/W^\perp$. By the Hahn-Banach theorem any $x \in W^\star$ can be extended to $y \in V^\star$ without increasing the norm, so h is surjective, and $\|h(y + W^\perp)| = \|x\| = \|y\| \geq \|y + W^\perp\|$ shows that h is also isometric. □

Theorem 2.4.26 *Let A be a bounded operator between real or complex Banach spaces. Then* im$A$ *is closed if and only if* imA^* *is closed if and only if it is w^*-closed.*

Proof Replacing $A: V \to W$ by the natural map $V/\ker A \to W$, we may by the previous proposition assume that A is injective. Replacing W by the closure of $A(V)$ we may further assume that im$A$ is dense in W. Then A^* is injective and $A^*(W^\star)$ is w^*-dense in $V^\star$ by Proposition 2.4.23.

Now the first forward implication is clear from Proposition 2.4.24, and so is the fact that im$A$ is closed whenever imA^* is w^*-closed.

Assume the latter set is only closed. Let $\{x_i\}$ be a net in both imA^* and the closed unital ball B of $V^\star$ that w^*-converges to $x \in V^\star$. Then $x \in B$, and by the open mapping theorem, the vectors $y_i = (A^*)^{-1}(x_i)$ form a bounded net, which by Alaoglu's theorem has a subnet w^*-convergent to some $y \in W^\star$. Then $A^*(y) = x$ by Proposition 2.4.22, so imA^* is w^*-closed by the Krein-Smulian theorem. □

The following result is clear from the theorem above and Proposition 2.4.23.

Corollary 2.4.27 *Let A be a bounded operator between real or complex Banach spaces. Then A is surjective if and only if A^* is a bounded isomorphism onto its image. The adjoint A^* is surjective if and only if A is a bounded isomorphism onto its image.*

2.5 Extreme Points

Recall that a subset E of a vector space is *convex* if $av + (1 - a)w \in E$ for any $v, w \in E$ and $a \in [0, 1]$. So a convex set contains the segment between any two vectors in it. Subspaces are obviously convex. For a subset to be convex it suffices that it contains the midpoint $(v + w)/2$ of any two members v and w.

Definition 2.5.1 A *face* of a convex subset A in a real or complex vector space is a non-empty convex subset B of A such that whenever $av + (1 - a)w \in B$ for $v, w \in A$ and $a \in \langle 0, 1\rangle$, then $v, w \in B$. An *extreme point* of A is a one-point face of A. The *extremal boundary* ∂A of A consists of all extreme points of A. The *convex hull* of a subset D in a real or complex vector space is the smallest convex set that contains D.

The convex hull of a subset D exists, being the intersection of all convex sets, including the whole vector space, that contains D. It is easy to see that it consists of all linear combinations of elements from D with non-negative coefficients that add up to one. Note that the face of a face of a convex set A is a face of A. Note also that an extreme point v of A is a point in A that cannot be written as a convex combination of two distinct points in A. In fact, it suffices to show that v is not a midpoint of two distinct points in A.

Example 2.5.2 A closed polygon in the plane has three types of faces; the polygon itself, its edges, and finally its vertices, which are then the extreme points. The convex hull of these points is the polygon. The closed unit disc has as faces the disc itself and every point on the boundary, there are no 1-dimensional faces here. Again the convex hull of the extreme points is the disc. The open disc has no extreme points, and only one face; the disc itself. ◇

The following important result is known as the *Krein-Milman theorem*.

Theorem 2.5.3 *The convex hull of the extremal boundary ∂A of a convex compact subset A in a locally convex topological vector space is dense in A.*

Proof Consider a closed face B of A. The family of closed faces of B, one of which is B itself, is inductively ordered under reverse inclusion since the intersection of all members of a totally ordered subfamily is a face, being non-empty as each of these members are compact, and this intersection is clearly a majorant of the subfamily. By Zorn's lemma B has therefore a face C that does not properly contain any smaller face.

Let x be a continuous functional on the vector space. Then the continuous function $v \mapsto \mathrm{Re}\, x(v)$ on the compact set C attains a minimum a. Hence the set $C_x \equiv \{v \in C \,|\, \mathrm{Re}\, x(v) = a\}$ is non-empty, so it is a face of C, and thus of B, and therefore $C_x = C$ by minimality of C. But the continuous functionals on the vector space separate points, so C cannot contain more than one point, which is then an extreme point of A. Thus $B \cap \partial A \neq \phi$.

The closure D of the convex hull of ∂A is clearly a convex subset of A. Suppose we have $v \in A \backslash D$. Then there is an open convex neighborhood of v which is disjoint from D. By the Hahn-Banach separation theorem there exist $x \in V^\star$ and $t \in \mathbb{R}$ such that $\mathrm{Re}\, x(v) < t \leq \mathrm{Re}\, x(D)$. Thus, if A_x denotes the set of elements in A for which $\mathrm{Re}\, x(\cdot)$ is minimal, then $A_x \cap D = \phi$, so $A_x \cap \partial A = \phi$. This contradicts the first part of the proof since A_x is clearly a closed face of A. So $D = A$. □

The following result is known as *Milman's theorem*, and concerns the location of extreme points.

Proposition 2.5.4 *Any closed set A of a locally convex topological vector space contains the extremal boundary of the closure B of the convex hull of A provided B is compact.*

Proof Let v be an extreme point of B. To show that $v \in A$ it suffices by closedness of A to show that $(v + C) \cap A \neq \phi$ for any neighborhood C of the origin. Since we may assume that the topology is weak, we can consider C of the form $\cap s^{-1}([0, r\rangle)$, where we intersect over finitely many seminorms s. Set $D = \frac{1}{2}C$. By closedness, and hence compactness, of A there are finitely many, say n, elements $v_i \in A$ such that the sets $v_i + D$ cover A. Thus the non-empty compact sets $A_i = (v_i + \overline{D}) \cap A$ cover A. The convex hull E of their union is actually compact. Indeed, it is the image of the set $\Delta \times \prod A_i$, where $\Delta \subset \mathbb{R}$ is the set of n non-negative real numbers that sum up to one. Now Δ is compact by the Heine-Borel theorem, so $\Delta \times \prod A_i$ is compact by Tychonoff's theorem, and E is the image of this set under the continuous map f given by $f(t_1, \ldots, t_n, w_1, \ldots, w_n) = \sum t_i w_i$. Now $A \subset E$ as each $A_i \subset E$, so $B \subset E$ by compactness and convexity of E. Thus $v \in E$, so we may write $v = \sum t_i w_i$ with the right-hand-side of the above form. Since all $w_i \in B$, and v is an extreme point of B, it must be one of the w_i's with $t_i > 0$. So $v \in A_i$ for some i. Hence $v - v_i \in \overline{D}$. Hence $(v - v_i + D) \cap D$ is non-empty, and if w is an element there, then $s(w) < r/2$ and $s(v - v_i - w) < r/2$, so $s(v - v_i) < r$ for the seminorms s defining C. Hence $v - v_i \in C$ and $(v + C) \cap A$ is non-empty. □

Say one wants to prove a property relating to a convex compact subset A of a real or complex vector space with weak topology induced by a separating set of functionals. If the property holds for the extreme points of A, and if the property is preserved under convex combinations and limits, then the Krein-Milman theorem tells us that it holds for the whole set A; the hope being then that the property is easier to prove for extreme points. With this strategy in mind we will supply examples both here and later on of extremal boundaries of convex compact sets, like closed unit balls in norm duals.

First observe that extreme points of any closed unit ball in a normed space are always unit vectors because if $\|v\| < 1$, there is $t \in \langle 0, 1\rangle$ such that $\|t^{-1}v\| \leq 1$ and $v = t(t^{-1}v) + (1 - t)0$.

Example 2.5.5 The closed unit ball in the Banach space $C(X)$ of continuous complex valued functions on a compact Hausdorff space X with pointwise operations and uniform norm is not compact when X is infinite. Even so, its extremal boundary consists of all functions f such that $|f| = 1$. Indeed, if $|f(x)| < 1$ for some x, then by continuity there is $r < 1$ and an open set A such that $|f| < r$ on A. So there is $g \in C(X)$ with support in A and such that $|g| \leq 1 - r$. Then $f = (f + g)/2 + (f - g)/2$ and both $f + g$ and $f - g$ belong to the closed unit ball. Hence the convex hull of the extremal boundary of the closed unit ball is uniformly dense in the ball. In contrast there are no extreme points of the unit ball in $C_0(X)$ when X is a non-compact locally compact Hausdorff space.

Also, if we instead considered real valued continuous functions on a connected compact Hausdorff space, the extremal boundary of the closed unit ball would consist of only two points, namely ± 1. ◇

We might not even have a single extreme point.

Example 2.5.6 If μ is the Lebesgue measure on a subset X of $\mathbb{R}^n$ with non-empty interior, then the closed unit ball in $L^1(\mu)$ has empty extremal boundary. Indeed, say $\|f\|_1 = 1$. Then there exists a measurable subset A of X such that $\int_A |f|\,d\mu \equiv a \in \langle 0, 1\rangle$. For instance, if $X = [0, 1] \subset \mathbb{R}$, then $x \mapsto \int_0^x |f(t)|\,dt$ is continuous with values 0 and 1 at 0 and 1, respectively, so there is $A = [0, x]$ such that $\int_A |f(t)|\,dt = 1/2$. For $X \subset \mathbb{R}^n$ reduce to a box and then use a similar argument for iterated integrals. Having such a set A, define measurable functions g and h on X to be $a^{-1}f$ on A and $(1 - a)^{-1}f$ on A^c, respectively, and otherwise zero. Then $f = ag + (1 - a)h$ and $\|g\|_1 = 1 = \|h\|_1$. By the Krein-Milman theorem the Banach space $L^1(\mu)$ thus cannot be the dual of another Banach space.

In $l^1(\mathbb{Z})$ the closed unit ball consists of all δ_n, where $\delta_n(m) = 1$ if $m = n$ and is otherwise zero. Indeed, if δ_n is the midpoint of two unit functions f_i in $l^1(\mathbb{Z})$, then $2 = f_1(n) + f_2(n)$, which forces $f_1 = f_2 = \delta_n$ as $\sum |f_i| = 1$. There are no other extreme points either because if f was non-zero for at least two distinct m and n, then there exists r such that $g_\pm$ defined to be $g_\pm(n) = f(n) \pm r$ and otherwise

$g_\pm = f$, would belong to the closed unit ball, and f would be their midpoint. Now $l^1(\mathbb{Z})$ is the dual of the Banach space $C_0(\mathbb{Z})$, which immediately implies that the convex hull of the extremal boundary of the closed unit ball in $l^1(\mathbb{Z})$ is w^*-dense in it, and here we see that it is dense even in norm. ◇

Example 2.5.7 If μ is any measure, the measurable functions of modulus one almost everywhere are clearly extreme points of the closed unit ball in $L^\infty(\mu)$, and there cannot be any other extreme points. Indeed, if f was, write $f = h|f|$ for a measurable h with $|h| = 1$. Then f is the midpoint of h and $2f - h$, and $\|2f - h\|_\infty = \|2|f| - 1\|_\infty \leq 1$ and $f \neq h$ on a set of positive measure. Recall also that $L^\infty(\mu)$ is the dual of the Banach space $L^1(\mu)$.

When μ is σ-finite, and $p \in \langle 1, \infty \rangle$, then the extremal boundary of the closed unit ball in the reflexive Banach spaces $L^p(\mu)$ consists of all the unit vectors, so here we have 'round' balls with no edges. ◇

Proposition 2.5.8 *The set $P(X)$ of unital functionals in the closed unit ball of the dual of the Banach space $C(X)$ is convex and w^*-compact. The extremal boundary of $P(X)$ consists of the Dirac measures δ_x.*

Proof Evaluating a functional at the constant function 1 is a w^*-continuous operation, so the convex set $P(X)$ is w^*-compact by Alaoglu's theorem.

We claim that any $\mu \in P(X)$ satisfies $|\mu(f)| \leq \mu(|f|)$ for $f \in C(X)$. Assume first that f is real, and write $\mu(f) = a + ib$ for $a, b \in \mathbb{R}$. Then

$$a^2 + b^2(1+n)^2 = |\mu(f + ibn)|^2 \leq \|f + ibn\|_u^2 = \|f\|_u^2 + b^2n^2$$

for each $n \in \mathbb{N}$, so $b = 0$. If $f \geq 0$, then $\|f\|_u - \mu(f) = \mu(\|f\|_u - f) \leq \|f\|_u$, so $\mu(f) \geq 0$. Hence $\operatorname{Re}\mu(f) = \mu(\operatorname{Re} f) \leq \mu(|f|)$ for any $f \in C(X)$. To prove the claim, we may assume that $\mu(f) \neq 0$. Then there is a complex number c of modulus one such that

$$|\mu(f)| = c\mu(f) = \mu(cf) = \operatorname{Re}\mu(cf) \leq \mu(\operatorname{Re}(cf)) \leq \mu(|cf|) = \mu(|f|),$$

which settles the claim.

If $\delta_x = (\mu + \nu)/2$ for $\mu, \nu \in P(X)$, then $|\mu(f)|/2 \leq \mu(|f|)/2 \leq \delta_x(|f|) = |f(x)|$, so $\ker \delta_x \subset \ker \mu$. But $f - f(x) \in \ker \delta_x$, so $\mu(f) = f(x)$ and $\mu = \delta_x$. Thus Dirac measures are extreme points in $P(X)$.

On the other hand, if μ is an extreme point in $P(X)$, then $d \equiv \mu(f) \in [0, 1]$ for $0 \leq f \leq 1$. If $d \in \langle 0, 1 \rangle$, then $\nu(g) = d^{-1}\mu(fg)$ and $\eta(g) = (1-d)^{-1}\mu((1-f)g)$ define unital functionals on $C(X)$, and $|\mu(fg)| \leq \mu(|fg|) \leq d\|g\|_u$, so $\|\nu\| \leq 1$, and similarly $\|\eta\| \leq 1$. Now $\mu = d\nu + (1-d)\eta$, so $\mu = \nu = \eta$ and $\mu(fg) = \mu(f)\mu(g)$. This also holds when $d = 0$ because $|\mu(fg)| \leq d\|g\|_u$ is true also in this case. Similarly $\mu(fg) = \mu(f)\mu(g)$ holds for $\mu(f) = 1$. But then it holds for any $f \in C(X)$ by linearity since f is a linear combination of positive functions with small norms. Finally, according to Example 5.1.10 characters on $C(X)$ are Dirac measures. □

Since the dual of the Banach space $C(X)$ is the space $M(X)$ of regular complex Borel measures on the compact Hausdorff space X, the result above tells us that every *probability measure* on X, that is, the positive elements in $M(X)$ that are one on X, can be approximated pointwise on $C(X)$ by measures with finite support.

2.6 Fixed Point Theorems

One of the simplest fixed point results is the following *contraction principle.*

Lemma 2.6.1 *A map T on a complete metric space has a unique fixed point if it is a contraction, that is, if there is $a < 1$ such that $d(T(x), T(y)) \leq ad(x, y)$ for all x and y.*

Proof Uniqueness holds since $d(x, y) \leq ad(x, y)$ is only possible if $d(x, y) = 0$.

Note that T is continuous. For any x_0 the sequence of elements $x_n \equiv T^n(x_0)$ is Cauchy since

$$d(x_n, x_m) \leq \sum_{k=m+1}^{n} d(x_k, x_{k-1}) \leq \sum_{k=m+1}^{n} a^{k-1} d(x_1, x_0) \leq a^m d(x_1, x_0)/(1-a)$$

for $n > m$. So it converges to some x, and $T(x) = \lim T(x_n) = \lim x_{n+1} = x$. □

Recall that a map from a convex set to a vector space is *affine* if it preserves convex combinations. Linear maps are obviously affine.

Proposition 2.6.2 *Let $f: A \to V$ be a continuous affine map from a compact convex subset of a locally convex topological vector space to another locally convex topological vector space. If v is an extreme point of the compact convex set $f(A)$, then there is an extreme point w of A such that $f(w) = v$. Moreover, if f maps A into $A \neq \phi$, then it has a fixed point; here f need not be continuous.*

Proof Pick an extreme point w of the compact convex set $f^{-1}(v)$. Then w is also an extreme point of A because if $w = (w_1 + w_2)/2$, then $v = (f(w_1) + f(w_2))/2$, so $v = f(w_i)$ since v is an extreme point of $f(A)$. But then $w = w_i$ as w is an extreme point of $f^{-1}(v)$.

For the last part, take any $v_0 \in A$. Then $v_n = \frac{1}{n} \sum_{i=0}^{n-1} f^i(v_0)$ form a sequence in A by convexity, which has a subnet converging to some point $v \in A$ by compactness. To show that v is a fixed point for f, it suffices to show that $xf(v) = x(v)$ for any $x \in V^\star$ due to the Hahn-Banach separation theorem. But this is now clear since the number $a_x = \sup |x(A)|$ is finite by compactness of A and continuity of x, and by affineness we get $|xf(v_n) - x(v_n)| = |x(\frac{1}{n} f^n(v_0) - \frac{1}{n} v_0)| \leq 2a_x/n$, which tends to 0 as $n \to \infty$. □

The previous fixed point result can be extended to a whole family of affine maps, resulting in the following *Markow-Kakutani fixed point theorem.*

Proposition 2.6.3 *Let A be a non-empty compact convex subset of a locally convex topological vector space. Then any family of continuous affine maps $A \to A$ has a common fixed point if the members of the family commute under composition.*

Proof Let $f^{(n)} = \frac{1}{n}\sum_{i=0}^{n-1} f^i : A \to A$ for f in the family F in the proposition. The collection of compact sets $f^{(n)}(A)$ as $n \in \mathbb{N}$ and $f \in F$ vary, has the finite intersection property by affineness and commutativity of the members of F, so there is some v which belongs to every $f^{(n)}(A)$, say $v = f^{(n)}(w)$ for $w \in A$ depending on f and n. By affineness of f we therefore get $f(v) - v = \frac{1}{n}(f^n(w) - w)$. Now $f^n(w) - w \in A - A$, which is compact and contains 0, and since the topology on a locally convex topological vector space is given by seminorms, there is $n \in \mathbb{N}$ to any neighborhood U of the origin such that $A - A \subset nU$. So $f(v) - v \in U$ for such an n, showing that $f(v) - v = 0$ by the Hausdorff property of the space. □

For a seminorm s on a vector space V, the *s-diameter* of a subset A of V is the extended number $\sup s(A - A)$.

Lemma 2.6.4 *Let A be a non-empty compact convex subset of a locally convex topological vector space with a continuous seminorm s. For $\varepsilon > 0$ there is a closed convex subset C of A different from A such that $A\backslash C$ has s-diameter less than ε.*

Proof Note that $S = s^{-1}([0, \varepsilon/4\rangle) + v$ is a neighborhood of $v \in \partial A$. Let A_1 and A_2 be the closed convex hulls of $\overline{\partial A}\backslash S$ and $\overline{\partial A} \cap S$, respectively, so $\partial A \subset A_1 \cup A_2$, and A is the convex hull of $A_1 \cup A_2$ by the Krein-Milman theorem; as noted in the proof of that theorem, we need not take closure here.

Let $r \in \langle 0, 1]$ and define a continuous map $f : A_1 \times A_2 \times [r, 1] \to A$ by $f(v_1, v_2, t) = tv_1 + (1 - t)v_2$, so its image C is compact. It is also convex as

$$\frac{1}{2}(tv_1 + (1-t)v_2) + \frac{1}{2}(t'v_1' + (1-t')v_2')$$
$$= \frac{t+t'}{2}(\frac{t}{t+t'}v_1 + \frac{t'}{t+t'}v_1') + (1 - \frac{t+t'}{2})(\frac{1-t}{2-t-t'}v_2 + \frac{1-t'}{2-t-t'}v_2').$$

If $C = A$ and $w \in \partial A$, then $w = tv_1 + (1 - t)v_2$ with $v_i \in A_i$ and $t \in [r, 1]$, so $w = v_1$ and $\partial A \subset A_1$, which means that $A = A_1$, but this is impossible as $\partial A \subset \overline{\partial A}\backslash S$ by Proposition 2.5.4, since this again implies the absurdity $S \cap \partial A = \phi$.

Hence there is $u \in A\backslash C$. By definition of C and since A is the convex hull of $A_1 \cup A_2$ we may write $u = tu_1 + (1 - t)u_2$ for $u_i \in A_i$ and $t \in [0, r\rangle$. Thus $s(u - u_2) = ts(u_1 - u_2) \le rd$, where d is the s-diameter of the compact set A. If also $u' \in A\backslash C$ with $u' = t'u_1' + (1 - t')u_2'$ for $u_i' \in A_i$ and $t' \in [0, r\rangle$, then

$$s(u - u') \le s(u - u_2) + s(u_2 - u_2') + s(u_2' - u') \le 2rd + \varepsilon/2$$

since $A_2 \subset \overline{S}$, which has s-diameter not greater than $\varepsilon/2$. Choose now any positive $r < \varepsilon/4d$ to get a $C \neq A$ with s-diameter less than ε. □

We are now in a position to prove the *Ryll-Nardzewski fixed point theorem.*

Theorem 2.6.5 *Let D be a non-empty compact convex subset of a locally convex topological vector space, and let X be a family of continuous affine maps $D \to D$ such that $0 \notin \{f(v) - f(w) \mid f \in X\}^-$ when $v \neq w$. Then X has a common fixed point.*

Proof We claim that every finite subset $f_1, \ldots, f_n$ of X has a fixed point. By the Markow-Kakutani theorem the map $h = (f_1 + \cdots f_n)/n : D \to D$ has a fixed point v. If v is not a fixed point of some f_k, say by renumbering, there is m such that $f_i(v) \neq v$ for $i \leq m$ and $f_i(v) = v$ for $i > m$. Then it is easy to see that v is a fixed point of $(f_1 + \cdots f_m)/m$. Thus we may assume that $f_i(v) \neq v$ for all i.

By assumption there is $\varepsilon > 0$ and a seminorm s such that $s(gf_i(v) - g(v)) > \varepsilon$ for all $g \in X$ and i. Let A be the closed convex hull of $X(v)$. By the lemma there is a closed convex subset C of A different from A such that $A \backslash C$ has s-diameter less than ε. Since C is convex and closed, there is $g \in X$ such that $g(v) \in A \backslash C$. Now

$$(gf_1(v) + \cdots + gf_n(v))/n = gh(v) = g(v) \in A \backslash C,$$

and since C is convex, we have $gf_i \in A \backslash C$ for some i. But then $s(gf_i(v) - g(v)) < \varepsilon$ violating the estimate above. This proves the claim.

Hence the set of fixed points under any finite subset of X is compact and convex and non-empty. The collection of such sets have clearly the finite intersection property, so the intersection of all of them is non-empty, and any element in this intersection is a fixed point for X. □

One might not only be interested in affine maps. The following *Brouwer fixed point theorem* really belongs to algebraic topology, and we need some results from homological algebra in the proof.

Proposition 2.6.6 *Any continuous function from the closed unit ball B^n in $\mathbb{R}^n$ to itself has a fixed point.*

Proof Let $f : B^n \to B^n$ be continuous. Let g be the continuous function from B^n to S^{n-1} that sends v to the point $g(v)$ that the segment from $f(v)$ through v hits. In order to always get a segment it is crucial that f has no fixed points. But the points of the sphere are fixed points of g, so if $h : S^{n-1} \to B^n$ is the inclusion map, we get $gh = \iota$, so we have homomorphisms $g_* : H_{n-1}(B^n, \mathbb{Z}) \to H_{n-1}(S^{n-1}, \mathbb{Z})$ and $h_* : H_{n-1}(S^{n-1}, \mathbb{Z}) \to H_{n-1}(B^n, \mathbb{Z})$ such that $g_* h_* = \iota$. This is impossible as $H_0(B^1, \mathbb{Z}) = \mathbb{Z}$ and $H_{n-1}(B^n, \mathbb{Z}) = \{0\}$ for $n \geq 2$, while $H_0(S^0, \mathbb{Z}) = \mathbb{Z} \oplus \mathbb{Z}$ and $H_{n-1}(S^{n-1}, \mathbb{Z}) = \mathbb{Z}$ for $n \geq 2$. □

The previous result remains true when the closed unit ball is replaced by a closed ball with center 0 and of any radius $r > 0$ because if f is a continuous map between such balls, then $r^{-1} f(r \cdot) : B^n \to B^n$ has a fixed point, say v, and then rv is a fixed point of f. We need not restrict to balls.

Corollary 2.6.7 *Any continuous function $f: A \to A$ for a non-empty compact convex subset A of a real or complex finite dimensional normed vector space has a fixed point.*

Proof We may assume that the vector space is real. By the Heine-Borel theorem we may pick a closed ball B with center 0 and of radius large enough for A to be contain in it. Let $g: B \to A$ be the continuous function which sends $v \in B$ to the unique vector $g(v) \in A$ with minimum distance to v, such a vector is attainable by compactness of A. Clearly g leaves each vector in A fixed. By the proposition above and the remark following it, the function $fg: B \to B$ has a fixed point v. But $v \in A$ as fg has range in A, so $g(v) = v$ and $f(v) = v$. □

The following extension of this result to infinite dimensional spaces is known as the *Schauder fixed point theorem.*

Corollary 2.6.8 *Any continuous map $f: A \to A$ for a non-empty compact convex subset A of a locally convex topological vector space has a fixed point.*

Proof Let s be a continuous seminorm on the vector space V in question. For each $n \in \mathbb{N}$ there is by compactness a finite subset A_n of A such that the open sets $\{v \in V \,|\, s(v - w) < 1/n\}$ cover A as w ranges over A_n. For w in A_n define a continuous function $g_w: A \to [0, 1/n]$ by $g_w(v) = 1/n - s(v - w)$ if $s(v - w) < 1/n$ and otherwise zero. Define a continuous function $g_n: A \to A$ by $g_n = \sum_{w \in A_n} w g_w / \sum_{w \in A_n} g_w$, which is well-defined as the denominator is a strictly positive function and A is convex. Note that

$$s(g_n(v) - v) \leq \sum_{w \in A_n} g_w(v) s(w - v) / \sum_{w \in A_n} g_w(v) < 1/n$$

as $s(w - v) < 1/n$ when $g_w(v) > 0$. Let $\text{span}A_n$ be the linear span of A_n. Then $A \cap \text{span}A_n$ is invariant under $g_n f$, and it is a compact convex subset of the finite dimensional vector space $\text{span}A_n$, which can be normed in such a way that $g_n f$ is continuous. By the corollary above there is $v_n \in A \cap \text{span}A_n$ fixed under $g_n f$.

By compactness of A the sequence $\{f(v_n)\}$ has a subnet $\{f(v_{n_i})\}$ converging to some $v \in A$. As

$$s(v_{n_i} - v) \leq s(g_{n_i} f(v_{n_i}) - f(v_{n_i})) + s(f(v_{n_i}) - v) < 1/n_i + s(f(v_{n_i}) - v)$$

the net $\{s(v_{n_i} - v)\}$ converges to zero. Thus $v_{n_i} \to v$ as the topology is the weak one induced by the seminorms s. So $f(v) = \lim f(v_{n_i}) = v$. □

We will need the following special case of the fixed point theorem.

Corollary 2.6.9 *Let E be a closed bounded convex subset of a real or complex normed space. Then any continuous map $f: E \to E$ has a fixed point provided its range has compact closure.*

Proof In the proof of the result above let s be the norm and let A be the closure of $f(E)$ to get g_n. Then $g_n f$ restricted to $E \cap \operatorname{span} A_n$ has a fixed point by the Heine-Borel theorem. The remaining part of the proof goes as before. □

2.7 The Eberlein-Krein-Smulian Theorems

The following result is known as the *Krein-Smulian theorem*, the converse is immediate from Alaoglu's theorem

Theorem 2.7.1 *Let $B^\star$ denote the closed unit ball in the dual of a complex Banach space V. Then a convex subset A of $V^\star$ is w^*-closed if the sets $A \cap rB^\star$ are w^*-compact for all $r > 0$.*

Proof The set A must be norm closed because any norm convergent sequence in A is contained in $rB^\star$ for some $r > 0$, and $A \cap rB^\star$ is certainly norm closed. (Since A is convex it is therefore also weakly closed, but since V might not be reflexive, we cannot at this stage conclude that it also is w^*-closed.)

In any case, since A is normed closed, if $x \notin A$, there is $r > 0$ such that $(A - x) \cap rB^\star = \phi$. Replacing A by the convex set $r^{-1}A - r^{-1}x$, we may assume that $A \cap B^\star = \phi$, and must then show that the origin does not belong to the w^*-closure of A.

For $v \in V$ let $P(v) = \{x \in V^\star \mid \operatorname{Re} x(v) \geq -1\}$, and extend this to any subset E of V by letting $P(E) = \cap_{v \in E} P(v)$, so $P(E)$ is a w^*-closed convex subset of $V^\star$.

Let $\overline{B}$ denote the closed unit ball in V. Clearly $r^{-1}B^\star \subset P(r\overline{B})$ as $|x(v)| \leq 1$ for $x \in B^\star$ and $v \in \overline{B}$. If $x \notin r^{-1}B^\star$, there is by the Hahn-Banach separation theorem, a vector $v \in V$ and $t \in \mathbb{R}$ such that

$$\operatorname{Re} x(v) < t \leq \operatorname{Re}(r^{-1}B^\star)(v) \leq -r^{-1}\|v\|.$$

Thus there is $v \in r\overline{B}$ such that $\operatorname{Re} x(v) < -1$, so $x \notin P(r\overline{B})$, and $r^{-1}B^\star = P(r\overline{B})$.

There exists a sequence of finite subsets E_n of V such that $E_1 = \{0\}$ and $E_n \subset (n-1)^{-1}\overline{B}$ and $A \cap nB^\star \cap P(E_1) \cap \cdots \cap P(E_n) = \phi$ for $n \geq 2$. Indeed, assuming this holds for n, the compact set $D = A \cap (n+1)B^\star \cap P(E_1) \cap \cdots \cap P(E_n)$ is disjoint from $nB^\star$, so $D \cap P(n^{-1}\overline{B}) = \phi$, so there is a finite set $E_{n+1} \subset n^{-1}\overline{B}$ such that $D \cap P(E_{n+1}) = \phi$, and the sequence exists by induction.

The elements in $\cup E_n$ can be ordered in a sequence $\{v_n\}_{n=1}^\infty$ that converges to 0 in norm. Then $T(x)(n) = x(v_n)$ defines a bounded operator $T\colon V^\star \to C_0(\mathbb{N})$. For $x \in A$ we have by construction that $\{x\} \cap mB^\star \cap P(\{v_n\}) = \phi$ for every m, and in particular for $m \geq \|x\|$, so $\inf \operatorname{Re} x(v_n) \leq -1$ and $\|T(x)\|_u \geq 1$. Thus the convex set $T(A)$ is disjoint from the open ball B in $C_0(\mathbb{N})$, and since the norm dual of $C_0(\mathbb{N})$ is $l^1(\mathbb{N})$, there is by the Hahn-Banach separation theorem an element $c = \{c_n\} \in l^1(\mathbb{N})$ and a scalar t such that $\operatorname{Re} c(B) < t \leq \operatorname{Re} c(T(A))$. The partial sums $s_k = \sum_{n=1}^k c_n v_n$ form a Cauchy sequence in the Banach space V, so they converge to an element $v \in V$. Then by normalizing $\|c\|_1 = 1$, we have

$$\operatorname{Re} x(v) = \sum \operatorname{Re} x(c_n v_n) = \operatorname{Re} c(T(x)) \geq 1$$

for every $x \in A$, so 0 is not in the w^*-closure of A. □

There are some immediate consequences.

Corollary 2.7.2 *A subspace A of the dual of a complex Banach space is w^*-closed if and only if $A \cap B^\star$ is w^*-closed.*

Corollary 2.7.3 *A linear functional on the dual of a complex Banach space V is w^*-continuous (and thus belongs to V) if and only if its restriction to the closed unit ball $B^\star$ is w^*-continuous.*

Proof In the corollary above let A be the kernel of the functional f in question. Then f is w^*-continuous if and only if A is w^*-closed if and only if $A \cap B^\star$ is w^*-closed if and only if f restricted $B^\star$ is w^*-continuous. □

If $\{v_n\}$ is a dense sequence in a real or complex Banach space V, then

$$d(x, y) = \sum_{n=1}^{\infty} 2^{-n} |x(v_n) - y(v_n)|$$

defines a metric on the closed unit ball $B^\star$ in $V^\star$ which induces the w^*-topology on $B^\star$. So the closed unit ball of the dual of a separable Banach space is metrizable in the w^*-topology. The converse is also true, namely if the w^*-topology on $B^\star$ is metrizable, then the Banach space V is separable. To see this pick a sequence of w^*-open sets A_n in $B^\star$ such that $\cap A_n = \{0\}$. For each n pick a finite subset V_n of V such that $x \in A_n$ whenever $x \in B^\star$ and $|x(V_n)| < 1$. Then $\cup V_n$ is dense in V.

Recall that a subset of a topological space is *sequentially closed* if every sequence in it has a limit in it. A subset of a topological space is *sequentially dense* if every point in the space is a limit of a sequence from the subset. The usual notions of closedness and denseness are characterized similarly using nets, so sequentially closedness is in general weaker than closedness, and sequentially denseness is in general stronger than denseness. Similarly, a map f between topological spaces is *sequentially continuous* if $\lim f(x_n) = f(x)$ for any sequence $\{x_n\}$ converging to x. This is weaker than ordinary continuity. In metric spaces sequences can clearly replace nets here.

Corollary 2.7.4 *A convex subset of the dual of a separable complex Banach space is w^*-closed if and only if it is sequentially w^*-closed. A linear functional on the dual of a separable complex Banach space is w^*-continuous if and only if it is sequentially w^*-continuous.*

Proof By separability we know that $rB^\star$ is w^*-metrizable for every $r > 0$, so if the convex subset A in question is sequentially w^*-closed, then $A \cap rB^\star$ is w^*-closed. Thus A is w^*-closed by the Krein-Smulian theorem.

If the functional f in question is sequentially w^*-continuous, then $\ker f$ is sequentially w^*-closed, so it is w^*-closed by the first part. Thus f is w^*-continuous. □

In a separable real or complex Banach space V there is dense sequence $\{v_n\}$ in the unit sphere $\{v \in V \; \|v\| = 1\}$. By the Hahn-Banach theorem there are $x_n \in V^\star$ of norm one such that $x_n(v_n) = 1$. Then $\|v\| = \sup_n |x_n(v)|$ for any $v \in V$, so $\{x_n\}$ is a countable family in $V^\star$ that separates points in V.

Lemma 2.7.5 *If there is a sequence of bounded functionals on a real or complex Banach space V that separates points in V, then any weakly compact subset of V is metrizable.*

Proof Let $\{x_n\}$ be such a sequence, removing possibly zero, let us assume they are normalized such that $\|x_n\| = 1$. Then $d(v, w) = \sum_{n=1}^{\infty} 2^{-n}|x_n(v - w)|$ is a metric on V. Suppose A is a weakly compact subset of V. Then for any $x \in V^\star$ the set $x(A)$ is compact and hence bounded by the Heine-Borel theorem. But then A is bounded by the principle of uniform boundedness since $V^\star$ is a Banach space. Hence the identity map from A with the weak topology to A with the metric topology is continuous, and therefore also has a continuous inverse map as A is weakly compact and the metric topology is Hausdorff. □

For any finite dimensional subspace W of the dual of a real or complex normed space, the unit sphere in W is compact by the Heine-Borel theorem. So there are finitely many norm one linear functionals x_i and unit vectors v_i with $|x_i(v_i)| > 3/4$ for all i, and $\|x - x_i\| \leq 1/4$ for any x and some i. Thus $\max_i |x(v_i)| \geq \|x\|/2$.

We can now prove the *Eberlein-Smulian theorem.*

Theorem 2.7.6 *The following statements are equivalent for a subset A of a real or complex Banach space:*

(i) *sequences in A have always weakly convergent subsequences;*
(ii) *sequences in A have always weak accumulation points;*
(iii) *the weak closure of A is weakly compact.*

These conditions would obviously be equivalent if the weak topology was metrizable.

Proof To show that (iii) implies (i), consider a sequence in the weakly compact set A. Since the norm closure W of the linear span of the sequence is weakly closed, the set $W \cap A$ is weakly compact in the separable Banach space W, so $W \cap A$ is metrizable by the lemma and the discussion prior to it. But then the sequence has a weakly convergent subsequence.

Assume (ii), and we prove (iii). For any bounded linear functional x, any sequence in $x(A)$ will clearly have an accumulation point, so $x(A)$ is compact and bounded. Thus A, viewed in the bidual, is bounded by the principle of uniform boundedness. Let f actually denote the canonical isometric isomorphism from the Banach space into its bidual. Then the w^*-closure $\overline{f(A)}$ of the bounded set $f(A)$ is w^*-compact by Alaoglu's theorem, so since the weak topology on the Banach space is the relative w^*-topology on $\mathrm{im}(f)$, it suffices to show that $\overline{f(A)}$ is contained in the range of f in order to conclude that A is weakly compact.

To this end we begin an inductive procedure. Let $w \in \overline{f(A)}$. To a functional x_1 of norm one, there is clearly $v_1 \in A$ such that $|(w - f(v_1)(x_1)| < 1$. By the discussion prior to this theorem, we may pick norm one linear functionals $x_2, \ldots, x_{n(2)}$ such that $\max_{i>1} |u(x_i)| \geq \|u\|/2$ for any u in the span of w and $w - f(v_1)$. Next pick $v_2 \in A$ such that $\max_i |(w - f(v_2)(x_i)| < 1/2$. Then find norm one functionals $x_{n(2)+1}, \ldots, x_{n(3)}$ such that $\max_{i>n(2)} |u(x_i)| \geq \|u\|/2$ for any u in the span of w and $w - f(v_1)$ and $w - f(v_2)$. Then pick $v_3 \in A$ such that $\max_i |(w - f(v_3)(x_i)| < 1/3$, and continue this way.

By hypothesis we may pick a weak accumulation point $v \in A$ of $\{v_n\}$. Now the normed closed linear span of the elements $w, w - f(v_1), w - f(v_2), \ldots$ is also weakly closed, so $w - f(v)$ belongs there and $\|w - f(v)\| \leq 2 \max_i |(w - f(v))(x_i)|$. But for any i pick k such that $n(k) > i$. Then

$$|(w - f(v))(x_i)| \leq |(w - f(v_m))(x_i)| + |f(v_m - v)(x_i)| < 1/k + |(v_m - v)(x_i)|$$

for $m > n(k)$, so $|(w - f(v))(x_i)| = 0$ as v is an accumulation point. Hence $w = f(v)$.

The remaining implication is trivial. □

The following result is now immediate.

Corollary 2.7.7 *A subset A of a real or complex Banach space is weakly compact if and only if $A \cap W$ is weakly compact for any separable closed subspace W.*

The following result is due to Krein-Smulian.

Corollary 2.7.8 *The weak closure of the convex hull of a weakly compact subset A of a real or complex Banach space is weakly compact.*

Proof Assume first that the Banach space V is separable. Consider A with the relative weak topology, and let $\nu \in M(A) \cong C(A)^\star$. Then the bounded linear functional f on $V^\star$ given by $f(x) = \int x \, d\nu$ is w^*-continuous. Indeed, by Corollary 2.7.4 it suffices to check sequentially w^*-continuity. So let $\{x_n\}$ be a sequence in $V^\star$ that w^*-converges to x. Then $\sup \|x_n\| < \infty$ by the principle of uniform boundedness, so $f(x_n) \to f(x)$ by Lebesgue's dominated convergence theorem. Hence there is a unique $v_\nu \in V$ such that $f(x) = x(v_\nu)$ for all $x \in V^\star$. Then $g(\nu) = v_\nu$ defines a linear map $g : C(A)^\star \to V$ that is clearly continuous with respect to the w^*-topology on $C(A)^\star$ and the weak topology on V. Thus $g(P(A))$ is convex and weakly compact by Proposition 2.5.8. But $g(\delta_v) = v$, so the weakly closed convex hull of A is contained in $g(P(A))$ and must be weakly compact.

If V is not separable, consider a sequence $\{v_n\}$ in the convex hull of A. Let W be the closure of the span of a countable set of elements in A expressing the elements v_n as convex combinations. Then $A \cap W$ is weakly compact in the separable Banach space W, and $\{v_n\}$ is contained in the weakly closed convex hull of it, which is weakly compact by the first part of the proof. By the theorem above the sequence $\{v_n\}$ has therefore a weakly convergent subsequence in the weakly closed convex hull of $A \cap W$. Applying the theorem once more we conclude that the weakly closed convex hull of A is weakly compact. □

The previous result was inspired by the following *Mazur's theorem.*

Proposition 2.7.9 *The closure of the convex hull of a compact subset A of a real or complex Banach space is compact.*

Proof By considering the closure of the linear span of A, we may assume that the Banach space V is separable. Since a norm compact set is weakly compact, the first paragraph of the proof above tells us that A is contained in the convex weakly compact set $g(P(A))$. So it suffices to show that the latter set is totally bounded. To $\varepsilon > 0$ there is a finite partition $\{A_i\}$ of A and $v_i \in V$ such that $\|v - s(v)\| < \varepsilon$ for every $v \in A$, where $s = \sum v_i \chi_{A_i}$. Define $h\colon M(A) \to V$ by $h(\nu) = \sum \nu(A_i) v_i$ for $\nu \in M(A)$. As $|x(g(\nu) - h(\nu))| = |\int x(\iota - s)\, d\nu| \le \|x\|\varepsilon$ for $\nu \in P(A)$ and $x \in V^\star$, we get $\|g(\nu) - h(\nu)\| < \varepsilon$. Since the bounded operator h has finite rank, the set $h(P(A))$ can be covered by finitely many balls of radius ε. By doubling these radii we get a covering of $g(P(A))$. □

We could also have proved directly that the closure of the convex hull of A is totally bounded.

The following result is known as the *Goldstine theorem.*

Proposition 2.7.10 *In a real or complex normed space V the closed unit ball is w^*-dense in the norm closed unit ball of the bidual $V^{\star\star}$.*

Proof The w^*-closure inside $V^{\star\star}$ of the closed unit ball in V is clearly contained in the closed unit ball in $V^{\star\star}$. If it is properly contained, pick w in the complement. The Hahn-Banach separation theorem for $V^{\star\star}$ with w^*-topology tells us that there is $x \in V^\star$ and $t \in \mathbb{R}$ such that $\operatorname{Re} w(x) < t \le \operatorname{Re} x(v)$ for all v in the closed unit ball of V. Since zero is there, we must have $t \le 0$, and we may assume that $t < 0$. Dividing by it and rescaling x, we get $\operatorname{Re} x(v) \le 1 < \operatorname{Re} w(x)$. Since we can multiply v by a scalar of modulus one and remain in the closed unit ball, we have $|x(v)| \le 1$, so $\|x\| \le 1$ and $1 < \operatorname{Re} w(x) \le |w(x)| \le 1$, which is absurd. □

Corollary 2.7.11 *A real or complex Banach space is reflexive if and only if the closed unit ball is weakly compact.*

Proof The forward implication is clear from Alaoglu's theorem, see also the proof of the Eberlein-Smulian theorem. In the converse direction, we observe that the unit ball, viewed in the bidual, is w^*-closed, and thus is the whole closed unit ball in the bidual by the proposition. □

Combining this corollary with the Eberlein-Smulian theorem we immediately get the following characterization of reflexivity.

Corollary 2.7.12 *A real or complex Banach space is reflexive if and only if every bounded sequence has a weakly convergent subsequence.*

2.8 Reflexivity and Functionals Attaining Extreme Values

Definition 2.8.1 We say a bounded linear functional x on a real or complex normed space *attains its norm* if $|x(v)| = \|x\|$ for some unit vector v.

For every unit vector v in a real or complex Banach space there is by the Hahn-Banach theorem, a bounded linear functional x such that $x(v) = 1 = \|x\|$. So on such spaces many functionals do attain their norm. We aim to show that they are dense in the dual Banach space.

Lemma 2.8.2 *Let x, y be two norm-one linear functionals on a real normed space and let $r > 0$. If $|y| \leq r/2$ on $\ker x$ intersected with the closed unit ball, then $\|x - y\| \leq r$ or $\|x + y\| \leq r$.*

Proof By the Hahn-Banach theorem there is a functional z that coincides with y on $\ker x$ and has $\|z\| \leq r/2$. Since $y - z$ vanishes on $\ker x$, there is a scalar a such that $y - z = ax$, so $\|y - ax\| \leq r/2$. If $a \geq 1$, then

$$\|y - x\| \leq \|(1 - a^{-1})y\| + \|a^{-1}(y - ax)\| \leq 1 - a^{-1} + r/2.$$

But $a = \|ax\| \leq 1 + \|y - ax\|$, so $a - 1 \leq a\|y - ax\|$ and $1 - a^{-1} \leq \|y - ax\|$. Thus $\|y - x\| \leq r$. If $a \in [0, 1\rangle$, then

$$\|y - x\| \leq \|y - ax\| + \|(1 - a)x\| \leq \|y - ax\| + \|y\| - \|ax\| \leq r.$$

When $a < 0$, replace f by $-f$ in the first part of the proof. □

The following result, or rather its corollary, is known as the *Bishop-Phelps theorem*.

Proposition 2.8.3 *The linear functionals that attain their norms on a real or complex Banach space are dense in the dual Banach space.*

Proof We must show that to $x \in V^\star$ and $r > 0$, there is $y \in V^\star$ such that $\|x - y\| \leq r$ and $|y(v)| = \|y\|$ for some unit vector $v \in V$. We may assume $\|x\| = 1$.

Consider the real case first. Let A be the convex hull in V of the union of the closed unit ball and $\overline{B_{2r^{-1}}(0)} \cap \ker x$. Suppose for the moment that there is $v \in \overline{A} \backslash A^o$ which also belongs to the closed unit ball. Since the interior A^o is both non-empty and convex, the Hahn-Banach separation theorem provides $y \in V^\star$ and $t \in \mathbb{R}$ with $y(A^o) < t \leq y(v)$. Hence $y(v) = \sup y(A)$. By rescaling y we may assume that it has norm one. Then $\|y\| \leq y(v) \leq \|y\|\|v\| \leq \|y\| = 1$. As $-A = A$ we get $\sup |y(A)| \leq 1$. In particular, for $w \in V$ with $x(w) = 0$ and $\|w\| \leq 1$, we know that $|y(2r^{-1}w)| \leq 1$, so replacing y by $-y$ if necessary, we are done by the lemma.

To show that there is indeed a $v \in \overline{A} \backslash A^o$ which is also in the closed unit ball, pick $u \in V$ in the closed unit ball such that $x(u) > 0$, and let $a = x(u)^{-1}(1+2r^{-1})$. We partially order the set of $w \in V$ in the closed unit ball such that $x(w) \geq x(u)$ by declaring that $w_2 \geq w_1$ if $x(w_2) \geq x(w_1)$ and $\|w_2 - w_1\| \leq ax(w_2 - w_1)$. If C is a chain, pick a sequence $\{w_n\}$ in it such that $x(w_n) \to \sup x(C)$. Then $\{w_n\}$

is Cauchy and converges to some element which is a majorant for C by continuity of x and the norm. By Zorn's lemma we have a maximal element $v \in V$, which by definition is in the closed unit ball, and hence in A.

Suppose $v \in A^o$. Pick $b > 0$ such that $v+bu \in A$. Since the closed unit ball in V and $B_{2r^{-1}}(0) \cap \ker x$ are convex sets, they have elements v_1 and v_2, respectively, such that $sv_1+(1-s)v_2 = v+bu$ for some $s \in [0, 1]$. Then $x(u) \leq x(v) < x(v+bu) = sx(v_1) \leq x(v_1)$, so v_1 belongs to the partially ordered set and $x(v) < x(v_1)$. Also

$$\|v_1-v\| = \|(1-s)(v_1-v_2)+bu\| \leq (1-s)(1+2r^{-1})+b \leq (1-s+b)(1+2r^{-1}),$$

while $x(v_1-v) = (1-s)x(v_1)+bx(u) > (1-s+b)x(u)$, so $\|v_1-v\| \leq ax(v_1-v)$ and $v_1 > v$, which contradicts maximality of v.

The complex case follows now from the real case by the usual decomposition into real and imaginary parts. □

By Alaoglu's theorem we see that for reflexive real or complex Banach spaces every bounded linear functional attains its norm since they are weakly continuous. Therefore we say that any real or complex normed space satisfying the conclusion of the Bishop-Phelps theorem is *subreflexive*. With this terminology every real or complex Banach space is subreflexive. However, not all spaces are subreflexive as the following example shows.

Example 2.8.4 Consider the vector space V of real polynomials on $[0, 1]$ with uniform norm. Since it is dense in the Banach space or real valued continuous functions on $[0, 1]$, we may identify $V^\star$ isometrically with the Banach space of regular real Borel measures on $[0, 1]$. Consider such a norm attaining member ν. Jordan decompose ν, so $\nu = \nu_+ - \nu_-$ and $|\nu| = \nu_+ + \nu_-$. If ν attains its norm at a polynomial f that is identically 1 or -1, then

$$|\nu_+([0, 1]) - \nu_-([0, 1])| = |\int f\, d\nu| = \|\nu\| = \nu_+([0, 1]) + \nu_-([0, 1]),$$

so ν is either ν_+ or ν_-.

If f is of norm one, but not constant, then the subset X of $[0, 1]$ where $|f| = 1$, is at least finite. If $|\nu|(X^c) > 0$, then $\|\nu\| \leq \int_{X^c} |f|\, d|\nu| + \int_X |f|\, d|\nu| < \|\nu\|$, which is absurd. So either ν is positive, negative or finitely supported.

Let m be the Lebesgue measure on $\mathbb{R}$. Then

$$\mu(A) = m(A \cap [0, 1/2]) - m(A \cap [1/2, 1])$$

defines a regular real Borel measure on $[0, 1]$ of norm one that it neither positive, negative or has finite support, so it does not attain its norm. Moreover, it is easy to check that $\|\mu - \nu\| \geq 1/2$, so the open ball of radius $1/2$ does not contain any norm attaining members of $V^\star$, which means that they are not dense in $V^\star$. ◇

A normed space need not be complete to be subreflexive.

Remark 2.8.5 Consider a dense subspaces W of a Hilbert space V, see the next chapter. Then the bounded linear functionals on V are of the form $(\cdot|v)$ with norm $\|v\|$ for $v \in V$. But then $|(w|v)| = \|v\|$ must hold for some unit vector $w \in W$ since otherwise $v \neq 0$ would be orthogonal to the closure of W. ◇

An easy adaptation of the proof of the proposition above gives the following result.

Corollary 2.8.6 *Let A be a non-empty bounded closed convex subset of a real or complex non-trivial Banach space V. Then the bounded linear functionals x on V such that* $\operatorname{Re} x(v) = \sup \operatorname{Re} x(A)$ *for some $v \in A$, are dense in $V^\star$.*

Any v such that $\operatorname{Re} x(v) = \sup \operatorname{Re} x(A)$ for some non-zero x is called a *support point* of a subset A of a real or complex normed space, and the bounded linear functional x is often referred to as the *support functional* for A *supporting* A at v. So the support functionals of A in the corollary are dense in the dual space. Any support point of a subset A in a normed space must be on the *boundary* $\overline{A} \backslash A^o$ of it, as is readily seen by the Hahn-Banach separation theorem.

Proposition 2.8.7 *The support points of a closed bounded convex subset A of a real or complex Banach space are dense in the boundary of A.*

Proof We may assume that the Banach space is real. To $v \in A \backslash A^o$ and $\varepsilon > 0$ there is $w \in A^c$ such that $\|w - v\| < \varepsilon/2$. By the Hahn-Banach separation theorem we may pick a linear functional x of norm one such that $x(w) > \sup x(A)$. Define a partial order on the vectors $u \in A$ satisfying $\|u - v\| \leq 2x(u - v)$ by declaring $u_2 \geq u_1$ if $\|u_2 - u_1\| \leq 2x(u_2 - u_1)$. As in the proof of Proposition 2.8.3 this partial order has a maximal element u. Using maximality it can be checked that x supports A at u and clearly $\|u - v\| \leq 2x(u - v) < 2x(w - v) < \varepsilon$. □

Definition 2.8.8 A bounded linear functional x on a real or complex normed space *attains its supremum* on a subset A of the space if $|x(v)| = \sup |x(A)|$ for some $v \in A$.

We have seen that the set of bounded functionals attaining their supremum on a bounded closed convex subset A of a real Banach space V is fairly rich. If A is weakly compact we get the entire dual Banach space $V^\star$, even when V is complex, since all bounded functionals are weakly continuous. We aim to prove the converse and need some preliminary results.

Lemma 2.8.9 *Suppose A is a non-empty subset of the closed unit ball in a real or complex normed space, and suppose we have a sequence of elements x_n in the closed unit ball of the dual space such that* $\sup |x(A)| \geq k \in \langle 0, 1\rangle$ *when x belongs to the convex hull of $\{x_n\}$. Then for any sequence of positive numbers b_n with sum one, there is $a \in [k, 1]$ and y_n in the convex hull of $\{x_j\}_{j \geq n}$ such that* $\sup |\sum_{j=1}^{\infty} b_j y_j(A)| = a$ *and* $\sup |\sum_{j=1}^{n} b_j y_j(A)| < a(1 - k\sum_{j=n+1}^{\infty} b_j)$ *for $n \in \mathbb{N}$.*

Proof Let s be the seminorm on the dual space given by $s(x) = \sup|x(A)|$. Pick a sequence of positive numbers r_n converging to zero in such a manner that $\sum_{n=1}^{\infty}(b_n r_n/(\sum_{j=n+1}^{\infty} b_j \sum_{j=n}^{\infty} b_j)) < 1 - k$. One can for instance let r_n be the denominator in the sum times half of $1 - k$.

We will inductively construct a sequence of numbers $a_n \geq \cdots \geq a_1 \geq k$ bounded by one, and a sequence of elements y_n in the convex hull of $\{x_j\}_{j \geq n}$ such that

$$s(\sum_{j=1}^{n-1} b_j y_j + (\sum_{j=n}^{\infty} b_j) y_n) < a_n(1 + r_n),$$

where a_n is the infimum of $s(\sum_{j=1}^{n-1} b_j y_j + (\sum_{j=n}^{\infty} b_j) y)$ as y vary over the convex hull of $\{x_j\}_{j \geq n}$. That we can find a_1 and y_1 as prescribed is clear from our assumptions. Assume that such required pairs have been found up till $n - 1$. For any y in the convex hull of $\{x_j\}_{j \geq n}$ we may then write

$$\sum_{j=1}^{n-2} b_j y_j + (\sum_{j=n-1}^{\infty} b_j)[(b_{n-1} y_{n-1} + \sum_{j=n}^{\infty} b_j y)/\sum_{j=n-1}^{\infty} b_j]$$

for the expression inside the seminorm s in the definition of a_n. The expression inside $[\cdots]$ is a convex combination of elements in the convex hull of $\{x_j\}_{j \geq n-1}$, so it belongs to this convex hull. So we are taking infinum over a smaller set of $y's$ than in the definition of a_{n-1}. Thus $a_n \geq a_{n-1}$, and there is y_n with the required properties to complete the induction step. Let $a = \lim a_n$, so $a \in [k, 1]$ and $a = s(\sum_{j=1}^{\infty} b_j y_j)$ as $a_n \leq s(\sum_{j=1}^{n-1} b_j y_j + (\sum_{j=n}^{\infty} b_j) y_n) < a_n(1 + r_n)$.

To prove the last property in the proposition, we get an upper bound of $s(\sum_{j=1}^{n} b_j y_j)$ in terms of $s(\sum_{j=1}^{n-1} b_j y_j)$ as follows:

$$\begin{aligned} s(\sum_{j=1}^{n} b_j y_j) &= s(b_n(\sum_{j=1}^{n-1} b_j y_j + (\sum_{j=n}^{\infty} b_j) y_n) + (\sum_{j=n+1}^{\infty} b_j) \sum_{j=1}^{n-1} b_j y_j)/\sum_{j=n}^{\infty} b_j \\ &< (b_n a_n(1 + r_n) + \sum_{j=n+1}^{\infty} b_j s(\sum_{j=n-1}^{\infty} b_j y_j))/\sum_{j=n}^{\infty} b_j \\ &= (\sum_{j=n+1}^{\infty} b_j)(\frac{b_n a_n(1 + r_n)}{\sum_{j=n+1}^{\infty} b_j \sum_{j=n}^{\infty} b_j} + \frac{s(\sum_{j=1}^{n-1} b_j y_j)}{\sum_{j=n}^{\infty} b_j}). \end{aligned}$$

Repeating this with n replaced by $n-1$, we get

$$\begin{aligned} s(\sum_{j=1}^{n} b_j y_j) &< (\sum_{j=n+1}^{\infty} b_j)\sum_{i=1}^{n}(\frac{b_i a_i(1+r_i)}{\sum_{j=i+1}^{\infty} b_j \sum_{j=i}^{\infty} b_j}) \\ &< a(\sum_{j=n+1}^{\infty} b_j)\sum_{i=1}^{n}(\frac{b_i(1+r_i)}{\sum_{j=i+1}^{\infty} b_j \sum_{j=i}^{\infty} b_j}) \\ &< a(\sum_{j=n+1}^{\infty} b_j)(\sum_{i=1}^{n}(\frac{1}{\sum_{j=i+1}^{\infty} b_j} - \frac{1}{\sum_{j=i}^{\infty} b_j}) + (1-k)), \end{aligned}$$

which equals $a(1 - k\sum_{j=n+1}^{\infty} b_j)$. □

Note that $y_n(v)$ above tends to zero for $v \in A$ when $x_n(v) \to 0$.

Lemma 2.8.10 *Let A be a non-empty separable, weakly closed subset of the closed unit ball of a real or complex Banach space. Then the following statements are equivalent:*

(i) *there is a bounded linear functional that does not attains its supremum on A;*
(ii) *the set A is not weakly compact;*
(iii) *there is a sequence $\{x_n\}$ as in the previous lemma with* $\lim x_n(v) = 0$ *for* $v \in A$.

Proof That (i) implies (ii) is clear. Assume (ii). Let V be the norm closed linear span of A, and consider the seminorm s on $V^\star$ given by $s(x) = \sup|x(A)|$. This is now a norm since a bounded linear functional that vanishes on A must also vanish on V. Denote the vector space $V^\star$ with the norm s by W, and define $f: A \to W^\star$ by $f(v)(x) = x(v)$. Then f has range in the closed unit ball of $W^\star$, and it is injective as $V^\star$ separates points in V. By the Hahn-Banach theorem a net $\{v_i\}$ in A converges weakly to $v \in V$ in the whole Banach space if and only if it converges weakly in V to v, and this happens if and only if $f(v_i) \to f(v)$ in the w^*-topology in $W^\star$. By Alaoglu's theorem $f(A)$ cannot be w^*-closed in $W^\star$, so there is g in the closure that does not belong to $f(A)$. We cannot have $v \in V$ such that $g(x) = x(v)$ for all $x \in W$, since otherwise v would be in the weak closure of A but not in A. So g cannot be zero and $s(g) > 0$, where s also denotes the dual norm on $W^\star$. Since A is contained in the closed unit ball, we see that g belongs to the bidual of V with norm $\|g\| \leq s(g)$. Viewing V in its bidual, pick $t \in \langle 0, r\rangle$, where $r = \inf\|g - V\|$. Pick a norm dense sequence $\{c_n\}$ in A. Then for $n \in \mathbb{N}$ and scalars d_i, we have

$$|d_1 t + \sum_{j=1}^{n} d_{j+1}0| \leq (t/r)\|d_1 g + \sum_{j=1}^{n} d_{j+1}c_j\|,$$

with the norm in the bidual of V. By Helly's theorem there are $z_n \in V^\star$ such that $\|z_n\| \leq (t/r) + (r-t)/2r < 1$ and $g(z_n) = t$ and $z_n(c_j) = 0$ for $j \leq n$. By the Hahn-Banach theorem we can extend z_n to x_n defined on the whole Banach space.

Clearly $\lim x_n(v) = 0$ for $v \in A$. And if x belongs to the convex hull of $\{x_n\}$, with restriction to V denoted by z, then $t = g(z) \leq s(g)s(z) = s(g)s(x)$. But we also have $0 < t = g(z_1) \leq \|g\|\|z_1\| < s(g)$, so $k \equiv t/s(g) \in \langle 0, 1\rangle$ and $\{x_n\}$ has the properties stated in (iii).

Assume (iii). Take any sequence of positive numbers b_n that sum up to one. Pick $\{y_n\}$ and a to $\{x_n\}$ and k and $\{b_n\}$ according to the previous lemma. Let $z = \sum_{j=1}^{\infty} b_j y_j$ and let $v \in A$. Choose $n \in \mathbb{N}$ with $|y_j(v)| < ak$ for $j > n$. Then

$$|z(v)| < \sup |\sum_{j=1}^{n} b_j y_j(A)| + ak \sum_{j=n+1}^{\infty} b_j < a(1 - k \sum_{j=n+1}^{\infty} b_j) + ak \sum_{j=n+1}^{\infty} = a,$$

so z does not attain its supremum on A. □

The following result is known as *James' theorem.*

Theorem 2.8.11 *A weakly closed subset A of a real or complex Banach space is weakly compact if and only if each bounded linear functional attains its supremum on A.*

Proof We will only give a proof for separable Banach spaces. The forward implication is anyway clear. For the opposite direction, note that the supremum of a bounded linear functional on any subset of a real or complex normed space clearly equals the supremum on the weakly closed convex hull of the subset. So we may assume that A is also convex. But it is also bounded because $\sup |x(A)| < \infty$ for each bounded linear functional x, and considering A inside the bidual under the canonical isometric isomorphism, we see that $\sup \|A\| < \infty$ by the principle of uniform boundedness. So we may assume that A is closed and convex and contained in the closed unit ball. Now the result is clear from the last lemma above. □

Having proved the theorem only in the separable case is not very restrictive due to Corollary 2.7.7. The reader should be warned that in the literature there are various faulty simplified proofs of James' result, so we stick to his own arguments. One immediate consequence of his theorem is the following striking result, some times referred to as *James' reflexivity theorem.*

Corollary 2.8.12 *A real or complex Banach space is reflexive if and only if every bounded linear functional attains its norm.*

Proof If all bounded linear functionals attain their norm, the closed unit ball is weakly compact by James' theorem, so we are done by Corollary 2.7.11. □

2.9 Compact Operators on Banach Spaces

Definition 2.9.1 A linear map $A\colon V \to W$ between Banach spaces over the same field is *compact* if the image $A(B)$ of the unit ball has compact closure. It has *finite rank* if $\dim A(V) < \infty$.

By the appendix we see that A is compact if and only if $A(B)$ is totally bounded. So when A is compact, then $A(V)$ is separable and $A \in B(V, W)$. If $A(V)$ is moreover closed, then A has finite rank since $A(V)$ is locally compact by the open mapping theorem. In particular, eigensubspaces of compact endomorphisms with non-zero eigenvalues are finite dimensional since the endomorphisms restrict to bijections on such closed subspaces. Denote the subspaces of $B(V, W)$ consisting of compact and finite rank operators by $B_0(V, W)$ and $B_f(V, W)$, respectively.

Proposition 2.9.2 *For Banach spaces V, W over the same field the linear space $B_0(V, W)$ is closed in $B(V, W)$.*

Proof If $A_n \to A \in B(V, W)$ with $A_n \in B_0(V, W)$, pick m to $\varepsilon > 0$ such that $\|A_m - A\| < \varepsilon/3$. Since $A_m(B)$ is totally bounded there are $v_1, \dots, v_n \in B$ such that for each $v \in B$, there is some i with $\|A_m(v) - A_m(v_i)\| < \varepsilon/3$. Then

$$\|A(v) - A(v_i)\| \leq \|A(v) - A_m(v)\| + \|A_m(v) - A_m(v_i)\| + \|A_m(v_i) - A(v_i)\| < \varepsilon.$$

□

Say V and W are real or complex Banach spaces. Then $B_f(V, W) \subset B_0(V, W)$ by the Heine-Borel theorem, so by the proposition above, the norm closure of $B_f(V, W)$ is a Banach subspace of $B_0(V, W)$. The image of a bounded set under $A \in B_0(V, W)$ has evidently compact closure, so compositions of A with linear maps to and from other Banach spaces produce compact operators. So $B_0(V) \equiv B_0(V, V)$ and $B_f(V)$ are ideals in $B(V)$, and $B_0(V) = B(V)$ if and only if I is compact if and only if $\dim V < \infty$ if and only if $B_0(V) = B_f(V)$.

Proposition 2.9.3 *A bounded operator A between real or complex Banach spaces is compact if and only if its adjoint A^* is compact.*

Proof Say $A \colon V \to W$ is compact. Any sequence of elements in the unit ball of $W^\star$ is equicontinuous, and by the Arzela-Ascoli theorem has a subsequence $\{y_n\}$ that converges uniformly on the compact set $\overline{A(V)}$. Hence $A^*(y_n) = y_n A$ form a Cauchy sequence that converges in the Banach space $V^\star$, so A^* is compact.

Conversely, if A^* is compact, then mapping V and W isometrically into their double duals, we see that $A(B)$ is a totally bounded subset inside the image of the unit ball in V^{**} under the operator A^{**}, which is compact by the first part of the proof. □

Given a bounded linear map $A \colon V \to W$ between real or complex Banach spaces. If the restriction of A^* to the closed unit ball $B^\star$ of $W^\star$ is continuous with respect to the relative w^*-topology on $B^\star$ and the norm topology on $V^\star$, then $\overline{A^*(B^\star)}$ is compact by Alaoglu's theorem, so A^* and A are compact.

Proposition 2.9.4 *Say $A \in \overline{B_f(V, W)}$ for real or complex Banach spaces V and W. Then A^* restricted to the closed unit ball $B^\star$ of $W^\star$ is continuous with respect to the relative w^*-topology on $B^\star$ and the norm topology on $V^\star$.*

Proof Say we have a net $\{x_i\}$ in $B^\star$ that converges to x in the w^*-topology. To $\varepsilon > 0$ pick $T \in B_f(V, W)$ with $\|A - T\| < \varepsilon/3$. Then

$$\|x_i A - xA\| \leq \|x_i T - xT\| + 2\varepsilon/3$$

and as $x_i T \to xT$ in the w^*-topology, and hence in norm on the finite dimensional space $T(V)$, we are done. □

This gives another proof that $\overline{B_f(V, W)}$ is a subspace of $B_0(V, W)$ for real or complex Banach spaces V and W.

Compact operators occur naturally in applications.

Example 2.9.5 Let X be a compact metric space with a Radon measure μ, and let $K \in C(X \times X)$. For $f \in C(X)$ and $x \in X$ define

$$A(f)(x) = \int K(x, \cdot) f \, d\mu.$$

As X is compact, the complex function K is uniformly continuous, so to $\varepsilon > 0$, there is $\delta > 0$ such that

$$|A(f)(x) - A(f)(y)| \leq \mu(X)\|K(x, \cdot) - K(y, \cdot)\|_u \cdot \|f\|_u < \varepsilon\mu(X) \cdot \|f\|_u$$

for $d(x, y) < \delta$. Thus $A(f) \in C(X)$. Let B be the unit ball in the Banach space $C(X)$ with respect to the uniform norm $\|\cdot\|_u$. The inequality above also shows that $A(B)$ is equicontinuous, and hence totally bounded by the Arzela-Ascoli theorem in the appendix since

$$\sup_{f \in B} |A(f)(x)| \leq \mu(X)\|K\|_u \cdot \|f\|_u.$$

So $A \in B(C(X))$ is a compact operator; an *integral operator* with *kernel* K. ◇

The next example displays a *Volterra integral operator*.

Example 2.9.6 For $f \in C([0, 1])$ define $A(f)(s) = \int_0^s f(t)\, dt$. As

$$|A(f)(s) - A(f)(r)| = |\int_r^s f(t)\, dt| \leq |s - r| \cdot \|f\|_u$$

we conclude again by the Arzela-Ascoli theorem that $A \in B(C([0, 1]))$ is compact. This operator has no eigenvalues because if $A(f) = \lambda f$, then $f(0) = 0$ and $\lambda f'(s) = f(s)$ with $f = 0$ as the only solution. ◇

We will push the analysis further in the context of real or complex Hilbert spaces.

2.10 Complemented and Invariant Subspaces

Definition 2.10.1 A closed subspace W of a real or complex normed space V is *complemented* in V if there is a closed subspace U of V such that $V = W \oplus U$, that is, if $V = W + U$ and $W \cap U = \{0\}$.

By extending a linear basis for an algebraic subspace of a vector space, we see that any subspace of a vector space is algebraically complemented, but we shall see that complementiveness does not always remain true in the non-algebraic sense.

Recall that an *idempotent* is an endomorphism E of a vector space such that $E^2 = I$. Then $I - E$ is also an idempotent and $\mathrm{im}E = \ker(I - E)$. Thus the image of a bounded idempotent E in a normed space V is closed and $V = \mathrm{im}E \oplus \ker E$, so the image is complemented in V. Conversely, if W is complemented in a real or complex Banach space V with $V = W \oplus U$ for U closed, then the canonical idempotent E with image W and kernel U is bounded by the closed graph theorem because if $v_n \to v$ and $E(v_n) \to w$ in V, then $v - w = \lim(I - E)(v_n) \in \ker E$, so $w = E(w) = E(v)$.

Proposition 2.10.2 *The closed subspace $C_0(\mathbb{N})$ of the Banach space $l^\infty(\mathbb{N})$ is not complemented in it.*

Proof If W was a closed subspace of $l^\infty(\mathbb{N})$ such that $l^\infty(\mathbb{N}) = W \oplus C_0(\mathbb{N})$, the quotient Banach space $l^\infty(\mathbb{N})/C_0(\mathbb{N}) = W$ would have a countable separating family in $W^\star$ since $\{\delta_n\}$ is such a family for $l^\infty(\mathbb{N})$.

Identify $\mathbb{N}$ with $\mathbb{Q}$. To each $a \in \mathbb{R}\backslash\mathbb{Q}$ pick by the axiom a choice a sequence of rational numbers that converges to a, and let $\chi_a \in l^\infty(\mathbb{N})$ be the characteristic function of that sequence. For distinct irrational numbers $a_1, \ldots, a_m$ and $n \in \mathbb{N}$ the Hahn-Banach theorem gives $x \in (l^\infty(\mathbb{N})/C_0(\mathbb{N}))^\star$ with $|x(\chi_{a_i} + C_0(\mathbb{N}))| \geq 1/n$. Pick scalars b_i of modulus one such that $b_i x(\chi_{a_i} + C_0(\mathbb{N})) = |x(\chi_{a_i} + C_0(\mathbb{N}))|$. Then $\sum b_i \chi_{a_i} \in l^\infty(\mathbb{N})$ has absolute value one on infinitely many members of $\mathbb{N}$ and exceeds that value for only finitely many members since sequences converging to different irrational numbers have only finitely many elements in common. Hence $\|x\| \geq |x(\sum b_i \chi_{a_i} + C_0(\mathbb{N}))| \geq m/n$, which shows that there can be only countably many irrational numbers a such that $x(\chi_a + C_0(\mathbb{N})) \neq 0$. Thus no countable family in $(l^\infty(\mathbb{N})/C_0(\mathbb{N}))^\star$ can separate the uncountable collection $\{\chi_a + C_0(\mathbb{N})\}$, and this contradicts the first paragraph. □

Corollary 2.10.3 *The canonical image of $C_0(\mathbb{N})$ in its bidual is not complemented there.*

Proof If it was, pick a bounded idempotent there with image $C_0(\mathbb{N})$. Invoking the isometric isomorphism $l^\infty(\mathbb{N}) \to C_0(\mathbb{N})^{\star\star}$, we get a bounded idempotent $l^\infty(\mathbb{N}) \to l^\infty(\mathbb{N})$ with image $C_0(\mathbb{N})$ contradicting the proposition. □

The next proposition then shows that $C_0(\mathbb{N})$ cannot be isomorphic to the dual of any normed space.

First notice that if $A: V \to W$ is a bounded map between two real or complex normed spaces and $v \in V$, then $A^{**}(v)$ is w^*-continuous and thus belongs to W. Moreover, the restriction of A^{**} to V is A.

Proposition 2.10.4 *If V is a real or complex normed space isomorphic to the dual of another normed space, then V canonically embedded in its bidual, is complemented in V^{**}.*

Proof Say we have an isomorphism $T: V \to W^*$ for some space W. Suppose $A: W \to W^{**}$ and $B: W^* \to W^{***}$ are the canonical embeddings, then $E = BA^*$ is a bounded idempotent onto W^*. By the remark prior to the proposition, the map $(T^{**})^{-1}ET^{**}$ is therefore a bounded idempotent with range V. □

Definition 2.10.5 An *invariant* subspace for a subset X of endomorphisms of a vector space is a subspace W such that $X(W) \subset W$. The subspace is *proper* if it is neither $\{0\}$ nor the whole space. A subspace is *hyperinvariant* for X if it is invariant for every operator in the *commutant* X' of X, that is, all bounded operators that commute with every element of X.

Note that a hyperinvariant subspace for a single operator is invariant for that operator.

Lemma 2.10.6 *Suppose X is a unital algebra of bounded operators on a real or complex normed space that has no proper closed invariant subspaces. For any non-zero compact operator A, there is $B \in X$ such that the kernel of $BA - I$ is non-trivial.*

Proof We may assume that $A: V \to V$ has norm one. Pick $v \in V$ such that $\|A(v)\| > 1$. Let B_v be the closed ball in V of radius one and center v. Then $0 \notin B_v$. And if $0 = \lim A(v_n)$ with $v_n \in B_v$, then $\|0 - A(v)\| = \lim \|A(v_n - v)\| \leq 1$, which is absurd, so 0 cannot belong to the closure of $A(B_v)$ either. For any non-zero $w \in V$ we have $\overline{X(w)} = V$ by hypothesis, so for every w in the closure of $A(B_v)$ there is $T \in X$ such that $\|T(w) - v\| < 1$. By compactness of A, there are finitely many $T_i \in X$ such that $\overline{A(B_v)} \subset \cup_i \{w \in V \mid \|T_i(w) - v\| < 1\}$. Define scalar valued functions f_i on $\overline{A(B_v)}$ by $f_i = \sup\{0, 1 - \|T_i(\cdot) - v\|\}$, so $\sum f_i > 0$ and $g_i = f_i / \sum f_i$ makes sense, and we get a continuous function $f: B_v \to V$ given by $f(w) = \sum g_i A(w) T_i A(w)$. In fact, we have $f(B_v) \subset B_v$ because B_v is convex and $\sum g_i A = 1$ on B_v and $T_i A(w) \in B_v$ for $w \in B_v$ when $g_i A(w) > 0$ as $f_i A(w) > 0$ then. Each operator $T_i A$ is compact, so the closure of the convex hull of all $T_i A(B_v)$ is compact by Mazur's theorem. Thus the closure of $f(B_v)$ is compact, so f has a fixed point $w \in B_v$ by Schauder's fixed point theorem. Then $B = \sum g_i A(w) T_i$ has the required properties. □

We can then prove *Lomonosov's theorem*.

Theorem 2.10.7 *Any bounded operator on a non-trivial complex Banach space that is not a multiple of the identity map and commutes with some non-zero compact operator has a proper closed hyperinvariant subspace.*

Proof Let X be the commutant of such an operator T, and let $A \in X$ be a non-zero compact operator. If X has no proper closed subspace, the lemma provides $B \in X$ such that $BA - I$ has non-trivial kernel W. Now BA is a compact operator that restricts to the identity map on W, so $\dim W < \infty$. Since T commutes with BA we see that $T(W) \subset W$, so T has an eigenvector in W, and since T is not a multiple of the identity map, we are done. □

Thus any compact operator on a non-trivial complex Banach has a proper closed invariant subspace. Moreover, any two commuting compact operators have a common proper closed invariant subspace.

Corollary 2.10.8 *Any bounded operator T on an infinite dimensional complex Banach space has a proper closed invariant subspace if $f(T)$ is compact for some polynomial f.*

Proof If $f(T) \neq 0$, the theorem applies to T. If $f(T) = 0$, then for $v \neq 0$, the linear span of $\{v, T(v), \ldots, T^n(v)\}$, where $n + 1$ is the degree of f, is clearly a proper closed invariant subspace for T. □

The following simple application of the Markow-Kakutani fixed point theorem can be seen as an invariant Hahn-Banach theorem.

Proposition 2.10.9 *Let X be a commuting family of bounded operators on a real or complex normed space V leaving a subspace W invariant. Then any $x \in W^\star$ such that $xT = x$ for all $T \in X$, has an extension $y \in V^\star$ with the same norm such that $yT = y$ for all $T \in X$.*

Proof Let A consist of all norm non-increasing extensions of x to V. By the Hahn-Banach theorem the set A is non-empty, and it is clearly convex. It is also w^*-compact by Alaoglu's theorem. Every $T \in X$ defines an affine map $y \mapsto yT$ from A to A which is clearly continuous with respect to the w^*-topology on A. The desired y is then given by the Markow-Kakutani fixed point theorem. □

2.11 An Approximation Property

Definition 2.11.1 A Banach space W has the *approximation property* if for every Banach space V, the set $B_f(V, W)$ is dense in $B_0(V, W)$.

We aim to show that this property is equivalent to a more intrinsic statement by Grothendieck.

Lemma 2.11.2 *Let X be a subset with compact closure of a real or complex Banach space V, and let W be the linear span of the closed convex hull B of all aX with $|a| \leq 1$. Then B is compact and contains X, and W is a Banach space for a norm m for which B is the closed unit ball, and the inclusion map from W with m-norm to V is a compact bounded operator.*

Proof That B is compact is clear from Mazur's theorem as B is the closed convex hull of the continuous image of $D \times \overline{X}$ in V given by scalar multiplication, where D is the closed unit disc. Clearly $X \subset B$.

It is easy to check that $m(w) = \inf\{t \in \langle 0, \infty \rangle \mid w \in tB\}$ defines a seminorm on W, see Lemma 2.4.3 for details. It is actually a norm since if $m(w) = 0$ for a non-zero w, there would be a sequence t_n converging to zero such that $t_n^{-1} w \in B$ contradicting boundedness of B in V. Clearly B is the closed unit ball for m, and the inclusion map from W with m-norm to V is therefore bounded and compact.

Finally, consider a non-convergent Cauchy sequence $\{v_n\}$ in B with respect to m. Then it will converge to some $v \in B$ with respect to the norm on V, so $w_n = v_n - v$ will form a non-convergent Cauchy sequence in B with respect to m, and $\|w_n\| \to 0$. So it has a subsequence of elements w_{n_i} with m-norm greater than some $r > 0$. Then $u_i = w_{n_i}/m(w_{n_i})$ is Cauchy with respect to m and $\|u_i\| \to 0$. For large enough i and j we have $m(u_i - u_j) < 1/2$, so $2(u_i - u_j) \in B$. Since $\|u_j\| \to 0$ and B is closed in V, we see that $2u_i \in B$, so $m(u_i) \leq 1/2$, which is impossible. □

Lemma 2.11.3 *Suppose $\{v_n\}$ is a sequence in a real or complex Banach space that converges to zero. Let D be the number one or a closed disc with center $\{0\}$. Then $A = \{\sum t_n a_n v_n \mid a_n \in D,\ t_n \geq 0,\ \sum t_n \leq 1\}$ is compact and coincides with the closed convex hull of $\cup_n D v_n$.*

Proof Notice that the sums in A are absolutely convergent in the Banach space, so A makes sense. Since all $a v_n \in A$ and A is contained in the closed convex hull of them, it suffices to show that A is closed and convex; compactness goes as in the proof of the previous lemma by using that $\{v_n\} \cup \{0\}$ is compact.

Convexity is clear as $|t t_n a_n + (1-t) s_n b_n| \leq (t t_n + (1-t) s_n) \sup_{a \in D} |a|$ for $t, t_n, s_n \in [0,1]$ and $a_n, b_n \in D$. As for closedness, consider a sequence $\{\sum t_{in} a_{in} v_n\}_i$ in A that converges in the Banach space. Now the sequence $\{\prod_n (t_{in}, a_{in})\}_i$ in the compact set $([0,1] \times D)^{\mathbb{N}}$ has a subsequence converging to some $\prod_n (t_n, a_n)$. Then $\sum t_n = \lim_i \sum t_{in} \leq 1$ by Lebesgue's dominated convergence theorem, and the vectors $\sum t_{in} a_{in} v_n$ converge to $\sum t_n a_n v_n \in A$ as $\|v_n\| \to 0$ while the coefficients are bounded. □

Lemma 2.11.4 *Any subset X with compact closure of a real or complex normed space is contained in the closed convex hull of some sequence converging to zero.*

Proof We may assume that X is non-empty. Pick members $v_1, \ldots, v_{n_1}$ of $2X$ such that $2X \subset \cup_{i=1}^{n_1} B_{2^{-1}}(v_i)$. Let $X_1 = \cup_{i=1}^{n_1} (2X \cap B_{2^{-1}}(v_i) - v_i) \subset B_{2^{-1}}(0)$. Then pick members $v_{n_1+1}, \ldots, v_{n_2}$ of $2X_1$ such that $2X_1 \subset \cup_{i=n_1+1}^{n_2} B_{2^{-2}}(v_i)$, and define $X_2 = \cup_{i=n_1+1}^{n_2} (2X_1 \cap B_{2^{-2}}(v_i) - v_i) \subset B_{2^{-2}}(0)$. Continuing this way gives a sequence $\{v_i\}$ that converges to zero, and if $v \in X$, there is i_1 between 1 and n_1 such that $2v - v_{i_1} \in X_1$. Then there is i_2 between $n_1 + 1$ and n_2 such that $4v - 2v_{i_1} - v_{i_2} \in X_2$, and so on, showing that $v = \sum 2^{-k} v_{i_k}$. □

For a real or complex Banach space every subset in the closed convex hull of some sequence converging to zero has of course compact closure.

Lemma 2.11.5 *If $\{v_n\}$ is a sequence converging to zero in a real or complex Banach space V, then there is a compact subset Y of V that contains the closed convex hull of $\{v_n\}$, and $m(v_n) \to 0$, where m is the norm associated to Y from the first lemma in this section.*

Proof We may assume the each v_n is non-zero. Let $w_n = \|v_n\|^{-1/2} v_n$ if $\|v_n\| < 1$ and otherwise let $w_n = v_n$. Then the closed convex hull Y of $\{w_n\}$ is compact as $\|w_n\| \to 0$, and Y contains the closed convex hull of $\{v_n\}$. When $\|v_n\| < 1$, then since $m(w_n) \leq 1$, we see that $m(v_n) \leq \|v_n\|^{1/2} \to 0$ as $n \to \infty$. □

Theorem 2.11.6 *A real or complex Banach space V has the approximation property if and only if the following holds: For any compact subset X of V and $\varepsilon > 0$ there is $T \in B_f(V)$ such that $\|T(v) - v\| < \varepsilon$ for $v \in X$.*

Proof Assume that the second property holds. If $T \in B_0(W, V)$ for some Banach space W, pick $T_n \in B_f(V)$ such that $\|T_n(v) - v\| < 1/n$ for $v \in T(B)$, where B is the closed unit ball in V. Thus $\|T_n T - T\| = \sup_{v \in T(B)} \|T_n(v) - v\| < 1/n$ and $T_n T \in B_f(W, V)$ shows that $B_f(W, V)$ is dense in $B_0(W, V)$.

Assume now that V has the approximation property. Let $\varepsilon > 0$ and let X be any compact subset of V. The previous two lemmas provide a sequence $\{v_i\}$ in V that converges to zero and with closed convex hull containing X and contained in another compact set Y such that $m(v_i) \to 0$, where m is the norm on a Banach space W constructed from Y according to the first lemma in this section. This lemma also tells us that the inclusion map $W \to V$ is a compact bounded operator, so $\|\cdot\| \leq am$ for some scalar a.

By hypothesis there is $S \in B_f(W, V)$ such that $\|S(v) - v\| < \varepsilon/2$ for every v in the bounded set X. Write $S = \sum x_i(\cdot)u_i$ for a linear basis $\{u_i\}_{i=1}^k$ of $S(W)$ and $x_i \in W^\star$ with respect to m. Let $r = \varepsilon/2k \sup_i \|u_i\|$. We claim there is $y \in V^\star$ to $x \in W^\star$ such that $|y(v) - x(v)| < r$ for $v \in X$. If this holds we are done because pick such y_i to each x_i, and then $T = \sum y_i(\cdot)v_i \in B_f(V)$ satisfies $\|T(v) - v\| \leq \|\sum y_i(v)u_i - \sum x_i(v)u_i\| + \|S(v) - v\| < \varepsilon$ for $v \in X$.

To prove our claim, let $x \in W^\star$ and pick n with $|x(v_i)| < r/2$ for $i > n$. Let A be the closed convex hull of $\cup_{i>n} 2r^{-1} D v_i$, where D is the closed unit disc. Looking at the alternative description of A given by Lemma 2.11.3, the sums converge absolutely with respect to both $\|\cdot\|$ and m, and the two possible limits one might get coincide because $\|\cdot\| \leq am$. So A is the same whether closure is with respect to $\|\cdot\|$ or m. Hence $|x(v)| < 1$ for $v \in A$.

Let B be the convex set consisting of all $v \in V$ in the linear span of $\{v_1, \ldots, v_n\}$ such that $\mathrm{Re}\, x(v) = 1$, so $|x(v)| \geq 1$ and $A \cap B = \phi$. As B is closed in a finite dimensional subspace of W, it is also closed in V, whereas A is compact in V by Lemma 2.11.3. If B is empty, then $x(v_i) = 0$ for $i \leq n$, and then we can pick $y = 0$ as $|x(v) - y(v)| < r/2$ for $v \in X$. While if B is non-empty, there is $y \in V^\star$ such that $\sup \mathrm{Re}\, y(A) < \inf \mathrm{Re}\, y(B)$ by the corollary to the Hahn-Banach separation theorem. If we had $v, w \in B$ such that $\mathrm{Re}\, y(v) \neq \mathrm{Re}\, y(w)$, then $v + t(v - w) \in B$ for $t \in \mathbb{R}$, so $\mathrm{Re}\, y(B) = \mathbb{R}$. Thus $\mathrm{Re}\, y$ is constant on B, and since $0 \in A$, we may rescale y by a positive scalar, still keeping the sup-inf-inequality, but now such that $\operatorname{Re} y = 1$ on B. We have thus two linear functionals on the linear span of $\{v_1, \ldots, v_n\}$ with equal real parts (both are one), so $y(v_i) = x(v_i)$ for $i \leq n$ since linear functionals are determined by their real parts. Pick a scalar b of modulus one to $v \in A$ such that $|y(v)| = \operatorname{Re} y(bv)$. Since $bv \in A$, we get $|y(A)| < 1$, so $|y(v_i)| < r/2$ for $i > n$.

By Lemma 2.11.3 write $v = \sum t_i v_i$ for $v \in X$, where $\sum t_i \leq 1$ and $t_i \geq 0$. From the previous two paragraphs we then see that

$$|y(v) - x(v)| = |\sum_{i=1}^{\infty} t_i(y(v_i) - x(v_i))| \leq \sum_{i=n+1}^{\infty} t_i(|y(v_i)| + |x(v_i)|) < r.$$

□

2.12 Weakly Compact Operators

Let us first introduce operators that lie between the bounded and compact ones.

Definition 2.12.1 An operator between real or complex Banach spaces is *Dunford-Pettis* if the image under every weakly compact set is compact.

Any Dunform-Pettis operator T is bounded because if we had a sequence of unit vectors v_n such that $\|T(v_n)\| > n^2$, then the image under T of the weakly compact set $\{n^{-1}v_n\} \cup \{0\}$ would not even by bounded.

Any compact operator S is Dunford-Pettis because any weakly compact set A is bounded (see the proof of James' theorem), so $S(A)$ has compact closure. But $S(A)$ is already closed because it is weakly compact, and thus weakly closed, as any bounded operator clearly is weak-to-weak continuous. Conversely, if A is Dunford-Pettis from a reflexive Banach space, then it is compact as the closed unit ball in the domain space is weakly compact in this case.

Proposition 2.12.2 *A operator is Dunford-Pettis if and only if it is sequentially weak-to-norm continuous.*

Proof Say T is Dunford-Pettis and that we have a sequence $\{v_n\}$ that converges weakly to zero, but $\|T(v_n)\| > r$ for some $r > 0$. As $\{v_n\} \cup \{0\}$ is weakly compact, the sequence $\{T(v_n)\}$ has a subsequence that converges in norm to some non-zero element. But it must then also converge weakly to this element, which contradicts the fact that T is weak-to-weak continuous.

The converse direction is clear from the Eberlein-Smulian theorem. □

Definition 2.12.3 A operator between real or complex Banach spaces is *weakly compact* if the weak closure of the image of a bounded set is weakly compact.

Again we see that compact operators are weakly compact and such operators are bounded. These inclusions are proper as a glance at the identity operators on $l^2(\mathbb{N})$ and $l^1(\mathbb{N})$ shows.

By Alaoglu's theorem we see that a bounded operator between real or complex Banach spaces is weakly compact if one of the Banach spaces is reflexive. The composition of two bounded operators between real or complex Banach spaces is weakly compact if one of the spaces is reflexive.

Also, an operator is weakly compact if and only if every bounded sequence has a subsequence with weakly convergent image.

Lemma 2.12.4 *Let A be a bounded convex balanced subset of a real or complex Banach space W. Then*

$$m_k(w) = \inf\{t \in \langle 0, \infty\rangle \mid t^{-1}w \in 2^k A + 2^{-k} B_1(0)^o\}$$

for $k \in \mathbb{N}$ defines a norm m_k on W equivalent to the original norm. The set V of $w \in W$ such that $s(w) \equiv (\sum_{k=1}^{\infty} m_k(w)^2)^{1/2} < \infty$ is a Banach space with norm s such that $s(A) < 1$. The inclusion map $\iota\colon V \to W$ is bounded with an injective biadjoint map having V as the inverse image of W. Moreover, the norm closure of A in W is weakly compact if and only if V is reflexive.

Proof It is easy to check, see Lemma 2.4.3, that m_k is a seminorm such that $m_k(w) < 1$ precisely when $w \in 2^k A + 2^{-k} B_1(0)^o$, from which it easily follows that m_k is a norm on W equivalent to the original one, and that $s(A) < 1$. Consider the Banach space W_2 of elements in the algebraic direct product $\prod_k W$ with finite $\|\cdot\|_2$-norm given by the Banach norms m_k, see Proposition 2.1.24. The embedding E of V into W_2 given by $E(v) = (v, v, \dots)$ is clearly an isometry with closed range, so V is a Banach space, and the inclusion $\iota\colon V \to W$ is bounded as it is the composition of E with the projection from W_2 onto say the first factor W. Since E is an isometry, its adjoint E^* is surjective because for $y \in V^\star$, define $x \in (\mathrm{im}\iota)^\star$ by $x\iota(v) = y(v)$, and extend x by the Hahn-Banach theorem to an element in $W^\star$ with image y under ι^*. But then E^{**} is evidently injective. Looking at the proof of reflexivity of $l^2(\mathbb{N})$, one sees that $(W_2)^{\star\star}$ is the algebraic direct product $\prod_k W^{\star\star}$ with finite $\|\cdot\|_2$-norm given by the bidual norms from m_k, and then $E^{**} = \prod_k \iota^{**}$, which shows that ι^{**} is also injective. Now if $\iota^{**}(v) \in W$ for some $v \in V^{\star\star}$, then $E^{**}(v) \in W_2$. By the Goldstine theorem there is a net $\{v_i\}$ in V that converges to v in the w^*-topology on $V^{\star\star}$, and as any adjoint map is w^*-to-$w^\star$ continuous, we see that $E^{**}(v) = \lim E^{**}(v_i) = \lim E(v_i)$ in the weak topology on W_2, and as norm closed convex sets are weakly closed, we conclude that $E^{**}(v) \in E(V)$, so V is the inverse image of W under E^{**}.

The image B of the closed unit ball under the bounded map ι is contained in the normed closed subset $2^k\overline{A} + 2^{-k}B_1(0)$ of W by the first sentence in this proof. If the norm closure $\overline{A}$ of A is also weakly compact, then the subset $2^k\overline{A}+2^{-k}B_1(0)$ of $W^{\star\star}$ is w^*-compact, where now $B_1(0)$ is the closed unit ball in $W^{\star\star}$. By Alaoglu's theorem and the Goldstine theorem we know that B is w^*-dense in the image under ι^{**} of the closed unit ball in $V^{\star\star}$, so this image is contained in $\cap_k(2^k\overline{A}+2^{-k}B_1(0))$, which again is contained in W. Thus V is reflexive by the previous paragraph. Conversely if V is reflexive, the closed unit ball is weakly compact, so its image under ι is weakly compact, and $\overline{A}$ is weakly compact as $s(A) < 1$. □

We can now prove the following decomposition result.

Theorem 2.12.5 *Any bounded map $T: U \to W$ between real or complex Banach spaces is weakly compact if and only if there is a reflexive Banach space V and bounded maps $R: V \to W$ and $S: U \to V$ such that $T = RS$.*

Proof The backward implication is trivial. For the forward one, let A be the image under T of the closed unit ball in U, and define V to A as in the lemma. Let R be the inclusion map $V \to W$. Note that $s(T(u)) < 1$ when $\|u\| \leq 1$ in U, so the linear map $S: U \to V$ defined by $S(u) = T(u)$ is bounded, and clearly $RS = T$. □

This gives the following *Gantmacher's theorem.*

Corollary 2.12.6 *Let $T: U \to W$ be a bounded operator between real or complex Banach spaces. Then T is weakly compact if and only if T^* is weakly compact if and only if $T^{**}(U^{\star\star}) \subset W$.*

Proof If T is weakly compact, write $T = RS$ as in the theorem. Then $T^* = S^*R^*$ is weakly compact by the theorem. Using that $V^{\star\star} = V$ we get $R^{**} = R$, so $T^{**} = RS^{**}$ has range in W. Conversely if $T^{**}(U^{\star\star}) \subset W$, then the image under T of the closed unit ball is contained in the w^*-compact image under T^{**} of the closed unit ball in $U^{\star\star}$, and this latter image is weakly compact, so T is weakly compact.

If T^* is weakly compact, then by the theorem there is a reflexive Banach space V_1 and bounded operators $R_1: V_1 \to U^\star$ and $S_1: W^\star \to V_1$ such that $T^* = R_1S_1$, so $T^{**} = S_1^*R_1^*$. The norm closure V_2 of $R_1^*(U)$ in the reflexive Banach space $V_1^\star$ is reflexive; this is true for any closed subspace of a reflexive Banach space by the corollary to the Goldstine theorem. Let S_2 be the restriction of R_1^* to U, and let R_2 be the restriction of S_1^* to V_2. Then $T = R_2S_2$ is weakly compact. □

Corollary 2.12.7 *The set of weakly compact operators between two fixed real or complex Banach spaces is a norm closed subspace of the Banach space of all the bounded operators between the spaces. In particular, the weakly compact operators on a real or complex Banach space is a closed two-sided ideal in the unital Banach algebra of all bounded operators on the space.*

Proof If we have a sequence of weakly compact operators $T_n: V \to W$ between the Banach spaces that converges in norm to a bounded operator T, then $T_n^{**}(V^{\star\star})$ has by the theorem its canonical image inside the norm closed subspace W of $W^{\star\star}$. But then so has $T^{**}(V^{\star\star})$, and T is weakly compact by the theorem. The remaining statements are clearly true. □

Clearly a bounded operator from a real or complex reflexive Banach space to another Banach space is Dunford-Pettis if and only if it is compact.

Definition 2.12.8 A real or complex Banach space V has the *Dunford-Pettis property* if every weakly compact operator from V into another Banach space is Dunford-Pettis.

No infinite dimensional real or complex reflexive Banach space has the Dunford-Pettis property since the identity operator is weakly compact but not Dunford-Pettis as the closed unit ball fails to be compact.

Proposition 2.12.9 *A real or complex Banach space V has the Dunford-Pettis property if and only if for every sequences $\{v_n\}$ in V and $\{x_n\}$ in $V^\star$ converging weakly to zero, the sequence $\{x_n(v_n)\}$ converges to zero.*

Proof If T is a weakly compact operator from V into another Banach space W that is not Dunford-Pettis, there is by Proposition 2.12.2 a sequence $\{v_n\}$ in V that converges weakly to zero but $\|T(v_n)\| \geq r > 0$ for all n. Pick by the Hahn-Banach theorem unit vectors $y_n \in W^\star$ such that $y_n T(v_n) = \|T(v_n)\|$. By Gantmacher's theorem T^* is weakly compact, so its image of the closed unit ball has weakly compact closure. By going to a subsequence we may assume by the Eberlein-Smulian theorem that $\{T^*(y_n)\}$ is weakly convergent to some $x \in V^\star$. Hence $\|T(v_n)\| = T^*(y_n)(v_n) \to 0$ as $x(v_n) \to 0$, which is absurd.

Conversely, if we have sequences $\{v_n\}$ in V and $\{x_n\}$ in $V^\star$ that converge weakly to zero, the formula $T(v) = \{x_n(v)\}$ defines a bounded operator T from V to the Banach space $C_0(\mathbb{N})$. Its adjoint operator sends the canonical basis element e_n of $l^1(\mathbb{N}) = C_0(\mathbb{N})^\star$ to x_n, so the image under T^* of the closed unit ball is contained in the weakly closed convex hull of $\{x_n\} \cup \{0\}$, which is weakly compact by Mazur's theorem. Thus T^* is weakly compact, and so is T by Gantmacher's theorem. By hypothesis T is therefore Dunford-Pettis, so $|x_n(v_n)| \leq \|T(v_n)\|_u \to 0$ due to Proposition 2.12.2. □

Exercises

For Sect. 2.1

1. Exhibit a vector space with countably many bounded operators on it.
2. Let A be the $\mathbb{Q}$-linear map on $\mathbb{Q} + \sqrt{2}\mathbb{Q}$ given by multiplication with $\sqrt{2}$. What does $\sum_{n=0}^{\infty} A^{-n}$ send $\sqrt{2}$ to? Is $B(\mathbb{Q} + \sqrt{2}\mathbb{Q})$ a Banach space?
3. Show that the Banach space l^∞ is non-separable, while the subspace c_0 of sequences tending to zero is closed and separable.
4. Prove that if U, V are normed spaces with $B(V, U)$ Banach, then U is Banach.
5. If W is a closed subspace of a Banach space V, does the quotient map $q : V \to V/W$ always take closed subsets of V to closed subsets of V/W?
6. Give an example of a non-commutative Banach algebra which has no non-trivial closed ideals.

For Sect. 2.2

1. Show that if on a Banach space one norm is smaller than another by an overall scalar, then the norms are equivalent.

2. Show that no non-trivial complete metric space is a countable union of closed sets with empty interiors. They are thus by definition not of Baire's first category, but of the remaining second category.
3. Let $P \in B(V)$ be an idempotent $P^2 = P$ for a Banach space V. Show that $P(V)$ and $\ker(P)$ are complementary subspaces of V in that they are closed, span V and have trivial intersection. Show that such subspaces are all of this form, and that every element of V can be written uniquely as a sum of two vectors with one in each subspace. Prove that $T \in B(V)$ with $TP = PT$ leaves the subspaces invariant.
4. Suppose U, V, W are Banach spaces and that $B : U \times V \to W$ is continuous and bilinear, that is, linear in each variable. Prove that there exists M such that $\|B(u, v)\| \leq M\|u\|\|v\|$ for all $u \in U$ and $v \in V$. Show that all such maps B form a Banach space in an obvious way.

For Sect. 2.3

1. Show that $c_0^\star$ is isometric isomorphic to l^1, and conclude that c_0 is not reflexive.
2. Prove that a real or complex normed space V is separable if $V^\star$ is.
3. Show that a real or complex Banach space V is reflexive if and only if $V^\star$ is reflexive.
4. Let V and W be real or complex Banach spaces, and let $BIL(V, W)$ be the space of bilinear continuous maps from $V \times W$ into the base field. Show that it is isometric isomorphic to the Banach spaces $B(V^\star, W)$ and $B(V, W^\star)$.
5. Let $BIL(V, W)$ be as in the previous exercise. Show that $(v, w) \mapsto v \otimes w$ with $(v \otimes w)(B) = B(v, w)$ defines a bounded bilinear map from $V \times W$ to $BIL(V, W)^\star$ such that $\|v \otimes w\| = \|v\|\|w\|$. The normed closed span $V \otimes W$ of all the elements $v \otimes w$ is called the *projective tensor product* of V and W. Show that $(V \otimes W)^\star$ is isometric isomorphic to $BIL(V, W)$. Show that for a continuous bilinear map $A : V \times W \to U$ into a Banach space, there is a unique $\tilde{A} \in B(V \otimes W, U)$ such that $\tilde{A}(v \otimes w) = A(v, w)$.

For Sect. 2.4

1. Show that whenever U is a closed subspace of a real or complex Banach space V, then V is reflexive if and only if U and V/U are reflexive.
2. Show that the smallest convex set that contains two compact convex subsets of a real or complex topological vector space is compact.
3. Let V be a real or complex normed space. Prove that a subspace of $V^\star$ is separating for V if and only if it is w^*-dense in $V^\star$.
4. Show that for a real or complex normed space, the closed unit ball is w^*-dense in the closed unit ball of the bidual space.
5. Show that a Banach space is reflexive if and only if the closed unit ball is weakly compact.

For Sect. 2.5

1. Prove that every element of a compact convex subset K of $\mathbb{R}^n$ is a convex combination of at most $n + 1$ extreme points of K.
2. Show that a w^*-closed ball in the dual of a real or complex normed space is the w^*-closed convex hull of its extreme points, and that no infinite dimensional real or complex normed space whose closed unit ball has only finitely many extreme points can be the dual of a normed space.
3. Show that a closed ball in a reflexive real or complex normed space is the closed convex hull of its extreme points.
4. Let Y be a non-empty bounded and w^*-closed subset in the dual of a real or complex normed space. Prove that the extreme points of the w^*-closed convex hull of Y lies in Y.
5. Consider a bounded sequence of complex continuous functions on a compact Hausdorff space that converges pointwise to a continuous function f. Show that f can be approximated uniformly by convex combinations of elements from the sequence.

For Sect. 2.6

1. Suppose X is a commutative subset of $B(V)$ for a real or complex normed space V with an X-invariant subspace U. We claim that any X-invariant element of $U^\star$ has an X-invariant extension with the same norm. Prove this claim using the Markow-Kakutani fixed point theorem by considering the set of all possible norm-preserving extensions.
2. We claim that for any abelian group G, there is a function μ from the power set of G to $[0, 1]$ that is one on G, that is translation invariant, and that takes any finite disjoint union to the sum of its values on the components. Prove this using the exercise above when G is infinite by considering those $f \in l^\infty(G)$ which have a limit at infinity.
3. Show that the map from the closed unit ball of $l^2(\mathbb{N})$ to itself that sends $v = (v_1, v_2, \ldots)$ to $((1 - \|v\|^2)^{1/2}, v_1, v_2, \ldots)$ is continuous but has no fixed points.
4. Use the Ryll-Nardzewski fixed point theorem to show that any compact group G has a unique Haar integral, that is, a unital complex valued $x \in C(G)^\star$ that is positive on non-zero positive functions, and such that $x(f(s\cdot)) = x(f) = x(f(\cdot s))$ for any $f \in C(G)$ and $s \in G$.

For Sect. 2.7

1. Let V be a separable complex Banach space. Show that the closed unit ball in $V^\star$ is metrizable in the w^*-topology, and that $V^\star$ therefore is separable in that topology.
2. Show that the set $\{0\} \cup \{n^{-1}\delta_n \mid n \in \mathbb{N}\}$ is compact in the complex normed space $C_c(\mathbb{N})$, while its convex hull is not weakly compact.
3. Prove that a subset of the complex Banach space l^1 is compact if and only if it is weakly compact.

4. Suppose V is a complex normed space such that the w^*-closed hyperplanes in $V^\star$ are the only hyperplanes having w^*-closed intersections with the closed unit ball. Show that V must then be a Banach space.
5. Show that a complex Banach space is reflexive if and only if every separable closed subspace is reflexive.
6. Show that a convex subset A of a complex normed space with closed unit ball B is closed if and only if $A \cap rB$ is closed for all $r > 0$. Show that the last equivalence also holds when norm closed is replaced by weakly closed. Here the normed space need not be complete.

For Sect. 2.8

1. Find a closed convex subset A of a real Banach space such that the set of support functionals for A is non-empty, but not dense in the dual space.
2. Suppose A is a non-empty closed subset of the dual of a real or complex Banach space V. Is it true that A is w^*-compact if and only if $\sup |A(v)|$ is attained whenever $v \in V$?
3. Find a non-empty closed subset A of a real or complex Banach space V such that A is not weakly compact even though $\sup |x(A)|$ is attained for any $x \in V^\star$.
4. Show that the non-trivial direction in James' theorem can fail when the real or complex normed space is not Banach.
5. Suppose A is a closed bounded non-weakly-compact convex subset of a real or complex Banach space. Prove that there is a decreasing sequence of non-empty closed convex subsets A_n of A such that $v + t(A - v)$ intersects only finitely many A_n for fixed $v \in A$ and $t \in [0, 1\rangle$.

For Sect. 2.9

1. Say T is a compact operator between real or complex Banach spaces. Show that $T(v_n) \to T(v)$ in norm if $v_n \to v$ weakly.
2. Show that bounded operators from c_0 to a reflexive complex normed space are automatically compact.
3. Suppose T is a compact operator on a real or complex Banach space with invertible $I - T$. Show that $I - T$ has closed range and that $I - (I - T)^{-1}$ is a compact operator.
4. Consider a shift and a multiplication operator on l^2 given by $S(f)(n) = f(n-1)$ for $n \geq 1$ and $S(f)(0) = 0$ and $M(f)(n) = (n+1)^{-1} f(n)$ for $n \geq 0$. Show that MS is a compact operator with no eigenvalues, while $MS - \lambda I$ is invertible for all complex numbers λ except for one value.

For Sect. 2.10

1. Characterize the compact idempotents in a real or complex Banach space among all idempotents.
2. Show that if W is a complementary subspace of a real or complex Banach space V, there is up to isomorphism only one closed subspace U with $W \oplus U = V$.
3. Show that if W and U are complementary subspaces in a real or complex Banach space, then $W^\perp$ and $U^\perp$ are complementary in the dual space.

4. Show that a bounded operator on a non-separable complex Banach space has a proper closed invariant subspace.
5. Show that the graph of a bounded operator T satisfying $TA_1 = A_2T$ for bounded operators A_i on a real or complex Banach space is invariant for $A_1 \oplus A_2$.

For Sect. 2.11

1. Prove that c_0 and l^p for $p \in [0, \infty\rangle$ have the approximation property.

For Sect. 2.12

1. Prove that the adjoint of an isometry between real or complex Banach spaces is surjective.
2. Show that an operator from a complex Banach space into l^1 is compact if and only if it is weakly compact.
3. Show that an operator from c_0 into a Banach space is compact if and only if it is weakly compact.
4. Show that a real or complex Banach space has the Dunford-Pettis property if its dual space has.
5. Use Gantmacher's theorem to prove that a real or complex Banach space is reflexive if and only if its dual is reflexive.

Chapter 3
Bases in Banach Spaces

A satisfactory account of functional analysis cannot be given without discussing some of the outcome of the Polish school centered around S. Banach. This community, which flourished between WWI and WWII, was sadly wiped out by the Nazis. A fundamental question in Banach space theory is whether two Banach spaces are isomorphic or not, or at least if one is isomorphic to a (complemented) subspace of the other. The Polish school understood that such questions, or even the desire to understand the differences or similarities between two spaces, can be effectively approached studying various bases in the them.

By a Schauder basis for a Banach space we mean a sequence of vectors such that any vector can be written as a unique 'infinite linear combination' of them. The unique scalar coefficients in the series are to be thought of as the coordinates of the vector in the Schauder basis coordinate system. The series converges in norm, so spaces with a Schauder basis are necessarily separable. Such bases are not to be confused with linear bases, where any vector is written uniquely as a (finite) linear combination of the basis. No countable linear basis exists for a separable Banach space.

The Schauder basis coordinate at the n-th place of a vector v can be singled out by the evaluation at v of the n-th element of a sequence of the dual space playing the role as dual coordinate functionals. The converse is also true, namely any sequence in a Banach space having such dual coordinate functionals must be a Schauder basis. Schauder bases of two Banach spaces are said to be equivalent if there is a bounded isomorphism from one space to another that takes the n-th basis element of one basis to the n-th basis element of the the other basis. This happens if the coordinates of a vector in one basis will always be the coordinates in the other basis of some vector, namely, for the image of the original vector under the isomorphism.

In the appendix we prove, using Zorn's lemma from set theory, that any vector space has a linear basis, even with unique cardinality, defined then as the dimension of the space. No such result exists for Schauder bases of separable Banach spaces. Yet, at the end of the 50s, new ways of constructing Schauder basic sequences, i.e.

L. Tuset, *Analysis and Quantum Groups*,
https://doi.org/10.1007/978-3-031-07246-8_3

sequences that are Schauder bases for their closures, rendered such bases subtle and effective tools in Banach space theory. Apart from reproducing profound classical results like the Eberlein-Smulian theorem, one obtained criteria for the existence of Schauder basic sequences. Note that the coordinate functionals for a Schauder basis form a Schauder basis sequence for the dual Banach space, then with the original Schauder basis as its coordinate functionals.

Having introduced the basic concepts and results for Schauder bases and basic sequences, and having supplied the fundamental examples of such bases, we introduce pertinent notions such as conditional-, (weakly) unconditional-, absolute-, subseries convergence of series in Banach spaces, and investigate the relation between these notions. We also study notions such as monotone-, shrinking- and boundedly complete Schauder bases, and that of block Schauder basic sequences. This gives a plethora of results, allowing us for instance to tell when a Banach space is reflexive.

We end the chapter with James' spectacular example of a Banach space that is isometrically isomorphic to its bidual without being reflexive, only quasi-reflexive.

3.1 Schauder Bases

Any real or complex Banach space V with a countable linear basis $\{v_i\}$ must be finite dimensional because it is the countable union of the span V_n of $\{v_1, \ldots, v_n\}$, so by Baire's theorem some V_n has non-empty interior, and therefore contains a small open ball of V and hence all of V. For infinite dimensional spaces we can in many cases nevertheless manage with sequences if we allow for limits of linear combinations.

Definition 3.1.1 A *Schauder basis* for a real or complex Banach space V is a sequence $\{v_n\}$ in V such that every element of V can be written as an infinite sum $\sum a_n v_n$ for unique scalars a_n. A sequence in V is a *Schauder basic sequence* for V if it is a Schauder basis for the closed span of the sequence.

The standard basis for $l^p(\mathbb{N})$ is a Schauder basis when $p \in [1, \infty\rangle$.

Any Banach space with a Schauder basis must be separable and infinite dimensional. By uniqueness the elements of any Schauder basis sequence are linearly independent. If we rescale the members of such a sequence by non-zero scalars, we get another Schauder basic sequence, so any Schauder basis can be normalized.

Proposition 3.1.2 *Given a Schauder basis $\{v_n\}$ for a real or complex Banach space V, the formula $\|\sum a_n v_n\|_s = \sup_m \|\sum_{n=1}^m a_n v_n\|$ defines a norm $\|\cdot\|_s$ on V that is equivalent to the original norm $\|\cdot\|$.*

Proof By continuity of $\|\cdot\|$ the formula above well-defines a function from V to $[0, \infty\rangle$ that majorizes $\|\cdot\|$, so $\|\cdot\|_s$ is indeed a norm on V. Since the identity map on V is a bounded operator with $\|\cdot\|_s$ on a domain space and $\|\cdot\|$ on the range space, we are done by the open mapping theorem provided we can show that V is complete with respect to $\|\cdot\|_s$.

So let $\{w_n\}$ be a $\|\cdot\|_s$-Cauchy sequence with k-th coordinate a_{nk}. Then

$$|a_{nk} - a_{mk}|\|v_k\| = \|\sum_{i=1}^{k}(a_{ni} - a_{mi})v_i - \sum_{i=1}^{k-1}(a_{ni} - a_{mi})v_i\| \leq 2\|w_n - w_m\|_s$$

shows that $\{a_{nk}\}_n$ is Cauchy and converges to a scalar a_k. By adding and subtracting terms we get the inequality

$$\|\sum_{k=i}^{j} a_k v_k\| \leq \sum_{k=i}^{j} |a_k - a_{nk}|\|v_k\| + 2\|w_n - w_m\|_s + \|\sum_{k=i}^{j} a_{mk} v_k\|.$$

To $\varepsilon > 0$ pick m and N such that $2\|w_n - w_m\|_s < \varepsilon/3$ and $\|\sum_{k=i}^{j} a_{mk} v_k\| < \varepsilon/3$ for $n > m$ and $j > i > N$. For any fixed such i, j pick n greater than m such that $\sum_{k=i}^{j} |a_k - a_{nk}|\|v_k\| < \varepsilon/3$. By the inequality above we get $\|\sum_{k=i}^{j} a_k v_k\| < \varepsilon$, so $w = \sum a_k v_k$ exists in V. Moreover, we have

$$\|w - w_m\|_s = \sup_j \|\sum_{k=1}^{j}(a_k - a_{km})v_k\| \leq \sup_j \sum_{k=1}^{j} |a_k - a_{kn}|\|v_k\| + \|w_n - w_m\|_s < 2\varepsilon/3,$$

so the $w_m \to w$ with respect to the $\|\cdot\|_s$-norm. □

Corollary 3.1.3 *The* k-th coordinate functional v_k^* *for a Schauder basis* $\{v_n\}$ *defined to be* a_k *at* $\sum a_n v_n$ *is a bounded linear functional on the Banach space.*

Proof The inequality $|a_k|\|v_k\| = \|\sum_{i=1}^{k} a_i v_i - \sum_{i=1}^{k-1} a_i v_i\| \leq 2\|\sum a_n v_n\|_s$ shows that $\|v_k^*\| \leq 2/\|v_k\|$ with respect to the $\|\cdot\|_s$-norm, and this norm is equivalent to the original norm on the Banach space, so the linear functional v_k^* is bounded. □

Hence a sequence $\{v_n\}$ in a real or complex Banach space V is a Schauder basis for V if and only if there is a sequence $\{v_n^*\}$ in $V^\star$ such that $v_n^*(v_m) = \delta_{nm}$ and $v = \sum v_n^*(v) v_n$ for all $v \in V$.

The *natural projection* P_m *associated to a Schauder basis* $\{v_n\}$, which sends $\sum a_n v_n$ to $\sum_{n=1}^{m} a_n v_n$, is a bounded idempotent. By the principle of uniform boundedness the *basis constant* $\sup_n \|P_n\|$ for $\{v_n\}$ is a non-negative real number K_b. The infinum over all Schauder bases of these constants for a fixed Banach space V is the *basis constant* for V.

A Schauder basis with basis constant one is *monotone*. Using the new norm in the proposition above, we see that any Banach space with a Schauder basis can be renormalized so that the basis becomes monotone. The standard bases for $C_0(\mathbb{N})$ and $l^p(\mathbb{N})$, for $p \in [1, \infty\rangle$, are clearly monotone.

The natural projections P_m in the definition above are commuting bounded idempotents with rank m such that $P_m P_n = P_m$ for $n \geq m$ and $\lim P_m(v) = v$ for all v. To such a collection of P_m's on a real or complex Banach space V, any sequence of non-zero $v_m \in P_m(V)$ with $P_{m-1}(v_m) = 0$ is a Schauder basis for V

with coordinate functionals v_m^* given by $v_m^*(v)v_m = P_m(v) - P_{m-1}(v)$, and with natural projections P_m.

Here is a useful criterion for deciding when dealing with a Schauder basis.

Proposition 3.1.4 *A sequence $\{v_n\}$ in a real or complex Banach space is a Schauder basis if and only if its linear span is dense and each member is non-zero and there is a number M such that $\| \sum_{i=1}^{n} a_i v_i \| \leq M \| \sum_{i=1}^{m} a_i v_i \|$ for all $m > n$ and scalars a_i.*

Proof The forward implication is clear from the remarks above.

For the converse, assume we have a sequence $\{v_n\}$ satisfying the properties above. If $\sum a_n v_n = \sum b_n v_n$, we have $|a_1 - b_1| \|v_1\| \leq M \| \sum a_n v_n - \sum b_n v_n \| = 0$. Then $a_1 = b_1$ as $v_1 \neq 0$, and $a_n = b_n$ by induction. So we have uniqueness of expansions, and $\{v_n\}$ is linearly independent. For existence of expansions, let P_m be the bounded extension with norm $\|P_m\| \leq M$ of the linear map that sends a linear combination $\sum a_n v_n$ to $\sum^m a_n v_n$. For any v in the Banach space and any w in the linear span of $\{v_n\}$, we have $\|P_m(v) - v\| \leq (M+1)\|v - w\| + \|P_m(w) - w\|$, so $\lim_m P_m(v) = v$ by denseness of the linear span. Hence $\{v_n\}$ is a Schauder basis for V with coordinate functionals v_m^* given by $v_m^*(v)v_m = P_m(v) - P_{m-1}(v)$. □

Example 3.1.5 Pick a dense sequence $\{r_i\}$ in $[0, 1]$ with $r_1 = 0$ and $r_2 = 1$. Define maps $P_m : C([0, 1]) \to C([0, 1])$ by $P_1(f) = f(r_1)$, and for $m \geq 2$, by letting $P_m(f)$ be the function with graph consisting of the points $(r_1, f(r_1)), \ldots (r_m, f(r_m))$ and the straight lines between the points with adjacent r_i's. This way we get a collection of commuting bounded idempotents P_m with rank m such that $P_m P_n = P_m$ for $n \geq m$ and $\lim P_m(v) = v$ for all v, and hence a Schauder basis for the Banach space $C([0, 1])$.

The same procedure works for $C(\mathbb{T})$ by identifying it will the subspace of functions f in $C([0, 2\pi])$ with *period* 2π, that is, satisfy $f(2\pi) = f(0)$. However, the exponential system $\{e_0, e_1, e_{-1}, e_2, e_{-2}, \ldots\}$ with $e_n(t) = e^{int}$ is not a Schauder basis for $C(\mathbb{T})$ under this identification despite the fact that this system is dense in $C(\mathbb{T})$ by the Stone-Wierstrass theorem.

To see that the exponential system is not a Schauder basis, consider the bounded functionals e_n^* on $C(\mathbb{T})$ with $e_n^*(f) = (1/2\pi) \int_{-\pi}^{\pi} f(t) e_{-n}(t)\, dt$ the *Fourier coefficients* of f. The natural projections given by $P_{2n+1}(f)(s) = \frac{1}{2\pi} \int f(s - t) D_n(t)\, dt$, where $D_n(t) = \sin((n + 1/2)t)/\sin(t/2)$ is the *Dirichlet kernel*, are not uniformly bounded. Indeed, as $\delta_0(P_{2n+1}(f)) = \frac{1}{2\pi} \int_{-\pi}^{\pi} D_n(t) f(-t)\, dt$ and $|\sin t| \leq |t|$, we have

$$\|P_{2n+1}\| \geq (1/2\pi) \int_{-\pi}^{\pi} |D_n(t)|\, dt \geq (2/\pi) \int_0^{(n+1/2)\pi} |\sin t|/t\, dt,$$

which tends to infinity by Fatou's lemma. ◇

The following example is also due to Schauder.

Example 3.1.6 The *Haar basis* $\{h_n\}$ of $L^p([0,1])$ for $p \in [0, \infty\rangle$ is defined to be $h_1 = 1$ and $h_n = \chi_I - \chi_J$ for $n \geq 2$, where $I = [(2n-2)/2^m - 1, (2n-1)/2^m - 1\rangle$ and $J = [(2n-1)/2^m - 1, 2n/2^m - 1\rangle$ for $m \in \mathbb{N}$ such that $2^{m-1} < n \leq 2^m$. It clearly spans $L^p([0,1])$, and if $\{a_n\}$ is a sequence of scalars, and a is the constant value of $\sum_{i=1}^{n-1} a_i h_i$ on $I \cup J$, then

$$\int |\sum_{i=1}^{n} a_i h_i|^p \, d\mu - \int |\sum_{i=1}^{n-1} a_i h_i|^p \, d\mu = \int_I |a + a_n|^p \, d\mu + \int_J |a - a_n|^p \, d\mu - \int_{I \cup J} |a|^p \, d\mu$$

equals the non-negative number $(|a+a_n|^p + |a-a_n|^p - 2|a|^p)/2^m$. By the previous proposition the Haar basis is thus a monotone Schauder basis for $L^p([0,1])$. ◇

The following result is perhaps not surprising.

Proposition 3.1.7 *Real or complex Banach spaces with a Schauder basis have the approximation property.*

Proof Say the Banach space has natural projections P_m associated to a Schauder basis with basis constant M. Let A be a compact subset of the Banach space and let $\varepsilon > 0$.

Pick finitely many members w_i of A such that every element of A is within distance $\varepsilon/(M+2)$ of some w_i. Pick also m such that $\|P_m(w_j) - w_j\| < \varepsilon/(M+2)$ for all j. To any $w \in A$ pick i such that $\|w - w_i\| < \varepsilon/(M+2)$. Then

$$\|P_m(w) - w\| \leq \|P_m(w - w_i)\| + \|P_m(w_i) - w_i\| + \|w_i - w\| < \varepsilon.$$

The result now follows from Theorem 2.11.6. □

We turn now to existence of Schauder basic sequences.

Lemma 3.1.8 *Let W be a non-trivial finite dimensional subspace of a real or complex Banach space V, and let $\varepsilon \in \langle 0, 1\rangle$. Cover the compact unit sphere of W with finitely many open $\varepsilon/4$-balls with centers $w_i \in W$ of norm one, and pick $x_i \in V^\star$ of norm one such that $x_i(w_i) = 1$. If $\{v_n\}$ is a sequence in V with $\inf\|v_n\| > 0$ and $\lim_n x_i(v_n) = 0$, there is $m \in \mathbb{N}$ greater than a given $k \in \mathbb{N}$ such that $\|w\| \leq (1+\varepsilon)\|w + av_m\|$ for any $w \in W$ and scalar a.*

Proof Pick $m > k$ such that $|x_i(v_m)| < \varepsilon \inf\|v_n\|/8$ for all i. Consider a unit vector $w \in W$ and a scalar a. We may assume that $|a| < 2/\|v_m\|$. Pick w_i such that $\|w - w_i\| < \varepsilon/4$. Then

$$\|w + av_m\| > \|w_i + av_m\| - \varepsilon/4 \geq |x_i(w_i + av_m)| - \varepsilon/4 > 1 - \varepsilon/2 > 1/(1+\varepsilon).$$

□

Lemma 3.1.9 *Let W be a finite dimensional subspace of an infinite dimensional real or complex Banach space V. For $\varepsilon > 0$ there is a unit vector $v \in V$ such that $\|w\| \leq \|(1+\varepsilon)\|w + av\|$ for any $w \in W$and scalar a.*

Proof We may pick w_i and x_i as in the lemma above. Since $\dim V = \infty$, we may pick a unit vector v in the kernels of all x_i. Letting $v_n = v$, we are done. □

Proposition 3.1.10 *An infinite dimensional real or complex Banach space has a normalized Schauder basic sequence with basis constant not greater than a given $M > 1$.*

Proof By the last lemma above we inductively pick a sequence $\{v_n\}$ of unit vectors such that $\| \sum_{n=1}^{m} a_n v_n \| \leq M^{1/2^m} \| \sum_{n=1}^{m+1} a_n v_n \|$ for any m and scalars a_n. Then

$$\| \sum_{n=1}^{m} a_n v_n \| \leq (\prod_{n=m}^{k-1} M^{1/2^n}) \| \sum_{n=1}^{k} a_n v_n \| \leq M \| \sum_{n=1}^{k} a_n v_n \|$$

for $k > m$, so the result follows from the proof of Proposition 3.1.4. □

Here is a criterion for when Schauder basic sequences can be located within another sequence.

Proposition 3.1.11 *Any sequence in a real or complex Banach space that converges to zero weakly but not in norm has a Schauder basic subsequence.*

Proof By going to a subsequence we may assume that our sequence $\{v_n\}$ satisfies $\inf \|v_n\| > 0$. By Lemma 3.1.8 it has a subsequence such that $\| \sum_{i=1}^{m} a_i v_{n_i} \| \leq M^{1/2^m} \| \sum_{i=1}^{m+1} a_i v_{n_i} \|$ for any m and scalars a_i. The proof is completed just as in the proof above. □

3.2 Unconditional Convergence

Definition 3.2.1 A series $\sum v_n$ in a real or complex normed space *converges unconditionally* if $\sum v_{\sigma(n)}$ converges for every permutation σ of $\mathbb{N}$. If $\sum v_n$ converges, but not unconditionally, the convergence is *conditional*. If $\sum \|v_n\|$ converges, we say that $\sum v_n$ *converges absolutely*.

Any absolutely convergent series converges if the normed space is complete, and the converse is clearly also true. The outstanding example of a series that converges conditionally is $\sum(-1)^n 1/n$. The *harmonic series* $\sum 1/n$ diverges by comparison with the integral $\int_1^\infty (1/x)\,dx$. That $\sum(-1)^n 1/n$ actually converges, is clear from the following *alternating test*.

Proposition 3.2.2 *If the set of partial sums of the complex series $\sum_{n=0}^{\infty} a_n$ is bounded, and $b_0, b_1, \ldots$ decreases towards zero, then $\sum a_n b_n$ converges.*

Proof Say the partial sums $s_n = \sum_{i=0}^n a_i$ are all bounded by M. To $\varepsilon > 0$ pick N with $b_N \le \varepsilon/(2M)$. For $q > p \ge N$ we then have

$$|\sum_{i=p}^{q} a_i b_i| = |\sum_{i=p}^{q-1} s_i(b_i - b_{i+1}) + s_q b_q - s_{p-1} b_p| \le M(\sum_{i=p}^{q-1}(b_i - b_{i+1}) + b_q + b_p) \le \varepsilon.$$

□

Proposition 3.2.3 *If $\sum v_n$ converges unconditionally in a normed space, then $\sum v_n = \sum v_{\sigma(n)}$ for every permutation σ of $\mathbb{N}$.*

Proof Say $r \equiv \|\sum v_n - \sum v_{\sigma(n)}\| > 0$. Pick a positive integer p_1 such that $\|\sum v_{\sigma(n)} - \sum_{n=1}^{p_1} v_{\sigma(n)}\| < r/3$. Then pick $q_1 \in \mathbb{N}$ such that $\sigma(\{1, \ldots, p_1\}) \subset \{1, \ldots, q_1\}$ and $\|\sum v_n - \sum_{n=1}^{q_1} v_n\| < r/3$. Then pick $p_2 \in \mathbb{N}$ such that $\{1, \ldots, q_1\} \subset \sigma(\{1, \ldots, p_2\})$ and $\|\sum v_{\sigma(n)} - \sum_{n=1}^{p_2} v_{\sigma(n)}\| < r/3$. Next pick q_2 to p_2 as above, and continue this way.

Let τ be the permutation of $\mathbb{N}$ by relisting it as $\sigma(1), \sigma(2), \ldots, \sigma(p_1)$, followed by the members of $1, \ldots, q_1$ not already listed, then $\sigma(1), \sigma(2), \ldots, \sigma(p_2)$ not already listed, then $1, \ldots, q_2$ not already listed, and so forth. The series $\sum v_{\tau(n)}$ cannot then converge since its partial sums swing within $\varepsilon/3$ between $\sum v_n$ and $\sum v_{\sigma(n)}$. □

Proposition 3.2.4 *A series $\sum v_n$ in a real or complex Banach space converges unconditionally if and only if each* subseries *$\sum_i v_{n_i}$ converges.*

Proof If $\sum_i v_{n_i}$ diverges, there are $r > 0$ and positive integers p_j, q_j such that $p_1 \le q_1 < p_2 \le q_2 < \cdots$ and $\|\sum_{i=p_k}^{q_k} v_{n_i}\| > r$ for every k. Then $\sum v_{\sigma(n)}$ diverges, where σ is the permutation of $\mathbb{N}$ into the order

$$n_{p_1}, n_{p_1+1}, \ldots, n_{q_1}, r_1, n_{p_2}, n_{p_2+1}, \ldots, n_{q_2}, r_2, \ldots$$

and $\{r_i\}$ is a sequence in ascending order of those elements $\mathbb{N}$ not found among the integers n_{p_j} and n_{q_j}.

Conversely, if $\sum v_{\tau(n)}$ diverges for some permutation τ of $\mathbb{N}$, there are $r > 0$ and integers p_j, q_j such that $0 < p_1 \le q_1 < p_2 \le q_2 < \cdots$ and $\|\sum_{i=p_k}^{q_k} v_{\tau(i)}\| > r$ for every k. We may also assume that $\sup \tau(\{p_k, \ldots, q_k\}) < \inf \tau(\{p_{k+1}, \ldots, q_{k+1}\})$ for each k. Let $\{n_i\}$ be the sequence of positive integers gotten by arranging $\tau(p_1), \ldots, \tau(q_1), \tau(p_2), \ldots, \tau(q_2), \ldots$ in ascending order. Then $\sum v_{n_i}$ diverges. □

Corollary 3.2.5 *If a series $\sum v_n$ in a real or complex Banach space converges unconditionally, then so does each of its subseries.*

It is also clear from the proposition above that if a series in a real or complex Banach space converges absolutely, then it converges unconditionally. The following result shows that the converse holds when the Banach space is just the real numbers.

Proposition 3.2.6 *Given numbers $A \leq B$ and a convergent real series $\sum a_n$ that diverge absolutely, there is a permutation σ of $\mathbb{N}$ with $\liminf_n \sum_{i=1}^n a_{\sigma(i)} = A$ and $\limsup_n \sum_{i=1}^n a_{\sigma(i)} = B$.*

Proof Pick numbers $A_n < B_n$ with $B_1 > 0$ such that $\lim A_n = A$ and $\lim B_n = B$. Let $p_1, p_2, \ldots$ be the non-negative terms of $\sum a_n$ in the order they occur, and let $q_1, q_2, \ldots$ be the absolute value of the negative terms of $\sum a_n$ in their original order. It is easy to see that both $\sum p_n$ and $\sum q_n$ diverge. Therefore there are smallest integers m_1, k_1 such that $s_1 \equiv p_1 + \cdots + p_{m_1} > B_1$ and $t_1 \equiv p_1 + \cdots + p_{m_1} - q_1 - \cdots - q_{k_1} < A_1$. Let m_2, k_2 be the smallest integers such that $s_2 \equiv p_1 + \cdots + p_{m_1} - q_1 - \cdots - q_{k_1} + p_{m_1+1} + \cdots + p_{m_2} > B_2$ and $t_2 \equiv p_1 + \cdots + p_{m_1} - q_1 - \cdots - q_{k_1} + p_{m_1+1} + \cdots + p_{m_2} - q_{k_1+1} - \cdots - q_{k_2} < A_2$, and so on. Then $|s_n - B_n| \leq p_{m_n}$ and $|t_n - A_n| \leq q_{k_n}$, so $s_n \to B$ while $t_n \to A$ as p_n and q_n converge to zero. □

By considering real and imaginary parts, we also see that an unconditionally convergent complex series is absolutely convergent. In fact, one checks that finite dimensional spaces are characterized among real or complex Banach spaces by the property that all unconditionally convergent series converge absolutely.

Lemma 3.2.7 *Given $\sum v_n$ in a normed space such that $\sum x(v_n)$ converges absolutely for every bounded linear functional x, there is a constant M such that $\sup_m \| \sum_{n=1}^m a_n v_n \| \leq M \|\{a_n\}\|_\infty$ for $\{a_n\} \in l^\infty(\mathbb{N})$.*

Proof The linear map T from the dual of the normed space to $l^1(\mathbb{N})$ given by $T(x) = \{x(v_n)\}$ is bounded by the closed graph theorem because if $x_i \to x$ and $T(x_i) \to \{b_n\} \in l^1(\mathbb{N})$, then $\sum |x_i(v_n) - b_n| \to 0$, so $x_i(v_n) \to b_n$. As also $x_i(v_n) \to x(v_n)$ we get $b_n = x(v_n)$. Hence

$$|x(\sum_{n=1}^{m} a_n v_n)| \leq \|\{a_n\}\|_\infty \|T(x)\|_1 \leq \|\{a_n\}\|_\infty \|T\| \|x\|$$

so the result follows from the Hahn-Banach theorem. □

The following result is known as the *bounded multiplier test*.

Proposition 3.2.8 *A series $\sum v_n$ in a real or complex Banach space is unconditionally convergent if and only if $\sum a_n v_n$ converges for any $\{a_n\} \in l^\infty(\mathbb{N})$.*

Proof The backward implication is clear from Proposition 3.2.4.

Conversely, let V be the subspace of $l^\infty(\mathbb{N})$ consisting of all $\{a_n\}$ such that $\sum a_n v_n$ converges. By Proposition 3.2.4 all $\{a_n\}$ with a_n either 0 or 1 belong to V. Let $\{b_n\}$ be of norm one in $l^\infty(\mathbb{N})$ and with only non-negative real entries. Let $0, s_{n1}s_{n2}\cdots$ be the binary expansion of b_n, so $\{b_n\} = \sum 2^{-i}\{s_{ni}\} \in V$ provided V is closed. So in this case $V = l^\infty(\mathbb{N})$, and it remains to prove closedness.

Suppose $\{a_{nj}\} \to \{a_n\}$ in $l^\infty(\mathbb{N})$ and $\{a_{nj}\} \in V$. As $\sum x(v_n)$ is absolutely convergent for any bounded linear functional x on the Banach space, there is M as in the lemma. Pick j and N such that $\|\{a_{nj}\} - \{a_n\}\|_\infty < \varepsilon$ and $\| \sum_{n=p}^q a_{nj} v_n \| < \varepsilon$ for $q > p \geq N$. Then $\| \sum_{n=p}^q a_n v_n \| < (1 + M)\varepsilon$ and $\{a_n\} \in V$. □

Corollary 3.2.9 *If a series $\sum v_n$ in a real or complex Banach space converges unconditionally, then so does $\sum a_n v_n$ for any $\{a_n\} \in l^\infty(\mathbb{N})$.*

Definition 3.2.10 A Schauder basis $\{v_n\}$ is *unconditional* if the expansion $\sum a_n v_n$ of each element in the Banach space converges unconditionally.

Thus $\{v_n\}$ is an unconditional Schauder basis if $\{\sigma(v_n)\}$ is a Schauder basis for every permutation σ of $\mathbb{N}$. Clearly the standard bases for $C_0(\mathbb{N})$ and $l^p(\mathbb{N})$ with $p \in [1, \infty\rangle$ are unconditional Schauder bases. Any unconditional Schauder basis can be normalized.

Proposition 3.2.11 *Given an unconditional Schauder basis $\{v_n\}$ for a Banach space V, the* bounded multiplier unconditional norm *$\|v\|_b$ obtained by taking the supremum of $\| \sum a_n v_n^*(v) v_n \|$ over the unit vectors $\{a_n\}$ in $l^\infty(\mathbb{N})$ is equivalent to the original norm on V and majorizes $\| \cdot \|_s$.*

Proof If $\|v\|_b = \infty$, then for n_1 pick $n_2 > n_1$ and scalars $a_1, \ldots, a_{n_2}$ of moduli not greater than one such that $\| \sum_{n=1}^{n_2} a_n v_n^*(v) v_n \| \geq 1 + \sum_{n=1}^{n_1} \|v_n^*(v) v_n\|$, so $\| \sum_{n=n_1+1}^{n_2} a_n v_n^*(v) v_n \| \geq 1$. We may thus construct $\{b_n\} \in l^\infty(\mathbb{N})$ such that the partial sums of $\sum b_n v_n^*(v) v_n$ is not Cauchy, contradicting unconditional convergence of $\sum v_n^*(v) v_n$ by the bounded multiplier test. So $\| \cdot \|_b$ is finite. Clearly it majorizes $\| \cdot \|_s$, so it is a norm, and as in the proof of Proposition 3.1.2, it remains to show completeness.

Let $\{w_i\}$ be Cauchy with respect to $\| \cdot \|_b$. Then it converges to w with respect to $\| \cdot \|$, and $v_n^*(w) = \lim v_n^*(w_i)$. To ε there are large i, j such that

$$\| \sum_{n=1}^{m} a_n v_n^*(w_i) v_n - \sum_{n=1}^{m} a_n v_n^*(w_j) v_n \| \leq \|w_i - w_j\|_b < \varepsilon$$

for every m and $\{a_n\} \in l^\infty(\mathbb{N})$. Letting first $j \to \infty$ and then $m \to \infty$, we get $\|w_i - w\|_b \leq \varepsilon$. □

The following result is now straightforward.

Corollary 3.2.12 *Consider the bounded multiplier unconditional norm $\| \cdot \|_b$ given by a normalized unconditional Schauder basis $\{v_n\}$ for V. Then V is a commutative non-unital Banach algebra with respect to the norm $\| \cdot \|_b$ and the product $(\sum a_n v_n)(\sum b_n v_n) = \sum a_n b_n v_n$.*

Recall that an *ordered vector space* is a real vector space with a partial order such that $v + u \leq w + u$ and $av \leq aw$ whenever $v \leq w$ and $a > 0$. It is a *vector lattice* if in addition every two vectors have a least upper bound. And it is a *normed lattice* if in addition it is a normed space such that $\|v\| \leq \|w\|$ when $|v| \leq |w|$, where the *absolute value* $|v|$ is the least upper bound of v and $-v$.

Proposition 3.2.13 *Any real Banach space with an unconditional basis $\{v_n\}$ is a $\| \cdot \|_b$-normed lattice by letting $v \leq w$ mean $v_n^*(v) \leq v_n^*(w)$ for all n. The least upper bound of v and w is $\sum \sup\{v_n^*(v), v_n^*(w)\} v_n$, and $|v| = \sum |v_n^*(v)| v_n$.*

Proof The result is clear from the bounded multiplier test and the easily established fact that $\|v\|_b \leq \|w\|_b$ when $|v_n^*(v)| \leq |v_n^*(w)|$ for all n. □

Both $C_0(\mathbb{N})$ and $l^p(\mathbb{N})$ for $p \in [0, \infty)$ are commutative non-unital Banach algebras with respect to $\| \cdot \|_b = \| \cdot \|$ and products $\{a_n\}\{b_n\} = \{a_n b_n\}$. In the real case they also become normed lattices by declaring $\{a_n\} \leq \{b_n\}$ when $a_n \leq b_n$ for every n.

The bounded multiplier unconditional norm can be used to prove boundedness of linear maps.

Proposition 3.2.14 *If $\{v_n\}$ is an unconditional Schauder basis for a Banach space V and $\{a_n\} \in l^\infty(\mathbb{N})$, the linear map T on V given by $T(v) = \sum a_n v_n^*(v) v_n$ is bounded, and the supremum of $\|T\|$ over all such $\{a_n\}$ of norm one, is finite.*

Proof Boundedness of T is clear from Proposition 3.2.11 since boundedness with respect to $\| \cdot \|_b$ follows straight from definitions.

The supremum of $\|T(v)\|$ over all $\{a_n\} \in l^\infty(\mathbb{N})$ of norm one is not greater than $\|v\|_b$, so we are done by the principle of uniform boundedness. □

Suppose $\{v_n\}$ is an unconditional Schauder basis for a Banach space V. We define the sum $\sum_{n \in A} v_n$ over $A \subset \mathbb{N}$ by listing the members of A in any order and adding the corresponding v_n's in that order; setting $\sum_{n \in \phi} v_n = 0$. This is well-defined by Corollary 3.2.5. Let P_A be the map T in the proposition above with $\{a_n\}$ the characteristic function of A, so P_A is a bounded projection on V onto the closed span of $\{v_n\}_{n \in A}$, and the *unconditional basis constant* $K_{ub} < \infty$ is the supremum of $\|P_A\|$ over all $A \subset \mathbb{N}$. Clearly K_{ub} is the least constant M such that $\|\sum_{n \in A} a_n v_n\| \leq M \|\sum_{n \in B} a_n v_n\|$ for all finite subsets $A \subset B$ and scalars a_n. This can evidently be used as a criterion to check whether a Schauder basis $\{v_n\}$ is unconditional.

The supremum of $\|T\|$ in the proposition above over all sequences $\{a_n\} \in l^\infty(\mathbb{N})$ of *signs*, meaning that all $a_n = \pm 1$, is called the *unconditional constant* $K_u < \infty$. It is straightforward to check that $1 \leq K_b \leq K_{ub} \leq (1 + K_u)/2 \leq K_u \leq 2K_{ub}$, so the constants are all one if $K_u \leq 1$, which for instance happens when K_u is calculated with respect to the $\| \cdot \|_b$-norm.

3.3 Equivalent Bases

Definition 3.3.1 Schauder bases $\{v_n\}$ and $\{w_n\}$ in two Banach spaces are *equivalent* if $\sum a_n v_n$ converges precisely when $\sum a_n w_n$ converges.

Proposition 3.3.2 *Two Schauder bases $\{v_n\}$ and $\{w_n\}$ in Banach spaces V and W, respectively, are equivalent if and only if there is a bounded isomorphism $T: V \to W$ such that $T(v_n) = w_n$ for all n.*

Proof For the forward implication consider the linear isomorphism given by $T(\sum a_n v_n) = \sum a_n w_n$. Say $u_i \to u$ and $T(u_i) \to v$. By continuity of the coordinate functionals $v_n^*(u_i) \to v_n^*(u)$, while $v_n^*(u_i) = w_n^* T(u_i) \to w_n^*(v)$, so $v_n^*(u) = w_n^*(v)$ and $T(u) = v$. Hence T is bounded by the closed graph theorem, and its inverse map is bounded by the open mapping theorem. The opposite direction is trivial. □

Corollary 3.3.3 *A Schauder basis equivalent to an unconditional Schauder basis is unconditional.*

Proposition 3.3.4 *Suppose $\{v_n\}$ is a Schauder basis with basis constant K_b, and that $\{w_n\}$ is another sequence in the Banach space with $\sum \|v_n\|^{-1}\|v_n - w_n\|$ less than $1/(2K_b)$. Then $\{w_n\}$ is a Schauder basis equivalent to $\{v_n\}$.*

Proof We may assume that $\{v_n\}$ is normalized. Clearly $\|v_n^*\| \leq 2K_b$ for all n, so $T(v) = \sum v_n^*(v)(v_n - w_n)$ is an operator on the Banach space V with $\|T\| < 1$. Thus $I - T$ is invertible in the Banach algebra $B(V)$, and $(I - T)(v_n) = w_n$. □

Corollary 3.3.5 *Any Schauder basis has an equivalent Schauder basis in a given dense subset of the Banach space.*

Let us focus on bases equivalent to the standard bases in $l^1(\mathbb{N})$ and $C_0(\mathbb{N})$.

Proposition 3.3.6 *A sequence $\{v_n\}$ in a real or complex Banach space is a Schauder basic sequence equivalent to the standard basis in $l^1(\mathbb{N})$ if and only if* $\sup \|v_n\| < \infty$ *and there is $M > 0$ such that $\sum^m |a_n| \leq M\|\sum^m a_n v_n\|$ for any m and scalars a_i.*

Proof For the forward implication, let T be a base preserving bounded isomorphism from the closed span of $\{v_n\}$ onto $l^1(\mathbb{N})$. Note that $\|v_n\| = \|T^{-1}(e_n)\| \leq \|T^{-1}\|$. We also get the desired inequality with $M = \|T\|$.

Conversely, that inequality for a general M shows that $M\|v_n\| \geq 1$, so $v_n \neq 0$. Moreover, for $m \leq n$, we have

$$\|\sum^m a_i v_i\| \leq (\sup \|v_j\|) \sum^m |a_i| \leq (\sup \|v_j\|) \sum^n |a_i| \leq M(\sup \|v_j\|)\|\sum^n a_i v_i\|$$

for any scalars a_i. Hence $\{v_n\}$ is a Schauder basic sequence, and

$$\sum_{i=m}^n |a_i| \leq M\|\sum_{i=m}^n a_i v_i\| \leq M(\sup \|v_j\|) \sum_{i=m}^n |a_i|$$

shows that it is equivalent to $\{e_n\}$. □

The following result is proved along the same lines.

Proposition 3.3.7 *A Schauder basic sequence $\{v_n\}$ in a real or complex Banach space is equivalent to the standard basis in $C_0(\mathbb{N})$ if and only if* $\inf\|v_n\| > 0$ *and there is $M > 0$ such that $\|\sum^m a_n v_n\| \leq M \sup |a_i|$ for any m and scalars a_i.*

Definition 3.3.8 A series $\sum v_n$ in a real or complex Banach space V is *weakly unconditional Cauchy* if $\sum_n |x(v_n)| < \infty$ for every $x \in V^\star$.

One should see the previous result in light of the following lemma.

Lemma 3.3.9 *A series $\sum v_n$ in a real or complex Banach space is weakly unconditionally Cauchy if and only if $\|\sum^m a_n v_n\| \leq M \sup|a_i|$ for some $M > 0$ and any m and scalars a_i.*

Proof The forward implication follows from Lemma 3.2.7.

Conversely, if $\sum x(v_n)$ is not absolutely convergent for some $x \in V^\star$, there are increasing integers $n_1 = 1, n_2, n_3, \ldots$ with $\sum_{n=n_i}^{n_{i+1}-1} |x(v_n)| \geq i$. Let b_i be the scalar of modulus $1/i$ such that $b_n x(v_n) = |x(v_n)|/i$ when $n \in [n_i, n_{i+1}\rangle$. Then $\{b_n\} \in C_0(\mathbb{N})$, but $\sum b_n x(v_n) = \infty$ and $\sum b_n v_n$ diverges. □

Lemma 3.3.10 *Suppose $\sum v_n$ is weakly unconditional Cauchy but not unconditionally convergent. Then we have increasing integers $n_1, n_2, \ldots$ and $m_1, m_2, \ldots$ such that $w_k = \sum_{i=m_k}^{m_{k+1}-1} v_{n_i}$ form a Schauder basic sequence equivalent to the standard basis in $C_0(\mathbb{N})$.*

Proof Pick $\{v_{n_i}\}$ that diverges, so there is $\varepsilon > 0$ and increasing integers $m_1, m_2, \ldots$ such that $\|w_k\| \geq \varepsilon$. Clearly $\sum w_k$ is weakly unconditionally convergent, so by Proposition 3.1.11 we may assume that $\{w_k\}$ is a Schauder basic sequence, which is equivalent to the standard basis in $C_0(\mathbb{N})$ by the previous proposition and lemma. □

Proposition 3.3.11 *The Banach space $C_0(\mathbb{N})$ cannot be embedded in a Banach space V if and only if all weakly unconditionally Cauchy series in V are convergent if and only if all weakly unconditionally Cauchy series in V are unconditionally convergent.*

Proof By the previous proposition and Lemma 3.3.9, whenever we have a bounded isomorphism $T: C_0(\mathbb{N}) \to V$, then $\sum T(e_n)$ diverges as $\inf\|T(e_n)\| > 0$, while $\sum T(e_n)$ is weakly unconditionally Cauchy.

The last lemma above settles the remaining non-trivial implication. □

The series $\sum e_n$ for the standard basis $\{e_n\}$ in $C_0(\mathbb{N})$ is weakly unconditionally Cauchy by Lemma 3.3.9 since $\|\sum^m e_n\|_u = 1$ for all m. However, it cannot converge weakly to any $v \in C_0(\mathbb{N})$ since $|e_m^*(v)| < 1/2$ for large m while $e_m^*(\sum^m e_n) = 1$.

Definition 3.3.12 A series $\sum v_n$ in a real or complex Banach space is *weakly subseries convergent* if each subseries of it is weakly convergent.

The following result is known as the *Orlicz-Pettis theorem.*

Proposition 3.3.13 *A series is weakly subseries convergent if and only if it is unconditionally convergent.*

Proof Given a weakly subseries convergent series that is not unconditionally convergent, we have by the previous lemma a basic Schauder sequence $\{w_k\}$ equivalent to the standard basis $\{e_k\}$ in $C_0(\mathbb{N})$. Let T be the bounded isomorphism from $C_0(\mathbb{N})$ to the closed span of $\{w_k\}$ that sends e_k to w_k. Then T^{-1} is bounded and thus weak-to-weak continuous, and since $\sum w_k$ converges weakly, so must $\sum e_n$, and this is false.

For the converse, note that any subseries of an unconditional convergent series is actually norm convergent. □

This result clearly implies that any weakly subseries convergent series $\sum v_n$ is *weakly reorder convergent*, in that, the series $\sum v_{\sigma(n)}$ is weakly convergent for any permutation σ of $\mathbb{N}$.

Definition 3.3.14 Given a Schauder basic sequence $\{v_n\}$ and integers p_i with $p_1 = 1 < p_2 < p_2 < \cdots$, then $w_n = \sum_{i=p_n}^{p_{n+1}-1} a_i v_i$ for scalars a_i not all zero, form a *block Schauder basic sequence* taken from $\{v_n\}$.

It is straightforward to check that any block Schauder basic sequence is a Schauder basic sequence with basis constant not greater than that of the original sequence.

Proposition 3.3.15 *Suppose $\{v_n\}$ is a Schauder basic sequence equivalent to the standard basis $\{e_n\}$ for $l^1(\mathbb{N})$. Then every permutation, subsequence or normalized block Schauder basic sequence taken from $\{v_n\}$ is equivalent to $\{e_n\}$. The same statement holds with $l^1(\mathbb{N})$ replaced by $C_0(\mathbb{N})$.*

Proof Let $\{w_k\}$ be a block Schauder basic sequence taken from $\{v_n\}$, and let T be the bounded isomorphism from $l^1(\mathbb{N})$ to the closed span of $\{v_n\}$ such that $T(e_n) = v_n$. Since

$$\sum^m |a_n| \leq \|T\| \sum^m |a_n| \|w_n\|^{-1} \|T^{-1}(w_n)\|_1 \leq \|T\| \|T^{-1}\| \| \sum^m a_n \|w_n\|^{-1} w_n \|$$

for any m and scalars a_i, the Schauder basic sequence $\{w_n/\|w_n\|\}$ is equivalent to $\{e_n\}$ by Proposition 3.3.6. The same proposition shows that permutations and subsequences remain equivalent to the standard basis for $l^1(\mathbb{N})$.

The proof for $C_0(\mathbb{N})$ is very similar, then referring to Proposition 3.3.7. □

We also include *The Bessaga-Pelczynski Selection Principle.*

Theorem 3.3.16 *Suppose $\{v_n\}$ is a Schauder basic sequence. Any sequence $\{w_n\}$ in the same Banach space that does not converge to zero in norm, but satisfies $\lim_m v_n^*(w_m) = 0$ for all n, has a subsequence equivalent to a block Schauder basic sequence taken from $\{v_n\}$.*

Proof By going to a subsequence, we may assume that $\inf\|w_m\| > 0$. Then $\lim_m v_n^*(w_m/\|w_m\|) = 0$ for all n, so we may assume that every w_m has norm one.

Let K be the basis constant for $\{v_n\}$. Set $q_1 = 1$. By assumption we can inductively produce increasing integers m_i and q_i such that both $\| \sum_{n=1}^{q_i} v_n^*(w_{m_i}) v_n \|$ and $\| \sum_{n=q_{i+1}+1}^{\infty} v_n^*(w_{m_i}) v_n \|$ are less than $K^{-1} 2^{-(i+3)}$. Set $u_i = \sum_{n=q_i+1}^{q_{i+1}} v_n^*(w_{m_i}) v_n$. Then

$$1 = \|w_{m_i}\| \leq \| \sum_{n=1}^{q_i} v_n^*(w_{m_i}) v_n \| + \|u_i\| + \| \sum_{n=q_{i+1}+1}^{\infty} v_n^*(w_{m_i}) v_n \|$$

shows that $\|u_i\| > 1/2$. So $\{u_n\}$ is a block Schauder basic sequence taken from $\{v_n\}$ with basis constant not exceeding K. Since

$$\sum_i \|u_i\|^{-1}\|u_i - w_{m_i}\| \leq 2\sum_i (\|\sum_{n=1}^{q_i} v_n^*(w_{m_i})v_n\| + \|\sum_{n=q_{i+1}+1}^{\infty} v_n^*(w_{m_i})v_n\|)$$

is less than $2\sum_i K^{-1}2^{-(i+2)} = 1/(2K)$, it follows then by Proposition 3.3.4 that $\{w_{m_n}\}$ is a Schauder basic sequence equivalent to $\{u_n\}$. □

3.4 Dual Bases

Under the isomorphism $l^p(\mathbb{N})^\star \cong l^q(\mathbb{N})$ for conjugate exponents $p, q \in \langle 1, \infty \rangle$, the coordinate functionals $\{e_n^*\}$ of the standard basis for $l^p(\mathbb{N})$ will be the standard basis for $l^q(\mathbb{N})$. Similarly $C_0(\mathbb{N})^\star \cong l^1(\mathbb{N})$ shows that the coordinate functionals of the standard basis in $C_0(\mathbb{N})$ is a Schauder basis for the dual space. The coordinate functionals of the standard basis in $l^1(\mathbb{N})$ is not a Schauder basis in the dual space $l^\infty(\mathbb{N})$; this space is not even separable as $\|\chi_A - \chi_B\|_\infty = 1$ for distinct subsets A and B of $\mathbb{N}$. We do however have the following result.

Proposition 3.4.1 *Let $\{v_n\}$ be a Schauder basis for a Banach space V with basis constant K. Then $\{v_n^*\}$ is a Schauder basic sequence in $V^\star$ with basis constant not greater than K, and its n-th coordinate functional is $v_n \in V \subset V^{\star\star}$ restricted to the closed span of $\{v_n^*\}$.*

Proof Given $r > 0$, integers $n \geq m > 0$ and scalars a_i, there are scalars b_j such that $\|\sum b_j v_j\| = 1$ and $|(\sum^m a_i v_i^*)(\sum b_j v_j)| \geq \|\sum^m a_i v_i^*\| - r$. One easily checks that $\|\sum^m a_i v_i^*\| \leq K\|\sum^n a_i v_i^*\| + r$. The rest is even easier. □

Proposition 3.4.2 *Let $\{v_n\}$ be a Schauder basis for V. Then $x \in V^\star$ is the w^*-limit of $\{\sum^n a_i v_i^*\}_n$ for unique a_i, and $v_m(x) = a_m$ for all m, where $V \subset V^{\star\star}$.*

Proof The formula $(\sum^n x(v_i)v_i^*)(\sum b_j v_j) = x(\sum^n b_j v_j)$ for any $\sum b_j v_j \in V$, shows that x is the w^*-limit of $\{\sum^n x(v_i)v_i^*\}_n$. The rest is clear. □

Lemma 3.4.3 *Let $\{v_n\}$ be a Schauder basis for V with basis constant K. The restriction $A(v)$ of $v \in V \subset V^{\star\star}$ to the closed span W of all v_n^* defines a norm decreasing linear map $A\colon V \to W^\star$ such that $\|A(v)\| \geq K^{-1}\|v\|$. It is an isometric embedding when $\{v_n\}$ is monotone.*

Proof The map A is in general a norm decreasing linear map since $V \subset V^{\star\star}$ is isometric and the restriction map from $V^{\star\star}$ to $W^\star$ is linear and norm decreasing. It remains to prove the inequality for v in the linear span of all v_n with $n-1$ less than a given integer m. To such v pick by the Hahn-Banach theorem a norm one functional

x on V such that $|x(v)| = \|v\|$. Let $y = \sum^m x(v)v_n^*$. Then

$$|y(\sum a_n v_n)| = |x(\sum^m a_n v_n)| \le K\|\sum a_n v_n\|$$

for $\sum a_n v_n$, so $\|v\| = |y(v)| = |A(v)(y)| \le \|y\|\|A(v)\| \le K\|A(v)\|$. □

Definition 3.4.4 A Schauder basis $\{v_n\}$ for a Banach space V is *shrinking* if the norm $\|x\|_{(m)}$ of the restriction of any $x \in V^\star$ to the closed span of $\{v_n\}_{n>m}$ tends to zero as $m \to 0$.

Proposition 3.4.5 *A Schauder basis $\{v_n\}$ for V is shrinking if and only if $\{v_n^*\}$ is a Schauder basis for $V^\star$.*

Proof Let K be the basis constant for $\{v_n\}$. For $\sum b_n v_n \in V$ and $x \in V^\star$, we get

$$|(x - \sum^m x(v_n)v_n^*)(\sum b_n v_n)| = |x(\sum_{n=m+1}^{\infty} b_n v_n)|$$

and

$$\|\sum_{n=m+1}^{\infty} b_n v_n\| \le \|\sum b_n v_n\| + \|\sum^m b_n v_n\| \le (1+K)\|\sum b_n v_n\|,$$

so $\|x - \sum^m x(v_n)v_n^*\| \le \|x\|_{(m)}(1+K)$, which proves the forward implication.

The converse is clear as $\|\sum a_n v_n^*\|_{(m)} \le \|\sum_{n=m+1}^{\infty} a_n v_n^*\|$ for $\sum a_n v_n^* \in V^\star$. □

Definition 3.4.6 A Schauder basic sequence $\{v_n\}$ is *boundedly complete* if $\sum a_n v_n$ converges whenever $\sup_m \|\sum^m a_n v_n\| < \infty$.

Proposition 3.4.7 *If $\{v_n\}$ is a shrinking Schauder basis for V, then $\{v_n^*\}$ is a boundedly complete Schauder basis for $V^\star$.*

Proof By the previous proposition we know that $\{v_n^*\}$ is a Schauder basis. If the sequence $\{\sum^m a_n v_n^*\}_m$ is bounded, it has by Alaoglu's theorem a subnet that converges to some $x \in V^\star$ in the w^*-topology, so $\sum a_n v_n^* = \sum x(v_n)v_n^* = x$. □

Summing up, we get the following result.

Proposition 3.4.8 *Let $\{v_n\}$ be a Schauder basis for V. Then $\{v_n^*\}$ is a Schauder basis for $V^\star$ if and only if the closed span of it is $V^\star$ if and only if $\{v_n\}$ is shrinking if and only if $\{v_n^*\}$ is boundedly complete.*

Proof If $\{v_n^*\}$ is boundedly complete, then $x \in V^\star$ is by Proposition 3.4.2 the w^*-limit of $\{\sum^m x(v_n)v_n^*\}_m$, which is bounded by the principle of uniform boundedness, so it converges to x in norm. Thus the closed span of $\{v_n^*\}$ is $V^\star$. The remaining implications follow from the last three propositions. □

Lemma 3.4.9 *The map A from Lemma 3.4.3 is surjective if and only if* $\{v_n\}$ *is boundedly complete.*

Proof For the forward implication, note that $\{A(v_n)\}$ is boundedly complete by the previous proposition. The same is then true for $\{v_n\}$.

Conversely, let x belong to the dual of the closed span of $\{v_n^*\}$ for a boundedly complete Schauder basis $\{v_n\}$ with basis constants K and L, respectively. For any convergent $\sum a_n v_n^*$ and $m \in \mathbb{N}$, we have

$$|A(\sum^m x(v_n^*)v_n)(\sum a_n v_n^*)| = |x(\sum^m a_n v_n^*)| \leq K\|x\|\|\sum a_n v_n^*\|,$$

so $\|\sum^m x(v_n^*)v_n\| \leq KL\|x\|$ by Lemma 3.4.3. Hence $\sum x(v_n^*)v_n$ converges. Since $A(\sum x(v_n^*)v_n)(v_k^*) = x(v_k^*)$, we get $A(\sum x(v_n^*)v_n) = x$, as desired. □

By Lemma 3.4.3 and the previous lemma and proposition, we see that a Banach space with a (monotone) boundedly complete Schauder basis is (isometrically) isomorphic to the dual of a Banach space with a shrinking Schauder basis. Also, a Schauder basis $\{v_n\}$ is boundedly complete if and only if the basic sequence $\{v_n^*\}$ is shrinking.

Theorem 3.4.10 *A Banach space with a Schauder basis is reflexive if and only if it has a Schauder basis that is both shrinking and boundedly complete if and only if every Schauder basis of it is shrinking and boundedly complete. In particular, the coordinate functionals of a Schauder basis in a reflexive Banach space V form a Schauder basis for* $V^\star$.

Proof If the Banach space V is reflexive with a Schauder basis $\{v_n\}$, then the closed span W of $\{v_n^*\}$ is a convex w^*-closed subset of $V^\star$, so $W = V^\star$ by Proposition 3.4.2. Thus $\{v_n\}$ is shrinking by the previous proposition. Similarly, we see that $\{v_n^*\}$ is shrinking, so $\{v_n\}$ is also boundedly complete.

The remaining implications follow from the previous lemma and proposition. □

Lemma 3.4.11 *The Banach space* $l_1(\mathbb{N})$ *is embedded in a real or complex Banach space V if* $C_0(\mathbb{N})$ *is embedded in* $V^\star$.

Proof Let $\{e_n\}$ and $\{f_n\}$ be the standard bases for $C_0(\mathbb{N})$ and $l^1(\mathbb{N})$, respectively. On occasion we will identify the Banach spaces $C_0(\mathbb{N})^\star$ and $l^1(\mathbb{N})$. By assumption we have a bounded isomorphism $A\colon C_0(\mathbb{N}) \to V^\star$ onto its image. By Corollary 2.4.27 its adjoint map is surjective and hence open by the open mapping theorem. So there is $r > 0$ such that all $f_n \in A^*(rB)$, where B is the closed unit ball in $V^{\star\star}$.

Consider now the neighborhood U_n of f_n consisting of those $x \in C_0(\mathbb{N})^\star$ such that $|x(e_n)| > 1/2$ but $|x(e_m)| < 1/n$ when $m < n$. By the Goldstine theorem the closed unit ball in V is w^*-dense in B, and since A^* is w^*-to-w^* continuous by Proposition 2.4.22, there is $v_n \in V$ with $\|v_n\| \leq r$ and $A^*(v_n) \in U_n$. By the Bessaga-Pelczynski selection principle there is a subsequence $\{v_{i_n}\}$ such that $A^*(v_{i_n})$ is a Schauder basic sequence equivalent to some block Schauder basic sequence $\{y_n\}$ taken from $\{f_n\} \subset C_0(\mathbb{N})$. By Proposition 3.3.15 the Schauder basic sequence $\{\|y_n\|^{-1}A^*(v_{i_n})\}$ is therefore equivalent to $\{f_n\}$, say with a identifying bounded isomorphism T from the closed span of the former sequence to the latter sequence. So there is t such that $t \geq \|\|y_n\|^{-1}A^*(v_{i_n})\| > 1/(2\|y_n\|)$. Hence, for $k \in \mathbb{N}$ and scalars a_i, we have

$$\sum^k |a_n| \leq 2t\|y_n\|\|\sum^k a_n f_n\|_1 = 2t\|TA^*(\sum^k a_n v_{i_n})\|_1 \leq 2t\|TA^*\|\|\sum^k a_n v_{i_n}\|,$$

so $\{v_{i_n}\}$ is a Schauder basic sequence equivalent to $\{f_n\}$, see Proposition 3.3.6. □

Lemma 3.4.12 *Any real or complex dual normed space with the Banach space $C_0(\mathbb{N})$ embedded in it cannot be separable.*

Proof We may assume that the dual space is $V^\star$ for a Banach space V. By the previous lemma we have an embedding of $l^1(\mathbb{N})$ in V. Its adjoint maps $V^\star$ onto the non-separable space $l^1(N)^\star$, so $V^\star$ cannot be separable. □

Theorem 3.4.13 *Let V be a real or complex Banach space with a Schauder basis $\{v_n\}$. Then the Banach space $l^1(\mathbb{N})$ cannot be embedded in V if $\{v_n\}$ is shrinking, and the Banach space $C_0(\mathbb{N})$ cannot be embedded in V if $\{v_n\}$ is boundedly complete. The converse statements hold when $\{v_n\}$ is unconditional.*

Proof If $\{v_n\}$ is boundedly complete, then Lemmas 3.4.3 and 3.4.9 tell us that the Banach space V is isomorphic to the dual of the closed span of $\{v_n^*\}$, and thus cannot have $C_0(\mathbb{N})$ embedded in it by the previous lemma.

If we have an embedding of $l^1(\mathbb{N})$ in V, its adjoint maps $V^\star$ onto the non-separabel space $l^1(N)^\star$, so V cannot have a shrinking basis by Proposition 3.4.8.

Suppose for the remaining part of the proof that $\{v_n\}$ is unconditional. We may and will also replace the norm with the equivalent bounded multiplier unconditional norm associated to $\{v_n\}$, using the same symbol $\|\cdot\|$.

If $\{v_n\}$ is not shrinking, there is $x \in V^\star$ such that $r = (\lim_m \|x\|_{(m)})/2 > 0$. Pick an increasing sequence $\{m_n\}$ of positive integers and a sequence of unit vectors w_n in the span of $v_{m_n}, \ldots, v_{m_{n+1}+1}$ such that $x(w_n) > r$. For $m \in \mathbb{N}$ and scalars a_i, we have by Proposition 3.2.13, that

$$r\sum^m |a_n| \leq x(\sum^m |a_n|w_n) \leq \|x\|\|\sum^m |a_n|w_n\| = \|x\|\|\sum^m a_n w_n\|.$$

Hence $\{w_n\}$ is a Schauder basic sequence equivalent to the standard basis of $l^1(\mathbb{N})$ by Proposition 3.3.6, so $l^1(\mathbb{N})$ is embedded in V.

Finally, if $\{v_n\}$ is not boundedly complete, there is $M > 0$ and scalars b_i such that $\sum b_n v_n$ diverges despite the fact that $\|\sum^m b_n v_n\| \leq M$ for all m. So there are $t > 0$ and integers p_i, q_i such that $0 < p_1 \leq q_1 < p_2 \leq q_2 < \cdots$ and $\|u_n\| \geq t$ for all n, where $u_n = \sum_{i=p_n}^{q_n} b_i v_i$. Again by Proposition 3.2.13, we get for $m \in \mathbb{N}$ and scalars a_i, that

$$\|\sum^m a_n u_n\| \le (\sup_i |a_i|)\|\sum^m u_n\| \le (\sup_i |a_i|)\|\sum^{q_m} b_n v_n\| \le M(\sup_i |a_i|).$$

Hence $\{u_n\}$ is a Schauder basic sequence equivalent to the standard basis for $C_0(\mathbb{N})$ by Proposition 3.3.7, so $C_0(\mathbb{N})$ is embedded in V. □

Corollary 3.4.14 *Every unconditional basis of a Banach space is shrinking (or boundedly complete), or none is. Any Banach space with an unconditional Schauder basis is reflexive if and only if neither the Banach space $C_0(\mathbb{N})$ nor $l^1(\mathbb{N})$ is embedded in it. In particular, any Banach space with an unconditional Schauder basic sequence has a reflexive subspace or a copy of $C_0(\mathbb{N})$ or a copy of $l^1(\mathbb{N})$.*

Proof This is immediate from the previous two theorems. □

3.5 The James Space *J*

Let $\{v_n\}$ be a Schauder basis for a Banach space. The set S of sequences $\{a_n\}$ of scalars such that $\|\{a_n\}\|_S \equiv \sup_m \|\sum^m a_n v_n\| < \infty$ is a Banach space under coordinate-wise operations and norm $\|\cdot\|_S$. The proof of this statement is analogous to that of showing that $\|\cdot\|_s$ was a complete norm.

Proposition 3.5.1 *Let $\{v_n\}$ be a shrinking Schauder basis for a Banach space V. Then the formula $f(v) = \{v(v_n^*)\}$ defines a bounded isomorphism $f : V^{\star\star} \to S$. When $\{v_n\}$ is also monotone, then f is isometric, and $\|v\|_S = \lim_m \|\sum^m v(v_n^*)v_n\|$ for $v \in V^{\star\star}$.*

Proof Replacing the norm on V by $\|\cdot\|_s$, we may assume that $\{v_n\}$ is monotone. Then $\{v_n^*\}$ is also a monotone Schauder basis for $V^\star$ by Propositions 3.4.8 and 3.4.1. Hence

$$|x(\sum^m v(v_n^*)v_n)| = |v(\sum^m x(v_n)v_n^*)| \le \|v\|\|x\|$$

for $m \in \mathbb{N}$ and $v \in V^{\star\star}$ and $x \in V^\star$. So $f(v) \in S$ with $\|f(v)\| \le \|v\|$ and f is clearly linear. If $f(v) = 0$, then $v(x) = v(\sum x(v_n)v_n^*) = \sum x(v_n)v(v_n^*) = 0$, so $v = 0$.

If $\{a_n\} \in S$ and $m > k$, then

$$|\sum_{n=k+1}^m a_n x(v_n)| \le \|x\|_{(k)}\|\sum_{n=k+1}^m a_n v_n\| \le 2\|x\|_{(k)}\|\{a_n\}\|_S$$

shows that $\sum a_n x(v_n)$ converges, and we can define a linear functional w on $V^\star$ by $w(x) = \sum a_n x(v_n)$. Clearly its norm does not exceed $\|\{a_n\}\|_S$, and $f(w) = \{a_n\}$.

Similarly, we see that $|v(x)| \le \|f(v)\|_S \|x\|$ for $v \in V^{\star\star}$ and $x \in V^\star$, so f is an isometric isomorphism. □

Thus, if $\{v_n\}$ is a monotone shrinking Schauder basis for a Banach space V, then V and $V^\star$ and $V^{\star\star}$ can be identified with the space of sequences $\{a_n\}$ such that $\sum^m a_n v_n$ converges and $\sum^m a_n v_n^*$ converges and $\sup_m \|\sum^m a_n v_n\| < \infty$, respectively, generalizing the usual way one identifies the dual $l^1(\mathbb{N})$ and bidual $l^\infty(\mathbb{N})$ of $C_0(\mathbb{N})$.

Definition 3.5.2 The *James space* J is the real subspace of $C_0(\mathbb{N})$ consisting of those sequences $\{a_n\}$ such that

$$\|\{a_n\}\|_a \equiv 2^{-1/2} \sup(\sum_{n=1}^{m-1} (a_{p_n} - a_{p_{n+1}})^2 + (a_{p_m} - a_{p_1})^2)^{1/2} < \infty,$$

where we sup over all $m \ge 2$ and $p_1 < \cdot < p_m$. Let $\|\{a_n\}\|_b$ be defined similarly, but with the term $(a_{p_m} - a_{p_1})^2$ skipped.

Proposition 3.5.3 *The James space J is a real Banach space with norm $\|\cdot\|_a$ equivalent to the norm $\|\cdot\|_b$.*

Proof That J is a normed real space is clear from Minkowski's inequality for sequences of numbers, which also shows that $\|\{a_n\}\|_a \le 2\|\{a_n\}\|_b$; the inequality $\|\{a_n\}\|_b \le \|\{a_n\}\|_a$ is obvious.

Note that

$$|a_{p_1} - a_{p_2}| = 2^{-1/2}((a_{p_1} - a_{p_2})^2 + (a_{p_2} - a_{p_1})^2)^{1/2} \le \|\{a_n\}\|_a,$$

so $|a_{p_1}| \le \|\{a_n\}\|_a$ as $p_2 \to \infty$. Hence, if $\{\{a_{ni}\}_n\}_i$ is Cauchy, then $\{a_{ni}\}_i$ is Cauchy and converges, say to a_n, for each n. As with $\|\cdot\|_s$ one shows that $\{a_n\} \in J$ and that $\|\{a_{ni}\} - \{a_n\}\|_a < \varepsilon$ for large enough i. □

Proposition 3.5.4 *The standard basis $\{e_n\}$ for $C_0(\mathbb{N})$ is a shrinking monotone Schauder basis for J with respect to $\|\cdot\|_a$, and $\{a_n\} = \sum a_n e_n$ for $\{a_n\} \in J$.*

Proof The standard basis belongs to J as $\|\{e_n\}\|_a = 1$. It is a monotone Schauder basic sequence in J by Proposition 3.1.4. Let $\{a_n\} \in J$. If $\|\{a_n\} - \sum^m a_n e_n\|_a$ does not tend to 0 as $m \to \infty$, there is $\varepsilon > 0$ and for each $i \in \mathbb{N}$, a collection $\{p_{1,i}, \ldots, p_{m(i),i}\}$ of integers such that $0 < p_{1,i} < \cdots < p_{m(i),i} < p_{1,i+1}$ and

$$\sum_{n=1}^{m(i)-1} (a_{p_{n,i}} - a_{p_{n+1,i}})^2 + (a_{p_{m(i),i}} - a_{p_{1,i}})^2 \ge \varepsilon.$$

As $\lim a_n = 0$, there is $j \in \mathbb{N}$ such that $\sum_{n=1}^{m(i)-1}(a_{p_{n,i}} - a_{p_{n+1,i}})^2 \geq \varepsilon/2$ for all $i \geq j$, which implies that $\|\{a_n\}\|_a$ is infinite.

It remains to show that $\{e_n\}$ is shrinking. If it is not, there is $x \in J^\star$ and $r > 0$ such that $x(v_n) \geq r$ for some $\|\cdot\|_b$-normalized block sequence $\{v_n\}$ taken from $\{e_n\}$, so $\sum n^{-1}v_n$ cannot converge. Say $v_i = \sum_{n=q_i}^{q_{i+1}-1} b_n e_n$ for integers $1 = q_1 < q_2 < \cdots$ and $b_n \in \mathbb{R}$ with $|b_n| \leq 1$.

Let $\{c_n\} = \sum_{n=m_1}^{m_2} n^{-1}v_n \in J$. For integers $m \geq 2$ and $0 < p_1 < \cdots < p_m$, the summands of $S \equiv \sum_{n=1}^{m-1}(c_{p_k} - c_{p_{k+1}})^2$ with $p_k, p_{k+1} \in [q_i, q_{i+1}\rangle$ are either zero or $i^{-2}(b_{p_k} - b_{p_{k+1}})^2$, so the sum S_1 of these does not exceed $2\sum_{n=m_1}^{m_2} n^{-2}$ as $\|v_n\|_b = 1$. For those summands in S with $p_k \in [q_i, q_{i+1}\rangle$ and $p_{k+1} \in [q_j, q_{j+1}\rangle$ we have

$$(c_{p_k} - c_{p_{k+1}})^2 \leq (i^{-1}b_{p_k})^2 + (j^{-1}b_{p_{k+1}})^2 + 2i^{-1}j^{-1}|b_{p_k}b_{p_{k+1}}| \leq 2(i^{-2} + j^{-2})$$

as $2i^{-1}j^{-1} \leq i^{-2} + j^{-2}$. Hence the sum S_2 of such summands in S cannot exceed $4\sum_{n=m_1}^{m_2} n^{-2}$. As $S = S_1 + S_2$, we get

$$\|\sum_{n=m_1}^{m_2} n^{-1}v_n\|_b = 2^{-1/2}S^{1/2} \leq 3^{1/2}(\sum_{n=m_1}^{m_2} n^{-2})^{1/2},$$

which is impossible. □

Using the standard basis $\{e_n\}$ for J in Proposition 3.5.1, we see that $\|\{a_n\}\|_S = \|\{a_n\}\|_a$ for $\{a_n\} \in J$. When $\{a_n\} \in J^{\star\star}$, it is easy to see that $\|\{a_n\}\|_S \geq \|\{a_n\}\|_a$. We also claim that $\lim a_n$ then exists, because otherwise there is $r > 0$ and integers p_n, q_n such that $1 \leq p_1 < q_1 < p_2 < q_2 < \cdots$ with $|a_{p_n} - a_{q_n}| \geq r$. So for $m \in \mathbb{N}$ we get the absurdity

$$\|\sum_{n=1}^{q_m} a_n e_n\|_a \geq 2^{-1/2}((a_{p_1}-a_{q_1})^2+\cdots+(a_{p_m}-a_{q_m})^2+(a_{q_m}-a_{p_1})^2)^{1/2} \geq r(m/2)^{1/2}.$$

Hence we can identify the elements of J inside $J^{\star\star}$ as those sequences $\{a_n\}$ with $\lim a_n = 0$.

Consider now the element $e_0 = \{1, 1, \ldots\} \in J^{\star\star}$ with $\|e_0\|_S = 1$. We claim that $\{e_0\} \cup \{e_n\}$ is a Schauder basis for $J^{\star\star}$. Indeed, we have $\{a_n\} = ae_0 + \sum_{n=1}^{\infty}(a_n - a)e_n$ with $a = \lim a_n$ for $\{a_n\} \in J^{\star\star}$ as the only possibility. So the vector space $J^{\star\star}/J$ is spanned by $e_0 + J$. We have proved the following result by James.

Theorem 3.5.5 *The James space J is not reflexive as it has codimension one in $J^{\star\star}$.*

Yet we have the theorem below. We need the following result, which is straightforward.

Lemma 3.5.6 *We have for $\{a_n\} \in J^{\star\star}$ that $\|\{a_n\}\|_a$ is given by the same formula as before except that in taking the supremum we also have to include the numbers $(\sum_{n=1}^{m-2}(a_{p_n} - a_{p_{n+1}})^2 + a_{p_{m-1}}^2 + a_{p_1}^2)^{1/2}$.*

Theorem 3.5.7 *The Banach space J is isometrically isomorphic to its bidual.*

Proof Let $\{a_n\} \in J^{\star\star}$ with $a = \lim a_n$. Then

$$T(\{a_n\}) = \{-a, a_1 - a, a_2 - a, \dots\}$$

defines a linear map $T \colon J^{\star\star} \to J$ such that $\|T(\{a_n\})\|_a = \|\{a_n\}\|_a$ by the lemma. Also, if $\{b_n\} \in J$, then $\{b_2, b_3, \dots\} - b_1 e_0 \in J^{\star\star}$ and its image under T is $\{b_n\}$. □

So Banach spaces with separable biduals are not always reflexive, and it does not always help that they are (non-canonically) isometrically isomorphic to their biduals. A Banach space with finite codimension in its bidual is called *quasi-reflexive*.

Exercises

For Sect. 3.1

1. Provide an infinite dimensional real or complex normed space with a vector space basis having only bounded coordinate functionals. What is the situation for Banach spaces?
2. Prove that the direct sum of two complex Banach spaces with Schauder bases have a Schauder basis.
3. Are the usual Schauder bases for $C([0, 1])$ monotone?
4. If $\{v_n\}$ is a Schauder basis for a real or complex Banach space, is it always true that $\sum a_n v_n$ converges when $\sum |a_n| < \infty$?
5. Let $\{v_n^*\}$ be the coordinate functionals for a Schauder basis $\{v_n\}$ in a real or complex Banach space. Prove that the numbers $\|v_n\|\|v_n^*\|$ all belong to $[1, M]$ for some scalar M.

For Sect. 3.2

1. Show that a sequence $\{v_n\}$ in a real or complex Banach space is an unconditional Schauder basis sequence if and only if there is a scalar M such that $\|\sum_{n\in A} a_n v_n\| \leq M\|\sum_{n\in B} a_n v_n\|$ for all finite sets $A \subset B$ and scalars a_i.
2. Prove that any permutation of an unconditional Schauder basis sequence in a real or complex Banach space is again an unconditional Schauder basis sequence.
3. Prove that the Haar basis for $L^1([0, 1])$ is a conditional Schauder basis.
4. Show that if any permutation of a sequence in a real or complex Banach space is a Schauder basis, then the original sequence is an unconditional Schauder basis.
5. Show that a series $\sum v_n$ in a real or complex Banach space is unconditionally convergent if and only if $\sum a_n v_n$ converges for all scalars a_n of moduli one if and only if the last series converges for all scalars $a_n = \pm 1$.

For Sect. 3.3

1. Show that the complex Banach spaces $C([0, 1])$ and $L^p([0, 1])$ for $p \in [1, \infty\rangle$ have Schauder bases consisting of polynomials.
2. Show that the normalized Haar basis for $L^1([0, 1])$ is not equivalent to the standard normalized Schauder basis for $l^1(\mathbb{N})$, but that a subsequence of the normalized Haar basis is.
3. Prove that a series $\sum v_n$ in a real or complex Banach space is weakly unconditional Cauchy if and only if there is a scalar majorizing $\| \sum_{n \in A} a_n v_n \|$ for any finite subset A and $a_n = \pm 1$.
4. Show that any block Schauder basic subsequence of a monotone Schauder basic sequence is again monotone.
5. Prove that $e_1 + \sum_{n=2}^{\infty} (e_n - e_{n-1})$ is weakly reordered convergent in $C_0(\mathbb{N})$, where $\{e_n\}$ is the normalized Schauder basis. Is this series weakly subseries convergent?

For Sect. 3.4

1. Show that the standard normalized Schauder basis for $l^p(\mathbb{N})$ is boundedly complete when $p \in [1, \infty\rangle$.
2. Prove that $C([0, 1])$ is not the dual of a normed space.
3. Suppose V is a real or complex Banach space with an unconditional Schauder basis and a separable dual space. Show that V is reflexive if and only if $C_0(\mathbb{N})$ cannot be embedded in V.
4. Say V is a real or complex Banach space with an unconditional Schauder basis. Show that V^{**} is separable if and only V is reflexive.

For Sect. 3.5

1. Show that J has no unconditional Schauder basis.
2. Show that J is not isomorphic to $J \oplus J$.
3. Prove that J cannot be the real part of a complex Banach space.

Chapter 4
Operators on Hilbert Spaces

Orthogonality in infinite dimensions is best studied in Hilbert spaces. These are Banach spaces with norms given by scalar products, so called inner products. The fundamental example of a Hilbert space is the L^2-space of the circle. A function, or vector, here can be thought of as a light signal, which through its Fourier series, is decomposed into orthogonal directions, each corresponding to a specific color identified with a certain frequency. This orthogonal decomposition into a basis of colors, which by the way, is a trivial example of a Schauder basis, can be seen as the mathematical counterpart of Newton's theory of colors; that white light is a mix of all colors, each of which can be detected by the use of a prism. Be aware that light waves of a certain color need not all have the same phase, which accounts for the negative part of the spectrum. Waves with the same frequency and phase superpose into a laser beam. In any case, the relevance of Hilbert spaces in physics is evident.

Thus in this chapter we devote two sections to Fourier series on the circle, and their continuous counterpart, the Fourier transforms on the real line. While we are at it, we also discuss pointwise convergence of Fourier series.

But first we introduce the basic notions and properties of Hilbert spaces, adapting the usual properties from Euclidean spaces. To each closed subspace of a Hilbert space there is an orthogonal projection, that is, a bounded operator on the Hilbert space that projects vectors orthogonally onto the subspace. Such operators are characterized as the idempotents that are self-adjoint; the adjoint of an operator corresponds to the coordinatewise conjugate of the transpose of a matrix. This gives a one-to-one correspondence between closed subspaces and orthogonal projections.

Using Zorn's lemma we show that every Hilbert space has an orthonormal basis, and that the cardinality of such a basis is unique, defined then as the Hilbert space dimension. In fact, for each cardinality there is up to isomorphism exactly one Hilbert space having this cardinality as its dimensions. Uniqueness is due to the fact that any operator sending basis elements to basis elements of two bases of the same cardinality is unitary, or an isomorphism, that is, a completely structure preserving

L. Tuset, *Analysis and Quantum Groups*,
https://doi.org/10.1007/978-3-031-07246-8_4

map. That all cardinalities are exhausted follows by considering l^2-spaces of sets of any cardinality.

In the remaining part of the chapter we take up the study of bounded operators on Hilbert spaces. We define what is meant by a positive operator, extending thus the idea of positive numbers. Then we show that one can take the square root of such operators, and in fact, decompose any bounded operator into one factor which is positive. The other factor is an operator which preserves the norm of the vectors in some closed subspace, while it kills the other vectors; a so called partial isometry. The positive factor serves the role of appropriately rescaling the resulting vectors. This is known as the polar decomposition of an operator, and has an evident analogue for complex numbers.

We then return to compact operators, now between Hilbert spaces. We characterize the compact operators which are normal, that is, those commuting with their adjoints, as the orthogonal diagonalizable ones with eigenvalues vanishing at infinity when counted with multiplicity.

We study the collection of compact operators as a whole, and show that they form a closed $*$-ideal $B_0(V)$ of the normed $*$-algebra $B(V)$ of all bounded operators on a Hilbert space V; in the separable case we furthermore show that no other proper closed ideal can occur. The Fredholm operators on V are those with invertible images in the Calkin algebra $B(V)/B_0(V)$. Such an operator has a definite index, defined to be the difference of the dimension of its kernel and the dimension of the kernel of its adjoint. The index is stable under pertubations by compact operators. In fact, it is a topological invariant, enumerating the connected components of the topological group of invertible elements in the Calkin algebra.

The trace of a compact positive operator is the infinite sum of its eigenvalues, and when this series converges, we say it is of traceclass. If this is true for the square of the operator, we say that it is Hilbert-Schmidt. We extend this to non-positive operators as well. An important class of Hilbert-Schmidt operators are given by integrating an L^2-function against a two-point kernel, which is an L^2-function on the product space. Properties of such integral operators are deduced in the general context of traceclass and Hilbert-Schmidt operators, which actually form ideals of the algebra of compact operators, and are of course in general not norm-closed. However, the trace allows turning the Hilbert-Schmidt operators into a Hilbert space, while it turns the traceclass operators into a Banach spaces which can be identified as the dual of the space of compact operators. The dual of the space of traceclass operators is in turn the space of all bounded operators. We will return to versions of this important duality several times in the sequel.

Recommended literature for this and the following two chapters is [7, 17, 34, 38, 46].

4.1 Hilbert Spaces

We study here a natural generalization of Euclidean spaces.

Definition 4.1.1 An *inner product* on a complex vector space W is a map $(\cdot|\cdot)\colon W\times W\to\mathbb{C}$ such that $(\cdot|w)=\overline{(w|\cdot)}$ is a linear functional which is strictly positive for every non-zero $w\in W$. The *associated norm* is given by $\|w\|\equiv(w|w)^{1/2}$.

The inequality

$$|a|^2\|v\|^2+2\operatorname{Re}a(v|w)+\|w\|^2=\|av+w\|^2\geq 0$$

for any vectors v, w and numbers a shows that the *Cauchy-Schwarz inequality*

$$|(v|w)|\leq\|v\|\,\|w\|$$

and the triangle inequality $\|v+w\|\leq\|v\|+\|w\|$ hold, so the associated norm of an inner product is indeed a norm. Note that equality in the Cauchy-Schwarz inequality holds if and only if v and w are linear dependent. Also, the proof of this inequality did not use the property that $(w|w)=0$ only holds for $w=0$.

Definition 4.1.2 An inner product (vector) space is a *Hilbert space* if it is a Banach space under the associated norm.

Example 4.1.3 The most important example of a Hilbert space is the space $L^2(\mu)$ of square integrable complex functions with respect to a measure μ with inner product $(f|g)=\int f\bar{g}\,d\mu$. The Hölder inequality for $p=q=1/2$ is a special case of the Cauchy-Schwarz inequality, and the counting measure on a finite set gives the complex Euclidean space with inner product $(v|w)=\sum v_i\bar{w}_i$. ◇

For any inner product on a complex vector space the *polarization identity*

$$4(v|w)=\sum_{n=0}^{3}i^n\|v+i^n w\|^2$$

and the *parallelogram law*

$$\|v+w\|^2+\|v-w\|^2=2\|v\|^2+2\|w\|^2$$

are easily seen to hold for the associated norm. Conversely, if the parallelogram law holds for a norm, then one checks that the first of these formulas defines an inner product having the norm as the associated one.

Example 4.1.4 The extreme points of the closed unit ball in a Hilbert space are precisely the unit vectors because if v, w and $(v + w)/2$ all have norm one, then $\|v - w\| = 0$ by the parallelogram law. Thus the closed unit ball is already the convex hull of its extreme boundary, and indeed, as we shall see later, it is weakly compact. ◇

Everything said so far holds also for real vector spaces except that the polarization identity must be replaced by $4(v|w) = \|v+w\|^2 - \|v-w\|^2$. The parallelogram law has an obvious geometric significance in the Euclidean case. Every statement made henceforth will hold in both the complex and real case unless otherwise stated. The polarization identities obviously hold without requiring that $(w|w) = 0$ only for $w = 0$.

Lemma 4.1.5 *A non-empty convex closed subset of a Hilbert space contains exactly one element of least norm.*

Proof By convexity and the parallelogram law, we see that

$$\|v - w\|^2 \leq 2\|v\|^2 + 2\|w\|^2 - 4a^2$$

for any v, w in the convex set E, where $a = \inf_{v \in E} \|v\|$. This establishes uniqueness.

It also shows that a sequence of $v_n \in E$ with $\|v_n\| \to a$ is Cauchy, which by completeness and closedness must converge to some $v \in E$ with $\|v\| = a$ since the norm is continuous. □

Definition 4.1.6 We say that v and w are *orthogonal* with respect to an inner product if $(v|w) = 0$. For a subset X of an inner product space denote by $X^\perp$ the subset of those vectors that are orthogonal to all the vectors of X. We also write $X \perp Y$ to say that every element of X is orthogonal to every element of Y.

By the Cauchy-Schwarz inequality the linear maps $(\cdot|w)$ are continuous, and as $X^\perp$ is the intersection of the kernels of those maps with $w \in X$, it must be a closed subspace.

Example 4.1.7 Given a family of Hilbert spaces V_i, we define an inner product on their algebraic direct sum by $(v|w) = \sum(\pi_i(v)|\pi_i(w))$, so the associated norm is the 2-norm with completion the *orthogonal sum* $\oplus V_i$ of the V_i's. This is clearly a Hilbert space and consists of those $v \in \prod V_i$ with finite 2-norm with respect to the counting measure on the index set. Each V_i is a closed subspace of $\oplus V_i$, and they are mutually orthogonal.

For norms associated to inner products *Pythagoras' identity*

$$\|v + w\|^2 = \|v\|^2 + \|w\|^2$$

holds for mutually orthogonal vectors v and w.

Proposition 4.1.8 *If V is a closed subspace of a Hilbert space W, then $W = V \oplus V^{\perp}$. In particular, there is always a non-zero vector orthogonal to a proper subspace of a Hilbert space.*

Proof By the lemma we know that to $w \in W$ there is a unique element $Q(w)$ with least norm in the non-empty closed convex subset $w + V$. So $P(w) \equiv w - Q(w)$ defines a map $P: W \to V$. The map Q has range $V^{\perp}$ because by the minimizing property we have

$$\|Q(w)\|^2 \leq \|Q(w) - (Q(w)|v)v\|^2$$

for any $v \in V$, and written out this gives $(Q(w)|v) = 0$. Pythagoras' identity shows that $P(w)$ is the nearest point in V to w. Linearity of both maps is clear by the evident uniqueness of the decomposition $w = P(w) + Q(w)$. □

Note that $E^{\perp\perp}$ is the least closed subspace of a subset E in a Hilbert space, so it is the norm closure of a subspace.

Corollary 4.1.9 *Every continuous linear map $W \to \mathbb{C}$ on a Hilbert space is of the form $(\cdot|w)$ for a unique $w \in W$.*

Proof Uniqueness is clear. As for existence, say x is a non-trivial continuous linear map $W \to \mathbb{C}$. Pick $w \in W$ with $\|w\| = 1$ orthogonal to the closed proper subspace $\ker x$. For any $v \in W$ we see that $x(v)w - x(w)v \in \ker x$, so it is orthogonal to w, and hence $x(v) = x(v)(w|w) = x(w)(v|w) = (v|\overline{x(w)}w)$. □

Definition 4.1.10 An *orthonormal basis* in a Hilbert space is a family of pairwise orthogonal vectors of norm one with dense span.

Given an uncountable family $\{a_i\}_{i \in I}$ in a topological vector space, then by $\sum_{i \in I} a_i$, or simply $\sum a_i$, we mean the limit, if it exists, of the net $\{\sum_{i \in J} a_i\}_J$ of finite sums, where the finite subsets of I are ordered under inclusion. This is a sort of unordered infinite summation.

If $\{v_i\}_{i \in I}$ is an orthonormal set in a Hilbert space V, then

$$\|v - \sum_{i \in J}(v|v_i)v_i\|^2 = \|v\|^2 - \sum_{i \in J}|(v|v_i)|^2$$

for finite $J \subset I$ shows that *Bessel's inequality* $\sum |(v|v_i)|^2 \leq \|v\|^2$ holds for $v \in V$. It shows in particular that $(v|v_i) \neq 0$ for only countably many i because the set of such i's is the union of the finite sets $\{i \in I \,|\, |(v|v_i)| > 1/m\}$ over $m \in \mathbb{N}$. When $\{v_i\}$ is an orthonormal basis we actually get equality in Bessel's inequality; then referred to as the *Parseval identity*. In this case, we can expanded $v \in V$ as $v = \sum(v|v_i)v_i$ in terms of the *Fourier coefficients* $(v|v_i)$ *of* v. To see this, let $\{i_k\}_{k=1}^{\infty}$ be an enumeration of those $i \in I$ with $(v|v_i) \neq 0$, and note that by Bessel's inequality, we get $\|\sum_{k=n}^{m}(v|v_{i_k})v_{i_k}\| \to 0$ as $n, m \to \infty$. Hence $v - \sum_{k=1}^{\infty}(v|v_{i_k})v_{i_k}$ converges as V is complete, and as it clearly is orthogonal to any v_i, it must be zero by denseness of the basis. Now to $\varepsilon > 0$, pick N such that $\|v - \sum_{k=1}^{N}(v|v_{i_k})v_{i_k}\| <$

$\varepsilon/2$ and $\|\sum_{k=N+1}^{n}(v|v_{i_k})v_{i_k}\| < \varepsilon/2$ for $n > N$. Let $J_0 = \{i_k\}_{k=1}^{N}$, and say $J \supset J_0$ is finite. Let N' be the greatest number of $N+1$ and the numbers k with $i_k \in J$. Then

$$\|v - \sum_{i\in J}(v|v_i)v_i\| \leq \|v - \sum_{k=1}^{N}(v|v_{i_k})v_{i_k}\| + \|\sum_{k=1}^{N}(v|v_{i_k})v_{i_k} - \sum_{i\in J}(v|v_i)v_i\| < \varepsilon,$$

as the latter norm expression is not greater than $\|\sum_{k=N+1}^{N'}(v|v_{i_k})v_{i_k}\|$. So our claim holds, and the Parseval identity is clear from the equality that led to Bessel's inequality. The reader will have noticed that these arguments resemble those proving completeness of $L^2(\mu)$ with respect to the counting measure μ on I. It is also worth noticing that V is the orthogonal sum of the 1-dimensional subspaces spanned by each v_i in the orthonormal basis.

Every Hilbert space has an orthonormal basis due to the following result.

Proposition 4.1.11 *Every family of orthogonal vectors of norm one in any Hilbert space can be enlarged to an orthonormal basis.*

Proof The collection of such families containing a given family is a partially ordered set under inclusion, which is moreover inductively ordered. By Zorn's lemma there is a maximal family, and this is an orthonormal basis since if the closure V of the subspace spanned by this family is not the whole Hilbert space, there is by the previous proposition a norm one vector in $V^{\perp}$, which violates maximality. □

A concrete way of producing an orthonormal basis $\{v_n\}$ from a countable basis $\{w_n\}$ in a Hilbert space, is the following *Gram-Schmidt orthonormalization process*, which sets $v_1 = w_1/\|w_1\|$ and defines inductively v_n to be the normalization of the component $w_n - \sum_{m=1}^{n-1}(w_n|v_m)v_m$ of w_n that is orthogonal to $\text{span}\{v_1, \ldots, v_{n-1}\} = \text{span}\{w_1, \ldots, w_{n-1}\}$.

Remark 4.1.12 An orthonormal basis for a separable Hilbert space V is a monotone Schauder basis for V as a Banach space because the obvious idempotent onto the first $n \in \mathbb{N}$ vectors has norm one. By the bounded multiplier test such a basis is also unconditional, and it is moreover shrinking and boundedly complete as V is reflexive. It basically has all the properties one can wish for in the context of Schauder bases.

We have already tacitly identified Hilbert spaces. The formal definition goes as follows.

Definition 4.1.13 Two Hilbert spaces V and W are *isomorphic*, written $V \cong W$, if there is a bijective linear map $A\colon V \to W$ such that $(A(v)|A(w)) = (v|w)$ for all $v, w \in V$.

Any isomorphism between Hilbert spaces is clearly isometric, and conversely, if $\|A(v)\| = \|v\|$ for all v, then $(A(v)|A(w)) = (v|w)$ for all v and w by the polarization identity. Two Hilbert spaces having orthonormal bases with the same cardinality are isomorphic since the linear map which sends basis elements to basis elements extends by the Parseval identity and Proposition 2.1.19 to a bijective isometry.

In particular, all infinite dimensional Hilbert spaces which are *separable*, i.e. having countable dense subsets, are pairwise isomorphic. Indeed, these are exactly the Hilbert spaces having countable orthonormal bases; the rational linear span of such a basis is countable and dense, and given a countable dense subset $\{w_n\}$, we may apply the Gram-Schmidt orthonormalization process to it discarding those v_n that might vanish due to possible linear dependence of the vectors w_n. So there is only one infinite dimensional separable Hilbert space up to isomorphism, namely $l^2(\mathbb{N})$. The finite dimensional Hilbert spaces are of course the Euclidean spaces.

The *dimension of a Hilbert space* is the cardinality of any orthonormal basis in it. This make sense as any two orthonormal bases have the same cardinality; a fact verified by modifying the proof of Theorem A.2.6 replacing linear combinations by series over sets F_j that must be countable by Parseval's identity.

Example 4.1.14 The 2-norm on the unit circle $\mathbb{T}$ is not greater than the uniform norm. Hence by the Stone-Weierstrass theorem and Lusin's theorem, the span of the functions $v_n : \mathbb{T} \to \mathbb{T}$ given by $v_n(z) = z^n$ as $n \in \mathbb{Z}$ is dense in $L^2(\mathbb{T})$. With inner product $(f|g) = \frac{1}{2\pi}\int_{-\pi}^{\pi} f(e^{it})\bar{g}(e^{it})\,dt$, we see that $(v_m|v_n) = \delta_{mn}$, so the *trigonometric system* $\{v_n\}$ is an orthonormal basis for $L^2(\mathbb{T})$. Set $\hat{f}(n) = (f|v_n)$. The *Fourier series* $\lim_{N\to\infty}\sum_{n=-N}^{N}\hat{f}(n)v_n$ of f converges to f in L^2-norm. By polarizing the Parseval identity we get

$$(f|g) = \sum_{n=-\infty}^{\infty} \hat{f}(n)\overline{\hat{g}(n)},$$

and $f \mapsto \hat{f}$ is a Hilbert space isomorphism from $L^2(\mathbb{T})$ to $l^2(\mathbb{Z})$. It is surjective because if $\{c_n\} \in l^2(\mathbb{Z})$ with only finitely many $c_n \neq 0$, then with $f = \sum c_n v_n$, we get $\hat{f}(n) = c_n$, and the sequences $\{c_n\}$ are dense in $l^2(\mathbb{Z})$. ◇

Example 4.1.15 Let A be a Borel subset of $\mathbb{R}$ with positive Lebesgue measure, and say $h > 0$ is a Borel function on A such that $\int_A t^{2n}h(t)dt < \infty$ for every non-negative integer n. The set V of Borel functions f on A such that $\int_A |f(t)|^2 h(t)dt < \infty$ is a Hilbert space with inner product $(f|g) = \int_A f(t)\overline{g(t)}h(t)dt$. By Lusin's theorem and the corollary to the Stone-Weierstrass theorem in the appendix, the sequence $1, t, t^2, \ldots$ of polynomials has dense span in V when A is locally compact Hausdorff. Applying the Gram-Schmidt orthonormalization process to these linear independent monomials gives then an orthonormal basis $\{v_n\}_{n=0}^{\infty}$ for V, where v_n is a polynomial of degree n.

When $A = [-1, 1]$ and $h(t) = (1-t)^a(1+t)^b$ for $a, b \in \langle -1, \infty\rangle$, then v_n is up to a scalar the *Jacobi polynomial* $P_n^{(a,b)}$. When $a = b = 0$ we get the *Legendre polynomial*

$$P_n(t) = (n+1/2)^{-1/2}v_n(t) = \frac{1}{2^n n!}\frac{d^n}{dt^n}(t^2-1)^n,$$

where the expression with the n-derivative is *Rodrigue's formula*. These are part of a broader subclass of Jacobi polynomials known as the *Gegenbauer polynomials* which includes the *Chebyshev polynomials* and the *Zernike polynomials*.

When $A = [0, \infty\rangle$ and $h(t) = t^a e^{-t}$, then v_n is up to a scalar the *Laguerre polynomial* L_n^a, whereas v_n is up to a scalar the *Hermite polynomial* H_n when $A = \mathbb{R}$ and $h(t) = e^{-t^2}$.

These are the main three sequences of polynomials from which the other *classical orthogonal polynomials* can be derived. ◇

Example 4.1.16 Let X and Y be measurable sets with σ-finite measures μ and ν. Let $\mu \times \nu$ be the product measure on $X \times Y$ with product topology. Suppose we have countable orthonormal bases $\{f_n\}$ and $\{g_m\}$ of $L^2(\mu)$ and $L^2(\nu)$, respectively. Define functions $f_n \otimes g_m \colon X \times Y \to \mathbb{C}$ by $(f_n \otimes g_m)(x, y) = f_n(x)g_m(y)$. We claim that $\{f_n \otimes g_m\}$ is an orthonormal basis of $L^2(\mu \times \nu)$. By the Fubini-Tonelli theorem they clearly form an orthonormal set in $L^2(\mu \times \nu)$. To see that its span is dense, consider any $h \in L^2(\mu \times \nu)$ such that $\int (f_n \otimes g_m)\bar{h}\, d(\mu \times \nu) = 0$. Since $\{f_n\}$ is an orthonormal basis, there is by the Fubini-Tonelli theorem a set X_n with $\mu(X_n) = 0$ such that $\int g_m \bar{h}(x, \cdot)\, d\nu = 0$ for $x \notin X_n$. Using now that $\{g_m\}$ is an orthonormal basis, we see that $h(x, \cdot) = 0$ almost everywhere for $x \notin \cup X_n$. Hence $h = 0$ except on $(\cup X_n) \times Y$ with product measure $\mu(\cup X_n)\nu(Y) = 0$.

The same is of course true for σ-finite Radon measures μ and ν with separable L^2-spaces and for the Radon product measure $\mu \otimes \nu$.

4.2 Fourier Transform Over the Reals

Let us begin with a definition.

Definition 4.2.1 A $*$-*algebra* is an algebra over the real or complex numbers which has a $*$-*operation*, that is, a map $a \mapsto a^*$ which is conjugate linear, antimultiplicative and involutive. We call a^* the *adjoint* of a. If $a^*a = aa^*$, then a is *normal*, and if $a^* = a$, it is *self-adjoint*. A complex Banach algebra is a *Banach* $*$-*algebra* if it is a $*$-algebra for an isometric $*$-operation.

In a complex $*$-algebra any element can be written uniquely as a linear combination of self-adjoint elements. In fact, we may write $a = b + ic$ for self-adjoint elements $b = (a + a^*)/2$ and $c = (a - a^*)/2i$. This representation is, moreover, unique since if $b + ic = 0$, then $0 = 0^* = b^* - ic^* = b - ic$, so $b = 0 = c$.

Consider the Banach $*$-algebra $L^1(\mathbb{R})$ with convolution product $(f * g)(s) = \int f(t)g(s - t)dt$ and involution $f^\star(s) = \overline{f(-s)}$. The *Fourier transform* of $f \in L^1(\mathbb{R})$ is the real function $\hat{f}$ given by $\hat{f}(t) = \int f(s)e^{-ist}\, ds$.

Proposition 4.2.2 *The Fourier transform $f \mapsto \hat{f}$ is a norm decreasing $*$-homomorphism from the Banach $*$-algebra $L^1(\mathbb{R})$ to $C_0(\mathbb{R})$, where the latter is also considered a Banach $*$-algebra with pointwise operations and uniform norm.*

Proof That $\hat{f}$ is continuous is immediate from Lebesgue's dominated convergence theorem, and the inequality $\|\hat{f}\|_u \leq \|f\|_1$ is also clear. Using $e^{\pi i} = -1$ and translation invariance of the Lebesgue measure on $\mathbb{R}$ we get

$$2|\hat{f}(t)| = |\int (f(s) - f(s - \pi/t))e^{-ist}\,ds| \leq \|f - f((\cdot) - \pi/t)\|_1 \to 0$$

as $t \to \pm\infty$ by Lemma A.10.3, so $\hat{f} \in C_0(\mathbb{R})$. Multiplicativity of the Fourier transform is a simple application of the Fubini-Tonelli theorem, and it is straightforward to check that it is $*$-preserving. □

Note that for $a \in \mathbb{R}$ and $f \in L^1(\mathbb{R})$ the Fourier transform of $s \mapsto f(s)e^{ias}$ is $t \mapsto \hat{f}(t - a)$, and conversely, the Fourier transform of $s \mapsto f(s - a)$ is $t \mapsto \hat{f}(t)e^{-iat}$. It clearly also sends $f((\cdot)/a)$ to $a\hat{f}(a(\cdot))$ when $a > 0$. If $g: s \mapsto -isf(s)$ also belongs to $L^1(\mathbb{R})$, then $\hat{f}$ is differentiable with derivative $\hat{g}$. This is clear from Lebesgue's dominated convergence theorem and the definition of the derivative.

Let $H(t) = e^{-|t|}$ and $h_a(s) = (2\pi)^{-1} \int H(at)e^{ist}\,dt$ for $a > 0$. Then $\|h_a\|_1 = 1$. From the proof of Proposition A.10.9 and Jensen's inequality we see that $f * h_a \to f$ in $L^p(\mathbb{R})$ as $a \to 0$ for $p \in [1, \infty\rangle$, and that $f * h_a \to f$ pointwise for $f \in L^\infty(\mathbb{R})$ at any point where f is continuous. We have the following *inversion theorem*.

Proposition 4.2.3 *If $f, \hat{f} \in L^1(\mathbb{R})$, then $s \mapsto (2\pi)^{-1} \int \hat{f}(t)e^{ist}\,dt$ belongs to $C_0(\mathbb{R})$ and equals f almost everywhere. In particular, if $\hat{f} = 0$, then $[f] = 0$.*

Proof The function g in question clearly belongs to $C_0(\mathbb{R})$. Since $H(at) \to 1$ as $a \to 0$, the Lebesgue dominated convergence theorem gives

$$g(s) = (2\pi)^{-1} \lim_{a \to 0} \int H(at)\hat{f}(t)e^{ist}\,dt = \lim(f * h_a)(s),$$

where the last equality follows from translation invariance and the Fubini-Tonelli theorem. But $f * h_a \to f$ in $L^1(\mathbb{R})$, and from the proof of the completeness of $L^1(\mathbb{R})$, we can thus find a sequence $\{a_n\}$ such that $f * h_{a_n} \to f$ almost everywhere. Thus $g = f$ almost everywhere. □

Theorem 4.2.4 *The* Plancherel transform *is the unitary operator F on $L^2(\mathbb{R})$ given by $F(f)(t) = (2\pi)^{-1/2} \lim_{n\to\infty} \int_{-n}^{n} f(s)e^{-ist}\,ds$ with inverse $F^{-1}(g)(s) = (2\pi)^{-1/2} \lim_{n\to\infty} \int_{-n}^{n} g(t)e^{ist}\,dt$. The Plancherel transform is $(2\pi)^{-1/2}$ times the Fourier transform on the dense subset $L^1(\mathbb{R}) \cap L^2(\mathbb{R})$.*

Proof Let $f \in L^1(\mathbb{R}) \cap L^2(\mathbb{R})$. Then $g = f * f^\star \in L^1(\mathbb{R})$ is continuous by Lemma A.10.3, and it is bounded with $\|g\|_u \leq \|f\|_2^2$ by left invariance and the Cauchy-Schwarz inequality. From the observations above we have

$$\|f\|_2^2 = g(0) = \lim_{a\to 0}(g * h_a)(0) = (2\pi)^{-1} \lim_{a\to 0} \int H(at)\hat{g}(t)\,dt = (2\pi)^{-1}\|\hat{f}\|_2^2,$$

where in the last step we used $\hat{g} = |\hat{f}|^2 \geq 0$ and the monotone convergence theorem as $H(at)$ increases towards 1 when a decreases to zero. Let F be the isometric extension to $L^2(\mathbb{R})$ of $(2\pi)^{-1/2}$ times the Fourier transform on the dense subset $L^1(\mathbb{R}) \cap L^2(\mathbb{R})$. Let V be the orthogonal complement of $F(L^1(\mathbb{R}) \cap L^2(\mathbb{R}))$. Clearly all possible translates of h_a for $a > 0$ belong to V. So if $h \in V^\perp$, then $(h_a * h)(s) = \int h_a(s-t)h(t)\,dt = 0$, so $h = 0$ in $L^2(\mathbb{R})$, and F is unitary.

Finally note that $(2\pi)^{-1/2}\int_{-n}^{n} e^{-ist} f(s)\,ds = F(f\chi_{[-n,n]})(t)$, and that

$$\|F(f) - F(f\chi_{[-n,n]})\|_2 = \|f - f\chi_{[-n,n]}\|_2 \to 0$$

as $n \to \infty$, and similarly for the inverse formula. □

The formula $(F(f)|F(g)) = (f|g)$ for $f, g \in L^2(\mathbb{R})$ is the *Parseval formula*. We also notice that $f(s) = (2\pi)^{-1/2}\int F(f)e^{ist}\,dt$ when $f \in L^2(\mathbb{R})$ and $F(f) \in L^1(\mathbb{R})$.

We include here some technical results that will be used later.

Lemma 4.2.5 *Let K be a compact subset of an open subset U of $\mathbb{R}$. Then there is $h \in L^1(\mathbb{R})$ such that $\hat{h}$ has range in $[0, 1]$, is 1 on K and has support in U.*

Proof Pick a neighborhood V of $0 \in \mathbb{R}$ with finite Lebesgue measure a such that $K + V - V \subset U$. Pick $f, g \in L^2(\mathbb{R})$ with $F(f) = (2\pi)^{-1/2}\chi_{K-V}$ and $F(g) = (2\pi)^{-1/2}\chi_V$. Then $h = fg/a$ does the trick as

$$\hat{h}(t) = a^{-1}(\chi_V * \chi_{K-V})(t) = a^{-1}\int_V \chi_{K-V}(t-s)\,ds.$$

□

Lemma 4.2.6 *Let $a \in \mathbb{R}$ and $f \in L^1(\mathbb{R})$ with $\hat{f}(a) = 0$, and let U be a neighborhood of a. Then to $\varepsilon > 0$ there is measurable h with $\|h\|_1 < 2$ and $\|f * h\|_1 < \varepsilon$ such that $\hat{h}$ has range in $[0, 1]$, is one on some neighborhood of a and has support in U.*

Proof We may assume that $a = 0$. Let $b = \varepsilon/4(1 + \|f\|_1)$ and pick a compact subset A of $\mathbb{R}$ such that $\int_{A^c} |f(s)|\,ds < b$. Pick $c > 0$ such that $[-3c, 3c] \subset U$ and $|1 - e^{ist}| < b$ for $s \in A$ and $t \in [-3c, 3c]$. Set $K = V = [-c, c]$ and pick $u, v \in L^2(\mathbb{R})$ such that $F(u) = (2\pi)^{-1/2}\chi_{K-V}$ and $F(v) = (2\pi)^{-1/2}\chi_V$. Set $h = uv/2c$. Then $K + V - V = [-3c, 3c] \subset U$, and as in the proof of the lemma above, we see that $\hat{h}$ has range in $[0, 1]$, is 1 on $\langle -c, c\rangle$ and has support in U. Also $\|h\|_1 \leq \|F(u)\|_2\|F(v)\|_2/2c < 2$.

Let $g_s(t) = g(t-s)$ for any function g on $\mathbb{R}$ and $s,t \in \mathbb{R}$. Since $(f*h)(t) = \int f(s)(h(t-s)-h(t))\,ds$ as $\hat{f}(0)=0$, we get $\|f*h\|_1 \leq \int |f(s)|\|h_s-h\|_1\,ds$. Splitting this integral up in one part over A and one part over A^c, we see that it suffices to show $\sup_{s\in A}\|h_s-h\|_1 \leq 4b$. Write $2c(h_s-h) = u(v_s-v)+(u_s-u)v_s$, and note that for $s \in \mathcal{A}$, we have $\|u_s-u\|_2^2 = \|F(u_s)-F(u)\|_2^2 = (2\pi)^{-1/2}\int_{-2c}^{2c}|1-e^{ist}|^2\,dt < 4cb^2$ and similarly $\|v_s-v\|_2^2 < 2cb^2$. Hence

$$\|h_s-h\|_1 \leq (2c)^{-1}(\|F(u)\|_2\|v_s-v\|_2 + \|u_s-u\|_2\|F(v_s)\|_2) \leq 4b$$

for $s \in A$. □

Proposition 4.2.7 *The set of $f \in L^1(\mathbb{R})$ such that $\hat{f}$ has compact support is dense in $L^1(\mathbb{R})$.*

Proof Since $C_c(\mathbb{R})$ is dense in $L^2(\mathbb{R})$, the set of $g \in L^2(\mathbb{R})$ with $\hat{g} \in C_c(\mathbb{R})$ is dense in $L^2(\mathbb{R})$ by the theorem above. Hence for any positive $h \in L^1(\mathbb{R})$ and $\varepsilon > 0$, there is $g \in L^2(\mathbb{R})$ with $\hat{g} \in C_c(\mathbb{R})$ such that $\|g-h^{1/2}\|_2 < \varepsilon$. Then

$$\|g^2-h\|_1 \leq \|h^{1/2}\|_2\|g-h^{1/2}\|_2 + \|g\|_2\|g-h^{1/2}\|_2 < \varepsilon(2\|h^{1/2}\|_2+\varepsilon)$$

and the Fourier transform of $g^2 \in L^1(\mathbb{R})$ is $\hat{g}*\hat{g} \in C_c(\mathbb{R})$. □

4.3 Fourier Series

Identifying L^2-functions on the circle $\mathbb{T}$ with 2π-periodic functions f on $\mathbb{R}$, meaning that $f(x+2\pi) = f(x)$, we have seen that any such square integrable function can be approximated by its Fourier series in L^2-norm. We include here some results on pointwise convergence of Fourier series on $\mathbb{R}$. We begin with a related result by Fejer.

Proposition 4.3.1 *Given any 2π-periodic continuous function f on $\mathbb{R}$, then as $N \to \infty$ we have $(N+1)^{-1}\sum_{n=0}^{N}\sum_{k=-n}^{n}\hat{f}(k)v_k \to f$ uniformly on $[\pi,\pi]$.*

Proof Let K_N be the expression on the left with $\hat{f}(k)$ replaced by 1. Then as $2(N+1)K_N(x) - e^{ix}(N+1)K_N(x) - e^{-ix}(N+1)K_N(x) = 2 - e^{i(N+1)x} - e^{-i(N+1)x}$ we get $K_N(x) = (N+1)^{-1}(1-\cos(N+1)x)(1-\cos x)^{-1}$ for $x \in [-\pi,\pi]\backslash\{0\}$, so K_N is non-negative and $\int_{-\pi}^{\pi}K_N(x)\,dx = 2\pi$. Letting F_N be the left hand side of the expression in the proposition, we get $F_N(x) = (2\pi)^{-1}\int_{-\pi}^{\pi}f(x-y)K_N(y)\,dy$ by periodicity. Since f is uniformly continuous on $[-\pi,\pi]$, to $\varepsilon > 0$, there is $\delta > 0$ such that $|f(x-y)-f(x)| < \varepsilon/2$ for $x \in [-\pi,\pi]$ and $|y| < \delta$. We also know that

$M \equiv \sup |f([-\pi, \pi])| < \infty$. Hence

$$
\begin{aligned}
|F_N(x) - f(x)| &\leq (2\pi)^{-1} \int_{-\pi}^{\pi} |f(x-y) - f(x)| \cdot K_N(y)\, dy \\
&\leq \frac{M}{\pi} \int_{-\pi}^{-\delta} K_N(y)\, dy + \frac{\varepsilon}{4\pi} \int_{-\delta}^{\delta} K_N(y)\, dy + \frac{M}{\pi} \int_{\delta}^{\pi} K_N(y)\, dy \\
&\leq \frac{4M}{\pi(N+1)(1-\cos\delta)} + \frac{\varepsilon}{2} < \varepsilon
\end{aligned}
$$

for large enough N. □

Pointwise convergence for Fourier series is more delicate. Let us recall some calculus. A *Riemann partition* for a bounded real-valued function f on $[a, b]$ is a finite sequence $P = \{x_i\}$ such that $a = x_0 < x_1 < \cdots < x_{n-1} < x_n = b$. Let $S = \sum M_i(x_i - x_{i-1})$ and $s = \sum m_i(x_i - x_{i-1})$, where M_i and m_i is the supremum and infinum, respectively, of f on $[x_{i-1}, x_i]$. Then f is *Riemann integrable* if $\inf_P S = \sup_P s$, and the *Riemann integral* $\int_a^b f(x)\, dx$ *of* f is this common value. We extend it by linearity to bounded complex-valued functions. Considering the obvious simple functions associated to each P, and using Lebesgue's dominated convergence theorem for a sequence of P's with $\max_i |x_i - x_{i-1}| \to 0$, we see that any Riemann integrable function on $[a, b]$ is Lebesgue integrable, and the two integrals do then coincide. Given an increasing right continuous function g on $[a, b]$, the *Riemann-Stieltjes integral* $\int_a^b f(x)\, dg(x)$ is defined as the Riemann integral except that $x_i - x_{i-1}$ is replaced by $g(x_i) - g(x_{i-1})$, so we recover the Riemann integral when g is the identity function. The Borel measure μ_g which we associate to $f \mapsto \int_a^b f\, dg$ by the Riesz representation theorem is uniquely determined by the property $\mu_g(\langle x, y]) = g(y) - g(x)$ for $x, y \in [a, b]$.

The *mean value theorem* says that for a real-valued continuous function f on $[a, b]$ that is differentiable on $\langle a, b\rangle$, there is $x \in \langle a, b\rangle$ such that $f(b) - f(a) = f'(x)(b - a)$. To see this, say g given by $g(y) = (f(b) - f(a))y - (b - a)f(y)$ on $[a, b]$ attains a local maximum or minimum at $x \in \langle a, b\rangle$. Then $g'(x) = 0$ by carefully considering the signs of the fraction defining the derivative. We record also the *fundamental theorem of calculus*, namely, if a real-valued Riemann integrable function f on $[a, b]$ has an antiderivative F, meaning that $F' = f$, then $\int_a^b f(x)\, dx = F(b) - F(a)$. Indeed, picking P such that $S - s < \varepsilon$, the mean value theorem provides $y_i \in [x_{i-1}, x_i]$ such that $F(x_i) - F(x_{i-1}) = f(y_i)(x_i - x_{i-1})$, so $\sum f(y_i)(x_i - x_{i-1}) = F(b) - F(a)$ and $|F(b) - F(a) - \int_a^b f(x)\, dx| < \varepsilon$.

We also have the following *integration by parts* result

$$
\int_{\langle a,b]} f\, d\mu_g + \int_{\langle a,b]} g\, d\mu_f = f(b)g(b) - f(a)g(a)
$$

for increasing right continuous f, g with one of them continuous. Indeed, if g is continuous, the Fubini-Tonelli theorem yields

$$\int_{(a,b]} f\,d\mu_g - f(a)(g(b)-g(a)) = \int_{(a,b]} (f(y)-f(a))\,d\mu_g(y) = \int d\mu_f d\mu_g$$

$$= \int d\mu_g d\mu_f = \int_{(a,b]} (g(b)-g(x))\,d\mu_f(x) = g(b)(f(b)-f(a)) - \int_{(a,b]} g\,d\mu_f.$$

Lemma 4.3.2 *Let f, g be real-valued functions on $[a, b]$ with f increasing and right continuous and g continuous. Then there is $c \in [a, b]$ such that*

$$\int_a^b f(x)g(x)\,dx = f(a)\int_a^c g(x)\,dx + f(b)\int_c^b g(x)\,dx.$$

Proof Replacing f by $f - f(a)$ we may assume that $f(a) = 0$. Consider G on $[a, b]$ given by $G(x) = \int_x^b g(y)\,dy$, so $G' = -g$ is continuous, and by considering the positive and negative parts of g, we can also assume that G is increasing. Using the integration by parts result from above we get $\int_a^b f(x)g(x)\,dx = \int_{(a,b]} G\,d\mu_f$ as $f(a) = 0 = G(b)$. Since $(f(b) - f(a))^{-1}\int_{(a,b]} G\,d\mu_f$ lies between $\inf G([a, b])$ and $\sup G([a, b])$, there must be $c \in [a, b]$ such that $G(c)$ equals this average. □

Let $D_N = \sum_{n=-N}^{N} v_n$ be the Dirichlet kernel. Then $(2\pi)^{-1}\int_{-\pi}^{\pi} D_N(x)\,dx = 1$, and since D_N is even, we get $(2\pi)^{-1}\int_{-\pi}^{0} D_N(x)\,dx = \frac{1}{2} = (2\pi)^{-1}\int_0^{\pi} D_N(x)\,dx$. Repeating the argument in the proof of the previous proposition we also see that $D_N(x) = (\sin(x/2))^{-1}\sin(N + \frac{1}{2})x$.

Lemma 4.3.3 *There is a constant d such that $|(2\pi)^{-1}\int_a^b D_N(x)\,dx| \leq d$ for every N and $-\pi \leq a \leq b \leq \pi$.*

Proof By the formula above

$$\int_a^b D_N(x)\,dx = \int_a^b \frac{\sin(N+\frac{1}{2})x}{(x/2)}\,dx + \int_a^b \sin(N+\frac{1}{2})x((\sin(x/2))^{-1} - (x/2)^{-1})\,dx.$$

Both factors inside the second integral are bounded on $[-\pi, \pi]$, whereas the first integral after a simple substitution, equals $2g((N + \frac{1}{2})b) - 2g((N + \frac{1}{2})a)$, where $g(x) = \int_0^x y^{-1}\sin y\,dy$. But g is also bounded since it is continuous and because it tends to a limit as $x \to \infty$. In fact, by the Fubini-Tonelli theorem this limit equals

$$\int_0^\infty \int_0^\infty e^{-xy}\sin y\,dxdy = \int_0^\infty \int_0^\infty e^{-xy}\sin y\,dydx = \int_0^\infty (1+x^2)^{-1}\,dx = \pi/2,$$

where we in the second step used partial integration twice, and then used the substitution $x = \tan u$ in the third step. □

Theorem 4.3.4 *If the real and imaginary parts of f are differences of two 2π-periodic increasing functions, then*

$$\lim_{N\to\infty}\sum_{n=-N}^{N}\hat{f}(n)v_n(x)=(f(x_+)+f(x_-))/2,$$

where $f(x_\pm)$ are the one-sided limits of f at x.

Proof One-sided limits of increasing functions clearly exist, and for this proof we might as well assume that f is increasing. We may also assume that f is right continuous since both sides of the formula we wish to prove are unaltered when we replace f by the function that is $f(y_+)$ at every y. By translation we can finally assume that $x=0$. Then $\sum_{n=-N}^{N}\hat{f}(n)v_n(x)-(f(x_+)+f(x_-))/2$ equals

$$\frac{1}{2\pi}\int_{-\pi}^{0}(f(x)-f(0_-))D_N(x)\,dx+\frac{1}{2\pi}\int_{0}^{\pi}(f(x)-f(0_+))D_N(x)\,dx.$$

We show that the second term tends to zero as $N\to\infty$. The first term tends to zero for similar reasons. Let d be as in the last lemma, and let $\varepsilon>0$. Chose $\delta>0$ such that $f(\delta)-f(0_+)<\varepsilon/d$. By the second last lemma there is $c\in[0,\delta]$ such that $(2\pi)^{-1}|\int_0^\delta(f(x)-f(0_+))D_N(x)\,dx|=|f(\delta)-f(0_+)|(2\pi)^{-1}\int_c^\delta D_N(x)\,dx|<\varepsilon$, whereas $|(2\pi)^{-1}\int_\delta^\pi(f(x)-f(0_+))D_N(x)\,dx|=\hat{g}_+(-N)-\hat{g}_-(N)$, where $g_\pm$ are the 2π-periodic functions on $[-\pi,\pi]$ given by

$$g_\pm(x)=(f(x)-f(0_+))(2i\sin(x/2))^{-1}e^{\pm i(x/2)}\chi_{[\delta,\pi\rangle}.$$

Clearly $g_\pm\in L^2(\mathbb{T})$, so $\hat{g}_\pm(\mp N)\to 0$ as $N\to\infty$ by Parseval's identity. □

In particular the Fourier series of f converges pointwise to f when it is also continuous at x. Any function with piecewise continuous derivative satisfies the hypothesis of the theorem, which is seen by considering the positive and negative parts of the derivative of the function. This special case was proved by Dirichlet, while the theorem above was by Jordan. Functions satisfying the hypothesis of the previous theorem are said to be of *bounded variation* on $[-\pi,\pi]$ since they are precisely those functions f on this interval such that the supremum of $V_P(f)\equiv\sum|f(x_i)-f(x_{i-1})|$ over all Riemann partitions P is finite. The forward implication of this claim is obvious, and for the converse write say the real part h of f as the difference of the two increasing functions $\sup_P V_P(h)$ and $\sup_P V_P(h)-h$, where $V_P(h)$ is a function of x gotten by replacing $\pi=b=x_n$ by x in P.

Example 4.3.5 Let $f(x)=x$ for $\langle-\pi,\pi\rangle$ and extend it to an odd 2π-periodic function f on $\mathbb{R}$, so $f(\pm\pi)=0$. Then $\hat{f}(0)=0$ and $\hat{f}(n)=i(-1)^n/n$ for $n\neq 0$ by partial integration. Thus $\sum_{n=-N}^{N}\hat{f}(n)v_n(x)=-2\sum_{n=1}^{N}\frac{(-1)^n}{n}\sin nx$ which tends to $(f(x_-)+f(x_+))/2$ by the theorem, and clearly matches for $x=0$ and $x=\pm\pi$. Setting $x=\pi/2$ gives $1-\frac{1}{3}+\frac{1}{5}-\frac{1}{7}+\cdots=\pi/4$ known as

Gregory's series. The Parseval identity yields $2\sum_{n=1}^{\infty}\frac{1}{n^2} = \sum_{n=-\infty}^{\infty}|\hat{f}(n)|^2 = \|f\|_2^2 = \frac{1}{\pi}\int_0^{\pi}x^2\,dx = \frac{\pi^2}{3}$, which solves Basel's problem. ◇

Notice that when a 2π-periodic integrable function f is real, then $\overline{\hat{f}(n)} = \hat{f}(-n)$ and $\sum_{n=-N}^{N}\hat{f}(n)v_n(x) = \sum_{n=0}^{N}((\hat{f}(n)+\overline{\hat{f}(n)})\cos nx + i(\hat{f}(n)-\overline{\hat{f}(n)})\sin nx)$ is real. It is moreover even with only cosines appearing when f is even, and it is odd with only sinus appearing when f is odd.

When a sequence $\{a_n\}$ has a limit, then its Cesaro limit $\lim_{N\to\infty}N^{-1}\sum_{n=1}^{N}a_n$ of arithmetic means also exists, and equals the same limit, but the converse is not always true as is seen by e.g. considering the sequence $a_n = (-1)^n$. This allows for the following even stronger convergence result in this average sense.

Proposition 4.3.6 *If f is a 2π-periodic Lebesgue integrable function on $\mathbb{R}$ with left and right limits $f(x_\pm)$ at x, then $(N+1)^{-1}\sum_{n=0}^{N}\sum_{k=-n}^{n}\hat{f}(k)v_k(x) \to (f(x_-)+f(x_+))/2$ as $N\to\infty$.*

Proof To $\varepsilon > 0$ pick $0 < \delta < \pi$ such that $|f(x+t)+f(x-t)-f(x_+)-f(x_-)| < \varepsilon$ for $t \in [0,\delta\rangle$. Using the expression for K_N in the proof of the previous proposition, we have $\sup_{t\in[\delta,\pi]}K_N(t) < \varepsilon$ for N large enough. Now $(N+1)^{-1}\sum_{n=0}^{N}\sum_{k=-n}^{n}\hat{f}(k)v_k(x)-(f(x_-)+f(x_+))/2$ equals $\frac{1}{\pi}\int_0^{\pi}K_N(t)(f(x+t)+f(x-t)-f(x_+)-f(x_-))\,dt$. Splitting up the latter integral at δ, the absolute value of the part from 0 to δ contributes with no more than ε for large enough N, whereas the absolute value of the part from δ to π contributes with no more than ε times $\|f-f(x_-)\|_1 + \|f-f(x_+)\|_1$ for such N. □

Hence if the Fourier series of such an f converges at such an x, then the limit must equal $(f(x_-)+f(x_+))/2$.

4.4 Polar Decomposition of Operators on Hilbert Spaces

We extend the notion of adjointness for matrices to operators on Hilbert spaces.

Proposition 4.4.1 *Let V be a Hilbert space. The Banach algebra $B(V)$ has an isometric $*$-operation $A \mapsto A^*$ such that $\|A^*A\| = \|A\|^2$ which is uniquely determined by $(A(v)|w) = (v|A^*(w))$ for all $v, w \in V$.*

Proof To any $w \in V$ there is by Corollary 4.1.9 a unique $A^*(w) \in V$ such that $(A(\cdot)|w) = (\cdot|A^*(w))$. The resulting map $A \mapsto A^*$ on $B(V)$ has plainly the required properties as $\|A(v)\|^2 \le \|A^*A\|\,\|v\|^2$. □

By polarization $A \in B(V)$ is normal if and only if $\|A^*(v)\| = \|A(v)\|$ for all $v \in V$. When V is a complex Hilbert space, then by polarization we see that $A \in B(V)$ is self-adjoint if and only if $(A(v)|v) \in \mathbb{R}$ for all $v \in V$.

Note that $\ker A^* = A(V)^{\perp}$ for any $A \in B(V)$, so $\mathrm{im}A$ is dense in V if and only if A^* is injective. By the open mapping theorem we therefore see that $A \in B(V)$

has $A^{-1} \in B(V)$ if and only if there is $\varepsilon > 0$ such that $\|A(v)\| \geq \varepsilon\|v\| \leq \|A^*(v)\|$ for all $v \in V$, saying that A and A^* are *bounded away from zero*.

Recall that a *sesquilinear form* on a complex vector space V is a map $V \times V \to \mathbb{C}$ that is linear in the first variable and conjugate linear in the second variable. The map $A \mapsto (A(\cdot)|\cdot)$ is an isomorphism from $B(V)$ to the vector space of sesquilinear forms B on V with norms

$$\|B\| \equiv \sup\{|B(v, w)| \mid \|v\| \leq 1, \ \|w\| \leq 1\} < \infty.$$

In particular, the map restricts to a bijective correspondence between the injective positive $A \in B(V)$ and the family of all possible inner products on V.

Definition 4.4.2 A bounded operator A on a Hilbert space V is *positive*, and we write $A \geq 0$, if $(A(v)|v) \geq 0$ for all $v \in V$. Let $A \leq B$ or $B \geq A$ mean that A and B are self-adjoint and that $B - A \geq 0$.

Proposition 4.4.3 *If A, B, S are bounded operators on a Hilbert space such that $A \leq B$, then $S^*AS \leq S^*BS$. If $A \geq 0$, then $\|A\| \leq \|B\|$.*

Proof Clearly S^*AS and S^*BS are self-adjoint and $((B - A)S(v)|S(v)) \geq 0$.

For the second assertion we use the Cauchy-Schwarz inequality to $(A(\cdot)|\cdot)$ and get

$$|(A(v)|w)|^2 \leq (A(v)|v)(A(w)|w) \leq (B(v)|v)(B(w)|w) \leq \|B\|^2\|v\|^2\|w\|^2,$$

so $\|A\|^2 \leq \|B\|^2$ by the considerations above. □

Proposition 4.4.4 *Every positive operator A on a Hilbert space has a unique positive operator with square A called the* square root *$A^{1/2}$ of A. It commutes with every operator that commutes with A.*

Proof We postpone the proof of the existence till we have established the continuous functional calculus in a later section. Then we also prove uniqueness. For another proof of uniqueness, see Corollary A.11.13. □

Proposition 4.4.5 *A positive operator A on a Hilbert space is invertible if and only if $A \geq \varepsilon I$ for some $\varepsilon > 0$. Then $A^{-1} \geq 0$ and $A^{1/2}$ is invertible with inverse $A^{-1/2} = (A^{-1})^{1/2}$, and if $B \geq A$, then $B^{-1} \leq A^{-1}$.*

Proof Since the product of two commuting positive operators is a positive operator by the two previous propositions, we get $A^2 \geq \varepsilon^2 I$ when $A \geq \varepsilon I$, so A is bounded away from zero and is invertible. Now $(A^{-1}A(v)|A(v)) \geq 0$, so $A^{-1} \geq 0$.

Conversely, an invertible positive operator A is bounded away from zero, so $(A^2(v)|v) \geq \varepsilon^2(v|v)$. Hence $A - \varepsilon I = (A + \varepsilon I)^{-1}(A^2 - \varepsilon^2 I) \geq 0$. Repeating this gives $A^{1/2} \geq \varepsilon^{1/2} I$, so $A^{1/2}$ has a positive inverse with square A^{-1}, and $(A^{1/2})^{-1} = (A^{-1})^{1/2}$ by uniqueness of postive square roots.

We have seen that $B^{-1/2}$ exists, so $B^{-1/2}AB^{-1/2} \leq I$ and

$$\|A^{1/2}B^{-1}A^{1/2}\| = \|B^{-1/2}A^{1/2}\|^2 = \|A^{1/2}B^{-1/2}\|^2 = \|B^{-1/2}AB^{-1/2}\| \leq 1$$

by Proposition 4.4.3, so $A^{1/2}B^{-1}A^{1/2} \leq I$ which again yields $B^{-1} \leq A^{-1}$. □

Definition 4.4.6 An isomorphism of a Hilbert space onto itself is *unitary*. An *orthogonal projection* is a self-adjoint idempotent from a Hilbert space to itself.

Hence A is unitary if and only if $A^*A = I = AA^*$, and in this case $\|A\| = 1$.

Example 4.4.7 The *unilateral shift* $A \in B(l^2(\mathbb{N}))$ given by $A(v_n) = v_{n+1}$ is an isometry $A^*A = I$ onto the proper subspace $\{v_1\}^\perp$. While AA^* is not the identity operator, it is an orthogonal projection. ◇

Orthogonal projections are positive. They are non-zero if and only if their norm is one. If $P \in B(V)$ is an orthogonal projection, then $I - P$ is also an orthogonal projection, and we have an orthogonal decomposition $V \cong P(V) \oplus (I - P)(V)$, where $P = \pi_1$ and $I - P = \pi_2$ if we identify the Hilbert spaces.

Definition 4.4.8 A bounded operator A on a Hilbert space is a *partial isometry* if A^*A is an idempotent.

Proposition 4.4.9 *Let A be a partial isometry on a Hilbert space V. Then $A = AA^*A$, so $P = A^*A$ and $Q = AA^*$ are orthogonal projections. The map A restricts to an isometry from it's* source space *$P(V) = A^*(V)$ onto it's* range space *$Q(V) = A(V)$ with the restriction of A^* as it's inverse map, while A vanishes on $P(V)^\perp$ and A^* vanishes on $Q(V)^\perp$. Conversely, any bounded map on a Hilbert spaces that is an isometry on a closed subspace W and vanishes on $W^\perp$ is a partial isometry with source space W. Given two closed subspaces that are isomorphic as Hilbert spaces, there exists a partial isometry with these as source and range spaces.*

Proof As $V = \overline{A(V)} \oplus \ker A^*$ we have

$$P(V) \subset A^*(V) \subset \overline{P(V)} = P(V).$$

Now $P(V)^\perp = A^*(V)^\perp = \ker A$ shows that $A(I - P) = 0$, or $A = AA^*A$. Since $(P(v)|v) = \|A(v)\|^2$ for $v \in V$, we see that A is a partial isometry with source space $P(V) = A^*(V)$ and range space $A(V)$, and this latter space equals $Q(V)$ by a similar argument.

Conversely, if A is a partial isometry on W with kernel $W^\perp$, then $(P(v)|v) = \|A(v)\|^2 = \|v\|^2$ for $v \in W$. Since equality in the Cauchy-Schwarz inequality holds, we get $P(v) = v$ for $v \in W$, and obviously $P(W^\perp) = \{0\}$, so P is an orthogonal projection onto W.

If W_1 and W_2 are two isomorphic subspaces of V with orthonormal bases $\{v_i\}$ and $\{w_i\}$, define a bounded map A on V by setting $A(v_i) = w_i$ for all i and letting A vanish on $W_1^\perp$. □

Proposition 4.4.10 *To each operator A on a Hilbert space $|A| \equiv (A^*A)^{1/2}$ is the unique positive operator such that $\||A|(v)\| = \|A(v)\|$ for all v. There is a unique partial isometry U such that $U|A| = A$ and $\ker U = \ker A$. It also satisfies $U^*U|A| = |A|$. When A is invertible, then U is unitary.*

Proof If $B \geq 0$ and $\|B(v)\| = \|A(v)\|$, then $(B^2(\cdot)|\cdot) = (A^*A(\cdot)|\cdot)$ by polarization, so $B^2 = |A|^2$ and $B = |A|$ by uniqueness of positive square roots.

From $\ker A = \ker |A| = \mathrm{im}|A|^\perp$, the required formulas well-define a unique partial isometry U with source projection U^*U on $\overline{\mathrm{im}|A|}$, and U is clearly unitary when A is invertible. □

We refer to $A = U|A|$ as the *polar decomposition of* A, see also the appendix. Then $U^*|A^*|$ with $|A^*| = U|A|U^*$ is the polar decomposition of A^*. The operators $|A|$ and U need not commute as the following example shows:

$$\begin{pmatrix} 0 & 1 \\ 2 & 0 \end{pmatrix} = \begin{pmatrix} 0 & 1 \\ 1 & 0 \end{pmatrix} \begin{pmatrix} 2 & 0 \\ 0 & 1 \end{pmatrix}.$$

Corollary 4.4.11 *For any normal operator A there is a unitary $\tilde{U}$ commuting with A and A^* and $|A|$ such that $A = \tilde{U}|A|$.*

Proof For the polar decomposition $A = U|A|$ we have $(U|A|U^*)^2 = AA^* = |A|^2$, so $U|A|U^* = |A|$ by uniqueness of positive square roots. Hence $U|A| = |A|U$.

Define the unitary operator $\tilde{U}$ to be U on $\overline{\mathrm{im}|A|}$ and I on $\ker |A| = \mathrm{im}|A|^\perp$. □

In the next result we need complex Hilbert spaces.

Proposition 4.4.12 *The* numerical radius

$$\|A\|_n = \sup_{\|v\| \leq 1} |(A(v)|v)|$$

of an operator A on a complex Hilbert space is a norm such that $\|\cdot\|/2 \leq \|\cdot\|_n \leq \|\cdot\|$ and $\|A^2\|_n \leq \|A\|_n^2$. Moreover, it coincides with the usual norm on normal operators.

Proof Obviously $\|\cdot\|_n$ is a seminorm dominated by $\|\cdot\|$.

From the identity

$$2(A(v)|w) + 2(A(w)|v) = (A(v+w)|v+w) - (A(v-w)|v-w)$$

and the parallelogram law, we get the estimate

$$\|A(v)\| + \|A(v)\|^{-1}\mathrm{Re}(A^2(v)|v) \leq 2\|A\|_n$$

for any unit vector v. For such vectors we thus get

$$\|A(v)\| + \|A(v)\|^{-1}|(A^2(v)|v)| \leq 2\|A\|_n$$

upon replacing A by aA for a suitable $a \in \mathbb{C}$ of modulus one. So $\|A\| \leq 2\|A\|_n$ and

$$0 \leq 2\|A\|_n\|A(v)\| - \|A(v)\|^2 - |(A^2(v)|v)| \leq \|A\|_n^2 - |(A^2(v)|v)|,$$

which shows that $\|A^2\|_n \leq \|A\|_n^2$.

Finally, if A is normal, then for any $m \in \mathbb{N}$, we get

$$\|A\|^{2^m} = \|(A^*A)^{2^m}\|^{1/2} = \|A^{*2^m}A^{2^m}\|^{1/2} = \|A^{2^m}\| \leq 2\|A^{2^m}\|_n \leq 2\|A\|_n^{2^m},$$

which can only happen if $\|A\| \leq \|A\|_n$. □

4.5 Compact Normal Operators

Say V is a Hilbert space. The compact operators $B_0(V)$ is a closed ideal in $B(V)$ which is *self-adjoint*, meaning that $A^* \in B_0(V)$ if $A \in B_0(V)$, since by polar decomposing A we see that both $|A| = U^*A$ and $A^* = |A|U^*$ are compact. The same argument shows that any closed ideal J in $B(V)$ is self-adjoint, and that $A \in J$ if and only if $|A| \in J$.

Another way to see that $|A|$ is compact is to observe that it is a norm-limit of polynomials in A^*A in the Banach algebra $B_0(V)$. If A is a finite rank operator on V, the orthogonal decomposition $V = A(V) \oplus \ker A^*$ shows that $A^*(V) = A^*A(V)$ is finite dimensional. So $B_f(V)$ and its closure are self-adjoint ideals in $B(V)$ which are contained in $B_0(V)$.

Corollary 4.1.9 provides an antilinear isometric bijection $f: V \to V^\star$ given by $f(v) = (\cdot|v)$. The Hilbert space adjoint A^* and the Banach space adjoint A^{*b} are then related by the formula $A^* = f^{-1}A^{*b}f$, which explains why the first operation is antilinear while the second operation is linear. This provides yet another proof that $B_0(V)$ is self-adjoint.

The w^*-topology on $V^\star$ transferred to V by f is the *weak topology*. So a net v_i converges weakly to v in V if $(v_i|w) \to (v|w)$ for all $w \in V$. Since we have adjoints, any $A \in B(V)$ is weak-weak continuous. Conversely, if a linear map $A: V \to V$ is weak-weak continuous, then $A \in B(V)$ by the closed graph theorem and the fact that the weak topology is Hausdorff.

By Alaoglu's theorem the closed unit ball $\overline{B}$ in V is compact in the weak topology. Hence by our Banach space results, the restriction of any $A \in \overline{B_f(V)}$ to the unit ball B is weak-norm continuous, and any $A \in B(V)$ with this continuity property is compact with $A(\overline{B}) = \overline{A(B)}$. For real or complex Banach spaces the finite rank operators are in general not dense in the compact operators, but for Hilbert spaces the result below shows that we do indeed have $\overline{B_f(V)} = B_0(V)$.

Definition 4.5.1 An *approximate unit* for a (not necessarily unital) Banach algebra is a net $\{w_i\}$ in the closed unit ball of the algebra such that $w_i w \to w$ and $w w_i \to w$ for every element w in the algebra.

Proposition 4.5.2 *The Banach algebra of compact operators on a Hilbert space has an approximate unit consisting of finite rank orthogonal projections.*

Proof Take any orthonormal basis $\{v_i\}$ for the Hilbert space V, and form the net of orthogonal projections P_J onto the subspace spanned by v_i for i in the finite set J, and partially order these sets under inclusion. Then $\|P_J v - v\| \to 0$ for any $v \in V$ by Parseval's identity. If $\{P_J A\}$ does not converge to $A \in B_0(V)$, then there is to $\varepsilon > 0$ a subnet $\{P_j\}$ with unit vectors w_j such that $\|(P_j A - A)w_j\| \geq \varepsilon$ and $A(w_j) \to w$ in norm for some w. Hence $\varepsilon \leq \|(I - P_j)(A(w_j) - w)\| + \|(I - P_j)w\| \to 0$. So $P_J A \to A$, and $A P_J \to A$ by taking adjoints. □

Consider the rank one operators $v \odot w \in B(V)$ given by $(v \odot w)(u) = (u|w)v$ for u, v, w in a Hilbert space V. By Corollary 4.1.9 every rank one operator is of this form, and for $v = w$ unit vectors one gets the rank one orthogonal projections.

For an orthonormal basis $\{v_i\}$ of V it is easily checked that $(v_i \odot v_j)^* = v_j \odot v_i$ and $(v_i \odot v_j)(v_k \odot v_l) = \delta_{jk} v_i \odot v_l$. Hence $B_f(V)$ is generated as an ideal by any single rank one operator. Since $(v_i \odot v_j)A = v_i \odot A^*(v_j)$ for any $A \in B(V)$, we see that $B_f(A)$ is contained in every non-zero ideal of $B(V)$, so $B_0(V)$ is contained in every non-zero closed ideal of $B(V)$. In particular, neither $B_f(V)$ nor $B_0(V)$ have proper ideals, respectively, proper closed ideals.

Any unitary $U \in B(V, W)$ induces a $*$-isomorphism from $B(W)$ to $B(V)$ by $\mathrm{Ad}_U(A) = U^* A U$ with the obvious definition of the adjoint $U^* \in B(W, V)$. If an orthogonal projection P on a separable Hilbert space V has infinite rank, then $P(V)$ is also separable, so we have a unitary $U \colon V \to P(V)$ such that $I = \mathrm{Ad}_U(P) = U^* P U$. Hence orthogonal projections in a proper closed ideal of $B(V)$ have finite rank when V is separable, and note in passing that $B_0(V)$ is then separable.

Proposition 4.5.3 *In a separable Hilbert space V the compact operators form the only proper closed ideal in $B(V)$.*

Proof This proof uses Borel functional calculus that will be included later. Let A be an element in a proper closed ideal J of $B(V)$. We may assume that $A \geq 0$ and $\|A\| \leq 1$. By the spectral theorem for Borel functions on $\mathrm{Sp}(A) \subset [0, 1]$, it suffices to show that each spectral projection $P(< r, s])$ of A belongs to J since we can then approximate A by finite rank operators. If $0 < r < s \leq 1$ and $\|v\| \leq 1$, then

$$\|((AP([r, 1]))^{1/n} - P([r, 1]))(v)\|^2 \leq \int_{\mathrm{Sp}(A)} |(\lambda \chi_{[r,1]}(\lambda))^{1/n} - \chi_{[r,1]}(\lambda)|^2 \, dP_{v,v}(\lambda) \to 0$$

as $n \to \infty$ since the function inside the integral is a bounded sequence that tends to zero pointwise. Hence $P([r, 1])$ is a norm limit of elements that belong to the closed ideal J. But then also $P(< r, s]) = P(< r, s])P([r, 1]) \in J$. □

The next result is useful also in the finite dimensional case.

Theorem 4.5.4 *For any Hilbert space isomorphism $U\colon V \to W$ the map Ad_U restricts to a $*$-isomorphism $B_0(W) \to B_0(V)$, and every such $*$-isomorphism is of this form.*

Proof Consider a $*$-isomorphism $f\colon B_0(W) \to B_0(V)$. Let $\{w_i\}$ be an orthonormal basis for W. Pick vectors $v_i \in V$ such that $v_i \odot v_i = f(w_i \odot w_i)$. Using Zorn's lemma and f^{-1} we see that $\{v_i\}$ is an orthonormal basis of V. It is also easy to see that there are scalars c_{ij} of modulus one such that $f(w_i \odot w_j) = c_{ij} v_i \odot v_j$. Fix j and define U by $U(v_i) = \overline{c_{ij}} w_i$ for all i. Then $\mathrm{Ad}_U(w_i \odot w_k) = f(w_i \odot w_k)$ for all i and k, so $\mathrm{Ad}_U = f$ on all rank one operators. Since the span of these is dense in the compact operators, we get $\mathrm{Ad}_U = f$ by continuity. □

Definition 4.5.5 We say $A \in B(V)$ is *orthogonally diagonalizable* if there is an orthonormal basis $\{v_i\}$ for V and scalars λ_i such that $A(v) = \sum \lambda_i (v|v_i) v_i$ for $v \in V$.

Then λ_i is an eigenvalue for A and $P_i = (\cdot|v_i)v_i$ is the orthogonal projection onto the subspace spanned by the eigenvector v_i. Since $A^*(v) = \sum \bar{\lambda}_i (v|v_i) v_i$, we see that orthogonally diagonalizable operators are normal.

Example 4.5.6 We assert that any normal complex square matrix A is orthogonally diagonalizable. Observe first that an upper triangular normal matrix B is orthogonally diagonalizable, actually diagonal, because the length $\|B(e_i)\|$ of its i-th column equals the length $\|B^*(e_i)\|$ of its i-th row. It remains to construct an upper triangular B and a unitary U with $AU = UB$ since B must then be normal. To this end let v_1 be a normalized eigenvector of A with eigenvalue λ_1. By a Gram-Schmidt process we can extend this to an orthonormal basis $v_1, \ldots, v_n$. Let U_1 be the unitary matrix with these vectors in the listed order as the columns from left to right. Then $AU_1 = U_1 B_1$, where B_1 is the matrix with λ_1 in the upper left corner and with only zeroes below it. Let A_1 be $(n-1) \times (n-1)$-matrix in the lower right corner of B_1. Next construct B_2 and U_2 to A_1 such that $A_1 U_2 = U_2 B_2$ as we did for A. Continuing inductively we get unitaries $V_k = \mathrm{diag}(I_{k-1}, U_k) \in M_n(\mathbb{C})$. By construction $U \equiv V_1 \cdots V_n$ will make $U^{-1} A U$ upper triangular. ◇

We will now extend this result to compact normal operators.

Lemma 4.5.7 *Let A be an orthogonally diagonalizable operator on a Hilbert space V, say $A(v) = \sum \lambda_i (v|v_i) v_i$ for an orthonormal basis $\{v_i\}$. Then A is compact if and only if there are only finitely many i to $\varepsilon > 0$ with $|\lambda_i| \geq \varepsilon$.*

Proof If such an index set J is infinite for compact A, then by Parseval's identity, the net $\{v_i\}_{i \in J}$ will converge weakly to zero, while $\|A(v_i)\| = |\lambda_i| \geq \varepsilon$, and this contradicts the fact that A is weak-norm continuous on the unit ball.

If every such J is finite, then A must be compact because $\|A - A_J\| \leq \varepsilon$ with $A_J = \sum_{i \in J} \lambda_i v_i \odot v_i \in B_f(V)$ as $\|(A - A_J)(v)\|^2 = \sum_{i \notin J} |\lambda_i|^2 |(v|v_i)|^2 \leq \varepsilon^2 \|v\|^2$. □

Let A be a normal operator on a Hilbert space. Any eigenvector v of it with eigenvalue λ is also an eigenvector for A^* with eigenvalue $\bar{\lambda}$ since $\|A^*(v) - \bar{\lambda} v\| = \|A(v) - \lambda v\| = 0$. Moreover, eigenvectors of distinct eigenvalues of A

are orthogonal because if also $A(w) = \mu w$ with $\lambda \neq \mu \neq 0$, then

$$(w|v) = \mu^{-1}(A(w)|v) = \mu^{-1}(w|A^*(v)) = \mu^{-1}\lambda(w|v).$$

In the next result the Hilbert spaces must be complex.

Lemma 4.5.8 *Every compact normal operator A on a complex Hilbert space has $\|A\|$ as the module of an eigenvalue.*

Proof If $v_i \to v$ weakly in the closed unit ball $\overline{B}$, then

$$|(A(v_i)|v_i) - (A(v)|v)| \leq \|A(v_i - v)\| + |(A(v)|v_i - v)| \to 0$$

since A is weak-norm continuous. So the numerical radius function $v \mapsto |(A(v)|v)|$ attains its maximum $\|A\|$ for some v in the weakly compact set $\overline{B}$. Equality in the Cauchy-Schwarz inequality is obtained only when vectors are proportional, so v is an eigenvector of A with eigenvalue having absolute value $\|A\|$. □

Theorem 4.5.9 *Operators on complex Hilbert spaces are compact normal if and only if they are orthogonally diagonalizable with eigenvalues vanishing at infininity when counted with multiplicity.*

Proof By the first lemma we need only show that a compact normal operator A on a complex Hilbert space V is diagonalizable. The family of sets consisting of orthonormal eigenvectors of A is inductively ordered under inclusion, so by Zorn's lemma, it has a maximal element $\{v_i\}$ with corresponding eigenvalues $\{\lambda_i\}$. Let $P = \sum(\cdot|v_i)v_i$ be the orthogonal projection onto the span of the vectors v_i. By the remarks prior to this proposition, we then have

$$AP = \sum(\cdot|v_i)\lambda_i v_i = \sum(\cdot|A^*(v_i))v_i = PA,$$

so $(I - P)A$ is normal and compact. If $I \neq P$, there is a unit vector in $(I - P)(V)$ that is an eigenvector for A by the last lemma, which contradicts maximality of $\{v_i\}$, so the latter must be an orthonormal basis for $P(V) = V$. □

Proposition 4.5.10 *For any compact operator A on a real or complex Hilbert space, and any scalar $\lambda \neq 0$, the operator $A - \lambda I$ has closed range.*

Proof Say $A \in B(V)$ and define an injective operator $T\colon W \to V$ by $T(v) = A(v) - \lambda v$ on the orthogonal complement W of $\ker(A - \lambda I)$. If suffices to show that there is some $r > 0$ such that $r\|v\| \leq \|T(v)\|$ for all $v \in W$. If this is not so, there is to every $r > 0$ a sequence of unit vectors $v_n \in W$ and $v \in V$ such that $T(v_n) \to 0$ and $A(v_n) \to v$ by compactness of A. Then $v = \lambda \lim v_n \in W$ and $T(v) = \lim \lambda T(v_n) = 0$, so $v = 0$, which contradicts $\|v\| = |\lambda| \lim \|v_n\| \neq 0$. □

Definition 4.5.11 The *spectrum* $\mathrm{Sp}(w)$ of an element w of a unital Banach algebra consists of all scalars λ such that $w - \lambda$ is not invertible.

There are two ways a bounded operator A on a Banach space fails to be boundedly invertible; either it is not injective or it is not surjective. Of course, these two conditions do not exclude one another. If $A - \lambda I$ is not injective, then λ is an eigenvalue for A.

We have just seen that a compact normal operator A on a complex Hilbert space V has an orthonormal basis $\{v_i\}$ of eigenvectors, say with corresponding eigenvalues $\{\lambda_i\}$. It cannot have more eigenvalues since any additional corresponding eigenvector would be orthogonal to the rest. If $A - \lambda I$ for $\lambda \neq 0$ is not surjective, then by the proposition above, we have $\ker(A^* - \bar{\lambda} I)^{\perp} = \text{im}(A - \lambda I) \neq V$, which shows that λ is an eigenvalue of A. If $\dim V = \infty$, then A cannot be invertible. So in this case $\text{Sp}(A) = \{\lambda_i\} \cup \{0\}$. Moreover, by the theorem above, we see that $\{\lambda_i\}$ is either finite, in which case A has finite rank, or it is a sequence that converges to zero; the λ_i's are indeed countable since those larger than $1/n$ would otherwise be uncountable for some $n \in \mathbb{N}$. In either case $\text{Sp}(A)$ is compact, see also the general statement in the appendix. We may write $A = \sum \lambda_i v_i \odot v_i$ with convergence in norm. By Parseval's identity the formula

$$f(A) = \sum f(\lambda_i) v_i \odot v_i$$

defines $f(A) \in B(V)$ for any complex continuous function f on $\text{Sp}(A)$. Then $f(A)$ is compact if and only if $f(0) = 0$. The *continuous spectral mapping* $f \mapsto f(A)$ will evidently be an isometric unital $*$-homomorphism $C(\text{Sp}(A)) \to B(V)$ that takes the identity function to A. We will generalize this without assuming compactness.

A closed $*$-subalgebra of $B(V)$ acts *irreducibly* on V if it has no proper closed invariant subspaces.

Proposition 4.5.12 *A closed $*$-subalgebra of $B(V)$ that acts irreducible on a complex Hilbert space V must contain $B_0(V)$ whenever it contains at least one non-zero compact operator.*

Proof We may assume that the non-zero compact operator A is self-adjoint, and then it must have a non-zero eigenvalue λ, and these are isolated, so $P = \delta_\lambda(A)$ is a non-zero orthogonal projection in the closed $*$-subalgebra W by the holomorphic functional calculus in the appendix, or by the continuous spectral mapping result above. Applying the spectral mapping to $(\iota - \lambda)\delta_\lambda = 0$, we get $(A - \lambda I)P = 0$, so P has finite rank since $\ker(A - \lambda I)$ has finite dimension as A is compact. We may assume that P has minimal rank in W.

By restricting to the finite dimensional space $P(V)$ we may consider PWP as a $*$-subalgebra of a matrix algebra. Every self-adjoint element of PWP will therefore be a sum of its spectral projections. Since they belong to PWP, it is spanned by its orthogonal projections. Since P has minimal rank, we must have $PWP = \mathbb{C}P$. Pick a unit vector $v \in P(V)$. Then the closure of $W(v)$ is invariant under the action of W, and so it must be the whole space V. If $u \in P(V)$, then $u = \lim A_n(v)$ for some $A_n \in W$. Hence $u = P(u) = \lim PA_nP(v) \in \mathbb{C}v$, so $P = v \odot v$. For any

$u \in V$ there are $A_n \in W$ such that $u = \lim A_n(v)$. Thus

$$u \odot u = \lim A_n(v) \odot A_n(v) = \lim A_n(v \odot v)A_n^* = \lim A_n P A_n^* \in W$$

and $B_0(V) \subset W$. □

4.6 Fredholm Operators

We have seen that the set $B_0(V)$ of compact operators on a Hilbert space V is a closed self-adjoint ideal of $B(V)$. Therefore the quotient $B(V)/B_0(V)$ is a unital Banach $*$-algebra, called the *Calkin algebra*, and the quotient map is a norm-decreasing unital $*$-epimorphism. To avoid trivialities we assume here that V is infinite dimensional.

Definition 4.6.1 A bounded operator on a Hilbert space is *Fredholm* if its image in the Calkin algebra is invertible. We denote the group of Fredholm operators on a Hilbert space V by $F(V)$.

By definition an operator A is Fredholm if there is another operator T such that AT and I and TA differ by compact operators. The following *Atkinson's theorem* shows that the notion is quite restrictive.

Theorem 4.6.2 *A bounded operator A on a Hilbert space is Fredholm if and only if* $\ker A$ *and* $\ker A^*$ *are finite dimensional and* $A(V)$ *is closed if and only if there is* $T \in B(V)$ *such that* TA *and* AT *are orthogonal projections on* $(\ker A)^\perp$ *and* $(\ker A^*)^\perp$, *respectively, of finite co-ranks.*

Proof Suppose A is Fredholm. For any sequence of orthonormal vectors v_n in $\ker A$, we get the absurdity $1 = \| - v_n\| = \|(TA - I)(v_n)\| \to 0$ as $v_n \to 0$ weakly by Parseval's identity. Replacing A by its adjoint shows that also $\ker A^*$ must have finite dimension. Next pick a finite rank operator R such that $\|TA - I - R\| \leq 1/2$. Then A restricted to $\ker R$ is bounded away from zero, so $A(\ker R)$ is closed. Let P be the orthogonal projection onto the orthogonal complement. Then

$$A(V) = A(\ker R) + A((\ker R)^\perp) = P^{-1}(PA(\mathrm{im} R^*))$$

is closed, being the inverse image under a continuous map of a finite dimensional space.

Assume now that $A \in B(V)$ satisfies the conditions in the second equivalence. By the open mapping theorem its restriction to $(\ker A)^\perp$ has a bounded inverse $T : \mathrm{im} A \to (\ker A)^\perp$, which we extend to the desired element of $B(V)$ by declaring it to vanish on $(\mathrm{im} A)^\perp$. □

In view of this theorem we may define the *index* of a Fredholm operator A by

$$\text{index}\, A = \dim\ker A - \dim\ker A^*.$$

With T a *partial inverse* associated to A as in Atkinson's theorem, and with the orthogonal projections $P = I - TA$ and $Q = I - AT$ of finite rank, we see that $-\text{index}\, T = \text{index}\, A = \text{rank}\, P - \text{rank}\, Q$. Going to adjoints also changes the sign of the index. Powers of the unilateral shift on $l^2(\mathbb{N})$ produces every negative integer as an index. Since every infinite set is a disjoint union of countable subsets, every infinite dimensional Hilbert space contains a copy of a separable Hilbert space where the unilateral shift acts. Extending this by using the identity operator on the remaining component, and observing that the index is additive on a finite direct sum of Fredholm operators, we see that every integer is attained as an index of a Fredholm operator on V. We partition $F(V)$ into non-empty subclasses $F_n(V)$ of index $n \in \mathbb{Z}$.

We turn to the amazing stability of the index under compact perturbations. Having in mind the analogue between finite rank (compact) operators and continuous functions with compact support (vanishing at infinity), we think of compact operators as describing quasi-local phenomena on the Hilbert space. The index is then stable with respect to such phenomena, and rather detects global topological features.

Lemma 4.6.3 *We have $I + S \in F_0(V)$ for any $S \in B_f(V)$.*

Proof Let T be a partial inverse of $I + S$, and let R be the orthogonal projection onto the finite dimensional subspace spanned by the images of P, Q, S and S^*. Considering the ordinary trace on the matrix algebra $B(R(V))$, we then get $\text{rank}\, P - \text{rank}\, Q = \text{Tr}(P - Q) = \text{Tr}(S(RTR) - (RTR)S) = 0$. □

Lemma 4.6.4 *If $A \in F_0(V)$ and $S \in B_0(V)$, then $A + S \in F_0(V)$.*

Proof Pick a partial isometry U with $\ker A$ and $\ker A^*$ as source and range spaces, respectively. Now $A + U$ is injective because if it vanishes on v, then $A(v) = -U(v) = 0$, so v vanishes as it belongs to both $\ker A$ and $(\ker A)^\perp$. It is also surjective as $(A + U)(V) = A(V) \oplus \ker A^* = V$, so it is invertible in $B(V)$ by the open mapping theorem. Pick $R \in B_f(V)$ such that $\|S - R\| < \|(A + U)^{-1}\|^{-1}$. Then $T \equiv (A+U)(I+(A+U)^{-1}(S-R))$ is invertible by the von Neumann series, and

$$\text{index}\,(A + S) = \text{index}\, T(I + T^{-1}(R - U)) = \text{index}\,(I + T^{-1}(R - U)) = 0$$

by the previous lemma. □

Theorem 4.6.5 *If A is Fredholm and S is compact, then*

$$\text{index}\,(A + S) = \text{index}\, A.$$

Proof By going to adjoints we may assume index $A = n > 0$ for $A \in F(V)$. Let R be the unilateral shift operator on $l^2(\mathbb{N})$. Then

$$(A + S) \oplus R^n = A \oplus R^n + S \oplus 0 \in F_0(V \oplus l^2(\mathbb{N}))$$

by the previous lemma. But then again $A + S \in F_n(V)$. □

The index respects composition of Fredholm operators.

Proposition 4.6.6 *If $A_k \in F_k(V)$, then $A_n A_m \in F_{n+m}(V)$.*

Proof We saw in the proof of the lemma above that $A_0 + U$ is invertible for some partial isometry $U \in B_f(V)$. Hence

$$\text{index}\, A_n = \text{index}\, A_n(A_0 + U) = \text{index}\,(A_n A_0 + A_0 U) = \text{index}\, A_n A_0$$

by the theorem above. By considering adjoints, we may assume $m > 0$. Then using the first part of the proof, together with the unilateral shift R on $l^2(\mathbb{N})$, we get

$$n = \text{index}\,(A_n \oplus I)(A_m \oplus R^m) = \text{index}\,(A_n A_m \oplus R^m) = \text{index}\, A_n A_m - m.$$

□

We will finally give a topological description of the index, and need some preliminary results.

Proposition 4.6.7 *The groups of respectively unitary and invertible bounded operators on a complex Hilbert space V are arcwise connected, and so are the Fredholm classes $F_n(V)$.*

Proof We will use the fact, to be proved later, that any unitary operator on a complex Hilbert space is of the form e^{iT} for a self-adjoint bounded operator T. Polar decompose $e^{iT}|A|$ a given invertible bounded operator A, and observe that $t \mapsto e^{itT}(t|A|+(1-t)I)$ defines a curve in the groups from I to A as $t|A|+(1-t)I$ is bounded away from zero.

To $B \in F_0(V)$ pick a partial isometry $S \in B_f(V)$ such that $B + S$ is invertible. Then $t \mapsto B + tS$ defines a curve in $F_0(V)$ from B to $B + S$. From $B + S$ to I use a curve in the invertibles. Using adjoints it suffices to show that $F_n(V)$ is arcwise connected for $n < 0$. Let $B \in F_n(V)$ and pick an extended unilateral shift operator R on V. Connect BR^{*n} to I by a curve f in $F_0(V)$. Then $f(\cdot)R^n$ is a curve in $F_n(V)$ between B and R^n. □

Since $F(V)$ is the inverse image under the quotient map of the open set of invertible elements in $B(V)/B_0(V)$, it is evidently open. More is true.

Lemma 4.6.8 *The Fredholm classes $F_n(V)$ are open in $B(V)$.*

Proof To $A \in F_0(V)$ pick a partial isometry $U \in B_f(V)$ such that $A + U$ is invertible. For $T \in B(V)$ close enough to A both $T + U$ and its adjoint are bounded away from zero, so $T + U$ is invertible, and index $T = \text{index}\,(T + U) = 0$.

If $A \in F_n(V)$ for $n > 0$, then $A \oplus R^n \in F_0(V \oplus l^2(\mathbb{N}))$. For $T \in B(V)$ close enough to A, the operator $T \oplus R^n$ must have vanishing index by the first part of the proof, so index $T = n$. To cover the case $n > 0$ remember that the adjoint operation is continuous with a continuous inverse. □

Let V be an infinite dimensional complex Hilbert space. Consider the open group G of invertible elements in $B(V)/B_0(V)$. By the above theorem and Proposition 4.6.6 we have a well-defined group epimorphism $f : G \to \mathbb{Z}$ given by

$$f([A]) = \text{index } A$$

for $A \in F(V)$. By the last proposition and lemma we see that the images $[F_n(V)]$ under the open quotient map $B(V) \to B(V)/B_0(V)$ are the connected components of G. The connected component of the identity operator $[F_0(V)]$ is obviously the kernel of f. By the fundamental isomorphism theorem for groups, the map f induces an isomorphism from the index group of the Calkin algebra to the integers. So we might view the index of a Fredholm operator as its image under the composition of two natural quotient maps.

We have already stated the following *Fredholm alternative* for compact normal operators.

Proposition 4.6.9 *Given a non-zero scalar λ and a compact operator A on an infinite dimensional complex Hilbert space. Then $A - \lambda I$ is injective if and only if it is surjective. The spectrum of A consists of the countably many eigenvalues of A together with the only accumulation point 0. Moreover, if λ is an eigenvalue of A, then $\bar{\lambda}$ is an eigenvalue of A^* with the same finite multiplicity.*

Proof As $A - \lambda I$ is Fredholm of index zero, its image is closed and its kernel if finite dimensional and of the same dimension as the kernel of its adjoint. When these dimensions are zero, then $A - \lambda I$ has image $\ker(A^* - \bar{\lambda} I)^{\perp}$ the whole space, so it has a bounded inverse by the open mapping theorem. □

4.7 Traceclass and Hilbert-Schmidt Operators

It is tempting to extend the trace on a matrix algebra to an 'integral' on the bounded operators on a Hilbert space, thinking of these as L^{∞}-functions.

Definition 4.7.1 Define the *trace* $\text{Tr}(A) \in [0, \infty]$ of a positive bounded operator A on a Hilbert space V by

$$\text{Tr}(A) = \sum (A(v_i)|v_i),$$

where $\{v_i\}$ is an orthonormal basis of V.

A priory the trace depends on the chosen orthonormal basis, but the next result shows that it actually does not.

Proposition 4.7.2 *We have* $\mathrm{Tr}(A^*A) = \mathrm{Tr}(AA^*)$ *for every bounded operator* A *on a complex Hilbert space. In particular, if* $B \geq 0$ *and* U *is unitary, then* $\mathrm{Tr}(UBU^*) = \mathrm{Tr}((UB^{1/2})(UB^{1/2})^*) = \mathrm{Tr}(B)$*, so the trace does not depend on the chosen orthonormal basis, and* $\|B\| \leq \mathrm{Tr}(B)$.

Proof Pick an orthonormal basis $\{v_i\}$ for V. Using continuity of the inner product and interchanging the order of (potentially uncountable) summation of positive terms, we get

$$\mathrm{Tr}(A^*A) = \sum_i (\sum_j (A(v_i)|v_j)v_j|A(v_i)) = \sum_j (\sum_i (A^*(v_j)|v_i)v_i|A^*(v_j)) = \mathrm{Tr}(AA^*).$$

To get $\|B\| \leq \mathrm{Tr}(B)$, pick as a first vector in an orthonormal basis one that realizes the numerical radius of B up to an epsilon. □

Definition 4.7.3 The set $B^1(V)$ of *trace class operators* on a complex Hilbert space V is the span of all positive bounded operators on V of finite trace.

We extend the trace to every $A \in B^1(V)$ by picking positive A_n with finite trace such that $A = \sum_{n=0}^3 i^n A_n$, and then set $\mathrm{Tr}(A) = \sum_{n=0}^3 i^n \mathrm{Tr}(A_n)$. This is independent of the decomposition because the trace is additive on positive operators. It also respects multiplication of such operators with non-negative scalars. It is easy to see that $\mathrm{Tr}(A) = \sum (A(v_i)|v_i)$ for any orthonormal basis $\{v_i\}$ of V, and moreover, with absolute convergence.

It is straightforward to check that for any elements a and b in any complex $*$-algebra, we have the following *polarization identity*

$$4a^*b = \sum_{n=0}^{3} i^n (a + i^n b)^*(a + i^n b).$$

For positive $A \in B^1(V)$ and $T \in B(V)$, the above identity with $a = A^{1/2}T$ and $b = A^{1/2}$ becomes

$$4AT = \sum_{n=0}^{3} i^n (T + i^n I)^* A (T + i^n I).$$

Combining this with the identity

$$\mathrm{Tr}(S^*AS) = \mathrm{Tr}(S^*A^{1/2}A^{1/2}S) = \mathrm{Tr}(A^{1/2}SS^*A^{1/2}) \leq \|SS^*\| \,\mathrm{Tr}(A)$$

for $S = T + i^n I$ shows that $AT \in B^1(V)$. So the self-adjoint subspace $B^1(V)$ of $B(V)$ is an ideal in $B(V)$ which, since $\mathrm{Tr}(v_i \odot v_j) = \delta_{ij}$, contains $B_f(V)$. We also have the description

$$B^1(V) = \{A \in B(V) \mid \mathrm{Tr}(|A|) < \infty\}$$

since if A belongs to the right hand side, then $|A| \in B^1(V)$, so $A = U|A| \in B^1(V)$, and conversely, if $A \in B^1(V)$, then $|A| = U^*A \in B^1(V)$, so $\mathrm{Tr}(|A|) < \infty$.

Definition 4.7.4 The set $B^2(V)$ of *Hilbert-Schmidt operators* on a complex Hilbert space V consists of all $A \in B(V)$ with $|A|^2 \in B^1(V)$.

From the *parallelogram formula*

$$(a+b)^*(a+b) + (a-b)^*(a-b) = 2(a^*a + b^*b)$$

in any $*$-algebra, we get $(A+T)^*(A+T) \leq 2(A^*A + T^*T)$ for bounded operators A, T on a Hilbert space V, so the Hilbert-Schmidt operators form a linear space, which by the proposition above is self-adjoint. If $A \in B^2(V)$ and $T \in B(V)$, then $|AT|^2 = T^*|A|^2T$ is trace class since such operators form an ideal in $B(V)$. So $B^2(V)$ is an ideal in $B(V)$, which moreover contains any trace class operator T because then $|T|$, and hence $|T|^2$, are trace class.

Furthermore, any $A \in B^2(V)$ is compact. To see this, pick an orthonormal basis $\{v_i\}$ of V. Then to any $\varepsilon > 0$, there is a finite set J such that $\sum_{i \in J}(|A|^2(v_i)|v_i) < \varepsilon$. If P is the orthogonal projection onto the span of v_i with $i \in J$, we get

$$\||A|(I-P)\|^2 = \|(I-P)|A|^2(I-P)\| \leq \mathrm{Tr}((I-P)|A|^2(I-P)) < \varepsilon$$

by the proposition above, so $|A|$ is compact, and so is A by polar decomposition.

The inclusions

$$B_f(V) \subset B^1(V) \subset B^2(V) \subset B_0(V) \subset B(V)$$

are proper exactly when the complex Hilbert space V is infinite dimensional, as the following example shows.

Example 4.7.5 By singling out a copy of a separable Hilbert space in a possibly larger complex Hilbert space, we reduce to the separabel case. Define a bounded operator on a countable orthonormal basis $\{v_n\}$ by $A(v_n) = n^{-s}v_n$ for $s \geq 0$. If $s = 0$, then A is bounded but not compact, and if $s = 1/2$, then A is compact but not Hilbert-Schmidt, and if $s = 1$, then A is Hilbert-Schmidt but not trace class, and if $s = 2$, then A is trace class but not of finite rank. ◇

In the infinite dimensional case the ideals $B^1(V)$ and $B^2(V)$ are therefore not closed in the operator norm. However, there are other norms on them under which they are closed; these we could think of as L^1-norms and L^2-norms, respectively.

Proposition 4.7.6 *Let V be a complex Hilbert space with an orthonormal basis $\{v_i\}$. Then $B^2(V)$ is a Hilbert space under the inner product $(S|A)_2 = \mathrm{Tr}(A^*S)$, which has $\{v_i \odot v_j\}$ as an orthonormal basis.*

Proof From the polarization identity in a $*$-algebra we see that A^*S is trace class, so the formula for $(\cdot|\cdot)_2$ makes sense, and we get an inner product because $\|A\|_2 \equiv \mathrm{Tr}(A^*A)^{1/2} \geq \|A\|$ by the previous proposition. This inequality also shows that any Cauchy sequence $\{A_n\}$ in $B^2(V)$ for the 2-norm will converge in operator norm to some $A \in B_0(V)$. For any orthogonal projection P on a finite dimensional subspace of V we have

$$\mathrm{Tr}(P(A-A_n)(A-A_n)^*P) = \lim \mathrm{Tr}(P(A_m-A_n)(A_m-A_n)^*P) \leq \limsup \|A_m-A_n\|_2^2,$$

so $\|A - A_n\|_2 \leq \limsup \|A_m - A_n\|_2$. For denseness in the last statement it suffices to observe that $(A|v_i \odot v_j)_2 = (A(v_j)|v_i)$ for $A \in B^2(V)$. □

Hilbert-Schmidt operators are important in applications mainly due to the following result.

Proposition 4.7.7 *Suppose μ is a σ-finite measure such that $L^2(\mu)$ is separabel. For $K \in L^2(\mu \times \mu)$ the integral operator T_K formally defined by*

$$T_K(f)(x) = \int K(x, \cdot)f\,d\mu$$

is a Hilbert-Schmidt operator on $L^2(\mu)$. The map $K \mapsto T_K$ is a $*$-preserving linear isometry from $L^2(\mu \times \mu)$ onto $B^2(L^2(\mu))$, where $K^*(x, y) = \overline{K(y, x)}$. Hence the kernel K *is conjugate symmetric if and only if T_K is self-adjoint, and in this case T_K is orthogonally diagonalizable with the eigenvalues λ_n tending to zero, in fact, we have $\sum |\lambda_n|^2 = \|K\|^2$ by Parseval's identity.*

Proof Let $f \in L^2(\mu)$. The Fubini-Tonelli theorem shows $K(x, \cdot) \in L^2(\mu)$ for almost all x. Then $K(x, \cdot)f$ belongs to $L^1(\mu)$ by the Cauchy-Schwarz inequality, so we may take its integral, and get an almost everywhere well-defined function $T_K(f)$, which can be regarded as measurable by the Fubini-Tonelli theorem. Using again the Cauchy-Schwarz inequality, we get

$$\|T_K(f)\|^2 \leq \int \|K(x, \cdot)\|^2 \|f\|^2\,d\mu(x) = \|f\|^2 \int\int |K(x, \cdot)|^2\,d\mu d\mu(x),$$

so by the Fubini-Tonelli theorem, we see that T_K is an operator on $L^2(\mu)$ with norm $\|T_K\| \leq \|K\| \equiv (\int |K|^2\,d(\mu \times \mu))^{1/2}$.

For any orthonormal basis $\{f_n\}$ of $L^2(\mu)$ we have $T_{f_n \otimes \bar{f}_m} = f_n \odot f_m$ for all m and n. We have seen that $\{f_n \otimes \bar{f}_m\}$ is an orthonormal basis of $L^2(\mu \times \mu)$, so by the proposition above, the map $K \mapsto T_K$ is an isometric surjection. That $T_{K^*} = T_K^*$ is easily checked for $K = f_n \otimes \bar{f}_m$, and then it holds for any $K \in L^2(\mu \times \mu)$ by conjugate linearity and continuity. □

Example 4.7.8 Suppose μ is a σ-finite measure such that $L^2(\mu)$ is separable, and consider the integral operator T_K with kernel $K^* = K \in L^2(\mu \times \nu)$. Then by the proposition above there is an orthonormal basis $\{f_n\}$ of $L^2(\mu)$ such that $T_K = \sum \lambda_n f_n \odot f_n$. The *Fredholm equation*

$$\int K(x, \cdot) f \, d\mu - \lambda f(x) = g(x)$$

with $g \in L^2(\mu)$ has then the unique solution $f = \sum(\lambda_n - \lambda)^{-1}(g|f_n)f_n$ provided the scalar λ differs from the eigenvalues λ_n. To see this observe that

$$\sum \lambda_n (f|f_n) f_n - \lambda \sum (f|f_n) f_n = T_K(f) - \lambda f = g = \sum (g|f_n) f_n,$$

which holds if and only if $(\lambda_n - \lambda) f = g$ for all n.

We return now to general trace class operators.

Lemma 4.7.9 *Let V be a complex Hilbert space. If $A \in B^1(V)$ and $S \in B(V)$, then* $|\operatorname{Tr}(SA)| \leq \|S\| \operatorname{Tr}(|A|)$ *and* $\operatorname{Tr}(SA) = \operatorname{Tr}(AS)$.

Proof Letting $A = U|A|$ be the polar decomposition of A, using that $|A|^{1/2}$ is Hilbert-Schmidt, that $B^2(V)$ is a self-adjoint ideal in $B(V)$, and the Cauchy-Schwarz inequality for the Hilbert-Schmidt inner product together with Proposition 4.7.2, we get

$$|\operatorname{Tr}(SA)|^2 = |(|A|^{1/2}|(SU|A|^{1/2})^*)_2|^2 \leq \operatorname{Tr}(|A|) \operatorname{Tr}(|A|^{1/2} U^* S^* SU|A|^{1/2}),$$

which gives the desired inequality as the expression in the last trace is not greater than $\|S\|^2|A|$.

For the next assertion assume first that $A, S \in B^2(V)$. By the polarization identity in a $*$-algebra and Proposition 4.7.2 and the fact that $B^2(V)$ is a self-adjoint linear space, we get

$$4 \operatorname{Tr}(A^* S) = \sum_{n=0}^{3} i^n \operatorname{Tr}((S + i^n A)^*(S + i^n A)) = 4 \operatorname{Tr}(SA^*)$$

since as we move i^n from one parenthesis inside the trace to the next one, we must conjugate it. We extend this to $A \in B^1(V)$ and $S \in B(V)$, assuming by linearity that A is positive, as follows

$$\operatorname{Tr}(SA) = \operatorname{Tr}(S|A|^{1/2}|A|^{1/2}) = \operatorname{Tr}(|A|^{1/2} S|A|^{1/2}) = \operatorname{Tr}(|A|^{1/2}|A|^{1/2} S) = \operatorname{Tr}(AS),$$

where we used that $B^2(V)$ is an ideal in $B(V)$. □

Proposition 4.7.10 *Let V be a complex Hilbert space. The ideal $B^1(V)$ of $B(V)$ is a Banach algebra under the norm* $\|A\|_1 = \operatorname{Tr}(|A|)$.

Proof Let $A, S \in B^1(V)$. Using the polar decomposition $A + S = U|A + S|$ and the previous lemma, we get

$$\|A + S\|_1 = \operatorname{Tr}(U^*(A + S)) \leq |\operatorname{Tr}(U^*A)| + |\operatorname{Tr}(U^*S)| \leq \|A\|_1 + \|S\|_1.$$

Using the polar decomposition $SA = R|SA|$ and Proposition 4.7.2, we get

$$\|SA\|_1 = \operatorname{Tr}(R^*SA) \leq \|S\| \operatorname{Tr}(|A|) \leq \||S|\| \operatorname{Tr}(|A|) \leq \|S\|_1 \|A\|_1,$$

so $\| \cdot \|_1$ is a submultiplicative norm on $B^1(V)$.

Any Cauchy sequence $\{A_n\}$ in $B^1(V)$ with respect to $\| \cdot \|_1$ must converge in operator norm to $A \in B_0(V)$. Using the polar decomposition $A - A_n = U|A - A_n|$, we have for each finite rank orthogonal projection P on V, that

$$\operatorname{Tr}(P|A - A_n|) = \operatorname{Tr}(PU^*(A - A_n)) = \lim \operatorname{Tr}(PU^*(A_m - A_n)) \leq \limsup \|A_m - A_n\|_1$$

by the previous lemma. Hence $\|A - A_n\|_1 \leq \limsup \|A_m - A_n\|_1$, which shows that $A \in B^1(V)$ and that $A_n \to A$ in $\| \cdot \|_1$-norm. □

We have the following remarkable duality result.

Theorem 4.7.11 *Let V be a complex Hilbert space. We have Banach space dualities $B_0(V)^\star = B^1(V)$ and $B^1(V)^\star = B(V)$ implemented by $(S, A) \mapsto \operatorname{Tr}(SA)$ in both cases.*

Proof Any $A \in B^1(V)$ defines a linear functional $\operatorname{Tr}(\cdot A)$ on $B_0(V)$ with $\|\operatorname{Tr}(\cdot A)\| \leq \|A\|_1$ by the lemma above. Using the polar decomposition $A = U|A|$, we get $\|A\|_1 = \operatorname{Tr}(U^*A) \leq \|\operatorname{Tr}(\cdot A)\|$, so the assignment $A \mapsto \operatorname{Tr}(\cdot A)$ from $B^1(V)$ to $B_0(V)^\star$ is linear and isometric. To see that it is surjective, take $x \in B_0(V)$. For $S \in B^2(V)$ we have $|x(S)| \leq \|x\|\|S\| \leq \|x\|\|S\|_2$, and as $B^2(V)$ is a Hilbert space, there is an element $A^* \in B^2(V)$ such that $x(S) = \operatorname{Tr}(AS) = \operatorname{Tr}(SA)$ by the lemma above. Using again polar decomposition of A, we get for every finite rank orthogonal projection P on V, that $|\operatorname{Tr}(P|A|)| = \operatorname{Tr}(PU^*A) = x(PU^*) \leq \|x\|$, so A is trace class.

For the next duality note that any $S \in B(V)$ defines a linear function $\operatorname{Tr}(S\cdot)$ on $B^1(V)$ with $\|\operatorname{Tr}(S\cdot)\| \leq \|S\|$ by the lemma above. To see that the linear map $S \mapsto \operatorname{Tr}(S\cdot)$ from $B(V)$ to $B^1(V)^\star$ is surjective and isometric, let $x \in B^1(V)^\star$, and define a sesquilinear form on V by $(v, w) \mapsto x(v \odot w)$. As

$$|v \odot w| = ((v \odot w)^*(v \odot w))^{1/2} = \|v\|\|w\|(\|w\|^{-1}w \odot \|w\|^{-1}w),$$

we get

$$|x(v \odot w)| \leq \|x\|\|v \odot w\|_1 = \|x\|\|v\|\|w\|,$$

so there is a unique $S \in B(V)$ such that $x(v \odot w) = (S(v)|w)$ and $\|S\| \leq \|x\|$. Any positive $A \in B^1(V)$ can be written as $A = \sum \lambda_i v_i \odot v_i$ for an orthonormal basis $\{v_i\}$ of V and scalars $\lambda_i > 0$ with $\sum \lambda_i = \|A\|_1$. Then as $\sum_{i \in J} \lambda_i v_i \odot v_i \to A$ in $\|\cdot\|_1$-norm for increasing finite sets J, we get

$$x(A) = \sum \lambda_i x(v_i \odot v_i) = \sum \lambda_i (S(v_i)|v_i) = \sum (SA(v_i)|v_i) = \mathrm{Tr}(SA).$$

We are now done since $B^1(V)$ is a linear span of positive operators with finite trace. □

We notice that $B(V)$ is therefore the bidual of $B_0(V)$.

Exercises

For Sect. 4.1

1. Show that the square integrable holomorphic complex functions on an open subset X of $\mathbb{C}$ is a Hilbert space $A^2(X)$ with the usual L^2-inner product.
2. Show that the Hardy space H^2, which is the closed span in $L^2(\mathbb{T})$ of half of the trigonometric system $\{v_n\}_{n \geq 0}$ is isomorphic to the Hilbert space $A^2(X)$, where X is the open unit disc in $\mathbb{C}$.
3. Prove that any infinite orthonormal sequence in a Hilbert space converges weakly to zero.
4. Show that any surjective function f between Hilbert spaces is linear if $(f(u)|f(v)) = (u, v)$ for all vectors u, v.
5. Show that $L^2([0, 1])$ is separable.

For Sect. 4.2

1. Show that for $0 < f \in L^1(\mathbb{R})$, one has $|\hat{f}(s)| < \hat{f}(0)$ for $s \neq 0$.
2. Find $f \in L^2(\mathbb{R})$ with $f \notin L^1(\mathbb{R})$ and $\hat{f} \in L^1(\mathbb{R})$.
3. Let S be the set of infinite differentiable real functions f on $\mathbb{R}$ such that for $n, m \in \{0, 1, \dots\}$ the expression $|t^n f^{(m)}(t)|$ is less then a constant independent of $t \in \mathbb{R}$, where $f^{(m)}$ is the m-th derivative of f. Prove that the Fourier transform maps S onto S.
4. Show that if $f \in L^p(\mathbb{R})$ and $g \in L^q(\mathbb{R})$ for conjugate exponents p and q, then $f * g$ is uniformly continuous.

For Sect. 4.3

1. Show that the Dirichlet kernels D_N tend to infinity in L^1-norm when N tends to infinity.
2. Show that the Fourier transform from $L^1(\mathbb{T})$ to $C_0(\mathbb{Z})$ is not surjective.
3. What can you say about Fourier analysis in $\mathbb{T}^n$ and $\mathbb{R}^n$?
4. Consider the function $f : \mathbb{R} \to \mathbb{R}$ that is an odd 2π-periodic extension of the identity function on $\langle -\pi, \pi \rangle$. Study oscillations of $f - \sum_{n=-N}^{N} \hat{f}(n) v_n$ at the

points of discontinuity of f by inspecting local extremas as $N \to \infty$. Deviations due to these sharp oscillations is known as the Gibbs phenomenon.

5. Prove Heisenberg's inequality $\int (t-a)^2|f(t)|^2\,dt \int (s-b)^2|\hat{f}(s)|^2\,ds \geq c\|f\|_2^4$ for some scalar c that works for all $a, b \in \mathbb{R}$ and $f \in L^2(\mathbb{R})$.

For Sect. 4.4

1. Given a Hilbert space V with inner product $(\cdot|\cdot)$, and suppose $(\cdot|\cdot)_1$ is another inner product on V such that $(v|v)_1 \leq (v|v)$ for $v \in V$. Show that there exists an injective $A \in B(V)$ such that $0 \leq A \leq I$ and $(A(v)|w) = (v|w)_1$ for $v, w \in V$.
2. A reflection operator is a self-adjoint bounded operator A on a Hilbert space V such that $A^2 = I$. Show that $P = (A + I)/2$ is an orthogonal projection, and that there is a closed subspace W of V with $v + A(v) \in W$ and $v - A(v) \in W^\perp$ for $v \in V$. Prove also a converse result.
3. Show that any map f on a real Hilbert space that preserves norm is of the form $f(v) = f(0) + A(v)$ for some isometry A.
4. Prove that the orthogonal projections on a Hilbert space are the extreme points of the positive operators in the closed unit ball.
5. Show that $A^2 \leq \|A\|A$ for a positive operator A on a Hilbert space.
6. Show that $0 \leq A \leq B$ implies $A^{1/2} \leq B^{1/2}$ for bounded operators A, B on a Hilbert space.

For Sect. 4.5

1. Show that the dimension of the image of a finite rank operator on a Hilbert space is the same as that of its adjoint.
2. Show that any non-zero multiplication operator on $L^2([0, 1])$ cannot be compact.
3. Show that a bounded operator A on a Hilbert space is compact if and only if $|A|$ is compact.
4. Show that weak-norm continuous operators on Hilbert spaces have finite rank.

For Sect. 4.6

1. Extend the index to all bounded operators on a Hilbert space under the convention $\infty - \infty = 0$. Show that the index can be non-zero for some compact operator.
2. On a separable Hilbert space show that the isomety in the polar decomposition of a bounded operator A can be taken to be unitary if and only if the extended index of A vanishes.
3. The Toeplitz operator of $f \in L^\infty(\mathbb{T})$ is the operator $T_f \in B(H^2)$ given by multiplication by f on $L^2(\mathbb{T})$ sandwiched by the orthogonal projection onto the Hardy space H^2. Show that $f \mapsto T_f$ is norm decreasing and $*$-preserving. Calculate $T_f(v_n)$ and identify T_{v_m} for the trigonometric system $\{v_n\}$, and show that T_f is compact only if $f = 0$ almost everywhere, and that T_f is Fredholm if f is continuous and invertible. In the latter case show that the index of T_f coincides with the index of the unitary in the polar decomposition of T_f.

4. Two functions $f, g \in C(\mathbb{T}, \mathbb{T})$ are homotopic if we have a continuous map $[0, 1] \times \mathbb{T} \to \mathbb{T}$, say $(t, s) \mapsto f_t(s)$, such that $f_0 = f$ and $f_1 = g$. Show that T_f and T_g are then Fredholm and have the same index.
5. For $f, g \in L^\infty(\mathbb{T})$ show that $T_f T_g - T_{fg}$ is compact if f or g is continuous. In the latter case relate the index of T_g to the winding number of g around 0.

For Sect. 4.7

1. For $p \in [1, \infty\rangle$ consider the set $B^p(V)$ of bounded operators A on a Hilbert space V such that $|A|^p$ is trace class. Show that this set of Schatten class operators is a self-adjoint ideal of $B(V)$ contained in $B_0(V)$, and that it is a Banach space under the norm $\|A\|_p = (\mathrm{Tr}(|A|^p))^{1/p}$.
2. Prove the inequality $|\,\mathrm{Tr}(SA)| \le \|S\|_p \|A\|_q$ for $S \in B^p(V)$ and $A \in B^q(V)$ and conjugate exponents $1 < p \le 2 \le q < \infty$. Use the trace to implement an isometric isomorphic duality between $B^p(V)$ and $B^q(V)$.
3. Consider the Volterra operator $A(f)(s) = \int_0^s f(t)\,dt$ on $L^2([0, 1])$. Find A^* and show that $A + A^*$ is the orthogonal projection onto the space of constant functions.
4. Find the eigenvalues of A^*A with A from the previous exercise, and prove that $\|A\| = 2/\pi$.

Chapter 5
Spectral Theory

In this chapter we develop functional calculi of operators. Given a bounded operator T on a Hilbert space V and a holomorphic function f with a power series that converges absolutely on a disc with radius larger than $\|T\|$. Replacing the variable in the power series by T we evidently get a convergent series in the Banach algebra $B(V)$ producing an operator denoted $f(T)$. The map $f \mapsto f(T)$ is called the holomorphic functional calculus at T. This way complex function theory is brought into the game. In the appendix we study holomorphic calculus for elements in general Banach algebras. There we also include basic complex function theory in the more general context of Banach space valued functions defined on domains in the complex plane, and perform integration of such functions, producing the holomorphic functional calculus as a special case.

Here we go beyond holomorphic functions. If the operator T is compact and normal, hence orthogonal diagonalizable, we can of course replace the i-th eigenvalue λ_i in the decomposition of T by $f(\lambda_i)$ for any complex function f, as long as the new series obtained converges, and this way get an operator $f(T)$. This generalization does not always work for more general operators since they might not have enough eigenvalues; in fact, they might not have any at all.

Instead of the set of eigenvalues, consider the spectrum of T. This consists of the complex numbers λ such that $T - \lambda$ is not invertible, which obviously holds for eigenvalues. The spectrum is never empty. In fact, the spectral radius, or the maximal distance of the spectral values from the origin, is determined by the norm of all non-negative powers of T. This holds when T is in any complex, unital Banach algebra. Moreover, when the algebra is commutative, the value at T of any character, that is, a unital, multiplicaltive linear functional of the algebra, belongs to the spectrum of T, and such evaluations exhaust the spectrum of T. This links to traditional algebra since the kernel of a character is a maximal ideal. The Gelfand theorem for a commutative unital complex Banach algebra says that its characters form a compact Hausdorff space when equipped with the w^*-topology, and that the evaluation of characters on elements defines a map, known as the Gelfand

L. Tuset, *Analysis and Quantum Groups*,
https://doi.org/10.1007/978-3-031-07246-8_5

transform, from the Banach algebra to the algebra of continuous functions on the character space. The uniform norm of the image of an element under the Gelfand transform is the spectral radius of the original element. The Gelfand transform on $L^1(\mathbb{R})$ is actually the Fourier transform. When the Banach algebra is a C*-algebra, the Gelfand transform is by the Stone-Weierstrass theorem from topology, an isomorphism. Returning to $T \in B(V)$, the link to commutative C*-algebras, and thus to the Gelfand isomorphism, is provided by normality since then the C*-algebra generated by T is commutative. Combining all this, we arrive at the continuous functional calculus for normal bounded operators. Having the continuous calculus in hand, we establish basic results of abstract C*-algebras, their ideals, hereditary subalgebras and homomorphisms. These are the Banach algebras with a $*$-operation such that the norm of any self-adjoint element is its spectral radius.

Next, we extend the functional calculus to bounded Borel functions at normal bounded operators. This is done by first applying a vector functional to $f(T)$ for a continuous function f, getting a bounded linear functional on the space of continuous functions on the spectrum of T. Such a functional is an integral with respect to some complex Borel measures on the spectrum, and we may now also integrate bounded Borel functions, giving the desired operators.

The image of the Borel functional calculus is contained in the bicommutant of $*$-algebra generated by T. The commutant X' of $X \subset B(V)$ are the operators that commute with all the operators in X, and the bicommutant X'' of X is $(X')'$. The bicommutant of any self-adjoint subset is a special kind of C*-algebra known as a von Neumann algebra; they coincide with their bicommutants and are pointwise weakly closed. Characteristic functions of Borel sets produce, under the functional calculus of any normal operator in a von Neumann algebra, an abundance of orthogonal projections. Such projections clearly generated the whole von Neumann algebra.

By extending the functional calculus to L^∞-functions one gets a $*$-isomorphism onto $\{T, T^*\}''$ that is spatial on each cyclic component, the so called maximal commutative ones. We explain the details around this in a separate section. Before this we establish the basic facts about von Neumann algebras. They offer the right framework for non-commutative measure theory. Utilizing the aforementioned type of duality, we show that a von Neumann algebra as a Banach space, is already the dual of another Banach space, known as its predual. The weak topology thus induced is called the σ-weak topology, which can be characterized differently. A von Neumann algebra is closed under this stronger and more canonical topology. We also include Kaplansky's density theorem which guarantees denseness of subsets of a C*-subalgebra W of $B(V)$ in corresponding subsets characterized within W''.

In the last section of the chapter, keeping in mind the Krein-Milman theorem, we describe the extreme points of some important subsets of C*-algebras.

5.1 Spectral Theory for Banach Algebras

Recall that a topological group G is a Hausdorff topological space that also is a group such that the map $G \times G \to G$ given by $(a, b) \mapsto ab^{-1}$ is continuous. The inverse map $a \mapsto a^{-1}$ in a topological group has a continuous inverse map since $(a^{-1})^{-1} = a$.

Say W is a complex unital Banach algebra. The *Neumann series* $\sum w^n$ converges to $(1 - w)^{-1}$ for $w \in W$ with $\|w\| < 1$. Hence if w is invertible, then $v = w(1 - w^{-1}(w - v)) \in W$ is invertible provided $\|w - v\| \leq 1/\|w^{-1}\|$. So the set $GL(W)$ of invertible elements in W form an open subset of W that is a topological group. Recall that the spectrum $\mathrm{Sp}(w)$ of w consists of those $a \in \mathbb{C}$ such that $a - w$ is not invertible. It is a closed subset of $\mathbb{C}$ with *spectral radius* $r(w) \equiv \sup |\mathrm{Sp}(w)| \leq \|w\|$. From the *first resolvent formula*

$$\frac{(a - w)^{-1} - (b - w)^{-1}}{b - a} = (a - w)^{-1}(b - w)^{-1}$$

and the fact that $GL(W)$ is a topological group, we see that the *resolvent map* $a \mapsto (a - w)^{-1}$ is holomorphic on $\mathrm{Sp}(w)^c$. Note that $(a - w)^{-1} \to 0$ as $|a| \to \infty$. So if $\mathrm{Sp}(w) = \phi$, the resolvent map would be holomorphic on the entire complex plane, and by Liouville's theorem it would therefore vanish, which is impossible. Using the Heine-Borel theorem we have proved the first part of the following result. The second part provides a more precise formula for the spectral radius.

Proposition 5.1.1 *The spectrum of an element w in a unital complex Banach algebra is a non-empty compact subset of $\mathbb{C}$ with $r(w) \leq \|w\|$. In fact, we have $r(w) = \lim \|w^n\|^{1/n}$.*

Proof We claim that $f(\mathrm{Sp}(w)) \subset \mathrm{Sp}(f(w))$ for any polynomial f. Indeed, any $\lambda \in \mathrm{Sp}(w)$ is a root of $f - f(\lambda)$, so there is another polynomial g such that $f(a) - f(\lambda) = (a - \lambda)g(a)$, and then $f(w) - f(\lambda) = (w - \lambda)g(w)$ cannot be invertible. Hence $\lambda^n \in \mathrm{Sp}(w^n)$ and $|\lambda|^n \leq r(w^n) \leq \|w\|^n$, so $r(w) \leq \inf \|w^n\|^{1/n}$.

For the opposite inequality, if $\lambda \in B(0, 1/r(w))$ with the convention $1/0 = \infty$, then by definition $1 - \lambda w$ is invertible, and $h(\lambda) = x((1 - \lambda w)^{-1})$ is analytic for any $x \in W^\star$. When $\|\lambda w\| < 1$ we have the Neumann series $(1 - \lambda w)^{-1} = \sum \lambda^n w^n$, so $h(\lambda) = \sum x(w^n)\lambda^n$ also when $\lambda \in B(0, 1/r(w))$. Thus $\{x(w^n)\lambda^n\}$ is certainly bounded, and there is positive M such that $\|w^n \lambda^n\| < M$ for all n by the principle of uniform boundedness. Hence $\limsup \|w^n\|^{1/n} \leq r(w)$. □

Non-emptiness of the spectrum provides the following classification result of complex Banach division algebras.

Corollary 5.1.2 *If 0 is the only non-invertible element in a complex unital Banach algebra W, then $W \cong \mathbb{C}$.*

Proof The canonical injection $\mathbb{C} \to W$ is surjective, because to any $w \in W$ there is $\lambda \in \mathbb{C}$ such that $\lambda - w$ is not invertible, so $w = \lambda$. □

Recall that a point in a topological space is a *boundary point* of a subset X if every neighborhood of it intersects both X and X^c.

Corollary 5.1.3 *If a sequence of invertible elements w_n in a complex unital Banach W converges to a boundary point of $GL(W)$, then $\|w_n^{-1}\| \to \infty$. If there exists M such that $\|v\|\|w\| \leq M\|vw\|$ for all $v, w \in W$, then $W \cong \mathbb{C}$.*

Proof If the first statement was false, there would be $a > 0$ and n and a boundary point w of $GL(W)$ such that $\|w_n^{-1}\| < a$ and $\|w_n - w\| < 1/a$. Then $\|1 - w_n^{-1}w\| = \|w_n^{-1}(w_n - w)\| < 1$, so $w = w_n(w_n^{-1}w) \in GL(W)$, which is absurd as $GL(W)$ is open.

For the second statement, consider any boundary point $w = \lim w_n$ of $GL(W)$ with $\|w_n^{-1}\| \to \infty$. As $\|w_n\|\|w_n^{-1}\| \leq M$, we get $\|w_n\| \to 0$, so $w = 0$. To see that the canonical map $\mathbb{C} \to W$ is surjective, observe that any boundary point of $\mathrm{Sp}(w)$ gives a boundary point $\lambda - w$ of $GL(W)$, so $w = \lambda$. □

Definition 5.1.4 A *character* on a unital algebra W over a field F is a non-trivial homomorphism $W \to F$, and we denote the set of all these by $\hat{W}$.

Characters are evidently surjective and unital. Let W be a unital complex Banach algebra. If $x \in \hat{W}$ and $w \in W$, then $x(w - x(w)) = 0$, so $x(w) \in \mathrm{Sp}(w)$ and $|x(w)| \leq r(w) \leq \|w\|$. Hence $\|x\| = 1$ and $\hat{W} \subset W^\star$. By continuity $\ker x$ is a closed ideal in W. It is obviously also maximal. Every maximal ideal V in W is closed since the closure $\overline{V}$ is an ideal and remains proper as $\|1 - w\| \geq 1$ for $w \in V$ not to be invertible.

Proposition 5.1.5 *Let W be a commutative unital complex Banach algebra. Then $x \mapsto \ker x$ is a bijection from $\hat{W}$ to the maximal ideals of W. Moreover, for any $w \in W$ we have $\mathrm{Sp}(w) = \{x(w) \mid x \in \hat{W}\}$.*

Proof The map is injective because if $\ker x = \ker y$, then $x(w - y(w)) = 0$ for all $w \in W$.

To see that it is surjective, consider a maximal ideal V of W. Then any non-zero element $w + V$ is invertible in the Banach algebra W/V because if not, the set $(w + V)(W/V)$ would be a proper ideal in the commutative algebra W/V, so $V + wW$ would be a proper ideal in W violating maximality. By Corollary 5.1.2 we conclude that the quotient map $W \to W/V$ is a character of W with kernel V.

We claim that any non-invertible $w \in W$ is contained in a maximal ideal. It is certainly contained in the proper ideal wW, and the non-empty family of all proper ideals of W that contain w is inductively ordered under inclusion, so it has a maximal element by Zorn's lemma, and this is a maximal ideal. Hence if $\lambda \in \mathrm{Sp}(w)$, then $w - \lambda \in \ker x$ for some $x \in \hat{W}$, so $\lambda = x(w)$. □

We have the following important *Gelfand theorem.*

Theorem 5.1.6 *Let W be a commutative unital complex Banach algebra. Then $\hat{W}$ equipped with the w^*-topology is compact Hausdorff, and $G\colon W \to C(\hat{W})$ given by $G(w)(x) = x(w)$ is a norm decreasing unital homomorphism. The image subalgebra $G(W)$ separates points in $\hat{W}$ and $\|G(w)\|_u = r(w)$.*

Proof If $x_n \to x \in W^\star$ and $x_n \in \hat{W}$, then

$$x(vw) = \lim x_n(vw) = \lim x_n(v) \lim x_n(w) = x(v)x(w),$$

so $\hat{W}$ is a closed subset of $B^\star$, which is compact by Alaoglu's theorem. The equality $\|G(w)\|_u = r(w)$ is immediate from the previous proposition. The rest of the statements are clear. □

The kernel of the *Gelfand transform* G is the radical of W, i.e. the intersection of all maximal ideals of W. So G is injective if and only if W is semisimple. In general the kernel consists of all $w \in W$ with $\lim \|w^n\|^{1/n} = r(w) = 0$, and such elements w are called *quasi-nilpotent.*

Corollary 5.1.7 *Any homomorphism from a unital complex Banach algebra into a semisimple commutative unital complex Banach algebra is continuous. Hence any algebra isomorphism between two semisimple commutative unital complex Banach algebras is continuous with a continuous inverse.*

Proof If $f : V \to W$ is such a map, and $v_n \to v$ in V and $f(v_n) \to w$, then by the closed graph theorem, it suffices to show that $f(v) = w$. For $x \in \hat{W}$, observe that $xf \in \hat{V}$, so

$$x(w) = \lim xf(v_n) = xf(\lim v_n) = xf(v)$$

and $w - f(v)$ belongs to the radical of W. □

Let us study some further properties of the Gelfand transform G.

Lemma 5.1.8 *Let W be a commutative unital complex Banach algebra, and set $s = \inf \|w^2\|/\|w\|^2 \leq 1$ and $t = \inf \|G(w)\|_u/\|w\| \leq 1$, where we have taken infinum over all non-zero $w \in W$. Then $t^2 \leq s \leq t$.*

Proof Clearly $\|w^2\| \geq \|G(w)\|_u^2 \geq t^2\|w\|^2$, so $t^2 \leq s$. For $m = 2^n$ with $n \in \mathbb{N}$, we get $\|w^m\| \geq s^{m-1}\|w\|^m$, so $\|G(w)\|_u = r(w) = \lim \|w^m\|^{1/m} \geq s\|w\|$ and $s \leq t$. □

Proposition 5.1.9 *Let W be a commutative unital complex Banach algebra. Then the Gelfand transform is an isometry if and only if $\|w^2\| = \|w\|^2$ for every $w \in W$. Its kernel is trivial and its image is closed if and only if there is a scalar M such that $\|w\|^2 \leq M\|w^2\|$ for all $w \in W$.*

Proof According to the lemma, the Gelfand transform G is an isometry if and only if $t = 1$ if and only if $s = 1$.

Clearly M exists if and only if $s > 0$ if and only if $t > 0$. If $t > 0$, then G is injective with continuous inverse from its image. So $G(W)$ is complete and closed. Conversely, if G is injective with closed image, then the open mapping theorem implies that $t > 0$. □

Example 5.1.10 Let X be a compact Hausdorff space. Let $\delta_x(f) = f(x)$ for $f \in C(X)$ and $x \in X$. We claim that $x \mapsto \delta_x$ is a continuous map from X onto $\widehat{C(X)}$ which has a continuous inverse map. Since $C(X)$ separates points in X, the map is clearly injective, and it is obviously continuous as each f is continuous. If it was not surjective, there would be a maximal ideal V of $C(X)$, which to each $x \in X$ contained a function f such that $f(x) \neq 0$. Hence by compactness there would be finitely many $f_i \in V$ such that the open sets $f_i^{-1}(\mathbb{C}\backslash\{0\})$ covered X. Then $\sum \bar{f_i} f_i \in V$ would be invertible, and this is impossible. A bijective continuous map from a compact space into a Hausdorff space takes closed sets to closed sets, so the inverse map is also continuous. Under this identification the Gelfand transform is clearly the identity map from $C(X)$ to itself. ◇

Example 5.1.11 Consider the Banach space $l^1(\mathbb{Z})$ of of doubly infinite summable sequences. Equipped with the convolution product it is a unital commutative Banach algebra with generator δ_1 as $w = \sum w_n \delta_1^n$ uniformly for $w \in l^1(\mathbb{Z})$. Since δ_1 is invertible with $\delta_1^{-1} = \delta_{-1}$, the continuous map $\widehat{l^1(\mathbb{Z})} \to \mathbb{T}$ given by $x \mapsto x(\delta_1)$ is surjective since to $z \in \mathbb{T}$ we can define a character by $x(w) = \sum w_n z^n$. Arguing as in the previous example the map has a continuous inverse map. Under this identification the Gelfand transform $G \colon l^1(\mathbb{Z}) \to C(\mathbb{T})$ is given by $G(w)(z) = \sum w_n z^n$. Identifying $\mathbb{T}$ with $\mathbb{R}/2\pi\mathbb{Z}$ we can think of the elements in $G(l^1(\mathbb{Z}))$ as the set of periodic functions on $\mathbb{R}$ with absolutely convergent Fourier series. If f is such a function which moreover is never zero, then we actually recover *Wiener's theorem*, which states that $1/f$ has also an absolutely convergent Fourier series. This is now easy because $f = G(w)$ for some $w \in l^1(\mathbb{Z})$ with $0 \notin \mathrm{Sp}(w)$ by assumption. Hence $w^{-1} \in l^1(\mathbb{Z})$ and $G(w^{-1}) = G(w)^{-1} = 1/f$. ◇

Characters on Banach algebras can be characterized without appealing directly to multiplicativity or continuity.

Lemma 5.1.12 *If f is a complex holomorphic function such that $f(0) = 1$ and $f'(0) = 0$ and $0 < |f(z)| \leq e^{|z|}$ for $z \in \mathbb{C}$, then $f = 1$.*

Proof Any power series has an anti-derivative power series. Pick holomorphic g such that $g' = f'/f$. Then the derivative of fe^{-g} is zero, so it is constant. Hence $f = e^g$ and $\mathrm{Re}(g(z)) \leq |z|$. For $s > 0$ the function $h(z) = s^2 g(z)/z^2(2s - g(z))$ is holomorphic in $B_{2s}(0)$ and $|h(z)| \leq 1$ for $|z| = s$. By the maximum modulus theorem we see that $|h| \leq 1$ on $B_s(0)$. Fixing z and letting $s \to \infty$ we conclude that $g(z) = 0$. □

Proposition 5.1.13 *Any unital linear functional on a unital complex Banach algebra that is non-zero on every invertible element is a character.*

Proof Let x be such a linear functional on a Banach algebra W, and set $V = \ker x$. Since V by assumption contains no invertible elements, we must have $\|1 - v\| \geq 1$ for $v \in V$. Hence $\|z - v\| \geq |z| = |x(z - v)|$ for $v \in V$ and $z \in \mathbb{C}$, which shows that x is continuous with norm 1.

Write $v, w \in W$ as $v = a + x(v)$ and $w = b + x(w)$ for $a, b \in V$. Then $x(vw) = x(v)x(w) + x(ab)$, so we need only show that $ab \in V$ for $a, b \in V$. It is actually enough to show that $a^2 \in V$ for $a \in V$. Indeed, then $x(v^2) = x(v)^2$, and replacing v by $v + w$, we get $x(vw + wv) = 2x(v)x(w)$, so $vw + wv \in V$ if $v \in V$ and $w \in W$. From the identity

$$(vw - wv)^2 + (vw + wv)^2 = 2(v(wvw) + (wvw)v)$$

we also get $(vw-wv)^2 \in V$. Hence $vw-wv \in V$, so $vw \in V$ as also $vw+wv \in V$.

For $a \in V$ with say $\|a\| \leq 1$, the holomorphic function given by $f(z) = x(e^{za})$ is never zero as $e^{za} \in W$ is invertible. But $f(0) = 1$ and $f'(0) = 0$ and $|f(z)| \leq e^{|z|}$ for $z \in C$, so by the lemma $f''(0) = 0$ and $x(a^2) = 0$. □

5.2 Spectral Theory for C*-Algebras

We study here a distinguished class of Banach algebras for which the spectral theory gains full force.

Definition 5.2.1 A *C*-algebra* is a Banach $*$-algebra such that $\|w^*w\| = \|w\|^2$.

Any complex Banach algebra with a $*$-operation satisfying $\|w\|^2 \leq \|w^*w\|$ is automatically a C*-algebra because $\|w\|^2 \leq \|w^*w\| \leq \|w^*\|\|w\|$, so $\|w\| \leq \|w^*\|$ and $\|w\| = \|w^*\|$ and $\|w\|^2 = \|w^*w\|$. Also, provided the algebra is non-trivial, any identity element in it will be self-adjoint and have norm one because $1^* = (1^*1)^* = 1$ and $\|1\|^2 = \|1^*1\| = \|1\|$.

Example 5.2.2 If V is a complex Hilbert space, then any closed $*$-subalgebra of $B(V)$ with the usual adjoint is a C*-algebra. We shall later see that up to isomorphism these are all. Among these the commutative ones turn out to be of the form $C_0(X)$ for a locally compact Hausdorff space X. ◇

For a self-adjoint element w in a unital C*-algebra W we have $r(w) = \|w\|$ by the previous section since $\|w^2\| = \|w\|^2$. There is only one C*-norm on W, namely $w \mapsto r(w^*w)^{1/2}$. In fact, we have the following more general result.

Proposition 5.2.3 *Any unital $*$-homomorphism from a Banach $*$-algebra to a C*-algebra is norm decreasing. In particular, all unital $*$-isomorphism between C*-algebras are isometric.*

Proof Say $f: V \to W$ is such a homomorphism. Then $\mathrm{Sp}(f(v)) \subset \mathrm{Sp}(v)$, so

$$\|f(v)\|^2 = r(f(v^*v)) \leq r(v^*v) \leq \|v^*v\| \leq \|v\|^2.$$

□

In a C*-algebra $\mathrm{Sp}(w^*) = \overline{\mathrm{Sp}(w)}$ as the inverse operation and the $*$-operation commute. An element u in a unital C*-algebra is unitary if $u^*u = 1 = uu^*$.

Proposition 5.2.4 *If u is a unitary element in a non-trivial unital C*-algebra, then* $\mathrm{Sp}(u) \subset \mathbb{T}$. *If the spectrum is not the whole circle, then $u = e^{iw}$ for a self-*

adjoint element w. This happens when $\|1-u\| < 2$. Self-adjoint elements have real spectrum. The spectral radius and the norm coincide on normal elements. Characters on C-algebras are $*$-preserving.*

Proof If $\lambda \in \mathrm{Sp}(u)$, then $\lambda^{-1} \in \mathrm{Sp}(u^{-1}) = \mathrm{Sp}(u^*)$, so neither $|\lambda|$ nor $|\lambda|^{-1}$ is greater than $r(u) \leq \|u\| = 1$. The second statement follows from the appendix. If v is normal, then

$$r(v)^2 \leq \|v\|^2 = \|v^*v\| = \lim \|(v^*v)^n\|^{1/n} \leq \lim \|(v^*)^n\|^{1/n} \lim \|v^n\|^{1/n} = r(v)^2.$$

If $r(1-u) = \|1-u\| < 2$, then $-1 \notin \mathrm{Sp}(u)$.

Given a self-adjoint element w and $t \in \mathrm{Sp}(w)$, the holomorphic spectral calculus in the appendix shows that $e^{it} \in \mathrm{Sp}(e^{iw}) \subset \mathbb{T}$, so $t \in \mathbb{R}$.

Any $x \in \hat{W}$ is $*$-preserving since we may write any element in W as $w_1 + iw_2$ with w_1 and w_2 self-adjoint, so $x(w_i) \in \mathrm{Sp}(w_i) \subset \mathbb{R}$, and then

$$x((w_1+iw_2)^*) = x(w_1 - iw_2) = x(w_1) - ix(w_2) = \overline{x(w_1) + ix(w_2)} = \overline{x(w_1+iw_2)}.$$

□

Gelfand's theorem gains full strength for C*-algebras.

Theorem 5.2.5 *For a commutative unital C*-algebra W the Gelfand transform $G\colon W \to C(\hat{W})$ is an isometric $*$-isomorphism.*

Proof By the previous proposition G is clearly $*$-preserving and isometric, and hence it is surjective by the Stone-Weierstrass theorem. □

We have seen that a function $g\colon X \to Y$ between compact Hausdorff spaces is continuous if and only if its *transpose* $T_g\colon C(Y) \to C(X)$ give by $T_g(f) = fg$ is a unital $*$-homomorphism. And g has also a continuous inverse map if and only if T_g is a $*$-isomorphism.

Corollary 5.2.6 *Let w be a normal element in a unital C*-algebra W, and let V denote the unital C*-subalgebra generated by w. There there is a unique isometric $*$-isomorphism $H\colon C(\mathrm{Sp}(w)) \to V$ such that $H(\iota) = w$.*

Proof Let $\mathrm{Sp}_V(w)$ denote the spectrum of w as an element of V, and observe that V is commutative since w is normal. We have seen that the map $g\colon \hat{V} \to \mathrm{Sp}_V(w)$ given by $g(x) = x(w)$ is continuous with a continuous inverse map. Composing its transpose $T_g\colon C(\mathrm{Sp}_V(w)) \to C(\hat{V})$ with the inverse of the Gelfand transform $G\colon V \to C(\hat{V})$, we get an isometric $*$-isomorphism $H = G^{-1}T_g$, which by the Stone-Weierstrass theorem is uniquely determined by $H(\iota) = w$.

Now $\mathrm{Sp}_V(w) = \mathrm{Sp}(w)$. Indeed, to $t \in \mathrm{Sp}_V(w)$ and $\varepsilon > 0$ pick $f \in C(\mathrm{Sp}_V(w))$ with $\|f\|_u = 1$ and $f(s) = 0$ when $|t-s| \geq \varepsilon$. Then $\|(w-t)H(f)\| = \|(\iota - t)f\| \leq \varepsilon$, so $t \in \mathrm{Sp}(w)$. The opposite inclusion is obvious. □

Since $H(\sum c_n t^n) = \sum c_n w^n$ whenever the series $\sum c_n t^n$ converges absolutely on $B_{\|w\|}(0)$, we write $f(w)$ for the normal element $H(f)$ even when $f \in C(\mathrm{Sp}(w))$. The map $f \mapsto f(w)$ is called the *continuous functional calculus* at w. By the uniqueness statement of the corollary above, we see that $f(x(w)) = x(f(w))$ for $x \in \hat{V}$. Hence $\mathrm{Sp}(f(w)) = \{x(f(w)) \mid x \in \hat{W}\} = f(\mathrm{Sp}(w))$. For the same reason, the functional calculus respects composition, in that $(gf)(w) = g(f(w))$ for $g \in C(\mathrm{Sp}(f(w)))$.

As already pointed out, the spectrum of an element depends from the outset what algebra it is calculated in, or where possible inverses live. For C*-algebras the situation is however particularly pleasant.

Corollary 5.2.7 *The spectrum of an element in a unital C*-algebra is unaltered when calculated in any unital C*-subalgebra that contains it.*

Proof Let U be a unital C*-subalgebra of W. When $u^* = u \in U$, then $\mathrm{Sp}_V(u) = \mathrm{Sp}(u) \subset \mathrm{Sp}_U(u) \subset \mathrm{Sp}_V(u)$ from the last part of the proof of the corollary above. So any self-adjoint element of U invertible in W, is also invertible in U.

Consider any $u \in U$ with inverse $w \in W$. Then uu^* has inverse $w^*w \in W$, so it is invertible in U, say $uu^*v = 1$ for $v \in U$. But then $w = u^*v \in U$. □

Remark 5.2.8 When a C*-algebra W is not unital, we may adjoin an identity element by considering the unital $*$-algebra $W \oplus \mathbb{C}$ with $(w, a)^* = (w^*, \bar{a})$. This contains W as a maximal self-adjoint ideal. However, the norm we used in the Banach algebra case does not in general give rise to a C*-algebra. We now rather use the legitimate norm given by

$$\|(w, a)\| = \sup_{\|v\|=1} \|wv + av\|.$$

When we speak of the spectrum of $w \in W$, we mean its spectrum as an element of the unital C*-algebra $W \oplus \mathbb{C}$, called the *unitization* of W. Then $0 \in \mathrm{Sp}(w)$ since W is an ideal of $W \oplus \mathbb{C}$.

Using this device together with the one-point compactification counterpart, simple modifications of proofs shows that many of the previous results hold also for non-unital C*-algebras, of course then with the appropriate formulation. For instance, in Gelfand's theorem and its corollaries replace 'compact Hausdorff' and C by 'locally compact Hausdorff' and C_0. The two propositions prior to Gelfand's theorem are also true if one keeps in mind that unitaries only make sense in unital $*$-algebras, and so forth. For simplicity we often continue to state results in the unital case although non-unital versions might well be true, and we might even apply these versions further down the road. ◇

Example 5.2.9 Consider the Banach $*$-algebra $L^1(\mathbb{R})$ with convolution product $(f * g)(s) = \int f(t)g(s - t)dt$ and involution $f^\star(s) = \overline{f(-s)}$ sitting as an ideal inside the Banach $*$-algebra $M(\mathbb{R})$ of regular complex Borel measures. Letting δ be the unit for $M(\mathbb{R})$, which under the duality with $C_0(\mathbb{R})$ corresponds to evaluation at 0, we consider the unital commutative Banach $*$-algebra $W \equiv L^1(\mathbb{R}) + \mathbb{C}\delta$. Define $F: \mathbb{R} \cup \{\infty\} \to \hat{W}$ by $F(t)(f + c\delta) = \int f(s)e^{-its}ds + c$ for $t \in \mathbb{R}$ and $F(\infty)(f + c\delta) = c$. Considering $\mathbb{R} \cup \{\infty\}$ as the one-point compactification of $\mathbb{R}$, and $\hat{W}$ with w^*-topology, we see that F is injective and continuous. It will thus be a homeomorphism if we can show that it is surjective. Consider $x \in \hat{W}$ different from $F(\infty)$, so its restriction to $L^1(\mathbb{R})$ does not vanish. By Theorem A.8.11 there is measurable h with $\|h\|_\infty = 1$ such that $x(f) = \int f(t)h(t)dt$. By Lemma A.10.3 the map $s \mapsto f((\cdot) - s)$ is continuous from $\mathbb{R}$ into $L^1(\mathbb{R})$. Hence by the Fubini-Tonelli theorem we get

$$\int x(f((\cdot)-s))g(s)ds = \int f(t-s)g(s)h(t)dsdt = x(f*g) = x(f)\int h(s)g(s)ds,$$

which shows that the identity $x(f((\cdot) - s)) = x(f)h(s)$ holds for almost all s. Picking f such that $x(f) \neq 0$, and recalling that the left-hand side of the identity depends continuously on s, we may assume that $h \in C_b(\mathbb{R})$. Using the identity repeatedly we also get $x(f)h(s+t) = x(f((\cdot) - t))h(s) = x(f)h(t)h(s)$, so $h(s+t) = h(s)h(t)$. Pick $r \in \mathbb{R}$ such that $h(1) = e^{-ir}$, so $h(s) = e^{-irs}$ for every rational s, and then by continuity, for every real s. Hence F is surjective. Identifying $\hat{W}$ with $\mathbb{R} \cup \{\infty\}$, we see that the Gelfand transform G on $L^1(\mathbb{R})$ is the Fourier transform $G(f)(t) = \int f(s)e^{-its}\,ds$ of f. Clearly $x(f^\star) = \overline{x(f)}$, so $G(L^1(\mathbb{R}))$ is uniformly dense in $C_0(\mathbb{R})$ by the Stone-Weierstrass theorem. ◇

Definition 5.2.10 We say an element w is a unital C*-algebra is positive, and write $w \geq 0$ if $w^* = w$ and $\mathrm{Sp}(w) \subset [0, \infty\rangle$. By $w \geq u$ or $u \leq w$ we mean that u, w are self-adjoint and that $w - u \geq 0$.

For a unital C*-subalgebra of $B(V)$ for a complex Hilbert space V, this notion of positivity coincides with the notion of positivity of an operator A on V introduced earlier. Indeed, if $A^* = A$ then $\mathrm{Sp}(A) \subset \mathbb{R}$ by the continuous functional calculus, and if $t \in \mathrm{Sp}(A)$, then $A - tI$ is not bounded away from zero, so there is a sequence of unit vectors $v_n \in V$ such that $\lim \|A(v_n) - tv_n\| = 0$. Thus if also $(A(v)|v) \geq 0$, then $\mathrm{Sp}(A) \subset [0, \infty\rangle$. On the other hand, if $A \geq 0$ in the C*-algebra sense, then $(A(v)|v) \geq 0$ for all $v \in V$ by the proposition below.

Let X be a compact Hausdorff space. Then the spectrum of $f \in C(X)$ is clearly its range $f(X)$. Hence $f \geq 0$ if and only if its range is positive. The following result is then immediate from the continuous functional calculus.

Proposition 5.2.11 *Any positive element w in a unital C*-algebra W has a unique positive square root $w^{1/2}$. Let $v^* = v \in W$. Define $|v|$ to be the positive square root of $v^2 \geq 0$. Then $v_+ = (|v| + v)/2$ and $v_- = (|v| - v)/2$ are the unique positive elements of W such that $v = v_+ - v_-$ and $v_+v_- = 0$. The elements $|v|, v_+, v_-$ commute with all elements that commute with v. We have that $v \geq 0$ if $\|v - t\| \leq t$ for some $t \in \mathbb{R}$. If $v \geq 0$ and $\|v\| \leq t$ for some $t \in \mathbb{R}$, then $\|v - t\| \leq t$.*

The norm characterization of positivity shows that $v+w \geq 0$ whenever $w, v \geq 0$ because

$$\|v + w - \|v\| - \|w\|\| \leq \|v - \|v\|\| + \|w - \|w\|\| \leq \|v\| + \|w\|.$$

We also notice that any element in a unital C*-algebra can be written as a sum of four positive elements.

The same is in fact true with 'positive' replaced by 'unitary' because if $w = w^*$ and $\|w\| \leq 1$, then $u = w + i(1 - w^2)^{1/2}$ and $v = w - i(1 - w^2)^{1/2}$ are evidently unitary and $w = (u + v)/2$.

Note that any positive element can be written in the form w^*w. The converse result is less obvious. It is based on the observation that if u, v are elements in a unital algebra, and $1 - uv$ has inverse w, then $1 - vu$ has inverse $1 + vwu$, so $\mathrm{Sp}(uv)\backslash\{0\} = \mathrm{Sp}(vu)\backslash\{0\}$.

Proposition 5.2.12 *In a unital C*-algebra any element of the form w^*w is positive. In particular, we have $w^*uw \leq w^*vw^*$ when $u \leq v$, or $v-u \geq 0$. When also $u \geq 0$, then $\|u\| \leq \|v\|$.*

Proof If $-w^*w \geq 0$, then $-ww^* \geq 0$ as $\mathrm{Sp}(-ww^*)\backslash\{0\} = \mathrm{Sp}(-w^*w)\backslash\{0\}$. Write $w = u + iv$ with u, v self-adjoint, so $w^*w = 2u^2 + 2v^2 - ww^* \geq 0$ and $w = 0$.

Writing $w^*w = (w^*w)_+ - (w^*w)_-$ for any w, then $-(w(w^*w)_-)^*(w(w^*w)_-) = ((w^*w)_-)^3 \geq 0$, so $(w^*w)_- = 0$ by the first part.

Hence $w^*vw - w^*uw = ((v-u)^{1/2}w)^*((v-u)^{1/2}w) \geq 0$. From $0 \leq u \leq v \leq \|v\|$, we get $\|u\| \leq \|v\|$, both from the functional calculus. □

We can thus extend the *absolute value* to any element w in a unital C*algebra by $|w| = (w^*w)^{1/2}$.

Proposition 5.2.13 *If $0 \leq v \leq w$ in a unital C*-algebra, then $v^{1/2} \leq w^{1/2}$, and if v and w are also invertible, then $w^{-1} \leq v^{-1}$.*

Proof If $1 \leq v^{-1/2}wv^{-1/2}$, then $v^{1/2}w^{-1}v^{1/2} = (v^{-1/2}wv^{-1/2})^{-1} \leq 1$ by the functional calculus, so $w^{-1} \leq v^{-1}$.

It remains to show that $v^2 \leq w^2$ implies $v \leq w$. For any positive real number t let a and b be the real and imaginary parts of $(t + w + v)(t + w - v)$, so

$$a = ((t+w+v)(t+w-v)+(t+w-v)(t+w+v))/2 = t^2+2tw+w^2-v^2 \geq t^2$$

is postive and invertible. Hence $a^{-1/2}(a+ib)a^{-1/2} = 1+ia^{-1/2}ba^{-1/2}$ is invertible, so $a + ib$ is invertible and $-t \notin \mathrm{Sp}(w - v)$. □

In view of the spectral theorem for compact normal operators, it is worth looking at normal operators with various spectra.

Example 5.2.14 Any compact subset X of $\mathbb{C}$ is the spectrum of some normal operator. Indeed, take any dense sequence $t_n \in X$ and consider the diagonal operator on $l^2(\mathbb{N})$ given by $A(v_n) = t_nv_n$ for some orthonormal basis $\{v_n\}$. Since the spectrum is closed, obviously $X \subset \mathrm{Sp}(A)$. Conversely, if $t \notin X$, then the linear map given by $(t - A)^{-1}(v_n) = (t - t_n)^{-1}v_n$ will be bounded since there is $\varepsilon > 0$ such that $|t - t_n| > \varepsilon$. ◇

Example 5.2.15 Let us exhibit a normal operator without a single eigenvalue in its spectrum. Take any compact subset X of $\mathbb{C}$ such that $X\backslash\overline{X}$ is dense in X, and consider the restriction m of the Lebesgue measure from $\mathbb{C}$. The normal operator A on $L^2(X)$ given by $A(f)(t) = tf(t)$ has spectrum X because if $s \notin X$, then $(s - A)^{-1}(f)(t) = (s - t)^{-1}f(t)$ defines a bounded operator $(s - A)^{-1}$, so $s \notin$

$\mathrm{Sp}(A)$. Conversely, if $s \in X$, then $(s - A)^{-1}$ is not bounded since to $\varepsilon > 0$, the normalization f of the characteristic function on $B_\varepsilon(s)$ satisfies $\|(s-A)(f)\|_2 \leq \varepsilon$, so $s \in \mathrm{Sp}(A)$. Yet A has no eigenvalues as $(t - s)g(t) = 0$ for all $t \in X$ yields $0 = [g] \in L^2(X)$.

Recall that an *isolated point* in a topological space is a point that forms an open set.

Proposition 5.2.16 *Any isolated point in the spectrum of a normal operator A on a complex Hilbert space is an eigenvalue of A.*

Proof Let t be an isolated point in the spectrum of A. Then $(t - \iota)\delta_t = 0$, so $\{0\} \neq \mathrm{im}(\delta_t(A)) \subset \ker(t - A)$. □

Clearly a complex square matrix A commutes with a normal matrix B if and only if it leaves the eigenspaces of B invariant. As B^* has the same eigenspaces as B, we conclude that A commutes with B^* when it commutes with B. The generalization of this result to infinite dimensions is known as *Fuglede's theorem*.

Proposition 5.2.17 *If A, T are bounded operators on a complex Hilbert space and T is normal, then $AT^* = T^*A$ when $AT = TA$.*

Proof For unit vectors v, w in the Hilbert space, the holomorphic complex function f on $\mathbb{C}$ given by

$$f(z) = (e^{-zT^*}Ae^{zT^*}(v)|w)$$

is bounded by $\|A\|$ as $e^{-zT^*}Ae^{zT^*} = U(-z)AU(z)$ with $U(z) = e^{zT^*-\bar{z}T}$ unitary. By Liouville's theorem f is constant, so $Ae^{zT^*} = e^{zT^*}A$, or $AT^* = T^*A$. □

Corollary 5.2.18 *Two normal operators on a complex Hilbert space are unitarily equivalent if they are similar.*

Proof Say we have an invertible bounded operator A such that $AT_1 = T_2A$ for normal operators T_i on a complex Hilbert space V. Define operators on $V \oplus V$ by $T = \mathrm{diag}(T_1, T_2)$ and S with $S_{ij} = \delta_{i2}\delta_{j1}A$. Then $ST = TS$, so $ST^* = T^*S$, or $AT_1^* = T_2^*A$, by the proposition. Hence $A^*AT_1 = T_1A^*A$ and $|A|$ commutes with T_1. Polar decomposing $A = U|A|$, we get $UT_1U^* = AT_1A^{-1} = T_2$. □

5.3 Ideals and Hereditary Subalgebras

Proposition 5.3.1 *Every C*-algebra has an approximate unit of positive elements in the unit ball.*

Proof The partially ordered set I of positive elements of norm less than one in a non-unital C*-algebra W is upward filtered. Indeed, the function from the positive elements of W to I sending v to $v(1 + v)^{-1} = 1 - (1 + v)^{-1}$ is increasing, so the image of $v(1-v)^{-1}+w(1-w)^{-1}$ is greater than or equal to the images $v, w \in I$ of the terms.

The net $I \to I$ sending i to $v_i = i$ is an approximate unit. For if $\varepsilon > 0$ and if f is the image of $v \in I$ under the Gelfand transform G of the C*-algebra generated by v, then Urysohn's lemma supplies a continuous function $g\colon \mathrm{Sp}(v)\backslash\{0\} \to [0, 1]$ with compact support that is one on $|f|^{-1}([\varepsilon, \infty\rangle)$. If $r \in \langle 0, 1\rangle$ with $1-r < \varepsilon$, we have $\|f - rgf\|_u \leq \varepsilon$, so $\|v - u_i v\| \leq \varepsilon$ when $i = G^{-1}(rg)$. For $j \geq i$, we then get

$$\|v - u_j v\|^2 \leq \|(1-u_j)^{1/2} v\|^2 = \|v(1-u_j)v\| \leq \|v(1-u_i)v\| \leq \varepsilon.$$

□

We refer to the approximate unit in the proof of this proposition as the *canonical* one. Every subnet of an approximate unit is another approximate unit, so there is in general no uniqueness. For a separable C*-algebra we can pick a subsequence of the canonical one; to a dense sequence $\{w_m\}$ choose an increasing sequence $\{u_n\}$ in the canonical approximate unit such that $\|w_m - u_n w_m\| < 1/n$ for $m \leq n$.

Proposition 5.3.2 *Any closed ideal W in a C*-algebra V is a C*-algebra and the Banach algebra V/W is a C*-algebra with ∗-operation* $(v + W)^* \equiv v^* + W$.

Proof Take the canonical approximate unit $\{u_i\}$ of the C*-algebra $W \cap W^*$, and let $w \in W$. As $\|w(1-u_i)\|^2 \leq \|w^* w(1-u_i)\|$, we see that $w^* = \lim u_i w^* \in W$. If $v \in V$ then $\|v - vu_i\| = \|(v+w)(1-u_i) - w + wu_i\| \leq \|v+w\| + \|w - wu_i\|$, so $\|v+W\| \equiv \inf_{w\in W} \|v+w\| = \lim \|v - vu_i\|$ and $\|v+W\|^2 \leq \lim \|v^* v(1-u_i)\| = \|(v^* + W)(v + W)\|$. □

The first argument in the proof above shows that any closed left ideal W in a C*-algebra has an increasing net $\{u_i\}$ of positive elements in the unit ball such that $\lim wu_i = w$ for every $w \in W$.

Corollary 5.3.3 *Any ∗-homomorphism between C*-algebras has closed range, and if is injective, it is isometric.*

Proof Consider an injective ∗-homomorphism f. We may assume that it is unital. Considering the unital C*-subalgebas generated by an element and its image, we may by Gelfand's theorem also assume that $f\colon C(X) \to C(Y)$ for compact Hausdorff spaces X and Y, so f is the transpose of a continuous surjection, and must be isometric.

Any ∗-homomorphism $h\colon V \to W$ between C*-algebras is norm decreasing, so its kernel is a closed ideal and the induced ∗-homomorphism $V/\ker h \to W$ is an injective ∗-homomorphism between C*-algebras. □

Corollary 5.3.4 *If U is a C*-subalgebra of a C*-algebra V, and W is a closed ideal of V, then U + W is a C*-subalgebra of V.*

Proof Note that $(U+W)/W$ is the image of the injective $*$-homomorphism $U/(U\cap W)\to V/W$ that sends $v+U\cap W$ to $v+W$. So it is a C*-algebra, and as W is complete, so is $U+W$. □

Any closed ideal W of a closed ideal U of a C*-algebra V is a closed ideal of V because if $v\in V$ and $0\le w\in W$, then $vw=\lim(vu_i w^{1/2})w^{1/2}\in W$ for the canonical approximate unit $\{u_i\}$ of U.

Definition 5.3.5 A C*-subalgebra W of a C*-algebra V is *hereditary* if $0\le v\le w$ implies $v\in W$ when $v\in V$ and $w\in W$.

Since the intersection of hereditary C*-subalgebras is again hereditary, it makes sense to define the hereditary C*-subalgebra *generated* by a subset as the smallest one containing the subset.

If p is an orthogonal projection in a C*-algebra W, then pWp is a hereditary C*-subalgebra of W. Indeed, if $0\le w\le pvp$ for $w,v\in W$, then $0\le(1-p)w(1-p)\le(1-p)pvp(1-v)=0$, so $\|w^{1/2}(1-p)\|^2=0$ and $w(1-p)=0$, or $w=pwp$.

Proposition 5.3.6 *The formula $L\mapsto L\cap L^*$ defines a map from the set of closed left ideals of a C*-algebra V to the set of hereditary C*-subalgebras of V with inverse map $W\mapsto\{v\in V\mid v^*v\in W\}$, and both maps preserve inclusions.*

Proof Clearly $L\cap L^*$ is a C*-subalgebra of V. Say $0\le w\le v$ with $w\in V$ and $v\in L\cap L^*$. Consider a net of positive u_i in the unit ball of L such that $\lim uu_i=u$ for $u\in L$. Then

$$\|w^{1/2}(1-u_i)\|^2=\|(1-u_i)w(1-u_i)\|\le\|(1-u_i)v(1-u_i)\|\le\|v(1-u_i)\|$$

shows that $w^{1/2}=\lim w^{1/2}u_i\in L$ and $w\in L\cap L^*$.

The inequalities $(v+w)^*(v+w)\le 2v^*v+2w^*w$ and $(vw)^*(vw)\le\|v\|^2w^*w$ show that the second map takes hereditary C*-subalgebras of V to closed left ideals of V. It is straightforward to check that the two maps are inverses of each other.

If $L_1\cap L_1^*\subset L_2\cap L_2^*$ for closed left ideals L_i of V, and $\{u_i\}$ is the canonical approximate unit for $L_2\cap L_2^*$, then $\|v(1-u_i)\|^2\le\|v^*v(1-u_i)\|$ for $v\in L_1$, shows that $v=\lim vu_i\in L_2$, so $L_1\subset L_2$. □

Corollary 5.3.7 *A C*-subalgebra W of a C*-algebra V is hereditary if and only if $WVW\subset W$. So closed ideals of C*-algebras are hereditary.*

Proof The forward implication is clear from the proposition. For the opposite direction, let $\{u_i\}$ be the canonical approximate unit for W, and let $0\le v\le w$ with $v\in V$ and $w\in W$. Then $\|v^{1/2}(1-u_i)\|^2\le\|w^{1/2}(1-u_i)\|^2$ shows that $v^{1/2}=\lim v^{1/2}u_i$ and $v=\lim u_i vu_i\in W$. □

If v is a positive element of a C*-algebra V, the closure of vVv is the hereditary C*-subalgebra of V generated by v. We note that v belongs to this subalgebra because $v^2=\lim vu_iv$, where $\{u_i\}$ is an approximate unit for V.

Proposition 5.3.8 *If W is a separable hereditary C*-subalgebra of a C*-algebra V, there is a positive element $w\in W$ such that W is the closure of wVw.*

Proof Pick a sequential positive approximate unit $\{u_n\}$ for W, and set $w = \sum 2^{-n} u_n$. By the previous corollary and the observation following it, we know that the closure of wVw belongs to W, and that u_n belongs to this hereditary C*-subalgebra of V. But then $v = \lim u_n v u_n$ also belongs to it for any $v \in W$. □

Separability is necessary here. Indeed, if the hereditary C*-subalgebra $B_0(V)$ of $B(V)$ was of such a form for some compact operator T, then for $v \in V$, we could write $v \odot v = \lim T S_n T$ for some $S_n \in B(V)$, so V would be the closure of the separable subspace $T(V)$.

Proposition 5.3.9 *If W is a hereditary C*-subalgebra of a unital C*-algebra V and $v \in V$ is positive, and if to $\varepsilon > 0$, there is positive $w \in W$ such that $v \leq w + \varepsilon$, then $v \in W$.*

Proof Pick positive $u \in W$ such that $v \leq u^2 + \varepsilon^2$. Then

$$\|v^{1/2}(1 - u(u+\varepsilon)^{-1})\|^2 \leq \varepsilon^2 \|v^{1/2}(u+\varepsilon)^{-1}\|^2 \leq \varepsilon^2$$

as $v \leq (u+\varepsilon)^2$. Hence $v = \lim (u+\varepsilon)^{-1} u v u (u+\varepsilon)^{-1} \in W$ as $\varepsilon \to 0$ by the observation prior to the previous proposition. □

The closed ideals of a hereditary C*-subalgebra of a C*-algebra V can be characterize in terms of closed ideals of V.

Proposition 5.3.10 *Closed ideals of a hereditary C*-subalgebra W of a C*-algebra V are of the form $W \cap I$ for closed ideals I of V.*

Proof If J is a closed ideal of W, then $I \equiv VJV$ is a closed ideal of V, and $W \cap I \subset WIW$. Hence $W \cap I = (WVJ)J(JVW)$, which is a subset of $WJW = J$ by the previous corollary. The reverse inclusion is obvious. □

Definition 5.3.11 A C*-algebra is *simple* if it contains no closed ideals other than itself or $\{0\}$.

So $B_0(V)$ is a simple C*-algebra for any complex Hilbert space V because $B_0(V)$ is the smallest possible non-trivial closed ideal of $B(V)$ when $V \neq \{0\}$. But C*-subalgebras of simple C*-algebras need not be simple as $\mathbb{C}P + \mathbb{C}Q \subset B_0(V)$ shows, where P and Q are rank one pairwise orthogonal projections.

Corollary 5.3.12 *Hereditary C*-subalgebras of simple C*-algebras are simple.*

Proof By the proposition any closed ideal of a hereditary C*-subalgebra W of simple C*-algebra V is of the form $W \cap I$ for a closed ideal I of V. Then I is either V or trivial $\{0\}$. □

5.4 The Borel Spectral Theorem

It is desirable to extend the continuous functional calculus to bounded Borel functions.

Definition 5.4.1 Let V be a complex Hilbert space. The *commutant* W' of a subset W of $B(V)$ consists of all elements in $B(V)$ that commute with all elements of W.

The commutant of any subset W of $B(V)$ is clearly a unital Banach subalgebra of $B(V)$, and if W is self-adjoint, then W' is a unital C*-subalgebra of $B(V)$. Clearly $W \subset W'' \equiv (W')'$ and since $U \subset W$ implies $W' \subset U'$, we get $W''' \subset W' \subset W'''$, or $W''' = W'$, so the process of taking commutants stabilizes.

Let $B_b(X)$ denote the commutative unital C*-algebra of bounded Borel functions on a locally compact Hausdorff space X with uniform norm.

Proposition 5.4.2 *Any $*$-homomorphism $H\colon C_0(X) \to B(V)$ can be extended to a unique $*$-homomorphism $\tilde{H}\colon B_b(X) \to H(C_0(X))'' \subset B(V)$ such that if $\{f_n\}$ is a bounded sequence in $B_b(X)$ that converges pointwise, then*

$$\lim(\tilde{H}(f_n)v|w) = (\tilde{H}(\lim f_n)v|w)$$

for every $v, w \in V$.

Proof According to the appendix $M(X)$ is isometric to the dual of the Banach space $C_0(X)$. So to $v, w \in V$ there is a unique $\mu_{v,w} \in M(X)$ such that

$$\int f \, d\mu_{v,w} = (H(f)v|w)$$

for all $f \in C_0(X)$, and $\|\mu_{v,w}\| \leq \|v\| \cdot \|w\|$. It also shows that $(v, w) \mapsto \mu_{v,w}$ is a sesquilinear map from $V \times V$ to $M(X)$. Let $f \in B_b(X)$. Then $(v, w) \mapsto \int f \, d\mu_{v,w}$ is therefore a bounded sesquilinear form on V with $|\int f \, d\mu_{v,w}| \leq \|f\|_u \|v\| \cdot \|w\|$. So there is a unique $\tilde{H}(f) \in B(V)$ such that

$$(\tilde{H}(f)v|w) = \int f \, d\mu_{v,w}$$

for all $v, w \in V$. This defines a linear map $\tilde{H}\colon B_b(X) \to B(V)$ that extends H and satisfies $\|\tilde{H}(f)\| \leq \|f\|_u$ for all $f \in B_b(X)$.

To prove that $\tilde{H}(\bar{f}) = \tilde{H}(f)^*$ for $f \in B_b(X)$, recall that there is $h \in L^1(|\mu_v|)$ of modulus one such that $d\mu_v = h \, d|\mu_v|$. As $\int fh \, d|\mu_v| = (H(f)v|v) \geq 0$ for positive $f \in C_0(X)$, we have $\int f \mathrm{Im}(h) \, d|\mu_v| = 0$ for all $f \in C_0(X)$. So for $f \in B_b(X)$, we get

$$(\tilde{H}(\bar{f})v|v) = \int \bar{f} \, d\mu_v = \overline{\int f \, d\mu_v} = \overline{(\tilde{H}(f)v|v)} = (\tilde{H}(f)^* v|v),$$

and $\tilde{H}(\bar{f}) = \tilde{H}(f)^*$ by the polarization identity.

To prove multiplicativity note that the equality $(\tilde{H}(fg)v|v) = (\tilde{H}(f)\tilde{H}(g)v|v)$ holds for all $f, g \in C_0(X)$ as $\tilde{H}$ is an extension of the homomorphism H. But $(\tilde{H}(fg)v|v) = \int gf \, d\mu_v$ and $(\tilde{H}(f)\tilde{H}(g)v|v) = \int g \, d\mu_{v,w}$ with $w = \tilde{H}(\bar{f})v$, so

$d\mu_{v,w} = f\,d\mu_v$, which means that the equality holds also when $g \in B_b(X)$. Taking complex conjugation on both sides of the equality and replacing f by $\bar{f}$ and g by $\bar{g}$, we get $(\tilde{H}(gf)v|v) = (\tilde{H}(g)\tilde{H}(f)v|v)$ for $f \in C_0(X)$ and $g \in B_b(X)$. But as above this means that $g\,d\mu_v = d\mu_{v,u}$ with $u = \tilde{H}(\bar{g})v$, so the equality must also hold when $f \in B_b(X)$.

By Lebesgue's dominated theorem we have

$$(\tilde{H}(\lim f_n)v|v) = \int \lim(f_n h)\,d|\mu_v| = \lim \int (f_n h)\,d|\mu_v| = \lim(\tilde{H}(f_n)v|v)$$

and again we are done by polarization.

If $f \in B_b(X)$ and $S \in B(V)$ satisfies $H(g)S = SH(g)$ for all $g \in C_0(X)$, then $\int g\,d\mu_{S(v),w} = \int g\,d\mu_{v,S^*(w)}$, so $\mu_{S(v),w} = \mu_{v,S^*(w)}$ and $\int f\,d\mu_{S(v),w} = \int f\,d\mu_{v,S^*(w)}$, or $(\tilde{H}(f)S(v)|w) = (S\tilde{H}(f)(v)|w)$, so $\tilde{H}(f)S = S\tilde{H}(f)$.

To $f \in B_b(X)$ there is by Lusin's theorem a bounded sequence of functions $f_n \in C_0(X)$ that converges pointwise to f almost everywhere with respect to $|\mu_{v,w}|$. Let $g_n = f_n$ where the sequence converges and otherwise let $g_n = f$. Then $g_n = f_n$ almost everywhere, and $\{g_n\}$ is a bounded sequence in $B_b(X)$ that converges pointwise to f. Thus

$$(\tilde{H}(f)v|w) = \lim(\tilde{H}(g_n)v|w) = \lim \int g_n\,d\mu_{v,w} = \lim \int f_n\,d\mu_{v,w} = \lim(H(f_n)v|w),$$

which proves uniqueness even without any restrictions on the range of $\tilde{H}$. □

Using that for any $f \in B_b(X)$ there is a bounded sequence in $C_0(X)$ that converges to f almost everywhere with respect to $|\mu_{v,w}|$, together with the continuity statement in the proposition above, gives a different proof that $\tilde{H}$ is a $*$-homomorphism with the described restriction of its range. The crucial step is however to get the measures $\mu_{v,w}$ since they allow for the extension to bounded Borel functions.

The following result, known as the *Borel functional calculus*, is now immediate.

Corollary 5.4.3 *Let T be a bounded normal operator on a complex Hilbert space V, and let $C^*(T)$ denote the unital C^*-subalgebra of $B(V)$ generated by T. Then the continuous functional calculus $H\colon C(\mathrm{Sp}(T)) \to C^*(T)$ has a unique extension to a unital $*$-homomorphism $\tilde{H}\colon B_b(\mathrm{Sp}(T)) \to C^*(T)''$ such that*

$$\lim(\tilde{H}(f_n)v|w) = (\tilde{H}(\lim f_n)v|w)$$

for every $v, w \in V$ and every pointwise convergent bounded sequence $\{f_n\}$ in $B_b(X)$.

We denote $\tilde{H}(f)$ by $f(T)$ for $f \in B_b(\mathrm{Sp}(T))$. We have the following *spectral mapping theorem* in this context.

Proposition 5.4.4 *Let T be a bounded normal operator on a complex Hilbert space V, and let $f \in B_b(\mathrm{Sp}(T))$. Then $(g \circ f)(T) = g(f(T))$ for $g \in C(\mathrm{Sp}(f(T)))$.*

Proof Clearly $g \circ f \in B_b(\mathrm{Sp}(T))$ and $f(T)$ is normal in $B(V)$, so each side of the formula makes sense. When $g \in C(\mathrm{Sp}(f(T)))$ we have equality in the formula because g is the pointwise limit of a bounded sequence $\{g_n\}$ of polynomials, in fact, even a uniform limit by the Stone-Weierstrass theorem, and for polynomials the validity of the formula just amounts to saying that the Borel functional calculus is a unital $*$-homomorphism that takes the identity function to $f(T)$. By the spectral mapping theorem for continuous functions we therefore get

$$((g \circ f)(T)v|w) = \lim((g_n \circ f)(T)v|w) = \lim(g_n(f(T))v|w) = (g(f(T))v|w).$$

□

Corollary 5.4.5 *Let U be a unitary operator in $B(V)$. Then there is a self-adjoint operator $T \in B(V)$ such that $U = e^{iT}$ and $\|T\| \leq 2\pi$.*

Proof The formula $f(t) = e^{it}$ defines a continuous function $f: [0, 2\pi\rangle \to \mathbb{T}$ with inverse $g = \bar{g} \in B_b(\mathbb{T})$, and $\mathrm{Sp}(U) \subset \mathbb{T}$. Then $T = g(U)$ is self-adjoint with $\|T\| \leq \|g\|_u \leq 2\pi$ and $U = (f \circ g)(U) = f(g(T)) = e^{iT}$ by the proposition. □

Definition 5.4.6 A *(regular) spectral measure* on a locally compact Hausdorff space X is a map P from the Borel sets of X to the set of orthogonal projections on a complex Hilbert space V such that $P_{v,w} \equiv (P(\cdot)v|w)$ is a (regular) complex Borel measure for every $v, w \in V$, and $P(\phi) = 0$ and $P(X) = I$ and $P(A \cap B) = P(A)P(B)$ for every Borel sets A and B.

There is a correspondence between spectral measures on X having orthogonal projections on V and unital $*$-homomorphism $B_b(X) \to B(V)$ satisfying a natural continuity property.

Proposition 5.4.7 *Suppose P is a spectral measure on a locally compact Hausdorff space X having orthogonal projections on a complex Hilbert space V. Then for $f \in B_b(X)$ the map $(v, w) \mapsto \int f\, dP_{v,w}$ is a bounded sesquilinear form, so there is a unique $H_P(f) \in B(V)$ such that $(H_P(f)v|w) = \int f\, dP_{v,w}$. Then $P \mapsto H_P$ is a bijection from the set of spectral measures on X with orthogonal projections on V to the set of unital $*$-homomorphisms $H: B_b(X) \to B(V)$ such that $\lim(H(f_n)v|w) = (H(\lim f_n)v|w)$ for every $v, w \in V$ and every bounded sequence $\{f_n\}$ in $B_b(X)$ that converges pointwise.*

Proof For $f = \chi_A$ we get a bounded sesquilinear form $(v, w) \mapsto (P(A)v|w)$. This remains true when f is a simple Borel function, and further when $f \in B_b(X)$ since f is a uniform limit of simple Borel functions f_n, and $|\int(f - f_n)\, dP_{v,w}| \leq \int |f - f_n|\, d|P_{v,w}| \leq \|f - f_n\|_u \|P_{v,w}\| \to 0$. So $(v, w) \mapsto \int f\, dP_{v,w} = \lim \int f_n\, dP_{v,w}$ is indeed a sesquilinear form. Also $|\int f\, dP_{v,v}| \leq \|f\|_u \|P_{v,v}\| = \|f\|_u P_{v,v}(X) = \|f\|_u \|v\|^2$ for $f \in B_b(X)$, so by the polarization identity, this sesquilinear form is certainly bounded by $4\|f\|_u$.

This gives us a map $H_P\colon B_b(X) \to B(V)$ such that $(H_P(f)v|w) = \int f\,dP_{v,w}$ and which evidently is unital and linear. Next note that $(H_P(\chi_A)v|w) = (P(A)v|w)$, so $H_P(\chi_A) = P(A)$. Thus $H_P(\chi_A\chi_B) = P(A \cap B) = P(A)P(B) = H_P(\chi_A)H_P(\chi_B)$ and $H_P(\overline{\chi_A}) = P(A) = P(A)^* = H_P(\chi_A)^*$, so H_P is multiplicative and $*$-preserving on simple Borel functions by linearity. Since $\|H_P(f)\| \leq 4\|f\|_u$ and we can approximate bounded Borel functions uniformly by simple ones, we conclude that H_P is a unital $*$-homomorphism. It has the prescribed continuity property due to Lebesgue's dominated convergence theorem applied to the measures $|P_{v,w}|$.

Since $H_P(\chi_A) = P(A)$ for every Borel set A, the map $P \mapsto H_P$ is injective. To see that it also is surjective, suppose we have a unital $*$-homomorphism $H\colon B_b(X) \to B(V)$. Then $P(A) = H(\chi_A)$ is evidently an orthogonal projection for every Borel set A, and $P(\phi) = 0$ and $P(X) = I$ and $P(A \cap B) = P(A)P(B)$. Also, if we have a countable partition $\{A_i\}$ of Borel sets, then

$$(P(\cup A_i)v|w) = (H(\chi_{\cup A_i})v|w) = \sum (H(\chi_{A_i})v|w) = \sum (P(A_i)v|w)$$

since $\{\sum_{i=1}^n \chi_{A_i}\}$ is a bounded sequence of Borel functions that converges pointwise to the characteristic function on $\cup A_i$. So $P_{v,w}$ is a complex measure, and $H_P = H$ since they coincide on characteristic functions and therefore also on any bounded Borel function by continuity. □

Write $\int f\,dP$ for $H_P(f)$. The spectral measure associated to $\tilde{H}$ from Proposition 5.4.2 is clearly regular.

We have the following *resolution of the identity* for a normal operator.

Corollary 5.4.8 *Let T be a bounded normal operator on a complex Hilbert space V. Then there is a unique regular spectral measure P on* $\mathrm{Sp}(T)$ *with orthogonal projections in $B(V)$ such that $T = \int \lambda\,dP(\lambda)$.*

Proof The regular spectral measure P provided by the proposition above when applied to $\tilde{H}$ from Corollary 5.4.3 has the required properties.

Suppose Q was another regular spectral measure on $\mathrm{Sp}(T)$ with orthogonal projections in $B(V)$ such that $T = \int \lambda\,dQ(\lambda)$. Since the unital $*$-subalgebra of $C(\mathrm{Sp}(T))$ generated by the identity function is norm dense by the Stone-Weierstrass theorem, we have $\int f\,dQ = H(f) = \int f\,dP$ for all $f \in C(\mathrm{Sp}(T))$, so $P_{v,w} = Q_{v,w}$ by regularity and duality, and thus $P = Q$. □

We refer to P above as the *spectral measure* of T. When $T = \int \lambda\,dP(\lambda)$, then $f(T) = \int f(\lambda)\,dP(\lambda)$ for $f \in B_b(\mathrm{Sp}(T))$ as both sides equal $H_P(f)$. By the spectral mapping theorem $g(f(T)) = \int g(f(\lambda))\,dP(\lambda)$ for $g \in C(\mathrm{Sp}(f(T)))$, and if a bounded sequence of Borel functions f_n converges pointwise to f, then $\int f(\lambda)\,dP_{v,w}(\lambda) = \lim \int f_n(\lambda)\,dP_{v,w}(\lambda)$ for all $v, w \in V$.

Proposition 5.4.9 *If λ is an eigenvalue for a normal operator T with eigenvector v, then $f(T)v = f(\lambda)v$ for $f \in B_b(\mathrm{Sp}(T))$.*

Proof This clearly holds when f is a polynomial, and hence also for f continuous as the continuous functional calculus is isometric and the polynomials are dense in $C(\mathrm{Sp}(T))$ by the Stone-Weierstrass theorem. Approaching $f \in B_b(\mathrm{Sp}(T))$ pointwise almost everywhere with respect to $|P_{v,w}|$ by a bounded sequence of continuous functions by Lusin's theorem, we thus get $(f(T)v|w) = f(\lambda)(v|w)$. □

We may use the Borel functional calculus to describe the spectrum further.

Proposition 5.4.10 *Let T be a bounded normal operator on a complex Hilbert space V with spectral measure P, and let $f \in B_b(\mathrm{Sp}(T))$. Then $\lambda \in \mathrm{Sp}(f(T))$ if and only if $P_r \equiv P(\{z \in \mathrm{Sp}(T) \,|\, |f(z) - \lambda| \leq r\}) \neq 0$ for all $r > 0$. And λ is an eigenvalue for $f(T)$ if and only if $P_0 \neq 0$, in which case P_0 is the orthogonal projection onto the subspace of eigenvectors of $f(T)$ with eigenvalue λ.*

Proof If $P_r = 0$ for some $r > 0$, let $g(z)$ be $(\lambda - f(z))^{-1}$ when $|f(z) - \lambda| \geq r$ and otherwise zero. Then $g \in B_b(\mathrm{Sp}(T))$ with $\|g\|_u \leq r^{-1}$ and $g(T)(\lambda - f(T)) = \int g(z)(\lambda - f(z))\, dP(z) = I$.

Conversely, if $P_r \neq 0$ for every $r > 0$, then for each unit vector $v \in P_r(V) \neq \phi$, we have $\|(\lambda - f(T))v\| \leq \|(\lambda - f(T))P_r\| \leq \|h\|_u \leq r$, where h is $\lambda - f$ restricted to $\{z \in \mathrm{Sp}(T) \,|\, |f(z) - \lambda| \leq r\}$.

If $P_0 \neq 0$, then $f(T)P_0 = \int f\chi_{f^{-1}(\lambda)}\, dP = \lambda P_0$. Conversely, if v is an eigenvector for $f(T)$ with eigenvalue λ, then the bounded sequence of $f_n \in B_b(\mathrm{Sp}(f(T)))$ given by $f_n = (\min\{1, |\lambda - f|\})^{1/n}$ converges pointwise to the characteristic function on $\mathrm{Sp}(T)\backslash f^{-1}(\lambda)$. Then $((I - P_0)v|w) = \lim(f_n(T)v|w) = 0$ by the previous proposition. □

In particular, for any $f \in B_b(\mathrm{Sp}(T))$ the inclusion $\mathrm{Sp}(f(T)) \subset \overline{f(\mathrm{Sp}(T))}$ holds. And as a special case of the second statement in the proposition we see that for each λ in the spectrum of a normal operator T with spectral measure P, the orthogonal projections $P(\{\lambda\})$ vanish unless λ is an eigenvalue, and then $P(\{\lambda\})$ is the orthogonal projection onto the subspace of eigenvectors of T with eigenvalue λ.

Example 5.4.11 Suppose T is a diagonalizable operator on a complex Hilbert space V. Then it is of course normal, and its spectral measure P vanishes on the complement A in $\mathrm{Sp}(T)$ of the eigenvalues of T because the orthogonal projection $P(A)$ commutes with T, so $\mathrm{im}P(A)$ is an invariant subspace for T that is not a subspace of eigenvectors. Thus the spectral measure is concentrated on the set of eigenvalues of T. Also, since P is non-zero on every closed ball with positive radius, each such ball must contain at least one eigenvalue, so the eigenvalues of T are dense in the closed set $\mathrm{Sp}(T)$.

Since $P(\{\lambda\})$ is the orthogonal projection onto the space of eigenvectors with eigenvalue λ, we may write $T(v) = \sum \lambda P(\{\lambda\})(v)$ for all $v \in V$, where we sum over all eigenvalues of T. When V is separable, or when T is compact, then there are at most countable many eigenvalues, and the spectral measure vanishes outside this set. ◇

Corollary 5.4.12 *Suppose T is a normal operator on a complex Hilbert space V. Then to $\varepsilon > 0$ there is a finite set $\{P_n\}$ of pairwise orthogonal projections with sum*

I, and complex scalars λ_n, such that $\|T - \sum \lambda_n P_n\| \leq \varepsilon$. In particular, the set of normal operators is the norm closure of the diagonalizable operators.

Proof Let $P_n = f_n(T)$, where f_n is the characteristic function of a sufficiently small half-open square of the plane, see the discussion in the appendix on the Lebesgue measure on $\mathbb{R}^2$. The scalars λ_n are chosen so that $\|\iota - \sum \lambda_n f_n\|_u \leq \varepsilon$ on $\mathrm{Sp}(T)$. Since such half-open squares form a partition of the plane, the projections P_n are pairwise orthogonal and sum up to one.

For the second statement, pick an orthonormal basis for each subspace $\mathrm{im} P_n$ and take the union of these to get an orthonormal basis for V for which $\sum \lambda_n P_n$ will be diagonal; the vectors in $\mathrm{im} P_n$ being the eigenvectors with eigenvalue λ_n. □

Corollary 5.4.13 *Let T be a positive operator on a complex Hilbert space such that $T \leq I$. Then there is a sequence of pairwise commuting orthogonal projections P_n such that $T = \sum 2^{-n} P_n$.*

Proof Note that $\mathrm{Sp}(T) \subset [0, 1]$. Let $f_n(T)$ be the orthogonal projections of the characteristic functions $f_1, f_2, f_3, \ldots$ of the Borel sets $\langle 1/2, 1]$ and $\langle 1/4, 2/4] \cup \langle 3/4, 1]$ and $\langle 1/8, 2/8] \cup \langle 3/8, 4/8] \cup \langle 5/8, 6/8] \cup \langle 7/8, 1]$, etc. Then $\iota = \sum 2^{-n} \chi_n$ in the uniform norm on $[0, 1]$. □

5.5 Von Neumann Algebras

Definition 5.5.1 Let V be a complex Hilbert space. The *strong topology* on $B(V)$ is the weak one induced by the separating family of seminorms $s(T) = \|T(v)\|$ for all $v \in V$. The *weak topology* on $B(V)$ is the weak one induced by the separating family of seminorms $s(T) = |(T(v)|w)|$ for all $v, w \in V$.

Both these topologies make $B(V)$ a locally convex topological vector space. A net $\{T_i\}$ in $B(V)$ converges strongly to T if and only if $\lim T_i(v) = T(v)$ for every $v \in V$, and the net converges weakly to T if and only if $\lim(T_i(v)|w) = (T(v)|w)$ for every $v, w \in V$. Noting that $(\cdot|w)$ are the bounded linear functionals on V, we could have done something similar if V was just a Banach space.

Since $|(T(v)|w)| \leq \|T(v)\|\|w\| \leq \|T\|\|v\|\|w\|$, we see that the weak topology is weaker than the strong one, which again is weaker than the norm topology. And one can be strictly weaker than the next. Indeed, as $|(T(v)|w)| = |(T^*(w)|v)|$, we see that the adjoint operation is weakly continuous. However, it is not always strongly continuous because if V is separable with an orthonormal basis $\{v_n\}$, then $\lim \|(v_1 \odot v_n)(v)\| = \lim |(v|v_n)| = 0$, while $\lim \|(v_1 \odot v_n)^*(v_1)\| = \lim |(v_1|v_1)| = 1$. Also, in this case the norm is not strongly continuous as $\|v_1 \odot v_n\| = 1 \neq 0$. The adjoint operation is strongly continuous on the set of normal operators.

Multiplication on $B(V)$ is neither weakly nor strongly continuous. For instance, if T is the unilateral shift operator on $l^2(\mathbb{N})$, then the sequences $\{T^n\}$ and $\{(T^*)^n\}$ both converge weakly to zero while $\{(T^*)^n T^n\}$ is the constant one. Multiplication is, however, if one of the factors is fixed. Even if the first factor remains bounded, multiplication is strongly continuous because

$$\|(ST - S_0T_0)(v)\| \leq \|(S - S_0)T_0(v)\| + \|S\|\|(T - T_0)(v)\|.$$

Given $n \in \mathbb{N}$ and a complex Hilbert space V, identify $B(V^n)$ and $M_n(B(V))$ as C^*-algebras . The *n-amplification* of $B(V)$ is the isometric unital $*$-homomorphism $f: B(V) \to B(V^n)$ with $f(T)$ the diagonal matrix with $T \in B(V)$ in the diagonal.

Proposition 5.5.2 *Let x be a linear functional on $B(V)$ for a complex Hilbert space V. Then x is strongly continuous if and only if it is weakly continuous if and only if $x = \sum(\cdot v_i|w_i)$ for finitely many $v_i, w_i \in V$.*

Proof If x is strongly continuous, there are $v_1, \ldots, v_n \in V$ such that $|x(T)| \leq 1$ when $\sup_i \|T(v_i)\| \leq 1$ for all $T \in B(V)$. So $|x(T)|^2 \leq \sum \|T(v_i)\|^2$. With $v = (v_i) \in V^n$ and f the n-amplification of $B(V)$, the linear functional y given by $y(f(T)v) = x(T)$ on the subspace $f(B(V))v$ of V^n is thus well-defined and bounded. Writing $y = (\cdot|w)$ for $w = (w_i) \in V^n$ gives $x = \sum(\cdot v_i|w_i)$. The remaining implications are clear. □

By the Hahn-Banach separation theorem we therefore get the following result.

Corollary 5.5.3 *Strongly closed convex subsets of $B(V)$ are weakly closed.*

The following result by von Neumann is known as the *bicommutant theorem*.

Theorem 5.5.4 *Any $*$-subalgebra W of $B(V)$ containing the identity map I on a complex Hilbert space V is strongly closed if and only if $W = W''$.*

Proof Assume that W is strongly closed. To $v \in V$ let P denote the orthogonal projection onto the norm closure of $W(v)$. Then $TP(V) \subset P(V)$, so $(I - P)TP = 0$, for $T \in W$. If in addition $T^* = T$, then $TP = PTP = PT$, so $P \in W'$. For $S \in W''$, we thus have $SP = PS$, and $S(v) = SP(v) \in P(V)$ as $I \in W$. Hence, for each $\varepsilon > 0$ there is $T \in W$ such that $\|(S - T)(v)\| < \varepsilon$.

Let now f be the n-amplification of $B(V)$, and note that a matrix in $B(V^n)$ commutes with $f(T)$ if and only if each of its entry commutes with T. Applying the result of the first paragraph to the amplified versions of V, W and S, we get $T \in W$ such that $\sum_{i=1}^n \|(S - T)(v_i)\|^2 = \|(f(S) - f(T))(v)\|^2 < \varepsilon^2$. Hence $S \in W$. The converse is obvious. □

So the strong or weak closure of a $*$-subalgebra W of $B(V)$ is W'' when $I \in W$.

Definition 5.5.5 Let V be a complex Hilbert space. A *von Neumann algebra* is a weakly closed $*$-algebra of $B(V)$ which contains the identity map on V.

A von Neumann algebra W equals its bicommutant, so it contains all its spectral projections and is the closed span of these. From the proof of Corollary 5.4.13 we see that the set of positive elements of W with norm not greater than one is the closed convex hull of the orthogonal projections in W. Note also that the commutant of any self-adjoint subset of $B(V)$ is a von Neumann algebra. The following result is known as *Vigier's theorem*.

Proposition 5.5.6 *A net of self-adjoint operators on a complex Hilbert space is strongly convergent if it is increasing and bounded above.*

Proof By truncating and translating the net $\{u_i\}$, we may assume that all $u_i \geq 0$ and $\|u_i\| \leq c < \infty$. Hence the net $\{(u_i(v)|v)\}$ converges for every v, and by the polarization identity, so does $\{(u_i(v)|w)\}$ for every v and w. Let u be the bounded operator given by $(u(v)|w) = \lim(u_i(v)|w)$. Clearly $u \geq u_i$, and

$$\|(u - u_i)(v)\|^2 \leq \|u - u_i\|\|(u - u_i)^{1/2}(v)\|^2 \leq 2c((u - u_i)(v)|v)$$

shows that $u_i \to u$ strongly. □

Combining this result with the fact that every C*-algebra has a canonical approximate unit, we conclude that its weak closure is unital. Also, every strongly closed ideal in W is of the form qW for a unique orthogonal projection $q \in W'$.

Remark 5.5.7 The identity element p of a weakly closed $*$-subalgebra W of $B(V)$ need not be the identity map on the complex Hilbert space V, but one can cut down to a subspace U of V, where it is. Indeed, the image W_p of W under the map $W \to B(p(V))$ given by $w \mapsto pw$ is isometrically $*$-isomorphic to W, and now $(W_p)'' = W_p$, see also the proposition below. It is worth noting that p is the orthogonal projection onto the closed linear span of $W(V)$. ◇

Proposition 5.5.8 *If W is a von Neumann algebra with an orthogonal projection p, then $W_p \equiv pWp$ and $W'_p \equiv W'p$ are von Neumann algebras on* im p. *We also have $(W_p)' = W'_p$.*

Proof Say W acts on the Hilbert space V. Clearly $W_p \subset (W'_p)'$. Equality holds because the extension of any element w in the latter set defined to vanish on $(1 - p)(V)$ evidently belongs to $W'' = W$, and this extension sandwiched by p gives w again. So $W_p = (W'_p)'$ is a von Neumann algebra on $p(V)$.

It remains to prove $(W_p)' = W'_p$, or that $w \in W'_p$ for any unitary $w \in (W_p)'$, to be fixed hereafter. Let $q \in W \cap W'$ be the orthogonal projection onto the closure of the span of $Wp(V)$. The formula $u(\sum w_i(v_i)) = \sum w_i w(v_i)$ for $v_i \in p(V)$ and $w_i \in W$ well-defines a linear map from the span of $Wp(V)$ to V because it is easily checked that $\| \sum w_i w(v_i)\|^2 = \| \sum w_i(v_i)\|^2$. By declaring u to be zero on $(1 - q)(V)$, it moreover extends to a partial isometry u on V with $u^*u = q$. By definition $u \in W'$ and $w = up \in W'_p$. □

We refer to W_p above as the *compression of W by p*. By abuse of language we sometimes write Wp for the von Neumann algebra W_p when p is *central*, that is, when p commutes with every element of W.

Definition 5.5.9 A C*-algebra W acts *non-degenerately* on a Hilbert space V if the closed linear span of $W(V)$ is V. It acts *irreducibly* on V if for any non-trivial closed subspace U of V invariant under the action of W, we must have $U = V$.

Equivalently, the action is non-degenerate if $v = 0$ when $W(v) = \{0\}$. Note that a weakly closed $*$-algebra W acts non-degenerately on the underlying Hilbert space if and only if $W'' = W$. Clearly non-trivial C*-algebras act non-degenerately if they act irreducibly.

Proposition 5.5.10 *A non-trivial C^*-algebra W acts irreducible on V if and only if $W' = \mathbb{C}$ if and only if W is strongly dense in $B(V)$.*

Proof Clearly an orthogonal projection p belongs to W' if and only if $p(V)$ is invariant under the action of W. The first equivalence in the proposition is then immediate as W' is the closed span of its orthogonal projections.

If W is strongly dense in $B(V)$, then $W' = B(V)' = \mathbb{C}'' = \mathbb{C}$. Conversely, if $W' = \mathbb{C}$, then $B(V) = \mathbb{C}' = W''$. □

Lemma 5.5.11 *If a sequence $\{f_n\}$ of increasing real-valued continuous functions on a compact space converges pointwise to a continuous function f, it converges uniformly to f.*

Proof Let $\varepsilon > 0$. Then $\{(f - f_n)^{-1}(\langle -\infty, \varepsilon\rangle)\}_n$ form an open cover of the compact set, so $f - f_m < \varepsilon$ for some m. □

Definition 5.5.12 The *range projection* of an operator T on a Hilbert space is the orthogonal projection onto the closure of $\mathrm{im}T$.

For any bounded operator T on a Hilbert space, we have $\ker T^* = \ker |T^*|$ by the polar decomposition of T^*, so the range projections of T and $(TT^*)^{1/2}$ coincide.

Proposition 5.5.13 *All von Neumann algebras contain the range projections of their elements.*

Proof By the discussion above it suffices to check this result for $T \geq 0$, and we may further assume $T \leq I$. By Proposition 5.5.6, the operators $T_n = T^{2^{-n}}$ converge strongly to a positive operator S in the von Neumann algebra.

We claim S is the range projection of T. The inequality $\|(S^2 - T_n^2)(v)\| \leq \|(S - T_n)(S(v))\| + \|(S - T_n)(v)\|$ and $T_n^2 = T_{n-1}$ show that $S^2 = S$. Clearly $\mathrm{im}S$ is contained in the closure of $\mathrm{im}T$. For the reversed inclusion, note that the functions $\mathrm{Sp}(T) \to \mathbb{R}$ sending $t \to t^{1+2^{-n}}$ converge to the identity function by the lemma, so $T = \lim T^{1+2^{-n}} = \lim TT_n = TS = ST$. □

Corollary 5.5.14 *If T belongs to a von Neumann algebra W, and $U|T|$ is the polar decomposition of T, then $U \in W$. If moreover T is normal and U is unitary, then $U = e^{iR}$ with $R^* = R \in W$.*

Proof If $A \in W'$ is unitary, then $A^*UA = U$ by uniqueness of the polar decomposition. Hence $U \in W'' = W$.

The last statement is clear from the Borel functional calculus and the corollary following that as $g(U) \in W'' = W$ for any bounded Borel function g. □

If an adjoint operator T^* has polar decomposition $U|T^*|$, then $T = |T^*|U^*$ and $|T^*| = (U^*U|T^*|)^* = (U^*T^*)^* = TU$, so $\mathrm{im}T = \mathrm{im}|T^*|$.

We remark also that if S is another bounded operator such that $SS^* \leq TT^*$, there is a bounded operator R such that $S = TR$ because $\|S^*(v)\|^2 \leq \|T^*(v)\|^2$, so we let R be the adjoint of a bounded extension of the well-defined linear map that sends $T^*(v)$ to $S^*(v)$.

Proposition 5.5.15 *A bounded operator T on a complex Hilbert space is compact if and only if its range has no infinite dimensional closed subspace.*

Proof If T is compact and P is an orthogonal projection onto a closed subspace of $\text{im}T$, the linear map $S\colon \ker(PT)^{\perp} \to \text{im}P$ that sends v to $PT(v)$ is compact and bijective, so $\text{im}P$ is finite dimensional by the open mapping theorem and the fact that the compact operators is an ideal in the algebra of bounded operators.

Suppose now $\text{im}T$ has no infinite dimensional closed subspace. By the first observation above, we may assume $0 \leq T \leq 1$. By Corollary 5.4.13 we can therefore write $T^2 = \sum 2^{-n} P_n$ for orthogonal projections P_n. Since $2^{-n} P_n \leq T^2$, there are by the second observation prior to this proposition, bounded operators R_n such that $P_n = TR_n$, so $\text{im}P_n$ is a closed subspace of $\text{im}T$ and must be finite dimensional, so T is compact. □

5.6 The σ-Weak Topology

Definition 5.6.1 For a complex Hilbert space V the *σ-weak topology* is the w^*-topology on $B(V) = B^1(V)^\star$.

Thus $T_i \to T$ in $B(V)$ if and only if $\text{Tr}(T_i S) \to \text{Tr}(TS)$ for all $S \in B^1(V)$. As $((T - T_i)(v)|w) = \text{Tr}((T - T_i)(v \odot w))$, we see that the weak topology is weaker than the σ-weak topology. However, restricted to the closed unit ball they coincide because by Alaoglu's theorem the ball is σ-weakly compact, and the identity map from the ball with the σ-weak topology to the ball with the weak topology is a continuous bijection from a compact space to a Hausdorff space. For the same reason the closed unit ball cannot be strongly compact in general as the adjoint operation restricted to the ball is not always strongly continuous.

Proposition 5.6.2 *A linear functional x on $B(V)$ for a complex Hilbert space V is σ-weakly continuous if and only if $x = \sum(\cdot v_n|w_n)$ for sequences $\{v_n\}$ and $\{w_n\}$ in V with $\sum \|v_n\|^2 < \infty$ and $\sum \|w_n\|^2 < \infty$. The vectors in the sequences can even by taken to be orthogonal, and if x is non-negative on the positive elements of $B(V)$, then we may in addition let all $v_n = w_n$.*

Proof Any σ-continuous functional x is of the form $\text{Tr}(\cdot T)$ for $T \in B^1(V)$. Polar decompose $T = U|T|$ and write $|T| = \sum \lambda_n u_n \odot u_n$ for a sequence $\{u_n\}$ of orthonormal vectors and $\lambda_n \geq 0$ with $\sum \lambda_n < \infty$. Set $v_n = \lambda_n^{1/2} U(u_n)$ and $w_n = \lambda_n^{1/2} u_n$. Then we get sequences of orthogonal vectors and

$$x(S) = \text{Tr}(ST) = \sum (ST(u_n)|u_n) = \sum (S(v_n)|w_n)$$

for $S \in B(V)$.

Conversely, let W be the closed span of all v_n and w_n, and let $\{u_n\}$ be an orthonormal basis for W. If W is finite dimensional, there are scalars c_{nm} such that $x(S) = \sum c_{nm}(S(u_n)|u_m) = \text{Tr}(ST)$, where T is the finite rank operator that vanishes on $W^\perp$ and is otherwise given by $T(u_m) = \sum c_{nm}u_n$. If W is infinite dimensional, let T_i be the Hilbert-Schmidt operators that vanish on $W^\perp$ and are otherwise given by $T_1(u_n) = v_n$ and $T_2(u_n) = w_n$. Then $T_1T_2^* \in B^1(V)$ and $\mathrm{Tr}(ST_1T_2^*) = \mathrm{Tr}(T_2^*ST_1) = \sum(S(v_n)|w_n)$ for $S \in B(V)$.

For the last statement observe that $(T(v)|v) = \mathrm{Tr}((v \odot v)T) = x(v \odot v) \geq 0$. □

We saw that the weak topology on $B(V)$ is the one induced by $B_f(V) \subset B^1(V)$.

Proposition 5.6.3 *Suppose W is a $*$-subalgebra of $B(V)$ for a complex Hilbert space V with $I \in W$. For $S \in W''$ and $\varepsilon > 0$ and each sequence $\{v_n\}$ in V with $\sum \|v_n\|^2 < \infty$, there is $T \in W$ with $\sum \|(S-T)(v_n)\|^2 < \varepsilon^2$. In particular, the σ-weak closure of W is W''.*

Proof Use the amplification $f\colon B(V) \to B(V^\infty)$ with the latter C*-algebra identified with those infinite matrices having entries in $B(V)$ giving bounded operators on V^∞. The first part of the proof of the bicommutant theorem gives then the first part of this proposition.

For the second part, let x be a σ-continuous functional on $B(V)$, and write $x = \sum(\cdot v_n|w_n)$ as in the previous proposition. Picking $T \in W$ to $S \in W''$ and $\{v_n\}$ and $\varepsilon > 0$ as above, and using the Cauchy-Schwarz inequality, we get

$$|x(S-T)| = |\sum((S-T)(v_n)|w_n)| \leq (\sum \|(S-T)(v_n)\|^2)^{1/2}(\sum \|w_n\|^2)^{1/2},$$

so S belongs to the σ-weak closure of W. □

The *σ-strong topology* on $B(V)$ for a complex Hilbert space V is the weak topology induced by the separating family of seminorms $T \mapsto (\sum_{n=1}^{\infty} \|T(v_n)\|^2)^{1/2}$ with $\sum \|v_n\|^2 < \infty$. It is clearly stronger than the strong and σ-weak topology, but weaker than the norm topology. The last proposition says that W'' is the σ-strong closure of W. The strong and σ-strong topologies coincide on bounded subsets. While the σ-strong and σ-weak topologies are in general distinct as the adjoint operation can fail to be σ-strongly continuous, they have the same continuous linear functionals, as is seen by applying the infinite version of the amplification described above in the proof of Proposition 5.5.2.

Theorem 5.6.4 *Let W be a von Neumann algebra on a Hilbert space V, and let $B(V) \to B^1(V)^\star$ be the isometric isomorphism $T \mapsto \mathrm{Tr}(\cdot T)$. Let $W^\perp$ be the closed subspace of $B^1(V)$ where $\mathrm{Tr}(\cdot T)$ vanishes for all $T \in W$. Then the linear functional f_T on $W_* \equiv B^1(V)/W^\perp$ which for $T \in W$ sends $S + W^\perp$ to $\mathrm{Tr}(ST)$ is bounded, and $T \mapsto f_T$ is an isometric isomorphism from W onto $W_*^\star$.*

Proof The map $T \mapsto f_T$ is clearly well-defined, injective and norm decreasing. As for surjectivity, let $x \in W_*^\star$. If $g\colon B^1(V) \to B^1(V)/W^\perp$ is the quotient map, there is $T \in B(V)$ such that $xg = \mathrm{Tr}(\cdot T)$. For $R \in W^\perp$ we have $\mathrm{Tr}(RT) = xg(R) = 0$, so $T \in W$ as W is weakly closed and thus σ-weakly closed. Clearly $x = f_T$. Moreover, for $\varepsilon > 0$ there is $S \in B^1(V)$ with $\|S\|_1 \leq 1$ such that $|\mathrm{Tr}(ST)| >$

$\| \operatorname{Tr}(\cdot T)\| - \varepsilon$, so $\|x\| \geq |xg(S)| > \|T\| - \varepsilon$, which shows that the bijection $T \mapsto f_T$ is isometric. □

We see that the w^*-topology on a von Neumann algebra W when viewed as the dual of the *predual* W_* is the relative σ-weak topology, and that any σ-weakly continuous linear functional on W is of the form $\operatorname{Tr}(\cdot T)$ for $T \in B^1(V)$. We will later see that the predual is unique, and that every C*-algebra dual to some normed space is a von Neumann algebra.

The *σ-strong* topology* on a von Neumann algebra W is the topology for which a net $\{w_i\}$ converges to 0 if $\sum(\|w_i(v_n)\|^2 + \|w_i^*(v_n)\|^2) \to 0$ for all square summable sequences $\{v_n\}$ in the Hilbert space of W. This is clearly the weakest topology on W that is stronger than the σ-strong topology and makes the adjoint operation continuous. As a locally convex topological space it has W_* as its dual space. The same can be said for the *strong* topology*, where we by definition replace the sequence $\{v_n\}$ by a single vector, and then skip σ in the statements.

5.7 The Kaplansky Density Theorem

Lemma 5.7.1 *Any* $f \in C_b(\mathbb{R})$ is strongly continuous, *meaning that* $f(T_i)$ *converges strongly to* $f(T)$ *whenever the net* $\{T_i\}$ *of bounded self-adjoint operators on any complex Hilbert space converges strongly to* T.

Proof The vector space W of strongly continuous functions under pointwise operations is closed under products when one of the factors is bounded. Hence $W \cap C_0(\mathbb{R})$ is a closed $*$-subalgebra of $C_0(\mathbb{R})$. Consider the functions f and g in the closed unit ball of $C_0(\mathbb{R})$ given by $f(z) = (1+z^2)^{-1}$ and $g(z) = zf(z)$. Since they separate points on $\mathbb{R}$ and do not vanish identically, it follows from the Stone-Weierstrass theorem that $C_0(\mathbb{R}) \subset W$ if we can show $f, g \in W$. But for self-adjoint bounded operators S and T on some Hilbert space, the inequality

$$\begin{aligned}\|g(S)(v) - g(T)(v)\| &= \|(1+S^2)^{-1}(S - T + S(T-S)T)(1+T^2)^{-1}(v)\| \\ &\leq \|(S-T)(1+T^2)^{-1}(v)\| + \|(T-S)T(1+T^2)^{-1}(v)\|\end{aligned}$$

for any vector v shows that g is strongly continuous, and so is f as $f(z) = 1 - zg(z)$. Then every $h \in C_b(\mathbb{R})$ belongs to W as $h(z) = h(z)f(z) + zh(z)g(z)$. □

The following result is known as *Kaplansky's density theorem*.

Theorem 5.7.2 *Let V be a complex Hilbert space and U a C*-subalgebra of $B(V)$ with strong closure W. Then the set of self-adjoint elements of U is strongly dense in the set of self-adjoint elements of W, and those self-adjoint ones in the closed unit ball of U are strongly dense in the set of those in the unit ball of W, and the unit ball of U is strongly dense in the unit ball of W. When $I \in U$, the unitaries of U are strongly dense in the set of unitaries of W.*

Proof For the first claim, if $T^* = T \in W$, there is a net $\{T_i\}$ in U that converges strongly to T. Then $T_i^* \to T^*$ weakly, so $\mathrm{Re}T_i \to T$ weakly, and T is in the strong closure of the convex set of self-adjoint elements of U.

If $T^* = T \in B$ has norm not greater than one, there is a net $\{T_i\}$ among the self-adjoint elements of U that is strongly convergent to T. The function $f \in C_0(\mathbb{R})$ given by $f(z) = z$ on $[-1, 1]$ and $f(z) = 1/z$ elsewhere is strongly continuous by the lemma, so $f(T_i) \to f(T)$ strongly. But $f(T_i)^* = f(T_i) \in U$ has norm not greater than one, and $f(T) = T$ as $\mathrm{Sp}(T) \subset [-1, 1]$.

If T is in the closed unit ball of W, then $S \equiv \begin{pmatrix} 0 & T \\ T^* & 0 \end{pmatrix} = S^*$ has norm not greater than one and belongs to the strong closure of $M_2(U)$, so there is a net $\{S_i\}$ of self-adjoint elements in the closed unit ball of $M_2(U)$ that converges strongly to S. But the net of 12-entries of $\{S_i\}$ is in the closed unit ball of U and converges strongly to T.

If T is a unitary in W, write $T = e^{iR}$ for $R^* = R \in W$. Pick a net $\{R_j\}$ of self-adjoint elements of U that converges strongly to R. Then $e^{iR_j} \to T$ strongly by the lemma. □

Isometric $*$-homorphisms between von Neumann algebras are not always weakly nor strongly continuous. For example, if V is a complex Hilbert space with an orthogonal sequence $\{v_n\}$ such that $\sum \|v_n\|^2 < \infty$, and $f\colon B(V) \to B(V^\infty)$ is the infinite dimensional version of the amplification, then $(\cdot v|v)$ with $v = \oplus n^{-1} v_n$ is a weakly continuous functional on $B(V^\infty)$ while $T \mapsto (f(T)(v)|v) = \sum n^{-2}(T(v_n)|v_n)$ is certainly not. Note however that it is σ-weakly continuous.

Proposition 5.7.3 *If W is a von Neumann algebra and $f\colon W \to B(V)$ is a weakly continuous unital $*$-homomorphism, then $f(W)$ is a von Neumann algebra.*

Proof Clearly $f(W)$ is a C*-algebra containing the identity map on V. Let $T \in W$ with polar decomposition $U|T|$, and assume that $f(T)$ has norm less than one, so $\|f(T)\| < c < 1$ for some scalar c. If P is the resolution of the identity for $|T|$, and A is the Borel set of points in the spectrum of $|T|$ not less than c, then $cP(A) \leq |T|P(A)$ and $cf(P(A)) \leq f(|T|)f(P(A))$, which shows that $c\|f(P(A))\| \leq \|f(|T|)\| < c$ as $|T| = U^*UT$. So $f(P(A)) = 0$ as it is an orthogonal projection with norm less than one. As $\|T(1-P(A))\| \leq c\|1-P(A)\| < 1$ and $f(T) = f(T(1 - P(A)))$, we see that f takes the open unit ball of W onto the open unit ball of $f(W)$.

The closed unit ball W_1 of W is the weak closure of the open unit ball, and as W_1 is weakly compact, so is $f(W_1)$. We claim it is the closed unit ball of $f(W)$. Indeed, if $f(S)$ belongs to the latter, and $r_n \in \langle 0, 1\rangle$ with $r_n \to 1$, then $r_n f(S)$ has norm less than one, and thus equals $f(S_n)$ for $\|S_n\| < 1$. As $r_n f(S) \to f(S)$ in norm, we conclude that $f(S) \in f(W_1)$, which proves our claim.

If $S \neq 0$ belongs to the weak closure of $f(W)$, then $S/\|S\|$ belongs to the weak closure of the closed unit ball of $f(W)$ by Kaplansky's density theorem. By the previous paragraph S then belongs to $f(W)$. □

5.8 Maximal Commutative Subalgebras

Suppose V is a separable complex Hilbert space. Any bounded subset B of $B(V)$ is metrizable in both the weak and strong topologies, for if $\{v_n\}$ is a dense sequence in the unit ball of V, then $d_s(S,T) = \sum 2^{-n}\|(S-T)(v_n)\|$ and $d_w(S,T) = \sum 2^{-n}((S-T)(v_n)|v_n)$ define metrics that induce the strong, respectively, weak topologies on B. Hence B and thus $B(V)$ are weakly separable. We claim that $B(V)$ is actually strongly separable. To see this first note that $B_f(V)$ is strongly dense in $B(V)$ as any strong neighborhood involves only finitely many vectors. Then use that $B_f(V)$ and thus $B_0(V)$ are norm separable, as elements in $B_f(V)$ can be approximated in norm by matrices (in a given countable orthonormal basis of V) having rational entries.

Definition 5.8.1 A commutative $*$-subalgebra of $B(V)$ for a complex Hilbert space V, is *maximal commutative* if it is not contained in strictly larger commutative $*$-subalgebras of $B(V)$.

Any $*$-subalgebra W of $B(V)$ is commutative if and only if $W \subset W'$. The algebra generated by W and $T^* = T \in W'$ will be another commutative $*$-subalgebra of $B(V)$, so W is maximal commutative if and only if it equals its commutant. So maximal commutative $*$-subalgebras of $B(V)$ are von Neumann algebras. Moreover, by Zorn's lemma every commutative von Neumann algebra is contained in a unique maximal commutative one.

Definition 5.8.2 Let W be a $*$-algebra of bounded operators on a complex Hilbert space V. A vector $v \in V$ is *cyclic* if $W(v)$ is norm dense in V, and it is *separating* for W if any $T \in W$ with $T(v) = 0$ must vanish. An orthogonal projection P is *cyclic relative to* W if $W(v)$ is norm dense in $P(V)$ for some $v \in V$.

So W above has a cyclic vector if and only if I is a cyclic orthogonal projection relative to W.

Lemma 5.8.3 *A vector in a complex Hilbert space V is cyclic for a $*$-subalgebra W of $B(V)$ if and only if it is separating for W'.*

Proof If v is cyclic for W, and $T \in W'$ with $T(v) = 0$, then $T(V)$ is contained in the norm closure of $TW(v) = WT(v) = \{0\}$, so $T = 0$.

If P is the cyclic orthogonal projection on the norm closure of $W(v)$ for a separating vector v for W', then $P = I$ as $(I-P)(v) = 0$ and $I - P \in W'$. □

Lemma 5.8.4 *For a $*$-algebra of bounded operators W on a complex Hilbert space V, there is a family of orthogonal projections $P_i \in W'$ that are cyclic relative to W and sum up to one. This family is countable when V is separable.*

Proof By Zorn's lemma there is a maximal family of unit vectors $v_i \in V$ such that $(W(v_i)|v_j) = 0$ when $i \neq j$. If P_i is the orthogonal projection onto the norm closure of $W(v_i)$, then $I = \sum P_i$ under strong convergence because otherwise there is a unit vector v orthogonal to every $W(v_i)$. □

Corollary 5.8.5 *Let V be a separable Hilbert space V. Then any commutative $*$-subalgebra W of $B(V)$ has a separating vector.*

Proof Pick vectors v_n to W as in the previous proof with corresponding cyclic projections $P_n \in W'$. Then $v = \sum 2^{-n} v_n$ is separating for W because if $T \in W$ with $T(v) = 0$, then $TP_n(V)$ is in the norm closure of $TW(v_n) = WP_nT(v) = \{0\}$, so $T = \sum TP_n = 0$. □

A von Neumann algebra is *σ-finite* if every orthogonal family of non-zero orthogonal projections must be countable. So von Neumann algebras acting on separable Hilbert spaces are σ-finite. Note also that a von Neumann algebra W is σ-finite if and only if it has a faithful normal state. Indeed, if x is such a state, and $\{p_i\}$ is an orthogonal family of orthogonal projections in W, then $\sum x(p_i) \leq x(\sum p_i) < \infty$, so only countably many $p_i \neq 0$. For the converse, decompose the Hilbert space into countably many components where W' is cyclic with cyclic unit vectors $\{v_n\}$, and notice that these are separating for W. Then $x = \sum 2^{-n}((\cdot)v_n|v_n)$ is a faithful normal state on W.

The unit ball is then metrizable in the σ-strong and the σ-strong* topologies with metrics d_1 and d_2 given by $d_1(w_1, w_2) = x((w_2 - w_1)^*(w_2 - w_1))^{1/2}$ and $d_2(w_1, w_2) = (x((w_2-w_1)^*(w_2-w_1))+x((w_2-w_1)(w_2-w_1)^*))^{1/2}$, respectively.

Proposition 5.8.6 *For a Radon measure μ on a σ-compact locally compact Hausdorff space X, let $M_f(g) = fg$ denote the* multiplication operator *of $f \in L^\infty(\mu)$ acting on $g \in L^2(\mu)$. Then $f \mapsto M_f$ is a $*$-isomorphism from $L^\infty(\mu)$ onto a maximal commutative $*$-subalgebra of $B(L^2(\mu))$ which contains $M_{C_c(X)}$ as a strongly dense $*$-subalgebra.*

Proof Clearly $f \mapsto M_f$ is a well-defined norm decreasing $*$-homomorphism from $L^\infty(\mu)$ into $B(L^2(\mu))$. It is also isometric, for to $\varepsilon > 0$ pick a compact set A with positive measure consisting only of points where $|f| \geq \|f\|_\infty - \varepsilon$. Letting $g = \bar{f}|f|^{-1}\chi_A$ and $h = \chi_A$ we get $(M_f(g)|h) = \int_A |f|\, d\mu \geq (\|f\|_\infty - \varepsilon)\|g\|_2\|h\|_2$, so $\|M_f\| \geq \|f\|_\infty - \varepsilon$.

Let $g_i, h_i \in C_c(X)$ with $\bar{h}_1 g_1 = \bar{h}_2 g_2$. Pick $e = \bar{e} \in C_c(X)$ that is one on the support on these four functions. For $T \in (M_{C_c(X)})'$ we have

$$(T(g_1)|h_1) = (T(g_1)|M_{h_1}e) = (T(\bar{h_1}g_1)|e) = (T(\bar{h_2}g_2)|e) = (T(g_2)|h_2),$$

so $\bar{h}g \mapsto (T(g)|h)$ is a well-defined function on $C_c(X)$. Picking to $g \in C_c(X)$ a function $e = \bar{e} \in C_c(X)$ that is one on the support of g, we can define a linear functional x on $C_c(X)$ by $x(g) = (T(g)|e)$. Then

$$|x(g)| = |(T(g|g|^{-1/2})||g|^{1/2})| \leq \|T\|\|g|g|^{-1/2}\|_2\||g|^{1/2}\|_2 = \|T\|\|g\|_1$$

with the understanding that $|g|^{-1/2}$ is only taken where $|g| \neq 0$. This shows that x can be extended by continuity to a bounded functional on $L^1(\mu)$, which then is of the form $x = \int \cdot f\, d\mu$ for $f \in L^\infty(\mu)$. As $(M_f(g)|h) = \int g\bar{h}f\, d\mu = (T(g)|h)$ for $g, h \in C_c(X)$, we conclude again by Lusin's theorem that $T = M_f$. We have shown that $(M_{C_c(X)})' \subset W \equiv M_{L^\infty(\mu)}$. Since also $M_{C_c(X)} \subset W \subset W'$, we thus get $W = W' = (M_{C_c(X)})''$. □

Observe that $d(f, g) = \sum 2^{-n}|f - g|(x_n)$ defines a metric on $C_0(X)$ for a locally compact Hausdorff space X with a dense sequence $\{x_n\}$.

Theorem 5.8.7 *Let W be a $*$-algebra of bounded operators on a separable Hilbert space V. Then W is maximal commutative if and only if W is a commutative von Neumann algebra with a cyclic vector if and only if there is a Radon measure μ on a second countable compact Hausdorff space and a surjective isometry $S\colon L^2(\mu) \to V$ such that $f \mapsto SM_fS^*$ is a $*$-isomorphism from $L^\infty(\mu)$ onto W.*

Proof The first forward implication is clear from Lemma 5.8.3 and the corollary above.

For the second forward implication, note that by the discussion at the beginning of this section we know that W is strongly separable. Let U be the C*-algebra generated by a strongly dense sequence of W. We may assume that $I \in U$. Let $G\colon U \to C(X)$ be the Gelfand transform, where now X is a second countable compact Hausdorff space. Let v be a cyclic vector for W, and let μ be the Radon measure representing the positive functional $f \mapsto (G^{-1}(f)(v)|v)$ on $C(X)$.

As v is also cyclic for the strongly dense subset U, the formula $S(f) = G^{-1}(f)(v)$ defines a surjective isometry $S\colon L^2(\mu) \to V$ such that $SM_f = G^{-1}(f)S$ on $L^2(\mu)$ for $f \in C(X)$ by Lusin's theorem. Since $T \mapsto STS^*$ is a strong homeomorphism from $B(L^2(\mu))$ to $B(V)$, we conclude from the previous proposition that $f \mapsto SM_fS^*$ takes $L^\infty(\mu)$ isometrically onto W.

The proof is now completed by using the previous proposition in conjunction with the observation that $T \mapsto STS^*$ takes maximal commutative $*$-subalgebras of $B(L^2(\mu))$ to maximal commutative $*$-subalgebras of $B(V)$. □

Corollary 5.8.8 *Any commutative von Neumann algebra on a separable Hilbert space is $*$-isomorphic to $L^\infty(\mu)$ for some Radon measure μ on a second countable compact Hausdorff space.*

Proof By the previous corollary the von Neumann algebra W has a separating vector v. Let $P \in W'$ be the orthogonal projection onto the closure of $W(v)$. Then $T \mapsto PTP$ is an $*$-isomorphism onto PWP, which is moreover weakly continuous. By proposition 5.7.3 its image is a von Neumann algebra that acts non-degenerately on $\mathrm{im}P$ and which has v as a cyclic vector. Then result now follows from the theorem above. □

Definition 5.8.9 We say that two von Neumann algebras W_i on Hilbert spaces V_i are *spatially isomorphic* if there is a surjective isomorphism $S\colon V_1 \to V_2$ such that $W_2 = SW_1S^*$.

Spatially isomorphism is in general strictly stronger than $*$-isomorphisms as the amplification shows.

Example 5.8.10 The matrices of the form $\begin{pmatrix} aI & bI \\ cI & dI \end{pmatrix}$, where $a, b, c, d \in \mathbb{C}$ and I is the identity map on $\mathbb{C}^3$, is a von Neumann algebra on $\mathbb{C}^6$ that is $*$-isomorphic to $M_2(\mathbb{C})$, but there is no surjective isometric map $\mathbb{C}^6 \to \mathbb{C}^2$ that implements a spatial isomorphism between the two algebras. In fact, the latter algebra has commutant $*$-isomorphic to $\mathbb{C}$ while the former has commutant $*$-isomorphic to $M_3(\mathbb{C})$. ◇

Corollary 5.8.11 *Any two maximal commutative $*$-subalgebras W_i of $B(V)$ for a separable complex Hilbert space V are spatially isomorphic if they are $*$-isomorphic.*

Proof By the theorem there are Radon measures μ_i on compact Hausdorff spaces and surjective isometries $S_i : L^2(\mu_i) \to V$ that implement the isomorphisms from $L^\infty(\mu_i)$ onto W_i. If $A: W_1 \to W_2$ is a $*$-isomorphism, we have a $*$-isomorphism $B: L^\infty(\mu_1) \to L^\infty(\mu_2)$ such that $A(S_1 M_f S_1^*) = S_2 M_{B(f)} S_2^*$ for $f \in L^\infty(\mu_1)$. By the Radon-Nikodym theorem there is a positive $h \in L^1(\mu_1)$ with $\int B(\cdot)\, d\mu_2 = \int \cdot h\, d\mu_1$. The formula $C(h^{1/2} f) = B(f)$ defines by Lusin's theorem a surjective isometry $C: L^2(\mu_1) \to L^2(\mu_2)$ such that $CM_f = M_{B(f)}C$ for $f \in L^\infty(\mu_1)$. Hence the unitary $S = S_2 C S_1^*$ satisfies $S(\cdot)S^* = A$. □

Definition 5.8.12 A normal bounded operator T on a separable Hilbert space is *multiplicity free* if the unital C*-algebra $C^*(T)$ it generates has a cyclic vector.

Since the von Neumann algebra $W^*(T)$ generated by T is the strong closure of $C^*(T)$, we see that T is multiplicity free if and only if $W^*(T)$ is maximal commutative. Then any eigenvalue λ of T has multiplicity one because if it has two orthogonal unit eigenvectors v and w, then the unitary operator S that takes v to w and w to $-v$ and leaves every element of $(\mathbb{C}v + \mathbb{C}w)^\perp$ fixed, belongs to $W^*(T)'$ as $ST = TS$ but is not in $W^*(T)$ as $(W^*(T)(v + w)|v - w) = \{0\}$ while $(S(v + w)|v - w) = -2$.

Letting $U = C^*(T)$ in the proof of the theorem above, we get the following *spatial version of the spectral theorem*.

Corollary 5.8.13 *For a multiplicity free operator T on a Hilbert space V, there is a Radon measure μ on $\mathrm{Sp}(T)$ and a surjective isometry $S: L^2(\mu) \to V$ such that $f \mapsto SM_f S^*$ induces a $*$-isomorphism $L^\infty(\mu) \to W^*(T)$ that extends the $*$-isomorphism $C(\mathrm{Sp}(T)) \to C^*(T)$ given by $f \mapsto f(T)$.*

Corollary 5.8.14 *Two multiplicity free operators T_i on the same Hilbert space are unitarily equivalent if and only if their spectrum coincide and there are cyclic vectors v_i for $C^*(T_i)$ such that the Radon measures μ_i representing $f \mapsto (f(T_i)v_i|v_i)$ for $f \in C(\mathrm{Sp}(T_i))$ are absolutely continuous with respect to each other if and only if there is a $*$-isomorphism $W^*(T_1) \to W^*(T_2)$ that takes T_1 to T_2.*

Proof The first forward implication is obvious. The second forward implication is immediate from the previous corollary as $L^\infty(\mu_1) = L^\infty(\mu_2)$. The proof is then completed using the second last corollary above. □

Corollary 5.8.15 *To a commuting family of normal operators T_i on a separable Hilbert space V, there is a Radon measure μ on a second countable compact Hausdorff space X and a surjective isometry $S\colon L^2(\mu) \to V$ such that $S^*T_iS = M_{f_i}$ for some $f_i \in L^\infty(\mu)$. If the family is countable, we may pick $f_i \in C(X)$.*

Proof Apply the theorem above, or its proof, to the maximal commutative $*$-subalgebra of $B(V)$ that contains all the operators T_i. □

Alternatively, one may cut up the Hilbert space into subspaces where the normal operator is multiplicity free. This leads to the *spatial spectral theorem.*

Theorem 5.8.16 *For a normal bounded operator T on a separable Hilbert space V, there is a sequence of pairwise orthogonal projections P_n that commute with T and has strong sum $\sum P_n = I$. Moreover, there are normalized Radon measures μ_n on $\mathrm{Sp}(T)$ and surjective isometries $S_n\colon L^2(\mu_n) \to P_n(V)$ such that $f \mapsto \sum S_n M_f S_n^*$, with strong sum, is a $*$-isomorphism $L^\infty(\sum 2^{-n}\mu_n) \to W^*(T)$ extending the $*$-isomorphism $C(\mathrm{Sp}(T)) \to C^*(T)$ that sends f to $f(T)$.*

Proof By Lemma 5.8.4 there is a sequence of pairwise orthogonal projections P_n that are cyclic relative to $W^*(T)$ and has strong sum I. If v_n is a cyclic unit vector for $W^*(TP_n)$ on $P_n(V)$, the theorem above provides the required surjective isometry S_n, where μ_n is the Radon measure representing the positive functional $f \mapsto (f(T)v_n|v_n)$ on $C(\mathrm{Sp}(T))$.

The map $A\colon f \mapsto \sum S_n M_f S_n^*$ is clearly a norm decreasing $*$-homomorphism from $L^\infty(\sum 2^{-n}\mu_n)$ to $B(V)$ such that $A(f) = \sum f(TP_n) = \sum f(T)P_n = f(T)$ for $f \in C(\mathrm{Sp}(T))$. It is also injective because if $M_f = 0$ on all $L^2(\mu_n)$, then f is zero almost everywhere with respect to $\mu \equiv \sum 2^{-n}\mu_n$.

Since $C(\mathrm{Sp}(T))$ is w^*-dense in $L^\infty(\mu) = L^1(\mu)^\star$, and $C^*(T)$ is σ-weakly dense in $W^*(T)$ by Proposition 5.6.3, it suffices to show that A is both continuous and open with respect to these topologies.

For continuity, let x be a σ-weakly continuous functional on $B(V)$ corresponding to a positive trace class operator. Pick by Proposition 5.6.2 an orthogonal sequence $\{w_n\}$ in V such that $x = \sum(\cdot(w_n)|w_n)$. With $g_{nm} = S_n^* P_n(w_m) \in L^2(\mu_n)$ and f any bounded Borel function on $\mathrm{Sp}(T)$, we have

$$xA(f) = \sum_{n,m}(S_n M_f S_n^*(w_m)|w_m) = \sum_{n,m}\int f|g_{nm}|^2\,d\mu_n.$$

As μ_n is absolutely continuous with respect to $2^n\mu$, there is by the Radon-Nikodym theorem a positive Borel function h_n on $\mathrm{Sp}(T)$ such that $\mu_n = h_n\mu$. Put $g = \sum|g_{nm}|^2h_n$. If in addition $f \geq 0$, then $xA(f) = \sum\int f|g_{nm}|^2h_n\,d\mu = \int fg\,d\mu$ by Lebesgue's monotone convergence theorem. For $f = 1$, we see that $g \in L^1(\mu)$ with $\|g\|_1 = x(I)$, so $xA(f) = \int fg\,d\mu$ for every $f \in L^\infty(\mu)$.

For openness, if $g \in L^1(\mathrm{Sp}(T))$ and $f \in L^\infty(\mu)$ are non-negative almost everywhere with respect to μ, then as $\sum 2^{-n}h_n = 1$, we get $u_n \equiv S_n((2^{-n}g)^{1/2}) \in P_n(V)$ and $\sum(A(f)u_n|u_n) = \sum 2^{-n}\int fgh_n\,d\mu = \int fg\,d\mu$. □

5.9 Unit Balls and Extremal Points in C*-Algebras

Proposition 5.9.1 *The closed unit ball of a unital C*-algebra is the closed convex hull of its unitaries.*

Proof Suppose an element w in a C*-algebra W with identity 1 has norm less than one. Since the spectrum of $1 - ww^*$ is strictly positive, the element $f(w, z) = (1 - ww^*)^{-1/2}(1 + zw)$ exists and is invertible in W for $z \in \mathbb{C}$ of modulus one. As $w^*(1 - ww^*)^{-1} = w^* \sum (ww^*)^n = (1 - w^*w)^{-1}w^*$, we see that

$$f(w,z)^* f(w,z) + 1 = (1 - ww^*)^{-1} + (1 - w^*w)^{-1}\bar{z}w^* + (1 - ww^*)^{-1}zw + (1 - w^*w)^{-1},$$

which is unaltered under $(w, z) \mapsto (w^*, \bar{z})$. Hence $u(z) \equiv f(w, z) f(w^*, \bar{z})^{-1}$ is unitary. Now

$$g(z) = (1 - ww^*)^{-1/2}(z + w)(1 + zw^*)^{-1}(1 - w^*w)^{1/2}$$

defines a function g on the closed unit disc that is holomorphic in the open unit disc, satisfies $g(z) = zu(\bar{z})$ on the boundary, and equals w at $z = 0$. Hence $w = (2\pi i)^{-1} \int_{|z|=1} g(z)z^{-1}\,dz = (2\pi)^{-1} \int_0^{2\pi} g(e^{it})\,dt$ by functional calculus. As the normalized Haar measure on the unit circle can be approximated by a convex combination of point measures, we see that w is contained in the closed convex hull of the unitary elements of W. □

Lemma 5.9.2 *Let u, v, w be elements of a C*-algebra W such that $u \geq 0$ and $v^*v \leq u^a$ and $ww^* \leq u^b$ for $a + b > 1$. Then the elements $c_n = v(n^{-1} + u)^{-1/2}w$ tend to an element in W with norm not greater than $\|u^{(a+b-1)/2}\|$.*

Proof Let $d_{nm} = (n^{-1} + u)^{-1/2} - (m^{-1} + u)^{-1/2}$. Then

$$\|c_n - c_m\|^2 = \|w^* d_{nm} v^* v d_{nm} w\| \leq \|u^{a/2} d_{nm} ww^* d_{nm} u^{a/2}\| \leq \|d_{nm} u^{(a+b)/2}\|^2.$$

By spectral theory and Lemma 5.5.11 the sequence $\{(n^{-1} + u)^{-1/2}u^{(a+b)/2}\}$ converges to $u^{(a+b)/2}$ in norm, so $\{c_n\}$ converges to an element in W with the required norm estimate since $\|c_n\| \leq \|u^{(a+b-1)/2}\|$ by the same type of argument as above. □

Proposition 5.9.3 *If w, v are elements of a C*-algebra W with $w^*w \leq v$ and $0 < a < 1/2$, there is $u \in W$ with $\|u\| \leq \|v^{1/2-a}\|$ and $w = uv^a$.*

Proof By the lemma $u_n = w(n^{-1} + v)^{-1/2}v^{1/2-a}$ tend to $u \in W$ with norm bounded by $\|v^{1/2-a}\|$. Also $\|w - u_n v^a\|^2 \leq \|v^{1/2}(1 - (n^{-1} + v)^{-1/2}v^{1/2})\|^2 \to 0$ by spectral theory and Lemma 5.5.11. □

Hence any element w in a C*-algebra admits a factorization $w = u|w|^a$ in the C*-algebra when $a \in \langle 0, 1\rangle$.

Theorem 5.9.4 *The extreme points in the positive part of the closed unit ball of a C*-algebra W are the orthogonal projections, whereas the extreme points of the closed unit ball of W consists of all w such that $(1 - ww^*)W(1 - w^*w) = \{0\}$. Then ww^* and w^*w are orthogonal projections and $1 = ww^* + w^*w - w(w^*)^2w$, so W is unital if and only if its closed unit ball has an extreme point.*

Proof A possible identity 1 in W is extreme in the closed unit ball because if $1 = (f + g)/2$ for $\|f\| \leq 1$ and $\|g\| \leq 1$, then f, g may be replaced by their real parts and as they generate a commutative C*-algebra, we may think of them as functions with range in $[-1, 1]$ where the scalar 1 is extreme, so $f = g = 1$.

If $p = (f + g)/2$ for an orthogonal projection $p \in W$ and positive elements v, w in the closed unit ball of W, then $f \leq 2p$ so $(1 - p)f^{1/2}((1 - p)f^{1/2})^* = 0$ and $f = pf = pfp$, and similarly $g = pgp$. But p is extreme, by the previous paragraph, as the identity among the positive elements in the closed unit ball of pWp, so $f = g = p$.

Conversely, if p is an extreme point in the positive part of the closed unit ball of W, by considering p as an extreme point in the C*-algebra generated by p, we may assume that $W = C_0(X)$ for a locally compact Hausdorff space X. If there is x such that $0 < p(x) < 1$, there is an open set A where $0 < p < 1$, so by Urysohn's lemma, there is a continuous function f with support in A and such that $0 \leq p \pm f \leq 1$ on A. Hence $p = ((p + f) + (p - f))/2$ is not extreme. So p is a characteristic function.

Next, if w is an extreme point of the closed unit ball of W, then $p = w^*w$ is an orthogonal projection. If not $w = (w(2 - p^{1/3}) + wp^{1/3})/2$ violates extremeness as $w^*wp^{1/3} \neq w^*w$ by spectral theory; both terms in the bracket are in the closed unit ball since writing $w = up^{1/3}$ with $\|u\| \leq 1$ gives $w(2 - p^{1/3}) = u(2p^{1/3} - p^{2/3})$. Then $q = ww^*$ is also an orthogonal projection and $w = wp = qw$. If $v = (1 - q)a(1 - p)$ is non-zero for $a \in W$ in the closed unit ball, then

$$\|w + v\|^2 = \|pw^*qwp + (1 - p)a^*(1 - q)a(1 - p)\| \in \langle 0, 1]$$

and $w = ((w + v) + (w - v))/2$ violates extremeness.

Conversely, if $(1 - ww^*)W(1 - w^*w) = \{0\}$ for $w \in W$, then $p = w^*w$ satisfies $p(1 - p)^2 = w^*(1 - ww^*)w(1 - w^*w) = 0$, so $p^2 = p$ by spectral theory. Say $w = (b + c)/2$ with $b, c \in W$ in the closed unit ball. Then $p = (pb^*bp + pb^*cp + pc^*bp + pc^*cp)/4$, and each term in the bracket is p as it is an identity, and therefore an extreme point, for pWp. Hence $p(b - c)^*(b - c)p = 0$, so $pb = pc$. Swapping the roles of w and w^*, we see that $q = ww^*$ is also an orthogonal projection such that $qb = qc$. Hence $b - c = (1 - q)(b - c)(1 - p) = 0$, so w is extreme.

Sticking the canonical approximate unit for W between $(1 - q)$ and $(1 - p)$ and taking the limit shows that $ww^* + w^*w - w(w^*)^2w$ is an identity for W. □

Exercises

For Sect. 5.1

1. Show that if U is a closed unital subalgebra of a complex unital Banach algebra W, then $G(U)$ is a union of components of $U \cap G(W)$. Show also that the boundary of $\mathrm{Sp}_U(u)$ lies in $\mathrm{Sp}_W(u)$ when $u \in U$.
2. Suppose w is a member of a complex unital Banach algebra W. Show that if $\mathrm{Sp}(w)$ belongs to an open subset of $\mathbb{C}$, then so does $\mathrm{Sp}(v)$ for any v in some δ-ball centered at w.
3. Use the identity $(vw)^n = v(vw)^{n-1}w$ to show that the elements vw and wv in a complex unital Banach algebra have the same spectral radii.
4. Show that commuting idempotents in a unital complex Banach algebra have distance not less than one.
5. Show that if w is an element of a unital complex Banach algebra W for which there is a sequence $\{v_n\}$ of unit elements such that $\lim \|v_n w\| = 0 = \lim \|v_n w\|$, a so called topological zerodivisor, then w cannot be invertible. Show that the boundary points of $G(W)$ are topological zerodivisors, and that any non-trivial $W \neq \mathbb{C}$ contains topological zerodivisors.
6. Let W be the set of complex polynomials of degree less than n, with product gotten by truncating the polynomials of degree larger than n. Show that W is a Banach algebra with L^1-norm, and that W has only one maximal ideal, which consists only of nilpotent elements.

For Sect. 5.2

1. Show that the set of piecewise linear continuous complex valued functions on a closed real interval J is dense in $C(J)$.
2. Show that any involution on a commutative semisimple complex unital Banach algebra is continuous.
3. Let A be a normal element in a unital C*-algebra, and let $\lambda \in \mathrm{Sp}(A)$. Show that there is a norm one functional x on the C*-algebra such that $x(A) = \lambda$ with x positive in that it is non-negative on positive elements.
4. Show that the spectrum of a product of two positive elements in a commutative unital C*-algebra belongs to $[0, \infty\rangle$.
5. Prove that an invertible element in a unital C*-algebra is unitary if and only if the norm of it and of its inverse is one.

For Sect. 5.3

1. Let X be a locally compact Hausdorff space. Show that $C_0(X)$ has a countable approximative unit consisting of orthogonal projections if and only if X is a union of a sequence of compact open sets.

   A positive element w in a C*-algebra W is said to be strictly positive if W is the closure of wWw.
2. Show that the strictly positive elements in a unital C*-algebra are the invertible positive ones.

3. Show that non-zero positive functionals on C*-algebras are strictly positive on strictly positive elements.
4. Show that a positive compact operator in the C*-algebra of compact operators on a complex Hilbert space is strictly positive if and only if it has dense range.

For Sect. 5.4

1. Suppose we have a sequence of normal bounded operators A_n on a complex Hilbert space that converges to A. Show that for $f \in C(X)$, where X is a compact subset of $\mathbb{C}$ that contains all $\mathrm{Sp}(A_n)$, we have $f(A_n) \to f(A)$.
2. Suppose A_i are bounded normal operators on a complex Hilbert space V with the same spectras X, and that there are $v_i \in V$ such that $C(X)v_1$ and $C(X)v_2$ are dense in V, and that $(f(A_1)v_1|v_1) = (f(A_2)v_2|v_2)$ for all $f \in C(X)$. Deduce that A_1 and A_2 are unitarily equivalent.
3. For a bounded operator A on a complex Hilbert space, we call the set $\Delta(A)$ of $(A(v)|v)$ as we vary over all unit vectors v, the numerical range of A. Show that $\Delta(A)$ is convex and that $\mathrm{Sp}(A)$ is contained in its closure.
4. Show that if A above is normal, then the convex hull of $\mathrm{Sp}(A)$ equals the closure of $\Delta(A)$.
5. A continuous function f on $[0, 1]$ is said to be operator monotone on $[0, 1]$ if $f(A) \leq f(B)$ for bounded operators A, B on a complex Hilbert space having spectra in $[0, 1]$ and such that $A \leq B$. Show that $f(t) = t(1 + at)^{-1}$ is operator monotone on $[0, 1]$ when $a \in \langle -1, 0]$.

For Sect. 5.5

1. Show that the weak and strong topologies coincide on the set of unitary operators on a complex Hilbert space.
2. Let X be a compact subset of $\mathbb{R}$ with $f \in C(X)$, and let V be a complex Hilbert space. Show that $A \mapsto f(A)$ is strongly continuous from the set of self-adjoint bounded operators on V with spectra in X to the set of normal bounded operators on V.
3. Show that the commutant of a union of von Neumann algebras on a complex Hilbert space is the intersection of their commutants.
4. Show that if A is an operator on a complex Hilbert space of norm less than one, then to any unitary B there are unitaries B_1 and C_1 such that $A + B = B_1 + C_1$.
5. Use the previous exercise inductively to show that if we have an operator A on a complex Hilbert space with $\|A\| < 1 - 2/n$ for some integer $n > 2$, there are unitaries B_i such that $nA = B_1 + \cdots + B_n$.

For Sect. 5.6

1. A positive functional x on a von Neumann algebra is completely additive if $x(\sum p_i) = \sum x(p_i)$ for pairwise orthogonal projections p_i with strongly convergent series. Show that x is then σ-weakly continuous.
2. Use the previous exercise to show that $*$-isomorphisms between von Neumann algebras are σ-weakly continuous.

3. Show that an element belongs to a von Neumann algebra if and only if it commutes with the orthogonal projections in the commutant.
4. Produce two normal operators such that the C*-algebra generated by their direct sum does not coincide with the direct sum of the C*-algebras generated by each one of them, while this holds on the von Neumann algebra level. Show that a decomposition in the C*-algebra case holds if and only if the spectra of the two operators do not intersect.

For Sect. 5.7

1. Show that the strong closure of the group of unitaries on a Hilbert space consists of the isometries.
2. Show that the weak closure of the group of unitaries on a Hilbert space is a self-adjoint semigroup containing the isometries and the orthogonal projections.

For Sect. 5.8

1. Show that a C*-algebra is contained in the C*-algebra generated by the spectral projections of its self-adjoint elements.
2. Let W be a self-adjoint unital commutative algebra of bounded operators on a complex Hilbert space with a cyclic vector. Show that the elements in the commutant W' are normal, so W' is commutative and W'' is maximal commutative.
3. Show that the weak and σ-weak topologies coincide on $L^\infty(\mu) \subset B(L^2(\mu))$ for a Radon measure μ on a locally compact σ-compact Hausdorff space.
4. Show that every commutative von Neumann algebra on a separable Hilbert space is isometrically $*$-isomorphic to $L^\infty(\mu)$ for some Radon measure μ on a compact second countable Hausdorff space.
5. Show that the center of a simple C*-algebra W is trivial. If W also contains a maximal abelian subalgebra, show that W acts irreducibly.
6. Prove that any normal bounded operator A is reflexive in that the bounded operators leaving invariant the closed invariant subspaces for A must belong to the von Neumann algebra generated by A; clearly any closed invariant subspace for A will be invariant for an operator in the latter algebra.
7. Show that a von Neumann algebra is σ-finite if its predual is separable.

For Sect. 5.9

1. Let p_i be orthogonal projections in a C*-algebra W. Show that the extreme points in the closed unit ball of the subspace $p_1 W p_2$ consists of those w in the subspace such that $(p_1 - ww^*)W(p_2 - w^*w) = \{0\}$.
2. Let u, v, w be elements of a C*-algebra such that $u^*u \leq vv^* + ww^*$. Show that $uu^* = aa^* + bb^*$ for elements a, b of the C*-algebra with $a^*a \leq v^*v$ and $b^*b \leq w^*w$.

Chapter 6
States and Representations

A positive linear functional on a C*-algebra is one that takes positive elements to non-negative numbers. They are automatically bounded, and among these, are exactly the ones that attain their norm at the unit, or in the non-unital case, attain it as the limit evaluated at any approximative unit; such 'units' always exist. Any bounded linear functional can be written as a difference of two positive functionals, which in the commutative case amounts to the Jordan decomposition of the corresponding complex measure.

A state is a unital positive linear functional. Any state produces a representation, which by definition is a $*$-homomorphism from the C*-algebra W to $B(V)$ for some Hilbert space V. This so called GNS-representation associated to the state is obtained by using the state to define an inner product on a maximal quotient of W. The representation is then given by left multiplication. By the Hahn-Banach theorem there are enough states to assure that the direct sum of the each GNS-representation, known as the universal representation, is injective. Its image is moreover closed, this is true for any $*$-homomorphism between C*-algebras, so we conclude that any abstract C*-algebra can be identified with a closed $*$-subalgebra of the bounded operators on some Hilbert space. Any state on W can in the associated GNS-representation be seen as a vector state with respect to a unit vector, which is cyclic, that is, has a dense orbit. In the second section we study this correspondence in detail, defining what is meant by (quasi-) equivalent representations.

The extreme points, apart from 0, of the convex w^*-compact set of the norm-decreasing positive linear functionals on a C*-algebra are the so called pure states, having the norm-decreasing positive functionals as their closed convex hull. The GNS-representations of the pure states are the irreducible ones, the ones that do not decompose into smaller bits, in other words, do not have proper closed invariant subspaces, or by the transitivity theorem, do not have any invariant subspaces at all. This tendency, of equivalences between topological and algebraic versions of various notions in a C*-algebra, is not surprising due to the spectral nature of

L. Tuset, *Analysis and Quantum Groups*,
https://doi.org/10.1007/978-3-031-07246-8_6

the norm. The interplay between positive linear functionals and representations is investigated further in the third section.

While characters and their maximal ideals were crucial for commutative C*-algebras, the focus in the non-commutative case shifts to irreducible representations and their kernels, characterized as so called primitive ideals. These are special cases of the more common notion of prime ideals among closed ideals. The map which sends an irreducible representation to its kernel descends to the space of unitary equivalence classes of irreducible representations, known as the spectrum of the C*-algebra. The spectrum is normally endowed with the weakest topology making the previous map continuous with respect to the so called Jacobson topology on the set of primitive ideals. In the unital case the Jacobson topology is compact and almost Hausdorff, but only fully Hausdorff in the commutative case. We study this correspondence in section four.

In the next section we study the relation between representations and compact operators, introducing the notions of (post-) liminal C*-algebras. On the way we characterize the finite dimensional C*-algebras as direct sums of full matrix algebras.

In the final section we show how infinite dimensional C*-algebras, the so called AF-algebras, can by constructed as sequential direct limits of finite dimensional ones. We include the very basic properties of AF-algebras, and study in more detail the more restricted class of UHF-algebras. There are uncountable many of them, and thanks to the presence of traces on full matrix algebras, they are simple and admit unique faithful tracial states. Taking the corresponding bicommutants yield the so called hyperfinite von Neumann algebras with trivial centers, so called factors, which moreover have unique faithful tracial states.

6.1 States

Definition 6.1.1 A linear map between C*-algebras is *positive* (and *faithful*) if it sends (non-zero) positive elements to (non-zero) positive elements.

Positive maps take self-adjoint elements to self-adjoint ones in an increasing manner. Clearly $*$-homomorphisms between C*-algebras are positive, and the faithful ones are the injective ones. The trace on $M_n(\mathbb{C})$ is a faithful positive linear functional that is not multiplicative when $n \geq 2$.

Proposition 6.1.2 *Positive linear functionals on C*-algebras are bounded.*

Proof Any positive linear functional x that is bounded on the positive elements in the unit ball of a C*-algebra, is bounded as the elements in the unit ball are sums of four positive elements in the closed unit ball. The hypothesis of the previous statement cannot fail, because if there was a sequence of positive elements v_n in the closed unit ball such that $x(v_n) \geq 2^n$, then $x(\sum 2^{-n} v_n)$ would not be finite. □

Positive linear functionals on commutative C*-algebras are represented by finite Radon measures on the corresponding character spaces.

If x is a positive functional on a C*-algebra V, then $(v, w) \mapsto x(w^*v)$ is a sesquilinear form on V, and $|x(w^*v)| \leq x(w^*w)^{1/2}x(v^*v)^{1/2}$. Observe that $y^* \equiv \overline{y((\cdot)^*)} \in V^\star$ for $y \in V^\star$ has norm $\|y\|$, and that $y = y_1 + iy_2$ with $y_1 = (y+y^*)/2$ and $y_2 = (y - y^*)/2i$ *self-adjoint functionals*, meaning $y_i^* = y_i$. So x is self-adjoint as $x(v^*) = \lim x(v^*u_i) = \lim \overline{x(u_i v)} = \overline{x(v)}$, where $\{u_i\}$ is the canonical approximate unit for V. Also $|x(v)|^2 = \lim |x(u_i v)|^2 \leq \sup_i x(u_i^2)x(v^*v) \leq \|x\|x(v^*v)$.

Proposition 6.1.3 *Let x be a bounded functional on a C*-algebra V. Then x is positive if and only if* $\lim x(u_i) = \|x\|$ *for every approximate unit $\{u_i\}$ of V with increasing positive elements if and only if* $\lim x(u_i) = \|x\|$ *for some such $\{u_i\}$. So the norm of positive functionals is additive, and when V is unital, then x is positive if and only if $x(1) = \|x\|$.*

Proof We may assume that $\|x\| = 1$. For the first forward implication, note that $\lim x(u_i) = \sup x(u_i) \leq 1$, and for v in the closed unit ball of V, we have $|x(u_i v)|^2 \leq x(u_i^2)x(v^*v) \leq \lim x(u_i)$, so $\lim x(u_i) = 1$.

If the last statement in the equivalences holds, and $v^* = v$ with $\|v\| \leq 1$ and $x(v) = a + ib$ for $a \in \mathbb{R}$ and $b \leq 0$, then $\|v - inu_i\|^2 \leq 1 + n^2 + n\|vu_i - u_i v\|$ for $n \in \mathbb{N}$ shows that $|a + ib - in|^2 \leq 1 + n^2$, or $-2bn \leq 1 - a^2 - b^2$, which is impossible unless $b = 0$. Hence for $v \geq 0$ with $\|v\| \leq 1$, the elements $u_i - v$ are self-adjoint and in the closed unit ball, so $1 - x(v) = \lim x(u_i - v) \leq 1$ and $x(v) \geq 0$. □

Definition 6.1.4 A *state* x on a C*-algebra W is a positive linear functional of norm one. If $W \subset B(V)$ for a complex Hilbert space V, and $x = (\cdot(v)|v)$ for a unit vector $v \in V$, then x is called a *vector state*.

So the states on a unital C*-algebra are the unital bounded linear functionals. The following result shows that states separate points in C*-algebras.

Proposition 6.1.5 *For any normal element v in a non-trivial C*-algebra, there is a state x such that $|x(v)| = \|v\|$.*

Proof Let W be the unital C*-subalgebra generated by $v \neq 0$ of the unitization of the C*-algebra V in question. Pick by Gelfand's theorem a character x on W such that $|x(v)| = \|v\|$ and extend it by the Hahn-Banach theorem to a norm one functional x on the unitization of V. The restriction to V then satisfies $|x(v)| = \|v\|$, so $\|x\| \geq 1$ also as a functional on V. □

Proposition 6.1.6 *Any positive linear functional x on a C*-subalgebra of a C*-algebra V has an extension to a positive linear functional on V with norm $\|x\|$.*

Proof If V is the unitization of the C*-subalgebra W where x is defined, then $y(w + a) = x(w) + a\|x\|$ for $w \in W$ and $a \in \mathbb{C}$ defines a positive extension y to V of x with norm $\|x\|$. Indeed, if $\{u_i\}$ is the canonical approximate unit for W, then $|y(w+a)| = |\lim x((w+a)u_i)| \leq \|x\|\|w+a\|$ shows that $\|y\| = \|x\| = y(1)$, so y is positive.

If W is any C*-subalgebra of V, we may by the previous paragraph assume that V has a unit 1 that lies in W. Now extend x on W by the Hahn-Banach theorem to a linear functional y on V with the same norm. As $y(1) = x(1) = \|x\| = \|y\|$, we see that y is positive. □

Corollary 6.1.7 *Any positive linear functional x on a hereditary subalgebra W of a C*-algebra V has a unique positive extension y with norm $\|x\|$. It is actually given by $y = \lim x(u_i \cdot u_i)$, where $\{u_i\}$ is an approximate unit for W consisting of increasing positive elements.*

Proof Existence of y is given by the proposition. For uniqueness note that $|y(v) - x(u_i v u_i)| \leq |y((1-u_i)v)| + |y(u_i v(1-u_i))|$ and applying the Cauchy-Schwartz inequality to the last two terms, gives the formula $y = \lim x(u_i \cdot u_i)$ as $\lim y(1 - u_i) = y(1) - \lim x(u_i) = y(1) - \|x\| = 0$. □

We extend the Jordan decomposition for finite real Borel measures to self-adjoint bounded functionals on C*-algebras.

Proposition 6.1.8 *For any self-adjoint bounded functional x on a C*-algebra, there are positive linear functionals $x_\pm$ with $x = x_+ - x_-$ and $\|x\| = \|x_+\| + \|x_-\|$.*

Proof By Alaoglu's theorem the positive functionals of norm not greater than one on the C*-algebra V in question form a w^*-compact Hausdorff space X, and $v \mapsto f(v)$ with $f(v)(y) = y(v)$ defines a real-linear order preserving map f from the space of self-adjoint elements of V to $C(X)$ which by Proposition 6.1.5 is an isometry.

By the Hahn-Banach theorem there is a real linear functional z on $C(X)$ such that $zf = x$ and $\|z\| = \|x\|$. By Jordan decomposing its representing measure, there are positive functionals $z_\pm$ on $C(X)$ with $z = z_+ - z_-$ and $\|z\| = \|z_+\| + \|z_-\|$. Thus $x_\pm = z_\pm f$ have the required properties. □

6.2 The GNS-Representation

Definition 6.2.1 A *representation* of a C*-algebra W on a complex Hilbert space V is a $*$-homomorphism $W \to B(V)$. It is *non-degenerate* (*cyclic with cyclic vector* v) if its image acts non-denerately on V (has a cyclic vector v in V), and it is *irreducible* if its image acts irreducibly.

Recall that the *direct sum* of a family of representations π_i of a C*-algebra W on V_i is the representation $\oplus\pi_i$ on $\oplus V_i$ given by $\oplus\pi_i(w)(\{v_i\}) = \{\pi_i(w)(v_i)\}$ for $w \in W$ and $v_i \in V_i$. Direct sums of non-degenerate representations are clearly non-degenerate. Conversely, if π is a representation of W on V and we have a family of pairwise orthogonal projections $P_i \in \pi(W)'$ that sum up strongly to the identity map on V, then we get representations π_i of W on $P_i(V)$ by restricting the operators $\pi(w)$ to $P_i(V)$, and π is a direct sum of these *subrepresentations* π_i. When π is non-degenerate, we can always find such P_i that are cyclic, so $\pi = \oplus\pi_i$ with π_i cyclic. If π from the outset failed to be non-degenerate, we could reduce to a non-degenerate representation by restricting the operators $\pi(w)$ to the closed span of $\pi(W)(V)$; a procedure which would not even alter their norm. Note also that if π is non-degenerate and $\{u_i\}$ is an approximate unit for W, then the net $\{\pi(u_i)\}$ converges strongly to the identity map on V.

The GNS-construction is a way to represent a C*-algebra W as operators on a complex Hilbert space V, and goes as follows:

Let x be a positive functional on W. By e.g. the Cauchy-Schwartz inequality the set $L = \{w \in W \mid x(w^*w) = 0\}$ is a closed left ideal of W, and $(v + L|w + L) = x(w^*v)$ is a well-defined inner product on W/L with a continuous extension to the Hilbert space completion V_x of W/L. The linear map $\pi_x(w)\colon W/L \to W/L$ given by $\pi_x(w)(v + L) = wv + L$ is well-defined with norm $\|\pi_x(w)\| \leq \|w\|$ since $\|\pi_x(w)(v + L)\|^2 = x(v^*w^*wv) \leq \|w^*w\|x(v^*v) = \|w^*w\|\|v + L\|^2$. It has therefore a continuous extension $V_x \to V_x$ also denoted by $\pi_x(w)$. Then π_x that sends $w \in W$ to $\pi_x(w)$ is a representation of W on V_x known as the *GNS-representation associated to* x. The *universal representation* of W is the direct sum $\oplus\pi_x$ of π_x over all states x of W.

Theorem 6.2.2 *The universal representation on any C*-algebra is faithful.*

Proof Suppose $\pi : W \to B(V)$ is a universal representation that vanishes on $w \in W$. Pick by Proposition 6.1.5 a state x on W such that $x(w^*w) = \|w^*w\|$, and put $u = (w^*w)^{1/4}$. Then $\|w\|^2 = x(u^4) = \|\pi_x(u)(u + L)\|^2 = 0$ as $\pi_x(u^4) = \pi_x(w^*w) = 0$ and $\pi_x(u) = 0$. □

Corollary 6.2.3 *Every C*-algebra is isometrically $*$-isomorphic to a C*-subalgebra of $B(V)$ for some complex Hilbert space V. When the C*-algebra is separabel, then V can also be taken to be separable.*

Proof The Hilbert spaces of the GNS-representations are clearly all separable, and to get a faithful representation it suffices to take a direct sum of only those π_{x_n} associated to states x_n with $x_n(a_n) = 1$ for any dense sequence $\{a_n\}$ in the subset of positive elements of unit length in the C*-algebra in question. □

Corollary 6.2.4 *Any self-adjoint element w of a C*-algebra W is positive if and only if $x(w) \geq 0$ for every positive functional x on W.*

Proof If $\pi : W \to B(V)$ is the universal representation, and $(\pi(w)(v)|v) \geq 0$ for all $v \in V$, then $\pi(w) \geq 0$, so $w \geq 0$ as π is a $*$-isomorphism onto its image. □

If μ is a finite Radon measure on a locally compact Hausdorff space X, then the GNS-representation π_x of $C_0(X)$ associated to $x = \int \cdot d\mu$ represents elements of $C_0(X)$ as multiplication operators on $L^2(\mu)$.

Proposition 6.2.5 *The GNS-representation π_x of a C*-algebra W associated to a state x has a cyclic unit vector $v_x \in V_x$ such that $x = (\pi_x(\cdot)(v_x)|v_x)$.*

Proof Extend the linear map $w + L \mapsto x(w)$ to a norm decreasing linear functional on V_x, which has the form $(\cdot|v_x)$ for $v_x \in V_x$. As $(w + L|\pi_x(u)(v_x)) = x(u^*w) = (w + L|u + L)$ for $u, w \in W$, we get $\pi_x(u)(v_x) = u + L$, so v_x is cyclic. If $\{u_i\}$ is an approximate unit for W, the net $\{\pi_x(u_i)\}$ converges strongly to the identity map on V_x, so $\|v_x\|^2 = \lim(\pi_x(u_i)(v_x)|v_x) = \lim x(u_i) = \|x\|$. □

So the universal representation of a C*-algebra is non-degenerate.

Proposition 6.2.6 *If x is a state on a C*-algebra W and y is a positive functional on W such that $y \leq x$, meaning that $x - y$ is positive, then there is a unique operator $T \in \pi_x(W)'$ such that $0 \leq T \leq 1$ and $y = (\pi_x(\cdot)T(v_x)|v_x)$.*

Proof The sesquilinear form on V_x sending $(v + L, w + L)$ to $y(w^*v)$ is well-defined with norm not greater than one, so there is $T \in B(V_x)$ with $\|T\| \leq 1$ such that $(T\pi_x(v)(v_x)|\pi_x(w)(v_x)) = y(w^*v)$ for $v, w \in W$, which incidentally also shows that $T \geq 0$ and $T \in \pi_x(W)'$. Replacing w by u_i, where $\{u_i\}$ is the canonical approximate unit for W, and taking the limit we also see that $y = (\pi_x(\cdot)T(v_x)|v_x)$. Plugging in w^*v we get uniqueness of T. □

The converse statement is obvious, namely that any vector state of the form above will be a positive linear functional less than x.

Definition 6.2.7 Two representations π_i on V_i of a C*-algebra are *unitarily equivalent* if there is a unitary $T: V_1 \to V_2$ such that $\pi_2 = T\pi_1(\cdot)T^*$. They are *quasi-equivalent* if there is a $*$-isomorphism $A: \pi_1(W)'' \to \pi_2(W)''$ such that $\pi_2 = A\pi_1$.

Clearly unitary equivalence is stronger than quasi-equivalence, and both are equivalence relations.

Proposition 6.2.8 *Two representations π_i of a C*-algebra on complex Hilbert spaces V_i with cyclic vectors v_i are unitarily equivalent under $T: V_1 \to V_2$ such that $T(v_1) = v_2$ if and only if $(\pi_1(\cdot)(v_1)|v_1) = (\pi_2(\cdot)(v_2)|v_2)$.*

Proof The easily verified identity $\|\pi_1(w)(v_1)\|^2 = \|\pi_2(w)(v_2)\|^2$ and cyclicity of v_i show that $\pi_1(w)(v_1) \mapsto \pi_2(w)(v_2)$ well-defines a linear map that extends by continuity to a unitary $T: V_1 \to V_2$. As $T\pi_1(v)\pi_1(w)(v_1) = \pi_2(vw)(v_1) = \pi_2(v)T\pi_1(w)(v_1)$, we get $T\pi_1(\cdot) = \pi_2(\cdot)T$, and $\pi_2(w)T(v_1) = \pi_2(w)(v_2)$ together with non-degeneracy of π_2 shows that $T(v_1) = v_2$. □

Corollary 6.2.9 *Any representation of a C*-algebra with cyclic unit vector v is unitarily equivalent to the GNS-representation associated to the state $(\pi(\cdot)(v)|v)$.*

Combining the last two propositions we see that if we have two linear functionals x and y on a C*-algebra W with $0 \leq y \leq x$, then the GNS-representation π_y associated to y is unitarily equivalent to the subrepresentation of π_x obtained from the orthogonal projection $P \in \pi_x(W)'$ onto the closure of $\pi_x(W)T^{1/2}(v_x)$.

The following result is immediate from Proposition 5.5.10.

Proposition 6.2.10 *A non-zero representation π of a C*-algebra is irreducible if and only if $(\text{im}\pi)' = \mathbb{C}$, and in this case every non-zero vector is cyclic.*

6.3 Pure States

Definition 6.3.1 A state x on a C*-algebra is *pure* if every positive linear functional it majorizes is of the form tx for $t \in [0, 1]$.

Proposition 6.3.2 *The GNS-representations associated to the pure states are the irreducible ones, and the pure states on commutative C*-algebras are the characters.*

Proof By Proposition 6.2.6 the positive linear functionals that a state x majorize are of the form $(\pi_x(\cdot)T(v_x)|v_x)$ for $T \in (\mathrm{im}\pi_x)'$ with $0 \leq T \leq 1$. Clearly $T = t$ for $t \in [0, 1]$ if and only if $tx = (\pi_x(\cdot)T(v_x)|v_x)$, so x is pure if and only if π_x is irreducible due to Proposition 6.2.10.

If the C*-algebra is commutative, and x is pure, then $\mathrm{im}\pi_x \subset (\mathrm{im}\pi_x)' = \mathbb{C}$, so $B(V_x) = \mathbb{C}$ and $\dim V_x = 1$, which shows that $x = (\pi_x(\cdot)(v_x)|v_x)$ is multiplicative.

Conversely, if x is a character on W majorizing a positive linear functional y, then $\ker x \subset \ker y$ because $|y(w)| \leq y(w^*w)^{1/2}$. So there is t with $y = tx$. Pick w with $x(w) = 1$. Then $0 \leq tx(w^*w) = t \leq x(w^*w) = 1$, so x is pure. □

Example 6.3.3 We claim that the pure states on $B_0(V)$ are the vector states on the non-trivial complex Hilbert space V. To this end first observe that $B_0(V)$ acts irreducible on V. Indeed, if $T \in B_0(V)'$, then $T(v \odot v) = (v \odot v)T$, so $T(v) = c_v v$ with $c_v \in \mathbb{C}$. But c_v must be independent of v since if w is not proportional to v, then $c_{v+w}(v + w) = T(v + w) = c_v v + c_w w$.

Hence any unit vector v in V is cyclic for $B_0(V)$ by Proposition 6.2.10, and by its previous corollary, the GNS-representation of the vector state x of v is unitary equivalent to the identity representation π of $B_0(V)$ on V, so x is pure by the previous proposition.

Conversely, if y is a pure state on $B_0(V)$, there is a positive trace class operator T on V such that $y = \mathrm{Tr}(\cdot T)$, so T is diagonalizable, say with $T(v_i) = \lambda_i v_i$ for an orthonormal basis $\{v_i\}$ of V. Hence $y \geq \lambda_i(\cdot(v_i)|v_i)$, which shows $y = (\cdot(v_i)|v_i)$ by pureness.

The corollary referred to above in this example therefore shows that any non-zero irreducible representation of $B_0(V)$ is unitarily equivalent to π. ◇

Proposition 6.3.4 *The convex w^*-compact set A of norm-decreasing positive linear functionals on a C*-algebra has as its extreme points the pure states and 0, so A is the closed convex hull of these functionals. When the C*-algebra is unital, the set of states is the closed convex hull of the pure states.*

Proof If $0 = z \equiv tx + (1 - t)y$ for $t \in [0, 1]$ and $x, y \in A$, then $0 \geq -tx(w) = (1 - t)y(w) \geq 0$ for positive w, so 0 is an extreme point of A. If instead z is a pure state, then $tx = sz$ for $s \in [0, 1]$. Since $1 = t\|x\| + (1 - t)\|y\|$ we get $\|x\| = \|y\| = 1$ and $t = s$, showing that $x = y = z$, so z is an extreme point of A.

Clearly any non-zero extreme point x of A is a state, and if it majorizes $y \in A$ with $y \neq x$, then $y = \|y\|x$ because $x = \|y\|(y/\|y\|) + (1 - \|y\|)(x - y)/\|x - y\|$.

The second statement is the Krein-Milman theorem. In the unital case clearly the states form a face of A, so the extreme points of this convex w^*-compact set are the pure states. □

Corollary 6.3.5 *To every positive element w of a non-trivial C*-algebra there is a pure state x such that $\|w\| = x(w) = \|\pi_x(w)\|$.*

Proof Let $w \neq 0$ be a positive element of the C*-algebra W. Then $y \mapsto y(w)$ is a w^*-continuous linear map from $W^\star$ to $\mathbb{C}$, and $\|w\| = \sup_{y \in A} y(w)$. By the Krein-Milman theorem the w^*-compact face $\{y \in A \mid y(w) = \|w\|\}$ of A has an extreme point x, which is also an extreme point of A, so x is a pure state. □

Proposition 6.3.6 *Any pure state of a (hereditary) C*-subalgebra of a C*-algebra extends to a (unique) pure state on the bigger algebra.*

Proof The set F of states on the bigger C*-algebra V extending a pure state on the C*-subalgebra W is a w^*-compact face of the set of norm decreasing positive functionals on V, and $F \neq \phi$ by Proposition 6.1.6 with F consisting of a single element when W is hereditary. By the Krein-Milman theorem F admits a non-zero extreme point, which is a pure state on V by Proposition 6.3.4. □

Proposition 6.3.7 *Suppose X is a set of states on a unital C*-algebra W such that when $w = w^* \in W$ and $x(w) \geq 0$ for all $x \in X$, then $w \geq 0$. Then the w^*-closed convex hull of X is the set of all states on W, and the w^*-closure of X contains the pure states on W.*

Proof Since the pure states are the extreme points of the set of all the states, it suffices by Milman's theorem to show that the w^*-closed convex hull Y of X is the set of all states on W. If we have only proper inclusion, there is a state $x \notin Y$. Now all w^*-continuous linear functionals on $W^\star$ correspond to elements of W under the canonical embedding $W \subset W^{\star\star}$, so by the Hahn-Banach separation theorem, there is $w \in W$ and $t \in \mathbb{R}$ such that $\mathrm{Re}\, x(w) < t \leq \mathrm{Re}\, y(w)$ for every $y \in Y$. Let $v = \mathrm{Re}\, w$. Then $y(t - v) \geq 0$, so by hypothesis $t \geq v$ and $t \geq x(v) = \mathrm{Re}\, x(w) < t$. □

Lemma 6.3.8 *Let v_i be n pairwise orthonormal vectors in a complex Hilbert space V, and let $w_i \in V$. Then there is $T \in B(V)$ with $\|T\| \leq \sqrt{2n} \sup \|w_i\|$ such that $T(v_i) = w_i$ for all i. If $S(v_i) = w_i$ for some $S = S^* \in B(V)$ and all i, we can choose T to be self-adjoint.*

Proof For the first statement use $T = \sum w_i \odot v_i$, and for the second statement use $T = SP + PS - PSP$, where P is the orthogonal projection $\sum v_i \odot v_i$. □

The following result is known as the *transitivity theorem*.

Theorem 6.3.9 *Let W be a non-trivial C*-algebras acting irreducible on a complex Hilbert space V, and consider $v_1, w_1, \ldots v_n, w_n \in V$. If $\{v_i\}$ are linear indepedent, there is $T \in W$ such that $T(v_i) = w_i$. We may choose T self-adjoint if there is $S = S^* \in B(V)$ with $S(v_i) = w_i$. And of the form e^{iR} for $R = R^* \in W$ if W is unital and there is a unitary $S \in B(V)$ such that $S(v_i) = w_i$.*

Proof To prove the second statement we may assume that $\{v_i\}$ is orthonormal because we may choose an orthonormal basis $\{u_i\}$ for the vector space U spanned by the v_i's and get $T = T^* \in W$ such that $T(u_i) = S(u_i)$, so $T = S$ on U and $T(v_i) = S(v_i) = w_i$. Clearly we may also assume that $\sup \|w_i\| \leq (2n)^{-1/2}$.

To $\varepsilon > 0$ consider $A_\varepsilon = \{T \in B(V) \mid \sup \|T(v_i)\| < \varepsilon\}$. By irreducibility and Kaplansky's density theorem there is to $T = T^* \in B(V)$ a self-adjoint element $T' \in W$ such that $T' - T \in A_\varepsilon$ and $\|T'\| \leq \|T\|$.

We claim there are self-adjoint elements $T_k \in W$ and $S_k \in B(V)$ in the closed ball of radius 2^{-k} with $T_k - S_k \in A_{2^{-k-1}(2n)^{-1/2}}$ and $S_k(v_i) = (S_{k-1} - T_{k-1})(v_i)$. Assuming this holds for 0 up till k, by the lemma there is $S_{k+1} = S_{k+1}^* \in B(V)$ such that $S_{k+1}(v_i) = (S_k - T_k)(v_i)$ and $\|S_{k+1}\| \leq \sqrt{2n} \sup_i \|S_k(v_i) - T_k(v_i)\| \leq 2^{-k-1}$. So there is $T_{k+1} = T_{k+1}^* \in W$ with $T_{k+1} - S_{k+1} \in A_{2^{-k-2}(2n)^{-1/2}}$ and $\|T_{k+1}\| \leq 2^{-k-1}$.

For the self-adjoint element $T = \sum_{k=0}^{\infty} T_k$ of W, we thus have $w_i - T(v_i) = \lim_m (w_i - \sum_{k=0}^{m} T_k(v_i)) = \lim_m S_{m+1}(v_i) = 0$.

To prove the first statement we cannot assume that S provided by the lemma is self-adjoint. We have seen that there are $\tilde{T}_i = \tilde{T}_i^* \in W$ with $\tilde{T}_1(v_i) = (\mathrm{Re}S)(v_i)$ and $\tilde{T}_2(v_i) = (\mathrm{Im}S)(v_i)$. With $T = T_1 + iT_2 \in W$ we then get $T(v_i) = S(v_i) = w_i$.

For the third statement, we may also assume that $\{v_i\}$ is orthonormal. If U is the linear span of $\{v_i\}$ and $\{w_i\}$, we may extend each of these to orthonormal bases of U denoted by the same symbols. Pick a unitary $S' \in B(U)$ such that $S'(v_i) = w_i$. Then $S'(u_j) = e^{is_j} u_j$ for an orthonormal base $\{u_j\}$ of U and $s_j \in \mathbb{R}$. Then $\sum s_i u_i \odot u_i$ is self-adjoint and sends u_i to $s_i u_i$. By the first part there is therefore $R = R^* \in W$ such that $R(u_i) = s_i u_i$ and $e^{iR}(v_j) = w_j$. □

Irreducibility for representations of C*-algebras in our topological sense is not weaker than irreducibility in a purely algebraic sense, as the following immediate consequence of the theorem shows.

Corollary 6.3.10 *Irreducible representations of C*-algebras have no proper invariant subspaces. In particular, for the GNS-representation of C*-algebra W associated to a pure state x, the Hilbert space is* $W/\{w \in W \mid x(w^*w) = 0\}$.

Let π be a representation of a C*-algebra W_1 on a complex Hilbert space V_1. If W_2 is a C*-subalgebra of W_1 and V_2 is a closed subspace of V_1 invariant under the action of $\pi(W_2)$, then π restricts to a representation of W_2 on V_2, which we denote by π_{W_2,V_2}. If V_2 is the closure of the span of $\pi(W_2)(V_1)$, we write π_{W_2} for π_{W_2,V_2} and call it the *restriction* of π to W_2.

Proposition 6.3.11 *Any non-degenerate representation of a C*-subalgebra* W_2 *of a C*-algebra* W_1 *is unitarily equivalent to* π_{W_2,V_2} *for some non-degenerate representation* π *of* W_1 *acting on a complex Hilbert space* V_1 *with a* $\pi(W_2)$*-invariant closed subspace* V_2*. If the original representation of* W_2 *is cyclic (irreducible), we may take* π *cyclic (irreducible) as well.*

Proof If the original representation is cyclic, extend the vector state on W_2 associated to a cyclic unit vector to a state x on W_1. If $\{u_i\}$ is the canonical approximate unit for W_2 with strong limit $P = P^* = P^2 = \lim \pi_x(u_i) \in \pi_x(W_2)'$, then $\pi = \pi_x$, and V_2 equalling the closure of $\pi(W_2)P(v_x)$, will have the required properties by Proposition 6.2.8.

The non-degenerate case follows now easily by decomposing into cyclic representations. Finally, if the original representation is irreducible, we may assume that x is a pure extension. □

Proposition 6.3.12 *Irreducible representations of C*-algebras restrict to irreducible representations on hereditary C*-subalgebras.*

Proof Consider a non-zero irreducible representation π on V of a C*-algebra with hereditary C*-subalgebra W. If $\{u_i\}$ is the canonical approximate unit for W with $P = \lim \pi(u_i)$ strongly, and $u, v \in P(V)$ with $u \neq 0$, the transitivity theorem provides w in the bigger C*-algebra such that $\pi(w)(u) = v$. Then $u_i w u_i \in W$ as W is hereditary, and

$$\|\pi(u_i w u_i)(u) - v\| \leq \|\pi(w)\pi(u_i)(u) - v\| + \|\pi(u_i)(v) - v\|$$

shows that $v = \lim \pi(u_i w u_i)(u)$, so π_W is irreducible. □

Corollary 6.3.13 *The restriction of a pure state on a C*-algebra to a non-zero hereditary C*-subalgebra W is of the form tx with* $t \in [0, 1]$ *and x a pure state on W.*

Proof We may assume that the restriction of the pure state y on the bigger C*-algebra is non-vanishing. Let $P \in \pi_y(W)'$ be the orthogonal projection of V_y onto the closed span of $\pi_y(W)(V_y)$. It is easily checked that $t = \|P(v_y)\|^2 \in \langle 0, 1]$ and $x = (\pi_y(\cdot)P(v_y)|P(v_y))/t$ have the required properties as π_y restricted to W is non-zero and irreducible by the proposition above. □

Proposition 6.3.14 *If W and V are C*-subalgebras of a C*-algebra with W hereditary and* $W \subset V$, *then* $W = V$ *provided the positive linear functionals that vanish on W also vanish on V.*

Proof We may assume that we have subalgebras of a unital C*-algebra U. Let $0 \leq v \in V$ and $\varepsilon > 0$. Let X be the w^*-compact subset of $U^\star$ of states x on U with $x(v) \geq \varepsilon$. By assymption there is then $w \in W$ with $x(w) \neq 0$. Consider the w^*-open neighborhood A of w of all $y \in U^\star$ such that $y(w) \neq 0$. Cover X by finitely many such neighborhoods and pick w_i in each one of them. Then the w^*-continuous functional $y \mapsto y(\sum w_i^* w_i)$ on $U^\star$ is not less on X than some $r > 0$. Let $u = (\|v\|/r)\sum w_i^* w_i \in W$. Then $x(u) \geq \|v\| \geq x(v)$ for $x \in X$. Hence every positive functional on U is non-negative on $u+\varepsilon-v$, so the latter element is positive by Corollary 6.2.4, and $v \in W$ by Proposition 5.3.9. □

Corollary 6.3.15 *If* L_i *are closed left ideals of a C*-algebra with* $L_1 \subset L_2$ *and such that every positive linear functional vanishes on* L_2 *if it vanishes on* L_1, *then* $L_1 = L_2$.

Proof By Proposition 5.3.6 the sets $L_i \cap L_i^*$ are hereditary C*-subalgebras, and if x is a positive functional vanishing on $L_1 \cap L_1^*$, then it vanishes on L_1 as $\|x(w)|^2 \leq \|x\|x(w^*w)$. So $L_1 \cap L_1^* = L_2 \cap L_2^*$ by the proposition above, and $L_1 = L_2$ by Proposition 5.3.6. □

Proposition 6.3.16 *Any proper closed left ideal L of a C*-algebra is the intersection of the closed left ideals containing L and associated to pure states.*

Proof Let A denote the set of norm-decreasing positive linear functionals on the C*-algebra, and let F be the non-empty w^*-compact face of A consisting of those having associated left ideal containing L. By Proposition 6.3.4 the extreme points of F are pure states since zero would contradict properness of L by the corollary above. The same corollary applied once more therefore gives the result. □

Proposition 6.3.17 *Let L be the closed left ideal associated to a state x on a C*-algebra. Then x is pure if and only if* $\ker x = \{v + w^* \mid v, w \in L\}$.

Proof The inclusion $\{v + w^* \mid v, w \in L\} \subset \ker x$ holds for any positive linear functional x, as is seen by e.g. using an approximative unit and the Cauchy-Schwarz inequality. If x is pure with GNS-representation π_x and cyclic vector v_x, and $w = w^* \in \ker x$ but $w \notin L$, then v_x and $w + L$ are orthogonal, so there is an orthogonal projection $P \in B(V_x)$ such that $P(v_x) = L$ and $P(w + L) = w + L$. By the transitivity theorem there is a self-adjoint element u such that $\pi(u)(v_x) = L$ and $\pi(u)(w + L) = w + L$. Hence $w = wu - (uw - w)^* \in \{v + w^* \mid v, w \in L\}$.

Conversely, if $\ker x = \{v + w^* \mid v, w \in L\}$ and y is a positive linear functional majorized by x, then $\ker x \subset \ker y$, so there is a scalar t such that $y = tx$. If $y \neq 0$ we see that $t > 0$, and if $\{u_i\}$ is the canonical approximate unit, then $t = \|y\| \leq \lim x(u_i) = 1$, so x is pure. □

Definition 6.3.18 A left ideal L of an algebra W is *modular* if there is $u \in W$ such that $v - vu \in L$ for all $v \in W$. If L happens to be a two-sided ideal and if also $v - uv \in L$, then it is *modular* as an ideal.

In a unital algebra all ideals are modular. The quotient algebra by an ideal is unital if and only if the ideal is modular. By Zorn's lemma every proper modular ideal is contained in a maximal ideal. If X is a locally compact Hausdorff space, the set of functions in $C_0(X)$ that vanish at a specific point x is a modular ideal in $C_0(X)$; pick as u any function in $C_0(X)$ that is one at x. This modular ideal is also maximal since it has codimension one. In a Banach algebra closures of proper modular left ideals are proper left ideals, and maximal modular left ideals are closed.

Theorem 6.3.19 *Let* L_x *denote the left ideal associated to a positive linear functional x on a non-trivial C*-algebra W. Then* $x \mapsto L_x$ *is a bijection from the set of pure states on W to the collection of maximal modular left ideals of W.*

Proof If x, y are pure states and $L_x \subset L_y$, then $\ker x \subset \ker y$ by the previous proposition, so $y = tx$ for a scalar t that must be one by evaluating both sides on an approximate unit. Incidentially this also shows that the map is injective.

To see that it has the correct range first observe that $V_x = W/L_x$ by the corollary to the transitivity theorem, so there is $u \in W$ with $v_x = u + L_x$. Then $w + L_x = \pi_x(w)(v_x) = wu + L_x$ for $w \in W$ shows that the left ideal L_x is modular.

If L is a proper left ideal containing L_x, then it is modular and its closure is a proper left ideal of W. By Proposition 6.3.16 this closure is contained in L_y for some pure state y, so $x = y$ by the first line in this proof, showing that L_x is maximal.

On the other hand, any maximal modular left ideal L is proper, so $L \subset L_x$ for some pure state x by Proposition 6.3.16, and $L = L_x$ by maximality. □

6.4 Primitive Ideals and Prime Ideals

Lemma 6.4.1 *Any modular left ideal L of an algebra W has a largest ideal J contained in it, namely $J = \{w \in W \mid wW \subset L\}$; the* ideal associated to L.

Proof Clearly J is an ideal of W. Pick $u \in W$ such that $w - wu \in L$ for all $w \in W$. If $w \in J$, then $wu, w - wu \in L$, so $w \in L$ and $J \subset L$. If J' is another ideal of W contained in L, then $wW \subset J' \subset L$ for $w \in J'$, so $J' \subset J$. □

Definition 6.4.2 The *primitive ideals* of an algebra are the ideals associated to the maximal modular left ideals of the algebra.

Proposition 6.4.3 *An ideal of a C*-algebra is primitive if and only if it is the kernel of a non-zero irreducible representation of the algebra.*

Proof By Theorem 6.3.19 any primitive ideal J of the C*-algebra W is associated to a maximal modular left ideal of the form L_x for some pure state x, so $J = \ker \pi_x$ as $\pi_x(w)(W/L_x) = 0$ means $wW \subset L_x$ for $w \in W$.

Conversely, by Proposition 6.2.10, the corollary prior to it, and Proposition 6.3.2, any non-zero irreducible representation π of W is unitarily equivalent to π_x, where x is any vector state of π, and $\ker \pi = \ker \pi_x$ is associated to the maximal modular left ideal L_x. □

Let hull(X) denote the set of primitive ideals of an algebra containing a subset X of the algebra. We denote by ker(Y) the intersection of a family Y of primitive ideals of an algebra with the convention that ker(ϕ) should be the whole algebra.

Proposition 6.4.4 *Any proper modular ideal of a C*-algebra is contained in a primitive ideal, while any proper closed ideal J is the intersection of the primitive ideals that contain it, so $J =$* ker(hull(J))*. In a C*-algebra modular maximal ideals are primitive.*

Proof By Zorn's lemma every proper modular ideal of the C*-algebra W is contained in a maximal modular left ideal L of W, so it is contained in the primitive ideal associated to L.

By Theorem 6.3.19 and Proposition 6.3.16, the intersection of the modular maximal left ideals of W containing J is J, and if J' is a primitive ideal of W associated to a left ideal L', then $J \subset J'$ if and only if $J \subset L'$. □

From Theorem 6.3.19 we see that $x \mapsto \ker x$ is a bijection from the set of characters on a commutative C*-algebra to the set of its maximal modular ideals, which are morover presicely the primitive ideals.

Definition 6.4.5 A C*-algebra is *primitive* if $\{0\}$ is a primitive ideal.

So a C*-algebra being primitive means that it has a faithful non-zero irreducible representation. Since the identity representation of $B(V)$ on the complex Hilbert space is irreducible, we conclude that $B(V)$ is primitive, but due to the compact operators, it is not simple when V is infinite dimensional. However, all simple C*-algebras are primitive since they admit pure states and hence irreducible representations with trivial kernels.

Definition 6.4.6 A closed ideal J of a C*-algebra is *prime* if $J_1 \subset J$ or $J_2 \subset J$ whenever $J_1 J_2 \subset J$ for closed ideals J_i. A C*algebra is *prime* if the closed ideal $\{0\}$ is prime.

If X_i are subsets of a C*-algebra W with $X_1 W X_2 \subset J$ for a prime ideal J, then X_1 or X_2 are contained in J. To see this just note that the closed ideal J_i equalling the closed span of $W X_i W$, contains X_i as W has an approximate unit.

That W is prime means that pairs of non-zero closed ideals have non-zero intersection because $J_1 \cap J_2$ equals the closed span of $J_1 J_2$ for closed ideals J_i of W. The following result shows that primitive C*-algebras are prime.

Proposition 6.4.7 *Primitive ideals of C*-algebras are prime.*

Proof If L_i are left ideals of a C*-algebra W with $L_1 L_2 \subset L_x$ but that L_2 is not contained in L_x for a pure state x, then there is $w \in L_2$ such that $\pi_x(w)(v_x) = v_x$ by the corollary of the transitivity theorem. If $v \in L_1$, we thus get $\pi_x(v)(v_x) = vw + L_x = L_x$, so $v \in L_x$ and $L_1 \subset L_x$. So, if J_i are closed ideals with $J_1 J_2 \subset \ker \pi_x \subset L_x$, then $J_1 \subset L_x$ or $J_2 \subset L_x$, so J_1 or J_2 is contained in $\ker \pi_x$. □

Theorem 6.4.8 *The* Jacobson topology *on the set* Prim(W) *of primitive ideals of a C*-algebra W is the topology with* hull(ker(X)) *the closure of* $X \subset$ Prim(W). *For every pair of distinct points in* Prim(W) *there is a neighborhood of one of them that does not contain the other point. And* $J \mapsto$ hull(J) *is an inclusion reversing bijection from the set of closed ideals of W to the set of closed subsets of* Prim(W). *When W is unital, then* Prim(W) *is compact.*

Proof An open set in the Jacobson topology is by definition the complement of hull(ker(X)) for $X \subset$ Prim(W). To see that it is a topology observe that the closure of ϕ is ϕ, that X is contained in its closure X and ker(X) = ker(hull(ker(X))), so taking closures twice is the same as taking closures once. That the closure operation respects arbitrary intersections follows straight from definitions, and that it respects finite unions holds if it respects the union of two subsets, which is immediate from definitions and the previous proposition.

The next two statements are immediate from Proposition 6.4.4, and the fact that distinct primitive ideals cannot be contained in each other.

Consider a collection of closed ideals J_i of a unital W such that $\{\text{hull}(J_i)\}$ has only non-empty finite intersections. Then finite sums of J_i are proper ideals and so is the union of all finite sums. By Proposition 6.4.4 this union contains a primitive ideal J, so $J_i \subset J$ and $J \in \text{hull}(J_i)$ for all i. □

Note that $\text{Prim}(B_0(V)) = \{0\}$ is compact, although $B_0(V)$ is non-unital when V is an infinite dimensional complex Hilbert space. In the Jacobson topology the closed points are the maximal ideals, so it is only Hausdorff when the C*-algebra is commutative. Then its 'spectrum', to be defined presently, consists of its characters with the locally compact Hausdorff w^*-topology.

Definition 6.4.9 The *spectrum* of a non-trivial C*-algebra W is the topological space $\hat{W}$ of unitary equivalence classes of non-zero irreducible representations of W with the weakest topology making the *canonical map* $\hat{W} \to \text{Prim}(W)$ given by $[\pi] \mapsto \ker \pi$ continuous.

Remark 6.4.10 Clearly the canonical map $f : \hat{W} \to \text{Prim}(W)$ is a well defined surjection which is automatically open. By the previous theorem the spectrum of a unital C*-algebra is compact. Moreover, it is easy to see that whenever two non-zero irreducible representations of a C*-algebra with the same kernels are unitarily equivalent, then f is a homeomorphism. This happens precisely when for every pair of distinct points in the spectrum, there is a neighbourhod of one point that does not contain the other point. Let $\text{hull}'(\cdot) = f^{-1}(\text{hull}(\cdot))$. Then $J \mapsto \text{hull}'(J)$ is a bijection from the set of closed ideals of W to the closed subsets of $\hat{W}$. ◇

Lemma 6.4.11 *If W is a non-zero hereditary C*-subalgebra of a C*-algebra V and J is a primitive ideal of V not containing W, then $J \cap W$ is a primitive ideal of W. And if K is a closed ideal of V such that $K \cap W \subset J$, then $K \subset J$.*

Proof By Proposition 6.4.3 there is a non-zero irreducible representation π of V such that $\ker \pi = J$, so $J \cap W = \ker \pi_W$, and π_W is non-zero as W is not contained in J. But π_W is irreducible by Proposition 6.3.12, so $J \cap W$ is primitive by Proposition 6.4.3.

Since W is hereditary, we have $WVW \subset W$, so $(WVK)(KVW) \subset K \cap W \subset J$ so $WVK \subset J$ or $KVW \subset J$ as J is prime. Thus $K \subset J$ or $W \subset J$, and the latter inclusion is by assumption ruled out. □

Theorem 6.4.12 *Let U be a non-zero hereditary C*-subalgebra of a C*-algebra W. Let f be the canonical map for U, and let f' be the restriction of the canonical map for W to $\hat{W} \backslash \text{hull}'(U)$. Then $g([\pi]) = [\pi_U]$ and $h(J) = J \cap U$ define homeomorphisms $g : \hat{W} \backslash \text{hull}'(U) \to \hat{U}$ and $h : \text{Prim}(W) \backslash \text{hull}(U) \to \text{Prim}(U)$ such that $fg = hf'$.*

Proof The maps g and h are well-defined by the lemma above, and their domains are non-empty by Propositions 6.3.12 and 6.4.3. It is also clear that $fg = hf'$.

Suppose $(\pi_2)_U = T(\pi_1)_U(\cdot)T^*$ for non-zero irreducible representations π_i of W on V_i and a unitary operator $T : V_1 \to V_2$. Pick unit vectors v_i in the closed linear span of $\pi_i(U)(V_i)$ with $T(v_1) = v_2$. Then $\pi_i(W)(v_i) = V_i$ by the transitivity theorem, and $x_i = (\pi_i(\cdot)(v_i)|v_i)$ are pure states on W such that π_i is unitarily equivalent to π_{x_i}. Clearly both x_i extend $((\pi_1)_U(\cdot)(v_1)|v_1) = ((\pi_2)_U(\cdot)(v_2)|v_2)$, so $x_1 = x_2$ by Corollary 6.1.7 and g is injective.

To see that g is surjective, we may by Proposition 6.3.11 assume that any element of $\hat{U}$ is of the form $[\pi_{U,V}]$ for an irreducible representation π of W and V is the closure of $\pi(U)(v)$ for some vector $v \in V$. Pick $u \in U$ such that $\pi(u)(v) = v$. To any vector v' in the representation space of π, pick $w \in W$ by the transitivity theorem such that $v' = \pi(w)(v)$. Then $\pi(U)(v') = \pi(Uwu)(v) \subset \pi(U)(v)$ as U is hereditary. Hence $\pi_{U,V} = \pi_U$ and $g([\pi]) = [\pi_{U,V}]$.

From $fg = hf'$ we thus see that h is surjective, and it is injective by the lemma above. It remains to show that h is a homeomorphism. As $h^{-1}(\mathrm{hull}_U(J)) = \mathrm{hull}(J) \cap (\mathrm{Prim}(W)\backslash\mathrm{hull}(U))$ for any closed ideal J of B, continuity of h is immediate from the theorem above. It also implies openness of h as surjectivity of h and the lemma above show that $h(\mathrm{hull}(K)\backslash\mathrm{hull}(U)) = \mathrm{hull}_U(K \cap U)$ for any proper closed ideal K of W. □

As $h(\{0\}) = \{0\}$ for h as in the theorem, we see that non-trivial hereditary C*-subalgebras of primitive C*-algebras are primitive. The primitive C*-algebra $B(V)$ contains commutative C*-subalgebras of dimension greater than one when the complex Hilbert space V has dimension greater than one, and these are not primitive.

Proposition 6.4.13 *Let $k\colon W \to V$ be a surjective $*$-homomorphism between C*-algebras, and let f be the canonical map for V and f' the restriction of the canonical map for W to* $\mathrm{hull}'(\ker k)$. *Then $g([\pi]) = [\pi k]$ and $h(J) = k^{-1}(J)$ define homeomorphisms $g\colon \hat{V} \to \mathrm{hull}'(\ker k)$ and $h\colon \mathrm{Prim}(V) \to \mathrm{hull}(\ker k)$ with $hf = f'g$.*

Proof The maps g, h are well-defined by Proposition 6.4.3 and surjectivity of k. It is easy to see that they are bijective and that $hf = f'g$, so it suffices to show that h is a homeomorphism. By Theorem 6.4.8 closed subsets of $\mathrm{hull}(\ker k)$ are of the form $\mathrm{hull}(J)$ for some closed ideal J of W containing $\ker k$. Then $h^{-1}(\mathrm{hull}(J)) = \mathrm{hull}(k(J))$ is closed by the same theorem, so h is continuous. Since $h(\mathrm{hull}(K)) = \mathrm{hull}(k^{-1}(K))$ for any closed ideal K of V, the same theorem shows that h is open. □

Let u be a unitary element in the unitization of a C*-algebra W. Then $x(u \cdot u^*)$ is evidently a pure state on W whenever x is. A set X of pure states on W is *unitarily invariant* if $x(u \cdot u^*) \in X$ for all $x \in X$ and unitaries u. Below we will see that its annihilator $X^{\perp}$ is then a closed ideal of W. If J is a closed ideal of W we write J_p for the set of pure states on W that kill J. Clearly J_p is a unitarily invariant w^*-closed subset of the set of pure states on W.

Lemma 6.4.14 *If x is a pure state on a C*-algebra with $\cap \ker \pi_i \subset \ker x$ for a family of representations π_i, then x belongs to the w^*-closure of the set consisting of the states $(\pi_i(\cdot)(v)|v)$ for all i and all unit vectors v.*

Proof By restriction we may assume that all π_i are non-degenerate. Passing to the quotient of W by the closed ideal $\cap \ker \pi_i$, we may also assume $\cap \ker \pi_i = \{0\}$. Further reflection shows that we may also assume that W is unital.

If $w = w^* \in W$ and all $(\pi_i(w)(v)|v) \geq 0$, then $\oplus \pi_i(w) \geq 0$, so $w \geq 0$, and the result follows from Proposition 6.3.7. □

Theorem 6.4.15 *Let W be a C*-algebra. Then the annihilator $X^\perp$ of a non-empty unitarily invariant set X of pure states on W is the intersection of $\ker \pi_x$ over $x \in X$. If X in addition is w^*-closed in the relative topology on the set of pure states on W, then $X = (X^\perp)_p$. For any closed ideal J of W we have $J = (J_p)^\perp$. So $X \mapsto X^\perp$ is a bijection from the family of unitarily invariant sets that are w^*-closed in the relative topology on the set of pure states on W to the family of closed ideals of W.*

Proof Clearly $\cap_{x\in X} \ker \pi_x \subset X^\perp$. If the inclusion is proper, there is $w \in X^\perp$ with $\pi_x(w) \neq 0$ for some $x \in X$. Pick a unit vector $v \in V_x$ with $(\pi_x(w)(v)|v) \neq 0$. Considering π_x as an irreducible representation of the unitization of W, there is by the transitivity theorem a unitary element u there such that $\pi_x(u)(v) = v_x$. Then $x(uwu^*) = (\pi_x(w)(v)|v) \neq 0$ and yet $x(u \cdot u^*) \in X$.

The inclusion $X \subset (X^\perp)_p$ in the second statement is obvious. For the reverse inclusion, note by the lemma above that any $y \in (X^\perp)_p$ is the w^*-limit of states of the form $(\pi_x(\cdot)(v)|v)$ for $x \in X$ and unit vectors $v \in V_x$, which belong to X by the previous paragraph. Hence $y \in X$ as the latter set is by assumption w^*-closed in the relative topology on the set of pure states on W.

For the third statement, by Proposition 6.3.16 a proper closed ideal J of W may be written as $J = \cap_{x\in X} L_x$ with X the set of pure states x on W with $J \subset L_x$. Note that X is non-empty unitarily invariant and w^*-closed in the relative topology on the set of pure states on W. As $\ker \pi_x$ is the largest ideal contained in L_x, we have $J \subset L_x$ if and only if $J \subset \ker \pi_x$, so $J = \cap_{x\in X} \ker \pi_x = X^\perp$. Hence $J_p = (X^\perp)_p = X$ and $(J_p)^\perp = X^\perp = J$.

The last statement is now clear. □

6.5 Postliminal C*-Algebras

Definition 6.5.1 A C*-algebra is *liminal* (*postliminal*) if the image of every non-zero irreducible representation is contained in (intersects non-trivially with) the compact operators.

Liminal C*-algebras are obviously postliminal, and proposition 4.5.12 shows that the image of a non-zero irreducible representation of a liminal (postliminal) C*-algebra equals (contains) the compact operators. Commutative C*-algebras are liminal as the image of any non-zero irreducible representation is $\mathbb{C}$.

Any finite dimensional C*-algebra W is liminal. Indeed, if π is a non-zero irreducible representation of W on V, then $V = \pi(W)(v)$ for any unit vector v, so V is finite dimensional, and $\pi(W) = B_0(V) = B(V)$. Combining this with the fact that the direct sum of the GNS-representations of W associated to the pure states is faithful by Corollary 6.3.5, we get the following result.

Proposition 6.5.2 *Every finite dimensional C*-algebra is $*$-isomorphic to a finite direct sum of full matrix algebras.*

From Example 6.3.3 we see that the compact operators is a liminal C*-algebra. However, the C*-algebra of all bounded operators is not liminal when the complex Hilbert space is infinite dimensional, since the identity representation is irreducible.

Proposition 6.5.3 *Quotient C*-algebras and C*-subalgebras of liminal C*-algebras are liminal.*

Proof Any non-zero irreducible representation of a C*-subalgebra W is by Proposition 6.3.11 unitarily equivalent to $\pi_{W,V}$ for a non-zero irreducible representation π of the bigger C*-algebra. Since $\pi(W)$ are all compact operators (as the latter form an ideal in the C*-algebra of bounded ones), so are their restrictions to V, so W is liminal.

If $f: W \to U$ is a surjective $*$-homomorphism between C*-algebras and π is a non-zero irreducible representation of U, then $\pi(U) = \pi f(W)$ consists only of compact operators as πf is a non-zero irreducible representation. □

Unital liminal C*-algebras have only finite dimensional irreducible representations because any non-zero irreducible representations is non-degenerate and take the identity element to the identity map on the Hilbert space, and the latter map can only be compact when the Hilbert space is finite dimensional. If V is an infinite dimensional complex Hilbert space, then the unital C*-algebra $W \equiv B_0(V) + \mathbb{C}$ is thus not liminal since the identity representation is irreducible. But $B_0(V)$ is a liminal closed ideal of it with liminal quotient $W/B_0(V) = \mathbb{C}$.

For postliminal C*-algebras we have a more symmetric situation.

Proposition 6.5.4 *Suppose J is a closed ideal of a C*-algebra W. Then W is postliminal if and only if J and W/J are postliminal.*

Proof As the compact operators form an ideal in the C*-algebra of bounded ones, to prove that J is postliminal when W is, we may by Proposition 6.3.11 assume that any non-zero irreducible representation of J is of the form $\pi_{J,V}$ for a non-zero irreducible representation π of W. Let v be a unit vector of V, and extend the orthogonal projection $P \in B(V)$ onto $\mathbb{C}v$ to an orthogonal projection P' on the bigger representation space that vanishes on $V^\perp$. We have $\pi_{J,V}(J)(v) = V$ by the transitivity theorem, so $v = \pi_{J,V}(u)(v)$ for some $u \in J$. Since P' is compact, it is $\pi(w)$ for some w in the postliminal C*-algebra W. Hence $P = \pi(u)P = \pi_{J,V}(uw)$ as $uw \in J$, so $\pi_{J,V}(J)$ contains the compact operators, showing that J is postliminal. That W/J is also postliminal, is obvious.

For the opposite direction, let π be a non-zero irreducible representation of W on a complex Hilbert space V. If $J \subset \ker\pi$, there is a $*$-homomorphism $f: W/J \to B(V)$ such that $\pi = fg$, where $g: W \to W/J$ is the quotient map. Since W/J is postliminal, we thus get $B_0(V) \subset f(W/J) = \pi(W)$. If $\ker\pi$ does not contain J, then π_J is non-zero and irreducible by Proposition 6.3.12, so $\pi_J(J)$ contains the compact operators. Hence in this case there is a non-zero element u of J such that $\pi(u)$ is compact. In either case $\pi(W)$ intersects non-trivially with the compact operators. □

So adjoining an identity element to a C*-algebra does not alter the fact that the algebra is postliminal or not. As any non-zero representation of a simple C*-

algebra W is faithful, we may identify W as a C*-algebra with its image. If W is postliminal, then since the compact operators form an ideal in the image of any non-zero irreducible representation of W, it is clear that postliminal simple C*-algebras are $*$-isomorphic to the compact operators on some complex Hilbert space. So infinite dimensional unital simple C*-algebras are not postliminal. By Proposition 4.5.3 the Calkin algebra of a separabel infinite dimensional complex Hilbert space V is simple and clearly unital, so it cannot be postliminal, and neither can $B(V)$ by the previous result.

Proposition 6.5.5 *Non-zero irreducible representations of a postliminal C*-algebra are unitarily equivalent if and only their kernels coincide.*

Proof If $\ker \pi_1 = \ker \pi_2$ for non-zero irreducible representations π_i of a postliminal C*-algebra W on V_i, the map $f: \pi_1(W) \to \pi_2(W)$ given by $f(\pi_1(w)) = \pi_2(w)$ is a well-defined $*$-isomorphism. As $B_0(V_i) \subset \pi_i(W)$ it makes sense to claim that $f(B_0(V_1)) \subset B_0(V_2)$, but if $P \in B(V_1)$ is a rank-one orthogonal projection, then $f(P)$ is also a rank-one orthogonal projection as $f(P)B_0(V_2)f(P) = \mathbb{C}f(P)$. By symmetry f restricts to a $*$-isomorphism $B_0(V_1) \to B_0(V_2)$. By Example 6.3.3 this irreducible representation is unitarily equivalent to the identity representation, so there is unitary $T: V_1 \to V_2$ such that $f = T \cdot T^*$ on $B_0(V_1)$. We complete the proof by showing that this holds also on $\pi_1(W)$. The elements in $B_0(V_2)$ are of the form $T\pi_1(u)T^*$ with $\pi_1(u) \in B_0(V_2)$ for $u \in W$. For $w \in W$ we therefore get $f(\pi_1(w))T\pi_1(u)T^* = f\pi_1(wu) = T\pi_1(w)T^*T\pi_1(u)T^*$, and $f(\pi_1(w)) = T\pi_1(w)T^*$ by letting $T\pi_1(u)T^*$ be rank-one operators. □

From this result and Remark 6.4.10 we see that for a non-trivial postliminal C*-algebra W the canonical map $\hat{W} \to \mathrm{Prim}(W)$ is a homeomorphism.

6.6 Direct Limits

Definition 6.6.1 A *direct system of C*-algebras* is a family of C*-algebras W_i indexed by an upward filtered ordered set and $*$-homomorphisms $f_{ij}: W_i \to W_j$ such that $f_{jk}f_{ij} = f_{ik}$ for $i < j < k$. Given such a system consider the $*$-subalgebra W of $\prod W_i$ consisting of the elements $w = \{w_i\}$ for which there is i such that $f_{jk}(w_j) = w_k$ when $i < j < k$. Then $\|w\| = \lim \|w_i\|$ is clearly a seminorm on W satisfying the C*-identity. Dividing out by the $*$-ideal where it vanishes and taking the completion gives a C*-algebra $\lim_{\to} W_i$ called the *direct limit* of the system. When the upward filtered ordered set is $\mathbb{N}$ we speak of a *direct sequence* of C*-algebras, and in this case the direct system can be constructed uniquely from the $*$-homomorphisms $f_{n,n+1}$.

Given a direct system $f_{ij}: W_i \to W_j$ of C*-algebras, the map that sends $w \in W_i$ to $\{w_k\}$, where $w_i = w$ and $w_k = f_{ik}(w)$ for $k > i$ and $w_k = 0$ for $k < i$, defines a $*$-homomorphism $f_i: W_i \to \lim_{\to} W_i$. These *natural maps* satisfy $f_j f_{ij} = f_i$ for $i < j$. It is easy to see that if V is another C*-algebra with $*$-homomorphisms $g_i: W_i \to V$ satisfying $g_i f_{ij} = g_j$, there is a unique $*$-

homomorphism $h\colon \lim_{\to} W_i \to V$ such that $hf_i = g_i$ for all i. Notice also that $\cup f_i(W_i)$ is dense in $\lim_{\to} W_i$, and that $\|f_i(w)\| = \lim_j \|f_{ij}(w)\|$.

Proposition 6.6.2 *Direct limits of simple C*-algebras are simple.*

Proof Observe that a C*-algebra is simple if and only if any non-zero $*$-homomorphism from it is injective and hence isometric. Restricting any non-zero $*$-homomorphism g from the direct limit $\lim_{\to} W_i$ to the images of the natural maps f_i we eventually get non-zero and hence isometric ones, rendering g isometric also on the closure, so the direct limit is simple. □

Definition 6.6.3 An *AF-algebra* is a direct limit of a direct sequence of finite dimensional C*-algebras.

Equivalently, an AF-algebra is a C*-algebra with a dense increasing countable union of finite direct sums of full matrix algebras. So they are closed linear spans of their orthogonal projections. The commutative ones are of the form $C_0(X)$ with X *totally disconnected*, that is, the non-empty connected subsets are the points.

Example 6.6.4 The compact operators on a separable complex Hilbert space V is AF. Indeed, pick an orthonormal basis $\{v_n\}$ and consider the approximate unit for $B_0(V)$ of orthogonal projections $P_n = \sum^n v_i \odot v_i$. Then $\cup P_n B_0(V) P_n$ is dense in $B_0(V)$, and $P_n T P_n = \sum (T(v_m)|v_i) v_i \odot v_m$ for $T \in B_0(V)$. ◇

The following result implies that a C*-algebra is AF if and only if its unitization is AF.

Proposition 6.6.5 *Any closed ideal of an AF-algebra is AF and so is the quotient C*-algebra.*

Proof Suppose J is a closed ideal of an AF-algebra W with a dense increasing countable union of finite dimensional C*-algebras W_n. Then their images under the quotient map $W \to W/J$ form a dense increasing union of finite dimensional C*-algebras in W/J, so the latter is AF.

Next, set $J_n = J \cap W_n$, so the closure $K \subset J$ of their union is a closed ideal of W. It remains to show that $K = J$, or that the well-defined $*$-homomorphism $f\colon W/K \to W/J$ given by $f(w + K) = w + J$ is isometric. Invoking the canonical $*$-isomorphisms $g\colon (W_n+J)/J \to W_n/(W_n \cap J)$ and $h\colon (W_n+K)/K \to W_n/(W_n \cap K)$ and using $W_n \cap K = W_n \cap J$, we see that the restriction of f to $(W_n + K)/K$ is the isometric $*$-homomorphism $g^{-1}h$. □

A positive functional x on a C*-algebra W is *tracial* if $x(w^*w) = x(ww^*)$ for all $w \in W$. By the polarization identity this is equivalent to requiring $x(vw) = x(wv)$ for all $v, w \in W$. Note that the left kernel L_x of x is a two-sided closed ideal of W, so if W is simple and x is non-zero, then x is faithful.

Example 6.6.6 The normalized trace on $M_n(\mathbb{C})$ is the only (faithful) tracial state because $M_n(\mathbb{C})$ is spanned by its rank one projections $v \odot v$, and if T is a unitary matrix, then $T(v) \odot T(v) = T(v \odot v)T^*$.

This argument shows that a tracial state x on $B_0(V)$ takes the same value $s > 0$ on all rank one projections as $B_0(V)$ is simple. So we get $1 \geq x(\sum_{i=1}^n v_i \odot v_i) = ns$

for an orthonormal basis $\{v_i\}$ of V, which becomes an absurdity when the complex Hilbert space V is infinite dimensional. In this case $B_0(V)$ does therefore not admit a tracial state. ◇

Proposition 6.6.7 *Suppose a unital C^*-algebra W has a dense increasing union of C^*-subalgebras W_i containing the same identity element as W. If each W_i admits a unique tracial state, then W has a unique tracial state.*

Proof Any tracial state x on W restricts to the unique tracial state x_i on W_i, so x is unique by denseness of $\cup W_i$ in W. We can also use this to consistently define a norm-decreasing unital functional x on $\cup W_i$, namely by setting $x(w) = x_i(w)$ if $w \in W_i$, and extend x by continuity to a tracial state on W. □

Definition 6.6.8 A *UHF-algebra* is a unital C*-algebra W with a dense increasing countable union of full matrix C^*-subalgebras containing the same identity element as W.

Equivalently, a UHF-algebra is an AF-algebra where the $*$-homomorphisms $f_n \colon M_n(C) \to M_m(\mathbb{C})$ in the direct sequence are unital. Note that this is only possible if n divides m. By Proposition 6.6.2 and the previous proposition and example, we see that UHF-algebras are simple C*-algebras that admit unique faithful tracial states. We shall see that there is an abundance of UHF-algebras.

Lemma 6.6.9 *Say p, q are orthogonal projections in a unital C^*-algebra W with $\|p - q\| < 1$. Then $w = 1 - p - q + 2qp$ is invertible, and the unitary element $u = w|w|^{-1} \in W$ satisfies $q = upu^*$ and $\|1 - u\| \leq \sqrt{2}\|p - q\|$.*

Proof The formula $w^*w = 1 - (p - q)^2$ shows that w is normal and invertible, so $|w|$ is invertible, while $wp = qp = qw$ shows that p commutes with $|w|$, so $qu = up$. Now $\mathrm{Re}\, w = 1 - (p - q)^2 = |w|^2$, so $\mathrm{Re}\, u = (\mathrm{Re}\, w)|w|^{-1} = |w|$ and

$$\|1-u\|^2 = \|(1-u^*)(1-u)\| = 2\|1-\mathrm{Re}\, u\| = 2\|1-|w|\| \leq 2\|1-|w|^2\| = 2\|p-q\|^2.$$

□

Lemma 6.6.10 *If w is a self-adjoint element of a C^*-algebra W such that $\|w - w^2\| < 1/4$, there is an orthogonal projection $p \in W$ with $\|w - p\| < 1/2$.*

Proof We may assume that $W = C_0(X)$ for a locally compact Hausdorff space X. If p is the characteristic function of the open compact set $|w|^{-1}(\langle 1/2, \infty\rangle)$, we get $|w(x) - p(x)| < 1/2$ for all $x \in X$, so $\|w - p\|_u < 1/2$. □

The following result by Glimm concerns the characterization of UHF-algebras by *supernatural numbers* (like f_s below).

Theorem 6.6.11 *Let S be the set of functions $s \colon \mathbb{N} \to \mathbb{N}$, and define $s! \in S$ by $s!(n) = s(1)s(2)\cdots s(n)$. Let W_s denote the UHF-algebra of the direct sequence of amplifications $M_{s!(n)}(\mathbb{C}) \to M_{s!(n+1)}(\mathbb{C})$. If W_s and W_t are $*$-isomorphic for $s, t \in S$, then $f_s = f_t$, where f_s is the function from the prime numbers to $\mathbb{N}\cup\{0, \infty\}$ that sends p to the largest exponent of p that divides $s!(n)$ for some n. In particular, there are uncountably many non-$*$-isomorphic such UHF-algebras.*

Proof Let $g\colon W_s \to W_t$ be a $*$-isomorphism, let x be the unique tracial state on W_t, and let $f_n^r\colon M_{r!(n)}(\mathbb{C}) \to W_r$ denote the natural maps for $r \in S$.

By symmetry it suffices to show $f_s \leq f_t$, or that for each n there is m such that $s!(n)$ divides $t!(m)$. Let P be a rank one orthogonal projection in $M_{s!(n)}(\mathbb{C})$, so $gf_n^s(P) \in W_t$ is an orthogonal projection. By denseness pick m and $T = T^* \in M_{t!(m)}(\mathbb{C})$ such that $\|gf_n^s(P) - f_m^t(T)\| < 1/8$ and $\|gf_n^s(P) - f_m^t(T^2)\| < 1/8$, so $\|T - T^2\| < 1/4$. The last lemma gives an orthogonal projection $Q \in M_{t!(m)}(\mathbb{C})$ such that $\|T - Q\| < 1/2$, so $\|gf_n^s(P) - f_m^t(Q)\| < 1$. By the second last lemma the orthogonal projections $gf_n^s(P)$ and $f_m^t(Q)$ are unitarily equivalent, so $xf_m^t(Q) = xgf_n^s(P) = 1/s!(n)$, while the left hand side is a multiple of $1/t!(m)$ as xf_m^t and xgf_n^s are tracial states.

For the last statement, consider a sequence of prime numbers p_n. To $s \in S$ define $\bar{s} \in S$ by $\bar{s}(n) = p_n^{s(n)}$, so $f_{\bar{s}}(p_n) = s(n)$. Then $s \mapsto [W_{\bar{s}}]$ is an injection from the uncountable set S to the $*$-isomorphism classes of C*-algebras $\{W_t\}_{t \in S}$. □

Definition 6.6.12 The *center* $Z(W)$ of a von Neumann algebra W is $W' \cap W$, and consists of the central elements in W. *Factors* are von Neumann algebras with trivial center. A von Neumann algebra is *hyperfinite* if it has a weakly dense C*-subalgebra which is a UHF-algebra.

Example 6.6.13 The C*-algebra $B(V)$ for a complex Hilbert space V is a factor. When V is separable, it is hyperfinite. Indeed, pick an infinite dimensional UHF-algebra W represented on a complex Hilbert space V by an irreducible representation. As any unit vector is cyclic and W is separable, we know that V is separable and infinite dimensional, and by irreducibility we know that $W' = \mathbb{C}$, so $B(V) = W''$. However, this hyperfinite factor has no faithful tracial state, just as $B_0(V)$ failed to have one. ◇

The following result shows that there are factors different than the bounded operators on a complex Hilbert space.

Proposition 6.6.14 *Let W be a UHF-algebra with unique faithful trace x. Then $\pi_x(W)''$ is a hyperfinite factor with a faithful tracial state.*

Proof Clearly $\pi_x(W)''$ is hyperfinite. By weak density and tracialness the functional $y = (\cdot v_x | v_x)$ on it is a tracial state. Again by tracialness we see that if $T(v_x) = 0$, then $\|TS(v_x)\|^2 = 0$ for $T \in \pi_x(W)''$ and $S \in \pi_x(W)$, so $T = 0$ as v_x is cyclic for $\pi_x(W)$. Hence y is faithful.

If P is an orthogonal projection in $\pi_x(W)' \cap \pi_x(W)''$, then $y(P\cdot)$ is a weakly continuous tracial positive functional on $\pi_x(W)''$, so there is by uniqueness a scalar t such that $y(P\cdot) = tx$ on $\pi_x(W)$, and hence $y(P\cdot) = ty$ by weak continuity. Thus $0 = y(P)y(1 - P)$, so $P = 0$ or $1 - P = 0$ by faithfulness. Finally recall that a von Neumann algebra is the weakly closed span of its orthogonal projections. □

Exercises

For Sect. 6.1

1. Show that the norm of a bounded linear functional x on a C*-algebra is given by $\sup |\text{Re}(x(w))|$, where we sup over w in the closed unit ball.
2. Let x, y be positive functionals on a unital C*-algebra W. Prove that $\|x - y\| = \|x\| + \|y\|$ if and only if to every $\varepsilon > 0$ there is a positive element w in the closed unit ball of W such that $x(1 - w) < \varepsilon$ and $y(w) < \varepsilon$.
3. Use the previous exercise to prove uniqueness in the Jordan desomposition of a self-adjoint functional.
4. Show that a bounded functional on a C*-algebra is positive if its norm can be achieved at a positive element in the closed unit ball.

For Sect. 6.2

1. What results in this and in the previous section hold for Banach $*$-algebras when formulated appropriately?
2. Show that an irreducible representation of a C*-algebra is either trivial or irreducible when restricted to a closed ideal.
3. Show that any cyclic representation of a separable C*-algebra W acts on a separable Hilbert space, and that W admits a faithful representation on a separable Hilbert space.
4. Show that any Banach $*$-algebra with a faithful representation admits sufficiently many irreducible representations to separate points.

For Sect. 6.3

1. Show that if $v \in V_x$ is the unit vector of a vector state in the GNS-representation of a pure state x on a C*-algebra, then $v = cv_x$ for a scalar c of modulus one.
2. Show that the vector states on $B(V)$ for a Hilbert space V, are all pure, but that these do not exhaust the pure states when V is separable and infinite dimensional.
3. Show that the map $x \mapsto \dim(V_x)$ from the set of positive linear functionals on a C*-algebra to $\{0, 1, \ldots, \infty\}$ is lower semicontinuous with respect to respectively the w^*-topology and discrete topology, and under the convention that all infinite dimensional Hilbert spaces have the same dimension ∞.
4. Let x be a state on a C*-algebra W. Prove that finitely many elements $w_i \in W$ in the GNS-space of x are linear independent if and only if the matrix with ij-entry $x(w_j^* w_i)$ has non-zero determinant.
5. Show that reflexive C*-algebras are finite dimensional.

For Sect. 6.4

1. Provide a primitive C*-algebra with a quotient that is not primitive.
2. Show that if J is a primitive ideal of a C*-algebra W, then $M_n(J)$ is a primitive ideal of $M_n(W)$. Conclude that the latter C*-algebra is primitive if W is primitive.
3. Prove that a C*-algebra W is prime if and only if $uWw = \{0\}$ is only possible if $u = 0$ or $w = 0$.

4. Suppose W is the closure of a union of a family of C*-subalgebras W_i that is upwards filtered ordered under inclusion. Show that W is prime if all W_i are prime.
5. Let J be a closed ideal of a C*-algebra W. Show that if $uw \in J$ for $u, w \in W$, then there are $a, b \in J$ such that $(u - a)(w - b) = 0$.

For Sect. 6.5

1. Suppose W is the closure of a union of a family of C*-subalgebras W_i that is upwards filtered ordered under inclusion. Show that W is postliminal if all W_i are postliminal.
2. Show that the sum $J_1 + J_2$ of two postliminal ideals J_i of a C*-algebra W is postliminal, hence that there is a largest postliminal ideal in W and that the quotient with respect to it has no non-zero postliminal ideals.
3. Let W be a finite dimensional C*-algebra written, say, as a direct sum of matrix algebras $M_{n_i}(\mathbb{C})$. Show that W is up to $*$-isomorphism uniquely determined by the sequence $\{n_i\}$ up to permutations.
4. Show that the identity representation of the commutant W' of W from the previous exercise is a direct sum of irreducible representations with multiplicities n_i.

For Sect. 6.6

1. Define morphisms between directed systems of C*-algebras, and show that when two systems are isomorphic, the corresponding direct limits are $*$-isomorphic.
2. Show that AF-algebras admit sequential approximative units consisting of orthogonal projections.
3. Given a normal bounded operator T on a Hilbert space V, produce a sequence of commuting orthogonal projections such that the C*-algebra they generate contains T. Construct then a C*-subalgebra of an AF-algebra that is not an AF-algebra.
4. Show that $M_n(W)$ is an AF-algebra when W is.

Chapter 7
Types of von Neumann Algebras

The classification of von Neumann algebras requires a good understanding of their orthogonal projections. We show that they form a complete lattice under the usual order given by positivity, so the supremum and infinum of any family of them belong to the von Neumann algebra W. We say two orthogonal projections $p, q \in W$ are equivalent if the corresponding reduced representations of the identity representation of W' associated to them are unitarily equivalent. This means that there is $w \in W$ such that $p = ww^*$ and $q = w^*w$. The set $D(W)$ of equivalence classes of such projections is then partially ordered, then given by comparing representatives, and it is a partial semigroup under addition of representatives whenever these are mutually orthogonal. The partial order on $D(W)$ becomes a total order when W is a factor.

Any von Neumann algebra W is uniquely a direct sum of von Neumann algebras of types I, II, III, and each of these types are characterized by the existence and properties of certain orthogonal projections. A factor of type I is isomorphic to $B(V)$ for some Hilbert space V. In the last section of this chapter we show that any factor on a separable Hilbert space possesses a certain trace, and evaluating this trace on representatives shows that $D(W)$ can be identified with $\{0, \ldots, n\}$ when W is of type I_n, and with $[0, 1]$ when W is of type II_1, and with $[0, \infty]$ when W is of type II_∞, and finally with $\{0, \infty\}$ when W is of type III. Moreover, examples of all these mutually exhausting types do exist. Thinking of this evaluation as a dimension function, since it has the axiomatic properties of such a function and since the type I_n factors are $M_n(\mathbb{C})$, one might wonder what the other types are.

The natural morphisms between von Neumann algebras are $*$-homomorphisms that are normal, that is, respect bounded increasing nets. Kernels and images of such maps are strongly closed, and any $*$-isomorphism is automatically normal. Positive maps between von Neumann algebras are normal if and only if they are σ-weakly continuous. The positive and negative parts of a normal bounded functional are again normal, and normal bounded functionals admit a polar type decomposition. Normal positive functionals correspond via the trace to positive

L. Tuset, *Analysis and Quantum Groups*,
https://doi.org/10.1007/978-3-031-07246-8_7

trace class operators. The normal bounded functionals of a von Neumann algebra W form a Banach space W_*, the predual, with dual W. In fact, we show that any C*-algebra which is the dual of a Banach space is $*$-isomorphic to a von Neumann algebra.

The bicommutant of any abstract C*-algebra W in the universal representations is called the enveloping von Neumann algebra of W, and as all bounded functionals on W are normal in the universal representation, this enveloping algebra is the bidual of W as a Banach space, leading to what is known as the Arens multiplication on the bidual. Any representation of W that is non-degenerate, meaning roughly that it acts sufficiently rich on the Hilbert space, extends uniquely to a normal representation of the enveloping von Neumann algebra. Taking the so called central cover of a non-degenerate representation, we obtain a bijection from the set of quasi-equivalence classes of non-degenerate representations of W to the orthogonal projections in the center of the enveloping von Neumann algebra of W. We study this correspondence in detail showing amongst other things that the bicommutant of the image of a representation of W is a factor when the corresponding central orthogonal projection is minimal.

In the remaining part of the chapter we focus on tracialness. The existence of certain trace-like maps on a von Neumann algebra is related to properties about their orthogonal projections, in particular, of the von Neumann algebra being finite and semifinite. This was used in the before mentioned classification. We also show that semifiniteness passes to commutants when the von Neumann algebras act on separable Hilbert spaces.

Remark 7.0.1 Relevant and recommended literature for this and the next two chapters is [6, 23, 31, 34, 40, 47]. $\diamond$

7.1 The Lattice of Projections

Notice that for orthogonal projections p, q in a C*-algebra we have $q \geq p$ if and only if $qp = p$. Indeed, if $qp = p$, then $q - p = (q - p)^*(q - p) \geq 0$. Conversely, if $q \geq p$, then $(qp - p)^*(qp - p) = p(p - q)p \leq 0$ which is only possible if $qp - p = 0$. In this case we say that q *majorizes* p, or that p is a *subprojection* of q.

Proposition 7.1.1 *The orthogonal projections in a von Neumann algebra form a* complete lattice *in the sense that any family* $\{p_i\}$ *has both a supremum* $\vee p_i$ *and an infinum* $\wedge p_i$ *in the von Neumann algebra.*

Proof If the von Neumann algebra W acts on a Hilbert space V, the orthogonal projection p onto $\cap p_i(V)$ commutes with $w \in W'$ as w leaves $p_i(V)$ invariant, so $p \in W'' = W$ and p is clearly the infinum of all p_i.

The orthogonal projection onto the closure of $\cup p_i(V)$ is $1 - \wedge(1 - p_i)$ by de Morgan's law, and this is clearly the supremum of all p_i. $\square$

Note that $q \vee p$ is the range projection of $p+q$, which is $p+q$ when $pq=0$.

Definition 7.1.2 Orthogonal projections q, p in a von Neumann algebra W are *(Murray-von Neumann) equivalent*, written $q \sim p$, if there is $w \in W$ such that $q = w^*w$ and $p = ww^*$.

So equivalent orthogonal projections are related by a partial isometry that lives in the von Neumann algebra W, and by polar decomposition in W, the range- and source projections of an element are always equivalent. To check that we actually get an equivalence relation, consider orthogonal projections $q, p, r \in W$, and note that $q = q^*q = qq^*$ shows $q \sim q$, while $q = w^*w$ and $p = ww^* = v^*v$ and $r = vv^*$ implies $(vw)^*(vw) = w^*pw = q$ and $(vw)(vw)^* = vpv^* = r$, so $q \sim r$, and $p \sim q$ by involutiveness of the $*$-operation. Also note that $p \sim 0$ is only possible when $p = 0$.

The equivalence classes of orthogonal projections on W is an abelian partial semigroup $D(W)$ under addition $[q]+[p]=[q+p]$ defined whenever there are pairwise orthogonal representatives q and p. Thus $D(M_n(\mathbb{C})) = \{0, 1, \ldots, n\}$ under the identification $[p] \mapsto \dim \operatorname{im} p$ with addition $k+m$ defined when $k+m \leq n$.

Proposition 7.1.3 *For orthogonal projections p, q in a von Neumann algebra we have $p \vee q - q \sim p - p \wedge q$.*

Proof Since the closed range of $q - q \wedge (1-p)$ is the closed range of qp, which again is $(\ker pq)^{\perp}$, and the closed range of $p - p \wedge (1-q)$ is the closed range of $pq = (qp)^*$, we get $q - q \wedge (1-p) \sim p - p \wedge (1-q)$ by polar decomposing pq. Replacing q by $1-q$ in this equivalence gives the required result. □

The following result shows that equivalence is preserved under arbitrary orthogonal summation of orthogonal projections.

Proposition 7.1.4 *If $\{p_i\}$ and $\{q_i\}$ are families of pairwise orthogonal orthogonal projections in a von Neumann algebra W with $q_i = w_i^*w_i \sim w_iw_i^* = p_i$ for $w_i \in W$, then $\sum q_i = w^*w \sim ww^* = \sum p_i$, where $w = \sum w_i \in W$ and the series converge strongly.*

Proof By Vigier's theorem $\sum p_i$ and $\sum q_i$ are orthogonal projections in W. As $\sum \|w_i(v)\|^2 = \sum (q_i(v)|v) = ((\sum q_i)(v)|v)$ and the vectors $w_i(v)$ are mutually orthogonal for a fixed vector v, it is clear that $\sum w_i$ converges strongly to some $w \in W$, and it is readily checked that $w^*w = \sum q_i$ and $ww^* = \sum p_i$ due to the obvious analogue identities for finite sums. □

Definition 7.1.5 For orthogonal projections q, p in a von Neumann algebra we write $q \prec p$ when q is equivalent to a projection majorized by p.

It is easy to see that $q \prec p$ if and only if p is equivalent to a projection that majorizes q. If also $p \prec r$, then clearly $q \prec r$, and evidently $p \prec p$. The following *Schröder-Bernstein property*, with proof inspired by the corresponding result in set theory, shows that $\prec$ induces a partial ordering on $D(W)$ for any von Neumann algebra W.

Proposition 7.1.6 *If $q \prec p$ and $p \prec q$ for orthogonal projections p, q in a von Neumann algebra, then $q \sim p$.*

Proof Pick orthogonal projections $r \sim p$ and $s \sim q$ with $r \leq s \leq p$. Choose u in the von Neumann algebra such that $uu^* = r$ and $u^*u = p$. Define decreasing orthogonal projections by $e_{2n} = u^n pu^{*n}$ and $e_{2n+1} = u^n su^{*n}$ and let $t = \inf e_n$, so $p = \sum_{n=0}^{\infty}(e_n - e_{n+1}) + t$ and $s = \sum_{n=1}^{\infty}(e_n - e_{n+1}) + t$. By induction we see that $e_m = (ue_m)^*(ue_m) \sim (ue_m)(ue_m)^* = e_{m+2}$, so $\sum_{n=0}^{\infty}(e_{2n} - e_{2n+1}) \sim \sum_{n=1}^{\infty}(e_{2n} - e_{2n+1})$ by the previous proposition. Adding $\sum_{n=0}^{\infty}(e_{2n+1} - e_{2n+2}) + t$ to both sides gives $p \sim s$. □

Lemma 7.1.7 *Suppose p and q are orthogonal projections in a von Neumann algebra W. If $qWp \neq \{0\}$, there are non-zero orthogonal projections $r \leq q$ and $s \leq p$ with $r \sim s$. While if $qWp = \{0\}$, there is a central orthogonal projection $t \geq p$ with $qt = 0$, so $W = tW \oplus (1-t)W$ with $p \in tW$ and $q \in (1-t)W$.*

Proof The range and source projections of a non-zero element qwp will do for the first statement. Indeed, for its polar decomposition $u|qwp|$ in W we have $|qwp| = p|qwp|p$ being a limit of polynomials in $(qwp)^*(qwp)$, and thus $u = qup$ by uniqueness of the decomposition. Hence $uu^* = qupu^*q = (uu^*)q$ and $u^*u = pu^*qup = (u^*u)p$.

For the second statement the orthogonal projection t onto the closed span V of the ranges of everything in Wp will do as V is invariant under both W and W'. □

Note that if $q \sim p$ in a von Neumann algebra with a central orthogonal projection t, then $qt \sim pt$ as $u^*ut = (ut)^*ut \sim ut(ut)^*) = uu^*t$.

We refer to the property in the next result as *generalized comparability*.

Proposition 7.1.8 *For orthogonal projections q, p in a von Neumann algebra, there is a central orthogonal projection t such that $qt \prec pt$ and $p(1-t) \prec q(1-t)$.*

Proof By Zorn's lemma there is a maximal family (p_i, q_i) of pairs of equivalent orthogonal projections in the von Neumann algebra W with p_i mutually orthogonal subprojections of p and q_i mutually orthogonal subprojections of q. Then $\sum p_i \sim \sum q_i$ by Proposition 7.1.4. Set $r = p - \sum p_i$ and $s = q - \sum q_i$. Then $rWs = \{0\}$ by the lemma above and maximality. By the same lemma there is a central projection t with $r \leq t$ and $st = 0$. Hence $qt = (\sum q_i)t \sim (\sum p_i)t \leq pt$ and similary $p(1-t) \prec q(1-t)$. □

Hence for a factor W we see that the partial order on $D(W)$ is a total order. When $W = M_n(\mathbb{C})$ we get the usual total order on $\{0, 1, \ldots, n\}$.

Definition 7.1.9 Let W be a von Neumann algebra. An orthogonal projection $p \in W$ is *abelian* if pWp is commutative. It is *finite* if the only orthogonal projection equivalent to p and majorized by p is p itself. We say W is of *type I* if the identity 1 is an orthogonal sum of abelian projections. It is *finite* if 1 is finite. It is *semifinite* if any non-zero central orthogonal projection majorizes a non-zero finite orthogonal projection, and W is *type II* if in addition it contains no non-zero abelian orthogonal projection. It is *type III* if it contains no non-zero finite orthogonal projections at all.

A von Neumann algebra is clearly of type I if and only if every non-zero central orthogonal projection in it has a non-zero abelian subprojection. Finiteness and abelianess of orthogonal projections are preserved under equivalence. Abelian orthogonal projections are finite. The center of W_p for an orthogonal projection p in a von Neumann algebra W is $pZ(W)$, so when p is abelian, then $pWp = pZ(W)$. If W is a factor, any abelian orthogonal projection p is minimal with $pWp = \mathbb{C}p$. Any minimal orthogonal projection has certainly this property and is therefore abelian. Subprojections and orthogonal sums of finite orthogonal projections remain finite; indeed, if $q \leq p$ and $r \leq q$ with $r \sim q$, then $r + (p - q) \sim p$, so $r + (p - q) = p$. Commutative von Neumann algebras are of type I, and by the lemma above a von Neumann algebra is of type I if and only if every non-zero orthogonal projection majorizes a non-zero abelian orthogonal projection.

Definition 7.1.10 The *central cover* of an orthogonal projection p in a von Neumann algebra W acting on a Hilbert space V is the central orthogonal projection $c(p) \geq p$ onto the closed span of $Wp(V)$.

Proposition 7.1.11 *Let p, q be orthogonal projections in a von Neumann algebra W. If $p \prec q$, then $c(p) \leq c(q)$, so $c(p) = c(q)$ if $p \sim q$. The central cover $c(p)$ of p is the minimal central projection majorizing p, and $c(p)$ is the supremum of the orthogonal projections equivalent to p. We have $c(\vee p_i) = \vee c(p_i)$ for orthogonal projections p_i. If t is a central projection, then $c(tp) = tc(p)$. We have $c(p)c(q) \neq 0$ if and only if $pWq \neq \{0\}$. The map $wp \mapsto wc(p)$ is a $*$-isomorphism from $W'p$ onto $W'c(p)$. It is a $*$-isomorphism onto W' if and only if $c(p) = 1$.*

Proof If $p = u^*u$ and $uu^* \leq q$ for $u \in W$ and W acts on V, then the closed linear span of $Wp(V)$ is contained in the closed span of $Wq(V)$, so $c(p) \leq c(q)$. If $t \geq p$ is a central orthogonal projection, then $t \geq c(p)$ as $tWp(V) = Wp(V)$. As the orthogonal projection $\vee\{q \mid q \sim p\}$ commutes with any unitary in W, it majorizes $c(p)$, and by what we have already said, it is clearly majorized by $c(p)$. Clearly $c(\vee p_i) \geq \vee c(p_i) \geq \vee p_i$ with the middle orthogonal projection central. We have $c(tp) = tc(p)$ for t a central orthogonal projection as $tWp(V) = Wtp(V)$. If $pWq = \{0\}$, then the closed spans of $Wp(V)$ and $Wq(V)$ are orthogonal, so $c(p)c(q) = 0$. Conversely, if this product vanishes, then $pWq = pc(p)Wqc(q) = 0$. The map in the proposition is a well-defined $*$-isomorphism as $wp = 0$ if and only if $wc(p) = 0$ when $w \in W'$. Finally, if it is a $*$-isomorphism onto W', then $c(p) = 1$ as $c(p) - 1 \in W'$ and $(c(p) - 1)p = 0$. □

It is easy to see that the central cover of an orthogonal projection p in a von Neumann algebra is the infinum of all central elements majorizing p.

Proposition 7.1.12 *For every orthogonal projection p in a von Neumann algebra W we have $c(p) = \vee u^*pu$, where u runs through the unitaries in W.*

Proof Any central element that majorizes p will also majorize u^*pu for any unitary $u \in W$, so $c(p) \geq q \equiv \vee u^*pu \in W$. But $u^*qu = q$, so q is central, and as $q \geq p$, we get $q = c(p)$. □

Proposition 7.1.13 *Any von Neumann algebra W is uniquely a direct sum $W_I \oplus W_{II} \oplus W_{III}$ of von Neumann algebras of types I, II and III.*

Proof Let t_3 be the supremum of all central orthogonal projections that do not majorize any non-zero finite orthogonal projections. Then the central orthogonal projection t_3 does not majorize any non-zero finite orthogonal projections either. Indeed, if $p \leq t_3$ is a finite orthogonal projection, and q is a central orthogonal projection without any non-zero finite subprojections, then pq is a finite subprojection of q, and must therefore vanish, so $p = pt_3 = 0$. Hence Wt_3 is of type III.

Let t_1 be the supremum of all central orthogonal projections t in W such that Wt is of type I. We claim that Wt_1 is also of type I. Indeed, suppose $p \in W$ is a non-zero central projection majorized by t_1. Then there is a central orthogonal projection $t \in W$ such that $pt \neq 0$ and Wt is of type I. Hence there is a non-zero abelian orthogonal projection $q \in Wt$ majorized by pt, so $q(Wt_1)q = q(Wt)q$ settles the claim.

Note that $t_1t_3 = 0$ as abelian orthogonal projections are finite. Letting $t_2 = 1 - t_1 - t_3$ we also get $t_1t_2 = 0 = t_2t_3$. By definition of t_3 we see that Wt_2 is semifinite. Suppose p is a non-zero abelian orthogonal projection of Wt_2, so $c(p) \leq t_2$. Then $Wc(p)$ is of type I. Indeed, any non-zero central orthogonal projection t of $Wc(p)$ has the non-zero abelian subprojection pt. Hence by definition of t_1, we get the contradiction $p \leq t_1$, so Wt_2 is of type II.

The uniqueness of the decomposition is clear by maximality of t_1 and t_3. □

A type II von Neumann algebra W is *type II_1* if it is finite, and it is *type II_∞* if it has no non-zero finite central orthogonal projections. The supremum t of all finite central orthogonal projections in W is a finite central orthogonal projection. Indeed, if $p \in W$ is an orthogonal projection equivalent to t and majorized by t, then for any finite central orthogonal projection q of W, we have $q = qt \sim qp \leq q$, so $qp = q$, or $q \leq p$ and $p = t$. Hence W is a unique direct sum of the type II_1 von Neumann algebra Wt and the type II_∞ von Neumann algebra $W(1 - t)$. Thus every factor falls into precisely one of the types I, II_1, II_∞ and III.

Proposition 7.1.14 *In a type II factor any orthogonal projection can be written as $p + q$ for equivalent orthogonal projections p and q such that $pq = 0$.*

Proof If r is the given orthogonal projection, then by resorting to an argument involving a maximal family of pairs (p_i, q_i) of non-zero equivalent orthogonal subprojections of r with all projections mutually orthogonal, it suffices to show that r contains such a pair. Now r cannot be minimal as it would then be abelian, so it has a non-zero orthogonal subprojection p. Then either $p \prec r - p$ or $r - p \prec p$, yielding in each case the required pair. □

For a set J and a Hilbert space V, let $l^2(J, V)$ denote the Hilbert space of J copies of V. To $v \in V$ and $f \in l^2(J)$, let $v \otimes f \in l^2(J, V)$ denote the function given by $(v \otimes f)(i) = f(i)v$. The span of such functions is dense in $l^2(J, V)$, and for $S \in B(V)$ and $T \in B(l^2(J))$, the formula $(S \otimes T)(v \otimes h) = S(v) \otimes T(h)$ defines by linearity and continuity an operator $S \otimes T \in B(l^2(J, V))$; we have anticipated the language of tensor products and refer the reader to that chapter for more details.

Theorem 7.1.15 *Any type I factor is $*$-isomorphic to the C^*-algebra of all bounded operators on some complex Hilbert space.*

Proof Say the factor W acts on V. As abelian orthogonal projections in a factor are minimal, and since W is of type I, we have an orthogonal family of minimal orthogonal projections p_i indexed by J such that $\sum p_i = 1$ strongly. Fix $k \in J$. By generalized comparison all minimal orthogonal projections are equivalent, so there are for each $i \in J$, elements $u_i \in W$ such that $p_i = u_i u_i^*$ and $p_k = u_i^* u_i$. Then $u_{ij} = u_i u_j^*$ form a *matrix unit* in W in the sense that $u_{ij}^* = u_{ji}$ and $\sum u_{ii} = 1$ strongly and $u_{ij}u_{mn} = \delta_{jm}u_{in}$. The latter property holds because $u_j^* = u_j^* u_j u_j^*$, so $u_{ij} = u_{ij}u_{jj}$ and for $j \neq m$, we have $u_{ij}u_{mn} = u_{ij}p_j p_m u_{mn} = 0$. Now $U(v \otimes \delta_i) = u_{ik}(v)$ for $v \in p_k(V)$ well-defines an isometry $U : l^2(J, p_k(V)) \to V$ as $(U(v \otimes \delta_i)|U(w \otimes \delta_j)) = (u_{ik}(v)|u_{jk}(w)) = (u_{jk}^* u_{ik}(v)|w) = \delta_{ij}(v|w)$. For $v \in V$ we see that $u_{ki}(v) \in p_k(V)$ and $U(u_{ki}(v) \otimes \delta_i) = p_i(v)$, so $v = \sum p_i(v)$ belongs to the image of U. Hence U is unitary with adjoint $U^* = \sum u_{ki}(\cdot) \otimes \delta_i$. Given a finite rank operator on $l^2(J)$ expanded as $T = \sum (T(\cdot)|\delta_n)\delta_n$, we get $U(p_k \otimes T)U^* = \sum (T(\delta_i)|\delta_n)u_{ni}$, which belongs to W. As $B_f(l^2(J))$ is strongly dense in $B(l^2(J))$ and the $*$-homomorphism $B(l^2(J)) \to B(V)$ given by $T \mapsto U(p_k \otimes T)U^*$ is strongly continuous, its image belongs to W. It is clearly injective. It is also surjective since for $w \in W$ and $v \in V$, we have $U^* w U(v \otimes \delta_i) = \sum u_{kj} w u_{ik}(v) \otimes \delta_j = (p_k \otimes c)(v \otimes \delta_i)$, with $c \in B(l^2(J))$ given by $c(\delta_i) = \sum c_{ij}\delta_j$ and c_{ij} given by $u_{kj} w u_{ik} = c_{ij} p_k$. □

Proposition 7.1.16 *Let p_i and q_i be orthogonal projections in a finite von Neumann algebra such that $p_1 \leq p_2$ and $q_1 \leq q_2$ and $p_i \sim q_i$. Then $p_2 - p_1 \sim q_2 - q_1$.*

Proof By generalized comparability there is a central projection t such that $(p_2 - p_1)t \prec (q_2 - q_1)t$ and $(q_2 - q_1)(1 - t) \prec (p_2 - p_1)(1 - t)$. Let $q \leq (q_2 - q_1)t$ with $q \sim (p_2 - p_1)t$. Then $q + q_1 t \leq q_2 t$ and $q + q_1 t \sim (p_2 - p_1)t + p_1 t \sim q_2 t$, so by finiteness $q + q_1 t = q_2 t$ and $(p_2 - p_1)t \sim (q_2 - q_1)t$. Similarly $(q_2 - q_1)(1 - t) \sim (p_2 - p_1)(1 - t)$ and the result follows. □

Corollary 7.1.17 *If we have an increasing sequence of orthogonal projections $p_n \prec p$ in a von Neumann algebra, then $\vee p_n \prec p$.*

Proof We claim there is an orthogonal family of orthogonal projections q_n majorized by p such that $q_n \sim p_n - p_{n-1}$ with $p_0 = 0$. This claim is clearly true for $n = 1$. Assume it holds up till n. Now $1 - p \prec 1 - p_{n+1}$ by the proposition, so

$$1 - (p_{n+1} - p_n) = 1 - p_{n+1} + \sum_{i=1}^{n} (p_i - p_{i-1}) \sim 1 - p_{n+1} + \sum_{i=1}^{n} q_i,$$

and again by the proposition, we get $p_{n+1} - p_n \prec p - \sum_{i=1}^{n} q_i$ allowing to pick q_{n+1} to fullfill the induction step. So $\vee p_n = \sum_{n=1}^{\infty} (p_n - p_{n-1}) \sim \sum_{n=1}^{\infty} q_n \leq p$. □

7.2 Normalcy

Von Neumann algebras can be characterized by their lattice property.

Lemma 7.2.1 *Let W be a C^*-subalgebra of $B(V)$ with 1 from $B(V)$ and an orthogonal projection p in its strong closure. For any sequence of unit vectors $v_i \in V$ there is $w \in B(V)$ that is the infinum of a decreasing sequence of elements in $B(V)$, each of which is the supremum of an increasing sequence of positive elements in the closed unit ball of W, such that $w(1-p)(v_i) = 0 = (1-w)p(v_i)$ for all i.*

Proof Pick by Kaplansky's density theorem positive $u_n \in W$ in the closed unit ball such that $\|(p - u_n p)(v_i)\| \leq 1/n$ and $\|u_n(1-p)(v_i)\| < 2^{-n}/n$ for $i \leq n$. For $n < m$ define positive elements $w_{nm} = (1 + \sum_{k=n}^{m} k u_k)^{-1} \sum_{k=n}^{m} k u_k$ in the unit ball of W. For $i \leq n$ we have $(w_{nm}(1-p)(v_i)|(1-p)(v_i)) \leq \sum_{k=n}^{m} 2^{-k} < 2^{-n+1}$ whereas for $i \leq m$ we have $((1-w_{nm})p(v_i)|p(v_i)) \leq 2(1+m)^{-1}$ as $w_{nm} \geq (1+mu_m)^{-1} m u_m$ implies $1 - w_{nm} \leq (1+mu_m)^{-1} \leq (1+m)^{-1}(1+m(1-u_m))$. Now for fixed n the elements w_{nm} increase to a supremum w_n, and these elements decreases in turn to an infinum w with the required properties. □

This leads to the *up-down theorem*.

Theorem 7.2.2 *Let W be a C^*-subalgebra of $B(V)$ with 1 from $B(V)$ and with V separable. Then any self-adjoint element (or a positive element of norm not greater than one) in the strong closure of W is the infinum of a decreasing sequence of elements in $B(V)$, each of which is the supremum of an increasing sequence of self-adjoint elements (or positive elements of norm not greater than one) in W.*

Proof Applying the lemma to a dense sequence of vectors in the closed unit ball of V we see that any orthogonal projection in the strong closure of W is the infinum of a decreasing sequence of elements in $B(V)$, each of which is the supremum of an increasing sequence of positive elements in the closed unit ball of W.

By Corollary 5.4.13 any positive element w in the closed unit ball of W is the norm limit of $\sum_{k=1}^{n} 2^{-k} p_k$ for spectral projections p_k. Pick elements w_{km} that decrease to p_k, where each is the supremum of a sequence of increasing positive elements in the closed unit ball of W. As the set of such supremums is convex the elements $w_n = \sum_{k=1}^{n} 2^{-k} w_{kn} + 2^{-n}$ also belong to this convex set. Moreover, they decrease to w as $w_n - w \leq \sum_{k=1}^{n} 2^{-k}(w_{kn} - p_k) + 2^{-m}$ for $n > m$.

The statement for general self-adjoint elements is also true since any such element in the strong closure of W is of the form $aw - b$ for positive scalars a, b and positive elements w in the closed unit ball of the strong closure of W. But aw is in the up-down extension of W by the above, while $-b$ is evidently there, and so is their sum. □

We have the following characterization of von Neumann algebras.

Corollary 7.2.3 *Any non-degenerate C^*-algebra W of bounded operators on a complex Hilbert space is a von Neumann algebra if and only if its real space of self-adjoint elements is closed under supremums among all bounded operators of increasing nets in W.*

Proof The forward direction is obvious. For the converse, consider an orthogonal projection p in the strong closure of W. By the lemma above there is to $u \in \operatorname{im} p$ and $v \in \operatorname{im}(1-p)$ a positive element $w \in W$ such that $w(u) = u$ and $w(v) = 0$. The range projection p_{uv} of w belongs to W as $\{(n^{-1} + w)^{-1}w\}$ increases towards it, and clearly $p_{uv}(u) = u$ while $p_{uv}(v) = 0$. Now $p_{uv_1} \wedge \cdots \wedge p_{uv_n}$ decrease towards $p_u \leq p$ as the v_i run through the finite subsets of $\operatorname{im}(1-p)$, while $p_{u_1} \vee \cdots \vee p_{u_n}$ increase towards p as the u_i run through the finite subsets of $\operatorname{im} p$. □

Next consider preservation of lattice structures between von Neumann algebras. Recall that *an increasing net in a von Neumann algebra* is a net $\{w_i\}$ in the von Neumann algebra such that all $w_k^* = w_k$ and $w_i \geq w_j$ when $i \geq j$.

Definition 7.2.4 A positive map f between von Neumann algebras is *normal* if $f(\vee w_i) = \vee f(w_i)$ for every bounded increasing net $\{w_i\}$.

By Vigier's theorem strongly continuous positive maps are normal. The converse fails because the infinite amplification associated to a separable complex Hilbert space is clearly normal but not strongly continuous. The amplification trick together with Vigier's theorem also shows that for every bounded increasing net $\{w_i\}$ in a von Neumann algebra we have σ-weak convergence $w_i \to \vee w_j$, so σ-weakly continuous positive maps between von Neumann algebras are normal.

Proposition 7.2.5 *Any $*$-isomorphism between von Neumann algebras is normal.*

Proof If $f\colon W \to V$ is such a map and $\{w_i\}$ increases towards w in W, then $f(w_i)$ increases towards $v \leq f(w)$. Similarly $f^{-1}(f(w_i))$ increases towards $w \leq f^{-1}(v)$, so $f(w) = v$. □

Proposition 7.2.6 *The kernel and image of a normal $*$-homomorphism between von Neumann algebras are strongly closed C*-subalgebras.*

Proof If $f\colon W \to V$ is such a map, by considering the strong closure of its image, we may assume that f preserves the identity elements. The C*-subalgebra $\ker f$ of W is strongly closed by Corollary 7.2.3. Applying the same corollary to the C*-subalgebra $f(W)$ of V, consider a bounded increasing net $\{v_i\}$ in $f(W)$ with supremum $v \in V$. By linear translation and rescaling we may assume that $0 \leq v_i \leq 1$. Let I denote the positive preimages of all v_i under f. Then I is an upward filtered set because if $f(w_i) = v_i$ with w_i positive, choose v_k larger than both v_i and v_j. Then $v_k - v_i = f(u^*u)$ for some $u \in W$ and $u_i \equiv u^*u + w_i$ has image v_k, and similarly we get u_j. Then $u_i + |u_j - u_i| \in W$ majorizes both w_i and w_j and has image v_k. Now for $\varepsilon > 0$ the bounded net $\{(1 + \varepsilon w)^{-1}w \mid w \in I\}$ increases towards $w' \in W$ and the image of the net under the normal map f increases towards $(1 + \varepsilon v)^{-1}v = f(w')$, which shows that v belongs to the C*-algebra $f(W)$. □

Proposition 7.2.7 *Let W be a von Neumann algebra. Every strongly closed hereditary C*-subalgebra (left ideal) of W is of the form pWp (Wp) for a unique orthogonal projection $p \in W$, which is central when the left ideal is two-sided. The image of W under a normal $*$-homomorphism is $*$-isomorphic to Wq for a unique central orthogonal projection q.*

Proof Any strongly closed hereditary C*-subalgebra V of W has a unique identity element p such that $pWp \subset V$ by Corollary 5.3.7, and clearly $V = pVp \subset pWp$. If L is a strongly closed left ideal of W, then L^* is by convexity also strongly closed, so $L \cap L^*$ is a strongly closed hereditary C*-subalgebra of W by Proposition 5.3.6, and is thus of the form pWp for an orthogonal projection $p \in L \cap L^*$. Hence $Wp \subset L$. If $v \in L$, then $v^*v \in pWp$, so by polar decomposition $v = vp$ and $L \subset Wp$. If L is two-sided, then $u^*pu \in L$ and as p is a the identity element for L, we get $u^*pu \leq p$ for any unitary $u \in W$, showing that $p \in Z(W)$. By the previous proposition the kernel of a normal $*$-homomorphism f from W is of the form Wp for an orthogonal projection $p \in Z(W)$. Then $W = Wp \oplus W(1-p)$ and f is a $*$-isomorphism from $W(1-p)$ onto $f(W)$. □

For functionals we extend the concept on normalcy.

Definition 7.2.8 A bounded functional x on a von Neumann algebra is *normal* if $\lim x(w_i) = x(\vee w_i)$ for every bounded increasing net $\{w_i\}$.

Proposition 7.2.9 *The normal bounded functionals on a von Neumann algebra W form a norm closed subspace of $W^\star$. The self-adjoint, positive and negative parts of a normal bounded functional x remain normal, and so do $x(\cdot v)$ and $x(v\cdot)$ for any $v \in W$.*

Proof If a sequence of bounded normal functionals x_n on W converge in norm to x, then the right hand side of $|x(w_i)| \leq \|w_i\|\|x - x_n\| + |x_n(w_i)|$ gets eventually below $\varepsilon > 0$ for large enough n and any bounded decreasing net $\{w_i\}$ in W with limit 0, so we get a normed closed subspace.

If x is a bounded normal functional on W, then $\overline{x((\cdot)^*)}$ and $x(v^* \cdot v)$ are clearly normal, and so are $x(\cdot v)$ and $x(v\cdot)$ by $4wv = \sum_{m=0}^{3} i^m(v+i^m)^*w(v+i^m)$.

For the Jordan decomposition $x = x_+ - x_-$ of a self-adjoint bounded normal functional, there is a self-adjoint element v in the unit ball of W and $\varepsilon > 0$ such that $x_+(1) + x_-(1) = \|x\| \leq (x_+ - x_-)(v) + \varepsilon$, so $x_+(1-v) + x_-(1+v) \leq \varepsilon$, showing that $x_+(1-u) \leq \varepsilon$ and $x_-(u) \leq \varepsilon$ for positive $u = (1+v)/2$ with $\|u\| \leq 1$. Hence for any $w \in W$, we get

$$|x_+(w) - x(uw)| \leq |x_+((1-u)w)| + |x_-(uw)| \leq 2\varepsilon^{1/2}\|x\|^{1/2}\|w\|$$

by the Cauchy-Schwarz inequality, so x_+, and thus x_- are normal. □

Lemma 7.2.10 *For any normal state x on a von Neumann algebra there is an orthogonal family of orthogonal projections p_i that sum strongly up to one and makes $x(\cdot p_i)$ weakly continuous.*

Proof Pick by Zorn's lemma a maximal orthogonal family of orthogonal projections p_i in the von Neumann algebra $W \subset B(V)$ such that $x(\cdot p_i)$ is weakly continuous. If $p = \sum p_i \neq 1$, pick a unit vector $v \in (1-p)(V)$ and consider the normal functional $y = 2(\cdot(v)|v)$ on W. By Zorn's lemma pick a maximal orthogonal family of orthogonal projections $q_i \in (1-p)W(1-p)$ such that $x(q_i) \geq y(q_i)$. By normalcy we then get $x(q) \geq y(q)$ for $q = \sum q_i$. Then $r = 1 - p - q \neq 0$ as

$y(1-p) = 2$ while $x(q) \leq 1$. As $x < y$ on every orthogonal projection majorized by r, we get $|x(wr)|^2 \leq x(rw^*wr) \leq y(rw^*wr) = 2\|wr(v)\|^2$ for $w \in W$, which shows that $x(\cdot r)$ is strongly and hence weakly continuous, contradicting maximality. □

Proposition 7.2.11 *Any positive map between von Neumann algebras is normal if and only if it is σ-weakly continuous.*

Proof If a net of increasing elements w_i in a von Neumann algebra has a bounded operator w as supremum, then with the infinite countable amplification f, the increasing net $\{f(w_i)\}$ has weak limit $f(w)$, so $w_i \to w$ in the σ-weak topology. Hence any σ-weakly continuous positive map is normal.

We claim that any normal bounded functional x on a von Neumann algebra W is σ-weakly continuous. By the previous proposition we may assume that x is a normal state, and by adding sufficiently many of the p_i's from the lemma, we get an orthogonal projection $p \in W$ such that $x(\cdot p)$ is weakly continuous and $x(1-p) < \varepsilon$. For any net $\{w_i\}$ in the closed unit ball of W that converges weakly to zero, we get $|x(w_i)| \leq |x(w_i p)| + \varepsilon^{1/2}$ by the Cauchy-Schwartz inequality, so $x(w_i) \to 0$. But the weak and σ-weak topologies coincide on the closed unit ball of W, so our claim holds by the Krein-Smulian theorem with the σ-weak topology as a w^*-topology.

Thus any positive normal map g between von Neumann algebras composed with a σ-weakly continuous bounded functional is σ-weakly continuous, so g is σ-weakly continuous. □

Proposition 7.2.12 *Any positive normal functional on a von Neumann algebra acting on V is of the form* $\mathrm{Tr}(\cdot w)$ *for a positive trace class operator w on V.*

Proof By Proposition 5.6.2 such a functional x is of the form $x = (f(\cdot)(u)|v)$, where f is the infinite countable amplification. On the positive part of the von Neumann algebra

$$(f(\cdot)(u+v)|u+v) \geq (f(\cdot)(u+v)|u+v) - (f(\cdot)(u-v)|u-v) = 4x,$$

so there is by Proposition 6.2.6 a positive element a in the commutant of $\operatorname{im} f$ such that $x = (f(\cdot)a^{1/2}(u+v)|a^{1/2}(u+v))$, which by Proposition 5.6.2 is of the required form. □

Definition 7.2.13 A positive linear functional x on a von Neumann algebra is *completely additive*completely additive if $x(\sum p_i) = \sum x(p_i)$ for any orthogonal family of orthogonal projections p_i.

Lemma 7.2.14 *Let x, y be two completely additive positive functionals on a von Neumann algebra with $x(p) \leq y(p)$ for a non-zero orthogonal projection p. Then there is a non-zero orthogonal projection $q \leq p$ such that $x(w) \leq y(w)$ for all positive $w \leq q$.*

Proof Let C be the collection of orthogonal families of non-zero orthogonal projections $p_i \leq p$ such that $x(p_i) > y(p_i)$. Partially order C under inclusion.

If $C \neq \phi$, there is by Zorn's lemma a maximal element $\{p_i\}$. Let $p_0 = \sum p_i$, so $0 \neq p_0 \leq p$ and $x(p_0) > y(p_0)$. Thus $0 \neq q = p - p_0 \leq p$, and if $r \leq q$ is an orthogonal projection, then $x(r) \leq y(r)$ by maximality. If $C = \phi$, then $q = p$ will also have the property that $x(r) \leq y(r)$ for any orthogonal projection $r \leq q$. In both cases we are done by Corollary 5.4.13. □

Proposition 7.2.15 *A positive linear functional on a von Neumann algebra is normal if and only if it is completely additive.*

Proof Say we have an orthogonal family of orthogonal projections p_i in a von Neumann algebra. Then the partial sums over finite sets ordered by inclusion form an increasing bounded net of positive elements that has $\sum p_i$ as supremum, so normalcy implies completely additivity.

Conversely, assume x is a completely additive positive functional on a von Neumann algebra W, and say without loss of generality that $x(1) = 1$. Given a non-zero orthogonal projection p, there is a vector v such that $x(p) \leq (p(v)|v)$, so by the lemma there is a non-zero orthogonal projection $q \leq p$ such that $x(w) \leq (w(v)|v)$ when $0 \leq w \leq q$. Hence $|x(uq)|^2 \leq x(qu^*uq) \leq \|u\|^2(qu^*uq(v)|v) = \|u\|^2\|uq(v)\|^2$ for any $u \in W$. Hence $x(\cdot q)$ is strongly continuous.

Take a maximal orthogonal family of orthogonal projections $q_i \in W$ such that $x(\cdot q_i)$ is strongly continuous, so $\sum x(q_i) = 1$ by the previous paragraph. Thus to $\varepsilon > 0$ there is a finite partial sum q of the projections q_i with $x(1 - q) < \varepsilon$. As $|x(w(1 - q))|^2 \leq x(w^*w)x(1 - q)$ for $w \in W$, we get $\|x - x(\cdot q)\| < \varepsilon^{1/2}$, so x is normal by Proposition 7.2.9. □

We have the following polar decomposition of normal bounded functionals.

Proposition 7.2.16 *For any normal bounded functional x on a von Neumann algebra W there is a unique positive normal functional $|x|$ with the same norm and such that $|x(w)|^2 \leq \|x\||x|(w^*w)$ for all $w \in W$. Furthermore, there is a partial isometry $u \in W$ such that $x = |x|(u\cdot)$ and $|x| = x(u^*\cdot)$.*

Proof We may assume that $\|x\| = 1$. Let u be the adjoint of an extreme point of the weakly closed face of elements in the closed unit ball of W that are one under x, so u exists by Alaoglu's theorem and the Krein-Milman theorem, and it is a partial isometry by Theorem 5.9.4. By Proposition 6.1.3 the functional $|x| = x(u^*\cdot)$ is positive and has norm one. Let $p = u^*u$ and $w \in W$. Then $\|nu^* + (1 - p)w\|^2 = \|n^2uu^* + w^*(1 - p)w\|$, so $|n + x((1 - p)w)|^2 \leq n^2 + \|w\|^2$, which implies $(\mathrm{Re}\, x)((1 - p)w) \leq 0$ and $x((1 - p)w) = 0$, so $x = |x|(u\cdot)$. The Cauchy-Schwarz inequality then shows that $|x(w)|^2 \leq \|x\||x|(w^*w)$.

If y is another positive functional of norm one satisfying $|x(w)|^2 \leq \|x\|y(w^*w)$, then $(|x|(1 + \varepsilon w))^2 \leq y((1 + \varepsilon w)^2)$ when $w^* = w$, so in this case $|x|(w) \leq y(w)$, which implies $y = |x|$. □

The left kernel of a normal state x on a von Neumann algebra W is σ-weakly closed and is thus of the form Wp for a unique orthogonal projection $p \in W$. Clearly p is the supremum of the orthogonal projections $q \in W$ with $x(q) = 0$. We call $1 - p$ the *support of* x and denote it by $S(x)$. So $x = x(S(x)\cdot) = x(\cdot S(x))$ by the Cauchy-Schwarz inequality. If $S(x)w = wS(x) = w \geq 0$ and $x(w) = 0$, then $w = 0$. Otherwise there is a spectral projection $q \neq 0$ with $x(q) = 0$ and $q \leq S(x)$.

Corollary 7.2.17 *For any normal bounded functional x on a von Neumann algebra there is a partial isometry v such that $x = |x|(v\cdot)$ and $vv^* = S(|x|)$, and this decomposition is unique in that both $|x|$ and v are uniquely determined.*

Proof Let u be as in the proposition. Set $v = S(|x|)u$. Then $x(v^*\cdot) = |x|$ and $vv^* = S(|x|)uu^*S(|x|) = S(|x|)$ as $|x|(1 - uu^*) = \|x\| - x(u^*uu^*) = \|x\| - x(u^*) = 0$.

Suppose $x = y(w\cdot)$ with $y \geq 0$ normal and $ww^* = S(y)$, so $y = x(w^*\cdot)$ and $\|y\| = \|x\|$, which we may assume is one. Then $1 = \|y\| = y(1) = |x|(vw^*)$ and $1 = |x|(1) = y(wv^*)$, so by the Cauchy-Schwarz inequality we get $1 = |x|(vw^*)^2 \leq |x|(wv^*vw^*) \leq 1$ and $|x|((vw^* - 1)u) = 0$ for any u in the von Neumann algebra. Hence $y = |x|$. Set $p = vw^*$. Then $S(|x|)pS(|x|) = S(|x|)pS(y) = p$. But $S(|x|) = pS(|x|)$ from what we already know. Hence $(w - v)w^* = p - p = 0$ and $(w - v)v^* = S(|x|)^* - S(|x|) = 0$, so $w = v$. □

Proposition 7.2.18 *Let W be a von Neumann algebra. If x is a normal bounded functional on W and $w \in W$, then $|x(w\cdot)| \leq \|w\||x|$. If a sequence of normal bounded functionals x_n tend to x in norm, then $|x_n| \to |x|$ in norm. If x is positive and majorizes another normal functional y, there is a sequence of elements $w_n \in W$ such that $x(w_n\cdot) \to y$ in norm.*

Proof By the previous proposition pick partial isometries $u, v \in W$ such that $x = |x|(u\cdot)$ and $|x(w\cdot)| = x(wv^*\cdot)$. Then $|x|(uwv^*\cdot) = |x|(\cdot vw^*u^*)$, so for positive $a \in W$, we have $|x|(uwv^*a)^2 \leq |x|((uwv^*)^2a)|x|(a)$ by the Cauchy-Schwarz inequality. By induction we therefore get $|x|(uwv^*a)^{2^n} \leq \|uwv^*\|^{2^n}\|a\|\|x\||x|(a)^{2^n-1}$, which can only hold for all $n \in \mathbb{N}$ if $|x|(uwv^*\cdot) \leq \|uwv^*\||x|$, so $|x(w\cdot)| \leq \|w\||x|$.

For the next statement pick by the previous proposition partial isometries $u_n, u \in W$ as in its proof such that $x_n = |x_n|(u_n\cdot)$ and $|x| = x(u^*\cdot)$. Then $|x_n|(u_nu^*) \to 1$ and $|x_n|((1 - u_nu^*)\cdot) \to 0$ in norm by the Cauchy-Schwarz inequality. Hence $|x_n| \to |x|$ in norm as $|x_n|(u_nu^*\cdot) = x_n(u^*\cdot) \to x(u^*\cdot) = |x|$ in norm.

Finally, pick $a \in \pi_x(W)'$ by Proposition 6.2.6 such that $y = (\pi_x(\cdot)a(v_x)|v_x)$. By cyclicity pick a sequence of elements $w_n \in W$ with $a(v_x) = \lim \pi_x(w_n^*)(v_x)$. Then $|y(w) - x(w_nw)| \leq \|w\|\|v_x\|\|a(v_x) - \pi_x(w_n^*)(v_x)\| \to 0$ for $w \in W$. □

We study now how normalcy maps out for representations.

Definition 7.2.19 The *enveloping von Neumann algebra* W'' of an abstract C*-algebra W is the von Neumann algebra $\pi(W)''$, where π is the universal representation of W.

Theorem 7.2.20 *Every non-degenerate representation π of a C*-algebra W has a unique extension to a normal unital $*$-epimorphism $\pi''\colon W'' \to \pi(W)''$. We have $W'' \cong W^{\star\star}$ as Banach spaces. The strong closure in W'' of any C*-subalgebra U of W is $*$-isomorphic to U''.*

Proof Assume first that $\pi\colon W \to B(V)$ has a cyclic unit vector v. Then π is unitarily equivalent to π_x with $x = (\cdot(v)|v)$. The orthogonal projection of the Hilbert space of the universal representation onto V_x belongs to the commutant of W'' as $\pi_x(w) = wp = pw$ for all $w \in W$. We get a normal $*$-homomorphism $\pi''\colon W'' \to B(V)$ by composing the map $w \mapsto wp$ with the spatial isomorphism $B(V_x) \to B(V)$, which clearly extends π and satisfies $\pi''(W'') = \pi(W)''$ by normality. When π is only non-degenerate, it is a direct sum of cyclic representations, and the direct sum π'' of the corresponding normal maps is normal. Uniqueness is clear from Corollary 7.2.3.

Every state on W is a vector state in the universal representation space, and hence normal on W'', so we have a linear map from $W^\star$ to the predual of W''. As W is σ-weakly dense in W'' the map will be isometric and, moreover, surjective as any element in the predual of W'' is the image of its restriction to W.

The double transpose of the inclusion map $U \to W$ is a linear isometry of U'' onto the weak closure of U in W''. Being a $*$-homomorphism when restricted to the weakly dense U in U'', we are done. □

Remark 7.2.21 Using the identification $W'' \cong W^{\star\star}$ in the theorem, we can transfer both the $*$-operation and the product from W'' to $W^{\star\star}$. This gives a $*$-operation and the so called *Arens multiplication* given by $a^*(x) = \overline{a(x^*)}$ and $(ab)(x) = a([b, x])$ with $x^*(w) = \overline{x(w^*)}$ and $[b, x](w) = b(x(w\cdot))$ for $a, b \in W^{\star\star}$ and $x \in W^\star$ and $w \in W$. One verifies this by invoking the isometric linear bijection $\pi_*\colon W''_* \to W^\star$ given by $\pi_*([S])(w) = \mathrm{Tr}(\pi(w)S)$, where π is the universal representation of W on V and we recall that the predual W''_* of W'' is the quotient of the trace class operators on V by the space of those annihilated by W''. The transpose of π_* is the isometric linear bijection $W^{\star\star} \to W''$ that extends π on $W \subset W^{\star\star}$. ◇

Proposition 7.2.22 *The GNS-representation of a von Neumann algebra associated to a normal positive functional is a normal $*$-homomorphism.*

Proof If x is a normal positive functional on the von Neumann algebra W, and $\{w_i\}$ is an increasing sequence in W with supremum $w \in W$, then $\{\pi_x(w_i)\}$ increases to a limit $u \in B(V_x)$ with $u \leq \pi_x(w)$. For $v \in W$ we have

$$(\pi_x(w)(v + L)|v + L) = x(v^*wv) = \lim x(v^*w_iv) = (u(v + L)|v + L),$$

so $(\pi_x(w) - u)(v + L) = 0$ and $\pi_x(w) = u$ by denseness. □

Definition 7.2.23 The *central cover* of a non-degenerate representation π of a C*-algebra W is the central orthogonal projection $c(\pi) \in W''$ such that $W''c(\pi)$ is $*$-isomorphic to $\pi(W)'' = \pi''(W'')$.

The existence and uniqueness of $c(\pi)$ is assured by Proposition 7.2.7.

Proposition 7.2.24 *The map $[\pi] \mapsto c(\pi)$ is a bijection from the set of quasi-equivalence classes of non-degenerate non-zero representations of a C*-algebra W to the set of non-zero central orthogonal projections in W''.*

Proof The map is clearly well-defined. It is surjective since for each non-zero central orthogonal projection $p \in W''$, the map $w \mapsto wp$ is clearly a non-degenerate $*$-representation of W on $\operatorname{im} p$ with central cover p and with the same formula yielding the normal extension to W''. It is injective because any non-degenerate non-zero representation π of W is quasi-equivalent to the representation given by $w \mapsto wc(\pi)$ on $\operatorname{im} c(\pi)$ with quasi-intertwiner the restriction of π'' to $W''c(\pi)$. □

Definition 7.2.25 A von Neumann algebra W is *σ-finite* if each orthogonal family of non-zero orthogonal projections in W is countable. An orthogonal projection $p \in W$ is *σ-finite* if W_p is σ-finite.

Clearly von Neumann algebras acting on separable Hilbert spaces are σ-finite.

Proposition 7.2.26 *A von Neumann algebra W has a faithful normal non-degenerate representation on a separable Hilbert space if and only if W is σ-finite and contains a strongly dense sequence. Any non-degenerate representation π of a separable C*-algebra is quasi-equivalent to a non-degenerate representation on a separable Hilbert space if and only if $c(\pi)$ is σ-finite. Non-degenerate representations of separable C*-algebras on separable Hilbert spaces are quasi-equivalent to cyclic representations.*

Proof If W acts on a separable Hilbert space V, then it is clearly σ-finite. Since the closed unit ball in $B(V)$ is strongly metrizable and separable, and hence second countable, it is clear that W is separable in the strong topology.

Suppose W is σ-finite and acts on a Hilbert space V, and let p_n be the orthogonal projection onto the closure of $W'(v_n)$ for some unit vector $v_n \in V$. By assumption any maximal orthogonal family of such projections is countable and $\sum p_n = 1$. Consider the normal state $x = \sum 2^{-n}(\cdot(v_n)|v_n)$ on W. If x vanishes on $w \geq 0$, then $w(v_n) = 0$ for all n, so $w = 0$ as $\sum p_n = 1$. Hence π_x is by Proposition 7.2.22 a faithful normal representation of W. If W also has a separable C*-subalgebra U that is strongly dense in W, then by Kaplansky's density theorem, there is to $w \in W$ a net $\{w_i\}$ in U that converges strongly to w and $\|w_i\| \leq \|w\|$. Then $\|(w+L)-(w_i+L)\|^2 = x((w-w_i)^*(w-w_i)) = \sum 2^{-n}\|(w-w_i)(v_n)\|^2$ tends to zero as $w_i \to w$ strongly. □

Proposition 7.2.27 *Suppose $f\colon W_1 \to W_2$ is a $*$-isomorphism between von Neumann algebras acting on Hilbert spaces V_i. Then they are both $*$-isomorphic to a von Neumann algebra W acting on V with orthogonal projections $p_i \in W'$ such that $c(p_i) = 1$ and partial isometries $u_i\colon V_i \to p_i(V)$ such that $u_i^* p_i W u_i = W_i$ and $f(u_1^* p_1 w u_1) = u_2^* p_2 w u_2$ for $w \in W$.*

Hence if π_1 and π_2 are quasi-equivalent non-degenerate representations of a C-algebra, there is a non-degenerate representation π quasi-equivalent to both and such that π_1 and π_2 are unitarily equivalent subrepresentations of π.*

Proof Choose by Zorn's lemma a maximal family of unit vectors $v_i \in V_2$ such that the orthogonal projections $p_i \in W_2'$ onto the closure of $W_2(v_i)$ are pairwise orthogonal. Then $\sum p_i = 1$. For each i pick a sequence of elements $v_{in} \in V_1$ such that $(f(\cdot)(v_i)|v_i) = \sum(\cdot(v_{in})|v_{in})$. Let V be the direct sum over all i, n of copies of V_1, and let $g: B(V_1) \to B(V)$ be the amplification with $W = g(W_1)$. Let $p_1 \in W'$ be the orthogonal projection from V onto one copy of V_1, so $c(p_1) = 1$, and let $u_1: V_1 \to V$ be the partial isometry that is the identity map on V_1, so $u_1^* p_1 W u_1 = W_1$.

Let $w_i = \oplus v_{in}$ and consider the sum p_2 of the orthogonal family of orthogonal projections $q_i \in W'$ onto the closure of $W(w_i)$. Now $(g(\cdot)(w_i)|w_i) = \sum(\cdot(v_{in})|v_{in}) = (f(\cdot)(v_i)|v_i)$, so by Proposition 6.2.8 there is an isometry u_i from $p_i(V_2)$ to $q_i(V)$ such that $u_i^* W q_i u_i = W_2 p_i$. Hence $u_2 = \sum u_i$ is an isometry from V_2 onto $p_2(V)$ such that $u_2^* W p_2 u_2 = W_2$, and $c(p_2) = 1$ as $u_2^* p_2 g(W_1) u_2 = f(W_1)$. □

Proposition 7.2.28 *Let $f: W_1 \to W_2$ be a $*$-isomorphism between von Neumann algebras. Then there are orthogonal projections $p_i \in W_i'$ with $c(p_i) = 1$ and a spatial $*$-isomorphism $g: p_1 W_1 \to p_2 W_2$ such that $p_2 f(w) = g(p_1 w)$ for $w \in W_1$.*

Hence, if π_1 and π_2 are quasi-equivalent non-degenerate representations of a C^*-algebra, they have unitarily equivalent subrepresentations that are quasi-equivalent to the original ones.

Proof By the previous proposition we may assume that $W_i = W q_i$ for a von Neumann algebra W and orthogonal projections $q_i \in W'$ with $c(q_i) = 1$. The closed unit ball in $q_2 W' q_1$ is σ-weakly compact and contains by the Krein-Milman theorem an extreme point u. Set $p_1 = u^* u$ and $p_2 = u u^*$. Then $p_i \in W_i'$ are orthogonal projections and $(q_2 - p_2) W' (q_1 - p_1) = 0$ by a corollary of Theorem 5.9.4 gotten by tracing the proof, and clearly $q_i - p_i$ are orthogonal projections. So for each unitary $w \in W'$ we have $w^*(q_2 - p_2) w (q_1 - p_1) = 0$, or $c(q_2 - p_2)(q_1 - p_1) = 0$ by Proposition 7.1.12 and thus $c(q_2 - p_2) c(q_1 - p_1) = 0$. So

$$1 \leq c(q_2 - p_2) + c(p_2) \leq 1 - c(q_1 - p_1) + c(p_2) \leq c(p_1) + c(p_2) = 2c(p_1)$$

and $c(p_i) = 1$. □

Proposition 7.2.29 *Let π_i be non-degenerate representations of a C^*-algebra W. Then $c(\pi_1) c(\pi_2) = 0$ if and only if $(\pi_1 \oplus \pi_2)(W)'' = \pi_1(W)'' \oplus \pi_2(W)''$ if and only if $(\pi_1 \oplus \pi_2)(W)' = \pi_1(W)' \oplus \pi_2(W)'$ if and only if there are no non-zero unitarily equivalent subrepresentations of π_1 and π_2.*

Proof If $c(\pi_1) c(\pi_2) = 0$, then

$$\ker(\pi_1 \oplus \pi_2)'' = \ker \pi_1'' \cap \ker \pi_2'' = W''(1 - c(\pi_1))(1 - c(\pi_2)) = W''(1 - c(\pi_1) - c(\pi_2)),$$

so $(\pi_1 \oplus \pi_2)''(W'')$ is isomorphic to $W''(c(\pi_1) + c(\pi_2)) = W'' c(\pi_1) \oplus W'' c(\pi_2)$, and $(\pi_1 \oplus \pi_2)(W)'' = \pi_1(W)'' \oplus \pi_2(W)''$. This last equality yields $(\pi_1 \oplus \pi_2)(W)' = \pi_1(W)' \oplus \pi_2(W)'$ by the bicommutant theorem.

If there are unitarily equivalent subrepresentations of $\pi_i: W \to B(V_i)$, there is a partial isometry $u: V_1 \to V_2$ such that $u^* u \in \pi_1(W)'$ and $u u^* \in \pi_2(W)'$ and

$u^*(\pi_2(\cdot)uu^*)u = \pi_1(\cdot)u^*u$. Thinking of $u \in B(V_1 \oplus V_2)$ we thus get

$$(\pi_1 \oplus \pi_2)(\cdot)u = \pi_2(\cdot)u = u\pi_1(\cdot) = u(\pi_1 \oplus \pi_2)(\cdot),$$

which gives $u = 0$ if $(\pi_1 \oplus \pi_2)(W)' = \pi_1(W)' \oplus \pi_2(W)'$.

The subrepresentations of π_1 and π_2 determined by the orthogonal projections $\pi_1''(c(\pi_2))$ and $\pi_2''(c(\pi_1))$ have central cover $c(\pi_1)c(\pi_2)$ and are quasi-equivalent. By the proposition above these subrepresentations have non-zero unitarily equivalent subrepresentations if $c(\pi_1)c(\pi_2) \neq 0$. □

Any two non-degenerate representations satisfying the equivalent conditions in the previous proposition are said to be *disjoint*. It is easy to see that for non-degenerate representations π_i of a C*-algebra W, there are central orthogonal projections $p_i \in \pi_i(W)'$ such that the subrepresentations on $\mathrm{im}\, p_1$ and $\mathrm{im}\, p_2$ are quasi-equivalent, while those on $\mathrm{im}(1 - p_1)$ and $\mathrm{im}(1 - p_2)$ are disjoint.

A non-degenerate representations π of a C*-algebra W is a *factor representation* if $\pi(W)''$ is a factor, and this happens precisely when $c(\pi)$ is minimal in the centre if W''. Hence two factor representations are either quasi-equivalent or disjoint.

We finally look at abstract characterizations of von Neumann algebras.

Proposition 7.2.30 *If W is a C*-algebra such that every bounded increasing net $\{w_i\}$ in W has a supremum $\vee w_i \in W$ and with a separating family of states x such that* $\sup x(w_i) = x(\vee w_i)$, *then there is a $*$-isomorphism f from W to a von Neumann algebra such that $\vee f(w_i) = f(\vee w_i)$.*

Proof The direct sum of the GNS-representation of the states x is clearly a faithful representation of W that is normal in the sense described for f above. We are then done by the proof of Proposition 7.2.6. □

The following characterization by Sakai is more striking.

Theorem 7.2.31 *Any C*-algebra linearly isometric to the dual of a Banach space V is $*$-isomorphic to a von Neumann algebra with predual V.*

Proof Let W be the C*-algebra in question, so $W \cong V^\star$ as Banach spaces. By Alaoglu's theorem the closed unit ball in W is w^*-compact, and has therefore by the Krein-Milman theorem an extremal point, so W is unital by Theorem 5.9.4.

We claim that the self-adjoint part of W is w^*-closed, and to show this it suffices by the Krein-Smulian theorem to show that if $\{x_j\}$ is a net in the closed unit ball of W that w^*-converges to $x + iy$ with x, y self-adjoint, then $y = 0$. Now $\|x_j + in\| \leq (1 + n^2)^{1/2}$ by using the triangle inequality on the imaginary part of $x_j + in$. But to $\delta > 0$ there is a unit vector $v \in V$ and x_j such that $\|x + i(y + n)\| \leq |(x + i(y + n))(v)| + \delta \leq |(x_j + in)(v)| + 2\delta$, so $(1 + n^2)^{1/2} \geq \|x + i(y + n)\| \geq \|y + n\|$. If $y \neq 0$ we may by passing to $\{-x_j\}$ assume there is a positive number t in the spectrum of y. Then $t + n \leq \|y + n\| \leq (1 + n^2)^{1/2}$, which cannot hold as $n \to \infty$.

But then also the set of positive elements in W is w^*-closed by the Krein-Smulian theorem, since the positive elements in the norm closed unit ball of W are of the form $(1 + u)/2$ for self-adjoint elements u in the norm closed unit ball of W.

But then the positive elements of $V \subset W^\star$ separate the points of W, since if $x \in W$ is self-adjoint with $-x$ non-positive, there is by the Hahn-Banach separation theorem a self-adjoint element $v \in V$ such that v is non-negative on the positive part of W and $-x(v) < 0$.

Also W contains with every bounded increasing net of self-adjoint elements $x_i \in W$ the supremum. Indeed, by Alaoglu's theorem it has a subnet $\{x_j\}$ that w^*-converges to $x = x^* \in W$. Then x majorizes all x_i as the positive elements in W form a w^*-closed set. For the same reason it is a least upper bound. Note also that every positive $v \in V$ has the property $\lim x_i(v) \leq x(v) = \lim x_j(v) \leq \lim x_i(v)$.

Hence the previous proposition provides a *-isomorphism from W onto a von Neumann algebra U, which by construction has a predual that contains the positive elements of V. Moreover, we know that for a non-zero $x = x^* \in U$, there is a positive $v \in V$ such that $x(v) \neq 0$. So the linear span of the positive elements in V is norm dense in V, and V is contained in the predual. But the w^*-topology and the σ-weak topology coincide on the norm closed unit ball of U as both are compact and Hausdorff. So V is the entire predual. □

7.3 Center Valued Traces

Proposition 7.3.1 *If x is a faithful positive tracial functional on a factor with orthogonal projections p, q, then $p \sim q$ if and only if $x(p) = x(q)$. Any von Neumann algebra with such a functional is finite.*

Proof The forward direction is obvious. For the opposite direction we may assume that $p \prec q$, say $p \sim r$ with $r \leq q$. Then $x(q - r) = x(q) - x(p) = 0$ and the orthogonal projection $q - r$ must vanish by faithfulness. If $p \sim 1$, then $x(1 - p) = 0$, so $p = 1$. □

A bounded linear functional x on a von Neumann algebra is tracial if it takes the same value on equivalent orthogonal projections. Indeed, if p is an orthogonal projection, then $x(u^* pu) = x(p)$ for any unitary u, and as the von Neumann algebra is the norm closed hull of its orthogonal projections, we get $x(u^* \cdot u) = x$.

Lemma 7.3.2 *Let W be a von Neumann algebra on a separable Hilbert space. Given a bounded set A of positive normal functionals on W such that for every sequence of orthogonal projections $p_n \in W$ and $\varepsilon > 0$, there is $m \in \mathbb{N}$ such that $x(p_n) \leq \varepsilon$ for all $x \in A$ and $n \geq m$. Then the weak closure of A in the predual of W is weakly compact.*

Proof It suffices to show that every element in the w^*-closure of A is normal, or completely additive, since such a bounded closure is compact by Alaoglu's theorem, and the weak topology on the predual of W is the relative topology of the w^*-topology on W^*.

To this end we claim that there is $k \in \mathbb{N}$ such that $x(p) - \sum_{i=1}^{n} x(p_i) < \varepsilon$ for all $x \in A$ and $n \geq k$, where $p = \sum p_i$. Indeed, if this was not so, there would be

$\varepsilon > 0$ such that for every k we would have $x(p) - \sum_{i=1}^{k} x(p_i) \geq \varepsilon$ for at least one $x \in A$. Pick such x_1 to $k_1 \in \mathbb{N}$. As x_1 is completely additive there is $k_2 \geq k_1 + 1$ with $x_1(p) - \sum_{i=1}^{k_2} x_1(p_i) < \varepsilon/2$, so $q_1 = \sum_{i=k_1+1}^{k_2} p_i$ is an orthogonal projection in W with $x_1(q_1) \geq \varepsilon/2$. Picking $x_2 \in A$ to k_2, and continuing this way, produces a sequence of orthogonal projections $q_j \in W$ and $x_j \in A$ such that $x_j(q_j) \geq \varepsilon/2$ for all j. This contradiction proves the claim.

Hence, if a net $\{x_i\}$ in A converges pointwise to $x \in W^*$, and if $\{p_n\}$ is an orthogonal sequence of orthogonal projections in W, and $\varepsilon > 0$, there is $k \in \mathbb{N}$ such that $x_i(p) - \sum_{j=1}^{n} x_i(p_j) < \varepsilon$ for all $n \geq k$ and i. To $n \geq k$ pick i such that $|x_i(\sum_{j=1}^{n} p_j) - x(\sum_{j=1}^{n} p_j)| < \varepsilon$ and $|x_i(p) - x(p)| < \varepsilon$. Then $x(p) - \sum_{j=1}^{n} x(p_j) < 3\varepsilon$ for all $n \geq k$ by the triangle inequality. By separability we are done. □

Lemma 7.3.3 *Let x belong to the predual of a finite von Neumann algebra W on a separable Hilbert space. Then the weakly closed convex hull of $x(u^* \cdot u)$ for all unitaries $u \in W$ is compact.*

Proof By the Jordan decomposition of bounded functionals we may assume that x is positive, and by Corollary 2.7.8 it suffices to show that the weak closure of all $x(u^* \cdot u)$ is compact. If it is not, we may by the previous proposition pick $\varepsilon > 0$ and a sequence of unitaries $u_n \in W$ and an orthogonal sequence of orthogonal projections $p_n \in W$ such that $x(q_n) \geq \varepsilon$ with $q_n = u_n^* p_n u_n$. Consider the decreasing sequences of orthogonal projections $r_n = p_n + p_{n+1} + \cdots$ and $s_n = q_n \vee q_{n+1} \vee \cdots$ and the increasing sequence of orthogonal projections $t_k = \sup\{q_i \mid n \leq i \leq n+k\}$. The latter is the range projection of $\sum_{i=n}^{n+k} q_i$, which is this projection itself as the q_i are mutually orthogonal. But $t_k \sim \sum_{i=n}^{n+k} p_i \leq r_n$ by Proposition 7.1.4, so $s_n \prec r_n$ by Corollary 7.1.17. By Proposition 7.1.16 we have $1 - r_n \prec 1 - s_n \leq 1 - \vee q_i$. Using Corollary 7.1.17 once more, we get $\vee(1 - r_n) \prec 1 - \vee q_i$. But $\vee(1 - r_n) = 1$ by separability, so $q_i \to 0$ strongly, which contradicts $x(q_i) \geq \varepsilon$ for all i. □

Lemma 7.3.4 *Let W be a finite von Neumann algebra on a separable Hilbert space with x in the predual of its center. Then there is a unique norm preserving extension y in the predual of W such that $y = y(u^* \cdot u)$ for all unitaries $u \in W$. When x is positive, so is y.*

Proof Since W with the σ-weak topology is a locally convex topological space, Proposition 2.4.11 provides an extension $y' \in W_*$ of x. The weakly closed convex hull D of $R_u(y') = y'(u^* \cdot u)$ for all unitaries $u \in W$ is compact by the lemma above, and clearly all $R_u(D) \subset D$. The Rull-Nardzewski fixed point theorem applies now to D and the family X of all R_u because each such map is evidently affine and continuous, and if 0 belonged to the closed convex hull of $X(\tilde{y})$ for some $\tilde{y} \in W_*$, there would be a unitary $u \in W$ such that $\|\tilde{y}\| = \|R_u(\tilde{y})\| < \varepsilon$, implying $\tilde{y} = 0$. It remains to show that the common fixed point $y \in W_*$ has norm $\|x\|$ and is unique. This will also show that the extension is positive when x is.

By Corollary 7.2.17 we may write $y = |y|(v \cdot)$ for $vv^* = S(|y|)$. For any unitary $u \in W$ we have $y = (R_u|y|)(uvu^* \cdot)$ and $S(R_u|y|) = uS(|y|)u^* = (uvu^*)(uvu^*)^*$, so $uvu^* = v$ by uniqueness of the decomposition. Hence v belongs to the center of W, and $\|y\| = y(v^*) = x(v^*) \leq \|x\|$, so $\|y\| = \|x\|$.

Uniqueness is now clear since the difference of two extensions extends the zero functional which has norm zero. □

Theorem 7.3.5 *Let W be a finite von Neumann algebra on a separable Hilbert space. Then there is a unique* finite center-valued trace *on W, by which we mean a norm decreasing σ-weakly linear map $g: W \to Z(W)$ that restricts to the identity map on $Z(W)$ and satisfies $g(u^* \cdot u) = g$ for all unitaries $u \in W$. It is moreover, positive, faithful and satisfies $g(vw) = vg(w)$ for $v \in Z(W)$ and $w \in W$.*

Proof The lemma gives a unique positive linear isometry $f: Z(W)_* \to W_*$ such that $f(x)$ restricts to x on $Z(W)$, and $R_u f(x) = f(x)$ for $x \in Z(W)_*$. Linearity is a consequence of uniqueness of extensions. The Banach space adjoint g of f is clearly a finite center-valued trace on W. It is the unique such one because if $h: W \to Z(W)$ was another, then $x \mapsto xh$ defines a map $Z(W)_* \to W_*$ which by uniqueness of f must be f, so $h = g$.

Clearly g is positive. For any unitary $v \in Z(W)$ the map $v^*g(v\cdot): W \to Z(W)$ must be g by uniqueness of finite center-valued traces, so $g(vw) = vg(w)$ for $w \in W$. As for faithfulness of g, it suffices to show that g is non-zero on the non-zero orthogonal projections in W. Suppose g vanishes on some non-zero orthogonal projection. Then by Zorn's lemma we have a maximal orthogonal family of orthogonal projections $p_i \neq 0$ such that $g(p_i) = 0$. Then g vanishes on $p = \sum p_i$ and $g(q) > 0$ for any non-zero orthogonal projection q majorized by $1-p$. Pick by generalized comparability a central orthogonal projection $t \in W$ such that $tp \prec t(1-p)$ and $(1-t)(1-p) \prec (1-t)p$. As $g(tp) = 0$ and tp is equivalent to an orthogonal projection majorized by $1-p$, we must have $tp = 0$ as g does not distinguish equivalent projections. Similarly $g((1-t)(1-p)) \leq g((1-t)p) = 0$ and $0 = (1-t)(1-p)$, so $p = 1-t$ and $0 = g(p) = 1-t = p \neq 0$. □

Any finite center-valued trace x has plainly the tracial property $x(wv) = x(vw)$ for all v, w. From this theorem and the proof of Proposition 7.3.1 we get the following result.

Corollary 7.3.6 *A von Neumann algebra on a separable Hilbert space is finite if and only if it has a faithful positive finite center-valued trace.*

On a factor any finite center-valued trace is of the form $x(\cdot)I$ for a tracial bounded linear functional x. So if the factor is finite and acts on a separable Hilbert space, it has a faithful tracial state. If it is not a factor it has a faithful family of tracial positive functionals, each of the form $\int g(\cdot) f \, d\mu$, where g is the finite center-valued trace and μ is a Radon measure on a second countable compact Hausdorff space gotten by identifying the center of the von Neumann algebra with $L^\infty(\mu)$, and $f \geq 0$ is a measurable function.

If we restrict g to the set P of orthogonal projections on the von Neumann algebra W, we get a *finite center-valued dimension function on W*, which is a map $d_Z: P \to Z(W)$ enjoying the following properties: it is positive and non-zero on non-zero orthogonal projections, additive on mutually orthogonal projections, does not distinguish equivalent projections, restricts to the identity map on the center, and satisfies $d_Z(pq) = pd_Z(q)$ for $p \in Z(W)$. Such a dimension function can only exist on a finite von Neumann algebra because if $p \sim q \leq p$, then $d_Z(q) = d_Z(p) = d_Z(p-q)+d_Z(q)$ implies $p-q = 0$. When W is a factor, we see that $d_Z = d(\cdot)I$, where d satisfies the basic properties in the following definition.

Definition 7.3.7 A *dimension function* on a factor W is a map $d\colon P \to [0, \infty]$ such that $d(p) = d(q)$ when $p \sim q$, and $d(p+q) = d(p) + d(q)$ when $pq = 0$, and $p = 0$ when $d(p) = 0$, and which takes finite value on some non-zero orthogonal projection if W contains a non-zero finite orthogonal projection.

7.4 Semifinite von Neumann Algebras

Definition 7.4.1 A *trace* on the positive part W_+ of a von Neumann algebra W is a map $x\colon W_+ \to [0, \infty]$ that respects addition $x(v + w) = x(v) + x(w)$ and scalar multiplication $x(av) = ax(v)$ for all $v, w \in W_+$ and $a \in [0, \infty\rangle$, and satisfies $x(v^*v) = x(vv^*)$ for all $v \in W$. It is *faithful* if $x(v) > 0$ for $v \neq 0$, and *semifinite* if to $v \neq 0$ there is non-zero $w \leq v$ such that $x(w) < \infty$, and it is *normal* if $x(\vee v_i) = \vee x(v_i)$ for any increasing bounded net $\{v_i\}$ in W_+.

Proposition 7.4.2 *Given a trace x on W_+, then $N = \{v \in W \,|\, x(v^*v) < \infty\}$ and the* definition ideal *M of x, being the linear span of N^2, are $*$-ideals of W. Also $M = N^2$ is the linear span of the positive part of M, which again consists of all positive $w \in W$ having finite trace. And x extends uniquely to a linear functional on M, also named x, such that $x(vw) = x(wv)$ for $v \in M$ and $w \in W$, or for $v, w \in N$, and $x(v) = x(u^*vu)$ for $v \in W_+$ and any unitary $u \in W$. The first of these conditions is equivalent to $x(w^*w) = x(ww^*)$ for $w \in W$ and any $x\colon W_+ \to [0, \infty]$ that respects addition and scalar multiplication.*

Proof The inequality $(v + w)^*(v + w) \leq 2v^*v + 2w^*w$ obtained from the parallelogram law shows that N is a vector subspace of W. The tracial property of x shows that N is $*$-closed, and it is clearly closed under multiplication by elements of W from the left, so N and M are $*$-ideals. If v is positive in W and of finite trace, then $v^{1/2} \in N$, so $v \in M$. If $v \in M$ is positive, then $v = \sum v_i^* w_i$ for finitely many $v_i, w_i \in N$, so $v = (v + v^*)/2 \leq \frac{1}{4}\sum(v_i + w_i)^*(v_i + w_i)$ by the polarization identity, showing that $x(v) < \infty$. The same identity shows that M is the span of the positive part of M. Polar decomposing $v = u|v|^{1/2}|v|^{1/2} \in M$ and using $|v| = u^*v \in M$, shows that $M = N^2$.

Since M is the span of its positive elements, we may uniquely extend x to a linear functional x on M. For unitary $u \in W$ and $v \in W_+$, we have $x(u^*vu) = x((v^{1/2}u)^*(v^{1/2}u)) = x((v^{1/2}u)(v^{1/2}u)^*) = x(v)$, so $x(v) = x(u^*vu)$ for $v \in M$, so $x(vw) = x(wv)$ for $v \in M$ and $w \in W$ as the unitaries span W. The polarization identity also shows that $x(vw) = x(wv)$ for $v, w \in N$.

For the forward implication of the last statement, let $v \in W$ with $x(v^*v) < \infty$ and polar decomposition $v = u|v|$. Defining N and M as above for x, we do get $*$-ideals with all the properties above, so $vv^* = uv^*vu^* \in M$ and $x(vv^*) < \infty$. In this case $x(vv^*) = x(uv^*vu^*) = x(u^*uv^*v) = x(v^*v)$ as $uv^*v \in M$ and $u^*uv^* = v^*$. □

For any trace x on the positive part of a von Neumann algebra W, the Cauchy-Schwarz inequality $|x(v^*w)|^2 \leq x(v^*v)x(w^*w)$ holds for $v, w \in N$, where on the left hand side x is the linear extension to the definition ideal M. Moreover, when $v \in M$ is positive and x is normal, then $x(v\cdot) = x(v^{1/2} \cdot v^{1/2})$ by the proposition above, shows that $x(v\cdot)$ is also normal.

Proposition 7.4.3 *A trace x on the positive part of a von Neumann algebra W is semifinite if and only if its definition ideal is σ-weakly dense in W. In this case any orthogonal projection $p \in W$ is the supremum of all orthogonal projections majorized by p having finite trace.*

Proof Say x is semifinite. The σ-weak closure of its definition ideal M is of the form Wt for a central orthogonal projection $t \in W$. If $1 - t \neq 0$, pick a non-zero positive $w \in W$ majorized by $1 - t$ with $x(w) < \infty$. Then w belongs to the positive part of M, which is impossible.

Conversely, if M is σ-weakly dense in W and $w \in W$ is non-zero and positive, there is by Kaplansky's density theorem a net $\{w_i\}$ in the closed unit ball of M that converges strongly to 1. So $0 \neq w^{1/2}w_i w w_i^* w^{1/2} \leq \|w\|w$ and x is semifinite.

The supremum in the proposition is an orthogonal projection $q \in W$ majorized by p. If $q \neq p$ we may absurdly pick a non-zero orthogonal projection with finite trace majorized by $p - q$. □

Remark 7.4.4 Given a faithful semifinite normal trace x on the positive part of a von Neumann algebra W, we may form a GNS-representation π_x of W associated to x. Indeed, with notation from Proposition 7.4.2 define an inner product on N by $(v|w) = x(w^*v)$, form the Hilbert space completion V_x, and define $\pi_x(w) \in B(V_x)$ for $w \in W$ as the continuous extension of the map $N \to N$ that sends v to wv. The resulting representation is faithful because if $wN = \{0\}$, then $w = 0$ as $M \subset N$ is σ-weakly dense in V by the previous proposition. Moreover, if $\{w_i\}$ is a bounded net that converges strongly to 0, then $(\pi_x(w_i)(v)|v) = x(v^*w_i^*v) = \overline{x(vv^*w_i)} \to 0$ for $v \in N$, so π_x is also normal.

Proposition 7.4.5 *Let x be a normal trace on the positive part of a von Neumann algebra W. The supremum t of all orthogonal projections in W killed by x is a central orthogonal projection in W. And Wt consists of all $w \in W$ with $x(w^*w) =$ 0, while x is faithful on $W(1 - t)$. We call $1 - t$ the* support *of the trace x and denote it by $s(x)$.*

Proof By traciality we see that $u^*tu = t$ for any unitary $u \in W$, so t is central. By Proposition 7.1.3 we know that $(p \vee q - p) \sim (q - p \wedge q)$ for orthogonal projections p and q, so if $x(p) = 0 = x(q)$ then $x(p \vee q) = 0$. This shows that the set of orthogonal projections in W killed by x is upward filtered under $\leq$, so the identity map on this set is a net, and $x(t) = 0$ by normality. Thus $x(W_+t) = \{0\}$. Any non-zero positive $w \in W(1-t)$ majorizes a positive scalar multiple of some spectral projection p, and $x(p) > 0$, so x is faithful on $W(1-t)$.

It is now clear that Wt consists of all $w \in W$ with $x(w^*w) = 0$. □

Theorem 7.4.6 *A von Neumann algebra W on a separabel Hilbert space is semifinite if and only if there is a separating family of semifinite normal traces on W_+.*

Proof If W is semifinite, it has a non-zero finite orthogonal projection p. Pick by Zorn's lemma a maximal orthogonal family of orthogonal projections $p_i \sim p$. By generalized comparability there is a central orthogonal projection t such that $p_0t \prec pt$ and $p(1-t) \prec p_0(1-t)$, where $p_0 = 1 - \sum p_i$ with an additional index 0. Then $pt \neq 0$, otherwise $p_0t = 0$ and $p \prec p_0$, contradicting maximality. Pick $u_i \in W$ such that $u_i^*u_i = pt$ and $u_iu_i^* = p_it$, including henceforth the index 0.

Given a normal tracial state x on W_{pt}, the formula $y = \sum x(u_i^* \cdot u_i)$ defines a normal trace y on the positive part of Wt because $u_i = u_iu_i^*u_i$ and $\sum u_iu_i^* = t$, so

$$y(w^*w) = \sum x(u_i^*w^*u_ju_j^*wu_i) = \sum x(u_j^*wu_iu_i^*w^*u_j) = y(ww^*)$$

for w in the positive part of Wt. Clearly p_it belongs to the positive part of the definition ideal M of y, and as $t = \sum p_it$, we see that M is σ-weakly dense in Wt, so y is semifinite by Proposition 7.4.3.

If $0 \neq w \in Wt$, some $wu_i \neq 0$, and $x(u_i^*w^*wu_i) \neq 0$ for some normal tracial state x on the finite von Neumann algebra W_{pt}, so $y(w^*w) \neq 0$ for the corresponding y. Since $W(1-t)$ is also semifinite, we may also find a separating family of semifinite normal traces on the positive part of $W(1-t)$. By Zorn's lemma, we thus get a separating family of semifinite normal traces on W_+.

Conversely, pick by Zorn's lemma a maximal orthogonal family of finite orthogonal projections p_i. If $p = 1 - \sum p_i \neq 0$, there is by Proposition 7.4.3 an orthogonal projection $q \in W$ majorized by p and a semifinite normal trace y on the positive part of W with $y(q) \in \langle 0, \infty \rangle$. So $y(q \cdot q)$ is a non-trivial normal finite trace on W_q. Its restriction to the compression $(W_q)s(y(q \cdot q))$ is moreover faithful, so the latter von Neumann algebra is finite by Proposition 7.3.1, and p has a non-zero finite orthogonal projection, which contradicts maximality. Thus $\sum p_i = 1$. Hence any central orthogonal projection $t \in W$ has the finite subprojections tp_i, which are non-zero for some i. □

Corollary 7.4.7 *A von Neumann algebra W on a separable Hilbert space is semifinite if and only if it has a faithful semifinite normal trace on W_+.*

Proof It W is semifinite it has by the theorem and Zorn's lemma a family of semifinite normal traces y_i on W_+ with orthogonal supports that add up to one. Then $\sum y_i$ is the required trace. □

From the proof of the theorem above it is clear that a von Neumann algebra on a separable Hilbert space is semifinite if and only if its identity element is the strong sum of an orthogonal family of finite orthogonal projections.

Proposition 7.4.8 *A von Neumann algebra is semifinite if and only if it has a finite orthogonal projection with central cover one.*

Proof Pick by Zorn's lemma a maximal family of finite orthogonal projections p_i with pairwise orthogonal central covers. Then the orthogonal projection $p = \sum p_i$ is finite because if q is an orthogonal projection majorized by p and equivalent to p, then $qc(p_i) \sim pc(p_i) = p_i$ and $qc(p_i) \leq pc(p_i) = p_i$, so all $qc(p_i) = p_i$, and hence $p = qp \sum c(p_i) = q$. Moreover, if $1 - c(p) \neq 0$, it has a finite orthogonal subprojection with central cover orthogonal to all $c(p_i)$, contradicting maximality. □

Lemma 7.4.9 *If x and $x(w\cdot)$ are positive linear functionals on a von Neumann algebra with $w \in W$, then $x(w\cdot) \leq \|w\|x$.*

Proof Let $x_0 = x$ and $x_{n+1} = x_0(w^{2^n}\cdot)$. For a positive element v in the von Neumann algebra, we have $0 \leq x_0(wv^{1/2}v^{1/2}) \leq x_0(v)^{1/2}x_0(wvw^*)^{1/2}$ and $x_2(v) = \overline{x_1(vw^*)} = x_0(wvw^*) \geq 0$ for any w in the von Neumann algebra. Proceeding inductively we thus get

$$0 \leq x_1(v) \leq x_0(v)^{\frac{1}{2}+\cdots+\frac{1}{2^n}} x_0(w^{2^n}v)^{1/2^n},$$

which gives the desired estimate in the limit. □

We have the following Radon-Nikodym theorem in this context.

Proposition 7.4.10 *If $x \leq y$ are normal positive linear functionals on a von Neumann algebra W, there is $w \in W$ with $0 \leq w \leq 1$ such that $x = y(w \cdot w)$.*

Proof By considering the compression $W_{s(y)}$, we may assume that y is faithful, and by considering its isometric normal GNS-representation π_y, we may further assume that y is of the form $(\cdot(v)|v)$ for a cyclic vector v of the Hilbert space V where W acts. By Proposition 6.2.6 write $x = (\cdot w'(v)|w'(v))$ for $w' \in W'$ with $0 \leq w' \leq 1$. By Corollary 7.2.17 there is a unique partial isometry $u \in W'$ such that $|(w' \cdot (v)|v)| = (w'u^* \cdot (v)|v)$ as normal bounded linear functionals on W'. As the latter functional is by the lemma above not greater than $\|w'u^*\|(\cdot(v)|v)$ on W', there is by Proposition 6.2.6 an element $w \in W$ with $0 \leq w \leq 1$ such that $(w'u^* \cdot (v)|v) = (w \cdot (v)|v)$ on W'. But if $p \in W$ is the orthogonal projection onto the closure of $W'(v)$, we get $((1-p)(v)|v) = 0$ and $1 - p = 0$, so v is also cyclic for W'. Hence $w(v) = uw'(v)$. According to our polar decomposition, we also have $(w' \cdot (v)|v) = (wu \cdot (v)|v)$ on W', so $w'(v) = u^*w(v)$. Hence $x = (\cdot u^*w(v)|w'(v)) = (\cdot w(v)|w(v))) = y(w \cdot w)$. □

Lemma 7.4.11 *Let W be a von Neumann algebra acting on a Hilbert space V with a cyclic and separating unit vector v. If its vector state is tracial, there is* *an involutive conjugate linear isometry $J: V \to V$ such that $w \mapsto JwJ$ defines a conjugate linear isometric $*$-preserving and multiplicative surjection from W to W'.*

Proof The formula $J(w(v)) = w^*(v)$ for $w \in W$ defines a unique extension to an involutive conjugate linear isometry J on V, and clearly $JWJ \subset W'$. If $w' \in W'$ with $0 \leq w' \leq 1$, then $0 \leq (\cdot w'(v)|v) \leq (\cdot(v)|v)$, so by the previous proposition there is positive $w \in W$ such that $(\cdot w'(v)|v) = (w \cdot w(v)|v)$ on W, and $w'(v) = w^2(v) = Jw^2J(v)$ by tracialness and cyclicity. Thus $w' = Jw^2J$ as v is also separating. □

Lemma 7.4.12 *Suppose W is a von Neumann algebra on a separable Hilbert space V with $v \in V$ and orthogonal projections $p \in W$ and $q \in W'$ onto the closures of $W'(v)$ and $W(v)$, respectively. Then p is finite if and only if q is finite.*

Proof By symmetry it suffices to show one of the implications. Say p is finite. Since W_{pq} and $(W_p)c(q)$ are $*$-isomorphic and the latter is finite, we conclude that W_{pq} is finite. Now $v \in pq(V)$ is cyclic for W_{pq} and its commutant, so it is also separating for W_{pq}. Composing the faithful vector state $(\cdot(v)|v)/\|v\|^2$ on the center of W_{pq} with the center-valued trace on W_{pq} from Corollary 7.3.6, we get a faithful normal tracial state x on W_{pq}. Now W_{pq} is $*$-isomorphic to the image U under the GNS-representation associated to x, so U is finite, and so is U' by the previous lemma. By Corollary 6.2.9 the von Neumann algebras W_{pq} and U are unitarily equivalent, so $(W_{pq})'$ is also finite. If $pqwqp = 0$ for $w \in W'$, then $0 = uqwqp(v) = qwqu(v)$ for $u \in W$, so $qwq = 0$ as the closure of $W(v)$ is $q(V)$. Hence $qwq \mapsto pqwqp$ defines a $*$-isomorphism from $qW'q$ onto $pqW'qp$, which is $(W_{pq})'$ by Proposition 5.5.8. Thus q is finite. □

Theorem 7.4.13 *The commutant of a semifinite von Neumann algebra on a separable Hilbert space is semifinite.*

Proof Suppose W is a semifinite von Neumann algebra on a separable Hilbert space V with a central orthogonal projection t such that $W't$ does not contain any non-zero finite orthogonal projection. Then as Wt is semifinite, we can from the outset assume that W is semifinite and that W' does not contain any non-zero finite orthogonal projection. By Proposition 7.4.8 there is a finite orthogonal projection $p \in W$ with central cover one, so $W = Wc(p)$ is $*$-isomorphic to W_p, and is thus finite, while W' is $*$-isomorphic to W_p' by Proposition 5.5.8, which therefore has no non-zero finite orthogonal projections. So we might as well assume that W is finite and that W' has no non-zero finite orthogonal projections. Let $p \in W$ and $q \in W'$ be the orthogonal projections associated as in the lemma above to $v \in V$. Since p is finite, then according to this lemma, so is q, which is a contradiction. □

Corollary 7.4.14 *A von Neumann algebra W on a separable Hilbert space is semifinite if and only if it is $*$-isomorphic to a von Neumann algebra on a separable Hilbert space with finite commutant.*

Proof For the forward implication, note that W' is semifinite by the theorem, so it has by Proposition 7.4.8 a finite orthogonal projection p with central cover one.

Then W is $*$-isomorphic to W_p and $(W_p)' = W'_p$. The opposite direction is clear from the theorem. □

7.5 Classification of Factors

Lemma 7.5.1 *Let v be a member of a von Neumann algebra such that $p = v^*v$ is an orthogonal projection with $vv^* < p$. Put $q_0 = p$ and $q_n = v^n v^{*n}$ for $n \in \mathbb{N}$. Then the elements $e_n = q_n - q_{n+1}$ form an orthogonal sequence of non-zero orthogonal projections that are pairwise equivalent, majorized by p, and which tends strongly to zero.*

Proof Now $(1-p)vv^*(1-p) = 0$, so $pv = v$ and $v^{*n}v^n = p$. Hence the q_n's are orthogonal projections such that $q_n q_{n+1} = v^n pvv^{*(n+1)} = q_{n+1}$, showing that they are decreasing. Thus the e_n's form an orthogonal sequence of orthogonal projections majorized by p which tends strongly to zero by Vigier's theorem. Now $(vq_n - vq_{n+1})^*(vq_n - vq_{n+1}) = e_n$ and $(vq_n - vq_{n+1})(vq_n - vq_{n+1})^* = e_{n+1}$ hold as $q_n p = q_n = v^n v^{*n}$, showing that the e_n's are pairwise equivalent and $e_0 \neq 0$ by assumption. □

Proposition 7.5.2 *Every non-zero central orthogonal projection in a von Neumann algebra W fails to be finite if and only if W has an orthogonal sequence of orthogonal projections p_n that are all equivalent to the identity element and add up to one if and only if W has an orthogonal projection p such that $p \sim (1-p) \sim 1$.*

Proof For the first forward implication, by using that 1 is not finite, there is $v \in W$ such that $v^*v = 1$ and $vv^* < 1$. Extend by Zorn's lemma the associated sequence of e_n's from the lemma above to a maximal orthogonal family $\{e_i\}_{i\in I}$ of orthogonal projections that are pairwise equivalent. Let $q = 1 - \sum_{i\in I} e_i$. By generalized comparability there is a non-zero central orthogonal projection $t \in W$ such that $qt \prec e_0 t$ and $e_0(1-t) \prec q(1-t)$. By Proposition A.2.4 write I as a countable disjoint union of sets I_m, each with cardinality I. The elements $r_1 = qt + \sum_{i\in I_1} e_i t$ and $r_m = \sum_{i\in I_m} e_i t$ for $m \geq 2$ form an orthogonal sequence of orthogonal projections that add up to t and which are all equivalent to t because $r_m \sim r_n \sim \sum_{i\in I} e_i t$ for all m, n by Proposition 7.1.4. Repeating this for the von Neumann algebra $W(1-t)$, we get an orthogonal family of non-zero central orthogonal projections $t_j \in W$ and for each j, and orthogonal sequence of orthogonal projections $r_{jn} \in W$ that add up to t_j and are all equivalent to t_j. Then the elements $p_n = \sum r_{jn}$ have clearly the required property. The second forward implication holds for $p = \sum_{n=0}^{\infty} p_{2n+1}$ by Proposition 7.1.4. The proof is completed by noting that $pt \sim (1-p)t \sim t$ for any non-zero central orthogonal projection $t \in W$. □

Corollary 7.5.3 *The supremum of two finite orthogonal projections in a von Neumann algebra is also finite.*

Proof We may assume for the finite orthogonal projections p, q that their supremum is one. Then $1 - p \sim q - p \wedge q \leq q$ by Proposition 7.1.3, so $1 - p$ is finite. Suppose t is a non-zero central orthogonal projection in the von Neumann algebra W such that every non-zero central orthogonal projection in Wt fails to be finite, then as pt and qt are finite with supremum t, we might as well assume that every non-zero central orthogonal projection in W is not finite. Pick by the proposition above an orthogonal projection $r \in W$ such that $r \sim 1 - r \sim 1$. By generalized comparability there is a central orthogonal projection $t \in W$ such that $rt \prec pt$ and $(1-r)(1-t) \prec (1-p)(1-t)$. Hence $t \sim rt$ and $1 - t \sim (1-r)(1-t)$ are finite, which is absurd. □

Lemma 7.5.4 *A von Neumann algebra W on a separable Hilbert space is semifinite if and only if W has an increasing net of finite orthogonal projections q_i with strong limit one.*

Proof If W is semifinite, pick an orthogonal family of finite orthogonal projections $p_i \in W$ that add up to one. The forward implication is now clear from the corollary above. The proof is completed by noting that for any non-zero central orthogonal projection $t \in W$, there is a q_i such that $q_i t \neq 0$. □

Proposition 7.5.5 *Let W be a factor on a separable Hilbert space and let x be a faithful normal trace on its positive part that is finite on some non-zero finite orthogonal projection whenever W contains a non-zero finite orthogonal projection. Then an orthogonal projection in W is finite if and only if its trace is finite, and $p \prec q$ if and only if $x(p) \leq x(q)$. If W contains a non-zero finite orthogonal projection, then x is semifinite. Finally x is unique up to multiplication by a positive scalar.*

Proof If p is a non-finite orthogonal projection of W, then as $(W_p)' = W'_p$ by Proposition 5.5.8 and W is a factor, we see that W_p has no non-zero central finite orthogonal projections, so by the previous proposition it has an orthogonal projection q such that $q \sim p - q \sim p$. Hence $x(p) = x(p-q) + x(q) = 2x(p)$ and $x(p) = \infty$ as x is faithful. If on the other hand $p \neq 0$ is finite, then by assumption there is an orthogonal projection $q \in W$ such that $x(q) < \infty$. Now $\prec$ is a total order in a factor, and if $p \prec q$, then $x(p) < \infty$. While if $q \prec p$, we may by Zorn's lemma pick a maximal orthogonal family of orthogonal subprojections p_i of p that are equivalent to q. Then $p - \sum p_i \prec q$ by maximality. Moreover, there can only be finitely many such projections p_i because otherwise we may pick out a sequence from them and take their sum r. By Corollary 7.1.17 we get $r \prec p$, so r should be finite and yet it is equivalent to any infinite subsum by Proposition 7.1.4. Now $x(p) - \sum x(p_i) \leq x(q)$ shows that $x(p) < \infty$.

The forward implication of the second statement is obvious. Conversely, if $x(p) \leq x(q)$ and q is equivalent to a proper orthogonal subprojection of p, then $x(q) < x(p)$. So the only option is that $p \prec q$ as we have a total order.

If W contains a non-zero finite orthogonal projection, then being a factor, it is semifinite. Picking to a non-zero $w \in W_+$ a non-zero orthogonal projection p and a scalar $c > 0$ with $cp \leq w$, then there is a non-zero finite orthogonal projection

$q \leq p$. Then $cq \leq w$ and $x(cq) \in \langle 0, \infty \rangle$ by the first paragraph above, so x is semifinite.

It remains to prove the last statement. If W contains no non-zero finite orthogonal projection, then x is infinite on every non-zero element of W, so x is uniquely determined. Say x, y are two faithful semifinite normal traces on W_+. Suppose first that W is finite. By the first proven claim x, y extend by Proposition 7.4.2 to faithful tracial normal positive functionals on W. By Proposition 7.4.10 there is positive $w \in W$ such that $x = (x + y)(w\cdot)$, and for any $v \in W$ we have $(x + y)((wv - vw)\cdot) = 0$ by tracialness, so $w \in Z(W) = \mathbb{C}$ by faithfulness. So $x = cy$ for a positive scalar c. For the more general case pick by the previous lemma an increasing net of finite orthogonal projections q_i with $\vee q_i = 1$. Since W_{q_i} is finite there are from above positive scalars c_i such that $x = c_i y$ on its positive part. But c_i is independent of i as the family $\{q_i\}$ is increasing. With $c = c_i$ all $x(q_i \cdot q_i) = cy(q_i \cdot q_i)$ on the positive part of W. For $u \in W_+$ we thus get $x(u) = \lim x(u^{1/2}q_i u^{1/2}) = \lim x(q_i u q_i) = c \lim y(q_i u q_i) = c \lim y(u^{1/2}q_i u^{1/2}) = cy(u)$ by normality. □

Theorem 7.5.6 *By Corollary 7.3.6 and Corollary 7.4.7 any factor on a separable Hilbert space possesses a trace x as prescribed under the hypothesis of the proposition above. Letting D denote the values of x on the orthogonal projections of W, then by appropriately rescaling x, the set D is $\{0, 1, \cdots, n\}$ when W is of type I, allowing also $n = \infty$, it is $[0, 1]$ when W is of type II_1, it is $[0, \infty]$ when W is of type II_∞, and it is $\{0, \infty\}$ when W is of type III.*

Proof The type III case is obvious, and the type I case is immediate from Theorem 7.1.15. For the type II_1 case, let x be the unique faithful normal tracial state on W. By Proposition 7.1.14 we may repeatedly half-en orthogonal projections into pairwise orthogonal ones that are all equivalent to the original one, so D must contain any number of the form $k2^n$ for $n = 0, 1, \ldots$ and k an integer in $[1, 2^n]$. For any $c \in [0, 1]$ we may therefore pick orthogonal projections $p_n \in W$ such that $\{x(p_n)\}$ increases towards c. By the proposition above $p_n \prec p_{n+1}$ for all n. Let $q_1 = p_1$ and $q_1 \sim q \leq p_2$. By Proposition 7.1.16 we have $1 - q_1 \sim 1 - q$, so $p_2 - q \prec 1 - q_1$. Let $p_2 - q \sim r \leq 1 - q_1$. Then $p_2 \sim q_2 \equiv q_1 + r \geq q_1$. Continuing this way we get an increasing sequence of orthogonal projections $q_n \sim p_n$. Then $x(\vee q_n) = \sup x(q_n) = \sup x(p_n) = c$, so $D = [0, 1]$ as D cannot be larger than the latter set. As for the type II_∞ case, pick an increasing net of finite orthogonal projections $p_i \in W$ with $\vee p_i = 1$. Then $\{x(p_i)\}$ increases without bound, and by the above, we know that $[0, x(p_i)] \subset D$, so $D = [0, \infty]$. □

In the type I case the factor is $*$-isomorphic to $B(V)$ for a separable Hilbert space V, and then $x(p) = \dim(p(V))$ for any orthogonal projection p in the factor, given x is scaled so that $x(1) = \dim(V) = n$. We say that the factor is of *type I_n*. The curious type II cases involve continuous dimensions, while the set D in the type III case suggests that tracialness is too limited a notion. We shall later see that factors of all types exist.

The trace x in the previous theorem restricted to the orthogonal projections on the factor W is clearly a dimension function on W, and the map $[p] \mapsto x(p)$ from

$D(W)$ to D identifies these spaces as totally ordered abelian partial semigroups with the obvious order and addition on D. We could have started defining a dimension function d on W, and then extended this to a trace of the desired type. For instance, when W is of type II_1 we can proceed as follows. First half-en the identity element $1 = p_1 + q_1$ with $p_1 \sim q_1$. By generalized comparison and finiteness, the orthogonal projection p_1 is unique up to equivalence. Next split $q_1 = p_2 + q_2$ with $p_2 \sim q_2$ unique up to equivalence. Continuing this way and using separability, we get a sequence of orthogonal projections $p_n \in W$ that add up to one and are uniquely constructed up to equivalence by repeated halving. Now given an orthogonal projection $p \in W$, we define $d(p)$ to be the binary expansion $0, c_1 c_2 \cdots$ where $c_1 = 1$ if $p_1 \prec p$ and $c_1 = 0$ otherwise, and $c_2 = 1$ if $c_1 p_1 + p_2 \prec p$ and $c_2 = 0$ otherwise, and so forth. Then $p \sim \sum c_n p_n$ and d is indeed a dimension function.

While it is not obvious how to extend such dimension functions to the required traces (if you do not already possess them), the following result shows that it is in principle possible, and this was the route taken by Murray and von Neumann. By the uniqueness statement for the trace in the theorem above, it also shows that any dimension function on a factor acting on a separable Hilbert space is unique up to multiplication by a positive scalar, a fact that can also be checked directly.

Proposition 7.5.7 *Any dimension function d on a factor W acting on a separable Hilbert space is the restriction of a trace x with the properties described in the previous proposition.*

Proof For a non-finite orthogonal projection p in the factor we have $d(p) = \infty$ because p is by Proposition 7.5.2 the sum of an orthogonal sequence of orthogonal projections, each of which is equivalent to p. So $d = x$ on the orthogonal projections of a factor without non-zero finite orthogonal projections.

Since minimal orthogonal projections in a type I factor are all equivalent, we may assume that d is one on each one of them, and as any orthogonal projection in this case is an orthogonal sum of minimal orthogonal projections, the image of d is $\{0, 1, \ldots, n\}$, and d is by the theorem above the restriction of the required trace.

In the type II_1 case, we may by the first paragraph above assume that $d(1) = 1$ and that x is a state. Given an orthogonal projection p in this case, we see from the proof of the theorem above that $d = x$ on orthogonal projections p_n such that $\{x(p_n)\}$ increase towards $x(p)$. Then $p_n \prec p$ and $d(p) \geq d(p_n) = x(p_n) \to x(p)$, so $d(p) \geq x(p)$ and similarly $d(1 - p) \geq x(1 - p)$, which yields $d(p) = x(p)$.

Finally, if the factor contains a non-zero orthogonal projection, then by assumption d is finite on some non-zero orthogonal projection p, which by the first paragraph of this proof must be finite. Scaling x appropriately we may therefore assume that $x(p) = d(p)$. It suffices to check equality on any finite orthogonal projection q. By the last corollary above $r = p \vee q$ is also finite, and by the two previous paragraphs we know that x and d coincide on the orthogonal projections in W_r, and hence also on q. □

Exercises

For Sect. 7.1

1. Say p_i, q_i are orthogonal projections in a von Neumann algebra with $p_i \prec q_i$. Show that $p_1 \vee p_2 \prec q_1 + q_2$ when $q_1 q_2 = 0$.
2. Prove that if $p \sim q$ are orthogonal projections in a finite von Neumann algebra, then $1 - p \sim 1 - q$.
3. Show that a von Neumann algebra W is finite if and only if to $p \sim q$ in W, there is a unitary $w \in W$ such that $wpw^* = q$.
4. Show that in a semifinite von Neumann algebra there is a pairwise orthogonal family of non-zero central orthogonal projections p_i such that $\sum p_i = 1$ strongly and $p_i = \sum q_{ij}$ strongly for a pairwise orthogonal family of mutually equivalent finite orthogonal projections q_{ij}.
5. Given finitely many orthogonal projections p_i, q_i in a von Neumann algebra such that $p_i \sim p_j$ and $q_i \sim q_j$ and $\sum p_i = 1 = \sum q_i$ strongly. Deduce $p_i \sim q_i$.

For Sect. 7.2

1. Show that for two norm closed ideals of a von Neumann factor one will be contained in the other.
2. Show that normal states on abelian von Neumann algebras are vector states.
3. Produce a factor acting on a separable Hilbert spaces with no separating vector.
4. Show that an orthogonal projection in a von Neumann algebra is the support of a normal state if and only if it is countably decomposable.
5. Suppose a C*-algebra W contains the strong limit of any bounded above increasing net in W. Show that W contains the range projections of its elements.

For Sect. 7.3

1. Suppose g is a finite center-valued trace on a finite von Neumann algebra W acting on a separable Hilbert space, and define $f: W \to W$ by $f(w) = w - g(w)$. Show that f is a bounded operator with $f^2 = f$. Find its range and kernel. Show that $\|f\| = 2$ when W is of type II_1, and that $\|f\| = 2(n-1)/n$ when W is a factor of type I_n with $n < \infty$.
2. Suppose g is a finite center-valued trace on a finite von Neumann algebra W acting on a separable Hilbert space V. Let X the set of vectors $v \in V$ with $u \in V$ such that $(\cdot v|v) \leq (g(\cdot)u|u)$ on the positive part of W. Show that X is separating for W.
3. Say f is a $*$-epimorphism from a von Neumann algebra W to a C*-algebra U. Show that $f(Z(W)) = Z(U)$ and that for fixed $a \in U$ the closed convex hull of the set of uau^* for all unitaries $u \in U$ meets $f(Z(W))$.
4. Suppose p and q are orthogonal projections in a von Neumann algebra W, and that q is a strong limit of an increasing family of orthogonal projections $q_i \in W$. Show that $q_i \vee p \to q \vee p$ strongly, that $\lim(q_i \wedge q) \leq q \wedge p$, and that equality can fail in the latter case when W is not finite.

5. Show that a von Neumann algebra W on a separable Hilbert space is finite if and only if for $w_i \in W$ with u_i contained in the intersection of $Z(W)$ with the closed convex hull of w_i, then $u_1 + u_2$ is contained in the intersection of $Z(W)$ with the closed convex hull of $w_1 + w_2$.

For Sect. 7.4

1. Suppose U, W are von Neumann algebras acting on the same separable Hilbert space, that $U \subset W$ and that W has a faithful normal tracial state x. Show that to $w \in W$ there is a unique element $f(w) \in U$ such that $x(wu) = x(f(w)u)$ for all $u \in U$. Prove that we thus get a σ-weakly continuous faithful bounded linear map $f\colon W \to U$ that reduces to the identity map on U.
2. Let g be a finite center-valued trace on a finite von Neumann algebra W on a separable Hilbert space. Show that $g(p) \leq c(p)$ and $c(g(p)) = c(p)$ for any orthogonal projection $p \in W$.
3. Show that a von Neumann algebra is finite if and only if it is $*$-isomorphic to a von Neumann algebra with abelian commutant.
4. Show that a normal trace on a factor is either faithful or identically zero.
5. Suppose W is a finite factor with finite commutant W', both acting on a separable Hilbert space, and let x and y be the tracial faithful normal states on W and W', respectively. Prove that there is a positive number c such that whenever $p \in W$ and $q \in W'$ are orthogonal projections, then $x(p) = cy(q)$ if and only if there is a vector v such that $\mathrm{im}(p)$ and $\mathrm{im}(q)$ are the closures of $W'(v)$ and $W(v)$, respectively.

For Sect. 7.5

1. Show that a von Neumann algebra on a separable Hilbert space is of type I, II, III if and only if its commutant is of the corresponding type.
2. Suppose W is a von Neumann algebra with a faithful semifinite normal trace x, and let U be a von Neumann subalgebra of W such that the restriction of x to it is semifinite. Show that there exists a faithful normal linear map $E\colon W \to U$ such that $E^2 = E$ and $xE = x$.
3. Prove that a semifinite normal trace on a von Neumann algebra is a sum of normal positive linear functionals.
4. Prove that a von Neumann algebra on a separable Hilbert space is finite if and only if the $*$-operation is σ-strongly continuous.
5. In the positive part of a von Neumann algebra W write $u \approx w$ if there is a family $\{v_i\}$ in W such that $u = \sum v_i^* v_i$ and $w = \sum v_i v_i^*$. Show that $\approx$ is an equivalence relation, that $\sum a_i \approx \sum b_i$ if all $a_i \approx b_i$, and that $p \approx q$ if $p \sim q$ and either p or q is finite as orthogonal projections.

Chapter 8
Tensor Products

Tensor products is the study of multilinear maps by linear maps, meaning that the multilinear maps from a space factor uniquely through a linear map defined on another vector space called the tensor product of the vector spaces occurring as direct products in the domain of the multilinear maps. Algebraic structures are easily passed on to the tensor product, but the issue of topology, while canonical for Hilbert spaces, is delicate already for C*-algebras; the algebraic tensor product can be given a whole series of norms leading to non-isomorphic C*-algebra completions, except when one of the C*-algebras in the product is so called nuclear. In the general case there is always a maximal and a minimal C*-norm, and the minimal, or spatial, one is gotten by pulling back the operator norm from the tensor product of injective representations. This result requires a careful study of the states of the C*-algebras involved, and this is also important in proving that any commutative C*-algebra is nuclear. The tensor product of commutative C*-algebras corresponds to the direct product of the locally compact Hausdorff character spaces. We also show that AF-algebras are nuclear, and that nuclearity is preserved under extensions.

The von Neumann tensor product of two von Neumann algebras is the weak closure of the algebraic tensor product acting on the Hilbert space tensor product of the original von Neumann algebras. The commutant of this tensor product is the von Neumann tensor product of the commutants. We investigate to what extend types of von Neumann algebras are preserved under formations of von Neumann tensor products and commutants. We also show that tensor products of normal $*$-epimorphisms are again such epimorphisms, and that such maps can be written as compositions of easier maps.

In the remaining part of the chapter we study completely positive linear maps, that is, linear maps between C*-algebras that extend to positive maps on all matrix algebras with elements of the C*-algebras as entries. Positive linear maps between C*-algebras are automatically bounded, and they are completely positive if their domain or range generate commutative algebras. Completely positive linear maps are shown to be of the form $T^*\pi(\cdot)T$ for some representation π and bounded linear

L. Tuset, *Analysis and Quantum Groups*,
https://doi.org/10.1007/978-3-031-07246-8_8

map T, and even more is known about this map. This allows inducing up completely positive linear maps from C*-subalgebras. The formation of tensor products of completely positive linear maps with mutually commuting images produces a map of the same kind. A special type of completely positive linear maps from a C*-algebra to itself are the so called conditional expectations, which behave almost like $*$-homomorphisms. These are exactly the norm one contractions that fix the elements in their image. Typical examples are states, center valued traces, and projections onto one component in a von Neumann tensor product; the latter are even σ-weakly continuous.

At the end we include a section on Hilbert modules, which generalize Hilbert spaces, and are non-commutative generalizations of sections on a vector bundle with Euclidean spaces as fibers. The 'inner product' for Hilbert modules takes values in a C*-algebra playing the role as the ground field. The bounded operators between Hilbert modules are taken over by maps having an adjoint map, and is endowed with the so called strict topology. The closed span by rank one type of operators is to be thought of as the space of compact operators. The C*-algebra of adjointable maps on a C*-algebra W is up to isomorphism the unique maximal one containing the 'compact operators' as an essential ideal. When W is represented non-degenerately on a Hilbert space, the adjointable maps form the so called multiplier algebra $M(W)$ of W, which consists of $T \in W''$ such that TW and WT are subsets of W. When $W = C_0(X)$, then $M(W) = C(Y)$, where Y is the Stone-Cech compactification of X. We discuss also extensions of maps in this context, and for the sake of completeness, include Kasparov's stabilization theorem.

8.1 Tensor Products of C*-Algebras

The *(algebraic) tensor product* of two real or complex vector spaces V, W is a real or complex vector space $V \otimes W$ and a bilinear map $h\colon V \times W \to V \otimes W$ such that any bilinear map $f\colon V \times W \to U$ into another real or complex vector space U factorizes as $f = gh$ for a unique linear map $g\colon V \otimes W \to U$. The tensor product is unique up to isomorphism of vector spaces because if U was another one we would get linear maps $V \otimes W \to U$ and $U \to V \otimes W$ that by uniqueness of the factorization would be inverses to each other. For existence consider the real or complex vector space with linear basis $V \times W$, in other words, the *free vector space* over $V \times W$. Divide this out by the subspace generated by elements of the type

$$(a_1v_1 + a_2v_2, b_1w_1 + b_2w_2) - a_1b_1(v_1, w_1) - a_1b_2(v_1, w_2) - a_2b_1(v_2, w_1) - a_2b_2(v_2, w_2)$$

for scalars a_i, b_i and $v_i \in V$ and $w_i \in W$, to get $V \otimes W$ with h as the quotient map restricted to $V \times W$. We denote $h(v, w)$ by $v \otimes w$ and call the latter an *elementary tensor*. Thus every element of $V \otimes W$ is a finite linear combination of elementary tensors. Such a linear combination is in general not unique.

Given linear maps $f_i : V_i \to W_i$ between vector spaces, we form their tensor product $f_1 \otimes f_2 : V_1 \otimes V_2 \to W_1 \otimes W_2$ as the unique map in the factorization of the bilinear map $V_1 \times V_2 \to W_1 \otimes W_2$ which sends (v_1, v_2) to $f_1(v_1) \otimes f_2(v_2)$ for $v_i \in V_i$. In particular, when the f_i's are linear functionals, we can use this to show that $\{v_i \otimes w_j\}$ is a linear basis of $V \otimes W$ whenever $\{v_i\}$ and $\{w_j\}$ are linear bases of V and W, respectively. Thus every element of $V \otimes W$ can be written as a finite sum $\sum u_j \otimes w_j$ for unique $u_j \in V$.

The *flip* is the linear isomorphism $V \otimes W \to W \otimes V$ given by the factorization of the bilinear map $(v, w) \mapsto w \otimes v$ for $v \in V$ and $w \in W$. In the same vain we have an isomorphism $V_1 \otimes (V_2 \otimes V_3) \to (V_1 \otimes V_2) \otimes V_3$ for vector spaces V_i.

By the *tensor product of Hilbert spaces* V, W we mean the Banach space completion of $V \otimes W$ with respect to the norm of the well-defined inner product given by $(v_1 \otimes w_1 | v_2 \otimes w_2) = (v_1|v_2)(w_1|w_2)$ for $v_i \in V$ and $w_i \in W$. We certainly get such a sesquilinear map, and if $(u|u) = 0$ then $u = 0$ because if $u = \sum v_i \otimes w_i$ with the w_i's orthonormal, then $\|u\|^2 = \sum \|v_i\|^2$, so all $v_i = 0$. We also denote the Hilbert space completion by $V \otimes W$; the context will make it clear what we mean. Note that $\|v \otimes w\| = \|v\|\|w\|$. If $\{v_i\}$ is an orthonormal basis of V and $\{w_j\}$ is an orthonormal basis of W, then $\{v_i \otimes w_j\}$ is an orthonormal basis of $V \otimes W$.

Example 8.1.1 Let X and Y be measurable sets with σ-finite measures μ and ν. It is clear from Example 4.1.16 that the map $L^2(\mu) \otimes L^2(\nu) \to L^2(\mu \times \nu)$ that sends $f \otimes g$ to the function on $X \times Y$ with the same symbol in that example is a unitary operator. So the Hilbert tensor product of the L^2-spaces corresponds to the L^2-space of the product measure. ◇

If $f \in B(V)$ and $g \in B(W)$ for Hilbert spaces V, W, then $f \otimes g$ has a continuous extension to a bounded linear map on the Hilbert space $V \otimes W$, again denoted by $f \otimes g$, such that $\|f \otimes g\| = \|f\|\|g\|$. For boundedness it suffices to assume that f and g are unitary, and on $u = \sum v_i \otimes w_i$ with the w_i's orthonormal, we then get $\|(f \otimes g)(u)\| = \|u\|$. For general f and g, write $f \otimes g$ as $(f \otimes \iota)(\iota \otimes g)$ and observe that $f \mapsto f \otimes \iota$ and $g \mapsto \iota \otimes g$ are injective $*$-homomorphism $B(V) \to B(V \otimes W)$ and $B(W) \to B(V \otimes W)$ respectively, and hence isometric maps, between C*-algebras. Thus $\|f \otimes g\| \leq \|f\|\|g\|$. Equality is obtained by finding unit vectors v, w to $\varepsilon > 0$ such that $\|f\| \geq \|f(v)\| - \varepsilon$ and $\|g\| \geq \|g(w)\| - \varepsilon$, so $\|(f \otimes g)(v \otimes w)\|$ approximates $\|f\|\|g\|$.

We also see that $(f_1 \otimes g_1)(f_2 \otimes g_2) = f_1 f_2 \otimes g_1 g_2$ and $(f_1 \otimes g_1)^* = f_1^* \otimes g_1^*$ for $f_i \in B(V)$ and $g_i \in B(W)$. If V and W are algebras, then the algebraic tensor product $V \otimes W$ will also be an algebra with product uniquely given by the same formula as above. The identity element will in this case be $1 \otimes 1$. If V and W are $*$-algebras, the last formula above turns $V \otimes W$ into a $*$-algebra. Indeed, if $\sum v_i \otimes w_i = 0$ and $w_i = \sum c_{ij} u_j$ for scalars c_{ij} and a linear basis $\{u_j\}$ of W, then $\sum c_{ij} v_i = 0$ for all j, so $\sum v_i^* \otimes w_i^* = \sum \bar{c}_{ij} v_i^* \otimes u_j^* = 0$, which shows that the $*$-operation on $V \otimes W$ is well-defined.

If U is a $*$-algebra and $f: V \to U$ and $g: W \to U$ are $*$-homomorphisms such that $f(V)$ and $g(W)$ commute with each other, then there is clearly a unique $*$-homomorphism $V \otimes W \to U$ such that $v \otimes w \mapsto f(v)g(w)$ for $v \in V$ and $w \in W$.

Lemma 8.1.2 *For representations $\pi_i: W_i \to B(V_i)$ of C*-algebras there is a unique representation $\pi_1 \otimes \pi_2$ of $W_1 \otimes W_2$ on $V_1 \otimes V_2$ such that $(\pi_1 \otimes \pi_2)(w_1 \otimes w_2) = \pi_1(w_1) \otimes \pi_2(w_2)$ for $w_i \in W_i$. If π_i are both injective, then so is $\pi_1 \otimes \pi_2$ on $W_1 \otimes W_2$.*

Proof Now $\pi_1 \otimes \pi_2$ is well-defined by the last paragraph before this lemma as $\pi_1(W_1) \otimes 1$ and $1 \otimes \pi_2(W_2)$ generate a commutative subalgebra of $B(V_1 \otimes V_2)$.

As for injectivity, if $(\pi_1 \otimes \pi_2)(\sum w_{1i} \otimes w_{2i}) = 0$ for $w_{1i} \in W_1$ and $\{w_{2i}\}$ a linear basis of W_2, then all $\pi_1(w_{1i}) = 0$ as the elements $\pi_2(w_{2i})$ are linearly independent by injectivity of π_2. Hence all $w_{1i} = 0$ by injectivity of π_1. □

Definition 8.1.3 The *spatial tensor product* $W_1 \otimes_{min} W_2$ of two C*-algebras W_i is the C*-algebra completion of the $*$-algebra $W_1 \otimes W_2$ with respect to the *spatial norm* $\| \cdot \|_{min}$, which is the norm pulled back from the image space of $\pi_1 \otimes \pi_2$ and where π_i is the universal representation of W_i.

Note that $\|w_1 \otimes w_2\|_{min} = \|w_1\| \|w_2\|$ for $w_i \in W_i$ with notation as above. In general there can be several C*-norms on the algebraic tensor product $V \otimes W$ of two C*-algebras. We denote the completion with respect to such a norm s by $V \otimes_s W$.

Proposition 8.1.4 *Given non-zero C*-algebras W_i and a C*-norm s on $W_1 \otimes W_2$ and a non-degenerate representation π of $W_1 \otimes_s W_2$ on V. Then there are unique representations π_i of W_i on V such that $\pi(w_1 \otimes w_2) = \pi_1(w_1)\pi_2(w_2) = \pi_2(w_2)\pi_1(w_1)$ for all $w_i \in W_i$.*

Proof Let $v = \sum \pi(w_{1i} \otimes w_{2i})(v_i)$ for $w_{ji} \in W_j$ and $v_i \in V$, and set $\pi_1(w)(v) = \lim_j \pi(w \otimes u_j)(v)$, where $\{u_j\}$ is the canonical approximate unit of W_2. That such a limit exists and equals $\sum_i \pi(ww_{1i} \otimes w_{2i})(v_i)$, so that $\pi_1(w)$ is a well-defined and bounded linear map and thus extends to a bounded linear map on V, is clear from the fact that $w_1 \mapsto w_1 \otimes w_2$ defines a bounded linear map from W_1 to $W_1 \otimes_s W_2$ for any $w_2 \in W_2$. To check this latter statement, say $r_n \to 0$ in W_1 and that $r_n \otimes w_2 \to b \in W_1 \otimes_s W_2$ with r_n and w_2 positive. Then for any positive linear functional x on $W_1 \otimes_s W_2$, the linear functional $x(\cdot \otimes w_2)$ on W_1 is positive and thus bounded, so $x(b) = \lim x(r_n \otimes w_2) = x(0 \otimes w_2) = 0$ and $b = 0$, allowing for the closed graph theorem to give the result. Hence we get a $*$-homomorphism $\pi_1: W_1 \to B(V)$, and similarly a $*$-homomorphism $\pi_2: W_2 \to B(V)$ such that $\pi(w_1 \otimes w_2) = \pi_1(w_1)\pi_2(w_2) = \pi_2(w_2)\pi_1(w_1)$ for all $w_i \in W_i$.

As for uniqueness, note first that any two representations satisfying the conclusion of the proposition are non-degenerate since if for instance $\pi_2(w_2)(v') = 0$ for all $w_2 \in W_2$, then $\pi(w_1 \otimes w_2)(v') = 0$ also for all $w_1 \in W_1$, so $v' = 0$ as π is non-degenerate. Hence $\pi_2(u_j) \to 1$ strongly, and $\{\pi(w \otimes u_j)\}$ converges strongly to any candidate for π_1 evaluated at w, so all such candidates coincide. The same is true for π_2. □

Corollary 8.1.5 *For any C*-seminorm s on the algebraic tensor product of two C*-algebras we have* $s(v \otimes w) \leq \|v\| \|w\|$.

Proof Letting t be the supremum of s and the spatial norm, and letting π be the universal representation of the C*-algebra completion of the algebraic tensor product, we have $\pi(v \otimes w) = \pi_1(v)\pi_2(w)$ by the theorem, so $s(v \otimes w) \leq t(v \otimes w) = \|\pi_1(v)\pi_2(w)\| \leq \|v\| \|w\|$. □

Definition 8.1.6 The *maximal C*-norm* $\| \cdot \|_{max}$ evaluated at an element of the tensor product $V \otimes W$ of two C*-algebras is the supremum of all C*-norms on $V \otimes W$ evaluated at the element, which is finite by the corollary above. Write $V \otimes_{max} W$ for the completion with respect to the maximal C*-norm on $V \otimes W$.

Any C*-seminorm on the algebraic tensor product of two C*-algebras is majorized by the maximal C*-norm.

Corollary 8.1.7 *If* $f : V \to U$ *and* $g : W \to U$ *are* $*$*-homomorphisms of C*-algebras with images generating a commutative subalgebra of* U*, the unique* $*$*-homomorphism* $h : V \otimes W \to U$ *such that* $v \otimes w \mapsto f(v)g(w)$ *for* $v \in V$ *and* $w \in W$*, extends by continuity to* $V \otimes_{max} W$.

Proof Considering the spatial norm we see that $\|h(\cdot)\|$ is majorized by the maximal C*-norm on $V \otimes W$. □

Definition 8.1.8 A C*-algebra V is *nuclear* if there is only one C*-norm on $V \otimes W$ for every C*-algebra W.

Due to the flip it is immaterial what order of V and W in the tensor product we use.

Proposition 8.1.9 *Given a non-empty collection of C*-subalgebras* V_i *of a C*-algebra* V *with dense union and which is upward filtered under inclusion. Then* V *is nuclear when all* V_i *are.*

Proof Let W be any C*-algebra and consider two C*-norms s and t on $V \otimes W$. Then $A = \cup_i V_i \otimes W$ is a $*$-subalgebra of $V \otimes W$ which is dense in both $V \otimes_s W$ and $V \otimes_t W$ by the argument in the first paragraph of the proof of the previous proposition. By assumption $s = t$ on A, so the identity map on A extends to a $*$-isomorphism $f : V \otimes_s W \to V \otimes_t W$. If $v \in V$ and $w \in W$ and $v_n \to v$ with $v_n \in \cup V_i$, then $v \otimes w = \lim v_n \otimes w$ with respect to both s and t, again by the argument in the first paragraph of the proof of the previous proposition. Hence $f(v \otimes w) = \lim f(v_n \otimes w) = \lim v_n \otimes w = v \otimes w$, so f is the identity map on $V \otimes W$. Thus $t = tf = s$ on $V \otimes W$ by uniqueness of C*-norms on $V \otimes_s W$. □

For any C*-algebra W and $n \in \mathbb{N}$, we may identify the $*$-algebras $M_n(\mathbb{C}) \otimes W$ and $M_n(W)$ by the unique map that sends $e_{ij} \otimes w$ to the W-valued matrix that has w as the ij-entry and otherwise only zeroes. Now $M_n(W)$ is already a C*-algebra. Indeed, if π is the universal representation of W on V, then we may consider $M_n(W)$ as a $*$-subalgebra of $B(V^n)$ by letting $a \in M_n(W)$ send $(v_1, \ldots, v_n)$ to $(\sum \pi(a_{1i})(v_i), \ldots, \sum \pi(a_{ni})(v_i))$. Then we see that $\|a_{ij}\| \leq \|a\|$, so $M_n(W)$ is actually closed in $B(V^n)$. So $M_n(\mathbb{C}) \otimes W$ is a C*-algebra and as there can be only one C*-norm on a C*-algebra, we see that $M_n(\mathbb{C})$ is nuclear. But every finite dimensional C*-algebra is a direct sum of full matrix algebras, so we have proved the first part of the following corollary. The second part is then clear from the proposition above.

Corollary 8.1.10 *Finite dimensional C*-algebras are nuclear, and so are the AF-algebras.*

Example 8.1.11 The C*-algebra of compact operators on a Hilbert space V is also nuclear. Indeed, if $\{v_i\}_{i\in I}$ is an orthonormal basis for V, then $p_F B_0(V) p_F$ with the orthogonal projection $p_F = \sum_{i\in F} v_i \odot v_i$ for each finite subset $F \subset I$, form an upward filtered collection of finite dimensional C*-subalgebras with dense union in $B_0(V)$ as the finite rank operators are dense in the compact operators. ◇

We turn now to states on tensor products.

For non-trivial C*-algebras V, W we have $\|\cdot\|_{min} = \sup \|(\pi_x \otimes \pi_y)(\cdot)\|$ on $V \otimes W$, where the supremum is taken over all states x of V and all states y of W. This is clear as the tensor product of the universal representations of V and W is unitarily equivalent to the direct sum of all the representations $\pi_x \otimes \pi_y$. As $x \otimes y$ is the vector state of $\pi_x \otimes \pi_y$, it is also clear that $x \otimes y$ is bounded with respect to the spatial norm on $V \otimes W$.

Proposition 8.1.12 *For any representations π_i of C*-algebras W_i we have $\|(\pi_1 \otimes \pi_2)(\cdot)\| \leq \|\cdot\|_{min}$ on $W_1 \otimes W_2$.*

Proof Firstly we may assume that the representations are non-degenerate, and hence are direct sums of cyclic ones. Such cyclic representations are unitarily equivalent to representations of the type in the last paragraph before this proposition. Since we are now taking the supremum over less states, the result is clear. □

Lemma 8.1.13 *For positive functionals x, y on C*-algebras V, W, respectively, the functional $x \otimes y$ on $V \otimes W$ satisfies $(x \otimes y)(a^*a) \geq 0$ for $a \in V \otimes W$.*

Proof If $a = \sum v_i \otimes w_i$, we have $(x \otimes y)(a^*a) = \sum x(v_i^* v_j) y(w_i^* w_j)$. As the matrix with ij-entries $y(w_i^* w_j)$ is positive, it can be diagonalized, say as a positive linear combination of rank one orthogonal projections p. It suffices to show that each $\sum x(v_i^* v_j) p_{ij} \geq 0$. But this is clear since $p = (\sum c_i e_i) \odot (\sum c_j e_j) = \sum c_i \bar{c}_j e_{ij}$, where $\{e_i\}$ is the standard basis and c_i are scalars, so $p_{ij} = c_i \bar{c}_j$. □

Proposition 8.1.14 *Given states x_i on C*-algebras W_i such that $x_1 \otimes x_2$ has a continuous extension to $W_1 \otimes_s W_2$ for a norm s on $W_1 \otimes W_2$. Then $x_1 \otimes x_2$ is a state and there is a unitary u from the Hilbert space $V_{x_1} \otimes V_{x_2}$ to the Hilbert space $V_{x_1 \otimes x_2}$ such that $\pi_{x_1 \otimes x_2} = u(\pi_{x_1} \otimes \pi_{x_2})(\cdot)u^*$ on $W_1 \otimes W_2$.*

Proof Pick a sequence of elements $a_n \in W_1 \otimes W_2$ converging to $a \in W_1 \otimes_s W_2$. Then $(x_1 \otimes x_2)(a^*a) = \lim (x_1 \otimes x_2)(a_n^* a_n) \geq 0$ by the lemma, so $x_1 \otimes x_2$ is positive. Let $\{u_{ij}\}_{j\in I_i}$ be approximate units for W_i. Then $\Lambda = I_1 \times I_2$ is upwards filtered under the product order and $\{\tilde{u}_{1\lambda} \otimes \tilde{u}_{2\lambda}\}_{\lambda\in\Lambda}$ with $\tilde{u}_{1(j,k)} = u_{1j}$ and $\tilde{u}_{2(j,k)} = u_{2k}$ is an approximate unit for $W_1 \otimes_s W_2$, and $\|x_1 \otimes x_2\| = \lim (x_1 \otimes x_2)(\tilde{u}_{1\lambda} \otimes \tilde{u}_{2\lambda}) = 1$, so $x_1 \otimes x_2$ is a state.

Clearly $(\pi_{x_1\otimes x_2}(\cdot)(v_{x_1\otimes x_2})|v_{x_1\otimes x_2}) = ((\pi_{x_1}\otimes\pi_{x_2})(\cdot)(v_{x_1}\otimes v_{x_2})|v_{x_1}\otimes v_{x_2})$ on $W_1\otimes W_2$, so $(\pi_{x_1}\otimes\pi_{x_2})(b)(v_{x_1}\otimes v_{x_2})\mapsto \pi_{x_1\otimes x_2}(b)(v_{x_1\otimes x_2})$ well-defines a linear isometry from $(\pi_{x_1}\otimes\pi_{x_2})(W_1\otimes W_2)(v_{x_1}\otimes v_{x_2})$ to $\pi_{x_1\otimes x_2}(W_1\otimes W_2)(v_{x_1\otimes x_2})$ that extends by continuity to the required unitary. □

Proposition 8.1.15 *Given C*-algebras W_i, the square of the spatial norm of $a\in W_1\otimes W_2$ is the supremum of $(x_1\otimes x_2)(b^*a^*ab)/(x_1\otimes x_2)(b^*b)$ over all states x_i of W_i and $b\in W_1\otimes W_2$ with $(x_1\otimes x_2)(b^*b)>0$. The spatial norm on $W_1\otimes\tilde{W}_2$, where $\tilde{W}_2$ is the unitization of W_2, restricts to the spatial norm on $W_1\otimes W_2$. If W_2 is non-unital, any C*-norm on $W_1\otimes W_2$ extends to a C*-norm on $W_1\otimes\tilde{W}_2$.*

Proof Now $\|a\|^2_{min}$ is the supremum of $\|(\pi_{x_1}\otimes\pi_{x_2})(a^*a)\|$ over all states x_i of W_i. By the previous proposition this last norm equals $\|\pi_y(a^*a)\|$, where the continuous extension y of $x_1\otimes x_2$ to $W_1\otimes_{min} W_2$ is indeed a state. And $\|\pi_y(a)\|^2$ is clearly the supremum of $y(b^*a^*ab)/y(b^*b)$ over all $b\in W_1\otimes W_2$ with $y(b^*b)>0$.

If s is the restriction to $W_1\otimes W_2$ of the spatial norm on $W_1\otimes\tilde{W}_2$, we get $\|a\|_{min}\le s(a)$ by the already proven assertion since any state on W_2 has a unique extension to a state on $\tilde{W}_2$. The opposite inequality follows from Proposition 8.1.12 by first restricting the universal representation of $\tilde{W}_2$ to W_2.

Say t is a C*-norm on $W_1\otimes W_2$ with W_2 non-unital. Let π be a faithful non-degenerate representation of $W_1\otimes_t W_2$ on V, and let π_i be the corresponding injective representations of W_i on V given by Proposition 8.1.4. The extension $\tilde{\pi}_2$ of π_2 to $\tilde{W}_2$ is also injective because if $\tilde{\pi}_2(w+c)=0$ for $w\in W_2$ and a non-zero scalar c, then we get the contradiction $\pi_2(-w/c)=1$. As the images of π_1 and $\tilde{\pi}_2$ generate a commutative algebra, we have a $*$-homomorphism $f\colon W_1\otimes\tilde{W}_2\to B(V)$ given by $f(w_1\otimes w_2)=\pi_1(w_1)\tilde{\pi}_2(w_2)$ for $w_1\in W_1$ and $w_2\in\tilde{W}_2$. Clearly $\|f(\cdot)\|$ would be the sought for norm extending $\|\pi(\cdot)\|=t$ provided f is injective. Now if $f(d)=0$, then by injectivity of π, we know that $d(W_1\otimes W_2)=\{0\}$, so $(\pi_1\otimes\tilde{\pi}_2)(d)$ vanishes on the span of $(\pi_1\otimes\tilde{\pi}_2)(W_1\otimes W_2)(V\otimes V)$, which is dense in $V\otimes V$, so $(\pi_1\otimes\tilde{\pi}_2)(d)=0$ and $\pi_1\otimes\tilde{\pi}_2$ is injective by Lemma 8.1.2, so $d=0$, indeed. □

Lemma 8.1.16 *If W_i are C*-algebras with unitaries v_i in their unitizations $\tilde{W}_i$, then the $*$-isomorphism on $W_1\otimes W_2$ given by $(v_1\otimes v_2)(\cdot)(v_1^*\otimes v_2^*)$ is isometric with respect to any C*-norm.*

Proof By unitarity we need only show that the map π in question is norm decreasing with respect to any C*-norm s on $W_1\otimes W_2$. Picking an approximate unit $\{u_{1\lambda}\otimes u_{2\lambda}\}_\lambda$ for $W_1\otimes_s W_2$ of the type described in the proof of Proposition 8.1.14, and using Corollary 8.1.5, we see that $\pi=\lim\pi((u_{1\lambda}\otimes u_{2\lambda})(\cdot)(u_{1\lambda}\otimes u_{2\lambda}))$, which gives the estimate $s\pi\le s$. □

Proposition 8.1.17 *Let W_i be C*-algebras with a C*-norm s on $W_1\otimes W_2$. Denote by X_s the set of pairs (x_1,x_2) of pure states x_i on W_i such that $x_1\otimes x_2$ is bounded with respect to s. Then X_s is closed in the product topology of the w^*-topologies on the set of pure states on W_i. If u_i are unitaries in the unitizations of W_i, then $(x_1(u_1\cdot u_1^*), x_2(u_2\cdot u_2^*))\in X_s$ whenever $(x_1,x_2)\in X_s$. If W_1 or W_2 is commutative, and $(x_1,x_2)\in X_s$, then $x_1\otimes x_2$ is a pure state on $W_1\otimes_s W_2$. If y is a pure state* *on $W_1 \otimes_s W_2$ with π_{yi} corresponding to π_y as in Proposition 8.1.4 such that y_1 is pure, where $y_i = (\pi_{yi}(\cdot)(v_y)|v_y)$, then $(y_1, y_2) \in X_s$ and $y = y_1 \otimes y_2$.*

Proof Say $(x_1, x_2) \in X_s$. Then $|(x_1 \otimes x_2)(\cdot)| \leq s$ by Proposition 8.1.14, so X_s is closed as described. Also $(x_1(u_1 \cdot u_1^*), x_2(u_2 \cdot u_2^*)) \in X_s$ from the lemma. Assume now that W_1 is commutative. Let π be the GNS-representation of the state $x_1 \otimes x_2$ on $W_1 \otimes_s W_2$, and let π_i denote the corresponding representation of W_i given by Proposition 8.1.4, also when restricted to the closure of $\pi_i(W_i)(v_{x_1 \otimes x_2})$. Then $(\pi_1(\cdot)(v_{x_1 \otimes x_2})|v_{x_1 \otimes x_2}) = \lim_\lambda(\pi_1(\cdot)\pi_2(u_\lambda)(v_{x_1 \otimes x_2})|v_{x_1 \otimes x_2})$, which in turn equals $\lim_\lambda(\pi(\cdot \otimes u_\lambda)(v_{x_1 \otimes x_2})|v_{x_1 \otimes x_2}) = \lim_\lambda(x_1 \otimes x_2)(\cdot \otimes u_\lambda) = x_1 = (\pi_{x_1}(\cdot)(v_{x_1})|v_{x_1})$ on W_1. Hence π_1 and π_{x_1} are unitarily equivalent, and the latter is irreducible as x_1 is pure. Hence $\pi_1(W_1) \subset \pi_1(W_1)' = \mathbb{C}$ and $\pi_1(w) = x_1(w)1$ for the restricted π_1. As the span of $\pi_1(W_1)\pi_2(W_2)(v_{x_1 \otimes x_2})$ is dense in $V_{x_1 \otimes x_2}$ for the non-restricted π_i's, we see that $\pi_1(w_1)\pi_2(w_2) = x_1(w_1)\pi_2(w_2)$, and $V_{x_1 \otimes x_2}$ is cyclic for π_2. Repeating the same argument for π_2 as for π_1, but now without restricting, we see that π_2 is irreducible, and $\pi(W_1 \otimes_s W_2)' = \mathbb{C}$. Thus $x_1 \otimes x_2$ is pure.

For the final statement, by repeating the arguments above, we see that the orthogonal projection p onto the closure of $\pi_{y1}(W_1)(v_y)$ belongs to $\pi_{y1}(W_1)'$ and that $p\pi_{y1}(W_1)p = \mathbb{C}p$. Since $\pi_{y2}(W_2) \subset \pi_{y1}(W_1)'$ and $p(v_y) = v_y$, we also see that $p\pi_{y2}(\cdot)p = y_2(\cdot)p$ and $y = y_1 \otimes y_2$ on $W_1 \otimes W_2$. To see that y_2 is a pure state, note that for a $\pi_{y_2}(W_2)$-invariant closed subspace V of V_{y_2}, the Hilbert space $V_{y_1} \otimes V$ is invariant for $\pi_{y_1} \otimes \pi_{y_2}$, which by Proposition 8.1.14 is unitarily equivalent to π_y, so π_{y_2} is irreducible. □

The next result is useful.

Theorem 8.1.18 *Commutative C^*-algebras are nuclear.*

Proof Let W_i be C*-algebras with a C*-norm s on $W_1 \otimes W_2$, and suppose W_1 is commutative. Let y be a pure state on $W_1 \otimes_s W_2$ with π_{yi} corresponding to π_y as in Proposition 8.1.4 and let $y_i = (\pi_{yi}(\cdot)(v_y)|v_y)$. Then $\pi_{y1}(W_1) \subset \pi_y(W_1 \otimes_s W_2)' = \mathbb{C}$ and $\pi_{y1} = y_1(\cdot)1$. So y_1 is a character and hence a pure state. By the proposition above y_2 is therefore also a pure state, and $y = y_1 \otimes y_2$ and $(y_1, y_2) \in X_s$. Since $\|\pi_y(\cdot)\| = \|(\pi_{y_1} \otimes \pi_{y_2})(\cdot)\|$ and $s(\cdot)$ is the supremum of $\|\pi_y(\cdot)\|$ over all pure state y on $W_1 \otimes_s W_2$, we conclude that $s(\cdot)$ is the supremum of $\|(\pi_{y_1} \otimes \pi_{y_2})(\cdot)\|$ over all pairs $(y_1, y_2) \in X_s$. Hence we are done if we can show that X_s consists of all possible pairs of pure states on W_1 and W_2.

Suppose it does not. By the previous proposition X_s is closed, so there are non-empty w^*-open sets A_i in the set of pure states of W_i with $A_1 \times A_2$ disjoint from X_s. Replacing A_i by the set of all $x(u \cdot u^*)$ with $x \in A_i$ and u unitary, we may by the proposition above also assume that A_i is unitarily invariant. The annihilator of the complement of A_i contains therefore by Theorem 6.4.15 a non-zero element w_i. Picking a pure state y of $W_1 \otimes_s W_2$ such that $s(w_1 \otimes w_2) = y(w_1 \otimes w_2)$ and letting y_i relate to y as in the first part of this proof, we get $y(w_1 \otimes w_2) = y_1(w_1)y_2(w_2) = 0$ since $y = y_1 \otimes y_2$ and $y_i \notin A_i$. Hence $w_1 \otimes w_2 = 0$, which is impossible. □

The following result justifies the notation $\|\cdot\|_{min}$ for the spatial C*-norm.

Theorem 8.1.19 *The spatial C*-norm on the algebraic tensor product of two C*-algebras is the least possible.*

Proof Let W_i be C*-algebras with a C*-norm s on $W_1 \otimes W_2$. By Proposition 8.1.15 we may assume that W_2 is unital. If X_s does not consist of all pairs of pure states on W_i, then as in the last part of the proof of the previous proposition, we know there are non-zero $w_i \in W_i$ such that $(x_1 \otimes x_2)(w_1 \otimes w_2) = 0$ for all $(x_1, x_2) \in X_s$. If V is the unital C*-algebra generated by w_2, then we may by the theorem above regard $W_1 \otimes_{min} V$ as a C*-subalgebra of $W_1 \otimes_s W_2$. Pick pure states x_1, x_2 on W_1, V that are positive on w_1, w_2, respectively, and extend $x_1 \otimes x_2$ to a pure state y on $W_1 \otimes_{min} V$ by the previous proposition, then extend this further to a pure state y on $W_1 \otimes_s W_2$. With notation as in the last statement of that same proposition, and noting that $y_1(w) = y(w \otimes 1) = x_1(w)$ for $w \in W_1$, we see that $(x_1, x_2) \in X_s$ and $y = x_1 \otimes x_2$, which is impossible. Hence X_s consists of all pairs of pure states on W_i. Using now that any state on a C*-algebra is the w^*-limit of a net of convex combinations of pure states and the zero functional, we see that the tensor product of any two states on W_i is bounded with respect to s. If x_i is any state on W_i, let x be the unique state extending $x_1 \otimes x_2$ on $W_1 \otimes_s W_2$ to its unitization W. If $b \in W$ satisfies $x(b^*b) > 0$, we have $s(a)^2 \geq x(b^*a^*ab)/x(b^*b)$ for $a \in W_1 \otimes W_2$ as $s(a^*a)1 \geq a^*a$ and $x(b^* \cdot b)$ is positive, so $s(\cdot) \geq \|\cdot\|_{min}$ by Proposition 8.1.15. □

Any C*-norm s on the algebraic tensor product of two C*-algebras W_i is *cross* in the sense that $s(w_1 \otimes w_2) = \|w_1\|\|w_2\|$ for $w_i \in W_i$. This holds by minimality of the spatial tensor product as $\|w_1\|\|w_2\| = \|w_1 \otimes w_2\|_{min} \leq s(w_1 \otimes w_2) \leq \|w_1\|\|w_2\|$.

Corollary 8.1.20 *For faithful representations π_i of C*-algebras W_i the spatial norm on $W_1 \otimes W_2$ equals $\|(\pi_1 \otimes \pi_2)(\cdot)\|$.*

Proof This is immediate from minimality of $\|\cdot\|_{min}$ and Proposition 8.1.12. □

Proposition 8.1.21 *The set $C_0(X, W)$ of continuous functions f from a locally compact Hausdorff space X to a C*-algebra W such that $\|f(\cdot)\|$ vanishes at infinity is a C*-algebra under pointwise operations and sup-norm. The linear map $C_0(X) \otimes W \to C_0(X, W)$ that sends $f \otimes w$ to $f(\cdot)w$ extends uniquely by continuity to a $*$-isomorphism $C_0(X) \otimes_{min} W \to C_0(X, W)$. For a locally compact Hausdorff space Y, the linear map $C_0(X) \otimes C_0(Y) \to C_0(X \otimes Y)$ that sends $f \otimes g$ to the function that evaluates (x, y) at $f(x)g(y)$ also extends by continuity to a $*$-isomorphism $C_0(X) \otimes_{min} C(Y) \to C_0(X \times Y)$.*

Proof The first statement is by now standard. As for the second one we see that the map h in question is a well-defined $*$-homomorphism. If it vanishes on $\sum f_i \otimes w_i$ with $\{w_i\}$ linear independent, we see that all $f_i = 0$, so h is injective. Hence $\|h(\cdot)\|$ is a C*-norm on $C_0(X) \otimes W$ that extends to $\|\cdot\|_{min}$ on $C_0(X) \otimes_{min} W$ by nuclearity of $C_0(X)$, so h extends by continuity to an isometric $*$-homomorphism. To see that it is surjective, let $f \in C_0(X, W)$ and $\varepsilon > 0$. Pick a compact set $K \subset X$ such that $\|f(\cdot)\| \leq \varepsilon$ outside K. By total boundedness of $f(K)$ there are finitely many $w_j \in f(K)$ such that $A_j = \{x \in X \mid \|f(x) - w_j\| < \varepsilon\}$ cover K. Pick a partition of unity $\{g_j\}$ on K associated to the set A_j. Then one easily checks that $\|f - \sum g_j(\cdot)w_j\| \leq \varepsilon$.

For the third statement, observe that the map $C_0(X, C_0(Y)) \to C_0(X, Y)$ that sends F to G, where $G(x, y) = (F(x))(y)$ for $x \in X$ and $y \in Y$ is a $*$-isomorphism. Composing this map with the $*$-isomorphism from the second statement when $W = C_0(Y)$ will clearly be the desired extension of the map in the third statement. □

In this proposition we could by nuclearity of commutative C*-algebras have used any C*-norm on the tensor products.

Corollary 8.1.22 *For $*$-homomorphisms $f_i : V_i \to W_i$ between C*-algebras, there is a unique $*$-homomorphism $f_1 \otimes f_2 \colon V_1 \otimes_{min} V_2 \to W_1 \otimes_{min} W_2$ such that $(f_1 \otimes f_2)(v_1 \otimes v_2) = f_1(v_1) \otimes f_2(v_2)$ for $v_i \in V_i$. Moreover, when each f_i is injective, so is $f_1 \otimes f_2$.*

Proof Pick faithful representations π_i of W_i, so $\pi_1 \otimes \pi_2$ is isometric for the spatial C*-norm, and $\|(f_1 \otimes f_2)(\cdot)\|_{\min} = \|(\pi_1 f_1 \otimes \pi_2 f_2)(\cdot)\| \leq \|\cdot\|_{min}$ on $V_1 \otimes V_2$.

If f_i are injective, so are $\pi_i f_i$ and in the inequality above we actually have equality, so $f_1 \otimes f_2$ is isometric. □

When V_i are C*-subalgebras of C*-algebras W_i, we can thus regard $V_1 \otimes_{min} V_2$ as a C*-subalgebra of $W_1 \otimes_{min} W_2$.

Definition 8.1.23 A sequence $0 \to U \xrightarrow{f} V \xrightarrow{g} W \to 0$ of $*$-homomorphisms between C*-algebras is *short exact* if f is injective and g is surjective and $\ker g = \operatorname{im} f$. We say then that V is an *extension of W by U*.

The quotient map g of a C*-algebra V by a closed ideal U form a short exact sequence with $W = V/U$ and f the inclusion map.

Lemma 8.1.24 *Given a short exact sequence $0 \to U \xrightarrow{f} V \xrightarrow{g} W \to 0$ of C*-algebras and a C*-algebra A such that $W \otimes A$ has a unique C*-norm, then*

$$0 \to U \otimes_{min} A \xrightarrow{f \otimes \iota} V \otimes_{min} A \xrightarrow{g \otimes \iota} W \otimes_{min} A \to 0$$

is also short exact.

Proof Clearly $g \otimes \iota$ is surjective and $f \otimes \iota$ is injective and its image is a closed ideal of $V \otimes_{min} A$. Let h be the quotient map onto the quotient C*-algebra B. Since $gf = 0$ there is a unique $*$-epimorphism $r \colon B \to W \otimes_{min} A$ such that $rh = g \otimes \iota$. It suffices to show that r is injective. Consider the well-defined unique $*$-homomorphism $s \colon W \otimes A \to B$ such that $s(g(v) \otimes a) = v \otimes a + \operatorname{im}(f \otimes \iota)$. Since the maximum of $\|s(\cdot)\|$ and $\|\cdot\|_{min}$ is a C*-norm on $W \otimes A$, and thus must be the spatial norm, we see that s is norm-decreasing with respect to the spatial norm, and its continuous extension is clearly a left inverse of r. □

A C*-algebra A for which the conclusion of the lemma above holds for every C*-algebra U, V, W is called *exact*. So nuclear C*- algebras are exact.

Proposition 8.1.25 *The extension of a nuclear C*-algebra by a nuclear C*-algebra is nuclear.*

Proof Suppose we have a short exact sequence $0 \to U \xrightarrow{f} V \xrightarrow{g} W \to 0$ of C*-algebras with U and W nuclear, and let A by any C*-algebra. If we can show that the identity map on $V \otimes A$ extended to $h\colon V \otimes_{max} A \to V \otimes_{min} A$ is injective, we are done. By the lemma above the sequence

$$0 \to U \otimes_{min} A \xrightarrow{f\otimes\iota} V \otimes_{min} A \xrightarrow{g\otimes\iota} W \otimes_{min} A \to 0$$

is short exact. Let r be the unique $*$-homomorphism from $U \otimes A$ to $V \otimes_{max} A$ such that $r(u \otimes a) = f(u) \otimes a$ for $u \in U$ and $a \in A$. By nuclearity of U the maximum of $\|r(\cdot)\|_{max}$ and the spatial norm must be the spatial norm, so r extends to $U \otimes_{min} A$, such that $f \otimes \iota = hr$. Similarly the unique $*$-homomorphism $s\colon V \otimes A \to W \otimes_{min} A$ such that $s(v \otimes a) = g(v) \otimes a$ extends to $V \otimes_{max} A$. Let t be the quotient map of $V \otimes_{max} A$ by the closed ideal $\operatorname{im} r$ onto the quotient C*-algebra B. As in the proof of the lemma above there is a unique $*$-homomorphism $k\colon W \otimes_{min} A \to B$ such that $k(g(v) \otimes a) = v \otimes a + \operatorname{im} r$ for $v \in V$ and $a \in A$.

Say $h(b) = 0$ for $b \in V \otimes_{max} A$. Then $0 = (g \otimes \iota)h(b) = s(b)$, so $0 = ks(b) = t(b)$ and $b = r(c)$ for some $c \in U \otimes_{min} A$. Hence $(f \otimes \iota)(c) = hr(c) = h(b) = 0$ and $c = 0$ by injectivity of f and the previous corollary. □

8.2 Von Neumann Tensor Products

Definition 8.2.1 The *von Neumann tensor product* $W_1 \bar{\otimes} W_2$ of two von Neumann algebras W_i acting on V_i is the weak closure of $W_1 \otimes W_2$ considered as a $*$-subalgebra of $B(V_1 \otimes V_2)$, so $W_1 \bar{\otimes} W_2 = (W_1 \otimes W_2)''$.

Any complex Hilbert space with inner product $(\cdot|\cdot)$ can be regarded as a real vector space with *real* inner product $\mathrm{Re}\,(\cdot|\cdot)$.

Lemma 8.2.2 *Given $*$-algebras U, W acting on a complex Hilbert space V with identity elements the identity map on V. Assume that $U \subset W'$ and that U has a cyclic vector $v \in V$. Let $\perp$ mean orthogonality with respect to real inner products, and let an index s on the symbol of a C*-algebra mean its self-adjoint part. Then $U_s(v)^\perp = i\overline{W_s(v)}$ if and only if $U_s(v) + iW_s(v)$ is dense in V if and only if $U' = W''$.*

Proof Clearly $i(U')_s(v)$ and $iW_s(v)$ are contained in $U_s(v)^\perp$, which proves the equivalence of the first two statements. To see that these two statements imply the last one, note that in this case $(U')_s(v) \subset \overline{W_s(v)}$. Using this in the inner product together with cyclicity of v, one checks that any element of $(U')_s$ belongs to W'', so $U' = W''$.

For the converse, let $r \in (U_s(v) + iW_s(v))^\perp$ and we aim at showing $r = 0$. Act on the Hilbert space $V \oplus V$ by $U_2 = \{\mathrm{diag}(u, u) \mid u \in U\}$. Consider the orthogonal projection $p = \begin{pmatrix} a & b \\ b^* & c \end{pmatrix} \in U_2'$ from $V \oplus V$ onto the closure of $U_2(\begin{pmatrix} v \\ r \end{pmatrix})$, so $a^* = a$ and $c^* = c$ and b all belong to U'. Since p fixes $(v, r)^T$, we have $a(v) + b(r) = v$. Since $r \in U_s(v)^\perp$ we get $(r|u(v)) = -(v|u(r))$ for $u \in U$, so p vanishes on $(r, v)^T$ and $a(r) + b(v) = 0$. Since $r \in iW_s(v)^\perp$, we get $(r|w(v)) = (v|w(r))$ for $w \in W'' = U'$. Combining this yields $(r|a(r)) = -(v|(1-a)(v))$. As $p^2 = p$ we get $a = a^2 + bb^*$ and $1 - a = (1-a)^2 + bb^*$, so $0 \leq \|a(r)\|^2 + \|b^*(r)\|^2 = -(\|(1-a)(v)\|^2 + \|b^*(v)\|^2) \leq 0$. So $a(r) = (1-a)(v) = 0$, and as v is separating for U', we get $a = 1$ and $r = 0$. □

Lemma 8.2.3 *Let V_j be complex Hilbert spaces with closed real subspaces U_j such that $U_j + iU_j$ are dense in V_j, and let $\perp$ mean orthogonality with respect to real inner products $\langle\cdot,\cdot\rangle$. Then $U_1 \otimes U_2 + iU_1^\perp \otimes U_2^\perp$ is dense in $V_1 \otimes V_2$.*

Proof Let $w \perp (U_1 \otimes U_2 + iU_1^\perp \otimes U_2^\perp)$ and define a bounded real linear map $f \colon V_1 \to V_2$ such that $\langle f(v_1)|v_2\rangle = \langle w|v_1 \otimes v_2\rangle$ for $v_j \in V_j$. One readily checks that $fi = -if$ and $f^*i = -if^*$ and $f(U_1) \subset U_2^\perp$ and $f^*(U_2) \subset U_1^\perp$ and $f(iU_1^\perp) \subset U_2$ and $f^*(U_2^\perp) \subset iU_1$. Hence $f^*f(U_1^\perp) \subset iU_1^\perp$ and $(f^*f)^2(U_1^\perp) \subset U_1^\perp$. Since f^*f is a norm limit of polynomials in $(f^*f)^2$, we therefore get $f^*f(U_1^\perp) \subset U_1^\perp \cap iU_1^\perp = \{0\}$ by denseness of $U_1 + iU_1$ in V_1. Thus $f(U_1^\perp) = \{0\}$. As $\langle f^*(U_2)|U_1^\perp\rangle = 0$, we have $f^*(U_2) = \{0\}$ and $f(U_1) \subset U_2^\perp \cap iU_2^\perp = \{0\}$ by denseness of $U_2 + iU_2$ in V_2. So $f = 0$ and $w = 0$. □

Theorem 8.2.4 *We have $(W_1 \bar{\otimes} W_2)' = W_1' \bar{\otimes} W_2'$ for any von Neumann algebras W_i.*

Proof Assume first that each W_j admits a cyclic vector $v_j \in V_j$. Let $U = W_1 \bar{\otimes} W_2$ and $W = W_1' \bar{\otimes} W_2'$. Since $v = v_1 \otimes v_2$ is cyclic for U it suffices by the first lemma above to show that $U_s(v) + iW_s(v)$ is dense in $V_1 \otimes V_2$. Let U_i be the closure of $(W_j)_s(v_j)$, so $U_1 \otimes U_2$ is contained in the closure of $U_s(v)$. By the same lemma applied to W_j and W_j', we know that $(W_j)_s(v_j)^\perp = \overline{i(W_j')_s(v_j)}$, so we also have $U_1^\perp \otimes U_2^\perp \subset \overline{W_s(v)}$. Thus we need only show $U_1 \otimes U_2 + iU_1^\perp \otimes U_2^\perp = V_1 \otimes V_2$, which by the last lemma above holds if $U_j + iU_j$ is dense in V_j. But this is clear as v_j is cyclic for W_j and the closure of $(W_j)_s(v_j) + i(W_j)_s(v_j)$ contains $W_j(v_j)$.

In general fix $v_j \in V_j$ and let p_j be the orthogonal projection onto the closure of $W_j(v_j)$. Then the von Neumann algebra $X_j = W_j p_j$ acts cyclically on $\mathrm{im} p_j$, so $(X_1 \otimes X_2)' = X_1' \bar{\otimes} X_2'$ on $\mathrm{im} p_1 \otimes \mathrm{im} p_2$ by the first paragraph. Let $p = p_1 \otimes p_2$. Then $(W_1 \bar{\otimes} W_2)p = X_1 \bar{\otimes} X_2$ and $p(W_1' \bar{\otimes} W_2')p = X_1' \bar{\otimes} X_2'$ by Proposition 5.5.8. Let $a \in (W_1 \bar{\otimes} W_2)'$ and $b \in (W_1' \bar{\otimes} W_2')'$. Then again by this proposition we see that $pbp = bp \in (W_1' \bar{\otimes} W_2')'p = (p(W_1' \bar{\otimes} W_2')p)' = (X_1' \bar{\otimes} X_2')' = X_1 \bar{\otimes} X_2$ while pap belongs to $p(W_1 \bar{\otimes} W_2)'p = ((W_1 \bar{\otimes} W_2)p)' = (X_1 \bar{\otimes} X_2)'$. Hence

$$(ab(v_1 \otimes v_2)|v_1 \otimes v_2) = (pappbp(v)|v) = (pbppap(v)|v) = (ba(v_1 \otimes v_2)|v_1 \otimes v_2),$$

which shows that $ab = ba$ and $(W_1 \bar{\otimes} W_2)' \subset (W_1' \bar{\otimes} W_2')' = W_1' \bar{\otimes} W_2'$, and the opposite inclusion is obvious. □

Corollary 8.2.5 *For von Neumann algebras W_i, U_i acting on V_i, we have*

$$((W_1\bar{\otimes}W_2)\cup(U_1\bar{\otimes}U_2))'' = (W_1\cup U_1)''\bar{\otimes}(W_2\cup U_2)''$$

and

$$(W_1\bar{\otimes}W_2)\cap(U_1\bar{\otimes}U_2) = (W_1\cap U_1\bar{\otimes}(W_2\cap U_2).$$

Proof The first identity is obvious. Replacing W_i, U_i by their commutants in it and then taking commutants of each side of that identity, we get the second identity by applying the theorem. □

Corollary 8.2.6 *We have $Z(W_1\bar{\otimes}W_2) = Z(W_1)\bar{\otimes}Z(W_2)$ for von Neumann algebras W_i. The von Neumann tensor product of two factors is a factor.*

Proof As $Z(W_1\bar{\otimes}W_2)$ is contained in $W_1\bar{\otimes}W_2$ and its commutant $W_1'\bar{\otimes}W_2'$, we get $Z(W_1\bar{\otimes}W_2)\subset(W_1\cap W_1')\bar{\otimes}(W_2\cap W_2') = Z(W_1)\bar{\otimes}Z(W_2)$ from the corollary above. The opposite inclusion is obvious. □

Example 8.2.7 For complex Hilbert spaces V_i we have by the theorem above that

$$B(V_1)\bar{\otimes}B(V_2) = (B(V_1)'\bar{\otimes}B(V_2)')' = (\mathbb{C}1\otimes\mathbb{C}1)' = B(V_1\otimes V_2)$$

as expected. ◇

Example 8.2.8 By Example 8.1.1 we have to σ-finite measures μ and ν a unitary operator $U: L^2(\mu)\otimes L^2(\nu)\to L^2(\mu\times\nu)$ such that $U(f\otimes g)(s,t) = f(s)g(t)$. For $a\in L^\infty(\mu)$ and $b\in L^\infty(\nu)$ we have $U(M_a\otimes M_b)U^* = M_{a\otimes b}$ for $a\in L^\infty(\mu)$ and $b\in L^\infty(\nu)$ as $(U(M_a\otimes M_b)(f\otimes g))(s,t) = a(s)f(s)b(t)g(t) = (M_{a\otimes b}U(f\otimes g))(s,t)$. Hence $h = U\cdot U^*$ is a spatial isomorphism from $L^\infty(\mu)\bar{\otimes}L^\infty(\nu)$ onto $L^\infty(\mu\times\nu)$.

Using the above in combination with Proposition 8.1.21 and Proposition 5.8.6, we see that a similar U according to the last paragraph in Example 4.1.16, implements a spatial isomorphism between $L^\infty(\mu)\bar{\otimes}L^\infty(\nu)$ and $L^\infty(\mu\otimes\nu)$ when μ and ν are Radon measures on σ-compact locally compact Hausdorff spaces and $\mu\otimes\nu$ is the Radon product measure of μ and ν; separability of the L^2-spaces is not needed here due to Proposition 8.1.21 and Lusin's theorem. ◇

Suppose W_i are von Neumann algebras on Hilbert spaces V_i. The linear map $u: V_1\otimes V_2\to V_2\otimes V_1$ which sends $v_1\otimes v_2$ to $v_2\otimes v_1$ is evidently unitary and $u\cdot u^*: W_1\bar{\otimes}W_2\to W_2\bar{\otimes}W_1$ is a $*$-isomorphic normal extension of the flip map. Since every permutation is a product of transpositions we can use this extended flip map to rearrange the order in any multiple von Neumann tensor product as we wish. For instance, the von Neumann algebras $W_1\bar{\otimes}W_2\bar{\otimes}W_3\bar{\otimes}W_4$ and $W_4\bar{\otimes}W_1\bar{\otimes}W_3\bar{\otimes}W_2$ are $*$-isomorphic under the unique normal extension of the linear map that sends $w_1\otimes w_2\otimes w_3\otimes w_4$ to $w_4\otimes w_1\otimes w_3\otimes w_2$.

Proposition 8.2.9 *Any normal $*$-epimorphism $W_1\to W_2$ between von Neumann algebras is of the form $f_3f_2f_1$, where $f_1(w) = w\otimes 1$ and 1 is the identity map on* *some Hilbert space* V*, and* $f_2(w \otimes 1) = (w \otimes 1)p$ *for some orthogonal projection* $p \in (W_1 \bar{\otimes} \mathbb{C}1)'$*, and* $f_3 \colon (W_1 \bar{\otimes} \mathbb{C}1)p \to W_2$ *is a spatial* $*$*-isomorphism.*

Proof Say W_i act on Hilbert spaces V_i, and let $f \colon W_1 \to W_2$ be a normal $*$-epimorphism. Assume first that W_2 has a cyclic vector v. Pick a sequence $\{v_n\}$ in V_1 with $\sum \|v_n\|^2 < \infty$ such that $(f(\cdot)(v)|v) = \sum((\cdot)(v_n)|v_n)$. Letting $V = l^2(\mathbb{N})$ and $h = \{v_n\} \in V_1 \otimes V$ with f_1 as in the proposition, we get $\sum((\cdot)(v_n)|v_n) = (f_1(\cdot)(h)|h)$, which again equals $(f_2 f_1(\cdot)(h)|h)$ with f_2 as above and where p is the orthogonal projection from $V_1 \otimes V$ onto the closure of $f_1(V_1)(h)$, so $p \in (W_1 \bar{\otimes} \mathbb{C}1)'$. Finally, letting u denote the unique unitary from the closure of $f(W_1)(v)$ to the space $p(V_1 \otimes V)$ such that $u(f(w)(v)) = f_2 f_1(w)(v) = \{w(v_n)\}$ for all $w \in W_1$, then $f_3 = u^* \cdot u$ yields $f = f_3 f_2 f_1$. The general case now follows by writing V_2 as a orthogonal sum of closed subspaces where W_2 acts cyclically, and composing f with the projection onto each such subspace, applying the first part of the proof to these maps, and then taking direct sums to get the desired functions f_i. □

Corollary 8.2.10 *Given normal* $*$*-epimorphisms* $f_i \colon U_i \to W_i$ *between von Neumann algebras, there is a unique normal* $*$*-epimorphism* $f \colon U_1 \bar{\otimes} U_2 \to W_1 \bar{\otimes} W_2$ *such that* $f(v_1 \otimes v_2) = f_1(v_1) \otimes f_2(v_2)$ *for* $v_i \in U_i$*. If* f_i *are (spatial) isomorphisms, then* f *is a (spatial) isomorphism.*

Proof By the proposition write $f_i(v_i) = u_i^*(v_i \otimes 1)p_i u_i$ with entities as described in that proposition. Letting $p = p_1 \otimes p_2$ and $u = u_1 \otimes u_2$, we see that $f = u^*((\cdot) \otimes 1 \otimes 1)pu$ has the required properties. If f_i are injective, so is the map given by $v_i \mapsto (v_i \otimes 1)p_i$, and $c(p_i) = 1$. Hence $c(p) = 1$ and f is injective. □

Proposition 8.2.11 *Let* x_i *be normal bounded linear functionals on von Neumann algebras* W_i*. Then the linear functional* $x_1 \otimes x_2$ *on the algebraic tensor product* $W_1 \otimes W_2$ *has a unique* σ*-weakly continuous extension* $x_1 \otimes x_2$ *to* $W_1 \bar{\otimes} W_2$ *with norm* $\|x_1\|\|x_2\|$*. Moreover, when* x_i *are states, so is the extension* $x_1 \otimes x_2$*.*

Proof Say W_i act on V_i and pick trace class operators w_i on V_i such that $x_i = \mathrm{Tr}(\cdot w_i)$. Clearly $w_1 \otimes w_2$ is trace class on $V_1 \otimes V_2$ with trace norm $\|w_1\|_1\|w_2\|_1$, and $\mathrm{Tr}(\cdot(w_1 \otimes w_2))$ is a normal σ-weakly continuous extension of $x_1 \otimes x_2$ with norm not less than $\|x_1\|\|x_2\|$, and yet the norm is not greater than $\|w_1\|_1\|w_2\|_1$. As for the second claim note that $\|x_1 \otimes x_2\| = \|x_1\|\|x_2\| = x_1(1)x_2(1) = (x_1 \otimes x_2)(1 \otimes 1)$. □

Given a normal bounded linear functional x on a von Neumann algebra U, and a second von Neumann algebra W with $a \in U \bar{\otimes} W$, the bounded linear functional on the predual of W which sends y to $(x \otimes y)(a)$ corresponds to an element in W with norm not greater than $\|x\|\|a\|$, which we denote by $(x \otimes \iota)(a)$. We call the linear map $U \bar{\otimes} W \to W$ with norm not greater than $\|x\|$ which sends a to $(x \otimes \iota)(a)$, the *(left) slice map* with respect to x.

Definition 8.2.12 Given normal bounded linear functionals x_i on von Neumann algebras W_i, then an element in $W_1 \bar{\otimes} W_2$ is *sliced from the left by* x_1*, or right by* x_2, when we apply $x_1 \otimes \iota$, or $\iota \otimes x_2$, respectively, to it.

Given complex Hilbert spaces V_j with an orthonormal basis $\{v_i\}$ of V_2. Then $V^i = V_1 \otimes \mathbb{C}v_i$ is a closed linear subspace of $V_1 \otimes V_2$ isomorphic to V_1 and $V_1 \otimes V_2$ and isomorphic to $\oplus V^i$. Let $u_i \colon V_1 \to V_1 \otimes V_2$ be the linear isometry given by $u_i(v) = v \otimes v_i$. Then $u_i^*(v \otimes v_j) = \delta_{ij} v$ and $u_i^* u_i$ is the identity map on V_1 and $p_i = u_i u_i^*$ is the orthogonal projection from $V_1 \otimes V_2$ onto V^i and $\sum p_i = 1$ strongly. The *operator matrix of* $a \in B(V_1 \otimes V_2)$ *with respect to the orthonormal basis* $\{v_i\}$ is the infinite matrix (a_{ij}) over $B(V_1)$ with $a_{ij} = u_i^* a u_j$ for all i, j. Since $a(w) = \sum u_i^* a(w) \otimes v_i$ for $w \in V_1 \otimes V_2$, we see that a is determined by (a_{ij}). The following result is straightforward.

Lemma 8.2.13 *Given $a, b \in B(V_1 \otimes V_2)$ with operator matrices (a_{ij}) and (b_{ij}) with respect to an orthonormal basis $\{v_i\}$ for V_2. Then $\sum a_{ik} b_{kj}$ converges strongly to $c_{ij} \in B(V_1)$ and (c_{ij}) is the operator matrix of ab with respect to the same basis. Given $a_i \in B(V_i)$, the operator matrix of $a_1 \otimes a_2 \in B(V_1 \otimes V_2)$ with respect to $\{v_i\}$ has ij-entry $(a_2(v_i)|v_j)a_1$, so $(\delta_{ij} a_1)$ is in particular the operator matrix of $a_1 \otimes 1$. Let $u_i : V_1 \to V_1 \otimes V_2$ be the map given by $u_i(v) = v \otimes v_i$. Then $B(V_1) \otimes \mathbb{C}$ is the commutant in $B(V_1 \otimes V_2)$ of all the elements $u_i u_j^*$. This latter set is a von Neumann algebra, so $B(V_1) \bar{\otimes} \mathbb{C} = B(V_1) \otimes \mathbb{C}$. If $W \subset B(V_1)$ is a von Neumann algebra, then $W \bar{\otimes} B(V_2)$ consists of all $a \in B(V_1 \otimes V_2)$ having operator matrices with respect to $\{v_i\}$ that have entries in W.*

Let us investigate how tensor products relate to types.

Proposition 8.2.14 *Two von Neumann algebras W_i on separabel Hilbert spaces are finite if and only if $W_1 \bar{\otimes} W_2$ is finite.*

Proof For the forward direction pick faithful normal tracial states x_i on W_i, say W_i acts on V_i, and write $x_i = \sum((\cdot)v_n^i|v_n^i)$ for v_n^i with $\sum \|v_n^i\|^2 < \infty$. Then the normal tracial state $x = \sum((\cdot)(v_m^1 \otimes v_n^2)|v_m^1 \otimes v_n^2)$ is faithful because the span of all $W_1'(v_m^1) \otimes W_2'(v_n^2)$ is dense in $V_1 \otimes V_2$. The opposite direction is clear as W_1 is $*$-isomorphic to $W_1 \otimes \mathbb{C}$ and W_2 is $*$-isomorphic to $\mathbb{C} \otimes W_2$. □

Lemma 8.2.15 *Let x be a semifinite normal trace on the positive part of a von Neumann algebra W. For $u \in Ws(x)$ with $x(u^*u) < \infty$ the map $w \mapsto uw^*$ is strongly continuous on any bounded ball of W.*

Proof Say $\{w_i\}$ is a net in the closed unit ball of W that converges strongly to zero. Since $w_i u^* u w_i^* \in Ws(x)$ we may assume that x is faithful. For the associated faithful normal GNS-representation π_x of W it suffices to show that $(\pi(w_i u^* u w_i^*)(v)|v) = x(v^* w_i u^* u w_i^* v)$ tends to zero for every $v \in N_x$. The latter expression equals $x(uw_i^* vv^* w_i u^*)$, which is not greater than $\|v\|^2 x(uw_i^* w_i u^*)$ and this expression does tend to zero. □

Proposition 8.2.16 *Given von Neumann algebras W_i on separabel Hilbert spaces, then $W_1 \bar{\otimes} W_2$ is of type III if and only if either W_1 or W_2 is of type III.*

Proof If W_i contain non-zero finite orthogonal projections, then so does $W_1 \bar{\otimes} W_2$ by Proposition 8.2.14. Conversely, suppose $W_1 \bar{\otimes} W_2$ is not of type III. Then by the type decomposition of von Neumann algebras, it has a central orthogonal projection

t such that $(W_1\bar{\otimes}W_2)t$ is semifinite, which possesses a faithful semifinite normal trace that clearly can be extended to a non-zero semifinite normal trace x on the positive part of $W_1\bar{\otimes}W_2$. Pick a non-zero element $a \in (W_1\bar{\otimes}W_2)_+ s(x)$ such that $x(a^2) < \infty$. Say W_i act on V_i and let $\{a_{kl}\}$ be the operator matrix of a with respect to an orthonormal basis of V_2. Pick k such that $a_{kk} > 0$. The chain

$$W_1 \xrightarrow{f} W_1\bar{\otimes}W_2 \xrightarrow{g} W_1\bar{\otimes}W_2 \xrightarrow{h} W_1$$

of maps given by $f(w) = w \otimes 1$ and $g(b) = ab^*$ and $h(b) = b_{kk}$ for $w \in W_1$ and $b \in W_1\bar{\otimes}W_2$ has a composition hgf which sends w to $a_{kk}w^*$ and is by the lemma strongly continuous on any bounded ball of W_1. Pick a non-zero orthogonal projection $p \in W_1$ and a positive scalar c such that $a_{kk}^2 \geq cp$. As $\|pw^*(v)\|^2 \leq c^{-1}\|a_{kk}w^*(v)\|^2$ for $w \in W_1$ and $v \in V_1$, we see that the map $w \mapsto pw^*$ is also strongly continuous on any bounded ball of W_1. So the $*$-operation is strongly continuous on any bounded ball of $W_1 p$. If p is not finite, then pick e_n to p as in Lemma 7.5.1, and then pick $w_n \in W_1$ such that $w_n^* w_n = e_n$ and $w_n w_n^* = e_1$. Then $\|w_n\| \leq 1$ and $w_n p = w_n$ and $w_n \to 0$ strongly while $\{w_n^*\}$ does not tend to zero strongly. Hence W_1 is not of type III, and by a similar argument, neither is W_2. □

Corollary 8.2.17 *Two von Neumann algebras W_i on separable Hilbert spaces are semifinite if and only if $W_1\bar{\otimes}W_2$ is semifinite.*

Proof The forward implication is immediate from Proposition 8.2.14 and Corollary 7.4.14 as $(W_1\bar{\otimes}W_2)' = W_1'\bar{\otimes}W_2'$, while the converse is immediate from the proposition and the type decomposition of von Neumann algebras. □

Proposition 8.2.18 *The von Neumann tensor product of two type I von Neumann algebras is of type I.*

Proof Given two type I von Neumann algebras with identities being orthogonal sums of abelian projections p_i and q_j, respectively, we see that $p_i \otimes q_j$ are pairwise orthogonal abelian projections with strong sum one by Vigier's theorem. □

In fact, the von Neumann tensor product of a type I_m factor and a type I_n factor is clearly a type I_{mn} factor.

Lemma 8.2.19 *A von Neumann algebra W on a separable Hilbert space is type I if and only if it has an abelian projection with central cover one if and only if W' is type I. If W has no non-zero abelian projections, neither does W'. So W is type II if and only if W' is type II.*

Proof For the first forward implication pick a maximal family of non-zero abelian projections p_i of W such that their central covers are pairwise orthogonal. Since W is type I, then $p = \sum p_i$ has by maximality central cover one and is abelian by Proposition 7.1.11.

For the second one, given an abelian projection $p \in W$ with $c(p) = 1$, then W_p is commutative and Proposition 7.1.11 tells us that its commutant W_p' is $*$-isomorphic to W'. Let q be any non-zero orthogonal projection of W'_p, let $0 \neq v \in \text{im} q$, and let $r \in W'_p$ be the orthogonal projection onto the closure of $W_p(v)$. As $(W_p)_r$ is commutative and admits a cyclic vector, Theorem 5.8.7 tells us that $(W_p)_r$ is maximal commutative, and $(W'_p)_r = (W_p)_r$ shows that r is an abelian subprojection of q, so W' is type I. The chain of implications is looped by replacing W with W'.

If W' contains a non-zero abelian projection p, then $W'c(p)$ is by what we have proved of type I and so is $(W'c(p))' = Wc(p)$. Thus W also contains a non-zero abelian projection. The last statement is now clear from Theorem 7.4.13. □

Proposition 8.2.20 *Let W_1 be type II and W_2 be semifinite von Neumann algebras on separable Hilbert spaces. Then $W_1 \bar{\otimes} W_2$ is of type II.*

Proof Pick by Proposition 7.4.8 finite orthogonal projections $p_1 \in W_1$ and $q \in W_2$ with central covers one. Repeated halving of p_1 produces orthogonal projections p_n, q_n such that $p_n = p_{n+1} + q_{n+1}$ with $p_{n+1}q_{n+1} = 0$ and $p_{n+1} \sim q_{n+1}$. By Proposition 8.2.14 the elements $e_n = p_n \otimes q \in W_1 \bar{\otimes} W_2$ form a decreasing sequence of finite projections such that $e_{n+1} \sim e_n - e_{n+1}$ and $c(e_1) = 1$. We claim that $(W_1 \bar{\otimes} W_2)_{e_1}$ contains no non-zero abelian projections. Then by the lemma, neither does $W_1 \bar{\otimes} W_2 = ((W_1 \bar{\otimes} W_2)'_{e_1})'$, and it is of type II by the previous corollary. To prove the claim we may assume that $e_1 = 1$, so $W_1 \bar{\otimes} W_2$ is finite, and thus possesses a faithful normal tracial state x by Corollary 7.3.6. Suppose we have an abelian projection $r \in W_1 \bar{\otimes} W_2$. By generalized comparability there are orthogonal projections r_n and central orthogonal projections t_n such that $e_n t_n \sim r_n \leq r t_n$ and $r(1 - t_n) \prec e_n(1 - t_n)$. Since $r t_n$ is an abelian projection, we get $r_n = c(r_n) r t_n$ because the central cover of r_n in the commutative von Neumann algebra $(W_1 \bar{\otimes} W_2)_{r t_n}$ is $c(r_n) r t_n$ by definition of the central cover. Thus $r_n = c(e_n t_n) r t_n = c(e_n) r t_n$. Now $c(e_n) \geq e_n$ and as $e_n \sim e_{n-1} - e_n$ we also get $c(e_n) \geq e_{n-1} - e_n$, so $2c(e_n) \geq e_{n-1}$ by spectral theory as $e_n e_{n-1} = e_n$, so $c(e_n) = 1$ as $e_1 = 1$. Hence $e_n t_n \sim r_n = r t_n$ and $r \prec e_n$. From $e_n = e_{n+1} + (e_n - e_{n+1})$ and $e_n - e_{n+1} \sim e_{n+1}$, we get $x(e_n) = 2x(e_{n+1})$, and as $x(e_1) = 1$, we get $x(r) \leq x(e_n) \leq 2^{1-n}$, so $x(r) = 0$ and $r = 0$ by faithfulness. □

In short one may say that the von Neumann tensor product of von Neumann algebras on separable Hilbert spaces is of a type that is the maximum of the types among the tensor factors involved.

Finally, note that any type II_∞-factor W is spatially isomorphic to $U \bar{\otimes} B(V)$ for some Hilbert space V and a type II_1-factor U. Indeed, pick a non-zero finite orthogonal projection $p \in W$, and let $\{p_i\}_{i \in J}$ be a maximal orthogonal family of orthogonal projection with $p_i \sim p$. Maximality forbids $1 - \sum p_i \geq q \sim p$ for some q, as otherwise $q(1 - \sum p_i) = q$ and all $qp_i = 0$, so by generalized comparability we must have $1 - \sum p_i \prec p$. Since the family is infinite, we have

$$1 = \sum p_i + 1 - \sum p_i \sim \sum_{i \neq j} p_i + 1 - \sum p_i \prec \sum_{i \neq j} p_i + p \sim \sum p_i \leq 1$$

for any j. Thus $\sum p_i = 1$ by the Schröder-Bernstein property. Using this and investigating the proof of Theorem 7.1.15 one sees that the choice $V = l^2(J)$ and $U = W_p$, which clearly is of type II_1, will do.

8.3 Completely Positive Maps

Definition 8.3.1 Recall that $W \otimes M_n(\mathbb{C}) \cong M_n(W)$ is a C*-algebra for any C*-algebra W and $n \in \mathbb{N}$, and that any linear map $f : W \to U$ to a C*-algebra extends to a linear map $f_n : M_n(W) \to M_n(U)$ such that $f_n((w_{ij})) = (f(w_{ij}))$. Then f is *n-positive* if f_n is positive, and f is *completely positive* if it is n-positive for all n.

The matrix in $M_n(W)$ with entries w_{ij} is positive if and only if $\sum w_i^* w_{ij} w_j \geq 0$ for all $w_i \in W$. Indeed, if the latter holds, pick a faithful family of cyclic representations $\pi_v : W \to B(V)$, and for $v_1, \ldots, v_n \in V$, pick $w_i(m) \in W$ such that $\pi_v(w_i(m))(v) \to v_i$. Then

$$(\pi_{vn}((w_{ij}))((v_i))|(v_i)) = \lim(\pi_v(\sum(w_i(m)^* w_{ij} w_j(m))(v)|v) \geq 0,$$

and (w_{ij}) is positive since we can consider $\oplus\pi_{vn}$. The forward direction is clear as the positive elements in a C*-algebra are of the form u^*u.

From this we see that if π is a representation of W on V and $T : V' \to V$ is a bounded linear map between Hilbert spaces, then $T^*\pi(\cdot)T$ is completely positive. Clearly also $*$-homomorphisms between C*-algebras are completely positive, and so are compositions of completely positive maps. Completely positive maps are also bounded. In fact, we have the following result.

Proposition 8.3.2 *Positive linear maps between C*-algebras are bounded.*

Proof Say $f : W \to V$ is a positive and linear map. Let $w_n \to 0$ in W and $f(w_n) \to v$. For any positive functional x on V, we know that xf is bounded, so $x(v) = \lim xf(w_n) = 0$ and the closed graph theorem applies. □

Proposition 8.3.3 *Let $f : W \to U$ be a positive linear map. If either W or U is commutative, then f is completely positive.*

Proof If $U = C_0(X)$ for a locally compact Hausdorff space X, and $w_i \in W$, $u_i \in U$, $x \in X$, then $(\sum u_i^* f(w_i^* w_j) u_j)(x) = f((\sum u_i(x) w_i)^* \sum u_i(x) w_i)(x) \geq 0$.

If $W = C_0(X)$ and $U \subset B(V)$ and $v_i \in V$, pick a complex Radon measure μ_{ij} on X such that $\int w \, du_{ij} = (f(w)(v_i)|v_j)$ for all $w \in W$. Let $\mu = \sum |\mu_{ij}|$ and pick $f_{ij} \in L^1(\mu)$ such that $\mu_{ij} = f_{ij}\mu$. For $c_i \in \mathbb{C}$ we have $\int |w|^2 d(\sum c_i \bar{c_j} \mu_{ij}) = (f(w^* w)(\sum c_i v_i)| \sum c_j v_j) \geq 0$, so $\sum c_i \bar{c_j} f_{ij} \geq 0$ almost everywhere with respect to μ. As a countable union of sets of measure zero has measure zero, we can therefore find a set A of measure zero such that $\sum c_i \bar{c_j} f_{ij} \geq 0$ on A^c for all complex rational numbers c_i, and thus all complex numbers c_i. So $\sum(f(w_i^* w_j)(v_i)|v_j) = \int \sum \overline{w_i} w_j f_{ij} \, d\mu \geq 0$. □

So positive linear functionals on C*-algebras are completely positive. We have the following *Stinespring result.*

Theorem 8.3.4 *Let $f: W \to B(V)$ be a completely positive linear map. Then there is a Hilbert space U, a representation π of W on U, a normal representation g of $f(W)'$ on U, a bounded linear map $T: V \to U$ such that $f = T^*\pi(\cdot)T$ and $g(\cdot)T = T(\cdot)$ and $\mathrm{im}g \subset \pi(W)'$ and U is the closed span of $\pi(W)T(V)$ and $\|T\| = \|f\|^{1/2}$. When W and f are unital, then T can be taken to be an isometry.*

Proof Turn the sesquilinear form on the algebraic tensor product $W \otimes V$ given by $(\sum w_i \otimes v_i, \sum w'_j \otimes v'_j) = (f((w'_j)^* w_i)(v_i)|v'_j)$ into an inner product $(\cdot|\cdot)$ on the Hilbert space U completion of the space of appropriate equivalence classes $[\cdot]$. Let $\pi(w)([\sum w_i \otimes v_i]) = \sum[ww_i \otimes v_i]$ and $g(v)([\sum w_i \otimes v_i]) = \sum[w_i \otimes vv_i]$. Then $\|\pi(w)([\sum^n w_i \otimes v_i])\|^2 = \sum^n(f(w_j^* w^* w w_i)(v_i)|v_j)$, which when written out using $n \times n$-matrices and n-vectors and then complete positivity, is seen to be less or equal to $\|w\|^2 \|\sum^n [w_i \otimes v_i]\|^2$. Hence $\pi(w)$ can be extended by continuity, yielding a representation π of W on U. Also $\|g(v)([\sum w_i \otimes v_i])\|^2 = \sum(v^* v f(w_i^* w_j)(v_j)|v_i) \leq \|v\|^2 \|\sum[w_i \otimes v_i]\|^2$, where the latter inequality is established by picking $w_{ij} \in f(W)'$ such that $(f(w_i^* w_j)) = (w_{ij}^*)(w_{ij})$ and writing out using $n \times n$-matrices and n-vectors. So we get a $*$-homomorphism $g: f(W)' \to \pi(W)'$. If $\{a_k\}$ is a bounded increasing net in $f(W)'_+$, then $\{g(a_k)\}$ is a bounded increasing net of $B(U)_+$, and $(g(a_k)\sum[w_i \otimes v_i]|\sum[w_j \otimes v_j]) = \sum(f(w_j^* w_i)a_k(v_i)|v_j)$, so $\sup g(a_k) = g(\sup a_k)$ and g is normal. If $\{u_k\}$ is the canonical approximate unit for W, then $\{f(u_k)\}$ is by Proposition 8.3.2 a bounded increasing net in $B(V)_+$, so $\sup f(u_k) = \lim f(u_k)$ strongly. The net of maps $T_k = [u_k \otimes (\cdot)]: V \to U$ converges strongly to $T: V \to U$ with norm not greater than $\|f\|^{1/2}$ and $T^*([w_i \otimes v_i]) = f(w_i)(v_i)$. Hence $f = T^*\pi(\cdot)T$, so $\|f\| \leq \|T\|^2$ and we have equality. One also checks that $\pi(w_i)T(v_i) = [w_i \otimes v_i]$, so U is the closed span of $\pi(W)T(V)$, and π is non-degenerate. Then $\pi(u_k) \to 1$ strongly by an ε-three argument. As $\pi(u_k)g(v)T(v_i) = \pi(u_k)Tv(v_i)$ we therefore get $g(v)T = Tv$. When f is unital, we may pick $u_k = 1$, so $T = [1 \otimes (\cdot)]$ and T is isometric. □

Corollary 8.3.5 *We have $f(w)^* f(w) \leq \|f\| f(w^* w)$ for any completely positive linear map f between C*-algebras.*

Proof Write $f = T^*\pi(\cdot)T$ as in the theorem and compute. □

Corollary 8.3.6 *Let U be a C*-subalgebra of a C*-algebra W. Then any completely positive linear map $f: U \to B(V)$ has a completely positive linear extension $W \to B(V)$.*

Proof Write $f = T^*\pi(\cdot)T$ as in the theorem. We extend π to a representation θ of W by decomposing π into cyclic representations π_x and extending each one of these to π_y, where y is an extension of the state x on U to a state on W, and then letting $\theta = \oplus\pi_y$. Let p be the orthogonal projection from the bigger Hilbert space onto V. Then $T^* p\theta(\cdot)pT$ is the required map. □

Corollary 8.3.7 *A positive linear map f from a C*-algebra with identity has norm $\|f(1)\|$.*

Proof We already know $f: W \to V$ is bounded, and by Proposition 5.9.1, it suffices to show that $\|f(u)\| \leq \|f(1)\|$ for any unitary $u \in A$. Since the C*-subalgebra of A generated by u is commutative, we may by Proposition 8.3.3 write $f = T^*\pi(\cdot)T$ as in the theorem above. Then $\|f(1)\| = \|T^*T\| = \|T\|^2 = \|f\|$. □

Lemma 8.3.8 *If $f_i: W_i \to U_i$ are completely positive linear maps between C*-algebras, then $\otimes_{i=1}^n f_i$ are completely positive maps between $\otimes W_i$ and $\otimes U_i$ when both are C*-algebras with spatial norms.*

Proof Write $f_i = T_i^*\pi_i(\cdot)T_i$ as in the theorem above, and note that $\otimes_{i=1}^n f_i = (\otimes T_i)^*(\otimes\pi_i)(\cdot)(\otimes T_i)$ which has the required properties. □

Lemma 8.3.9 *Let W be a C*-algebra and suppose $S(k) \in M_n(W)_+$ with mutually commuting matrix entries. Then the matrix with ij-entry $S(1)_{ij} \cdots S(m)_{ij}$ is positive in $M_n(W)$.*

Proof We may assume that $m = 2$. The C*-algebras $W(k)$ generated by all $S(k)_{ij}$ are commutative and have elements $v(k)_{rs}$ such that $S(k)_{ij} = \sum v(k)_{ri}^* v(k)_{rj}$. Hence $(S(1)_{ij}S(2)_{ij}) = (\sum(v(1)_{ri}v(2)_{si})^* v(1)_{rj}v(2)_{sj}) \geq 0$. □

Proposition 8.3.10 *Let $f_i: W_i \to U$ be completely positive linear maps with mutually commuting images. Then there is a completely positive linear map f from $\otimes W_i$ with maximal norm into U such that $f(\otimes w_i) = f_1(w_1)\cdots f_n(w_n)$.*

Proof The well-defined map f from the algebraic tensor product is positive because with $w = \sum_j \otimes w(i)_j$ we have $f(w^*w) = \sum b_{jk}$, where

$$b_{jk} = f_1(w(1)_j^* w(1)_k)\cdots f_n(w(n)_j^* w(n)_k),$$

and $(b_{jk}) \geq 0$, so $\sum u_l b_{jk} u_l \in U_+$ for the canonical approximate unit $\{u_l\}$ for U. Hence $f(w^*w) \in U_+$. Define a linear map $g: U^\star \to (\otimes W_i)^\star$ by letting $g(x)$ be the continuous extension of $x \circ f$. If $x_k \to 0$ in $U^\star$ and $x_k \circ f \to y$ in $(\otimes W_i)^\star$, then $y(\otimes w_i) = \lim x_k(f_1(w_1)\cdots f_n(w_n)) = 0$, so $y = 0$ and g is bounded by the closed graph theorem. For $w \in \otimes W_i$ we therefore get $\|f(w)\| = \sup_{\|x\|\leq 1}|x \circ f(w)| \leq \|g\|\|w\|_{max}$, so f has a continuous extension to $\otimes W_i$ which we still denote by f. For complete positivity it suffices to show that $\sum b_i^* f(v_i^* v_j) b_j \in U_+$ for $v_i = \sum_k \otimes w(s)_{ik}$ with $w(s)_{ik} \in W_s$ and $b_i \in U$. Since f_s is completely positive we know that $(f_s(w(s)_{ik}^* w(s)_{jl}))$ is a positive U-valued matrix, and so is $(f_1(w(1)_{ik}^* w(1)_{jl})\cdots f_n(w(n)_{ik}^* w(n)_{jl}))$ by the lemma. □

Definition 8.3.11 Let U be a C*-subalgebra of W with the same identity as W. A linear surjective contraction $P: W \to U$ that fixes the elements of U is a *projection of norm one* from W onto U. A completely positive unital linear map $E: W \to W$ is a *conditional expectation* if $E(E(w)v) = E(w)E(v) = E(wE(v))$ and $E(w)^*E(w) \leq E(w^*w)$ for $v, w \in W$.

Conditional expectations are positive and $*$-preserving.

Lemma 8.3.12 *Let p be an orthogonal projection on a Hilbert space V. Then* $\|pS(1-p)+(1-p)Tp\| = \max\{\|pS(1-p)\|, \|(1-p)Tp\|\}$ *for* $S, T \in B(V)$.

Proof We have $\|pS(1-p)+(1-p)Tp\|^2 = \|a+b\| = \max\{\|a\|, \|b\|\}$ by spectral theory as $a \equiv (1-p)S^*pS(1-p)$ and $b \equiv pT^*(1-p)Tp$ have vanishing product. □

Theorem 8.3.13 *Every conditional expectation is a projection of norm one onto its image, and the converse is also true.*

Proof If $E\colon W \to W$ is a conditional expectation, then it has norm one by Corollary 8.3.7, and $E^2 = E$. Since $E(w)E(v) = E(E(w)1E(v))$ and E is $*$-preserving, we see that $E(W)$ is a normed $*$-algebra with the same identity as W. It is also closed because if $E(w_n) \to w \in W$, then $E(w_n) = E^2(w_n) \to E(w)$, so $w = E(w)$.

Let P be a norm one projection from W onto U. It is positive since by picking a faithful unital representation π of U, every functional $x = (\pi P(\cdot)(v)|v)$ satisfies $x(1) = \|x\|$ and is thus positive, and thus $*$-preserving.

By Theorem 7.2.20 the bi-transpose P^{**} of P is a projection of norm one from W'' onto U'' that extends P and is σ-weakly continuous. So we may assume that U is the closed span of its orthogonal projections p. We claim that $P(pw) = pP(w)$ for $w \in W$, which suffices to show $P(P(w)v) = P(w)P(v) = P(wP(v))$ for all $v, w \in W$. If $w \geq 0$ and $\|w\| \leq 1$ we have $p \geq pwp$, so $p = P(p) \geq P(pwp)$. Hence $pP(pwp)p = P(pwp)$ holds for all $w \in W$. Say $\|w\| \leq 1$. Then $\|pw(1-p) \pm np\| = \|pw(1-p)w^*p + n^2p\|^{1/2} \leq (1+n^2)^{1/2}$. Let $a = P(pw(1-p))$ and $b = (pap + pa^*p)/2$. If $b \neq 0$ there is a non-zero real $\lambda \in \mathrm{Sp}(b)$. Then

$$\lambda \pm n \leq \|b \pm np\| \leq \|pap \pm np\| \leq \|a \pm np\| \leq (1+n^2)^{1/2},$$

which is impossible for large $|n|$. Hence $b = 0$, and replacing n by in in the argument above, we also get $pap - pa^*p = 0$. Hence $pap = 0$. Since $a^* = P((1-p)w^*p)$, we similarly get $(1-p)a^*(1-p) = 0$, so $(1-p)a(1-p) = 0$. Combining this with the lemma we get $\|a + n(1-p)ap\| = (n+1)\|(1-p)ap\|$ for large enough n. If we in addition use that $(1-p)ap$ is fixed under P, we get $\|a+n(1-p)ap\| \leq n\|(1-p)ap\|$ for large enough n, and arrive at a contradiction unless $(1-p)ap = 0$. Combining previous identities we see that $a = pa(1-p)$, or $P(pw(1-p)) = pP(pw(1-p))(1-p)$. Replacing in addition p by $1-p$ in this identity and in $pP(pwp)p = P(pwp)$ gives four identities, which when combined with

$$P(w) = P(pwp) + P(pw(1-p)) + P((1-p)wp) + P((1-p)w(1-p)),$$

proves our claim.

If $w_i \in W$ and $u_i \in U$, then $\sum u_i^* P(w_i^* w_j) u_j = P((\sum w_i u_i)^* \sum w_j u_j) \geq 0$, so P is completely positive. Finally, for any $w \in W$, we have

$$P(w^* w) - P(w)^* P(w) = P((w - P(w))^* (w - P(w))) \geq 0.$$

□

States and center valued traces on von Neumann algebras are conditional expectations.

Proposition 8.3.14 *Letting W, V be von Neumann algebras, then there is a σ-weakly continuous projection P of norm one from $W \bar{\otimes} V$ onto $\mathbb{C} \otimes V \cong V$.*

Proof Fix a normal state x on W. By Proposition 8.2.11 we can define P by $P(a)(y) = (x \otimes y)(a)$ for $a \in W \bar{\otimes} V$ and $y \in V_*$. □

8.4 Hilbert Modules

Recall that a *right module* over a complex algebra W is a complex vector space V with a right action of W on it that is $\mathbb{C}$-linear.

Definition 8.4.1 A *pre-Hilbert module* V over a C*-algebra W is a right module over W together with a map $\langle \cdot | \cdot \rangle : V \times V \to W$ that is linear in the second variable and satisfies $\langle v_1 | v_2 w \rangle = \langle v_1 | v_2 \rangle w$ and $\langle v_1 | v_2 \rangle^* = \langle v_2 | v_1 \rangle$ for $v_i \in V$ and $w \in W$, and such that $\langle v_1 | v_1 \rangle$ is positive and non-zero for $v_1 \neq 0$.

For v_i in a pre-Hilbert module V, the formula $\|v_1\| = \|\langle v_1 | v_1 \rangle\|^{1/2}$ defines a norm on V, and $\langle v_1 | v_2 \rangle^* \langle v_1 | v_2 \rangle \leq \|v_1\|^2 \langle v_2 | v_2 \rangle$, which is is seen by setting $\|v_1\| = 1$ and $a = \langle v_1 | v_2 \rangle$ and observing that $0 \leq \langle v_2 - v_1 a | v_2 - v_1 a \rangle \leq \langle v_2 | v_2 \rangle - a^* a$ as $a^* \langle v_1 | v_1 \rangle a \leq a^* \|v_1\|^2 a$. Hence the *Cauchy-Schwarz inequality* $\|\langle v_1 | v_2 \rangle\| \leq \|v_1\| \|v_2\|$ holds in V. We say V is a *Hilbert module* if it is complete as a normed space with respect to the previously defined norm. The Cauchy-Schwarz inequality shows that the Banach space completion of a pre-Hilbert module is a Hilbert module with the module inner product extended by continuity.

Submodules of Hilbert modules are themselves Hilbert modules if they are closed as normed subspaces. The *direct sum* $\oplus V_i$ of Hilbert modules V_i over W consists of all $v \equiv \{v_i\}$ with $v_i \in V_i$ such that $\sum \langle v_i | v_i \rangle$ converges in W. It is a Hilbert module over W with $vw = \{v_i w\}$ and $\langle v | u \rangle = \sum \langle v_i | u_i \rangle$.

Example 8.4.2 Let V be a Hilbert space and W a C*-algebra. The algebraic tensor product $W \otimes V$ is a pre-Hilbert module over W with right action determined by $(w_1 \otimes v_1) w = w_1 w \otimes v_1$ and module inner product $\langle w_1 \otimes v_1 | w_2 \otimes v_2 \rangle = \langle v_1 | v_2 \rangle w_1^* w_2$ for $v_i \in V$ and $w_i, w \in W$. We denote its completion by $W \otimes V$. By picking an orthonormal basis for V indexed by I, it can be identified with the direct sum Hilbert module $\oplus_{i \in I} W$. When V is separable we also write V_W for $W \otimes V$. ◇

Definition 8.4.3 A map $T\colon V_1 \to V_2$ between Hilbert modules over the same C*-algebra is *adjointable* if there exists a map $T^*\colon V_2 \to V_1$ such that $\langle T(v_1)|v_2\rangle = \langle v_1|T^*(v_2)\rangle$ for $v_i \in V_i$.

Any adjointable map $T\colon V_1 \to V_2$ is automatically a linear module map, and since the set of $\langle T(v_1)|v_2\rangle$ for $\|v_1\| \leq 1$ is bounded for each v_2, the map T is bounded by the principle of uniform boundedness. So the set $L(V_1, V_2)$ of adjointable maps $T\colon V_1 \to V_2$ is a closed subspace of $B(V_1, V_2)$ with $T^* \in L(V_2, V_1)$ and $\|T^*\| = \|T\|$. Since $ST \in L(V_1, V_3)$ when $S\colon V_2 \to V_3$, we see that $L(V_1) \equiv L(V_1, V_1)$ is a unital C*-algebra because $\|\langle T(v_1)|T(v_1)\rangle\| \leq \|T^*T\|$ when $\|v_1\| \leq 1$, so $\|T\|^2 = \|T^*T\|$.

For v_i in Hilbert modules V_i define $v_2 \odot v_1\colon V_1 \to V_2$ by $v_2 \odot v_1 = v_2\langle v_1|\cdot\rangle$. Then $(v_2 \odot v_1)^* = v_1 \odot v_2$ and $(v_2 \odot v_1)(u_1 \odot u_0) = v_2\langle v_1|u_1\rangle \odot u_0$ and $T(v_2 \odot v_1) = T(v_2) \odot v_1$ and $(v_2 \odot v_1)S = v_2 \odot S^*(v_1)$ for $u_i \in V_i$ and $T \in L(V_2, V_0)$ and $S \in L(V_0, V_1)$. The closed span $K(V_1, V_2)$ of all $v_2 \odot v_1$ is thus a closed subspace of $L(V_1, V_2)$, and $K(V_1) \equiv K(V_1, V_1)$ is a closed two-sided ideal in $L(V_1)$, referred to as the *compact operators* on V_1, being precisely that when V_1 is a Hilbert module over $\mathbb{C}$. In the latter case one checks that $K(W \otimes V_1)$ is $*$-isomorphic to $W \otimes_{min} B_0(V_1)$.

Definition 8.4.4 The *strict topology* on $L(V_1, V_2)$ for Hilbert modules V_i over the same C*-algebra is the weak topology induced by the seminorms $T \mapsto \|T(v_1)\|$ and $T \mapsto \|T^*(v_2)\|$ for all $v_i \in V_i$.

Proposition 8.4.5 *The unit ball of $K(V_1, V_2)$ is strictly dense in the unit ball of $L(V_1, V_2)$ for Hilbert modules V_i.*

Proof Say V_i are Hilbert modules over W. The closed span U of $\langle V_1|V_1\rangle$ is a two sided ideal of W, and letting $\{u_i\}$ be the canonical approximate unit for the C*-algebra U, then $\langle v - vu_i|v - vu_i\rangle \to 0$ for $v \in V_1$, so $V_1\langle V_1|V_1\rangle$ is dense in V_1. Let now $\{T_i\}$ be the canonical approximate unit for $K(V_1)$. Then $T_i v\langle v_1|v_1'\rangle = T_i(v \odot v_1)(v_1') \to v\langle v_1|v_1'\rangle$ for $v, v_1, v_1' \in V_1$, so $T_i v \to v$ by boundedness of the approximate unit and by the previous claimed denseness. For $T \in L(V_1, V_2)$ we thus see that $TT_i \to T$ strictly, and we are done by the formulas in the last paragraph prior to the definition above. □

When V is the C^*-algebra W itself, then $K(W)$ is $*$-isomorphic to W under the map which sends $w_1 \odot w_2$ to left multiplication on W by $w_1 w_2^*$. When W is unital, also $L(W) \cong W$ where $T \in L(W)$ corresponds to left multiplication on W by $T(1)$. We refer to $L(W)$ as the *multiplier algebra* of W, and sometimes denote it by $M(W)$.

Definition 8.4.6 A closed ideal W in a C*-algebra U is *essential* if no non-zero ideal of U has trivial intersection with W, i.e. if $u \in U$ vanishes when $uW = \{0\}$.

An essential ideal is never a non-zero direct summand. If W is a C*-algebra, then $K(W)$ is essential in $L(W)$. Indeed, if $\{v_i\}$ is the canonical approximate unit for W and $T \in L(W)$, then $T(w \odot v_i)(v_i) = T(wv_i^2) \to T(w)$ for $w \in W$.

Definition 8.4.7 Let V be a Hilbert module over some C*-algebra, and let W be a C*-algebra. A $*$-homomorphism $f : W \to L(V)$ is *non-degenerate* if V is the closed span of $f(W)V$.

From the proof of the previous proposition we see that for a Hilbert module V, the inclusion $K(V) \subset L(V)$ is non-degenerate.

Proposition 8.4.8 *Let W be a closed ideal of a C*-algebra U, and let V be a Hilbert module over some C*-algebra. Any non-degenerate $*$-homomorphism $f : W \to L(V)$ extends uniquely to a $*$-homomorphism $U \to L(V)$ which is injective whenever f is injective and W is essential in U.*

Proof Using the canonical approximate unit of W, we see that the assignment $\sum f(w_n)v_n \mapsto \sum f(uw_n)v_n$ for $v_n \in V, w_n \in W$ and $u \in U$ extend by continuity to a well-defined $g(u) \in L(V)$ with adjoint $g(u^*)$. We thus get a unique $*$-homomorphic extension $g : U \to L(V)$ of f. When f is injective, the ideal $\ker g$ of U has trivial intersection with W, so g is injective whenever W is essential in U. □

Corollary 8.4.9 *Let W be a C*-algebra. Then $L(W)$ is up to isomorphism the unique maximal C*-algebra that contains $K(W)$ as an essential ideal.*

Proof The maximality is immediate from the proposition. As for uniqueness, if U is a maximal essential extension of W, then as W is an essential ideal of $L(W)$ there is by maximality a $*$-monomorphism $h : L(W) \to U$ extending the inclusion $W \subset U$. The existence result in the proposition tells us that $W \subset L(W)$ has an injective extension $g : U \to L(W)$, while the uniqueness result, tells us that $gh = \iota$, so g is surjective and $U \cong L(W)$. □

Corollary 8.4.10 *Let V be a Hilbert module over some C*-algebra, and let W be another C*-algebra. If we have a non-degenerate inclusion $W \subset L(V)$, then $L(W)$ is $*$-isomorphic to the unital C*-algebra U consisting of all $u \in L(V)$ such that $uW \subset W$ and $Wu \subset W$.*

Proof Clearly W is an ideal of U which is essential by non-degeneracy. By the corollary it suffices to show that U is maximal with this property. But if W is an essential ideal in a C*-algebra U_1, then $W \subset L(V)$ extends by the proposition above to an inclusion $U_1 \subset L(V)$, and $U_1 \subset U$ as W is an ideal of U_1. □

When W is a non-degenerate C*-subalgebra of $B(V)$ for a Hilbert space V, the set of $T \in B(V)$ such that $Tw, wT \in W$ for all $w \in W$ is $*$-isomorphic to $M(W)$ by the corollary above, and $M(W)$ is the unique maximal C*-algebra containing W as an essential ideal. Since $TW \subset W$ implies $TW'' \subset W''$ and W'' is unital, we see that $M(W) \subset W''$.

Example 8.4.11 Let X be a locally compact Hausdorff space. If X is also σ-compact, represent $C_0(X)$ non-degenerately as multiplication operators on some $L^2(\mu)$. Then, as $C_0(X)'' = L^\infty(\mu)$, the previous corollary shows that $M(C_0(X)) = C_b(X)$. This holds also when we do not assume σ-compactness. One proof goes as follows: The inclusion $C_0(X) \subset M(C_0(X))$ extends to a $*$-monomorphism

$F: C_b(X) \to M(C_0(X))$ as $C_0(X)$ is an essential ideal of $C_b(X)$. To see that F is surjective, consider a positive $f \in M(C_0(X))$. Considering the canonical approximate unit $\{u_i\}$ for $C_0(X)$, then for each $x \in X$, the bounded increasing net $\{fu_i(x)\}$ converges to $h(x)$, producing a bounded function h on X such that $hg = fg$ for any $g \in C_0(X)$. To see that h is continuous, consider $x_i \to x$ in X. We may assume that this happens inside a compact subset Y of X. Pick a function $g \in C_0(X)$ that is one on Y. Then $hg = fg \in C_0(X)$, so $h(x_i) = hg(x_i) \to hg(x) = h(x)$. As $F(h)g = fg$ for all $g \in C_0(X)$, we get $F(h) = f$.

Given a compact Hausdorff space Y. Then $C_0(X)$ is an ideal of $C(Y)$ if and only if X is open in Y, and it is essential if and only if X is dense in Y. In this case we say that Y is a *Hausdorff compactification* of X. The smallest one is the one-point compactification, and then $C(Y)$ is just $C_0(X)$ with an identity adjoined. The maximal compactfication is called the *Stone-Cech compactification*, for which $C(Y) = M(C_0(X)) = C_b(X)$. The space Y consists then of the non-zero characters on $C_b(X)$ with w^*-topology. We may think of the multiplier algebra of a C*-algebra W as a sort of non-commutative Stone-Cech compactification of the underlying 'space' of W.

Given a locally compact Hausdorff space Z and a continuous map $p: X \to Z$, the *transpose* of p is the non-degenerate $*$-homomorphism $P: C_0(Z) \to C_b(X) = M(C_0(X))$ given by $P(f) = f \circ p$. It is easy to see that the image of P is in $C_0(X)$ if and only if p is *proper*, meaning that inverse images of compact sets are compact. Conversely, given a non-degenerate $*$-homomorphism $P: C_0(Z) \to M(C_0(X))$, we set $p(x) = \tilde{x} \circ P$ for x a character on $C_0(X)$; these characters form a topological space under w^*-topology which we by Gelfand's theorem identify with X. By $\tilde{x}$ we mean the extension of the non-degenerate $*$-homomorphism x to $M(C_0(X))$. Now $p(x) \neq 0$ as P is non-degenerate and $\tilde{x}$ is multiplicative, so $p(x)$ is indeed a character on $C_0(Z)$, and thus belongs to Z. Now if $x_i \to x$ in the w^*-topology, then $\tilde{x}_i(g)x_i(g') = x_i(gg') \to x(gg') = \tilde{x}(g)x(g')$ for $g \in M(C_0(X))$ and $g' \in C_0(X)$. Pick g' such that $x(g') = 1$. Then

$$|\tilde{x}_i(g) - \tilde{x}(g)| \leq |\tilde{x}_i(g)| \cdot |x(g') - x_i(g')| + |\tilde{x}_i(g)x_i(g') - \tilde{x}(g)x(g')| \to 0$$

as $|\tilde{x}_i(g)| \leq \|g\|$. Hence $p(x_i)(f) = \tilde{x}_i(P(f)) \to \tilde{x}(P(f))$ and p is continuous, and by definition of the Gelfand transform, we see that P is the transpose of $p: X \to Z$.

This shows that a *morphism* from one C^*-algebra V to another W should be a non-degenerate $*$-homomorphism $V \to M(W)$. ◇

Corollary 8.4.12 *For any Hilbert module V we have* $L(V) \cong M(K(V))$.

Proof The inclusion $K(V) \subset L(V)$ is non-degenerate, so by the previous corollary it extends to a $*$-isomorphism $M(K(V)) \to L(V)$. □

A positive element w of a C*-algebra W is *strictly positive* if $x(w) > 0$ for all states x on W.

Lemma 8.4.13 *A positive element w of a C*-algebra W is strictly positive if and only if the right ideal generated by w is W. In particular, if V is a Hilbert module, then a positive element T of $K(V)$ is strictly positive if and only if T has dense range.*

Proof If $\overline{wW} = W$ and $x(w) = 0$ for a state x, then $x = 0$ by the Cauchy-Schwarz inequality. Conversely, if $\overline{wW} \neq W$, pick a state x on W vanishing on $\overline{wW}$. With the canonical approximate unit $\{u_i\}$ of W, we get $x(w) = \lim x(wu_i) = 0$.

If T is strictly positive, then $TK(V)$ is dense in $K(V)$ by the paragraph above, so $T(V)$ is dense in V by the proof of Proposition 8.4.5. If $T(V)$ is dense in V, then to $v \in V$, there is $\{v_n\} \subset V$ such that $T(v_n) \to v$. Then $v \odot w = \lim T(v_n \odot w)$ for $w \in V$, so T is strictly positive by the paragraph above. □

The following result is known as *Kasparov's stabilization theorem.*

Theorem 8.4.14 *We have $U \oplus V_W \cong V_W$ for any countably generated Hilbert module U over W.*

Proof Let $\tilde{W}$ be C*-algebra obtained by adjoining an identity to W, and regard U as a Hilbert module over $\tilde{W}$, see the proof of Proposition 8.4.5. Since $V_W \cong \overline{V_{\tilde{W}} W}$ and $U \oplus V_W \cong \overline{(U \oplus V_{\tilde{W}})W}$, we may assume that W is unital.

Let $\{u_n\}$ be a sequence of unit vectors in U where each element of a generating set occurs infinitely often. Let $\{v_n\}$ be an orthonormal basis for V. Set $w_n = 1 \otimes v_n$, so $\{w_n\}$ is generating for V_W. Define $T = \sum_{n=1}^{\infty}(2^{-n}u_n \odot w_n + 4^{-n}w_n \odot w_n)$ in $K(V_W, U \oplus V_W)$, so $T(w_n) = 2^{-n}u_n + 4^{-n}w_n$. So whenever $u_m = u_n$ we have $T(2^n w_m) = u_n + 2^{-m}w_m$, and since there are infinitely many such m, we see that $u_n \in \overline{T(V_W)}$. Also $w_n = T(4^n w_n) - 2^n u_n \in \overline{T(V_W)}$, so T has dense range. Now

$$T^*T = \sum_{n,m} 2^{-(n+m)} w_n(\langle u_n | u_m \rangle + \langle 2^{-n}w_n | 2^{-m}w_m \rangle) \odot w_n \geq \sum 4^{-2n} w_n \odot w_n$$

and the latter operator is positive with dense range, so it and hence T^*T is strictly positive by the previous lemma. Again by the lemma $|T| = (T^*T)^{1/2}$ has dense range. The desired well-defined isomorphism sends $|T|(v)$ to $T(v)$ for $v \in V_W$.

□

Exercises

For Sect. 8.1

1. Show that the C*-algebra of compact operators on a Hilbert space V is nuclear, while the C*-algebra of all bounded operators on V is never nuclear when V is infinite dimensional.
2. Show that the spatial tensor product of two AF-algebras is an AF-algebra.

3. Show that $(W_1 \otimes W_2) \otimes W_3$ is $*$-isomorphic as a $*$-algebra to $W_1 \otimes (W_2 \otimes W_3)$ whenever W_i are $*$-algebras, and that this also holds for spatial tensor products of C*-algebras. Prove that if a C*-algebra W is nuclear, then so is $M_n(W)$.
4. Show that the spatial tensor product of simple C*-algebras is again simple.
5. Show that postliminal C*-algebras are nuclear.
6. Show that a closed ideal J of a nuclear C*-algebra W is nuclear, and that W/J is nuclear. Show that nuclearity is inherited by hereditary C*-subalgebras.

For Sect. 8.2

1. Prove that the von Neumann tensor product of maximal abelian von Neumann algebras is again maximal abelian.
2. Show that if $\pi_i\colon W_i \to B(V_i)$ are faithful normal representations of von Neumann algebras, then the representation $\pi_1 \otimes \pi_2$ of the spatial tensor product extends uniquely to a faithful normal representation of $W_1 \bar{\otimes} W_2$ with range $\pi_1(W_1)\bar{\otimes}\pi_2(W_2)$.
3. If T is the transpose map of $M_2(\mathbb{C})$, show that $\iota \otimes T$ on $M_2(\mathbb{C}) \otimes M_2(\mathbb{C})$ is neither an isometry nor of norm one.
4. Suppose W_i are subfactors of a von Neumann factor W, such that W_1 and W_2 mutually commute and generate W. Show that the map $w_1 \otimes w_2 \mapsto w_1 w_2$ extends to a $*$-isomorphims $W_1 \bar{\otimes} W_2 \to W$ if and only if there is a non-zero normal bounded functional x on W such that $x(w_1 \otimes w_2) = x(w_1)x(w_2)$ for $w_i \in W_i$.

For Sect. 8.3

1. Suppose $W \subset B(V)$ is a von Neumann algebra with a commutative von Neumann subalgebra U. Show that there exist norm one projections from $B(V)$ onto U and from $B(V)$ onto U' and from W onto U and from W onto $W \cap U'$.
2. Say W is a von Neumann algebra with a type III von Neumann subalgebra U, and that there is a separating family of norm one projections from W onto U that are weakly continuous on the closed unit ball. Show that W is then also of type III.

For Sect. 8.4

1. A double centraliser for a C*-algebra W is a pair (L, R) of bounded maps on W such that $L(uw) = L(u)w$ and $R(uw) = uR(w)$ and $R(u)w = uL(w)$ for $u, w \in W$. Show that $\|L\| = \|R\|$, that the set X of double centralisers is a closed subspace of $B(W) \oplus B(W)$, and that it as a C*-algebra with product $(L_1, R_1)(L_2, R_2) = (L_1L_2, R_2R_1)$ and $*$-operation $(L, R)^* = (R^*, L^*)$, where $f^* \equiv f((\cdot)^*)^*$ for a linear map f on W, is $*$-isomorphic to $M(W)$. Describe the inclusion $W \subset X$ here.
2. Show that $M(W \otimes_{min} U)$ contains a canonical copy of $M(W) \otimes_{min} M(U)$ for C*-algebras U, W.
3. Prove that the multiplier algebra of a non-unital C*-algebra is never separable.
4. Let X be a locally compact Hausdorff space. Show that the strict topology on $M(C_0(X))$ is the topology of uniform convergence on compact subsets of X.

5. Let X be a locally compact Hausdorff space, and let W be a C*-algebra. Show that $M(C_0(X) \otimes_{min} W)$ is the set of strictly continuous functions from the Stone-Cech compactification of X to $M(W)$.
6. Show that separable C*-algebras always contain strictly positive elements.
7. Show that a Hilbert module V over a C*-algebra W is countably generated if and only if $K(V)$ has a strictly positive element.

Chapter 9
Unbounded Operators

Most operators occurring in applications, like differential operators or multiplication operators, are unbounded. Quantum physics requires that such operators should act on some Hilbert spaces, so that one can talk about self-adjointness which is needed to give expectation values that are real numbers; such values are supposed to be the outcome of a measurement. Now a self-adjoint operator defined on the whole Hilbert space must be bounded, so unbounded ones can only be densely defined, rendering the business of dealing with unbounded operators a minefield, especially in composing operators having different ranges and domains, which might well have empty intersections, etc.

Fortunately, the relevant operators are not too wild, since they often have closed graphs, so topology comes into play already at the Hilbert space level. In fact, self-adjoint operators are always closed. Often one can also focus on operators restricted to common cores, thus having the original graphs as their closures. We discuss these basic notions in the first section, concluding with the result that any operator bounded below has a self-adjoint extension with the same bound.

In the next section we invoke the Cayley transform from complex analysis. We use it to set up a correspondence between the set of densely defined symmetric operators and a certain class of isometries. This is then used to find conditions, like equality of so called deficiency indices, telling us when the operators are self-adjoint.

We study also the concept of affiliation, that a densely defined closed operator is affiliated to a von Neumann algebra acting on the same Hilbert space. This will become important in the next section on modular theory, where the unbounded (conjugate linear) operator in question is the $*$-operation in the GNS-representation of a C*-algebra.

We are now at the core of the topic, namely the spectral theorem for self-adjoint operators. The proof of this hinges on the spectral theorem for normal bounded operators, and a clever utilization of the Caley transform. We discuss various

L. Tuset, *Analysis and Quantum Groups*,
https://doi.org/10.1007/978-3-031-07246-8_9

aspects of this, including quadratic forms, which are intimately linked to unbounded operators.

Having the spectral machinery at hand, we polar decompose densely defined closed operators, and we establish Stone's theorem, which tells us that any strongly continuous one-parameter unitary group of operators on a Hilbert space has a generator, that is, a self-adjoint operator A such that $\{e^{itA}\}_{t\in\mathbb{R}}$ is the one-parameter group. Using this we give a sensible definition of what it means that self-adjoint operators commute, and show that this happens exactly when their one-parameter groups mutually commute. We also study tensor products of unbounded operators.

In the final section we study convergence of sequences of unbounded operators, ending with familiar versions of formulas for the exponential map from Lie theory, like Trotter's product formula.

9.1 Definitions and Basic Properties

Definition 9.1.1 An *operator in* a complex Hilbert space V is a linear map $T\colon D(T) \to V$ having *domain* $D(T)$ a subspace of V. It is *densely defined* if $D(T)$ is dense in V. Its *range* $R(T)$ is $T(D(T))$. If S is another operator in V that coincide with T on $D(S) \subset D(T)$, then T *extends* S, and we write $S \subset T$.

The sum $S + T$ and composition ST of two operators S, T in a Hilbert space are the usual ones, then with domains $D(S) \cap D(T)$ and $D(T) \cap T^{-1}(D(S))$, respectively. The distributive law need not hold. We define the *inverse* of an injective operator T to be the operator such that $T^{-1}T(v) = v$ for all v in its domain $D(T^{-1}) \equiv R(T)$. So $R(T^{-1}) = D(T)$ and TT^{-1} is the identity map on $R(T)$. If also S is injective, then $(ST)^{-1} = T^{-1}S^{-1}$.

Definition 9.1.2 The *adjoint* of a densely defined operator T in a Hilbert space V is the operator T^* with domain $D(T^*)$ consisting of all $v \in V$ such that $(T(\cdot)|v)$ is bounded on $D(T)$. Then there is a unique $T^*(v) \in V$ such that $(\cdot|T^*(v)) = (T(\cdot)|v)$ on $D(T)$. It is *self-adjoint* if $T^* = T$.

By definition $\ker T^* = R(T)^{\perp}$, so the eigenvectors of T^* with eigenvalue $c \in \mathbb{C}$ form the vector space $R(T - \bar{c}I)^{\perp}$. The adjoint need not be densely defined. Note that $S^* + T^* \subset (S+T)^*$ and $T^*S^* \subset (ST)^*$ whenever $S, T, S+T, ST$ are densely defined. If $S \subset T$ with S densely defined, then $T^* \subset S^*$.

A densely defined operator T in V is *symmetric* if $T \subset T^*$, which happens if and only if $(T(v)|v) \in \mathbb{R}$ for all $v \in D(T)$. When T is symmetric and $a, b \in \mathbb{R}$, we have $\|(T - (a+ib)I)(v)\|^2 = \|(T - aI)(v)\|^2 + b^2\|v\|^2$ for $v \in V$. So $T - (a+ib)I$ is injective when $b \neq 0$, and its inverse is then bounded by $|b|^{-1}$. A symmetric operator T is *maximal symmetric* if $T = S$ whenever $S \supset T$ is symmetric. Self-adjoint operators are maximally symmetric because if $T^* = T$ and $S \supset T$ is symmetric, then $S \subset S^* \subset T^* = T$.

Definition 9.1.3 An operator T in a Hilbert space V is *closed* if it has a closed *graph* $G(T) \equiv \{(v, T(v)) \mid v \in D(T)\}$. We say T is *closable* if the closure of its graph is the graph of another operator, which would then be the minimal closed operator extending T, called the *closure* of T and denoted by $\overline{T}$.

So T is closed if whenever $\{v_n\}$ is a sequence in $D(T)$ converging to v and $T(v_n) \to w$, then $v \in D(T)$ and $T(v) = w$. An operator T is closable if and only if for every sequence $\{v_n\}$ in $D(T)$ converging to zero, the only accumulation point of $\{T(v_n)\}$ is zero. By the closed graph theorem everywhere defined closed operators are bounded, and so is an operator whose domain and adjoint domain is the whole Hilbert space. Thus everywhere defined symmetric operators are bounded.

Definition 9.1.4 A *core* of a closed operator T is any subspace D of $D(T)$ such that the closure of the graph of T when restricted to D is $G(T)$.

Projecting dense subspaces of $G(T)$ onto the first factor provide cores for any closed operator T. Dealing with several operators it is useful to find a common core to avoid handling several domains in computations.

Proposition 9.1.5 *Let T be a densely defined operator in a Hilbert space V. Then T^* is closed and $V \oplus V = \overline{G(T)} \oplus U(G(T^*))$, where U is the unitary operator on $V \oplus V$ given by $U((v, w)) = (-w, v)$. Also T is closable if and only if T^* is densely defined, and T^{**} is the closure of T.*

Proof When $(v, w) \in G(T)^\perp$ we have $0 = ((v, w)|(u, T(u))) = (v|u) + (w|T(u))$ for $u \in D(T)$, so $w \in D(T^*)$ and $T^*(w) = -v$. Hence $G(T)^\perp = U(G(T^*))$, and T^* is closed by unitarity, and we also get the required decomposition.

If $v \in D(T^*)^\perp$, then since $U^* = -U$ and $U(W^\perp) = U^*(W)^\perp$ for any subspace W of V, we get

$$(0, v) = U((v, 0)) \in U(G(T^*)^\perp) = U^*(G(T^*))^\perp = U(G(T^*))^\perp = \overline{G(T)}.$$

Hence if T is closable, then $v = 0$ and T^* is densely defined. Conversely, if T^* is densely defined, apply the decomposition to T^{**}. Then

$$G(T^{**}) = U^*(U(G(T^{**}))) = U^*(G(T^*)^\perp) = U(G(T^*))^\perp = \overline{G(T)},$$

so T^{**} is the minimal closed extension of T. □

Every symmetric operator T is closable. We say T is *essentially self-adjoint* if its closure is self-adjoint, meaning that $\overline{T} = T^*$. The following result is then straightforward from definitions.

Proposition 9.1.6 *If T is a symmetric operator in a Hilbert space V, and if $D \subset D(T)$ is a dense subspace of V such that T when restricted to D is essentially self-adjoint, then T is essentially self-adjoint and its closure is the closure of T when restricted to D.*

Proposition 9.1.7 *If T is an injective densely defined closed operator with dense range, the so are T^{-1} and T^*, and $(T^*)^{-1} = (T^{-1})^*$. In particular, injective self-adjoint operators have self-adjoint inverses.*

Proof If the Hilbert space for T is V and S is the unitary operator on $V \oplus V$ given by $S((v, w)) = (w, v)$, then $S(G(T)) = G(T^{-1})$, so T^{-1} is closed and is evidently densely defined with dense range. By Proposition 9.1.5 the operator T^* is closed and densely defined, so $\ker T^* = R(T)^\perp = \{0\}$. But $T^{**} = T$ is also injective and $R(T^*)^\perp = \ker T^{**} = \{0\}$, so T^* has dense range. With U from that proposition

$$US(G((T^*)^{-1})) = U(G(T^*)) = G(T)^\perp = S(G(T^{-1}))^\perp = S^*(G(T^{-1})^\perp),$$

which equals $S^*U(G((T^{-1})^*))$, so $(T^*)^{-1} = (T^{-1})^*$ as $S^* = S$ and $US = -SU$. □

Proposition 9.1.8 *Let T be a densely defined closed operator in a Hilbert space V. Then T^*T is self-adjoint and $D(T^*T)$ is a core for T. And $T^*T + I$ is a bijection from $D(T^*T)$ onto V with $0 \leq (T^*T + I)^{-1} \leq I$. The closure of $(TT^* + I)^{-1}T$ is $T(T^*T + I)^{-1}$, which has norm not greater than one, and $(n^{-2}T^*T + I)^{-1} \to I$ strongly as $n \to \infty$.*

Proof The decomposition in Proposition 9.1.5 gives to $v \in V$ unique vectors $S(v) \in D(T)$ and $R(v) \in D(T^*)$ such that $(v, 0) = (S(v), TS(v)) \oplus (T^*R(v), -R(v))$. Hence $\|v\|^2 = \|S(v)\|^2 + \|TS(v)\|^2 + \|T^*R(v)\|^2 + \|R(v)\|^2$, so $\|S(v)\| \leq \|v\|$ and $\|R(v)\| \leq \|v\|$, showing that the linear assignments $v \mapsto S(v)$ and $v \mapsto R(v)$ define operators $S, R \in B(V)$ with norm not greater than one. By the decomposition above we also see that $R = TS$ and $(I + T^*T)S = I$, which again shows that S is injective and that $(S^*(v)|v) = (S^*(T^*T + I)S(v)|v) = \|R(v)\|^2 + \|S(v)\|^2 \geq 0$, so $S \geq 0$. By the previous proposition S^{-1} is also self-adjoint and thus maximal symmetric, and since $T^*T + I \supset S^{-1}$ is also symmetric, we get $T^*T + I = S^{-1}$. So T^*T is self-adjoint and $(T^*T + I)^{-1} = (S^{-1})^{-1} = S$ and $R = T(T^*T + I)^{-1}$.

Using the same arguments for T^* instead of T, we get operators $TT^* + I$ and $(TT^* + I)^{-1}$ such that for $v \in D(T)$ and $w \in D(TT^*)$ we have

$$((TT^* + I)^{-1}T(v)|(TT^* + I)(w)) = (T(T^*T + I)^{-1}(v)|(TT^* + I)(w)),$$

and since $TT^* + I$ is surjective, this means that $(TT^* + I)^{-1}T \subset R$.

Let $S_n = (n^{-2}T^*T + I)^{-1}$. As $R(S_n) = D(T^*T)$ the considerations above yield

$$n^2(I - S_n) = T^*T(n^{-2}T^*T + I)^{-1} \supset T^*(n^{-2}TT^* + I)^{-1}T \supset (n^{-2}T^*T + I)^{-1}T^*T,$$

so $\|v - S_n(v)\| \leq n^{-2}\|T^*T(v)\| \to 0$ for $v \in D(T^*T)$, and everywhere as $\|S_n\| \leq 1$.

To see that $D(T^*T)$ is a core for T consider $v \in D(T)$. Then $S_n(v) \in D(T^*T)$ and $S_n(v) \to v$. But $(n^{-2}TT^* + I)^{-1} \to I$ strongly as $n \to \infty$, so from the above $TS_n(v) = (n^{-2}TT^* + I)^{-1}T(v) \to T(v)$, and the graph of the restriction of T to $D(T^*T)$ is dense in $G(T)$. □

Any operator satisfying the equivalent statements in the result below is said to be *normal*.

Proposition 9.1.9 *Let T be a densely defined closed operator in a Hilbert space V. Then $D(T) = D(T^*)$ and $\|T(\cdot)\| = \|T^*(\cdot)\|$ on $D(T)$ if and only if $T^*T = TT^*$ if and only if there are self-adjoint operators A, B in V such that $T = A + iB$ and $T^* = A - iB$ and $\|T(\cdot)\|^2 = \|A(\cdot)\|^2 + \|B(\cdot)\|^2$ on $D(T)$. Then A and B would be the closures of $(T + T^*)/2$ and $i(T^* - T)/2$, respectively.*

Proof For the first forward implication we have

$$4(T^*T(v)|w) = \sum i^k \|T(v+i^k w)\|^2 = \sum i^k \|T^*(v+i^k w)\|^2 = 4(T^*(v)|T^*(w))$$

for $v \in D(T^*T)$ and $w \in D(T)$, so $T^*(v) \in D(T^{**}) = D(T)$ and $TT^*(v) = T^*T(v)$. Hence $T^*T \subset TT^*$ and equality is obtained by symmetry.

In the opposite direction we clearly get $\|T(v)\| = \|T^*(v)\|$ for $v \in D(T^*T)$. If $v \in D(T)$, then $S_n(v) \to v$ and $TS_n(v) \to T(v)$ with S_n as in the previous proof. So $\{T^*S_n(v)\}$ is Cauchy and converges to $w \in V$ with $\|w\| = \|T(v)\|$. As T^* is closed we thus see that $v \in D(T^*)$ and $T^*(v) = w$. Hence $D(T) \subset D(T^*)$ and $\|T^*(\cdot)\| = \|T(\cdot)\|$ on $D(T)$. Again by symmetry, we get $D(T) = D(T^*)$.

Assuming now the first two conditions hold, then $A_0 \equiv (T + T^*)/2 \subset A_0^*$, and if $v \in D(A_0)$, then for $w \in D(T)$, we have $(S_n A_0^*(v)|w) = (v|A_0 S_n(w)) = (v|S_n A_0(w)) = (A_0 S_n(v)|w)$ with S_n as above, so $S_n A_0^*(v) = A_0 S_n(v)$. As $S_n \to I$ strongly we thus see that $\overline{A_0} = A_0^* \equiv A$. Similarly, we let B be the adjoint, or closure, of $i(T^* - T)/2$, so $B^* = B$. Now $D(T) \subset D(A)$ and $D(T) \subset D(B)$, and the restriction of $A+iB$ to $D(T)$ is T. On the other hand, if $v \in D(A) \cap D(B)$, then as above $(A + iB)(v) = \lim S_n(A + iB)(v) = \lim(A + iB)S_n(v) = \lim TS_n(v)$. As T is closed, we get $v \in D(T)$ and $T(v) = (A + iB)(v)$. Similarly, we get $A - iB = T^*$. Clearly $D(T^*T)$ is contained in both $D(A^2)$ and $D(B^2)$ and $A^2 + B^2 = T^*T$ on $D(T^*T)$. But $A^2 + B^2$ is symmetric and T^*T is self-adjoint, and whence maximal symmetric, so $T^*T = A^2 + B^2$. Invoking S_n once again, we thus see that $\|T(v)\|^2 = \|A(v)\|^2 + \|B(v)\|^2$ for $v \in D(T)$.

In the opposite direction we clearly have $D(T) = D(A) \cap D(B) = D(T^*)$, and for $v \in D(T)$ we have $\|T(v)\|^2 - \|T^*(v)\|^2 = 4\,\mathrm{Im}(A(v)|B(v) = 0$ as $\|(A + iB)(v)\|^2 = \|A(v)\|^2 + \|B(v)\|^2$. □

Definition 9.1.10 A densely defined symmetric operator T in a Hilbert space is *bounded from below* if for some $c \in \mathbb{R}$ we have $(T(v)|v) \geq c\|v\|^2$ for $v \in D(T)$, and we then write $T \geq cI$. We say T is *positive* if $T \geq 0$.

By Proposition 9.1.8 we know that $T^*T \geq 0$ for any densely defined closed operator T in a Hilbert space. Clearly a densely defined symmetric operator S in a Hilbert space is bounded if and only if both S and $-S$ are bounded from below.

Theorem 9.1.11 *Every operator in a Hilbert space that is bounded below has a self-adjoint extension with the same lower bound.*

Proof Say T is such an operator in a Hilbert space V. By appropriately adding a scalar times the identity map on V, we may assume that $T \geq I$. Then $(T(\cdot)|\cdot)$ is an inner product s on $D(T)$ with associated norm greater than $\|\cdot\|$. The identity map on

$D(T)$ extends by continuity to an operator A with $\|A\| \leq 1$ from the Hilbert space completion W of $D(T)$ with respect to s into V. If $w_n \to w \in W$ with $w_n \in D(T)$, then $A(w_n) \to A(w)$ and $(T(v)|A(w)) = \lim(T(v)|A(w_n)) = \lim s(v, w_n) = s(v, w)$ for $v \in D(T)$. So A is injective and we regard W as a subspace of V. Also, if $A(w) \in D(T^*)$, then $(T^*A(w)|A(w)) = \lim(T^*A(w)|A(w_n))$ equals $\lim(A(w)|TA(w_n)) = \lim s(w, w_n) = s(w)^2$, which is not less than $\|A(w)\|^2$. Hence the restriction S of T^* to $A(W) \cap D(T^*)$ is symmetric and $S \geq I$. For more general $w \in D(T)$ we have $(T(w)|AA^*(v)) = s(w, A^*(v)) = (A(w)|v) = (w|v)$, so $AA^*(v) \in D(T^*)$ and $T^*(AA^*(v)) = v$, which means that $AA^*(v) \in D(S)$ and $SAA^* = I$. Hence S extends $(AA^*)^{-1}$, which is self-adjoint by Proposition 9.1.7, and as S is symmetric we must therefore have $S = (AA^*)^{-1}$, which is the desired self-adjoint extension of T. □

The extension we constructed in the proof of this theorem is known as the *Friedrichs extension.* An operator in a Hilbert space that is not essentially self-adjoint may have several self-adjoint extensions.

Example 9.1.12 Let μ be a Radon measure on a locally compact Hausdorff space X. For every Borel function f on X, define the multiplication operator M_f in $L^2(\mu)$ by $M_f(g) = fg$ for $g \in D(M_f) \equiv \{g \in L^2(\mu) \mid fg \in L^2(\mu)\}$. Clearly $M_f = 0$ if and only if f vanishes almost everywhere. It has $C_c(X)$ as a core, so it is densely defined. As $D(M_f^*) = D(M_f)$ and $M_f^* = M_{\bar{f}}$ it is closed and normal. If $\bar{f} = f$ almost everywhere, then it is self-adjoint, and if $f \in L^\infty(\mu)$, then it is the usual bounded multiplication operator. For a more concrete unbounded self-adjoint operator, consider the case when f is the identity map on $\mathbb{R}$ with Lebesgue measure. ◇

For later purposes we need the following result.

Lemma 9.1.13 *Let μ be a finite Radon measure and M_f be the multiplication operator of a real $f \in L^p(\mu)$ for $p \in \langle 2, \infty\rangle$. Then any dense subspace of $L^q(\mu)$, where $q^{-1} + p^{-1} = 1/2$, is a core for M_f.*

Proof By Hölder's inequality we see that $L^q(\mu) \subset D(M_f)$. Let $g \in D(M_f)$. Letting g_n be the function that vanishes when $|g| > n$ and coincides with g otherwise, Lebesgue's dominated convergence theorem tells us that $g_n \to g$ and $fg_n \to fg$ in L^2-norm. Since all g_n belong to $L^q(\mu)$, it must therefore be a core for M_f. If D is a dense subspace of $L^q(\mu)$, so there are $h_n \in D$ that converge to g in L^q-norm, then Hölder's inequality shows again that g belongs to the domain of the closure of M_f restricted to D, so D is a core for M_f. □

Example 9.1.14 Let X be a non-empty open subset of $\mathbb{R}^n$ with restricted Lebesgue measure μ. The *Laplace operator* is the operator Δ in $L^2(\mu)$ with dense domain all complex valued smooth functions on X with compact support, and which is given by $\Delta(f) = \sum \partial_i^2(f)$ for $f \in D(\Delta)$, where ∂_i is the partial derivative with respect to the i-th coordinate. As $f \in D(\Delta)$ vanishes near the boundary of X, partial integration gives $(\Delta(f)|f) \leq 0$, so $-\Delta$ is a positive symmetric operator, and its Friedrichs extension will be a self-adjoint positive operator. ◇

9.2 The Cayley Transform

The Möbius transform $K(z) = (z - i)/(z + i)$ of the Riemann sphere is known as the Cayley transform, and it takes the one-point compactification of $\mathbb{R}$ homeomorphically onto the circle $\mathbb{T}$ with $K(\infty) = 1$ and inverse transform $z \mapsto i(1 + z)/(1 - z)$. We will use this to extend the spectral analysis to the unbounded self-adjoint case.

Let T be a densely defined symmetric operator in a Hilbert space V. Then $(T + iI)^{-1}$ is a bounded operator with domain $(T+iI)D(T)$ and range $D(T)$. The *Cayley transform of* T is the operator $K(T) \equiv (T - iI)(T + iI)^{-1}$ in V.

Lemma 9.2.1 *The Cayley transform $K(T)$ is an isometry from $(T + iI)D(T)$ onto $(T - iI)D(T)$ such that $I - K(T)$ is injective and has range $D(T)$. Moreover, we have $i(I + K(T))(I - K(T))^{-1} = T$.*

Proof For $v \in D(T)$ we have

$$\|(T + iI)(v)\|^2 = \|T(v)\|^2 + \|v\|^2 = \|(T - iI)(v)\|^2,$$

and since $K(T)(T + iI)(v) = (T - iI)(v)$, we see that $K(T)$ is isometric and that $I - K(T)$ is injective.

As $I - K(T) = (T+iI)(T+iI)^{-1} - K(T) = 2i(T+iI)^{-1}$, it has range $D(T)$. Similarly, we have $I + K(T) = 2T(T+iI)^{-1}$, so $i(I+K(T))(I-K(T))^{-1} = T$. □

Theorem 9.2.2 *The Cayley transform K is an order-preserving isomorphism from the set of densely defined symmetric operators in a Hilbert space V onto the set of isometries A on V such that $I - A$ has dense range. Moreover for a densely defined operator T in V either T, $K(T)$, $R(T + iI)$, $R(T - iI)$ are all closed, or none are.*

Proof Clearly $S \subset T$ implies $K(S) \subset K(T)$, and K is injective by the previous lemma. To prove surjectivity, let A be as described in the theorem. By polarization $(A(v)|A(w)) = (v|w)$ for $v, w \in D(A)$, so whenever $A(v) = v$, then $(v|(I - A)(w)) = 0$ and $v = 0$ by denseness. So $I - A$ is injective and we can define an operator T in V with domain $R(I - A)$ by $T((I - A)(v)) = i(I + A)(v)$. As $(T((I - A)(v))|(I - A)(v)) = 2\mathrm{Im}(v|A(v))$, we see that T is a densely defined symmetric operator, and as $T = i(I + A)(I - A)^{-1}$, the lemma gives $A = K(T)$.

If T is closed, then $T + iI$ is closed, and since $\|(T + iI)^{-1}\| \leq 1$, we conclude that $R(T + iI)$ is closed because if $(T + iI)(v_n) \to w$, then $\{v_n\}$ converges to some v, so $(v, w) \in G(T + iI)$, or $(T + iI)(v) = w$. By the lemma the spaces $R(T + iI)$ and $R(T - iI)$ are isometrically isomorphic, so they are simultaneously closed, and then $K(T)$ is closed by the lemma. On the other hand, if $K(T)$ is closed, and a sequence $\{v_n\}$ in $D(T)$ converges to v while $T(v_n) \to w$, then $(T \pm iI)(v_n) \to w \pm iv$, so $(w + iv, w - iv) \in G(K(T))$. Hence there is $u \in D(T)$ such that $w \pm iv = (T \pm iI)(u)$, so $v = u$ and $w = T(u)$, and T is closed. □

Proposition 9.2.3 *Let T be a densely defined symmetric operator. Then $T^* = T$ if and only if $T \pm iI$ are surjective if and only if $K(T)^* = K(T)^{-1}$.*

Proof If $T^* = T$, then T is closed and $R(T \pm iI)$ are closed by the theorem, and they are also dense as $\pm i$ are not eigenvalues for T. Conversely, if to $v \in D(T^*)$ there is $w \in D(T)$ such that $(T + iI)(w) = (T^* + iI)(v)$, then as $T \subset T^*$, we get $v - w \in \ker(T^* + iI) = R(T - iI)^{\perp} = \{0\}$, so $D(T^*) = D(T)$. The last equivalence in the proposition is immediate from the lemma. □

Definition 9.2.4 The *deficiency indices* $\Delta_{\pm}$ of a densely defined symmetric operator T in a Hilbert space are defined to be $\dim R(T \pm iI)^{\perp}$, where by dimension here we mean the cardinality of an orthonormal basis in the Hilbert space.

So the deficiency indices of T are the co-dimensions of $D(K(T))$ and $R(K(T))$, alternatively, we may view $\Delta_{\pm}$ as the multiplicities of the eigenvalues $\pm i$ of T^*.

Proposition 9.2.5 *A densely defined symmetric operator T in a Hilbert space admits a self-adjoint extension if and only if $\Delta_+ = \Delta_-$.*

Proof By the previous results T in V admits a self-adjoint extension exactly when there is a unitary A on V that restricts to $K(T)$ on $R(T + iI)$, and by extending orthonormal bases, such a unitary exists exactly when the deficiency indices of T are equal. □

When T is closed, so that $K(T)$ is already a partial isometry on the Hilbert space, i.e. no extension by continuity, then $T^* = T$ if and only if $\text{index}K(T) = 0$

Example 9.2.6 Let A be an isometry from a Hilbert space onto a proper subspace, for example, the unilateral shift. By the previous theorem we know that $T = K^{-1}(A)$ is a closed symmetric operator that is maximal symmetric because if $T \subset S$, then $A = K(T) \subset K(S)$, so $A = K(S)$ and $T = S$. But T is not self-adjoint by Proposition 9.2.3, and indeed, the deficiency indices are 0 and $\dim(\text{im}A)^{\perp}$.

For the record we note that a densely defined symmetric operator is maximal symmetric if and only if one of the deficiency indices is zero, and it is self-adjoint if both deficiency indices vanish. ◇

Here is another condition that guarantees self-adjoint extensions.

Proposition 9.2.7 *Let T be a densely defined symmetric operator in a separable Hilbert space V. If there is a continuous involutive conjugate linear map $J: V \to V$ such that $J(D(T)) \subset D(T)$ and $TJ = JT$ on $D(T)$, then T has a self-adjoint extension.*

Proof Since $J^2 = I$ we get $J(D(T)) = D(T)$. To $v \in V$ there is a unique $J^*(v) \in V$ such that $(w|J^*(v)) = (v|J(w))$ for all $w \in V$. This defines an involutive conjugate linear map $J^*: V \to V$. For $v \in D(T^*)$ and $w \in D(T)$ we have $(J^*T^*(v)|w) = (TJ(w)|v) = (JT(w)|v) = (J^*(v)|T(w))$, so $J^*(v) \in D(T^*)$ and $T^*J^*(v) = J^*T^*(v)$, and as $J^{*2} = I$, we get $T^*J^* = J^*T^*$ on $D(T^*) = J^*(D(T^*))$. Hence $J^*(\ker(T^* \pm iI)) = \ker(T^* \mp iI)$ and the deficiency indices of T are equal. □

Suppose S is an operator in a Hilbert space V and that $T \in B(V)$. Then S and T *commute* if $TS \subset ST$, which happens precisely when $(T \oplus T)G(S) \subset G(S)$. Thus the operators commuting with S form a subalgebra A_S of $B(V)$ which is strongly closed when S is closed. When S is also densely defined, then T^* commutes with S^* whenever T commutes with S, so in this case $A_S \cap A_{S^*}$ is a von Neumann algebra on V. We denote its commutant by $W^*(S)$.

Definition 9.2.8 A densely defined closed operator S in a Hilbert space V is *affiliated with a von Neumann algebra* W on V, written $S\eta W$, if $W^*(S) \subset W$.

Clearly $S\eta W$ if and only if $W' \subset A_S \cap A_{S^*}$.

Lemma 9.2.9 *Let S be a self-adjoint operator in a Hilbert space. Then any bounded operator commutes with S if and only if it commutes with $K(S)$. So $W^*(S)$ is the von Neumann algebra generated by $K(S)$, and S is affiliated with a von Neumann algebra W if and only if $K(S) \in W$.*

Proof Let T be a bounded operator. If $TS \subset ST$, then $T(S \pm iI) \subset (S \pm iI)T$ and $T(S \pm iI)^{-1} = (S \pm iI)^{-1}T$ by Proposition 9.2.3, so $TK(S) = K(S)T$.

On the other hand, if this holds, then by Lemma 9.2.1, we get

$$TS(I-K(S)) = iT(I+K(S)) = i(I+K(S))T = S(I-K(S))T = ST(I-K(S)),$$

so $TS \subset ST$. □

Lemma 9.2.10 *If a normal operator T in a Hilbert space is affiliated with a von Neumann algebra, then so are the self-adjoint operators A and B in the decomposition $T = A + iB$ described in Proposition 9.1.9.*

Proof Say $T\eta W$, so if $S \in W' \subset A_T \cap A_{T^*}$, we get $S(T + T^*) \subset (T + T^*)S$, and $S^*A \subset AS^*$ by definition of A. Hence $A\eta W$, and similarly $B\eta W$. □

Definition 9.2.11 The *resolvent set* of an operator T in a Hilbert space V consists of the complex numbers λ such that $\lambda I - T$ is a bijection from $D(T)$ onto V and $\|(\lambda I - T)(\cdot)\| \geq t\| \cdot \|$ on $D(T)$ for some $t > 0$. The *spectrum* $\mathrm{Sp}(T)$ of T is the complement in $\mathbb{C}$ of the resolvent set of T, and the *resolvent function* is the function $R\colon \mathrm{Sp}(T)^c \to B(V)$ given by $R(\lambda) = (\lambda I - T)^{-1}$.

That the spectrum of an operator in a Hilbert space is closed, is clear from the following result.

Proposition 9.2.12 *Let T be an operator in a Hilbert space. If $\lambda \notin \mathrm{Sp}(T)$, then for any complex number z in the open disc with center at the origin and radius $\|R(\lambda)\|^{-1}$ we have $\lambda - z \notin \mathrm{Sp}(T)$ and $R(\lambda - z) = \sum_{n=0}^{\infty} R(\lambda)^{n+1} z^n$.*

Proof Using the Neumann series in the computation

$$((\lambda I - T)(I - zR(\lambda)))^{-1} = (I - zR(\lambda))^{-1}R(\lambda) = \sum_{n=0}^{\infty} R(\lambda)^{n+1} z^n,$$

we see that $\lambda - z$ is in the resolvent set, and we obtain the formula for $R(\lambda - z)$. □

The multiplication operator of the identity function on $\mathbb{C}$ in Example 9.1.12 has clearly spectrum $\mathbb{C}$. The spectrum can also be empty. For instance, if T is an injective bounded operator with dense range and $\text{Sp}(T) = \{0\}$, then T^{-1} is a densely defined operator such that for every $\lambda \in \mathbb{C}$ we have $(\lambda I - T^{-1})^{-1} = -\sum_{n=0}^{\infty} \lambda^n T^{n+1}$, which is bounded because $\|T^n\| \leq |2\lambda|^{-n}$ for large enough n. So $\text{Sp}(T^{-1}) = \phi$.

Proposition 9.2.13 *Let T be a self-adjoint operator in a Hilbert space. Then* $\text{Sp}(T)$ *is a non-empty subset of* $\mathbb{R}$, *and* $\lambda \in \text{Sp}(T)$ *if and only if* $K(\lambda) \in \text{Sp}(K(T))$, *and* λ *is an eigenvalue for T if and only if $K(\lambda)$ is an eigenvalue for $K(T)$.*

Proof If $\lambda \in \mathbb{C}\backslash\mathbb{R}$, then $(T - \lambda I)$ is surjective by Proposition 9.2.3, and $\|(T - \lambda I)^{-1}\| \leq |\text{Im}\lambda|^{-1}$, so $\lambda \notin \text{Sp}(T)$. Hence $\text{Sp}(T) \subset \mathbb{R}$. The same proposition shows that $\text{Sp}(K(T)) \subset \mathbb{T}$.

By Lemma 9.2.1 we have $\lambda I - T = (\lambda + i)(K(\lambda)I - K(T))(I - K(T))^{-1}$ for any real λ. Hence any $v \in D(T)$ is an eigenvector for T with eigenvalue λ if and only if $v = (I - K(T))(w)$ for some eigenvector w for $K(T)$ with eigenvalue $K(\lambda)$.

We also see that if $K(\lambda) \notin \text{Sp}(K(T))$, then $\lambda \notin \text{Sp}(T)$ because $(\lambda I - T)^{-1}$ equals $(I - K(T))(K(\lambda)I - K(T))^{-1}(\lambda + i)^{-1}$.

If $K(\lambda) \in \text{Sp}(K(T))$, pick a sequence of unit vectors v_n in the Hilbert space such that $\|K(\lambda)v_n - K(T)(v_n)\| \to 0$. Now $w_n = (I - K(T))(v_n) \in D(T)$ by Lemma 9.2.1, and also $\lim \|w_n\| = \lim \|v_n - K(T)(v_n)\| = |1 - K(\lambda)| > 0$, while $(\lambda I - T)(w_n) = (\lambda + i)(K(\lambda)I - K(T))(v_n) \to 0$, so $\lambda \in \text{Sp}(T)$. □

9.3 Sprectral Theory for Unbounded Operators

Let μ be a Radon measure on a σ-compact locally compact Hausdorff space X. Then the unital $*$-algebra $L(\mu)$ consisting of all classes of measurable complex-valued functions on X coincide with the $*$-algebra $B(\mu)$ of classes of Borel functions. That any measurable function is Borel almost everywhere follows from a straightforward argument, where one uses σ-compactness to reduce the problem to integrable functions, follows up with Lusin's theorem and wraps things up by using the simple fact that sequential pointwise limits of Borel functions are again Borel.

Definition 9.3.1 A map h from the algebra of (classes of) measurable functions on a measure space to the set of normal operators in a Hilbert space is an *essential homomorphism* if $h(fg), h(f+g)$ and $h(cf)$ are the closures of the operators $h(f)h(g), h(f) + h(g)$ and $ch(f)$, respectively, for any measurable functions f, g and scalar c.

Note that $ch(g)$ is automatically closed.

Proposition 9.3.2 *Let μ be a Radon measure on a σ-compact locally compact Hausdorff space X. Then $f \mapsto M_f$ with M_f as in Example 9.1.12 is a $*$-preserving essential isomorphism from $L(\mu)$ onto the set of normal operators in $L^2(\mu)$ affiliated with $L^\infty(\mu)$.*

Proof Let $h_n = (1 + n^{-1}(|f| + |g|))^{-1}h \in D(M_f) \cap D(M_g)$ for $f, g \in L(\mu)$ and $h \in D(M_{f+g})$. By Lebesgue's monotone convergence theorem we see that $h_n \to h$ and $(M_f + M_g)h_n \to M_{f+g}h$ in L^2-norm, so $G(M_{f+g})$ is the closure of $G(M_f + M_g)$. The case for the product is similar. Hence our map is a $*$-preserving essential monomorphism from $L(X)$ to a commutative algebra of normal operators in $L^2(X)$. As $M_g M_f \subset M_f M_g$ when $g \in L^\infty(\mu)$, we see that M_f is affiliated with $L^\infty(\mu)$ by maximal commutativity. As for surjectivity, if $T = T^* \eta L^\infty(\mu)$, then $K(T) \in L^\infty(\mu)$ by Lemma 9.2.9, so there is a Borel function h of modulus one such that $M_h = K(T)$, and $h \neq 1$ almost everywhere. Then $M_{K^{-1}(h)} = i(I + K(T))(I - K(T))^{-1} = T$. Lemma 9.2.10 gives the general normal case. □

Theorem 9.3.3 *Let T be a self-adjoint operator in a separable Hilbert space V with multiplicity free $K(T)$. Then there is a finite Radon measure μ on* Sp(T) *and an isometry $U: L^2(\mu) \to V$ such that the map $f \mapsto f(T) \equiv UM_fU^*$ is a unital $*$-preserving essential isomorphism from $L(\mu)$ to the set of normal operators in V affiliated with $W^*(T)$. Moreover, we have $\iota(T) = T$, and the two meanings of $K(T)$ coincide.*

Proof By assumption we may pick a unit vector $v \in V$ that is cyclic for the C*-algebra generated by $K(T)$. Now $f \circ K^{-1} \in C_0(\mathbb{T}\backslash\{0\})$ for $f \in C_0(\mathbb{R})$. Hence the linear functional $f \mapsto (f \circ K^{-1}(K(T))(v)|v)$ on $C_0(\mathrm{Sp}(T))$ is the integral of f over a finite Radon measure μ on Sp(T). The map $f \mapsto f \circ K^{-1}(K(T))(v)$ on $C_0(\mathbb{R})$ extends by continuity to an isometry $U: L^2(\mu) \to V$. By the previous proposition the map $f \mapsto f(T)$ is a $*$-preserving essential isomorphism from $L(\mu)$ onto the set of normal operators in V affiliated with the maximal commutative von Neumann algebra $W = \{f(T) \mid f \in L^\infty(\mu)\}$. For $f \in C_0(\mathbb{R})$ we have $UM_K(f) = U(Kf) = K(T)U(f)$, so $UM_KU^* = K(T)$, and as $M_\iota = i(I + M_K)(I - M_K)^{-1}$ we get $\iota(T) = T$. As f ranges over the bounded Borel functions on $\mathbb{R}$, then $f \circ K^{-1}$ ranges over the bounded Borel functions on $\mathbb{T}\backslash\{1\}$. Since 1 is not an eigenvalue of $K(T)$, we therefore see that W consists of all $f(K(T))$ as f ranges over the bounded Borel functions on Sp$(K(T))$. Hence W is by Corollary 5.8.13 the von Neumann algebra generated by $K(T)$, which is $W^*(T)$ by Lemma 9.2.9. □

Definition 9.3.4 A self-adjoint operator T in a Hilbert space V is *multiplicity free* if there is $v \in \cap D(T^n)$ such that the closed span of $\{T^n(v)\}$ is V.

Proposition 9.3.5 *A self-adjoint operator T in a Hilbert space is multiplicity free if and only if $K(T)$ is multiplicity free.*

Proof If the closed span of $\{T^n(v)\}$ is the Hilbert space V for $v \in \cap D(T^n)$, let P be the orthogonal projection onto the closure of the C*-algebra W generated by $K(T)$ applied to v. Then $P(v) = v$ and $P \in W'$, so $P \in A_T$ by Lemma 9.2.9. Thus $(I - P)T^n(v) = T^n(I - P)(v) = 0$ and $P = I$, so $K(T)$ is multiplicity free.

If on the other hand this holds, then with U from the theorem above, we see that $U(g) \in \cap D(T^n)$ when $g(t) = e^{-t^2}$, and the closed span of $\{T^n(U(g))\}$ is V. □

The following preliminary result is straightforward.

Lemma 9.3.6 *Given a sequence of self-adjoint operators T_n in Hilbert spaces V_n, there is a self-adjoint operator T in $\oplus V_n$ that restricts to T_n on $D(T_n)$ and has as core the span of all $D(T_n)$. Its domain $D(T)$ consists of the vectors $(v_n) \in \oplus V_n$ such that $v_n \in D(T_n)$ and $\sum \|T_n(v_n)\|^2 < \infty$.*

Theorem 9.3.7 *For a self-adjoint operator T in a separable Hilbert space there is a finite Radon measure on* $\mathrm{Sp}(T)$ *and a unital $*$-preserving essential isomorphism $f \mapsto f(T)$ from $L(\mu)$ onto the set of normal operators affiliated with $W^*(T)$ such that $\iota(T) = T$ and that the two meanings of $K(T)$ coincide.*

Proof By Lemma 5.8.4 there is a sequence of pairwise orthogonal projections P_n that are cyclic relative to $W^*(T)$ and has strong sum I. So $K(T)P_n$ is multiplicity free on $\mathrm{im} P_n$. By Lemma 9.2.9 the projections P_n commute with T, and regarding TP_n as a self-adjoint operator in $\mathrm{im} P_n$, we have $K(TP_n) = K(T)P_n$. Considering $\mathrm{Sp}(TP_n)$ as a closed subset of $\mathrm{Sp}(T)$, the theorem above supplies a normalized Radon measure μ_n on $\mathrm{Sp}(T)$, and an isometry $U_n \colon L^2(\mu_n) \to \mathrm{im} P_n$ and a $*$-preserving essential isomorphism $f \mapsto U_n M_f U_n^*$ from $L(\mu_n)$ onto the set of normal operators in $\mathrm{im} P_n$ affiliated with $W^*(TP_n)$.

Consider now the normalized Radon measure $\mu = \sum 2^{-n}\mu_n$ on $\mathrm{Sp}(T)$, and for $f \in L(\mu)$ let $f(T)$ be the normal operator associated as in the lemma to the sequence of operators $U_n M_f U_n^*$. The corresponding map $f \to f(T)$ has the required properties; the only thing one might doubt is surjectivity, which we now check.

If S is a self-adjoint operator affiliated with $W^*(T)$, then by Lemma 9.2.9 the operator $K(S)$ belongs to the von Neumann algebra generated by $K(T)$. Pick by Theorem 5.8.16 a measurable function h on $\mathrm{Sp}(T)$ of modulus one such that $K(S) = h(K(T))$. Identifying $K(S)P_n$ with $K(SP_n)$ and letting $g = h \circ K$, we get $K(SP_n) = h(K(T))P_n = h(K(TP_n)) = U_n M_g U_n^*$ by the previous theorem, so $SP_n = U_n M_f U_n^*)$ with $f = K^{-1} \circ g$. By the lemma S and $f(T)$ coincide on a common core, so $S = f(T)$. The general case is clear from Lemma 9.2.10. □

We discuss continuity properties of the spectral map.

Proposition 9.3.8 *For any self-adjoint operator T in a separable Hilbert space V and every $v \in V$ there is a finite Radon measure μ_v on* $\mathrm{Sp}(T)$ *that is absolutely continuous with respect to the measure μ from the theorem above and such that $\int f\, d\mu_v = (f(T)(v)|v)$ for ever positive Borel function f on* $\mathrm{Sp}(T)$.

Proof The formula $\int g \circ K\, d\mu_v = (g(K(T))(v)|v)$ defines a Radon measure μ_v on $\mathrm{Sp}(T)$ with $\mu_v(\mathrm{Sp}(T)) = \|v\|^2$ since $g \circ K$ ranges over $C_0(\mathbb{R})$ as g ranges over $C_0(\mathbb{T}\backslash\{0\})$. Writing $v = \sum v_n$ with $v_n = \sum P_n(v)$ and P_n from the proof above, we get

$$\int f\, d\mu_v = (f(T)(v)|v) = \sum((f(T)(v_n)|v_n) = \sum \int f\, d\mu_{v_n}$$

for $f \in C_0(\mathrm{Sp}(T))$. Hence $\mu_v = \sum \mu_{v_n}$ and the integral identity in the proposition, which holds for each v_n by Theorem 9.3.3, will clearly also hold for v. A measurable function f on $\mathrm{Sp}(T)$ vanishes almost everywhere with respect to μ if and only if $f(T) = 0$, so $\int f\, d\mu_v = 0$ and $\mu_v \ll \mu$. □

Corollary 9.3.9 *If T is a self-adjoint operator in a separable Hilbert space and f, g, f_n are Borel functions on $\mathrm{Sp}(T)$ such that $f_n \to f$ pointwise and $|f_n| \leq g$, then $f_n(T)(v) \to f(T)(v)$ for every $v \in D(g(T))$.*

Proof Since $\int g^2\, d\mu_v < \infty$ with μ_v from the proposition, Lebesgue's dominated convergence theorem yields $\|f(T)(v) - f_n(T)(v)\|^2 = \int |f - f_n|^2\, d\mu_v \to 0$. □

The following spectral measure point of view is immediate.

Theorem 9.3.10 *Let T be a self-adjoint operator in a separable Hilbert space V. Consider the unital $*$-preserving essential homomorphism $f \mapsto f(T)$ from Theorem 9.3.7. Restricting it to the bounded Borel functions on $\mathrm{Sp}(T)$), we get by Proposition 5.4.7 a unique regular spectral measure P on $\mathrm{Sp}(T)$) with orthogonal projections in $B(V)$. By continuity we have $(f(T)(v)|v) = \int f\, dP_{v,v}$ for any $v \in V$ and any bounded Borel function f on $\mathrm{Sp}(T)$. The same identity holds when f is unbounded provided we only consider v in the domain $\{v \in V \mid \int |f|^2\, dP_{v,v} < \infty\}$ of $f(T)$, sometimes also written $\int f(\lambda)\, dP(\lambda)$, and referring to the case $f = \iota$ as* a resolution of the identity.

Definition 9.3.11 Two self-adjoint operators A and B in a separabel Hilbert space *commute* if their spectral projections pairwise commute.

By the theorem above $K(A)$ and $K(B)$ commute if A and B are commuting self-adjoint operators in a separabel Hilbert space. Conversely, if $K(A)K(B) = K(B)K(A)$, then as the spectral projections of A belong to $W^*(A) = W^*(K(A))$, they belong to the commutant of $W^*(K(B)) = W^*(B)$, and thus must commute with the spectral projections of B. So A and B commute.

Remark 9.3.12 Just as in the bounded case Theorem 9.3.3 shows that for any commuting family $\{T_i\}$ of self-adjoint operators on a separable Hilbert space V one may embed $\{K(T_i)\}$ into a maximal commutative $*$-subalgebra of $B(V)$ to get a representation of the operators T_i as a commuting family of multiplication operators M_{f_i}. ◇

The following result, known as *Stone's formula*, relates resolvents and spectral projections.

Corollary 9.3.13 *Given a self-adjoint operator T in a separable Hilbert space we have the strong limit*

$$\lim_{t \to 0} (\pi i)^{-1} \int_a^b ((T - \lambda I - itI)^{-1} - (T - \lambda I + itI)^{-1})\, dP(\lambda) = P([a, b]) + P(\langle a, b \rangle)$$

for real scalars $b > a$ as t decreases towards zero.

Proof This is clear from the theorem as the functions on $\mathbb{R}$ given by

$$x \mapsto (\pi i)^{-1} \int_a^b ((x - \lambda - it)^{-1} - (x - \lambda + it)^{-1})\, d\lambda$$

are uniformly bounded as t varies, and tend pointwise to 0 when $x \notin [a, b]$, to 1 when $x = a$ or $x = b$, and to 2 when $x \in \langle a, b \rangle$. □

Definition 9.3.14 A *quadratic form* is a sesquilinear map q on a dense subspace $D(q)$, called its *form domain*, of a complex Hilbert space. It is *bounded from below* by $M \geq 0$ if $q(v, v) \geq -M\|v\|^2$ for all $v \in D(q)$. When M can be picked to be 0, we say q is *positive*.

Any quadratic form q bounded from below is *symmetric*, i.e. $q(v, w) = \overline{q(w, v)}$.

Example 9.3.15 The quadratic form q_T *associated to* a self-adjoint operator T in a separabel Hilbert space V has $D(q_T) = \{v \in V \mid \int |\lambda| \, dP_{v,v}(\lambda) < \infty\}$ and is given by $q_T(v, w) = \int \lambda \, dP_{v,w}(\lambda)$, where P is the spectral measure of T. ◇

It is easy to see that an operator T in a Hilbert space is closed if its domain $D(T)$ is complete with respect to the *graph norm* given by $\| \cdot \|_T = \|T(\cdot)\| + \| \cdot \|$. In view of this we give the following definition.

Definition 9.3.16 A quadratic form q bounded from below by M is *closed* if $D(q)$ is complete with respect to the norm $\|v\|_q = (q(v, v) + (M + 1)\|v\|^2)^{1/2}$, and then any dense subspace of $D(q)$ is a *form core* for q.

The domain of a closed quadratic form q bounded from below by M is clearly a Hilbert space with inner product $(v|w)_q = q(v, w)+(M+1)(v|w)$ for $v, w \in D(q)$. Any quadratic form q bounded from below is closed if and only if whenever $v_n \to v$ with $v_n \in D(q)$ and $q(v_n - v_m, v_n - v_m) \to 0$ as $n, m \to 0$, then $v \in D(q)$ and $q(v - v_n, v - v_n) \to 0$. Combining this with Lebesgue's dominated convergence theorem, one readily sees that the quadratic form q_T associated to a self-adjoint operator T is closed when T is bounded from below, and then any core for T is a form core for q_T.

We know that densely defined symmetric operators always have closures, but that these need not be self-adjoint. The following result shows that whenever a bounded from below quadratic form on a separable Hilbert space has a closed extension, which it might well not have, then it is associated to a self-adjoint operator.

Proposition 9.3.17 *Every closed quadratic form q with dense domain in a separable Hilbert space V is the quadratic form q_T associated to a unique self-adjoint operator T in V.*

Proof Let V_q denote the Hilbert space $D(q)$ with respect to $(\cdot|\cdot)_q$, so we have an inclusion $V_q \subset V$. Let j denote the injective linear map from V into the space V_{-q} of continuous conjugate linear functionals on V_q given by $j(v) = (v|\cdot)$. Let $S \colon V_q \to V_{-q}$ denote the isometric isomorphism such that $(S(v))(w) = (v|w)_q$. It is straightforward to check that $j^{-1}S$ is densely defined and symmetric. Now $S^{-1}j$ is everywhere defined and symmetric, and is thus bounded and self-adjoint. As it is also injective, spectral theory then shows that its inverse $j^{-1}S$ is in fact self-adjoint. Then $T = j^{-1}S - I$ is self-adjoint, and $(T(v)|w) = q(v, w)$ for $v, w \in D(T)$. As $D(T)$ is dense in V_q we conclude that $q_T = q$. Uniqueness of T is even easier. □

Polar decomposition can be extended to unbounded operators. Let T be a densely defined closed operator in a separable Hilbert space V. Then T^*T is self-adjoint and $(\lambda I + T^*T)^{-1} \in B(V)$ for $\lambda .> 0$, so $\mathrm{Sp}(T^*T)$ consists only of non-negative numbers. The *absolute value* of T is then defined to be $|T| \equiv (T^*T)^{1/2}$. The square root is unique because if S is a positive self-adjoint operator in V with $S^2 = T^*T$, then both S and $|T|$ are affiliated with $W^*(S)$, so $S = |T|$ by Theorem 9.3.7 as multiplication operators clearly have unique positive square roots.

Proposition 9.3.18 *Let T be a densely defined closed operator in a separable Hilbert space. Then $\||T|(\cdot)\| = \|T(\cdot)\|$ on $D(|T|) = D(T)$, and there is a unique partial isometry U with* $\ker U = \ker T$ *such that $T = U|T|$. We also have $U^*U|T| = |T|$ and $U^*T = |T|$ and $UU^*T = T$.*

Proof The norm identity clearly holds for $v \in D(T^*T)$, which by Proposition 9.1.8 is a core for both $D(T)$ and $D(|T|)$, so $D(|T|) = D(T)$ and the same norm identity holds for $v \in D(T)$ by Proposition 9.1.9. Letting $U(|T|(v)) = T(v)$ for $v \in D(T)$ and extending U by continuity to the norm closure of $R(|T|)$, and even further, by declaring it to vanish on $R(|T|)^\perp = \ker T$, we get a partial isometry such that $T = U|T|$. The remaining statements are proved as in the bounded case. □

From the proof above we see that the partial isometry in the polar decomposition of T is a unitary if and only if T and T^* are injective. So U is unitary when T is invertible. Arguing as in the bounded case we also see that when T is normal, so $\ker T^* = \ker T$, then the partial isometry U in the polar decomposition of T is also normal and commutes with $|T|$. We may therefore extend it to a unitary, also denoted by U, such that $T = U|T| = |T|U$.

Remark 9.3.19 Let T be a normal operator in a separable Hilbert space with polar decomposition $T = U|T|$. Then $L(T) \equiv U|T|(I + |T|)^{-1}$ is a normal operator such that $\|L(T)\| \leq 1$ and $I - |L(T)|$ is injective. Conversely, given any normal operator S with $\|S\| \leq 1$ and $I - |S|$ injective, then $T = (I - |S|)^{-1}S$ is normal in the Hilbert space, and $L(T) = S$. Using the correspondence L we may extend the spectral theory to normal operators. In fact, define h on the open unit disc by $h(\lambda) = \lambda(1 - |\lambda|)^{-1}$, and set $f(T) = f \circ h(L(T))$ for every bounded Borel function f on $\mathrm{Sp}(T) = h(\mathrm{Sp}(L(T)))$. The resulting spectral theory is then completely analogues to the bounded case, and can with caution, also be extended to include unbounded Borel functions, thus generalizing Theorem 9.3.7. ◇

Definition 9.3.20 A *strongly continuous one-parameter unitary group* of operators on a Hilbert space V is a strongly continuous function $U : \mathbb{R} \to B(V)$ taking values in the unitaries such that $U_{s+t} = U_s U_t$ holds for $s, t \in \mathbb{R}$.

Proposition 9.3.21 *Let A be a self-adjoint operator in a separable Hilbert space V. Then $t \mapsto U_t \equiv e^{itA}$ is a strongly continuous one-parameter unitary group. Moreover, we can recover the* infinitesimal generator *of the one-parameter group from $A(v) = \lim_{t\to 0}(U_t(v) - v)/it$ for $v \in D(A)$. In fact, any $v \in V$ for which this limit exists belongs to $D(A)$, and then the limit is $A(v)$.*

Proof The first statement makes sense and is correct due to the theorem above and its previous corollary. Let $v \in D(A)$. The square function is integrable with respect to μ_v from Proposition 9.3.8, and $\|(it)^{-1}(U_t(v) - v) - A(v)\|^2 = \int |f_t|^2 \, d\mu_v$, where $f_t(s) = (it)^{-1}(e^{ist} - 1 - ist)$. As $f_t \to 0$ pointwise when $t \to 0$, and $|f_t(s)| \leq a|s| + b$ for suitable constants a, b and real s, t with $|t| \leq 1$, Lebesgue's dominated convergence theorem shows that $\int |f_t|^2 \, d\mu_v \to 0$ as $t \to 0$.

On the other hand, with v as in the last statement and with $T(v)$ the corresponding limit, we get a symmetric operator T that extends $A^* = A$, so $T = A$. □

Example 9.3.22 The formula $(U_t(f))(s) = e^{its} f(s)$ for $f \in L^2(\mathbb{R})$ defines a strongly continuous one-parameter unitary group. The self-adjoint multiplication operator M_ι in $L^2(\mathbb{R})$ is an infinitesimal generator because $U_t = e^{itM_\iota}$. ◇

The following converse result is known as *Stone's theorem.*

Theorem 9.3.23 *Any strongly continuous one-parameter unitary group of operators on a separable Hilbert space V is of the form $t \mapsto e^{itA}$ for a self-adjoint operator A in V.*

Proof By Example 5.2.9 we know that the Gelfand transform G of $L^1(\mathbb{R})$ is a norm-decreasing $*$-homomorphism from $L^1(\mathbb{R})$ onto a dense $*$-subalgebra of $C_0(\mathbb{R})$. Let $\{U_t\}$ be a strongly continuous one-parameter unitary group of operators on V. For $f \in L^1(\mathbb{R})$ the formula $(v, w) \mapsto \int (U_t(v)|w) f(t) dt$ defines a sesquilinear form on V bounded by $\|f\|_1$, so there is a unique $S_{G(f)} \in B(V)$ such that $(S_{G(f)}(v)|w) = \int (U_t(v)|w) f(t) dt$ for all $v, w \in V$. Clearly $f \mapsto S_{G(f)}$ is a norm decreasing $*$-homomorphism from $L^1(\mathbb{R})$ to $B(V)$. Hence $\|S_{G(f)}\| = r(S_{G(f)}) \leq r(f) = \|G(f)\|_u$, and since $G(L^1(\mathbb{R}))$ is dense in $C_0(\mathbb{R})$, the map $G(f) \mapsto S_{G(f)}$ extends by continuity to $C_0(\mathbb{R})$. For $v \in V$ consider the finite Radon measure μ_v on $\mathbb{R}$ given by $\int h \, d\mu_v = (S_h(v)|v)$ for $h \in C_0(\mathbb{R})$. So $\int G(f) \, d\mu_v = \int (U_t(v)|v) f(t) dt$ for $f \in L^1(\mathbb{R})$. Taking non-negative f's with $\|f\|_1 = 1$ and support increasingly concentrated at 0, we get $\mu_v(\mathbb{R}) = \|v\|^2$ by strong continuity of $t \mapsto U_t$. Also, for $g \in L^1(\mathbb{R})$ we have

$$(U_{-t} S_{G(g)}(v)|v) = (S_{G(g)}(v)|U_t(v)) = \int (U_s(v)|v) g(s+t) ds = \int e_t G(g) \, d\mu_v,$$

where $e_t(s) = e^{its}$. Arguing as above by picking the g's appropriately, we see that $(U_{-t}(v)|v) = \int e_t \, d\mu_v$. We extend $h \mapsto S_h$ to $C(\mathbb{R} \cup \{\infty\})$ by letting $S_1 = I$. Consider the unitary operator S_K on V given by the Cayley transform K. If $S_K^*(v) = v$, then $0 = ((I - S_K)(v)|v) = \int (1 - K) \, d\mu_v$, and since $\operatorname{Re}(1 - K) > 0$ on $\mathbb{R}$, we get $\mu_v = 0$ and $v = 0$. Hence $A \equiv -K^{-1}(S_K)$ is by Theorem 9.2.2 and Proposition 9.2.3 a self-adjoint operator in V. With the strongly continuous one-parameter unitary group $t \mapsto e^{itA}$ from the previous proposition, we get

$$(e^{itA}(v)|v) = (e^{-itK^{-1}(S_K)}(v)|v) = \int e_{-t} \, d\mu_v = (U_t(v)|v)$$

for all v, so $e^{itA} = U_t$. □

Proposition 9.3.24 *Two self-adjoint operators A and B in a separable Hilbert space commute if and only their non-real resolvents pairwise commute if and only if $e^{isA}e^{itB} = e^{itB}e^{isA}$ for all $s, t \in \mathbb{R}$.*

Proof If $R^A(\lambda)R^B(\mu) = R^B(\mu)R^A(\lambda)$ for non-real λ, μ, then A and B commute by Stone's formula and the fact that the strong limits $\lim_{t\to 0} itR^A(a + it)$ and $\lim_{t\to 0} itR^B(b + it)$ are the spectral projections of A and B at $\{a\}$ and $\{b\}$, respectively.

Assume $e^{isA}e^{itB} = e^{itB}e^{isA}$ for all $s, t \in \mathbb{R}$. Let $\hat{f}(t) = \int f(s)e^{-its}\,ds$ for $f \in L^1(\mathbb{R})$ and $t \in \mathbb{R}$. Then

$$\int f(s)(e^{-isA}(v)|v)\,ds = \int f(s)(\int e^{-ist}\,dP^A(t))(v)|v)\,ds = (\hat{f}(A)(v)|v)$$

for all v by the Fubini-Tonelli theorem. Applying it once more, then by hypothesis we get $(\hat{f}(A)\hat{g}(B)(v)|v) = (\hat{g}(B)\hat{f}(A)(v)|v)$. By Example 5.2.9 the collection of all $\hat{f}$ are dense in $C_0(\mathbb{R})$. Writing characteristic functions on closed intervals as pointwise limits of uniformly bounded sequences from this collection, and using spectral theory, the corresponding spectral projections must also commute.

The remaining implications are immediate from spectral theory. □

Remark 9.3.25 With trivial modifications the proof of Stone's theorem shows that for any strongly continuous function $t \mapsto U_t$ from $\mathbb{R}^n$ into the unitaries on a separable Hilbert space V such that $U_{s+t} = U_sU_t$ for $s, t \in \mathbb{R}^n$, there are commuting self-adjoint operators A_k in V such that $U_t = e^{it_1A_1}\cdots e^{it_nA_n}$, where t_i are the coordinates of t. ◇

Given densely defined operators S and T on Hilbert spaces V and W, respectively. The operator A in the Hilbert space $V \otimes W$ given by $v \otimes w \mapsto S(v) \otimes T(w)$ on the algebraic tensor product $D(S) \otimes D(T)$ is then closable whenever S and T are closable. Indeed, by definition of the tensor product inner product we see that $D(S^*) \otimes D(T^*) \subset D(A^*)$, and as $D(S^*)$ and $D(T^*)$ are dense in V and W, respectively, then $D(A^*)$ is dense in $V \otimes W$. The *tensor product* of S and T is the closure of A, and is denote by $S \otimes T$. Similarly, the closure of $S \otimes I + I \otimes T$ exists and is denoted by $S + T$ on account of the fact that we also consider S and T as $S \otimes I$ and $I \otimes T$, respectively. When S and T are bounded, then $S \otimes T$ has the usual meaning, and $\|S \otimes T\| = \|S\|\|T\|$. For any finite number of densely defined closable operators we have the obvious analogue definitions of tensor products and sums.

Proposition 9.3.26 *For densely defined closed operators T_i in Hilbert spaces V_i the adjoint of $T_1 \otimes T_2$ is $T_1^* \otimes T_2^*$.*

Proof By polar decomposition if suffices to prove that the algebraic tensor product $T_1 \otimes T_2$ on $D(T_1) \otimes D(T_2)$ is essentially self-adjoint when T_i are positive self-adjoint. Let $S_\pm = i(I + T_1)^{-1} \otimes (I + T_2)^{-1} \pm T_1(I + T_1)^{-1} \otimes T_2(I + T_2)^{-1}$. Then $S_\pm((I + T_1)(D(T_1)) \otimes (I + T_2)(D(T_2))) = (iI \otimes I \pm T_1 \otimes T_2)(D(T_1) \otimes$

$D(T_2))$. As $S_\pm$ are bounded with dense range we see that the latter set is dense in $V_1 \otimes V_2$. □

Theorem 9.3.27 *Consider finitely many self-adjoint operators T_k in separable Hilbert spaces V_k, let p be a real polynomial of degree n_k in the k-th variable, and let D_k^e be a domain of essential self-adjointness for $T_k^{n_k}$. Then the operator $p(T_1, T_2, \ldots)$ in $\otimes V_k$ is essentially self-adjoint on $\otimes D_k^e$, and the spectrum of the closure of it is the closure of $p(\prod \mathrm{Sp}(T_k))$.*

Proof We first prove essential self-adjointness on $D \equiv \otimes D(T_k^{n_k})$. By spectral theory we can regard T_k as multiplication operators M_{f_k} on $L^2(\mu_k)$. By replacing μ_k with $e^{-f_k^2}\mu_k$ and utilizing the obvious unitary operator $L^2(\mu_k) \to L^2(e^{-f_k^2}\mu_k)$, we may assume that $f_k \in L^q(\mu_k)$ for all $q \in [1, \infty\rangle$ while μ_k remains finite. By Example 8.1.1 we identify the Hilbert space $\otimes L^2(\mu_k)$ with $L^2(\otimes\mu_k)$, where $p(T_1, T_2, \ldots)$ becomes $M_{p(f_1, f_2, \ldots)}$ with domain all finite linear combinations of functions $g_1 g_2 \cdots$ such that $f_k^{n_k} g_k \in L^2(\mu_k)$. Clearly $p(f_1, f_2, \ldots) \in L^4(\otimes\mu_k)$, and by self-adjointness of $M_{f_k^{n_k}}$ its domain contains the characteristic functions of the measurable sets, so D is dense in $L^4(\otimes\mu_k)$. Hence Lemma 9.1.13 tells us that $p(T_1, T_2, \ldots)$ is essentially self-adjoint on D. To show that $P \equiv p(T_1, T_2, \ldots)$ is essentially self-adjoint on $\otimes D_k^e$, by Proposition 9.1.6 it suffices to show that the closure of the restriction of P to $\otimes D_k^e$ extends the restriction of P to D, and this is straightforward.

If λ is not in the closure of $p(\prod \mathrm{Sp}(T_k))$, then $(\lambda I - p(f_1, f_2, \ldots))^{-1}$ is bounded, so λ is in the resolvent of the closure of $p(T_1, T_2, \ldots)$. Conversely, if λ belongs to the closure of $p(\prod \mathrm{Sp}(T_k))$, and also belongs to an open interval J, then $p^{-1}(J)$ contains a product $\prod J_k$ of open intervals such that $J_k \cap \mathrm{Sp}(T_k) \neq \phi$. Hence $\mu_k((f_k^{n_k})^{-1}(J_k)) \neq 0$, so $(\otimes\mu_k)(p(f_1, f_2, \ldots)^{-1}(J)) \neq 0$ and λ belongs to the spectrum of the closure of $p(T_1, T_2, \ldots)$. □

Of course, when all T_k are bounded, then $p(\prod \mathrm{Sp}(T_k))$ is in fact compact.

Corollary 9.3.28 *Suppose $T_1, \ldots, T_N$ are self-adjoint operators in separable Hilbert spaces, and let D_k^e be domains of essential self-adjointness for T_k. Then the operators $T_1 \otimes \cdots \otimes T_N$ and $T_1 + \cdots + T_N$ are essentially self-adjoint on $\otimes D_k^e$, and $\mathrm{Sp}(T_1 \otimes \cdots \otimes T_N)$ is the closure of $\mathrm{Sp}(T_1) \cdots \mathrm{Sp}(T_N)$, while $\mathrm{Sp}(T_1 + \cdots + T_N)$ is the closure of $\sum \mathrm{Sp}(T_k)$.*

9.4 Generalized Convergence of Unbounded Operators

To avoid trouble with domains we give the following definition of generalized convergence, recalling that the spectrum of a self-adjoint operator is real.

Definition 9.4.1 A sequence of self-adjoint operators T_n converge to a self-adjoint operator T in the *strong (norm) resolvent sense* if $R^{T_n}(\lambda) \to R^T(\lambda)$ strongly (in norm) whenever λ is non-real.

When T, T_n are uniformly bounded, then $T_n \to T$ in norm resolvent sense if and only if $T_n \to T$ in norm. Indeed, if the latter holds and λ is non-real, then

$$(T_n - \lambda I)^{-1} = (T - \lambda I)^{-1}(I + (T_n - T)(T - \lambda I)^{-1})^{-1} \to (T - \lambda I)^{-1}.$$

The opposite direction is clear from $\sup \|T_n\| < \infty$ and the identity

$$T_n - T = (T_n - iI)((T - iI)^{-1} - (T_n - iI)^{-1})(T - iI).$$

Generalized convergence is only needed at one point off the real axis.

Proposition 9.4.2 *Convergence in norm (strong) resolvent sense holds if it holds for one non-real λ_0.*

Proof If $\|R^{T_n}(\lambda_0) - R^T(\lambda_0)\| \to 0$, then $R(\lambda) = \sum_{n=0}^{\infty}(\lambda_0 - \lambda)^n R(\lambda_0)^{n+1}$ which converges in norm for $|\lambda - \lambda_0| < |\lambda_0|^{-1}$, shows that $R^{T_n}(\lambda) \to R^T(\lambda)$ for such λ. Repeating this process we cover the open half plane containing λ_0. As $\|R^{T_n}(\bar{\lambda}) - R^T(\bar{\lambda})\| = \|R^{T_n}(\lambda) - R^T(\lambda)\|$, we also cover the other open half-plane.

In the strongly convergent case we cover the open half-plane as above, but as the $*$-operation is not strongly continuous, to cover the other open half-plane, we use that $(T_n - iI)^{-1} - (T - iI)^{-1}$ equals

$$((T_n + iI)(T_n - iI)^{-1})((T_n + iI)^{-1} - (T + iI)^{-1})((T + iI)(T - iI)^{-1}).$$

□

Proposition 9.4.3 *If $T_n \to T$ in norm resolvent sense, then $f(T_n) \to f(T)$ in norm for $f \in C_0(\mathbb{R})$. If $T_n \to T$ in strong resolvent sense, then $g(T_n) \to g(T)$ strongly for $g \in C_b(\mathbb{R})$. Here the underlying Hilbert space is assumed to be separable.*

Proof By the Stone-Weierstrass theorem we can approximate $f \in C_0(\mathbb{R})$ uniformly by polynomials with variables $x \pm i$. The first statement is then clear from spectral theory. For the same reason the second statement holds when $g \in C_0(\mathbb{R})$. The functions $g_n \in \mathbb{C}_0(\mathbb{R})$ given by $g_n(t) = e^{-t^2/n}$ tend to 1 pointwise, so $g_n(A) \to I$ strongly, and $g_n g \in C_0(\mathbb{R})$ when $g \in C_b(\mathbb{R})$, so the second statement must then also hold for such g. □

Hence if $T_n \to T$ in norm resolvent sense among self-adjoint positive operators in a separabel Hilbert space, then $e^{-sT_n} \to e^{-sT}$ in norm for $s \geq 0$.

Proposition 9.4.4 *As operators in a separable Hilbert space $A_n \to A$ in the strong resolvent sense if and only if $e^{itA_n} \to e^{itA}$ strongly for each real t.*

Proof The forward direction is clear from the previous proposition. For the opposite direction, if $\mathrm{Im}\lambda < 0$, then the formula $(\lambda - \mu)^{-1} = i\int_0^\infty e^{-i\lambda t}e^{i\mu t}\,dt$ combined with spectral theory and the Fubini-Tonelli theorem give the formula $R^A(\lambda)(v) = i\int_0^\infty e^{-i\lambda t}e^{itA}(v)\,dt$. Hence by Lebesgue's dominated convergence theorem, we get the required convergence on the lower open half-plane. The case for the upper open half-plane is done similarly. □

Proposition 9.4.5 *Let $\{T_n\}$ be a sequence of self-adjoint operators in a separable Hilbert space. If the sequence converge in the strong resolvent sense at one point in each open half-plane to operators with dense range, then there is a self-adjoint operator T such that $T_n \to T$ in the strong resolvent sense.*

Proof If $S(\lambda_0)$ is the strong limit operator of the resolvents of T_n at the given point λ_0 in the upper open half-plane, then as $\|R^{T_n}(\lambda_0)\| \leq |\mathrm{Im}\lambda_0|^{-1}$, the operator $S(\lambda) = \sum_{m=0}^{\infty}(\lambda_0-\lambda)^m S(\lambda_0)^{m+1}$ is well-defined when $|\lambda-\lambda_0| \leq |\mathrm{Im}\lambda_0|^{-1}$ and $R^{T_n}(\lambda) \to S(\lambda)$ strongly for such λ's. Repeating this and using the given point in the lower open half-plane, we get $S(\lambda)$ with the same strong convergence for every λ off the real axis. As $S(\lambda)^* = S(\bar{\lambda})$ mutually commute and satisfy the first resolvent formula, they must all have the same range, which is moreover dense, so each $S(\lambda)$ is injective when λ is non-real. The first resolvent formula shows that $A \equiv \lambda I - S(\lambda)^{-1}$ is independent of such λ's, and as the range of $A \pm iI = S(\mp i)^{-1}$ is the whole Hilbert space, the deficiency indices of A are equal. □

We needed convergence at points in both open half-planes here since we could not assume that the limit operator from the outset was self-adjoint. Let us study how spectra relate under generalized convergence.

Proposition 9.4.6 *Say $T_n \to T$ in norm resolvent sense on a separable Hilbert space. If $\lambda \notin \mathrm{Sp}(T)$, then $\lambda \notin \mathrm{Sp}(T_n)$ for large enough n, and $R^{T_n}(\lambda) \to R^T(\lambda)$ in norm. If $a < b$ and $a, b \notin \mathrm{Sp}(T)$, then $\|P^T(\langle a,b\rangle) - P^{T_n}(\langle a,b\rangle)\| \to 0$.*

Proof We need only consider real λ. Since the spectrum of T is closed, there is $t > 0$ such that $\langle\lambda-t, \lambda+t\rangle$ does not intersect $\mathrm{Sp}(T)$. By spectral theory $\|R^T(\lambda+it/3)\| < t^{-1}$ and $\|R^{T_n}(\lambda+it/3)\| < 2t^{-1}$ for large enough n. Then the power series for $R^{T_n}(\mu)$ about $\lambda+it/3$ has radius of convergence at least $t/2$.

As for the second statement, pick $0 < s < (b-a)/2$ so that the norms of $(T_n - aI)^{-1}$ and $(T_n - bI)^{-1}$ are not greater than $1/s$ for large enough n. Then $\mathrm{Sp}(T_n) \cap \langle a,b\rangle \subset \langle a+s, b-s\rangle$ by spectral theory. Pick by Urysohn's lemma a continuous function f that is one on $\langle a+s, b-s\rangle$ and zero outside $\langle a,b\rangle$, so $f(T)$ and $f(T_n)$ are the spectral projections on $\langle a,b\rangle$ of T and T_n, respectively. Now use Proposition 9.4.3. □

Proposition 9.4.7 *Suppose $T_n \to T$ in the strong resolvent sense on a separable Hilbert space. If an open interval does not intersect the spectrum of any T_n, neither does it intersect the spectrum of T, so any $\lambda \in \mathrm{Sp}(T)$ can be approached by $\lambda_n \in \mathrm{Sp}(T_n)$. And $P^{T_n}(\langle a,b\rangle) \to P^T(\langle a,b\rangle)$ strongly if $a < b$ are not eigenvalues of T.*

Proof Now $\langle a,b\rangle \cap \mathrm{Sp}(T) = \phi$ if and only if $\|(T-\lambda I)^{-1}\| \leq 2^{1/2}/(b-a)$ with $2\lambda = (a+b)+i(b-a)$. This, and the fact that $\|(T-\lambda I)^{-1}\| \leq \limsup\|(T_n-\lambda I)^{-1}\|$ when $(T_n-\lambda I)^{-1} \to (T-\lambda I)^{-1}$ strongly, proves the first statement.

As for the second statement, pick uniformly bounded sequences of continuous functions f_n (and g_n) that increase (decrease) pointwise to the characteristic functions on $\langle a,b\rangle$ and $[a,b]$, respectively. Then $f_n(T) \to P^T(\langle a,b\rangle)$ and

$g_n(T) \to P^T([a,b]) = P^T(\langle a,b\rangle)$ strongly by spectral theory. Combining this with Proposition 9.4.3 completes the proof. □

Thus the spectrum of the limiting operator in strong resolvent sense on a separable Hilbert space cannot suddenly expand. So if $T_n \to T$ in this sense and $T_n \geq 0$, then $T \geq 0$. While Proposition 9.4.6 tells us that the spectrum of the limiting operator in norm resolvent sense on a separable Hilbert space cannot suddenly contract, in the strongly convergent case, this might well happen. For instance, self-adjoint operators T_n in $L^2(\mathbb{R})$ given by multiplication with ι/n, converge in the strong resolvent sense to 0, which has zero spectrum while $\mathrm{Sp}(T_n) = \mathbb{R}$.

The following result avoids assumptions on the often inaccessible resolvents.

Proposition 9.4.8 *Suppose T, T_n are self-adjoint operators with common core D. If $T_n \to T$ strongly on D, then $T_n \to T$ in the strong resolvent sense. If $\sup_{\|v\|_T=1} \|(T_n - T)(v)\| \to 0$, then $T_n \to T$ in the norm resolvent sense. Finally, if T, T_n are self-adjoint and positive with a common form domain such that the supremum of $|q_{T-T_n}(v,v)|/q_{(T+I)}(v,v)$ over all non-zero v in the form domain, tends to zero as $n \to \infty$, then $T_n \to T$ in the norm resolvent sense.*

Proof For the first statement, the identity

$$((T_n + iI)^{-1} - (T + iI)^{-1})(T + iI)(v) = (T_n + iI)^{-1}(T - T_n)(v)$$

for $v \in D$ shows that $(T_n + iI)^{-1} \to (T + iI)^{-1}$ strongly. A similar result holds with $-iI$, and we may apply Proposition 9.4.5.

The assumption in the second statement says that $(T_n - T)(T + iI)^{-1} \to 0$ in ordinary norm. Hence

$$(T_n + iI)^{-1} = (T + iI)^{-1}(I + (T_n - T)(T + iI)^{-1})^{-1} \to (T + iI)^{-1}$$

in norm, and similarly $(T_n - iI)^{-1} \to (T - iI)^{-1}$ in norm, and again $T_n \to T$ in the norm resolvent sense.

From the assumption in the third statement $(T+I)^{-1/2}(T_n - T)(T+I)^{-1/2} \to 0$ in ordinary norm. Then

$$(T_n + I)^{-1} = (T + I)^{-1/2}(I + (T + I)^{-1/2}(T_n - T)(T + I)^{-1/2})^{-1}(T + I)^{-1/2}$$

shows that $(T_n + I)^{-1} \to (T + I)^{-1}$ in norm, and one then extends by power series from $\lambda = 1$ to non-real λ's in the upper and lower open half-planes. □

Definition 9.4.9 We say $(v, w) \in V \times V$ is in the *strong graph limit* of operators T_n in a Hilbert space V if there are $v_n \in D(T_n)$ such that $v_n \to v$ and $T_n(v_n) \to w$. We denote the set of such pairs by G^s. If G^s is the graph of an operator T, then T is the *strong graph limit* of T_n, and we write $T = \mathrm{glim}\, T_n$.

Proposition 9.4.10 *We have $T_n \to T$ in the strong resolvent sense if and only if $T = \mathrm{glim}\, T_n$ is self-adjoint.*

Proof For the forward direction, note that $v_n \equiv (T_n + iI)^{-1}(T + iI)(v) \to v$ and $T_n(v) = (T + iI)(v) - iv_n \to T(v)$ for $v \in D(T)$, so $(v, T(v)) \in G^s$ and $G(T) \subset G^s$. On the other hand, if $v_n \in D(T_n)$ and $v_n \to v$ and $T_n(v_n) \to w$, then $w_n \equiv (T + iI)^{-1}(T_n + iI)(v_n) \in D(T)$ and $\|w_n - v_n\| \to 0$ as $n \to \infty$. Hence $w_n \to v$ and $T(w_n) \to w$, and since T is closed, we get $(v, w) \in G(T)$ and $G(T) = G^s$.

For the converse, if $T = \text{glim}\, T_n$ is self-adjoint and $v \in D(T)$, then there are $v_n \in D(T_n)$ such that $v_n \to v$ and $T_n(v_n) \to T(v)$. Hence

$$((T_n+iI)^{-1}-(T+iI)^{-1})(T+iI)(v)=(T_n+iI)^{-1}((T+iI)(v)-(T_n+iI)(v_n))-(v-v_n),$$

which tend to zero in norm. As $T + iI$ has full range, the strong convergence of $(T_n + iI)^{-1} \to (T + iI)^{-1}$ follows. The remaining bit is by now standard. □

Proposition 9.4.11 *Let G^s be the strong graph limit of symmetric operators T_n in a Hilbert space V. If $D \equiv \{v \in V \mid (v, w) \in G^s$ for some $w \in V\}$ is dense in V, then G^s is the graph of an operator T that is symmetric and closed.*

Proof Suppose $v_n, v_n' \in D(T_n)$ such that $v_n \to v$ and $T_n(v_n) \to w$ and $v_n' \to v$ and $T_n(v_n') \to w'$. Let $u \in D$, so that there are $u_n \in D(T_n)$ such that $u_n \to u$ and $\{T_n(v_n)\}$ converges. Then

$$(w - w'|u) = \lim(T_n(v_n - v_n')|u_n) = \lim(v_n - v_n'|T_n(u_n)) = 0$$

and $w = w'$ by denseness. Hence we may well-define an operator T with the required properties. □

Recall the following Lie product formula and its useful proof.

Proposition 9.4.12 *We have $e^{A+B} = \lim(e^{A/n}e^{B/n})^n$ for complex square matrices A and B.*

Proof The formula $S_n^n - T_n^n = \sum_{m=0}^{n-1} S_n^m(S_n - T_n)T_n^{n-1-m}$ with $S_n = e^{(A+B)/n}$ and $T_n = e^{A/n}e^{B/n}$ gives the estimate $\|S_n^n - T_n^n\| \leq n\|S_n - T_n\|e^{\|A\|+\|B\|}$, while expansions of the exponential functions involved gives $\|S_n - T_n\| \leq C/n^2$ with C independent of n, so $\|T_n^n - S_n^n\| \to 0$. □

The following generalization is known as *Trotter's product formula.*

Proposition 9.4.13 *Let A, B and $A + B$ be self-adjoint operators in a separable Hilbert space V. Then $(e^{itA/n}e^{itB/n})^n \to e^{it(A+B)}$ strongly as $n \to \infty$.*

Proof Since $A + B$ is closed its domain D is a Banach space with graph norm. By Proposition 9.3.21 we see that $L(t)(v) = t^{-1}(e^{itA}e^{itB} - e^{it(A+B)})(v) \to 0$ for each $v \in D$ as $t \to 0$, and also when $t \to \infty$. As each map $L(t)\colon D \to V$ is bounded, the principle of uniform boundedness tells us that $L(t)$ are uniformly bounded, so on compact subsets of D we know that $L(t)(v) \to 0$ uniformly. As $A + B$ is self-adjoint its strongly continuous one-parameter unitary group leaves D invariant. It is moreover continuous with respect to the graph norm on D. So

$\{e^{is(A+B)}(v) \mid s \in [-1,1]\}$ is compact in D for each $v \in D$, and

$$t^{-1}(e^{itA}e^{itB} - e^{it(A+B)})e^{is(A+B)}(v) \to 0$$

uniformly for $s \in [-1,1]$. Write $((e^{itA/n}e^{itB/n})^n - (e^{it(A+B)/n})^n)(v)$ as

$$\sum_{m=0}^{n-1}(e^{itA/n}e^{itB/n})^m(e^{itA/n}e^{itB/n} - e^{it(A+B)/n})(e^{it(A+B)/n})^{n-1-m}(v)$$

and observe that the norm of this is not greater than

$$|t|\max_{|s|<t}\|(t/n)^{-1}(e^{itA/n}e^{itB/n} - e^{it(A+B)/n})e^{is(A+B)}(v)\|,$$

which tends to zero as $n \to \infty$. Convergence extends to all $v \in V$ by denseness. □

From the proof we see that on a fixed vector convergence is uniform for t on any compact subset of $\mathbb{R}$. Arguing as in the proof above one also sees that $(e^{-tA/n}e^{-tB/n})^n \to e^{-t(A+B)}$ strongly as $n \to \infty$ whenever A and B in addition are bounded from below.

Remark 9.4.14 Much of the theory in this chapter can easily be adapted to work for not everywhere defined linear maps between different Hilbert spaces, even between Banach spaces. This allows also to treat not everywhere defined conjugate linear maps on a Hilbert space. The reason for this is that for any Hilbert space V with inner product $(\cdot|\cdot)$, we can define its *conjugate Hilbert space* $\overline{V}$ as the same set V with the same addition of vectors but with scalar multiplication given by $c \bullet v = \bar{c}v$ for $v \in V$ and $c \in \mathbb{C}$. The inner product is changed accordingly, and is given by $(v, w) \mapsto (w|v)$. A conjugate linear map on V can then be regarded as a linear map $V \to \overline{V}$. For instance, the polar decomposition of a densely defined closed conjugate linear operator T on a separabel Hilbert space goes as before, but now the partial isometry U will be conjugate linear. Then $(U(v)|w) = (U^*(w)|v)$. ◇

Exercises

For Sect. 9.1

1. Show that $R(T)^\perp = \ker T^*$ for a densely defined operator T in a Hilbert space, and prove that $R(T^*)^\perp = \ker T$ when in addition T is closed.
2. An operator T in a Hilbert space V is boundedly invertible if there exists a bounded operator S on V such that $TS = 1$ and $ST \subset 1$. Prove that this happens precisely when $\ker T = \{0\}$ and $R(T) = V$ and $G(T)$ is closed.
3. Show that there exists an operator T in any infinite dimensional Hilbert space with dense graph. What does this say about $D(T^*)$?

4. Consider $L^2([0,1])$ with respect to the Lebesgue measure, and consider the Volterra operator T on this Hilbert space given by $T(f)(t) = i\int_0^t f(s)\,ds$. Show that T is injective, and prove that $D_1 \equiv \text{im}(T) + \mathbb{C}$ and $D_2 \equiv \{g \in D_1 \mid g(0) = g(1)\}$ and $D_3 \equiv \{g \in D_2 \mid g(0) = 0\}$ are all dense subsets of $L^2([0,1])$. Finally, show that the operators S_k in $L^2([0,1])$ given by $S_k(g) = f$ for $g = T(f) + c \in D_k$, satisfy $S_1^* = S_3$ and $S_2^* = S_2$ and $S_3^* = S_1$.

For Sect. 9.2

1. Show that for a self-adjoint operator T in a Hilbert space, we have $\lambda \in \text{Sp}(T)$ if and only if $T - \lambda I$ is surjective.
2. Prove that eigenvectors of distinct eigenvalues of a symmetric operator in a Hilbert space are mutually orthogonal.
3. Show that each deficiency index can take as value any natural number.
4. Let $\{v_n\}$ be an orthonormal basis of a separable Hilbert space V. Given complex numbers c_n, consider the operator T in V given by $T(v) = \sum c_n(v|v_n)v_n$ for $v \in D(T) \subset V$ which by definition satisfies $\sum |c_n(v|v_n)|^2 < \infty$. Show that T is a densely defined operator with T^* given by $T^*(v) = \sum \overline{c_n}(v|v_n)v_n$ for $v \in D(T^*) = D(T)$. Calculate the Cayley transform $K(T)$ of T when the numbers c_n are real.
5. What is the Cayley transform of a multiplication operator (with the usual domain) by a real Borel function acting in the L^2-space with respect to a Radon measure?

For Sect. 9.3

1. Prove that symmetric normal operators in Hilbert space are self-adjoint.
2. Show that if S is a bounded operator on a Hilbert space V and T_i are normal operators in V such that $ST_1 \subset T_2S$, then $ST_1^* \subset T_2^*S$.
3. Prove that isometric operators in a Hilbert space have closed isometric extensions.
4. Show that any non-empty closed subset of $\mathbb{C}$ is the spectrum of a normal operator in an infinite dimensional Hilbert space.
5. Show that for a positive self-adjoint operator T we have
$$T^s(v) = \pi^{-1}sin(\pi s)\int_0^\infty t^{s-1}(T + tI)^{-1}\,dt\,T(v)$$
for $v \in D(T)$ and $s \in \langle 0, 1\rangle$.
6. Suppose S, T are unbounded self-adjoint operators affiliated with a finite von Neumann algebra W. Show that ST and $S + T$ are densely defined with unique closed extensions affiliated with W.
7. Show that we may replace strongly continuous by weakly continuous in the definition of a strongly continuous one-parameter unitary group of operators without altering the substance of the definition.

For Sect. 9.4

1. Prove that $T_n \to T$ in the strong resolvent sense if the resolvents converge weakly for non-real scalars.
2. Show that $T_n \to T$ in the strong resolvent sense for positive operators if and only if $(T_n + I)^{-1} \to (T + I)^{-1}$ strongly, and that this happens if $e^{-tT_n} \to e^{-tT}$ strongly for $t > 0$.
3. Prove that $tT \to sT$ in norm resolvent sense if $t \to s \neq 0$, and that $e^{itT} \to e^{isT}$ in norm as $t \to s$ if and only if T is bounded.
4. Show that nI has a strong graph limit as $n \to \infty$, but that it cannot be the graph of an operator.
5. Let $\{T_n\}$ be a uniformly bounded sequence of self-adjoint operators, and let T be a bounded operator, all on the same Hilbert space. Show that $T_n \to T$ weakly if and only if T is the so called weak graph limit of T_n, where we now mean $T_n(v_n) \to w$ weakly and $v_n \to v$ in norm in the definition of the strong graph limit.

Chapter 10
Tomita-Takesaki Theory

This chapter, dealing with modular theory for (operator valued) weights, is particularly important for the second half of the book, where we study quantum groups and their various crossed products. It is also fundamental for to the finer understanding of operator algebras. Modular theory was discovered in the 1960s. One then revealed hidden dynamics on C*-algebras entirely due to the non-commutativity of the algebras. This manifested itself through the $*$-operation, which in the non-commutative case is of course antimultiplicative but never multiplicative.

A weight is a 'linear functional' on the positive part of a von Neumann algebra, respecting then only non-negative scalars, since its values should be non-negative, and we allow it to take infinite values. We say it is semifinite when the elements where it is finite generate the von Neumann algebra. We think of weights as generalizations of positive linear functionals, of traces, and of integrals. It is as generalizations of 'Haar integrals' they later will become important for us. In weight theory one focuses on faithful semifinite normal weights. We show that any von Neumann algebra has one. We also include Haagerup's fundamental theorem characterizing the normal weights as those that are upward limits of positive normal functionals.

A modification of the GNS-construction works for weights, yielding what we call a semicyclic representation. More precisely, given a weight x on a von Neumann algebra W, we consider an analogue of the Hilbert-Schmidt operators, namely the left ideal $N_x = \{w \in W \,|\, x(w^*w) < \infty\}$ of W. Then x can be extended to a genuine linear functional on the $*$-algebra $N_x^* N_x$ called the definition domain of the weight. Dividing N_x out by the left ideal L_x of $w \in W$ with $x(w^*w) = 0$, and completing to a Hilbert space V_x, we get a quotient map $q_x \colon N_x \to N_x/L_x \subset V_x$, and the semicyclic representation π_x of W on V_x is given by left multiplication extended by continuity. It is called semicyclic because it almost has a cyclic vector, only approximately so.

When x is faithful, semifinite and normal, then π_x is faithful and normal. Then $A_x \equiv q_x(N_x \cap N_x^*)$ is a $*$-algebra in an obvious way; it is dense in the Hilbert

L. Tuset, *Analysis and Quantum Groups*,
https://doi.org/10.1007/978-3-031-07246-8_10

space V_x, and the $*$-operation is actually a closable operator in V_x, and the left multiplication represents A_x non-degenerately as bounded operators on V_x. We have an example of what we abstractly axiomatize as a full left Hilbert algebra. We can polar decompose the closure of the $*$-operation as $J_x\Delta_x^{1/2}$, where J_x is a conjugate linear involutive isometry, the modular conjugation, and Δ_x is a self-adjoint positive generally unbounded operator known as the modular operator. One proves then that $J_x\pi_x(W)J_x = \pi_x(W)'$ and $\sigma_t^x\pi_x(w) \equiv \Delta^{it}\pi_x(w)\Delta^{-it} \in \pi_x(W)$ for $w \in W$. This can be done in any abstract full left Hilbert algebra, and from any such Hilbert algebra one can construct a faithful, semifinite and normal weight on the bicommutant of the bounded operators given by left multiplication. In fact, starting with weights or left Hilbert algebras amounts to the same thing, we have equivalent pictures, and in this text we actually introduce left Hilbert algebras first. All of this can also be done for abstract C*-algebras; one then talks about lower semicontinuity of the weights rather than normality.

The one-parameter group $\{\sigma_t^x\}$ of automorphisms on $\pi_x(W)$ is the sought for hidden dynamics. This so called modular automorphism group associated with x is the unique one satisfying the so called KMS-condition, which says that there exists a bounded continuous function f on the horizontal strip between $\mathbb{R}$ and $\mathbb{R}+i$, that is moreover holomorphic in the interior of the strip, and satisfies $f(t) = x(\sigma_t^x(v)w)$ and $f(t+i) = x(w\sigma_t^x(v))$ for $v, w \in \pi_x(W)$ and $t \in \mathbb{R}$. Note the change of order in v, w on the two horizontal boarders; the modular group governs the degree of tracialness. This also tells us that traces are too limited to study type III von Neumann algebras. A more general weight is only tracial on its centralizer, that is, where the associated modular group acts as the identity map.

From the outset the modular automorphism group depends on x, but not too much. Connes' cocycle theorem says it is unique up to conjugation by a one-parameter family of unitaries u_t satisfying a cocycle condition $u_{s+t} = u_s\sigma_s^x(u_t)$; this family depends on two choices of weights, and we write $(Dx : Dy)_t$ for u_t calling the family the cocycle derivative of x with respect to y. So dividing out by the inner automorphisms, we get an intrinsic dynamics due to non-commutativity. Conversely, we show that given such an abstract family of u_t's, there exists a unique y yielding the family as the cocycle derivative of x with respect to y. In fact, a von Neumann algebra is semifinite if and only if the intrinsic dynamics is trivial, and then for a fixed tracial faithful semifinite normal weight x, any faithful semifinite normal weight y is morally of the form $x(h\cdot)$ for some positive self-adjoint operator h affiliated with the von Neumann algebra. We show that such a Radon-Nikodym derivative h also exists in the case of general von Neumann algebras; if x is a faithful semifinite normal weight that is invariant under the modular automorphism group of another faithful semifinite normal weight y, then y is morally of the form $x(h\cdot)$, where h is affiliated with the centralizer of x. In a separate section we show that this roughly holds when the invariance is slightly weaker involving a scalar.

In the semicyclic representation $\pi_x(W)$ of a faithful semifinite normal weight x, a von Neumann algebra W is in a so called standard form. It has then a certain self-dual convex cone in V_x whose elements are fixed under the modular conjugation

J_x. In the standard form any normal state is a vector state for a unique vector in the cone. The standard forms of a von Neumann algebra associated to two weights are equivalent by a unique unitary. Axiomatizing what is meant by the standard form of a von Neumann algebra, a similar result holds for the standard forms of isomorphic von Neumann algebras. As a result every $*$-automorphism of a von Neumann algebra in standard form is inner; in fact, any automorphism group, seen as a topological group, has a unique unitary representation implementing the automorphisms as inner ones.

For a von Neumann algebra $W \subset B(V)$, and faithful semifinite normal weights x and y on W and W', respectively, there is a positive operator dx/dy in V, called the spatial derivative of x with respect to y, that implements the modular group of x, and with inverse dy/dx implementing the modular group of y. Moreover, any implementation of a modular group of a weight is the spatial derivative of it with respect to some weight on the commutant. A certain chain rule involving the cocycle derivative also exists.

In the next section we show that a faithful semifinite normal weight x on a von Neumann algebra restricts to a similar type of weight on a von Neumann subalgebra U that is invariant under the modular group of x if and only there exists a faithful normal conditional expectation $E: W \to U$ such that $xE = x$. We also show that such an E must be unique.

Towards the end of this chapter, in the course of generalizing, we go one step further by studying operator valued weights; they generalize both weights and conditional expectations. The trick is to invoke the predual of a von Neumann algebra $W \subset B(V)$ by introducing its extended positive part $\bar{W}_+$ consisting of weight-like objects on the positive part of the predual W_*. The positive elements of W can under evaluation be seen as such 'weights', and new 'weights' are formed from old ones by applying $w^*(\cdot)w$ for $w \in W$. By spectral theory $\bar{W}_+$ can be identified with the positive self-adjoint operators in V affiliated with compressions of W.

An operator valued weight E from W to a von Neumann subalgebra U is a map $W_+ \to \bar{U}_+$ behaving like a weight and like a conditional expectation onto $U \subset \bar{U}_+$. It can be extended to $\bar{W}_+$ and composed with any operator valued weight into W, or from U. One can also form tensor products of operator valued weights, and the usual properties for weights make sense for operator valued weights, including modular automorphism groups and cocycle derivatives. In fact, we show that given faithful semifinite normal weights x and x' on U and W, respectively, with modular groups coinciding on U, there exists a unique faithful semifinite normal operator valued weight E from W to U such that $x' = xE$. Conversely, for any such operator valued weight E, the weights xE and x will have automorphism groups coinciding on U. A similar statement holds for cocycle derivatives.

Finally, we show that there is a unique faithful semifinite normal operator valued weight E' from U' to W' such that $dx/d(yE') = d(xE)/dy$ for all faithful semifinite normal weights x and y on U and W', respectively. This way spatial derivatives provide a beautiful one-to-one correspondence from well-behaved operator valued weights between von Neumann algebras in an inclusion to well-behaved operator valued weights between the commutants of the von Neumann algebras.

Recommended literature for this chapter is [19–21, 45, 48, 50].

10.1 Left and Right Hilbert Algebras

Definition 10.1.1 A *left Hilbert algebra* is a $*$-algebra A with an inner product such that if V is its Hilbert space completion, left multiplication by elements of A on A extend by continuity to a non-degenerate $*$-homomorphism $\pi_l\colon A \to B(V)$, and the involution is a closable operator in V. Its *left von Neumann algebra* is $\pi_l(A)''$.

Note that non-degeneracy in this definition is saying that the linear span A^2 of all products of two elements in A is dense in A for the inner product norm.

Right Hilbert algebra and *right von Neumann algebras* are defined as above, but now multiplication from the right gives a $*$-antihomomorphism $A \to B(V)$. In both cases A^2 is dense in V. If the involution in a left Hilbert algebra is an isometry, then it is also a right Hilbert algebra; we say then it is a *Hilbert algebra*.

Example 10.1.2 Let W be a von Neumann algebra with a separating and cyclic vector v. Then $W(v)$ is a unital left Hilbert algebra with inner product given by restriction, with multiplication and involution given by $u(v) \cdot w(v) = uw(v)$ and $u(v)^* = u^*(v)$ for $u, w \in W$. Clearly $\|\pi_l(u(v))\| \leq \|u\|$ and $(\pi_l(u(v))(w(v))|w(v)) = (w(v)|\pi_l(u(v)^*)(w(v)))$ hold, and while closability of the involution is true, it is by no means obvious, and we presently skip the proof of this.

The vector functional of $v/\|v\|$ is a faithful normal state on W, and conversely, the cyclic vector in the GNS-representation of any faithful normal state on a von Neumann algebra is separating. ◇

Let S denote the closure of the involution in a left Hilbert algebra A with Hilbert space completion V. Note that $v \in V$ belongs to $D(S)$ if and only if there is a sequence of vectors $v_n \in A$ that converges to v and such that $\{v_n^*\}$ Cauchy in V. Then $S(v) = \lim v_n^*$. By involutiveness of $*$ we therefore see that S is a bijection from $D(S)$ onto $D(S)$. So $S = S^{-1}$. Let $S = J\Delta^{1/2}$ be the polar decomposition, where the *modular operator* $\Delta = S^*S$ is a linear self-adjoint positive operator in V with square root having domain $D(S)$ and the *modular conjugation* J is a conjugate linear partial isometry on V. Since S is invertible, so are $\Delta^{1/2}$ and J, and as $S = S^{-1}$, uniqueness in the polar decomposition gives $J = J^{-1} = J^*$ and $J\Delta^{1/2} = \Delta^{-1/2}J$ and $S^* = \Delta^{1/2}J$.

We say $v \in V$ is *right bounded* if there is $\pi_r(v) \in B(V)$ with $\pi_r(v) = \pi_l(\cdot)(v)$ on A. The set B' of all such v is a subspace of V and $\pi_r\colon B' \to B(V)$ is linear. It is easy to check that $\pi_l(A)'(B') \subset B'$, that $\pi_r(B')$ is a left ideal of $\pi_l(A)'$, and that $\pi_r(w(v)) = w\pi_r(v)$ for $w \in \pi_l(A)'$ and $v \in B'$. Let $A' = B' \cap D(S^*)$. Then one easily checks that $\pi_r(B')^*(B') \subset A'$ and $S^*(\pi_r(v_1)^*(v_2)) = \pi_r(v_2)^*(v_1)$ for $v_i \in B'$.

Lemma 10.1.3 *Let S denote the closure of the involution in a left Hilbert algebra A. Let $v \in D(S^*)$. Then $a \equiv \pi_l(\cdot)(v)$ and $b \equiv \pi_l(\cdot)(S^*(v))$ are closable with $a \subset b^*$ and $b \subset a^*$, and $\pi_r(v) \equiv a^{**}$ and $\pi_r S^*(v) \equiv b^{**}$ are affiliated with $\pi_l(A)'$. Letting $U|\pi_r(v)|$ be the polar decomposition of $\pi_r(v)$, then for $f \in C_c(\langle 0, \infty\rangle)$, both $f(|\pi_r(v)|)(S^*(v))$ and $f(U|\pi_r(v)|U^*)(v)$ are right bounded, and their images under π_r are $|\pi_r(v)|f(|\pi_r(v)|)U^*$ and $U|\pi_r(v)|U^* f(U|\pi_r(v)|U^*)U$, respectively.*

Proof It suffices to show that a^* and b^* are affiliated to $\pi_l(A)'$, and this and the remaining part of the proof is straightforward. □

Lemma 10.1.4 *Let S be the closure of the involution in a left Hilbert algebra A. Then A' and $\pi_r(A')(A')$ are dense in $D(S^*)$ with respect to the graph norm of S^*.*

Proof Let $v \in D(S^*)$. By the previous lemma

$$|\pi_r(v)|f(|\pi_r(v)|) = U^*U|\pi_r(v)|U^* f(U|\pi_r(v)|U^*)U = \pi_r(U^* f(U|\pi_r(v)|U^*)(v))$$

and

$$U|\pi_r(v)|U^* f(U|\pi_r(v)|U^*) = U|\pi_r(v)|f(|\pi_r(v)|)U^* = \pi_r(Uf(|\pi_r(v)|)(S^*(v)))$$

both belong to $\pi_r(B')$. Pick $g \in C_c(\langle 0, \infty\rangle)$ such that $f(c) = g(c)cf(c)$ for $c > 0$. Then $f(|\pi_r(v)|), f(U|\pi_r(v)|U^*) \in \pi_r(B')$. Actually they belong to $\pi_r(B')^*\pi_r(B')$ as is seen by picking $f_i \in C_c(\langle 0, \infty\rangle)$ such that $f = \bar{f}_1 f_2$. Hence $f(U|\pi_r(v)|U^*)(v)$ and $f(|\pi_r(v)|)(S^*(v))$ both belong to A' and to $\pi_r(A')(A')$ and

$$S^*(f(U|\pi_r(v)|U^*)(v)) = f(|\pi_r(v)|)^*(S^*(v)).$$

Consider the range projections p and q of $|\pi_r(v)|$ and $U|\pi_r(v)|U^*$, respectively. Pick a net of vectors $v_i \in A$ such that $\pi_l(v_i) \to 1$ strongly. Then $v = \lim \pi_r(v)(v_i) \in q(V)$ and $S^*(v) = \lim \pi_r(v)^*(v_i) \in p(V)$, so $q(v) = v$ and $p(S^*(v)) = S^*(v)$. Pick an increasing sequence of positive functions $g_n \in C_c(\langle 0, \infty\rangle)$ such that $g_n \to 1$ pointwise. Then $g_n(|\pi_r(v)|) \to p$ and $g_n(U|\pi_r(v)|U^*) \to q$ strongly. Since p is also the range projection of $\pi_r(v)^*$, and q is also the range projection of $\pi_r(v)$, we see that $g_n(U|\pi_r(v)|U^*)(v) \to v$ while $g_n(|\pi_r(v)|)(S^*(v)) \to S^*(v)$. So $\{g_n(U|\pi_r(v)|U^*)(v)\}$ converges to v with respect to the graph norm of S^*. □

The following result is now immediate.

Corollary 10.1.5 *Let S denote the closure of the involution in a left Hilbert algebra A with Hilbert space completion V. Then the set A' of right bounded vectors in $D(S^*)$ is a right Hilbert algebra with Hilbert space completion V, multiplication $(v_1, v_2) \mapsto \pi_r(v_2)(v_1)$ and involution $v_1 \mapsto S^*(v_1)$ for $v_i \in A'$.*

Note that $\pi_l(A)' = \pi_r(A')''$ since by the lemma above the $*$-subalgebra $\pi_r(A')$ of $\pi_l(A)'$ is non-degenerate, so that for $w \in \pi_l(A)'$ there are $w_i \in \pi_r(A')$ with $w = \lim w_i^* w w_i$ and $w_i^* w w_i \in \pi_r(B')^*\pi_r(B') \subset \pi_r(A')$.

Lemma 10.1.6 *Let S denote the closure of the involution in a left Hilbert algebra A with Hilbert space completion V. Then A^2 is dense in $D(S)$ with respect to the graph norm of S. Thus if $v_i \in V$ satisfy $(w_1^* w_2|v_1) = (v_2|w_2^* w_1)$ for all $w_i \in A$, then $v_1 \in D(S^*)$ and $v_2 = S^*(v_1)$. We also have $\pi_r(A') = \pi_r(B') \cap \pi_r(B')^*$.*

Proof Pick $u_i \in A'$ such that $\pi_r(u_i) \to 1$ strongly. Let $v \in A$. Then $v = \lim \pi_l(v)(u_i)$ belongs to the image of the range projection of $\pi_l(v)\pi_l(v)^*$. Rescale v so that $\|\pi_l(v)\| \leq 1$. By spectral theory, we thus see that if $p_n(t) = 1 - (1-t)^n$ for $t \in \mathbb{R}$, then $v = \lim p_n(\pi_l(v)\pi_l(v)^*)(v) = \lim p_n(vv^*)v$, while $S(v) = v^* = \lim p_n(\pi_l(v^*)\pi_l(v^*)^*)(v^*) = \lim p_n(v^*v)v^* = \lim v^* p_n(vv^*) = \lim(p_n(vv^*)v)^*$.

As for the last statement, if $\pi_r(v_1)^* = \pi_r(v_2)$ for $v_i \in B'$, then $(w_1^* w_2|v_1) = (v_2|w_2^* w_1)$, so $v_2 = S^*(v_1) \in A'$. □

We can repeat the whole process above, but starting now with the right Hilbert algebra A' having Hilbert space completion V. Then the adjoint operator of the involution S^* on A' coincides with the original involution S on A by Lemma 10.1.4. The linear space B of *left bounded vectors* consisting of those $v \in V$ such that $\pi_l(v) = \pi_r(\cdot)v$ on A' is invariant under $\pi_l(A)''$, and $\pi_l(B)$ is a left ideal of $\pi_l(A)''$ and $\pi_l(wv) = w\pi_l(v)$ for $w \in \pi_l(A)''$ and $v \in B$. We also have the obvious analogue of the first part of Lemma 10.1.3. Now the set $A'' = B \cap D(S)$ is a left Hilbert algebra with Hilbert space completion V, multiplication $(v_1, v_2) \mapsto \pi_l(v_1)(v_2)$ and involution $v_1 \mapsto S(v_1)$ for $v_i \in A''$. Also $\pi_l(A'') = \pi_l(B) \cap \pi_l(B)^*$. Clearly $A \subset A''$ and $\pi_l(A'')'' = \pi_l(A)''$.

Repeating this process once more gives nothing new as $A''' = A'$. We say A is *full* if $A'' = A$. Since in general A and A'' have the same left von Neumann algebra, which is our point of interest, we might as well assume that A is full from the outset.

Lemma 10.1.7 *Let A be a full left Hilbert algebra. For $z \in \mathbb{C}$ off the positive real axis let $h(z) = (2(|z| - \mathrm{Re} z))^{-1/2}$. Then $(\Delta - z)^{-1}(A') \subset A$ and $\|\pi_l((\Delta - z)^{-1}(\cdot))\| \leq h(z)\|\pi_r(\cdot)\|$ on A', and similarly $(\Delta^{-1} - z)^{-1}(A) \subset A'$ and $\|\pi_r((\Delta^{-1} - z)^{-1}(\cdot))\| \leq h(z)\|\pi_l(\cdot)\|$ on A.*

Proof Let $v \in A'$, so $w \equiv (\Delta - z)^{-1}(v) \in D(\Delta) \subset D(S)$. Let $U|\pi_l(w)|$ be the polar decomposition of $\pi_l(w)$, and set $a = U|\pi_l(w)|U^*$. From the proof of the dual version of Lemma 10.1.4 we know that $f(a)(w) \in A$ for $f \in C_c(\langle 0, \infty \rangle)$ and $S(f(a)(w)) = f(|\pi_l(w)|)^*(S(w))$. Now

$$
\begin{aligned}
h(z)^{-2}\||\pi_l(w)|f(|\pi_l(w)|)(S(w))\|^2 &= h(z)^{-2}(\Delta(w)|af(a)^* af(a)(w)) \\
&\leq 2|z|\|af(a)\Delta(w)\|\|af(a)(w)\| - 2\mathrm{Re}\, z(af(a)\Delta(w)|af(a)(w)) \\
&\leq \|af(a)(v)\|^2 = \|\pi_r(v)Uf(|\pi_l(w)|)(S(w))\|^2 \leq \|\pi_r(v)\|^2\|f(|\pi_l(w)|)(S(w))\|^2.
\end{aligned}
$$

With the resolution of the identity $|\pi_l(w)| = \int_0^\infty \lambda \, dP(\lambda)$, this shows that

$$
\int \lambda^2 |f(\lambda)|^2 d(P(\lambda)(S(w))|S(w)) \leq c^2 \int |f(\lambda)|^2 d(P(\lambda)(S(w))|S(w))
$$

for $f \in C_c(\langle 0, \infty\rangle)$, where $c = h(z)\|\pi_r(v)\|$. So the measure $(P(\cdot)(S(w))|S(w))$ is supported on $[0, c]$, and $P(c)(S(w)) = S(w)$. Hence $\pi_l(w)^* = P(c)\pi_l(w)^*$, which has norm not greater than c, so w^*, and thus w, is left bounded, and $\|\pi_l(w)\| \leq c$. The proof of the second statement is similar. □

Lemma 10.1.8 *Let A be a full left Hilbert algebra, and let $w = (\Delta + t)^{-1}(v)$ for $v \in A'$ and $t > 0$. Then*

$$(\pi_r(v)(v_1)|v_2) = (J\pi_l(w)^* J\Delta^{-1/2}(v_1)|\Delta^{1/2}(v_2)) + t(J\pi_l(w)^* J\Delta^{1/2}(v_1)|\Delta^{-1/2}(v_2))$$

for $v_i \in D(\Delta^{1/2}) \cap D(\Delta^{-1/2})$.

Proof The equality holds for $v_i \in A \cap D(\Delta^{-1/2})$ because

$$(\pi_r(v)(v_1)|v_2) = (v_1|S(wS(v_2))) + t(S(S(w)S(v_1))|v_2).$$

It thus suffices to show that to $u \in D(\Delta^{1/2}) \cap D(\Delta^{-1/2})$, there is a sequence of vectors $u_n \in A \cap D(\Delta^{-1/2})$ such that $u_n \to u$ and $\Delta^{\pm 1/2}(u_n) \to \Delta^{\pm 1/2}(u)$. As $\Delta^{-1/2}(A') = JS^*(A') = J(A')$ is dense in the Hilbert space completion of A, there is a sequence of vectors $b_n \in A'$ such that $(\Delta^{1/2} + \Delta^{-1/2})(u) = \lim \Delta^{-1/2}(b_n)$. Since $\Delta^{-1/2}(\Delta^{1/2} + \Delta^{-1/2})^{-1} = (1 + \Delta)^{-1}$ we have $u_n \equiv (1 + \Delta)^{-1}(b_n) \in A \cap D(\Delta^{-1/2})$ by the previous lemma, and

$$u = (\Delta^{1/2} + \Delta^{-1/2})^{-1} \lim \Delta^{-1/2}(b_n) = \lim(\Delta^{1/2} + \Delta^{-1/2})^{-1}\Delta^{-1/2}(b_n) = \lim u_n$$

while $\Delta^{\pm 1/2}(u_n) \to \Delta^{\pm 1/2}(u)$ by boundedness of $(\Delta + 1)^{-1}$ and $\Delta(\Delta + 1)^{-1}$. □

Lemma 10.1.9 *Let $u : \mathbb{C} \to W$ be a holomorphic map into the invertible members of a unital Banach algebra W such that* $\sup \|u(\mathbb{R})\| < \infty$ *and $u(z_1 + z_2) = u(z_1)u(z_2)$ for $z_i \in \mathbb{C}$. Then for $s \in \mathbb{R}$ the element $e^{-s/2}u(-i/2) + e^{s/2}u(i/2)$ is invertible with inverse $\int_{-\infty}^{\infty} e^{-ist}(e^{\pi t} + e^{-\pi t})^{-1}u(t)\,dt$.*

Proof The function $h : \mathbb{C} \to W$ given by $h(z) = e^{isz}(e^{\pi z} + e^{-\pi z})^{-1}u(z)$ has simple poles at $\mathbb{Z}i$, and $\|h(z)\| \leq \sup \|u(\mathbb{R})\| e^{-sr} |e^{\pi z} + e^{-\pi z}|^{-1} \|u(ir)\|$, where $z = t + ir$. Let C be the counterclockwise path along the boundary of a rectangle in the complex plane with center at the origin, length $2R$ and height one. Then

$$I = \lim_{R \to \infty} \int_C h(z)\,dz = \int_{-\infty}^{\infty} (h(t - i/2) - h(i + i/2))\,dt$$

by the above estimate, while $I = 2\pi i \lim_{z \to 0} zh(z) = i$. Now plug in h. □

Lemma 10.1.10 *Let A be a full left Hilbert algebra, and let $s \in \mathbb{R}$. Then*

$$e^{s/2}\Delta^{1/2}(\Delta + e^s)^{-1} = \int_{-\infty}^{\infty} e^{-ist}(e^{\pi t} + e^{-\pi t})^{-1}\Delta^{it}\,dt.$$

Proof Consider $\Delta = \int_0^\infty \lambda\, dP(\lambda)$ and let $P_r = P(r) - P(r^{-1})$ for $r > 1$. The previous lemma applied to $W = B(\mathrm{im} P_r)$ and $u(z) = (\Delta P_r)^{iz}$ gives

$$e^{s/2}\Delta^{1/2}(\Delta + e^s)^{-1}P_r = \int_{-\infty}^{\infty} e^{-ist}(e^{\pi t} + e^{-\pi t})^{-1}\Delta^{it}P_r\, dt.$$

Then take the limit as $r \to \infty$. □

Lemma 10.1.11 *Let A be a full left Hilbert algebra with Hilbert space completion V, and let $s \in \mathbb{R}$. If $(a(v_1)|v_2) = (b\Delta^{-1/2}(v_1)|\Delta^{1/2}(v_2)) + e^s(b\Delta^{1/2}(v_1)|\Delta^{-1/2}(v_2))$ for $a, b \in B(V)$ and $v_i \in D(\Delta^{1/2}) \cap D(\Delta^{-1/2})$, then*

$$e^{s/2}b = \int_{-\infty}^{\infty} e^{-ist}(e^{\pi t} + e^{-\pi t})^{-1}\Delta^{it}a\Delta^{-it}\, dt.$$

Proof Write $\Delta = \int_0^\infty \lambda\, dP(\lambda)$ and $P_r = P(r) - P(r^{-1})$ for $r > 1$. Let $W = B(B(P_r(V)))$ and $u(z) = \Delta^{iz}(\cdot)\Delta^{-iz}$ on $B(P_r(V))$. By assumption $P_r a P_r = e^{s/2}(e^{-s/2}u(-i/2) + e^{s/2}u(i/2))(P_r b P_r)$, and Lemma 10.1.9 gives

$$e^{s/2}P_r b P_r = \int_{-\infty}^{\infty} e^{-ist}(e^{\pi t} + e^{-\pi t})^{-1}\Delta^{it}P_r a P_r \Delta^{-it}\, dt.$$

It remains to take the limit as $r \to \infty$. □

Theorem 10.1.12 *Let A be a left Hilbert algebra with modular operator Δ and modular conjugation J, and let $t \in \mathbb{R}$. Then Δ^{it} acts on A' and A'' as automorphisms, and $J(A') = A''$ and $J(v_1 v_2) = J(v_2)J(v_1)$ for $v_i \in A'$. Hence $J\pi_l(A)''J = \pi_l(A)'$ and $\Delta^{it}\pi_l(A)''\Delta^{-it} = \pi_l(A)''$.*

Proof The previous lemma and Lemma 10.1.8 show that

$$e^{s/2}J\pi_l((\Delta + e^s)^{-1}(v))^*J = \int_{-\infty}^{\infty} e^{-ist}(e^{\pi t} + e^{-\pi t})^{-1}\Delta^{it}\pi_r(v)\Delta^{-it}\, dt$$

for $v \in A'$ and $s \in \mathbb{R}$. Sandwiching the right hand side with J and evaluating at $u \in A'$, we thus get $e^{s/2}\pi_r(u)J\Delta^{1/2}(\Delta + e^s)^{-1}(v)$. Invoking Lemma 10.1.10 we deduce $J\Delta^{it}\pi_r(v)\Delta^{-it}J(u) = \pi_r(u)J\Delta^{it}(v)$. So $J\Delta^{it}(v)$ is left bounded and $\pi_l(J\Delta^{it}(v)) = J\Delta^{it}\pi_r(v)\Delta^{-it}J$. As $J(D(S^*)) = J(D(\Delta^{-1/2})) = D(\Delta^{1/2}) = D(S)$, we thus see that $J\Delta^{it}(v) \in A''$. Setting $t = 0$ gives $J(A') \subset A''$ and $\pi_l(J(v)) = J\pi_r(v)J$, which shows that J is antimultiplicative on A'. Similarly $J\Delta^{it}(w) \in A'$ and $\pi_r(J(w)) = J\pi_l(w)J$ for $w \in A''$. Hence $J(A') = A''$ and $\Delta^{it}(A') = A'$ and $\Delta^{it}(A'') = A''$. Moreover, we have

$$\pi_l(\Delta^{it}(w)) = \pi_l(J\Delta^{-it}(J(w))) = J\Delta^{-it}\pi_r(J(w))\Delta^{it}J = \Delta^{it}\pi_l(w)\Delta^{-it},$$

which shows that Δ^{it} is multiplicative on A''. It is also multiplicative on A'. □

Proposition 10.1.13 *Let A be a left Hilbert algebra. Then $a \in \pi_l(A)' \cap \pi_l(A)''$ leaves $D(S)$ and $D(S^*)$ invariant, and $S(av) = a^*S(v)$ and $S^*(aw) = a^*S^*(w)$ and $JaJ = a^*$ and $\Delta^{it}a\Delta^{-it} = a$ for $v \in D(S)$, $w \in D(S^*)$ and $t \in \mathbb{R}$.*

Proof We can assume A is full. We know a leaves $\pi_r(B') \cap \pi_r(B')^* = \pi_r(A')$ and $\pi_l(B) \cap \pi_l(B)^* = \pi_l(A)$ invariant, so $aA' \subset A'$ and $aA \subset A$. Clearly also $S(av) = a^*S(v)$ and $S^*(aw) = a^*S^*(w)$ for $v \in A$ and $w \in A'$. The statements for $D(S)$ and $D(S^*)$ follow now by denseness.

If a is unitary, then $aA = A$ and $aA' = A'$, and $Sa(v) = a^*S(v)$ for $v \in A$. Hence $J\Delta^{1/2} = S = aSa = aJaa^*\Delta^{1/2}a$ and $J = aJa$ and $\Delta^{1/2} = a^*\Delta^{1/2}a$. As C*-algebras are spanned by their unitaries, we are done. □

10.2 Weight Theory

Definition 10.2.1 Let W be a von Neumann algebra. A *weight* on W is a map $x: W_+ \to [0, \infty]$ such that $x(c_1w_1+c_2w_2) = c_1x(w_1)+c_2x(w_2)$ for $w_i \in W_+$ and scalars $c_i \geq 0$. It is *semifinite* if the elements of finite weight generate W, and *faithful* if x vanishes only at 0, and *normal* if $x(\sup w_i) = \sup x(w_i)$ for every bounded increasing net $\{w_i\}$ in W_+, and *completely additive* if $x(\sum v_i) = \sum x(v_i)$ for every σ-strongly summable net $\{v_i\} \subset W_+$.

Normal weights are clearly completely additive. In fact, we have the following fundamental result known as *Haagerup's Theorem*.

Theorem 10.2.2 *Let x be a weight on a von Neumann algebra W. Then*

$$x = \sup\{y \in W_* \mid 0 \leq y \leq x \text{ on } W_+\}$$

pointwise if and only if x is σ-weakly lower semicontinuous if and only if x is normal if and only if x is completely additive.

The forward implications are trivial, but in order to enclose the circle of statements, we need a good deal of preparation.

Given a weight x on W, and let $N_x = \{w \in W \mid x(w^*w) < \infty\}$. Its *definition domain* M_x is $N_x^*N_x$. As $(vw)^*(vw) \leq \|v\|^2w^*w$ and $(v \pm w)^*(v \pm w) \leq 2(v^*v + w^*w)$ for $v, w \in W$, we see that N_x is a left ideal of W. So M_x is a $*$-subalgebra of W, and by the polarization identity we know that it is spanned by the elements of finite weight. In fact, every positive element of M_x, say of the form $w = \sum v_n^*w_n$ with $v_n, w_n \in N_x$, is majorized by $\sum(v_n + w_n)^*(v_n + w_n)$ and hence has finite weight. We extend x to a linear functional on M_x denoted by the same symbol.

The set L_x of all $w \in W$ such that $x(w^*w) = 0$ is a left ideal of W contained in N_x. Let $q: N_x \to N_x/L_x$ be the quotient map, and let V_x denote the Hilbert space completion of N_x/L_x with respect to the inner product given by $(q(w_1)|q(w_2)) = x(w_2^*w_1)$ for $w_i \in N_x$. Extend left multiplication of $w \in W$ on elements of N_x/L_x by continuity to $\pi_x(w) \in B(V_x)$. This gives a unital $*$-representation π_x of W on V_x called the *semicyclic representation* of W associated to x. When x is normal, so is π_x. If x is faithful and semifinite, then π_x is injective. Indeed, if $\pi_x(w) = 0$ for $w \in W$, then $0 = \|\pi_x(w)q(v)\|^2 = \|q(wv)\|^2 = x((wv)^*(wv))$, so $wv = 0$ for all $v \in N_x$, and hence for all positive elements of finite weight, so $w = 0$.

Lemma 10.2.3 *Let W be a von Neumann algebra on V. If $w_1^* w_1 \leq w_2^* w_2$ for $w_i \in W$, there is a unique $u \in W$ that vanishes on $w_2(V)^\perp$ and satisfies $w_1 = uw_2$. Also $\|u\| \leq 1$. If we have a family of $w_i \in W$ such that $a = \lim w_i^* w_i$ strongly, and $u_i \in W$ is the operator that vanishes on $a(V)^\perp$ and satisfies $w_i = u_i a^{1/2}$, then $\sum u_i^* u_i$ converges strongly to the range projection p of a. If $\{b_i\}$ is a bounded increasing net in W_+ with $b = \sup b_i$, and if $s_i \in W$ vanishes on $b(V)^\perp$ and satisfies $b_i = s_i b^{1/2}$, then $\{s_i^* s_i\}$ increases towards the range projection q of b, and $q = \lim s_i$ strongly.*

Proof Let p_2 be the orthogonal projection onto the closure of $w_2(V)$. For $v \in V$ we have $\|w_1(v)\| \leq \|w_2(v)\|$, so $s(w_2(v)) = w_1(v)$ well-defines an operator s from $\operatorname{im} p$ to V. Then $u = sp_2$ is the required operator, and we have no other choice. Since w_1 and w_2 commute with the unitaries in W', so does u by uniqueness of the decomposition, and thus $u \in W$. Clearly $\|u\| \leq 1$.

For the second statement, let $v = a^{1/2}(v')$ for $v' \in V$. Then

$$(\sum_J u_i^* u_i(v)|v) = (\sum_J w_i^* w_i(v')|v') \leq (a(v')|v') = \|v\|^2$$

for finite J. Let $p_J = \sum u_i^* u_i$, so $(p_J(r)|r) \leq (p(r)|r)$ for $r \in a^{1/2}(V) + (1 - p)(V)$, and $p_J \leq p$ by continuity. By Vigier's theorem the net $\{p_j\}$ converges strongly to $p' \in W$ majorized by p. As $(p'(v)|v) = \|v\|^2$, we get $p' = p$.

The proof of the remaining part is similar. □

Lemma 10.2.4 *Let x be a weight on a von Neumann algebra W with associated semicyclic representation π_x. Then there exists a unique completely positive linear map $f_x : M_x \to (\pi_x(W)')_*$ such that $(f_x(w^*v))(a) = (aq(v)|q(w))$ for $v, w \in N_x$ and $a \in \pi_x(W)'$.*

Proof Uniqueness is clear from the definition of M_x. If $w^*w = v^*v$, there is by the previous lemma a partial isometry $u \in W$ such that $v = uw$ and $w = u^*v$. So $(aq(w)|q(w)) = (aq(v)|q(v))$ and we can define a map $f_x : (M_x)_+ \to (\pi_x(W)')_*$ such that $(f_x(w^*w))(a) = (aq(w)|q(w))$ for $w \in N_x$. Clearly $f_x(c(\cdot)) = cf_x(\cdot)$ for $c > 0$. If $w = w_1 + w_2$ for positive $w_i \in M_x$, choose $u_i \in M$ by the lemma such that $w_i^{1/2} = u_i w^{1/2}$ and $p = u_1^* u_1 + u_2^* u_2$ is the range projection of w. Then

$$(aq(w^{1/2})|q(w^{1/2})) = (a\pi_x(p)q(w^{1/2})|q(w^{1/2})) = (f_x(w_1) + f_x(w_2))(a),$$

so $f_x(w) = f_x(w_1) + f_x(w_2)$. We can now extend f_x by linearity to M_x. Complete positivity is clear from

$$(f_x(w_i^* w_j))(a_i^* a_j) = \|\sum a_k q(w_k)\|^2 \geq 0$$

for $w_i \in N_x$ and $a_i \in \pi_x(W)'$. □

Lemma 10.2.5 *Let x be a weight on a von Neumann algebra W with f_x as in the previous lemma. Then $\|f_x(w)\| = \inf\{x(w_1)+x(w_2) \mid w = w_1-w_2,\ w_i \in (M_x)_+\}$ for any self-adjoint $w \in M_x$.*

Proof Let $f(w)$ be the right hand side of the identity above. Then we get a function f on the self-adjoint part of M_x such that $f(cw) = |c|f(w) \geq 0$ for $c \in \mathbb{R}$. Let $w_i = w_i^* \in M_x$ and $\varepsilon > 0$, and pick positive $v_i, u_i \in M_x$ such that $w_i = v_i - u_i$ and $x(v_i) + x(u_i) < f(w_i) + \varepsilon$. Then

$$f(w_1 + w_2) \leq x(v_1 + v_2) + x(u_1 + u_2) \leq f(w_1) + f(w_2) + 2\varepsilon,$$

so f is a seminorm on the self-adjoint part of M_x that coincides with x on $(M_x)_+$.

Fix $w = w^* \in M_x$. Pick by the Hahn-Banach theorem a real valued linear functional g on the self-adjoint part of M_x such that $g(w) = f(w)$ and $|g(\cdot)| \leq f(\cdot)$. Extending g to a self-adjoint linear functional on M_x, again denoted by g, we get $-x(v^*v) = -f(v^*v) \leq g(v^*v) \leq f(v^*v) = x(v^*v)$ for $v \in N_x$, which shows that there is $a \in B(V_x)$ with $\|a\| \leq 1$ such that $(aq(v)|q(v)) = g(v^*v)$. Clearly $a = a^* \in \pi_x(W)'$. Consider the polar decomposition $u|w| = |w|u$ of w, and note that $|w|^{1/2} \in N_x$. Then by polarization we get

$$f(w) = g(w) = (aq(|w|^{1/2}u)|q(|w|^{1/2})) = (f_x(|w|u))(a) \leq \|f_x(|w|u)\| = \|f_x(w)\|.$$

For the reverse inequality, note that if $w \in (M_x)_+$, then $f_x(w)$ is positive, so $\|f_x(w)\| = (f_x(w))(1) = x(w) = f(w)$, and $\|f_x(\cdot)\| \leq f(\cdot)$ on the self-adjoint part of M_x. □

Lemma 10.2.6 *Let x be a normal weight on a von Neumann algebra W with f_x as in the previous lemma, and let $\{w_n\}$ be a bounded sequence in $(M_x)_+$ such that $\{f_x(w_n)\}$ converges in norm. If $w_n \to w \in W$ σ-strongly, then $w \in (M_x)_+$, and if $w = 0$, then $f_x(w_n) \to 0$.*

Proof Let $g = \lim f_x(w_n)$ and $\varepsilon > 0$. By going to a subsequence we may assume that $\|g - f_x(w_n)\| < \varepsilon/2^{n+1}$, so $\|f_x(w_{n+1}) - f_x(w_n)\| < \varepsilon/2^n$. Pick by the previous lemma $a_n, b_n \in (M_x)_+$ with $w_{n+1} - w_n = a_n - b_n$ and $x(a_n) + x(b_n) < \varepsilon/2^n$, and note that $w_{n+1} \leq w_1 + \sum_{k=1}^n a_k$ and $w_1 - w_{n+1} \leq \sum_{k=1}^n b_k$. Let $c > 0$ and define $h_c(t) = t(1+ct)^{-1}$ for $t \in \langle -1/c, \infty \rangle$. Then $h_c(w_{n+1}) \leq h_c(w_1 + \sum_{k=1}^n a_k) \equiv d_n \leq 1/c$ and the bounded increasing sequence $\{d_n\}$ converges σ-strongly to $d \in W_+$.

If $w_n \to w \in W$ σ-strongly, then $h_c(w) = \lim h_c(w_n)$ σ-strongly by the proof of Lemma 5.7.1, and $h_c(w) \leq d$. By normality we thus get

$$x(h_c(w)) \leq \lim x(d_n) \leq \lim x(w_1 + \sum_{k=1}^{n} a_k) \leq x(w_1) + \sum_{k=1}^{\infty} \varepsilon/2^k = x(w_1) + \varepsilon.$$

But $h_c(w)$ increases strongly towards w as c decreases towards zero, so by normality $x(w) = \lim x(h_c(w)) \leq x(w_1) + \varepsilon < \infty$ and $w \in (M_x)_+$.

Suppose now that $w = 0$, and set $K = \sup \|w_n\|$. Then $w_1 - w_{n+1} \geq -K$, so $h_c(w_1 - w_{n+1}) \leq h_c(\sum_{k=1}^{n} b_n)$ when $0 < c < 1/K$. Then by the same lemma $h_c(w_1) = \lim h_c(w_1 - w_{n+1}) \leq \lim h_c(\sum_{k=1}^{n} b_k) \equiv r_c$, and by normality

$$x(h_c(w_1)) \leq x(r_c) = \lim x(h_c(\sum_{k=1}^{n} b_k)) \leq \lim x(\sum_{k=1}^{n} b_k) \leq \sum_{k=1}^{\infty} \varepsilon/2^k = \varepsilon,$$

and again $x(w_1) = \lim x(h_c(w_1)) \leq \varepsilon$. Hence

$$\|g\| \leq \|g - f_x(w_1)\| + \|f_x(w_1)\| < \varepsilon/2 + x(w_1) < 2\varepsilon$$

and $g = 0$. □

Lemma 10.2.7 *If x is a normal weight on a σ-finite von Neumann algebra W, the closed unit ball of $G \equiv \{(w, q(w)) \mid w \in N_x\}$ in the Banach space $W \times V_x$ is w^*-compact.*

Proof Let B be the closed unit ball of $W \times V_x$. Since $B \cap G$ is convex, we only need to check closedness under any locally convex topology on $W \times V_x$ that has $W_* \times V_x^*$ as its dual space. Pick the product topology of the σ-strong* topology on W and the norm topology on V_x. Since $B \cap G$ is metrizable, if (w, v) is a limit point of $B \cap G$, there is a sequence of $w_n \in N_x$ such that $w_n \to w$ σ-strongly* and $q(w_n) \to v$ in norm, and $\|w_n\| \leq 1$ and $\|q(w_n)\| \leq 1$. Hence $w_n^* w_n \to w^* w$ σ-strongly, while $f_x(w_n^* w_n) = ((\cdot)q(w_n)|q(w_n))$ converges to $((\cdot)v|v)$ in norm. By the previous lemma $w \in N_x$. Then $\{(w_n - w)^*(w_n - w)\}$ tends to zero σ-strongly, while $f_x((w_n - w)^*(w_n - w)) = ((\cdot)q(w_n - w)|q(w_n - w))$ converges to $((\cdot)(v - q(w))|v - q(w))$ in norm. Hence by the previous lemma, the latter is zero on $\pi_x(W)'$, so $\|v - q(w)\|^2 = 0$. □

Lemma 10.2.8 *The three last conditions in the previous theorem are equivalent when the von Neumann algebra is σ-finite.*

Proof Any σ-finite von Neumann algebra W admits a faithful normal state y. Suppose x is a completely additive weight on W, and that $w = \sup w_i$ for a bounded increasing net $\{w_i\} \subset W$. Pick a subsequence $\{v_n\}$ such that $y(v_n) > y(w) - 1/n$, which by Vigier's theorem converges strongly, and hence σ-strongly, to $v \leq w$. Since $y(w) - 1/n < y(v_n) \leq y(v) \leq y(w)$, we get $v = w$ by faithfulness. Letting $u_n = v_n - v_{n-1}$ with $v_0 = 0$, we therefore get $w = \sum u_n$, and as x is completely additive, we finally get $x(w) = \sum x(u_n) = \lim x(v_n) \leq \lim x(w_i) \leq x(w)$. So x is normal.

Assume now that x is a normal weight on W. We show that it is σ-weakly lower semicontinuous. For $r, s > 0$ let $B_{r,s} = \{(w, v) \in W \times V_x \mid \|w\| \leq r, \ \|v\| \leq s\}$, and let G be as in the previous lemma. Then $G \cap B_{r,s}$ is w^*-compact, so the image $C_{r,s} = \{w \in W \mid \|w\| \leq r, \ x(w^*w) \leq s^2\}$ of the projection on the von Neumann algebra side is σ-weakly compact. Set $E_r = \{w \in W \mid x(w^*w) \leq s^2\}$, and let B be the closed unit ball of W. Then $C_{r,s} = E_r \cap rB$, and E_r is σ-weakly closed by the Krein-Smulian theorem. Now what we want to prove is that $F_s = \{w \in W_+ \mid x(w) \leq s^2\}$ is σ-weakly closed. Again by the Krein-Smulian theorem it suffices to show that $F_s \cap rB$ is σ-weakly closed, so say $w_i \to w \in W_+$ σ-weakly with $w_i \in F_s \cap rB$. Then the net $\{w_i^{1/2}\}$ in E_s converges σ-strongly to $w^{1/2}$, so $w^{1/2} \in E_s$ and $w \in F_s$. □

Lemma 10.2.9 *Let S be the collection of orthogonal projections p in a von Neumann algebra W such that pWp is σ-finite. Suppose we have a convex set $F \subset (\cup_{p \in S} pWp)_+$ such that $0 \leq v \leq w \in F$ implies $v \in F$ when $v \in W$. Then F is σ-weakly closed in $\cup_{p \in S} pWp$ if and only if $F \cap pWp$ is σ-weakly closed for all $p \in S$.*

Proof The forward implication is clear. For the converse, let B be the closed unit ball in W, and note that the σ-strong topology on $pWp \cap B$ is metrizable.

Now $E = \{w \in W \mid w^*w \in F\}$ is convex because if $w_i \in E$ and $c \in [0, 1]$, then

$$0 \leq (cw_1 + (1-c)w_2)^*(cw_1 + (1-c)w_2) \leq cw_1^*w_1 + (1-c)w_2^*w_2 \in F$$

as $w_1^*w_2 + w_2^*w_1 \leq w_1^*w_1 + w_2^*w_2$. Also if $a \in B$, then $aE \subset E$ as $(aw)^*aw \leq w^*w$.

The set $E \cap pWp \cap rB$ is σ-weakly closed as it is the inverse image of the σ-weakly closed set $F \cap pWp$ under the σ-strongly* continuous map $w \mapsto w^*w$ from $rB \cap pWp$ to $r^2B \cap pWp$. By the Krein-Smulian theorem we thus conclude that $E \cap pWp$ is σ-weakly closed.

We claim that pE is σ-weakly closed, or equivalently, by the Krein-Smulian theorem, that $rB \cap pE$ is σ-strongly closed. Any w in this σ-strong closure can be approximated σ-strongly, and thus σ-weakly, by a sequence of $w_n \in rB \cap pE$. Hence there is an orthogonal projection $q \in S$ with $w_n = qw_nq \in E \cap qWq$, which σ-weakly closed be the previous paragraph, so $w \in E \cap qWq$. But $pw = w$ and $w \in rB$, so $w \in rB \cap pE$.

Let $\overline{F}$ be the σ-strong closure of F, and say $v \in \overline{F} \cap (\cup_{p \in S} pWp)$. If $\{v_i\}$ is a net in F that converges σ-strongly to v, then also $v_i^{1/2} \to v^{1/2}$ σ-strongly. But the range projection p of v belongs to S as it is majorized by some projection in S, and $pv_i^{1/2} \to pv^{1/2}$ σ-strongly. Hence $v^{1/2} \in pE$ by the previous paragraph, and as $pE \subset E$, we get $v \in F$. □

Proof (Of the Equivalences of the Three Last Statements of the Previous Theorem) So let x be a completely additive weight on W. By Lemma 10.2.8 it is σ-weakly lower semicontinuous on pWp for $p \in S$ with S as in the previous lemma. Letting $F = \{w \in W_+ \,|\, x(w) \leq 1\}$, that lemma shows that F is σ-weakly closed in $\cup_{p\in S} pWp$. Pick by Zorn's lemma a maximal orthogonal family $\{p_i\}$ in S indexed say by I. The orthogonal projections $q_J = \sum_{i\in J} p_i \in S$ for finite $J \subset I$ increases towards the identity. Suppose $\{w_i\}$ in $F \cap rB$ converges σ-strongly to $w \in W$. Then $w_i^{1/2} q_J w_i^{1/2} \to w^{1/2} g_J w^{1/2}$ σ-strongly, and the convergence happens in $\cup_{p\in S} pWp$ because the latter is an ideal in W. Indeed, if v belongs there and $u \in W$, then the range projections of vu and $(vu)^*$ belong to S, and so does their supremum q, and clearly $q(vu)q = vu$. Similarly $uv \in \cup_{p\in S} pWp$. Now $w_i^{1/2} q_J w_i^{1/2} \in F$, so the previous lemma tells us that $w^{1/2} g_J w^{1/2} \in F$, or $x(w^{1/2} g_J w^{1/2}) \leq 1$. Complete additivity then gives $x(w) = x(\lim_J w^{1/2} q_J w^{1/2}) = \lim_J x(w^{1/2} q_J w^{1/2}) \leq 1$ and $w \in F$. As $\{w \in W_+ \,|\, x(w) \leq s\} = sF$ for $s > 0$, we thus see that x is σ-weakly lower semicontinuous. □

Let A be an ordered locally convex topological vector space over $\mathbb{R}$ with positive cone A_+ such that $A = A_+ - A_+$. Then the dual space $A^\star$ is an ordered vector space with positive cone $(A^\star)_+$ consisting of all $x \in A^\star$ such that $x(A_+) \subset [0, \infty\rangle$. For any subset F of A, let $F^\vee = \{x \in A^\star \,|\, x(F) \subset \langle -\infty, 1]\}$ and $F^\wedge = F^\vee \cap (A^\star)_+$. For a subset E of $A^\star$ we define the subsets $E^\vee$ and $E^\wedge$ of A analogously. We say a subset F of A_+ is *hereditary* if $0 \leq x \leq y$ implies $x \in F$ whenever $y \in F$.

Lemma 10.2.10 *Let A be an ordered locally convex topological real vector space with positive cone A_+ such that $A = A_+ - A_+$. Then the following are equivalent:*

(i) *We have $F = \overline{(F - A_+)} \cap A_+$ for every hereditary convex closed $F \subset A_+$.*
(ii) *We have $F^{\wedge\wedge} = F$ for every hereditary convex closed $F \subset A_+$.*
(iii) *Any additive increasing lower semicontinuous function $x \colon A_+ \to [0, \infty]$ such that $x(\lambda(\cdot)) = \lambda x(\cdot)$ for $\lambda > 0$ is of the form $x = \sup\{y \in (A^\star)_+ \mid y \leq x\}$.*

Proof To show $(i) \Rightarrow (ii)$, note that $F - A_+$ is the convex hull of $F \cup (-A_+)$, so $(F - A_+)^\vee = (F \cup (-A_+))^\vee = F^\vee \cap (A^\star)_+ = F^\wedge$. The Hahn-Banach separation theorem then yields

$$F = \overline{(F - A_+)} \cap A_+ = (F - A_+)^{\vee\vee} \cap A_+ = (F^\wedge)^\vee \cap A_+ = F^{\wedge\wedge}.$$

To show $(ii) \Rightarrow (iii)$, let x satisfy the assumptions of (iii). Consider the hereditary convex closed subset $F = \{a \in A_+ \,|x(a) \leq 1\}$ of A_+. Let $x'(a) = \sup\{y(a) \,|\, y \in F^\wedge\}$ for $a \in A_+$. Clearly $x' \leq x$ on A_+, and if strict inequality holds for some $b \in A_+$, by rescaling b we may assume that $x'(b) < 1 < x(b)$, so $b \notin F$, and yet

$$F = F^{\wedge\wedge} = \{a \in A_+ \,|y(a) \leq 1 \text{ for } y \in F^\wedge\} = \{a \in A_+ \,|\, x'(a) \leq 1\}.$$

Hence x equals x' which is of the desired form.

As for $(iii) \Rightarrow (i)$, let F be a hereditary convex closed subset of A_+. Then $x \colon A_+ \to [0, \infty]$ given by $x(a) = \inf\{r > 0 \,|\, a \in rF\}$ for $a \in A_+$ satisfies the assumptions of (iii). Since F is closed it equals $\{a \in A_+ \,|\, x(a) \leq 1\}$. Now letting

$H = \{y \in (A_+)^\star \mid y \leq x \text{ on } A_+\}$, we get $F = H^\vee \cap A_+$, and $H = F^\vee \cap (A_+)^\star = (F - A_+)^\vee$ so $H^\vee = \overline{F - A_+}$. □

Proof (Completing the Proof of the Previous Theorem) Clearly it suffices to show that the self-adjoint part A of the von Neumann algebra W with the σ-weak topology satisfies (i) of the previous lemma. Let F be a σ-strongly closed hereditary convex subset of A_+, and consider h_c from the proof of Lemma 10.2.6. Let $G = \{w \in A \mid h_c(w) \in F - A_+,\ 0 < c < c_w\}$ with $c_w = \sup\{c > 0 \mid w \geq -1/c\}$. For $r > 0$ consider a net of $w_i \in G \cap rB$ that converges σ-strongly to $w \in A$. By assumption $h_c(w_i) \in F - A_+$ for $0 < c < 1/2r$, so there are $v_i \in F$ such that $h_c(w_i) \leq v_i$. Thus $h_{2c}(w_i) = h_c(h_c(w_i)) \leq h_c(v_i) \leq 1/c$ and $h_{2c}(w_i) \to h_{2c}(w)$ σ-strongly. By going to a subnet we may also assume that the bounded net $\{h_c(v_i)\}$ converges σ-weakly to $v \in A$. Since $0 \leq h_c(v_i) \leq v_i$ we have $h_c(v_i) \in F$ as F is hereditary, and thus $v \in F$ as F is σ-weakly closed. Now $v - h_{2c}(w) = \lim(h_c(v_i) - h_{2c}(w_i)) \geq 0$, so $h_{2c}(w) \in F - A_+$, or $h_c(w) \in F - A_+$ for $0 < c < 1/r$. Since $h_{c_1} \geq h_{c_2}$ for $0 < c_1 < c_2$, we thus see that $h_b(w) \in F - A_+$ when $0 < b < c_w$. Hence $w \in G$ and G is σ-strongly closed by the Krein-Smulian theorem.

We claim that $G \cap rB = \overline{((F - A_+) \cap sB)} \cap rB$ for $s > r$, where bar means σ-strong closure. If $w \in G \cap rB$, then $h_c(w) \in F - A_+$ for $0 < c < c_w$. For sufficiently small $c > 0$ we have $h_c(w) \in sB$, and as $\{h_c(w)\}$ increases towards w as c decreases towards zero, we get $w \in \overline{((F - A_+) \cap sB)}$ for $s > r$. On the other hand, since $h_c(w) \leq w$ for $0 < c < c_w$, we have $G \supset F - A_+$, so $G \cap sB \supset (F - A_+) \cap sB$ and $G \cap sB \supset \overline{(F - A_+) \cap sB}$ by the previous paragraph. This proves our claim. Hence $G \cap rB$ is convex, and therefore so is G, which together with the previous paragraph also shows that G is σ-weakly closed. Since $F - A_+ \subset G$ and as G clearly sits inside the σ-weak closure of $F - A_+$, we get $G = \overline{F - A_+}$, where bar now means σ-weak closure. If $w \in G \cap A_+$, then as $0 \leq h_c(w) \leq w$ we get $h_c(w) \in F - A_+$, so $h_c(w) \in F$ by heredity of F. Hence $w = \lim h_c(w) \in F$ and $F = G \cap A_+$. □

Let x be a normal weight on a von Neumann algebra W, and consider its associated semicyclic representation π_x on V_x with quotient map $q_x : N_x \to V_x$. Let E_x consist of all $cy \in W_*$ such that $0 \leq y \leq x$ on W_+ and $c \geq 0$. Let $y \in E_x$. Then there exists a unique positive $h_y \in \pi_x(W)'$ such that $(h_y q_x(w)|q_x(w)) = y(w^*w)$ for $w \in N_x$. Let π_y be the representation of W on V_y with cyclic vector v_y, and let t_y be the unique bounded operator from V_x to V_y such that $t_y(q_x(w)) = \pi_y(w)(v_y)$ for $w \in N_x$. One checks that $h_y = t_y^* t_y$, so the polar decomposition of t_y is $u_y h_y^{1/2}$. As $t_y \pi_x(w) = \pi_y(w) t_y$ for $w \in W$, the same holds for u_y. Let $w_y = u_y^*(v_y)$, so $\pi_x(v)(w_y) = h_y^{1/2} q_y(v)$ for $v \in N_x$. The *opposite weight* x' of x is defined as follows.

Proposition 10.2.11 *Let x be a normal weight on a von Neumann algebra W with notation as in the previous paragraph. Letting $x'(w) = \|y\|$ when $w = h_y$, and $x'(w) = \infty$ otherwise, we get a faithful semifinite normal weight x' on $\pi_x(W)'$.*

Proof The collection of all h_y is evidently a convex subcone F of $(\pi_x(W)')_+$. If $0 \leq w \leq h_y$, then by Lemma 10.2.3 there is $u \in \pi_x(W)'$ with $\|u\| \leq 1$ such that $w^{1/2} = uh_y^{1/2}$. Setting $z = (\pi_x(\cdot)u(v_y)|u(v_y))$ on W, we get $z(a^*a) = (w(q_x(a))|q_x(a))$ for $a \in N_x$, so $0 \leq z \leq y$. Hence F is hereditary and $x'(w) = \|z\| \leq x'(h_y)$. By construction x' is faithful.

Say $w = \sum w_i$ σ-strongly in $(\pi_x(W)')_+$. If $\sum x'(w_i) < \infty$, each $w_i = h_{y_i}$ for some $y_i \in E_x$, and $\sum \|y_i\| < \infty$, so $y = \sum y_i \in (W_*)_+$ in norm. But $y(v^*v) = (wq_x(v)|q_x(v)) \leq \|w\|x(v^*v)$ for $v \in N_x$, so $y \in E_x$ and $h_y = w$. Hence $x'(w) = \|y\| = \sum \|y_i\| = \sum x'(w_i)$. On the other hand, if $x'(w) < \infty$, then $w = h_y$ for some $y \in E_x$, and $w_i \in F$ as F is hereditary. But $\sum x'(w_i) \leq x'(w)$, so $x'(w) = \sum x'(w_i)$ by the previous argument. It remains to show semifiniteness. As F is the positive part of $M_{x'}$, we see that $\|q_x(w)\|^2 = \sup\{\|vq_x(w)\|^2 \mid v \in N_{x'}$ with $0 \leq v \leq 1\}$. Hence the open unit ball of $(N_{x'} \cap N_{x'}^*)_+$ is upward filtered and converges σ-strongly to the identity map on V_x. As $b\pi_x(W)'b \subset N_{x'} \cap N_{x'}^*$ for $b \in N_{x'}$, we see that $\pi_x(W)'$ is the σ-strong closure of $N_{x'} \cap N_{x'}^*$. □

Given a normal weight x on a von Neumann algebra W, let $p \in W$ be the orthogonal projection such that Wp is the σ-strong closure of N_x. Also, let $q \in W$ be the orthogonal projection such that $Wq = \{w \in W \mid x(w^*w) = 0\}$. Then x is semifinite on pWp and faithful on $(1-q)W(1-q)$. The *support* $s(x)$ of x is $p-q$.

Proposition 10.2.12 *Every von Neumann algebra admits a faithful semifinite normal weight.*

Proof Consider a von Neumann algebra W. By Zorn's lemma there is a maximal family of normal positive linear functionals x_i on W with pairwise orthogonal support projections $s(x_i)$. By maximality $\sum s(x_i) = 1$. Set $x(w) = \sum x_i(w)$ for $w \in W_+$. This defines a normal weight x that is semifinite since any finite sum of $s(x_i)$ belongs to the positive part of M_x. If $x(w) = 0$, then all $x_i(w) = 0$, so $w^{1/2}s(x_i) = 0$ and $w^{1/2} = \sum w^{1/2}s(x_i) = 0$. □

10.3 Weights and Left Hilbert Algebras

Lemma 10.3.1 *Let A be a full left Hilbert algebra with left von Neumann algebra W and set B of left bounded vectors. The positive part $(M_l)_+$ of $M_l \equiv \pi_l(B)^*\pi_l(B)$ is a hereditary convex subcone of W_+, and $\pi_l(B) = \{w \in W \mid w^*w \in (M_l)_+\}$.*

Proof Clearly $(M_l)_+$ is a convex subcone of W_+. Suppose the finite sum $a = \sum w_i^* v_i$ with $v_i, w_i \in \pi_l(B)$ dominates $b \in W_+$. Since $a = (a + a^*)/2 \leq \sum(w_i^* w_i + v_i^* v_i) \in (M_l)_+$, we may assume $a = \sum w_i^* w_i$. Say $w_i = \pi_l(u_i)$ for $u_i \in B$. By Lemma 10.2.3 there are $s_i \in W$ with $\|s_i\| \leq 1$ such that $w_i^{1/2} = s_i a^{1/2}$ and with $p = \sum s_i^* s_i$ the range projection of a. Set $u = \sum s_i^*(u_i)$. Then $u \in B$ and $\pi_l(u) = a^{1/2}$. Again by Lemma 10.2.3 pick $s \in W$ such that $b^{1/2} = sa^{1/2}$. Then $b^{1/2} = \pi_l(su) \in \pi_l(B)$, so $(M_l)_+$ is hereditary.

If $w \in \pi_l(B)$, then $w^*w \in (M_l)_+$. Conversely, if $w^*w \in (M_l)_+$, then $|w| = \pi_l(v)$ for some $v \in B$ by the previous paragraph. By the polar decomposition of w, we thus get $w \in \pi_l(B)$. □

The same conclusion clearly holds with W, B and the index l replaced by W', B' and r. The following result is proved along the same lines as Lemma 10.2.4.

Lemma 10.3.2 *Let A be a full left Hilbert algebra with left von Neumann algebra W. Let $(W')_1$ and $(W')_1^o$ denote the closed and open unit balls of W'. Then there exists a completely positive map $f\colon M_r \to W_*$ such that $f(\pi_r(v)^*\pi_r(v)) = ((\cdot)v|v)$ for $v \in B'$, and $f((M_r)_+ \cap (W')_1) = \{((\cdot)v|v) \mid v \in B',\ \|\pi_r(v)\| \leq 1\}$ and $f((M_r)_+ \cap (W')_1^o) = \{((\cdot)v|v) \mid v \in B',\ \|\pi_r(v)\| < 1\}$ with $M_r = \pi_r(B')^*\pi_r(B')$.*

The same conclusion holds with W and B replaced by W' and B', and with r and l swopped. Note that for instance $K_l \equiv \{((\cdot)v|v) \mid v \in B',\ \|\pi_r(v)\| < 1\}$ is a hereditary convex subset of the positive part of W_*. Also recall that $\pi_l(A) = \pi_l(B) \cap \pi_l(B)^*$ and $\pi_r(A') = \pi_r(B') \cap \pi_r(B')^*$ for a full left Hilbert algebra A.

Lemma 10.3.3 *Let A be a full left Hilbert algebra with left von Neumann algebra W. Define $x_l\colon W_+ \to [0,\infty]$ by letting $x_l(w) = \|v\|^2$ if $w^{1/2} = \pi_l(v)$ and otherwise $x_l(w) = \infty$. Then $x_l = \sup\{y \mid y \in K_l\}$ pointwise with K_l as above.*

Proof Let $x = \sup\{y \mid y \in K_l\}$ pointwise. Then x is clearly a normal weight on W. By Lemma 10.3.1 and $\pi_l(A) = \pi_l(B) \cap \pi_l(B)^*$, for $w \in (M_l)_+$ there is $v \in A$ such that $w^{1/2} = \pi_l(v)$. Since $(M_l)_+ \cap (W')_1^o$ is upward filtered and converges strongly to the identity, the previous lemma yields

$$x(w) = \sup\{\|\pi_r(u)(v)\|^2 \mid u \in B',\ \|\pi_r(u)\| < 1\} = \|v\|^2 = x_l(w).$$

Suppose on the other hand that $x(w) < \infty$ for $w \in W_+$. Consider the positive linear functional $y = f(\cdot)(w)$ on M_r, where f is as in the previous lemma. As $\|y\| = x(w)$, we can extend y to the norm closure of M_r. Denote the positive linear functional on this C*-algebra also by y. By the Cauchy-Schwarz inequality for y we have $|y(\pi_r(v))| \leq \|y\|^{1/2}y(\pi_r(v)^*\pi_r(v))^{1/2} = \|y\|^{1/2}\|w^{1/2}(v)\|$ for $v \in B'$, where in the last equality we used the previous lemma. By the Riesz representation theorem there is u in the closed image V of $w^{1/2}$ such that $y(\pi_r(v')) = (w^{1/2}(v')|u)$ for $v' \in B'$. Hence $\overline{(w^{1/2}(v')|w^{1/2}(v))} = (w^{1/2}(v')|\pi_r(v)(u))$ for $v, v' \in B'$. Since $\pi_r(v)(u) \in \pi_r(v)(\overline{w^{1/2}(V)}) \subset \overline{w^{1/2}(V)}$, we conclude that $w^{1/2}(v) = \pi_r(v)(u)$ for $v \in B'$. So u is left bounded and $w^{1/2} = \pi_l(u) \in \pi_l(B)$. Hence $w \in (M_l)_+$ and $x_l(w) = \|u\|^2 < \infty$. The first part of the proof then shows that $x_l = x$. □

The same conclusion holds with W, B' and l replaced by W', B and r, respectively.

Theorem 10.3.4 *Let A be a full left Hilbert algebra with left von Neumann algebra W and Hilbert space completion V. Then the normal weights x_l and x_r on W and W', respectively, are faithful and semifinite. Let π_{x_l} be the semicyclic representation of x_l with quotient map q_l. Then $U(v) = q_l(\pi_l(v))$ for $v \in B$ defines a unique isometric bijection $U\colon V \to V_{x_l}$ that intertwines π_{x_l} and the representation given* *by the action of W on V. It identifies x_r with the opposite weight of x_l. Also $N_{x_l} = \pi_l(B)$ and $x_l(\pi_l(v)^*\pi_l(w)) = (w|v)$ for $v, w \in B$, and similarly for r.*

Proof Semifiniteness of x_l is clear as the von Neumann algebra generated by $(M_l)_+$ is W, while normality was already proved in the previous lemma. By the definition of x_l and the polarization identity, we get $(q_l(\pi_l(w))|q_l(\pi_l(v))) = x_l(\pi_l(v)^*\pi_l(w)) = (w|v)$ for $v, w \in B$. So U extends by continuity to a surjective isometry. Upon identifying V and V_{x_l} the maps π_l and q_l are inverses of each other. The relation to the opposite weight is clear from Lemma 10.3.2. □

In the reverse direction we have the following result.

Theorem 10.3.5 *Let x be a faithful semifinite normal weight on a von Neumann algebra W, and let π_x be the associated semicyclic representation on V_x with quotient map $q_x\colon N_x \to V_x$. Then $A_x = q_x(N_x \cap N_x^*)$ is a full left Hilbert algebra with Hilbert space completion V_x, product $q_x(w_1)q_x(w_2) = q_x(w_1w_2)$ and involution $q_x(w_1)^* = q_x(w_1^*)$ for $w_i \in N_x \cap N_x^*$. Its left von Neumann algebra $\pi_l(A_x)''$ is $\pi_x(W)$, and $x = x_l\pi_x$, where x_l is the weight associated to A_x.*

Proof As $\pi_l(q_x(w)) = \pi_x(w)$ for $w \in N_x$, we get $\pi_l(A_x) = \pi_x(N_x \cap N_x^*)$ and $\pi_l(A_x)'' = \pi_x(W)$. As $\pi_l(A_x)$ acts non-degenerately on V_x and has an increasing net converging strongly to the identity, we know that A_x^2 is dense in V_x. We show that the involution on A_x is closable. Let $K = \{y \in (W_*)_+ \mid (1+\varepsilon)y \leq x$ for some $\varepsilon > 0\}$. Arguing as we did when we defined the opposite weight in the previous section, we see that to $y \in K$ there is h_y in the positive part of $\pi_x(W)'$ with $\|h_y\| < 1$ and $w_y \in V_x$ with $\pi_x(v)(w_y) = h_y^{1/2}(q_x(v))$ for $v \in N_x$, such that $y = ((\cdot)w_y|w_y)$ on W. For $y_i \in K$ and $u \in \pi_x(W)'$ and $v \in N_x \cap N_x^*$ we thus get $(q_x(v^*)|h_{y_1}^{1/2}uw_{y_2}) = (h_{y_2}^{1/2}u^*w_{y_1}|q_x(v))$. Hence if we have a sequence of elements $v_n \in N_x \cap N_x^*$ such that $q_x(v_n) \to 0$ and $q_x(v_n^*) \to v \in V$ in norm, then $(v|h_{y_1}^{1/2}uw_{y_2}) = 0$. So to show $v = 0$, it suffices to show that the collection of all $h_{y_1}^{1/2}uw_{y_2}$ is dense in V_x. By Haagerup's theorem, we have

$$\|q_x(w)\|^2 = \sup\{y(w^*w) \mid y \in K\} = \sup\{(h_y(q_x(w))|q_x(w)) \mid y \in K\}$$

for $w \in N_x \cap N_x^*$, so since K is convex, we see that the identity is in the strong closure of $\{h_y \mid y \in K\}$. Thus is suffices to show that the closed span R of all uw_{y_2} is V_x. Let p be the orthogonal projection onto R, so $p \in \pi_x(W)$ and $(1-p)(w_y) = 0$ for $y \in K$. So there is an orthogonal projection $q \in W$ such that $\pi_x(q) = 1 - p$. Then $x(q) = \sup\{y(q) \mid y \in K\} = \sup\{((1-p)(w_y)|w_y) \mid y \in K\} = 0$ and $q = 0$. To see that A_x is full, it follows from the above that w_y corresponding to $y \in K$ is right bounded and $\pi_r(w_y) = h_y^{1/2}$. Hence $w_y \in (A_x)'$ by Lemma 10.1.6. For any left bounded $v \in V_x$ we have

$$x(\pi_l(v)^*\pi_l(v)) = \sup\{y(\pi_l(v)^*\pi_l(v)) \mid y \in K\} = \sup\{\|h_y^{1/2}q_x(v)\| \mid y \in K\} = \|v\|^2$$

as the identity is in the strong closure of $\{h_y \mid y \in K\}$. So $\pi_l(B_x) \subset \pi_x(N_x)$, where B_x are the left bounded vectors in V_x. Hence

$$\pi_l(A_x) \subset \pi_l((A_x)'') = \pi_l(B_x) \cap \pi_l(B_x)^* \subset \pi_x(N_x \cap N_x^*) = \pi_l(A_x),$$

which means that $(A_x)'' = A_x$. Finally, if $u|w|$ is the polar decomposition of $w \in N_x$, then $|w| = u^*w \in N_x \cap N_x^*$, so $q_x(w) = \pi_x(u)q_x(|w|) \subset \pi_x(u)(A_x) \subset B_x$ and $\pi_l(q_x(w)) = \pi_x(w)$. Hence

$$x(w^*w) = \|q_x(w)\|^2 = x_l(\pi_l(q_x(w))^*\pi_l(q_x(w))) = x_l(\pi_x(w)^*\pi_x(w)),$$

so $x = x_l\pi_x$. □

10.4 Weights on C*-Algebras

Weights on C-algebras* are defined just as for von Neumann algebras.

Proposition 10.4.1 *Let x be a weight on a C*-algebra W that is lower semi-continuous with respect to the norm-topology. Then the associated semicyclic representation π_x is non-degenerate, and $x = \sup\{y \in W^\star \mid 0 \leq y \leq x\}$ pointwise.*

Proof Let $\{u_i\}$ be the canonical approximate unit for W. For $w \in N_x$, we then have

$$\lim \|q_x(w) - \pi_x(u_i)(q_x(w))\|^2 = \lim(x(w^*w) - 2x(w^*u_iw) + x(w^*u_i^*u_iw)) = 0$$

by lower semicontinuity, so π_x is non-degenerate.

As for the second statement, we show that the self-adjoint part A of W with norm-topology satisfies (i) in Lemma 10.2.10. Let F be a hereditary convex norm closed subset of A_+. Let A'' be the enveloping von Neumann algebra of A, and let $\tilde{E}$ denote the σ-weak closure in A'' of any subset E of A. So $\tilde{F} \cap A = F$ by the Hahn-Banach separation theorem. From the proof of Haagerup's theorem we know that $\tilde{F} = (\tilde{F} - \tilde{A}_+)^{\sim} \cap \tilde{A}_+$ provided $\tilde{F}$ is hereditary, and in this case $F \subset \overline{(F - A_+)} \cap A_+ \subset \tilde{F} \cap A = F$. To show that $\tilde{F}$ is hereditary, let $0 \leq v \leq w \in \tilde{F}$. As usual we may pick $u \in \tilde{A}$ with $\|u\| \leq 1$ such that $v^{1/2} = uw^{1/2}$. Pick elements u_i in the unit ball of A and $w_i \in F$ such that $u_i \to u$ and $w_i \to w$ σ-weakly. The boundedness of $\{u_i\}$ thus implies that $u_iw_i^{1/2} \to uw^{1/2} = v^{1/2}$ σ-strongly. So $w_i^{1/2}u_i^*u_iw_i^{1/2} \to v$ σ-weakly, and $w_i^{1/2}u_i^*u_iw_i^{1/2} \in F$ by heredity of F as it is majorized by w_i. Hence $v \in \tilde{F}$. □

Let x be a lower semicontinuous weight on a C*-algebra W. As in the previous section one shows that $K = \{y \in (W^\star)_+ \mid (1 + \varepsilon)y \leq x \text{ for some } \varepsilon > 0\}$ is upward filtered and $x = \sup K$. Now any positive functional y on a C*-algebra is a vector functional $\tilde{y}$ on its (universal) enveloping von Neumann algebra. Hence x can be extended to a normal weight $\tilde{x}$ on the enveloping von Neumann algebra W''

by letting $\tilde{x} = \sup\{\tilde{y} \mid y \in K\}$. Viewing $\tilde{x}$ as a normal weight on $\pi_{\tilde{x}}(W'')$, we can construct its opposite weight on the commutant, which is then faithful, semifinite and normal.

10.5 The Modular Automorphism

Definition 10.5.1 A lower semicontinuous weight x on a C*-algebra W satisfies the *KMS-condition* for a one parameter group $\{\sigma_t\}$ of automorphisms on W if $x = x\sigma_t$, and if for every $v, w \in N_x \cap N_x^*$ there is bounded continuous function f on the closed horizontal strip H bounded by $\mathbb{R}$ and $\mathbb{R}+i$ that is holomorphic on the interior H^o and satisfies $f(t) = x(\sigma_t(v)w)$ and $f(t+i) = x(w\sigma_t(v))$ for $t \in \mathbb{R}$.

Lemma 10.5.2 *Let T be an invertible positive self-adjoint operator on a Hilbert space V, let $a \in \mathbb{R}$ and let H_a denote the closed horizontal strip in the complex plane bounded by $\mathbb{R}$ and $\mathbb{R} - ia$. If $v \in D(T^a)$ the function $\mathbb{R} \to V$ sending $t \in \mathbb{R}$ to $T^{it}(v)$ extends to a bounded continuous function $H_a \to V$ that is holomorphic on H_a^o. Moreover, any $v \in V$ satisfying this conclusion must belong to $D(T^a)$.*

Proof Turning to T^{-1} if needed, we may assume $a > 0$. If $z = t - is \in H_a^o$ with $s, t \in \mathbb{R}$, then $D(T^{iz}) = D(T^s)$. The inequality $\|T^{iz}(v)\| \leq \|(I+T)^a(v)\|$ shows that $g(z) = T^{iz}(v)$ defines a bounded continuous function g on H_a. Write $T = \int_0^\infty \lambda\, dP(\lambda)$. For $w \in M \equiv \cup_{n=1}^\infty (P(n) - P(1/n))(V)$, the function $h\colon \mathbb{C} \to V$ given by $h = T^{i(\cdot)}(w)$ is holomorphic as $(T^{i(\cdot)}(w)|u) = \int_{1/n}^n \lambda^{i(\cdot)}\, d(P(\lambda)(w)|u)$ for $u \in V$ and sufficiently large n. As $(g(z)|w) = \overline{(h(-iz)|v)}$, we see that $(g(\cdot)|w)$ is holomorphic on H_a^o, and remains so for every $w \in V$ as M is dense in V.

For the converse, extend f given by $f(t) = T^{it}(v)$ to H_a as prescribed. By the paragraph above each $w \in D(T^a)$ gives a bounded continuous function $z \mapsto T^{iz}(w)$ on H_a that is holomorphic on H_a^o. The functions $z \mapsto (f(z)|w)$ and $z \mapsto (v|T^{-i\bar{z}}(w))$ agree on $\mathbb{R}$ and hence on H_a, so $(f(-ia)|w) = (v|T^a(w))$ and $v \in D(T^a)$ with $T^a(v) = f(-ia)$. □

Theorem 10.5.3 *Every faithful semifinite normal weight x on a von Neumann algebra W satisfies the KMS-condition for a unique one parameter group $\{\sigma_t\}$ of automorphisms on W called the* modular automorphism group *associated with x.*

Proof Let π_x be the semicyclic representation associated to x, and let $A = q_x(N_x \cap N_x^*)$ be the full left Hilbert algebra corresponding to x, and identify W with $\pi_x(W) = \pi_l(A)''$. Using the modular operator Δ of A we get a one parameter group $\{\sigma_t\}$ of automorphisms on W given by $\sigma_t(w) = \Delta^{it} w \Delta^{-it}$. Since Δ^{it} leaves A invariant, it leaves $N_x \cap N_x^*$ and hence the definition domain M_x, invariant. Moreover, for $v_i \in A$ we have

$$x(\sigma_t((\pi_l(v_1)^*\pi_l(v_2))) = x(\pi_l(\Delta^{it}(v_1))^*\pi_l(\Delta^{it}(v_2))) = (\Delta^{it}(v_2)|\Delta^{it}(v_1)) = (v_2|v_1),$$

which equals $x((\pi_l(v_1)^*\pi_l(v_2))$, so $x\sigma_t = x$. Also, since $\Delta^{-i(\cdot)/2}(v_i)$ by the lemma are bounded continuous functions on H that are holomorphic on H^o, so is the function f given by $f(z) = (\Delta^{-iz/2}(v_1)|\Delta^{i\bar{z}/2}(v_2))$, and

$$f(t) = (v_1|\Delta^{it}(v_2)) = x(\pi_l(\Delta^{it}(v_2))^*\pi_l(v_1)) = x(\sigma_t(\pi_l(v_2))\pi_l(v_1))$$

while

$$f(t+i) = (J\Delta^{1/2}\Delta^{it}(v_2)|J\Delta^{1/2}(v_1)) = (S(\Delta^{it}(v_2))|S(v_1)) = x(\pi_l(v_1)\pi_l(\Delta^{it}(v_2))^*)$$

which equals $x(\pi_l(v_1)\sigma_t(\pi_l(v_2)))$ for $t \in \mathbb{R}$.

Suppose x satisfies the KMS-condition for another one parameter group $\{\beta_t\}$ of automorphisms on W. Since $x\beta_t = x$ the formula $U_t(q_x(w)) = q_x(\beta_t(w))$ for $w \in N_x$ and $t \in \mathbb{R}$, defines a one parameter unitary group $\{U_t\}$ on V_x that is strongly continuous by continuity of the corresponding f's in the KMS-condition. By Stone's theorem there is a self-adjoint operator T in V_x such that $U_t = e^{itT}$.

Since β_t are $*$-preserving, we have $SU_t = U_t S$. Hence $\|\Delta^{1/2}(v)\| = \|U_t S(v)\| = \|\Delta^{1/2}U_t(v)\|$ for $v \in D(S)$, so $\Delta U_t = U_t\Delta$ and $JU_t = U_t J$.

For $v, w \in D(S)$ there are $v_n, w_n \in A$ such that $v_n \to v$ and $w_n \to w$ in graph norm. Let f_n be the corresponding functions on H such that $f_n(t) = x(\beta_t(\pi_l(v_n)^*)\pi_l(w_n)) = (w_n|U_t(v_n))$ and $f_n(t+i) = x(\pi_l(w_n)\beta_t(\pi_l(v_n)^*))$ which equals $(U_t(S(v_n))|S(w_n))$. Thus $f_n(t) \to (w|U_t(v))$ and $f_n(t+i) \to (U_t(S(v))|S(w))$ uniformly in $t \in \mathbb{R}$. By the Phragmen-Lindelöf theorem they converge uniformly on H to a bounded continuous function g that is holomorphic in the interior and that satisfies $g(t) = (w|U_t(v))$ and $g(t+i) = (U_t(S(v))|S(w)) = (\Delta^{1/2}(v)|\Delta^{1/2}U_t(w))$ for $t \in \mathbb{R}$.

Write $T = \int \lambda\, dP(\lambda)$ and $P_n = P(n) - P(-n)$. Then $P_n(D(S)) \subset D(S)$ and $D = \cup_{n=1}^{\infty} P_n(D(S))$ is a core for $\Delta^{1/2}$ as $\lim(I + \Delta^{1/2})P_n(w) = \lim P_n(I + \Delta^{1/2})(w)$ for $w \in D(S)$. If $v \in D$, then $g(t) = \int_{-n}^{n} e^{-i\lambda t}\, d(w|P(\lambda)(v))$ for sufficiently large n. Hence g is holomorphic and $(\Delta^{1/2}(v)|\Delta^{1/2}U_t(w)) = g(t+i) = (w|e^T U_t(v))$. Setting $t = 0$ we get $(\Delta^{1/2}(v)|\Delta^{1/2}(w)) = (w|e^T(v))$, so $\Delta^{1/2}(D) \subset D(S)$ and $e^T(v) = \Delta(v)$ for $v \in D$. Now $(1 + e^T)(D) = \cup_{n=1}^{\infty}(1 + e^T)P_n(D(S))$ and $(1 + e^T)P_n(D(S))$ is dense in $P_n(V_x)$, so D is dense in $D(e^T)$ with respect to the graph norm. Thus D is a common core for e^T and Δ, where they coincide, so $e^T = \Delta$. Hence $q_x(\beta_t(w)) = U_t(q_x(w)) = \Delta^{it}(q_x(w)) = q_x(\sigma_t(w))$ for $w \in N_x$, and $\beta_t = \sigma_t$ for $t \in \mathbb{R}$ as N_x generates W as a von Neumann algebra. □

By uniqueness of the modular automorphism group associated to a weight, we immediately see that whenever $g\colon U \to W$ is an isomorphism of von Neumann algebras, and x is a faithful semifinite normal weight on W with modular automorphism group $\{\sigma_t\}$, then the modular automorphism group on U associated to the weight xg is $\{g^{-1} \circ \sigma_t \circ g\}$.

Proposition 10.5.4 *Let x be a faithful semifinite lower semicontinuous weight on a C*-algebra W satisfying the KMS-condition for a one parameter automorphism group $\{\sigma_t\}$. Then the corresponding normal weight $\tilde{x}$ on $\pi_x(W)''$ considered in the* *previous section is faithful and semifinite, and its modular automorphism group* $\{\beta_t\}$ *satisfies* $\beta_t \circ \pi_x = \pi_x \circ \sigma_t$ *for* $t \in \mathbb{R}$.

Proof We claim that $A = q_x(N_x \cap N_x^*)$ is a full left Hilbert algebra with the usual product, involution and inner product. Indeed, suppose we have a sequence $\{v_n\}$ in $N_x \cap N_x^*$ such that $x(v_n^* v_n) \to 0$ and $x((v_n - v_m)(v_n - v_m)^*) \to 0$ and $\lim q_x(v_n^*) = v \in V_x$. For $w \in N_x \cap N_x^*$ there are bounded continuous functions f_n on H that are holomorphic in H^o and satisfy $f_n(t) = x(\sigma_t(w^*)v_n^*)$ and $f_n(t+i) = x(v_n^*\sigma_t(w^*))$ for $t \in \mathbb{R}$. Now $f_n(t) \to (v|q_x(\sigma_t(w)))$ while by the Cauchy-Schwarz inequality $f_n(t+i) \to 0$, both uniformly in $t \in \mathbb{R}$. Hence $\{f_n\}$ converges uniformly on H to a bounded continuous function f on H that is holomorphic on H^o and satisfies $f(t+i) = 0$ for $t \in \mathbb{R}$. So $f = 0$ and $(v|q_x(\sigma_t(w))) = 0$ and $v = 0$. Thus the involution is closable, and A is a full left Hilbert algebra. Now $\pi_l(A)'' = W$ and the associated weight of A is $\tilde{x}$. The identity $\beta_t \circ \pi_x = \pi_x \circ \sigma_t$ for $t \in \mathbb{R}$ follows from the uniqueness argument in the previous theorem. □

Let W_i be von Neumann algebras with faithful semifinite normal weights x_i and associated full left Hilbert algebras A_i with involutions S_i and Hilbert space completions V_i. Then the algebraic tensor product $A_1 \otimes A_2$ is a full left Hilbert algebra with involution $S_1 \otimes S_2$ in $V_1 \otimes V_2$. The faithful semifinite normal weight $x_1 \otimes x_2$ on $W_1 \bar{\otimes} W_2$ associated to this left Hilbert algebra is the *tensor product* of the weights x_i. Note that $\pi_l(A_1 \otimes A_2)'' = \pi_l(A_1)'' \bar{\otimes} \pi_l(A_2)''$. By Proposition 9.3.26 we see that the modular automorphism group of $x_1 \otimes x_2$ is $\{\sigma_t^1 \otimes \sigma_t^2\}$, where $\{\sigma_t^i\}$ is the modular automorphism group of x_i.

In the non-faithful case consider the support projections $s(x_i)$ of x_i, and form the semifinite normal weight $x_1 \otimes x_2$ on $W_1 \bar{\otimes} W_2$ with support $s(x_1) \otimes s(x_2)$ such that it on $(s(x_1) \otimes s(x_2))(W_1 \bar{\otimes} W_2)(s(x_1) \otimes s(x_2))$ is the tensor product of the faithful semifinite normal weights x_i on $s(x_i)W_i s(x_i)$.

10.6 Centralizers of Weights

Definition 10.6.1 Let x be a faithful semifinite normal weight on a von Neumann algebra W with associated modular automorphism group $\{\sigma_t\}$. The von Neumann subalgebra $W_x \equiv \{w \in W \mid \sigma_t(w) = w, \ \forall t \in \mathbb{R}\}$ is the *centralizer* of x.

A one parameter group $\{\sigma_t\}$ of automorphism on a von Neumann algebra W is *continuous* if $t \mapsto y(\sigma_t(w))$ is continuous for all $w \in W$ and $y \in W_*$. We say w is *entire* if $t \mapsto \sigma_t(w)$ can be extended to a holomorphic function on all of $\mathbb{C}$ in the Banach space sense. Its value at $z \in \mathbb{C}$ is denoted by $\sigma_z(w)$, and the set of all entire elements is denoted by W_e.

Lemma 10.6.2 *Let B be a compact convex subset of a locally convex topological vector space A. Suppose $f : \mathbb{R} \to A$ is continuous and that* $\operatorname{im} f \subset B$. *Then* $f_r = (r/\pi)^{1/2} \int e^{-rt^2} f(t)\, dt \in B$ *for* $r > 0$ *and* $\lim_{r\to\infty} f_r \to f(0)$.

Proof As $(r/\pi)^{1/2}\int e^{-rt^2}\,dt = 1$, convexity and compactness show $f_r \in B$. If s is a continuous seminorm on A, then

$$s((r/\pi)^{1/2}\int e^{-rt^2} f(t)\,dt - f(0)) \le (r/\pi)^{1/2}\int e^{-rt^2} s(f(t) - f(0))\,dt$$

and the latter tends to $s(f(0) - f(0))$ as $r \to \infty$. □

Lemma 10.6.3 *Let $\{\sigma_t\}$ be a continuous one parameter group of automorphism on a von Neumann algebra W. Then W_e is a σ-weakly dense unital $*$-subalgebra of W. We also have $\sigma_a(wv) = \sigma_a(w)\sigma_a(v)$ and $\sigma_{a+b}(w) = \sigma_a(\sigma_b(w))$ and $\sigma_{\bar{a}}(w) = \sigma_a(w^*)^*$ for $a, b \in \mathbb{C}$ and $v, w \in W_e$.*

Proof Let $w \in W$, and set $f_r(z) = (r/\pi)^{1/2}\int e^{r(t-z)^2}\sigma_t(w)\,dt$ for $r > 0$ and $z \in \mathbb{C}$. Let $y \in W_*$. Boundedness of $t \mapsto y(\sigma_t(w))$ shows that $z \mapsto y(f_r(z))$ is holomorphic on all of $\mathbb{C}$, so f_r is holomorphic on all of $\mathbb{C}$ and it clearly extends $t \mapsto \sigma_t(f_r(0))$ for real t, so $f_r(0) \in W_e$. By the lemma $\lim_{r\to\infty} f_r(0) = w$ σ-weakly.

If $v, w \in W_e$ the derivative of $z \mapsto \sigma_z(v)\sigma_z(w)$ with respect to the norm is $\sigma_z'(v)\sigma_z(w) + \sigma_z(v)\sigma_z'(w)$, where $'$ means norm derivative, which do exist by Taylor expansions. So $vw \in W_e$ and $\sigma_z(vw) = \sigma_z(v)\sigma_z(w)$ by uniqueness of holomorphic extensions. The remaining parts are even easier. □

Lemma 10.6.4 *Let x be a faithful semifinite normal weight on a von Neumann algebra W with associated modular automorphism group $\{\sigma_t\}$, and let A be the corresponding left Hilbert algebra, so $\pi_l(A) = N_x \cap N_x^*$. Let*

$$A_0 = \{v \in \cap_{n\in\mathbb{Z}} D(\Delta^n) \mid \Delta^n(v) \in A,\ \forall n\}.$$

Then $\pi_l(A_0) \subset W_e$ and $\sigma_z(\pi_l(v)) = \pi_l(\Delta^{iz}(v))$ for $v \in A_0$ and $z \in \mathbb{C}$. Both $\pi_l(A)$ and M_x are W_e-bimodules, and $\pi_l(A_0)$ is an ideal of W_e.

Proof The first claim is immediate from Lemma 10.5.2. To see that $\pi_l(A)$ is a W_e-bimodule, it suffices to show $W_e\pi_l(A) \subset N_x^*$ as $\pi_l(A)$ and W_e are $*$-algebras and N_x is a left ideal in W. Let $w \in W_e$ and $v \in \pi_l(A)$. Then $wv \in N_x^*$ if and only if $q_x(wv) \in D(S) = D(\Delta^{1/2})$. Set $g(t) = \Delta^{it}(q_x(wv))$ for $t \in \mathbb{R}$. Then $g(t) = q_x(\sigma_t(wv)) = \sigma_t(w)\Delta^{it}(q_x(v))$ shows that g can be extended to a bounded continuous function on $H_{1/2}$ that is holomorphic on $H_{1/2}^o$, so $q_x(wv) = g(0) \in D(\Delta^{1/2})$ by Lemma 10.5.2. To see that M_x is a W_e-bimodule, it clearly suffices to show $W_e(M_x)_+ \subset M_x$, so let $w \in W_e$ and $v \in (M_x)_+$. Then $v^{1/2} \in N_x \cap N_x^*$, so $wv = (wv^{1/2})v^{1/2} \in M_x$. The last statement is proved similarly. □

Lemma 10.6.5 *Let x be a faithful semifinite normal weight on a von Neumann algebra W with associated modular automorphism group $\{\sigma_t\}$. If $a \in W$ satisfies $aM_x \subset M_x$ and $M_x a \subset M_x$, then for $v, w \in N_x \cap N_x^*$, there is a holomorphic function f on all of $\mathbb{C}$ that is bounded on the strip H and satisfies $f(t) = x(\sigma_t(a)vw^*)$ and $f(t+i) = x(vw^*\sigma_t(a))$ for $t \in \mathbb{R}$. If $a \in W_e$ and $u \in M_x$,* *then the function defined by* $g_u(z) = x(\sigma_z(a)u)$ *is holomorphic on all of* $\mathbb{C}$ *and bounded on* H *and satisfies* $g_u(t) = x(\sigma_t(a)u)$ *and* $g_u(t+i) = x(u\sigma_t(a))$ *for real* t.

Proof The function $f(z) = (a\Delta^{-iz}(q_x(v))|\Delta^{-i\bar{z}+1}(q_x(w)))$ settles the first claim as $f(t) = (\sigma_t(a)q_x(v)|\Delta(q_x(w))) = (S(q_x(w))|S(q_x(\sigma_t(a)v))) = x(\sigma_t(a)vw^*)$ and $f(t+i) = (\Delta^{1/2}(q_x(v))|\Delta^{1/2}(\sigma_t(a^*)q_x(w))) = x(vw^*\sigma_t(a))$.

For the second statement we may assume that $u = vw^*$ with $v, w \in N_x \cap N_x^*$. By the previous lemma $q_x(av) \in D(\Delta^{1/2})$ and $\sigma_z(a)\Delta^{iz}(q_x(v)) = \Delta^{iz}(q_x(av))$ for $z \in H_{1/2}$. Replacing z by $-i/2$ and a by $\sigma_z(a)$ for $z \in \mathbb{C}$, we get $\Delta^{1/2}(\sigma_z(a)q_x(v)) = \sigma_{z-i/2}(a)\Delta^{1/2}(q_x(v))$, so

$$g_u(z) = (S(q_x(w))|S(q_x(\sigma_z(a)v))) = (\sigma_{z-i/2}(a)\Delta^{1/2}(q_x(v))|\Delta^{1/2}(q_x(w)))$$

and g_u has the required properties with $g_u(t+i) = x(u\sigma_t(a))$ for real t. □

Theorem 10.6.6 *Let* x *be a faithful semifinite normal weight on a von Neumann algebra* W, *and let* $a \in W$. *Then* $a \in W_x$ *if and only if* $aM_x \subset M_x$ *and* $M_x a \subset M_x$ *and* $x(au) = x(ua)$ *for* $u \in M_x$.

Proof If $a \in W_x$, then $z \mapsto \sigma_z(a) = a$ is holomorphic on all of $\mathbb{C}$, so $a \in W_e$ and aM_x and $M_x a$ are subsets of M_x by Lemma 10.6.4. Each $u \in M_x$ gives g_u with the properties of the second statement of the last lemma. But $g_u(t) = x(au)$ for real t, so g_u is constant on $\mathbb{C}$, and $x(au) = x(ua)$ holds.

For the converse, note that each $v, w \in N_x \cap N_x^*$ gives f with the properties in the first statement of the previous lemma. Since $x\sigma_t = x$, we thus get $f(t) = f(t+i)$, so f is bounded on $\mathbb{C}$ and is constant by Liouville's theorem. By the formula for f in the proof of the previous lemma $((\sigma_t(a) - a)q_x(v)|\Delta(q_x(w))) = 0$, so $a \in W_x$ as $q_x(\pi_l(A_0)) = A_0$ and $\Delta(A_0) = A_0$ are dense in V_x. □

Lemma 10.6.7 *Let* x *be a faithful semifinite normal weight on a von Neumann algebra* W. *For positive* $h \in W_x$ *the function* $x_h = x(h^{1/2} \cdot h^{1/2})$ *is a semifinite normal weight on* W, *and* $h \mapsto x_h$ *is additive, preserves positive constants, and is pointwise increasing.*

Proof By Theorem 10.6.6 the normal weight x_h is finite on $(M_x)_+$, so it is also semifinite. We show additivity of $h \mapsto x_h$. To $h_i \in (W_x)_+$ pick $u_i \in W_x$ with $\|u_i\| \leq 1$ such that $h_i^{1/2} = u_i(h_1 + h_2)^{1/2}$ and such that $u_1^*u_1 + u_2^*u_2$ is the range projection of $h_1 + h_2$. Suppose $x_{h_1+h_2}(w) < \infty$ for $w \in W_+$, so $v = (h_1 + h_2)^{1/2}w(h_1 + h_2)^{1/2} \in (M_x)_+$. By Theorem 10.6.6 we have $u_i v u_i^* \in M_x$ and

$$\sum x_{h_i}(w) = x(\sum u_i v u_i^*) = x((\sum u_i^* u_i)v) = x(v) = x_{h_1+h_2}(w).$$

If $x_{h_i}(w) < \infty$, then

$$(h_1+h_2)^{1/2}w(h_1+h_2)^{1/2} = \lim_{\varepsilon \to 0}(h_1+h_2+\varepsilon)^{-1/2}(h_1+h_2)w(h_1+h_2)(h_1+h_2+\varepsilon)^{-1/2}$$

which cannot be greater than

$$2\lim_{\varepsilon\to 0}(h_1+h_2+\varepsilon)^{-1/2}(h_1wh_1+h_2wh_2)(h_1+h_2+\varepsilon)^{-1/2}=2\sum u_i^*h_i^{1/2}wh_i^{1/2}u_i,$$

so $x_{h_1+h_2}(w)<\infty$ by Theorem 10.6.6. The first part of the proof shows then that $x_{h_1+h_2}(w)=x_{h_1}(w)+x_{h_2}(w)$. If $h_1\leq h_2$, then $h_2=h_1+(h_2-h_1)$ and $x_{h_2}=x_{h_1}+x_{h_2-h_1}\geq x_{h_1}$. □

Lemma 10.6.8 *Let h be a positive self-adjoint operator affiliated with the centralizer W_x of a faithful semifinite normal weight x on a von Neumann algebra W, and consider the bounded positive operator $h_\varepsilon=h(1+\varepsilon h)^{-1}$ for $\varepsilon>0$. Then $x_h=\lim_{\varepsilon\to 0}x(h_\varepsilon^{1/2}\cdot h_\varepsilon^{1/2})$ defines a semifinite normal weight x_h on W which is faithful if and only if h is invertible.*

Proof As $h_\varepsilon\in W_x$ the previous lemma shows that x_{h_ε} is a semifinite normal weight on W. The same lemma shows that x_{h_ε} increase pointwise towards x_h on W_+ as ε decreases towards zero, so $x_h=\sup x_{h_\varepsilon}$ makes sense and is a normal weight on W. Let P_n be the spectral projection of h corresponding to $[0,n]$ for $n\in\mathbb{N}$. Then $\cup P_nM_xP_n$ is σ-weakly dense in W. For positive $w\in M_x$ we have $x_{h_\varepsilon}(P_nwP_n)=x_{h_\varepsilon P_n}(w)\leq x_{hP_n}(P_nwP_n)<\infty$ by the previous lemma and Theorem 10.6.6. Thus x_h is semifinite. If q is the range projection of h, then $x_h(1-q)=0$, so x_h is faithful if and only if $q=1$, which happens exactly when h is invertible. □

Theorem 10.6.9 *Let h be an invertible positive self-adjoint operator affiliated with the centralizer W_x of a faithful semifinite normal weight x on a von Neumann algebra W, and let $\{\sigma_t\}$ be the modular automorphism group of x. Then the modular automorphism group $\{\beta_t\}$ of x_h from the lemma above is given by $\beta_t=h^{it}\sigma_t(\cdot)h^{-it}$ for $t\in\mathbb{R}$.*

Proof We first restrict to $h\in W_x$. By Theorem 10.6.6 we have $M_{x_h}=h^{-1/2}M_xh^{-1/2}\subset M_x$ and $M_x=h^{-1/2}h^{1/2}M_xh^{1/2}h^{-1/2}\subset h^{-1/2}M_xh^{-1/2}=M_{x_h}$, so $M_{x_h}=M_x$ and $N_{x_h}=N_x$. Let $v,w\in N_{x_h}\cap N_{x_h}^*$. Let A be the left Hilbert algebra with involution S associated to x. Pick a sequence of $v_n\in\pi_l(A_0)$ such that $q_x(v_n)\to q_x(v)$ in the graph norm of S. Now

$$f_n(z)=(h^{iz+1}\Delta^{iz+1}(q_x(v_n))|S(h^{-iz}q_x(w)))$$

clearly defines a function f_n that is holomorphic on all of $\mathbb{C}$ that is bounded on the strip H and satisfies

$$f_n(t)=x(hh^{it}\sigma_t(v_n)h^{-it}w)=(q_x(h^{-it}wh^{it}h)|\Delta^{it}(q_x(v_n^*)))$$

and

$$f_n(t+i)=x(hwh^{it}\sigma_t(v_n)h^{-it})=(\Delta^{it}(q_x(v_n))|q_x(h^{-it}w^*h^{it}h))$$

for real t. As $\lim_{n\to\infty} \|\Delta^{it}(q_x(v) - q_x(v_n))\|_S = 0$ uniformly in $t \in \mathbb{R}$, the sequence $\{f_n\}$ converges uniformly on the boundary of H. By the Phragmen-Lindelöf theorem, it thus converges to a bounded continuous function f on H that is holomorphic on H^o and satisfies $f(t) = x(hh^{it}\sigma_t(v)h^{-it}w) = x_h(h^{it}\sigma_t(v)h^{-it}w)$ while $f(t+i) = x(hwh^{it}\sigma_t(v)h^{-it}) = x_h(wh^{it}\sigma_t(v)h^{-it})$, where we also used the previous theorem. So $\beta_t = h^{it}\sigma_t(\cdot)h^{-it}$ for $t \in \mathbb{R}$ by uniqueness of the KMS-condition.

In the general unbounded case, let P_n be the spectral projection of h corresponding to $[1/n, n]$ for $n \in \mathbb{N}$. The restriction of x to P_nWP_n is a faithful normal semifinite weight with modular automorphism group the restriction of $\{\sigma_t\}$. By the paragraph above we have for $w \in P_nWP_n$ that $\beta_t(w) = (hP_n)^{it}\sigma_t(w)(hP_n)^{-it} = h^{it}\sigma_t(w)h^{-it}$ for real t as $P_n \in W_x$. It remains to observe that $\cup P_nWP_n$ is σ-weakly dense in W. □

10.7 Cocycle Derivatives

Let x, y be faithful semifinite normal weights on a von Neumann algebra W. Consider

$$w = \begin{pmatrix} w_{11} & w_{12} \\ w_{21} & w_{22} \end{pmatrix} = \sum w_{ij} \otimes e_{ij} \in U \equiv M_2(W) = W \otimes M_2(\mathbb{C}).$$

Since $(w^*w)_{11} = w_{11}^*w_{11} + w_{21}^*w_{21}$ and $(w^*w)_{22} = w_{12}^*w_{12} + w_{22}^*w_{22}$, the formula $z(w) = x(w_{11}) + y(w_{22})$ defines a faithful weight z on U such that

$$N_z = \{w \in U \mid w_{11}, w_{21} \in N_x \text{ and } w_{12}, w_{22} \in N_y\}.$$

It is normal as $z(w) = \sup\{x'(w_{11}) + y'(w_{22})\}$, where the supremum is taken over all positive normal linear functionals x', y' on W such that $x' \leq x$ and $y' \leq y$ pointwise. Picking increasing nets $\{v_i\} \subset (M_x)_+$ and $\{w_i\} \subset (M_y)_+$ that converge strongly to 1, we see that $v_i \otimes e_{11} + w_i \otimes e_{22} \to 1$ strongly, so z is semifinite. We call z the *balanced weight* of x and y, and denote it sometimes by $x \oplus y$.

Consider the full left Hilbert algebra $A_z = q_z(N_z \cap N_z^*)$ associated to z. Note that

$$N_z \cap N_z^* = \{w \in U \mid w_{11} \in N_x \cap N_x^*,\ w_{12} \in N_x^* \cap N_y,\ w_{21} \in N_x \cap N_y^*,\ w_{22} \in N_y \cap N_y^*\}.$$

We also have the orthogonal decomposition $V_z = V_1 \oplus V_2 \oplus V_3 \oplus V_4$, where V_1, V_2, V_3, V_4 are the closures of $q_z(N_x \cap N_x^*) \otimes \mathbb{C}e_{11}$, $q_z(N_x \cap N_y^*) \otimes \mathbb{C}e_{21}$, $q_z(N_x^* \cap N_y) \otimes \mathbb{C}e_{12}$ and $q_z(N_y \cap N_y^*) \otimes \mathbb{C}e_{22}$, respectively. With respect to this decomposition the involution S in V_z has the form

$$S = \begin{pmatrix} S_x & 0 & 0 & 0 \\ 0 & 0 & S_{xy} & 0 \\ 0 & S_{yx} & 0 & 0 \\ 0 & 0 & 0 & S_y \end{pmatrix}$$

and similarly for J, while

$$\Delta = \begin{pmatrix} \Delta_x & 0 & 0 & 0 \\ 0 & \Delta_{yx} & 0 & 0 \\ 0 & 0 & \Delta_{xy} & 0 \\ 0 & 0 & 0 & \Delta_y \end{pmatrix} = \begin{pmatrix} S_x^* S_x & 0 & 0 & 0 \\ 0 & S_{yx}^* S_{yx} & 0 & 0 \\ 0 & 0 & S_{xy}^* S_{xy} & 0 \\ 0 & 0 & 0 & S_y^* S_y \end{pmatrix},$$

whereas the semicyclic representation π_z is of the form

$$\pi_z(w) = \begin{pmatrix} \pi_x(w_{11}) & \pi_x(w_{12}) & 0 & 0 \\ \pi_x(w_{21}) & \pi_x(w_{22}) & 0 & 0 \\ 0 & 0 & \pi_y(w_{11}) & \pi_y(w_{12}) \\ 0 & 0 & \pi_y(w_{21}) & \pi_y(w_{22}) \end{pmatrix}$$

for $w \in U$. Hence

$$J\pi_z(\begin{pmatrix} 0 & 1 \\ 1 & 0 \end{pmatrix})J = \begin{pmatrix} 0 & 0 & J_x J_{xy} & 0 \\ 0 & 0 & 0 & J_{xy} J_y \\ J_{yx} J_x & 0 & 0 & 0 \\ 0 & J_y J_{yx} & 0 & 0 \end{pmatrix},$$

which belongs to $\pi_z(U)'$, so

$$J_x J_{xy} \pi_y(w_{11}) = (J\pi_z(\begin{pmatrix} 0 & 1 \\ 1 & 0 \end{pmatrix})J\pi_z(w))_{13} = \pi_x(w_{11}) J_x J_{xy}$$

for $w \in U$. Thus $J_x J_{xy}$ intertwines π_x and π_y, and we may identify these representations and suppress them in further calculations. Writing $\sigma^{xy} = \Delta_x^{it}(\cdot)\Delta_{yx}^{-it}$ and $\sigma^{yx} = \Delta_{yx}^{it}(\cdot)\Delta_x^{-it}$ the modular automorphism group $\{\sigma_t\}$ of z is given by

$$\sigma_t(w) = \begin{pmatrix} \sigma_t^x(w_{11}) & \sigma_t^{xy}(w_{12}) \\ \sigma_t^{yx}(w_{21}) & \sigma_t^y(w_{22}) \end{pmatrix}$$

for $w \in U$, where $\{\sigma_t^x\}$ and $\{\sigma_t^y\}$ are the modular automorphism groups of x and y, respectively. Set $u_t = \sigma_t^{xy}(1)$. Since

$$\begin{pmatrix} a & 0 \\ 0 & 0 \end{pmatrix} = \begin{pmatrix} 0 & 1 \\ 0 & 0 \end{pmatrix} \begin{pmatrix} 0 & 0 \\ 0 & a \end{pmatrix} \begin{pmatrix} 0 & 0 \\ 1 & 0 \end{pmatrix} \text{ and } \begin{pmatrix} 0 & ab \\ 0 & 0 \end{pmatrix} = \begin{pmatrix} 0 & a \\ 0 & 0 \end{pmatrix} \begin{pmatrix} 0 & 0 \\ 0 & b \end{pmatrix}$$

for $a, b \in W$ we get $\sigma_t^x = u_t\sigma_t^y(\cdot)u_t^*$ and $\sigma_t^{xy}(ab) = \sigma_t^{xy}(a)\sigma_t^y(b)$, so $u_{s+t} = u_s\sigma_s^y(u_t)$.

We have arrived at *Connes' cocycle theorem.*

Theorem 10.7.1 *Let x, y be faithful semifinite normal weights on a von Neumann algebra W with modular automorphism groups $\{\sigma_t^x\}$ and $\{\sigma_t^y\}$, respectively. There exists a σ-strongly continuous one parameter family of unitaries $u_t \in W$ such that $u_{s+t} = u_s\sigma_s^y(u_t)$ and $\sigma_t^x = u_t\sigma_t^y(\cdot)u_t^*$ for $s, t \in \mathbb{R}$. Moreover, for $a \in N_x \cap N_y^*$ and $b \in N_x^* \cap N_y$ there exists a bounded continuous function f on the strip H that is holomorphic on H^o and satisfies $f(t) = x(u_t\sigma_t^y(b)a)$ and $f(t+i) = y(au_t\sigma_t^y(b))$ for $t \in \mathbb{R}$. The family $\{u_t\}$ is uniquely determined by this.*

Proof We have already proved the first statement, and the existence of f is immediate from the KMS-condition of the weight z. To prove the uniqueness part, assume $\{v_t\}$ plays the same role as $\{u_t\}$. Then it suffices to show that $b(t) = v_t\sigma_t^y(b)$ satisfies $b(t) = \sigma_t^{xy}(b(0))$ for $t \in \mathbb{R}$. To achieve this it suffices to consider z instead of x in the following statement: If $b \colon \mathbb{R} \to N_x \cap N_x^*$ is a σ-weakly continuous function such that $t \mapsto x(b(t)^*b(t)) + x(b(t)b(t)^*)$ is bounded and if there exists a bounded continuous function f on H that is holomorphic on H^o and satisfies $f(t) = x(b(t)a)$ and $f(t+i) = x(ab(t))$ for real t, then $b(t) = \sigma_t^x(b(0))$. To prove this statement, consider the full left Hilbert algebra A associated to x with A_0 as in Lemma 10.6.4. Let $v(t) = q_x(b(t))$. Using J, then to each $w \in q_x(A_0)$, there is by assumption a bounded continuous function f_w on H that is holomorphic on H^o and satisfies $f_w(t) = (\Delta^{1/2}(v(t))|w)$ and $f_w(t+i) = (v(t)|\Delta^{-1/2}(w))$ for real t. Since $\|v(\cdot)\|_S$ is bounded on $\mathbb{R}$, also by assumption, the Phragmen-Lindelöf theorem shows that for each $c \in H$ the map which sends w to $f_w(c)$ extends to an antilinear bounded functional on $D(S^{-1})$ with respect to the graph norm of S^{-1}. Since $D(S^{-1})$ is a Hilbert space with respect to this norm, as S^{-1} is closed, we get by the Riesz representation theorem a vector $d(c) \in D(S^{-1})$ such that $f_w(c) = (d(c)|w)_{S^{-1}} = (d(c)|(1+\Delta^{-1})(w))$. Now $g(c, r) = (d(c)|(1+\Delta^{-1})\Delta^{i\bar{r}}(w))$ for $r \in \mathbb{C}$ is holomorphic on $H^o \times \mathbb{C}$. As $d(t) \in D(\Delta^{-1})$ and $(1+\Delta^{-1})(d(t)) = \Delta^{1/2}(v(t))$ for real t, we get $g(t, r) = (v(t)|\Delta^{i\cdot\overline{(r+i/2)}}(w))$ so $g(t+i, r) = g(t, r-i)$. Putting $g(r) = g(r, r)$ for $r \in H$, then since $g(t) = g(t+i)$ for real t, it extends to a bounded function that is holomorphic on all of $\mathbb{C}$, so it must be constant by Liouville's theorem. Hence $(\Delta^{1/2}(v(0))|w) = g(0) = g(t) = (\Delta^{1/2}\Delta^{-it}(v(t))|w)$, which shows that $v(t) = \Delta^{it}(v(0))$, and thus $b(t) = \pi_l(v(t)) = \sigma_t^x(b(0))$. □

Remark 10.7.2 We define the centralizer and modular automorphism group of a normal weight x on a von Neumann algebra W to be the ones defined by the corresponding faithful semifinite normal weight on the reduced algebra $W_{s(x)}$. If x in this theorem fails to be faithful, we see that the theorem still holds, except that u_t will form a σ-strongly $*$-continuous one-parameter group of partial isometries such that $u_t u_t^* = s(x)$ and $u_t^* u_t = \sigma_t^x(s(x))$.

Corollary 10.7.3 *Let x, y be faithful semifinite normal weights on a von Neumann algebra W with modular automorphism groups $\{\sigma_t^x\}$ and $\{\sigma_t^y\}$, respectively. Then $y \circ \sigma_t^x = y$ for real t if and only if $y = x_h$ for x_h as in Theorem 10.6.9 if and only if $x \circ \sigma_t^y = x$ for real t.*

Proof If $y = x_h$, then $y \circ \sigma_t^x = (x\sigma_t^x)_h = x_h$ by the argument in the proof of Theorem 10.6.9. Conversely, if $y \circ \sigma_t^x = y$ for real t, then by uniqueness of the $\{u_t\}$ in the theorem above, and the properties $x \circ \sigma_t^x = x$ and $y \circ \sigma_t^y = y$ for real t, we see that all $u_t \in W_x$. By Stone's theorem there exists an invertible positive self-adjoint operator h affiliated with W_x such that $u_t = h^{it}$ for real t. Hence $y = x_h$ by the KMS-type condition in the theorem above. The second equivalence is similar. □

Definition 10.7.4 The family $\{u_t\}$ in the previous theorem is called the *cocycle derivative* of x with respect to y, and we write $u_t = (Dx : Dy)_t$ for $t \in \mathbb{R}$.

We have the following *chain rule*.

Proposition 10.7.5 *If x_i are faithful semifinite normal weights on a von Neumann algebra W, then $(Dx_1 : Dx_3)_t = (Dx_1 : Dx_2)_t (Dx_2 : Dx_3)_t$ for real t.*

Proof On $W \otimes M_3(\mathbb{C})$ considered the faithful semifinite normal weight z given by $z(\sum w_{ij} \otimes e_{ij}) = \sum x_i(w_{ii})$. Then the proof is immediate from $e_{13} = e_{12}e_{23}$ and $(Dx_i : Dx_j)_t \otimes e_{ij} = \sigma_t^z(1 \otimes e_{ij})$ for real t. □

Note also that if $(Dx_1 : Dx_2)_t = (Dx_1 : Dx_3)_t$ for real t, then $x_2 = x_3$.

We also have a converse to the cocycle theorem.

Theorem 10.7.6 *Let x be a faithful semifinite normal weight on a von Neumann algebra W with modular automorphism group $\{\sigma_t\}$. Given a σ-strongly continuous family $\{u_t\}$ of unitaries in W such that $u_{s+t} = u_s \sigma_s(u_t)$ for $s, t \in \mathbb{R}$, there exists a faithful semifinite normal weight y on W with $(Dy : Dx)_t = u_t$ for real t.*

Proof Let $\{\sigma_t\}$ be the modular automorphism group of x with entire elements W_e, and consider the σ-weakly one parameter group of isometric linear transformations $\alpha_t = u_t \sigma_t(\cdot)$ with entire elements $\tilde{W}$ defined in the obvious way. Since $\alpha_t(w)^* \alpha_t(w) = \sigma_t(w^* w)$ for $w \in W$, we get $\alpha_t(N_x) \subset N_x$ and $\tilde{W}^* \tilde{W} \subset M_x$. By Stone's theorem there is an invertible positive self-adjoint operator T in V_x such that $u_t = T^{it} \Delta_x^{-it}$ for real t, so $T^{it}(q_x(w)) = q_x(\alpha_t(w))$ for $w \in N_x$. The formula $q(\sum v_n w_n^*) = \sum \pi_x(v_n) J_x T^{1/2}(q_x(w_n))$ for $v_n, w_n \in \tilde{W}$ well-defines a linear injection $q \colon \tilde{W}^* \tilde{W} \to V_x$. Indeed, if $\sum v_n w_n^* = 0$ for $v_n, w_n \in \tilde{W}$, then

$$\|q(\sum v_n w_n^*)\|^2 = \sum (J_x \pi_x(v_n^* v_m) J_x q_x(\alpha_{-i/2}(w_m)) | T^{1/2}(q_x(w_n)))$$

vanishes as $J_x\pi_x(v_n^*v_m)J_xq_x(\alpha_{-i/2}(w_m)) = \alpha_{-i/2}(w_m)J_x(q_x(v_m^*v_n))$ and since we have $v_m^*v_n \in W_x$ and $J_x(q_x(v_n^*v_m)) = q_x(\sigma_{-1/2}(v_m^*v_n))$ and finally the equality $\alpha_{-i/2}(w_m)\sigma_{-i/2}(v_m^*v_n) = \sigma_{-i/2}(w_m v_m^* v_n)$. So q is well-defined and linear.

A similar calculation shows that when $q(\sum v_m w_m^*) = 0$, then we have the equality $(T^{1/2}(q_x((\sum v_m w_m^*)^*v))|T^{1/2}(q_x(w))) = 0$ for $v, w \in \tilde{W}$. As $T^{1/2}(q_x(w_n)) = q_x(\alpha_{-i/2}(w_n))$ and $\alpha_{-i/2}(\tilde{W}) = \tilde{W}$ and $q_x(\tilde{W})$ is dense in V_x, it is clear that elements of the type in the second entry of the inner product are dense in V_x, so $(\sum v_m w_m^*)^*v = 0$. By taking the limit $v \to 1$ we see that q is injective. Using that $\tilde{W}$ is strongly dense in W, we also see that q has dense range in V_x. Let $\beta_t = u_t\sigma_t(\cdot)u_t^*$. As $\beta_t(vw^*) = \alpha_t(v)\alpha_t(w)^*$ for real t, the elements in $\tilde{W}\tilde{W}^*$ are entire with respect to $\{\beta_t\}$, and the set is β-invariant. The map that sends $z \in \mathbb{C}$ to $\Delta(z)(q(vw^*)) \equiv q(\beta_{-iz}(vw^*)) = \alpha_{-iz}(v)J_xT^{1/2-\bar{z}}(q_x(w))$ for $v, w \in \tilde{W}$ is holomorphic on all of $\mathbb{C}$. So both $z \mapsto (\Delta(z)(q(v_1w_1^*))|q(v_2w_2^*))$ and $z \mapsto (q(v_1w_1^*)|\Delta(\bar{z})(q(v_2w_2^*)))$ for $v_i, w_i \in \tilde{W}$ are holomorphic everywhere, and by a calculation similar to the one above, they coincide for $z = it$ and real t, and hence for all $z \in \mathbb{C}$. Hence $(\Delta(1)(q(v_1w_1^*))|q(v_2w_2^*)) = (\Delta(\frac{1}{2})(q(v_1w_1^*))|\Delta(\frac{1}{2})q(v_2w_2^*)) = (q(w_2v_2^*)|q(w_1v_1^*))$. Now $A \equiv q(\tilde{W}\tilde{W}^*)$ is a left Hilbert algebra with inner product from V_x, with multiplication $q(v)q(w) = q(vw)$ and involution $q(v) \mapsto q(v^*)$, which is closable since $(q(w_2v_2^*)|q(v_1w_1^*)) = (q(\beta_{-i}((v_1w_1^*)^*))|q(v_2w_2^*))$ by the property of $\Delta(1)$ shown above. Since $\tilde{W}$ is σ-weakly dense in W, we have $A^2 = A$, and W is the left von Neumann algebra of A. Let y be the faithful semifinite normal weight on W associated to the full left Hilbert algebra A''. By construction $\{\beta_t\}$ is the modular automorphism group on W of y. By definition of q we have $q_x(vw^*) = \pi_y(v)J_{yx}T^{1/2}(q_x(w))$ for $v, w \in \tilde{W}$, so $\tilde{W}^* \subset N_x$ and $\Delta_{yx}^{1/2}(q_x(w)) = T^{1/2}(q_x(w))$. As $q_x(\tilde{W})$ is a core for $T^{1/2}$ we get $\Delta_{yx}^{1/2} = T^{1/2}$ since both sides are self-adjoint operators. Hence $u_t = \Delta_{yx}^{it}\Delta_x^{-it} = (Dy : Dx)_t$. □

Remark 10.7.7 If u_t in this theorem form a σ-strongly $*$-continuous one-parameter group of partial isometries with $u_tu_t^* = e$ and $u_t^*u_t = \sigma_t^x(e)$ for some orthogonal projection $e \in W$, the theorem clearly still holds, but now y is not faithful and $s(y) = e$.

The modular automorphism group $\{\sigma_t\}$ of a weight on a von Neumann algebra W is *inner* if $\sigma_t = u_t \cdot u_t^*$ for some continuous one parameter group of unitaries $u_t \in W$.

Proposition 10.7.8 *A von Neumann algebra W is semifinite if and only if the modular automorphism group of every faithful semifinite normal weight on W is inner if and only if this is true for at least one such weight. Then every faithful semifinite normal weight on W is of the form τ_h for a fixed faithful semifinite normal trace τ on W and some positive self-adjoint operator h affiliated with W.*

Proof If τ is a faithful semifinite normal trace on W, then every faithful semifinite normal weight on W is by Corollary 10.7.3 of the form τ_h for an invertible positive self-adjoint operator h affiliated with W. The modular automorphism group is thus inner by Theorem 10.6.9.

If on the other hand the modular automorphism group $\{\sigma_t\}$ of some faithful semifinite normal weight x on W is inner, say $\sigma_t = u_t \cdot u_t^*$, then as N_x is a left ideal of W, we have $N_x u_t^* = \sigma_t(N_x) = N_x$ as $x \circ \sigma_t = x$. Hence $u_t \in W_x$ by Theorem 10.6.6. Stone's theorem thus provides an invertible positive self-adjoint operator h affiliated with W_x such that $u_t = h^{-it}$ for real t. By Theorem 10.6.9 the modular automorphism group of the faithful semifinite normal weight x_h is trivial, so x_h is a trace. □

Proposition 10.7.9 *Let x be a faithful semifinite normal weight on a von Neumann algebra W with modular automorphism group $\{\sigma_t\}$. If y is a σ-invariant semifinite normal weight on W that coincide with x on a σ-weakly dense σ-invariant $*$-subalgebra of M_x, then $y = x$.*

Proof Let M be the σ-weakly dense σ-invariant $*$-subalgebra of M_x, and let $\{a_j\}$ be a net of positive elements of M that increases σ-strongly to 1. Let $w_j = (\pi)^{-1/2} \int e^{-t^2} \sigma_t(a_j)\, dt$. As M is σ-invariant and x, y are normal, the functionals $x(w_j \cdot w_j)$ and $y(w_j \cdot w_j)$ coincide on M and hence on W. Each w_j is entire with respect to $\{\sigma_t\}$ with $\sigma_z(w_j) = (\pi)^{-1/2} \int e^{-(t-z)^2} \sigma_t(a_j)\, dt$ for $z \in \mathbb{C}$. Also, for each $v \in V_x$, the continuous functions f_j on $\mathbb{R}$ given by $f_j(t) = \|\sigma_t(1-a_j)(v)\|$ decrease to zero uniformly on compact sets by Dini's theorem. Hence $\{\|\sigma_z(1 - w_j)(v)\|\}$ decrease to zero as $j \to \infty$. By σ-weakly continuity of y and entireness of w_j we thus get

$$y(w) \leq \lim y(w_j w w_j) = \lim x(w_j w w_j) = \lim \|\sigma_{-i/2}(w_j) J(q_x(w^{1/2}))\|^2 = x(w)$$

for $w \in (M_x)_+$, so $y \leq x$. Let $p = 1 - s(y)$. For positive $w \in M_x$ with $w \leq p$ we have

$$\begin{aligned} x(w) &\leq \lim x(w_j w w_j) = \lim y(w_j w w_j) \leq 2 \lim x((1 - w_j) w (1 - w_j)) \\ &\leq 2 \lim y((1 - w_j) w (1 - w_j)) = 2 \lim \|(1 - \sigma_{-i/2}(w_j))(q_x(w^{1/2}))\| = 0, \end{aligned}$$

so $w = 0$. Since $p \in W_x$, we have $p(M_x)_+ p \subset (M_x)_+$, so $p(M_x)_+ p = \{0\}$ and $p = 0$ and y is faithful. Since x is invariant under the modular automorphism group of y, we can swop the roles of x and y above, to get $x \leq y$. □

Corollary 10.7.10 *If two faithful semifinite normal weights x, y with the same modular automorphism group agree on a σ-weakly dense $*$-subalgebra of M_x, then $x = y$.*

Proof Now the element h in Corollary 10.7.3 is affiliated with the center of the von Neumann algebra W. If $h \neq 1$, there is a central orthogonal projection q such that $(1+\varepsilon)q \leq hq$ or $(1-\varepsilon)q \geq hq$. In the first case we get for any a in the given dense algebra, the inequality

$$(1+\varepsilon)x(a^*qa) \leq x(a^*hqa) = y(a^*qa) = x(a^*qa).$$

The second case leads to an equally absurd conclusion, so $h = 1$ and $y = x$. □

Lemma 10.7.11 *Let x be a faithful semifinite normal weight on a von Neumann algebra W with modular automorphism group $\{\sigma_t\}$. Fix $a \in W$. Then $x(a \cdot a^*) \leq x$ on W_+ if and only if $t \mapsto \sigma_t(a)$ can be extended to a bounded continuous function on $H_{1/2}$ that is holomorphic on $H^o_{1/2}$ and satisfies $\|\sigma_{-i/2}(a)\| \leq 1$. If also $aa^* \in W_x$, then $x(aa^*w) = x(\sigma_{-i/2}(a)^* w \sigma_{-i/2}(a))$ for $w \in M_x$.*

Proof Suppose $t \mapsto \sigma_t(a)$ can be extended to a bounded continuous function on $H_{1/2}$ that is holomorphic on $H^o_{1/2}$ and satisfies $\|\sigma_{-i/2}(a)\| \leq 1$. For $v \in D(S) = D(\Delta^{1/2})$ the elements $\sigma_t(a)(v)$ belong to $D(S)$, so $\sigma_{-i/2}(a)\Delta^{1/2}(v) = \Delta^{1/2}a(v)$ by Lemma 10.5.2. Thus for $w \in N_x \cap N_x^*$ we get $Sa(q_x(w^*)) = J\sigma_{-i/2}(a)J(q_x(w))$, so $x(awa^*) = \|q_x(w^{1/2}a^*)\|^2 \leq \|q_x(w^{1/2})\|^2 = x(w)$ when $w \in (M_x)_+$.

For the forward implication note that $q_x(w) \mapsto q_x(wa^*)$ for $w \in N_x$ extends to a bounded operator b on V_x. When $w \in N_x \cap N_x^*$ we have $wa \in N_x \cap N_x^*$ and $J\Delta^{1/2}(q_x(wa^*)) = aq_x(w)$, so $Jb(q_x(w)) = \Delta^{1/2}a\Delta^{-1/2}J(q_x(w))$. Hence for $v_i \in D(S)$ we get $|(a\Delta^{-iz}(v_1)|\Delta^{-i\bar{z}}(v_2))| \leq \|v_1\|\|v_2\|$.

For the last statement let $c = \sigma_{-i/2}(a)$. Then $t \mapsto \sigma_t(c^*)$ extends to a bounded continuous function on $H_{1/2}$ that is holomorphic on $H^o_{1/2}$ and satisfies $\sigma_{-i/2}(c^*) = a^*$. From the first paragraph we see that $q_x(wa^*) = J\sigma_{-i/2}(a)J(q_x(w))$ for $w \in N_x$, so $q_x(wc) = Ja^*J(q_x(w))$ and $\|q_x(wc)\|^2 = (Jaa^*J(q_x(w))|q_x(w)) = x(aa^*w^*w)$ by Theorem 10.6.6. □

Theorem 10.7.12 *Let x, y be faithful semifinite normal weights on a von Neumann algebra W. Then there exists $M > 0$ such that $y \leq Mx$ on W_+ if and only if $t \mapsto (Dy : Dx)_t$ extends to a bounded continuous function on $H_{1/2}$ that is holomorphic on $H^o_{1/2}$ and satisfies $\|(Dy : Dx)_{-i/2}\| \leq M^{1/2}$. In this case $y = x((Dy : Dx)^*_{-i/2} \cdot (Dy : Dx)_{-i/2})$ on M_y.*

Proof Replacing x by $M^{-1}x$ we may assume $M = 1$. Consider the weight $z = y \oplus x$ on $W \otimes M_2(\mathbb{C})$. For $a = 1 \otimes e_{12}$ we have $\sigma^z_t(a) = (Dy : Dx)_t \otimes e_{12}$ and $z(a(\sum w_{ij} \otimes e_{ij})a^*) = y(w_{11})$, which is not greater than $z(\sum w_{ij} \otimes e_{ij})$ when the argument is positive. The equivalence in the theorem is now clear from the lemma applied to z. For the final statement note that $aa^* = 1 + e_{11} \in W_z$, so the lemma applied to z now gives with $u = (Dy : Dx)_{-i/2}$ the identity

$$y(w) = z((1+e_{11})^2(w \otimes e_{11})) = z((u \otimes e_{12})^*(w \otimes e_{11})(u \otimes e_{12})) = x(u^*wu)$$

for $w \in (M_y)_+$. □

Let $\{\sigma_t\}$ be a pointwise σ-weakly continuous one-parameter group of isometries on a von Neumann algebra W. Let W^σ denote the elements $w \in W$ of *exponential type*, which means that $t \mapsto \sigma_t(w)$ admits an extension to $\mathbb{C}$ that is entire, and bounded on every horizontal line, and such that $\sup_{s,t\in\mathbb{R}}\|\sigma_{s+it}(w)\|\exp(-r|t|) < \infty$ for some $r > 0$. In general, we define σ_{s+it} on those $w \in W$, where the evaluation at w is bounded and continuous on H_t and holomorphic on H_t^o.

Lemma 10.7.13 *The linear space W^σ defined above is σ-weakly dense in W.*

Proof If $x \in W_*$ annihilates W^σ, then $\int \hat{f}(t)x\sigma_t(w)\,dt = 0$ for any $w \in W$ and smooth function f on $\mathbb{R}$ with compact support, where $\hat{f}$ is the Fourier transform of f. Since such $\hat{f}$ form a dense subspace in $L^1(\mathbb{R})$, we see that $x\sigma_t(w)$ vanishes for almost every t, and hence for every t by continuity, showing that $x = 0$. □

Proposition 10.7.14 *If $\{\sigma_t\}$ and $\{\sigma_t'\}$ are pointwise σ-weakly continuous one-parameter group of isometries on a von Neumann algebra W, and σ_i is a restriction of σ_i', then $\sigma_t' = \sigma_t$ for all $t \in \mathbb{R}$.*

Proof We claim that $W^\sigma \subset W^{\sigma'}$. Indeed, say there are $M \geq 0$ and $r > 0$ such that $\|\sigma_{s+it}(w)\| \leq Me^{r|t|}$ for all $s, t \in \mathbb{R}$. By assumption $\sigma_{it}' = \sigma_{it}$ when $t = n \in \mathbb{Z}$. The Phragmen-Lindelöf Theorem shows that for $t = pn$ with $p \in [0, 1]$, we have $\|\sigma_{s+it}'(w)\| \leq Me^{rp|n|} = Me^{r|t|}$, so the claim holds.

Let $x \in W_*$ and $w \in W^\sigma$. Then $z \mapsto x(\sigma_z(w)-\sigma_z'(w))$ defines an entire function on $\mathbb{C}$ of exponential type that vanishes when $z \in i\mathbb{Z}$. Hence it vanishes for all $z \in \mathbb{C}$ by Corollary A.13.2, and the previous lemma then gives the result. □

Proposition 10.7.15 *Consider faithful semifinite normal weights x, y on a von Neumann algebra W, and let $a, b \in W$. Then $b = \sigma_{-i}^{yx}(a)$ if and only if $aN_x^* \subset N_y^*$ and $N_y b \subset N_x$ and $y(ac) = x(cb)$ for $c \in N_x^* N_y$.*

Proof Considering the balanced weight $x \oplus y$ and using the usual 2×2-matrix trick, we may assume that $y = x$.

For the forward direction note that $\sigma_{-i/2}^x(a) = \sigma_{i/2}^x(b)$ and $\sigma_{i/2}^x(b)^* = \sigma_{-i/2}^x(b^*)$. Then Lemma 10.7.11 implies $aN_x^* \subset N_x^*$ and $N_x b \subset N_x$, so $x(ac)$ and $x(cb)$ make sense, and $(q_x(v)|q_x(wa^*))$ equals

$$(q_x(v)|J_x\sigma_{-i/2}^x(a)J_xq_x(w)) = (J_x\sigma_{-i/2}^x(b^*)J_xq_x(v)|q_x(w)) = (q_x(vb)|q_x(w)),$$

for $v, w \in N_x$.

Conversely, consider the full left Hilbert algebra associated to x with A_0 as in Lemma 10.6.4. For $z \in \mathbb{C}$ and $v, w \in A_0$, set $B_z(v, w) = (a\Delta^{-iz}v|\Delta^{-i\bar{z}}w)$. Then $B_{-i}(v, w) = (bv|w)$ by assumption. Also, the function $z \mapsto B_z(v, w)$ is entire, and is bounded on $\mathbb{R} + ip$ for $p \in [s, t]$ and real $s < t$. It equals $(\sigma_t^x(a)v|w)$ for $z = t$ and $(\sigma_t^x(b)v|w)$ for $z = t - i$, and by the Phragmen-Lindelöf Theorem, we get $|B_z(v, w)| \leq \sup\{\|a\|, \|b\|\}\|v\| \cdot \|w\|$ when $z \in H_1$. Therefore, for such z, there exists $g(z) \in B(V_x)$ such that $(g(z)v|w) = (a\Delta^{-iz}v|\Delta^{-i\bar{z}}w)$. Hence g is continuous on H_1 and holomorphic on H_1^o, and $g(t) = \sigma_t^x(a)$ while $g(t - i) = \sigma_t^x(b)$, so $g(z) \in W$ and $b = \sigma_{-i}^x(a)$. □

Hence for normal faithful states x, y on a von Neumann algebra W with elements a, b, we see that $b = \sigma_{-i}^{yx}(a)$ if and only if $y(ac) = x(cb)$ for $c \in W$.

10.8 A Generalized Radon-Nikodym Theorem

An unbounded self-adjoint operator on a Hilbert space is *strictly positive* if it is positive and has dense range. For later purposes we need the following generalization of Corollary 10.7.3.

Theorem 10.8.1 *Let x and y be faithful semifinite normal weights on a von Neumann algebra W, and let ν be a positive constant such that $x\sigma_t^y = \nu^t x$ for all real t. Then there exists a unique strictly positive operator h affiliated with W such that $\sigma_t^x(h) = \nu^t h$ for real t, and $y = x_h$, where now x_h is the faithful semifinite normal weight on W associated to the left Hilbert algebra $q(N \cap N^*) \subset V_x$ with inner product from V_x, where N consists of those $a \in W$ such that $ah^{1/2}$ has bounded closure in N_x and $q: N \to V_x$ is the map that sends a to $q_x(ah^{1/2})$, and where the $*$-algebra structure is given by $q(a)q(b) = q(ab)$ and $q(a)^* = q(a^*)$. Here we have denoted the bounded closure by the same symbol.*

Proof Let $u_t \equiv (Dy : Dx)_t$. As $x(u_t^*(\cdot)u_t) = x\sigma_{-t}^x(u_t^*(\cdot)u_t) = x\sigma_{-t}^y = \nu^{-t}x$ for real t, both $N_x u_t$ and $u_t^* N_x$ are subsets of N_x, and we have $x(wu_t) = \nu^{-t}x(u_t w)$ for $w \in M_x$. Adapting the proof of Theorem 10.6.6 we thus see that $\sigma_s^x(u_t) = \nu^{ist}u_t$ for real s, t. But $t \mapsto \nu^{-it^2/2}u_t \in W$ is a strongly continuous one-parameter group of unitaries, so there is a strictly positive operator h affiliated to W such that $h^{it} = \nu^{-it^2/2}u_t$ for real t. It remains to show that $(Dx_h : Dx)_t = u_t$ for real t, so that $x_h = y$, and that we really have a left Hilbert algebra as described with associated von Neumann algebra W, so that x_h actually exists. Clearly left multiplication is bounded, and the adjoint is the Hilbert space adjoint. As for denseness consider the self-adjoint elements $v_n \in W$ given by

$$v_n = 2n^2\Gamma(1/2)^{-1}\Gamma(1/4)^{-1}\int_{-\infty}^{\infty} e^{-n^2s^2-n^4t^4}\nu^{is}h^{it}ds\,dt,$$

where Γ is the gamma function, so $\Gamma(z) = \int_0^\infty t^{z-1}e^{-t}dt$ for complex z with positive real part. Then $(z_1, z_2, z_3) \mapsto h^{z_1}\nu^{z_2}\sigma_{z_3}^x(v_n)$ is analytic from $\mathbb{C}^3$ to W and $\{\sigma_{z_1}^x(v_n)\}$ is bounded and converges to 1 in the σ-strong* topology. Hence for $a \in N_x \cap N_x^*$ we see that $v_n a v_n, v_n a h^{-1/2}v_n \in N \cap N^*$ and $v_n a v_n \to a$ σ-strongly*, while

$$q(v_n a h^{-1/2}v_n) = q_x(v_n a v_n) = J_x\sigma_{i/2}^x(v_n)^* J_x v_n q_x(a) \to q_x(a),$$

so $N \cap N^*$ is σ-strongly* dense in W and $q(N \cap N^*)^2$ is dense in V_x. To show that the map $q(a) \mapsto q(a^*)$ is closable, it suffices to show that for b, b' in the Tomita algebra for x, the element $q_x(v_n bb' v_n)$ belongs to the domain of its adjoint map. But this is evident from the easily checked identity

$$(q_x(h^{1/2}\sigma_{-i}(v_n)\sigma_{-i}(b'^* b^* h^{-1/2}v_n))|q(a)) = (q(a^*)|q_x(v_n bb' v_n))$$

for $a \in N \cap N^*$. So we have indeed a left Hilbert algebra with associated von Neumann algebra W, say with closure S of the adjoint map, and a corresponding weight x_h with modular automorphism group $\{\sigma_t\}$. Invoking the elements v_n and arguing with analyticity, it is straightforward to check that the modular operator Δ of x_h is the closure of $J_x h^{-1} J_x h \Delta_x$ and the modular conjugation J is $\nu^{i/4} J_x$, so $S = J\Delta^{1/2}$, and one further checks that $\sigma_t = h^{it}\sigma_t^x(\cdot)h^{-it}$ for real t. Then one already has $u_{s+t} = u_s\sigma_s^x(u_t)$ and $\sigma_t = u_t\sigma_t^x(\cdot)u_t^*$ for real s, t, and it remains only to check the KMS-type condition in Connes' cocycle theorem. To this end consider the strictly positive operator k in V_x such that $k^{is} = \nu^{is^2/2}h^{is}\Delta_x^{is}$ by verifying that the right hand side defines a strongly continuous one-parameter group in the real parameter s, so that Stone's theorem applies. Then one verifies that $J_x\nu^{-i/8}k^{1/2}q_x(a) = q_{x_h}(a)$ for $a \in N_x \cap N_{x_h}^*$. This follows by an analyticity argument that hinges on the fact that all $v_n a \in N_x \cap N_x^*$, which is obtained by first approximating 1 σ-strongly* by a net of elements in W that are analytic with respect to σ_t, and then sandwiching them between v_n's to get elements in $N \cap N^*$ with the desired approximation property. Using this we can for $b \in N_x^* \cap N_{x_h}$ well-define a map $f \colon H \to \mathbb{C}$ that is given by $f(z) = (k^{iz+1}q_x(a)|q_x(b^*))$ for $\mathrm{Im}(z) \geq 1/2$ and $f(z) = (J_x q_{x_h}(a^*)|k^{-i\bar{z}}J_x q_{x_h}(b))$ for $\mathrm{Im}(z) \leq 1/2$. Then it is verified that f is continuous, analytic in the interior, and that $f(t) = x_h(u_t\sigma_t^x(a)b)$ while $f(t+i) = x(bu_t\sigma_t^x(a))$ for real t. □

Note from the proof that the σ-strong* closure of q, denoted again by q, and the left multiplication, extended by boundedness, is just the GNS-construction q_{x_h} and π_{x_h} for the weight x_h. We also note that $\sigma_t^{x_h} = h^{it}\sigma_t^x(\cdot)h^{-it}$ and $\sigma_t^{x_h}(h) = \nu^t h$ for real t. Studying the proof above it is also fairly straightforward to show a converse of this theorem, namely, whenever we have a strictly positive operator h affiliated to W such that $\sigma_t^x(h) = \nu^t h$ for real t, and with $x_h = y$, then $x\sigma_t^y = \nu^t x$ for real t. For similar reasons this is again equivalent to the property $y\sigma_t^x = \nu^{-t}y$ for real t.

10.9 Standard Form

The *dual cone* of a convex cone C in a Hilbert space V consists of those $v \in V$ such that $(v|w) \geq 0$ for all $w \in C$. When C equals its dual cone, it is *self-dual*.

Definition 10.9.1 A *standard form* of a von Neumann algebra W on a Hilbert space V consists of an antilinear self-adjoint involution J on V and a self-dual convex cone C in V whose elements are fixed under J and such that $JWJ = W'$ and $wJwJ(C) \subset C$ for $w \in W$, and $JwJ = w^*$ when $w \in W \cap W'$.

Consider a von Neumann algebra W with a faithful semifinite normal weight x and associated full left and right Hilbert algebras A and A', respectively. Let A_0 be those $v \in \cap_{n\in\mathbb{Z}} D(\Delta^n)$ with all $\Delta^n(v) \in A$. As $v = (1-\Delta^{-1})^{-1}(1-\Delta^{-1})(v) \in A'$ by Lemma 10.1.7, we have $A_0 \subset A \cap A'$ and $J(A_0) = A_0$. Let C_x be the closure of $\{wJ(w) \,|\, w \in A_0\}$. By Theorem 10.1.12 every element of C_x is fixed under J. For $v, w \in A_0$ we thus have $\pi_l(v)J\pi_l(v)J(wJ(w)) = vwJ(vw) \in C_x$. By

Kaplansky's density theorem we get $wJwJ(C_x) \subset C_x$ for $w \in \pi_l(A)''$. We also know that $J\pi_l(A)''J = \pi_l(A)'$ and $JwJ = w^*$ when $w \in \pi_l(A)'' \cap \pi_l(A)'$ by Proposition 10.1.13. Hence J and C_x is a standard form for $\pi_l(A)'' \cong W$ provided we can show that C_x is a self-dual convex cone in V_x. We do this in our next result.

Lemma 10.9.2 *Let x be a faithful semifinite normal weight on a von Neumann algebra with associated full left- and right Hilbert algebras A and A', respectively. Then $C_l = \{vv^* \mid v \in A\}^-$ and $C_r = \{\pi_r(S^*(w))(w) \mid w \in A'\}^-$ are mutual dual convex cones in V_x, and $C_x = (\Delta^{1/4}(C_l))^- = (\Delta^{-1/4}(C_r))^-$, and C_x is a self-dual convex cone in V_x such that $\Delta^{it}(C_x) = C_x$ for $t \in \mathbb{R}$.*

Proof Each cone is contained in the dual of the other because A^2 and $(A')^2$ are dense in V_x and $(\pi_l(v^*)(v)|\pi_r(S^*(w))(w)) = \|\pi_l(v)(w)\| \geq 0$ for $v \in A$, $w \in A'$. If $v \in V_x$ and $(v|\pi_r(S^*(w))(w)) \geq 0$ for $w \in A'$, again by polarization and the right hand version of Lemma 10.1.6, we see that $\pi_l(v)$ is a positive symmetric operator affiliated with $\pi_l(A)'' = W$. Let P be the spectral measure of its Friedrichs extension. Then the elements $P([0, n])v$ in A converge to v as $n \to \infty$. Letting Q be the spectral measure of the positive bounded operator $P([0, n])v$, we note that $\lambda(1 - Q([0, \lambda])) \leq P([0, n])v$ for $\lambda > 0$, so $x(1 - Q([0, \lambda])) < \infty$ and $v_m = q_x((1 - Q([0, 1/m]))P([0, n])v^{1/2}) \in A$. But $v_m^* v_m = (1 - Q([0, 1/m])P([0, n])v \to P([0, n])v$ as $m \to \infty$, so all $P([0, n])v \in C_l$ and thus $v \in C_l$. Hence C_l is the dual of C_r, and similarly, we see that C_r is the dual of C_l.

For the second statement, note that $C_l \subset D(S) = D(\Delta^{1/2})$, so $\Delta^{1/4}(C_l)$ makes sense. By Theorem 10.1.12, the Phragmen-Lindelöf theorem and uniqueness of analytic extensions, we know that $A_0 \subset D(\Delta^z)$ and that Δ^z is multiplicative on A_0 for $z \in \mathbb{C}$. Since $vJ(v) = \Delta^{1/4}(\Delta^{-1/4}(v)\Delta^{-1/4}(v)^*)$ for $v \in A_0$, we get the inclusion $C_x \subset (\Delta^{1/4}(C_l))^-$. Let $v \in A$. Then $v_s = (s/\pi)\int e^{-st^2}\Delta^{it}(v)\,dt \in D(\Delta^{iz})$ for $z \in \mathbb{C}$, and $\Delta^{iz}(v_s) = (s/\pi)\int e^{-s(t-z)^2}\Delta^{it}(v)\,dt$. For $w \in A'$ we have $\pi_r(w)\Delta^{iz}(v_s) = (s/\pi)(\int e^{-s(t-z)^2}\Delta^{it}\pi_l(v)\Delta^{-it}\,dt)(w)$, so $\Delta^{iz}(v_s)$ is left bounded by Lemma 10.6.2. Thus $\Delta^{iz}(v_s) \in A$ and $v_s \in A_0$. The same lemma also shows that $\lim v_s \to v$ and $\lim v_s^* \to v^*$ in norm. Hence $C_l = \{vv^* \mid v \in A_0\}^-$. If $v \in C_l$, pick v_n of the form ww^* with $w \in A_0$ such that $v_n \to v$ in norm. Since $S(v) = v$ and $S(v_n) = v_n$, we get $\Delta^{1/2}(v_n) = J(v_n) \to J(v) = \Delta^{1/2}(v)$, so $\|\Delta^{1/4}(v) - \Delta^{1/4}(v_n)\| = (\Delta^{1/2}(v - v_n)|v - v_n) \to 0$ and $\Delta^{1/4}(v) \in C_x$. Thus $C_x = (\Delta^{1/4}(C_l))^-$, and similarly $C_r = \{wS^*(w) \mid w \in A_0\}^-$ and $C_x = (\Delta^{-1/4}(C_r))^-$.

For the third statement, first note that C_x is contained in its dual because for $v \in C_l$ and $w \in C_r$ it suffices by the second statement to show $(\Delta^{1/4}(v)|\Delta^{-1/4}(w)) = (v|w) \geq 0$, which again holds by the first statement. Let $v \in V_x$ satisfy $(v|w) \geq 0$ for $w \in C_x$, and let v_s be as in the previous paragraph. As $J\Delta^{it} = \Delta^{it}J$ for real t, we have $\Delta^{it}(wJ(w)) = \Delta^{it}(w)J(\Delta^{it}(w)) \in C_x$ by Theorem 10.1.12, which incidentally proves the last statement, and shows that

$$(v_s|wJ(w)) = (s/\pi)\int e^{-st^2}(v|\Delta^{-it}(w)J(\Delta^{-it}(w))\,dt \geq 0.$$

Hence $(\Delta^{-1/4}(v_s)|wS^*(w)) = (v_s|\Delta^{-1/4}(w)J(\Delta^{-1/4}(w))) \geq 0$, so $\Delta^{-1/4}(v_s) \in C_l$ and $v_s \in C_x$ by the last line of the previous paragraph. Thus $v = \lim v_s \in C_x$, and C_x is self-dual. □

Observe that A_0 is dense in A by the proof above, so A_0^2 is dense in V_x.

Lemma 10.9.3 *Let x, y be faithful semifinite normal weights on a von Neumann algebra W with semicyclic representations π_x, π_y and standard forms with self-dual cones C_x, C_y, respectively. Then there is a unique unitary $u_{xy}: V_x \to V_y$ such that $u_{xy}\pi_y(\cdot) = \pi_x(\cdot)u_{xy}$ and $u_{xy}(C_y) = C_x$.*

Proof Using notation from that proof of Connes' cocycle theorem, consider the unitary intertwiner $u_{xy} = J_x J_{xy} = J_{xy} J_y$ of π_x and π_y. For $v \in N_x \cap N_x^*$ and $w \in N_y \cap N_y^*$ we have $w^*v \in N_x \cap N_y$ and a straightforward calculation gives

$$(\pi_x(v)J_x(q_x(v))|u_{xy}\pi_y(w)J_y(q_y(w))) = (\Delta_{xy}^{1/2}q_x(w^*v)|q_x(w^*v)) \geq 0,$$

so self-duality of C_x yields $u_{xy}(C_y) \subset C_x$ and equality is obtained using $u_{xy}^* = u_{yx}$.

If u was another unitary intertwiner of π_x and π_y such that $u(C_y) = C_x$, then $U = u_{xy}u^* \in \pi_x(W)'$ and $U(C_x) = C_x$. Since element of C_x are fixed under J_x, we get $U = J_x U J_x$ by polarization and denseness, so U is central in $\pi_x(W)$ and $U = U^* = U^{-1}$. By spectral theory, letting p and q be the central orthogonal projections of $\pi_x(W)$ corresponding to the characteristic functions that are one where the function corresponding to U is positive and negative, we get $U = p - q$ and $pq = 0$ and $p + q = 1$. Since $qJ_x = J_x q$ as U commutes with J_x, and q is central and $q^2 = q$, we get $q(vJ_x(v)) = q(v)J_x(q(v)) \in C_x$ for $v \in A_0$, so $q(C_x) \subset C_x$. If $q(w) = w$ for $w \in C_x$, then $(U(w)|w) = -\|q(w)\|^2 \leq 0$, so $q(w) = 0$ as C_x is self-dual, and thus by polarization and denseness of A_0^2 in V_x, we get $q = 0$, so $U = 1$, and uniqueness is established. □

Lemma 10.9.4 *Let x be a faithful semifinite normal weight on a von Neumann algebra W with semicyclic representation π_x and standard form with self-dual cone C_x. For $v \in C_x$ consider $(\cdot(v)|v)$ on W with support projection p, and denote the restriction of the vector functional to W_p by y, so y is a faithful positive normal bounded functional on W_p with cyclic representation π_y having cyclic vector v_y. Then the formula $u_{xy}(\pi_y(w)(v_y)) = \pi_x(w)(v)$ for $w \in W_p$ defines a unique isometry $u_{xy}: V_y \to \overline{W(v) \cap W'(v)}$ such that $u_{xy}\pi_y(w) = \pi_x(w)u_{xy}$ for $w \in W_p$ and $u_{xy}(C_y) = C_x \cap \overline{W(v)} \cap \overline{W'(v)}$.*

Proof Note that $p \in W$ is the orthogonal projection onto $\overline{W'(v)}$ and that $J_x p J_x \in W'$ is the orthogonal projection onto $\overline{W(v)}$, so the orthogonal projection $pJ_x pJ_x$ commutes with J_x and u_{xy} is an isometry from V_x onto $pJ_x pJ_x(V_x)$ such that $u_{xy}\pi_y(w) = \pi_x(w)u_{xy}$ for $w \in W_p$. Then $J = u_{xy}^* J_x u_{xy}$ fixes v_y and $J\pi_y(W_p)J = \pi_y(W_p)'$. For $w \in W_p$ we have

$$(JS_y(\pi_y(w)(v_y))|\pi_y(w)(v_y)) = (J\pi_y(w^*)J\pi_y(w^*)(v_y)|v_y) \geq 0,$$

so $JS_y = \Delta_y^{1/2}$ and $J = J_y$ by uniqueness of the polar decomposition of S_y. Thus $J_x u_{xy} = u_{xy} J_y$. Now $\{J_y \pi_y(w) J_y \pi_y(w)(v_y) \mid w \in W_p\}^- \subset C_y$ and as both are self-dual convex cones, we actually have equality, and $u_{xy} J_y \pi_y(w) J_y \pi_y(w)(v_y) = J_x \pi_x(w) J_x \pi_x(w)(v)$ for $w \in W_p$, so $u_{xy}(C_y) \subset C_x$. By self-duality of C_y we get $u_{xy}^*(C_x \cap \overline{W(v)} \cap \overline{W'(v)}) \subset C_y$. The proof of uniqueness is similar to that in the proof above. □

Theorem 10.9.5 *Let x be a faithful semifinite normal weight on a von Neumann algebra W with semicyclic representation π_x and standard form with self-dual cone C_x. For each positive $y \in W_*$ there exists a unique $v \in C_x$ such that $y = (\pi_x(\cdot)(v)|v) \equiv y_v$. Moreover, we have the inequalities*

$$\|v - w\|^2 \leq \|y_v - y_w\| \leq \|v - w\|\|v + w\|$$

for $v, w \in C_x$.

Proof Assume first that $y \leq x$ and $y\sigma_t = y$ for all real t, where $\{\sigma_t\}$ is the modular automorphism group of x. From the proof of the theorem where the left Hilbert algebra is constructed from x, we know there is positive $h \in \pi_x(W)'$ and $v \in V_x$ such that $h^{1/2}(q_x(\cdot)) = \pi_x(\cdot)(v)$ and $y = (\pi_x(\cdot)(v)|v)$. Hence $v \in C_r$ from Lemma 10.9.2, and as $\Delta_x^{it} q_y = q_y$, we get $v \in \Delta_x^{-1/4}(C_r) \subset C_x$.

In the general case pick a semifinite normal weight y' on W with support projection $1 - s(y)$, and then replace x by $y + y'$ in the paragraph above to get $v \in C_{y+y'}$ such that $y = (\pi_{y+y'}(\cdot)(v)|v)$. Then the vector $u_{x(y+y')}(v)$ will do the trick, where $u_{x(y+y')}$ is as in Lemma 10.9.3.

If $v_i \in C_x$ and $y = (\pi_x(\cdot)(v_i)|v_i)$, there is a partial isometry $u \in W'$ such that $u^*(v_2) = v_1$, so $\overline{W'(v_1)} = \overline{W'(v_2)}$ and $\overline{W(v_1)} = J_x(\overline{W'(v_1)}) = \overline{W(v_2)}$. Hence if we use v_i instead of v in the previous lemma, the uniqueness result tells us that we get the same isometry u_{xy} for v_1 and v_2, so $v_1 = v_2$.

Since $y_v - y_w = ((\pi_x(\cdot)(v + w)|v - w) + (\pi_x(\cdot)(v - w)|v + w))/2$ we get $\|y_v - y_w\| \leq \|v - w\|\|v + w\|$ for $v, w \in C_x$. For the other inequality, write $v - w = v_+ - v_-$, where $v_\pm \in C_x$ are mutually orthogonal. This is indeed possible because let v_+ be the vector in the closed convex cone C_x with least distance to $v - w$, so v_+ is orthogonal to $v - w - v_+$, and if $0 \neq u \in C_x$, then $\|v - w - v_+\|^2 < \|v - w - (v_+ + cu)\|^2$ for $c > 0$, which is only possible if $(v - w - v_+|u) \leq 0$, so $v_- \equiv v_+ - v + w \in C_x$ by self-duality. By Lemma 10.9.6 we know that $s(y_{v_+}) \perp s(y_{v_-})$, so with $p = s(y_{v_+}) - s(y_{v_-})$ we get $\|y_v - y_w\| \geq |(y_v - y_w)(p)|$ and

$$|(y_v - y_w)(p)| = (v+w|v_+ + v_-) \geq (v|v_+) - (v|v_-) - (w|v_+) + (w|v_-) = \|v - w\|^2$$

as $p(v_\pm) = \pm v_\pm$. □

Lemma 10.9.6 *Let x be a faithful semifinite normal weight on a von Neumann algebra W with semicyclic representation π_x and standard form having self-dual cone C_x. Let p_i be the support projection of the functional $(\pi_x(\cdot)(v_i)|v_i)$ on W for $v_i \in C_x$. Then $p_1 p_2 = 0$ if $v_1 \perp v_2$.*

Proof From the first part of the previous theorem and the previous lemma, we may for $v \in C_x$ and $y_v \in W_*$, by using u_{xy_v}, identify the pair (V_x, v_x) with $(\overline{W(v) \cap W'(v)}, v)$, so $C_v \equiv C_{y_v} \subset C_x$ and π_x restricted to W_p is π_{y_v}, where p is the support projection in W of y_v. We claim that $C_v = \{(\mathbb{R}_+ v - C_x) \cap C_x\}^-$ for $v \in C_x$, where $\mathbb{R}_+ = \langle 0, \infty\rangle$. The inclusion $C_v \subset \{(\mathbb{R}_+ v - C_x) \cap C_x\}^-$ if clear from the identity $\Delta_v^{1/4}\pi_x(w)(v) = \|w\|v - \Delta_v^{1/4}\pi_x(\|w\|p - w)(v)$ for positive $w \in W_p$. For the reverse inclusion, say $v_1 - v_2 \in C_x$, and pick a semifinite normal weight y' on W with support projection $1 - p_1$, and let $y = y' + y_{v_1}$. By Lemma 10.9.3 we may therefore assume that $y_{v_1}\sigma_t = y_{v_1}$ for all real t, where $\{\sigma_t\}$ is the modular automorphism group of x. So $\Delta_x(v_1) = v_1$. For $s > 0$ let $(v_i)_s$ and $(v_1 - v_2)_s$ be as in the proof of Lemma 10.9.2. Then $v_1 = (v_1)_s = (v_2)_s + (v_1 - v_2)_s$ for all $s > 0$. Staying with the notation from this lemma, we have $v_1 = \Delta^{-1/4}((v_2)_s) + \Delta^{-1/4}((v_1 - v_2)_s) \in C_l$ and with the same true for each term. Since v_1 is left- and right bounded, we conclude that each of these terms are left bounded with $\pi_l(\Delta_x^{-1/4}((v_2)_s)) \leq \pi_l(v_1) \leq p_1$ and $\pi_l(\Delta_x^{-1/4}((v_1 - v_2)_s)) \leq p_1$. Since also p_1 and Δ_x commute, we get $p_1((v_2)_s) = (v_2)_s$ and $p_1((v_1 - v_2)_s) = (v_1 - v_2)_s$. As these approach v_2 and $v_1 - v_2$, respectively, as $s \to \infty$, we get $v_2 = p_1(v_2) = J_x p_1 J_x p_1(v_2) \in C_{v_1}$ and $v_1 - v_2 = p_1(v_1 - v_2) = J_x p_1 J_x p_1(v_1 - v_2) \in C_{v_1}$, so $(\mathbb{R}_+ v_1 - C_x) \cap C_x \subset C_{v_1}$ and the claim holds.

If $v_1 \perp v_2$ for $v_i \in C_x$, then $(\mathbb{R}_+ v_1 - C_x) \cap C_x \perp (\mathbb{R}_+ v_2 - C_x) \cap C_x$. Indeed, if $w_i \in C_x$ and $c_i > 0$ with $c_i v_i - w_i \in C_x$, then $0 \leq (w_1|w_2) \leq c_1(v_1|w_2) \leq c_1c_2(v_1|v_2) = 0$. Hence $C_{v_1} \perp C_{v_2}$. It remains to observe that the closed span of C_{v_i} is $\overline{W(v_i) \cap W'(v_i)}$ to conclude that $p_1 p_2 = 0$. □

Theorem 10.9.7 *If (J_i, C_i) are standard forms of the von Neumann algebra W_i on V_i, and $f: W_1 \to W_2$ is a $*$-isomorphism, then there is a unique unitary $u: V_1 \to V_2$ such that $f = u \cdot u^*$ and $J_2 = uJ_1u^*$ and $C_2 = u(C_1)$.*

Proof Given a standard form (J, C) of a von Neumann algebra W on V, we show that there exists a unitary $u: V \to V_x$ such that $\pi_x = u \cdot u^*$ and $J_x = uJu^*$ and $C_x = u(C)$. Uniqueness of u is then clear from Lemma 10.9.3.

For $v \in C$ let $V_v = \overline{W(v) \cap W'(v)}$ and let p_v be the support projection in W of $(\cdot(v)|v)$. Now $JS_v = \Delta_v^{1/2}$ and $J_v = J$ by the argument in the proof of Lemma 10.9.4. Since $C_v = \{wJw(v) \mid w \in M_{p_v}\}^-$, we get $C_v \subset C$, and the self-duality of C_v in V_v gives $C_v = C \cap V_v$.

Assume that W is σ-finite. By Lemma 10.9.6 any maximal orthogonal family of vectors $v_i \in C$ is then countable, and we may assume that $\sum \|v_i\|^2 < \infty$. The same lemma then shows that $(\cdot(v)|v)$ with $v = \sum v_i$ is faithful, so $V_v = V$ and $C_v = C$.

In general for each orthogonal projection $p \in W$ set $V(p) = p(V) \cap J(p(V))$. If p is σ-finite, then $C \cap V(p) = C_v$ for some $v \in C$, and there exists a unique unitary $u_p: V(p) \to \pi_x(p)J_x\pi_x(p)(V_x)$ such that $u_p\pi_x(w) = \pi_x(w)u_p$ for $w \in W_p$, and $u_p(C_v) = C_x \cap u_p(V(p))$. If $q \leq p$, then u_p extends u_q by uniqueness of u_q. The family of σ-finite orthogonal projections in W is upward filtered with supremum 1, so there is a common extension u of all u_p. Since $C = \cup C_v$ this u has the required properties. □

Definition 10.9.8 Let W be a von Neumann algebra on V, and let G be a group of $*$-automorphism of W. A *unitary implementation* of G is a representation $g \mapsto u_g$ of G on V as unitary operators such that $g(w) = u_g w u_g^*$ for $g \in G$ and $w \in W$.

Theorem 10.9.9 *Let (J, C) be a standard form of a von Neumann algebra W on V. Then the whole group* $\mathrm{Aut}(W)$ *of $*$-automorphism of W has a unique unitary implementation $g \mapsto u_g$ such that $u_g J u_g^* = J$ and $u_g(C) = C$ for $g \in$* $\mathrm{Aut}(W)$. *Moreover, the implementation is a homeomorphism from* $\mathrm{Aut}(W)$ *onto a closed subgroup of the unitary group on V with strong (weak) operator topology, where the topology on* $\mathrm{Aut}(W)$, *turning it into a topological group, is induced from the seminorms $g \mapsto \|x \circ g\|$ for $x \in W_*$.*

Proof The previous theorem shows that there to $g \in \mathrm{Aut}(W)$ exists a unique unitary $u_g \in B(V)$ such that $g = u_g \cdot u_g^*$ and $u_g J u_g^* = J$ and $u_g(C) = C$. The uniqueness part also shows that $u_{gh} = u_g u_h$ for $g, h \in \mathrm{Aut}(W)$, so we have a unitary implementation.

The given seminorms certainly turn $\mathrm{Aut}(W)$ into a topological group, and Theorem 10.9.5 with its inequalities show that $v \mapsto y_v$ is a homeomorphism from C onto the positive part of W_*. Hence $g \mapsto u_g$ is a homeomorphism onto its range, which consists of the unitaries $u \in B(V)$ satisfying $uWu^* = W$ and $uJu^* = J$ and $u(C) = C$, and this set is strongly closed relative to the unitary group of $B(V)$. □

The unitary implementation of $\mathrm{Aut}(W)$ is the *canonical implementation*.

10.10 Spatial Derivative

Let x be a faithful semifinite normal weight on a von Neumann algebra $W \subset B(V)$. Then $v \in V$ is *x-bounded* if the linear map $q_x(w) \to w(v)$ for $w \in N_x$ extends to a bounded map $R_v^x \colon V_x \to V$. The space D_x of x-bounded vectors is dense in V. Indeed, writing $x = \sum(\cdot v_i | v_i)$ for $v_i \in V$, we clearly have $v_i \in D_x$, and since D_x is W'-invariant, the orthogonal projection p onto the closure of D_x belongs to W. As $pv_i = v_i$ we get $x(1-p) = 0$, so $p = 1$. Clearly $R_{w(v)}^x = wR_v^x$ for $w \in W'$, while $wR_v^x = R_v^x \pi_x(w)$ for $w \in W$. Picking a net $\{w_i\}$ in N_x such that $w_i \to 1$ strongly, then $R_v^x(w_i) \to v$, so $v \in \overline{R_v^x(V_x)}$. The subspace F_x of $B(V)$ spanned by $R_v^x(R_u^x)^*$ for $u, v \in D_x$ is a strongly*-dense ideal of W'. By the above we only need to check denseness. But to a non-zero $w \in W'$, we may pick $v \in D_x$ with $w(v) \neq 0$, so $0 \neq R_{w(v)}^x(R_{w(v)}^x)^* \leq \|R_v^x\|^2 ww^*$, which tells us that every positive element of W' majorizes non-zero elements of the ideal F_x, so the latter must be strongly*-dense in W'. Moreover, every positive element b of F_x is a finite sum of elements $R_v^x(R_v^x)^*$. Indeed, say $b = \sum R_{v_k}^x(R_{u_k}^x)^*$. Then $b = (b+b^*)/2 \leq c \equiv 2^{-1}\sum R_{v_k+u_k}^x(R_{v_k+u_k}^x)^*$, and picking $w \in W'$ with $wcw^* = b$, we get $b = 2^{-1}\sum R_{w(v_k+u_k)}^x(R_{w(v_k+u_k)}^x)^*$. This also shows that there exists a family of $v_i \in D_x$ such that $\sum R_{v_i}^x(R_{v_i}^x)^* \to 1$ in the strong*-topology.

Recall that when $W = \pi_x(W)$, then $R_v^x = \pi_r(v)$, and the weight x_r on W' associated to $x = x_l$ is given by $x_r(\pi_r(v)^*\pi_r(v)) = (v|v)$. In general we claim that $x_r((R_v^x)^*R_v^x) = (v|v)$ for $v \in D_x$. Indeed, the partial isometry $U: V \to V_x$ in the polar decomposition of $(R_v^x)^*$ satisfies $U^*U(v) = v$ and $\pi_x(w)U = Uw$ for $w \in W$, so $\pi_r U(v) = UR_v^x$, and $x_r((R_v^x)^*R_v^x) = x_r(\pi_r U(v)^*\pi_r U(v)) = (U(v)|U(v)) = (v|v)$.

Let y be a faithful semifinite normal weight on W'. Then the function f on D_y sending v to $x(R_v^y(R_v^y)^*) \in [0, \infty]$ is lower semicontinuous since writing $x = \sum(\cdot v_i|v_i)$, we get $f(v) = \sum \|(R_v^y)^*v_i\|^2 = \sum \sup|(w^*v_i|v)|$, where we sup over $w \in N_y$ of norm not greater than one. Also, since N_x^* is strongly dense in W as x is semifinite, and $f(wv) \leq \|R_v^y\|^2 x(ww^*) < \infty$, we see that f restricted to those v with finite image, has an extension to a closed positive quadratic form which is minimal as such. Let T be the associated greatest positive self-adjoint operator in V as given by Proposition 9.3.17. We denote T by dx/dy and call it the *spatial derivative* of x with respect to y. So $((dx/dy)v|v) = x(R_v^y(R_v^y)^*)$ for $v \in D_x$.

Theorem 10.10.1 *Let $W \subset B(V)$ be a von Neumann algebra, let x and y be faithful semifinite normal weights on W and W', respectively. Then dx/dy is strictly positive with $(dx/dy)^{-1} = dy/dx$, and $\sigma_t^x = (dx/dy)^{it}(\cdot)(dx/dy)^{-it}$ while $\sigma_t^y = (dx/dy)^{-it}(\cdot)(dx/dy)^{it}$ for real t. For any faithful semifinite normal weight z on W we have $(dz/dy)^{it} = (Dz : Dx)_t(dx/dy)^{it}$ for real t.*

Proof Consider the faithful semifinite normal weight $y_r = y(J_y(\cdot)^*J_y)$ on $\pi_y(W')'$, so $y_r(\pi_r(v)^*\pi_r(v)) = (v|v)$ for $v \in V_y$ with $\pi_r(v)$ bounded. The semicyclic representation of y_r is unitarily equivalent to the identity representation with $S_{y_r} = S_y^*$ and $\Delta_{y_r} = \Delta_y^{-1}$ and $J_{y_r} = J_y$. Let π be the representation of W' on $V \oplus V_{y_r}$ given by $\pi(w) = \begin{pmatrix} w & 0 \\ 0 & \pi_{y_r}(w) \end{pmatrix}$, and let z be the balanced weight on $\pi(W')' \subset B(V \oplus V_{y_r})$ of x and y_r. Let $\{S_t^x\}$ be the one-parameter group of isometries on the space of intertwiners $a_{12}: V_y \to V$ of π_y and the identity representation of W' that satisfy

$$\sigma_t^z\begin{pmatrix} 0 & a_{12} \\ 0 & 0 \end{pmatrix} = \begin{pmatrix} 0 & S_t^x(a_{12}) \\ 0 & 0 \end{pmatrix}$$

for real t. We claim that there exists a strongly continuous one-parameter group of unitaries $u_t^x \in B(V)$ uniquely determined by $\sigma_t^x = u_t^x(\cdot)u_{-t}^x$ and $\sigma_t^y = u_{-t}^x(\cdot)u_t^x$ and $S_t^x(R_v^y) = R_{u_t^x(v)}^y$ for $v \in D_y$. Uniqueness is clear from the last formula. As for existence, the argument goes in three steps. Firstly, if the claim holds for one specific weight x_0, then it holds for any x, as is seen by setting $u_t^x = (Dx : Dx_0)_t u_t^{x_0}$ and using $R_{u_t^x(v)}^y = (Dx : Dx_0)_t R_{u_t^{x_0}(v)}^y$ in combination with the fact that for the balanced weight z_0 of x_0 and y_r, we have $(Dz : Dz_0)_t = \begin{pmatrix} (Dx : Dx_0)_t & 0 \\ 0 & 0 \end{pmatrix}$. Secondly, if the claim holds in one realization of W', then it holds also in any other realization. To verify this, note that by Proposition 8.2.9 any $*$-isomorphism is a composition of an amplification, an injective induction and a spatial isomorphism, so we need only check it for each these cases, and for spatial isomorphisms it is clear. As for amplifications, say the claim holds for $W' \subset B(V)$, so we must check it for $W' \otimes \mathbb{C} \subset B(V)\bar{\otimes}B(l^2(I))$ for an arbitrary set I. In this case $(W' \otimes \mathbb{C})' = W\bar{\otimes}B(l^2(I)) \subset B(V)\bar{\otimes}B(l^2(I))$. But $v = \{v_i\} \in V \otimes l^2(I)$ is y-bounded when $v_i \in V$ is y-bounded and $\sum \|R^y_{v_i}\|^2 < \infty$, and then $R^y_v(u)_i = R^y_{v_i}(u)$ for $u \in V$. Replacing x by $x \otimes \mathrm{Tr}$ on $W\bar{\otimes}B(l^2(I))$, then as $\sigma_t^{x\otimes\mathrm{Tr}} = \sigma_t^x \otimes \iota$, we see that $u_t^{x\otimes\mathrm{Tr}} = u_t^x \otimes 1$ will do for the amplification. Hence by the first step above, the amplification case holds. For injective inductions, we need to prove that if the claim holds for $W' \subset B(V)$, then it holds for $W'p \subset B(pV)$, where $p \in W$ is an orthogonal projection with full central support. We may assume that $p \in W_x$ by choosing x correctly. Indeed, if W is semifinite, we may use any tracial x, and if W is of type III, then $W \cong M_2(pWp)$ and we may take x to be the balanced weight of any faithful semifinite normal weight on pWp with itself. When $p \in W_x$, then x restricts to a faithful semifinite normal weight on pWp, with modular group given by restriction. Then $u_t^x p = pu_t^x$ will do the job since pD_y will be the y-bounded vectors in pV. Again, by step one, the specific choice of x is irrelevant, and we are done for injective inductions. The third step consists of assuming $W' = \pi_y(W')$, so $W = \pi_y(W')'$, and then letting $y_r = x$. By the first two steps above, we are permitted to do this. Then $u_t^x = \Delta_y^{-it}$ will do as $S_t^x = \sigma_t^x$ and $S_t^x(R_v^y) = \sigma_t^x(\pi_r(v)) = R^y_{u_t^x(v)}$. We have settled the claim.

By Stone's theorem there exists a unique strictly positive operator A in V such that $A^{it} = u_t^x$ for real t. We show that $A = dx/dy$. Indeed, consider again the balanced weight z, write $\Delta_z = \int \lambda\, dP(\lambda)$, and to $a \in N_z \subset \pi(W')'$, consider the regular positive Borel measure μ on $\langle 0, \infty\rangle$ given by $d\mu(\lambda) = (dP(\lambda)\pi_z(a)|\pi_z(a))$. Then $z(a^*\sigma_t^z(a)) = (\Delta_z^{it}\pi_z(a)|\pi_z(a)) = \int \lambda^{it}\, d\mu(\lambda)$, while $z(aa^*) = \|\pi_z(a)^*\|^2 = \|S_z(\pi_z(a))\| = \|\Delta_z^{1/2}\pi_z(a)\|^2 = \int \lambda\, d\mu(\lambda)$ for real t. Note also that μ is finite as $D(S_z) \cap N_z = N_z \cap N_z^*$. Pick a with $a_{12} = R_v^y$ for $v \in D_y$ as the only non-zero entry. Then $\int \lambda^{it}\, d\mu(\lambda) = y_r((R_v^y)^* S_t^y(R_v^y)) = (A^{it}v|v)$, so μ is the spectral measure associated to A and v, and $\|A^{1/2}v\|^2 = \int \lambda\, d\mu(\lambda) = x(R_v^y(R_v^y)^*) = \|(dx/dy)^{1/2}v\|^2$. But thanks to the proved claim, the automorphism on $B(V)$ implemented by u_t^x fixes W, W', x, y, and thus also dx/dy, so the latter operator commutes with A, and thus must equal A.

To prove the last part of the theorem, and the property $(dx/dy)^{-1} = dy/dx$, consider the balanced weight z' on $M_2(W')$ of faithful semifinite normal weights y_1 and y_2 on W'. Let $\tilde{x}$ be the weight on $M_2(W')'$ corresponding to x on the isomorphic W. Identify $V_{z'}$ with $V_{y_1} \oplus V_{y_2} \oplus V_{y_1} \oplus V_{y_2}$. Then $v = (v_1, v_2) \in V \oplus V$ is z'-bounded if and only if $v_i \in D_{y_i}$, and then $R_v^{z'} = \begin{pmatrix} R_{v_1}^{y_1} & R_{v_2}^{y_2} & 0 & 0 \\ 0 & 0 & R_{v_1}^{y_1} & R_{v_2}^{y_2} \end{pmatrix}$, so $\tilde{x}(R_v^{z'}(R_v^{z'})^*) = x(R_{v_1}^{y_1}(R_{v_1}^{y_1})^*) + x(R_{v_2}^{y_2}(R_{v_2}^{y_2})^*)$, and $d\tilde{x}/dz' = \mathrm{diag}(dx/dy_1, dx/dy_2)$. Hence, by what we have already proved, we see that $\begin{pmatrix} 0 & 0 \\ (Dy_2 : Dy_1)_t & 0 \end{pmatrix}$ equals

$\sigma_t^{z'}\begin{pmatrix}0&0\\1&0\end{pmatrix} = \begin{pmatrix}(dx/dy_1)^{-it}&0\\0&(dx/dy_2)^{-it}\end{pmatrix}\begin{pmatrix}0&0\\1&0\end{pmatrix}\begin{pmatrix}(dx/dy_1)^{it}&0\\0&(dx/dy_2)^{it}\end{pmatrix}$, so $(dx/dy_2)^{-it}(dx/dy_1)^{it} = (Dy_2 : Dy_1)_t$. Combining this with the arguments in step one and two above, we get $(dx/dy)^{-1} = dy/dx$, since when $W' = \pi_y(W')$ and $x = y_r$ it is trivially true. This also settles the last part of the theorem. □

Corollary 10.10.2 *Let y be a faithful semifinite normal weight on the commutant of a von Neumann algebra $W \subset B(V)$. For any strictly positive operator A in V we have $\sigma_{-t}^{y} = A^{it}(\cdot)A^{-it}$ for real t if and only if there is a faithful semifinite normal weight x on W such that $dx/dy = A$.*

Proof For the forward direction, if z is any faithful semifinite normal weight on W, the map $t \mapsto A^{it}(dz/dy)^{-it}$ is a σ_t^z-cocycle, so by Connes' cocycle theorem there is a unique faithful semifinite normal weight x on W such that $(Dx : Dz)_t = A^{it}(dz/dy)^{-it}$. By the last statement in the theorem above, we thus get $(dx/dy)^{it} = A^{it}$ and $dx/dy = A$. The converse direction is immediate from the theorem. □

10.11 Weights and Conditional Expectations

Theorem 10.11.1 *Let V be a von Neumann subalgebra of a von Neumann algebra W with a faithful semifinite normal weight x. Then the restriction of x to V is semifinite and V is invariant under the modular automorphism group of x if and only if there is a faithful normal conditional expectation $E : W \to V$ such that $xE = x$. Any such conditional expectation is unique.*

Proof Given such an E, even without assuming faithfulness, we see that the restriction y of x to V is a faithful semifinite normal weight. The condition $xE = x$ viewed in the semicyclic representation of y clearly assures that E is unique. Let A_x and A_y be the left Hilbert algebras associated to the weights x and y with Hilbert spaces V_x and V_y for the semicyclic representations π_x and π_y of W and V, respectively. Let P be the orthonormal projection of V_x onto V_y viewed as a closed subspace of V_x such that $q_y(v) = q_x(v)$ for $v \in N_y = N_x \cap V$, so $P \in \pi_x(V)'$, and $\pi_x(v)$ restricted to V_y is $\pi_y(v)$. For $w \in N_x$ we have $(q_x(w)|q_x(v)) = xE(v^*w) = x(v^*E(w)) = (q_x(E(w))|q_x(v))$, so $Pq_x(w) = q_x(E(w))$. Since E is $*$-preserving we thus get $PS_x \subset S_xP$, where S_x is the closure of the involution in A_x. Hence $(1-2P)S_x \subset S_x(1-2P)$ and $(1-2P)S_x(1-2P) \subset S_x$ by multiplying with $1-2P$ from the right, so $S_x \subset (1-2P)^2S_x(1-2P)^2 \subset (1-2P)S_x(1-2P) \subset S_x$ and we obtain $S_x = (1-2P)S_x(1-2P)$. By taking the adjoint, this equality remains true when S_x is replaced by S_x^* and hence also by the modular operator $\Delta_x = S_x^*S_x$. So P and Δ_x commute, and as $PA_x = A_y = A_x \cap V$, we get $\Delta_x^{it}A_y = \Delta_x^{it}PA_x = P\Delta_x^{it}A_x = PA_x = A_y$ for all real t. Then $q_x(\sigma_t^x(v)) = \Delta_x^{it}q_x(v)$ for $v \in N_x \cap N_x^* \cap V$ shows that the modular group $\{\sigma_t^x\}$ of x leaves $N_x \cap N_x^* \cap V$ invariant, and the latter is dense in V, so V is invariant.

Conversely, if $\sigma_t^x(V) \subset V$ for all real t, and y is semifinite, then y is certainly a faithful semifinite normal weight on V. Now $q_x(\sigma_t^x(v)) = \Delta_x^{it} q_x(v)$ for $v \in N_x \cap N_x^*$ shows that $\Delta_x^{it} A_y = A_y$, so $\Delta_x^{it} P = P\Delta_x^{it}$, and Δ_y^{it} is the restriction of Δ_x^{it} to V_y. The same holds for $\Delta_y^{\pm 1/2}$ and S_y and S_y^* and the modular conjugation J_y of y. By definition $A_y \subset A_x \cap V_y$ and $A_x' \cap V_y \subset A_y'$, and by applying J_x twice we see that these inclusions are not proper. For $w \in A_x$ and $u \in A_y'$ we thus have $\pi_l^y(Pw)u = \pi_r^y(u)Pw = P\pi_r^x(u)w = P\pi_l^x(w)u$, so $A_y = PA_x$ and $\pi_l^y P(w) = P\pi_l^x(w)$. Hence $\pi_y(V) = P\pi_x(W)P$. It is now straightforward to check that $\pi_y(E(w)) = P\pi_x(w)P$ defines a faithful normal conditional expectation $E\colon W \to V$ such that $E\sigma_t^x = \sigma_t^y E$ for all real t, and such that $yE(v^*wv) = x(v^*wv)$ for $w \in W$ and $v \in N_y$. As $N_y^* W N_y$ is a σ-weakly dense and σ^x-invariant $*$-subalgebra of the definition domain M_x of x, we get $yE = x$ from Corollary 10.7.10. □

The conditional expectation E in the theorem is given by $\pi_y(E(w)) = P\pi_x(w)P$ for $w \in W$, where P is the orthogonal projection onto $V_y \subset V_x$.

Corollary 10.11.2 *Let W be a von Neumann algebra with a von Neumann subalgebra V such that $V' \cap W$ is the center of V. Then any normal conditional expectation $E\colon W \to V$ is faithful and unique. In fact, if y is any faithful normal semifinite weight on V, then E is the conditional expectation associated to the weight yE on W as described in the theorem above.*

Proof The weight yE is normal and semifinite, and we claim it is also faithful. The left kernel of it is $W(1 - s(yE))$, where $s(yE) \in W$ is the support of yE. Since the left kernel consists of all $w \in W$ such that $E(w^*w) = 0$, and E is a conditional expectation onto V, we see that $W(1-s(yE))$ is invariant under right multiplication by elements of V, so $s(yE) \in V' \cap W$ which by assumption is the center of V. But $y = yE$ on V, so $s(yE) = 1$.

If $F\colon W \to V$ is another normal conditional expectation, then the modular automorphism groups of yF and yE restricted to V coincide with the one of y, so the cocycle derivative $(DyF : DyE)_t$ belongs to the center of V, where it acts trivially, so by Stone's theorem there exists a strictly positive operator h affiliated with the center such that $(DyF : DyE)_t = h^{it}$ for real t. Hence $yF = yE(h\cdot)$, and as $yF = yE$ on V, we get $h = 1$, so $F = E$ by the theorem above. □

10.12 The Extended Positive Part of a von Neumann Algebra

Definition 10.12.1 The *extended positive part* $\bar{W}_+$ of a von Neumann algebra W is the set of additive lower semicontinuous maps $(W_*)_+ \to [0, \infty]$ that respect multiplication by non-negative scalars.

Evaluation of positive normal functionals at positive elements of a von Neumann algebra W shows that W_+ can be considered a subset of $\bar{W}_+$. The extended positive part is clearly closed under pointwise addition and multiplication by non-negative scalars and increasing limits, and also under the operation that sends $m \in \bar{W}_+$

to $a^*ma \equiv m(a(\cdot)a^*)$ for $a \in W$, where $a\omega a^* = \omega(a^*(\cdot)a)$ for $\omega \in W_*$. The following alternative description is useful.

Proposition 10.12.2 *Let $W \subset B(V)$ be a von Neumann algebra with an orthogonal projection $e \in W$. Let $A = \int \lambda\, dP(\lambda)$ be a positive self-adjoint operator in eV affiliated with the von Neumann algebra eWe. Then $m_A(x) = \int \lambda\, dx(P(\lambda)) + \infty x(1-e)$ for positive $x \in W_*$ defines $m_A \in \bar{W}_+$ with the property that $m_A(x_v)$ equals $\|A^{1/2}v\|^2$ when $v \in D(A^{1/2})$ and is infinite otherwise, and $A \mapsto m_A$ defines a one-to-one correspondence between the set of such (e, A) and $\bar{W}_+$.*

Proof As $m_A(x) = \sup_{n\in\mathbb{N}} x(\int^n \lambda\, dP(\lambda)) + \infty x(1-e)$, we see that m_A, which clearly respects addition and multiplication by non-negative scalars, is lower semicontinuous. The formulas for $m_A(x_v)$ are obvious. Injectivity of $A \mapsto m_A$ is clear from uniqueness in the spectral decomposition of A. As for surjectivity, suppose $m \in \bar{W}_+$. Then the map $V \to [0, \infty]$ which sends v to $m(x_v)$ is clearly lower semicontinuous, satisfies the polarization identity, sends λv to $|\lambda|^2 m(x_v)$ for $\lambda \in \mathbb{C}$, and sends uv to $m(x_v)$ for any unitary $u \in W'$. Hence the set of those $v \in V$ such that $m(x_v) < \infty$ is a subspace U_0 of V, and there is a positive sesquilinear form $\langle\cdot|\cdot\rangle$ on U_0 such that $\langle v|v\rangle = m(x_v)$ for $v \in U_0$. Let T be the quotient map from U_0 into its normification, and viewing it as an unbounded operator from the closure U of U_0, we claim it is closed. Say $v_n \to v \in U$ and $Tv_n \to v'$ for a sequence $\{v_n\}$ in U_0. Then $m(x_v) \leq \sup m(x_{v_n}) = \|Tv_n\|^2 < \infty$, so $v \in U$, and $v' = Tv$, again by lower semicontinuity. Thus $A = T^*T$ is a positive operator in U and $m(x_v) = \|Tv\|^2 = \|A^{1/2}v\|^2 = m_A(x_v)$ for any $v \in U_0$ by the spectral decomposition of A, so $m = m_A$ as normal states on W are infinite sums of vector states. The orthogonal projection e onto U belongs to W, and A is affiliated to W, both due to the property above involving $u \in W'$. □

Remark 10.12.3 From the above result it is clear that if $m \in \bar{W}_+$ never reaches infinity, there is $A \in W$ such that $m(x) = x(A)$ for positive $x \in W_*$. Note that the spectral projections $P(\lambda)$ in the proposition above increase towards e, and the function $\lambda \mapsto P(\lambda)$ is strongly continuous from the right. Working in the amplification of W, so that every normal state is a vector state, we note that $P(0) = 0$ if and only if A is injective if and only if m is *faithful*, i.e. is non-zero on non-zero positive elements of W_*. Also, we have that $e = 1$ if and only if m is *semifinite*, that is, if the positive elements $x \in W_*$ with $m(x) < \infty$ form a dense subset of $(W_*)_+$. For the forward implication, note that $D(A^{1/2})$ is then dense, so for $v \in V$ the domain contains a sequence $\{v_n\}$ such that $x_{v_n} \to x_v$. The opposite direction is clear as $m(x) < \infty$ implies $x(1-e) = 0$ for positive $x \in W_*$. Elements of $\bar{W}_+$ are pointwise limits of increasing sequences in W_+. Indeed, we have $m_A(x) = \lim x(w_n)$ with $w_n = \int^n \lambda\, dP(\lambda) + n(1-e)$. Finally, any normal weight y on W has a unique extension y to $\bar{W}_+$ that respects increasing limits, addition and multiplication by non-negative scalars. Uniqueness is clear, and define the extension by $y(m_A) = \lim y(w_n)$. Writing the weight on W as $y = \sum x_i$ for normal functionals x_i on W, we get $y(m_A) = \sum x_i(m_A)$, which shows that the extension has the claimed properties.

The converse of the following result is obvious.

Proposition 10.12.4 *Let W be a von Neumann algebra. Then any countably additive map $(W_*)_+ \to [0, \infty]$ that respects multiplication by non-negative scalars belongs to $\bar{W}_+$.*

Proof Consider such a map m, and say $W \subset B(V)$. Then $\tilde{m}(x) = m(x|W)$ for positive $x \in B(V)_*$ defines a countable additive map $\tilde{m}: (B(V)_*)_+ \to [0, \infty]$ that respects multiplication by non-negative scalars. Also, if $\tilde{m}$ is lower semicontinuous, so $\tilde{m} = \tilde{m}_A$ for some operator A and orthogonal projection $e \in B(V)$ provided by the previous proposition, then as $u^*\tilde{m}u$ for $u \in W'$, we see that e and A are affiliated with W, so m is lower semicontinuous by the same proposition. Therefore we may assume that $W = B(V)$, and by replacing m by $m + 1$, we can also assume that $m(x) \geq \|x\|$ for positive $x \in B(V)_*$. Define a weight y on $B(V)$ by $y(T) = m(\mathrm{Tr}(\cdot T))$ when T is a positive trace class operator on V, and $y(T) = \infty$ when T is positive but not trace class. Inspection shows that y is not only countably additive but completely additive, hence normal by Haagerup's theorem, and then m must be lower semicontinuous. □

10.13 Operator Valued Weights

Definition 10.13.1 An *operator valued weight* E from a von Neumann algebra W to a von Neumann subalgebra V is an additive map $W_+ \to \bar{V}_+$ that respects multiplication by non-negative scalars, and satisfies $E(v^*(\cdot)v) = v^*E(\cdot)v$ for $v \in V$. It is *normal* if it respects increasing limits, and *faithful* if is non-zero on non-zero elements, and *semifinite* if the V-bimodule $N_E \equiv \{w \in W \mid E(w^*w) \in V_+\}$ is σ-weakly dense in W.

Any operator valued weight E from W to V has a unique extension to a linear V-bimodule map $\tilde{E}: M_E \equiv \mathrm{span}\{w \in W_+ \mid E(w) \in V_+\} = N_E^* N_E \to V$, which is a conditional expectation when $E(1) = 1$. When E is normal, then just as we did for ordinary weights, we can extend E uniquely to a normal map $\bar{E}: \bar{W}_+ \to \bar{V}_+$ which respects increasing limits, addition and multiplication by non-negative scalars, and satisfies $\bar{E}(v^*(\cdot)v) = v^*\bar{E}(\cdot)v$ for $v \in V$. Hence we may compose operator valued weights in the obvious way, and the composition $E'E$ is semifinite if both E and E' are because if $w \in E$, and $a_i \in N_{E'}$ with $a_i \to 1$ σ-strongly, then $E'E(a_i^*w^*wa_i) = E'(a_i^*E(w^*w)a_i) \leq \|E(w^*w)\|E'(a_i^*a_i)$, so $wa_i \in N_{E'E}$, and $N_{E'E}$ is σ-strongly dense in W as N_E is. Here we used the same symbol for E' and its extension, a convention we will stick to in the sequel.

Lemma 10.13.2 *Let y be a semifinite normal weight on a von Neumann algebra W with $a_n \in W_{s(y)}$. Then $y = \sum y(a_n(\cdot)a_n^*)$ on W_+ if and only if $s(y) = \sum \sigma_{-i/2}^y(a_n)^*\sigma_{-i/2}^y(a_n)$.*

Proof Working on $W_{s(y)}$ we may assume that y is faithful. For the forward implication, by Theorem 10.7.12 and the lemma (with its proof) prior to it, note that

$$\|q_y(w)\|^2 = \sum y(a_n w^* w a_n^*) = (J_y \sum \sigma^y_{-i/2}(a_n)^* \sigma^y_{-i/2}(a_n) J_y q_y(w) | q_y(w))$$

for $w \in N_y$, and then just use the polarization identity. To prove the converse, trace backwards. □

Theorem 10.13.3 *Let W be a von Neumann algebra with a von Neumann subalgebra V. If there are faithful semifinite normal weights x and x' on V and W, respectively, such that $\sigma_t^x = \sigma_t^{x'}$ on V for $t \in \mathbb{R}$, then there is a unique faithful semifinite normal operator valued weight E from W to V such that $x' = xE$. Conversely, given a faithful semifinite normal operator valued weight E from W to V and a faithful semifinite normal weight x on V, then $\sigma_t^{xE} = \sigma_t^x$ on V for $t \in \mathbb{R}$. Moreover, if y is another faithful semifinite normal weight on V, then $(DxE : DyE)_t = (Dx : Dy)_t$, so $\sigma_t^{xE,yE} = \sigma_t^{xy}$ for $t \in \mathbb{R}$.*

Proof By the converse theorem of Connes' cocycle theorem, for any semifinite normal weight y on V, there is a unique semifinite normal weight y' on W associated to x, x' given as above, such that $s(y') = s(y)$ and $(Dy' : Dx')_t = (Dy : Dx)_t$ for all real t. Let $E(w)(y) = y'(w) \in [0, \infty]$ for $w \in W_+$ and positive $y \in V_*$. As

$$(D(\lambda y) : Dx)_t = \lambda^{it}(Dy : Dx)_t = \lambda^{it}(Dy' : Dx')_t = (D(\lambda y') : Dx')_t$$

for $\lambda \geq 0$, we see that $(\lambda y)' = \lambda y'$, and $E(w)$ respects multiplication by λ. We claim $E(w)$ is countably additive. Indeed, if $y = \sum y_n \in V_*$ for positive $y_n \in V_*$, then by Theorem 10.7.12 we have $y_n = y((Dy : Dy_n)^*_{-i/2} \cdot (Dy : Dy_n)_{-i/2})$ on $V_{s(y)}$. Using $(Dy'_n : Dy')_{-i/2} = (Dy_n : Dy)_{-i/2}$, which follows by the chain rule for cocycle derivatives, we thus get $y' = \sum y'_n$ by applying the previous lemma in both directions. Hence by Proposition 10.12.4 we have defined an additive map $E \colon W_+ \to \bar{V}_+$ that respects multiplication by non-negative scalars, is normal, and evidently satisfies $x' = xE$, which also shows that E is faithful. For any unitary u in V, we have that $(D(uyu^*)' : Dx')_t$ equals

$$(Duyu^* : Dx)_t = u(Dy : Dx)_t \sigma_t^x(u^*) = u(Dy' : Dx')_t \sigma_t^{x'}(u^*)_t = (Duy'u^* : Dx')_t,$$

where we used that $\sigma_t^x = \sigma_t^{x'}$ on V. So $(uyu^*)' = uy'u^*$. Next consider a positive element a in the centraliser of y. Then it is clear that

$$(D(aya)' : Dy')_t = (Daya : Dy)_t = a^{2it} = (Day'a : Dy')_t,$$

so $(aya)' = ay'a$. To get the result for general $a \in V$, we first observe that $(y \otimes \mathrm{Tr})' = y' \otimes \mathrm{Tr}$ on $(W \otimes M_2(\mathbb{C}))_+$. There is of course no loss in generality assuming $\|a\| \leq 1$. Then

$$\begin{pmatrix} a & (1 - aa^*)^{1/2} \\ -(1 - aa^*)^{1/2} & a^* \end{pmatrix} \in V \otimes M_2(\mathbb{C})$$

is unitary, so $(u^*(y \otimes \mathrm{Tr})u)' = u^*(y' \otimes \mathrm{Tr})u$ by what we have shown. Since $1 \otimes e_{11}$ is a positive element in the centralizer of $y \otimes \mathrm{Tr}$, and $a \otimes e_{11} = (1 \otimes e_{11})u(1 \otimes e_{11})$, we therefore get $((a^* \otimes e_{11})(y \otimes \mathrm{Tr})(a \otimes e_{11}))' = (a^* \otimes e_{11})(y' \otimes \mathrm{Tr})(a \otimes e_{11})$, which shows that $(a^*ya)' = a^*y'a$ and $E(a^*(\cdot)a) = a^*E(\cdot)a$, so E is an operator valued weight. To show that E is semifinite, we know that $x' = xE$ is semifinite, so the positive elements w of $M_{x'}$ form a σ-weakly dense subset of W. Writing $E(w) = \int \lambda\, P(\lambda) + \infty(1-e)$, then as x is faithful and $x'(w) < \infty$, we must have $e = 1$, so $P(\lambda)wP(\lambda) \to w$ in the σ-weak topology as $\lambda \to \infty$, and the net is in M_E as $E(P(\lambda)wP(\lambda)) = P(\lambda)E(w)P(\lambda) \in V_+$. Thus M_E is σ-weakly dense in W. Before we prove uniqueness of E, we turn to the second half of the theorem, and show $\sigma_t^{xE} = \sigma_t^x$ on V for given x and E. By Proposition 10.7.14 it suffices to show $b = \sigma_{-i}^{xE}(a)$ when $b = \sigma_{-i}^x(a)$. Then $a, b^* \in D(\sigma_{-i/2}^x)$, so there is $M \geq 0$ such that $x(ava^*) \leq M^2x(v)$ and $x(b^*vb) \leq M^2x(v)$ for $v \in V_+$ by Lemma 10.7.11. Taking increasing limits, these inequalities still hold for any $v \in \bar{V}_+$. Hence $xE(awa^*) \leq M^2xE(w)$ and $xE(b^*wb) \leq M^2xE(w)$ for positive $w \in W$, and $\|q_{xE}(wa^*)\| \leq M\|q_{xE}(w)\|$ and $\|q_{xE}(wb)\| \leq M\|q_{xE}(w)\|$ when $w \in N_{xE}$. By Proposition 10.7.15 we need only verify $xE(aw) = xE(wb)$ for $w \in M_{xE}$. Consider first $w = u^*v$ with $u, v \in N_{xE} \cap N_E$. By assumption $xE(aw) = x(aE(w)) = x(E(w)b) = xE(wb)$. Considering then the general case, i.e. when $u, v \in N_{xE}$, we may write $E(u^*u) = \int \lambda\, dP_u(\lambda)$ as $xE(u^*u) < \infty$. Then $\|q_{xE}(uP_u(\lambda) - u)\|^2 = x(\int_\lambda^\infty r\, dP_u(r)) \to 0$ as $\lambda \to \infty$. Similarly, we get $\|q_{xE}(vP_v(\lambda) - v)\|^2 = x(\int_\lambda^\infty r\, dP_v(r)) \to 0$ as $\lambda \to \infty$. Combining this with the last two inequalities, we get $\lim \|q_{xE}(uP_u(\lambda)a^* - ua^*)\| = 0 = \lim \|q_{xE}(vP_v(\lambda)b - vb)\|$. Hence

$$
\begin{aligned}
xE(aw) &= (q_{xE}(v)|q_{xE}(ua^*)) = \lim(q_{xE}(vP_v(\lambda))|q_{xE}(uP_u(\lambda)a^*)) \\
&= \lim xE(aP_u(\lambda)u^*vP_v(\lambda))
\end{aligned}
$$

which then equals $\lim xE(P_u(\lambda)u^*vP_v(\lambda)b) = xE(bw)$ by the previous calculation as $uP_u(\lambda), vP_v(\lambda) \in N_{xE} \cap N_E$. To show the last assertion in the theorem, by what we have just proved, we see that $\sigma_t^{y\otimes\mathrm{Tr}} = \sigma_t^y \otimes \iota = \sigma_t^{yE} \otimes \iota = \sigma_t^{yE\otimes\mathrm{Tr}}$ on $W \otimes M_2(\mathbb{C})$, so by what we have proved so far, there is a faithful semifinite normal operator valued weight E' from $W \otimes M_2(\mathbb{C})$ to $V \otimes M_2(\mathbb{C})$ such that $(y \otimes \mathrm{Tr})E' = yE \otimes \mathrm{Tr}$. Considering the balanced weight $y \oplus x$ on $V \otimes M_2(\mathbb{C})$, note that $(y \oplus x)E' = yE \oplus xE$, so $(DxE : DyE)_t = \sigma_t^{yE\oplus xE}(1 \otimes e_{21}) = \sigma_t^{y\oplus x}(1 \otimes e_{21}) = (Dx : Dy)_t$ for real t, again by what we have already proved. Turning to uniqueness, if F is another faithful semifinite normal operator valued weight from $W \to V$ such that $x' = xF$, then by what we have just proved, we have for any faithful semifinite normal weight y on V that $(DyE : Dx')_t = (Dy : Dx)_t = (DyF : Dx')_t$, so $yE = yF$. Given positive $z \in V_*$, pick y such that $y(s(z)(\cdot)s(z)) = z$. Then $E(w)(z) = zE(w) = yE(s(z)ws(z)) = F(w)(z)$ for $w \in W_+$ as the support projection $s(z)$ of z belongs to V. □

Corollary 10.13.4 *Given an inclusion of von Neumann algebras $U \subset W$ and faithful semifinite normal operator valued weights E, E' from W to U, and faithful semifinite normal weights x, y on U, then $(DxE : DxE')_t = (DyE : DyE')_t \in U' \cap W$. In particular, if $U' \cap W = \mathbb{C}$, then E and E' are proportional.*

Proof Consider the faithful semifinite normal operator valued weight E'' from $W \otimes M_2(\mathbb{C})$ to $U \otimes \mathbb{C} = U$ given by $E''(\sum w_{ij} \otimes e_{ij}) = (E(w_{11}) + E'(w_{22})) \otimes 1$, so $xE'' = xE \oplus xE'$. As $1 \otimes e_{12} \in (U \otimes \mathbb{C})' \cap (W \otimes M_2(\mathbb{C}))$, and $\sigma_t^{xE''}$ leaves $U \otimes \mathbb{C}$ invariant by the theorem, we get $(DxE : DxE')_t \otimes e_{12} = \sigma_t^{xE''}(1 \otimes e_{12}) \in (U' \cap W) \otimes M_2(\mathbb{C})$, so $(DxE : DxE')_t \in U' \cap W$. Using this together with the chain rule for cocycle derivatives and the last statement in the theorem above, we get the required equality involving y. For the last statement, use Connes' cocycle theorem for xE and xE' to conclude that they are proportional, since the cocycle derivatives are proportional. □

Since σ_t^{xE} restricts to an automorphism on $U' \cap W$ and is independent of x, we may speak of it as the *modular automorphism* σ_t^E of E. We have also seen that $(DxE : DxE')_t$ is independent of x, so we may speak of it as the *cocycle derivative* $(DE : DE')_t$ of E relative to E'.

The following corollary defines the *tensor product* $E_1 \otimes E_2$ of operator valued weights E_i.

Corollary 10.13.5 *Consider faithful semifinite normal operator valued weights E_i from a von Neumann algebra W_i to a von Neumann subalgebra V_i. Then there exists a unique faithful semifinite normal operator valued weight $E_1 \otimes E_2$ from $W_1 \bar{\otimes} W_2$ to $V_1 \bar{\otimes} V_2$ such that $(x_1 \otimes x_2)(E_1 \otimes E_2) = x_1E_1 \otimes x_2E_2$ for all faithful semifinite normal weights x_i on V_i.*

Proof Fixing x_i, and observing that $\sigma_t^{x_1E_1 \otimes x_2E_2} = \sigma_t^{x_1 \otimes x_2}$ by definition of tensor products of weights and by the last part of the theorem above, the first part of the theorem establishes a unique $E_1 \otimes E_2$ as described for the fixed x_i. That the same $E_1 \otimes E_2$ also works for other x_i, say y_i, follows from the last part of the theorem, since

$$(D(y_1 \otimes y_2)(E_1 \otimes E_2) : D(x_1 \otimes x_2)(E_1 \otimes E_2))_t = (D(y_1 \otimes y_2) : D(x_1 \otimes x_2))_t$$

$$= (Dy_1 : Dx_1)_t \otimes (Dy_2 : Dx_2)_t = (Dy_1E_1 : Dx_1E_1)_t \otimes (Dy_2E_2 : Dx_2E_2)_t$$

$$= (D(y_1E_1 \otimes y_2E_2) : D(x_1E_1 \otimes x_2E_2))_t = (D(y_1E_1 \otimes y_2E_2) : D(x_1 \otimes x_2)(E_1 \otimes E_2))_t$$

for real t implies $(y_1 \otimes y_2)(E_1 \otimes E_2) = y_1E_2 \otimes y_2E_2$. □

From the proof we see that uniqueness only requires $(x_1 \otimes x_2)(E_1 \otimes E_2) = x_1E_1 \otimes x_2E_2$ to hold for some faithful semifinite normal weights x_i. Note that $(E_1 \otimes E_2)(w_1 \otimes w_2) = E_1(w_1) \otimes E_2(w_2)$ for all $w_i \in M_{E_i}$. Later we shall consider the operator valued weights $\iota \otimes x$ and $x \otimes \iota$ for a weight x.

The following result relates operator valued weights to spatial derivatives of weights.

Theorem 10.13.6 *Let U be a von Neumann subalgebra of a von Neumann algebra W. Then there is a bijection $E \mapsto E'$ from the set of faithful semifinite normal operator valued weights $W \to U$ to the set of faithful semifinite normal operator valued weights $U' \to W'$ uniquely determined by $dy/d(xE') = d(yE)/dx$ for all faithful semifinite normal weights y on U and x on W'.*

Proof Fix x and y as prescribed. By Theorem 10.10.1 we know that $\sigma_t^{yE} = (d(yE)/dx)^{it}(\cdot)(d(yE)/dx)^{-it}$ while $\sigma_t^x = (d(yE)/dx)^{-it}(\cdot)(d(yE)/dx)^{it}$. By the previous theorem we have that σ_t^y and σ_t^{yE} coincide on U. By Theorem 10.10.1 and its corollary, there exists a unique faithful semifinite normal weight z on U' such that $(dy/dz)^{it} = (d(yE)/dx)^{it}$, so $\sigma_t^z = \sigma_t^x$ on $W' \subset U'$. By the previous theorem there is a unique E' from U' to W' such that $z = xE'$, and thus E' satisfies the property of the theorem for the given pair x and y.

Now $dy/d(xE') = d(yE)/dx$ determines xE' uniquely by normality since every positive element of U is a strong*-limit of elements in F_y. But then E' is uniquely determined as well. Similarly one constructs a map $E' \to E$ satisfying the same relation, which therefore must be the inverse map. Using the last part of Theorem 10.10.1 together with the last part of the previous theorem, one sees that the same map $E \to E'$ yields the relation $dy/d(xE') = d(yE)/dx$ for any other x and y. □

Exercises

For Sect. 10.1

1. Let A be a full left Hilbert algebra. Prove that A is a Banach $*$-algebra with respect to the norm given by $\|a\|_n \equiv \sup\{\|a\|, \|S(a)\|, \|\pi_l(a)\|\}$ for $a \in A$.
2. Show that a left Hilbert algebra A is full if and only if the set of $a \in A$ with $\|\pi_l(a)\| \leq 1$ is closed in $D(S)$.
3. Let V be a Hilbert space, and let $\bar{v} \equiv (\cdot|v) \in V^\star$. Given an invertible positive self-adjoint operator h in V, show that $D(h) \otimes \overline{D(h^{-1})}$ is a left Hilbert algebra with product and $*$-operation given by $(v_1 \otimes \bar{u}_1)(v_2 \otimes \bar{u}_2) = \overline{h^{-1}(u_1)(v_2)} v_1 \otimes \bar{u}_2$ and $(v_1 \otimes \bar{u}_1)^* = h^{-1}(u_1) \otimes \overline{h(v_1)}$ for $v_i \in D(h)$ and $u_i \in D(h^{-1})$. Prove that the completion of this left Hilbert algebra is $V \otimes \bar{V}$, that its left von Neumann algebra is $B(V) \otimes \mathbb{C}$, and that $J(v_1 \otimes \bar{u}_1) = u_1 \otimes \bar{v}_1$ and $\Delta^{1/2}(v_1 \otimes \bar{u}_1) = h(v_1) \otimes \overline{h^{-1}(u_1)}$.
4. Consider a full left Hilbert algebra. Show that the map π_l is closed with respect to the Hilbert space topology on $D(S)$ and the σ-strong* topology on the left von Neumann algebra.

5. Let A be a left Hilbert algebra. Show that to $v \in A''$ there is a sequence $\{v_n\}$ in A with $\|\pi_l(v_n)\| \leq \|\pi_l(v)\|$ that approaches v in the graph norm of S. Conclude that $\{\pi_l(v_n)\}$ converges to $\pi_l(v)$ in the strong*-topology.

For Sect. 10.2

1. Given a normal weight x on a von Neumann algebra W. Show that $x(u \cdot w)$ extends to a normal bounded linear functional on W when $u, w \in N_x$.
2. Let x be the weight on the von Neumann algebra $l^\infty(\mathbb{N})$ given by $x(\{v_n\}) = \sum v_n$ if only finitely many v_n's are non-zero, and let x be infinite otherwise. Show that x is completely additive on orthogonal projections, but that it is not normal.
3. Given weights $x \leq y$ on a von Neumann algebra W, show that there exists a unique $a \in \pi_y(W)'$ with $0 \leq a \leq 1$ and $x(u^*w) = (a(q_y(w))|a(q_y(u)))$ for $u, w \in N_y$.
4. Show that the integral on $L^\infty(\mu)$ with respect to a Radon measure μ on a σ-compact locally compact Hausdorff space is a semifinite normal weight.

For Sect. 10.3

1. Let x be a faithful semifinite normal weight on a von Neumann algebra W, let A_x be the associated left Hilbert algebra, and identify W with $\pi_x(W)$. Prove that $q_x : N_x \cap N_x^* \to A_x$ is a closed linear map with respect to the σ-strong* topology on W and the norm topology on V_x. Show that its inverse map is $\pi_l : A_x \to W$.
2. Let x be a faithful semifinite normal weight on a von Neumann algebra W, let A_x be the associated left Hilbert algebra, identify W with $\pi_x(W)$, and let B be the algebra of left bounded vectors in V_x. Show that $\pi_l : B \to W$ is a closed linear map with respect to the norm topology in V_x and the σ-strong topology in W. Prove that its inverse map is $q_x : N_x \to B$.
3. Let x be a semifinite normal weight on a von Neumann algebra W, and suppose we have a σ-weakly continuous function $f : \mathbb{R} \to W_+$ with $\int \|f(t)\|\, dt < \infty$. Show that $x(\int f(t)\, dt) = \int x(f(t))\, dt$.

For Sect. 10.4

1. Let W be a C*-algebra with $w \in W_+$ and with a unitary u in the unitization of W. Show that $u^*w = 4^{-1} \sum_{n=0}^{3} i^n (1 + i^n u)^* w (1 + i^n u)$, and use this to show that M_x and N_x are ideals of a weight x that is tracial, meaning that $x(u^*wu) = x(w)$ for all u, w as above. Deduce $x(wu) = x(uw)$. When x is also lower semicontinuous, show that $x(v^*v) = x(vv^*)$ for $v \in W$, and that $x(ab) = x(ba)$ for $a, b \in N_x$.
2. Let W be a C*-algebra and define an equivalence relation $\approx$ on W_+ by letting $a \approx b$ mean that there are finitely many elements $w_i \in W$ such that $a = \sum w_i^* w_i$ and $b = \sum w_i w_i^*$. Show that if x is a lower semicontinuous trace on W, then $x(a) = x(b)$, and that the converse also holds when x is the usual trace on $B_0(V)$ for any Hilbert space V and a, b of finite rank.

3. Suppose $\sum a_n^* a_n = \sum b_m b_m^*$ for finitely many a_n, b_m in a C*-algebra. Show that the limits c_{nm} in the C*-algebra of the sequences $\{a_n(k^{-1} + \sum a_i^* a_i)^{-1/2} b_m\}_k$ satisfy $\sum c_{nm} c_{nm}^* = a_n a_n^*$ and $\sum c_{nm}^* c_{nm} = b_m^* b_m$.
4. Combining the previous three exercises, define $a \precsim b$ in the positive part of a C*-algebra W to mean $a \approx c \leq b$ for some positive $c \in W$. If x is a lower semicontinuous weight on a hereditary C*-subalgebra U of W, define

$$y(w) = \sup\{x(u) \mid u \in U_+,\ u \precsim w\}$$

for $w \in W_+$. Show that y is a lower semicontinuous trace on W, and that y restricted to U_+ is the smallest trace dominating x.

For Sect. 10.5

1. Show that in the KMS-condition of a finite lower semicontinuous weight the requirement of its invariance under the modular automorphism group is redundant.
2. Let h be a positive invertible operator on a finite dimensional Hilbert space V. Show that the modular automorphism group of the weight $Tr(\cdot h)$ on $B(V)$ is given by $\sigma_t = h^{it} \cdot h^{-it}$ for real t.
3. Suppose W is a unital C*-algebra with a one-parameter group of automorphism that is pointwise-norm continuous. Prove that the set X of states on W satisfying the KMS-condition with respect to this group is a closed convex subset of the set of states on W. Show that a state x of W is an extreme point of X if and only if $\pi_x(W)''$ is a factor.
4. Show that the σ-strong closure of the unitary group of a von Neumann algebra W is the semigroup of isometries in W.
5. Show that if x is a faithful normal state on a von Neumann algebra W with bounded modular operator Δ, then the convex closure of the set of $x(u \cdot u^*)$ for all unitaries $u \in W$ is weakly compact. Using some fixed point theorem, prove then that W admits a finite faithful normal trace if it has a faithful normal state.

For Sect. 10.6

1. Suppose W is a von Neumann algebra on a Hilbert space V. Show that a positive self-adjoint operator T in V is affiliated to W if and only if $(1 + \varepsilon T)^{-1} T \in W$ for some $\varepsilon > 0$, and thus for every $\varepsilon > 0$.
2. A faithful semifinite normal weight x on a von Neumann algebra W is called strictly semifinite if it is a sum of normal positive linear functionals with mutually orthogonal support projections. Show that x is strictly semifinite if and only if its restriction to its centralizer is strictly semifinite if and only if $z\sigma_{\mathbb{R}}^x$ is weakly compact in W_* for each $z \in W_*$.
3. Suppose x is a faithful semifinite normal weight on a von Neumann algebra W and $w \in W$. Prove that the W-valued function $t \mapsto \sigma_t^x(w)$ can be extended to a bounded continuous function on the strip H that is holomorphic in the interior if and only if $N_x w \subset N_x$ and there exists $u \in W$ such that $N_x u^* \subset N_x$ and $x(\cdot w) = x(u \cdot)$ on W_x.

For Sect. 10.7

1. Show that for two faithful semifinite normal weights with commuting modular automorphism groups, their sum is a semifinite weight.
2. Suppose we have two faithful semifinite normal weights and that one of them is finite. Show that their modular automorphism groups commute if and only if one of the weights is invariant under the other weights modular automorphism group.
3. Consider positive bounded normal linear functionals x, y on a von Neumann algebra with x faithful. Show that $y\sigma_t^x = y$ if and only if $|x + iy| = |x - iy|$.
4. Suppose x is a faithful positive bounded normal linear functional on a factor W, and that any automorphism α on W with $x\alpha = x$ has only the scalars as fixed points. Prove that x is then either a trace or that W is of type *III*.
5. Let x be a faithful semifinite normal weight on a von Neumann algebra W. Show that W admits a semifinite normal trace if and only if for every non-zero orthogonal projection p of W, there is a non-zero orthogonal projection q of W_p such that the modular operator for the restriction of x to W_q is bounded.

For Sect. 10.8

1. Show that if we have two faithful semifinite normal weights x, y on a factor with commuting modular automorphism groups, then there is a real number r such that $y\sigma_t^x = e^{rt}y$ and $x\sigma_t^y = e^{-rt}x$ for $t \in \mathbb{R}$.

For Sect. 10.9

1. Let x be a faithful semifinite normal weight on a von Neumann algebra W with semicyclic representation π_x and standard form with self-dual cone C_x. Consider the canonical implementation $g \mapsto u_g$ of $\mathrm{Aut}(W)$. Show that $u_g q_x = q_{xg^{-1}} g$ on N_x.
2. Let $W \subset B(V)$ be a von Neumann algebra in standard form with self-dual cone C, and let $v(y) \in C$ be such that $y = y_{v(y)}$ for positive $y \in W_*$. Strictly speaking one should work on von Neumann algebras reduced by support projections to insure faithfulness. Define an order on V by $u \geq v$ if $u - v \in C$. Show that $y \mapsto v(y)$ is order preserving and satisfies $v(cy) = c^{1/2}v(y)$ for scalars $c \geq 0$.
3. Prove that if W is a von Neumann algebra and $x, y \in W_*$ are positive with $x \leq cy$ for some scalar c, then $v(x) = (Dx : Dy)_{-i/2}v(y)$, with notation as in the exercise above.
4. Let $W \subset B(V)$ be a von Neumann algebra with semicyclic representation and standard form with involution J and self-dual cone C. Define a right action of elements $w \in W$ on elements $v \in V$ by $vw = Jw^*J(v)$. Again, let $v(y) \in C$ be such that $y = y_{v(y)}$ for positive $y \in W_*$. Show that $w(v(y)) = v(y)w$ if and only if $y(w\cdot) = y(\cdot w)$. Prove that $\|y(w\cdot) - y(\cdot w)\| \leq 2\|y\|^{1/2}\|w(v(y)) - v(y)w\|$.
5. With notation as in the previous exercise show that there is a scalar $c > 0$ such that $\|(\Delta_y^{it} - 1)w(v(y))\| \leq c(1 + |t|)\|w(v(y)) - v(y)w\|$ when y is faithful.

For Sect. 10.10

1. Compute the spatial derivative dx/dy, where $x = \mathrm{Tr}(h\cdot)$ on $B(V)$ for an invertible positive self-adjoint operator h on a Hilbert space V, and where y is the identity weight on $\mathbb{C} = B(V)'$.
2. Let x be a faithful semifinite normal weight on a von Neumann algebra W, and let y be a faithful semifinite normal weight on W'. Show that $|(v|u)|^2 \leq x(R_u^y(R_u^y)^*)y(R_v^x(R_v^x)^*)$ for $v \in D_x$ and $u \in D_y$.
3. Let $W \subset B(V)$ be a von Neumann algebra and let $t \mapsto u_t$ be a strongly continuous one-parameter group of unitaries on V. Show that there exists a faithful semifinite normal weight on W with modular automorphism group $u_t(\cdot)u_t^*$ if and only if there exists a faithful semifinite normal weight on W' with modular automorphism group $u_t^*(\cdot)u_t$.
4. Suppose x, y are faithful semifinite normal weights on a von Neumann algebra W and that z is a faithful semifinite normal weight on W'. Show that $x \leq y$ if and only if $dx/dz \leq dy/dz$, and that $dx(h \cdot h^*)/dz = h^*(dx/dz)h$ for any invertible $h \in W$.
5. Let y be a faithful semifinite normal weight on a von Neumann algebra W, let z be a faithful semifinite normal weight on W', and let $x_n \to x$ among positive faithful elements in W_*. Show that in the strong* topology we have the following convergences, namely, that $\sigma_t^{x_n}(w) \to \sigma_t^x(w)$ and $(Dx_n : Dy)_t \to (Dx : Dy)_t$ and $(dx_n/dz)^{it} \to (dx/dz)^{it}$ for $w \in W$ and real t. Prove that the convergence is uniform for $|t| \leq c$.

For Sect. 10.11

1. Let E be the faithful normal conditional expectation from a von Neumann algebra W onto a von Neumann subalgebra V such that $xE = x$ for a faithful semifinite normal weight x on W. Let α be an automorphism of W. Show that $\alpha^{-1}E\alpha$ is the faithful normal conditional expectation F onto $\alpha^{-1}(V)$ such that $x\alpha F = x\alpha$.
2. Suppose E is a faithful normal conditional expectation from a von Neumann algebra W onto a von Neumann subalgebra V, and let y be a faithful semifinite normal weight on V. Show that $\sigma_t^y E = \sigma_t^{yE} E = E\sigma_t^{yE}$ on W.
3. Suppose E is a faithful normal conditional expectation from a von Neumann algebra W onto a von Neumann subalgebra V, and let y_i be faithful semifinite normal weights on V. Prove that $(Dy_1E : Dy_2E)_t = (Dy_1 : Dy_2)_t \in V$.
4. Let $\alpha\colon G \to \mathrm{Aut}(W)$ be a pointwise strongly continuous homomorphism from a locally compact group to the group of automorphisms on a von Neumann algebra W. Let W^α be the elements $w \in W$ such that $\alpha_t(w) = w$ for all $t \in G$. Show that there exists a faithful normal conditional expectation $E\colon W \to W^\alpha$ with $E\alpha_t = E$ for all t if and only if for each non-zero $w \in W$, there is a normal state x on W with $x(w) \neq 0$ and $x\alpha_t = x$ for all $t \in G$.

For Sect. 10.12

1. Write $a + \infty(1 - e)$ for $m = m_A$ in the extended positive part of a von Neumann algebra W. Put $m_0 = (1 + a)^{-1}e$ and $m_\varepsilon = a(1 + \varepsilon a)^{-1}e + \varepsilon^{-1}(1 - e)$ for a scalar $\varepsilon > 0$. Show that in the extended positive part of W we have that $m \leq n$ if and only if $m_0 \geq n_0$ if and only if $m_\varepsilon \leq n_\varepsilon$ for all $\varepsilon > 0$. Prove also that m_i increases towards m if and only if $(m_i)_0$ decreases towards m_0 if and only if $(m_i)_\varepsilon$ increases towards m_ε for all $\varepsilon > 0$. Note that m_0 and m_ε are bounded.
2. Let x be a faithful semifinite normal weight on a von Neumann algebra W. For m in the extended positive part of the centralizer W_x of the weight x, let $x_m(w) = \lim_{\varepsilon \to 0} x(m_\varepsilon^{1/2} w m_\varepsilon^{1/2})$ for positive $w \in W$, where we use notation from the previous exercise. Show that this defines an order preserving bijection $m \mapsto x_m$ onto the set of σ_t^x-invariant normal weights on W such that if m_i increases towards m, then x_{m_i} increases pointwise towards x_m.

For Sect. 10.13

1. Show that the centralizer W_x of a faithful semifinite normal weight x on a von Neumann algebra W is semifinite if and only if there exists a σ_t^x-invariant faithful semifinite normal operator valued weight from W to W_x.
2. Let W be a von Neumann algebra on a Hilbert space V. Show that there is a unique bijection $x \mapsto E_x$ from the set of faithful semifinite normal weights on W' to the set of faithful semifinite normal operator valued weights from $B(V)$ to W such that $yE_x = \mathrm{Tr}_{dy/dx}$ for any faithful semifinite normal weight y on W. Moreover, prove that $E_x(v \odot v) = R_v^x (R_v^x)^*$ for $v \in D_x$.
3. Assume we have a faithful semifinite normal operator valued weight E from W to U. Show that $F \mapsto (DF : DE)$ is a bijection from the set of faithful semifinite normal operator valued weights from W to U to the set of unitary σ_t^E-cocycles in the von Neumann algebra $U' \cap W$.
4. Let $U \subset W$ be an inclusion of von Neumann algebras. Suppose we have a map $x \mapsto \tilde{x}$ from the set of faithful semifinite normal weights on U to the set of faithful semifinite normal weights on W such that $\sigma_t^x = \sigma_t^{\tilde{x}}$ on U and $(D\tilde{y} : D\tilde{x})_t = (Dy : Dx)_t$ for real t. Show that there exists a unique faithful semifinite normal operator valued weight E from W to U such that $xE = \tilde{x}$ for all faithful semifinite normal weights x on U.
5. Consider the Lebesgue integral $\int$ on $L^\infty(\mathbb{R}) \subset B(L^2(\mathbb{R}))$. Show that the operator valued weight E from $B(L^2(\mathbb{R}))$ to $L^\infty(\mathbb{R})$ such that $\int E = \mathrm{Tr}$ is given by $E(w) = \int u(t) w u(t)^* \, dt$, where $u(t)$ is the multiplication operator on $L^2(\mathbb{R})$ of the function $s \mapsto e^{ist}$.

Chapter 11
Spectra and Type III Factors

In this section we study some useful invariants especially of type III von Neumann algebras bringing our classification program to a certain level of completion. From the outset these invariants are associated with dynamical systems.

By a von Neumann dynamical system of $\mathbb{R}$ we mean an action of $\mathbb{R}$ on a von Neumann algebra W by $*$-automorphisms α_t such that $t \mapsto x(\alpha_t(w))$ is continuous, where $w \in W$, $x \in W_*$. The Arveson spectrum of this dynamical system is the closed subset $\text{sp}(\alpha)$ of $\mathbb{R}$ consisting of all $t \in \mathbb{R}$ vanishing at the Fourier transforms of those $f \in L^1(\mathbb{R})$ with $\int x(\alpha_s(w)) f(s)\, ds = 0$ for all $w \in W$ and $x \in W_*$. This spectrum is homeomorphic to the character space with w^*-topology on the abelian Banach subalgebra of $B(W)$ generated by $\alpha(L^1(\mathbb{R}))$, where $\alpha(f)$ for $f \in L^1(\mathbb{R})$ is the operator on W that sends w to the element of W defined by the integral above for any x. Hence the map $t \mapsto \alpha_t$ from $\mathbb{R}$ to $B(W)$ is norm continuous if and only if $\text{sp}(\alpha)$ is compact. In this case we show that the dynamics is inner.

In general, for any orthogonal projection p in the fixed point algebra of all the automorphisms α_t, let α^p denote the von Neumann dynamical system of $\mathbb{R}$ obtained by restricting the automorphisms α_t to pWp. The Connes spectrum $\Gamma(\alpha)$ is the intersection of $\text{sp}(\alpha^p)$ over all such p. We then show that $\Gamma(\alpha)$ is a closed subgroup of $\mathbb{R}$, and these are just $\mathbb{R}$ and $c\mathbb{Z}$ for $c \geq 0$. Moreover, the Connes spectrum coincides for cocycle equivalent dynamical systems of $\mathbb{R}$. Thus, by considering the modular group of a faithful semifinite normal weight x on a von Neumann algebra W as a von Neumann dynamical system, its Connes spectrum is indeed an invariant $\Gamma(W)$ for W. In this case its exponential is the spectrum of the modular operator Δ_x except 0. Using this we can decide when a factor on a separable Hilbert space is semifinite or of type III. Depending on the different subgroups above, we talk about factors of type III_λ for $\lambda \in [0, 1]$. There are factors of all these sub-types. We show that one needs only intersect over the orthogonal projections in the center of the fixed point algebra of the dynamical system, and we also consider the case of several non-faithful weights x.

Recommended literature for this chapter is [8].

L. Tuset, *Analysis and Quantum Groups*,
https://doi.org/10.1007/978-3-031-07246-8_11

11.1 The Arveson Spectrum

Definition 11.1.1 For a closed ideal J of $L^1(\mathbb{R})$ denote by $J^\perp$ the closed subset $\{t \in \mathbb{R} | \hat{J}(t) = \{0\}\}$ of $\mathbb{R}$. For a closed subset A of $\mathbb{R}$ denote by $J(A)$ the closed ideal $\{f \in L^1(\mathbb{R}) \mid \hat{f}(A) = \{0\}\}$ of $L^1(\mathbb{R})$.

Lemma 11.1.2 *If A is a closed subset of $\mathbb{R}$, then $J(A)^\perp = A$. If A is compact, then $J(A)$ admits $h \in L^1(\mathbb{R})$ such that $f - f * h \in J(A)$ for all $f \in L^1(\mathbb{R})$.*

Proof If $t \notin A$, then by Lemma 4.2.5 there is $f \in L^1(\mathbb{R})$ such that $\hat{f}$ vanishes on A and equals one at t, so $f \in J(A)$ and $t \notin J(A)^\perp$.

If A is compact, the same lemma provides $h \in L^1(\mathbb{R})$ with $\hat{h}$ that is one on A, so $g - g * h \in J(A)$ for $g \in L^1(\mathbb{R})$. □

Lemma 11.1.3 *Let U_i be open in $\mathbb{R}$, and $K \subset U_2$ compact. Given $f \in L^1(\mathbb{R})$ and a closed ideal J of $L^1(\mathbb{R})$ with $f_i \in J$ such that $\hat{f_i} = \hat{f}$ on U_i. Then there is $g \in J$ such that $\hat{g} = \hat{f}$ on $U_1 \cap K$.*

Proof By Lemma 4.2.5 pick $h \in L^1(\mathbb{R})$ such that $\hat{h}$ is one on K and has support in U_2. Then $g = f_1 - f_1 * h + f_2 * h$ does the trick. □

Proposition 11.1.4 *Let J be a closed ideal of $L^1(\mathbb{R})$. If $f \in L^1(\mathbb{R})$ and $\hat{f}$ vanishes on an open set containing $J^\perp$, then $f \in J$.*

Proof We claim that if $t \notin J^\perp$, there is $g \in J$ such that $\hat{g} = \hat{f}$ on some neighborhood of t. Pick a neighborhood of t with compact closure K disjoint from $J^\perp$. By the proof of the second last lemma above $J(K)$ admits $h \in L^1(\mathbb{R})$ such that $\hat{h}$ is one on K and $f_1 - f_1 * h \in J(K)$ for $f_1 \in L^1(\mathbb{R})$. If $J + J(K)$ is proper, it is contained in a maximal ideal of $L^1(\mathbb{R})$, which by Example 5.2.9 is of the form $\{g \in L^1(\mathbb{R}) \mid \hat{g}(s) = 0\}$ for some $s \in \mathbb{R}$. But $s \in K \cap J^\perp = \phi$ by the second last lemma above, so $J + J(K) = L^1(\mathbb{R})$ and $h = k + h_1$ with $k \in J$ and $h_1 \in J(K)$. As $\hat{h}_1$ vanishes on K, we see that $g = f * k$ settles the claim.

Suppose first that the support A of $\hat{f}$ is compact. By assumption $A \cap J^\perp = \phi$, so by the claim above, each $t \in A$ has a neighborhood U_t and $g_t \in J$ such that $\hat{g}_t = \hat{f}$ on U_t. Then pick a neighborhood V_t of t with compact closure inside U_t. By compactness of A, there are $t_1, \dots t_n \in A$ such that the union of $V_i = V_{t_i}$ contains A and $g_i = g_{t_i} \in J$ satisfy $\hat{g}_i = \hat{f}$ on $U_i = U_{t_i}$. Let $U_{n+1} = A^c$ and $g_{n+1} = 0$. By the previous lemma applied repeatedly, first to the V_i's and then finally to $\cup V_i$ and U_{n+1}, there is $g \in J$ such that $\hat{g} = \hat{f}$ on $U_{n+1} \cup A = \mathbb{R}$, so $f = g$ by the inversion theorem.

In the general case pick $u \in L^1(\mathbb{R})$ to $\varepsilon > 0$ such that $\|f - f * u\|_1 < \varepsilon/2$. By Proposition 4.2.7 there is $v \in L^1(\mathbb{R})$ with $\|v - u\|_1 < \varepsilon/2\|f\|_1$ such that $\hat{v}$ has compact support. Then $\|f - f * v\|_1 \leq \|f - f * u\|_1 + \|f\|_1\|v - u\|_1 < \varepsilon$. The Fourier transform of $f * v$ is $\hat{f}\hat{v}$, so it has support inside the support of v, which is compact, and it vanishes on a neighborhood of $J^\perp$, so $f * v \in J$ by the previous paragraph. Thus $f \in J$ by closedness. □

Recall that $L^1(\mathbb{R})$ is a closed two-sided $*$-ideal of the abelian Banach algebra $M(\mathbb{R}) \cong C_0(\mathbb{R})^*$ of bounded complex Borel measures on $\mathbb{R}$ with convolution product $(\mu * \nu)(f) = \int f(s+t)\, d\mu(s) d\nu(t)$ and identity $\delta(f) = f(0)$ for $f \in C_0(\mathbb{R})$. The Fourier transform can be extended to $M(\mathbb{R})$ by $\hat{\mu}(t) = \int e^{-ist}\, d\mu(s)$, and then the Fourier transform of $f * \mu$ is easily seen to be $\hat{f}\hat{\mu}$.

Definition 11.1.5 By a *von Neumann dynamical system* (W, G, α) we mean an action of a locally compact group G on a von Neumann algebra W by $*$-automorphisms α_s such that $s \mapsto x(\alpha_s(w))$ is continuous on G for $w \in W$, $x \in W_*$.

Proposition 11.1.6 *Let $(W, \mathbb{R}, \sigma)$ be a von Neumann dynamical system. Then to $\mu \in M(\mathbb{R})$ there is a σ-weakly continuous linear operator $\sigma(\mu)$ on W with $\|\sigma(\mu)\| \leq \|\mu\|$ and $x\sigma(\mu)(w) = \int x(\sigma_t(w))\, d\mu(t)$ for $x \in W_*$ and $w \in W$. For $f \in L^1(\mathbb{R})$ we write $\sigma(f)$ for $\sigma(f(t)dt)$.*

Proof Consider positive $x \in W_*$ and $\mu \in M(\mathbb{R})$, and a bounded increasing net of positive elements $w_i \in W$ such that $w = \vee w_i$. Then $\{x(\sigma_t(w_i))\}$ increases towards $x(\sigma_t(w))$ uniformly for t in any compact set by Dini's theorem. Combined with regularity of μ and boundedness of $\{w_i\}$, we get $x\sigma(\mu)(w) = \sup x\sigma(\mu)(w_i)$. □

Definition 11.1.7 Let $(W, \mathbb{R}, \sigma)$ be a von Neumann dynamical system. The *Arveson spectrum* $\mathrm{sp}(\sigma)$ of σ is the closed subset $\{f \in L^1(\mathbb{R}) \mid \sigma(f) = 0\}^\perp$ of $\mathbb{R}$. Denote by $\mathrm{sp}_\sigma(w)$ the closed subset $\{f \in L^1(\mathbb{R}) \mid \sigma(f)(w) = 0\}^\perp$ of $\mathrm{sp}(\sigma)$ for $w \in W$. The *associated spectral subspace* $M(\sigma, A)$ of a closed subset A of $\mathbb{R}$ is the subset $\{w \in W \mid \mathrm{sp}_\sigma(w) \subset A\}$ of W.

Since $\sigma(f)(1) = \hat{f}(0)$ for $f \in L^1(\mathbb{R})$, we see that $0 \in \mathrm{sp}(\sigma)$. By Lemma 4.2.5 we get $\mathrm{sp}_\sigma(0) = \phi$ and $0 \in M(\sigma, A)$.

Proposition 11.1.8 *Let $(W, \mathbb{R}, \sigma)$ be a von Neumann dynamical system. Then:*

(i) $\mathrm{sp}_\sigma(w^*) = -\mathrm{sp}_\sigma(w)$ *and* $\mathrm{sp}_\sigma(\sigma_t(w)) = \mathrm{sp}_\sigma(w)$ *and* $\sigma_t(M(\sigma, A)) = M(\sigma, A)$ *for a closed $A \subset \mathbb{R}$ and $w \in W$ and $t \in \mathbb{R}$;*
(ii) *$w \in M(\sigma, A)$ if and only if $\sigma(f)(w) = 0$ for all $f \in L^1(\mathbb{R})$ with $\hat{f} = 0$ on some open $U \supset A$. So $M(\sigma, A)$ is a σ-weakly closed subspace of W;*
(iii) *if* $\mathrm{sp}_\sigma(w) = \{0\}$, *then* $w = 0$;
(iv) $\mathrm{sp}(\sigma)$ *is the closure of* $\cup_{w \in W} \mathrm{sp}_\sigma(w)$;
(v) $\mathrm{sp}_\sigma(\sigma(f)(w))$ *is contained in the intersection of* $\mathrm{sp}_\sigma(w)$ *and the support of* $\hat{f}$ *for $f \in L^1(\mathbb{R})$ and $w \in W$;*
(vi) *if $w \in W$ and $\mu \in M(\mathbb{R})$ with $\hat{\mu} = 0$ on some open set containing* $\mathrm{sp}_\sigma(w)$, *we have $\sigma(\mu)(w) = 0$;*
(vii) *if $f \in L^1(\mathbb{R})$ with $\hat{f} = 0$ or $\hat{f} = 1$ on some open set containing* $\mathrm{sp}_\sigma(w)$, *either $\sigma(f)(w) = 0$ or $\sigma(f)(w) = w$, respectively.*

Proof The first equality in (i) is clear from $\sigma(f)(w)^* = \sigma(\bar{f})(w^*)$, and the second equality is clear from $\sigma(f)(\sigma_t(w)) = \sigma(f_t)(w)$, and $\hat{f}_t(s) = e^{-ist}\hat{f}(s)$, and that $\hat{f}_t = 0$ if and only if $\hat{f} = 0$, where $f_t(s) = f(s-t)$. The last equality is clear.

For (ii) let $J = \{g \in L^1(\mathbb{R}) \mid \sigma(g)(w) = 0\}$. Then $\mathrm{sp}_\sigma(w) = J^\perp \subset A$, and $f \in J$ by Proposition 11.1.4, so $\sigma(f)(w) = 0$. Conversely, say $\sigma(f)(w) = 0$ for all $f \in L^1(\mathbb{R})$ with $\hat{f} = 0$ on some open $U \supset A$. If $s \in \mathrm{sp}_\sigma(w)\backslash A$, then Lemma 4.2.5 yields $k \in L^1(\mathbb{R})$ with $\hat{k}$ that is one at s and has support in some neighborhood U of s such that $\overline{U} \cap A = \phi$, so $\sigma(k)(w) = 0$ by assumption, and yet $\hat{k}(s) \neq 0$. Thus $\mathrm{sp}_\sigma(w) \subset A$ and $w \in M(\sigma, A)$. Hence $M(\sigma, A)$ is a subspace of W and it is σ-weakly closed by the previous proposition.

As for (iii) we get $\sigma(f)(w) = 0$ for all $f \in L^1(\mathbb{R})$ by (ii). Hence $w = 0$ by the previous proposition.

As for (iv), clearly the closure B of $\cup_{w\in W}\mathrm{sp}_\sigma(w)$ is contained in $\mathrm{sp}(\sigma)$. If $s \notin B$, then by Lemma 4.2.5 there is $k \in L^1(\mathbb{R})$ with $\hat{k}$ that is one at s and vanishes on some open set containing B. Then $\sigma(k)(w) = 0$ by (iii) for all $w \in W$, so $\sigma(k) = 0$ and yet $\hat{k}(s) \neq 0$. Thus $s \notin \mathrm{sp}(\sigma)$ and $\mathrm{sp}(\sigma) = B$.

To prove (v), note that $\mathrm{sp}_\sigma(w) \supset \mathrm{sp}_\sigma(\sigma(f)(w))$ because

$$\{g \in L^1(\mathbb{R}) \mid \sigma(g)(w) = 0\} \subset \{h \in L^1(\mathbb{R}) \mid \sigma(h)(\sigma(f)(w)) = 0\}$$

as $\sigma(g)(\sigma(f)(w)) = \sigma(g * f)(w)$. If $s \in \mathbb{R}$ does not belong to the support of $\hat{f}$, then by Lemma 4.2.5 there is $k \in L^1(\mathbb{R})$ with $\hat{k}$ that is one at s and has support disjoint from the support of $\hat{f}$. Thus $\hat{k}\hat{f} = 0$, so $k * f = 0$ and $\sigma(k)(\sigma(f)(w)) = 0$ and yet $\hat{k}(s) \neq 0$, so $s \notin \mathrm{sp}_\sigma(\sigma(f)(w))$.

To get (vi), note that $f * \mu = 0$ while its Fourier transform $\hat{f}\hat{\mu} = 0$ on some open set containing $\mathrm{sp}_\sigma(w)$, so $\sigma(f)(\sigma(\mu)(w)) = 0$ for $f \in L^1(\mathbb{R})$ by (iii). Hence $\sigma(\mu)(w) = 0$ by the previous proposition.

Finally, suppose we have $f \in L^{(}\mathbb{R})$ with $\hat{f}$ that is one on some open set containing $\mathrm{sp}_\sigma(w)$. Now $\sigma(\delta)(w) = w$ and $\hat{\delta}(s) = 1$ for all $s \in \mathbb{R}$. Hence the Fourier transform of $f - \delta$ vanishes on some open set containing $\mathrm{sp}_\sigma(w)$. By (vi) we thus get $0 = \sigma(f-\delta)(w) = \sigma(f)(w) - w$. □

Proposition 11.1.9 *Let* $(W, \mathbb{R}, \sigma)$ *be a von Neumann dynamical system. If* A_i *are closed subsets of* $\mathbb{R}$ *and* $w_i \in M(\sigma, A_i)$, *then* $w_1 w_2 \in M(\sigma, \overline{A_1 + A_2})$. *In particular, the set* $\mathrm{sp}_\sigma(w_1 w_2)$ *is contained in the closure of* $\mathrm{sp}_\sigma(w_1) + \mathrm{sp}_\sigma(w_2)$.

Proof Assume first that $\mathrm{sp}_\sigma(w_i)$ is compact. Replacing A_i by $\mathrm{sp}_\sigma(w_i)$, we may then assume that A_i are compact, so $A = \overline{A_1 + A_2}$ is compact. Let $f \in L^1(\mathbb{R})$ have $\hat{f}$ that vanishes on $A + \overline{U} + \overline{U}$, where U is a neighborhood of 0. By Lemma 4.2.5 pick $f_i \in L^1(\mathbb{R})$ such that $\hat{f}_i$ is one on some open set containing A_i and with support in $A_i + U$. By (vii) in the previous proposition, we have $\sigma(f_i)(w_i) = w_i$. Let $x \in W_*$.

By the Fubini-Tonelli theorem and translation-invariance write

$$
\begin{aligned}
x(\sigma(f)(w_1w_2)) &= \int f(s)x(\sigma_s(\sigma(f_1)(w_1)\sigma(f_2)(w_2)))\,ds \\
&= \int f(s)f_1(t_1)f_2(t_2)x(\sigma_{s+t_1}(w_1)\sigma_{s+t_2}(w_2))\,dsdt_1dt_2 \\
&= \int k(s_1,s_2)x(\sigma_{s_1}(w_1)\sigma_{s_1+s_2}(w_2))ds_1ds_2,
\end{aligned}
$$

where $k(s_1, s_2) = \int f(s)f_1(s_1 - s)f_2(s_2 + s_1 - s)\,ds$. Taking Fourier transform with respect to s_2 gives $\hat{k}(t, s_2) = \hat{f}(t)(\hat{f}_1 * \hat{g})(t)$, where $g(s) = f_2(s + s_2)$. As $\hat{g}(t) = e^{is_2t}\hat{f}_2(t)$, we see that $\hat{g}$ and $\hat{f}_2$ have the same support, so the support of $\hat{f}_1 * \hat{g}$ is contained in $A + U + U$, and as $\hat{f}$ is zero on this set, we see that $\hat{k}(t, s_2) = 0$ for all $t \in \mathbb{R}$ and hence $k(s_1, s_2) = 0$ for almost all $s_1 \in \mathbb{R}$. The Fubini-Tonelli theorem shows then that $x(\sigma(f)(w_1w_2)) = 0$ and $w_1w_2 \in M(\sigma, \overline{A_1 + A_2})$ by (ii) in the previous proposition.

Pick an approximate unit $\{u_i\}$ in the unit ball of $L^1(\mathbb{R})$. As $\{\sigma(u_i)(w)\}$ converges σ-weakly to w, any $w \in W$ belongs to the σ-weak closure of the set $\{\sigma(f)(w) \mid f \in L^1(\mathbb{R})$ with $\|f\|_1 \leq 1\}$. By Proposition 11.1.6 and Proposition 4.2.7 we may even assume that all $\hat{f}$ have compact support. As the σ-weak topology coincides with the strong topology on convex sets, and as products of bounded nets are continuous in the latter topology, we see that w_1w_2 belongs to the σ-weak closure of the set of all $\sigma(f_1)(w_1)\sigma(f_2)(w_2)$, where $f_i \in L^1(\mathbb{R})$ have $\|f_i\| \leq 1$ and $\hat{f}_i$ have compact support. But such $\sigma(f_1)(w_1)\sigma(f_2)(w_2) \in A$ by the previous paragraph and (v) in the previous proposition. As $M(\sigma, A)$ is σ-weakly closed by the same proposition, we are therefore done. □

Lemma 11.1.10 *Let $(W, \mathbb{R}, \sigma)$ be a von Neumann dynamical system, let K be a compact subset of $\mathbb{R}$, and let $\varepsilon > 0$. Then there is a neighborhood U of $t \in \mathbb{R}$ with compact closure such that $\|\sigma_s(w) - e^{-ist}w\| \leq \varepsilon\|w\|$ for $w \in M(\sigma, \overline{U})$ and $s \in K$.*

Proof By Lemma 4.2.5 pick a neighborhood W_1 of t with compact closure and $f \in L^1(\mathbb{R})$ such that $\hat{f}$ has compact support and is one on $\overline{W_1}$. For $s \in K$, let $f^s(r) = f(r - s) - e^{-ist}f(r)$, so $\hat{f^s}(t) = 0$. By Lemma 4.2.6 there is $k^s \in L^1(\mathbb{R})$ and a neighborhood W^s of t such that $\hat{k^s}$ is one on W^s and $\|f^s * k^s\|_1 < \varepsilon$. Since K is compact and $s \mapsto f^s$ is continuous from $\mathbb{R}$ to $L^1(\mathbb{R})$, there is a neighborhood W_2 of t with compact closure such that for each $s \in K$, there is $k \in L^1(\mathbb{R})$ with $\hat{k}$ one on $\overline{W_2}$ and $\|f^s * k\|_1 < \varepsilon$. Let $U \subset W_1 \cap W_2$ be a neighborhood of t with compact closure and w in $M(\sigma, \overline{U})$. Pick k to $s \in K$ as above. Then $\sigma(f * k)(w) = w$ by (vii) in Proposition 11.1.8, so $\|\sigma_s(w) - e^{-ist}w\| = \|\sigma(f^s * k)(w)\| < \varepsilon\|w\|$. □

Proposition 11.1.11 *Let $(W, \mathbb{R}, \sigma)$ be a von Neumann dynamical system. Then $t \in$ $\mathrm{sp}(\sigma)$ if and only if $M(\sigma, \overline{U}) \neq \{0\}$ for every neighborhood U of t if and only if there is a net $\{w_j\}$ of unit vectors in W such that $\|\sigma_s(w_j) - e^{-ist}w_j\| \to 0$ uniformly in s on any compact $K \subset \mathbb{R}$ if and only if $|\hat{f}(t)| \leq \|\sigma(f)\|$ for all $f \in L^1(\mathbb{R})$.*

Proof If $t \in \mathrm{sp}(\sigma)$ and U is a neighborhood of t such that $M(\sigma, \overline{U})$ is trivial, pick $f \in L^1(\mathbb{R})$ by Lemma 4.2.5 with $\hat{f}(t) = 1$ and support of $\hat{f}$ in U. Then by (v) in Proposition 11.1.8 we have $\mathrm{sp}_\sigma(\sigma(f)(w)) \subset U$ for $w \in W$, so $\sigma(f) = 0$, and yet $\hat{f}(t) \neq 0$. For the second forward implication, pick $w_U \in M(\sigma, \overline{U})$ with $\|w_U\| = 1$ for every neighborhood U of t, and apply the lemma. For the third forward implication, use

$$\|\sigma(f)\| = \| \int \sigma_s(w_j) f(s)\, ds\| \geq \|\hat{f}(t) w_j\| - \int \|\sigma_s(w_j) - e^{-ist} w_j\| \cdot |f(s)|\, ds$$

for $f \in L^1(\mathbb{R})$. The converse is now immediate. □

The following result motivates the definition of the Arveson spectrum.

Proposition 11.1.12 *Let $(W, \mathbb{R}, \sigma)$ be a von Neumann dynamical system, and let V be the abelian Banach subalgebra of $B(W)$ generated by $\sigma(L^1(\mathbb{R}))$. Then* $\mathrm{sp}(\sigma)$ *is homeomorphic to $\hat{V}$ with w^*-topology.*

Proof Note that $\sigma(f)(\sigma(g)(w)) = \sigma(f * g)(w)$ and $\sigma(f)\sigma(g) = \sigma(f * g)$ for $f, g \in L^1(\mathbb{R})$ and $w \in W$, so by Example 5.2.9, there is $h(x) \equiv t \in \mathbb{R}$ to each $x \in \hat{V}$ such that $x(\sigma(f)) = \hat{f}(t)$ for all $f \in L^1(\mathbb{R})$. As $\hat{f}(t) = 0$ when $\sigma(f) = 0$, we see that $t \in \mathrm{sp}(\sigma)$, so $h \colon \hat{V} \to \mathrm{sp}(\sigma)$. Clearly h is injective, and it is surjective by Proposition 11.1.6. As each element of $\hat{V}$ is continuous, so is h, and it is open since the Fourier transform of $L^1(\mathbb{R})$ is dense in $C_0(\mathbb{R})$ by Example 5.2.9. □

Proposition 11.1.13 *Let $(W, \mathbb{R}, \sigma)$ be a von Neumann dynamical system. Then the map $t \mapsto \sigma_t$ from $\mathbb{R}$ to $B(W)$ is norm continuous if and only if* $\mathrm{sp}(\sigma)$ *is compact.*

Proof If $\mathrm{sp}(\sigma)$ is compact, then by Lemma 4.2.5 there is $f \in L^1(\mathbb{R})$ such that $\hat{f}$ has compact support and is one on some open set containing $\mathrm{sp}(\sigma)$. Thus $\sigma(f)(w) = w$ for $w \in W$ by (vii) in Proposition 11.1.8. Hence

$$\|\sigma_t(w) - w\| = \|\sigma(\delta_t * f)(w) - \sigma(f)(w)\| \leq \|f_t - f\|_1 \|w\|,$$

so $\|\sigma_t - \iota\| \to 0$ as $t \to 0$. Conversely, if u_n has support in $[-1/n, 1/n]$ and form an approximate unit for $L^1(\mathbb{R})$, then

$$\|\sigma(u_n)(w) - w\| \leq \int \|\sigma_t(w) - w\| u_n(t)\, dt \leq \|w\| \sup_{|t| \leq 1/n} \|\sigma_t - \iota\|,$$

so the Banach algebra generated by $\sigma(L^1(\mathbb{R}))$ is unital, and $\mathrm{sp}(\sigma)$ is compact by the previous proposition. □

Theorem 11.1.14 *Let $(W, \mathbb{R}, \sigma)$ be a von Neumann dynamical system such that the map $t \mapsto \sigma_t$ from $\mathbb{R}$ to $B(W)$ is norm continuous. Then there is a self-adjoint operator a affiliated to W such that $\sigma_t = e^{ita} \cdot e^{-ita}$ for $t \in \mathbb{R}$.*

Proof For $\lambda \in \mathbb{R}$ let $p_\lambda \in W$ be the supremum of all orthogonal projections p in W such that $p\sigma(f)(w) = 0$ for all $w \in W$ and $f \in L^1(\mathbb{R})$ with $\hat{f}$ having compact support in $\langle \lambda, \infty \rangle$. Clearly $1 - p_\lambda$ is the minimal orthogonal projection in W that fixes every such $\sigma(f)(w)$, and $(1 - p_\lambda)W$ is the σ-weak closure of all such elements multiplied with W from the right. The function $\lambda \mapsto p_\lambda$ is increasing, and is one when λ is larger than the maximum of the compact set $\mathrm{sp}(\sigma)$, while it vanishes whenever $\lambda < \min(\mathrm{sp}(\sigma))$. Indeed, for $\lambda > \max(\mathrm{sp}(\sigma))$, any f as above has vanishing $\hat{f}$ on a open set containing $\mathrm{sp}(\sigma)$, so $\sigma(f) = 0$ by (vii) in Proposition 11.1.8, and $p_\lambda = 1$. A similar argument applies when λ is below $\mathrm{sp}(\sigma)$. The map $\lambda \mapsto p_\lambda$ is also strongly continuous from the right. Indeed, if $\lambda_n \to \lambda_+$ then $\wedge p_{\lambda_n} \geq p_\lambda$. And for any f as above, there is m such that $\hat{f}$ has support in $\langle \lambda_n, \infty \rangle$ for $n \geq m$. Hence $\wedge p_{\lambda_n}\sigma(f)(w) = 0$ and $p_\lambda = \wedge p_{\lambda_n}$.

It makes sense therefore to define a self-adjoint operator $a = -\int \lambda\, p_\lambda$ affiliated to W, and $u_t \equiv e^{ita} = \int e^{-it\lambda}\, dp_\lambda$. Then $\int f(t)u_t\, dt = \int \hat{f}(s)\, dp_s$ for $f \in L^1(\mathbb{R})$, and $p_\lambda \int f_1(t)u_t\, dt = \int f_1(t)u_t\, dt$ for $f_1 \in L^1(\mathbb{R})$ with $\hat{f_1}$ having compact support in $\langle -\infty, \lambda]$, while $(1 - p_\lambda)\int g_1(t)u_t\, dt = \int g_1(t)u_t\, dt$ for $g_1 \in L^1(\mathbb{R})$ with $\hat{g_1}$ having compact support in $\langle \lambda, \infty \rangle$.

Let $\lambda_0 \in \mathbb{R}$ and $w \in W$ with $\mathrm{sp}_\sigma(w) \subset \langle -\infty, \lambda_0]$, and consider f to λ as in the first paragraph. By Proposition 11.1.9 and (i) in Proposition 11.1.8, we get

$$\mathrm{sp}_\sigma(w^*\sigma(f)(v)) \subset \overline{[-\lambda_0, \infty\rangle + [\lambda + \varepsilon, \infty\rangle} \subset \langle \lambda - \lambda_0, \infty \rangle,$$

where $v \in W$ and $\varepsilon > 0$ is such that the support of $\hat{f}$ is contained in $[\lambda + \varepsilon, \infty\rangle$. Pick $g \in L^1(\mathbb{R})$ with $\hat{g}$ having compact support in $\langle \lambda - \lambda_0, \infty \rangle$ and that is one on an open set containing the compact set $\mathrm{sp}_\sigma(w^*\sigma(f)(v))$. By (vii) in Proposition 11.1.8, we then get $\sigma(g)(w^*\sigma(f)(v)) = w^*\sigma(f)(v)$, so $p_{\lambda-\lambda_0}w^*\sigma(f)(v) = 0$ by the first paragraph above, which furthermore shows that $(1 - p_\lambda)wp_{\lambda-\lambda_0} = 0$.

Pick $f, g \in L^1(\mathbb{R})$ with compact supports in $\langle -\infty, 0\rangle$ and $\langle 0, \infty \rangle$, respectively. Let $f_1(t) = f(t)e^{it(\lambda-\lambda_0)}$ and $g_1(t) = g(t)e^{it\lambda}$. Then f_1, g_1 have compact supports in $\langle -\infty, \lambda - \lambda_0\rangle$ and $\langle \lambda, \infty \rangle$, respectively, so by the second paragraph we see that $p_{\lambda-\lambda_0}\int f_1(s)u_s\, ds = \int f_1(s)u_s\, ds$ and $(1 - p_\lambda)\int g_1(s)u_s\, ds = \int g_1(s)u_s\, ds$.

Suppose W acts on the Hilbert space V, and let $v_i \in V$. Then the integral $(w\int f_1(s)u_s\, ds(v_1)|\int g_1 ts)u_t\, dt(v_2))$ vanishes by the third paragraph, while calculating the left hand side yields

$$\int (u_t^* w u_t u_{s-t}(v_1), v_2) f_1(s)\overline{g_1(t)}\, dsdt = \int (\beta_t(w)u_s(v_1), v_2) f_1(s-t)\overline{g_1(-t)}\, dsdt,$$

which equals $\hat{h}(\lambda_0 - \lambda)$, where $\beta_t = u_t \cdot u_t^*$ and

$$h(s) = \int (\beta_t(w)u_s(v_1), v_2) f(s-t)\overline{g(-t)}e^{it\lambda_0}\, dt.$$

Hence $h = 0$ almost everywhere, and thus everywhere, by continuity, so $0 = h(0) = (\beta(k)(w)(v_1), v_2)$, where $k(t) = f(-t)\overline{g(-t)}e^{it\lambda_0}$. Thus $\beta(k)(w) = 0$ and $\mathrm{sp}_\sigma(w)$

is contained in the set where $\hat{k}$ vanishes. Now $\hat{k}(s) = \int \hat{g}(r)\hat{f}(\lambda_0 - s + r)\,dr$, and we may even limit the integration range over $\langle \varepsilon, \infty \rangle$, where $\varepsilon > 0$ is such that the support of $\hat{g}$ is inside $\langle \varepsilon, \infty \rangle$. Since the support of $\hat{f}$ is inside $\langle -\infty, 0 \rangle$, we get $\hat{k}(s) = 0$ when $s < \lambda_0 + \varepsilon$, and as $\mathrm{sp}_\beta(w)$ is closed, it is thus contained in $\langle -\infty, \lambda_0]$. Hence $M(\sigma, \langle -\infty, \lambda_0]) \subset M(\beta, \langle -\infty, \lambda_0])$, as well as $M(\sigma, [\lambda_0, \infty\rangle) \subset M(\beta, [\lambda_0 \infty\rangle)$ by (i) in Proposition 11.1.8.

Let $f, g \in L^1(\mathbb{R})$ with support in $\langle -\infty, 0 \rangle$ and $\langle 0, \infty \rangle$, respectively. Let $f_1(t) = f(t)e^{i\lambda t}$ and $g_1(t) = g(t)e^{i\lambda t}$. Then $f_1, g_1 \in L^1(\mathbb{R})$ have support in $\langle -\infty, \lambda \rangle$ and $\langle \lambda, \infty \rangle$, respectively. Since $\sigma(f_1)(w) \in M(\sigma, \langle -\infty, \lambda]) \subset M(\beta, \langle -\infty, \lambda])$, it follows that $\hat{g_1}$ vanishes on an open set containing $\mathrm{sp}_\beta(\sigma(f_1)(w))$. Thus by (vii) in Proposition 11.1.8 we get from a calculation similar to the one in the previous paragraph that $0 = \beta(g_1)(\sigma(f_1)(w)) = \int \beta_t(\sigma_s(w))g_1(t)f_1(s)\,dsdt = \hat{h}(-\lambda)$, where now $h(s) = \int \gamma_t(\sigma_s(w))f(s-t)g(t)\,dt$ and $\gamma_t = \beta_t\sigma_{-t}$. Again $h = 0$ and $0 = h(0) = \gamma(k)(w)$, where now $k(t) = f(-t)g(t)$. Hence $\mathrm{sp}_\gamma(w)$ is contained in the set where $\hat{k}$ vanishes. But $\hat{k}(s) = \int_\varepsilon^\infty \hat{g}(r)\hat{f}(r-s)\,dr$, where now $\varepsilon > 0$ is such that $\hat{g}$ has support in $\langle \varepsilon, \infty \rangle$. Thus $\hat{k}(s) = 0$ when $s < \varepsilon$, and as $\mathrm{sp}_\gamma(w)$ is closed, it is contained in $\langle -\infty, 0]$. Similarly, one shows that $\mathrm{sp}_\gamma(w)$ is contained in $[0, \infty\rangle$. Since this holds for all $w \in W$, we see that $\gamma_t = \iota$ for all $t \in \mathbb{R}$. □

11.2 The Connes Spectrum

Definition 11.2.1 Let $(W, \mathbb{R}, \sigma)$ be a von Neumann dynamical system with *fixed point algebra* $W^\sigma = \{w \in W \mid \sigma_t(w) = w \,\forall t \in \mathbb{R}\}$. For an orthogonal projection $p \in W^\sigma$, consider the von Neumann dynamical system $(W_p, \mathbb{R}, \sigma^p)$, where σ_t^p is the restriction of σ_t to $W_p = pWp$ for $t \in \mathbb{R}$. The *Connes spectrum* $\Gamma(\sigma)$ Connes spectrum $\Gamma(\sigma)$ is the intersection of $\mathrm{sp}(\sigma^p)$ over all such p.

So the Connes spectrum is a closed subset of $\mathbb{R}$ that contains 0.

Proposition 11.2.2 *Let $(W, \mathbb{R}, \sigma)$ be a von Neumann dynamical system, and let $p \in W^\sigma$ be an orthogonal projection. Then $M(\sigma^p, A) = M(\sigma, A) \cap W_p$ for any closed subset A of $\mathbb{R}$. Moreover, we have $\Gamma(\sigma) + \mathrm{sp}(\sigma) = \mathrm{sp}(\sigma)$, and $\Gamma(\sigma)$ is a closed subgroup of $\mathbb{R}$.*

Proof As $\sigma^p(f)(w) = \sigma(f)(w)$ for $w \in W_p$ and $f \in L^1(\mathbb{R})$, we get $\mathrm{sp}_{\sigma^p}(w) = \mathrm{sp}_\sigma(w)$, which proves the first statement.

The inclusion $\mathrm{sp}(\sigma) \subset \Gamma(\sigma) + \mathrm{sp}(\sigma)$ is clear as $0 \in \Gamma(\sigma)$. For the opposite inclusion, let $c_1 \in \Gamma(\sigma)$ and $c_2 \in \mathrm{sp}(\sigma)$. By Proposition 11.1.11 it suffices to show that $M(\sigma, U) \neq \{0\}$ for every neighborhood of $c \equiv c_1 + c_2$ with compact closure U. Pick neighborhoods of c_i with compact closure U_i such that $U_1 + U_2 \subset U$. Then there is non-zero $w_2 \in M(\sigma, U_2)$ by Proposition 11.1.11. Let $u_t h_t$ be the polar decomposition of $\sigma_t(w_2^*)$, and let $p_t = u_t u_t^*$. Then $p = \vee p_t$ is a non-zero orthogonal projection in W^σ because if W acts on V, then $p(V)$ is the closed span of $\sigma_{\mathbb{R}}(w_2^*)(V)$, so $\sigma_t(p)(V) = p(V)$ and $\sigma_t(p) = p$ for all $t \in \mathbb{R}$. So $c_1 \in$

$\mathrm{sp}(\sigma^p)$ and there is non-zero $w_1 \in M(\sigma, U_1) \cap W_p$ by the first paragraph and Proposition 11.1.11. So there are $t \in \mathbb{R}$ and $v_i \in V$ such that $0 \neq (p_t w_1(v_1)|v_2) = (w_1(v_1)|p_t(v_2))$. As $p_t(v_2)$ belongs to the closure of $\sigma_t(w_2^*)(V)$, there is $v \in V$ such that $(w_1(v_1)|\sigma_t(w_2^*)(v)) \neq 0$. Thus $\sigma_t(w_2)w_1$ is non-zero, and moreover belongs to $M(\sigma, U)$ by Proposition 11.1.9 and (i) in Proposition 11.1.8.

For the final statement, let $q \in W^\sigma$ be an orthogonal projection. Using the obvious inclusions $\Gamma(\sigma) \subset \Gamma(\sigma^q)$ and $\Gamma(\sigma) \subset \mathrm{sp}(\sigma^q)$, we get $\Gamma(\sigma) + \Gamma(\sigma) \subset \Gamma(\sigma^q) + \mathrm{sp}(\sigma^q) = \mathrm{sp}(\sigma^q)$, so $\Gamma(\sigma) + \Gamma(\sigma) \subset \Gamma(\sigma)$. By (i) and (iv) in Proposition 11.1.8 we have $\mathrm{sp}(\sigma^q) = -\mathrm{sp}(\sigma^q)$, so $\Gamma(\sigma) = -\Gamma(\sigma)$, and $\Gamma(\sigma)$ is a closed subgroup of $\mathbb{R}$. □

Proposition 11.2.3 *The proper closed subgroups of $\mathbb{R}$ are $c\mathbb{Z}$ for $c \geq 0$.*

Proof Consider such a subgroup G. If there are positive $c_n \in G$ such that $c_n \to 0$, then to $a \in \mathbb{R}$ there is $m_n \in \mathbb{N}$ such that $m_n c_n \leq a \leq (m_n + 1)c_n$, so a is within distance c_n from G. As G is closed we thus get $a \in G$, which is a contradiction. Hence there is non-negative $c \in G$ such that $G \cap \langle 0, c\rangle = \phi$. If there is positive $b \in G \backslash c\mathbb{Z}$, there is n such that $nc < b < (n+1)c$. Then $0 < b - nc < c$ and $b - nc \in G$, which is absurd, so $G = c\mathbb{Z}$. □

From the two previous results we see that the multiplicative subgroup $e^{\Gamma(\sigma)}$ of $\langle 0, \infty\rangle$ is either $\{1\}$ or $\langle 0, \infty\rangle$ or $a^{\mathbb{Z}}$ for $a \in \langle 0, 1\rangle$.

Lemma 11.2.4 *Let $(W, \mathbb{R}, \sigma)$ be a von Neumann dynamical system. Given an open cover $\{U_i\}$ of $\mathbb{R}$ and a non-zero $w \in W$, there is $f \in L^1(\mathbb{R})$ such that $\sigma(f)(w) \neq 0$ and with $\hat{f}$ having support in some U_i.*

Proof Let J be the closure in $L^1(\mathbb{R})$ of the span of all $f \in L^1(\mathbb{R})$ such that $\hat{f}$ has compact support in some U_i, so J is a closed ideal of $L^1(\mathbb{R})$. Any $t \in \mathbb{R}$ belongs to some U_i, and by Lemma 4.2.5 there is $f \in L^1(\mathbb{R})$ such that $\hat{f}(t) = 1$ and with $\hat{f}$ having compact support in U_i. So $f \in J$ and $t \notin J^\perp$. Thus $J^\perp = \phi$ and $J = L^1(\mathbb{R})$ by Proposition 11.1.4. Picking an approximate unit $\{u_j\}$ for $L^1(\mathbb{R})$, there is some u_j such that $\sigma(u_j)(w) \neq 0$. By the above we can then approximate u_j with linear combinations of functions f, each with $\hat{f}$ having support in the U_i's, and then $\sigma(f)(w) \neq 0$ for some such function. □

Lemma 11.2.5 *Let $(W, \mathbb{R}, \sigma)$ be a von Neumann dynamical system, and let p_i be non-zero orthogonal projections in W^σ. If $p_1 \sim p_2$ in W, then $\Gamma(\sigma^{p_1}) = \Gamma(\sigma^{p_2})$.*

Proof Let $0 \neq c \in \Gamma(\sigma^{p_1})$. By Propositions 11.2.2 and 11.1.11 it suffices to show that for a neighborhood of c with compact closure A and a non-zero orthogonal projection $q \in W^\sigma$ with $q \leq p_2$, we have $M(\sigma, A) \cap W_q \neq \{0\}$. Let A_1 and A_2 be the compact closures of neighborhoods of c and 0, respectively, and such that $A_1 + A_2 \subset A$. Pick an open cover $\{U_i\}$ of $\mathbb{R}$ such that all $U_j - U_j \subset A_2$. Pick $u \in W$ with $p_1 = u^*u$ and $p_2 = uu^*$, so $qu \neq 0$. The lemma provides $g \in L^1(\mathbb{R})$ such that $a \equiv \sigma(g)(qu) \neq 0$ and with $\hat{g}$ having support in some U_j. Clearly $a = qap_1$. And $\mathrm{sp}_\sigma(a)$ is in the support of $\hat{g}$ by (v) in Proposition 11.1.8, so $\mathrm{sp}_\sigma(a) - \mathrm{sp}_\sigma(a) \subset A_2$. Say W acts on V. Let $q_1 \in W^\sigma$ be the non-zero orthogonal projection onto the closed span of $\sigma_{\mathbb{R}}(a^*)(V)$, so $q_1 \leq p_1$ as $p_1\sigma_t(a^*) = \sigma_t(a^*)$. As $0 \neq c \in A_1 \cap \mathrm{sp}(\sigma^{q_1})$, there is non-zero $b \in M(\sigma, A_1) \cap W_{q_1}$ by Propositions 11.2.2 and 11.1.11 and (iv) in Proposition 11.1.8. From the proof of Proposition 11.2.2 we also see that $\sigma_{t_1}(a)bq_1 \neq 0$ for some $t_1 \in \mathbb{R}$. By definition of q_1 there is then $t_2 \in \mathbb{R}$ such that $d \equiv \sigma_{t_1}(a)b\sigma_{t_2}(a^*) \in W_q$ is non-zero. By Proposition 11.1.9 and (i) in Proposition 11.1.8 and $\mathrm{sp}_\sigma(b) \subset A_1$ we have $\mathrm{sp}_\sigma(d) \subset A$. □

Definition 11.2.6 Two actions σ, β of $\mathbb{R}$ by automorphisms on a von Neumann algebra W are *outer equivalent* if there is a strongly continuous one-parameter group of unitaries $u_t \in W$ with $u_{s+t} = u_s\sigma_s(u_t)$ and $\beta_t = u_t\sigma_t(\cdot)u_t^*$ for $s, t \in \mathbb{R}$.

The following result is due to Connes.

Theorem 11.2.7 *Let $(W, \mathbb{R}, \sigma)$ and $(W, \mathbb{R}, \beta)$ be von Neumann dynamical systems. If σ and β are outer equivalent, then $\Gamma(\sigma) = \Gamma(\beta)$.*

Proof Let $u_t \in W$ be as in the previous definition, and consider the von Neumann dynamical system $(W \otimes M_2(\mathbb{C}), \mathbb{R}, \gamma)$, where

$$\gamma_t(\sum w_{ij}\otimes e_{ij}) = \sigma_t(w_{11})\otimes e_{11}+\sigma_t(w_{12})u_t^*\otimes e_{12}+u_t\sigma_t(w_{21})\otimes e_{21}+\beta_t(w_{22})\otimes e_{22}$$

for $w_{ij} \in W$ and $t \in \mathbb{R}$. Let $v = 1 \otimes e_{21}$. Then $v^*v = 1 \otimes e_{11}$ and $vv^* = 1 \otimes e_{22}$ are fixed points for γ. Hence $\Gamma(\gamma^{1\otimes e_{11}}) = \Gamma(\gamma^{1\otimes e_{22}})$ by the previous lemma, and $\gamma_t^{1\otimes e_{11}}$ restricted to $W_{1\otimes e_{11}} \cong W$ is σ_t, while $\gamma_t^{1\otimes e_{22}}$ restricted to $W_{1\otimes e_{22}} \cong W$ is β_t for $t \in \mathbb{R}$. □

The following result is an immediate consequence of the theorem above and the cocycle theorem of Connes.

Corollary 11.2.8 *Let x be a faithful semifinite normal weight on a von Neumann algebra W with modular automorphism group $\{\sigma_t^x\}$. Then $\Gamma(W) = \Gamma(\sigma^x)$ is an invariant for W which is independent of the chosen x.*

11.3 Classification of Type III Factors

Since we are only going to consider factors on separable Hilbert spaces, we could here have considered states rather than weights, but since we have developed the weight theory for later purposes, we might as well use it.

Just like in the bounded case one checks that the spectrum of an unbounded self-adjoint positive operator is a closed subset of the non-negative real line, and it can well contain zero even if the operator is invertible as an unbounded operator.

Proposition 11.3.1 *Let x be a faithful semifinite normal weight on a von Neumann algebra W, let $\{\sigma_t\}$ be the modular automorphism group of x, and let Δ be the corresponding modular operator. Then $e^{\mathrm{sp}(\sigma)} = \mathrm{Sp}(\Delta)\backslash\{0\}$.*

Proof Let $\Delta = \int \lambda \, dP(\lambda)$ be a resolution of the identity for Δ. For $f \in L^1(\mathbb{R})$ and $w \in N_x$, the Fubini-Tonelli theorem gives

$$q_x(\sigma(f)(w)) = \int f(t) q_x(\sigma_t(w)) \, dt = \int f(t) \Delta^{it} q_x(w) \, dt$$

$$= \int \lambda^{it} f(t) \, dP(\lambda) dt q_x(w) = \int \hat{f}(-\ln \lambda) \, dP(\lambda) q_x(w),$$

so by injectivity of q_x and denseness of N_x in W, we see that $\sigma(f) = 0$ if and only if $\hat{f}(-\ln \lambda) = 0$ for all positive $\lambda \in \Delta$. Thus $\mathrm{sp}(\sigma) = \{\ln \lambda \mid \forall 0 < \lambda \in \Delta\}$ as $\mathrm{sp}(\sigma)$ is symmetric about the origin by Proposition 11.1.8. □

Let x be a semifinite normal weight on a von Neumann algebra W, and let p be its support projection. By the *modular automorphism group* σ^x and the *modular operator* Δ_x of x we mean the ones given by the faithful semifinite normal weight on W_p given by restricting x.

Proposition 11.3.2 *Let W be a von Neumann algebra, and define $S(W)$ to be the intersection of all* $\mathrm{Sp}(\Delta_x)$ *over all semifinite normal weights x on W. Then*

$$e^{\Gamma(W)} = S(W) \cap \langle 0, \infty \rangle = \cap\{e^{\mathrm{sp}(\sigma^x)} \mid x \text{ is a semifinite normal weight on } W\}.$$

Proof Let x be a semifinite normal weight on W with support projection p, and pick a semifinite normal weight y on W with support projection $1 - p$, so $x + y$ is a faithful semifinite normal weight on W, say with modular automorphism group $\{\sigma_t\}$, while $\{\sigma_t^x\}$ is the modular automorphism group of x on W_p. By the KMS-condition and uniqueness of such groups, the automorphism σ_t restricted to W_p is σ_t^x for each $t \in \mathbb{R}$. Clearly $(x + y)(wp) = (x + y)(pw)$ for $w \in M_{x+y}$, so $p \in W^\sigma$ by Theorem 10.6.6. Hence $\Gamma(W) = \Gamma(\sigma) \subset \Gamma(\sigma^x) \subset \mathrm{sp}(\sigma^x) = \{\ln \lambda \mid 0 < \lambda \in \mathrm{Sp}(\Delta_x\}$ by the previous proposition, and $e^{\Gamma(W)} \subset S(W) \cap \langle 0, \infty \rangle$.

Conversely, suppose $e^s \in S(W)$ for $s \in \mathbb{R}$. Let x be a faithful semifinite normal weight on W with modular automorphism group $\{\sigma_t\}$. For any non-zero orthogonal projection $q \in W^\sigma$, the restriction of σ_t to W_q gives as above the modular automorphism group of y, where y is now x restricted to W_q. Thus $e^s \in \mathrm{Sp}(\Delta_y)$, and $s \in \mathrm{sp}(\sigma^y) = \mathrm{sp}(\sigma^q)$ by the previous proposition. Hence $s \in \Gamma(\sigma) = \Gamma(W)$. □

From the previous section we know that $\Gamma(W)$ for any von Neumann algebra W is either $\langle 0, \infty \rangle$ or $\{1\}$ or $\{\lambda^{\mathbb{Z}}\}$ for some $\lambda \in \langle 0, 1 \rangle$.

Proposition 11.3.3 *Let W be a factor on a separable Hilbert space. Then W is semifinite if and only if $S(W) = \{1\}$ if and only if $0 \notin S(W)$. So the type III factors W on separable Hilbert spaces are exactly those with $0 \in S(W)$.*

Proof Suppose W is semifinite. Let x be a faithful semifinite normal trace on W_+. Then its modular operator $\Delta_x = 1$ because on the corresponding left Hilbert algebra $S_x^* = S_x = S_x^{-1}$ due to the tracialness of the weight x. So $S(W) = \{1\}$ and $0 \notin S(W)$. Assuming the latter, there is by definition a semifinite normal weight y on

W such that $0 \notin \mathrm{Sp}(\Delta_y)$. But then Δ_y^{-1} is bounded away from zero, so both it and Δ_y are bounded operators. By for instance considering the expansion series of Δ_y^{it}, one sees that $t \mapsto \sigma_t$ is norm continuous, and by Theorem 11.1.14 it is inner on W_p, where p is the support projection of y. By Proposition 10.7.8 we conclude that W_p is semifinite, but then W is also semifinite as W is a factor. □

We thus give the following definition to distinguish type III factors.

Definition 11.3.4 A factor W on a separabel Hilbert space is of *type* III_0, III_1 *or* III_λ if $S(W)$ equals resp. $\{0, 1\}$, $[0, \infty\rangle$, or $\{\lambda^{\mathbb{Z}}\} \cup \{0\}$ for $\lambda \in \langle 0, 1\rangle$.

In calculating the Connes spectrum it suffices to consider central projections in the fixed point algebra.

Proposition 11.3.5 *Let* $(W, \mathbb{R}, \sigma)$ *be a von Neumann dynamical system. Then* $\Gamma(\sigma) = \cap \mathrm{sp}(\sigma^p)$ *where we intersect over all orthogonal projections* $0 \neq p \in Z(W^\sigma)$.

Proof The inclusion $\subset$ is obvious, so it suffices to show that for any non-zero orthogonal projection $p \in W^\sigma$, there is a non-zero orthogonal projection $q \in Z(W^\sigma)$ such that $\sigma^q = \sigma^p$. Let q be the supremum of upu^* over all unitaries $u \in W^\sigma$. Clearly q is a non-zero orthogonal projection of $Z(W^\sigma)$, so by Propositions 11.1.11 and 11.2.2, we need only show that $M(\sigma, A) \cap W_p \neq \{0\}$ if an only if $M(\sigma, A) \cap W_q \neq \{0\}$ for any closed subset A of $\mathbb{R}$. The forward implication is obvious as $p \leq q$. If $0 \neq w = qwq \in M(\sigma, A)$, there are unitaries $u, v \in W^\sigma$ such that $(upu^*)w(vpv^*) \neq 0$. Then $0 \neq a \equiv pu^*wvp \in W_p$. As $\sigma(f)(b) = \hat{f}(0)b$ for $b \in W^\sigma$, we see that $\mathrm{sp}_\sigma(b) = \{0\}$ for all $b \in W^\sigma \backslash \{0\}$. By Proposition 11.1.9 we thus see that $\mathrm{sp}_\sigma(a) = \mathrm{sp}_\sigma(w) \subset A$, and $M(\sigma, A) \cap W_p \neq \{0\}$. □

Corollary 11.3.6 *Let* $\{\sigma_t\}$ *be the modular automorphism group of a faithful semifinite normal weight on a von Neumann algebra* W. *Then* $\Gamma(W) = \cap \mathrm{sp}(\sigma^p)$ *where we intersect over all non-zero orthogonal projections* $p \in Z(W^\sigma)$.

Proposition 11.3.7 *Let* W *be a factor on a separable Hilbert space, let* x *be a faithful semifinite normal weight on* W *with modular automorphism group* $\{\sigma_t\}$, *and let* Δ_p *be the modular operator for the restriction of* x *to* W_p, *where* p *is any non-zero orthogonal projection in* $Z(W^\sigma)$. *Then* $S(W) = \cap_p \mathrm{sp}(\Delta_p)$.

Proof By the corollary, Propositions 11.3.1 and 11.3.2, we get

$$e^{\Gamma(W)} = S(W) \backslash \{0\} = \cap_p \mathrm{sp}(\Delta_p) \backslash \{0\}.$$

If W is type III, so is W_p, and $0 \in S(W)$ and $0 \in \mathrm{sp}(\Delta_p)$ by Proposition 11.3.3. If W is semifinite, the same proposition tells us that $0 \notin S(W)$, and we must find a non-zero orthogonal projection $p \in Z(W^\sigma)$ such that $0 \notin \mathrm{sp}(\Delta_p)$. From the proof of Proposition 10.7.8, we have $\sigma_t = h^{-it} \cdot h^{it}$ for an invertible positive self-adjoint operator h in the Hilbert space where W acts, and the spectral projections $P(\lambda)$ of h clearly belong to $Z(W^\sigma)$. Set $p = \int_{1/n}^n dP(\lambda)$. Then $y = x_{hp}$ on W_p is a trace by Theorem 10.6.9 and the KMS-condition. So $x(ww^*) \leq ny(ww^*) = ny(w^*w) \leq$

$n^2x(w^*w)$ for $w \in W_p$ and

$$\|\Delta_p(q_x(w))\|^2 = \|J\Delta_p(q_x(w))\|^2 = \|q_x(w^*)\|^2 = x(ww^*) \leq n^2\|q_x(w)\|^2.$$

when also $w \in N_x$, so Δ_p is bounded. □

Exercises

For Sect. 11.1

1. Generalize the Arveson spectrum to von Neumann dynamical systems (W, G, α), where G is an abelian locally compact group. Generalize some of the results relating to the Arveson spectrum.
2. Let v be an element of a unital Banach algebra V. Show that $f(t) = e^{tv}$ for $t \in \mathbb{R}$ is a norm continuous one-parameter group of invertible elements of V that is differentiable in norm. Find its derivative.
3. Prove that a norm continuous one-parameter group of invertible elements of a unital Banach algebra V is of the form $t \mapsto e^{tv}$ for $v \in V$, and show that the spectrum of v is purely imaginary if $\|e^{tv}\| = 1$ for all $t \in \mathbb{R}$.
4. Given a von Neumann dynamical systems $(W, \mathbb{R}, \sigma)$, let h be an unbounded operator $D \subset W \to W$ with domain D consisting of those $w \in W$ with $t \mapsto x(\sigma_t(w))$ differentiable for every $x \in W_*$, and such that $x \mapsto dx(\sigma_t(w))/dt|_{t=0}$ is bounded and is thus of the form $x \mapsto x(h(w))$. Prove that D is σ-weak* dense in W, and that h is σ-weakly* closed. Prove that its spectrum is purely imaginary.

For Sect. 11.2

1. Define the Connes spectrum of von Neumann dynamical systems (W, G, α), where G is an abelian locally compact group. Generalize some of the results relating to the Connes spectrum.
2. Let $(W, \mathbb{R}, \sigma)$ be a von Neumann dynamical system. Show that $\Gamma(\sigma) = \mathrm{sp}(\sigma)$ when W^σ is a factor.
3. Let $(W, \mathbb{R}, \sigma)$ be a von Neumann dynamical system with a non-zero orthogonal projection $p \in W^\sigma$ with central support q. Show that for each closed subset A of $\mathbb{R}$, we have $M(\sigma, A) \cap W_p \neq 0$ if and only if $M(\sigma, A) \cap W_q \neq 0$.
4. Let $(B(L^2(\mathbb{R})), \mathbb{R}^2, \alpha)$ be the von Neumann dynamical system with $\alpha_{(s,t)} = u(s)v(t)(\cdot)v(t)^*u(s)^*$, where $u(s)f(r) = f(r+s)$ and $v(t)f(r) = e^{irt}f(r)$, $f \in L^2(\mathbb{R})$. Show $\Gamma(\alpha) = \mathbb{R}^2$, and that $H \mapsto \{w \in B(L^2(\mathbb{R})) \,|\, \alpha_{(s,t)}(w) = w \text{ for } (s,t) \in H\}$ is an injection from the closed subgroups of $\mathbb{R}^2$ to von Neumann subalgebras of $B(L^2(\mathbb{R}))$.

For Sect. 11.3

1. Given a faithful state x on $M_k(\mathbb{C})$ written as $x = \mathrm{Tr}(h\cdot)$ for a positive invertible matrix $h \in M_k(\mathbb{C})$ with the sum of its eigenvalues equaling one. Show that the list $0 < \lambda_k \leq \cdots \leq \lambda_1$ of eigenvalues of h is an invariant for x. Assuming that h is diagonal with λ_n in the n-th row, show that the modular automorphism group of x is given by $\sigma_t^x = h^{it}(\cdot)h^{-it}$, and that $\sigma_t^x(e_{mn}) = (\lambda_m/\lambda_n)^{it} e_{mn}$ for the matrix unit e_{mn}.
2. Given a sequence (W_n, x_n) of full matrix algebras W_n and faithful states x_n on W_n, consider the embedding $W_1 \otimes \cdots \otimes W_n \subset W_1 \otimes \cdots \otimes W_{n+1}$ which sends $w \to 1 \otimes w$, and consider the state x on the corresponding inductive limit C*-algebra V given by $x(w_1 \otimes \cdots \otimes w_n \otimes 1 \otimes \cdots) = x_1(w_1) \cdots x_n(w_n)$ for $w_i \in W_i$. Show that x is faithful. Consider the von Neumann algebra W in the GNS-representation of x, and denote the corresponding normal state by y. Show that the modular group of y is given by $\sigma_t^y(w_1 \otimes \cdots \otimes w_n \otimes 1 \otimes \cdots) = \sigma_t^{x_1}(w_1) \otimes \cdots \otimes \sigma_t^{x_n}(w_n) \otimes 1 \otimes \cdots$.
3. In the previous exercise, let $0 < \lambda_{n,m_n} \leq \cdots \leq \lambda_{n,1}$ be the eigenvalue list of x_n as described in exercise 1. Show that the von Neumann algebra W in exercise 2 is of type I if and only if $\sum_n |1 - \lambda_{n,1}| < \infty$, that W is of type II_1 if and only if $\sum_{n,j} |(m_n)^{-1/2} - (\lambda_{n,j})^{1/2}|^2 < \infty$. Prove that if all $\lambda_{n,1}$ are greater than a positive scalar, then W is of type III if and only if $\sum_{n,j} \inf\{|(\lambda_{n,1}/\lambda_{n,j}) - 1|^2, r\} = \infty$ for some $r > 0$.
4. In exercise 2 put all $W_n = M_2(\mathbb{C})$ and $x_n((a_{ij})) = s a_{11} + (1 - s) a_{22}$ for a fixed $s \in \langle 0, 1/2 \rangle$. Show that the corresponding W in that exercise is a factor of type III_λ with $\lambda = s/(1 - s)$, known as the Powers factor.

Chapter 12
Quantum Groups and Duality

The basic notion in this chapter and the second half of the book, is that of a locally compact quantum group. By definition it consists of a von Neumann algebra W with a so called coproduct, that is, a normal $*$-homomorphism $\Delta\colon W \to W\bar{\otimes}W$ satisfying what we call coassociativity $(\Delta\otimes\iota)\Delta = (\iota\otimes\Delta)\Delta$, and we impose the existence of two faithful semifinite normal weights, one, x, that satisfies so called left invariance $(\iota\otimes x)\Delta = x$, and the other one, y, is right invariant, meaning that $(y\otimes\iota)\Delta = y$. In due time we will prove that these weights are unique up to multiplication with positive constants.

Locally compact quantum groups generalize locally compact groups. Namely, when W is commutative, it is of the form $L^\infty(G)$ for a unique locally compact group G, and the product of the group is encoded in the coproduct in the sense that $\Delta(f)(s,t) = f(st)$ for $f \in L^\infty(G)$ and $s, t \in G$. The weights x and y are then the left- and right invariant Haar integrals.

In another direction, when x and y are states, they must coincide, and we speak of a compact locally compact quantum group. Finally, when W is finite dimensional, we get a Hopf $*$-algebra. This is by definition a unital $*$-algebra W with a (non-normal) coproduct Δ, and moreover, it should have a so called counit, i.e. a character ε on W satisfying $(\varepsilon\otimes\iota)\Delta = \iota = (\iota\otimes\varepsilon)\Delta$, and a so called coinverse, i.e. an invertible linear map S on W satisfying $m(S\otimes\iota)\Delta = \varepsilon = m(\iota\otimes S)\Delta$, where $m\colon W\otimes W \to W$ is the linear extension of the multiplication on W. Conversely, when we have a Hopf $*$-algebra with W a finite dimensional C^*-algebra, or a finite direct sum of full matrix algebras, then it is a locally compact quantum group. And when W in addition is commutative, then $W = L^\infty(G) = C_0(G)$ for a finite group G. Then the unit e in G is encoded in the counit as $\varepsilon(f) = f(e)$, while the inverse operation on G is encoded in the coinverse as $S(f)(t) = f(t^{-1})$. Hopf $*$-algebras also occur when the group is no longer finite, but then the algebras are either only dense in some C*-algebra, or they contain unbounded elements.

Starting with a locally compact quantum group we will study to what extent it behaves like a Hopf $*$-algebra. We will do so by invoking the modular theory of

L. Tuset, *Analysis and Quantum Groups*,
https://doi.org/10.1007/978-3-031-07246-8_12

the left- and right invariant weights, and derive a series of properties of numerous intrinsic quantities naturally associated with the quantum group. For the sake of expedience, and because our fundamental objects of study are locally compact quantum groups, we postpone discussing in detail the relation to the classical situation of groups till the next chapter when we have developed enough abstract theory.

In section one we include the basic algebraic theory of Hopf $*$-algebras. In section two we go through the theory of compact quantum groups, where we prove existence of an invariant state from axioms inspired by the cancellation properties for groups. Working with a generalization of the right regular representation of a group, and decomposing this into its irreducibles, we obtain a generalization of the Peter-Weyl theory. The modular properties are governed by a one-parameter family $\{f_z\}_{z\in\mathbb{C}}$ of functionals on a dense Hopf $*$-algebra playing the role as the regular functions on the group. In this setting one works from the outset with C*-algebras playing the role as $C_0(G)$; the relevant von Neumann algebra is the bicommutant of the Hopf $*$-algebra in the GNS-representation of the state.

A striking feature is that already in the compact case neither the counit nor the coinverse can be expected to be bounded; the coinverse is in general no longer involutive. However, we can always polar decompose the coinverse $S = R\tau_{-i/2}$ into the so called unitary coinverse R and a generally unbounded part controlled by a one-parameter group $\{\tau_t\}$, the so called scaling group. When the scaling group is trivial, then $S = R$ is involutive, and we have what is known as a Kac algebra. The theory for them was developed long before locally compact quantum groups entered the arena, so the new obstacle was really the scaling group. In the compact case the scaling group can be expressed in terms of the functionals f_z.

These phenomena are already manifest in the historic example of the compact quantum group $SU_q(2)$, which is a deformation of the group of unitary two-by-two complex matrices with determinant one obtained when the deformation parameter $q \in [-1, 1]\backslash\{0\}$ is set to one. Quantum $SU_q(2)$ belongs to a large class of quantum groups defined in terms of generators and relations, and is better studied from the Lie algebra point of view; every semisimple compact Lie group has a q-deformation corresponding to each of its Poisson Lie group structures, and the non-compact case is a chapter for itself hitherto not much explored. It should be noted that due to well-known no-go theorems, these deformations do not merely mean parametrising the stucture constants of the Lie algebras; the deformation is an entirely different one, voyaging beyond the universal enveloping algebras of Lie algebras and, on the dual side, into a non-commutative regime closely related to quantum physics, hence the name quantum, thus calling for an appropriate analytical machinery.

In section three we define locally compact quantum groups and show that they possess a regular corepresentation, where we use the prefix co- to indicate that we have 'transposed' the notion of, in this case, a representation. Actually we rather focus on the closely related notion of a multiplicative unitary on a Hilbert space V. By definition this is any unitary $M \in B(V \otimes V)$ satisfying the so called pentagonal equation $M_{12}M_{13}M_{23} = M_{23}M_{12}$ in $B(V \otimes V \otimes V)$, where we have spread the legs of M according to the indices. The regular corepresentation corresponds then to the so called fundamental multiplicative unitary $M \in W\bar{\otimes}B(V_x)$ on V_x uniquely determined by $M^*(q_x(v) \otimes q_x(w)) = (q_x \otimes q_x)(\Delta(w)(v \otimes 1))$ for $v, w \in N_x$, so $\Delta(\cdot) = M^*(1 \otimes (\cdot))M$ and $(\Delta \otimes \iota)(M) = M_{13}M_{23}$ as the set of slices $(\iota \otimes \omega)(M)$ of M by all $\omega \in B(V_x)_*$ is weakly dense in W. This density condition, and another one stating that W is the weakly closed span of the set of elements $(\omega \otimes \iota)\Delta(w)$ for $\omega \in W_*$ and $w \in W$, are proved invoking the densely defined closed conjugate linear involution G in V_x that sends a typical core element $q_x(y \otimes \iota)(\Delta(a^*)(b \otimes 1))$ to $q_x(y \otimes \iota)(\Delta(b^*)(a \otimes 1))$ for $a, b \in N_x^* N_x$. Polardecomposing $G = I|G|$ yields $\tau_t = |G|^{-2it}(\cdot)|G|^{2it}$ and $R = I(\cdot)^*I$ and $(\sigma_t^y \otimes \tau_t)\Delta = \Delta\sigma_t^y$ for real t, so $(\tau_t \otimes \tau_t)\Delta = \Delta\tau_t$ while $f(R \otimes R)\Delta = \Delta R$, where f is the flip on $W\bar{\otimes}W$. We prove what we call strong left invariance of x, which means that $S(\iota\otimes x)(\Delta(c^*)(1\otimes d)) = (\iota\otimes x)((1\otimes c^*)\Delta(d))$ for $c, d \in N_x$, and we prove a similar strong right invariance of y. Using this we show that the coinverse, scaling group and unitary coinverse are independent of x and y. When $W = C(G)$ for a finite group G, then with $s \in G$ seen as characters on $C(G)$, the map $s \mapsto (s \otimes \iota)(M)$ is a unitary representation of G on $V_x = l^2(G)$, namely the (left) regular one.

In general, we show what we call relative invariance of x, meaning that if another faithful semifinite normal weight is invariant under σ_t^x up to multiplication by the t-th power of a positive constant, then that weight coincides with x up to multiplication by a positive scalar. We show that this holds for the weight $x\tau_s$, so there is a unique scalar $\nu > 0$ called the scaling constant satisfying $x\tau_s = \nu^{-s}x$. Applying the generalized Radon-Nikodym theorem from Chap. 9, we prove then that there exists a unique strictly positive operator h affiliated with W such that x_h equals the right invariant weight xR and $\sigma_t(h) = \nu^t h$. Note that due to the property of R involving the flip, the faithful semifinite normal weight xR is indeed right invariant. We call the group-like element h the modular element of the quantum group thanks to its role in the classical case. Using it together with relative invariance we show finally that x and y are unique up to multiplication with positive scalars, and that h, ν depend only on the quantum group.

We proceed then to discuss modularity and manageability of multiplicative unitaries, reduced, or topological, versions of quantum groups, and duality, establishing a fairly complete list of fundamental relations between the various quantities thus at hand. We close the chapter discussing discrete quantum groups, being by definition, duals of compact ones. The discreteness exposes itself in that the von Neumann algebra is a direct product of full matrix algebras corresponding to the irreducible unitary corepresentations of the compact dual quantum group.

Just a few words about duality. It generalizes Pontryagin duality for locally compact abelian groups. The fundamental multiplicative unitary $\hat{M}$ for the dual locally compact quantum group $(\hat{W}, \hat{\Delta})$ is M_{21}^*, and $\hat{\Delta} = \hat{M}^*(1 \otimes (\cdot))\hat{M}$, where $\hat{W}$ is the von Neumann algebra generated by the slices $(\iota \otimes \omega)(\hat{M})$ for $\omega \in B(V_x)_*$, so $M \in W\bar{\otimes}\hat{W}$. One also has the identification $V_{\hat{x}} = V_x$, where $\hat{x}$ is an appropriately scaled left invariant faithful semifinite normal weight on the dual quantum group. Clearly the bidual quantum group coincides with the original quantum group. Let us also point out that the C*-algebra generated by the slices $(\iota\otimes\omega)(M)$ with coproduct given by restriction yields the so called reduced C*-algebra version of the quantum group. The left invariant lower semicontinuous weight given by restriction is locally finite, which is the C*-algebra version of semifinitenes. It satisfies a KMS-type property, and clearly it is faithful.

Recommended literature for this chapter is [15, 24–27, 30, 32, 35, 43, 44, 49, 54].

12.1 Hopf Algebras

Definition 12.1.1 A *bialgebra* (W, Δ) consists of a complex unital algebra W and unital homomorphisms $\Delta\colon W \to W \otimes W$ and $\varepsilon\colon W \to \mathbb{C}$ such that $(\Delta \otimes \iota)\Delta = (\iota\otimes\Delta)\Delta$ and $(\varepsilon\otimes\iota)\Delta = \iota = (\iota\otimes\varepsilon)\Delta$. It is a *Hopf algebra* if it has an invertible linear map $S\colon W \to W$ such that $m(S\otimes\iota)\Delta = 1\varepsilon(\cdot) = m(\iota\otimes S)\Delta$, where $m\colon W\otimes W \to W$ is the multiplication on W extended to the tensor product. If W is also a $*$-algebra and Δ is $*$-preserving, then (W, Δ) is a *Hopf $*$-algebra*. We call Δ, ε and S the *comultiplication, counit* and *coinverse*, respectively.

We adapt the *Sweedler notation* $\Delta(w) = w_{(0)} \otimes w_{(1)}$, and omit parenthesis in $(\Delta \otimes \iota)\Delta(w) = w_{(0)} \otimes w_{(1)} \otimes w_{(2)}$ due to *coassociativity* $(\Delta \otimes \iota)\Delta = (\iota \otimes \Delta)\Delta$. In a Hopf algebra $\varepsilon(w_{(0)})w_{(1)} = w = w_{(0)}\varepsilon(w_{(1)})$ and $S(w_{(0)})w_{(1)} = \varepsilon(w)1 = w_{(0)}S(w_{(1)})$.

Counits and coinverses are unique. Indeed, if ε' and S' are other ones, then $\varepsilon' = (\varepsilon' \otimes \varepsilon)\Delta = \varepsilon$ and $S'(w) = S'(w_{(0)})w_{(1)}S(w_{(2)}) = S(w)$. Note that $(A, \sigma\Delta)$ is a Hopf algebra with counit ε and coinverse S^{-1}, where σ is the flip on $W \otimes W$.

Proposition 12.1.2 *Let (W, Δ) be a Hopf algebra. Its coinverse S is a unital antimultiplicative map such that $\sigma(S \otimes S)\Delta = \Delta S$. If (W, Δ) is a Hopf $*$-algebra, then ε is $*$-preserving and $S(S(w^*)^*) = w$ for $w \in W$.*

Proof We have $S(v)S(w) = S(w_{(0)}v_{(0)})w_{(1)}v_{(1)}S(v_{(2)})S(w_{(2)}) = S(wv)$ for $v, w \in W$ as the counit is multiplicative. Also

$$S(v_{(1)}) \otimes S(v_{(0)}) = (S(v_{(1)}) \otimes S(v_{(0)}))\Delta(v_{(2)}S(v_{(3)})) = \Delta S(v)$$

as Δ is multiplicative. In the Hopf $*$-algebra case, note that $w \mapsto \overline{\varepsilon(w^*)}$ and $w \mapsto S(w^*)^*$ clearly satisfy the axioms for the counit and the coinverse for $(A, \sigma\Delta)$, so we are done by uniqueness of these maps. □

So the coinverse is $*$-preserving if and only if it is involutive. Here is an alternative characterization of Hopf algebras.

Proposition 12.1.3 *A bialgebra (W, Δ) without a counit is a Hopf algebra if and only if the linear maps on $W\otimes W$ given by $T_1(v\otimes w) = v_{(0)}\otimes v_{(1)}w$ and $T_2(v\otimes w) = vw_{(0)} \otimes w_{(1)}$ are invertible.*

Proof For the forward direction it is readily checked that the inverses are given by $T_1^{-1}(v \otimes w) = v_{(0)} \otimes S(v_{(1)})w$ and $T_2^{-1}(v \otimes w) = vS(w_{(0)}) \otimes w_{(1)}$.

Conversely, consider the linear map E on W given by $E(w) = mT_1^{-1}(w \otimes 1)$ for $w \in W$. By surjectivity of T_1, write $w \otimes 1 = \sum \Delta(v^i)(1 \otimes w^i)$, and note that

$$(\iota \otimes E)\Delta(w) = w_{(0)} \otimes mT_1^{-1}(w_{(1)} \otimes 1) = (\iota \otimes mT_1^{-1})(\Delta \otimes \iota)(w \otimes 1)$$

$$= \sum (\iota \otimes mT_1^{-1})(v^i_{(0)} \otimes v^i_{(1)} \otimes v^i_{(2)} w^i) = \sum (\iota \otimes m)(v^i_{(0)} \otimes v^i_{(1)} \otimes w^i) = w \otimes 1.$$

Multiplying from the left by $v \otimes 1$ gives $(\iota \otimes E)((v \otimes 1)\Delta(w)) = vw \otimes 1$, and surjectivity of T_2 shows that E is scalar valued. Hence $\varepsilon(w)1 = E(w)$ defines a linear functional ε on W such that $(\iota \otimes \varepsilon)\Delta = \iota$. Similarly we get $(\varepsilon \otimes \iota)\Delta = \iota$. The coinverse is now given by $S(w) = (\varepsilon \otimes \iota)T_1^{-1}(w \otimes 1)$ for $w \in W$, and one checks in a similar manner that it actually satisfies the axioms for a coinverse. □

The proof did not use multiplicativity of the counit.

Given a finite dimensional Hopf algebra (W, Δ) with counit ε and coinverse S, we turn the vector space W' of linear functionals on W into a *dual* Hopf algebra with product, identity, coproduct $\hat{\Delta}$, counit $\hat{\varepsilon}$ and coinverse $\hat{S}$ given by $xy = (x \otimes y)\Delta$ and ε and $\hat{\Delta}(x) = xm \in (W \otimes W)' = W' \otimes W'$ and $\hat{\varepsilon}(x) = x(1)$ and $\hat{S}(x) = xS$. In the Hopf $*$-algebra case the $*$-operation on W' is given by $x^* = \overline{x(S(\cdot)^*)}$. Upon dualizing once more the natural vector space isomorphims $W \to W''$ will be an isomorphism of Hopf $*$-algebras. In the infinite dimensional case one settles for *dual pairings*, where the bilinear pairing $(w, x) \mapsto x(w)$ is only assumed to be non-degenerate.

Example 12.1.4 Let G be a finite group. Then $(C(G), \Delta)$ is a Hopf $*$-algebra with comultiplication, counit and coinverse given by $\Delta(f)(s, t) = f(st)$ and $\varepsilon(f) = f(e)$ and $S(f)(t) = f(t^{-1})$ for $f \in C(G)$ and $s, t \in G$ and the unit e in G. The group axioms are reflected in the Hopf algebra axioms in a natural way.

Another Hopf $*$-algebra $(\mathbb{C}[G], \hat{\Delta})$ has comultiplication, counit and coinverse given by $\hat{\Delta}(t) = t \otimes t$ and $\varepsilon(t) = 1$ and $S(t) = t^{-1} = t^*$ for $t \in G$. Recall that the product on the group algebra $\mathbb{C}[G]$ is a linearization of the group multiplication. This Hopf $*$-algebra is isomorphic to the dual of the previous one under the identification $\mathbb{C}[G] \to C(G)'$ which sends $t \in G$ to δ_t. While $(C(G), \Delta)$ is commutative, its dual is *cocommutative*, in that $\sigma\hat{\Delta} = \hat{\Delta}$, and is only commutative when G is abelian, in which case the former Hopf $*$-algebra is also cocommutative.

The formula $x(f) = \sum_{t \in G} f(t)$ defines a faithful tracial positive linear functional x on $C(G)$ such that $(x \otimes \iota)\Delta = x(\cdot)1 = (\iota \otimes x)\Delta$, reflecting right- and left invariance of the counting measure on G. Similarly, the formula $\hat{x}(t) = \delta_{t,e}$ for $t \in G$ defines a faithful tracial positive linear functional $\hat{x}$ on $\mathbb{C}[G]$ that is right- and left invariant with respect to $\hat{\Delta}$. ◇

Another example of a cocommutative Hopf algebra $(U(\mathfrak{g}), \hat{\Delta})$ is the universal enveloping Lie algebra $U(\mathfrak{g})$ of a Lie algebra $\mathfrak{g}$ with coproduct $\hat{\Delta}(X) = X \otimes 1 + 1 \otimes X$, counit $\hat{\varepsilon}(X) = 0$ and coinverse $\hat{S}(X) = -X$ for $X \in \mathfrak{g}$. These maps extend to $U(\mathfrak{g})$ by universality, and clearly obey the axioms.

12.2 Compact Quantum Groups

When G is a compact group, the formulas for Δ, ε and S from the example in the previous section works provided one identifies $C(G\times G)$ with the C*-tensor product $C(G) \otimes C(G)$ and extend the necessary maps on the tensor product by continuity. This way one gets a Hopf $*$-algebra type of notion from which the group can be recovered with continuous product $\delta_s\delta_t = (\delta_s \otimes \delta_t)\Delta$ for $s, t \in C(G)$. The left- and right invariant faithful positive functional x on $C(G)$ is then the integral with respect to the Haar measure on G.

Extending this to arbitrary unital C*-algebras must be done with caution; new examples show that the coinverse shold not be considered bounded, neither should this be assumed for the counit, and finally, the product map does not extend to the C*-tensor product.

Definition 12.2.1 A *compact quantum group* (W, Δ) consists of a unital C*-algebra W and a unital $*$-homomorphism $\Delta\colon W \to W \otimes W$ (with the minimal C*-norm) such that $W \otimes W$ is the closed span of $(W \otimes 1)\Delta(W)$ and the closed span of $(1 \otimes W)\Delta(W)$.

When G is a compact group, the example $(C(G), \Delta)$ in the first paragraph is a compact quantum group by the Stone-Weierstrass theorem as the span of $(C(G) \otimes 1)\Delta(C(G))$ is the unital $*$-subalgebra of $C(G \times G)$ spanned by functions of the form $(s, t) \mapsto f(s)g(st)$ for $f, g \in C(G)$ and $s, t \in G$, and similarly for the span of $(1 \otimes C(G))\Delta(C(G))$.

Proposition 12.2.2 *Any commutative compact quantum group* (W, Δ) *is of the form* $(C(G), \Delta)$ *for some compact group* G.

Proof By Gelfand's theorem write $W = C(G)$ for a compact Hausdorff space G. Defining the continuous product of characters on $C(G)$ by $xy = (x \otimes y)\Delta$, coassociativity of Δ gives $f(x(yz)) = f((xy)z)$ for $f \in C(G)$, so G becomes a compact semigroup. If $xy = xz$, then $f(x)g(xy) = f(x)g(xz)$, so $h(x, y) = h(x, z)$ for $h \in C(G \times G)$ by the first closed span requirement, so $y = z$. Similarly $yx = zx$ implies $y = z$. So we have a compact semigroup with cancellation.

Then it has to be a group. Indeed, the closed subgroup H of G generated by an element $t \in G$ is abelian; it is the closure of the subset of all t^n for $n \in \{0\} \cup \mathbb{N}$. By an ideal in a semigroup we mean in the remaning part of this proof, any subset that is invariant under left and right multiplication by elements in the semigroup. The intersection $I \cap J$ of two closed ideals of H is non-empty as it contains $IJ = JI$. By compactness the intersection I_t of all closed ideals of H is therefore a non-empty closed ideal of H. Since sI_t for $s \in I_t$ is a closed ideal of H contained in I_t, it equals I_t. So there is $e \in I_t$ such that $se = s$. Then $seu = su$ for $u \in G$, and $eu = u$ by cancellation. Similarly $ue = u$. As $e \in I_t = tI_t = I_t t$, we see that t is invertible. So G is a group. Since the continuous bijection on $G \times G$ given by $(s, t) \mapsto (s, st)$ has a continuous inverse by compactness, we see that $t \mapsto t^{-1}$ is continuous. □

The existence of a *Haar state* is automatic.

Proposition 12.2.3 *Every compact quantum group (W, Δ) admits a unique state x on W such that $(x \otimes \iota)\Delta = x(\cdot)1 = (\iota \otimes x)\Delta$.*

Proof If x' was another one, then $x' = (x \otimes x')\Delta = x$.

To a state y, take a w^*-accumulation point x of the states $n^{-1} \sum_{k=1}^{n} y^k$. Clearly $xy = x = yx$. We claim that if z is a positive functional on W such that $z \leq y$ pointwise, then $xz = z(1)x = zx$. Indeed, for $w \in W$ put $v = (\iota \otimes x)\Delta(w)$. Then $(x \otimes y)((\Delta(v) - v \otimes 1)^*(\Delta(v) - v \otimes 1)) = 0$ as $(\iota \otimes y)\Delta(v) = v$ by coassociativity. Thus $(x \otimes z)((u \otimes 1)(\Delta(v) - v \otimes 1)) = 0$ by the Cauchy-Schwarz inequality, and

$$(x \otimes zx)((u \otimes 1)\Delta(w)) = (x \otimes z)((u \otimes 1)\Delta(v)) = z(1)(x \otimes x)((u \otimes 1)\Delta(w)).$$

From the denseness criteria in the axioms of a compact quantum group we thus get $zx = z(1)x$, and similarly $xz = z(1)x$.

For a finite set F of states y_i on W, take their average y and form x_F to y as above. Then $y_i x_F = x_F = x_F y_i$ by our proved claim. So any w^*-accumulation point x of the increasing net $\{x_F\}$ will do. □

Let us develop the representation theory in this context. A *right corepresentation* of (W, Δ) on a Hilbert space V is an invertible element N of $M(B_0(V) \otimes W)$ such that $(\iota \otimes \Delta)(N) = N_{12}N_{13}$. When the corepresentation is *finite dimensional* with a finite orthonormal basis $\{e_i\}$ of V, and matrix units m_{ij} sending e_j to e_i and all others to zero, we can write $N = \sum m_{ij} \otimes u_{ij}$, where the *matrix elements* u_{ij} satisfy $\Delta(u_{ij}) = \sum u_{ik} \otimes u_{kj}$. An operator $T: V \to V'$ *intertwiners* right corepresentations N and N' if $(T \otimes 1)N = N'(T \otimes 1)$. Write $\mathrm{Mor}(N, N')$ for the normed space of such intertwiners. If it contains an (unitary) invertible element, then N and N' are *(unitarily) equivalent*. If $\mathrm{End}(N) \equiv \mathrm{Mor}(N, N) = \mathbb{C}$, we say N is *irreducible*. Since $T^* \in \mathrm{Mor}(N', N)$, we immediately get Schur's lemma telling us that two irreducible (unitary) right corepresentations N and N' are either (unitarily) equivalent and $\mathrm{Mor}(N, N')$ is one-dimensional, or $\mathrm{Mor}(N, N') = \{0\}$. Any right corepresentation N is equivalent to a unitary representation. Indeed, since $N^*N \geq \varepsilon$ for some $\varepsilon > 0$, then $T = (\iota \otimes x)(N^*N) \geq \varepsilon$ is invertible and satisfies $T \otimes 1 = N^*(T \otimes 1)N$ by left invariance of x, so $N' = (T^{1/2} \otimes 1)N(T^{-1/2} \otimes 1)$ is a unitary right corepresentation, and $T^{1/2} \in \mathrm{Mor}(N, N')$. Every unitary right corepresentation N on V *decomposes* into a direct sum of finite dimensional irreducible unitary right corepresentations, meaning that there are minimal finite rank projections $p_i \in \mathrm{End}(N)$ such that $V = \oplus p_i V$. Indeed, pick a net $\{q_j\}$ of finite rank projections that increases towards 1 strongly. Then the net $\{(\iota \otimes x)(N^*(q_j \otimes 1)N)\}$ in $\mathrm{End}(N) \cap B_0(V)$ increases strongly to 1.

We now define the right *regular* corepresentation.

Proposition 12.2.4 *Let (W, Δ) be a compact quantum group, say with the C^*-algebra W universally represented on V. Then $M(q_x(a) \otimes v) = \Delta(a)(q_x(1) \otimes v)$ for $a \in W$ and $v \in V$ defines a unitary right corepresentation M of the quantum group on the GNS-space V_x of the Haar state x. Moreover, we have $(\pi_x \otimes \iota)\Delta(a) = M(\pi_x(a) \otimes 1)M^*$, and that $\{(z \otimes \iota)(M) \mid z \in B_0(V_x)^*\}$ is dense in W.*

Proof We have $\| \sum \Delta(a_k)(q_x(1) \otimes v_k)\|^2 = \| \sum q_x(a_k) \otimes v_k\|^2$ for finite sums due to right invariance of x, so M is a well-defined isometry, which is surjective as the closed span of $\Delta(W)(1 \otimes W)$ is $W \otimes W$. Approximating $\Delta(a)$ by finite sums of elementary tensors, one sees that $M(T \otimes 1) \in B_0(V_x) \otimes W$ for any rank one operator T with range in $q_x(W)$, and approximating $a \otimes 1$ by elements in the span of $\Delta(W)(1 \otimes W)$ we also see that $M^*(T \otimes 1) \in B_0(V_x) \otimes W$, so $M \in M(B_0(V_x) \otimes W)$. Since every state on W is a vector state in the universal representation, we get $(\iota \otimes \omega)(M)q_x(a) = (\iota \otimes \omega)\Delta(a)q_x(1)$ for any $\omega \in W^\star$. Therefore $(\iota \otimes \Delta)(M) = M_{12}M_{13}$ by applying $(\iota \otimes \omega \otimes \omega')$ to both sides and using coassociativity. Since $(\omega_{q_x(a),q_x(b)} \otimes \iota)(M) = (x \otimes \iota)((b^* \otimes 1)\Delta(a))$, denseness is clear, and the remaining identity is obvious. □

We have the following generalization of the Peter-Weyl theorem for compact groups.

Theorem 12.2.5 *Let (W, Δ) be a compact quantum group. There is a dense Hopf $*$-subalgebra W_0 of W with comultiplication given by restricting Δ, and with a linear basis $\{u^i_{mn} \,|\, m, n \in F_i$ finite$\}$ such that $\Delta(u^i_{mn}) = \sum_k u^i_{mk} \otimes u^i_{kn}$ and with the counit and coinverse given by $\varepsilon(u^i_{mn}) = \delta_{mn}$ and $S(u^i_{mn}) = (u^i_{nm})^*$. For the Haar state x one has the orthogonality relations $x((u^i_{mn})^* u^j_{kl}) = \delta_{ij}\delta_{nl} f_{-1}(u^i_{km})/d_i$ and $x(u^i_{mn}(u^j_{kl})^*) = \delta_{ij}\delta_{mk} f_1(u^i_{ln})/d_i$, where $d_i = \sum f_1(u^i_{rr})$ and $\{f_z\}_{z\in\mathbb{C}}$ consists of pointwise entire unital multiplicative functionals on W_0 with $S^2 = (f_1 \otimes \iota \otimes f_{-1})\Delta$ and $f_0 = \varepsilon$ and $f_{z_1} S = f_{-z_1}$ and $f_{z_1} f_{z_2} = f_{z_1+z_2}$ and $\overline{f_{z_1}((\cdot)^*)} = f_{-\bar{z_1}}$ for $z_i \in \mathbb{C}$. The family $\{f_z\}$ is the unique one with these properties.*

Proof Let N be a unitary right corepresentation of (W, Δ) on a finite dimensional space V. Let $B \equiv \{F(a) \equiv (\iota \otimes x)(N(1 \otimes a)) \,|\, a \in W\}$. Then $N^*(F(a) \otimes 1) = (\iota \otimes \iota \otimes x)(N_{13}(1 \otimes \Delta(a)))$ and $F(b)^*F(a) = (\iota \otimes x \otimes x)(N_{13}(1 \otimes (b^* \otimes 1)\Delta(a)))$, so B is a $*$-subalgebra of $B(V)$ that is non-degenerate as N is unitary, and hence $N^* \in B \otimes W$. As End(N) is the commutant of B, we thus see that $B = B(V)$ if and only if N is irreducible.

Pick an conjugate linear involution $J = J^*$ on V, and let $j = J(\cdot)^*J$. Then $N^c = (j \otimes \iota)(N^{-1})$ clearly satisfies $(\iota \otimes \Delta)(N^c) = N^c_{12}N^c_{13}$, and we claim it is invertible. Indeed, the positive operators $T_l = (\iota \otimes x)(N^c(N^c)^*)$ and $T_r = (\iota \otimes x)((N^c)^*N^c)$ satisfy $T_l \otimes 1 = N^c(T_l \otimes 1)(N^c)^*$ and $T_r \otimes 1 = (N^c)^*(T_r \otimes 1)N^c$ and their traces equal dim(V). Letting $p \in B(V)$ be the orthogonal projection onto the kernel of T_l, we get $(j(p) \otimes 1)N(j(T_l^{1/2}) \otimes 1) = 0$, so $j(p)Bj(T_l^{1/2}) = 0$, so $p = 0$ and T_l is invertible if N is irreducible. In this case also T_r is invertible for a similar reason. Unitarizing and decomposing into irreducibles, we conclude that N^c is always a right corepresentation. Multiplying the first morphism identity from above on the left by $(N^c)^{-1}$ and applying $j \otimes \iota$, we get $j(T_l) \in$ Mor(N, N^{cc}), and similarly $j(T_r) \in$ Mor(N^{cc}, N), and T_lT_r is a scalar when N is irreducible. In this case let $\rho \in$ Mor(N, N^{cc}) be the unique positive operator with Tr$(\rho) =$ Tr(ρ^{-1}). By Schur's lemma $(\iota \otimes x)(N(T \otimes 1)N^*)$ is a scalar for $T \in B(V)$, so it equals Tr$(T\rho_r)$ for all such T and a unique positive $\rho_r \in B(V)$. Taking traces on both sides we then see that $\rho_r = j(T_r)/$dim(V), which again equals $\rho/$Tr(ρ). Inserting $T = m_{ij}$

thus yields $x(u_{kl}u_{ij}^*) = \delta_{ki}\rho_{jl}/\mathrm{Tr}(\rho)$. Similarly $x(u_{ij}^*u_{kl}) = \delta_{jl}\rho_{ki}^{-1}/\mathrm{Tr}(\rho)$. If $N' = \sum m_{ij} \otimes v_{ij}$ is an irreducible unitary right corepresentation on V' not equivalent to N, then $(\iota\otimes x)(N'(T\otimes 1)N^*) \in \mathrm{Mor}(N, N') = \{0\}$ for any linear map $T: V \to V'$, yielding $x(v_{kl}u_{ij}^*) = 0$, and similarly $x(u_{ij}^*v_{kl}) = 0$.

Let u_{mn}^i be the matrix coefficients of a complete family $\{N_i\}$ of pairwise inequivalent irreducible unitary right corepresentations of the quantum group. From the orthogonality relations from the last paragraph, we see that these coefficients are linear independent, and by decomposing the right regular corepresentation M from the previous proposition into its irreducible constituents, we also see that their span W_0 is dense in W. Considering the matrix coefficients of N^c, we see that W_0 is $*$-preserving, and considering the coefficients of the irreducible constituents of the right corepresentation $N \times N' \equiv N_{13}N'_{23}$ on $V \otimes V'$ and of $1 \otimes 1$ on $\mathbb{C}$, we see that W_0 is actually a unital $*$-subalgebra of W having $\{u_{mn}^i\}$ as a linear basis. Now we just define the counit and the coinverse by the formulas in the theorem, and easily verify that (W_0, Δ) is a Hopf $*$-algebra. Next define entire unital functionals f_z on W_0 for $z \in \mathbb{C}$ by $f_z(u_{mn}^i) = (\rho_i^z)_{mn}$, so $d_i = \mathrm{Tr}(\rho_i)$ with ρ_i associated to N_i. This gives the orthogonality relations in the theorem. The only non-trivial algebraic property to check for the family $\{f_z\}$ is their individual multiplicativity. To this end we extend the definition of ρ to the case when N is not irreducible by decomposing it as say $\oplus(1 \otimes N_i)$ and then letting $\rho = \oplus(1 \otimes \rho_i)$. Then ρ is uniquely characterized by its usual properties because another one would be of the form $\oplus(T_i \otimes \rho_i)$ for positive invertible matrices T_i satisfying $\mathrm{Tr}(\cdot T_i) = \mathrm{Tr}(\cdot T_i^{-1})$, yielding $T_i = 1$. Now multiplicativity of f_z would be clear from $\rho_{N\times N'} = \rho_N \otimes \rho_{N'}$ for irreducible N and N'. To prove this, consider the states $f_N = \mathrm{Tr}(\cdot\rho_N^{-1})/\mathrm{Tr}(\rho_N)$ and $g_N = \mathrm{Tr}(\cdot\rho_N)/\mathrm{Tr}(\rho_N)$ on $B(V)$. From the last paragraph we have $g_{N'}(T') = (\iota \otimes x)(N'(T' \otimes 1)(N')^*)$ for $T' \in B(V')$, giving $g_{N'}(T') = (\iota \otimes g_{N'})(N'(T' \otimes 1)(N')^*)$ by right invariance of x. Thus for $T \in \mathrm{End}(N \times N')$, we get $N((\iota \otimes g_{N'})(T) \otimes 1)N^* = (\iota \otimes g_{N'})(T) \otimes 1$, or $(\iota\otimes g_{N'})(T) \in \mathrm{End}(N)$. Similarly, we get $(f_N\otimes\iota)(\mathrm{End}(N\times N')) \subset \mathrm{End}(N')$. By the unique characterization of $\rho_{N\times N'}$, we thus only need to check that $(g_N\otimes g_{N'})(T) = (f_N \otimes f_{N'})(T)$, but since $g_N = f_N$ on $\mathrm{End}(N)$, and $g_{N'} = f_{N'}$ on $\mathrm{End}(N')$, this clearly holds. Uniqueness of the family $\{f_z\}$ is a standard argument in modular theory. □

It is clear form the last paragraph of the proof above that the *quantum dimension* $d_N = \mathrm{Tr}(\rho_N)$ of a finite dimensional unitary right corepresentation N on V of a compact quantum group is a multiplicative and additive on the set of such corepresentations with *direct sums* $N \oplus N'$ and *tensor products* $N \times N'$.

The orthogonality relations give $x(u_{mn}^i) = 0$ except when N^i is the *trivial right corepresentation* $1\otimes 1$. They also tell us that $x(vw) = x(w(f_1\otimes\iota\otimes f_1)(\Delta\otimes\iota)\Delta(v))$ for $v, w \in W_0$, so x satisfies the KMS-condition in its GNS-representation with modular automorphism given by the formula $\sigma_t^x = (f_{it} \otimes \iota \otimes f_{it})(\Delta \otimes \iota)\Delta$. This immediately shows that x can be extended to a faithful normal state on the von Neumann algebra of W in this representation.

Let $\tau_z = (f_{iz} \otimes \iota \otimes f_{-iz})(\Delta \otimes \iota)\Delta$ for $z \in \mathbb{C}$. The one-parameter group $\{\tau_t\}_{t\in\mathbb{R}}$ of $*$-automorphisms on W_0 is called the *scaling group* of the compact quantum group. Using the theorem above one checks that the *unitary coinverse* $R = S\tau_{i/2}$ is an involutive $*$-antiautomorphism on W_0 such that $\Delta R = \sigma(R \otimes R)\Delta$. This map is bounded in the GNS-representation of W with respect to x. Note that x is tracial if and only if all $f_z = \varepsilon$ if and only if $S^2 = \iota$ if and only if S is $*$-preserving.

Interesting examples of compact quantum groups are given in terms of generators and relations. The *universal* unital $*$-algebra U_0 generated by $g_1, \ldots, g_n$ satisfying finitely many polynomial relations $r_j \equiv p_j(g_1, g_1^*, \ldots, g_n, g_n^*) = 0$ is the quotient of the group algebra of the free group on the generators g_i and their adjoints g_i^* by the ideal generated by the elements r_j. Whenever a unital $*$-algebra V is generated by generators h_i satisfying the same relations $p_j(h_1, h_1^*, \ldots, h_n, h_n^*) = 0$, there is a unique surjective unital $*$-homomorphism $f : U_0 \to V$ such that $f(g_i) = h_i$.

Let $\|\cdot\| = \sup \|\pi(\cdot)\|$ with supremum taken over all unital $*$-homomorphisms from U_0 into the bounded operators on Hilbert spaces. Whenever there is something to take supremum over, and if $\|u\| < \infty$ for all $u \in U_0$, then $\|\cdot\|$ is a seminorm on U_0, and we get a unital C*-algebra U by first taking the quotient of U_0 by the $*$-ideal of the elements with zero seminorm, and then completing with respect to the induced norm on the quotient. Whenever $\|\cdot\|$ is already a norm on U_0, then U_0 will sit as a dense unital $*$-subalgebra of U. In general U has the same universal property as U_0 does, except that now the V above is in addition assumed to be a C*-algebra. We call U the *universal unital C*-algebra* generated by g_i satisfying the relations $r_j = 0$. Note that the form of the relations r_j determine whether $\|u\| < \infty$ for all $u \in U_0$, and whether there is anything to take supremum over, preferably even enough to produce a norm already on U_0.

Example 12.2.6 Let $q \in [-1, 1]\backslash\{0\}$. The compact quantum group $SU_q(2)$ is defined as follows. The universal unital $*$-algebra W_0 generated by a and b satisfying $a^*a + b^*b = 1 = aa^* + q^2bb^*, b^*b = bb^*, ab = qba, ab^* = qb^*a$ is a Hopf $*$-algebra with comultiplication, counit and coinverse defined with the right properties by universality through the formulas $\Delta(a) = a \otimes a - qb^* \otimes b$ and $\Delta(b) = b \otimes a + a^* \otimes b$ and $\varepsilon(a) = 1$ and $\varepsilon(b) = 0$ and $S(a) = a^*$ and $S(b) = -qb$.

The existence of ε and the relation $a^*a + b^*b = 1$ show that we may also form the unital universal C*-algebra W. By universality we have a unital $*$-homomorphism $\Delta : W \to W \otimes W$ satisfying coassociativity, and (W, Δ) is indeed a compact quantum group since a quotient of W_0 is dense in W and we may invoke Proposition 12.1.3.

We can do better than this. Using the defining relations it is easy to see that W_0 is spanned by $B \equiv \{a^r b^l (b^*)^m \mid l, m \in \mathbb{N} \cup \{0\},\ r \in \mathbb{Z}\}$, where $a^r \equiv (a^*)^{-r}$ when $r < 0$. We claim B is a basis for W_0. In fact, we will see that the image of the elements in B under the quotient map $W_0 \to W$ are also linear independent, which immediately implies that W_0 is embedded densely in W. To this end we introduce some representations that seem rather ad hoc. When $|q| < 1$ define a unital $*$-representation $\pi : W_0 \to B(V)$ by $\pi(a)v_{nk} = (1 - q^{2n})^{1/2} v_{n-1,k}$ and $\pi(b)v_{nk} = q^n v_{n,k+1}$, where $\{v_{nk} \mid n \in \mathbb{N} \cup \{0\},\ k \in \mathbb{Z}\}$ is an orthonormal basis for V. In the case $|q| = 1$ we define for $t \in [0, 1]$ a unital $*$-representation $\pi_t : W_0 \to B(V_t)$ by $\pi_t(a)u_{nk} = (1 - t^2)^{1/2}u_{n-1,k}$ and $\pi_t(b)u_{nk} = tq^n u_{n,k+1}$, where $\{u_{nk} \mid k, n \in \mathbb{Z}\}$ is an orthonormal basis for V_t.

Suppose $c \equiv \sum c_{rlm}a^r b^l (b^*)^m$ vanishes under π when $|q| < 1$ and under all π_t when $|q| = 1$, but that the complex coefficients are not all zero. Let s be the smallest number such that $s = l + m$ with $c_{rlm} \neq 0$. Then $c_{r,s-m,m}$ equals $\lim_{p\to\infty} q^{-sp}(v_{p+r,s-2m}|\pi(c)v_{p0})$ and $\lim_{t\to 0} t^{-s}(u_{r,s-2m}|\pi_t(c)u_{00})$ when $|q|$ is respectively less or equal to one. Both cases are absurd, and our claim is proved.

When $q = 1$ the C^*-algebra W is commutative. We claim it is $*$-isomorphic to $C(SU(2))$, where $SU(2)$ is the compact group of unitary complex 2×2-matrices with determinant one. An element u of $SU(2)$ is typically of the form

$$u = \begin{pmatrix} \tilde{a} & -\tilde{b}^* \\ \tilde{b} & \tilde{a}^* \end{pmatrix}$$

for complex numbers $\tilde{a}$ and $\tilde{b}$ such that $|\tilde{a}|^2 + |\tilde{b}|^2 = 1$. Considering the coordinate functions a' and b' on $SU(2)$ given by $a'(u) = \tilde{a}$ and $b'(u) = \tilde{b}$, we see that they satisfy the same relations as a and b do, so we have a surjective unital $*$-homomorphism f from W_0 onto the unital $*$-subalgebra $\mathrm{Pol}(SU(2))$ of $C(SU(2))$ generated by a' and b' such that $f(a) = a'$ and $f(b) = b'$. The identity

$$f(w)(\begin{pmatrix} y(a) & -y(b^*) \\ y(b) & y(a^*) \end{pmatrix}) = y(w)$$

for $w \in W_0$ and y a $*$-character on W, shows that $f(w) = 0$ if and only if $y(w) = 0$ for all y, so then $w = 0$ as W is commutative. Hence f identifies W_0 and $\mathrm{Pol}(SU(2))$, and extends by continuity to a unital $*$-isomorphism from W onto $C(SU(2))$ by the Stone-Weierstrass theorem. So when $q \neq 1$ we can think of $SU_q(2)$ as a deformation of $SU(2)$. The Haar state x on W_0 is then given by

$$x = (1 - q^2) \sum q^{2n} (\pi(\cdot)v_{n0}|v_{n0}),$$

while the modular functional f_1 on W_0 is given by $f_1(a) = |q|^{-1}$ and $f_1(b) = 0$. $\diamond$

12.3 Locally Compact Quantum Groups

When G is a locally compact group the formula $\Delta(f)(s,t) = f(st)$ defines a $*$-homomorphism $\Delta\colon C_0(G) \to C_b(G \times G) = M(C_0(G) \otimes C_0(G))$, and of course, the Haar measure is no longer unimodular. For non-unital C*-algebras W with coassociative $\Delta\colon W \to M(W \otimes W)$ satisfying an axiom corresponding to cancellation, there exists presently no proof of the existence of 'Haar integrals'.

Their existence has to be imposed, and as weight theory really belongs to von Neumann algebras, it is natural to phrase everything in that context. The resulting theory relies heavily on the previously developed theory of Kac algebras obtained by considering bounded coinverses.

We will employ the *leg numbering notation*. Say we have two von Neumann algebras V and W. We define an injective normal unital $*$-homomorphism f_{ij} from $V\bar{\otimes}W$ into the tensor product of a number of von Neumann algebra with V and W at the i-th and j-th place, respectively, such that $f_{ij}(v \otimes w)$ is the elementary tensor with v and w at the i-th and j-th place, respectively, and otherwise only 1's. With $X \in V\bar{\otimes}W$ set $X_{ij} = f_{ij}(X)$.

Definition 12.3.1 A *multiplicative unitary* on a Hilbert space V is a unitary $M \in B(V \otimes V) = B(V)\bar{\otimes}B(V)$ such that $M_{12}M_{13}M_{23} = M_{23}M_{12}$ in $B(V \otimes V \otimes V)$.

The flip F on $V \otimes V$ is a self-adjoint involution in $B(V \otimes V)$, and with $f(\cdot) = F(\cdot)F$, we get $M_{13} = (f \otimes \iota)(1 \otimes M)$. Note that $f(M^*)$ is a multiplicative unitary on V, called the *dual* multiplicative unitary of M.

Say x is a faithful semifinite normal weight on a von Neumann algebra W, and let G_x be the upward filtered set of positive $y \in W_*$ such that $(1 + \varepsilon)y \leq x$ pointwise on W_+. For a von Neumann algebra V, let $M^+_{x\otimes\iota}$ be those $X \in (W\bar{\otimes}V)_+$ such that the net $\{(y \otimes \iota)(X)\}_{y\in G_x}$ converges to some element, say $(x \otimes \iota)(X) \in V_+$, in the σ-strong* topology. Then $N_{x\otimes\iota} = \{X \in W\bar{\otimes}V \mid X^*X \in M^+_{x\otimes\iota}\}$ is a left ideal of $W\bar{\otimes}V$ such that $N^*_{x\otimes\iota}N_{x\otimes\iota}$ is the linear span $M_{x\otimes\iota}$ of $M^+_{x\otimes\iota}$. This defines an increasing homogeneous additive map $x \otimes \iota$ on $M^+_{x\otimes\iota}$ that extends linearly to $M_{x\otimes\iota}$ and is denoted by the same symbol. We define $\iota \otimes x$ similarly.

Remark 12.3.2 What we are dealing with here are really tensor products of operator valued weights, but we have deliberately avoided this formalism as it is strictly not necessary at this stage. The definition we have given of $x \otimes \iota$ above evidently coincides on $M^+_{x\otimes\iota}$ with the tensor product $x \otimes \iota\colon (W\bar{\otimes}V)_+ \to \bar{V}_+$ of the faithful semifinite normal operator valued weights $x\colon W_+ \to [0, \infty] = \bar{\mathbb{C}}_+$ and $\iota\colon V_+ \to V_+ \subset \bar{V}_+$, and the definitions of $M_{x\otimes\iota}$ and $N_{x\otimes\iota}$ clearly also coincide with the ones for operator valued weights. We will return to operator valued weights in the context of quantum groups when we discuss strong left invariance and crossed products by coactions.

Lemma 12.3.3 *For $X, Y \in N_{\iota\otimes x}$ we have*

$$(\iota \otimes x)(X^*Y)^*(\iota \otimes x)(X^*Y) \leq \|(\iota \otimes x)(X^*X)\|(\iota \otimes x)(Y^*Y).$$

Proof We need only prove the identity with x replaced by $y \in G_x$, and working in the semicyclic representation π of y with cyclic vector v, we can write

$$(\iota \otimes y)(X^*Y) = ((\iota \otimes \pi)(X)(1 \otimes f_v))^*((\iota \otimes \pi)(Y)(1 \otimes f_v)),$$

where $f_v(c) = cv$ for $c \in \mathbb{C}$. Using this and the C*-identities $a^*b^*ba \leq \|b^*b\|a^*a$ and $\|b^*b\| = \|b\|^2 = \|bb^*\|$, we are done. □

Definition 12.3.4 A *locally compact quantum group* (W, Δ) consists of a von Neumann algebra W, a normal unital $*$-homomorphism $\Delta\colon W \to W\bar{\otimes}W$ such that $(\Delta \otimes \iota)\Delta = (\iota \otimes \Delta)\Delta$, and normal semifinite faithful weights x, y on W such that $(\iota \otimes x)\Delta = x$ and $(y \otimes \iota)\Delta = y$ on the positive parts of M_x and M_y, respectively.

We say x and y are *left-* and *right invariant* weights, respectively. In this section we prove the following result.

Theorem 12.3.5 *Let (W, Δ) be a locally compact quantum group with left- and right invariant faithful semifinite normal weights x and y. There is a unique multiplicative unitary M on V_x such that $M^*(q_x(v)\otimes q_x(w)) = (q_x\otimes q_x)(\Delta(w)(v\otimes 1))$ for $v, w \in N_x$, called the* fundamental multiplicative unitary *of (W, Δ).*

Certainly $\Delta(w)(v \otimes 1) \in N_{x\otimes x}$ by left invariance. Note that the linear map $q_x\colon N_x \to V_x$ is σ-strong* closed. We assume that W is represented by the semicyclic representation of x as bounded operators on V_x, so it is in a standard form. Then every element of W_* is of the form $\omega_{v,w} \equiv (\cdot(v)|w)$ for $v, w \in V_x$. When using vector functionals for vectors in other representation spaces, we often suppress the representation in the notation.

By Theorem 8.3.13 we have for any state $z \in W_*$ that $(z \otimes \iota)(X)^*(z \otimes \iota)(X) \leq (z \otimes \iota)(X^*X)$ for $X \in W\bar{\otimes}W$, so $(z \otimes \iota)\Delta(w) \in N_x$ and $\|q_x((z \otimes \iota)\Delta(w))\| \leq \|q_x(w)\|$ for $w \in N_x$. By the right version of the lemma above and right invariance, we get

$$(y\otimes\iota)(\Delta(u^*)X(v\otimes1))^*(y\otimes\iota)(\Delta(u^*)X(v\otimes1)) \leq y(u^*u)(y\otimes\iota)((v^*\otimes1)X^*X(v\otimes1))$$

for $X \in W\bar{\otimes}W$ and $u, v \in N_y$. Combining this with left invariance of x, we see that $(y \otimes \iota)(\Delta(u^*w)(v \otimes 1)) \in N_x$ for $w \in N_x$ and

$$\|q_x((y \otimes \iota)(\Delta(u^*w)(v \otimes 1)))\| \leq \|q_y(v)\|\|q_y(u)\|\|q_x(w)\|.$$

Lemma 12.3.6 *The closed span of the above elements $q_x((z \otimes \iota)\Delta(w))$ coincides with V_x, and so does the closed span of the elements $q_x((y \otimes \iota)(\Delta(u^*w)(v \otimes 1)))$.*

Proof Let V denote the latter closed span. We show that it coincides with the first closed span. Let X denote the isometry on $V_y \otimes V_x$ given by $X(q_y(a) \otimes q_x(b)) = (q_y \otimes q_x)(\Delta(a)(1 \otimes b))$ for $a \in N_y$ and $b \in N_x$. Then $(\omega_{q_y(a),q_y(c)} \otimes \iota)(X^*) = (y \otimes \iota)(\Delta(c^*)(a \otimes 1))$ for $a, c \in N_y$ as is easily seen by evaluating both sides on a vector functional with vectors in $q_x(N_x)$. Letting A_0 be the *Tomita algebra* from Lemma 10.6.4 associated to the weight y, and combining the last inequality before this lemma with closedness of q_x and denseness of A_0 in $\pi_y(M)''$ and denseness of $q_y(A_0)$ in V_y and $\sigma^y_{-i}(A_0) = A_0$, we see that V is the closed span of the elements $q_x((\omega_{q_y(a\sigma^y_{-i}(b^*)),q_y(w^*u)}\otimes\iota)(X^*))$, where $a, b \in A_0$. But the latter expression equals $q_x(\omega_{q_y(a),q_y(b)} \otimes \iota)\Delta(u^*w))$ by the KMS-condition for y. Using denseness of N_y in W and of $q_y(A_0)$ in V_y, we thus see that V is the required closed span.

It remains to show that $V = V_x$. Consider the isometry Y on $V_y \otimes V_x$ given by $Y((q_y(a)\otimes q_x(b)) = (q_y\otimes q_x)(\Delta(b)(a\otimes 1))$ for $a \in N_y$ and $b \in N_x$. Regarding V

as the latter closed span we get $Y(V_y \otimes V_x) \subset V_y \otimes V$ from Lemma 12.3.7 below. Regarding on the other hand V as the first closed span we get $V_y \otimes V \subset Y(V_y \otimes V)$ from Lemma 12.3.8 below. Hence $Y(V_y \otimes V_x) = Y(V_y \otimes V)$, so $V = V_x$. □

Lemma 12.3.7 *We have the equalities* $\sum \|q_x((\omega_{v,v_i} \otimes \iota)\Delta(b)\|^2 = \|v\|^2 x(b^*b)$ *and* $Y(v \otimes q_x(b)) = \sum v_i \otimes q_x((\omega_{v,v_i} \otimes \iota)\Delta(b))$ *for* $v \in V_y$ *and* $b \in N_x$ *and any orthonormal basis* $\{v_i\}$ *of* V_y.

Proof Let $u \in V_x$. Evaluating $\omega_{u,u}$ on $\sum((\omega_{v,v_i} \otimes \iota)\Delta(b))^*(\omega_{v,v_i} \otimes \iota)\Delta(b)$, one sees that it converges to $(\omega_{v,v} \otimes \iota)\Delta(b^*b)$ since $\omega_{u,u}(c^*c) = \sum(c(u)|u_i)(u_i|c(u))$, where $\{u_i\}$ is an orthonormal basis of V_x and $c \in W$. The first equality in the lemma follows now from normality and left invariance of x. So the right hand side of last equality in the lemma makes sense. Letting $a, d \in N_y$ and $e \in N_x$, we get

$$(\sum v_i \otimes q_x((\omega_{q_y(a),v_i} \otimes \iota)\Delta(b))|q_y(d) \otimes q_x(e)) = (Y(q_y(a) \otimes q_x(b))|q_y(d) \otimes q_x(e)).$$

□

Lemma 12.3.8 *Let* $a, u \in N_y$ *and* $w \in N_x$ *and* $v \in V_{y'}$ *and* $\{v_i\}$ *be an orthonormal basis of* $V_{y'}$*, where* y' *is a faithful semifinite normal weight on* W*. Then* $\sum \|q_x((y \otimes \iota)(\Delta(u^*w)((\iota \otimes \omega_{v,v_i})\Delta(a) \otimes 1)))\|^2 < \infty$ *and*

$$Y(\sum v_i \otimes q_x((y \otimes \iota)(\Delta(u^*w)((\iota \otimes \omega_{v,v_i})\Delta(a) \otimes 1)))) = v \otimes q_x((y \otimes \iota)\Delta(u^*w)(a \otimes 1))).$$

Proof By the last inequality before Lemma 12.3.6, and the proof of Lemma 12.3.7, we see that the left hand side of the inequality in this lemma is not greater than $\|v\|^2\|q_y(u)\|^2\|q_y(a)\|^2\|q_x(w)\|^2$, which is finite.

Just after the proof of Haagerup's theorem we saw that for any positive $z \in W_*$ with $z \leq x$ pointwise, there is $h_z \in W'$ with $0 \leq h_z \leq 1$ and unique $w_z \in V_x$ such that $(h_z q_x(w)|q_x(w)) = z(w^*w)$ and $\pi_x(w)(w_z) = h_z^{1/2} q_x(w)$ for $w \in N_x$. Note that this did not use left invariance of x.

We claim that $((1 \otimes h_z^{1/2})Y(v \otimes q_x(w))|r \otimes s) = (\omega_{v,r} \otimes \omega_{w_z,s})\Delta(w)$ for $r \in V_{y'}$ and $s \in V_x$. Pick positive $z' \in W_*$ such that $z' \leq y'$ pointwise, and consider $w_{z'}$ associated to y'. Clearly $(h_{z'}^{1/2} \otimes h_z^{1/2})Y(q_{y'}(u') \otimes q_x(w)) = \Delta(w)(u' \otimes 1)(w_{z'} \otimes w_z)$ for $u' \in N_{y'}$, so $((h_{z'}^{1/2} \otimes h_z^{1/2})Y(q_{y'}(u') \otimes q_x(w))|r \otimes s) = (\Delta(w)(h_{z'}^{1/2} q_{y'}(u') \otimes w_z)|r \otimes s)$, and the claim follows as $\{h_{z'}^{1/2}\}_{z' \in G_{y'}}$ converges strongly to the identity.

By the proved claim above, we see that applying $1 \otimes h_z^{1/2}$ to the left hand side of the last identity in the lemma, and taking the inner product with $r \otimes s$, yields

$$\sum y((\iota \otimes \omega_{v_i,r})\Delta(\iota \otimes \omega_{w_z,s})\Delta(u^*w)(\iota \otimes \omega_{v,v_i})\Delta(a)),$$

which equals $y((\iota \otimes \omega_{v,r})(\Delta(\iota \otimes \omega_{w_z,s})\Delta(u^*w))\Delta(a))$ by the proof of the previous lemma. Using right invariance of y this in turn equals $(v|r)y((\iota \otimes \omega_{w_z,s})\Delta(u^*w)a)$, which again equals $(v, r)(h_z^{1/2} q_x((y \otimes \iota)(\Delta(u^* w)(a \otimes 1)))|s)$. Using that $\{h_z^{1/2}\}_{z \in G_x}$ converges strongly to the identity, we are therefore done. □

Proof of the Previous Theorem That we get a well defined unitary element M of $B(V_x \otimes V_x)$ is immediate from Lemma 12.3.8 and Lemma 12.3.6. By definition we see that $\Delta(w) = M^*(1 \otimes w)M$ for $w \in W$. To check the pentagonal equation, let $z_i \in B(V_x)_*$. Then as

$$((\omega_{q_x(a),q_x(b)} \otimes \iota)(M^*)q_x(c)|q_x(d)) = (q_x((\omega_{q_x(a),q_x(b)} \otimes \iota)\Delta(c))|q_x(d))$$

for $a, b, c, d \in N_x$, we get

$$\begin{aligned}(z_1 \otimes z_2 \otimes \iota)(M_{23}^* M_{13}^*)q_x(a) &= (z_2 \otimes \iota)(M^*)(z_1 \otimes \iota)(M^*)q_x(a)\\ &= (z_2 \otimes \iota)(M^*)q_x((z_1 \otimes \iota)\Delta(a))\\ &= q_x((z_2 \otimes \iota)\Delta(z_1 \otimes \iota)\Delta(a))\\ &= q_x(((z_1 \otimes z_2)\Delta \otimes \iota)\Delta(a))\\ &= ((z_1 \otimes z_2)\Delta \otimes \iota)(M^*)q_x(a)\\ &= (z_1 \otimes z_2 \otimes \iota)(M_{12}^* M_{23}^* M_{12})q_x(a).\end{aligned}$$

□

As $(\omega_{q_x(a),q_x(b)} \otimes \iota)(M^*)q_x(c) = q_x(\omega_{q_x(a),q_x(b)} \otimes \iota)(\Delta(c))$ for $a, b, c \in N_x$, we see that $M \in W \bar{\otimes} B(V_x)$. Similarly, one shows that there is a unitary $N \in B(V_x) \bar{\otimes} W$ such that $N(q_y(a) \otimes q_y(b)) = (y \otimes \iota)(\Delta(a)(1 \otimes b))$ for $a, b \in N_y$. In this case we get $(\omega_{q_y(a),q_y(b)} \otimes \iota)(N^*) = (y \otimes \iota)(\Delta(b^*)(a \otimes 1))$ and $\Delta(w) = N(w \otimes 1)N^*$ for $w \in W$.

We also need the following result in the sequel.

Lemma 12.3.9 *With notation as in the previous lemma, we have for $u' \in N_{y'}$ and $X = (y \otimes \iota \otimes \iota)(\Delta(u^* w)_{13}\Delta(a)_{12})$ that $X(u' \otimes 1) \in N_{y' \otimes x}$ and*

$$Y(q_{y'} \otimes q_x)(X(u' \otimes 1)) = q_{y'}(u') \otimes q_x((y \otimes \iota)(\Delta(u^* w)(a \otimes 1))).$$

Proof By the second last inequality before Lemma 12.3.6, we have

$$X^* X \leq y(u^* u)(y \otimes \iota \otimes \iota)(\Delta(a^*)_{12}\Delta(w^* w)_{13}\Delta(a)_{12}),$$

so $X(u' \otimes 1) \in N_{y' \otimes x}$ and

$$(q_{y'} \otimes q_x)(X(u' \otimes 1)) = \sum v_i \otimes q_x((y \otimes \iota)(\Delta(u^* w)((\iota \otimes \omega_{q_{y'}(u'),v_i})(\Delta(a)) \otimes 1)))$$

by the proof of Lemma 12.3.7. The result is now clear from the previous lemma. □

12.4 A Fundamental Involution

Proposition 12.4.1 *Let (W, Δ) be a locally compact quantum group with left- and right invariant semifinite normal faithful weights x, y. Then there is a unique densely defined closed conjugate linear involutive operator G in V_x such that the span of the elements $q_x((y \otimes \iota)(\Delta(a^*)(b \otimes 1)))$ with $a, b \in N_x^* N_y$ is a core for G and*

$$G(q_x((y \otimes \iota)(\Delta(a^*)(b \otimes 1)))) = q_x((y \otimes \iota)(\Delta(b^*)(a \otimes 1))).$$

Proof By Lemma 12.3.6 and the inequality just before it, we see that the elements $q_x((y \otimes \iota)(\Delta(a^*)(b \otimes 1)))$ for $a, b \in N_x^* N_y$ span a dense subset of V_x. It remains to show that if $\sum_{i=1}^{m_n} q_x((y \otimes \iota)(\Delta(a_{ni}^*)(b_{ni} \otimes 1))) \to 0$ as $n \to \infty$ for $a_{ni}, b_{ni} \in N_x^* N_y$, while $\sum_{i=1}^{m_n} q_x((y \otimes \iota)(\Delta(b_{ni}^*)(a_{ni} \otimes 1))) \to v \in V_x$, then $v = 0$. By the previous lemma we thus see that

$$\sum_{i=1}^{m_n} (q_x \otimes q_x)((1 \otimes d^*)(y \otimes \iota \otimes \iota)(\Delta(a_{ni}^*)_{13}(\Delta(b_{ni})_{12}))(c \otimes 1)) \to 0$$

while

$$\sum_{i=1}^{m_n} (q_x \otimes q_x)((1 \otimes c^*)(y \otimes \iota \otimes \iota)(\Delta(b_{ni}^*)_{13}(\Delta(a_{ni})_{12}))(d \otimes 1)) \to M\Delta(c^*)(q_x(d) \otimes v)$$

for $c, d \in N_x$. Using the flip map to the expression in the first convergence above gives $(d^* \otimes 1)(y \otimes \iota \otimes \iota)(\Delta(a_{ni}^*)_{12}(\Delta(b_{ni})_{13}))(1 \otimes c) \in N_{x \otimes x} \cap N_{x \otimes x}^*$, and $\sum_{i=1}^{m_n} (q_x \otimes q_x)$ applied to it converges to zero. The second convergence above tells us that

$$S_{x \otimes x} \sum_{i=1}^{m_n} (q_x \otimes q_x)((d^* \otimes 1)(y \otimes \iota \otimes \iota)(\Delta(a_{ni}^*)_{12}(\Delta(b_{ni})_{13}))(1 \otimes c)) \to M\Delta(c^*)(q_x(d) \otimes v),$$

where $S_{x \otimes x}$ is the closure in $V_x \otimes V_x$ of the adjoint operator on W given by modular theory. Thus $M\Delta(c^*)(q_x(d) \otimes v) = 0$. By Lemma 12.3.6 we may write $q_x(d) \otimes v = M^*(X)$ for $X \in V_x \otimes V_x$. Then $0 = (1 \otimes c^*)X$ and $X = 0$ by denseness of N_x in W. So $q_x(d) \otimes v = 0$ and $v = 0$. □

Considering the polar decomposition $I|G|$ of G in the proposition above, we see that I is conjugate linear with $I^* = I$ and $I^2 = 1$ and $I|G|I = |G|^{-1}$.

Proposition 12.4.2 *Let (W, Δ) be a locally compact quantum group with left- and right invariant semifinite normal faithful weights x, y. With N as in the previous section, and G as in the proposition above, we have $(\omega_{v,w} \otimes \iota)(N^*)G \subset G(\omega_{w,v} \otimes \iota)(N^*)$ and $(\omega_{v,w} \otimes \iota)(N)G^* \subset G^*(\omega_{w,v} \otimes \iota)(N)$ for $v, w \in V_y$. In particular, we have $N(\Delta_y \otimes |G|^2) = (\Delta_y \otimes |G|^2)N$.*

Proof Pick an orthonormal basis $\{e_i\}$ for V_y and let $f_i = (\cdot|e_i)$. Note that $P_J = \sum_{i\in J} f_i^* f_i$ for finite J is the orthogonal projection onto the closed subspace of V_y spanned by $\{e_i\}_{i\in J}$. To $c \in N_x^* N_y$ define $c_i \in N_x^*$ by $z(c_i) = f_i q_y((\iota \otimes z)\Delta(c))$ for all $z \in W_*$, and let $\bar{z} = \overline{z((\cdot)^*)}$ in what follows. We have

$$\sum (e_i|q_y(r)) f_i q_y(\iota \otimes z_1(s^*\cdot))(a' \otimes a'') z_2(t^*) = y(\iota \otimes z_1 \otimes z_2)((r \otimes s \otimes t)^*((a' \otimes a'') \otimes 1))$$

for $z_j \in W_*$ and $a', a'', r, s, t \in N_y$ since the left hand side, when summed over a finite set J, is $(P_J q_y(a')|q_y(r)) z_1(s^* a'') z_2(t^*)$. Both sides are continuous and linear with respect to $a' \otimes a''$ and continuous and conjugate linear with respect to $r \otimes s \otimes t$. So we may replace $a' \otimes a''$ by $\Delta(d)$, where $d \in N_x^* N_y$, and maintain equality. Now the left hand side equals

$$\sum \overline{f_i q_y(\iota \otimes \overline{z_1(\cdot d_i)} \otimes \overline{z_2})(r \otimes s \otimes t)},$$

which is also continuous and conjugate linear in $r \otimes s \otimes t$, so equality is maintained when we replace $r \otimes s \otimes t$ by $(\iota \otimes \Delta)\Delta(c)$ for $c \in N_x^* N_y$, and we obtain

$$y(\iota \otimes z_1 \otimes z_2)((\iota \otimes \Delta)\Delta(c)^* \Delta(d)_{12}) = \sum \overline{f_i q_y(\iota \otimes \overline{z_1(\cdot d_i)} \otimes \overline{z_2})(\iota \otimes \Delta)\Delta(c)}.$$

By right invariance the left hand side equals $(z_1 \otimes z_2)(1 \otimes (y \otimes \iota)(\Delta(c^*)(d \otimes 1))$, while the right hand side equals $(z_1 \otimes z_2)(\sum \Delta(c_i^*)(d_i \otimes 1))$, so we get

$$\sum \Delta(c_i^*)(d_i \otimes 1) = 1 \otimes (y \otimes \iota)(\Delta(c^*)(d \otimes 1).$$

Hence

$$\sum (y \otimes \iota)(\Delta(a^* c_i^*)(d_i b \otimes 1)) = (y \otimes \iota)(\Delta(a^*)(b \otimes 1))(y \otimes \iota)(\Delta(c^*)(d \otimes 1))$$

for $a, b \in N_y$. On the other hand $\sum q_x(y \otimes \iota)(\Delta(a^* c_i^*)(d_i b \otimes 1))$ is also convergent because by reasoning as above we get

$$\|\sum_{i\in J} q_x(y \otimes \iota)(\Delta(a^* c_i^*)(d_i b \otimes 1))\|^2 \leq y(l^* l) y(a^* a) x(k^* k) \|(P_J \otimes 1)(q_y \otimes q_y)\Delta(d)(1 \otimes b)\|^2$$

when $c = k^* l$ with $k \in N_x$ and $l \in N_y$, so the Cauchy criterion applies. But as q_x is closed, the net $\{\sum_{i\in J} q_x(y \otimes \iota)(\Delta(a^* c_i^*)(d_i b \otimes 1))\}_J$ thus converges to the element $\pi_x((y \otimes \iota)(\Delta(a^*)(b \otimes 1))) q_x(y \otimes \iota)(\Delta(c^*)(d \otimes 1))$. Similarly

$$\sum q_x(y \otimes \iota)(\Delta(b^* d_i^*)(c_i a \otimes 1)) = \pi_x((y \otimes \iota)(\Delta(b^*)(a \otimes 1))) q_x(y \otimes \iota)(\Delta(d^*)(c \otimes 1)).$$

By definition of G we have

$$G(\sum_{i\in J} q_x(y\otimes\iota)(\Delta(a^*c_i^*)(d_ib\otimes 1))) = \sum_{i\in J} q_x(y\otimes\iota)(\Delta(b^*d_i^*)(c_ia\otimes 1)),$$

and as G is closed with core spanned by elements of the form $q_x(y\otimes\iota)(\Delta(c^*)(d\otimes 1))$, we therefore get

$$\pi_x((y\otimes\iota)(\Delta(b^*)(a\otimes 1)))G \subset G\pi_x((y\otimes\iota)(\Delta(a^*)(b\otimes 1))).$$

Approximating v, w by elements from $q_y(N_y)$ and using closedness of G once more, we therefore get the first inclusion in the proposition. The second one follows by taking the adjoint of the first one.

As for the final identity, we claim that

$$(\omega_{q_y(a),q_y(b)}\otimes\iota)(N)^* = (\omega_{q_y(a^*),q_y(\sigma^y_{-i}(b^*))}\otimes\iota)(N)$$

for a, b in the Tomita algebra A_0. Indeed, from the proof of Lemma 12.3.6 we get

$$(\omega_{q_y(a),q_y(b\sigma^y_{-i}(c^*))}\otimes\iota)(N)^* = (\omega_{q_y(b),q_y(c)}\otimes\iota)\Delta(a^*) = (\omega_{q_y(a^*),q_y(c\sigma^y_{-i}(b^*))}\otimes\iota)(N)$$

for $c \in A_0$. Pick a net $\{e_j\}$ in $N_y \cap N_y^*$ that converges to 1 in the strong*-topology. Then clearly $c_j = (\pi)^{-1/2}\int e^{-t^2}\sigma_t^y(e_j)dt \in A_0$ and the nets $\{c_j\}$ and $\{\sigma^y_{i/2}(c_j)\}$ are bounded and converge to 1 in the strong*-topology. Then $q_y(b\sigma^y_{-i}(c_j^*)) = J_y\pi_y(\sigma^y_{i/2}(c_j))J_yq_y(b) \to q_y(b)$, so plugging in c_j for c above, and taking the limit, the claim is settled. Using this claim twice in combination with the previous inclusions we thus get

$$(\omega_{q_y(a),q_y(b)}\otimes\iota)(N)G^*G \subset G^*G(\omega_{q_y(\sigma_i^y(a)),q_y(\sigma^y_{-i}(b))}\otimes\iota)(N).$$

As $q_y(\sigma_i^y(a)) = \Delta_y^{-1}q_y(a)$ and $q_y(\sigma^y_{-i}(b)) = \Delta_yq_y(b)$ and $q_y(A_0)$ is a core for $\Delta_y^{\pm 1}$, we therefore get

$$(\omega_{v,w}\otimes\iota)(N)G^*G \subset G^*G(\omega_{\Delta_y^{-1}(v),\Delta_y(w)}\otimes\iota)(N)$$

for $v \in D(\Delta_y^{-1})$ and $w \in D(\Delta_y)$. Now $N(\Delta_y\otimes|G|^2) = (\Delta_y\otimes|G|^2)N$ is immediate from the lemma below. □

Lemma 12.4.3 *Let X be a unitary on $V\otimes W$ for Hilbert spaces V, W, and let S_i and T_i be strictly positive operators in V and W, respectively, such that $(\omega_{v,w}\otimes\iota)(X)T_1 \subset T_2(\omega_{S_1^{-1}(v),S_2(w)}\otimes\iota)(X)$ for $v \in D(S_1^{-1})$ and $w \in D(S_2)$. Then $X(S_1\otimes T_1) = (S_2\otimes T_2)X$.*

Proof By assumption

$$(X(p_1 \otimes q_1)|(S_2 \otimes T_2)(p_2 \otimes q_2)) = (X(S_1(p_1) \otimes T_1(q_1))|p_2 \otimes q_2)$$

for $p_i \in D(S_i)$ and $q_i \in D(T_i)$. As the span of tensors of the form $p_2 \otimes q_2$ is a core for $S_2 \otimes T_2$, we may replace $p_2 \otimes q_2$ in the identity above by any element in $D(S_2 \otimes T_2)$, and using self-adjointness of $S_2 \otimes T_2$ we get $(S_2 \otimes T_2)X(p_1 \otimes q_1) = X(S_1(p_1) \otimes T_1(q_1))$. Using closedness of $S_2 \otimes T_2$ we thus get $X(S_1 \otimes T_1) \subset (S_2 \otimes T_2)X$. Taking the adjoint of this inclusion and using unitarity of X, we are done. □

12.5 Density Conditions

In the sequel we will need this density result.

Theorem 12.5.1 *Let (W, Δ) be a locally compact quantum group. Then W is the σ-strong*-closed span of the elements $(\omega \otimes \iota)\Delta(w)$, or of $(\iota \otimes \omega)\Delta(w)$, for $\omega \in W_*$ and $w \in W$, or of $(\eta \otimes \iota)(N)$ for $\eta \in B(V_x)_*$ with N from the previous section.*

Proof Let A_0 be the Tomita algebra of the right invariant faithful semifinite normal weight y on W. Then the last closed span in the theorem equals the closed span of the elements $(y \otimes \iota)((a^* \otimes 1)\Delta(b)(\sigma^y_{-i}(c) \otimes 1))$ for $a, b \in N_y$ and $c \in A_0$. And this equals the first closed span W_r of the theorem. Now W_r is a von Neumann algebra being evidently self-adjoint and furthermore an algebra acting non-degenerately on V_x as N is a multiplicative unitary of the quantum group. Considering the quantum group with opposite comultiplication we conclude that also the second closed span W_l is a von Neumann subalgebra of W, and by the commutant theorem for tensor products we see that $\Delta(W) \subset W_l \bar{\otimes} W_r$. From the previous proposition the formula $\tau_t(w) = |G|^{-2it} w |G|^{2it}$ for $w \in W_r$ and real t defines a one-parameter group $\{\tau_t\}$ of automorphism of W_r such that $\Delta\sigma^y_t = (\sigma^y_t \otimes \tau_{-t})\Delta$ as $\Delta(w) = N(w \otimes 1)N^*$ for $w \in W$. Hence $\sigma^y_t(W_l) = W_l$, which is the closed span of the elements $(\iota \otimes \omega)\Delta(w)$ for $w \in W$ and $\omega \in (W_r)_*$. By right invariance the restriction of y to W_l is semifinite. Thus by Theorem 10.11.1 there is a unique normal faithful conditional expectation $E\colon W \to W_l$ such that $yE = y$, and $E(\cdot)P = P \cdot P$, where P is the orthogonal projection onto $q_y(N_y \cap W_l)$, which in this case contains the elements $q_y((\iota \otimes \omega)\Delta(w))$ for $\omega \in W_*$ and $w \in N_y$. Hence $P = 1$ and $W_l = W$. Using the opposite coproduct, we get similarly $W_r = W$. □

12.6 The Coinverse

From the proof of the previous theorem, we have a strongly continuous one-parameter group $\{\tau_t\}$ of $*$-automorphisms on W given by $\tau_t(\cdot) = |G|^{-2it} \cdot |G|^{2it}$

such that $(\sigma_t^y \otimes \tau_{-t})\Delta = \Delta\sigma_t^y$. We call this one-parameter group the *scaling group* of the quantum group (W, Δ). The theorem also gives $\Delta\tau_t = (\tau_t \otimes \tau_t)\Delta$ because using the relation with σ_t^y and coassociativity we have $(\sigma_t^y \otimes \Delta\tau_{-t})\Delta = (\sigma_t^y \otimes (\tau_{-t} \otimes \tau_{-t})\Delta)\Delta$. Using the polar decomposition $G = I|G|$ of the fundamental involution G of the quantum group, we also define the *unitary coinverse* of (W, Δ) as the involutive $*$-antiautomorphism R of W given by $R(\cdot) = I(\cdot)^*I$. We define the *coinverse* of the quantum group as the σ-strongly* closed map $S = R\tau_{-i/2}$ with dense domain and dense range. Note that S commutes with both R and τ_t for real t, and has an inverse map $S^{-1} = R\tau_{i/2}$, so $S^2 = \tau_{-i}$. Clearly S is antimultiplicative with $S(\cdot)^*$ involutive in the sense that $S(ab) = S(b)S(a)$ and $S(S(a)^*)^* = a$ for $a, b \in D(S)$.

The last formula in the proposition below is known as *strong right invariance*.

Proposition 12.6.1 *Let (W, Δ) be a locally compact quantum group with right invariant faithful semifinite normal weight y. Let a, b belong to the Tomita algebra of y. Then*

$$(y \otimes \iota)((b^* \otimes 1)\Delta(a))|G|^{2z} \subset |G|^{2z}(y \otimes \iota)((\sigma_{iz}^y(b^*) \otimes 1)\Delta(\sigma_{iz}^y(a))$$

for $z \in \mathbb{C}$, where G is the fundamental involution of the quantum group. Moreover, the span of the elements $(y \otimes \iota)((a^* \otimes 1)\Delta(b))$ is a core for both R and S, and

$$R((y \otimes \iota)((a^* \otimes 1)\Delta(b))) = (y \otimes \iota)(\Delta\sigma_{-i/2}^y(a^*)(\sigma_{-i/2}^y(b) \otimes 1)),$$

while

$$S((y \otimes \iota)((a^* \otimes 1)\Delta(b))) = (y \otimes \iota)(\Delta(a^*)(b \otimes 1)).$$

Proof The entire function $z \mapsto (y \otimes \iota)((\sigma_{iz}^y(b^*) \otimes 1)\Delta\sigma_{iz}^y(a))$ attains at it for real t the value $(\omega_{\Delta_{-it}^y q_y(a), \Delta_{-it}^y q_y(b)} \otimes \iota)(N)$, which by Proposition 12.4.2 equals $|G|^{-2it}(y \otimes \iota)((b^* \otimes 1)\Delta(a))|G|^{2it}$. This gives the inclusion in the proposition.

Combining this inclusion for $z = 1/2$ together with one of the inclusions

$$(y \otimes \iota)((b^* \otimes 1)\Delta(a))|G| \subset |G|I(y \otimes \iota)((a^* \otimes 1)\Delta(b))I$$

from Proposition 12.4.2 and using that $|G|$ is injective and densely defined, we get the formula for R. Using this inclusion once more, we get the formula for S. The dense space in question is clearly a core for S as it is invariant under τ_t by its relation with σ_t^y. □

Proposition 12.6.2 *For a locally compact quantum group (W, Δ) we have $f(R \otimes R)\Delta = \Delta R$, where R is the unitary coinverse and f is the flip on $W \bar{\otimes} W$.*

Proof Given $\omega \in W_*$ and $n \in \mathbb{N}$, then $\omega_n = n(\pi)^{-1/2} \int e^{-n^2t^2}\omega\tau_t(\cdot)dt \in W_*$ has the property that the map from $\mathbb{R}$ to W_* given by $t \mapsto \omega_n\tau_t$ has an extension to an analytic function on $\mathbb{C}$, and the sequence $\{\omega_n\}$ converges to ω in W_*. Hence it suffices to prove the desired formula by checking that it holds under evaluation by $\omega \otimes \eta$ for ω, η satisfying such extension properties from the outset.

Then $(\omega \otimes \eta)\Delta S = (\eta S \otimes \omega S)\Delta$ on $D(S)$, where ηS and ωS denote the extensions. Indeed, for $a, b \in N_y$ apply the right hand side above to $(y \otimes \iota)((a^* \otimes 1)\Delta(b))$ and use the formula for S in the previous proposition twice arriving at the left hand side applied to the same element, then note that the span of such elements is a core for S.

Now as $\Delta\tau_t = (\tau_t \otimes \tau_t)\Delta$ for real t, we thus get

$$(\omega \otimes \eta)(R \otimes R)\Delta S = (\eta\tau_{-i/2} \otimes \omega\tau_{-i/2})\Delta = (\eta \otimes \omega)\Delta RS,$$

and then use that S has dense image. □

A *Kac algebra* is a locally compact quantum group with trivial scaling group, so that the coinverse is an involutive $*$-preseving antiautomorphism.

Proposition 12.6.3 *Consider a locally compact quantum group (W, Δ). If $\Delta(w) = 1 \otimes w$ or $\Delta(w) = w \otimes 1$ for $w \in W$, then $w \in \mathbb{C}1$.*

Proof Let y be a right invariant faithful semifinite normal weight on W, and say $\Delta(w) = 1 \otimes w$. The elements $a_n = n(\pi)^{-1/2} \int e^{-n^2t^2} \sigma_t^y(w)dt$ in W are analytic with respect to σ^y. Define $b_n = n(\pi)^{-1/2} \int e^{-n^2t^2} \tau_{-t}(w)dt \in W$. Since $(\sigma_t^y \otimes \tau_{-t})\Delta = \Delta\sigma_t^y$ for real t, we get $\Delta(a_n) = 1 \otimes b_n$. Pick a positive $c \in M_y$ with $y(c) = 1$. Then for $\omega \in W_*$ we have

$$y(ca_n)\omega(1) = y(\iota \otimes \omega)\Delta(ca_n) = y(\iota \otimes \omega(\cdot b_n))\Delta(c) = \omega(b_n),$$

and as $b_n \to w$ strongly, the sequence $\{y(ca_n)\}$ is Cauchy and converges to a number, so $w \in \mathbb{C}1$. If $\Delta(w) = w \otimes 1$ consider the quantum group with the opposite comultiplication. □

We aim now to establish *strong left invariance*.

Lemma 12.6.4 *Let (W, Δ) be a locally compact quantum group. Suppose we have $a, b \in W$ such that $a \otimes 1 = \sum \Delta(p_i)(1 \otimes q_i)$ and $b \otimes 1 = \sum(1 \otimes p_i)\Delta(q_i)$ for nets $\{p_i\}$ and $\{q_i\}$ in W such that $\sum p_i p_i^*$ and $\sum q_i q_i^*$ converge in the σ-strong* topology. Then $a \in D(S)$ and $S(a) = b$.*

Proof Using Proposition 12.6.2 and antimultiplicativity of R, we may as well show that when $1 \otimes a = \sum(p_i \otimes 1)\Delta(q_i)$ and $1 \otimes b = \sum \Delta(p_i)(q_i \otimes 1)$, then $S(a) = b$. First one easily checks that the infinite sums converge and that any finite subsum is bounded. Moreover, for finite sums

$$S(\sum(y \otimes \iota)((c^* p_i \otimes 1)\Delta(q_i d))) = \sum(y \otimes \iota)(\Delta(c^* p_i)(q_i d \otimes 1))$$

for $c, d \in N_y$ by strong right invariance. As S is closed we thus get

$$S(a(y \otimes \iota)((c^* \otimes 1)\Delta(d))) = (y \otimes \iota)(\Delta(c^*)(d \otimes 1))b,$$

so $S(aw) = S(w)b$ for $w \in D(S)$, again by strong right invariance. This gives the conclusion since for any bounded net $\{u_i\}$ in W that converges σ-strongly* to 1, then $w_i = (\pi)^{-1/2} \int e^{-t^2} \tau_t(u_i)dt$ gives a net $\{w_i\}$ in $D(S)$ that converges to 1 and such that $\{S(w_i)\}$ also converges to 1, and we can again use closedness of S. □

Proposition 12.6.5 *Let (W, Δ) be a locally compact quantum group with a left invariant faithful semifinite normal weight x on W. Then the span of the elements $(\iota \otimes x)(\Delta(a^*)(1 \otimes b))$ for $a, b \in N_x$ is a core for S and*

$$S((\iota \otimes x)(\Delta(a^*)(1 \otimes b))) = (\iota \otimes x)((1 \otimes a^*)\Delta(b)).$$

Proof Pick an orthonormal basis $\{e_i\}$ for V_x, and let $f_i = (\cdot|e_i)$. Then define $p_i, q_i \in W$ by $z(p_i^*) = f_i q_x(z \otimes \iota)\Delta(a)$ and $z(q_i) = f_i q_x(z \otimes \iota)\Delta(b)$. Arguing as in Proposition 12.4.2 we then see that $\sum p_i p_i^*$ and $\sum q_i^* q_i$ converge, and

$$\sum \Delta(p_i)(1 \otimes q_i) = (\iota \otimes \iota \otimes x)((\Delta \otimes \iota)\Delta(a^*)(1 \otimes \Delta(b))) = (\iota \otimes x)(\Delta(a^*)(1 \otimes b)) \otimes 1$$

and $\sum(1 \otimes p_i)\Delta(q_i) = (\iota \otimes x)((1 \otimes a^*)\Delta(b)) \otimes 1$ by left invariance. The formula for S is now immediate from the lemma. That we get a core is clear from the next result. □

Proposition 12.6.6 *Let (W, Δ) be a locally compact quantum group with left invariant faithful semifinite normal weight x and a scaling group $\{\tau_t\}$ defined by a right invariant faithful semifinite normal weight. Then $(I \otimes J_x)M = M^*(I \otimes J_x)$ and $(|G|^{-2} \otimes \Delta_x)M = M(|G|^{-2} \otimes \Delta_x)$ and $(\tau_t \otimes \sigma_t^x)\Delta = \Delta\sigma_t^x$ for real t, and*

$$(\omega_{v,w} \otimes \iota)(M)^* = (\omega_{I|G|^{-1}v, I|G|w} \otimes \iota)(M)$$

for $v \in D(|G|^{-1})$ and $w \in D(|G|)$. Here M is the fundamental multiplicative unitary while G is the fundamental involution of the quantum group.

Proof Using strong left invariance and $RS = \tau_{-i/2}$ and the definition of R, we get $(\iota \otimes \omega_{q_x(a), q_x(b)})(M)|G| \subset |G|I(\iota \otimes \omega_{q_x(a), q_x(b)})(M)I$ for $a, b \in N_x$, so

$$((\omega_{I|G|^{-1}v, I|G|w} \otimes \iota)(M)q_x(b)|q_x(a)) = ((\omega_{v,w} \otimes \iota)(M)^* q_x(b)|q_x(a)),$$

which proves the last assertion in the proposition.

Now for $c \in N_x \cap N_x^*$ we have

$$S_x(\omega_{v,w} \otimes \iota)(M^*)q_x(c) = q_x((\omega_{w,v} \otimes \iota)\Delta(c)^*) = (\omega_{w,v} \otimes \iota)(M^*)S_x q_x(c),$$

so $(\omega_{w,v} \otimes \iota)(M^*)S_x \subset S_x(\omega_{v,w} \otimes \iota)(M^*)$ and $(\omega_{w,v} \otimes \iota)(M)S_x^* \subset S_x^*(\omega_{v,w} \otimes \iota)(M)$ by taking the adjoint. Using these two inclusions together with the last assertion of the proposition, we get $(\omega_{v,w} \otimes \iota)(M)\Delta_x \subset \Delta_x(\omega_{|G|^2 v, |G|^{-2} w} \otimes \iota)(M)$ when $v \in D(|G|^2)$ and $w \in D(|G|^{-2})$. Lemma 12.4.3 yields the second formula in the proposition, and then the third formula also follows. For the first one, let now $v \in$

$D(|G|^{-1})$ and $w \in D(|G|)$. Then from $M^*(|G|^{-1} \otimes \Delta_x^{1/2}) = (|G|^{-1} \otimes \Delta_x^{1/2})M^*$, we get

$$(\omega_{Iw,Iv} \otimes \iota)(M)\Delta_x^{1/2} = (\omega_{|G|^{-1}v,|G|w} \otimes \iota)(M^*)\Delta_x^{1/2} \subset \Delta_x^{1/2}(\omega_{v,w} \otimes \iota)(M^*)$$

by the last assertion in the proposition. Comparing this with the inclusion

$$J_x(\omega_{w,v} \otimes \iota)(M^*)J_x\Delta_x^{1/2} = J_x(\omega_{w,v} \otimes \iota)(M^*)S_x \subset \Delta_x^{1/2}(\omega_{v,w} \otimes \iota)(M^*)$$

and using that $\Delta_x^{1/2}$ has dense range, we get $(\omega_{Iw,Iv}\otimes\iota)(M) = J_x(\omega_{w,v}\otimes\iota)(M^*)J_x$. □

By strong left invariance we see that the coinverse S does not depend on the right invariant weight y, and neither does the scaling group $\{\tau_t\}$ by using the formula $(\tau_t \otimes \sigma_t^x)\Delta = \Delta\sigma_t^x$ together with the density conditions. Thus the unitary coinverse R is also independent of y. Similarly, using right invariance and the analogous formula for the scaling group with the modular group of y together with the density conditions, the triple $(S, R, \{\tau_t\})$ is also independent of the left invariant weight x. It only depends on W and Δ, which will also be evident later on when we discuss manageability. In any case our notation is justified, leaving no indices y or x on any of the entries in the triple.

12.7 Relative Invariance

Proposition 12.7.1 *Given left invariant faithful semifinite normal weights $x_1 \leq x_2$ on a locally compact quantum group, there is $r \in \langle 0, 1]$ such that $x_1 = rx_2$. In particular, if to a left invariant faithful semifinite normal weight x, there is a positive constant λ such that $x_1\sigma_t^x = \lambda^t x_1$ for all real t, then there is a positive constant s such that $x_1 = sx$.*

Proof For the quantum group (W, Δ) note that the unitary X on $V_x \otimes V_{x_2}$ defined by $X(q_x(a) \otimes q_{x_2}(b)) = (q_x \otimes q_{x_2})(\Delta(b)(a \otimes 1))$ for $a \in N_x$ and $b \in N_{x_2}$ satisfies $(\iota \otimes \pi_{x_2})\Delta(w) = X(1 \otimes \pi_{x_2}(w))X^*$ for $w \in W$. Define $F \in B(V_{x_2}, V_{x_1})$ by $Fq_{x_2}(b) = q_{x_1}(b)$, so $T = F^*F \in \pi_{x_2}(W)'$ satisfies $x_1(b^*c) = (Tq_{x_2}(c)|q_{x_2}(b))$ for $c \in N_{x_2}$. Picking an orthonormal basis $\{e_i\}$ of V_x we may write $X(q_x(a) \otimes q_{x_2}(b)) = \sum e_i \otimes q_{x_2}((\omega_{q_x(a),e_i} \otimes\iota)\Delta(b))$, so $((1\otimes T)X(q_x(a)\otimes q_{x_2}(b))|X(q_x(a)\otimes q_{x_2}(b)))$ equals

$$\sum x_1((\omega_{q_x(a),e_i} \otimes \iota)\Delta(b)^*(\omega_{q_x(a),e_i} \otimes \iota)\Delta(b)) = x_1(\omega_{q_x(a),q_x(a)} \otimes \iota)\Delta(b^*b),$$

which in turn equals $((1 \otimes T)(q_x(a) \otimes q_{x_2}(b))|q_x(a) \otimes q_{x_2}(b))$ by left invariance. Hence $X^*(1 \otimes T)X = 1 \otimes T$ by polarization. A trivial modification of the proof of the previous proposition yields $X^*(I \otimes J_{x_2}) = (I \otimes J_{x_2})X$, and combining these two identities with the fact that $J_{x_2}TJ_{x_2} \in \pi_{x_2}(W)$, we get $\Delta\pi_{x_2}^{-1}(J_{x_2}TJ_{x_2}) = 1 \otimes \pi_{x_2}^{-1}(J_{x_2}TJ_{x_2})$. Then Proposition 12.6.3 shows that $\pi_{x_2}^{-1}(J_{x_2}TJ_{x_2}) = r1$ for a scalar r, so $x_1 = rx_2$ and $r \in \langle 0, 1]$.

For the last statement consider the left invariant faithful semifinite normal weight $x_1 + x$ that majorizes x. Indeed, to see this we clearly only need to show that $N_{x_1} \cap N_x$ is a core for both q_{x_1} and q_x. Define an injective positive operator Λ in V_{x_1} by $\Lambda^{it}q_{x_1}(a) = \lambda^{-t/2}q_{x_1}(\sigma_t^x(a))$ for $a \in N_{x_1}$ and real t, and note that $a_n = n(\pi)^{-1/2}\int e^{-n^2t^2}\sigma_t^x(a)dt \in N_x \cap D(\sigma_{i/2}^x)$ converges to a, while the elements $q_{x_1}(a_n) = n(\pi)^{-1/2}\int e^{-n^2t^2}\lambda^{t/2}\Lambda^{it}q_{x_1}(a)dt$ tend to $q_{x_1}(a)$. Picking a bounded net $\{u_i\}$ in N_x that converges to 1 σ-strongly*, then the net $\{a_nu_i\}_{(n,i)}$ is contained in $N_x \cap N_{x_1}$ and converges to a, while q_{x_1} applied to it converges to $q_{x_1}(a)$, so $N_{x_1} \cap N_x$ is a core for x_1. It is also a core for x because if instead $a \in N_x$, and we pick a bounded net $\{u_j\}$ in N_{x_1} that converges to 1, then $v_j = (\pi)^{-1/2}\int e^{-t^2}\sigma_t^x(u_j)dt \in N_{x_1} \cap D(\sigma_{i/2}^x)$ and the elements $\sigma_{i/2}^x(v_j) = (\pi)^{-1/2}\int e^{-(t-i/2)^2}q_{x_1}\sigma_t^x(u_j)dt$ form bounded nets that tend to 1. So $av_j \to a$ and $q_x(av_j) = J_x\sigma_{i/2}^x(v_j)^*J_xq_x(a) \to q_x(a)$ and $av_j \in N_{x_1} \cap N_x$.

By the first part of this proof, there is $r \in \langle 0, 1]$ such that $x = r(x_1 + x)$. But then $s = (1-r)/r > 0$ and $x_1 = sx$ on the intersection of the positive parts of M_{x_1} and M_x. Referring to the paragraph above we are done. □

In the next section we prove that any left invariant faithful semifinite normal weight satisfies such a relative invariance property with respect to a fixed x.

Proposition 12.7.2 *Let (W, Δ) be a locally compact quantum group with left invariant faithful semifinite normal weight x. Consider the right invariant faithful semifinite normal weight $y = xR$ with modular group $\sigma_t^y = R\sigma_{-t}^xR$ for real t. Then the one-parameter groups $\{\sigma_t^x\}$ and $\{\sigma_t^y\}$ and $\{\tau_t\}$ pairwise commute, and $\Delta\tau_t = (\sigma_t^x \otimes \sigma_{-t}^y)\Delta$, and there exists a unique positive scalar ν such that $x\sigma_t^y = \nu^tx = x\tau_{-t}$ and $y\sigma_t^x = \nu^{-t}y = y\tau_t$ for real t.*

Proof For a positive element a in the definition domain of y we have

$$y\sigma_t^x\tau_{-t}(a)1 = (y \otimes \iota)\Delta\sigma_t^x\tau_{-t}(a) = (y \otimes \iota)(\iota \otimes \sigma_t^x\tau_{-t})\Delta(a) = y(a)1,$$

so $\sigma_s^y\sigma_t^x\tau_{-t} = \sigma_t^x\tau_{-t}\sigma_s^y$ for real s, t. Hence $(\iota \otimes \sigma_{-t}^x\tau_t)\Delta = (\iota \otimes \tau_s\sigma_{-t}^x\tau_t\tau_{-s})\Delta$ and thus $\sigma_{-t}^x = \tau_s\sigma_{-t}^x\tau_{-s}$ by the density conditions. Therefore the left invariant faithful semifinite normal weight $x_1 = x\tau_s$ is invariant under $\{\sigma_t^x\}$, so there is by the previous result a unique $\nu > 0$ with $x\tau_s = \nu^{-s}x$. Invoking R gives $y\tau_s = \nu^{-s}y$, so $\{\sigma_t^y\}$ and $\{\tau_t\}$ commute. But as $y\sigma_t^x = \nu^{-t}x\tau_{-t}\sigma_t^x = \nu^{-t}x$ from our arguments above, we see that also $\{\sigma_t^y\}$ and $\{\sigma_t^x\}$ commute, and $x\sigma_t^y = y\sigma_{-t}^xR = \nu^tx$ as R is involutive. Hence $S\sigma_t^y = \sigma_{-t}^xS$. By strong left invariance combined with the formulas we have established so far, we get

$$S\sigma_t^y(\iota \otimes x)(\Delta(a^*)(1 \otimes b)) = (\iota \otimes x)((1 \otimes a^*)(\iota \otimes \sigma_{-t}^y)\Delta\tau_{-t}(b))$$

and $\sigma^x_{-t} S(\iota \otimes x)(\Delta(a^*)(1 \otimes b)) = (\iota \otimes x)((1 \otimes a^*)(\sigma^x_{-t} \otimes \iota)\Delta(b))$ for $a, b \in N_x$, so $\Delta\tau_{-t} = (\sigma^x_{-t} \otimes \sigma^y_t)\Delta$ by faithfulness of x and the density conditions. □

We call ν the *scaling constant* of the quantum group.

12.8 Invariance and the Modular Element

Let (W, Δ) be a locally compact quantum group with left invariant faithful semifinite normal weight x, and let $y = xR$, where R is the unitary coinverse. By Theorem 10.8.1 there exists a unique strictly positive operator h affiliated with W such that $x_h = y$ and $\sigma^x_t(h) = \nu^t h$ for real t. Note that $\sigma^y_t = h^{it}\sigma^x_t(\cdot)h^{-it}$ for real t. We also consider the semicyclic representation map $q \colon N \to V_x$ of x_h defined by $q(a) = q_x(ah^{1/2})$, where N consists of those $a \in W$ with $ah^{1/2} \in N_x$. We call h the *modular element* of (W, Δ), and by the end of this section it will be clear that both h and the scaling constant ν depend only on W and Δ.

Proposition 12.8.1 *The modular element h of a locally compact quantum group (W, Δ) is* group like, *that is, it satisfies $\Delta(h) = h \otimes h$. Moreover, we have $\tau_t(h) = h = R(h^{-1})$ and $S(h^{it}) = h^{-it}$ for real t.*

Proof Retaining the notation from the paragraph above, we introduce elements $v_n = n(\pi)^{-1/2} \int e^{-n^2t^2} h^{it}\, dt$ in W that are analytic with respect to $\{\sigma^y_t\}$. Let C be the linear span of the elements $v_n w$ with w in the Tomita algebra A_0 of y, so $C \subset A_0$. Let $c \in C$ and $d \in A_0$. We see that $h^{-1/2}\sigma^y_{i/2}(c) \in N^*_y$. As $x = y_{h^{-1}}$, we know that $\sigma^y_{i/2}(c)^* \in N_x$ and

$$x(\sigma^y_{i/2}(c)\sigma^y_{i/2}(c)^*) = y((\sigma^y_{i/2}(c)^*h^{-1/2})^*(\sigma^y_{i/2}(c)^*h^{-1/2})) = y((h^{-1/2}c)^*(h^{-1/2}c)),$$

where we used $h^{-1/2}c \in N_y$ and $\sigma^y_{i/2}(h^{-1/2}c) = \nu^{-i/4}h^{-1/2}\sigma^y_{i/2}(c)$ by the property of the v_n's, together with the KMS-condition of y. Hence by the KMS-condition of y and Proposition 12.6.1 and left invariance of x we get

$$\begin{aligned} y(y(c^*(\cdot)c) \otimes \iota)\Delta(d^*d) &= xR(y \otimes \iota)((\sigma^y_i(c)c^* \otimes 1)\Delta(d^*d)) \\ &= x(y \otimes \iota)(\Delta\sigma^y_{-i/2}(\sigma^y_i(c)c^*)(\sigma^y_{-i/2}(d^*d) \otimes 1)) \\ &= y(d^*d)y((h^{1/2}c)^*(h^{1/2}c)), \end{aligned}$$

so $y(\omega_{q(c),q(c)} \otimes \iota)\Delta(d^*d) = y(d^*d)(h^{-1/2}q(c)|h^{-1/2}q(c))$. But as $q(v_n w) = v_n q(w)$ and $q(A_0)$ is dense in V_x, we see that $q(C)$ is a core for $h^{-1/2}$. Thus the lemma below shows that we may replace $q(c)$ by $v \in D(h^{-1/2})$, and using the lemma once more we can assume that $d \in N_y$ in $y(\omega_{v,v} \otimes \iota)\Delta(d^*d) = y(d^*d)(h^{-1/2}v|h^{-1/2}v)$; an identity which we will be using repeatedly in the remaining part of the proof.

Using it twice we see that $y(\omega_{p_1\otimes p_2,p_1\otimes p_2}\otimes\iota)(\Delta\otimes\iota)\Delta(a)$ equals

$$y(a)((h^{-1/2}\otimes h^{-1/2})(p_1\otimes p_2)|(h^{-1/2}\otimes h^{-1/2})(p_1\otimes p_2))$$

for $p_i\in D(h^{-1/2})$ and positive $a\in M_y$. As $D(h^{-1/2})\otimes D(h^{-1/2})$ is a core for $h^{-1/2}\otimes h^{-1/2}$, we may replace $p_1\otimes p_2$ by $v\in D(h^{-1/2}\otimes h^{-1/2})$ due to the lemma below, yielding $y(\omega_{v,v}\otimes\iota)(\Delta\otimes\iota)\Delta(a)=y(a)\|(h^{-1/2}\otimes h^{-1/2})v\|^2$.

Let $\{e_i\}$ be an orthonormal basis of V_x and let $f_i=(\cdot|e_i)$ and $w_i=(f_i\otimes 1)M(v)$ for $v\in D(\Delta(h^{-1/2}))$, where M is the fundamental multiplicative unitary of (W,Δ). As $(f_i\otimes 1)(1\otimes h^{-1/2})\subset h^{-1/2}(f_i\otimes 1)$ and $\Delta(h^{-1/2})=M^*(1\otimes h^{-1/2})M$, we have $(f_i\otimes 1)(1\otimes h^{-1/2})M(v)=h^{-1/2}w_i$ and $\omega_{v,v}\Delta=\sum\omega_{w_i,w_i}$ pointwise. Hence by normality of y and the identity from the first paragraph, we have

$$\begin{aligned}y(\omega_{v,v}\otimes\iota)(\Delta\otimes\iota)\Delta(a)&=\sum y(\omega_{w_i,w_i}\otimes\iota)\Delta(a)\\&=\sum y(a)\|h^{-1/2}w_i\|^2=y(a)\|\Delta(h^{-1/2})v\|^2\end{aligned}$$

for positive $a\in M_y$.

Combining the results of the last two paragraphs we get $\|(h^{-1/2}\otimes h^{-1/2})v\|=\|\Delta(h^{-1/2})v\|$ when v is in the intersection of the respective domains. But the two domains coincide, and $\Delta(h)=h\otimes h$, as we have a common core, and this is true because $\Delta(h)$ and $h\otimes h$ commute, which again holds thanks to Proposition 12.7.2 and the other identities between Δ and the scaling group, which implies $(\sigma^y_{-t}\sigma^x_t\otimes\sigma^y_{-t}\sigma^x_t)\Delta=\Delta\sigma^y_{-t}\sigma^x_t$, so $(h^{it}\otimes h^{it})\Delta(h^{is})(h^{-it}\otimes h^{-it})=\Delta(h^{is})$ for $s,t\in\mathbb{R}$.

As for the final statement, observe that

$$\tau_t(h^{is})\otimes\sigma^x_t(h^{is})=\Delta\sigma^x_t(h^{is})=\nu^{is}h^{is}\otimes h^{is}=h^{is}\otimes\sigma^x_t(h^{is}),$$

so $\tau_t(h^{is})=h^{is}$ for real s,t. Since $1\otimes h^{it}=(h^{-it}\otimes 1)\Delta(h^{it})$ and $1\otimes h^{-it}=\Delta(h^{-it})(h^{it}\otimes 1)$, we get $S(h^{it})=h^{-it}$ by Lemma 12.6.4. The formula for R is now clear from $S=R\tau_{-i/2}$. □

Let us include the lemma which we used several times in the proof above.

Lemma 12.8.2 *Suppose we have a linear map $Q\colon V\to X$ from a vector space into a normed space, and a sesquilinear map $T\colon V\times V\to W$ to a von Neumann algebra with a faithful semifinite normal weight x such that $T(v,v)\geq 0$ for $v\in V$, and say there is a constant $k\geq 0$ such that $xT(u,u)=k\|Q(u)\|^2$ for all u in a subspace U of V. If to every $v\in V$, there is a net $\{u_i\}$ in U such that $Q(u_i)\to Q(v)$ and $T(u_i,u_i)\to T(v,v)$ σ-strongly*, then $xT(v,v)=k\|Q(v)\|^2$.*

Proof Using a net $\{u_i\}$ as prescribed to $v\in V$ yields $xT(v,v)\leq k\|Q(v)\|^2$ by normality of x Hence $v\mapsto xT(v,v)^{1/2}$ is a seminorm on V, and

$$|xT(v,v)^{1/2}-xT(v',v')^{1/2}|\leq xT(v-v',v-v')\leq k\|Q(v)-Q(v')\|$$

for $v'\in V$. This shows that $\{xT(u_i,u_i)\}$ converges both to $xT(v,v)$ and $k\|Q(v)\|^2$. □

Theorem 12.8.3 *Up to multiplication with a positive scalar there is only one left invariant faithful semifinite normal weight on a locally compact quantum group.*

Proof Say we have two such weights x, x' on the quantum group (W, Δ), and say x' has modular group $\{\sigma_t'\}$. By a trivial adaptation of the proof of Proposition 12.6.6 we then have $(\tau_t \otimes \sigma_t')\Delta = \Delta\sigma_t'$ for real t, so by Proposition 12.7.2, we get $(\iota \otimes \sigma_{-s}^x\sigma_s')\Delta = \Delta\sigma_{-s}^x\sigma_s'$ for real s. Arguing as in the proof of that proposition we moreover see that $y \equiv xR$ is invariant under $\sigma_{-s}^x\sigma_s'$, so $\{\sigma_t'\}$ commutes with $\{\sigma_t^x\}$ by Proposition 12.7.2. Combining all this then gives $\Delta(\sigma_t'(h^{is})h^{-is}) = 1 \otimes \sigma_t'(h^{is})h^{-is}$, so by Proposition 12.6.3 there is a complex number z of modulus one such that $\sigma_t'(h^{is}) = zh^{is}$ for real s. Hence $\{\sigma_t'\}$ and $\{\sigma_t^x\}$ commute. Thus for each real t we conclude that $x\sigma_t'$ is a left invariant faithful semifinite normal weight that is invariant under $\{\sigma_t^x\}$. Using Proposition 12.7.1 twice, we are therefore done. □

Thanks to the property $f(R \otimes R)\Delta = \Delta R$, where f is the flip on $W\bar{\otimes}W$, we deduce that up to multiplication by a positive constant, there is only one right invariant faithful semifinite normal weight on a locally compact quantum group (W, Δ) as well.

12.9 Modularity and Manageability

Quantum groups can be approached starting with multiplicative unitaries satisfying additional constrains, and we can pursuit this in a pure C*-algebra setting. For any Hilbert space V fix a conjugate linear self-adjoint involution J, and define the *transpose* of a closed operator A in V as the closed operator $T(A) = JA^*J$ in V with domain $JD(A^*)$. We introduce the following concepts.

Definition 12.9.1 A multiplicative unitary M on $V \otimes V$ is *modular* if there are strictly positive operators $Q, \hat{Q}$ in V and a unitary M_0 on $V \otimes V$ such that M commutes with $\hat{Q} \otimes Q$ and $(M(c \otimes d)|a \otimes b) = (M_0(Ja \otimes Q^{-1}d)|Jc \otimes Qb)$ for $a, c \in V$ and $b \in D(Q)$ and $d \in D(Q^{-1})$. We say M is *manageable* if $\hat{Q} = Q$.

Recall that the dual of a multiplicative unitary $M \in B(V \otimes V)$ is the multiplicative unitary $\hat{M} \equiv FM^*F$, where F is the flip on $V \otimes V$. So $\hat{\hat{M}} = M$.

Proposition 12.9.2 *A multiplicative unitary is modular if and only if its dual is modular. In particular it is manageable if and only if its dual is manageable.*

Proof One checks easily that whenever $Q, \hat{Q}$ and M_0 are the operators associated to a modular multiplicative unitary M, then $\hat{Q}, Q$ and $(J \otimes J)FM_0F(J \otimes J)$ are the corresponding operators associated to $\hat{M}$, and in doing this one first shows that M_0 commutes with $T(\hat{Q}) \otimes Q^{-1}$. □

Theorem 12.9.3 *Let M be a modular multiplicative unitary on $V \otimes V$. Then the norm closure A of $(\omega \otimes \iota)(M)$ and the closure $\hat{A}$ of $(\iota \otimes \omega)(M^*)$ as ω ranges over $B(V)_*$, are non-degenerate C*-subalgebras of $B(V)$ such that $M \in M(\hat{A} \otimes A)$. The formula $\Delta(a) = M(a \otimes 1)M^*$ for $a \in A$ defines a non-degenerate *-*

homomorphism $\Delta\colon A \to M(A \otimes A)$ *such that* $(\Delta \otimes \iota)\Delta = (\iota \otimes \Delta)\Delta$ *and such that linear spans of* $(A \otimes 1)\Delta(A)$ *and of* $\Delta(A)(1 \otimes A)$ *are both dense in* $A \otimes A$. *There is a unique closed linear operator* S *on the Banach space* A *with core* $\{(\omega \otimes \iota)(M) \mid \omega \in B(V)_*\}$ *and* $S(\omega \otimes \iota)(M) = (\omega \otimes \iota)(M^*)$, *and* $S(ab) = S(b)S(a)$ *and* $S(S(a^*)^*) = a$ *for* $a, b \in D(S)$. *Finally, there is a unique* σ*-weakly continuous involutive* $*$*-antiautomorphism* R *of* $A \subset B(V)$ *and a unique* σ*-weakly continuous one-parameter group of* $*$*-automorphisms* τ_t *of* A *such that* $S = R\tau_{-i/2}$ *and* $R\tau_t = \tau_t R$ *for real* t.

Proof Assume first that M is manageable with associated Q and M_0. Consider a Hilbert space U and a unitary N on $U \otimes V$ such that $M_{23}N_{12} = N_{12}N_{13}M_{23}$. Consider then the unitary $\tilde{N} = (T\otimes T\otimes\iota)(N_{12})(1\otimes M_0)(T\otimes T\otimes\iota)(N_{12}^*)(1\otimes M_0^*)$ on $U\otimes V\otimes V$. Using the defining property of M_0 twice together with its commutativity with $T(Q) \otimes Q^{-1}$ and the unitarity of M_0, it is straightforward to verify

$$Je \otimes Q(\omega_{a,c} \otimes \iota)(N)^* b = (\omega_{Jc,Ja} \otimes \iota \otimes \iota)(\tilde{N})^*(Je \otimes Qb)$$

for $a, c \in U$ and $b, e \in D(Q)$. So it makes sense to define a unitary N_0 on $U \otimes V$ with $(N_0)_{13} = \tilde{N}$. Then $(N(c \otimes d)|a \otimes b) = (N_0(Ja \otimes Q^{-1}d)|Jc \otimes Qb)$ for $d \in D(Q^{-1})$.

Let $B(V)_Q$ be the normed space of $a \in B(V)$ with $\|a\|_Q = \mathrm{Tr}(a^*Q^{-2}a)^{1/2} < \infty$. Consider the linear surjective isometry $f\colon B(V)_Q \to V \otimes V$ such that $(a(v)|u) = (f(a)|J(v) \otimes Q(u))$ for $v \in V$ and $u \in D(Q)$. Let $g_v\colon \mathbb{C} \to V$ be the linear function that sends 1 to $v \in V$. For such v and $u \in D(Q^{-1})$ we see that $f(g_u g_v^*) = Jv \otimes Q^{-1}u$, so the span of all elements E_1 of the form $g_u g_v^*$ is dense in $B(V)_Q$, and so is the span of the elements E_2 of the form $(g_v^* \otimes 1)M(1 \otimes g_u)$ as their image under f is $M_0(Jv \otimes Q^{-1}u)$, and the span of the elements E_3 of the form $(1 \otimes g_v^*)M^*(g_u \otimes 1)$ is also dense in $B(V)_Q$ as their image under f is $(J \otimes J)FM_0F(J \otimes J)(Jv \otimes Q^{-1}u)$.

We claim that $(1 \otimes g_u^*)N(1 \otimes a)N^*(1 \otimes g_v) = (\iota \otimes \omega)(N^*)$ for $v \in V$ and $u \in D(Q)$, where $\omega = (M_0(T(\cdot) \otimes 1)M_0^* f(a)|Jv \otimes Qu) \in B(V)_*$. Writing any $X \in U \otimes V$ as $X = \sum x_i \otimes e_i$, where $\{e_i\}$ is an orthonormal basis of V, and $x_i \in U$ satisfy $\sum \|x_i\|^2 = \|X\|^2$, we get

$$((1\otimes g_u^*)N(1\otimes a)X|w) = (N(1\otimes a)X|w\otimes u) = \sum(N_0(Jw\otimes Q^{-1}ae_i)|Jx_i\otimes Qu)$$

for $w \in U$, so

$$((1 \otimes g_u^*)N(1 \otimes a)X|w) \leq \sum \|x_i\|\|Qu\|\|w\|\|Q^{-1}ae_i\| \leq \|w\|\|Qu\|\|X\|\|a\|_Q,$$

and the norms of both sides of the claimed identity are bounded by $\|Qu\|\|v\|\|a\|_Q$. By linearity and continuity if suffices therefore to check the claimed identity under $\omega_{s,r}$ for $r, s \in U$ when $a \in E_2$, and there it holds due to $M_{23}N_{12} = N_{12}N_{13}M_{23}$.

Inserting elements of the type in E_1 in the now proved identity, we get

$$(\iota \otimes \omega')(N)(\iota \otimes \omega'')(N^*) = (\iota \otimes \omega)(N^*),$$

where $\omega, \omega', \omega''$ are all functionals of types that span dense subsets of $B(V)_*$. Letting B be the norm closure of $(\iota \otimes \eta)(N^*)$ for all $\eta \in B(V)_*$, we thus deduce that B^*B spans a dense subset of B, so $B^* = B$ is a C*-subalgebra of $B(U)$. If $0 \neq r \in U$ and $0 \neq v \in V$, there are $s \in U$ and $u \in V$ such that $((\iota \otimes \omega_{v,u})(N^*)r|s) = (N^*(r \otimes v)|s \otimes u) \neq 0$, so B acts non-degenerately on U.

The span of elements of the form $(1 \otimes g_v g_u^*)N(1 \otimes g_r g_s^*) = (\iota \otimes \omega_{r,u})(N) \otimes g_v g_s^*$ with $u \in D(Q)$ and $r \in D(Q^{-1})$ and $v, s \in V$ is clearly dense in $B \otimes B_0(V)$. By the estimates above we can replace $g_r g_s^* \in E_1$ by $(1 \otimes g_s^*)M^*(g_r \otimes 1) \in E_3$, showing that elements of the form

$$(1 \otimes g_v g_u^* \otimes g_s^*)N_{12}M_{23}^*(1 \otimes g_r \otimes 1)$$

span a dense subspace of $B \otimes B_0(V)$. Multiplying such an element from the right by N^* and using $M_{23}N_{12} = N_{12}N_{13}M_{23}$, we get $(1 \otimes g_v)(1 \otimes g^*_{M(u\otimes s)})(N \otimes 1)(1 \otimes g_r \otimes 1)$, and this equals $(\iota \otimes \omega_{r,u})(N) \otimes g_v g_s^*$ when we replace $M(u \otimes s)$ by $u \otimes s$, so right multiplication by N^* leaves us again with a dense span in $B \otimes B_0(V)$. Hence $(B \otimes B_0(V))N^* = B \otimes B_0(V)$. Multiplying this from the right by N, using unitarity and taking adjoints, we see that $N \in M(B \otimes B_0(V))$. When $N = \hat{M}$, we get $B = A$ and $M = F\hat{M}^*F \in M(B_0(V) \otimes A)$. Thus $N_{13} = N_{12}^*M_{23}N_{12}M_{23}^*$ belongs to $M(B \otimes B_0(V) \otimes A)$, so $N \in M(B \otimes A)$. But now when $N = M$, then $B = \hat{A}$ and $M \in M(\hat{A} \otimes A)$.

It is fairly easy to see that $\{\omega(\cdot c) \,|\, c \in B_0(V),\ \omega \in B(V)_*\} = B(V)_*$. To show that the closed span of $\Delta(A)(1 \otimes A)$ is $A \otimes A$, observe that the closed span of elements of the form $\Delta(\omega(\cdot c) \otimes \iota)(M)(1 \otimes a) = (\omega \otimes \iota \otimes \iota)(M_{12}(M(c \otimes a))_{13})$ for $a \in A$ and $c \in B_0(V)$ equals the closed span of the elements above with the last M removed, and these are $(\omega(\cdot c) \otimes \iota)(M) \otimes a$, so we get $A \otimes A$. The closed span of $(A \otimes 1)\Delta(A)$ is similarly $A \otimes A$. Hence we get a non-degenerate $*$-homomorphism $\Delta\colon A \to M(A \otimes A)$, and $(\Delta \otimes \iota)\Delta = (\iota \otimes \Delta)\Delta$, again by the pentagonal equation.

Now $\tau_t = Q^{-2it}(\cdot)Q^{2it}$ defines a one-parameter group of σ-weakly continuous $*$-automorphism on A since $\tau_t(\omega \otimes \iota)(M) = (\omega(Q^{2it}(\cdot)Q^{-2it}) \otimes \iota)(M)$ for $\omega \in B(V)_*$. As $(\tau_{-i/2}(\omega_{r,u} \otimes \iota)(M)v|s) = (M(Q^{-1}r \otimes v)|Qu \otimes s) = ((\omega_{r,u}T \otimes \iota)(M_0)v|s)$ for $s, v \in V$ and $u \in D(Q)$ and $r \in D(Q^{-1})$, and as the operator $\tau_{-i/2}$ is closed, we see that $\tau_{-i/2}(\omega \otimes \iota)(M) = (\omega T \otimes \iota)(M_0)$ for $\omega \in B(V)_*$. The elements $(\omega \otimes \iota)M$ form a core for $\tau_{-i/2}$ since each $R_n = n(\pi)^{-1/2} \int e^{-n^2t^2}\tau_{-t/2}dt$ leaves this set invariant and $R_n(A) \subset D(\tau_{-i/2})$.

Taking the conjugate of $z((\iota\otimes\omega')(N)(\iota\otimes\omega'')(N^*)) = z(\iota\otimes\omega)(N^*)$ for $z\in B(U)_*$ amounts to replacing z by $\bar{z}$ and ω' by $\overline{\omega''}$. This gives the first step in

$$
\begin{aligned}
&(M_0(T(z\otimes\iota)(N^*)\otimes 1)M_0^*(Jv\otimes Q^{-1}r)|Ju\otimes Q(s))\\
&= (M_0(T(z\otimes\iota)(N)\otimes 1)M_0^*(Js\otimes Q(u))|Jr\otimes Q^{-1}v)\\
&= (M_0(T(zT\otimes\iota)(N_0)\otimes 1)M_0^*(JQs\otimes u)|JQ^{-1}r\otimes v)
\end{aligned}
$$

for $u,s\in D(Q)$ and $r,v\in D(Q^{-1})$, where in the last step we used the commutativity of M_0 with $T(Q)\otimes Q^{-1}$ together with $(zT\otimes\iota)(N_0)Q^{-1}\subset Q^{-1}(z\otimes\iota)(N)$, which is immediate from the first paragraph in this proof. Hence by density we get $\tilde{J}(1\otimes(z\otimes\iota)(N^*))\tilde{J} = 1\otimes(zT\otimes\iota)(N_0)^*$, where $\tilde{J}$ is the self-adjoint conjugate linear involution $(J\otimes J)FM_0^*F(J\otimes J)FM_0F(J\otimes J)$. Letting $N=M$ we see that $R(\omega\otimes\iota)(M^*) = (\omega T\otimes\iota)(M_0)$ for $\omega\in B(V)_*$ defines a σ-weakly continuous involutive $*$-preserving conjugate linear map R on A such that $R\tau_t = \tau_t R$ for real t as $\tilde{J}(T(Q)\otimes Q^{-1})\tilde{J} = (T(Q)\otimes Q^{-1})^{-1}$. Defining $S = R\tau_{-i/2}$ we have thus proved the theorem in the case when M is manageable.

Suppose now that M is only modular with associated M_0 and Q and $\hat{Q}$. Let q be the self-adjoint multiplication operator on $L^2(\mathbb{R})$ by the identity function, while p is the positive self-adjoint generator of translation, so $p^{it}h(t') = h(t'-t)$ for $h\in L^2(\mathbb{R})$. In this Schrödinger representation we have $p^{it}qp^{-it} = q-t$ for real t. Adapt the notation A_k to denote the operator acting on multitensors as the identity at every factor except at the k-th place, where it acts as A. Then $X\equiv(1\otimes Q)^{i(q\otimes 1)}(1\otimes\hat{Q})^{-i(q\otimes 1)} = Q_2^{iq_1}\hat{Q}_2^{-iq_1}$, and $X^*(p\otimes Q)X = p\otimes\hat{Q}$ as

$$
(p\otimes Q)^{it}X = p_1^{it}Q_2^{i(q_1+t)}\hat{Q}_2^{-iq_1} = Q_2^{iq_1}p_1^{it}\hat{Q}_2^{-iq_1} = Q_2^{iq_1}\hat{Q}_2^{-i(q_1-t)}p_1^{it} = X(p\otimes Q)^{it}
$$

for real t. Similarly $Q_2^{it}MQ_2^{-it} = \hat{Q}_1^{-it}M\hat{Q}_1^{it}$ for real t by definition of M. Let $V_m = L^2(\mathbb{R})\otimes V$ and define σ-weakly $*$-homomorphisms $\alpha,\beta\colon B(V)\to B(V_m)$ by $\alpha(a) = X(1\otimes a)X^*$ and $\beta(a) = 1\otimes a$. Using the formulas in this paragraph together with the definition of M as a modular multiplicative unitary, it is now straightforward, thought tedious, to check that $M_m = (\alpha\otimes\beta)(M)$ is a manageable multiplicative unitary on V_m with associated $(M_m)_0 = T(X)_{12}^*(M_0)_{24}T(X)_{12}$ and $Q_m = p\otimes Q$ with core $D(p)\otimes D(Q)$. By the first part of the proof we can thus apply the theorem to M_m obtaining the desired quantities which we denote as before but with a subscript m. To $\omega\in B(V)_*$ there are $\eta,\eta'\in B(V_m)_*$ such that $\omega = \eta\alpha = \eta'\beta$, and since $\beta(\omega\otimes\iota)(M) = (\eta\otimes\iota)(M_m)$, we see that β restricts to a $*$-isomorphism $\beta\colon A\to A_m$, and similarly α restricts to a $*$-isomorphism $\alpha\colon\hat{A}\to\hat{A}_m$. Then $\Delta = (\beta^{-1}\otimes\beta^{-1})\Delta_m\beta$ and $S = \beta^{-1}S_m\beta$ and $R = \beta^{-1}R_m\beta$ and $\tau_t = \beta^{-1}(\tau_m)_t\beta$ for real t, and all the required properties clearly hold. □

In the proof we showed that given a manageable multiplicative unitary M on $V\otimes V$ and a unitary N on $U\otimes V$ such that $M_{23}N_{12} = N_{12}N_{13}M_{23}$, then $N\in M(B\otimes A)$, where B is the norm closure of $\{(\iota\otimes\omega)(N)\mid\omega\in B(V)_*\}$, so $(\iota\otimes\Delta)(N) = N_{12}N_{13}$, and $S(\eta \otimes \iota)(N) = (\eta \otimes \iota)(N^*)$ for $\eta \in B(U)_*$. It is also easy see that $\{\tau_t\}$ is pointwise norm-continuous.

Proposition 12.9.4 *The fundamental multiplicative unitary of any locally compact quantum group is manageable.*

Proof Let (W, Δ) be a locally compact quantum group with left invariant faithful semifinite normal weight x and fundamental multiplicative unitary M. Since $x\tau_t = \nu^{-t}x$ for real t, we may define a strictly positive operator Q in V_x by $Q^{it}q_x(a) = \nu^{t/4}q_x(\tau_{t/2}(a))$ for $a \in N_x$, so $\tau_t = Q^{2it}(\cdot)Q^{-2it}$. By Proposition 12.6.6 we can therefore write $(I(\iota \otimes \omega_{r,s})(M)I(v)|w) = ((\iota \otimes \omega_{s,r})(M)Q^{-1}(v)|Q(w))$ for $v \in D(Q^{-1})$ and $w \in D(Q)$ and $r, s \in V_x$. The same proposition thus yields

$$(\hat{M}(q \otimes v)|p \otimes w) = (FMF(J_x p \otimes Q^{-1}v)|J_x q \otimes Qw),$$

where F is the flip on $V_x \otimes V_x$. Now M is manageable if $\hat{M}$ is manageable and this happens if it commutes with $Q \otimes Q$. But by definition

$$M^*(Q^{it} \otimes Q^{it})(q_x(a) \otimes q_x(b)) = (Q^{it} \otimes Q^{it})M^*(q_x(a) \otimes q_x(b))$$

for $a, b \in N_x$. □

Note that if M is the fundamental multiplicative unitary of a locally compact quantum group (W, Δ), then by convention the weak extensions of the C*-algebra A and the coproduct associated by the theorem above to the manageable multiplicative unitary $\hat{M}$ is actually the locally compact quantum group $(W, F\Delta(\cdot)F)$. This explains the discrepancy between the implementation of τ_t in the proof above and in the proof of the previous theorem.

Remark 12.9.5 The theorem offers a passage from the von Neumann setting to a C*-algebra setting for quantum groups. Indeed, if (W, Δ) is a locally compact quantum group with left invariant faithful semifinite normal weight x and with fundamental multiplicative unitary $M \in B(V_x \otimes V_x)$, then the norm closure W_r of $\{(\iota \otimes \omega)(M) \mid \omega \in B(V_x)_*\}$ is a C*-algebra and Δ restricts to a non-degenerate $*$-homomorphism $\Delta_r \colon W_r \to M(W_r \otimes W_r)$ such that $\Delta_r(w) = M^*(1 \otimes w)M$ for $w \in W_r$, and $(\Delta_r \otimes \iota)\Delta_r = (\iota \otimes \Delta_r)\Delta_r$, and the closed spans of $\Delta(W_r)(W_r \otimes 1)$ and $\Delta(W_r)(1 \otimes W_r)$ are dense in $W_r \otimes W_r$. Let y be the right invariant weight xR. The formulas $R(\iota \otimes \omega_{u,v})(M) = (\iota \otimes \omega_{J_x v, J_x u})(M)$ and $\tau(\iota \otimes \omega_{u,v})(M) = (\iota \otimes \omega_{\Delta_x^{it}u, \Delta_x^{it}v})(M)$ and $\sigma_t^x(\iota \otimes \omega_{u,v})\Delta = (\iota \otimes \omega_{u,v}\sigma_t^y)\Delta\tau_t$ for $u, v \in V_x$ and real t, show that R, τ_t, σ_t^x restrict to norm continuous maps on W_r. The weight x restricts to a faithful lower semicontinuous weight x_r on W_r which is *locally finite*, that is, with N_{x_r} dense, as $(y \otimes \iota)(\Delta(b^*c)(a \otimes 1)) \in N_{x_r}$ for $a, b \in N_y$ and $c \in N_x$, and it clearly also satisfies the KMS-condition. Let z be a positive functional on W_r with GNS-representation π_z, so $z = \omega_v\pi_z$ for $v \in V_z$. For positive $\eta \in B(V_x)_*$ with $\eta(1) = 1$ define positive $\theta \in B(V_x)_*$ by $\theta(w) = (\omega_v \otimes \eta)((\pi_z \otimes \iota)(M)^*(1 \otimes w)(\pi_z \otimes \iota)(M))$.

Then $(\eta\otimes\iota)\Delta(z\otimes\iota)\Delta_r=(\theta\otimes\iota)\Delta$ on W_r. Using left invariance of x twice we get

$$x_r(z\otimes\iota)\Delta_r=x(z\otimes\iota)\Delta_r=x(\eta\otimes\iota)\Delta(z\otimes\iota)\Delta_r=\theta(1)x=z(1)x_r$$

on the positive domain of x_r. We also conclude that y_r is right invariant in a similar C*-algebraic sense. We have arrived at the reduced C*-algebraic counterpart of (W,Δ). Departing from such an object with faithful lower semicontinuous locally finite KMS-weights having these invariance properties, together with a coproduct Δ_r with the cancellation properties as described above, we can produce, even more easily, a multiplicative unitary implementing the coproduct, and extending this by continuity to a normal map from the von Neumann algebra closure, alongside the invariant weights producing faithful semifinite normal weights, we get a locally compact quantum group (in the von Neumann setting). ◇

The advantage starting from a modular multiplicative unitary is that we are not from the outset assuming the existence of invariant weights. The closest we are to a proof of their existence is the result below.

Proposition 12.9.6 *Let M be a modular multiplicative unitary with associated $M_0,Q,\hat{Q}$, and consider the faithful lower semicontinuous weight $y=\mathrm{Tr}(\hat{Q}(\cdot)\hat{Q})$ on A with (A,Δ) as in the theorem above. Then y is right invariant in the sense that $y(\iota\otimes z)\Delta=y$ for any state z on A.*

Proof Say $M\in B(V\otimes V)$. By $\mathrm{Tr}(\hat{Q}(\cdot)\hat{Q})$ we mean of course $\sum\omega_{\hat{Q}e_i}$ for any orthonormal basis $\{e_i\}$ of V with $e_i\in D(\hat{Q})$. Writing $z=\omega_v\pi_z$ as above, we get for $a\in A$ that

$$y(\iota\otimes z)\Delta(a^*a)=\sum\|(a\otimes 1)(\iota\otimes\pi_z)(M)^*(\hat{Q}e_i\otimes v)\|^2,$$

which equals $\sum|((\iota\otimes\pi_z)(M)(a^*e_i\otimes f_j)|\hat{Q}e_i\otimes v)|^2$ for any orthonormal basis $\{f_j\}$ of V_z. Whenever $y(a^*a)<\infty$ we thus get

$$y(\iota\otimes z)\Delta(a^*a)=\sum|((\iota\otimes\pi_z)(M_0)(Je_i\otimes f_j)|J\hat{Q}a^*e_i\otimes v)|^2=\sum\|\hat{Q}a^*e_i\|^2=y(a^*a).$$

□

If y is locally finite, we immediately get a locally compact quantum group. In examples y is often locally finite. Then S,R,τ are, as we have seen, automatically unique, and as the following result shows, this is nevertheless always true.

Proposition 12.9.7 *Suppose $M\in B(V\otimes V)$ is a modular multiplicative unitary, and construct S,R and $\{\tau_t\}$ to A and Δ as in the theorem above. Then these quantities and the σ-weak topology on $A\subset B(V)$ do not depend on the choice of M. Up to isomorphism, neither do $\hat{A}$ and $\hat{\Delta}$ associated to $\hat{M}$, nor M considered as an element of $M(\hat{A}\otimes A)$.*

Proof Say $N\in B(V\otimes V)$ is another modular multiplicative unitary producing the same A and Δ. The unitary $X=(T\otimes R)(M)FN^*MF(T\otimes R)(M^*)$ satisfies

$X_{12}M_{23}X_{12}^* = N_{23}$. Hence $X(1 \otimes (\iota \otimes \omega)(M^*))X^* = 1 \otimes f(\iota \otimes \omega)(M^*)$ for $\omega \in B(V)_*$ defines a $*$-isomorphism $f: \hat{A} \to \hat{B}$ that is a σ-weakly homeomorphism such that $(f \otimes \iota)(M) = N$, where $\hat{B}$ is the norm closure of $\{(\iota \otimes \omega)(N^*) \mid \omega \in B(V)_*\}$. Hence the coproduct associated to $\hat{N}$ is $(f \otimes f)\hat{\Delta} f^{-1}$. Applying this to $\hat{\hat{M}} = M$, we get the second statement of the proposition. As S is σ-weakly closed with the formula on the core involving M, we see that S is also independent of the choice of M, and so is R, with a similar formula on the core involving both M and M_0, and whence $\tau_{-i/2}$ does also not depend on the choice of M. □

12.10 The Dual Quantum Group

Let (W, Δ) be a locally compact quantum group with left invariant faithful semifinite normal weight x and fundamental multiplicative unitary M. Consider the multiplicative unitary $\hat{M} = FM^*F$. From the previous section we know that the σ-strong* closed span $\hat{W}$ of elements of the form $(\iota \otimes \omega)(\hat{M})$ for $\omega \in B(V_x)_*$ is a von Neumann algebra, and $\hat{\Delta}(w) = \hat{M}^*(1 \otimes w)\hat{M}$ defines a normal $*$-homomorphism $\hat{\Delta}: \hat{W} \to \hat{W}\bar{\otimes}\hat{W}$ such that $(\hat{\Delta} \otimes \iota)\hat{\Delta} = (\iota \otimes \hat{\Delta})\hat{\Delta}$ and such that the closed spans of $\hat{\Delta}(\hat{W})(\hat{W} \otimes 1)$ and $\hat{\Delta}(\hat{W})(1 \otimes \hat{W})$ are dense in $\hat{W}\bar{\otimes}\hat{W}$. The pentagonal equation for $M \in W\bar{\otimes}\hat{W}$ clearly yields $(\Delta \otimes \iota)(M) = M_{13}M_{23}$ and $(\iota \otimes \hat{\Delta})(M) = M_{13}M_{12}$.

Aiming to construct invariant weights to $\hat{\Delta}$, regard W_* as a Banach algebra with product $\omega\eta = (\omega \otimes \eta)\Delta$, and define an injective algebra homomorphism $\lambda: W_* \to \hat{W}$ by $\lambda(\omega) = (\omega \otimes \iota)(M)$. Let K consist of those $\omega \in W_*$ for which there is a positive scalar c such that $|\omega(w^*)| \leq c\|q_x(w)\|$ for $w \in N_x$. By the Riesz theorem for Hilbert spaces there is to $\omega \in K$ a unique $p(\omega) \in V_x$ such that $\omega(w^*) = (p(\omega)|q_x(w))$ for $w \in N_x$, yielding a closed linear map $p: K \to V_x$. Clearly K is a left ideal in W_* with $p(\eta\omega) = \lambda(\eta)p(\omega)$ for $\eta \in W_*$, which is dense in W_* since $p(x(b^* \cdot a)) = q_x(a\sigma_i^x(b)^*)$ for a, b in the Tomita algebra associated to x.

Let h be the modular element of (W, Δ) with weight $x_h = xR = y$ and associated semicyclic representation map $q: N \to V_x$. The formula $\rho_t(\omega)(w) = \omega(h^{-it}\tau_{-t}(w))$ for $w \in W$ and real t evidently defines a continuous one-parameter group $\{\rho_t\}$ of automorphisms of W_*. Let Q denote the operator associated to M as a manageable multiplicative unitary, so $Q^{it}q_x(a) = \nu^{t/4}q_x\tau_{t/2}(a)$ for $a \in N_x$. As $J_x h^{it} J_x q_x(a) = \nu^{-t/2}q_x(ah^{-it})$ and $\tau_t(h) = h$, we see that Q and $J_x h J_x$ commute. One now checks that $p(\rho_t(\omega)) = Q^{2it}J_x h^{it} J_x p(\omega)$ for $\omega \in K$, so $\hat{\sigma}_t = Q^{2it}J_x h^{it} J_x(\cdot)(Q^{2it}J_x h^{it} J_x)^*$ defines a continuous one-parameter group $\{\hat{\sigma}_t\}$ of automorphism of $\hat{W}$ uniquely determined by $\hat{\sigma}_t(\lambda(\omega)) = \lambda(\rho_t(\omega))$ for $\omega \in W_*$.

Lemma 12.10.1 *With notation as above, the formula $Uq(a) = q_x R(a^*)$ for $a \in N_y$ defines a conjugate linear map $U: V_x \to V_x$ such that $U^*U = UU^* = 1$ and $p(\omega\eta) = U^*\lambda\rho_{i/2}(\eta)^*Up(\omega)$ for $\omega \in K$ and $\eta \in D(\rho_{i/2})$.*

Proof Only the last formula requires a proof. The formulas $h_t^*(\omega) = \omega(h^{it}\cdot)$ and $\tau_t^*(\omega) = \omega\tau_t$ for $\omega \in W_*$ and real t, define commuting continuous one-parameter groups $\{h^*_t\}$ and $\{\tau^*_t\}$ of endomorphisms of W_* with $\rho_t = \tau^*_{-t}h^*_{-t}$, so $\tau^*_{-i/2}h^*_{-i/2}$ has closure $\rho_{i/2}$, so we may assume $\eta \in D(\overline{\tau^*_{-i/2}h^*_{-i/2}})$. Pick v_n as in the proof of Proposition 12.8.1, and let $b \in N_x$. Then $(\iota \otimes \overline{(h^*_{-i/2})}(\eta))\Delta(bh^{-1/2}v_n) \in N_y$ by right invariance, and a by now standard calculation shows that its image under q is $U^*(\rho_{i/2}(\eta) \otimes \iota)(W)Uq_x(bv_n)$. But $(\iota \otimes \overline{(h^*_{-i/2})}(\eta))\Delta(bh^{-1/2}v_n) = (\iota \otimes \bar{\eta})\Delta(bv_n)h^{-1/2}$ by analyticity as this identity clearly holds with $h^{-1/2}$ replaced by h^{it} for real t. Hence $q(\iota \otimes \bar{\eta})\Delta(bv_n) = U^*\lambda\rho_{i/2}(\eta)Uq_x(bv_n)$, which in turn gives

$$(\omega\eta)((bv_n)^*) = (U^*\lambda\rho_{i/2}(\eta)^*Up(\omega)|q_x(bv_n)),$$

and letting n tend to infinity we get the desired formula with $\omega\eta \in K$. □

Lemma 12.10.2 *Keeping notation as above, there is a unique σ-strong* closed linear map $\hat{q}_x$ into V_x such that $\hat{q}_x\lambda(\omega) = p(\omega)$ on a core $\lambda(K)$ which is σ-strong* dense in $\hat{W}$.*

Proof Say we have a net $\{\omega_j\}$ in K such that $\lambda(\omega_j) \to 0$ and $p(\omega_j) \to v \in V_x$. Let $\eta \in D(\rho_{i/2}) \cap K$. Then $\lambda(\omega_j)p(\eta) = U^*\lambda\rho_{i/2}(\eta)^*Up(\omega_j)$ by the previous lemma. So $0 = U^*\lambda\rho_{i/2}(\eta)^*Uv$, and $v = 0$ as $\rho_{i/2}(D(\rho_{i/2}) \cap K)$ is dense in W_* which in turn is clear from

$$p(n(\pi)^{-1/2}\int e^{-n^2(t+i/2)^2}\rho_t(\omega)\,dt) = n(\pi)^{-1/2}\int e^{-n^2(t+i/2)^2}Q^{2it}J_xh^{it}J_xp(\omega)\,dt$$

and

$$\rho_{i/2}(n(\pi)^{-1/2}\int e^{-n^2(t+i/2)^2}\rho_t(\omega)\,dt) = n(\pi)^{-1/2}\int e^{-n^2t^2}\rho_t(\omega)\,dt$$

for $\omega \in K$. □

Letting S be the coinverse of (W, Δ), set $\omega^* = \bar{\omega}S$ for $\omega \in W_*$, and consider the set $W^\sharp$ of those ω such that ω^* has a continuous extension to an element of W_*, which we again denote by ω^*. From the manageability of M and the proof of Proposition 12.6.2 we see that $W^\sharp$ is a subalgebra of W_* which is $*$-preserving for the antimultiplicative involution $\omega \mapsto \omega^*$. Since $(\iota \otimes \eta)(M)$ for $\eta \in B(V_x)_*$ form a σ-strong* core for S and $S(\iota \otimes \eta)(M) = (\iota \otimes \eta)(M^*)$, we also see that $W^\sharp$ consists of those $\omega \in W_*$ such that $\lambda(\omega)^* \in \lambda(W_*)$, and then $\lambda(\omega)^* = \lambda(\omega^*)$.

Lemma 12.10.3 *The spaces $K \cap W^\sharp$ and $(K \cap W^\sharp)^*$ are dense in W_*, and $\lambda(K \cap W^\sharp)$ is a σ-strong* core for $\hat{q}_x$.*

Proof Let $\omega \in K$ and define $\omega(n, z) = n(\pi)^{-1/2}\int e^{-n^2(t+z)^2}\omega\tau_t\,dt \in W_*$ for $n \in \mathbb{N}$ and $z \in \mathbb{C}$. Then $\omega(n, z)^* = n(\pi)^{-1/2}\int e^{-n^2(t+\bar{z}+i/2)^2}\bar{\omega}R\tau_t\,dt \in W^\sharp$ and $\omega(n, z) \in K$ with $p(\omega(n, z)) = n(\pi)^{-1/2}\int e^{-n^2(t+z)^2}\nu^{-t/2}Q^{-2it}p(\omega)\,dt$. Letting $z = 0$ and $n \to \infty$, and using that K is dense in W_* with a core $\lambda(K)$ for $\hat{q}_x$, we see that $K \cap W^\sharp$ is dense in W_*, and that $\lambda(K \cap W^\sharp)$ is a σ-strong* core for $\hat{q}_x$. As $\omega(n, i/2)^* \to \bar{\omega}R$, we also see that $(K \cap W^\sharp)^*$ is dense in W_*. □

Lemma 12.10.4 *The set* $K^\sharp = (K \cap W^\sharp) \cap (K \cap W^\sharp)^*$ *is a* $*$*subalgebra of* $W^\sharp$ *which is dense in* W_*, *and* $\lambda(K^\sharp)$ *is a* σ*-strong* core for* $\hat{q}_x$.

Proof Clearly $K^\sharp$ is a $*$-subalgebra of $W^\sharp$, and since K is a left ideal of W_*, we get $(K \cap W^\sharp)^*(K \cap W^\sharp) \subset K^\sharp$, so to show that $K^\sharp$ is dense in W_*, it suffices by the previous lemma to show that $W_* W_*$ is dense in W_*. But this follows easily from the fact that $(\Delta(\cdot)M^*(v \otimes v_1)|M^*(v \otimes v_2)) = \omega_{v_1,v_2}$ for $v, v_i \in V_x$ with $\|v\| = 1$.

Since $K \cap W^\sharp$ is dense in W_* by the previous lemma, we know that 1 belongs to the σ-strong* closure of $\lambda(K \cap W^\sharp)^*$. Combining this with $\lambda(K \cap W^\sharp)^*\lambda(K \cap W^\sharp) \subset \lambda(K^\sharp)$ and the previous lemma, we get the final result in the lemma. □

We can now construct the invariant weights by going via left Hilbert algebras.

Lemma 12.10.5 *By requiring* $p \colon K^\sharp \to V_x$ *to be a* $*$*-homomorphism, its image is a left Hilbert algebra with Hilbert space* V_x *and left von Neumann algebra* $\hat{W}$. *The associated faithful semifinite normal weight* $\hat{x}$ *has* $q_{\hat{x}} = \hat{q}_x$ *with the identity representation of* $\hat{W}$ *on* $V_{\hat{x}} = V_x$.

Proof To see that we get the desired left Hilbert algebra, in view of the previous proposition, and the fact that $p(\eta\omega) = \lambda(\eta)p(\omega)$ for $\omega, \eta \in K^\sharp$, we only need to check that the involution is closable. To this end observe that the set of $a \in N_x \cap D(S^{-1})$ with $S^{-1}(a)^* \in N_x$ is dense in V_x since by strong right invariance of y, it contains the usual core of the fundamental involution G of (W, Δ). On this set define an unbounded operator T in V_x by $Tq_x(a) = q_x(S^{-1}(a)^*)$, so $(p(\omega)^*|q_x(a)) = (Tq_x(a)|p(\omega))$ for $\omega \in K^\sharp$. This shows that T is a restriction of a densely defined adjoint of the involution, which therefore must be closable. The remaining statements are now immediate. □

Theorem 12.10.6 *Let* (W, Δ) *be a locally compact quantum group. Then* $(\hat{W}, \hat{\Delta})$ *defined in the first paragraph of this section is a locally compact quantum group with left invariant faithful semifinite normal weight* $\hat{x}$ *defined in the previous lemma. We call this quantum group the* dual *of* (W, Δ).

Proof Let $\omega \in K^\sharp$ and $\eta \in \hat{W}_*$. Since $(\iota \otimes \hat{\Delta})(M) = M_{13}M_{12}$, we get $(\eta \otimes \iota)\hat{\Delta}(\lambda(\omega)) = \lambda(\omega(\cdot(\iota \otimes \eta)(M))$. As $p(\omega(\cdot a)) = ap(\omega)$ for $a \in W$, we thus get $\hat{q}_x(\eta\otimes\iota)\hat{\Delta}(\lambda(\omega)) = (\iota\otimes\eta)(M)\hat{q}_x(\lambda(\omega))$, and by closedness of $\hat{q}_x$ this latter identity holds even when $\lambda(\omega)$ is replaced by any element w of $N_{\hat{x}}$. Pick an orthonormal basis $\{v_i\}$ of V_x and let $v \in V_x$ with $\|v\| = 1$. Then using this identity twice we get

$$\begin{aligned}\hat{x}(\omega_{v,v} \otimes \iota)\hat{\Delta}(w^*w) &= \sum \hat{x}((\omega_{v,v_i} \otimes \iota)\hat{\Delta}(w)^*(\omega_{v,v_i} \otimes \iota)\hat{\Delta}(w)) \\ &= \sum ((\iota \otimes \omega_{v,v_i})(M)^*(\iota \otimes \omega_{v,v_i})(M)\hat{q}_x(w)|\hat{q}_x(w)),\end{aligned}$$

which is $\hat{x}(w^*w)$ by unitarity of M. So $\hat{x}$ is left invariant since every normal state is of the form $\omega_{v,v}$ as we are working in a standard representation.

Let $\hat{R} = J_x(\cdot)^* J_x$. Since $(I \otimes J_x)M^*(I \otimes J_x) = M$, we get $\hat{R}(\lambda(\omega)) = \lambda(\omega R)$ for $\omega \in W_*$, so we have an involutive antimultiplicative $*$-preserving map $\hat{R}\colon \hat{W} \to \hat{W}$ such that $(R \otimes \hat{R})(M) = M$. Using $(\iota \otimes \hat{\Delta})(M) = M_{13}M_{12}$ and introducing the flip f on $\hat{W}\bar{\otimes}\hat{W}$, we thus get $f(\hat{R} \otimes \hat{R})\hat{\Delta} = \hat{\Delta}\hat{R}$ on elements of the form $\lambda(\omega)$ for $\omega \in W_*$, and thus on any element in $\hat{W}$. Hence $\hat{x}\hat{R}$ is a right invariant faithful semifinite normal weight on $\hat{W}$. □

When $\{v_j\}$ is an orthonormal basis for V_x, then as $(\iota \otimes \hat{\Delta})(M) = M_{13}M_{12}$, we get $\hat{\Delta}(\lambda(\omega_{u,v})) = \sum \lambda(\omega_{u,v_j}) \otimes \lambda(\omega_{v_j,v})$ for $u, v \in V_x$, which can be interpreted as $\hat{\Delta}(\omega_{u,v})(a \otimes b) = \omega_{u,v}(ba)$ for $a, b \in W$. Notice that $\hat{\Delta}$ is an extension of the Hopf algebraic dual of W with the opposite product on W.

The following result is a vast generalization of Pontryagin's duality theorem.

Theorem 12.10.7 *Any locally compact quantum group (W, Δ) is isomorphic to its double dual $(\hat{\hat{W}}, \hat{\hat{\Delta}})$, and their left invariant faithful semifinite normal weights x and $\hat{\hat{x}}$, respectively, coincide.*

Proof If M is the fundamental multiplicative unitary of (W, Δ), then since $\hat{q}_x(\eta \otimes \iota)\hat{\Delta}(w) = (\iota \otimes \eta)(M)\hat{q}_x(w)$ for $w \in N_{\hat{x}}$ and $\eta \in B(V_x)_*$ from the previous proof, we see that $\hat{M} \equiv FM^*F$ is the fundamental multiplicative unitary of $(\hat{W}, \hat{\Delta})$. Hence $M = \hat{\hat{M}}$ is the fundamental multiplicative unitary of the double dual, which therefore must be isomorphic to the original quantum group, then with the identity map as the isomorphism.

The usual representation of $(\hat{W})_*$ on V_x is given by $\hat{\lambda}(\omega) = (\omega \otimes \iota)(\hat{M})$. As $\hat{\hat{x}}$ and x are proportional by uniqueness, we get $N_{\hat{\hat{x}}} = N_x$, so for $\omega \in \hat{K}$, we get

$$\omega(\lambda(\eta)^*) = \overline{\eta(\hat{\lambda}(\omega)^*)} = \overline{(p(\eta)|q_x(\hat{\lambda}(\omega)))} = (q_x(\hat{\lambda}(\omega))|\hat{q}_x(\lambda(\eta)))$$

for $\eta \in K$. In this identity we may replace the typical element $\lambda(\eta)$ in the core for $\hat{q}_x$ by any element in $N_{\hat{x}}$. Thus $\hat{\hat{q}}_x(\hat{\lambda}(\omega)) = q_x(\hat{\lambda}(\omega))$. Since $\hat{\lambda}(\hat{K})$ is a core for $\hat{\hat{q}}_x$ and q_x is closed, we thus get $\hat{\hat{q}}_x = q_x$ and $\hat{\hat{x}} = x$. □

In fact, we see that K consists of those $\eta \in W_*$ such that $\lambda(\eta) \in N_{\hat{x}}$. Indeed, the above proof shows that $(q_x(\hat{\lambda}(\omega))|\hat{q}_x(\lambda(\eta))) = \omega(\lambda(\eta)^*)$ for $\omega \in \hat{K}$. Rewriting this identity, and using that $\hat{\lambda}(\hat{K})$ is a core for q_x, we get $(\hat{q}_x(\lambda(\eta))|q_x(a)) = \eta(a^*)$ for $a \in N_x$, so $\eta \in K$.

We can push the analysis further to obtain more relations between the various operators associated to quantum groups. We follow an approach that can be adapted more easily to the C*-algebra setting, though we could have worked more directly with left Hilbert algebras as was done in the Kac algebra case.

Lemma 12.10.8 *The modular group of $\hat{x}$ from the previous lemma is $\{\hat{\sigma}_t\}$ defined above. Hence $\Delta_{\hat{x}}^{it} = Q^{2it} J_x h^{it} J_x$ for real t.*

Proof Using Lemma 12.10.1 and closedness of $\hat{q}_x$, we see that its domain D is a left ideal of $\hat{W}$ such that $\hat{q}_x(wa) = w\hat{q}_x(a)$ and $\hat{q}_x\hat{\sigma}_t(a) = Q^{2it} J_x h^{it} J_x \hat{q}_x(a)$ and

$\hat{q}_x(ab) = U^*\hat{\sigma}_{i/2}(b)U\hat{q}_x(a)$ for $w \in \hat{W}$ and $a \in D$ and $b \in D(\hat{\sigma}_{i/2})$. Now $D \cap D^*$ is a core for $\hat{q}_x$ since it contains D^*D, and using the usual regularizing integrals, we conclude that the $*$-subalgebra C of $\hat{W}$ consisting of the analytic elements $a \in D \cap D^*$ such that $\hat{\sigma}_z(a) \in D \cap D^*$ for complex z, is also a core for $\hat{q}_x$, and clearly $C \subset D \cap D(\hat{\sigma}_{i/2})$ and $\hat{\sigma}_{i/2}(C)^* \subset C$. Calculating carefully we then get $(c\hat{q}_x(a)|\hat{q}_x(a)) = (Uc^*U^*\hat{q}_x(\hat{\sigma}_{i/2}(a)^*)|\hat{q}_x(\hat{\sigma}_{i/2}(a)^*))$ for $a, c \in C$, so we may replace c by 1, and using the usual approximation techniques involving regularizing integrals, we get this equality (with $c = 1$) for every $a \in D \cap D(\hat{\sigma}_{i/2})$. By straightforward left Hilbert algebra theory this yields the KMS-condition for $\hat{x}$ with respect to $\{\hat{\sigma}_t\}$, so it must be the modular group by uniqueness. □

From the proof of Lemma 12.10.5 and the definition of $\hat{q}_x$, we see that the closure of T is $\Delta_{\hat{x}}^{1/2} J_{\hat{x}}$. That proof also shows that this closure extends the fundamental involution G of (W, Δ), and since

$$\tau_t(y \otimes \iota)(\Delta(a^*)(b \otimes 1))h^{-it} = \nu^t(y \otimes \iota)(\Delta((h^{it}\tau_t(a))^*)(h^{it}\tau_t(b) \otimes 1))$$

for $a, b \in N_x^* N_y$ and real t, the usual core of G is actually invariant under $\Delta_{\hat{x}}^{1/2} J_{\hat{x}}$, so it is also a core for $\Delta_{\hat{x}}^{1/2} J_{\hat{x}}$, which then must coincide with G. Thus $|G| = \Delta_{\hat{x}}^{1/2}$ and $I = J_{\hat{x}}$ by uniqueness of the polar decomposition. We also record the fact that the set of all $q_x(c)$ with $c \in N_x \cap D(S^{-1})$ and $S^{-1}(a)^* \in N_x$ is a core for $\Delta_{\hat{x}}^{1/2} J_{\hat{x}}$, and that $\Delta_{\hat{x}}^{1/2} J_{\hat{x}} q_x(c) = q_x(S^{-1}(c)^*)$.

We claim that $J_{\hat{x}}$ equals U from Lemma 12.10.1. Indeed, let c belong to the core just described. Let v_n be as in the proof of Proposition 12.8.1. Then proceeding carefully we see that

$$\Delta_{\hat{x}}^{-1/2} q_x(cv_n) = q_x\rho_{i/2}(cv_n) = q(\rho_{i/2}(cv_n)h^{-1/2}) = q(\tau_{i/2}(c)v_n)$$

holds, so $\Delta_{\hat{x}}^{-1/2} q_x(c) = q\tau_{i/2}(c)$. Thus $U\Delta_{\hat{x}}^{-1/2} q_x(c) = q_x(S^{-1}(c)^*) = J_{\hat{x}}\Delta_{\hat{x}}^{-1/2} q_x(c)$, and $U = J_{\hat{x}}$.

Proposition 12.10.9 *We have* $J_{\hat{x}} J_x = \nu^{i/4} J_x J_{\hat{x}}$.

Proof From the proof of Theorem 10.8.1 we know that the modular conjugation J_y of $y = xR$ with respect to the semicyclic representation map q is $\nu^{i/4} J_x$. For $a \in N_y \cap D(\sigma_{i/2}^y)$ we therefore get

$$\begin{aligned} J_{\hat{x}} J_x q(a) &= \nu^{i/4} J_{\hat{x}} q(\sigma_{i/2}^y(a)^*) = \nu^{i/4} q_x(R(\sigma_{i/2}^y(a)^*)^*) \\ &= \nu^{i/4} q_x(\sigma_{i/2}^x(R(a)^*)^*) = \nu^{i/4} J_x q_x(R(a)^*) = \nu^{i/4} J_x J_{\hat{x}} q(a), \end{aligned}$$

where we used $J_{\hat{x}} = U$ twice and $R\sigma_t^y = \sigma_{-t}^x R$ for real t. □

For real t we clearly have $p(\tau_t^*(\omega)) = \nu^{-t/2} Q^{-2it} p(\omega)$ when $\omega \in K$. Since τ_t^* is multiplicative, there is therefore a unique continuous one-parameter group $\{\hat{\tau}_t\}$ of automorphisms on $\hat{W}$ such that $\hat{\tau}_t\lambda(\omega) = \lambda(\omega\tau_{-t})$, and we have moreover that

$\hat{\tau}_t = Q^{2it}(\cdot)Q^{-2it}$. As we evidently have $\hat{q}_x(\hat{\tau}_t(a)) = \nu^{t/2}Q^{2it}\hat{q}_x(a)$ for $a \in N_{\hat{x}}$, we get $\hat{x}\hat{\tau}_t = \nu^t\hat{x}$ for real t. Also, we see that $(\tau_t \otimes \hat{\tau}_t)(M) = M$.

Proposition 12.10.10 *The dual quantum group $(\hat{W}, \hat{\Delta})$ has unitary coinverse $\hat{R}$ and scaling group $\{\hat{\tau}_t\}$ defined as above. Hence its scaling constant is ν^{-1}. Moreover, we have $\tau_t = \Delta_{\hat{x}}^{it}(\cdot)\Delta_{\hat{x}}^{-it}$.*

Proof By the previous lemma we have $\hat{\sigma}_t = Q^{2it}J_xh^{it}J_x(\cdot)Q^{2it}J_xh^{it}J_x$ for real t. Using $M(Q^{2it} \otimes Q^{2it}) = (Q^{2it} \otimes Q^{2it})M$ and $J_xh^{it}J_x \in W'$ and $\hat{\Delta}(a) = \hat{M}^*(1 \otimes a)\hat{M}$ for $a \in \hat{W}$, we get $\hat{\Delta}\hat{\sigma}_t = (\hat{\tau}_t \otimes \hat{\sigma}_t)\hat{\Delta}$, which is the usual identity for the scaling group. The density conditions thus show that $\{\hat{\tau}_t\}$ is indeed the scaling group for $(\hat{W}, \hat{\Delta})$.

For $\omega \in D(\tau^*_{-i/2})$ we have

$$\hat{R}(\omega \otimes \iota)(M) = (\omega R \otimes \iota)(M^*) = (\tau^*_{-i/2}(\omega) \otimes \iota)(M) = \hat{\tau}_{i/2}(\omega \otimes \iota)(M),$$

where we used the definition of $\hat{S}$, and analyticity combined with the relations $(R \otimes \hat{R})(M) = M$ and $(\tau_t \otimes \hat{\tau}_t)(M) = M$ for real t.

That the scaling constant is ν^{-1} is now clear from $\hat{x}\hat{\tau}_t = \nu^t\hat{x}$, while the formula for τ_t is clear from $J_xh^{it}J_x \in W'$. □

By duality we also get $\hat{\tau}_t = \Delta_x^{it}(\cdot)\Delta_x^{-it}$ and $R = J_{\hat{x}}(\cdot)^*J_{\hat{x}}$, so $M(\Delta_{\hat{x}} \otimes \Delta_x) = (\Delta_{\hat{x}} \otimes \Delta_x)M$ and $M(J_{\hat{x}} \otimes J_x) = (J_{\hat{x}} \otimes J_x)M$ and $M(Q^2 \otimes \Delta_x) = (Q^2 \otimes \Delta_x)M$ and $M(\Delta_{\hat{x}} \otimes Q^2) = (\Delta_{\hat{x}} \otimes Q^2)M$.

Recall that q appears in the semicyclic representation of $y = xR$ using the modular element h of (W, Δ), and let $\hat{q}$ be the corresponding map used in the representation of $\hat{y} = \hat{x}\hat{R}$ involving the modular element $\hat{h}$ of $(\hat{W}, \hat{\Delta})$. Then one easily checks that $\Delta_y^{it}q_x(a) = \nu^{-t/2}q_x(\sigma_t^y(a))$ for $a \in N_x$, and $Q^{2it}q(b) = \nu^{t/2}q\tau_t(b)$ for $b \in N_y$. Using that $\{\sigma_t^x\}$, $\{\sigma_t^y\}$ and $\{\tau_t\}$ commute, that $\tau_t(h) = h$, $R(h) = h^{-1}$ and $\sigma_t^x(h^{is}) = \nu^{its}h^{is} = \sigma_t^y(h^{is})$, that $\sigma_t^x R = R\sigma_{-t}^y$, that $\nu^{i/4}J_x$ is the modular conjugation of y, and combining this with modular theory and duality and some more of the familiar formulas stated in this section, it is straightforward to prove the following result, where we for convenience have gathered some further formulas used in quantum group theory.

Proposition 12.10.11 *With notation as above, we get the relations*

$$\Delta_{\hat{x}}^{it}\Delta_x^{is} = \nu^{ist}\Delta_x^{is}\Delta_{\hat{x}}^{it},\ \Delta_{\hat{y}}^{it}\Delta_y^{is} = \nu^{ist}\Delta_y^{is}\Delta_{\hat{y}}^{it},\ \Delta_{\hat{x}}^{it}\Delta_y^{is} = \nu^{ist}\Delta_y^{is}\Delta_{\hat{x}}^{it},\ \Delta_x^{it}\Delta_y^{is} = \Delta_y^{is}\Delta_x^{it},$$

$$J_{\hat{x}}\Delta_x J_{\hat{x}} = \Delta_y,\ J_x\Delta_x J_x = \Delta_x^{-1},\ J_x\Delta_y J_x = \Delta_y^{-1},\ J_{\hat{x}}QJ_{\hat{x}} = Q^{-1},\ J_{\hat{x}}hJ_{\hat{x}} = h^{-1},$$

$$Q^{is}\Delta_x^{it} = \Delta_x^{it}Q^{is},\ Q^{is}\Delta_y^{it} = \Delta_y^{it}Q^{is},\ Q^{is}h^{it} = h^{it}Q^{is},$$

$$\Delta_x^{it}h^{is} = \nu^{ist}h^{is}\Delta_x^{it},\ \Delta_y^{it}h^{is} = \nu^{ist}h^{is}\Delta_y^{it},\ \Delta_{\hat{x}}^{it}h^{is} = h^{is}\Delta_{\hat{x}}^{it},\ \Delta_{\hat{y}}^{it}h^{is} = h^{is}\Delta_{\hat{y}}^{it}$$

for real s, t.

By duality we also obtain valid relations by adding hats, canceling double hats, replacing ν by ν^{-1}, and using $\hat{Q} = Q$.

Recall that $N(q(a) \otimes q(b)) = (q \otimes q)(\Delta(a)(1 \otimes b))$ for $a, b \in N_y$ defines a multiplicative unitary N such that $\Delta(w) = N(w \otimes 1)N^*$ for $w \in W$, and that a similar formula with $\hat{q}, \hat{\Delta}$ instead of q, Δ, defines a multiplicative unitary $\tilde{N}$ such that $\hat{\Delta}(w) = \tilde{N}(w \otimes 1)\tilde{N}^*$ when $w \in \hat{W}$. In fact, we have $N = (J_{\hat{x}} \otimes J_{\hat{x}})\hat{M}(J_{\hat{x}} \otimes J_{\hat{x}})$ since $M^* F(J_{\hat{x}} \otimes J_{\hat{x}})(q(a) \otimes q(b)) = F(J_{\hat{x}} \otimes J_{\hat{x}})N(q(a) \otimes q(b))$ as $f(R \otimes R)\Delta = \Delta R$ and $q(a) = J_{\hat{x}} q_x(R(a)^*)$ from above. Hence $N \in \hat{W}' \bar{\otimes} W$. By duality we thus get $\tilde{N} = (J_x \otimes J_x)M(J_x \otimes J_x) \in W' \bar{\otimes} \hat{W}$.

Let us say a few words on quantum groups by opposites and commutants.

Definition 12.10.12 Let (W, Δ) be a locally compact quantum group. Its *opposite* quantum group $(W, \Delta)^{\mathrm{op}}$ has the same von Neumann algebra W but the opposite coproduct $f\Delta$, where f is the flip on $W \bar{\otimes} W$, whereas its *commutant* quantum group $(W, \Delta)'$ has as von Neumann algebra W' and coproduct given by $(J_x \otimes J_x)\Delta(J_x(\cdot)J_x)(J_x \otimes J_x)$, where J_x is the modular conjugation of the left invariant faithful semifinite normal weight x of (W, Δ).

If R is the unitary coinverse of (W, Δ), then clearly xR is the left invariant faithful semifinite normal weight on $(W, \Delta)^{\mathrm{op}}$ with q in the semicyclic representation, while $x(J_x(\cdot)J_x)$ is the left invariant faithful semifinite normal weight on $(W, \Delta)'$ with $J_x q_x(J_x(\cdot)J_x)$ in the semicyclic representation. The fundamental multiplicative unitaries are then $\hat{N}$ and $\tilde{N}$, respectively, whereas the unitary coinverses are R and $J_x R(J_x(\cdot)J_x)J_x$, so the corresponding modular elements are h^{-1} and $J_x h J_x$, where h is the modular element of (W, Δ), and the scaling groups are $\{\tau_{-t}\}$ and $\{J_x \tau_t(J_x(\cdot)J_x)J_x\}$, where $\{\tau_t\}$ is the scaling group of (W, Δ).

Using $(R \bar{\otimes} R)\Delta^{\mathrm{op}} = \Delta R$ and $R = J_{\hat{x}}(\cdot)^* J_{\hat{x}}$, we see that $J_{\hat{x}} J_x(\cdot) J_x J_{\hat{x}} \colon W \to W'$ is an isomorphism between the quantum groups (W, Δ) and $(W, \Delta)'^{\mathrm{op}}$.

Proposition 12.10.13 *Let (W, Δ) be a locally compact quantum group, letting $(W, \Delta)^\wedge$ denote the dual quantum group. Then*

$$(W, \Delta)^{\mathrm{op}\wedge} = (W, \Delta)^{\wedge\prime}, \quad (W, \Delta)'^{\wedge} = (W, \Delta)^{\wedge \mathrm{op}}, \quad (W, \Delta)'^{\mathrm{op}} = (W, \Delta)^{\mathrm{op}\prime}.$$

Proof The von Neumann algebra of $(W, \Delta)^{\mathrm{op}\wedge}$ is generated by elements of the form $(\iota \otimes \omega)(N^*)$ for $\omega \in B(V_x)_*$, so it is the commutant of the von Neumann algebra of $(W, \Delta)^\wedge$. As $N = (J_{\hat{x}} \otimes J_{\hat{x}})\hat{M}(J_{\hat{x}} \otimes J_{\hat{x}})$, we moreover see that the coproducts of $(W, \Delta)^{\mathrm{op}\wedge}$ and $(W, \Delta)^{\wedge\prime}$ coincide, so they are the same quantum groups. Replacing (W, Δ) by $(W, \Delta)^\wedge$ in this identification and using duality twice, we get the second equality in the proposition. As for the final equality it suffices to observe that the modular conjugation of the left invariant weight xR of $(W, \Delta)^{\mathrm{op}}$ is $\nu^{i/4} J_x$, where ν is the scaling constant of (W, Δ). □

Note that $W \cap \hat{W} = \mathbb{C}$. This is clear from the first equality in the previous proposition, and the fact that $W \cap (\hat{W})' = \mathbb{C}$, which is immediate from the implementation of Δ using $M \in W \bar{\otimes} \hat{W}$. Similarly, we get $W' \cap \hat{W} = \mathbb{C} = W' \cap (\hat{W})'$.

Example 12.10.14 In the discussion after Theorem 12.2.5 we saw that any compact quantum group gives a *compact* locally compact quantum group (W, Δ) with a norm dense $*$-subalgebra W_0 of the C*-algebra W_r of the reduced quantum group that turns into a Hopf $*$-algebra when restricting Δ to W_0. The family $\{u^i_{kl}\}$ in the theorem can be regarded as matrix coefficients of irreducible unitary corepresentations $N_i = \sum u^i_{kl} \otimes m^i_{kl}$ of (W_r, Δ_r) in $B(V_i)$, where m^i_{kl} are the usual matrix units with respect to some chosen orthonormal basis in V_i. Since $xR = x$ the scaling constant ν is one, and the modular element h is the identity map.

The dual quantum group $(\hat{W}, \hat{\Delta})$ is a *discrete* locally compact quantum group. Recall that for the left invariant faithful semifinite normal weight $\hat{x}$ on it, we have $q_{\hat{x}}\lambda(\omega) = p(\omega)$, where $\omega(a^*) = (p(\omega)|q_x(a))$ for $\omega \in K^\sharp \subset W_*$ and $a \in N_x$. Since the identity element belongs to $W_0 \subset N_x$, we have $\hat{b} \equiv x(\cdot b) \in K^\sharp$ for $b \in W_0$, and $(p(\hat{b})|q_x(a)) = x(a^*b) = (q_x(b)|q_x(a))$, so $q_{\hat{x}}\lambda(\hat{b}) = q_x(b)$. Thinking of $b \mapsto \hat{b}$ as a generalized Fourier transform, we thus get the Plancherel isomorphism in the form $\hat{x}\lambda(\hat{a}^*\hat{b}) = x(a^*b)$ for $a, b \in W_0$. For the matrix coefficients u^i_{kl} of pairwise inequivalent irreducible unitary corepresentations $N_i = \sum u^i_{kl} \otimes m^i_{kl}$ of (W, Δ) on V_i we get

$$\hat{u}^i_{kl}\hat{u}^j_{rs} = d_i^{-1} f_{-1}(u^i_{rl})\hat{u}^i_{ks}\delta_{ij} \text{ and } (\hat{u}^i_{kl})^* = \hat{u}^i_{lk},$$

which is easily seen by evaluating each one of the identities on similar matrix coefficients and using the orthogonality relations for x. Combining these by setting $a = \hat{u}^i_{kl}$ and $b = \hat{u}^j_{rs}$ in the Plancherel isomorphism and using the orthogonality relations for x once more, we get $\hat{x}\lambda(\hat{u}^i_{ls}) = \delta_{ls}$ since $f_{-1}(u^i_{rk}) \neq 0$ for some r, k. By linearity we therefore get $\hat{x}\lambda(\hat{a}) = \varepsilon(a)$ for $a \in W_0$. By faithfulness of x the Plancherel isomorphism is injective, so we could have used this formula to define $\hat{x}\lambda$ on the $*$-subalgebra $\{\hat{a} \,|\, a \in W_0\}$ of $K^\sharp$, and then shown that one gets a faithful right invariant functional with respect to the Hopf algebraic dual of (W_0, Δ), all within a purely algebraic framework that allows an algebraic version of Pontryagin duality. The challenge then would be to extend everything to an analytic level. In the compact-discrete case this is easy since $\hat{W} = \prod B(V_i)$ due to the decomposition $M = \oplus N_i = \sum u^i_{kl} \otimes m^i_{kl}$ of the fundamental multiplicative unitary of (W, Δ). We will actually find a formula for $\hat{x}$ that respects this decomposition and immediately yields the analytic extension. First notice that $\sigma^{\hat{x}}_t = \hat{\tau}_t$ and that $\hat{\tau}_t\lambda(\omega) = \lambda(\omega\tau_{-t})$ for $\omega \in K^\sharp$. As $\tau_{-t} = (f_{-it} \otimes \iota \otimes f_{it})(\Delta \otimes \iota)\Delta$, we therefore get $\hat{\tau}_t = \lambda(f_{-it})(\cdot)\lambda(f_{it})$ for real t, so $\hat{S}^2 = \lambda(f_{-1})(\cdot)\lambda(f_1)$ and $\hat{x}\lambda(\hat{a}\hat{b}) = \hat{x}\lambda(\hat{b} f_{-1}\hat{a} f_1)$ for $a, b \in W_0$, which could also have been verified algebraically using the counit formula for $\hat{x}\lambda$. The spoke of formula is $\hat{x}\lambda(\hat{a}) = \oplus_i d_i \mathrm{Tr}_i \lambda(p_i \hat{a} f_{-1})$ for $a \in W_0$, where Tr_i is the non-normalized trace on the matrix algebra $B(V_i)$, and where $p_i = d_i \sum f_1(u^i_{lk})\hat{u}^i_{kl}$ has the property that $\lambda(p_i)$ is the identity element in $B(V_i)$. The latter statement is clear form the fact that p_i commutes with all $\hat{u}^i_{kl}$, which is again clear from the identities above for these elements. The functional κ on $\{\hat{a} \,|\, a \in W_0\}$ given by $\kappa = d_i^{-1}\hat{x}\lambda(p_i(\cdot)f_1)$ is clearly tracial, and as $p_i f_1 = d_i \sum \hat{u}^i_{kk}$, which is checked using the orthogonality relations for x, we see that $\kappa(p_i) = \dim(V_i)$. By uniqueness of the trace on $B(V_i)$ we conclude that $\kappa = \mathrm{Tr}_i \lambda$, so the claimed formula holds. We also note that thanks to the identities for $\hat{u}^i_{kl}$ above the $*$-algebra $\{\lambda(\hat{a}) \mid a \in W_0\}$ is the algebraic direct sum $\oplus B(V_i)$. $\diamond$

Exercises

For Sect. 12.1

1. Prove that the comultiplication of a Hopf algebra is injective, and that the coinverse composed with the counit is just the counit.
2. Prove that the coinverse of a Hopf algebra is involutive if the Hopf algebra is commutative or cocommutative.
3. Let (W, Δ) be a bialgebra with product $m: W \otimes W \to W$. Show that the vector space of linear maps $W \to W$ is an algebra under the product $f * g = m(f \otimes g)\Delta$. Prove that it is a unital algebra such that the identity map on W has a bijective inverse for $*$ if and only if (W, Δ) is a Hopf algebra.
4. Given Hopf algebras (W_i, Δ_i), show that the tensor product algebra $W_1 \otimes W_2$ is a Hopf algebra with coproduct $(\iota \otimes \sigma \otimes \iota)(\Delta_1 \otimes \Delta_2)$, where $\sigma: W_1 \otimes W_2 \to W_2 \otimes W_1$ is the flip map.
5. Produce a Hopf algebra that is neither commutative nor cocommutative and with an involutive coinverse.
6. Given a bialgebra (W, Δ) with W a finite dimensional C*-algebra and Δ a $*$-homomorphism. Suppose we have a faithful positive linear functional x on W such that $(x \otimes \iota)\Delta = x(\cdot)1 = (\iota \otimes x)\Delta$. Prove that (W, Δ) is a Hopf $*$-algebra.

For Sect. 12.2

1. Prove that the Haar state on a compact quantum group is faithful on the dense Hopf $*$-algebra spanned by the matrix elements of a maximal family of pairwise inequivalent finite dimensional irreducible unitary right corepresentations.
2. Prove that there is only one dense Hopf $*$-subalgebra of a compact quantum group with coproduct given by restriction.
3. Show that a compact quantum group is finite dimensional if and only if it is a Hopf $*$-algebra for the coproduct.
4. Prove that the quantum dimension is greater than the classical dimension.
5. Show that the quantum dimension can take any real number in $\{0, 1\} \cup [2, \infty\rangle$.
6. Describe the classical group underlying a compact quantum group with a dense Hopf $*$-algebra generated by one finite dimensional unitary right corepresentation.
7. We say a vector space V is a right comodule of a Hopf $*$-algebra (M, Δ) with counit ε if we have a linear map $f: V \to V \otimes M$ such that $(f \otimes \iota)f = (\iota \otimes \Delta)f$ and $(\iota \otimes \varepsilon)f = \iota$. Given a finite dimensional right corepresentation N of (M, Δ) on V, in that $(\iota \otimes \Delta)(N) = N_{12}N_{13}$ and $(\iota \otimes \varepsilon)(N) = 1$, show that V is a finite

dimensional right comodule with $f(v) = N(v \otimes 1)$ of (M, Δ), and that this gives a 1-1 correspondence between such corepresentations and comodules.
8. Given a Hopf $*$-algebra generated as an algebra by the matrix coefficients of (the right corepresentations associated to) its finite dimensional right comodules. Show that it can be completed in an obvious fashion to a compact quantum group.

For Sect. 12.3

1. Let $M \in B(V)\bar{\otimes}B(V)$ be unitary, and let $\Delta(a) = M^*(1 \otimes a)M$ for $a \in B(V)$. Show that M is a multiplicative unitary if and only if $(\Delta \otimes \iota)(M) = M_{13}M_{23}$. Provided this holds, can you get $(\iota \otimes \hat{\Delta})(M) = M_{12}M_{13}$ for some injective normal $*$-homomorphism $\hat{\Delta}\colon B(V) \to B(V)\bar{\otimes}B(V)$?
2. Describe the fundamental multiplicative unitary of the locally compact quantum group $(L^\infty(G), \Delta)$ with $\Delta(f)(s,t) = f(st)$, where G is a locally compact group, and where the weights are integration with respect to left and right invariant Haar measures.
3. Exhibit a non-zero element $M \in B(V)\bar{\otimes}B(V)$ such that $M_{12}M_{13}M_{23} = M_{23}M_{12}$ and which is not unitary.

For Sect. 12.4

1. Simplify the proof of the main results in this section when the locally compact quantum group is the von Neumann completion of a compact quantum group in the GNS-representation of the Haar state.
2. Describe the fundamental involution for the locally compact quantum group $(L^\infty(G), \Delta)$ with $\Delta(f)(s,t) = f(st)$, where G is a locally compact group, and where the weights are integration with respect to left and right invariant Haar measures.

For Sect. 12.5

1. Let (W, Δ) be a compact quantum group. Prove norm density of the span of $(\iota \otimes \omega)\Delta(W)$ and of the span of $(\omega \otimes \iota)\Delta(W)$ in W as ω runs through the bounded functionals on W.
2. Provide a multiplicative unitary M on V such that the σ-strong*-closed span W of $\{(\iota \otimes \omega)(M) \mid \omega \in W_*\}$ does not coincide with the σ-strong*-closed span of $\{(\iota \otimes \omega)(M^*(\mathbb{C} \otimes W)M) \mid \omega \in W_*\}$.

For Sect. 12.6

1. In a Hopf $*$-algebra (W, Δ) with $w \in W$, find finitely many $p_i, q_i \in W$ such that $w \otimes 1 = \sum \Delta(p_i)(1 \otimes q_i)$. Then show that $S(w) \otimes 1 = \sum (1 \otimes p_i)\Delta(q_i)$, where S is the coinverse.
2. Given locally compact quantum groups (W_i, Δ_i) with coinverses S_i, unitary coinverses R_i and scaling groups $\{\tau_t^i\}$. Prove that for a $*$-isomorphism $f\colon W_1 \to W_2$ such that $\Delta_2 f = (f \otimes f)\Delta_1$, we have $S_2 f = f S_1$ and $R_2 f = f R_1$ and $\tau_t^2 f = f \tau_t^1$ for real t.

3. Given a locally compact quantum group (W, Δ) with unitary coinverse R, show that the invertible elements $w \in W$ with $\Delta(w) = w \otimes w$ form a subgroup of the group of invertible elements in W that consists only of unitaries, and such that $R(w) = w^*$.
4. Let (W, Δ) be a locally compact quantum group with unitary coinverse R. Given an element $w = R(w)$ in the center of W such that $\Delta(w) \geq w \otimes w$, show that $\Delta(w)(w \otimes 1) = w \otimes w = \Delta(w)(1 \otimes w)$.

For Sect. 12.7

1. Suppose we have a locally compact quantum group (W, Δ) and an orthogonal projection $p \in W$. Show that $p \in \{0, 1\}$ if $\Delta(p) \leq p \otimes 1$ or $\Delta(p) \leq 1 \otimes p$.
2. Find the scaling constant of a Kac algebra.

For Sect. 12.8

1. Compare the usual proof of uniqueness of the Haar integral with the proof of the uniqueness of the Haar weight when the locally compact quantum group is of the form $(L^\infty(G), \Delta)$ with $\Delta(f)(s,t) = f(st)$ for a locally compact group G. Is the modular element trivial in this case?
2. Suppose (W, Δ) is a locally compact quantum group with modular element h. If f is the flip on $W \bar{\otimes} W$, then $(W, f\Delta)$ is again a locally compact quantum group. Express its modular element in terms of h.

For Sect. 12.9

1. Show that the fundamental multiplicative unitary of a Kac algebra is manageable with associated operator $Q = 1$.

For Sect. 12.10

1. Show that if f is an isomorphism between locally compact quantum groups (W_i, Δ_i), in that $f : W_1 \to W_2$ is a $*$-isomorphism satisfying $\Delta_2 f = (f \bar{\otimes} f)\Delta_1$, then the corresponding dual quantum groups are isomorphic.
2. Prove that the dual of a Kac algebra is a Kac algebra.
3. We say that a Kac algebra is unimodular of the left invariant weight is tracial. Show that the dual Kac algebra is then also unimodular. Is this true for locally compact quantum groups? If G is a locally compact group, show that the dual of the locally compact quantum group $(L^\infty(G), \Delta)$ with $\Delta(f)(s,t) = f(st)$ is unimodular if and only if G is unimodular.
4. Consider the locally compact quantum group $(L^\infty(G), \Delta)$ with $\Delta(f)(s,t) = f(st)$ for a locally compact group G. Show that $\hat{\Delta}(a) = a \otimes a$ if and only if $a \in \lambda(G)$.

Chapter 13
Special Cases

Let (W, Δ) be a locally compact quantum group with coinverse S and fundamental multiplicative unitary M. The predual W_* of W is a Banach algebra with product $\omega\omega' = (\omega \otimes \omega')\Delta$. Restricting to where the operation $\omega \mapsto \omega^* \equiv \overline{\omega(S(\cdot)^*)}$ makes sense, we obtain a Banach $*$-algebra with norm $\|\omega\|_* = \sup\{\|\omega\|, \|\omega^*\|\}$ and universal C*-completion $\hat{W}_u$. Non-degenerate representations of this $\hat{W}_u$ correspond to unitary corepresentations of (W, Δ). Letting $\hat{\pi} : \hat{W}_u \to \hat{W}_r$ be the $*$-epimorphism given by lifting λ, we get a corepresentation $\hat{\mathcal{M}}$ such that $(\iota \otimes \hat{\pi})(\hat{\mathcal{M}}) = M$. Repeating this for the dual quantum group, and using $\hat{\mathcal{M}}$, we get a C*-algebra W_u with coproduct Δ_u satisfying cancellation properties and with a bounded counit ε_u. We also obtain the so called universal corepresentation $M_u \equiv \mathcal{M}_{12}^* \hat{\mathcal{M}}_{23} \mathcal{M}_{12} \hat{\mathcal{M}}_{23}^* \in M(W_u \otimes \hat{W}_u)$ such that $(\pi \otimes \hat{\pi})(M_u) = M$. Essentially all the quantum group entities we have produced so far lift to the universal level. We have arrived at the universal quantum group and its dual. They are not really quantum groups since they are living in the C*-algebra world, and the invariant weights are in general not faithful because π nor $\hat{\pi}$ need not be injective. They are however convenient vehicles that keep track of representations and corepresentations.

In the case when (W, Δ) is cocommutative, then W_u is the universal group C*-algebra $C_u^*(G)$ for a locally compact group G, whereas $\hat{W}_u = \hat{W}_r = C_0(G)$. We discuss all aspects of the commutative and cocommutative cases in section two. We know that $C_u^*(G)$ equals the reduced group C*-algebra $C_r^*(G)$ exactly when G is amenable, that is, has an invariant mean. We discuss this in the general context of locally compact quantum groups, and refer to amenability when we have an invariant mean, while the fact that the universal and reduced quantum groups coincide, is referred to as coamenability, succinctly expressed using the counit. Weak containment is also considered. We show that a locally compact quantum group is amenable if its dual is coamenable. The converse direction is only known in the setting of discrete and compact quantum groups. This also holds when we talk

L. Tuset, *Analysis and Quantum Groups*,
https://doi.org/10.1007/978-3-031-07246-8_13

of amenability and coamenability of corepresentations. We also discuss the relation to injectivity and nuclearity.

Recommended literature for this chapter is [3, 5, 29, 37].

13.1 The Universal Quantum Group

Let (W, Δ) be a locally compact quantum group with fundamental multiplicative unitary M and coinverse S. Consider $W^\sharp$ as a Banach $*$-algebra under the norm given by $\|\omega\|_* = \sup\{\|\omega\|, \|\omega^*\|\}$. Since λ is an injective $*$-representation of $W^\sharp$, the supremum $\|\omega\|_u$ of $\|f(\omega)\|$ over all $*$-representations f of $W^\sharp$ defines a C*-norm on $W^\sharp$ with C*-completion $\hat{W}_u \subset B(V_u)$ and embedding $\lambda_u \colon W^\sharp \to \hat{W}_u$. Clearly, if $w \in D(S)$, then $\omega(w\cdot), \omega(\cdot w) \in W^\sharp$ with $\omega(w\cdot)^* = \omega^*(S(w)^*\cdot)$ and $\omega(\cdot w)^* = \omega^*(\cdot S(w)^*)$, so $\|\omega(w\cdot)\|_*$ and $\|\omega(\cdot w)\|_*$ are not greater than $\|\omega\|_* \sup\{\|w\|, \|S(w)\|\}$. Set $\lambda^*(z) = (z \otimes \iota)(M)$ for $z \in B(V_x)_*$, so $z \mapsto \omega(\lambda^*(z)\cdot)^* = \omega(\lambda^*(\bar{z})\cdot)$ is continuous.

Proposition 13.1.1 *With notation as above, there is a unique $*$-representation f of $W^\sharp$ on $V_x \otimes V_u$ such that $(f(\omega)(v_1 \otimes w_1)|v_2 \otimes w_2) = (\lambda_u(\omega(\lambda^*(\omega_{v_1,v_2})\cdot))w_1|w_2)$ for all $v_i \in V_x$ and $w_i \in V_u$, called the* Kronecker product *of λ and λ_u.*

Proof Let $\{e_i\}$ be an orthonormal basis for V_x, and set $a_{kl} = \lambda^*(\omega_{e_l,e_k})$. Thanks to the pentagonal equation we get $\sum(\omega^*(a_{l'k}\cdot)(\omega(a_{kl}\cdot) = (\omega^*\omega)(a_{l'l}\cdot)$ with convergence in $W^\sharp$, so $\sum \| \sum_{l\in L} \lambda(\omega(a_{kl}\cdot)w_l\|^2 = \sum_{l,l'\in L}(\lambda_u((\omega^*\omega)(a_{l'l}\cdot))w_l|w_{l'})$ for finite L. Hence $g(\omega) = \sum_{l,l'\in L}(\lambda_u(\omega(a_{l,l'}\cdot))w_l|w_{l'})$ defines a linear functional g on $W^\sharp$ that is positive and satisfies $|g(\omega)|^2 \leq (\sum_{l\in L}\|w_l\|^2)g(\omega^*\omega)$ by the Cauchy-Schwarz inequality. By the lemma below g is continuous with norm bounded by $\sum_{l\in L}\|w_l\|^2$. This shows that there is a contractive linear map $f \colon W^\sharp \to B(V_x \otimes V_u)$ satisfying $(f(\omega)(e_l \otimes w_1)|e_k \otimes w_2) = (\lambda_u(\omega(\lambda^*(\omega_{e_l,e_k})\cdot))w_1|w_2)$ and hence also the identity in this proposition. It is now straightforward to check that f is $*$-preserving, and then multiplicativity follows from polarization. □

Lemma 13.1.2 *If a non-zero positive linear functional g on a Banach $*$-algebra satisfies $|g(a)|^2 \leq cg(a^*a)$ for some $c > 0$ and all a, it is continuous with $\|g\| \leq c$.*

Proof The obvious extension of g to the Banach $*$-algebra with a unit added satisfies $g((a+k)^*(a+k)) \geq (g(a^*a)^{1/2} - c^{1/2}|k|)^2$ for $a \in A$ and $k \in \mathbb{C}$, so g is positive. We claim that every positive linear functional g on a unital Banach $*$-algebra is continuous with $\|g\| = g(1)$, and this now settles the lemma. To prove the claim note that if $a^* = a$ has norm less than one, then the spectrum of $1 - a$ is contained in the open unit disc around 1. Let h be the analytic extension to this disc of $\sqrt{\cdot}$ on the open unit interval. Then $b \equiv h(1-a)$ is self-adjoint as h has a power series around 1 with real coefficients, and we have $b^2 = 1 - a$ by holomorphic functional calculus. Hence $g(1) - g(a) = g(b^2) \geq 0$, so $|g(a)|^2 \leq g(1)g(a^*a) \leq g(1)^2$ by the Cauchy-Schwarz inequality and the fact that $\|a^*a\| \leq \|a\|^2 \leq 1$. □

Let $\{\tau_t\}$ be the scaling group of (W, Δ). Then $\omega \mapsto \omega\tau_t$ is a $*$-automorphism of $W^\sharp$, so by universality there is a unique norm continuous one-parameter group of $*$-automorphisms $\hat{\tau}_t^u$ on $\hat{W}_u$ such that $\hat{\tau}_t^u(\lambda_u(\omega)) = \lambda_u(\omega\tau_t)$ for $\omega \in W^\sharp$ and real t.

Lemma 13.1.3 *For $z \in \hat{W}_u$ and $n \in \mathbb{N}$, define a bounded linear functional on $\hat{W}_u$ by $z_n = n(\pi)^{1/2} \int e^{-n^2t^2} z\hat{\tau}_t^u \, dt$. Then there is $w \in W$ such that $z_n\lambda_u(\omega) = \omega(w)$ for $\omega \in W^\sharp$. The collection of functionals with this latter property is therefore separating for $\hat{W}_u$.*

Proof Now $h(\omega) = n(\pi)^{-1/2} \int e^{-n^2t^2} \omega\tau_{-t} \, dt$ defines a continuous function h on $W^\sharp$ as $h(\omega)^* = n(\pi)^{-1/2} \int e^{-n^2(t-i/2)^2} \bar{\omega}R\tau_{-t} \, dt$, but then we are done since $z_n\lambda_u = z\lambda_u h$ extends to a bounded functional on the predual of W. □

Lemma 13.1.4 *By universality there are $*$-homomorphisms $s \colon \hat{W}_u \to \hat{W}$ and $s' \colon \hat{W}_u \to B(V_x \otimes V_u)$ such that $s\lambda_u = \lambda$ and $s'\lambda_u = f$ with f from the proposition above. Then* $\ker s \subset \ker s'$. *Let P be the orthogonal projection onto the closed span of $f(W^\sharp)(V_x \otimes V_u)$. Then there is a unique $U \in M(W_r \otimes B_0(V_x \otimes V_u))$ such that $U^*U = UU^* = 1 \otimes P$ and $f(\omega) = (\omega \otimes \iota)(U)$ for $\omega \in W^\sharp$.*

Proof Suppose $s(r) = 0$. Let $a, b \in N_x$ and $c \in N_{xR}$, and let $v = J_x q_x(c^*a)$ and $w = J_x q_x(b)$, and note that $\lambda^*(\omega_{v,w}) = R(\iota \otimes x)(\Delta(a^*c)(1 \otimes b)) \in N_x^*$. Pick $z \in (\hat{W}_u)^\star$ with $d \in W$ such that $z\lambda_u(\omega) = \omega(d)$ for $\omega \in W^\sharp$. Then for $\eta \in K \cap W^\sharp$ with K as in the previous section, the proposition above yields

$$z(\omega_{v,w} \otimes \iota)(s'(\lambda_u(\omega)) f(\eta)) = (\omega\eta)(\lambda^*(\omega_{v,w})d) = (s(\lambda_u(\omega))p(\eta)|d^* q_x(\lambda^*(\omega_{v,w})^*))$$

with p as in the previous section. Replacing $\lambda_u(\omega)$ by r in this identity, we get $(\omega_{v,w} \otimes \iota)(s'(r)f(\eta)) = 0$ by the previous lemma. Hence $s'(r) = 0$.

We can now define a unique $*$-homomorphism $g \colon \hat{W}_r \to B(V_x \otimes V_u)$ such that $gs = s'$. Then $U = (\iota \otimes g)(M)$ does the job. □

Theorem 13.1.5 *Let $W, \Delta)$ be a locally compact quantum group. Then there is a unique element $\hat{\mathcal{M}} \in M(W_r \otimes B_0(V_u))$ such that $\lambda_u(\omega) = (\omega \otimes \iota)(\hat{\mathcal{M}})$ for $\omega \in W^\sharp$, and also $\hat{\mathcal{M}}$ is a unitary element of $M(W_r \otimes \hat{W}_u)$ satisfying $(\Delta \otimes \iota)(\hat{\mathcal{M}}) = \hat{\mathcal{M}}_{13}\hat{\mathcal{M}}_{23}$.*

Proof Pick z and d as in the proof above. With U as above, the previous proposition yields $z(\omega \otimes \eta \otimes \iota)(U) = (\omega \otimes \eta)(M(d \otimes 1))$ for $\omega \in W^\sharp$ and $\eta \in B(V_x)_*$. Replacing $\omega \otimes \eta$ by $(\omega \otimes \eta)(M^* \cdot)$ gives $(\omega \otimes \iota \otimes \iota)(M_{12}^* U) = 1 \otimes \lambda_u(\omega)$ by Lemma 13.1.3. Hence $M_{12}^* U \in (\mathbb{C} \otimes B(V_x) \otimes \mathbb{C})'$, so there is $\hat{\mathcal{M}} \in M(W_r \otimes B_0(V_u))$ such that $\hat{\mathcal{M}}_{13} = M_{12}^* U$. Thus $\lambda_u(\omega) = (\omega \otimes \iota)(\hat{\mathcal{M}})$ and $\hat{\mathcal{M}}^*\hat{\mathcal{M}} = 1 \otimes P' = \hat{\mathcal{M}}\hat{\mathcal{M}}^*$ for an orthogonal projection $P' \in B(V_u)$ such that $1 \otimes P' = P$. If $P'v = 0$ for $v \in V_u$, then $\hat{\mathcal{M}}(w \otimes v) = 0$ for $w \in V_x$, so $\lambda_u(\omega)v = 0$ for $\omega \in W^\sharp$, and $v = 0$, which shows that $P' = 1$. Multiplicativity of λ shows that $(\Delta \otimes \iota)(\hat{\mathcal{M}}) = \hat{\mathcal{M}}_{13}\hat{\mathcal{M}}_{23}$, so we are done by the proof of Theorem 12.9.3. □

Recall that to each non-degenerate $*$-homomorphims $g \colon W^\sharp \to M(B)$ for a C*-algebra B, there is a unique non-degenerate $*$-homomorphims $f \colon \hat{W}_u \to M(B)$

such that $f\lambda_u = g$, and vice-versa. A *unitary corepresentation* of (W, Δ) *in B* is a unitary element $N \in M(W_r \otimes B)$ such that $(\Delta \otimes \iota)(N) = N_{13}N_{23}$.

Corollary 13.1.6 *Given a C*-algebra B, then $f \mapsto (\iota \otimes f)(\hat{\mathcal{M}})$ is a bijection from the set of non-degenerate ∗-homomorphims $f : \hat{W}_u \to M(B)$ to the set of unitary corepresentations of (W, Δ) in B.*

Proof The map is clearly injective. As for surjectivity, given a unitary corepresentation N of (W, Δ) in B, consider the multiplicative linear map $g : W^\sharp \to M(B)$ given by $g(\omega) = (\omega \otimes \iota)(N)$. Let $\xi(a) = (z \otimes \eta)(N(a \otimes 1)N^*)$ for $a \in B_0(V_x)$ and $z \in B_0(V_x)^\star$ and $\eta \in B^\star$. Then

$$S((\iota\otimes z)(M)(\iota\otimes\eta)(N)) = S(\iota\otimes\xi)(M) = (\iota\otimes\xi)(M^*) = (\iota\otimes\eta)(N^*)S(\iota\otimes z)(M),$$

so $S(\iota \otimes \eta)(N) = (\iota \otimes \eta)(N^*)$, and $g(\omega^*) = g(\omega)^*$ as is now seen by applying η to both sides of this equality. But g is also non-degenerate since the closed span of elements of the form $(\omega\otimes\iota)(N(1\otimes b))$ with $b \in B$, or of the form $(\omega\otimes\iota)(N(w\otimes b)$ with $w \in W_r$, or of the form $(\omega \otimes \iota)(w \otimes b)$, all coincide with B. Picking f such that $f\lambda_u = g$ then gives $(\iota \otimes f)(\hat{\mathcal{M}}) = N$. □

Let $\hat{\pi} : \hat{W}_u \to \hat{W}_r$ be the surjective ∗-homomorphism such that $\hat{\pi}\lambda_u = \lambda$, so we clearly have $(\iota \otimes \hat{\pi})(\hat{\mathcal{M}}) = M$.

We can of course repeat all this for the dual of (W, Δ), and taking into account the generalized Pontryagin duality, we denote by $\pi : W_u \to W_r$ the surjective ∗-homomorphism such that $\pi\hat{\lambda}_u = \hat{\lambda}$, where $\hat{\lambda}(\omega) = (\iota \otimes \omega)(M^*)$ for $\omega \in (\hat{W})_*$. We then have a unique unitary element $\mathcal{M} \in M(W_u \otimes \hat{W}_r)$ such that $\hat{\lambda}_u(\omega) = (\iota \otimes \omega)(\mathcal{M}^*)$ for $\omega \in (\hat{W})^\sharp$, which moreover satisfies $(\iota \otimes \hat{\Delta})(\mathcal{M}) = \mathcal{M}_{13}\mathcal{M}_{12}$ and $(\pi \otimes \iota)(\mathcal{M}) = M$. In this case $f \mapsto (f \otimes \iota)(\mathcal{M})$ will be a bijection from the set of non-degenerate ∗-homomorphisms $W_u \to M(B)$ to the set of unitaries $N \in M(B \otimes \hat{W}_r)$ such that $(\iota \otimes \hat{\Delta})(N) = N_{13}N_{12}$. Since $\hat{\lambda}$ restricts to a ∗-isomorphism from $(\hat{W})^\sharp$ onto $\hat{\lambda}((\hat{W})^\sharp) \cap \hat{\lambda}((\hat{W})^\sharp)^*$, we see that W_u is the universal enveloping C*-algebra of the ∗-algebra $\mathcal{W} \cap \mathcal{W}^*$, where $\mathcal{W}$ is the algebra consisting of the elements $(\iota \otimes z)(M)$ with $z \in B_0(V_x)^\star$. Similarly $\hat{W}_u$ is the universal enveloping C*-algebra of the ∗-algebra $\hat{\mathcal{W}} \cap \hat{\mathcal{W}}^*$, where $\hat{\mathcal{W}}$ consists of the elements of the type $(z \otimes \iota)(M)$.

We next extend the various quantum group entities to the universal level.

Proposition 13.1.7 *To any locally compact quantum group (W, Δ) we have a unique non-degenerate ∗-homomorphism $\Delta_u : W_u \to M(W_u \otimes W_u)$ such that $(\Delta_u\otimes\iota)(\mathcal{M}) = \mathcal{M}_{13}\mathcal{M}_{23}$. The closed spans of $\Delta_u(W_u)(W_u\otimes\mathbb{C})$ and $\Delta_u(W_u)(\mathbb{C}\otimes W_u)$ are both $W_u \otimes W_u$, and $(\Delta_u \otimes \iota)\Delta_u = (\iota \otimes \Delta_u)\Delta_u$. There is a unique ∗-homomorphism $\varepsilon_u : W_u \to \mathbb{C}$ such that $(\varepsilon_u \otimes \iota)(\mathcal{M}) = 1$, and $(\varepsilon_u \otimes \iota)\Delta_u = \iota = (\iota\otimes\varepsilon_u)\Delta_u$. We also have $(\pi\otimes\pi)\Delta_u = \Delta\pi$ and $(\iota\otimes\pi)\Delta_u(w) = \mathcal{M}^*(1\otimes\pi(w))\mathcal{M}$ for $w \in W_u$.*

Proof Existence of such Δ_u and ε_u is immediate from the bijective correspondence described in the paragraph just before the proposition since $\mathcal{M}_{13}\mathcal{M}_{23}$ and $1 \otimes 1$ satisfy the required algebraic relations. The coassociativity relations and the counit relations are easily checked on the dense set of elements of the form $(\iota \otimes z)(\mathcal{M})$ with $z \in B_0(V_x)^\star$. The closed span of $\Delta_u(W_u)(\mathbb{C} \otimes W_u)$ is the closed span of the elements of the form $\Delta_u(\iota \otimes z)(\mathcal{M})(1 \otimes a)$ with $a \in W_u$, or of the form $(\iota \otimes \iota \otimes z)(\mathcal{M}_{13}\mathcal{M}_{23}(1 \otimes a \otimes c))$ with $c \in B_0(V_x)$, or of the form $(\iota \otimes z)(\mathcal{M}) \otimes a)$, which is $W_u \otimes W_u$. The proof of the other density condition is similar. Using $(\pi \otimes \iota)(\mathcal{M}) = M$ and the easily verified identity $(\iota \otimes \pi)\Delta_u(\iota \otimes z)(\mathcal{M}) = \mathcal{M}^*(1 \otimes \pi(\iota \otimes z)(\mathcal{M}))\mathcal{M}$, the final two identities in the proposition are clear. □

We get a similar result on the dual side with $(\hat{\pi} \otimes \iota) f \hat{\Delta}_u(w) = \hat{\mathcal{M}}(\hat{\pi}(w) \otimes 1)\hat{\mathcal{M}}^*$ for $w \in \hat{W}_u$, where f is the flip on $\hat{W}_u \otimes \hat{W}_u$ extended to the multiplier algebra. Note also that Δ_u and $\hat{\Delta}_u$ are always injective.

Proposition 13.1.8 *We call the unitary element* $M_u = \mathcal{M}_{12}^*\hat{\mathcal{M}}_{23}\mathcal{M}_{12}\hat{\mathcal{M}}_{23}^*$ *in* $M(W_u \otimes \mathbb{C} \otimes \hat{W}_u) \cong M(W_u \otimes \hat{W}_u)$ *the* universal corepresentation *of* (W_u, Δ_u). *It satisfies* $(\Delta_u \otimes \iota)(M_u) = (M_u)_{13}(M_u)_{23}$ *and* $(\iota \otimes \hat{\Delta}_u)(M_u) = (M_u)_{13}(M_u)_{12}$, *and* $(\iota \otimes \hat{\pi})(M_u) = \mathcal{M}$ *and* $(\pi \otimes \iota)(M_u) = \hat{\mathcal{M}}$ *and* $(\pi \otimes \hat{\pi})(M_u) = M$.

Proof Clearly $M_u \in M(W_u \otimes W_r \otimes \hat{W}_u) \cap M(W_u \otimes \hat{W}_r \otimes \hat{W}_u) \cong M(W_u \otimes \hat{W}_u)$ as $W_r \cap \hat{W}_r = \mathbb{C}$. The element $\mathcal{M}_{12}(M_u)_{13} = \hat{\mathcal{M}}_{23}\mathcal{M}_{12}\hat{\mathcal{M}}_{23}^*$ satisfies

$$(\Delta_u \otimes \iota \otimes \iota)(\mathcal{M}_{12}(M_u)_{13}) = (\Delta_u \otimes \iota \otimes \iota)(\mathcal{M}_{12})(M_u)_{14}(M_u)_{24},$$

so $(\Delta_u \otimes \iota)(M_u) = (M_u)_{13}(M_u)_{23}$, and similarly $(\iota \otimes \hat{\Delta}_u)(M_u) = (M_u)_{13}(M_u)_{12}$. Finally, we have

$$(\iota \otimes \iota \otimes \hat{\pi})((M_u)_{13}) = \mathcal{M}_{12}^* M_{23}\mathcal{M}_{12}M_{23}^* = \mathcal{M}_{12}^*(\iota \otimes f\hat{\Delta})(\mathcal{M}) = \mathcal{M}_{13},$$

and similarly $(\pi \otimes \iota)(M_u) = \hat{\mathcal{M}}$, so $(\pi \otimes \hat{\pi})(M_u) = M$. □

Hence W_u is the closed span of the elements of the form $(\iota \otimes \omega)(M_u)$ with $\omega \in \hat{W}_u^\star$, while $\hat{W}_u$ is the closed span of the elements of the form $(\omega \otimes \iota)(M_u)$, where now ω ranges over $W_u^\star$.

As before $f \mapsto (\iota \otimes f)(M_u)$ is a bijection from the set of non-degenerate $*$-homomorphisms $f\colon \hat{W}_u \to M(B)$ to the set of unitary corepresentations of (W_u, Δ_u) in a C*-algebra B, where for the latter case one replaces W_r by W_u in the definition of a corepresentation. The following result shows that (W, Δ) and (W_u, Δ_u) have the same corepresentation theory, and this goes now without a proof.

Corollary 13.1.9 *The map which associates to every unitary corepresentation N of (W_u, Δ_u) in a C*-algebra B the unitary corepresentation $(\pi \otimes \iota)(N)$ of (W, Δ) in B is a bijection between the respective sets.*

Proposition 13.1.10 *There is a unique $*$-antiautomorphism R_u on W_u such that $(R_u \otimes \hat{R})(\mathcal{M}) = \mathcal{M}$. Also $R_u^2 = \iota$ and $\pi R_u = R\pi$ and $f(R_u \otimes R_u)\Delta = \Delta R_u$, where f is the flip on $W_u \otimes W_u$.*

Proof Let g be the obvious $*$-antiautomorphism from W_u to its opposite C*-algebra W_u^o. Then $(g \otimes \hat{R})(\mathcal{M}) \in M(W_u^o \otimes \hat{W}_r)$ is a unitary and

$$(\iota\otimes\hat{\Delta})(g\otimes\hat{R})(\mathcal{M}) = (\iota\otimes f)(g\otimes\hat{R}\otimes\hat{R})(\mathcal{M}_{13}\mathcal{M}_{12}) = (g\otimes\hat{R})(\mathcal{M})_{13}(g\otimes\hat{R})(\mathcal{M})_{12},$$

so there is a non-degenerate $*$-homomorphism $h\colon W_u \to M(W_u^o)$ which satisfies $(h \otimes \iota)(\mathcal{M}) = (g \otimes \hat{R})(\mathcal{M})$. Then $R_u = g^{-1}h$ uniquely satisfies the first part of the proposition, so $R_u(W_u) = W_u$ and $(R_u^2 \otimes \iota)(\mathcal{M}) = \mathcal{M}$ and R_u is involutive. Clearly $f(R_u \otimes R_u)\Delta = \Delta R_u$, and $(\pi R_u \otimes \hat{R})(\mathcal{M}) = (\pi \otimes \iota)(\mathcal{M}) = M = (R \otimes \hat{R})(M) = (R\pi \otimes \hat{R})(\mathcal{M})$, so $\pi R_u = R\pi$. □

Suppose (W, Δ) is a locally compact quantum group with left- and right invariant faithful semifinite normal weights x and y. Now $W_u^\star$ is a Banach algebra under $\omega\omega' = (\omega \otimes \omega')\Delta_u$ having ε_u as an identity element, and $\pi^*\colon W_r^\star \to W_u^\star$ given by $\pi^*(\eta) = \eta\pi$ is an isometric homomorphism. And $\pi^*(W_r^\star)$ is a two-sided ideal of $W_u^\star$ since $\omega\pi^*(\eta) = (\omega \otimes \eta)(\mathcal{M}^*(1 \otimes \pi(\cdot))\mathcal{M})$ and $f(R_u \otimes R_u)\Delta_u = \Delta_u R_u$.

Since $\pi(\iota \otimes \omega_{q_x(a),q_x(c^*b)})(\mathcal{M}) = (\iota \otimes x)(\Delta(b^*c)(1 \otimes a))$ for $a, b \in N_x$ and $c \in N_y$, we see that $y_u = y\pi$ is a non-zero lower semicontinuous weight on (W_u, Δ_u) with GNS-representation π having $q_u = q\pi$. While $x_u = x\pi$ is a non-zero lower semicontinuous weight on (W_u, Δ_u) with GNS-representation π having $q_{xu} = q_x\pi$. Note that $x_u = y_u R_u$. Thanks to $(\pi \otimes \pi)\Delta_u = \Delta\pi$ and left invariance of x, we get $x_u(\pi^*(\eta) \otimes \iota)\Delta_u = \pi^*(\eta)(1)x_u$ for positive η on the positive definition domain of x_u. Replacing $\pi^*(\eta)$ by $\omega\pi^*(\eta)$ with ω positive and η satisfying $\eta(1) = 1$, we get

$$\omega(1)x_u = x_u(\omega\pi^*(\eta) \otimes \iota)\Delta_u = x_u(\pi^*(\eta) \otimes \iota)\Delta_u(\omega \otimes \iota)\Delta_u = x_u(\omega \otimes \iota)\Delta_u,$$

so x_u is left invariant, and thus y_u is right invariant. One checks that $\mathcal{M}$ satisfies

$$(\omega_{q_{xu}(b)} \otimes \iota)(\mathcal{M}^*)q_{xu}(a) = q_{xu}(\omega_{q_{xu}(b)} \otimes \iota)\Delta_u(a)$$

and

$$(\iota \otimes \omega_{q_x(a),q_x(b)})(\mathcal{M}) = (\iota \otimes x_u)(\Delta_u(b^*)(1 \otimes a))$$

for $\omega \in W_u^\star$ and $a, b \in N_{x_u}$. To get the remaining entities at the universal level we need the following result.

Lemma 13.1.11 *Given $*$-automorphisms f, g on W_r satisfying $(f \otimes g)\Delta = \Delta f$. Then $(g \otimes g)\Delta = \Delta g$, and there is a scalar $r > 0$ such that $xf = rx = xg$. The unitary operators u, v on V_x given by $uq_x(a) = r^{-1/2}q_x f(a)$ and $vq_x(a) = r^{-1/2}q_x g(a)$ for $a \in N_x$ satisfy $f(\iota\otimes\eta)(M) = (\iota\otimes\eta(u^*\cdot v))(M)$ and $g(\iota\otimes\eta)(M) = (\iota \otimes \eta(v^* \cdot v))(M)$ for $\eta \in \hat{W}_*$. There are unique $*$-automorphisms f', g' on W_u with*

$$f'(\iota \otimes \eta)(\mathcal{M}) = (\iota \otimes \eta(u^* \cdot v))(\mathcal{M}) \quad \textit{and} \quad g'(\iota \otimes \eta)(\mathcal{M}) = (\iota \otimes \eta(v^* \cdot v))(\mathcal{M}).$$

Moreover, we have $\pi f' = f\pi$ and $\pi g' = g\pi$ and $(f' \otimes g')\Delta_u = \Delta_u f'$.

Proof Coassociativity yields $(f \otimes (g \otimes g)\Delta)\Delta = (f \otimes \Delta g)\Delta$, so $(g \otimes g)\Delta = \Delta g$ by density. Hence $xg = rx$ for some $r > 0$ by uniqueness of left invariant faithful semifinite normal weights on W. Pick a normal state ω on W_r. Then $xf(b) = xg(\omega f \otimes \iota)\Delta(b) = r\omega f(1)x(b) = rx(b)$ for positive $b \in M_x$. The identity

$$(\omega f \otimes \iota)(M^*)q_x(a) = q_x(\omega f \otimes \iota)\Delta(a) = r^{-1/2}v^*q_x(\omega \otimes \iota)\Delta f(a) = v^*(\omega \otimes \iota)(M^*)uq_x(a)$$

gives $f(\iota \otimes \eta)(M) = (\iota \otimes \eta(u^* \cdot v))(M)$, and similarly $g(\iota \otimes \eta)(M) = (\iota \otimes \eta(v^* \cdot v))(M)$. The first of these formulas shows that $\tilde{f}(\eta) = \eta(v^* \cdot u)$ defines an endomorphism on $\hat{W}_*$ satisfying $f\hat{\lambda} = \hat{\lambda}\tilde{f}$ and which restricts to a $*$-homomorphism on $\hat{W}^\sharp$. By universality there is therefore a $*$-homomorphism $f'\colon W_u \to W_u$ such that $f'\lambda_u = \lambda_u\tilde{f}$. By treating f^{-1} the same way, we thus see that f' is bijective, and clearly $\pi f'(\iota \otimes \eta)(\mathcal{M}) = f\pi(\iota \otimes \eta)(\mathcal{M})$, so $\pi f' = f\pi$. The case g' is done similarly, and uniqueness of f' and g' is obvious. The last identity in the lemma clearly holds on elements of the type $(\iota \otimes \eta)(\mathcal{M})$. □

Since $(\sigma_t^y \otimes \tau_{-t})\Delta = \Delta\sigma_t^y$ for real t, there is according to the lemma above a unique norm continuous one-parameter group of automorphisms σ_t^{yu} on W_u such that $\sigma_t^{yu}(\iota \otimes \eta)(\mathcal{M}) = (\iota \otimes \eta(\Delta_y^{-it} \cdot Q^{-2it}))(\mathcal{M})$ for $\eta \in \hat{W}_*$, and $\pi\sigma_t^{yu} = \sigma_t^y\pi$ for real t. So y_u is a KMS-weight on W_u with modular group $\{\sigma_t^{yu}\}$. Also x_u is a KMS-weight on W_u with modular group given by $\sigma_t^{xu} = R_u\sigma_t^{yu}R_u$, and $\pi\sigma_t^{xu} = \sigma_t^x\pi$. They are the unique left (right) invariant non-zero lower semicontinuous weights on (W_u, Δ_u) up to multiplication by positive scalars. Note that the weights are in general not faithful, so the modular groups are in general not unique.

By the previous lemma there is a unique norm continuous one-parameter group of automorphisms τ_t^u on W_u such that $\tau_t^u(\iota \otimes \eta)(\mathcal{M}) = (\iota \otimes \eta(Q^{-2it} \cdot Q^{2it}))(\mathcal{M})$ for $\eta \in \hat{W}_*$, and $\pi\tau_t^u = \tau_t\pi$ for real t. Note that $(\tau_t^u \otimes \hat{\tau}_t)(\mathcal{M}) = \mathcal{M}$, so $\tau_t^u R_u = R_u\tau_t^u$.

Proposition 13.1.12 *The automorphism groups* $\{\sigma_t^{xu}\}$, $\{\sigma_t^{yu}\}$ *and* $\{\tau_t^u\}$ *pairwise commute, and*

$$\Delta_u\sigma_t^{xu} = (\tau_t^u \otimes \sigma_t^{xu})\Delta_u, \quad \Delta_u\sigma_t^{yu} = (\sigma_t^{yu} \otimes \tau_{-t}^u)\Delta_u, \quad \Delta_u\tau_t^u = (\tau_t^u \otimes \tau_t^u)\Delta_u$$

and $x_u\sigma_t^{yu} = \nu^t x_u$, $y_u\sigma_t^{xu} = \nu^{-t}y_u$, $x_u\tau_t^u = \nu^{-t}x_u$ *and* $y_u\tau_t^u = \nu^{-t}y_u$ *for real* t.

Proof The last four identities are immediate from the corresponding result in the reduced case. The last two identities involving Δ_u are clear from the previous lemma, and the first of them follows from $f(R_u \otimes R_u)\Delta_u = \Delta_u R_u$. Since $\{\sigma_t^{yu}\}$ and $\{\tau_t^u\}$ commute, so does Δ_y and Q, and $(\tau_t^u\sigma_t^{yu} \otimes \iota)(\mathcal{M}) = (\sigma_t^{yu}\tau_t^u \otimes \iota)(\mathcal{M})$. Hence also $\{\sigma_t^{xu}\}$ and $\{\tau_t^u\}$ commute, and now $\Delta_u\sigma_s^{yu}\sigma_t^{xu} = \Delta_u\sigma_t^{xu}\sigma_s^{yu}$, so we get the final commutation relation by injectivity of Δ_u. □

The coinverse $S_u = R_u\tau_{-i/2}^u$ of (W_u, Δ_u) is a densely defined closed operator in W_u with dense range and with $S_u^{-1} = R_u\tau_{i/2}^u$. It is antimultiplicative and satisfies $S_u(S_u(a)^*)^* = a$ for $a \in D(S_u)$, it commutes with both R_u and τ_t^u, and $\pi S_u \subset S\pi$.

We also have strong left invariance in the following sense (with a similar result for strong right invariance).

Proposition 13.1.13 *The elements* $(\iota \otimes x_u)(\Delta_u(b^*)(1 \otimes a))$ *for* $a, b \in N_{x_u}$ *form a core for* S_u *and* $S_u(\iota \otimes x_u)(\Delta_u(b^*)(1 \otimes a)) = (\iota \otimes x_u)((1 \otimes b^*)\Delta_u(a))$. *The elements* $(\iota \otimes \omega)(\mathcal{M})$ *for* $\omega \in B_0(V_x)^\star$ *form a core for* S_u *and* $S_u(\iota \otimes \omega)(\mathcal{M}) = (\iota \otimes \omega)(\mathcal{M}^*)$.

Proof By the previous proposition we have for c, d in the Tomita algebra (in the C*-algebra W_u) of x_u that

$$\tau_t^u(\iota \otimes \omega_{q_{xu}(c),q_{xu}(d)})(\mathcal{M}) = \tau_t^u(\iota \otimes x_u)(\Delta_u(d^*)(1 \otimes c)) = (\iota \otimes \omega_{\Delta_{xu}^{it} q_{xu}(c),\Delta_{xu}^{it} q_{xu}(d)})(\mathcal{M})$$

for real t. Replacing t by $-i/2$ in this, we get

$$S_u(\iota \otimes \omega_{q_{xu}(c),q_{xu}(d)})(\mathcal{M}) = R_u(\iota \otimes \omega_{\Delta_{xu}^{1/2} q_{xu}(c),\Delta_{xu}^{-1/2} q_{xu}(d)})(\mathcal{M}),$$

which equals $(\iota \otimes \omega_{q_{xu}(c),q_{xu}(d)})(\mathcal{M}^*)$ as $(R_u \otimes \hat{R})(\mathcal{M}) = \mathcal{M}$ and $\hat{R}$ is implemented by J_x, and x_u is a KMS-weight. We are thus done by closedness of S_u. □

The modular element h of (W, Δ) actually belongs to $M(W_r)$ because it is group-like, so $(\iota \otimes \omega)(M)h^{it} = (\iota \otimes \omega(h^{it} \cdot h^{-it}))(M) \in W_r$ for $\omega \in B_0(V_x)^\star$. Hence there is a unique element $u_t \in M(W_u)$ such that $\Delta_u(u_t) = u_t \otimes u_t$ and $\pi(u_t) = h^{it}$, and by Stones theorem there is a unique strictly positive $h_u \in B(V_u)$ such that $h_u^{it} = u_t$. Then $h_u^{it}(\iota \otimes \omega)(\mathcal{M}) = (\iota \otimes \omega(h^{it} \cdot h^{-it}))(\mathcal{M})$ for $\omega \in B_0(V_x)^\star$. Uniqueness of h_u implies $\tau_t^u(h_u^{is}) = h_u^{is}$ and $R_u(h_u^{is}) = h_u^{-is}$ and $S_u(h_u^{is}) = h_u^{-is}$, which by its group-like nature, yields $\sigma_t^{xu}(h_u^{is}) = \nu^{ist}\sigma_t^{yu}(h_u^{is})$ for real t, s. Moreover, this *universal modular element* h_u satisfies the property below.

Proposition 13.1.14 *We have* $\sigma_t^{yu}(a) = h_u^{it}\sigma_t^{xu}(a)h_u^{-it}$ *for real* t *and* $a \in W_u$. *Hence* $(\sigma_t^{xu} \otimes \sigma_{-t}^{yu})\Delta_u = \Delta_u\tau_t^u$ *for real* t.

Proof The $*$-automorphism $g = h^{it}\tau_t(\cdot)h^{-it}$ on W_r satisfies $(g \otimes g)\Delta = \Delta g$ and $xg = x$, so there is a unitary u on V_x satisfying $uq_x(a) = q_x g(a)$ for $a \in N_x$. Thus $M(u \otimes u) = (u \otimes u)M$, and $(g \otimes \sigma_t^y)\Delta = \Delta\sigma_t^y$ gives $M(u \otimes \Delta_y^{it}) = (u \otimes \Delta_y^{it})M$, so $u^*wu = \Delta_y^{it} w \Delta_y^{-it}$ for $w \in \hat{W}_r$. By the previous lemma there is therefore a $*$-automorphism g_u on W_u such that $g_u(\iota \otimes \omega)(\mathcal{M}) = (\iota \otimes \omega(\Delta_y^{-it} \cdot \Delta_y^{it}))(\mathcal{M})$ for $\omega \in B_0(V_x)^\star$. Combining this with $\sigma_t^{yu}(\iota \otimes \omega)(\mathcal{M}) = (\iota \otimes \omega(\Delta_y^{-it} \cdot Q^{-2it}))(\mathcal{M})$ and arguing as in the proof of the lemma, gives $((g_u \otimes \sigma_t^{yu})\Delta_u \otimes \iota)(\mathcal{M}) = (\Delta_u\sigma_t^{yu} \otimes \iota)(\mathcal{M})$, so $(g_u \otimes \sigma_t^{yu})\Delta_u = \Delta_u\sigma_t^{yu}$. Hence the $*$-automorphisms $m = h_u^{-it} g_u(\cdot)h_u^{it}$ and $n = h_u^{-it}\sigma_t^{yu}(\cdot)h_u^{it}$ on W_u satisfy $(m \otimes n)\Delta_u = \Delta_u n$ and $\pi n = \pi\sigma_t^{xu}$. But since $(\tau_t^u \otimes \sigma_t^{xu})\Delta_u = \Delta_u\sigma_t^{xu}$, we thus get $\sigma_t^{xu} = n$ by the lemma below. The second identity in the proposition is now a straightforward calculation. □

Lemma 13.1.15 *If* $*$*-automorphisms* f_i, g_i *on* W_u *satisfy* $(g_i \otimes f_i)\Delta_u = \Delta_u f_i$ *and* $\pi f_1 = \pi f_2$, *then* $f_1 = f_2$ *and* $g_1 = g_2$.

Proof By using $f(R_u \otimes R_u)\Delta_u = \Delta_u R_u$, we can assume $(f_i \otimes g_i)\Delta_u = \Delta_u f_i$ instead. As is clear from the proof, the previous lemma holds with M replaced by

$\mathcal{M}$ and by sticking an index u to the various relevant quantities involved. So there are scalars $r_i > 0$ and unitaries u_i, v_i on V_u and unique $*$-automorphism f_i', g_i' on W_r such that $x_u f_i = x_u g_i = r_i x_u$, and $u_i q_{xu}(a) = r_i^{-1/2} q_{xu} f_i(a)$ and $v_i q_{xu}(a) = r_i^{-1/2} q_{xu} g_i(a)$ for $a \in N_{xu}$, and $f_i' \pi = \pi f_i$ and $g_i' \pi = \pi g_i$. Then $x f_i' = r_i x = x g_i'$ and $u_i q_x(b) = r_i^{-1/2} q_x f_i'(b)$ while $v_i q_x(b) = r_i^{-1/2} q_x g_i'(b)$ for $b \in N_x$. As $f_1' = f_2'$ by hypothesis, we thus get $u_1 = u_2$, and as $(f_1' \otimes g_1')\Delta = \Delta f_1' = \Delta f_2' = (f_2' \otimes g_2')\Delta = (f_1' \otimes g_2')\Delta$, so $g_1' = g_2'$ by density, we also get $v_1 = v_2$. Hence $f_1(\iota \otimes \omega)(\mathcal{M}) = (\iota \otimes \omega(u_1^* \cdot v_1))(\mathcal{M}) = f_2(\iota \otimes \omega)(\mathcal{M})$, and similarly $g_1 = g_2$. □

Remark 13.1.16 Note that (W_u, Δ_u) is not a locally compact quantum group in our sense since we have formulated everything in the C*-algebra context, and more importantly, because the invariant weights in general are not faithful as π need not be injective. But we have seen that basically all the quantum group entities can be lifted to the universal level, so it makes sense to consider (W_u, Δ_u) as some sort of quantum group which keeps track of all relevant representations and corepresentations. ◇

13.2 Commutative and Cocommutative Quantum Groups

We discuss in some detail two cases that are important in harmonic analysis and served as motivation for the more general definition of a quantum group.

Let G be a locally compact group with (left invariant) Haar measure ds. By Proposition A.10.12 we know that $L^\infty(G) = L^1(G)^\star$ as Banach spaces, and as is clear from the proof of Proposition 5.8.6, we may represent $L^\infty(G)$ on $L^2(G)$ as a maximal commutative von Neumann algebra of multiplication operators with predual $L^1(G)$ and strongly dense $*$-subalgebra $C_c(G)$. Recall that $L^1(G)$ is actually a Banach $*$-algebra with convolution product $f * g = \int f(t)g(t^{-1}\cdot)\,dt$ and $*$-operation $f^\star(t) = \overline{f(t^{-1})}\Delta(t^{-1})$, where the modular function $\Delta\colon G \to \langle 0, \infty\rangle$ is not to be confused with the same symbol introduced below. We denote by $K(G)$ the dense $*$-subalgebra of compactly supported functions.

By Proposition 8.1.21, which identifies $C_0(G) \otimes C_0(G)$ in any C*-tensor norm, with the C*-algebra $C_0(G \otimes G)$, and Lusin's theorem, we can further identify the Hilbert tensor product $L^2(G) \otimes L^2(G)$ with the Hilbert space $L^2(G \times G)$, and the von Neumann tensor product $L^\infty(G)\bar{\otimes}L^\infty(G)$ with the von Neumann algebra $L^\infty(G \times G)$, which in turn yields $L^\infty(G)\bar{\otimes}L^\infty(G)\bar{\otimes}L^\infty(G) = L^\infty(G \times G \times G)$.

Define a unital $*$-monomorphism $\Delta\colon L^\infty(G) \to L^\infty(G \times G) = L^\infty(G)\bar{\otimes}L^\infty(G)$ by $\Delta(f)(s,t) = f(st)$. Now

$$\int \Delta(f)(s,t)(g \otimes h)(s,t)\,dsdt = \int f(st)g(s)h(t)\,dsdt = \int f(t)(g * h)(t)\,dt$$

for $g, h \in L^1(G)$ by left invariance of ds and by the Fubini-Tonelli theorem. So Δ is normal as the span of functions of the form $g \otimes h$ is w^*-dense in $L^\infty(G \times G)$, while $(\Delta \otimes \iota)\Delta(f)(r, s, t) = f((rs)t) = f(r(st)) = (\iota \otimes \Delta)\Delta(f)(r, s, t)$ shows that Δ is coassociative. The small calculation above also shows that the product on the predual $L^1(G)$ induced by Δ is indeed the convolution product.

Now $x(f) = \int f(t)\, dt$ and $y(f) = \int f(t^{-1})\, dt$ define faithful (by definition of $L^\infty(G)$), semifinite (as $C_c(G)$ is w^*-dense in $L^\infty(G)$), normal tracial weights on $L^\infty(G)$ that are left and right invariant, respectively, since for $z = \int g(s)(\cdot)\, ds$ and g positive in $L^1(G)$, we have

$$x(z \otimes \iota)\Delta(f) = \int g(s) f(st)\, ds dt = \int g(s)(\int f(st)\, dt)\, ds = z(1)x(f)$$

by left invariance of ds and the Fubini-Tonelli theorem, and similarly for right invariance, whenever f is positive and belongs to the definition domains of the weights. In fact $V_x = L^2(G)$, $N_x = L^2(G) \cap L^\infty(G)$, $M_x = L^1(G) \cap L^\infty(G)$, and x is the weight associated to the full left Hilbert algebra $L^2(G) \cap L^\infty(G)$ with associated von Neumann algebra $L^\infty(G)$, while the modular operator Δ_x is the identity map on $L^2(G)$.

Hence $(L^\infty(G), \Delta)$ is a locally compact quantum group. Its modular element h is the inverse of the modular function Δ on G since $y = x(h^{1/2} \cdot h^{1/2})$ translates as $\int f(t^{-1})\, dt = \int h(t) f(t)\, dt$ for $f \in C_c(G)$. The fundamental multiplicative unitary M of this quantum group is given by $M(f)(s, t) = f(s, s^{-1}t)$ with inverse given by $M^*(f)(s, t) = f(s, st)$ for $f \in L^2(G \times G)$ because

$$M^*(q_x(f_1) \otimes q_x(f_2))(s, t) = (q_x \otimes q_x)(\Delta(f_2)(f_1 \otimes 1))(s, t) = (q_x \otimes q_x)(f_1 \otimes f_2)(s, st)$$

for $f_i \in C_c(G) \subset N_x$ and $s, t \in G$. The $*$-isomorphism S on $L^\infty(G)$ given by $S(g)(t) = g(t^{-1})$ for $g \in L^\infty(G)$, which is then automatically normal, must by strong left invariance be the coinverse, and hence also the unitary coinverse R, and the scaling group must be trivial, with a trivial scaling constant. In fact, in this case we get the Hopf algebra type of identities $(\varepsilon_r \otimes \iota)\Delta_r = \iota = (\iota \otimes \varepsilon_r)\Delta_r$ on $C_0(G)$ and $m_r(S_r \otimes \iota)\Delta_r = \varepsilon_r(\cdot)1 = m_r(\iota \otimes S_r)\Delta_r$ on $C_b(G)$, where $\varepsilon_r(f) = f(e)$ for the unit $e \in G$, and $m_r \colon M(C_0(G) \otimes C_0(G)) \to M(C_0(G))$ is the extension of the non-degenerate $*$-homomorphism $C_0(G) \otimes C_0(G) \to C_0(G)$ which sends $g_1 \otimes g_2$ to $g_1 g_2$ for $g_i \in C_0(G)$. It is the transpose of the continuous *diagonal map* $G \to G \times G$ which sends t to (t, t).

Proposition 13.2.1 *Any commutative locally compact quantum group is isomorphic to $(L^\infty(G), \Delta)$ for a locally compact group G unique up to isomorphism.*

Proof Let (W, Δ) be a commutative locally compact quantum group with left invariant faithful semifinite normal weight x. Then $\Delta_r \colon W_r \to M(W_r \otimes W_r)$ is a non-degenerate $*$-homomorphism thanks to the cancellation properties. Identifying by Gelfand's theorem the commutative C*-algebra W_r with $C_0(G)$, where G is the locally compact Hausdorff space of characters on W_r with w^*-topology, we see that

$\Delta_r : C_0(G) \to M(C_0(G) \otimes C_0(G)) = C_b(G \otimes G)$ is the transpose of a continuous map $p : G \times G \to G$, meaning $p(f)(s,t) = (s \otimes t)\Delta(f)$, which is associative as Δ is coassociative. This semigroup has the cancellation properties, meaning that $st_1 = st_2$ or $t_1 s = t_2 s$ for all s implies $t_1 = t_2$, which is immediate from the cancellation properties of the quantum group (W_r, Δ_r). By strong left invariance and commutativity of W, we have

$$S^2(\iota \otimes x)(\Delta(a^*)(1 \otimes b)) = S(\iota \otimes x)((1 \otimes a^*)\Delta(b)) = (\iota \otimes x)(\Delta(a^*)(1 \otimes b))$$

for $a, b \in N_x$, so $\tau_{-i} = S^2 = \iota$ and $S_r = R_r$. We may view the $*$-preserving isomorphism S_r as the transpose of an involutive continuous map on G which sends t to say t^{-1}, and which satisfies $(st)^{-1} = t^{-1}s^{-1}$ as $f(S_r \otimes S_r)\Delta_r = \Delta_r S_r$. Now $m(\iota \otimes S_r)\Delta_r : W_r \to W_r$ makes sense and is a continuous unital $*$-homomorphism. But $m(\iota \otimes S_r)\Delta(\iota \otimes x)(\Delta(a^*)(1 \otimes b))$ equals

$$m(\iota \otimes S_r)(\iota \otimes \iota \otimes x)((\iota \otimes \Delta)\Delta(a^*)(1 \otimes 1 \otimes b)) = m(\iota \otimes \iota \otimes x)(\Delta(a^*)_{13}\Delta(b)_{23}) = x(a^*b)1,$$

so we have a character ε_r on W_r such that $m(\iota \otimes S_r)\Delta_r(\cdot) = \varepsilon_r(\cdot)1$. Applying S_r to this identity and using $S_r^2 = \iota$, we also get $m(S_r \otimes \iota)\Delta_r(\cdot) = \varepsilon_r(\cdot)1$. Now ε_r is given by evaluation at some $e \in G$. Hence $tt^{-1} = e = t^{-1}t$ for all t. In particular, we get $e^{-1} = (ee^{-1})^{-1} = ee^{-1} = e$, so $e^2 = e$ and $te^2 = te$, which by cancellation gives $te = t$ and similarly $et = t$. So e is a unit for G and t^{-1} is the inverse of t. Thus G is a locally compact group, and since the Haar measure on G provides a left invariant faithful semifinite normal weight on W, it must be x up to a non-zero scalar by uniqueness of invariant weights for quantum groups. So the Hilbert spaces are both $L^2(G)$ and we get the same weak closures W and $L^\infty(G)$, and by construction the same coproducts. Uniqueness is clear. □

Given the formula written above for the fundamental multiplicative unitary $M \in M(C_0(G) \otimes \hat{L^\infty(G)}_r)$ of $(L^\infty(G), \Delta)$, then for a character $\delta_t \in C_0(G)^\star$ associated to the point measure of $t \in G$, we see that $\lambda(t) \equiv (\delta_t \otimes \iota)(M) \in B(L^2(G))$ is given by $\lambda(t)(f) = f(t^{-1}\cdot)$. In other words, it is the usual strongly continuous unitary *left regular representation* of G on $L^2(G)$. From the quantum group point of view multiplicativity comes from $(\Delta_r \otimes \iota)(M) = M_{13}M_{23}$ and $\delta_{st} = \delta_s * \delta_t$, where the convolution product of measures corresponding to $z_i \in C_0(G)^\star$ as described in Theorem A.10.10 is given by $z_1 * z_2 = (z_1 \otimes z_2)\Delta$. Unitarity is clear from $(S_r \otimes \iota)(M) = M^*$ and $\delta_t^\star = \delta_{t^{-1}}$, where the adjoint of the measure corresponding to $z \in C_0(G)^\star$ is given by $z^\star = \overline{z(S_r(\cdot)^*)}$.

The unique extension of λ to a unital representation of the unital Banach $*$-algebra $M(G) \cong C_0(G)^\star$ given by $\lambda(z) = (z \otimes \iota)(M)$ restricts to a non-degenerate representation to the two-sided $*$-ideal $L^1(G)$ and is given by $\lambda(f) = \int f(t)\lambda(t)\,dt$. This correspondence is generic.

Proposition 13.2.2 *The map $u \mapsto \int (\cdot)u(t)\,dt$ is a bijection from the set of strongly (or equivalently, weakly) continuously unitary representations on a Hilbert space V*

of a locally compact group G to the set of non-degenerate representations on V of the Banach $*$-algebra $L^1(G)$.

Proof We need only observe that any map $u\colon G \to B(V)$ is a strongly continuous representation if and only if the linear map $N\colon C_0(G) \otimes B_0(V) \to C_0(G) \otimes B_0(V) \cong C_0(G, B_0(V))$ given by $N(f)(t) = u(t)(f(t))$ is a unitary element of $M(C_0(G) \otimes B_0(V))$ such that $(\Delta_r \otimes \iota)(N) = N_{13}N_{23}$. Then u extends uniquely to $M(G) = C_0(G)^\star$ as $z \mapsto (z \otimes \iota)(N)$. □

The correspondence is clearly functorial, and the proof above obviously also served as motivation for the definition of a unitary corepresentation of a quantum group.

Definition 13.2.3 The *group von Neumann algebra* $W^*(G) \subset B(L^2(G))$ of a locally compact group G is the weak closed span of $\lambda(G)$. The *reduced group C*-algebra* $C_r^*(G)$ of G is the norm closure of $\lambda(L^1(G))$, while the *universal group C*-algebra* $C_u^*(G)$ of G is the completion of $L^1(G)$ with respect to the universal C*-norm $\|\cdot\|_u = \sup_\pi \|\pi(\cdot)\|$, where we sup over all non-degenerate $*$-representations π of the Banach $*$-algebra $L^1(G)$.

The universal norm makes sense, and is less than the L^1-norm of an L^1-function, as $\|\pi(f)\| \leq \|f\|_1$ by the remark below.

Remark 13.2.4 Any $*$-homomorphism $A \to B$ from a Banach $*$-algebra A with a self-adjoint bounded approximate unit $\{a_i\}$, into a C^*-algebra B, is norm decreasing due to the spectral radius formula and the fact that the image of an invertible element under a unital homomorphism is always invertible. Any positive linear functional z on A is automatically bounded, as is seen by performing a GNS-construction with respect to z, so $\|z\| = \lim z(a_i)$. We can obviously define a universal C*-norm $\|\cdot\|_u$ for any such A as well, so $\|a\|_u \leq \|a\|$ for $a \in A$, and the C*-algebra A_u is defined to be the normification of A. When we have a faithful non-degenerate $*$-representation of A, which is evidently the case for $L^1(G)$ as λ will do, then $A \subset A_u$ and $\{a_i\}$ is an approximate unit for A_u with respect to the C*-norm, and A and A_u have the same state spaces. ◇

The von Neumann algebra $W^*(G)$, which is $\lambda(G)''$ or the weak closure of $\lambda(M(G))$ or $\lambda(L^1(G))$, is by definition $L^\infty\hat{}(G)$, while $C_r^*(G) = L^\infty\hat{}(G)_r$ and $C_u^*(G) = L^\infty\hat{}(G)_u$ as C*-algebras. In general $\lambda(t) \in M(L^\infty\hat{}(G)_r) \subset L^\infty\hat{}(G)$, but $\lambda(t) \notin C_r^*(G)$ unless G is discrete. We have a simple description of the coproduct on $W^*(G)$, namely, as δ_t is multiplicative, we have

$$\hat{\Delta}(\lambda(t)) = (\delta_t \otimes \iota \otimes \iota)(\iota \otimes \hat{\Delta})(M) = (\delta_t \otimes \iota \otimes \iota)(M_{13}M_{12}) = \lambda(t) \otimes \lambda(t)$$

for $t \in G$, which uniquely determines $\hat{\Delta}$ and shows that the dual of the commutative quantum group $(L^\infty(G), \Delta)$ is cocommutative. The converse is also true, namely, the dual of a cocommutative quantum group (W, Δ) is commutative. Indeed, if $z_i \in W_*$, and M is the fundamental multiplicative unitary of (W, Δ), then

$$(z_1 \otimes \iota)(M)(z_2 \otimes \iota)(M) = (z_1 \otimes z_2 \otimes \iota)(M_{13}M_{23}) = (z_1 \otimes z_2 \otimes \iota)(\Delta \otimes \iota)(M)$$

equals $(z_2 \otimes \iota)(M)(z_1 \otimes \iota)(M)$ as $(z_1 \otimes z_2)\Delta = (z_2 \otimes z_1)\Delta$. So by generalized Pontryagin duality a quantum group is commutative if and only if its dual is cocommutative.

The GNS-map q of the right invariant faithful semifinite normal weight $y = xS$ of $(L^\infty(G), \Delta)$ is given by $q(a) = q_x(ah^{1/2})$ for $a \in L^\infty(G)$ with $ah^{1/2} \in N_x$. For the associated multiplicative unitary $N \in \hat{L^\infty(G)}' \bar{\otimes} L^\infty(G)$ we therefore have

$$N(q_x(a) \otimes q_x(b)) = N(q(ah^{-1/2}) \otimes q(bh^{-1/2}) = (q \otimes q)(\Delta(ah^{-1/2})(1 \otimes bh^{-1/2}))$$

which equals

$$(q_x \otimes q_x)(\Delta(ah^{-1/2})(1 \otimes bh^{-1/2})(h^{1/2} \otimes h^{1/2})) = (q_x \otimes q_x)(\Delta(a)(1 \otimes bh^{-1/2})$$

as h is group-like. So the *right* regular representation is the strongly continuous unitary representation of G given by $\rho(t) \equiv (\iota \otimes \delta_t)(N)$, so $\rho(t)(f) = \Delta(t)^{1/2} f(\cdot t)$ for $f \in L^2(G)$. The *right* group von Neumann algebra $\rho(G)'' \subset B(L^2(G))$ of G is therefore the commutant of $W^*(G)$. Also, the unitary T on $L^2(G)$ given by $T(a)(t) = \Delta(t)^{-1/2} a(t^{-1})$ satisfies $T\lambda(t) = \rho(t)T$, so the (left) regular and right regular representations of G are unitarily equivalent. Combining these representations we get the *left-right* regular representation of $G \times G$ on $L^2(G)$ which sends (s, t) to $\lambda(s)\rho(t) = \rho(t)\lambda(s)$. Its restriction to the diagonal subgroup consisting of all (t, t), is called the *conjugate* representation.

Now $K(G)$ is a left Hilbert algebra with the convolution product and $*$-operation described earlier, and with the usual L^2-inner product. Its associated von Neumann algebra is $W^*(G)$. The faithful semifinite normal weight associated to the this left Hilbert algebra is called the *Plancherel weight*, which we denote suggestively by $\hat{x}$. By definition it satisfies the following Plancherel identity

$$\hat{x}(\lambda(a)^*\lambda(b)) = \hat{x}\pi_l(a^\star * b) = (b|a) = \int \overline{a(t)} b(t)\, dt$$

for left bounded $a, b \in L^2(G)$. As the coinverse of $(L^\infty(G), \Delta)$ is bounded and normal we see that $K^\sharp = K \cap K^*$ consists of all f such that $f, f^\star \in L^1(G) \cap L^2(G)$, and the elements in $K(G)$ certainly belong there. The dual weight of $(L^\infty(G), \Delta)$ is associated to the left Hilbert algebra gotten as the image of $p \colon K^\sharp \to V_x = L^2(G)$ uniquely determined by $\omega(w^*) = (p(\omega)|q_x(w))$ for w in $L^\infty(G) \cap L^2(G)$, and with operations gotten by forcing this p to be a $*$-homomorphism. This left Hilbert algebra clearly contains the dense left Hilbert $K(G)$, so $\hat{x}$ is indeed the dual left invariant faithful semifinite normal weight of the Haar integral x, and its GNS-map satisfies $q_{\hat{x}}\lambda(\omega) = p(\omega)$. Hence $(W^*(G), \hat{\Delta})$ with the Plancherel weight $\hat{x}$ is the locally quantum group dual to $(L^\infty(G), \Delta)$ with the Haar integral x. Thanks to the formula $\hat{\tau}_t = \Delta_x^{it} \cdot \Delta_x^{-it}$ for the scaling group $\{\hat{\tau}_t\}$, the coinverse $\hat{S}$ on the dual side is the unitary coinverse $\hat{R}$ uniquely determined by $\hat{S}(\lambda(t)) = \lambda(t^{-1})$ for $t \in G$. The scaling constant is of course one, and since $\hat{\Delta} = \hat{\Delta}^{\mathrm{op}}$ the weight $\hat{x}$

is also right invariant with modular element $\hat{h} = 1$. The modular operator $\Delta_{\hat{\chi}}$ for the Plancherel weight is just multiplication by the modular function Δ, while its modular conjugation is given by $J_{\hat{\chi}}(f) = f^\star$.

Combining Proposition 13.2.1 with generalized Pontryagin duality we conclude that any cocommutative locally compact quantum group is isomorphic to a quantum group of the above type $(W^*(G), \hat{\Delta})$ for a locally compact group G unique up to isomorphism.

There is another way of recovering the group G from a cocommutative locally compact quantum group $(\hat{W}, \hat{\Delta})$ which begins by noting that the elements $\lambda(t)$ are group-like unitaries in $B(L^2(G))$ when we are in the setting $\hat{W} = W^*(G)$. Indeed, the *intrinsic group* of any locally compact quantum group $(\hat{W}, \hat{\Delta})$ consists of the invertible $a \in B(V_x)$ such that $M(a \otimes 1)M^* = a \otimes a$. For $b \in \hat{W}'$ we have

$$a \otimes ba = (1 \otimes b)M(a \otimes 1)M^* = M(a \otimes 1)M^*(1 \otimes b) = a \otimes ab,$$

so $a \in \hat{W}'' = \hat{W}$ and $\hat{\Delta}(a) = a \otimes a$. Writing $a = u|a|$, we get $(u \otimes u)(|a| \otimes |a|) = \hat{\Delta}(u)\hat{\Delta}(|a|)$, which by uniqueness of the polar decomposition, shows that both u and $|a|$ are group-like. As $\hat{\Delta}$ is injective, we get $\||a|\| = \|\Delta(|a|)\| = \||a| \otimes |a|\| = \||a|\|^2$, so $|a|$ has norm one, and as $|a|^{-1}$ is also group-like, it has also norm one, so $|a| = 1$ by spectral theory. Hence a is unitary. So the intrinsic group of $(\hat{W}, \hat{\Delta})$ is a locally compact group of the unitary group-like elements of $\hat{W}$ under, say, the σ-weak topology. Now $a = (\iota \otimes z)\hat{\Delta}(a) \in M(\mathbb{C} \otimes \hat{W}_r)$ for any $z \in \hat{W}_*$ with $z(a) = 1$, and in Sect. 13.1 we established a bijection $f \mapsto (f \otimes \iota)(\mathcal{M})$ from the characters of W_u to the set of such elements a, which is moreover a homeomorphism when we use the w^*-topology on the character space turning the latter into a locally compact topological group with a product $z_1 z_2 = (z_1 \otimes z_2)\Delta_u$ for characters z_i on W_u which corresponds to the product in the intrinsic group. The unit for the group of characters is then the counit ε_u, while the inverse of z_1 is $z_1 S_u$, which is bounded despite the fact that the coinverse S_u need not be, and indeed $z_1 S_u(\iota \otimes \omega)(\mathcal{M}) = z_1(\iota \otimes \omega)(\mathcal{M}^*)$ for $\omega \in B_0(V_x)^\star$.

When W_u is commutative, then by Gelfand's theorem it is a C*-algebra of the form $C_0(G)$, and by what we have seen, identifying $t \in G$ with the character δ_t on $C_0(G)$, we know that G is a locally compact group such that $\Delta_u(f)(s, t) = f(st)$. Now $W_u = W_r$ by the properties of the Haar integral and since the uniform norm is the universal one because the (equivalence classes of the) non-degenerate representations of a commutative C*-algebra are just the characters. Hence $\mathcal{M} = M$ and $(\delta_t \otimes \iota)(\mathcal{M}) = \lambda(t)$. So in this case the intrinsic group is $\lambda(G)$ as expected.

Let G be an abelian locally compact group. Then the locally compact quantum group $(C_u^*(G) = C_r^*(G), \hat{\Delta}_u)$ is both commutative and cocommutative. By the Gelfand transform $C_u^*(G)$ is isomorphic to $C_0(\hat{G})$ for the locally compact group $\hat{G}$ of characters on $C_u^*(G)$ with w^*-topology. These 1-dimensional non-degenerate representations are in one-to-one correspondence with the 1-dimensional strongly continuous unitary representations u on G with corresponding characters $\int (\cdot)u(t)\,dt$ on the dense $*$-subalgebra $L^1(G)$ of $C_u^*(G)$. We may therefore view $\hat{G}$

as the group of continuous homomorphisms $G \to \mathbb{T}$ into the circle with pointwise product. A net of such *group characters* converges if it converges uniformly on any compact subset of G. The locally compact group $\hat{G}$ viewed this way, is what we mean by the *dual group* of G. The Plancherel weight on $W^*(G) \supset C_u^*(G) \cong C_0(\hat{G})$ is finite on $C_c(\hat{G})$ and thus by the Riesz representation theorem, corresponds to a Radon measure on $\hat{G}$, which is up to a scalar the unique Haar measure $d'u$ as the Plancherel weight is biinvariant with respect to the coproduct $\hat{\Delta}$ which on $C_0(\hat{G})$ is the transpose of the product on $\hat{G}$. We call this measure $d'u$ on $\hat{G}$ the *Plancherel measure*. Here we have merely exploited the correspondence between characters on $C_u^*(G)$ and unitary group-like elements $u \in M(C_0(\hat{G})_r) = C_b(\hat{G})$ with respect to $\hat{\Delta}$.

The Gelfand transform $C_u^*(G) \to C_0(\hat{G})$ restricts to the *Fourier transform* F on $L^1(G)$, so $F(f)(u) = \int f(t)u(t)\,dt$ for a group character u on G. By the Plancherel identity for $\hat{x}$, the restriction of F to $L^1(G) \cap L^2(G)$ extends uniquely by continuity to a unitary $L^2(G) \to L^2(\hat{G})$ with inverse given by $F^*(g)(t) = \int g(u)\overline{u(t)}\,d'u$ for $g \in L^1(\hat{G}) \cap L^2(\hat{G})$.

Every $t \in G$ defines a character $u \mapsto u(t)$ on $\hat{G}$ producing an isomorphism $G \mapsto \hat{\hat{G}}$ of topological groups known as *Pontryagin duality* for abelian groups, and this is just generalized Pontryagin duality restricted to the intrinsic groups.

Returning to a general locally compact group G, any $f \in L^\infty(G) = L^1(G)^\star$ is *positive definite* if it is positive as a bounded functional on the Banach $*$-algebra $L^1(G)$. By the Fubini-Tonelli theorem and the property of the modular function this happens if and only if $\int \overline{g(t)}g(s)f(t^{-1}s)\,dtds \geq 0$ for all $g \in K(G)$ since this integral is the evaluation of the functional at $g^\star * g$ and $K(G)$ is dense in $L^1(G)$. We may approximate the integral by Riemann sums $\sum \overline{g(t_i)}g(t_j)f(t_i^{-1}t_j)\mu(A_i)\mu(A_j)$, where $\mu(A_i)$ is the Haar measure of the neighborhood A_i of $t_i \in G$. Hence f is positive definite if the matrix with ij-entry $f(t_i^{-1}t_j)$ is positive for any finite set $\{t_i\}$ of G. To see that the converse is also true first note that the cyclic representation (π, v) of $L^1(G)$ associated to a positive definite f corresponds to a unitary representation u of G via $\int g(t)f(t)\,dt = (\pi(g)v|v) = \int g(t)(u(t)v|v)\,dt$, so $f = (u(\cdot)v|v)$ almost everywhere, and every function of the form $(u(\cdot)v|v)$ is by the same identity, positive definite. But then $\sum \overline{a_i}a_i f(t_i^{-1}t_j) = \|\sum a_i u(t_i)v\|^2 \geq 0$ for finitely many $a_i \in \mathbb{C}$, so the matrix is indeed positive. In passing note that any positive definite f has a continuous representative in $L^\infty(G)$. Considering its positive 2×2-matrix associated to the finite set $\{e, t\}$, which thus have positive determinant, we get $|f(t)| \leq f(e)$.

The predual of $W^*(G)$ is a Banach $*$-algebra with product and $*$-operation dual to $(W^*(G), \hat{\Delta})$; keep in mind that the coinverse of this quantum group is the unitary coinverse. To $z \in W^*(G)_*$ we define the function $z\lambda$ on G, and by transferring operations, the $*$-algebra operations we get on the set of all such functions $z\lambda$ are the usual pointwise ones. These functions are clearly continuous, and they vanish at infinity since as we are in the standard representation, we may approximate $z\lambda$ uniformly by functions of the form $(\lambda(\cdot)f|g)$ with $f, g \in C_c(G)$ as $C_c(G)$ is dense in $L^2(G)$, and $t \mapsto (\lambda(t)f|g) = \int f(t^{-1}s)\overline{g(s)}\,ds$ has again compact support. By the Stone-Weierstrass theorem this $*$-algebra $A(G)$ is moreover dense in the C*-algebra $C_0(G)$, and is known as the *Fourier algebra*. The fact that $A(G) \subset C_0(G)$ shows that the regular representation λ of any non-compact locally compact group cannot have a finite dimensional subrepresentation u since $\sum |(u(t)v_i|v_j)|^2 = 1$ for any orthonormal basis $\{v_i\}$ of the finite dimensional space for u.

The *Fourier-Stieltjes algebra* $B(G)$ of a locally compact group G is the $*$-subalgebra of $C_b(G)$ consisting of all matrix coefficients $(u(\cdot)v_1|v_2)$ of strongly continuous unitary representations u of G. By polarization it is spanned by the positive definite functions on G. It can also be identified with $C_u^*(G)^\star$ with operations dual to those of the universal quantum group $(C_u^*(G), \hat{\Delta}_u)$. This identification is clear since $C_u^*(G)^\star$ is spanned by the states on $C_u^*(G)$ and each such produces a cyclic representation which again is associated to a unitary representation of G. So $B(G)$ is indeed a $*$-subalgebra of $C_b(G)$, and with the norm transferred from $C_u^*(G)^\star$ it is even a Banach $*$-algebra, although this norm need not be the uniform norm. The Fourier algebra $A(G)$ is nevertheless an ideal of $B(G)$. This is clear from the discussion following Proposition 13.1.10 since states of $C_u^*(G)$ are vector states. Note also that the w^*-topology on $C_u^*(G)^\star$ corresponds to the topology on $B(G)$ of uniform convergence on compact subsets of G, and that the states on $C_u^*(G)$ are the positive definite functions that are one on the unit of G.

13.3 Amenability

A locally compact group G is *amenable* if it has a *left invariant mean*, that is, a state m on the C*-algebra $L^\infty(G)$ such that $m(g(s\cdot)) = m(g)$ for $s \in G$ and $g \in L^\infty(G)$. Such a mean restricts to a left invariant mean on the C*-subalgebra $UCB(G)$ of *left uniformly continuous* on G, that is, those $g \in C_b(G)$ with $\|g(s\cdot)-g\|_u \to$ as $s \to e$. From the discussion following Corollary A.10.7, we see that $L^1(G) * L^\infty(G) \subset UCB(G)$ and that $\|f * g\|_u \leq \|f\|_1\|g\|_\infty$ for $f \in L^1(G)$ and $g \in L^\infty(G)$. Then $\lim \|f * f_i * g - f * g\|_u = 0$ for any approximate unit $\{f_i\}$ of $L^1(G)$. Since the integral $f * g = \int f(s)g(s^{-1}\cdot)ds$ is $\|\cdot\|_u$-convergent whenever $g \in UCB(G)$, we get $m(f*g) = m(g)$ for such g and positive $f \in L^1(G)$ with $\|f\|_1 = 1$. Fixing any such f, we therefore see that the state $\tilde{m}$ on $L^\infty(G)$ given by $g \mapsto m(f*g)$ has the invariant property $\tilde{m}(h*g) = \tilde{m}(g)$ for $h \in L^1(G)$ and $g \in L^\infty(G)$. Considering the normal state $\omega = \int h^\star(t) \cdot dt$ on $L^\infty(G)$ corresponding to the positive $h^\star \in L^1(G)$ with $\|h\|_1 = 1$, we can rephrase this as $\tilde{m}(\omega \otimes \iota)\Delta(g) = \omega(1)\tilde{m}(g)$. According to the definition below this means that the quantum group $(L^\infty(G), \Delta)$ is amenable. Such an invariant state $\tilde{m}$ is in fact a left invariant mean for G because $\tilde{m}(g(s^{-1}\cdot)) = \tilde{m}(k_s * g) = \tilde{m}(h * k_s * g) = \tilde{m}(g)\int(h * k_s)(t)\,dt = \tilde{m}(g)$. However, when G is non-discrete a left invariant mean for G might not be invariant for $(L^\infty(G), \Delta)$; to get an invariant mean for the quantum group one invariable has to go through the procedure described above.

Let (W, Δ) be a locally compact quantum group. In Sect. 13.1 we did set up a bijection $N \mapsto \pi_N$ from the set $C(\hat{W}, \hat{\Delta})$ of unitaries N in $M(\hat{W}_r \otimes B_0(V_N))$ such that $(\hat{\Delta} \otimes \iota)(N) = N_{13}N_{23}$ to the set $R(W_u, \Delta_u)$ of non-degenerate $*$-representations π_N of W_u on a Hilbert space V_N which is defined such that $(\iota \otimes \pi_N)(F\mathcal{M}^*F^*) = N$, where $F\colon V_u \otimes V_x \to V_x \otimes V_u$ sends $u \otimes v$ to $v \otimes u$. So $\pi_{1\otimes 1} = \varepsilon_u$ and $\pi_{\hat{M}} = \pi$.

By *weak containment* $N \prec N'$ for $N, N' \in C(\hat{W}, \hat{\Delta})$ we mean $\pi_N \prec \pi_{N'}$, or $\ker \pi_{N'} \subset \ker \pi_N$. This means there is a unique $*$-epimorphism $f\colon \pi_{N'}(W_u) \to \pi_N(W_u)$ such that $f\pi_{N'} = \pi_N$. Clearly $\prec$ is both transitive and reflexive.

Definition 13.3.1 Let (W, Δ) be a locally compact quantum group. It is *coamenable* if there is a state ε on W_r such that $(\iota \otimes \varepsilon)\Delta_r = \iota$. It is *amenable* if it has a state m on W such that $m(\omega \otimes \iota)\Delta = \omega(1)m$ for $\omega \in W_*$; we call m a *(left) invariant mean*. We say $N \in C(\hat{W}, \hat{\Delta})$ is *amenable* if there is a state m_N on $B(V_N)$ such that $m_N(\omega \otimes \iota)(N(1 \otimes (\cdot))N^*) = \omega(1)m_N$ for $\omega \in \hat{W}_*$. It has the *weak containment property* if $1 \otimes 1 \prec N$.

Thanks to the presence of the unitary coinverse, we also get $(\varepsilon \otimes \iota)\Delta_r = \iota$, so ε is a counit for (W_r, Δ_r). Applying $\iota \otimes \varepsilon \otimes \iota$ to both sides of $(\Delta_r \otimes \iota)(M) = M_{13}M_{23}$ gives $(\varepsilon \otimes \iota)(M) = 1$. Thus

$$\varepsilon((\iota \otimes \omega)(M)(\iota \otimes \eta)(M)) = \varepsilon(\iota \otimes \omega \otimes \eta)(M_{23}M_{13}M_{23}^*) = \omega(1)\eta(1),$$

which shows that ε is multiplicative, and thus a $*$-character. Hence $\varepsilon\pi = \varepsilon_u$, so coamenability of (W, Δ) means that $\hat{M} \in C(\hat{W}, \hat{\Delta})$ has the weak containment property. Any state ε on W_r such that $(\varepsilon \otimes \iota)(M) = 1$ will of course satisfy $(\varepsilon \otimes \iota)\Delta_r = \iota$, and (W, Δ) is then coamenable. For coamenability to hold it suffices actually that we have a $*$-character z on W_r. Indeed, from the pentagonal equation for M one readily checks that the state $\varepsilon = z((z \otimes \iota)(M) \cdot (z \otimes \iota)(M)^*)$ satisfies $(\varepsilon \otimes \iota)\Delta_r = \iota$.

If $(\hat{W}, \hat{\Delta})$ has an invariant mean m, any $N \in C(\hat{W}, \hat{\Delta})$ is amenable. Indeed, pick a normal state z on $B(V_N)$. Then $m_N = m\hat{R}(\iota \otimes z)(N(1 \otimes (\cdot))N^*)$ satisfies $m_N(\omega \otimes \iota)(N(1 \otimes (\cdot))N^*) = \omega(1)m_N$ for $\omega \in \hat{W}_*$ since $m\hat{R}(\iota \otimes \omega)\hat{\Delta} = \omega(1)m\hat{R}$. The converse is also true since $(\hat{W}, \hat{\Delta})$ is amenable when $\hat{M} \in C(\hat{W}, \hat{\Delta})$ is amenable.

Note that (W, Δ) is compact if and only if it has a normal invariant mean.

Proposition 13.3.2 *A locally compact quantum group (W, Δ) is coamenable if and only if $\pi : W_u \to W_r$ is injective.*

Proof We must show $\|(\iota \otimes \omega)(N)\| \leq \|(\iota \otimes \omega)(M)\|$ for $N \in C(\hat{W}, \hat{\Delta})$ and $\omega \in (\hat{W})^\sharp$, given that we know $|\eta(1)| \leq \|(\iota \otimes \eta)(M)\|$ for $\eta \in \hat{W}_*$. Pick a normal state z on $B(V_N)$, and let $\eta = (z \otimes \omega)(N(1 \otimes (\cdot)))$. Then

$$|z(\iota\otimes\omega)(N)| \leq \|(\iota\otimes\eta)(M)\| = \|(\iota\otimes z)(N_{21}^*((\iota\otimes\omega)(M)\otimes 1)N_{21})\| \leq \|(\iota\otimes\omega)(M)\|.$$

□

Thus $(W_u, \Delta_u) \cong (W_r, \Delta_r)$ in this case.

Let $T_\omega(m) = m(\omega \otimes \iota)\Delta$ for $\omega \in W_*$ and any state m on W. Then the collection $\{T_\omega\}_{\omega\in W_*}$ of affine maps acting on the state space of W is a commuting family exactly when (W, Δ) is cocommutative. In this case the Kakutani fixed point theorem provides a fixed point which obviously is an invariant mean. Hence every cocommutative quantum group is amenable. In particular, considering $W = L^\infty(G)$, we see that every abelian locally compact group is amenable. That $(W^*(G), \hat{\Delta})$ is amenable for a general locally compact group G is less interesting, and is also immediate from the proposition below since we saw in the previous section that $(L^\infty(G), \Delta)$ is always coamenable. In any case we see that in the group case, the concepts of amenability and coamenability are completely dual. Such a complete duality for quantum groups is so far only known in the discrete and compact cases. Before we move on let us include the outstanding example of a non-amenable group.

Example 13.3.3 Let F_2 be the free group on two generators $\{a, b\}$ considered then as a discrete group. Denote by χ_c the characteristic function on F_2 on the subset beginning with $c \in \{e, a, a^{-1}, b, b^{-1}\}$. Any invariant mean m on F_2 would then give the contradiction that $1 = m(\chi_e)+m(\chi_a)+m(\chi_{a^{-1}})+m(\chi_b)+m(\chi_{b^{-1}})$ should equal $m(\chi_e)+m(\chi_a)+m(\chi_{a^{-1}}(a\cdot))+m(\chi_b)+m(\chi_{b^{-1}}(b\cdot)) = m(\chi_e)+1+1$. ◇

Proposition 13.3.4 *The dual of a coamenable locally compact quantum group is amenable.*

Proof Say (W_r, Δ_r) has a counit, which we have extended to a state ε on $B(V_x)$, and let m denote the restriction of ε to $\hat{W}$. For any normal state ω on $\hat{W}$, we clearly have $\varepsilon(\omega \otimes \iota)(\hat{M}) = 1 = \varepsilon(\omega \otimes \iota)(\hat{M}^*)$, so m is an invariant mean for $(\hat{W}, \hat{\Delta})$ by the lemma below as $\hat{\Delta}(a) = \hat{M}^*(1 \otimes a)\hat{M}$ for $a \in \hat{W}$. □

Lemma 13.3.5 *If $f : A \to B$ is a conditional expectation between C^*-algebras, and $f(w^*w) = f(w)^*f(w)$ for $w \in A$, then $f(a^*w) = f(a)^*f(w)$ and $f(w^*a) = f(w)^*f(a)$ for all $a \in A$. In particular, if w is unitary and $f(w) = 1$, then $f(aw) = f(a) = f(wa)$ for all $a \in A$.*

Proof Since $f(a^*a) - f(a)^*f(a) = f((f(a)-a)^*(f(a)-a)) \geq 0$, the expression $t(f(w)^*f(a)+f(a)^*f(w)) = f(tw+a)^*f(tw+a) - t^2f(w)^*f(w) - f(a)^*f(a)$ is not greater than $tf(w^*a+a^*w)+f(a^*a)-f(a)^*f(a)$ for real t. Dividing by $\pm t$ and letting $|t| \to \infty$ gives $f(w)^*f(a)+f(a)^*f(w) = f(w^*a)+f(a^*w)$, which combined with the equation gotten by replacing w with $-iw$, gives the result. □

Proposition 13.3.6 *An element $N \in C(\hat{W}, \hat{\Delta})$ has the weak containment property if and only if there is a state z on $B(V_N)$ such that $z(\omega \otimes \iota)(N) = 1$ for each normal state ω on $\hat{W}$ if and only if there is a net $\{v_i\}$ of unit vectors in V_N such that $\lim \|N(v \otimes v_i) - v \otimes v_i\| \to 0$ for $v \in V_x$.*

Proof Assume we have a character z on $\pi_N(W_u)$ such that $z\pi_N = \varepsilon_u$. Extend it to a state z on $B(V_N)$. Then $z(\omega \otimes \iota)(N) = z\pi_N\hat{\lambda}_u(\omega) = 1$ for $\omega \in (\hat{W})^\sharp$. To any such state z pick a net $\{v_i\}$ of unit vectors in V_N such that $z(a) = \lim \omega_{v_i}(a)$ for $a \in \pi_N(W_u)$. Then $\lim(N(v \otimes v_i)|v \otimes v_i) = z(\omega_v \otimes \iota)(N) = 1$ for a unit vector

$v \in V_x$, which gives the last statement in the proposition. Conversely given a net as in this last statement, then using Alaogu's theorem and passing to a subnet $\{v_i\}$, we get a state z on $B(V_N)$ such that $z(a) = \lim \omega_{v_i}(a)$ for $a \in B(V_N)$. To $b \in \ker \pi_N$ pick a sequence $\{\omega_n\}$ in $(\hat{W})^\sharp$ with $b = \lim \hat{\lambda}_u(\omega_n)$. Then

$$\varepsilon_u(b) = \lim \varepsilon_u \hat{\lambda}_u(\omega_n) = \lim \omega_n(1) = \lim z(\omega_n \otimes \iota)(N) = z\pi_N(b) = 0.$$

□

Hence (W, Δ) is coamenable if and only if there is a net $\{v_i\}$ of unit vectors in V_x such that $\lim \|M(v_i \otimes v) - v_i \otimes v\| = 0$ for $v \in V_x$. The restriction ω_i to W of ω_{v_i} will clearly satisfy

$$\|\omega_i \omega_v - \omega_v\| \leq 2\|v\| \|M(v_i \otimes v) - v_i \otimes v\| \to 0,$$

so $\{\omega_i\}$ is a bounded left approximate unit for W_*, and $\{\omega_i R\}$ is similarly a bounded right approximate unit for W_*, while $\{\omega_i + \omega_i R - (\omega_i R)\omega_i\}$ is a two-sided one. Passing to a subnet, any of these approximate units will converge weakly to a counit rendering (W, Δ) coamenable. Using the lemma as in the proof just before it, we also see that $N \in C(\hat{W}, \hat{\Delta})$ is amenable if it has the weak containment property.

We focus now on the converse direction. Suppose G is an amenable locally compact group with left invariant mean m on $(L^\infty(G), \Delta)$. Since the normal states on $L^\infty(G)$ are w^*-dense in the set of all states, there is a net of positive norm one elements $g_i \in L^1(G)$ such that $(f|g_i) \to m(f)$ for $f \in L^\infty(G)$. Then $(f|h*g_i - g_i) = (h^\star * f|g_i) - (f|g_i) \to m(h^\star * f) - m(f) = 0$ for a positive norm one element $h \in L^1(G)$. For such $\{h_1, \ldots, h_n\}$ the convex set in the n-fold product of $L^1(G)$ with coordinates $h_k * g - g$ as g varies, has therefore by Lemma 13.3.15 a weak limit point, which by the Hahn-Banach theorem must be a norm limit point. So there is a g with $\sup_k \|h_k * g - g\|_1 < \varepsilon$, and we get a net of positive norm one elements $g_j \in L^1(G)$ such that $\|h * g_j - g_j\|_1 \to 0$. Then

$$\|(h * g_j)(s\cdot) - h * g_j\|_1 \leq \|h(s\cdot) * g_j - g_j\|_1 + \|h * g_j - g_j\|_1 \to 0$$

for each $s \in G$, and by Lemma A.10.3, we see that for any compact subset K of G, there is a positive norm one $g \in L^1(G)$ with $\|g(s\cdot) - g\|_1 < \varepsilon$ for all $s \in K$. Then $|(g^{1/2}(s\cdot)|g^{1/2}) - 1|^2 \leq \|g^{1/2}(s\cdot) - g^{1/2}\|_2^2 \leq \|g(s\cdot) - g\|_1 < \varepsilon$ for all $s \in K$. So the function in $A(G)$ given by $s \mapsto (g^{1/2}(s^{-1}\cdot)|g^{1/2})$ converges to the constant function 1 uniformly on compacts, and since $A(G)$ is an ideal in $B(G)$, we conclude that it is dense in $B(G)$. So $W^*(G)_*$ is w^*-dense in $C_u^*(G)^\star$ and the canonical map $C_u^*(G) \to C_r^*(G)$ is injective. Hence G is amenable if and only if the canonical map $C_u^*(G) \to C_r^*(G)$ is a C*-algebra isomorphism. This also shows that the chain of implications in the argument we have just been through are in fact equivalences. For instance, the group G is amenable if and only if there is a net of positive norm one elements $h_i \in L^2(G)$ such that $(h_i(s\cdot)|h_i) \to 1$ uniformly for s on any compact subset of G, a statement known as the *Godement condition*.

Recall that a von Neumann algebra $W \subset B(V)$ is *injective* if there is a linear norm one projection $P: B(V) \to W$. Such a P is automatically a conditional expectation. Clearly direct sums of injective von Neumann algebras are injective.

Proposition 13.3.7 *If $(\hat{W}, \hat{\Delta})$ is amenable, the universal enveloping von Neumann of W_u is injective, so W is in particular injective.*

Proof In view of the lemma below it suffices to show that $\pi_N(W_u)'$ is injective for any $N \in C(\hat{W}, \hat{\Delta})$. The unital $*$-homomorphism $\alpha: B(V_N) \to \hat{W}\bar{\otimes}B(V_N)$ given by $\alpha = N^*(1 \otimes (\cdot))N$ satisfies $(\iota \otimes \alpha)\alpha = (\hat{\Delta} \otimes \iota)\alpha$, and the *fixed point algebra* $B(V_N)^\alpha = \{w \in B(V_N) \,|\, \alpha(w) = 1 \otimes w\}$ equals $\pi_N(W_u)'$ as $(\hat{W})^\sharp$ is dense in $\hat{W}_*$. We construct a norm one projection $P: B(V_N) \to B(V_N)^\alpha$. Indeed, if we have an invariant mean m, then since $|m(\iota \otimes \eta)\alpha(w)| \leq \|\eta\|\|w\|$ for $\eta \in B(V_N)_*$ and $w \in B(V_N)$, we have a linear contraction $P: B(V_N) \to B(V_N)$ such that $\eta P = m(\iota \otimes \eta)\alpha$, which clearly fixes elements in the fixed point algebra, and which projects onto it as $(\omega \otimes \eta)\alpha P(w) = m(\iota \otimes \omega \otimes \eta)(\hat{\Delta} \otimes \iota)\alpha(w) = (\omega \otimes \eta)(1 \otimes P(w))$ for $\omega \in \hat{W}_*$ and $w \in B(V_N)$. □

Lemma 13.3.8 *A von Neumann algebra is injective if and only if its commutant is injective.*

Proof If $W \subset B(V)$ is in standard form with conjugate linear involution J on V, and if it is injective with norm one projection $P: B(V) \to W$, then the linear map $JP(J \cdot J)J: B(V) \to W'$ will obviously be a norm one projection, so W' is injective. Any von Neumann W is $*$-isomorphic to one in standard form. Hence the result follows from Proposition 8.2.9, and the claim that $W\bar{\otimes}B(U)$ is injective for any Hilbert space U, whenever W is injective, say with a norm one projection $P: B(V) \to W$. To prove the claim, first note that when U is finite dimensional, then $P \otimes \iota: B(V \otimes U) \to W \otimes B(U)$ is a norm one projection. When U is infinite dimensional, pick a net $\{p_i\}$ of finite rank orthogonal projections in U that converges σ-weakly to the identity map on U. Then $P'(T) = \lim(P \otimes \iota)((1 \otimes p_i)T(1 \otimes p_i))$ for $T \in B(V \otimes U)$ defines a norm one projection $P': B(V \otimes U) \to W\bar{\otimes}B(U)$. □

Proposition 13.3.9 *The universal enveloping von Neumann algebra of a nuclear C*-algebra is injective.*

Proof By the lemma above it suffices to show that $\pi_z(W)'$ is injective for any unital nuclear C*-algebra W and any state z on W. By nuclearity the formula $(w, w') \mapsto ww'$ defines a $*$-representation f of $\pi_z(W) \otimes \pi_z(W)'$ on V_z, where we are using the minimal tensor product to produce a C*-subalgebra of $\pi_z(W) \otimes B(V_z)$ again with the minimal tensor product. Extending $\omega_{v_z} f$ to a state y on the latter C*-algebra, and consider the orthogonal projection p from V_y onto the closure of $\pi_y(\pi_z(W) \otimes \pi_z(W)')v_y$, which we identify with V_z. Then $P(a) = p\pi_y(1 \otimes a)$ restricted to V_z defines a norm one projection $P: B(V_z) \to \pi_z(W)'$. □

Let (W, Δ) be a compact locally compact quantum group with dense Hopf $*$-algebra W_0. Define a self-adjoint unitary operator u on V_x by the formula $uq_x(b) = q_x((S \otimes f_1)\Delta(b))$, so $\mathrm{Ad}(u)(a)q_x(b) = q_x(b(S \otimes f_1)\Delta(a))$ for $a, b \in W_0$, and $\mathrm{Ad}(u)(W) \subset W'$. Then $C(a) = a_{(1)}\mathrm{Ad}(u)(a_{(2)})$ is a non-degenerate $*$-representation of W_0 on V_x, and thus extends to a unique $*$-representation C_u of W_u on V_x. When (W, Δ) in addition is a Kac algebra, then $(C_u(\cdot)q_x(1)|q_x(1)) = \varepsilon_u$, so $\varepsilon_u \prec C_u$.

Proposition 13.3.10 *Let (W, Δ) be a compact locally compact quantum group. If $m(w \otimes w') = ww'$ defines by linearity and continuity a $*$-homomorphism m from $W \otimes W'$, with minimal tensor product, to $B(V_x)$, then $C_u \prec \pi$, so in the Kac algebra case (W, Δ) is coamenable.*

Proof One checks that $C_u = m(\iota \otimes \mathrm{Ad}(u))\Delta_r\pi$. □

When (W, Δ) is compact and W is injective, the orthogonality relations show that the state $\kappa = (m(\iota \otimes \mathrm{Ad}(u))\Delta_r(\cdot)q_x(1)|q_x(1))$ on W_r has the value $\dim(V_i)d_i^{-1}\delta_{kl}$ on the matrix coefficients u^i_{kl} of any irreducible corepresentation.

In the remaining part of this section we adapt the notation in Example 12.10.14.

Lemma 13.3.11 *Let (W, Δ) be a compact locally compact quantum group. There exists a unique injective unital $*$-homomorphism $g\colon W_0 \to \hat{W}_*$ with dense image such that $g(a)\lambda(\hat{b}) = x(S^{-1}(a)b)$ for $a, b \in W_0$.*

Proof Pick $c \in W_0$ to $a \in W_0$ such that $\hat{c}\widehat{S(a^*)} = \widehat{S(a^*)}$. Then by the Plancherel isomorphism we get

$$g(a)\lambda(\hat{b}) = x(S(a^*)^*b) = \hat{x}(\widehat{S(a^*)}^*\hat{c}^*\hat{b}) = \omega_{q_{\hat{x}}\lambda(\hat{a}),q_{\hat{x}}\lambda(\hat{c})}\lambda(\hat{b})$$

by the KMS-property of $\hat{x}$, so $g(a)$ extends to a normal functional on $\hat{W}$, denoted by the same symbol, and we get a linear map $g\colon W_0 \to \hat{W}_*$ which is injective by faithfulness of x. A similar argument shows that if we to $d, e \in W_0$ pick $f \in W_0$ such that $(S^{-1}(f)^*)^\wedge = \hat{e}(f_1\hat{d}f_{-1})^*$, then $\omega_{q_{\hat{x}}\lambda(\hat{d}),q_{\hat{x}}\lambda(\hat{e})} = g(f)$, so g has dense image. That g is multiplicative is clear from

$$g(a_1)g(a_2)\lambda(\hat{b}) = (g(a_1) \otimes g(a_2))\Delta\lambda(\hat{b}) = x(S^{-1}(a_2)S^{-1}(a_1)b)$$

for $a_i \in W_0$, and antimultiplicativity of S^{-1}. That $g(1)$ is the identity is then clear from denseness. Finally, that g is $*$-preserving means that $g(a^*)\lambda(\hat{b}) = \overline{g(a)}(\hat{S}\lambda(\hat{b}))$, which is readily checked. □

Lemma 13.3.12 *Let (W, Δ) be a compact locally compact quantum group. Then $N \in C(\hat{W}, \hat{\Delta})$ has the weak containment property if there is a net $\{v_i\}$ of unit vectors in V_N such that $\lim \|N(q_{\hat{x}}\lambda(p_k) \otimes v_i) - q_{\hat{x}}\lambda(p_k) \otimes v_i\| = 0$ for all k.*

Proof Arguing as in the proof of Proposition 13.3.6 we get a state z on $B(V_N)$ such that $z(\omega_{q_{\hat{x}}\lambda(p_k)} \otimes \iota)(N) = \omega_{q_{\hat{x}}\lambda(p_k)}(1)$. If this holds with $\omega_{q_{\hat{x}}\lambda(p_k)}$ replaced by $g(a)$ for $a \in W_0$, where g is as in the previous lemma, then we are done by Proposition 13.3.6 and denseness of $g(W_0)$ in $\hat{W}_*$. Now $g(u^k_{ij})(1) = \delta_{ij}$, so we need only show that the state $\kappa = z\pi_N\hat{\lambda}_u g = z(g(\cdot) \otimes \iota)(N)$ on W_u satisfies $\kappa(u^k_{ij}) = \delta_{ij}$

for all i, j, k. Let $b_k = d_k \sum f_1(u^k_{ij})u^k_{ji}$, so $p_k = \hat{b}_k$. Then we get $\omega_{q_{\hat{x}}\lambda(p_k)}\lambda(\hat{c}) = \hat{x}(p^*_k\hat{c}) = x(b^*_k c) = g(b_k)\lambda(\hat{c})$ for $c \in W_0$ since $S(b^*_k) = b_k$. Hence $\kappa(b_k) = \omega_{q_{\hat{x}}\lambda(p_k)}(1) = \hat{x}(p_k) = d_k$ by the first line in the proof and the formula for $\hat{x}$ in terms of traces, so $(\kappa f_1)(d) = f_1(d)$ with $d = \sum_i u^k_{ii}$. Let us introduce $X_{ij} = (\iota \otimes f_{1/2})\Delta(u^k_{ij}) - f_{1/2}(u^k_{ij})$. Then a straightforward calculation, using the properties of the f_z's, shows that $\kappa(\sum X^*_{ij}X_{ij}) = 2(f_1(d) - \mathrm{Re}(\kappa f_1)(d))$. The latter expression vanishes as $(\kappa f_1)(d) = f_1(d)$ is real. So $\kappa(X^*_{ij}X_{ij}) = 0$ by positivity of κ, and $\kappa(X_{ij}) = 0$ by the Cauchy-Schwarz inequality. Thus $(\kappa f_{1/2})(u^k_{ij}) = f_{1/2}(u^k_{ij})$, or $\kappa f_{1/2} = f_{1/2}$, which gives $\kappa(u^k_{ij}) = \varepsilon(u^k_{ij}) = \delta_{ij}$. □

Let N be a unitary corepresentation of a locally compact quantum group (W, Δ) on V_N. Consider the normal $*$-preserving linear maps $l, r \colon B(V_N) \to B(B^2(V_N))$ given by $l(x)y = xy$ and $r(x)y = yx$, where $B^2(V_N)$ is the Hilbert space of Hilbert-Schmidt operators on V_N. So l is multiplicative and r is antimultiplicative. The *Hilbert-Schmidt corepresentation* N_{hs} associated to N is the unitary corepresentation $(\iota \otimes l)(N)(\iota \otimes r)(N)$ of (W, Δ) on $B^2(V_N)$. That it really is a unitary corepresentation is easy to check.

Theorem 13.3.13 *Let (W, Δ) be a compact locally compact quantum group. Then $N \in C(\hat{W}, \hat{\Delta})$ is amenable if and only its associated Hilbert-Schmidt corepresentation N_{hs} has the weak containment property.*

Proof In this proof we denote $\lambda(p_k)$ and $\lambda(f_t)$ by p_k and f_t, respectively. Let $y \in B^2(V_N)$, and consider n such that $\hat{S}(p_k) = p_n$, so $\hat{R}(p_k) = f_{-1/2}\hat{S}(p_k)f_{1/2} = p_n$. Now $(\hat{R} \otimes r)(N)(p_k \otimes 1) = (\hat{R} \otimes r)((p_n \otimes 1)N)$ by continuity and linearity since it holds when N is an elementary tensor. But $(p_n \otimes 1)N \in B(V_n) \otimes B(V_N)$, so we can use algebraic tensor product of linear maps, which we here denote by $\odot$ for clearity. Thus $(\hat{R} \otimes r)(N)(p_k \otimes 1) = (\iota \odot r)(\hat{R} \odot \iota)((p_n \otimes 1)N)$, and $(\hat{R} \odot \iota)((p_n \otimes 1)N) = (f_{-1/2} \otimes 1)N^*(p_k f_{1/2} \otimes 1)$, which is checked by slicing with $\iota \otimes \omega$ on both sides and using $\hat{S}(\iota \otimes \omega)(N) = (\iota \otimes \omega)(N^*)$. Combining this, we see that

$$N_{hs}(q_{\hat{x}}(p_k) \otimes y) = (\iota \otimes l)(N)(\hat{R} \otimes r)(N)(p_k \otimes 1)(q_{\hat{x}}(p_k) \otimes y)$$

which equals $(\iota \odot l)(N(p_k \otimes 1))(\iota \odot r)((f_{-1/2} \otimes 1)N^*)(p_k f_{1/2} \otimes 1))(q_{\hat{x}}(p_k) \otimes y)$ which again equals $(q_{\hat{x}} \odot \iota)(N(p_k f_{-1/2} \otimes y)N^*(f_{1/2} \otimes 1))$, so by the formula for $\hat{x}$ in terms of traces, we get

$$\|N_{hs}(q_{\hat{x}}(p_k) \otimes y) - q_{\hat{x}}(p_k) \otimes y\| = d_k^{1/2}\|N(p_k f_{-1/2} \otimes y)N^* - p_k f_{-1/2} \otimes y\|_k,$$

where $\|\cdot\|_k$ is the Hilbert-Schmidt norm on $B(V_k) \otimes B^2(V_x)$.

Now assume that N is amenable. By the previous lemma it suffices therefore to produce a net y_i of positive elements in $B^2(V_N)$ of Hilbert-Schmidt norm one such that when we replace y by y_i in the last norm expression above, and take the limit, we get zero. As the normal states are w^*-dense in the space of states on $B(V_N)$, we may pick a net of positive operators s_i on V_N of trace class norm one such that an invariant mean for N is a w^*-limit point of $\{\mathrm{Tr}(\cdot s_i)\}$. Set $y_i = s_i^{1/2}$. To $e, e' \in \oplus B(V_j) \subset \hat{W}_r$, set $c_n = f_1 p_n e' f_{-1} e^*$. Also write $(p_n \otimes 1)N = \sum a_r \otimes b_r$. Then using the formula for $\hat{x}$ in terms of traces, we get for $b \in B(V_N)$ that

$$\begin{aligned} 0 &= \lim_i \omega_{q_{\hat{x}}(p_n e'), q_{\hat{x}}(e)}(1)\mathrm{Tr}(b s_i) - \lim_i \mathrm{Tr}((\omega_{q_{\hat{x}}(p_n e'), q_{\hat{x}}(e)} \otimes \iota)(N(1 \otimes b)N^*)s_i) \\ &= \lim_i (\hat{x} \odot \mathrm{Tr})(c_n \otimes b y_i^2 - (c_n \otimes y_i^2)N(1 \otimes b)N^*) \\ &= \lim_i d_n(\mathrm{Tr}_n \odot \mathrm{Tr})((c_n \otimes b)(p_n f_{-1} \otimes y_i^2 - \sum a_r p_n a_s^* f_{-1} \otimes b_s^* y_i^2 b_r)). \end{aligned}$$

As e, e' vary, the elements c_n will exhaust $B(V_n)$, and the span of the elements $c_n \otimes b$ as also b varies, will be $B(V_n) \otimes B(V_N) = B(V_n \otimes V_N)$. But this latter space, under that trace as prescribed, is the dual of $B^1(V_n \otimes V_N)$. To $z \in B^1(V_N)$ let $T_n(z) = p_n f_{-1} \otimes z - \sum a_r p_n a_s^* f_{-1} \otimes b_s^* z b_r$. Then we have just shown that $T_n(y_i^2) \to 0$ weakly in the Banach space $B^1(V_n \otimes V_N)$. Consider the locally convex topological vector space $\prod_n B^1(V_n \otimes V_N)$ with the product topology of the $\|\cdot\|_{1,n}$-norms on each block. The weak topology on $\prod_n B^1(V_n \otimes V_N)$ is by the first lemma below, the product topology of the weak topologies on each block. So $\lim_i (T_n(y_i^2))_n = 0$ weakly in $\prod_n B^1(V_n \otimes V_N)$. Now the set E of all $(T_n(z))_n$ as z varies over all positive operators on V_N of trace class norm one, is evidently a convex subset of $\prod_n B^1(V_n \otimes V_N)$, and we have shown that 0 belongs to the weak closure of E, and by Proposition 2.4.14, it belongs the closure of E in the product topology of the norms. Hence there is some net in E that converges to 0 in the latter topology. Writing the i-th member of this net as say $(T_n(y_i^2))_n$ for a positive operator y_i on V_n of Hilbert-Schmidt norm one, we therefore get

$$\lim_i \|p_n f_{-1} \otimes y_i^2 - \sum a_r p_n a_s^* f_{-1} \otimes b_s^* y_i^2 b_r\|_{1,n} = 0$$

for each n. Since $B(V_n)$ is a matrix algebra we can apply the linear map $\hat{S}(\cdot f_2)$ on the first tensor factors and still keep convergence. If we also use $(\hat{S} \odot \iota)(N^*(p_n \otimes 1)) = (f_1 \otimes 1)N(p_k f_{-1} \otimes 1)$, we then get

$$\lim_i \|p_k f_{-1} \otimes y_i^2 - N(p_k f_{-1} \otimes y_i^2)N^*\|_{1,k} = 0.$$

By the second lemma below we therefore get the desired limit.

For the converse direction, it is immediate from Proposition 13.3.6 and the proof of Proposition 13.3.4 that N_{hs} is amenable, say with an invariant mean $m_{N_{hs}}$. Fix a conjugate linear map J on V_N such that $J^2 = J = J^*$, and consider the unitary $A\colon V_N \otimes V_N \to B^2(V_N)$ given by $A(u \otimes v)(v') = (v'|J(u))v$ for any $u, v, v' \in V_N$. Then $m_N(x) = m_{N_{hs}}(A(1 \otimes x)A^*)$ for $x \in B(V_N)$ defines an invariant mean m_N for N since $A(1 \otimes x)A^* = l(x)$ and $A(x \otimes 1)A^* = r(Jx^*J)$, so that for $\omega \in \hat{W}_*$

we see that $m_N(\omega \otimes \iota)(N(1 \otimes x)N^*)$ equals

$$m_{N_{hs}}(\omega \otimes \iota)(A_{23}N_{13}N_{12}A_{23}^*A_{23}(1 \otimes 1 \otimes x)A_{23}^*A_{23}N_{12}^*N_{13}^*A_{23}),$$

which again equals $m_{N_{hs}}(\omega \otimes \iota)(N_{hs}(1 \otimes l(x))N_{hs}^*) = \omega(1)m_{N_{hs}}(l(x)) = \omega(1)m_N(x)$. □

It is clear from the proof that the converse direction in the theorem is true for any locally compact quantum group. It is an open question whether the more elaborate forward direction holds in full generality.

Corollary 13.3.14 *A compact locally compact quantum group is coamenable if its dual if amenable.*

Proof Say we have a compact locally compact quantum group (W, Δ) with $(\hat{W}, \hat{\Delta})$ amenable. Then $\hat{M}$ is amenable, so its associated Hilbert-Schmidt corepresentation $\hat{M}_{hs}$ has the weak containment property by the theorem above, say with a state z on $B(B^2(V_x))$ such that $z(\omega \otimes \iota)(\hat{M}_{hs}) = \omega(1)$ for $\omega \in \hat{W}_*$. Consider the unitary $A\colon V_x \otimes V_x \to B^2(V_x)$ from the last part of the proof of the theorem above, now with $J = J_{\hat{x}}$ for convenience. Then the state z' on $B(V_x)$ given by $z'(x) = z(A\hat{M}_{21}(1 \otimes x)\hat{M}_{21}A^*)$ satifies

$$z'(\omega \otimes \iota)(\hat{M}) = z(A(\omega \otimes \iota \otimes \iota)(\hat{M}_{32}\hat{M}_{13}\hat{M}_{32}^*)A^*) = z(\omega \otimes \iota)(\hat{M}_{hs}) = 1$$

for $\omega \in \hat{W}_*$ because $\hat{M}_{32}\hat{M}_{13}\hat{M}_{32}^* = \hat{M}_{13}\hat{M}_{12}$ by the pentagonal equation and because of the properties of A in the last proof combined with $(\hat{R} \otimes R)(\hat{M}) = \hat{M}$. Hence (W, Δ) is coamenable. □

Lemma 13.3.15 *Let V_λ be normed spaces and let $V = \prod V_\lambda$ be the algebraic product with product topology. Then the weak topology on V is the same as the product topology of the weak topologies on the normed spaces V_λ.*

Proof Let $\pi_\lambda\colon V \to V_\lambda$ and $I_\lambda\colon V_\lambda \to V$ denote the projections and injections associated to the direct product. Suppose a net $\{v_i\}$ converges weakly to v in V, so $x(v_i) \to x(v)$ for every $x \in V^\star$. Such a functional x has the property that $x(u_j) \to x(u)$ when $u_j \to u$ in the original topology on V, that is, when each $\pi_\lambda(u_j) \to \pi_\lambda(u)$ in norm. Then $v_i \to v$ in the product topology of the weak topologies, meaning that $y\pi_\lambda(v_i) \to y\pi_\lambda(v)$ for each λ and $y \in V_\lambda^\star$, and this holds simply because $y\pi_\lambda \in V^\star$. Conversely, assuming this holds for each y and λ, we will see that $x(v_i) \to x(v)$ for all $x \in V^\star$ by characterizing such $x \neq 0$. Since $\ker x \neq V$, we may by Theorem 2.4.4 assume there is a point $w' \in V$ and a balanced neighborhood W of 0 such that $w' + W$ is disjoint from the closed subset $\ker x$ of V. Then $x(W)$ must be bounded. Otherwise $x(W) = \mathbb{C}$, and for suitable $w'' \in W$ we would get the contradiction $x(w' + w'') = 0$. Given the nature of the product topology on V, we may assume that W is the set of all w in the product such that, for each λ in a fixed finite set A of indices, each w_λ belongs to a given neighborhood W_λ of zero in V_λ. If $\pi_\lambda(w) = 0$ for all $\lambda \in A$, then every scalar multiple of w

also belongs to W, so $x(w) = 0$ since otherwise $x(W)$ would be unbounded. Hence $0 = x(v' - \oplus_{\lambda\in\Lambda} I_\lambda \pi_\lambda(v'))$ for all $v' \in V$, so $x = \oplus_{\lambda\in\Lambda} x I_\lambda \pi_\lambda$. Now just note that each linear functional xI_λ is continuous since every norm convergent sequence in V_λ can be extended to a norm convergent sequence in V that is zero on all the other blocks. □

Lemma 13.3.16 *We have* $\mathrm{Tr}(|a - b|) \geq \mathrm{Tr}((a^{1/2} - b^{1/2})^2)$ *for positive Hilbert-Schmidt operators a and b.*

Proof Let $c = a^{1/2} - b^{1/2}$ and $d = a^{1/2} + b^{1/2}$, so $a - b = (cd + dc)/2$ is self-adjoint and $d \geq \pm c$. Let $\{v_i\}$ be an orthonormal basis of eigenvectors for the compact self-adjoint operator c with corresponding eigenvalues $\{t_i\}$. Since $|a - b| \geq \pm(a - b)$, we therefore get

$$\mathrm{Tr}(|a-b|) \geq \sum |((a-b)(v_i)|v_i)| = \sum |t_i(d(v_i)|v_i)| \geq \sum t_i^2 = \sum (c^2(v_i)|v_i).$$

□

Exercises

For Sect. 13.1

1. Investigate to what extend the main result in this section can be extended to C*-algebras with coproducts associated to manageable multiplicative unitaries.
2. Let (W, Δ) be a locally compact quantum group with associated $\mathcal{M}$ as described in this section. Show that $f \mapsto (f \otimes \iota)(\mathcal{M})$ is a bijection from the set of non-degenerate $*$-homomorphisms $W_u \to L(V)$ to the set of unitaries N in $L(V \otimes \hat{W}_r)$ such that $(\iota \otimes \hat{\Delta})(N) = N_{13}N_{12}$, where V is a Hilbert module over a C*-algebra and $L(V)$ are the adjointable maps on V.

For Sect. 13.2

1. Let G be a locally compact group. Show that any automorphism of $L^\infty(G)$ that commutes with the right translations on G is a left translation.
2. Show that a closed subspace of $L^1(G)$ of a locally compact group G is an ideal if and only if it is translation invariant.
3. The Bohr compactification of an abelian locally compact group G is the dual $\tilde{G}$ of $\hat{G}$ considered as a discrete group. Show that the natural inclusion of G in $\tilde{G}$ is continuous onto a dense subgroup of $\tilde{G}$.
4. Let G be an abelian compact group. Show that if the Fourier transforms of the members of a closed subalgebra B of $L^1(G)$ separate the points in $\hat{G}$, then B is either $L^1(G)$ or a maximal ideal of $L^1(G)$.

For Sect. 13.3

1. Prove that any discrete group that contains the free group F_2 on two generators as a closed subgroup is not amenable.
2. Show that $SU(2)$ considered as a discrete group is not amenable.
3. Show that the compact quantum group $SU_q(2)$ with $q \in \langle 0, 1]$ is coamenable.
4. Exhibit a locally compact quantum that is neither amenable nor coamenable.

Chapter 14
Classical Crossed Products

Since in the remaining part of this text we will study various crossed products associated to quantum groups, it makes sense to begin with a small chapter on crossed products of C*-dynamical systems of locally compact groups. Such systems are defined just as von Neumann dynamical systems except for the fact that we consider continuity with respect to norm. A crossed product is then a certain C*-algebra describing the dynamical system. This interplay between dynamical systems and C*-algebras is useful for studying such systems by C*-algebra techniques, and conversely, in constructing new C*-algebras from old ones by introducing such systems.

The universal crossed product of a C*-dynamical system generalizes the universal group C*-algebra of the group; the latter C*-algebra occurs when the action is the trivial one on $\mathbb{C}$. There is a correspondence between non-degenerate representations of the universal crossed product of a dynamical system and so called covariant representations of the system; these are the representations of the C*-algebra in the system where the group action appears as inner by a strongly continuous unitary representation of the group. Hence, by invoking the regular representation of the group, any non-degenerate representation of the C*-algebra in the dynamical system induces a representation of the universal crossed product of the system. Taking the supremum over all such representations one obtains a C*-norm on a dense $*$-subalgebra of the universal crossed product playing the role of L^1-functions on the group. The completed C*-algebra thus obtained is the so called reduced crossed product of the dynamical system; when the action is the trivial one on $\mathbb{C}$, we get the reduced group C*-algebra of the group. We show that the group is amenable if and only if the $*$-epimorphism from the universal crossed product to the reduced one is injective. In general, both the group and the original C*-algebra of the dynamical system can be regarded as sitting inside the multiplier algebra of the universal crossed product of the system.

When the group is amenable there is an action of the Pontryagin dual of the group on the universal crossed product of the original C*-dynamical system; note

L. Tuset, *Analysis and Quantum Groups*,
https://doi.org/10.1007/978-3-031-07246-8_14

that abelian groups are amenable. Taking the crossed product of this dual system one obtains the C*-algebra in the original system modulo tensoring with the C*-algebra of compact operators. This is known as Takesaki-Takai duality. A similar result in the von Nuemann setting is also proved.

In the third section we study so called G-products. This is a way of characterizing the C*-algebras arising uniquely as crossed products by C*-dynamical systems of abelian groups. Again we also prove a von Neumann version of this result. We also study simple and prime C*-algebras in this context.

In the last section we study some examples of crossed products of dynamical systems. We show that the C*-algebra of compact operators on a Hilbert space is a G-product, and that the von Neumann algebra closure of all bounded operators is a von Neumann G-product. We also see how semidirect products of groups give rise to crossed products, and finally we study Weyl systems as a historical approach to crossed products.

Recommended literature for this chapter is [10, 39].

14.1 Crossed Products of Actions

Definition 14.1.1 A C*-dynamical system (W, G, α) consists of a C*-algebra W and a homomorphism $t \mapsto \alpha_t$ from a locally compact group G to the group $\mathrm{Aut}(W)$ of automorphisms of W such that $t \mapsto \alpha_t(w)$ is continuous for $w \in W$.

Given a C*-dynamical system (W, G, α) we denote by $L^1(G, \alpha)$ the linear space of measurable functions $f : G \to W$ with finite norm $\|f\|_1 \equiv \int \|f(t)\|\, dt$. Under this norm it is a Banach $*$-algebra with product and $*$-operation given by $(f*g)(t) = \int f(s)\alpha_s(g(s^{-1}t))\, ds$ and $f^\star(t) = \Delta(t)^{-1}\alpha_t(f(t^{-1}))^*$. The continuous functions $G \to W$ with compact support form a $*$-subalgebra with a subspace $K(G) \odot W$ which is dense in $L^1(G, \alpha)$. Note that $L^1(G, \alpha) = L^1(G)$ when $W = \mathbb{C}$ and $\alpha = \iota$.

Proposition 14.1.2 *Given a C*-dynamical system (W, G, α) with an approximate unit $\{f_i\}_{i\in I} \subset K(G)$ for $L^1(G)$ and an approximate unit $\{u_j\}_{j\in J}$ for W, then $t \mapsto f_i(t)\alpha_t(u_j)$ with product order on $I \times J$ is an approximate unit for $L^1(G, \alpha)$, and $\{f_i u_j\}$ is also an approximate unit. The Banach $*$-algebra $L^1(G, \alpha)$ admits a faithful representation.*

Proof For $w \in W$ and $f \in L^1(G)$ we have

$$\|(fw)*f_i(\cdot)\alpha_{(\cdot)}(u_j) - fw\|_1 \leq \|w\| \cdot \|f * f_i - f\|_1 + \int |(f * f_i)(t)| \cdot \|w\alpha_t(u_j) - w\|\, dt.$$

The latter integral tends to zero as $(i, j) \to \infty$ since the first factor in it will be small outside some compact subset of G whereas the second factor converges uniformly to zero on any compact set. Similarly we get the required behavior when we convolute with fw from the right. Since functions of the form $t \mapsto f(t)\alpha_t(w)$ span a dense subspace of $L^1(G, \alpha)$, we see that $\{f_i u_j\}$ is also an approximate unit.

Assuming $W \subset B(V)$ for some Hilbert space V, we define a representation π of $L^1(G, \alpha)$ on $L^2(G, V) = L^2(G) \otimes V$ by $(\pi(f)\eta)(t) = \int \alpha_{t^{-1}}(f(s))\eta(s^{-1}t)\, ds$. If $\pi(f) = 0$, then taking inner product we see that for $u, v \in V$ and $g \in K(G)$, the function $h \colon t \mapsto \int (\alpha_{t^{-1}}(f(s))u|v)g(s^{-1}t)\, ds$ vanishes almost everywhere. But it must actually vanish everywhere because it is continuous since $|h(t) - h(r)|$ is not greater than

$$\int \|f(s)\| \cdot \|u\| \cdot \|v\| \cdot |g(s^{-1}t) - g(s^{-1}r)|\, ds + 2\|u\| \cdot \|v\| \cdot \|g\|_u \cdot \|f - \sum g_i \otimes w_i\|$$

plus $\sum \int |((\alpha_{t^{-1}} - \alpha_{r^{-1}})(w_i)u|v)| \cdot |g_i(s)g(s^{-1}t)|\, ds$. In particular it must vanish for $t = e$, so $(f(s)u|v) = 0$ almost everywhere. Letting V be the Hilbert space for the universal representation of W, we conclude that $zf(s) = 0$ almost everywhere and for every $z \in W^\star$. Since there are $k, n_k \in \mathbb{N}$ and $w_{kn} \in W$ and $h_{kn} \in L^1(G)$ such that $\|f - \sum_{k=1}^{n_k} h_{kn} \otimes w_{kn}\| \leq 1/n_k$, we see that $f(G)$ is contained in the closed span of $\{w_{nk}\}$. So we may assume that W is separable. Pick a w^*-dense sequence $\{z_n\}$ in the closed unit ball of $W^\star$. Let B_n be the Borel subset of measure zero where $z_n f \neq 0$. Then $B = \cup B_n$ has measure zero and all $z_n f = 0$ outside B. But then also $f = 0$ outside B. □

Definition 14.1.3 The *universal crossed product* $G \ltimes_\alpha W$ of a C*-dynamical system (W, G, α) is the universal enveloping C*-algebra of $L^1(G, \alpha)$.

By the previous proposition and Remark 13.2.4 this crossed product contains $L^1(G, \alpha)$ as a dense $*$-subalgebra and $\| \cdot \|_u \leq \| \cdot \|_1$ there with $K(G) \odot W$ dense. They have the same state spaces and $\{f_i u_j\}$ is an approximate unit for $G \ltimes_\alpha W$. When $\alpha = \iota$ the crossed product equals the maximal tensor product $C_u^*(G) \otimes_{max} W$.

A pair (π, u) is a *covariant representation* of a C*-dynamical system (W, G, α) if π is a non-degenerate representation of W on a Hilbert space V and u is a strongly continuous unitary representation of G on V such that $\pi(\alpha_t(w)) = u_t \pi(w) u_t^*$.

Proposition 14.1.4 *The map $(\pi, u) \mapsto \pi \times u$ with $(\pi \times u)(f) = \int \pi(f(t))u_t\, dt$ is a bijection from the class of covariant representations of a C*-dynamical system (W, G, α) to the class of non-degenerate representations of $L^1(G, \alpha)$.*

Proof It is straightforward to check that $\pi \times u$ is a non-degenerate representation of $L^1(G, \alpha)$. Conversely, given such a representation θ, pick an approximate unit $\{v_i\}$ for $L^1(G, \alpha)$ and let $\pi(w)$ and u_t be the strong limits of $\theta(wv_i)$ and $\theta(\alpha_t(v_i(t^{-1}\cdot)))$ respectively. Then one readily checks that (π, u) is a covariant representation of the dynamical system and $\pi \times u = \theta$. □

Note in passing that the weak closure of $(\pi \times u)(G \ltimes_\alpha W)$ is the bicommutant of $\pi(W) \cup u_G$.

The function $f \colon G \to W^\star$ given by $t \mapsto (\pi(\cdot)u_t v|v)$ for $v \in V$ is w^*-continuous and *positive definite* in the sense that $\sum f(t_i^{-1}t_j)(\alpha_{t_i^{-1}}(w_i^* w_j)) \geq 0$ for any finite collection of $t_i \in G$ and $w_j \in W$. For any positive definite function f of the dynamical system, the complex valued function on G given by $t \mapsto f(t)(w^*\alpha_t(w))$ with $w \in W$ is positive definite. So $f(e) \geq 0$. By polarization and by using an approximate unit for W, we see that $\|f(t)\| \leq 2\|f(e)\|$ for $t \in G$. It is also clear that if g is positive definite on G, then the function $G \to W^\star$ given by $t \mapsto g(t)f(t)$ is positive definite. The following result is now easily checked.

Proposition 14.1.5 *Let (W, G, α) be a C*-dynamical system. The formula $\tilde{g}(f) = \int g(t)(f(t))\,dt$ for a positive definite w^*-continuous function $g\colon G \to W^\star$, defines a positive linear functional $\tilde{g}$ on $L^1(G,\alpha)$, and $g \mapsto \tilde{g}$ is a bijection between these two sets, which takes those g with $\|g(e)\| = 1$ to the states on $L^1(G,\alpha)$. The inverse image of a positive linear functional z on $L^1(G,\alpha)$ is the positive definite w^*-continuous function associated (as in the previous paragraph with $v = v_z$) to the covariant representation of the GNS-representation of z.*

Definition 14.1.6 The *regular representation* $\mathrm{Ind}\,\pi$ of $G \ltimes_\alpha W$ induced by a non-degenerate representation π of W on V is the representation $\pi' \times (\lambda \otimes \iota)$ associated to the covariant representation $(\pi', \lambda \otimes \iota)$ on $L^2(G, V)$ of the C*-dynamical system (W, G, α), where $(\pi'(w)\eta)(t) = \pi(\alpha_{t^{-1}}(w))\eta(t)$.

Hence $(\mathrm{Ind}\,\pi(f)\eta)(t) = \int \pi(\alpha_{t^{-1}}(f(s)))\eta(s^{-1}t)\,dt$ for $f \in L^1(G,\alpha)$. Using the fact that the Haar measure is a Radon measure, it is not difficult to see that whenever v_i are cyclic vectors for $\pi(W)$ and f_j are cyclic vectors for $\lambda(L^1(G))$, then $f_j \otimes v_i$ are cyclic vectors for $\mathrm{Ind}\,\pi(G \ltimes_\alpha W)$. Thus by invoking an approximate unit for $L^1(G)$ with compact support, we see that if z is a positive linear functional on W with GNS-representation (π_z, V_z, v_z), the linear span of $\mathrm{Ind}\,\pi_z(g)(f \otimes v_z)$ with $g \in K(G, W)$ and $f \in K(G)$ is dense in $L^2(G, V_z)$. Therefore $\|\mathrm{Ind}\,\pi_z(h)\|$ for $h \in L^1(G,\alpha)$ is the supremum of $\|\mathrm{Ind}\,\pi_z(h*g)(f \otimes v_z)\|$ over $g \in K(G, W)$ and $f \in K(G)$ with $\|\mathrm{Ind}\,\pi_z(g)(f \otimes v_z)\| = 1$. Since $(\mathrm{Ind}\,\pi_z(h)(f_1 \otimes v_z)|f_2 \otimes v_z) = \int z\alpha_{t^{-1}}(h(s)) f_1(s^{-1}t)\overline{f_2(t)}\,dsdt \equiv z_{f_1 f_2}(h)$, we can also say that $\|\mathrm{Ind}\,\pi_z(h)\|$ is the supremum of $z_{ff}(g^\star * h^\star * h * g)^{1/2}$ over f and g with $z_{ff}(g^\star * g) = 1$.

Lemma 14.1.7 *Given a C*-dynamical system (W, G, α), consider the regular representation $\mathrm{Ind}\,\pi$ of $G \ltimes_\alpha W$ induced by a non-degenerate representation π of W on V, and let $h \in L^1(G,\alpha)$. Then $\|\mathrm{Ind}\,\pi(h)\|$ is the supremum of the fraction $z_{ff}(g^\star * h^\star * h * g)^{1/2}/z_{ff}(g^\star * g)^{1/2}$ over $f \in K(G)$ and $g \in K(G, W)$ and positive linear functionals z of W such that $z_{ff}(g^\star * g) > 0$ and $\ker\pi \subset \ker\pi_z$.*

Proof Such a supremum certainly majorizes $\|\mathrm{Ind}\,\pi(h)\|$ by the discussion prior to the lemma since $\pi = \oplus_{z\in I}\pi_z$ and $\mathrm{Ind}\,\pi = \oplus_{z\in I}\mathrm{Ind}\,\pi_z$ for some I. We are done if we can show that this norm is never strictly less than $\|\mathrm{Ind}\,\pi_z(h)\|$ for any of the z's we are taking supremum over. When $z = \omega_{v,v}\pi$ for some $v \in V$, consider v a cyclic vector for a subrepresentation of π among a family of other cyclic vectors of subrepresentations π_i of π on orthogonal subspaces of V. Again as Ind respects such a decomposition, then $\|\mathrm{Ind}\,\pi(h)\|$ majorizes each $\|\mathrm{Ind}\,\pi_i(h)\|$ and hence also $\|\mathrm{Ind}\,\pi_z(h)\|$. If z is a sum of such vector states z^j with $j = 1, \ldots, n$ we claim that $\|\mathrm{Ind}\,\pi_z(h)\| \leq \sup \|\mathrm{Ind}\,\pi_{z^j}(h)\|$, so again we get the desired inequality. The claim holds because if $a_j = z^j_{ff}(g^\star * g)$ and $b_j = z^j_{ff}(g^\star * h^\star * h * g)$, then

$$z_{ff}(g^\star * h^\star * h * g)/z_{ff}(g^\star * g) = \frac{b_1 + \cdots + b_n}{a_1 + \cdots + a_n} \leq \sup \frac{b_i}{a_i}.$$

For a general state z among those we are taking supremum over, consider it a state on $W/\ker\pi$ and extend it to a state on $B(V)$, which by the Hahn-Banach separation theorem is a w^*-limit of a net of convex sums z_i of vector states z^j of the above type. As we have w^*-convergence $z_i\alpha_t \to z\alpha_t$ uniformly on any compact subset of G, we see that $(z_i)_{ff} \to z_{ff}$ pointwise on $K(G) \odot W$, and since $\|(z_i)_{ff}\| \leq \|f\|_\infty \|f\|_1$, we get even pointwise convergence on $L^1(G,\alpha)$. Hence $\|\mathrm{Ind}\,\pi_z(h)\| \leq \sup \|\mathrm{Ind}\,\pi_{z_i}(h)\| \leq \|\mathrm{Ind}\,\pi(h)\|$. □

Definition 14.1.8 The *reduced crossed product* $G \ltimes_{\alpha r} W$ of a C*-dynamical system (W, G, α) is the completion of $L^1(G,\alpha)$ with respect to the C*-norm $\|h\|_r$ being the supremum of $\|\mathrm{Ind}\,\pi(h)\|$ over the non-degenerate representations π of W.

Since every C*-algebra W has a faithful non-degenerate representation π, and that $\mathrm{Ind}\,\pi$ is then automatically faithful by the proof of Proposition 14.1.2, the reduced norm is indeed a norm. By the lemma above $\|\mathrm{Ind}\,\pi(h)\|$ is the supremum of the fraction $z_{ff}(g^\star * h^\star * h * g)^{1/2}/z_{ff}(g^\star * g)^{1/2}$ over $f \in K(G)$and $g \in K(G,W)$ and positive linear functionals z of W such that $z_{ff}(g^\star * g) > 0$.

Proposition 14.1.9 *Let (W, G, α) be a C*-dynamical system and let π be a non-degenerate representation of W on V. Then $\oplus_{t\in G}\pi\alpha_t$ is a faithful representation of W if and only if $\|\mathrm{Ind}\,\pi(h)\| = \|h\|_r$ for $h \in L^1(G,\alpha)$. So if π is faithful, then* Ind π *extends uniquely to a faithful representation of $G \ltimes_{\alpha r} W$.*

Proof If $\|\mathrm{Ind}\,\pi(h)\| = \|h\|_r$ and $\pi\alpha_t(w) = 0$, then $\|g \otimes w\|_r = 0$ for $g \in K(G)$, so $w = 0$. Conversely, assume that $\oplus_{t\in G}\pi\alpha_t$ is faithful. Note that $(U_t f)(s) = \Delta(t)^{1/2} f(st)$ defines a unitary U_t on $L^2(G,V)$ such that $U_t^*\mathrm{Ind}\,\pi(\cdot)U_t = \mathrm{Ind}\,\pi\alpha_t(\cdot)$. By the previous lemma and the discussion before it $\|\mathrm{Ind}\,\pi(h)\|$ is therefore the supremum of $z_{ff}(g^\star * h^\star * h * g)^{1/2}/z_{ff}(g^\star * g)^{1/2}$ over $f \in K(G)$ and $g \in K(G,W)$ and positive linear functionals z of W such that $z_{ff}(g^\star * g) > 0$ and $\ker\pi\alpha_t \subset \ker\pi_z$. But as in the proof of the lemma, any state on W is the w^*-limit of states that are convex sums of vector states of the representations $\pi\alpha_t$, so we can drop the requirement $\ker\pi\alpha_t \subset \ker\pi_z$, and the supremum will simply yield $\|h\|_r$. □

Proposition 14.1.10 *Let (W, G, α) be a C*-dynamical system and let U be C*-subalgebra of W that is invariant under the action α. Then $G \ltimes_{\alpha r} U \subset G \ltimes_{\alpha r} W$.*

Proof If π is a non-degenerate faithful representation of U on V, then since it is a direct sum of GNS-representations of states on U that can be extended to states on W, we can clearly extend π to a representation π' of W on V' such that π' restricted to V is π. Hence for $h \in L^1(G,\alpha|U)$ we have $\mathrm{Ind}\,\pi'(h) = \mathrm{Ind}\,\pi(h)$ on $L^2(G,V) \subset L^2(G,V')$. Thus the norm of h in $G \ltimes_{\alpha r} W$ is greater than $\|\mathrm{Ind}\,\pi'(h)\| \geq \|\mathrm{Ind}\,\pi(h)\| = \|h\|_r$ with the latter norm taken in $G \ltimes_{\alpha r} U$. The converse inequality is obvious since every representation of the former crossed product restricts to a representation of the latter crossed product. □

As a corollary of the theorem below we have a similar inclusion of the universal crossed products when G is amenable. When $\alpha = \iota$ and $W \subset B(V)$, then $G \ltimes_{\alpha r} W$

is the minimal C*-tensor product $C_r^*(G) \otimes_{min} W$. The forward implication in the following theorem is therefore clear.

Theorem 14.1.11 *We have $G \ltimes_\alpha W = G \ltimes_{\alpha r} W$ for any C*-dynamical system (W, G, α) if and only if G is amenable.*

Proof If G is amenable there is by Godement's condition a net of unit vectors $g_i \in L^2(G)$ such that $(g_i(t^{-1}\cdot)|g_i) \to 1$ uniformly on compacts. To any state z on $G \ltimes_\alpha W$ write $\pi_z = \pi \times u$ for a covariant representation (π, u) of $L^1(G, \alpha)$. Let $f_i(s) = g_i(s)u_{s^{-1}}v_z$, so f_i are unit vectors in $L^2(G, V_z)$. For $h \in L^1(G, \alpha)$ the number $z(h^\star * h)$ equals

$$\lim \int (g_i(t^{-1}\cdot)|g_i)(\pi((h^\star * h)(t))u_t v_z|v_z)\, dt = \lim(\mathrm{Ind}\,\pi(h^\star * h) f_i | f_i) \leq \|h\|_r^2,$$

so the universal norm of h is not greater than $\|h\|_r$. □

Let (W, G, α) be a C*-dynamical system with a non-degenerate faithful representation of $G \ltimes_\alpha W$, say $\pi \times u$ for a covariant representation (π, u) of $L^1(G, \alpha)$, so π is also non-degenerate and faithful. The map T_w on $L^1(G, \alpha)$ given by $T_w(f)(t) = wf(t)$ for $w \in W$ is bounded by $\|w\|$ and satisfies $(\pi \times u)(T_w(f)) = \pi(w)(\pi \times u)(f)$, so for the norm in the crossed product $\|T_w(f)\| = \|(\pi \times u)(T_w(f))\| \leq \|w\|\|f\|$ and T_w can be extended to a linear map on $G \ltimes_\alpha W$ that is bounded in norm by $\|w\|$. It is easy to see that T_w is adjointable, so $T_w \in M(G \ltimes_\alpha W)$ and $w = \lim T_w(f_i)$ for an approximate unit $\{f_i\}$ in $L^1(G, \alpha)$, where the limit is the strong one in any faithful non-degenerate representation of the crossed product. Hence $w \mapsto T_w$ is a $*$-monomorphism from W to $M(G \ltimes_\alpha W)$. Similarly the map T_s on $L^1(G, \alpha)$ given by $T_s(f)(t) = \alpha_s(f(s^{-1}t))$ for $s \in G$ extend by continuity to a bounded adjointable linear map T_s on $G \ltimes_\alpha W$ such that $(\pi \times u)(T_s(f)) = u_s(\pi \times u)(f)$ for $f \in L^1(G, \alpha)$, and invoking $\{f_i\}$, we see that the map $s \mapsto T_s$ is a monomorphism from G to the group of unitary elements in $M(G \ltimes_\alpha W)$. This map is moreover continuous with respect to the strict topology on the multiplier algebra since $\|T_s(f) - f\| \leq \int \|f(s^{-1}t) - f(t)\|\, dt + \int \|(\alpha_s - \iota)(f(t))\|\, dt$. Using these monomorphisms we embed the group G and the C*-algebra W into $M(G \ltimes_\alpha W)$, and regarding $\pi \times u$ as a faithful representation of $M(G \ltimes_\alpha W)$ we have $(\pi \times u)(w) = \pi(w)$ and $(\pi \times u)(s) = u_s$ and $\alpha_s(w) = sws^{-1}$ for $w \in W$ and $s \in G$.

14.2 Takesaki-Takai Duality

Two C*-dynamical systems (W, G, α) and (V, G, β) are isomorphic if there is a $*$-isomorphism $f \colon W \to V$ such that $f\alpha_t = \beta_t f$ for $t \in G$. Then $G \ltimes_\alpha W$ and $G \ltimes_\beta V$ are clearly isomorphic. The *tensor product* of the two systems is the C*-dynamical system $(W \otimes V, G, \alpha \otimes \beta)$, where we use the spatial tensor product of the C*-algebras and where $(\alpha \otimes \beta)_t(w \otimes v) = \alpha_t(w) \otimes \beta_t(v)$. Letting ρ be the right regular representation of G, we form the C*-dynamical system $(B_0(L^2(G)), G, \mathrm{Ad}\,\rho)$, where $(\mathrm{Ad}\,\rho)_t(a) = \rho(t)a\rho(t)^*$ acts on the ideal of compact operators a on $L^2(G)$. That it is pointwise norm continuous is easily checked on rank one operators. For $t \in G$ let ρ'_t and λ_t be the $*$-isomorphism on $C_0(G)$ given by $\rho'_t(g) = g(\cdot t)$ and $\lambda_t(g) = g(t^{-1}\cdot)$. Notice that $C_0(G)$ and $B_0(L^2(G))$ are nuclear.

Proposition 14.2.1 *Let (W, G, α) be a C*-dynamical system. Then we get a C*-dynamical system $(G \ltimes_{(\lambda\otimes\alpha)r} (C_0(G) \otimes W), G, \rho' \otimes \iota)$ which is isomorphic to the C*-dynamical system $(W \otimes B_0(L^2(G)), G, \alpha \otimes \mathrm{Ad}\,\rho)$.*

Proof Since $(\lambda \otimes \alpha)_s$ and $(\rho' \otimes \iota)_t$ commute, the latter is a $*$-isomorphism of $L^1(G, \lambda\otimes\alpha)$ that extends to a $*$-isomorphism of $G\ltimes_{(\lambda\otimes\alpha)r}(C_0(G)\otimes W)$, and $\rho'\otimes\iota$ is pointwise norm continuous as is checked on the dense subspace $L^1(G)\odot K(G, W)$.

Consider a faithful non-degenerate representation π of W on a Hilbert space V. Then $\mathrm{Ind}(\iota \otimes \pi)$ is a faithful non-degenerate representation of $G \ltimes_{(\lambda\otimes\alpha)r} (C_0(G) \otimes W)$ on $L^2(G \times G, V)$ such that

$$\mathrm{Ind}(\iota \otimes \pi)(a)d(s, t) = \int \pi\alpha_{s^{-1}}(a(u, st))d(u^{-1}s, t)\, du$$

for $a \in K(G, C_0(G)\otimes W)$ and $d \in L^2(G\times G, V)$ and $s, t \in G$. Now $(Nd)(s, t) = \Delta(t)^{1/2}d(st, t)$ defines a unitary operator N on $L^2(G \times G, V)$. It satisfies

$$N^*\mathrm{Ind}(\iota \otimes \pi)(a)N = (\tilde{\pi} \otimes \iota)(w \otimes (\cdot|f')g)$$

for a given by $a(s, t) = \alpha_t(w)f(s^{-1}t)g(t)$, where $f, g \in K(G)$ and $f' = \overline{f/\Delta}$ and $w \in W$ and where $(\tilde{\pi}(w)c)(t) = \pi\alpha_t(w)c(t)$ defines a faithful non-degenerate representation of W on $L^2(G, V)$. Now elements of the form a span a dense subspace of $G\ltimes_{(\lambda\otimes\alpha)r}(C_0(G)\otimes W)$, whereas elements of the form $w\otimes(\cdot|f')g$ span a dense subspace of $W \otimes B_0(L^2(G))$. Moreover, if one replaces a by $(\rho' \otimes \iota)_t(a)$ the right hand side above becomes $(\tilde{\pi} \otimes\iota)(\alpha \otimes \mathrm{Ad}\,\rho)_t(w \otimes (\cdot|f')g)$, so the systems are isomorphic. □

Definition 14.2.2 The *dual system* of a C*-dynamical system (W, G, α) with G abelian is the C*-dynamical system $(G \ltimes_\alpha W, \hat{G}, \hat{\alpha})$, where $\hat{\alpha}_u(f)(t) = \overline{u(t)}f(t)$ for $f \in L^1(G, \alpha)$ and $u \in \hat{G}$ and $t \in G$.

It is easy to check that $\hat{\alpha}_u$ extends to a $*$-isomorphism of $G \ltimes_\alpha W$, and that $u \mapsto \hat{\alpha}_u$ is a homomorphism which is pointwise norm continuous with respect to the topology of uniform convergence on compacts on the dual group $\hat{G}$ of G. One refers to $\hat{\alpha}$ as the *dual action*. Clearly two isomorphic C*-dynamical systems have isomorphic dual systems. Recall that we have Pontryagin duality for abelian locally compact groups, and that such groups are amenable. The following result shows that the bidual system is isomorphic to the original dynamical system modulo tensoring with a fairly simple action on the compact operators.

Theorem 14.2.3 *Let (W, G, α) be a C*-dynamical system with G abelian. Then the C*-dynamical system $(\hat{G} \ltimes_{\hat{\alpha}} (G \ltimes_{\alpha} W), G, \hat{\hat{\alpha}})$ is isomorphic to any of the two C*-dynamical systems in the previous proposition.*

Proof Clearly $\hat{G} \ltimes_{\iota} W = C^*_u(\hat{G}) \otimes W$. Letting $\beta_t(f)(u) = u(t)\alpha_t(f(u))$ for $f \in L^1(\hat{G}, \iota)$ and $t \in G$ one gets a C*-dynamical system $(\hat{G} \ltimes_{\iota} W, G, \beta)$. Consider the Fourier transform $F : C^*_u(\hat{G}) \to C_0(G)$ given by $F(g)(s) = \int \overline{u(s)} g(u)\, du$ for g in $L^1(\hat{G})$. Then $F \otimes \iota$ is an isomorphism from $(\hat{G} \ltimes_{\iota} W, G, \beta)$ to $(C_0(G) \otimes W, G, \lambda \otimes \alpha)$. Hence $G \ltimes_{\beta} (\hat{G} \ltimes_{\iota} W) \cong G \ltimes_{\lambda \otimes \alpha} (C_0(G) \otimes W)$.

Pick a faithful non-degenerate representation π of W on V, and define a unitary A on $L^2(\hat{G} \times G, V)$ by $(Ab)(u, t) = \overline{u(t)} b(u, t)$. Now $(\iota \otimes \pi) \times (\lambda \otimes \iota)$ is a faithful representation of $G \ltimes_{\iota} W$ on $L^2(\hat{G}, V)$, so $\pi' \equiv \text{Ind}((\iota \otimes \pi) \times (\lambda \otimes \iota))$ is a faithful representation of $G \ltimes_{\beta} (\hat{G} \ltimes_{\iota} W)$ on $L^2(\hat{G} \times G, V)$, and is given by $(\pi'(f)b)(u, t) = \int v(t) \pi \alpha_{t^{-1}}(f(v, s)) b(v^{-1}u, s^{-1}t)\, dv ds$ for $f \in K(\hat{G} \times G, W)$. Another faithful representation on the same Hilbert space is $\text{Ind}(\overline{\text{Ind}\,\pi})$ and it is easy to check that $\text{Ind}(\text{Ind}\,\pi)(f) = A^* \pi'(f') A$, where $f'(u, t) = \overline{u(t)} f(u, t)$ for f as above. Since $K(\hat{G} \times G, W)$ is dense in both $G \ltimes_{\beta} (\hat{G} \ltimes_{\iota} W)$ and $\hat{G} \ltimes_{\hat{\alpha}} (G \ltimes_{\alpha} W)$, these C*-algebras are therefore isomorphic. Combining this isomorphism with $F \otimes \iota$ from the first paragraph, we get an isomorphism between $G \ltimes_{\lambda \otimes \alpha} (C_0(G) \otimes W)$ and $\hat{G} \ltimes_{\hat{\alpha}} (G \ltimes_{\alpha} W)$ which moreover respects the actions $\rho' \otimes \iota$ and $\hat{\hat{\alpha}}$. □

A *von Neumann dynamical system* is a triple (W, G, α), where $W \subset B(V)$ is a von Neumann algebra and G is a locally compact group acting by $*$-automorphisms α_t on W such that $t \mapsto \alpha_t$ is a pointwise weakly continuous homomorphism. Then $t \mapsto \alpha_t$ is pointwise strongly continuous since for any unitary $w \in W$ we have $\|\alpha_{t_i}(w)v - \alpha_t(w)v\|^2 = 2\|v\|^2 - (\alpha_{t_i}(w)v|\alpha_t(w)v) - (\alpha_t(w)v|\alpha_{t_i}(w)v) \to 0$ as $t_i \to t$, so the map is actually pointwise σ-weakly continuous.

Write π for the representation of $W \subset B(V)$. In defining the regular representation induced by π observe that $(\pi', \lambda \otimes \iota)$ is a covariant representation on $L^2(G, V)$ of the system (W, G, α), and the faithful representation π' is normal since for $v \in V$ we see by Dini's theorem that $(\alpha_{t^{-1}}(w_i)v|v)$ increases towards $(\alpha_{t^{-1}}(w)v|v)$ uniformly on any compact subset of G as w_i increases towards w in W_+. The *von Neumann crossed product* of W by the action α of G is the von Neumann algebra $G \bar{\ltimes}_{\alpha} W$ generated by $\pi'(M)$ and $(\lambda \otimes \iota)(G)$. Since $(\lambda \otimes \iota)(t)\pi'(w)(\lambda \otimes \iota)(t)^* = \pi'\alpha_t(w)$, we see that $G \bar{\ltimes}_{\alpha} W$ is the σ-weak closure of the span of the elements $\pi'(w)(\lambda \otimes \iota)(t)$ for $w \in W$ and $t \in G$. In fact, if U is a σ-weakly dense C*-subalgebra of W invariant under α such that (U, G, α) is a C*-dynamical system, then the σ-weak closure of $G \ltimes_{\alpha r} U$ is $G \bar{\ltimes}_{\alpha} W$as is seen by invoking the regular representation induced by π. For the existence of such a C*-dynamical system (U, G, α) let U consist of those $w \in W$ such that $t \mapsto \alpha_t(w)$ is norm continuous. Then U is clearly a C*-subalgebra of W invariant under α. And U is σ-weakly dense in W because it contains the elements $\alpha_f(w) \equiv \int f(s)\alpha_s(w)\, ds$ for any $w \in W$ and $f \in L^1(G)$ as $\alpha_t(\alpha_f(w)) = \int \alpha_s(w) f(t^{-1}s)\, ds$, so $\|\alpha_t(\alpha_f(w)) - \alpha_f(w)\| \to 0$ as $t \to e$, and if z is a normal state on W, then

$|z(w - \alpha_{f_i}(w))| \leq \int |z(w - \alpha_s(w))| f_i(s)\, ds \to 0$, where $\{f_i\}$ is an approximate unit for $L^1(G)$.

We can use this correspondence between C*-dynamical systems and von Neumann dynamical systems to provide a von Neumann version of the theorem above. Clearly the crossed products of two isomorphic von Neumann dynamical systems will be isomorphic, and by considering faithful covariant representations of von Neumann dynamical systems, we can also form the tensor product von Neumann dynamical system. The *von Neumann dual* of a von Neumann dynamical system (W, G, α) with G abelian is the von Neumann dynamical system $(G \bar{\ltimes}_\alpha W, G, \hat{\alpha})$, where the *dual action* is given by $\hat{\alpha}_u = A_u(\cdot)A_u^* \otimes \iota$, and where A_u is the unitary on $L^2(G)$ given by $A_u(f)(t) = \overline{u(t)} f(t)$ for $f \in L^2(G)$. One checks that $A_u \lambda(t) A_u^* = \overline{u(t)} \lambda(t)$ and $(A_u(\cdot)A_u^* \otimes \iota)\pi'(w) = \pi'(w)$, so $\hat{\alpha}_u$ leaves the crossed product invariant. Dual systems of isomorphic von Neumann dynamical system will of course remain isomorphic. Since the weak closure of $B_0(L^2(G))$ is $B(L^2(G))$, the following result is now immediate from the theorem above.

Corollary 14.2.4 *Let (W, G, α) be a von Neumann dynamical system with G abelian. Then the bidual von Neumann dynamical system $(\hat{G} \bar{\ltimes}_{\hat{\alpha}} (G \bar{\ltimes}_\alpha W), G, \hat{\hat{\alpha}})$ is isomorphic to the von Neumann dynamical system $(W \bar{\otimes} B(L^2(G)), G, \alpha \otimes \mathrm{Ad}\, \rho)$.*

14.3 Landstad Theory

Can we characterize those C*-algebras or von Neumann algebras that are crossed products by abelian groups? Let G be an abelian locally compact group with Pontryagin dual $\hat{G}$. A C*-algebra W is a *G-product* if we have a C*-dynamical system $(W, \hat{G}, \hat{\alpha})$ and a pointwise norm continuous homomorphism $\lambda' : G \to U(M(W))$ such that $\hat{\alpha}_u \lambda'(t) = \overline{u(t)} \lambda'(t)$ for $u \in \hat{G}$ and $t \in G$. We say $w \in M(W)$ satisfies *Landstad's conditions* if it is a fixed point for $\hat{\alpha}$ and $t \mapsto \lambda'(t) w \lambda'(t^{-1})$ is norm continuous and $\lambda'(f)w, w\lambda'(f) \in W$ for $f \in L^1(G)$, where $\lambda'(f) = \int \lambda'(t) f(t)\, dt$. Clearly any C*-algebra of the form $G \ltimes_\alpha W$ for G abelian is a G-product for the dual action $\hat{\alpha}$ and with $\lambda' = \lambda \otimes \iota$ acting on $L^2(G, V)$, where $W \subset B(V)$. The elements of W then satisfy Landstad's conditions. To prove the converse, we need integrability.

Let W be a G-product. Then $w \in M(W)_+$ is *$\hat{\alpha}$-integrable* if there is a positive element $I(w)$ in $M(W)$ such that $z(I(w)) = \int z\hat{\alpha}_u(w)\, du$ for $z \in M(W)^\star$. Linear combinations of such elements are also called $\hat{\alpha}$-integrable. If $0 \leq w' \leq w$ for w' in $M(W)$, then since both functionals $z \mapsto \int z\hat{\alpha}_u(w')\, du$ and $z \mapsto \int z\hat{\alpha}_u(w - w')\, du$ are evidently lower w^*-semicontinuous and their sum is w^*-continuous, they are both w^*-continuous and are thus evaluation at positive elements of $M(W)$. So the integrable elements form a hereditary $*$-subalgebra of $M(W)$. If w^*w and w'^*w' are $\hat{\alpha}$-integrable, then so is w^*w' by polarization, and $|(I(w^*w')v'|v)| \leq \int \|\hat{\alpha}_u(w')v'\| \cdot \|\hat{\alpha}_u(w)v\|\, du \leq (I(w^*w)v|v)^{1/2}(I(w'^*w')v'|v')^{1/2}$, where $v, v' \in V$ for $M(W) \subset$

$B(V)$, so $\|I(w^*w')\| \leq \|I(w^*w)\| \cdot \|I(w'^*w')\|$. Finally notice that $I(w)$ is $\hat{G}$-invariant.

Lemma 14.3.1 *Let W be a G-product. Elements of the form $w' \equiv \lambda'(f)^*w\lambda'(g)$ for $f, g \in L^1(G) \cap L^2(G)$ and $w \in M(W)$ are $\hat{\alpha}$-integrable, and $t \mapsto I(w'\lambda(t))$ is norm continuous, and $w' \to w$ when $f = g \in L^2(G)$ ranges over an approximate unit for $L^1(G)$. When $f = g$, then $I(w')$ satisfies Landstad's conditions, and w' is a norm limit of elements of the form $\int I(w'\lambda'(t^{-1}))\lambda'(t)f_i(t)\,dt$, where $\{f_i\} \subset L^1(G)$ and $\{F(f_i)\}$ is an approximate unit for $L^1(\hat{G})$.*

Proof For the first statement, by the previous paragraph, it suffices to show that $\lambda(f)^*\lambda(f)$ is $\hat{\alpha}$-integrable. Letting $F\colon C_u^*(G) \to C_0(\hat{G})$ be the Gelfand transform, and extending λ' from $L^1(G)$ to $C_u^*(G)$, and going to the multiplier algebras, we get a $*$-homomorphism $\lambda'F^{-1}\colon C_b(\hat{G}) \to M(W)$. Then $\lambda'F^{-1}(k_u * F(f)) = \lambda'(u^{-1}f) = \int \overline{u(t)}\lambda'(t)f(t)\,dt = \hat{\alpha}_u(\lambda'F^{-1}F(f))$ for $f \in L^1(G)$ and by continuity $\lambda'F^{-1}(k_u * f') = \hat{\alpha}_u(\lambda'F^{-1}(f'))$ for $f' \in C_b(\hat{G})$. Therefore $\hat{\alpha}_u(\lambda'(f)^*\lambda'(f)) = \hat{\alpha}_u(\lambda'F^{-1}(|F(f)|^2)) = \lambda'F^{-1}(k_u * |F(f)|^2)$, and $I(\lambda'(f)^*\lambda'(f))$ is $\hat{\alpha}$-integrable, being simple the scalar $\|F(f)\|^2 = \|f\|^2$.

The second statement is clear as $\|I(\lambda'(f)^*w\lambda'(g)(\lambda'(s) - \lambda'(t)))\|^2$ is less than $\|I(\lambda'(f)^*w^*w\lambda'(f))\| \cdot \|I((\lambda'(s) - \lambda'(t))^*\lambda'(g)^*\lambda'(g)(\lambda'(s) - \lambda'(t)))\|$, which is again less than $\leq \|w\|^2\|f\|_2^2\|g * (k_s - k_t)\|_2^2$, whereas the third statement is clear from

$$\|\lambda'(f)^*w\lambda'(f) - w\| \leq \int \|\lambda'(t)w - w\| f^\star(t)\,dt + \int \|w\lambda'(s) - w\| f(s)\,ds$$

for positive f with $\|f\|_1 = 1$.

For the very last statement, notice that

$$\int I(w'\lambda'(t^{-1}))\lambda'(t)f_i(t)\,dt = \int \hat{\alpha}_u(w')F(f_i)(u)\,du \in W$$

converges to w' in norm as $\{F(f_i)\}$ is an approximate unit for $L^1(\hat{G})$. Now $I(w')$ is clearly a fixed point for $\hat{\alpha}$ By the last paragraph above chose to $g \in L^1(G)$ a neighborhood B of the unit e in G such that $\|I(w'\lambda'(t)) - I(w')\| < \varepsilon$ and $\|g - k_t * g\| < \varepsilon$ for $t \in B$. Pick positive f with support in B and $\|f\|_1 = 1$. Then $\|I(w')\lambda'(g) - \int I(w'\lambda'(t^{-1}))\lambda'(t)f(t)\,dt\lambda'(g)\|$ is not greater than

$$\|I(w')\|\cdot\|g - f * g\|_1 + \int \|I(w') - I(w'\lambda'(t^{-1})\| f(t)\,dt\|g\|_1 \leq (\|I(w')\| + \|g\|_1)\varepsilon,$$

so $I(w')\lambda'(g) \in W$ by the previous observation. Similarly $\lambda'(g)I(w') \in W$. Finally, the map $t \mapsto \lambda'(t)I(w')\lambda'(t^{-1}) = I(\lambda'(t)w'\lambda'(t^{-1}))$ is continuous by the previous paragraph, so Landstad's conditions hold for $I(w')$. □

Theorem 14.3.2 *Every G-product is of the form $G \ltimes_\alpha W$ for a C*-dynamical system (W, G, α) which is unique up to isomorphism. In fact, the C*-algebra W consists of those elements that satisfies Landstad's conditions, whereas the action is given by $\alpha_t = \lambda'(t)(\cdot)\lambda'(t^{-1})$, where λ' defined the G-product.*

Proof Let U be a G-product with C*-dynamical system $(U, \hat{G}, \hat{\alpha})$ and continuous homomorphism $\lambda' \colon G \to U(M(U))$ such that $\hat{\alpha}_u\lambda'(t) = \overline{u(t)}\lambda'(t)$, and let W be the C*-subalgebra of $M(U)$ of elements that satisfy Landstad's conditions. Then (W, G, α) is evidently a C*-dynamical system. Consider the universal representation π of W on the same Hilbert space V as that of the universal representation of $M(U)$. The regular representation $\mathrm{Ind}\,\pi$ of $G \ltimes_\alpha W$ on $L^2(G, V)$ induced by π is faithful. We have also the representation π' of $M(U)$ on $L^2(\hat{G}, V)$ given by $(\pi'(w)f)(u) = \hat{\alpha}_{u^{-1}}(w)f(u)$ for $w \in M(U)$ and $f \in L^2(\hat{G}, V)$ and $u \in \hat{G}$. Let $A \colon L^2(G, V) \to L^2(\hat{G}, V)$ be the surjective isometry given by $A(f)(u) = \int \lambda'(t)(f)u(t)\,dt$. Then one readily checks that $w \otimes g = A^*\pi'(wg)A$ for $w \in W$ and $g \in K(G)$, so $G \ltimes_\alpha W \subset A^*\pi'(U)A$. For the opposite inclusion, consider the $\hat{\alpha}$-integrable element $w' = \lambda'(g)^*w\lambda'(g) \in U$ with $g \in L^1(G) \cap L^2(G)$ and $w \in U$. From the above we have $I(w'\lambda'(t^{-1})) \otimes (k_t * f) = A^*\pi'(I(w'\lambda'(t^{-1}))\lambda'(t)\lambda'(f))A$ for $f \in K(G)$ and $t \in G$. Let $w_i' = \int I(w'\lambda'(t^{-1}))\lambda'(t)f_i(t)\,dt \in U$ for $f_i \in L^1(G)$ such that $\{F(f_i)\}$ is an approximate unit for $L^1(\hat{G})$. By the previous lemma the map from G to $G \ltimes_\alpha W$ given by $t \mapsto I(w'\lambda'(t^{-1})) \otimes (k_t * f)$ is norm continuous, so $A^*\pi'(w_i'\lambda'(f))A = \int (I(w'\lambda'(t^{-1})) \otimes (k_t * f))f_i(t)\,dt \in G \ltimes_\alpha W$. Hence $A^*\pi'(U)A \subset G \ltimes_\alpha W$ by the same lemma.

As for uniqueness, let W' denote the elements in $M(G \ltimes_\alpha W)$ that satisfy Landstad's conditions. Clearly $W \subset W'$ covariantly, and since $G \ltimes_\alpha W = G \ltimes_\alpha W'$ from what we have already proved, we get $W = W'$ since clearly strict inclusion in Proposition 14.1.10 would lead to strict inclusion of crossed products. □

Proposition 14.3.3 *Let $W, G, \alpha)$ be a C*-dynamical system with G abelian. Consider the dual action $\hat{\alpha}$ and $\lambda \otimes \iota \colon G \to M(G \ltimes_\alpha W)$ and $M(W) \subset M(G \ltimes_\alpha W)$. Then $w \in M(W)$ if and only if w is $\hat{G}$-invariant and $t \mapsto (\lambda \otimes \iota)(t)w(\lambda \otimes \iota)(t^{-1})w'$ is norm continuous for every $w' \in W$.*

Proof The forward implication is obvious. For the converse, a straightforward calculation, using the previous lemma, shows that $\int \alpha_{t^{-1}}(ww')\overline{f(t)}g(t)\,dt = I((\lambda \otimes \iota)(f)^*ww'(\lambda \otimes \iota)(g))$ for $f, g \in L^1(G) \cap L^2(G)$ and $F(g) \in L^1(\hat{G})$. Landstad's conditions show therefore that the integral belongs to W when f, g runs over an approximate unit for $L^1(G)$, and the integral then converges to $ww' \in W$. Similarly $w'w \in W$, so $w \in M(W)$. □

We include also the von Nemann versions of these two results. A von Neumann algebra W is a *G-product* with G an abelian locally compact group if there is a von Nemann dynamical system $(W, \hat{G}, \hat{\alpha})$ and a strongly continuous unitary representation λ' of G with $\lambda'(G) \subset W$ and such that $\hat{\alpha}_u(\lambda'(t)) = \overline{u(t)}\lambda'(t)$ for $u \in \hat{G}$ and $t \in G$. Evidently any von Neumann algebra which is a von Neumann crossed product by an abelian group G is a G-product under the dual action, and members of the original von Neumann algebra are clearly fixed under this dual action.

Corollary 14.3.4 *Any von Neumann G-product is of the form $G\bar{\ltimes}_\alpha W$ for a von Neumann dynamical system (W, G, α), which is unique up to isomorphism, and W is the fixed point algebra for the dual action of $\hat{G}$ on the crossed product, while $\alpha_t = \lambda'(t)(\cdot)\lambda'(t^{-1})$, where λ' defines the G-product.*

Proof If $(U, \hat{G}, \hat{\alpha})$ is the von Neumann dynamical system with homomorphism λ' defining the G-product, consider the C*-dynamical system $(U^c, \hat{G}, \hat{\alpha})$ consisting of $w \in U$ such that $u \mapsto \hat{\alpha}_u(w)$ is norm continuous. Then $\lambda'(t)w \in U^c$ since $\hat{\alpha}_u(\lambda'(t)) = \overline{u(t)}\lambda'(t)$, so $\lambda'(t) \in M(U^c)$. Let B be the C*-subalgebra of U^c consisting of those $w \in U$ with $t \mapsto \lambda'(t)w$ and $t \mapsto w\lambda'(t)$ norm continuous. Since it contains the elements $\lambda'(f)w\lambda'(g)$ for any $f, g \in L^1(G)$ and $w \in U^c$, it is σ-weakly dense in U^c, which is again dense in U. Clearly B is a G-product for $\hat{\alpha}$ and λ' appropriately restricted, so by the theorem above $B \cong G \ltimes_\alpha A$ for the prescribed action α, where A is the C*-algebra of elements in $M(B)$ that satisfy Landstad's conditions, and where, upon regarding $U \subset B(V)$, the isomorphism is spatial. So $U = (G \ltimes_\alpha A)'' = G\bar{\ltimes}_\alpha W$, where $W = A'' \subset U$.

Evidently the elements in W are $\hat{G}$-invariant. Let $\tilde{U}$ denote the von Nemann algebra of $\hat{G}$-invariant elements in U. Note that $I(\lambda'(f)(\cdot)\lambda'(f)) \colon U \to \tilde{U}$ with $f \in L^1(G) \cap L^2(G)$ is a positive normal linear map, and from the proof of the previous proposition we have $I(\lambda'(f)^* w\lambda'(f)) = \int \alpha_{-t}(w)|f(t)|^2\, dt$ for $w \in \tilde{U}$ when $F(f) \in L^1(\hat{G})$. Pick a net $\{w_i\}$ in B which converges σ-weakly to $w \in \tilde{U}$. Then $I(\lambda'(f)^* w_i\lambda'(f)) \to \alpha_g(w)$ σ-weakly, where $g(t) = |f(t^{-1})|^2$. Hence $\alpha_g(w) \in W$, and letting f range over an approximate unit for $L^1(G)$, we $\alpha_g(w) \to w$ σ-weakly, so $w \in W$, and $\tilde{U} = W$. Uniqueness is now also clear. □

Let (W, G, α) be a C*-dynamical system. Then W is *G-simple* if it has no non-trivial closed G-invariant ideals. It is *G-prime* if any two non-trivial closed G-invariant ideals intersect non-trivially. Clearly this happens precisely whenever $\alpha_t(w)W\alpha_s(w') = \{0\}$ for all $s, t \in G$ can only occur when $w = 0$ or $w' = 0$.

Proposition 14.3.5 *Let (W, G, α) be a C*-dynamical system with G abelian. Then W is G-simple if and only if $G \ltimes_\alpha W$ is $\hat{G}$-simple. And W is G-prime if and only if $G \ltimes_\alpha W$ is $\hat{G}$-prime.*

Proof Observe first that the image under the map I on $M(G \ltimes_\alpha W)$ defined above of the $\hat{\alpha}$-integrable elements of a $\hat{G}$-invariant $*$-ideal of $G \ltimes_\alpha W$ is a G-invariant $*$-ideal of W. This is clear since $\alpha_t I = I((\lambda\otimes\iota)(t)(\cdot)(\lambda\otimes\iota)(t^{-1}))$ and $I(w(\cdot)w') = wI(\cdot)w'$ for $w, w' \in W$.

If J is a non-trivial closed $\hat{G}$-invariant ideal of $G \ltimes_\alpha W$, and z is a state on the crossed product that annihilates J with a unique extension to $M(G\ltimes_\alpha W)$ also called z, then since an approximate unit for W will be an approximate unit for $G \ltimes_\alpha W$, we must have $z(J) = 0$ by the Cauchy-Schwarz inequality. But if $w \in J$ is $\hat{\alpha}$-integrable then $zI(w) = \int z\hat{\alpha}_u(w)\, du = 0$ as J is $\hat{G}$-invariant. So the corresponding ideal of W given by the first paragraph is not dense in W, and yet it is not $\{0\}$ as

$I((\lambda\otimes\iota)(f)^*J(\lambda\otimes\iota)(f))$ is contained in it for $f\in L^1(G)\cap L^2(G)$. So its closure is non-trivial, and $G\ltimes_\alpha W$ is $\hat{G}$-simple whenever W is G-simple. The converse also holds since $W\otimes B_0(L^2(G))$ is then G-simple by what we have already proved and by Takesaki-Takai duality and because if J was a non-trivial closed G-invariant ideal of W, then $J\otimes B_0(L^2(G))$ would be a similar ideal in $W\otimes B_0(L^2(G))$.

If J_i are non-trivial closed $\hat{G}$-invariant ideals of $G\ltimes_\alpha W$ with $J_1\cap J_2=\{0\}$, then their corresponding images as described in the first paragraph are again non-zero G-invariant ideals, and yet their product vanishes as $z(I(w)I(w'))=\int z\hat{\alpha}_u(w)z\hat{\alpha}_{u'}(w')\,du du'=0$ for any state z on $G\ltimes_\alpha W$ and $\hat{\alpha}$-integrable elements $w\in J_1$ and $w'\in J_2$. So $G\ltimes_\alpha W$ is $\hat{G}$-prime when W is G-prime. Conversely, one concludes as above that $W\otimes B_0(L^2(G))$ is G-prime whenever $G\ltimes_\alpha W$ is $\hat{G}$-prime, and again, then W must also be G-prime. □

14.4 Examples of Crossed Products

Here is a simple version of Takasaki-Takai duality.

Example 14.4.1 Let G be a locally compact group. Consider $C_0(G)$ as multiplication operators on $L^2(G)$ and let $\alpha_t(f)(s)=f(t^{-1}s)$ for $f\in C_0(G)$. Then $(C_0(G),G,\alpha)$ is a C*-dynamical system and $G\ltimes_{\alpha r}C_0(G)$ is σ-weakly dense in $G\bar{\ltimes}_\alpha L^\infty(G)$, where the von Neumann dynamical system $(L^\infty(G),G,\alpha)$ is defined with the same formula for α. The latter crossed product is just the bicommutant of $\pi(L^\infty(G))\cup(\lambda\otimes\iota)(G)$, where $(\pi(f)v)(s,t)=f(st)v(s,t)$ for $v\in L^2(G\times G)$. Letting N be the unitary operator on $L^2(G\times G)$ given by $(Nv)(s,t)=\Delta(t)^{1/2}v(st,t)$ one checks that $N^*\pi(f)N=f\otimes 1$ and $N^*(\lambda\otimes\iota)(t)N=\lambda(t)\otimes 1$, so as $L^\infty(G)$ is maximally commutative in $B(L^2(G))$, we see that $G\bar{\ltimes}_\alpha L^\infty(G)$ is spatially isomorphic to $B(L^2(G))$. Similarly, with $h(s,t)=f(s^{-1}t)g(t)$ for $f,g\in K(G)$, one gets $N^*(\pi\times(\lambda\otimes\iota))(h)N=(\cdot|f')g\otimes 1$, where $f'=\bar{f}/\Delta$. So $G\ltimes_{\alpha r}C_0(G)\cong B_0(L^2(G))$.

This shows that $B_0(V)$ and $B(V)$ for any Hilbert space V are G-products in the C*-algebra and von Neumann algebra sense, respectively, for a discrete abelian group G of appropriate cardinality. ◇

Crossed products generalize semidirect products in the following sense.

Example 14.4.2 Let H,K be locally compact groups with a homomorphism $\rho\colon K\to\mathrm{Aut}(H)$ such that the map from $H\times K$ to H given by $(h,k)\mapsto\rho(k)h$ is continuous. Then the *semidirect product* of (H,K,ρ) is the locally compact group $K\times_\rho H$ which is $H\times K$ as a topological space (with product topology) and with multiplication given by $(h,k)(h',k')=(h\rho(k)h',kk')$. To each $k\in K$, there is by uniqueness of the Haar measure on H a positive constant $\delta(k)$ such that $d\rho(k)h=\delta(k)dh$ for $h\in H$. Clearly δ is multiplicative and equals one on the unit of K, and the Haar measure on $K\times_\rho H$ is $d(h,k)=\delta(k)^{-1}dhdk$, and $\Delta_{K\times_\rho H}(h,k)=\delta(k)^{-1}\Delta_H(h)\Delta_K(k)$ and $\Delta_H(\rho(k)h)=\Delta_H(h)$.

The formula $(\alpha_k f)(h) = \delta(k)^{-1} f(\rho(k)^{-1}h)$ for $f \in L^1(H)$ evidently defines a C*-dynamical system $(C_u^*(H), K, \alpha)$. We claim that $C_u^*(K \times_\rho H) \cong K \ltimes_\alpha C_u^*(H)$. Indeed, the formula $B(g)(k,h) = \delta(k)^{-1} g(h,k)$ defines a $*$-isomorphism B from $L^1(K \times_\rho H)$ onto $L^1(K, L^1(H), \alpha)$, so we need only check that every non-degenerate representation π' of $L^1(K, L^1(H), \alpha)$ extends to one on $L^1(K, C_u^*(H), \alpha)$. If $\{g_i\}$ is an approximate unit for $L^1(K \times_\rho H)$, then $\{B(g_i)\}$ is an approximate unit for $L^1(K, C_u^*(H), \alpha)$ by denseness, so we can define $\pi(f)$ and u_k for $f \in L^1(H)$ and $k \in K$ to be the respective strong limits of $\rho(f B(g_i))$ and $\rho(\alpha_k(B(g_i)(k^{-1}, \cdot)))$. Then $\pi' = \pi \times u$ and we can clearly extend π to $C_u^*(H)$ and get a corresponding extension of π'.

A concrete example of this is $C_u^*(G \times_\rho G) \cong G \ltimes_\alpha C_u^*(G)$, where G is a locally compact group acting on itself by $\rho(s)t = sts^{-1}$ with corresponding $(\alpha_s f)(t) = \Delta(s) f(s^{-1}ts)$ for $f \in L^1(G)$ and $s, t \in G$. ◇

Definition 14.4.3 Let G be an abelian locally compact group. Two strongly continuous unitary representations A and B of G and $\hat{G}$, respectively, acting on the same Hilbert space, satisfy the *Weyl relations* if $A(u)B(t) = \overline{u(t)} B(t) A(u)$ for all $u \in \hat{G}$ and $t \in G$.

Here is the historically motivating example for studying crossed products.

Example 14.4.4 Consider the trivial action α on $\mathbb{C}$ of an abelian locally compact group G. Its von Neumann dual action is given by $\hat{\alpha}_u = A_u(\cdot)A_u^*$, where A is the strongly continuous unitary representation of $\hat{G}$ on $L^2(G)$ given by $(A_u f)(t) = \overline{u(t)} f(t)$. The left regular representation λ of G acts on the same Hilbert space, and we have seen that $A_u \lambda(t) A_u^* = \overline{u(t)} \lambda(t)$, so A and λ satisfy the Weyl relations. Takesaki-Takai duality tells us that the von Neumann dynamical systems $(\hat{G} \bar{\ltimes}_{\hat{\alpha}} (G \bar{\ltimes}_\alpha \mathbb{C}), G, \hat{\hat{\alpha}})$ and $(B(L^2(G)), G, \mathrm{Ad}\rho)$ are spatially isomorphic. Hence $(A_{\hat{G}} \cup \lambda(G))'' = \hat{G} \bar{\ltimes}_{\hat{\alpha}} (G \bar{\ltimes}_\alpha \mathbb{C}) = B(L^2(G))$, and the *Weyl system* (A, λ) is irreducible, generating a factor of type I.

We are to some extend repeating ourselves here because $G \ltimes_\alpha \mathbb{C} = C_u^*(G)$ which is isomorphic to $C_0(\hat{G})$ under the Gelfand transform F, and then $\hat{\alpha}$ is simply the action described in the first example above, with G replaced by $\hat{G}$ there, thus obtaining the factor $B(L^2(\hat{G}))$, which is spatially isomorphic $B(L^2(G))$ under the Fourier transform between the L^2-spaces.

Now given any Weyl system (A, B) with unitary representations on a Hilbert space V, denote by π the unital representation of $\mathbb{C}$ on V, and note that (π, B) is a covariant representation of $(\mathbb{C}, G, \alpha)$, so $(\pi \times B, A)$ is a covariant representation of $(G \ltimes_\alpha \mathbb{C}, \hat{G}, \hat{\alpha})$, and $(\pi \times B) \times A$ is a non-degenerate representation of the crossed product $\hat{G} \ltimes_{\hat{\alpha}} (G \ltimes_\alpha \mathbb{C}) = B_0(L^2(G))$, and is thus a multiple copy of the representation gotten by using the irreducible Weyl system (A, λ) from above. We have proved the *Stone-von Neumann theorem*, which states that every Weyl system is a multiple copy of the irreducible one (A, λ) described above.

When $G = \mathbb{R} = \hat{G}$ under the pairing $\langle a, b \rangle = e^{-iab}$, the system (A, λ) is called the *Schrödinger representation*. The strongly continuous one-parameter groups A and λ are those associated by Stone's theorem to the unbounded self-adjoint operators on $L^2(\mathbb{R})$ representing the position q and momentum p operators from quantum mechanics given as multiplication and differential operators, respectively, and which satisfy *Heisenberg's commutation relation* $[p, q] = i$ on some appropriate smooth core, which in turn yields the famous *Heisenberg uncertainty relation* $\Delta p \Delta q \geq 1/2$ for the root-mean square deviations of these observables. This relation says that the position and momentum of a single particle moving in one dimension cannot both be measured accurately simultaneously. One may view the Schrödinger representation as the irreducible strongly continuous representation of the *Heisenberg group* of invertible upper triangular real 3×3-matrices given by

$$\begin{pmatrix} 1 & a & c \\ 0 & 1 & b \\ 0 & 0 & 1 \end{pmatrix} \mapsto e^{ic} A_a \lambda(b).$$

Alternatively, one can say that any Weyl system (A, B) on a Hilbert space V gives a *projective representation* of $\hat{G} \times G$, in that, the map $(u, t) \mapsto A_u B(t)$ followed up the quotient map $U(B(V)) \to U(B(V))/\mathbb{T}$ is a group homomorphism. ◇

Exercises

For Sect. 14.1

1. Establish a correspondence between C*-dynamical systems of a locally compact group G on commutative C*-algebras and actions of G on locally compact Hausdorff spaces; here an action on X is a continuous map $G \times X \to X$; $(s, x) \mapsto sx$ such that $ex = x$ and $(st)x = s(tx)$ for $s, t \in G$ and $x \in X$ and where e denotes the unit of G. Investigate how isomorphisms of C*-dynamical systems transfer to corresponding actions.
2. Simplify the formulas in this section when the group is finite.
3. Show that the crossed product $G \ltimes_\alpha W$ of a C*-dynamical system (W, G, α) contains canonical copies of G and W exactly when G is discrete and W is unital.
4. Let (W, G, α) be a C*-dynamical system and let U be a C*-subalgebra of W invariant under α. Show that $G \ltimes_{\alpha r} U \subset G \ltimes_{\alpha r} W$ is proper when $U \neq W$.

For Sect. 14.2

1. Let G be an abelian group of order n, and consider the dual action α of G on $W \equiv C^*_u(\hat{G})$. Show that the fixed point algebra $\{w \in W \,|\, \alpha_t(w) = w,\ t \in G\}$ of α is $\mathbb{C}$, and that the universal crossed product of the C*-dynamical system (W, G, α) is isomorphic to $M_n(\mathbb{C})$.

2. Consider the cyclic group $\mathbb{Z}_n$ acting on the circle S^1 by rotation through multiples of $2\pi/n$. Show that the corresponding universal crossed product is isomorphic to $M_n(C(S^1))$.
3. Let θ be an irrational number between 0 and 1, and consider the discrete group $\mathbb{Z}$ of integers acting via α on the circle S^1 by rotation through multiples of $2\pi\theta$. The irrational rotation algebra A_θ is the universal crossed product $\mathbb{Z} \ltimes_\alpha C(S^1)$. Show that A_θ is isomorphic to the universal C*-algebra generated by unitaries u, v satisfying $uv = e^{2\pi i\theta} vu$. What would we have gotten when $\theta = 0$?
4. Consider a von Neumann dynamical system (W, G, α) with G discrete and W acting on a separable Hilbert space. Show that the von Neumann algebra $W \bar{\ltimes}_\alpha G$ is finite if and only if W has a tracial faithful normal state $x = x\alpha_t$ for $t \in G$.

For Sect. 14.3

1. Let (W, G, α) be a von Neumann dynamical system. An α-cocycle is a strongly continuous map $u \colon G \to U(W)$ such that $u_{st} = u_t\alpha_{t^{-1}}(u_s)$ for $s, t \in G$. Show that (W, G, α^u) with $\alpha^u_s = u_s\alpha_{s^{-1}}(\cdot)u^*_s$ for $s \in G$ is another von Neumann dynamical system. Any von Neumann dynamical system (W, G, β) isomorphic to this is said to be cocycle conjugate to (W, G, α), and we write $\alpha \sim \beta$. Show that $\sim$ is an equivalence relation, and that $G \bar{\ltimes}_\alpha W$ is isomorphic to $G \bar{\ltimes}_\beta W$ when $\alpha \sim \beta$.
2. We say an action $G \times X \to X$ of a locally compact group G on a locally compact Hausdorff space X is topologically free if $sx = x$ implies $s = e$. It is minimal if Gx is dense in X for every $x \in X$. Consider the induced action α of G on $C_0(X)$, and assume that G is amenable and acts freely on X. Show that the action is minimal if and only if the C*-algebra $G \ltimes_\alpha C_0(X)$ is simple.
3. Prove that the irrational rotation algebra A_θ is simple and has a unique tracial state, is nuclear, and that $A_{\theta'}$ is not isomorphic to A_θ when $\theta' \notin \{\theta, 1-\theta\}$.

For Sect. 14.4

1. Let (W, G, α) be a von Neumann dynamical system with W of type III. Pick a faithful semifinite normal weight x on W, and consider its modular group as a von Neumann dynamical system $(W, \mathbb{R}, \sigma^x)$ with dual action θ of $\hat{\mathbb{R}}$, seen as the multiplicative group $\mathbb{R}^*_+$, on $U \equiv \mathbb{R} \bar{\ltimes}_{\sigma^x} W$. Use Connes' cocycle theorem to show that the von Neumann dynamical system $(U, \mathbb{R}^*_+, \theta)$ is independent of the choice x up to cocycle conjugacy and isomorphism. Use the biduality theorem for crossed products to show that $W \cong \mathbb{R}^*_+ \bar{\ltimes}_\theta U$. Prove that U is of type II_∞ with a faithful semifinite normal trace y satisfying $y\theta_\lambda = \lambda y$ for $\lambda \in \mathbb{R}^*_+$. One thinks of this as a continuous decomposition of type III von Neumann algebras into type II_∞ ones. The restriction of θ_s to $Z(U)$ is an invariant of W known as the flow of weights. Show that the Connes' spectra $S(W)$ and $T(W)$ can be recovered from the flow of weights in the sense that $S(W) \cap \mathbb{R}^*_+ = \{\lambda \in \mathbb{R}^*_+ \mid \theta_\lambda = \iota\}$ and $T(W) = \{t \in \mathbb{R} \mid \exists 0 \neq w \in Z(U) \text{ with } \theta_\lambda(w) = \lambda^{it}(w)\, \forall \lambda \in \mathbb{R}^*_+\}$. Prove that

W is a factor if and only if θ acts ergodically on $Z(U)$ in the sense that $\theta_s(w) = w$ for all $s \in \mathbb{R}_+^*$ implies $w \in \mathbb{C}$. Conclude that U is a factor when W is a factor of type III_1. Show, conversely, that if we have a von Neumann dynamical system $(U, \mathbb{R}_+^*, \theta)$ with the above property for every such y, then $\mathbb{R}_+^* \bar{\ltimes}_\theta U$ will be of type III. What can you say about factor correspondences when $\lambda \in \langle 0, 1 \rangle$?

Chapter 15
Crossed Products for Quantum Groups

Crossed products are best dealt with in the framework of locally compact quantum groups since generalized Pontryagin duality holds there; the purely operator algebraic setup is also much cleaner. Von Neumann dynamical systems are then replaced by the more general concept of a coaction of a locally compact quantum group (W, Δ) on a von Neumann algebra U, that is, a normal unital $*$-monomorphism $\alpha\colon U \to W\bar{\otimes}U$ such that $(\iota \otimes \alpha)\alpha = (\Delta \otimes \iota)\alpha$. The (von Neumann) crossed product of α is the von Neumann algebra $W \ltimes_\alpha U$ generated by $\alpha(U)$ and $\hat{W} \otimes \mathbb{C}$. The dual coaction $\hat{\alpha}$ is a coaction of $(\hat{W}, \hat{\Delta}^{\mathrm{op}})$ on $W \ltimes_\alpha U$ uniquely determined by $\hat{\alpha}\alpha = 1 \otimes \alpha$ and $\hat{\alpha}(a \otimes 1) = \hat{\Delta}^{\mathrm{op}}(a) \otimes 1$ for $a \in \hat{W}$. The biduality theorem in this context says that the bidual $\hat{\hat{\alpha}}$ is a generalized cocycle equivalent to α modulo performing a flip and tensoring with $B(V_x)$.

Given any faithful semifinite normal weight y on U there is a faithful semifinite normal weight $\tilde{y}$ on the crossed product called the dual weight of y. Then $K \equiv J_{\tilde{y}}(J_{\hat{x}} \otimes J_y)$ is a unitary corepresentation of $(W, \Delta^{\mathrm{op}})$ such that $\alpha = K(1 \otimes (\cdot))K^*$. It is called the unitary implementation of α since it does not depend too much on the chosen weight y. This fundamental result for coactions is a far reaching generalization of the standard implementation of an action on a von Neumann algebra; the proof uses Connes' relative modular theory and a stronger form of left invariance, which we first prove, for the left invariant weights on quantum groups.

Recommended literature for this chapter is [13, 16, 51].

15.1 Complete Left Invariance for Locally Compact Quantum Groups

Crossed products can be defined for 'actions' of quantum groups. In order to develope a full theory we need a stronger form of left invariance of the 'Haar weights' on locally compact quantum groups.

L. Tuset, *Analysis and Quantum Groups*,
https://doi.org/10.1007/978-3-031-07246-8_15

Proposition 15.1.1 *Let (Δ, W) be a locally compact quantum group with a left invariant faithful semifinite normal weight x, and consider a positive element $X \in U\bar{\otimes}W$ for a von Neumann algebra U. Then $(\iota\otimes\iota\otimes x)(\iota\otimes\Delta)(X) = (\iota\otimes x)(X)\otimes 1$ as an equality in the extended positive part of $U\bar{\otimes}W$, so both $\iota\otimes x$ and $\iota\otimes\iota\otimes x$ are considered as tensor products of operator valued weights. In particular, we have $(\iota\otimes x)\Delta = x(\cdot)1$ on W_+, coined* complete left invariance *of x.*

Proof By Pontryagin duality we may prove the proposition for the dual quantum group $(\hat{W}, \hat{\Delta})$ with left invariant weight $\hat{x}$. As $U \subset B(V)$, we may also assume that $U = B(V)$. Consider $\tilde{N} = (J_x \otimes J_x)M(J_x \otimes J_x) \in W'\bar{\otimes}\hat{W}$ defined just after Proposition 12.10.11, and define for any positive $Y \in B(V \otimes V_x)$ the element $E(Y) \in \overline{B(V \otimes V_x)_+}$ by $E(Y) = (\iota\otimes\iota\otimes\hat{x})((1\otimes\tilde{N})(Y\otimes 1)(1\otimes\tilde{N}^*))$. Extend a unit vector $u \in V \otimes V_x$ to an orthonormal basis $\{e_i\}$, and let $P = (\cdot|u)u$. For $v \in V \otimes V_x$ we then have

$$\begin{aligned}
E(P)(\omega_v) &= \hat{x}(\omega_v \otimes \iota)((1\otimes\tilde{N})(P\otimes 1)(1\otimes\tilde{N}^*)) \\
&= \sum \hat{x}((\omega_{v,e_i}\otimes\iota)((P\otimes 1)(1\otimes\tilde{N}^*))^*(\omega_{v,e_i}\otimes\iota)((P\otimes 1)(1\otimes\tilde{N}^*))) \\
&= \hat{x}((\omega_{v,u}\otimes\iota)(1\otimes\tilde{N}^*)^*(\omega_{v,u}\otimes\iota)(1\otimes\tilde{N}^*)) \\
&= \hat{x}((\omega_{(1\otimes w)v,(1\otimes w)u}\otimes\iota)(1\otimes M)^*(\omega_{(1\otimes w)v,(1\otimes w)u}\otimes\iota)(1\otimes M)),
\end{aligned}$$

where $w = J_{\hat{x}}J_x$. In defining $\hat{x}$ we introduced a map $p\colon K \subset W_* \to V_x$, and $\omega \in W_*$ satisfies $(\omega\otimes\iota)(M) \in N_{\hat{x}}$ if and only if $\omega \in K$. Hence $E(P)(\omega_v) < \infty$ if and only if $\omega_{(1\otimes w)v,(1\otimes w)u}(1\otimes(\cdot)) \in K$, and then $E(P)(\omega_v) = \|p(\omega_{(1\otimes w)v,(1\otimes w)u}(1\otimes(\cdot)))\|^2$ by definition of $\hat{x}$. In this case we easily see that

$$p(\omega_{(1\otimes w)(1\otimes J_x w' J_x)v,(1\otimes w)u}(1\otimes(\cdot))) = R((w')^*)p(\omega_{(1\otimes w)v,(1\otimes w)u}(1\otimes(\cdot)))$$

for any unitary $w' \in W$, so for such w' we have $E(P)(\omega_v) = E(P)(\omega_{(1\otimes J_x w' J_x)v})$. We may thus consider $E(P) \in \overline{(B(V)\bar{\otimes}W)_+}$ by first extending normal functionals on $B(V)\bar{\otimes}W$ to $B(V\otimes V_x)$ and then apply the original $E(P)$ to them; we have shown that the element we get in $[0,\infty]$ does not depend on the chosen normal extension. This still holds when we replace P by any positive $Y \in B(V\otimes V_x)$ because E is lower semicontinuous and $Y = \sum(\cdot|Y^{1/2}e_i)Y^{1/2}e_i$. So when $Y \in B(V)\bar{\otimes}\hat{W}$, we get

$$E(Y) = (\iota\otimes\iota\otimes\hat{x})(\iota\otimes\hat{\Delta})(Y) \in \overline{B(V)\bar{\otimes}W_+} \cap \overline{B(V)\bar{\otimes}\hat{W}_+}.$$

Writing $E(Y) = \int \lambda\, dP(\lambda) + \infty(1-e)$, we get $P(\lambda), e \in B(V)\otimes\mathbb{C}$ as $W\cap\hat{W} = \mathbb{C}$. Define $T \in \overline{B(V)_+}$ by $\int\lambda\, dQ(\lambda) + \infty(1-f)$ with $e = f\otimes 1$ and $P(\lambda) = Q(\lambda)\otimes 1$, so $T\otimes 1 = (\iota\otimes\iota\otimes\hat{x})(\iota\otimes\hat{\Delta})(Y)$. When $V = \mathbb{C}$, then $T \in [0,\infty]$ and $(\iota\otimes\hat{x})\Delta(Y) = T1$. If there exists a non-zero positive $\omega \in \hat{W}_*$ with $(\omega\otimes)\hat{\Delta}(Y) \in M_{\hat{x}}$, then $T < \infty$, in which case $\hat{x}(\omega'\otimes\iota)\Delta(Y) = T\omega'(1) < \infty$ for any positive $\omega' \in \hat{W}_*$. But then $Y \in M_{\hat{x}}$ and $T = \hat{x}(Y)$, again by left invariance of $\hat{x}$. In the other case we must have $T = \infty$, and by left invariance, we cannot have $Y \in M_{\hat{x}}$, so $\hat{x}(Y) = \infty = T$. In either case $(\iota \otimes \hat{x})\hat{\Delta}(Y) = \hat{x}(Y)1$. For general V, by the special case already proved, we have $T(\omega)\omega'(1) = \hat{x}(\omega \otimes \omega' \otimes \iota)(\iota \otimes \Delta)(Y) = \omega'(1)(\iota \otimes \hat{x})(Y)(\omega)$ for positive $\omega \in B(V)_*$ and $\omega' \in \hat{W}_*$. So $T = (\iota \otimes \hat{x})(Y)$. □

15.2 Coactions and Integrability

Definition 15.2.1 A *(left) coaction* of a locally compact quantum group (W, Δ) on a von Neumann algebra U is a injective normal unital $*$-homomorphism $\alpha : U \to W\bar{\otimes}U$ such that $(\iota \otimes \alpha)\alpha = (\Delta \otimes \iota)\alpha$.

This generalizes the notion of an action of a group on a von Neumann algebra in the following sense.

Proposition 15.2.2 *Let (W, α, G) be a von Neumann dynamical system. Then the map $\tilde{\alpha} : W \to L^\infty(G, W) \cong L^\infty(G)\bar{\otimes}W$ defined by $\tilde{\alpha}(w)(t) = \alpha_{t^{-1}}(w)$ is a coaction of $(L^\infty(G), \Delta)$ on W for the usual coproduct Δ on $L^\infty(G)$. Furthermore, the map $\alpha \mapsto \tilde{\alpha}$ provides a one-to-one correspondence between these notions.*

Proof That we get a coaction is clear from

$$(\iota \otimes \tilde{\alpha})\tilde{\alpha}(w)(s, t) = \tilde{\alpha}(\alpha_{s^{-1}}(w))(t) = \alpha_{(st)^{-1}}(w) = \tilde{\alpha}(w)(st) = (\Delta \otimes \iota)\tilde{\alpha}(w)(s, t).$$

To show that the map in the final statement is surjective, say we are given a coaction β of $(L^\infty(G), \Delta)$ on W. Using $\Delta = M^*(1 \otimes (\cdot))M$ and $\lambda(t) = (\delta_t \otimes \iota)(M)$ and $(\iota \otimes \beta)\beta = (\Delta \otimes \iota)\beta$, we see that $(\lambda(t) \otimes 1)\beta(w)(\lambda(t)^* \otimes 1) \in \beta(W)$, so we may consider its inverse image $\alpha_t(w)$ under β in order to define α. Then (W, α, G) will be a von Neumann dynamical system, and $(\iota \otimes \beta)\tilde{\alpha} = (\iota \otimes \beta)\beta$, so $\tilde{\alpha} = \beta$. □

Given a coaction α of (W, Δ) on U. A unitary $X \in W\bar{\otimes}U$ is an *α-cocycle* if $(\Delta \otimes \iota)(X) = X_{23}(\iota \otimes \alpha)(X)$. In this case $\beta = X\alpha(\cdot)X^*$ is clearly a coaction of (W, Δ) on U, and X^* is a β-cocycle. We say α and β are *cocycle equivalent* and write $\alpha \sim \beta$. That $\sim$ is an equivalence relation is straightforward; transitivity is obtained by multiplying together the corresponding cocycles. Notice that when α is trivial, the cocycle satisfies $(\Delta \otimes \iota)(X) = X_{23}X_{13}$. When we have a von Neumann dynamical system (U, α, G) with $\tilde{\alpha}$ as in the previous proposition, and if we have a strongly continuous one-parameter family $\{X_t\}$ of unitaries in U satisfying $X_{st} = X_t\alpha_{t^{-1}}(X_s)$ for $s, t \in G$, and let $\beta_t = X_t\alpha_{t^{-1}}(\cdot)X_t^*$, the von Neumann dynamical systems (U, β, G) and (U, α, G) are *cocycle equivalent* in that $t \mapsto X_t$, considered as an element in $L^\infty(G)\bar{\otimes}U$, is an $\tilde{\alpha}$-cocycle and $\tilde{\alpha} \sim \tilde{\beta}$. Again one checks that every $\tilde{\alpha}$-cocycle stems from such a unique family.

We want to study to what extend a coaction of a quantum group is implemented by an appropriate unitary.

Let α be a coaction on U of a locally compact compact quantum group (W, Δ), and let k be a self-adjoint strictly positive operator affiliated with W. A faithful semifinite normal weight y on U is *k-invariant* if $y(\omega_v \otimes \iota)\alpha = \|k^{1/2}v\|^2 y$ on $(M_y)_+$ for $v \in D(k^{1/2})$.

Proposition 15.2.3 *Let α be a coaction on U of a locally compact compact quantum group (W, Δ) with left invariant faithful semifinite normal weight x and modular element h. Suppose we are given an h^{-1}-invariant faithful semifinite normal weight y on U, and assume in addition that*

$$y(\omega_{h^{\frac{1}{2}}v_1, h^{\frac{1}{2}}v_2} \otimes \iota)(\alpha(u_1^*)(1 \otimes u_2)) = y(\omega_{\Delta_{\hat{x}}^{\frac{1}{2}} h^{\frac{1}{2}} J_{\hat{x}} v_2, \Delta_{\hat{x}}^{-\frac{1}{2}} h^{-\frac{1}{2}} J_{\hat{x}} v_1} \otimes \iota)((1 \otimes u_1^*)\alpha(u_2))$$

for $u_i \in N_y$ and $v_i \in D \equiv \cup_{s,t\in\mathbb{R}} D(h^s \Delta_{\hat{x}}^t)$, where by $h^s \Delta_{\hat{x}}^t$ we here mean the closure of the composition. Then there exists a unique unitary $K \in W\bar{\otimes}B(V_y)$ such that $(\omega_{v,v'} \otimes \iota)(K)q_y = q_y(\omega_{h^{1/2}v,v'} \otimes \iota)\alpha$ on N_y for $v \in D(h^{1/2})$ and $v' \in V_x$. Moreover, it satisfies $(\Delta \otimes \iota)(K) = K_{23}K_{13}$ and $(\iota \otimes \pi_y)\alpha(u) = K(1 \otimes \pi_y(u))K^*$ for $u \in U$, and K commutes with $h\Delta_{\hat{x}} \otimes \Delta_y$, and finally $K^* = (J_{\hat{x}} \otimes J_y)K(J_{\hat{x}} \otimes J_y)$.

Proof First note that since h is fixed under the scaling group, which is implemented by $\Delta_{\hat{x}}$, these unbounded operators commute strongly, so the compositions of any of their real powers are closable with D as a dense core containing $q_x(N_x)$. Now h^{-1}-invariance clearly shows that $K(h^{-1/2}q_x(w) \otimes q_y(u)) = q_{x\otimes y}(\alpha(u)(w \otimes 1))$ for $w \in N_x$ and $u \in N_y$ defines an isometry K on $V_x \otimes V_y$ that satisfies the first condition in the proposition, and is evidently uniquely determined by it. Since h is affiliated with W, and $\alpha(U) \subset W\bar{\otimes}U$, we get for $w' \in W'$, that

$$(\omega_{w'v,v'} \otimes \iota)(K)q_y = q_y(\omega_{h^{1/2}v,w'^*v'} \otimes \iota)\alpha = (\omega_{v,w'^*v'} \otimes \iota)(K)q_y$$

on N_y, so $K(w' \otimes 1) = (w' \otimes 1)K$, and $K \in (W'\bar{\otimes}B(V_y)')' = W\bar{\otimes}B(V_y)$. The property $(\iota \otimes \pi_y)\alpha(u)K = K(1 \otimes \pi_y(u))$ for $u \in U$, is clear from the formula for K given at the beginning of this proof together with multiplicativity of α. That K implements α in the representation π_y follows then from unitarity of K, which we prove below. Letting S_y as usual denote the closure of $q_y(u) \mapsto q_y(u^*)$, we get

$$(\omega_{h^{-1/2}v_1,v_2} \otimes \iota)(K)S_y q_y(u) = q_y(\omega_{v_1,v_2} \otimes \iota)\alpha(u^*) = S_y(\omega_{h^{-1/2}v_2,v_1} \otimes \iota)(K)q_y(u),$$

which also implies $(\omega_{v_1,h^{-1/2}v_2} \otimes \iota)(K^*)S_y^* \subset S_y^*(\omega_{v_2,h^{-1/2}v_1} \otimes \iota)(K^*)$ by taking adjoints. On the other hand, we have

$$(\omega_{h^{1/2}v_1,v_2} \otimes \iota)(K^*) = (\omega_{\Delta_{\hat{x}}^{1/2} J_{\hat{x}} v_2, \Delta_{\hat{x}}^{-1/2} h^{-1/2} J_{\hat{x}} v_1} \otimes \iota)(K),$$

which is checked by evaluating both sides on $\omega_{q_y(u_1),q_y(u_2)}$ and using the second assumption in the proposition with $u_i \in N_y$ together with the condition uniquely determining K. Combining this with $\Delta_y = S_y^* S_y$ and the properties just proved for S_y and S_y^*, we get $(\omega_{v_1,h^{-1/2}v_2} \otimes \iota)(K^*)\Delta_y = \Delta_y(\omega_{\Delta_{\hat{x}}^{-1}h^{-1}v_1, \Delta_{\hat{x}}^{1}h^{1/2}v_2} \otimes \iota)(K^*)$, and hence $K(h\Delta_{\hat{x}} \otimes \Delta_y) = (h\Delta_{\hat{x}} \otimes \Delta_y)K$. Using this last identity and $J_y = S_y\Delta_y^{-1/2}$

together with the previous identities involving K, we get

$$J_y(\omega_{v_1,h^{-1/2}\Delta_{\hat{x}}^{-1/2}v_2} \otimes \iota)(K)J_y = (\omega_{J_{\hat{x}}v_1,J_{\hat{x}}\Delta_{\hat{x}}^{-1/2}h^{-1/2}v_2} \otimes \iota)(K^*),$$

thus $K^* = (J_y \otimes J_{\hat{x}})K(J_y \otimes J_{\hat{x}})$. Hence

$$KK^* = (J_y\otimes J_{\hat{x}})K^*(J_y\otimes J_{\hat{x}})(J_y\otimes J_{\hat{x}})K(J_y\otimes J_{\hat{x}}) = (J_y\otimes J_{\hat{x}})K^*K(J_y\otimes J_{\hat{x}}) = I$$

as $K^*K = I$. To prove the corepresentation property, observe that

$$(\omega_{v,v'} \otimes \omega_{w,w'} \otimes \iota)(K_{23}K_{13})q_y = (\omega_{w,w'} \otimes \iota)(K)q_y(\omega_{h^{1/2}v,v'} \otimes \iota)\alpha$$

equals $q_y((\omega_{h^{1/2}v,v'} \otimes \omega_{h^{1/2}w,w'})\Delta \otimes \iota)\alpha$ as $(\iota \otimes \alpha)\alpha = (\Delta \otimes \iota)\alpha$. Using $\Delta(h^{1/2}) = h^{1/2}\otimes h^{1/2}$ and the determining condition for K, one shows that the latter expression again equals $(\omega_{v,v'} \otimes \omega_{w,w'} \otimes \iota)(\Delta \otimes \iota)(K)q_y$ by using the by now standard techniques for quantum groups and weights in a completely rigorous argument. □

We shall later see that this result, generalizing the usual theory for actions, is true under minimal assumptions. In fact, that the second assumption in the previous result is redundant, can already be seen by tracing our earlier proof of strong left invariance for the Haar-weight, adjusting it slightly by taking into account that the comultiplication must be replaced by a coaction, and by modifying the invariance condition in the first assumption by h. However, instead of following this instructive but lengthy and technical route, we need the above result only for the specific case considered in the next result, and here the second assumption above can be proved directly with much less fuss.

Proposition 15.2.4 *Let $\alpha: U \to W\bar{\otimes}U$ be a coaction of a locally compact quantum group (W, Δ) with modular element h and right invariant faithful semifinite normal weight y. The* fixed point algebra *of α is the von Neumann subalgebra $U^\alpha \equiv \{u \in U \mid \alpha(u) = 1 \otimes u\}$ of U. Consider the tensor product operator valued weight $y \otimes \iota: (W\bar{\otimes}U)_+ \to \overline{U}_+$. For $u \in U_+$, we have $(y \otimes \iota)\alpha(u) \in \overline{U}_+$, and we define $E(u) \in \overline{U^\alpha}_+$ by $E(u)(z) \equiv (y \otimes \iota)\alpha(u)(\tilde{z})$ for positive $z \in (U^\alpha)_*$, where $\tilde{z}$ is any normal positive extension of z to U. Then E is a faithful normal operator valued weight from U to U^α. Moreover, when α is* integrable, *meaning that E is semifinite, then $z'E$ is an h^{-1}-invariant faithful semifinite normal weight on U for any faithful semifinite normal weight z' on U^α, and $z'E$ satisfies furthermore the second assumption of the previous proposition.*

Proof Let F be positive in $(W\bar{\otimes}U)_*$. By definition of tensor product operator valued weights, we have

$$((y \otimes \iota)\alpha(u))(F\alpha) = y(\iota \otimes F\alpha)\alpha(u) = ((y \otimes \iota \otimes \iota)(\Delta \otimes \iota)\alpha(u))(F),$$

which upon applying the right invariant version of Proposition 15.1.1, is equal to $(1 \otimes (y \otimes \iota)\alpha(u))(F)$. Hence $\tilde{\alpha}((y \otimes \iota)\alpha(u)) = 1 \otimes (y \otimes \iota)\alpha(u)$. Thus $(y \otimes \iota)\alpha(u)$

can be written as a limit of elements belonging to U^α, and by normality of $\tilde{z}$, this shows that $E(u)(z)$ is independent of the extension $\tilde{z}$. Also, we get

$$E(aua^*)(z) = ((y\otimes\iota)((1\otimes a)\alpha(u)(1\otimes a^*)))(\tilde{z}) = y(\iota\otimes a^*\tilde{z}a)\alpha(u) = E(u)(a^*za)$$

for $a \in U^\alpha$, so E is indeed an operator valued weight. It is normal and faithful as both α and $y \otimes \iota$ are.

When α is integrable, clearly $z'E$ is a faithful semifinite normal weight on U. To show h^{-1}-invariance, we may by uniqueness of right invariant weights, assume that $y = x(h^{1/2} \cdot h^{1/2})$, and then, for $v \in D(h^{-1/2})$, we get by Proposition 15.1.1 that

$$y(\omega_v\otimes\iota)\Delta = x(\omega_{h^{-1/2}v}\otimes\iota)\Delta(h^{1/2}\cdot h^{1/2}) = \omega_{h^{-1/2}v}(1)x(h^{1/2}\cdot h^{1/2}) = \|h^{-1/2}v\|^2 y$$

as $h^{1/2}$ is group-like. By tensor product of weights we thus get $(\omega_v\otimes y\otimes z')(\Delta\otimes\iota) = \|h^{-1/2}v\|^2 y \otimes z'$. Combining this with $z'E(\omega_v \otimes \iota) = (\omega_v \otimes z')(\iota \otimes (y \otimes \iota)\alpha)$ and $(\iota \otimes \alpha)\alpha = (\Delta \otimes \iota)\alpha$, we get $z'E(\omega_v \otimes \iota)\alpha = \|h^{-1/2}v\|^2 z'E$. The second condition in the previous proposition we get from

$$z'E(\omega_{h^{\frac{1}{2}}v_1,h^{\frac{1}{2}}v_2}\otimes\iota)(\alpha(u_1^*)(1\otimes u_2)) = (\omega_{h^{\frac{1}{2}}v_1,h^{\frac{1}{2}}v_2}\otimes y\otimes z')((\iota\otimes\alpha)\alpha(u_1^*)(1\otimes\alpha(u_2))$$

by using $(\iota \otimes \alpha)\alpha = (\Delta \otimes \iota)\alpha$ twice combined with

$$(\omega_{h^{\frac{1}{2}}v_1,h^{\frac{1}{2}}v_2}\otimes\iota)(\Delta(a_1^*)(1\otimes a_2)) = y(\omega_{\Delta_{\hat{x}}^{\frac{1}{2}}h^{\frac{1}{2}}J_{\hat{x}}v_2,\Delta_{\hat{x}}^{-\frac{1}{2}}h^{-\frac{1}{2}}J_{\hat{x}}v_1}\otimes\iota)((1\otimes a_1^*)\Delta(a_2))$$

for $a_i \in N_y$. To prove this latter identity, use $y = x(h^{1/2} \cdot h^{1/2})$ that $h^{1/2}$ is group-like, then strong left invariance of x, and then the property $\omega_{h^{1/2}v_1,v_2}S^{-1} = \omega_{\Delta_{\hat{x}}^{1/2}J_{\hat{x}}v_2,\Delta_{\hat{x}}^{-1/2}J_{\hat{x}}h^{1/2}v_1}$, which follows from $S^{-1} = R\tau_{i/2}$ and $\tau_{i/2} = \Delta_{\hat{x}}^{-1/2} \cdot \Delta_{\hat{x}}^{1/2}$ and $R = J_{\hat{x}}(\cdot)^* J_{\hat{x}}$. Then reverse the route. □

Clearly α restricts to a trivial action on U^α. We say that α is *ergodic* if $U^\alpha = \mathbb{C}$. Of course there always exists a weight z' as prescribed in the previous proposition. Hence the second last proposition tells us that any integrable α is implemented in the semicyclic representation of $z'E$ by a unitary corepresentation K of (W, Δ) that in addition has nice modular properties.

15.3 Crossed Products of Coactions

Definition 15.3.1 The *crossed product* of a coaction α of a locally compact quantum group (W, Δ) on a von Neumann algebra U is the von Neumann subalgebra $W \ltimes_\alpha U$ of $B(V_x)\bar{\otimes}U$ generated by $\alpha(U)$ and $\hat{W} \otimes \mathbb{C}$.

Proposition 15.3.2 *Suppose we are given a coaction* α *on a von Neumann algebra* U *of a locally compact quantum group* (W, Δ) *with left invariant faithful semifinite normal weight* x *and fundamental multiplicative unitary* M, *and let* $N' = (J_x \otimes J_x)M_{21}(J_x \otimes J_x)$. *Then* $\hat{\alpha} = (N' \otimes 1)(1 \otimes (\cdot))(N'^* \otimes 1)$ *defines a coaction of* $(\hat{W}, \hat{\Delta}^{\rm op})$ *on* $W \ltimes_\alpha U$ *that is uniquely determined by the properties* $\hat{\alpha}\alpha = 1 \otimes \alpha$ *and* $\hat{\alpha}(a \otimes 1) = \hat{\Delta}^{\rm op}(a) \otimes 1$ *for* $a \in \hat{W}$. *This* dual coaction of α *is moreover integrable.*

Proof Clearly $\hat{\alpha}$ is a faithful normal unital $*$-homomorphism. As the second leg of N' lies in W', the property $\hat{\alpha}\alpha = 1 \otimes \alpha$ is obvious, while the property $\hat{\alpha}(a \otimes 1) = \hat{\Delta}^{\rm op}(a) \otimes 1$ is clear from $\hat{\Delta} = M_{21}(1 \otimes (\cdot))M_{21}^*$ and $\hat{R} = J_x(\cdot)^* J_x$ and $(\hat{R} \otimes \hat{R})\hat{\Delta}\hat{R} = \hat{\Delta}^{\rm op}$. These two properties clearly determines $\hat{\alpha}$ uniquely. As

$$(\iota \otimes \hat{\alpha})\hat{\alpha}(a \otimes 1) = (\iota \otimes \hat{\Delta}^{\rm op} \otimes \iota)(\hat{\Delta}^{\rm op}(a) \otimes 1) = (\hat{\Delta}^{\rm op} \otimes \iota)\hat{\alpha}(a \otimes 1)$$

by coassociativity, we thus conclude that $\hat{\alpha}$ is a coaction of $(\hat{W}, \hat{\Delta}^{\rm op})$ on $W \ltimes_\alpha U$.

To see that it is integrable, let $E = (\hat{x} \otimes \iota)\hat{\alpha}$ be the faithful normal operator valued weight from $W \ltimes_\alpha U$ to $(W \ltimes_\alpha U)^{\hat{\alpha}}$ associated to $\hat{\alpha}$. As $E(a^*a \otimes 1) = (\hat{x} \otimes \iota)(\hat{\Delta}^{\rm op}(a^*a) \otimes 1) = \hat{x}(a^*a)1$ for $a \in N_{\hat{x}}$, we see that N_E is dense in $W \ltimes_\alpha U$. □

We have the following *biduality theorem for crossed products* of quantum groups.

Theorem 15.3.3 *Suppose we are given a coaction* α *on a von Neumann algebra* U *of a locally compact quantum group* (W, Δ) *with left invariant faithful semifinite normal weight* x *and fundamental multiplicative unitary* M. *Consider the fundamental multiplicative unitary* N_{21}^* *for* $(W, \Delta)^{\rm op}$ *with* $N = (J_{\hat{x}} \otimes J_{\hat{x}})\hat{M}(J_{\hat{x}} \otimes J_{\hat{x}})$, *see Proposition 12.10.13, and recall that* $g = J_{\hat{x}} J_x(\cdot) J_x J_{\hat{x}}$ *is an isomorphism from the quantum group* (W, Δ) *to* $(W, \Delta)'^{\rm op} = (W, \Delta)^{\wedge{\rm op}\wedge{\rm op}}$. *Let* F *denote the flip on* $B(V_x \otimes V_x)$, *and let* $Z = (\alpha(U) \cup (B(V_x) \otimes \mathbb{C}))'' \subset B(V_x)\bar{\otimes}U$.

Then the map $M_{12}(\iota \otimes \alpha)(\cdot)M_{12}^*\colon B(V_x)\bar{\otimes}V \to B(V_x \otimes V_x)\bar{\otimes}U$ *sends* $\alpha(u), a, b$ *to* $\alpha(u)_{23}, \hat{\Delta}^{\rm op}(a)_{12}, b \otimes 1 \otimes 1$ *for* $u \in U, a \in \hat{W}, b \in W'$, *respectively. It sends* $W \ltimes_\alpha U$ *onto* $\hat{\alpha}(W \ltimes_\alpha U)$, *and restricts to a* $*$-*isomorphism* $f\colon Z \to \hat{W} \ltimes_{\hat{\alpha}} (W \ltimes_\alpha U)$. *The map* $(F \otimes \iota)(\iota \otimes \alpha)\colon B(V_x)\bar{\otimes}U \to W\bar{\otimes}B(V_x)\bar{\otimes}U$ *restricts to a coaction* β *of* (W, Δ) *on* Z, *and* $N_{21}^* \otimes 1$ *is a unitary* β-*cocycle. The corresponding coaction* $\gamma = N_{21}^*\beta(\cdot)N_{21}$ *is isomorphic to the bidual coaction* $\hat{\hat{\alpha}}$ *of* $(W, \Delta)^{\wedge{\rm op}\wedge{\rm op}}$ *on* $\hat{W} \ltimes_{\hat{\alpha}} (W \ltimes_\alpha U)$ *in the sense that* $\hat{\hat{\alpha}}f = (g \otimes f)\gamma$.

Proof That the restriction defines a $*$-isomorphism f is clear from how the map evidently operates on the elements u, a, b, and the fact that $W \cap (\hat{W})' = \mathbb{C}$, which by taking commutants, shows that $B(V_x)$ as a von Neumann algebra, is generated by W' and $\hat{W}$. That $(F \otimes \iota)(\iota \otimes \alpha)$ is a coaction of (W, Δ), is straightforward, and to see that it restricts to a coaction on Z, observe that

$$(F \otimes \iota)(\iota \otimes \alpha)(Z) = (N_{21}\alpha(U)_{23}N_{21}^* \cup (\mathbb{C} \otimes B(V_x) \otimes \mathbb{C}))'' \subset W\bar{\otimes}U$$

as $\Delta^{\rm op} = N_{21}(1 \otimes (\cdot))N_{21}^*$ and $N_{21} \in W\bar{\otimes}\hat{W}'$. This also shows that $N_{21}^* \otimes 1$ belongs to $W\bar{\otimes}Z$, and the cocycle property now boils down to the pentagonal equation for N, which holds. The final formula needs only be checked on the generating elements $\alpha(u), a, b$, and that is by now straightforward. □

In the above theorem the ominous Z can actually be replaced by $B(V_x)\bar{\otimes}U$. Indeed, let $\alpha\colon U \to M\bar{\otimes}U$ be a coaction of a locally compact quantum group (W, Δ) on a von Neumann algebra U. We say that α satisfies A if $(\alpha(U) \cup (B(V_x) \otimes \mathbb{C}))'' = B(V_x)\bar{\otimes}U$, that α satisfies B if the inclusion $\alpha(U) \subset (W \ltimes_\alpha U)^{\hat{\alpha}}$ is an equality, that α satisfies C if $\alpha(U) \subset \{X \in W\bar{\otimes}U \mid (\iota \otimes \alpha)(X) = (\Delta \otimes \iota)(X)\}$ is an equality, that α satisfies D if there exists a faithful normal $*$-representation π of U and a unitary $K \in W\bar{\otimes}B(V_\pi)$ with $K^* = (J_{\hat{x}} \otimes J)K(J_{\hat{x}} \otimes J)$ for some antilinear J on V_π with $J^* = J = J^2$ and $J\pi(U)J = \pi(U)'$, and such that $(\Delta \otimes \iota)(K) = K_{23}K_{13}$ and $(\iota \otimes \pi)\alpha = K(1 \otimes \pi(\cdot))K^*$. We need a sequence of lemmas, and will in the end prove that all coactions satisfy properties A, B, C, D. In the lemmas below we assume that α is a coaction of a locally compact quantum group (W, Δ) on a von Neumann algebra U.

Lemma 15.3.4 *Any integrable coaction satisfies D, and any dual coaction satisfies A and D.*

Proof The first claim was proved in the last section. For the second we have

$$(\hat{\alpha}(W \ltimes_\alpha U) \cup (B(V_x) \otimes \mathbb{C} \otimes \mathbb{C}))'' = ((\mathbb{C} \otimes \alpha(U)) \cup ((\hat{\Delta}^{\rm op}(\hat{W}) \cup (B(V_x) \otimes \mathbb{C}))'' \otimes \mathbb{C}))'',$$

which equals $B(V_x)\bar{\otimes}(W \ltimes_\alpha U)$ as $((\hat{\Delta}^{\rm op}(\hat{W}) \cup (B(V_x) \otimes \mathbb{C}))'' = B(V_x)\bar{\otimes}\hat{W}$. The third claim follows now from the previous proposition. □

If α satisfies D, then $\alpha' = K^*(1 \otimes (\cdot))K$ defines a coaction of $(W, \Delta^{\rm op})$ on U' since taking the commutant of $K(\mathbb{C} \otimes U)K^* \subset B(V_x)\bar{\otimes}U$ yields $K^*(\mathbb{C} \otimes U')K \subset W\bar{\otimes}U'$, and $(\iota \otimes \alpha')\alpha'(b) = K_{23}^*K_{13}^*(1 \otimes 1 \otimes b)K_{13}K_{23} = (\Delta^{\rm op} \otimes \iota)\alpha'(b)$ for $b \in U'$. Here we have for simplicity identified U with $\pi(U)$.

Lemma 15.3.5 *Suppose α satisfies D. Then α satisfies C if and only if α' satisfies A. In this case α also satisfies B.*

Proof Let $X \in W\bar{\otimes}U$. Write $(\Delta^{\rm op} \otimes \iota)(K) = K_{13}K_{23}$ and consider the fundamental multiplicative unitary $N_{21}^* \in W\bar{\otimes}\hat{W}'$ for $(W, \Delta)^{\rm op}$. This gives $N_{12}^*K_{23} = K_{13}N_{12}^*K_{13}^*$, and the left hand side commutes with X_{13} if and only if $(\iota \otimes \alpha)(X) = (\Delta \otimes \iota)(X)$, as is seen by flipping the first two legs in the latter identity and using that N and K implement $\Delta^{\rm op}$ and α, respectively. By the right hand side of the identity we see that this happens if and only if $X \in K(\hat{W}\bar{\otimes}B(V_\pi))K^*$. As $K \in W\bar{\otimes}B(V_\pi)$ and $W \cap \hat{W} = \mathbb{C}$, we thus see that the set of $X \in W\bar{\otimes}U$ such that $(\iota \otimes \alpha)(X) = (\Delta \otimes \iota)(X)$ coincides with the set $(B(V_x)\bar{\otimes}U) \cap K(\mathbb{C} \otimes B(V_\pi))K^*$. This latter set equals $K(\mathbb{C} \otimes U)K^*$ if and only if α satisfies C. Taking commutants on both sides of that equality shows that its validity is equivalent to α' satisfying A. To see that α then also satisfies B, it now suffices to show that $(W \ltimes_\alpha U)^{\hat{\alpha}} \subset (B(V_x)\bar{\otimes}U) \cap K(\mathbb{C} \otimes B(V_\pi))K^*$. We claim that $K^*(\hat{W} \otimes \mathbb{C})K \subset \hat{W}\bar{\otimes}B(V_\pi)$.

Indeed, the von Neumann algebra $B(V_x)\bar{\otimes}\hat{W}$ is generated by M^* and $B(V_x)\otimes\mathbb{C}$, so

$$K_{23}^*(B(V_x)\bar{\otimes}\hat{W}\otimes\mathbb{C})K_{23}=((B(V_x)\otimes\mathbb{C}\otimes\mathbb{C})\cup\{K_{23}^*M_{12}^*K_{23}\})''.$$

Now $M_{12}^*K_{23}M_{12}=(\Delta\otimes\iota)(K)=K_{23}K_{13}$, so $K_{23}^*M_{12}^*K_{23}=K_{13}M_{12}^*$, which belongs to $W\bar{\otimes}\hat{W}\bar{\otimes}B(V_\pi)$, and the claim follows. Hence

$$W\ltimes_\alpha U=(K(\mathbb{C}\otimes U)K^*\cup(\hat{W}\otimes\mathbb{C}))''\subset K(\hat{W}\bar{\otimes}B(V_\pi))K^*.$$

Now $X\in W\ltimes_\alpha U$ belongs to the fixed point algebra of $\hat{\alpha}$ if and only if X_{23} and $\tilde{M}_{12}\in\hat{W}\bar{\otimes}W'$ commute, in other words, presicely when $X\in W\bar{\otimes}U$. So

$$(W\ltimes_\alpha U)^{\hat{\alpha}}=(W\ltimes_\alpha U)\cap(W\bar{\otimes}U)\subset K((\hat{W}\bar{\otimes}B(V_\pi))\cap(W\bar{\otimes}U))K^*$$

and the latter set equals $K(\mathbb{C}\otimes B(V_\pi))K^*$ as $\hat{W}\cap W=\mathbb{C}$. □

We record the useful identity $(W\ltimes_\alpha U)^{\hat{\alpha}}=(W\ltimes_\alpha U)\cap(W\bar{\otimes}B(V))$, valid for any coaction α with $U\subset B(V)$, which we showed in the proof just above.

Lemma 15.3.6 *Any coaction that satisfies A and D also satisfies B. So dual coactions satisfy B.*

Proof If α satisfies A and D, then

$$((K(\mathbb{C}\otimes U)K^*)\cup(B(V_x)\otimes\mathbb{C}))''=B(V_x)\bar{\otimes}U.$$

Using $K^*=(J_{\hat{x}}\otimes J)K(J_{\hat{x}}\otimes J)$ and $J\pi(U)J=\pi(U)'$, we see that α' satisfies A. But then α satisfies B by the last lemma. The second last lemma shows that dual coactions satisfy B. □

Lemma 15.3.7 *We have*

$$W\bar{\otimes}(W\ltimes_\alpha U)=((\iota\otimes\alpha)\alpha(U)\cup(W\bar{\otimes}\hat{W}\otimes\mathbb{C}))''.$$

Proof The inclusion $\supset$ is obvious. The inclusion $\subset$ is clear from $(\iota\otimes\alpha)\alpha=(\Delta\otimes\iota)\alpha$ and that Δ is implemented by $M\in W\bar{\otimes}\hat{W}$. □

Theorem 15.3.8 *Cocations of locally compact quantum groups satisfy A, B, C.*

Proof Say α is a coaction of a locally compact quantum group (W,Δ) on a von Neumann algebra $U\subset B(V)$. By Lemma 15.3.6 the bidual coaction satisfy B, and thus, so does β in the biduality theorem. But

$$\begin{aligned}M\ltimes_\beta Z&=((F\otimes\iota)(\iota\otimes\alpha)(\alpha(U)\cup(B(V_x)\otimes\mathbb{C}))\cup(\hat{W}\otimes\mathbb{C}\otimes\mathbb{C}))''\\&=(F\otimes\iota)((\iota\otimes\alpha)\alpha(U)\cup(W\bar{\otimes}\hat{W}\otimes\mathbb{C})\cup(B(V_x)\otimes\mathbb{C}\otimes\mathbb{C}))''\\&=(F\otimes\iota)((W\bar{\otimes}(W\ltimes_\alpha U))\cup(B(V_x)\otimes\mathbb{C}\otimes\mathbb{C}))''\end{aligned}$$

by Lemma 15.3.7, so by the remark just after Lemma 15.3.5, we have

$$\begin{aligned}(M \ltimes_\beta Z)^{\hat{\beta}} &= (M \ltimes_\beta Z) \cap (W \bar{\otimes} B(V_x \otimes V)) \\ &= (F \otimes \iota)((B(V_x) \bar{\otimes} (W \ltimes_\alpha U)) \cap (W \bar{\otimes} B(V_x \otimes V)) \\ &= (F \otimes \iota)((B(V_x) \bar{\otimes} (W \ltimes_\alpha U)) \cap (B(V_x) \bar{\otimes} W \bar{\otimes} B(V))) \\ &= (F \otimes \iota)(B(V_x) \bar{\otimes} (W \ltimes_\alpha U)^{\hat{\alpha}}).\end{aligned}$$

Hence as $(M \ltimes_\beta Z)^{\hat{\beta}} = \beta(Z)$, we get

$$B(V_x) \bar{\otimes} (W \ltimes_\alpha U)^{\hat{\alpha}} = (\iota \otimes \alpha)(\alpha(U) \cup (B(V_x) \otimes \mathbb{C}))''.$$

Now the right hand side is evidently contained in $B(V_x) \bar{\otimes} \alpha(U)$, and thus α satisfies B. But inserting $(W \ltimes_\alpha U)^{\hat{\alpha}} = \alpha(U)$, and using injectivity of $\iota \otimes \alpha$, we also see that α satisfies A. If α satisfies D, then as α' satisfies A, we see that α satisfies C by Lemma 15.3.5. For general α, consider the coaction $\iota \otimes \alpha$ of (W, Δ) on $\alpha(U)$. As $(\iota \otimes \alpha)\alpha = (\Delta \otimes \iota)\alpha$ and Δ is implemented by its fundamental multiplicative unitary, we see that the coaction $\iota \otimes \alpha$ satisfies D, and thus also C. Hence

$$(\iota \otimes \alpha)\alpha(U) = (\iota \otimes \alpha)(\{X \in W \bar{\otimes} U \mid (\iota \otimes \iota \otimes \alpha)(\iota \otimes \alpha)(X) = (\Delta \otimes \iota \otimes \iota)(\iota \otimes \alpha)(X)\}),$$

which by injectivity of $\iota \otimes \iota \otimes \alpha$ and $\iota \otimes \alpha$, says that α satisfies C. □

In the next section we prove that all coations also satisfy D.

15.4 Corepresentation Implementation of Coactions

Let α be a coaction on a von Neumann algebra U of a locally compact group group (W, Δ) with a left invariant faithful semifinite normal weight x. Consider the operator valued weight $E = (\hat{x} \otimes \iota \otimes \iota)\hat{\alpha}$ from $W \ltimes_\alpha U$ to $(W \ltimes_\alpha U)^{\hat{\alpha}} = \alpha(U)$ associated to the dual coaction $\hat{\alpha}$ on $W \ltimes_\alpha U$ of the quantum group $(\hat{W}, \hat{\Delta})^{\mathrm{op}}$. As dual actions are integrable, we know that E is normal faithful and semifinite. The *dual weight* $\tilde{y}$ associated to α of a faithful semifinite normal weight y on U is the faithful semifinite normal weight $y\alpha^{-1}E$ on $W \ltimes_\alpha U$.

Lemma 15.4.1 *We have* $\tilde{y}(\alpha(u^*)(a^*a \otimes 1)\alpha(u)) = y(u^*u)\hat{x}(a^*a)$, $u \in U, a \in N_{\hat{x}}$.

Proof We have $\hat{\alpha}(\alpha(u^*)(a^*a \otimes 1)\alpha(u)) = (1 \otimes \alpha(u^*))(\hat{\Delta}^{\mathrm{op}}(a^*a) \otimes 1)(1 \otimes \alpha(u))$, so if we set $z' = z(\alpha(u^*)((\cdot) \otimes 1)\alpha(u))$ on $\hat{W}$ for positive $z \in (W \ltimes_\alpha U)_*$, we get

$$E(\alpha(u^*)(a^*a \otimes 1)\alpha(u))(z) = \hat{x}(\iota \otimes z')\hat{\Delta}^{\mathrm{op}}(a^*a) = z\alpha(u^*u)\hat{x}(a^*a).$$

□

We need the following quantum group result, and the next weight theory result.

Lemma 15.4.2 *We have $q_{\hat{x}}\lambda(\omega_{v,q_x(b)}) = J_x\sigma^x_{i/2}(b)J_x v$ for $v \in V_x$ and b in the Tomita algebra of x. The elements $\lambda(\omega_{v,q_x(b)})$ span a σ-strong*-norm core for $q_{\hat{x}}$.*

Proof The first statement follows from the definition of $\hat{x}$ and the fact that $\omega_{v,q_x(b)}(w^*) = (w^*v|q_x(b)) = (J_x\sigma^x_{i/2}(b)J_x v|q(w))$ for $w \in N_x$. Next, let L be the set of $a \in N_x$ such that $x(\cdot a)$ defined on N_x^* has a normal bounded (unique) extension ax to W. Clearly $q_{\hat{x}}\lambda(ax) = q_x(a)$. We claim that the set D of all $\lambda(ax)$ is a σ-strong*-norm core for $q_{\hat{x}}$. Let $\bar{D}$ denote the σ-strong*-norm closure of the restriction of $q_{\hat{x}}$ to D. Adapting the notation prior to Lemma 12.10.1, we see that $b = \tau_t(a)h^{-it} \in N_x$ and

$$x(wb) = \nu^t x(h^{-it}w\tau_t(a)) = x(h^{-it}\tau_{-t}(w)a) = (\rho_t(ax))(w)$$

for $w \in N_x^*$, so $bx = \rho_t(ax)$ and $\hat{\sigma}_t\lambda(ax) = \lambda(bx) \in D$ for real t. Hence D, and thus also $\bar{D}$, is invariant under $\hat{\sigma}_t$. Let $z \in W_*$ be such that there is $z' \in W_*$ with $z' = zS^{-1}$ on $D(S^{-1})$. Then $b \equiv (z'\otimes\iota)\Delta(a) \in N_x$ and $x(wb) = x((z\otimes\iota)\Delta(w)a)$ by strong left invariance of x. So $bx = (z \otimes ax)\Delta$ and $\lambda(z)\lambda(ax) = \lambda(bx) \in D$. As such $\lambda(z)$ form a σ-strong* dense subset of $\hat{W}$, we conclude that $\bar{D} \subset N_{\hat{x}}$ is a $\hat{\sigma}$-invariant σ-strong* dense left ideal of $\hat{W}$, and the claim follows. But then the span of the elements $\lambda(\omega_{q_x(a),q_x(b)}) = \lambda(abx)$ with $a \in L$ and b in the Tomita algebra, is a σ-strong*-norm core for $q_{\hat{x}}$. □

Lemma 15.4.3 *Let y be a faithful semifinite normal weight on a von Neumann algebra U. Suppose $D \subset N_y$ is a weakly dense left ideal in U, and $q: D \to V$ is a linear map with dense range in a Hilbert space V, and that q is σ-strongly*-norm closed, and that π is a normal representation of U on V such that $\pi(u)q(w) = q(uw)$ for $u \in U$ and $w \in D$, and that $X: V \to V_y$ is an isometry with $Xq = q_y$ on D. Then there is a unique faithful semifinite normal weight z on U with $N_z = D$ and (q, π) as a semicyclic representation. So z is a restriction of y in the sense that $z(w) = y(w)$ for positive $w \in M_z \subset M_y$.*

Proof Now q is injective as X is an isometry. So $q(D \cap D^*)$ is a $*$-algebra in an obvious fashion, and it is dense in V. In fact, it is a left Hilbert algebra since if we have a sequence of elements $w_n \in D \cap D^*$ with $q(x_n) \to 0$ and $q(x_n^*) \to v$, then using X, we get $q_y(w_n) \to 0$ and $q_y(w_n^*) \to Xv$, so $Xv = 0$ and $v = 0$. The von Neumann algebra associated with the left Hilbert algebra is clearly $\pi(U)$, and π is injective as q is. Hence the weight on $\pi(U)$ associated with the left Hilbert algebra, will by composition with π, yield a faithful semifinite normal weight z on U. By definition of z every $w \in D \cap D^*$ belongs to N_z and $q_z(w) = q(w)$. If only $w \in D$, take a net of elements $w_i \in D$ that converges to the unit σ-strongly*. Then $w_i^*w \to w$ σ-strongly* and $w_i^*w \in D \cap D^*$ and $q_z(w_i^*w) = q(w_i^*w) = \pi(w_i^*)q(w) \to q(w)$, so $w \in N_z$ and $q_z(w) = q(w)$ as q_z is σ-strongly*-norm closed. Conversely if $w \in N_z$, by denseness of left Hilbert algebras, there is a net of elements $w_i \in D \cap D^*$ of norm less than $\|w\|$ such that $w_i \to w$ σ-strongly* and $q(w_i) = q_z(w_i) \to q_z(w)$. Hence $w \in D$ and $q(w) = q_z(w)$, now by σ-strongly*-norm closedness of q. □

Lemma 15.4.4 *Let y be a faithful semifinite normal weight on a von Neumann algebra U. By the first lemma above we have an isometry $X: V_x \otimes V_y \to V_{\tilde{y}}$ such that $X(q_{\hat{x}}(a) \otimes q_y(u)) = q_{\tilde{y}}((a \otimes 1)\alpha(u))$ for $a \in N_{\hat{x}}$ and $u \in N_y$. Letting D be the span of the elements $(a \otimes 1)\alpha(u)$, we therefore have a linear map $q_0: D \to V_x \otimes V_y$ that sends $(a \otimes 1)\alpha(u)$ to $q_{\hat{x}}(a) \otimes q_y(u)$. Since $q_{\tilde{y}}$ is σ-strongly*-norm closed, we can close q_0 for these topologies and obtain a linear map $q: \bar{D} \to V_x \otimes V_y$ such that $\bar{D} \subset N_{\tilde{y}}$ and $Xq = q_{\tilde{y}}$. Then $\bar{D}$ is a weakly dense left ideal in $W \ltimes_\alpha U$, and $q(cd) = cq(d)$ for $c \in W \ltimes_\alpha U$ and $d \in \bar{D}$.*

Proof Chose an orthonormal basis $\{v_i\}_{i \in I}$ for V_x, let $v \in V_x$ and $u \in U$, and let b be in the Tomita algebra of x. As M implements Δ, and $(\Delta \otimes \iota)\alpha = (\iota \otimes \alpha)\alpha$, we get

$$\alpha(u)(\lambda(\omega_{v,q_x(b)}) \otimes 1) = \sum(\lambda(\omega_{v_i,q_x(b)}) \otimes 1)\alpha(\omega_{v,v_i} \otimes \iota)\alpha(u)$$

in the σ-strongly* topology. By the quantum group lemma above, for any finite subset F of I, the element $d_F = \sum_{i \in F}(\lambda(\omega_{v_i,q_x(b)}) \otimes 1)\alpha((\omega_{v,v_i} \otimes \iota)(\alpha(u)u')$ belongs to D for $u' \in U$, and $q_0(d_F) = (J_x\sigma^x_{i/2}J_x \otimes 1)(P_F \otimes 1)\alpha(u)(v \otimes q_y(u'))$, where P_F is the orthogonal projection onto the closed span of those v_i's with $i \in F$. Hence $d_F \to d \equiv \alpha(u)(\lambda(\omega_{v,q(b)}) \otimes 1)\alpha(u')$ in the σ-strong* topology, while $\{q(d_F)\}$ converges in norm to

$$(J_x\sigma^x_{i/2}J_x \otimes 1)\alpha(u)(v \otimes q_y(u')) = \alpha(u)q((\lambda(\omega_{v,q(b)}) \otimes 1)\alpha(u')),$$

so $d \in \bar{D}$ and $q(d)$ equals the right hand side. Since the elements $\lambda(\omega_{v,q(b)})$ form a σ-strong*-norm core for q be the aforementioned lemma, we conclude that $\alpha(u)d \in \bar{D}$ and $q(\alpha(u)d) = \alpha(u)q(d)$ for any $d \in \bar{D}$. For $a \in \hat{W}$, it is clear that $(a \otimes 1)d \in \bar{D}$ and $q((a \otimes 1)d) = (a \otimes 1)q(d)$. □

Combining the two previous lemmas, there is to any faithful semifinite normal weight y on a von Neumann algebra U subject to a coaction α of (W, Δ), a unique faithful semifinite normal weight z on $W \ltimes_\alpha U$ such that $N_z = D$ and such that (q, ι) is a semicyclic representation for z. We aim to prove that $z = \tilde{y}$.

Proposition 15.4.5 *The unitary $K = J_z(J_{\hat{x}} \otimes J_y) \in B(V_x \otimes V_y)$, called the* unitary implementation *of α, satisfies $\alpha = K(1 \otimes (\cdot))K^*$ and $K(J_{\hat{x}} \otimes J_y) = (J_{\hat{x}} \otimes J_y)K^*$, and $\sigma^z_t\alpha = \alpha\sigma^y_t$ for real t.*

Proof Let $u \in D(\sigma^y_{i/2})$ and $a \in N_{\hat{x}}$ and $u' \in N_y$. Then

$$q((a \otimes 1)\alpha(u')\alpha(u)) = q_{\hat{x}}(a) \otimes q_y(u'u) = (1 \otimes J_y\sigma^y_{i/2}(u)^*J_y)q((a \otimes 1)\alpha(u')),$$

and by continuity we can replace $(a \otimes 1)\alpha(u') \in D$ by a general element d in N_z, yielding

$$J_z\sigma^z_{i/2}\alpha(u)^*J_zq(d) = q(d\alpha(u)) = (1 \otimes J_y\sigma^y_{i/2}(u)^*J_y)q(d),$$

so $J_z\sigma^z_{i/2}\alpha(u)^*J_z = 1\otimes J_y\sigma^y_{i/2}(u)^*J_y$. Using $\sigma^y_{i/2}(u)^* = \sigma^y_{-i/2}(u^*)$, we therefore get $\sigma^z_{-i/2}\alpha(u) = K(1\otimes\sigma^y_{-i/2}(u))K^*$ when $u\in D(\sigma^y_{-i/2})$. Taking adjoints we may replace $-i/2$ by $i/2$. As $\sigma^y_{-i}(u)\in D(\sigma^y_{i/2})$ when $u\in D(\sigma^y_{-i})$, we then get $\sigma^z_{i/2}\alpha\sigma^y_{-i}(u) = K(1\otimes\sigma^y_{-i/2}(u))K^*$. In this case also the first formula with K can be applied as $D(\sigma^y_{-i})\subset D(\sigma^y_{-i/2})$, yielding the same result, so $\sigma^z_{-i/2}\alpha(u) = \sigma^z_{i/2}\alpha\sigma^y_{-i}(u)$, or $\sigma^z_{-i}\alpha(u) = \alpha\sigma^y_{-i}(u)$. By Proposition 10.7.14 we may replace $-i$ by any real number t. Replacing then t by $i/2$, and using the corresponding formula above involving K, we get $\alpha = K(1\otimes(\cdot))K^*$ by denseness of the domain of $\sigma^y_{-i/2}$. The formula with K and J's is obvious. □

Lemma 15.4.6 *We have $(Dy_2 : Dy_1)_t\in U^\alpha$ for $t\in\mathbb{R}$ and h^{-1}-invariant faithful semifinite normal weights y_i on U.*

Proof The balanced weight $y_1\oplus y_2$ is clearly h^{-1}-invariant for the coaction $\alpha\otimes\iota$ of (W,Δ) on $U\otimes M_2(\mathbb{C})$. By Proposition 15.2.3 we have $(\alpha\otimes\iota)\sigma_t^{y_1\oplus y_2} = (\sigma_t^{xR}\sigma^x_{-t}\tau_t\otimes\sigma_t^{y_1\oplus y_2})(\alpha\otimes\iota)$ for real t. Hence

$$\alpha((Dy_2 : Dy_1)_t)\otimes e_{21} = (\alpha\otimes\iota)\sigma_t^{y_1\oplus y_2}(1\otimes e_{21}) = (\sigma_t^{xR}\sigma^x_{-t}\tau_t\otimes\sigma_t^{y_1\oplus y_2})(1\otimes 1\otimes e_{21})$$

equals $1\otimes(Dy_2 : Dy_1)_t\otimes e_{21}$. □

Proposition 15.4.7 *We have $z = \tilde{y}$.*

Proof As $\hat{h}^{-1}$ is the modular element of $(\hat{W},\hat{\Delta}^{\mathrm{op}})$, by Proposition 15.2.4, we see that $\tilde{y}$ is $\hat{h}$-invariant for $\hat{\alpha}$. We claim that z is also $\hat{h}$-invariant. Let $a\in N_{\hat{x}}$ and $u\in N_y$ and $v\in D(\hat{h}^{1/2})$ and $v'\in V_x$ and

$$d\equiv(\omega_{v,v'}\otimes\iota\otimes\iota)\hat{\alpha}((a\otimes 1)\alpha(u)) = ((\omega_{v,v'}\otimes\iota)\hat{\Delta}^{\mathrm{op}}(a)\otimes 1)\alpha(u).$$

From the proof of the aforementioned proposition, we see that $q_{\hat{x}}(\omega_{v,v'}\otimes\iota)\hat{\Delta}^{\mathrm{op}}(a) = (\iota\otimes\omega_{\hat{h}^{1/2}v,v'})(\hat{N})q_{\hat{x}}(a)$, where $\hat{N} = (J_x\otimes J_x)M(J_x\otimes J_x)$ is the fundamental multiplicative unitary of $(W,\Delta)^{\wedge\mathrm{op}}$. Hence $q(z) = ((\iota\otimes\omega_{\hat{h}^{1/2}v,v'})(\hat{N})\otimes 1)q((a\otimes 1)\alpha(u))$. By denseness of D, we therefore get $q(\omega_{v,v'}\otimes\iota\otimes\iota)\hat{\alpha}(d') = ((\iota\otimes\omega_{\hat{h}^{1/2}v,v'})(\hat{N})\otimes 1)q(d')$ for $d'\in N_z$, so our claim holds by unitarity of $\hat{N}$. From the previous lemma we get $(Dz : D\tilde{y})_t\in(W\ltimes_\alpha U)^{\hat{\alpha}} = \alpha(U)$, so pick unitaries $u_t\in U$ such that $\alpha(u_t) = (Dz : D\tilde{y})_t$ for real t. By the theory of operator valued weights, we conclude that $\sigma_t^{\tilde{y}}\alpha = \alpha\sigma_t^y$, so $\{u_t\}$ is a σ_t^y-cocycle as $\{(Dz : D\tilde{y})_t\}$ is a $\sigma_t^{\tilde{y}}$-cocycle. Let y_0 be the faithful semifinite normal weight on U such that $(Dy_0 : Dy)_t = u_t$. From the theory of operator valued weights we thus get

$$(D\tilde{y}_0 : D\tilde{y})_t = \alpha((Dy_0 : Dy)_t) = (Dz : D\tilde{y})_t,$$

and $\tilde{y}_0 = z$, so $\tilde{y}_0$ is a restriction of $\tilde{y}$. Fix positive $a \in M_{\hat{x}}$ with $\hat{x}(a) = 1$, and let $u \in N_{y_0}$. Then by the first lemma in this section, we have

$$y(u^*u) = \tilde{y}(\alpha(u^*)(a^*a \otimes 1)\alpha(u)) = \tilde{y}_0(\alpha(u^*)(a^*a \otimes 1)\alpha(u)) = y_0(u^*u),$$

so y_0 is a restriction of y. Using the theory of operator valued weights in the first equality, and the previous proposition in the last, we get $\alpha\sigma_t^{y_0} = \sigma_t^{\tilde{y}_0}\alpha = \sigma_t^z\alpha = \alpha\sigma_t^y$, and by injectivity of α, we get $\sigma_t^{y_0} = \sigma_t^y$. Hence $y_0 = y$, and thus $z = \tilde{y}_0 = \tilde{y}$. □

We focus now on properties of the unitary implementation K of α.

Proposition 15.4.8 *We have $K \in M \otimes B(V_y)$.*

Proof Consider the one-parameter group of automorphisms of W given by $f_t = J_x\Delta_{\hat{x}}^{it}J_x(\cdot)J_x\Delta_{\hat{x}}^{-it}J_x$ for real t, so $J_xwJ_x\Delta_{\hat{x}}^{1/2} \subset \Delta_{\hat{x}}^{1/2}J_xf_{-i/2}(w)J_x$ for w in $D(f_{-i/2})$, and one checks that $f_tR = Rf_t$. Let $w \in D(f_{i/2})$ and $u, u' \in N_y$ and $v \in D(S_{\hat{x}})$. Then $(J_xwJ_x \otimes 1)S_{\tilde{y}}\alpha(u^*)(v \otimes q_y(u')) = (J_xwJ_x \otimes 1)\alpha(u'^*)(S_{\hat{x}}v \otimes q_y(u))$ equals $\alpha(u'^*)(J_{\hat{x}}J_xR(w^*)J_x\Delta_{\hat{x}}^{1/2}v \otimes q_y(u))$. From above we have $J_xR(w^*)J_x\Delta_{\hat{x}}^{1/2} \subset \Delta_{\hat{x}}^{1/2}J_xR(f_{i/2}(w)^*)J_x$, and inserting this in the latter expression, and using that the elements $\alpha(u^*)(v \otimes q_y(u'))$ span a core for $S_{\tilde{y}}$, we get

$$(J_xwJ_x \otimes 1)S_{\tilde{y}} \subset S_{\tilde{y}}(J_xR(f_{i/2}(w)^*)J_x \otimes 1).$$

Using this and the one obtained by taking the adjoint, together with $\Delta_{\tilde{y}} = S_{\tilde{y}}^*S_{\tilde{y}}$ gives $(J_xwJ_x \otimes 1)\Delta_{\tilde{y}} \subset \Delta_{\tilde{y}}(J_xf_{-i}(w)J_x \otimes 1)$ for $w \in D(f_{-i})$, or $g_{-i}(J_xwJ_x \otimes 1) = J_xf_{-i}(w)J_x \otimes 1$, where $g_t = \Delta_{\tilde{y}}^{-it}(\cdot)\Delta_{\tilde{y}}^{it}$ defines a one-parameter group on $B(V_x \otimes V_y)$. By Proposition 10.7.14 we may replace $-i$ by any real number t. Replacing then t by $i/2$ gives $(J_xwJ_x \otimes 1)\Delta_{\tilde{y}}^{-1/2} \subset \Delta_{\tilde{y}}^{-1/2}(J_xf_{-i}(w)J_x \otimes 1)$ for $w \in D(f_{i/2})$. Combining this with a previous inclusion gives

$$(J_xwJ_x \otimes 1)S_{\tilde{y}}\Delta_{\tilde{y}}^{-1/2} \subset S_{\tilde{y}}(J_xR(f_{i/2}(w)^*)J_x \otimes 1)\Delta_{\tilde{y}}^{-1/2} \subset J_{\tilde{y}}(J_{\hat{x}}J_xwJ_xJ_{\hat{x}} \otimes 1).$$

Hence $(J_xwJ_x \otimes 1)J_{\tilde{y}} = J_{\tilde{y}}(J_{\hat{x}}J_xwJ_xJ_{\hat{x}} \otimes 1)$ for $w \in W$, or $(J_xwJ_x \otimes 1)K = K(J_xwJ_x \otimes 1)$, and $K \in W\bar{\otimes}B(V_y)$. □

The following result shows how the unitary implementation varies with y.

Proposition 15.4.9 *Let y_i be faithful semifinite normal weights on U subject to a coaction α of (W, Δ), and consider the unitary implementations K_i of α with respect to y_i. By Lemma 10.9.3 there is a unique unitary $u_{y_1y_2}$ that intertwines π_{y_i} and relates the self-dual cones. Then $1 \otimes u_{y_1y_2}$ intertwines the representations $\pi_{\tilde{y}_i} = \iota \otimes \pi_{y_i}$ of $W \ltimes_\alpha U$, and $K_2 = (1 \otimes u_{y_1y_2})K_1(1 \otimes u_{y_1y_2}^*)$.*

Proof Let $a \in N_{\hat{x}}$ and $u' \in N_{y_1}$ and $u \in \sigma^{y_1 y_2}_{-i/2}$. By standard modular theory, we have $q_{y_2}(u'u^*) = J_{y_1 y_2}\pi_{y_1}\sigma^{y_1 y_2}_{-i/2}(u)J_{y_1}q_{y_1}(u')$, so

$$q_{\tilde{y}_2}((a \otimes 1)\alpha(u')\alpha(u)^*) = q_{\hat{x}}(a) \otimes q_{y_2}(u'u^*)$$

equals $(1 \otimes J_{y_1 y_2}\pi_{y_1}\sigma^{y_1 y_2}_{-i/2}(u)J_{y_1})q_{\tilde{y}_1}((a \otimes 1)\alpha(u'))$. By denseness and continuity we may replace $(a \otimes 1)\alpha(u')$ above by any element in $N_{\tilde{y}_1}$, so we get

$$J_{\tilde{y}_1\tilde{y}_2}(\iota \otimes \pi_{y_1})\sigma^{\tilde{y}_1\tilde{y}_2}_{-i/2}\alpha(u)J_{\tilde{y}_1} = (1 \otimes J_{y_1 y_2}\pi_{y_1}\sigma^{y_1 y_2}_{-i/2}(u)J_{y_1})$$

by invoking the first formula in this proof. As $(D\tilde{y}_2 : D\tilde{y}_1)_t = \alpha((Dy_2 : Dy_1)_t)$ for real t, we get $\sigma^{\tilde{y}_1\tilde{y}_2}_t\alpha = \alpha\sigma^{y_1 y_2}_t$, and inserting the version for $t = -i/2$ into the previous formula gives us

$$(\iota \otimes \pi_{y_1})\alpha\sigma^{y_1 y_2}_{-i/2}(u) = J^*_{\tilde{y}_1\tilde{y}_2}(J_{\hat{x}} \otimes J_{y_1 y_2})(1 \otimes \pi_{y_1}\sigma^{y_1 y_2}_{-i/2}(u))K_1^*.$$

Since K_1 implements α, we therefore get $K_1 = J^*_{\tilde{y}_1\tilde{y}_2}(J_{\hat{x}} \otimes J_{y_1 y_2})$. Using $K_1 K_1^* = 1$ and $K_1 = (J_{\hat{x}} \otimes J_{y_1})J_{\tilde{y}_1}$ and $u_{y_1 y_2} = J_{y_1 y_2}J_{y_1}$, we thus get $J_{\tilde{y}_1\tilde{y}_2}J_{\tilde{y}_1} = 1 \otimes u_{y_1 y_2}$. Lemma 10.9.3 settles therefore the first statement, and also the second statement as $u_{y_1 y_2}$ relates the J_{y_i}'s while $1 \otimes u_{y_1 y_2}$ relates the $J_{\tilde{y}_i}$'s. □

Next we investigate how the unitary implementation changes when we deforme α with a α-cocycle.

Proposition 15.4.10 *Let y be a faithful semifinite normal weight on U subject to the coaction α of (W, Δ), consider the coaction $\beta = X\alpha(\cdot)X^*$ for an α-cocycle X, and let $K(\alpha)$ and $K(\beta)$ be the unitary implementations of α and β with respect to y. Then $K(\beta) = XK(\alpha)(J_{\hat{x}} \otimes J_y)X^*(J_{\hat{x}} \otimes J_y)$, and $(\Delta \otimes \iota)(K(\alpha)) = K(\alpha)_{23}K(\alpha)_{13}$ if and only if $(\Delta \otimes \iota)(K(\beta)) = K(\beta)_{23}K(\beta)_{13}$.*

Proof Let $v, v' \in V_x$ and let $\{v_i\}_{i \in I}$ be an orthonormal basis of V_x. Applying $\omega_{v,v'} \otimes \iota \otimes \iota$ to the cocycle identity $X^*_{23}M_{12}X_{23} = M_{12}(\iota \otimes \alpha)(X^*)$ gives

$$X^*(\lambda(\omega_{v,v'}) \otimes 1)X = \sum(\lambda(\omega_{v_i,v'}) \otimes 1)\alpha(\omega_{v,v_i} \otimes \iota)(X^*)$$

with convergence in the σ-strong* topology. Hence $f \equiv X^*(\cdot)X \colon W \ltimes_\beta U \to W \ltimes_\alpha U$ is a well-defined $*$-isomorphism. Let $u \in N_y$ and let w belong to the Tomita algebra of x. Then

$$X^*(\lambda(\omega_{v,q(w)}) \otimes 1)\beta(u)X = \sum(\lambda(\omega_{v_i,q(w)}) \otimes 1)\alpha((\omega_{v,v_i} \otimes \iota)(X^*)u).$$

Let d be the left hand side, and let d_F be the right hand side summing only over a finite subset F of I. By Lemma 15.4.2 we get

$$q_{\tilde{y}}^{\alpha}(d_F) = \sum_{i\in F} q_{\hat{x}}\lambda(\omega_{v_i,q(w)}) \otimes (\omega_{v,v_i} \otimes \iota)(X^*)q_y(u) = (J_x\sigma^x_{i/2}(w)J_x P \otimes 1)X^*(v \otimes q_y(u))$$

for the orthogonal projection P onto the closed span of v_i with $i \in F$. Then $d_F \to d$ σ-strongly*, while

$$q_{\tilde{y}}^{\alpha}(d_F) \to (J_x\sigma^x_{i/2}(w)J_x \otimes 1)X^*(v \otimes q_y(u)) = X^* q_{\tilde{y}}^{\beta}\lambda(\omega_{v,q(w)}) \otimes 1)\beta(u)$$

in norm, so the latter limit equals $q_{\tilde{y}}^{\alpha}(d)$. As the elements $\lambda(\omega_{v,q(w)}) \otimes 1)\beta(u)$ span a core for $q_{\tilde{y}}^{\beta}$, we thus get $q_{\tilde{y}}^{\alpha} f = X^* q_{\tilde{y}}^{\beta}$. Hence $J_{\tilde{y}}^{\beta} = XJ_{\tilde{y}}^{\alpha}X^*$ and

$$K(\beta) = J_{\tilde{y}}^{\beta}(J_{\hat{x}} \otimes J_y) = XJ_{\tilde{y}}^{\alpha}X^*(J_{\hat{x}} \otimes J_y) = XK(\alpha)(J_{\hat{x}} \otimes J_y)X^*(J_{\hat{x}} \otimes J_y).$$

Let $L_y = J_y(\cdot)^*J_y$ on U. If $(\Delta \otimes \iota)(K(\alpha)) = K(\alpha)_{23}K(\alpha)_{13}$, then by using the above formula for $K(\beta)$ together with the cocycle property of X, that $K(\alpha)$ implements α with its property relating to $J_{\hat{x}} \otimes J_y$, and that $\Delta R = (R \otimes R)\Delta^{\mathrm{op}}$, an easy calculation shows that $(\Delta \otimes \iota)(K(\beta)) = K(\beta)_{23}K(\beta)_{13}$. □

The unitary implementation is the expected one when y is h^{-1}-invariant.

Proposition 15.4.11 *If the faithful semifinite normal weight y on U, subject to a coaction α of (W, Δ), is h^{-1}-invariant, then the unitary implementation of α coincides with the K defined in Proposition 15.2.3.*

Proof Let $u \in N_y \cap N_y^*$, let u' belong to the Tomita algebra of y, let $v' \in V_x$ and $v \in D$ with D from Proposition 15.2.3, and set $f_{v'}(c) = cv'$ for $c \in \mathbb{C}$. Then

$$\begin{aligned}(f_{v'}^* \otimes 1)S_{\tilde{y}}\alpha(u'^*)(v \otimes q_y(u)) &= (f_{v'}^* \otimes 1)\alpha(u^*)(S_{\hat{x}}v \otimes q_y(u'))\\ &= q_y((\omega_{S_{\hat{x}}v,v'} \otimes \iota)\alpha(u^*)u')\\ &= J_y\sigma^y_{i/2}(u')^*J_yq_y(\omega_{S_{\hat{x}}v,v'} \otimes \iota)\alpha(u^*)\\ &= J_y\sigma^y_{i/2}(u')^*J_y(\omega_{h^{-1/2}S_{\hat{x}}v,v'} \otimes \iota)(K)q_y(u^*)\end{aligned}$$

equals $(f_{v'}^* \otimes 1)(1 \otimes J_y\sigma^y_{i/2}(u')^*J_y)K(h^{-1/2}S_{\hat{x}}v \otimes q_y(u^*))$. Using the easy identity $h^{-1/2}S_{\hat{x}}v = J_{\hat{x}}(h\Delta_x)^{1/2}v$, we thus get from closedness of $S_{\tilde{y}}$, that

$$S_{\tilde{y}}(v \otimes q_y(u)) = K(J_{\hat{x}} \otimes J_y)((h\Delta_x)^{1/2} \otimes \Delta_y^{1/2})(v \otimes q_y(u)).$$

As elements of the form v and $q_y(u)$ form cores for $(h\Delta_x)^{1/2}$ and $\Delta_y^{1/2}$, respectively, we therefore get $K(J_{\hat{x}} \otimes J_y)((h\Delta_x)^{1/2} \otimes \Delta_y^{1/2}) \subset S_{\tilde{y}}$. Hovever, the domain of $(h\Delta_x)^{1/2} \otimes \Delta_y^{1/2}$ is a core for $S_{\tilde{y}}$, and thus we get equality above, provided $S_{\tilde{y}}$

and $(h\Delta_x)^{it} \otimes \Delta_y^{it}$ commute for real t, and we claim this holds. In this case, by uniqueness in the polar decomposition, we then get $K(J_{\hat{x}} \otimes J_y) = J_{\tilde{y}}$ and $K = K'$. As for the claim, if $u, u' \in N_y$ and $v \in D(S_{\hat{x}})$, then by the aforementioned proposition, we get

$$S_{\tilde{y}}\alpha(u'^*)(v \otimes q_y(u)) = \alpha(u^*)(S_{\hat{x}}v \otimes q_y(u')) = K(1 \otimes u^*)K^*(S_{\hat{x}}v \otimes q_y(u')).$$

Applying $(h\Delta_x)^{it} \otimes \Delta_y^{it}$ on both sides, using that it commutes with K, and that $(h\Delta_x)^{it}$ commutes with $S_{\hat{x}}$, we get

$$((h\Delta_x)^{it} \otimes \Delta_y^{it})S_{\tilde{y}}\alpha(u'^*)(v \otimes q_y(u)) = S_{\tilde{y}}\alpha\sigma_t^y(u'^*)((h\Delta_x)^{it}v \otimes q_y\sigma_t^y(u)),$$

which equals $S_{\tilde{y}}((h\Delta_x)^{it} \otimes \Delta_y^{it})\alpha(u'^*)(v \otimes q_y(u))$. □

Theorem 15.4.12 *The unitary implementation K of a coaction α of (W, Δ) on a von Neumann algebra U satisfies $(\Delta \otimes \iota)(K) = K_{23}K_{13}$.*

Proof Let y be a faithful semifinite normal weight on U that provides K. By Proposition 15.2.4 the weight $\tilde{\tilde{y}}$ is $J_x h J_x$-invariant with respect to $\hat{\hat{\alpha}}$. Pick f and γ as in the biduality theorem for crossed products. Then $f\tilde{\tilde{y}}$ is a h^{-1}-invariant faithful semifinite normal weight on $B(V_x)\bar{\otimes}U$ for the coaction γ of (W, Δ) on $B(V_x)\bar{\otimes}U$, as α satisfies property A. Hence by the previous proposition and Proposition 15.2.3, the unitary implementation of γ with respect to $f\tilde{\tilde{y}}$ satisfies the corepresentation property, and by the proposition before that, so does the unitary implementation of $\beta = (F \otimes \iota)(\iota \otimes \alpha)$ from the biduality theorem for crossed products. By Proposition 15.4.9 this is also true for the unitary implementation $K(\beta)$ of β with respect to the weight $\mathrm{Tr} \otimes y$, rather than $f\tilde{\tilde{y}}$, where Tr is the usual trace on $B(V_x)$. Then $\iota \otimes \pi_{\mathrm{Tr}} \otimes \iota$ is the semicyclic representation of $(\mathrm{Tr} \otimes y)^{\sim}$, and one easily checks that $\Sigma_{12}S_{(\mathrm{Tr}\otimes y)^{\sim}}\Sigma_{12} = J_{\mathrm{Tr}} \otimes S_{\tilde{y}}$, where Σ is the flip on $V_x \otimes V_{\mathrm{Tr}}$. But then $\Sigma_{12}J_{(\mathrm{Tr}\otimes y)^{\sim}}\Sigma_{12} = J_{\mathrm{Tr}} \otimes J_{\tilde{y}}$ by uniqueness in the polar decomposition, and thus $\Sigma_{12}K(\beta)\Sigma_{12} = 1 \otimes K$. The corepresentation property of K follows now from the corepresentation property of $K(\beta)$. □

In this section we have therefore shown that property D holds for any coaction of a locally compact quantum group on a von Neumann algebra.

Exercises

For Sect. 15.1

1. What does the stronger form of left invariance mean classically, that is, when the locally compact quantum group stems from a locally compact group G and when $X \in L^\infty(\mu)\bar{\otimes}L^\infty(G)$ is positive for a Radon measure μ on a σ-compact locally compact Hausdorff space?

For Sect. 15.2

1. Show that the comultiplication of a locally compact quantum group (W, Δ) is an ergodic coaction on W, that is, with trivial fixed point algebra. Can Δ restrict to ergodic coactions of (W, Δ) on von Neumann subalgebras of W?
2. Given a locally compact quantum group (W, Δ) with fundamental multiplicative unitary M. Show that $a \mapsto M^*(1 \otimes a)M$ defines a coaction of (W, Δ) on the Neumann algebra $\hat{W}$ of the dual quantum group. Can you describe it in concrete simple cases?
3. Can you actually prove that the discussed 'second condition' is redundant.

For Sect. 15.3

1. Show that the definition of the crossed product of coactions of locally compact quantum groups is consistent with the definition of the crossed product for actions of locally compact groups.
2. Find the dual coaction of the coaction Δ on W of a locally compact quantum group (W, Δ).

For Sect. 15.4

1. What is the dual weight of the left invariant faithful semifinite normal weight x on (W, Δ) associated to the coaction Δ on W, and what is the canonical unitary implementation in this case?
2. What is the relation between the unitary implementation of an action of a locally compact group G on a von Neumann U and the canonical unitary implementation of the associated coaction on U of the corresponding locally compact quantum group $(L^\infty(G), \Delta)$?
3. Can you define tensor products of coactions?
4. Let $q \in \langle 0, 1\rangle$ and consider the compact quantum group $SU_q(2)$ with reduced C*-algebra W_r universally generated by a, b satisfying certain relations, and with a coproduct Δ defined on these generators. Extend the $*$-characters f_{it} for $t \in \mathbb{R}$ to W_r, and consider the C*-subalgebra $U_r = \{w \in W_r \mid (\iota \otimes f_{it})\Delta(w) = w \, \forall t \in \mathbb{R}\}$ of W_r. Show that $\Delta(U_r) \subset W_r \otimes U_r$ with a similar inclusion for the von Neumann algebras $W = W_r''$ and $U = U_r''$. Prove that the coproduct Δ on the von Neumann level is a coaction of the compact locally compact quantum group (W, Δ) on U. What is the canonical unitary implementation of the normal Haar state associated to this coaction? What does all this correspond to in the classical case when $q = 1$, and what is the underlying space for U_r?

Chapter 16
Generalized and Continuous Crossed Products

The classical notion of twisted actions of groups on algebras can be rephrased as so called cocycle coactions of locally compact quantum groups on von Neumann algebras generalizing both twisted actions and coactions. The added difficulty is that only $(\iota\otimes\alpha)\alpha = X(\Delta\otimes\iota)\alpha(\cdot)X^*$ holds for a unitary $X \in W\bar{\otimes}W\bar{\otimes}U$ satisfying a cocycle type of condition. One can then define the cocycle crossed product $W \ltimes_{\alpha,X} U$ of the cocycle coaction. Basically everything in the previous chapter can then be carefully generalized to this setting. We devote the first section to this.

In the second section we study a special case of cocycle crossed products, namely the cocycle bicrossed product of a pair of locally compact quantum groups (W_1, Δ_1) and (W_2, Δ_2) subject to a cocycle matching (r, X, Y). This means that $r: W_1\bar{\otimes}W_2 \to W_2\bar{\otimes}W_1$ is a normal $*$-monomorphism such that $(\alpha \equiv r(1\otimes(\cdot)), X)$ is a cocycle coaction of (W_1, Δ_1) on W_2, with a second cocycle coaction obtained by switching legs and indices, and finally some compatibility conditions should hold. Consider then the cocycle crossed product $W \equiv W_1 \ltimes_{\alpha,X} W_2$. Using the fundamental multiplicative unitaries of (W_i, Δ_i) together with the matching one can define a multiplicative unitary, ultimately playing the role of the fundamental one, and thus a coproduct Δ on W, of a locally compact quantum group (W, Δ) called the cocycle bicrossed product of (W_1, Δ_1) and (W_2, Δ_2), suppressing the cocycle matching. By invoking the dual weight construction of the left invariant weights of the constituent quantum groups one gets a left invariant faithful semifinite normal weight of the cocycle bicrossed product. The dual quantum group is obtained simply by swapping the roles of the constituent quantum groups when forming the cocycle bicrossed product.

The cocycle bicrossed product is a generalization of a matched pair of, say finite, groups G_1 and G_2. This consists of a group G, a homomorphism $i: G_1 \to G$ and an antihomomorphism $j: G_2 \to G$ such that $(g, s) \mapsto i(g)j(s)$ is a bijection $G_1 \times G_2 \to G$. Let $r(f)(g, s) = f(\beta_s(g), \alpha_g(s))$, where $j(\alpha_g(s))i(\beta_s(g)) = i(g)j(s)$ for $g \in G_1$ and $s \in G_2$. Then $(r, 1,1)$ is a cocycle matching for the quantum groups $(L^\infty(G_1), \Delta_1)$ and $(L^\infty(G_2), \Delta_2)$. This can also be done for

L. Tuset, *Analysis and Quantum Groups*,
https://doi.org/10.1007/978-3-031-07246-8_16

locally compact groups. As an example we define a cocycle bicrossed product which can be seen as a q-deformation of the locally compact group $ax + b$.

In the last section we carry through basically everything in the previous chapter in the context of continuous coactions, which are to be thought of as C*-algebra versions of coactions, mainly of reduced locally compact quantum groups. We limit ourselves to so called regular quantum groups, i.e. those with regular fundamental multiplicative unitaries. We say a multiplicative unitary N on V is regular if the space of slices $(\iota \otimes \omega)(FN)$ for all $\omega \in B(V)_*$ is norm dense in $B_0(V)$, where F is the flip on $V \otimes V$. This holds for all Kac algebras and compact and discrete quantum groups, but not for all locally compact quantum groups. It is however questionable if the notion of a continuous coaction really makes sense for non-regular quantum groups.

Recommended literature for this chapter is [4, 52].

16.1 Cocycle Crossed Products

Crossproducts of coactions can be generalized.

Definition 16.1.1 A *cocycle coaction* of a locally compact quantum group (W, Δ) on a von Neumann algebra U is a pair (α, X) such that $\alpha: U \to W\bar{\otimes}U$ is an injective normal unital $*$-homomorphism and $X \in W\bar{\otimes}W\bar{\otimes}U$ is a unitary satisfying $(\iota\otimes\alpha)\alpha = X(\Delta\otimes\iota)\alpha(\cdot)X^*$ and $(\iota\otimes\iota\otimes\alpha)(X)(\Delta\otimes\iota\otimes\iota)(X) = X_{234}(\iota\otimes\Delta\otimes\iota)(X)$. The *cocycle crossed product* of (α, X) is the von Neumann subalgebra $W \ltimes_{\alpha,X} U$ of $B(V_x)\bar{\otimes}U$ generated by $\alpha(U)$ and $(\omega\otimes\iota\otimes\iota)(\tilde{M})$ for $\omega \in W_*$, where $\tilde{M} \equiv M_{12}X^*$.

Example 16.1.2 By a *twisted action* of a separable locally compact group G on a σ-finite von Neumann algebra U, we mean Borel maps $G \to \mathrm{Aut}U : s \to \alpha_s$ and $f: G \times G \to U$, such that f takes values in the unitaries of U and satisfies $\alpha_s\alpha_t = f(s,t)\alpha_{st}(\cdot)f(s,t)^*$ and $\alpha_r(f(s,t))f(r,st) = f(r,s)f(rs,t)$ almost everywhere. Identifying $W\bar{\otimes}U$ with $L^\infty(G, U)$ etc., we see that $(\tilde{\alpha}, X)$ is a cocycle coaction of $(L^\infty(G), \Delta)$ on U, where $\tilde{\alpha}(u)(t) = \alpha_{t^{-1}}(u)$ and $X(s,t) = f(t^{-1}, s^{-1})$. This is by now clearly a one-to-one correspondence. ◇

We may also define dual coactions of cocycle coactions.

Proposition 16.1.3 *Given a cocycle coaction (α, X) of (W, Δ) on U. Then $\hat{\alpha} = N'_{12}(1\otimes(\cdot))N'^*_{12}$ with $N' = (J_x\otimes J_x)M_{21}(J_x\otimes J_x)$ defines a coaction of $(W, \Delta)^{\wedge\mathrm{op}}$ on $W \ltimes_{\alpha,X} U$ uniquely determined by $\hat{\alpha}\alpha = 1\otimes\alpha$ and $(\iota\otimes\hat{\alpha})(\tilde{M}) = M_{12}\tilde{M}_{134}$. We also have $\hat{\alpha} = \tilde{M}(\iota\otimes\alpha)(\cdot)\tilde{M}^*$. So $u \in U$ satisfies $\alpha(u) = 1\otimes u$ if and only if $\alpha(u)$ commutes with the elements $(\omega\otimes\iota\otimes\iota)(\tilde{M})$ for $\omega \in B(V_x)_*$.*

Proof The determining properties of $\hat{\alpha}$ are clear from $N' \in \hat{W}\bar{\otimes}W'$ and $\hat{\Delta}^{\mathrm{op}} = N'(1\otimes(\cdot))N'^*$ and $(\iota\otimes\hat{\Delta}^{\mathrm{op}})(M) = M_{12}M_{13}$. Hence $\hat{\alpha}$ will be a well-defined normal unital $*$-monomorphism from $W \ltimes_{\alpha,X} U$ to $\hat{W}\bar{\otimes}(W \ltimes_{\alpha,X} U)$. The property $(\iota\otimes\hat{\alpha})\hat{\alpha} = (\hat{\Delta}^{\mathrm{op}}\otimes\iota)\hat{\alpha}$ is clear on $\alpha(U)$, while on elements of the form $(\omega\otimes\iota\otimes\iota)(\tilde{M})$

for $\omega \in B(V_x)_*$, it follows from

$$(\iota \otimes \iota \otimes \hat{\alpha})(\iota \otimes \hat{\alpha})(\tilde{M}) = M_{12}M_{13}\tilde{M}_{145} = (\iota \otimes \hat{\Delta}^{\mathrm{op}} \otimes \iota)(\iota \otimes \hat{\alpha})(\tilde{M}).$$

The defining properties of cocycle coactions and of their dual coactions immediately yields $\hat{\alpha}\alpha = \tilde{M}(\iota \otimes \alpha)\alpha(\cdot)\tilde{M}^*$, and

$$\tilde{M}_{234}(\iota \otimes \iota \otimes \alpha)(\tilde{M})\tilde{M}_{234}^* = \tilde{M}_{234}M_{12}(\Delta \otimes \iota \otimes \iota)(X)(\iota \otimes \Delta \otimes \iota)(X^*)X_{234}^*\tilde{M}_{234}^*,$$

which equals $M_{23}M_{12}M_{23}^*X_{134}^* = (\iota \otimes \hat{\alpha})(\tilde{M})$ thanks to the pentagonal equation for M. The final statement follows from $(\iota \otimes \alpha)\alpha(u) = \tilde{M}^*(1 \otimes \alpha(u))\tilde{M}$, since this shows that $\alpha(u)$ commutes with the elements $(\omega \otimes \iota \otimes \iota)(\tilde{M})$ if and only if $(\iota \otimes \alpha)\alpha(u) = 1 \otimes \alpha(u)$, and one now uses injectivity of $\iota \otimes \alpha$. □

Definition 16.1.4 A cocycle coaction (α, X) of (W, Δ) on U is *Y-stabilizable* if there is a unitary $Y \in W\bar{\otimes}U$ satisfying $Y_{23}(\iota \otimes \alpha)(Y) = (\Delta \otimes \iota)(Y)X^*$.

Proposition 16.1.5 *Given a* Y-stabilizable *cocycle coaction* (α, X) *of* (W, Δ) *on* U, *then* $\beta = Y\alpha(\cdot)Y^*$ *defines a coaction of* (W, Δ) *on* U, *and* $Y^*(\cdot)Y$ *reduces to a* $*$*-isomorphism* f *from* $W \ltimes_\beta U$ *to* $W \ltimes_{\alpha,X} U$ *such that* $\hat{\alpha} f = (\iota \otimes f)\hat{\beta}$.

Proof The properties $(\iota \otimes \beta)\beta = (\Delta \otimes \iota)\beta$ and $f\beta = \alpha$ and $(\iota \otimes f)(M_{12}) = \tilde{M}(\iota \otimes \alpha)(Y^*)$ are immediate from definitions. So f restricts appropriately, and the last statement is clear since $1 \otimes Y$ and $N' \otimes 1$ commute. □

Proposition 16.1.6 *For any cocycle coaction* (α, X) *of* (W, Δ) *on* U, *the cocycle coaction* $(\alpha \otimes \iota, X \otimes 1)$ *of* (W, Δ) *on* $U\bar{\otimes}B(V_x)$ *is* $(N_{31}^*X_{312}^*)$*-stabilizable.*

Proof Clearly $Y \equiv N_{31}^*X_{312}^* \in W\bar{\otimes}U\bar{\otimes}B(V_x)$ is unitary. The property $Y_{234}(\iota \otimes \alpha \otimes \iota)(Y) = (\Delta \otimes \iota \otimes \iota)(Y)X_{123}^*$, or

$$Y_{341}(\iota \otimes \iota \otimes \alpha)(Y_{231}) = (\iota \otimes \Delta \otimes \iota)(Y_{231})X_{234}^*,$$

follows as the left hand side of the latter identity equals

$$N_{13}^*X_{134}^*N_{12}^*(\Delta \otimes \iota \otimes \iota)(X)(\iota \otimes \Delta \otimes \iota)(X^*)X_{234}^* = N_{13}^*X_{134}^*X_{134}N_{12}^*(\iota \otimes \Delta \otimes \iota)(X^*)X_{234}^*,$$

which again equals $(\iota \otimes \Delta)(N^*)_{123}(\iota \otimes \Delta \otimes \iota)(X^*)X_{234}^*$. □

Proposition 16.1.7 *Let* (α, X) *be a cocycle coaction of* (W, Δ) *on* U. *Then* $(W \ltimes_{\alpha,X} U)^{\hat{\alpha}} = \alpha(U)$ *and* $((W \ltimes_{\alpha,X} U) \cup (W' \otimes \mathbb{C}))'' = B(V_x)\bar{\otimes}U$ *and* $W \ltimes_{\alpha,X} U$ *is the* σ*-strong* closed span of the elements* $(\omega \otimes \iota \otimes \iota)(\tilde{M})\alpha(u)$ *for* $u \in U$ *and* $\omega \in W_*$.

Proof The first two assertions hold by the previous two propositions since ordinary coactions have properties A and B. With β, Y and f from the first of these propositions, we know from the proof of Lemma 15.4.4, that $W \ltimes_\beta U$ is the closed span of $((\omega \otimes \iota)(M) \otimes 1)\beta(u)$. Applying $f \otimes \iota$ and using $(\iota \otimes f)(M_{12}) =$

$\tilde{M}(\iota \otimes \alpha)(Y^*)$, we thus see that $W \ltimes_{\alpha,X} U$ is the closed span of the elements $((\omega \otimes \iota \otimes \iota)(\tilde{M}(\iota \otimes \alpha)(Y^*))\alpha(u)$. Letting L be the closed span in the proposition, these latter elements belong to L because if $\{v_i\}$ is an orthonormal basis of V_x and $v, v' \in V_x$, we have

$$(\omega_{v,v'} \otimes \iota \otimes \iota)(\tilde{M}(\iota \otimes \alpha)(Y^*))\alpha(u) = \sum(\omega_{v_i,v'} \otimes \iota \otimes \iota)(\tilde{M})\alpha((\omega_{v,v_i} \otimes \iota)(Y^*)u).$$

The opposite inclusion is obtained using the same trick again, after having inserted $1 = (\iota \otimes \alpha)(Y^*Y)$ in the middle of the typical elements in L. □

Proposition 16.1.8 *Let (α, X) be a cocycle coaction of (W, Δ) on U. Then $E = (\hat{x} \otimes \iota \otimes \iota)\hat{\alpha}$ is a faithful semifinite normal operator valued weight from $W \ltimes_{\alpha,X} U$ to $(W \ltimes_{\alpha,X} U)^{\hat{\alpha}} = \alpha(U)$. The* dual weight *of a faithful semifinite normal weight y on U is the faithful semifinite normal weight $\tilde{y} = y\alpha^{-1}E$ on $W \ltimes_{\alpha,X} U$. It satisfies*

$$\tilde{y}(\alpha(u^*)(\omega \otimes \iota \otimes \iota)(\tilde{M})^*(\omega \otimes \iota \otimes \iota)(\tilde{M})\alpha(u)) = \|p(\omega) \otimes q_y(u)\|^2$$

for $u \in N_y$ and $\omega \in K$, where $p\colon K \to V_x$ is as in Lemma 12.10.2.

Proof To prove semifiniteness of E, and to settle the formula for $\tilde{y}$ in the proposition, for the argument d of $\tilde{y}$ in the formula, we may, for $v \in V_x \otimes V_y$ and an orthonormal basis $\{v_i\}$ for $V_x \otimes V_y$, write

$$(\iota\otimes\omega_{v,v})\hat{\alpha}(d) = (\iota\otimes\omega_{\alpha(u)v,\alpha(u)v})((\omega\otimes\iota\otimes\iota\otimes\iota)(M_{12}\tilde{M}_{134})^*(\omega\otimes\iota\otimes\iota\otimes\iota)(M_{12}\tilde{M}_{134})),$$

which equals

$$\sum((\iota \otimes \omega_{\alpha(u)v,v_i})(\tilde{M})\omega \otimes \iota)(M)^*((\iota \otimes \omega_{\alpha(u)v,v_i})(\tilde{M})\omega \otimes \iota)(M),$$

which by normality of $\hat{x}$, yields

$$\hat{x}(\iota\otimes\omega_{v,v})\hat{\alpha}(d) = \sum \|p((\iota\otimes\omega_{\alpha(u)v,v_i})(\tilde{M})\omega)\|^2 = \sum \|(\iota\otimes\omega_{\alpha(u)v,v_i})(\tilde{M})p(\omega)\|^2,$$

which finally equals $\|p(\omega) \otimes \alpha(u)v\|^2 = \|p(\omega)\|^2\omega_{v,v}\alpha(u^*u)$. □

Proposition 16.1.9 *Let (α, X) be a cocycle coaction of (W, Δ) on U with a faithful semifinite normal weight y. Then there is a semicyclic representation $(\tilde{q}_y, \iota \otimes \pi_y)$ of $\tilde{y}$ such that the elements $(\omega_{v,q_x(a)} \otimes \iota \otimes \iota)(\tilde{M})\alpha(u)$ for $v \in V_x$ and $u \in U$ and a in the Tomita algebra of x, span a σ-strong*-norm core for $\tilde{q}_y\colon N_{\tilde{y}} \to V_x \otimes V_y$, and that $\tilde{q}_y((\omega \otimes \iota \otimes \iota)(\tilde{M})\alpha(u)) = p(\omega) \otimes q_y(u)$ for $\omega \in K$.*

Proof We may assume that $\pi_y = \iota$. Suppose that (α, X) is Y-stabilizable, and pick f and β as in Proposition 16.1.5. As $\iota \otimes f)\hat{\beta} = \hat{\alpha}f$, we see that $\tilde{y}f$ equals the dual weight $\tilde{y}_\beta$ of y with respect to β. Then $(\tilde{q}_y = Y^*q_{\tilde{y}_\beta}f^{-1}, \iota \otimes \iota)$ is a semicyclic representation of $\tilde{y}$. From the previous section and the properties of f, we see that the elements $(\omega_{v,q_x(a)}\otimes\iota\otimes\iota)(\tilde{M}(\iota\otimes\alpha)(Y^*))\alpha(u)$ span a σ-strong*-norm core for $\tilde{q}_y$, and that $\tilde{q}_y((\omega\otimes\iota\otimes\iota)(\tilde{M}(\iota\otimes\alpha)(Y^*))\alpha(u)) = Y^*(p(\omega)\otimes q_y(u))$. By the previous proposition we have an isometry Z on $V_x\otimes V_y$ such that $Z(p(\omega)\otimes q_y(u)) = \tilde{q}_y((\omega\otimes\iota\otimes\iota)(\tilde{M})\alpha(u))$. Let D be the span of the elements in this proposition, and let $\overline{D}$ be the domain of the σ-strong*-norm closure of the restriction of $\tilde{q}_y$ to D. Using Lemma 15.4.2 and again the trick with an orthonormal basis as in the proof of Proposition 16.1.7, we see that $d\equiv(\omega\otimes\iota\otimes\iota)(\tilde{M}(\iota\otimes\alpha)(Y^*))\alpha(u)\in\overline{D}$ and $\tilde{q}_y(d) = ZY^*(p(\omega_{v,q_x(a)})\otimes q_y(u))$. Hence D is a σ-strong*-norm core for $\tilde{q}_y$ and $Z=1$. If (α, X) is not Y-stabilizable, we know that $(\alpha\otimes\iota, X\otimes 1)$ is a cocycle coaction of (W,Δ) on $U\bar{\otimes}B(V_x)$ that is Y-stabilizable for some Y. Clearly $W\ltimes_{\alpha\otimes\iota,X\otimes 1}(U\bar{\otimes}B(V_x)) = (W\ltimes_{\alpha,X}U)\bar{\otimes}B(V_x)$ and $\hat{\alpha}\otimes\iota$ is the dual coaction of $(\alpha\otimes\iota, X\otimes 1)$, while the dual weight of $y\otimes\mathrm{Tr}$ is $\tilde{y}\otimes\mathrm{Tr}$. From the first part of the proof we get $\tilde{q}_{\tilde{y}\otimes\mathrm{Tr}}((\omega\otimes\iota\otimes\iota)(\tilde{M})\alpha(u)\otimes w) = p(\omega)\otimes q_y(u)\otimes q_{\mathrm{Tr}}(w)$ for $w\in N_{\mathrm{Tr}}$, and the elements $(\omega_{v,q_x(a)}\otimes\iota\otimes\iota)(\tilde{M})\alpha(u)\otimes w$ span a σ-strong*-norm core for $\tilde{q}_{\tilde{y}\otimes\mathrm{Tr}}$. Let D be the span of the elements $(\omega_{v,q_x(a)}\otimes\iota\otimes\iota)(\tilde{M})\alpha(u)$, let $d\in N_{\tilde{q}}$, and pick w', w'' in the Tomita algebra of Tr such that $\mathrm{Tr}(w'^*ww'')=1$, assuming that $w\neq 0$. Pick a net of elements $e_j\in D\odot N_{\mathrm{Tr}}$ such that $e_j\to d\otimes w$ σ-strongly* and $(q_{\tilde{y}}\otimes q_{\mathrm{Tr}})(e_j)\to q_{\tilde{y}}(d)\otimes q_{\mathrm{Tr}}(w)$ in norm. But then $(\iota\otimes\omega_{q_{\mathrm{Tr}}(w''),q_{\mathrm{Tr}}(w')})(e_j)\to d$ σ-strongly* and the net of elements

$$q_{\tilde{y}}(\iota\otimes\omega_{q_{\mathrm{Tr}(w'')},q_{\mathrm{Tr}(w')}})(e_j) = (1\otimes f^*_{q_{\mathrm{Tr}}(w'\sigma_i^{\mathrm{Tr}}(w'')^*)})(q_{\tilde{y}}\otimes q_{\mathrm{Tr}})(e_j)$$

converges to $(1\otimes f^*_{q_{\mathrm{Tr}}(w'\sigma_i^{\mathrm{Tr}}(w'')^*)})(q_{\tilde{y}}(d)\otimes q_{\mathrm{Tr}}(w)) = q_{\tilde{y}}(d)$. So D is a σ-strong*-norm core for $q_{\tilde{y}}$. Hence by the previous proposition we have a unitary $Z\colon V_x\otimes V_y\to V_{\tilde{y}}$ such that $Z(p(\omega)\otimes q_y(u)) = q_{\tilde{y}}((\omega\otimes\iota\otimes\iota)(\tilde{M})\alpha(u))$. Now define $\pi = Z^*\pi_{\tilde{y}}(\cdot)Z$ and $\tilde{q}_y = Z^*q_{\tilde{y}}$. Then D is a core for $\tilde{q}_y$ and $\tilde{q}_y((\omega\otimes\iota\otimes\iota)(\tilde{M})\alpha(u)) = p(\omega)\otimes q_y(u)$. Since $\tilde{q}_y\otimes q_{\mathrm{Tr}}$ and $\tilde{q}_{\tilde{y}\otimes\mathrm{Tr}}$ agree on a common core, they must be equal, so $\pi\otimes\pi_{\mathrm{Tr}} = \iota\otimes\iota\otimes\pi_{\mathrm{Tr}}$ and $\pi=\iota$. □

Letting $\tilde{J}$ be the modular conjugation of the dual weight $\tilde{y}$ in the semicyclic representation $(\tilde{q}_y, \iota\otimes\pi_y)$ from the previous proposition, we introduce the *unitary implementation* of (α, X) by $K = \tilde{J}(J_{\hat{x}}\otimes J_y)$. Tracing the proofs of the results of the previous two sections using Proposition 16.1.5 and Proposition 16.1.6, as we for instance did in the previous proposition, we get the next two theorems for free.

Theorem 16.1.10 *Let (α, X) be a cocycle coaction of a locally compact quantum group (W,Δ) on a von Neumann algebra U with a faithful semifinite normal weight y and a unitary implementation K. Then $(\iota\otimes\pi_y)\alpha = K(1\otimes\pi_y(\cdot))K^*$, and $K\in W\bar{\otimes}B(V_y)$ satisfies $(J_{\hat{x}}\otimes J_y)K = K^*(J_{\hat{x}}\otimes J_y)$ and $(\Delta\otimes\iota)(K) = (\iota\otimes\iota\otimes\pi_y)(X^*)K_{23}K_{13}(J_{\hat{x}}\otimes J_{\hat{x}}\otimes J_y)(\iota\otimes\iota\otimes\pi_y)(X_{213})(J_{\hat{x}}\otimes J_{\hat{x}}\otimes J_y)$.*

The unitary implementation K does not depend on the chosen weight y, because if y' was another and if $u\colon V_y\to V_{y'}$ is the unitary operator that intertwines the corresponding semicyclic representations, then the unitary implementation with respect to y' is $(1 \otimes u)K(1 \otimes u^*)$. The biduality theorem in this context goes as follows.

Theorem 16.1.11 *Let (α, X) be a cocycle coaction of a locally compact quantum group (W, Δ) on a von Neumann algebra U. Then*

$$f = \tilde{M}(\iota \otimes \alpha)(\cdot)\tilde{M}^* : B(V_x)\bar{\otimes}U \to \hat{W} \ltimes_{\hat{\alpha}} (W \ltimes_{\alpha,X} U)$$

is a $*$-isomorphism determined by $f = \hat{\alpha}$ on $W \ltimes_{\alpha,X} U$ and $f(w' \otimes 1) = w' \otimes 1 \otimes 1$ for $w' \in W'$. Defining the coaction γ of (W, Δ) on $B(V_x)\bar{\otimes}U$ by

$$\gamma = (\Sigma N^* \otimes 1)X^*(\iota \otimes \alpha)(\cdot)X(N\Sigma \otimes 1) : B(V_x)\bar{\otimes}U \to W\bar{\otimes}B(V_x)\bar{\otimes}U,$$

where Σ is the flip on $V_x \otimes V_x$, we have $(g \otimes f)\gamma = \hat{\hat{\alpha}}f$, where $g = J_{\hat{x}}J_x(\cdot)J_xJ_{\hat{x}}$ is the isomorphism from (W, Δ) to $(W, \Delta)'^{\mathrm{op}}$. Moreover, we have

$$W \ltimes_{\alpha,X} U = \{d \in B(V_x)\bar{\otimes}U \mid N_{12}^*X^*(\iota \otimes \alpha)(d)XN_{12} = d_{13}\}.$$

Landstad theory can also be phrased in this context.

Theorem 16.1.12 *Let (W, Δ) be a locally compact quantum group, and let β be a coaction of $(\hat{W}, \hat{\Delta}^{\mathrm{op}})$ on a von Neumann algebra U. Then there is a cocycle coaction (α, X) of (W, Δ) on U^β such that $\hat{\alpha}$ is isomorphic to β if and only there is a unitary $Z \in W\bar{\otimes}U$ such that $(\iota \otimes \beta)(Z) = M_{12}Z_{13}$. If we have such a unitary Z, then $\alpha = Z^*(1\otimes(\cdot))Z$ and $X = Z_{23}^*Z_{13}^*(\Delta\otimes\iota)(Z)$ defines the required cocycle coaction (α, X), and $f = Z^*\beta(\cdot)Z : U \to W \ltimes_{\alpha,X} U^\beta$ is the required $*$-isomorphism satisfying $\hat{\alpha}f = (\iota \otimes f)\beta$, and moreover $(\iota \otimes f)(Z) = M_{12}X^*$ and $f = \alpha$ on U^β.*

Proof For the forward implication use $Z = M_{12}X^*$. For the converse, one easily shows that $(\iota \otimes \beta)\alpha(u) = \alpha(u)_{13}$ for $u \in U^\beta$, so $\alpha(U^\beta) \subset W\bar{\otimes}U^\beta$. Similarly, one shows that $(\iota \otimes \iota \otimes \beta)(X) = X_{124}$, so $X \in W\bar{\otimes}W\bar{\otimes}U^\beta$. It is now straightforward to check that (α, X) is a cocycle coaction of (W, Δ) on U^β. Using the characterization of the cocycle crossed product at the end of the previous theorem, a standard computation shows that $f(U) \subset W \ltimes_{\alpha,X} U$. The easily verified last two formulas for f in the theorem, show that the opposite inclusion holds. The formula $\hat{\alpha}f = (\iota \otimes f)\beta$ is clear from an appropriate implementation of $\hat{\Delta}^{\mathrm{op}}$. □

16.2 Cocycle Bicrossed Products

We say a triple (r, X, Y) is a *cocycle matching* for a pair of locally compact quantum groups (W_i, Δ_i) if $r : W_1\bar{\otimes}W_2 \to W_2\bar{\otimes}W_1$ is a normal $*$-monomorphism such that $(\alpha = r(1 \otimes (\cdot)), X)$ is a cocycle coaction of (W_1, Δ_1) on W_2 and $(F\beta, X_{321})$ is a cocycle coaction of (W_2, Δ_2) on W_1, where $\beta = r((\cdot) \otimes 1)$ and $F: W_1 \otimes W_2 \to W_2 \otimes W_1$ is the flip, and moreover if the following compatibility conditions hold:

$$r_{13}(\alpha \otimes \iota)\Delta_2 = Y_{132}(\iota \otimes \Delta_2)\alpha(\cdot)Y_{132}^*, \quad r_{23}F_{23}(\beta \otimes \iota)\Delta_1 = X(\Delta_1 \otimes \iota)\beta(\cdot)X^*$$

and

$$(\Delta_1 \otimes \iota \otimes \iota)(Y)(\iota \otimes \iota \otimes \Delta_2^{\rm op})(X^*) = X_{12}^*(\iota \otimes rF \otimes \iota)((\beta \otimes \iota \otimes \iota)(X^*)(\iota \otimes \iota \otimes \alpha)(Y))Y_{23}.$$

Note that $(F\beta, X_{321})$ being a cocycle coaction simply means that $(\beta \otimes \iota)\beta = Y(\iota \otimes \Delta_2^{\rm op})\beta(\cdot)Y^*$ and $(\beta \otimes \iota \otimes \iota)(Y)(\iota \otimes \iota \otimes \Delta_2^{\rm op})(Y) = Y_{12}(\iota \otimes \Delta_2^{\rm op} \otimes \iota)(Y)$.

Given such a triple (r, X, Y) for (W_i, Δ_i) with fundamental multiplicative unitaries M_i on V_{x_i}, we introduce unitaries $M = \hat{M}_{21}^*$ and

$$\hat{M} = (\beta \otimes \iota \otimes \iota)((M_1 \otimes 1)X^*)(\iota \otimes \iota \otimes \alpha)(Y(1 \otimes \hat{M}_2))$$

on the Hilbert space $V = V_{x_1} \otimes V_{x_2}$ and a $*$-monomorphism $\Delta = M^*(1 \otimes (\cdot))M: W \to B(V \otimes V)$, where $W = W_1 \ltimes_{\alpha,X} W_2$.

Definition 16.2.1 With notation from the previous paragraph we call (W, Δ) the *cocycle bicrossed product* of (W_1, Δ_1) and (W_2, Δ_2).

Of course we will prove that (W, Δ) is a locally compact quantum group with fundamental multiplicative unitary M with respect to some left invariant faithful semifinite normal weight. We will also describe the dual quantum group. In the sequel we shall denote $(M_1 \otimes 1)X^*$ by $\tilde{M}$. The following result is obvious.

Lemma 16.2.2 *We have $W \subset B(V_{x_1})\bar{\otimes} W_2$ and*

$$\Delta^{\rm op} = (\beta \otimes \iota \otimes \iota)(\tilde{M})(\iota \otimes \iota \otimes \alpha)(Y(\iota \otimes \Delta_2^{\rm op})(\cdot)Y^*)(\beta \otimes \iota \otimes \iota)(\tilde{M}^*).$$

The next result shows that $\Delta(W) \subset W\bar{\otimes}W$.

Proposition 16.2.3 *We have $\Delta\alpha = (\alpha \otimes \alpha)\Delta_2$ and*

$$(\iota \otimes \Delta^{\rm op})(\tilde{M}) = (\tilde{M} \otimes 1 \otimes 1)((\iota \otimes \alpha)\beta \otimes \iota \otimes \iota)(\tilde{M})(\iota \otimes \alpha \otimes \alpha)(Y).$$

Proof By the first compatibility identity, we get

$$(\iota \otimes \iota \otimes \alpha)(Y(\iota \otimes \Delta_2^{\rm op})\alpha(\cdot)Y^*) = (r \otimes \iota \otimes \iota)((\iota \otimes \alpha \otimes \iota)(\alpha \otimes \iota)\Delta_2()_{1342}),$$

which by the (α, X)-cocycle property and the definition of α and β in terms of r, equals $(\beta \otimes \iota \otimes \iota)(\tilde{M}^*)(\alpha \otimes \alpha)\Delta_2^{\rm op}(\cdot)(\beta \otimes \iota \otimes \iota)(\tilde{M})$, so $\Delta^{\rm op}\alpha = (\alpha \otimes \alpha)\Delta_2^{\rm op}$ by the previous lemma.

For the next statement, the previous lemma shows that $(\iota \otimes \Delta^{\rm op})(\tilde{M})$ equals

$$(1 \otimes (\beta \otimes \iota \otimes \iota)(\tilde{M}))(\iota \otimes \iota \otimes \iota \otimes \alpha)(C)(1 \otimes (\beta \otimes \iota \otimes \iota)(\tilde{M}^*)),$$

where $C \equiv Y_{234}(\iota \otimes \iota \otimes \Delta_2^{\mathrm{op}})(\tilde{M})Y_{234}^*$ equals

$$(M_1 \otimes 1 \otimes 1)X_{234}^*(\iota \otimes rF \otimes \iota)((\beta \otimes \iota \otimes \iota)(X^*)(\iota \otimes \iota \otimes \alpha)(Y))$$

by the definition of $\tilde{M}$, that $\Delta_1 = M_1^*(1 \otimes (\cdot))M_1$, and the compatibility between X and Y. Hence by definition of $\tilde{M}$ and $\Delta_1 = M_1^*(1 \otimes (\cdot))M_1$, the expression $(\iota \otimes \iota \otimes \iota \otimes \alpha)(C)$ equals

$$(\tilde{M}\otimes 1\otimes 1)r_{23}F_{23}(\beta\otimes\iota\otimes\iota\otimes\iota)((\iota\otimes\iota\otimes\alpha)(X^*)(1\otimes\tilde{M}^*))(\iota\otimes\alpha\otimes\alpha)(Y)(1\otimes(\beta\otimes\iota\otimes\iota)(\tilde{M})).$$

So by the formula at the top of this paragraph, we see that $(\iota \otimes \Delta^{\mathrm{op}})(\tilde{M})$ equals

$$(1\otimes(\beta\otimes\iota\otimes\iota)(\tilde{M}))(\tilde{M}\otimes 1\otimes 1)r_{23}F_{23}(\beta\otimes\iota\otimes\iota\otimes\iota)((\iota\otimes\iota\otimes\alpha)(X^*)(1\otimes\tilde{M}^*))(\iota\otimes\alpha\otimes\alpha)(Y).$$

But the expression consisting of the first two factors in this expression equals

$$(\tilde{M} \otimes 1 \otimes 1)r_{23}F_{23}(\beta \otimes \iota \otimes \iota \otimes \iota)(\Delta_1 \otimes \iota \otimes \iota)(\tilde{M})$$

by the second compatibility relation, so $(\iota \otimes \Delta^{\mathrm{op}})(\tilde{M})$ equals

$$(\tilde{M}\otimes 1\otimes 1)r_{23}F_{23}(\beta\otimes\iota\otimes\iota\otimes\iota)((\Delta_1\otimes\iota\otimes\iota)(\tilde{M})(\iota\otimes\iota\otimes\alpha)(X^*)(1\otimes\tilde{M}^*))(\iota\otimes\alpha\otimes\alpha)(Y).$$

The argument of $r_{23}F_{23}(\beta \otimes \iota \otimes \iota \otimes \iota)$ in this expression equals $\tilde{M}_{134}$ as is seen by using the definition of $\tilde{M}$, then $(\Delta_1\otimes\iota)(M_1) = (M_1)_{13}(M_1)_{23}$ and the compatibility between X and Y, then the fact that M_1 implements Δ_1, and finally the definition of $\tilde{M}$. Using this and the definition of α involving r, completes the proof. □

Proposition 16.2.4 *We have* $(\iota \otimes \Delta^{\mathrm{op}})(\hat{M}) = \hat{M}_{12}\hat{M}_{13}$.

Proof By definition of $\hat{M}$ and the previous lemma $(\iota \otimes \iota \otimes \Delta^{\mathrm{op}})(\hat{M})$ equals

$$(\beta \otimes \iota \otimes \iota)(\tilde{M})_{1234}((\beta \otimes \alpha)\beta \otimes \iota \otimes \iota)(\tilde{M})(\beta \otimes \alpha \otimes \alpha)(Y)$$

times $(\iota\otimes\iota\otimes\alpha\otimes\alpha)((\iota\otimes\iota\otimes\Delta_2^{\mathrm{op}})(Y)(\hat{M}_2)_{23}(\hat{M}_2)_{24})$. Using the $(F\beta, Y_{321})$ cocycle property and $\Delta_2^{\mathrm{op}} = \hat{M}_2((\cdot) \otimes 1)\hat{M}_2^*$, gives

$$(\beta \otimes \iota \otimes \iota)(Y)(\iota \otimes \iota \otimes \Delta_2^{\mathrm{op}})(Y)(\hat{M}_2)_{23}(\hat{M}_2)_{24} = Y_{123}(\hat{M}_2)_{23}Y_{124}(\hat{M}_2)_{24},$$

so $(\iota \otimes \iota \otimes \Delta^{\mathrm{op}})(\hat{M})$ equals

$$(\beta \otimes \iota \otimes \iota)(\tilde{M})_{1234}((\beta \otimes \alpha)\beta \otimes \iota \otimes \iota)(\tilde{M})$$

times $(\iota \otimes \iota \otimes \alpha)(Y(1 \otimes \hat{M}_2))_{1234}(\iota \otimes \iota \otimes \alpha)(Y(1 \otimes \hat{M}_2))_{1256}$. Using the $(F\beta, Y_{321})$ cocycle property and $\Delta_2^{\mathrm{op}} = \hat{M}_2((\cdot) \otimes 1)\hat{M}_2^*$ on the factor $((\beta \otimes \alpha)\beta \otimes \iota \otimes \iota)(\tilde{M})$, then gives the desired result. □

Hence $(\Delta \otimes \iota)(M) = M_{13}M_{23}$, so $\Delta \colon W \to W \bar{\otimes} W$ is coassociative and both M and $\hat{M}$ are multiplicative unitaries. Now let $x = x_2\alpha^{-1}(\hat{x}_1 \otimes \iota \otimes \iota)\hat{\alpha}$ be the dual weight on W of the left invariant faithful semifinite normal weight x_2 of (W_2, Δ_2). According to Proposition 16.1.9 we have a semicyclic representation (q_x, ι) of M such that the elements $(\omega_{v,q_{x_1}(a)} \otimes \iota \otimes \iota)(\tilde{M})\alpha(u)$ for $v \in V_{x_1}$ and $u \in N_{x_2}$ and a in the Tomita algebra of x_1, span a σ-strong*-norm core for $q_x : N_x \to V_{x_1} \otimes V_{x_2}$, and that $q_x((\omega \otimes \iota \otimes \iota)(\tilde{M})\alpha(u)) = p_1(\omega) \otimes q_{x_2}(u)$ for $\omega \in K_1$.

Proposition 16.2.5 *The weight x defined above is left invariant. In fact, we have $q_x(\omega \otimes \iota)\Delta(w) = (\omega \otimes \iota)(M^*)q_x(w)$ for $w \in N_x$ and $\omega \in W_*$.*

Proof By Proposition 16.2.3 we see that $\Delta^{\rm op}((\omega_{v,q_{x_1}(a)} \otimes \iota \otimes \iota)(\tilde{M})\alpha(u))$ equals the σ-strongly*-converging series over k of $(\omega_{v_k,q_{x_1}(a)} \otimes \iota \otimes \iota)(\tilde{M}) \otimes 1 \otimes 1$ times $(\alpha \otimes \iota \otimes \iota)((\omega_{v,v_k} \otimes \iota \otimes \iota \otimes \iota)((\beta \otimes \iota \otimes \iota)(\tilde{M})(\iota \otimes \iota \otimes \alpha)(Y))(\iota \otimes \alpha)\Delta_2^{\rm op}(u))$ for any orthonormal basis $\{v_k\}$ for V_{x_1}. For any $v', v'' \in V$ and any orthonormal basis $\{u_j\}$ of V, we therefore get

$$(\iota \otimes \iota \otimes \omega_{v'v''})\Delta^{\rm op}((\omega_{v,q_{x_1}(a)} \otimes \iota \otimes \iota)(\tilde{M})\alpha(u)) = \sum_{kj}(\omega_{v_k,q_{x_1}(a)} \otimes \iota \otimes \iota)(\tilde{M})\alpha(L),$$

where $L = (\omega_{v,v_k} \otimes \iota \otimes \omega_{u_j,v''})((\beta \otimes \iota \otimes \iota)(\tilde{M})(\iota \otimes \iota \otimes \alpha)(Y))(\iota \otimes \omega_{v',u_j}\alpha)\Delta_2^{\rm op}(u)$. Hence q_x of a finite such double sum is the finite double sum of $J_{x_1}\sigma_{i/2}^{x_1}(a)J_{x_1}v_k \otimes L'q_{x_2}(u)$, where L' is L with $\Delta_2^{\rm op}(u)$ replaced by $\tilde{M}_2$. Since q_x is σ-strongly*-norm continuous, we can first take the limit over j and then the limit over k, thus concluding that with $w' = (J_{x_1}\sigma_{i/2}^{x_1}(a)J_{x_1}v_i \otimes 1)\Delta^{\rm op}((\omega_{v,q_{x_1}(a)} \otimes \iota \otimes \iota)(\tilde{M})\alpha(u))$, then $q_x(w')$ equals $(J_{x_1}\sigma_{i/2}^{x_1}(a)J_{x_1}v_k \otimes 1)(\iota \otimes \iota \otimes \omega_{v',v''})(\hat{M})(v \otimes q_{x_2}(u))$. Then $\hat{M} \in W_1\bar{\otimes}B(V_{x_2} \otimes V)$ gives $q_x(w') = (\iota \otimes \iota \otimes \omega_{v',v''})(\hat{M})(p_1(\omega_{v,q_{x_1}(a)}) \otimes q_{x_2}(u))$, so $q_x(\iota \otimes \omega_{v',v''})\Delta^{\rm op}(w) = (\iota \otimes \omega_{v',v''})(\hat{M})q_x(w)$, and we get the formula in the proposition. □

To get the right invariant weight, we go via R, or $J_{\hat{x}}$, for some dual version $\hat{x}$ of x. As (FrF, Y_{321}, X_{321}) is a cocycle matching for (W_2, Δ_2) and (W_1, Δ_1), our results so far give, by using the flip $V_{x_2} \otimes V_{x_1} \to V_{x_1} \otimes V_{x_2}$, the following dual result.

Proposition 16.2.6 *Let $\hat{W}$ be the von Neumann subalgebra of $W_1\bar{\otimes}B(V_{x_2})$ generated by $\beta(W_1)$ and $(\iota \otimes \iota \otimes \omega)(Y(1 \otimes \hat{M}_2))$ for $\omega \in (W_2)_*$. Let $\hat{\Delta} = \hat{M}^*(1 \otimes (\cdot))\hat{M}$. Then $\hat{\Delta}$ is a normal $*$-monomorphism such that $(\iota \otimes \hat{\Delta})\hat{\Delta} = (\hat{\Delta} \otimes \iota)\hat{\Delta}$. There exists a faithful normal semifinite weight $\hat{x}$ on $\hat{W}$ and a semicyclic representation $(q_{\hat{x}}, \iota)$ of $\hat{M}$ such that the elements $(\iota \otimes \iota \otimes \omega_{v,q_{x_1}(a)})((1 \otimes \hat{M}_2^*)Y^*)\beta(u)$ for $v \in V_{x_2}$ and $u \in N_{x_1}$ and a in the Tomita algebra of x_2, span a σ-strong*-norm core for $q_{\hat{x}} : N_{\hat{x}} \to V_{x_1} \otimes V_{x_2}$, and that $q_{\hat{x}}((\iota \otimes \iota \otimes \omega)((1 \otimes \hat{M}_2^*)Y^*)\beta(u)) = q_{x_1}(u) \otimes p_2(\omega)$ for $\omega \in K_2$, and $\hat{x}$ is left invariant for $(\hat{M}, \hat{\Delta})$, and $q_{\hat{x}}(\omega \otimes \iota)\hat{\Delta}(w) = (\iota \otimes \omega)(M)q_{\hat{x}}$ on $N_{\hat{x}}$ for $\omega \in \hat{W}_*$.*

Of course this turns out be the dual quantum group of (W, Δ).

Lemma 16.2.7 *The σ-strong*-closure of $\{(\iota \otimes \omega)(M) \mid \omega \in B(V)_*\}$ is W. We have $(J_{\hat{x}} \otimes J_x)M = M^*(J_{\hat{x}} \otimes J_x)$ and $\Delta R = (R \otimes R)\Delta^{\text{op}}$, where $R = J_{\hat{x}}(\cdot)^* J_{\hat{x}}$ is a $*$-antiautomorphism on W.*

Proof By Proposition 16.2.5 we have $(\bar{\omega} \otimes \iota)(M^*)q_x(w) = q_x(\bar{\omega} \otimes \iota)\Delta(w)$ for $\omega \in B(V)_*$ and $w \in N_x \cap N_x^*$. Hence the left hand side belongs to $D(\Delta_x^{1/2})$ and $J_x \Delta_x^{1/2}(\bar{\omega} \otimes \iota)(M^*)q_x(w) = q_x(\omega \otimes \iota)\Delta(w^*) = (\omega \otimes \iota)(M^*)J_x \Delta_x^{1/2} q_x(w)$. As the elements $q_x(w)$ form a core for $\Delta_x^{1/2}$ we get $(\omega \otimes \iota)(M^*)J_x \Delta_x^{1/2} \subset J_x \Delta_x^{1/2}(\bar{\omega} \otimes \iota)(M^*)$. Similarly we get $(\iota \otimes \omega)(M)J_{\hat{x}}\Delta_{\hat{x}}^{1/2} \subset J_{\hat{x}}\Delta_{\hat{x}}^{1/2}(\iota \otimes \bar{\omega})(M)$ by the previous proposition. Combining these two inclusions on vectors and replacing ω by vector functionals, all vectors being in appropriate domains, we see that M and $\Delta_{\hat{x}} \otimes \Delta_x$ commute, hence M and $\Delta_{\hat{x}}^{1/2} \otimes \Delta_x^{1/2}$ commute, and using this in a second round with the two inclusions, we get $(J_{\hat{x}} \otimes J_x)M = M^*(J_{\hat{x}} \otimes J_x)$.

Let L_1 be the closed span in the proposition, and let L_2 be the σ-strong*-closed span of the elements $(\iota \otimes \omega)\Delta(w)$ for $\omega \in W_*$ and $w \in W$. Here we can assume that $w \in N_x$ and $\omega = \omega_{q_x(a),q_x(b)}$ for a, b in the Tomita algebra of x. Therefore, as $(\iota \otimes \omega_{q_x(a),q_x(b)})\Delta(w) = (\iota \otimes \omega_{q_x(w),q_x(b\sigma^x_{-i}(a^*))})(M^*)$, we get $L_1 = L_2$. Now $L_2^* = L_2$, while L_1 is an algebra as M is multiplicative. Also, as $(\iota \otimes \omega)\Delta\alpha = \alpha(\iota \otimes \omega\alpha)\Delta_2$ by Proposition 16.2.3, and α is injective, we have $\alpha(W_2) \subset L_2$, so L_2 is unital, and L_1 is a von Neumann algebra. Now $(\iota \otimes \Delta^{\text{op}})(W_1 \bar{\otimes} W) \subset W_1 \bar{\otimes} W \bar{\otimes} L_1$ by Takesaki's commutant theorem, so $(\iota \otimes \Delta^{\text{op}})(\tilde{M}) \in W_1 \bar{\otimes} W \bar{\otimes} L_1$. By Proposition 16.2.3 and $\alpha(W_2) \subset L_1$, we therefore see that $((\iota \otimes \alpha)\beta \otimes \iota \otimes \iota)(\tilde{M}) \in W_1 \bar{\otimes} W \bar{\otimes} L_1$, and as $(\iota \otimes \alpha)\beta$ is injective, we see that $(\omega \otimes \iota \otimes \iota)(\tilde{M}) \in L_1$ for $\omega \in (W_1)_*$. Hence $L_1 = W$.

From the first paragraph we have $R(\iota \otimes \omega_{u,v})(M) = (\iota \otimes \omega_{J_x v, J_x u})(M)$ for $u, v \in V$. By the second paragraph above, we thus see that $R(W) \subset W$, and R is a $*$-antiautomorphism on W. Letting $\{v_i\}$ be an orthonormal basis of V, we have $\Delta(\iota \otimes \omega_{u,v})(M) = \sum(\iota \otimes \omega_{v_i,u})(M) \otimes (\iota \otimes \omega_{v,v_i})(M)$ by multiplicativity of M, so

$$(R \otimes R)\Delta^{\text{op}}(w) = \sum(\iota \otimes \omega_{J_x v_i, J_x u})(M) \otimes (\iota \otimes \omega_{J_x v, J_x v_i})(M) = \Delta R(w),$$

for $w = (\iota \otimes \omega_{u,v})(M)$, so the second paragraph gives $\Delta R = (R \otimes R)\Delta^{\text{op}}$. □

Theorem 16.2.8 *The cocycle bicrossed product (W, Δ) is a locally compact quantum group with fundamental multiplicative unitary M associated to the left invariant faithful semifinite normal weight x on W, and its dual locally compact quantum group is $(\hat{W}, \hat{\Delta})$.*

Proof Since the weight xR is right invariant, we know that (W, Δ) is a locally compact quantum group, and M is its fundamental multiplicative unitary, again by Proposition 16.2.5. Hence the von Neumann algebra of the dual quantum group is the σ-strong*-closed span of the elements $(\omega \otimes \iota)(M)$ for $\omega \in M_*$, and by the proposition above applied to the cocycle bicrossed product given by the cocycle matching (FrF, Y_{321}, X_{321}) for (W_2, Δ_2) and (W_1, Δ_1), we thus see that $(\hat{W}, \hat{\Delta})$ is indeed the dual quantum group. □

Since $x(1 \otimes 1) = \hat{x}_1(1)x_2(1)$, the cocycle bicrossed product (W, Δ) is compact if and only if (W_1, Δ_1) is discrete and (W_2, Δ_2) is compact. Since $(\hat{W}, \hat{\Delta})$ is the cocycle bicrossed product of (W_2, Δ_2) and (W_1, Δ_1), we therefore see that $(\hat{W}, \hat{\Delta})$ is discrete if and only if (W_1, Δ_1) is compact and (W_2, Δ_2) is discrete.

Definition 16.2.9 A *matched pair* of locally compact groups G_1 and G_2 consists of a locally compact group G and a homomorphism $i: G_1 \to G$ and an antihomomorphism $j: G_2 \to G$ that are homeomorphisms onto their images, which are assumed to be closed, and $u: G_1 \times G_2 \to G$ given by $u(g, s) = i(g)j(s)$ must be a homeomorphism onto an open subset H of G with complement of measure zero.

Consider the homeomorphism $v: G_1 \times G_2 \to H^{-1}$ given by $v(g, s) = j(s)i(g)$. Then $v^{-1}u$ is a homeomorphism from the open subset $u^{-1}(H \cap H^{-1})$ of $G_1 \times G_2$ onto the open subset $v^{-1}(H \cap H^{-1})$ of $G_1 \times G_2$. Define $\beta_s(g) \in G_1$ and $\alpha_g(s) \in G_2$ by $v^{-1}u(g, s) = (\beta_s(g), \alpha_g(s))$ for $(g, s) \in u^{-1}(H \cap H^{-1})$, so $j(\alpha_g(s))i(\beta_s(g)) = i(g)j(s)$. We may normalize the Haar measures on G and G_i such that $\int f(k)\,dk = \int f(i(g)j(s))\Delta(j(s))\,dgds$ for positive $f \in C_c(G)$. So a Borel subset of $G_1 \times G_2$ has measure zero if and only if its image under u has measure zero in G. Since $H \cap H^{-1}$ has complement of measure zero, so have $u^{-1}(H \cap H^{-1})$ and $v^{-1}(H \cap H^{-1})$, and we can define a $*$-isomorphism $r: L^\infty(G_1)\bar{\otimes}L^\infty(G_2) \to L^\infty(G_2)\bar{\otimes}L^\infty(G_1)$ by $r(f)(g, s) = f(\beta_s(g), \alpha_g(s))$ for $(g, s) \in u^{-1}(H \cap H^{-1})$. Then it is straightforward to check that $(r, 1, 1)$ is a cocycle matching for $(L^\infty(G_1), \Delta_1)$ and $(L^\infty(G_2), \Delta_2)$. If one wishes, one can also easily supply measurable maps $X: G_1 \times G_1 \times G_2 \to \mathbb{T}$ and $Y: G_1 \times G_2 \times G_2 \to \mathbb{T}$ satisfying obvious relations almost everywhere, to get a cocycle matching (r, X, Y) with non-trivial cocycles. In either case one gets cocycle bicrossed products this way that can be further studied, even finite dimensional ones. Here is an example of a cocycle bicrossed product that is neither compact nor discrete.

Example 16.2.10 Consider the $ax + b$ group $G = \{(a, b) \mid a \in \mathbb{R}\backslash\{0\},\ b \in \mathbb{R}\}$ with $(a, b)(c, d) = (ac, bc+d)$. Consider $G_i = \mathbb{R}\backslash\{0\}$ as groups with multiplication, and define i, j by $i(g) = (g, g-1)$ and $j(s) = (s, 0)$. Then G_1 and G_2 is a matched pair of groups. One checks that $\alpha_g(s) = gs(s(g-1)+1)^{-1}$ and $\beta_s(g) = s(g-1)+1$ for $g, s \in \mathbb{R}\backslash\{0\}$ with $s(g-1) \neq -1$. Let (W, Δ) be the cocycle bicrossed product with respect to the cocycle matching $(r, 1, 1)$ described above. Since $(\hat{W}, \hat{\Delta})$ is obtained from the matched pair with the roles of α and β switched, and since $S(\beta_s(g)) = \alpha_{S^{-1}(s)}S(g)$ with $S: G_1 \to G_2$ the inverse operation, we see that (W, Δ) is self-dual. For $q > 0$ let f_q be the automorphism of G_1 given by $f_q(g) = g^q$ with $g^q \equiv -(-g)^q$ when $g < 0$. Letting $i_q = if_q$ and keeping j as above, we get another matched pair of groups, now with $\alpha_g^q(s) = g^q s(s(g^q - 1) + 1)^{-1}$ and $\beta_s^q(g) = (s(g^q - 1) + 1)^{1/q}$. One checks that the associated cocycle bicrossed products will all be isomorphic, say with multiplicative unitaries M^q. Let P be the orthogonal projection from $L^2(G_1)$ onto the space of vectors $h \in L^2(G_1)$ satisfying $h(g) = 0$ whenever $g < 0$. Now $\alpha_g^q(s) \to s$ and $\beta_s^q(g) \to g^s$ for $g > 0$ as $q \to 0$. Hence $(P\otimes 1\otimes P\otimes 1)M^q(P\otimes 1\otimes P\otimes 1)$ converges σ-strongly* to a multiplicative

unitary that is isomorphic, via the Fourier transform, to the multiplicative unitary of G. In this sense our cocycle bicrossed product is a deformation of the $ax + b$ group.

16.3 Continuous Coactions and Regularity

Let $[X]$ denote the closed linear span of a subset X in a normed space.

Definition 16.3.1 Let (W, Δ) be a locally compact quantum group with reduced C*-quantum group (W_r, Δ_r). A *continuous (left) coaction* of (W, Δ) on a C*-algebra U is a non-degenerate $*$-monomorphism $\alpha: U \to M(W_r \otimes U)$ satisfying $(\iota \otimes \alpha)\alpha = (\Delta_r \otimes \iota)\alpha$ and $W_r \otimes U = [(W_r \otimes \mathbb{C})\alpha(U)]$. The *(reduced) crossed product* is the C*-subalgebra $W_r \ltimes_\alpha U$ of $M(B_0(V_x) \otimes U)$ generated by $\hat{W}_r \otimes \mathbb{C}$ and $\alpha(U)$.

Remark 16.3.2 One could also have defined a universal version of a continuous coaction α by replacing (W_r, Δ_r) with the universal C*-quantum group (W_u, Δ_u) of (W, Δ), leading to a universal crossed product, but we won't enter that here. ◇

From Lemma 15.4.4, we saw that the *crossed product* of a coaction α of a locally compact quantum group (W, Δ) on a von Neumann algebra U is the σ-strong*-closed span of elements of the form $(a \otimes 1)\alpha(u)$ for $a \in \hat{W}$ and $u \in U$. A similar result holds in the pure context of C*-algebras.

Proposition 16.3.3 *The* crossed product *of a continuous coaction α of a locally compact quantum group (W, Δ) on a C*-algebra U is $[(\hat{W}_r \otimes \mathbb{C})\alpha(U)]$.*

Proof We only need to show that $\alpha(u)((\omega_{v,v'} \otimes \iota)(M^*) \otimes 1)$ belongs to the closed span L described above, where M is the fundamental multiplicative unitary of (W, Δ) and $v, v' \in V_x$. The element under question equals

$$(f_v^* \otimes 1 \otimes 1)(1 \otimes \alpha(u))(M^* \otimes 1)(f_{v'} \otimes 1 \otimes 1),$$

where $f_v: \mathbb{C} \to V_x$ sends c to cv. Since $\hat{W}_r$ acts non-degenerately on V_x we may replace f_v^* by $f_v^* a$ for $a \in \hat{W}_r$, and then replacing $a \otimes w$ by $(a \otimes 1)\alpha(u)$, it suffices to check that elements of the form

$$(f_v^* \otimes 1 \otimes 1)(\iota \otimes \alpha)(a \otimes 1)\alpha(u))(M^* \otimes 1)(f_{v'} \otimes 1 \otimes 1)$$

belong to L. But this element equals $((\omega_{a^*v,v'} \otimes \iota)(M^*) \otimes 1)\alpha(u)$ as is seen by using $(\iota \otimes \alpha)\alpha = (\Delta_r \otimes \iota)\alpha$ and that M implements Δ. □

Let α be a continuous coaction of a locally compact quantum group (W, Δ) on a C*-algebra U. If M is the fundamental multiplicative unitary of (W, Δ), then $(J_x \otimes J_x)\hat{M}(J_x \otimes J_x)$ is the fundamental multiplicative unitary of $(W, \Delta)^{\wedge\mathrm{op}}$, and $\hat{\alpha} = ((J_x \otimes J_x)\hat{M}(J_x \otimes J_x))_{12}^*(1 \otimes (\cdot))((J_x \otimes J_x)\hat{M}(J_x \otimes J_x))_{12}$ is a continuous coaction of $(W, \Delta)^{\wedge\mathrm{op}}$ on $W_r \ltimes_\alpha U$, known as the *dual coaction*, and which is uniquely determined by $\hat{\alpha} = 1 \otimes \alpha(\cdot)$ and $\hat{\alpha}(a \otimes 1) = \hat{\Delta}_r^{\mathrm{op}}(a) \otimes 1$ for $a \in \hat{W}_r$. Then

$$\hat{W}_r \ltimes_{\hat{\alpha}} (W_r \ltimes_\alpha U) = [(J_x W_r J_x \otimes \mathbb{C} \otimes \mathbb{C})(\hat{\Delta}^{\mathrm{op}}(\hat{W}_r) \otimes \mathbb{C})(\mathbb{C} \otimes \alpha(U))].$$

Since $\hat{\Delta}_r^{\mathrm{op}} = M((\cdot) \otimes 1)M^*$ and $(\Delta_r \otimes \iota)\alpha = (\iota \otimes \alpha)\alpha$, the $*$-isomorphism given by $(M^* \otimes 1)(\cdot)(M \otimes 1)$ maps $\hat{W}_r \ltimes_{\hat{\alpha}} (W_r \ltimes_\alpha U)$ onto $[(J_x W_r J_x \hat{W}_r \otimes \mathbb{C} \otimes \mathbb{C})(\iota \otimes \alpha)\alpha(U)]$. We have thus arrived at the following *continuous bicrossed product theorem.*

Theorem 16.3.4 *Let α be a continuous coaction of a locally compact quantum group (W, Δ) on a C*-algebra U. If $[J_x W_r J_x \hat{W}_r] = B_0(V_x)$, then*

$$\hat{W}_r \ltimes_{\hat{\alpha}} (W_r \ltimes_\alpha U) \cong B_0(V_x) \otimes \alpha(U) \cong B_0(V_x) \otimes U.$$

Proof In view of the discussion prior to the theorem, we need only show that $[(B_0(V_x) \otimes \mathbb{C} \otimes \mathbb{C})(\iota \otimes \alpha)\alpha(U)] = B_0(V_x) \otimes \alpha(U)$, but this is clear as

$$[(B_0(V_x) \otimes \mathbb{C})\alpha(U)] = [(B_0(V_x) \otimes \mathbb{C})(W_r \otimes \mathbb{C})\alpha(U)] = [B_0(V_x) W_r \otimes U]$$

by the non-degenerate action of W_r on V_x. □

We shall soon see that quantum groups fulfilling the assumption in the previous theorem are presicely those satisfying the following property.

Definition 16.3.5 A multiplicative unitary N on a Hilbert space V is *regular* whenever $[\{(\iota \otimes \omega)(FN) \mid \omega \in B(V)_*\}] = B_0(V)$, where F is the flip on $V \otimes V$. A locally compact quantum group is *regular* if its fundamental multiplicative unitary is regular.

As $(\iota \otimes \omega)(F(FN^*F)) = ((\iota \otimes \bar{\omega})(FN))^*$, we see that FN^*F is regular when N is regular. So duals of regular locally compact quantum groups are regular. Note that the span of the elements $(\iota \otimes \omega)(FN)$ is always an algebra. Indeed, simple manipulations show that $(\iota \otimes \omega)(FN)(\iota \otimes \omega')(FN) = (\iota \otimes \eta)(FN)$, where $\eta = (\omega' \otimes \omega)(NF(1 \otimes (\cdot))N)$.

Example 16.3.6 Recall that an antiunitary J on a Hilbert space V with inner product $(\cdot|\cdot)$ can be seen as a unitary from V to the conjugate Hilbert space $\bar{V}$. As usual $u \odot v = (\cdot|v)u$, and by Proposition 4.7.6, the map $v \otimes u \mapsto u \odot v$ defines a unitary $\bar{V} \otimes V \to B^2(V)$, and we know that $B^2(V)$ is norm dense in the C*-algebra $B_0(V)$ of compact operators on V. Consider now a Kac algebra, that is, a locally compact quantum group (W, Δ) with trivial scaling group. Then $R(\iota \otimes \omega)(M) = (\iota \otimes \omega)(M^*)$ for $\omega \in B(V_x)_*$, where $R = J_{\hat{x}}(\cdot)^* J_{\hat{x}}$ is the unitary coinverse. Picking ω to be a vector functional and evaluating the expression on a vector functional gives $((\iota \otimes \omega_{u,v})(FM)u'|v') = (M^*(J_{\hat{x}}v \otimes u)|J_{\hat{x}}u' \otimes v')$ for $u, v, u', v' \in V_x$. Viewing $L \equiv (J_{\hat{x}} \otimes 1)M^*(J_{\hat{x}} \otimes 1)$ as a unitary on $\bar{V}_x \otimes V_x$, we thus see that $(\iota \otimes \omega_{u,v})(FM) \in B^2(V_x)$ corresponds to $L(v \otimes u) \in \bar{V}_x \otimes V$. As L is unitary and $\omega_{u,v} = \mathrm{Tr}((u \odot v)\cdot)$, we therefore see that M is regular, so Kac algebras are regular. This example initiated the study of both regular and manageable multiplicative unitaries, where for manageable multiplicative unitaries one needs to consider S instead of R. ◇

Example 16.3.7 Let (W, Δ) be a compact quantum group. Then as W_r is unital and $\hat{M} \in M(B_0(V_x) \otimes W_r)$, we get $(a \otimes 1)\hat{M} \in B_0(V_x) \otimes W_r$ for $a \in B_0(V_x)$, and thus $(1 \otimes a)(F\hat{M})(1 \otimes b) = F(a \otimes 1)\hat{M}(1 \otimes b) \in B_0(V_x) \otimes B_0(V_x)$ for $b \in B_0(V_x)$, so $(\iota \otimes \omega)(F\hat{M}) \in B_0(V_x)$ for $\omega \in B(V_x)_* \cong B_0(V_x)^\star$ as $B_0(V_x) = B_0(V_x)B_0(V_x)$. To see that $B \equiv [\{(\iota \otimes \omega)(F\hat{M}) \mid \omega \in B(V_x)_*\}]$ is all of $B_0(V_x)$, it suffices to show that for each non-zero $v \in V_x$, there is $b \in B$ such that $(v_x|bv) \neq 0$, because then $B^* v_x$ is dense in V_x, and $u \odot (b^* v_x) = (\iota \otimes \omega_{u,v_x})(F\hat{M})b \in BB \subset B$. Say on the contrary that $(v_x|bv) = 0$ for all $b \in B$. Then $((\omega_{v,u} \otimes \iota)(\hat{M})^* v_x|u') = (v_x|(\iota \otimes \omega_{u',u})(F\hat{M})v) = 0$, so $(\omega_{v,u} \otimes \iota)(\hat{M})v_x = 0$ and $\hat{M}(v \otimes v_x) = 0$ and $v = 0$. So compact quantum groups are regular, and thus, so are their duals, the discrete quantum groups. ◇

In conclusion, the category of regular locally compact quantum groups is self-dual and contains all compact quantum groups and Kac algebras, and thus also all locally compact groups. We need various reformulations of regularity.

Proposition 16.3.8 *A multiplicative unitary N on V is regular if and only if $[(B_0(V) \otimes \mathbb{C})N(\mathbb{C} \otimes B_0(V))] = B_0(V) \otimes B_0(V)$.*

Proof Combine $(1 \otimes a)FN(1 \otimes b) = F(a \otimes 1)N(1 \otimes b)$ for $a, b \in B_0(V)$ with $(1 \otimes (v \odot v'))FN(1 \otimes (u \odot u')) = (\iota \otimes \omega_{v',u})(FN) \otimes (v \odot u')$. □

Lemma 16.3.9 *Let $N \in B(V \otimes V)$ be a multiplicative unitary and C an involutive unitary on V such that $N' = F(C \otimes 1)N(C \otimes 1)F$ and $N'' = (C \otimes C)N'(C \otimes C)$ are multiplicative unitaries. Then*

$$N_{12}(1 \otimes C \otimes 1)N_{23}(1 \otimes C \otimes 1) = (1 \otimes C \otimes 1)N_{23}(1 \otimes C \otimes 1)N_{13}N_{12}$$

and $N'_{23}N_{12}N_{13} = N_{13}N'_{23}$ and $N''_{12}N_{13} = N_{13}N_{23}N''_{12}$ and $[N_{12}, (FN'F)_{23}] = 0$. If in addition $[N_{12}, N''_{23}] = 0 = [N'_{12}, N_{23}]$ and

$$\{(\omega \otimes \iota)(N), (\iota \otimes \omega)(N) \mid \omega \in B(V)_*\}' = \mathbb{C} = \{(\omega \otimes \iota)(N), (\omega \otimes \iota)(N') \mid \omega \in B(V)_*\}',$$

then $(1 \otimes C)FN'NN''$ is a scalar.

Proof One checks that $F_{13}N'_{12}F_{13} = (1 \otimes C \otimes 1)N_{23}(1 \otimes C \otimes 1)$ and $F_{13}N'_{13}F_{13}$ equals $(C \otimes 1 \otimes 1)N_{13}(C \otimes 1 \otimes 1)$ and $F_{13}N'_{23}F_{13} = (C \otimes 1 \otimes 1)N_{12}(C \otimes 1 \otimes 1)$. Since N' is multiplicative and $[C \otimes 1 \otimes 1, N_{23}] = 0$, one gets the first formula in the lemma. Multiplying this formula from the right and left with F_{23}, gives the second formula in the lemma, while the third formula is obtained by multiplying the first formula from the right and left with F_{12}. From the second formula and multiplicativity of N one gets $N'_{23}N_{23}N_{12} = N_{13}N'_{23}N_{23}$, which shows that $[F_{23}N'_{23}F_{23}, N_{12}] = 0$, showing that the fourth formula in the lemma holds. Letting $M \equiv (1 \otimes C)FN'NN''$, the fourth formula in the lemma and $[N_{12}, N''_{23}] = 0$, yield $[N_{12}, M_{23}] = 0$. Replacing N by N'' in the lemma, then the fourth formula in the lemma gives $[N''_{12}, F_{23}N_{23}N''_{23}] = 0$. Since $M = (C \otimes C)N(C \otimes 1)FNN''$, and $[N'_{12}, N_{23}] = 0$, which gives $[N''_{12}, (1 \otimes C \otimes C)N_{23}(1 \otimes C \otimes 1)] = 0$, we also get$[N''_{12}, M_{23}] = 0$. Hence M commutes with $a \otimes 1$ for a under the commutator on the left hand side in the lemma. Replacing N by N' in this last argument, one sees that $(C\otimes C)N'(C\otimes 1)FN'N$ commutes with $b\otimes 1$ for b under the commutator on the right hand side in the lemma. But $(C\otimes C)N'(C\otimes 1)FN'N = F(C\otimes 1)M(C\otimes 1)F$, so M commutes with $1 \otimes b$. Hence M is a scalar by assumption and because of Takesaki's commutant theorem for tensor products. □

Proposition 16.3.10 *A locally compact quantum group* (W, Δ) *is regular if and only if* $[J_{\hat{x}}\hat{W}_r J_{\hat{x}} W_r] = B_0(V_x)$.

Proof Consider the multiplicative unitary $N = (J_{\hat{x}} \otimes J_{\hat{x}})M(J_{\hat{x}} \otimes J_{\hat{x}})$ satisfying $\Delta = N((\cdot)\otimes 1)N^*$. Clearly N is regular if and only if the fundamental multiplicative unitary M is regular. With $C = J_x J_{\hat{x}}$ the lemma above shows that $(1\otimes C)FN'NN''$ is a scalar, so FN equals $(C \otimes 1)N^*(1 \otimes C)FN^*F(C \otimes 1)$ up to a multiplicative scalar. Now $(W, \Delta)^{\wedge'}$ is a locally compact quantum group with associated reduced C*-algebra $A \equiv J_{\hat{x}}\hat{W}_r J_{\hat{x}}$ since A is gotten by slicing the image of the coproduct with normal functionals. Using $[(A\otimes\mathbb{C})\hat{\Delta}'(A)] = A\otimes A$ and $(J_x\otimes J_{\hat{x}})N(J_x\otimes J_{\hat{x}}) = N^*$, $\hat{\Delta}' = N^*(1 \otimes (\cdot))N$, with $B \equiv \{(\iota \otimes \omega)(FN) \mid \omega \in B(V_x)_*\}$ we get

$$[B] = [\{(\iota \otimes \omega)((1 \otimes J_x a J_x)FN(1 \otimes J_{\hat{x}} b J_{\hat{x}})) \mid a, b \in A, \omega \in B(V_x)\}]$$

which again equals

$$[\{(\iota\otimes\omega)(F(J_x\otimes J_{\hat{x}})(a\otimes 1)\hat{\Delta}'(b)N^*(J_x\otimes J_{\hat{x}})) \mid a, b \in A, \omega \in B(V_x)_*\}] = [J_{\hat{x}} A J_{\hat{x}} B].$$

As $J_x A J_x = A$ we thus get $[CBC] = [ACBC]$. Combining this equality with the identity we got from the lemma above, and using $N \in M(A \otimes B_0(V_x))$, we get

$$[CBC] = [\{(\iota\otimes\omega)((a\otimes b)N^*(1\otimes C)FN^*F) \mid a \in A, b \in B_0(V_x), \omega \in B(V_x)_*\}],$$

which equals $[AW_r]$. □

Remark 16.3.11 Dualizing and using that $V_{\hat{x}} = V_x$ shows that a locally quantum group is regular if and only if $B_0(V_x) = [J_x W_r J_x \hat{W}_r]$. This shows that the biduality theorem above for crossed products of continuous actions of regular locally compact quantum groups hold. In the general von Neumann case, we have $B(V_x) = (W \cap \hat{W}')' = (W' \cup \hat{W})''$ and the latter equals $[W'\hat{W}]$ by the remark at the beginning of this section. So regularity in this von Neumann sense always holds. Using this, the proof of the biduality theorem in the continuous case also works for the von Neumann case. ◇

Remark 16.3.12 For any locally compact quantum group (W, Δ) we can consider Δ_r^{op} as a continuous coaction of $(W_r, \Delta_r)^{\text{op}}$ on W_r, and then

$$W_r \ltimes_{\Delta_r^{\text{op}}} W_r = [((J_{\hat{x}} \hat{W}_r J_{\hat{x}}) \otimes 1)\Delta_r^{\text{op}}(W_r)]$$

by Proposition 16.3.3 and the fact that $(W_r, \Delta_r)^{\text{op}\wedge\text{op}}$ has $J_{\hat{x}} \hat{W}_r J_{\hat{x}}$ as associated reduced C*-algebra. Using $\Delta_r^{\text{op}} = M_{21}^*((\cdot) \otimes 1)M_{21}$, we thus see that $M_{21}(\cdot)M_{21}^*$ is a $*$-isomorphism from $W_r \ltimes_{\Delta_r^{\text{op}}} W_r$ to $[J_{\hat{x}} \hat{W}_r J_{\hat{x}} W_r]$. ◇

Proposition 16.3.13 *For a regular locally compact group (W, Δ) we have $[(B_0(V_x) \otimes \mathbb{C})N(\mathbb{C} \otimes W_r)] = W_r \otimes B_0(V_x)$, where $N = (J_{\hat{x}} \otimes J_{\hat{x}})M(J_{\hat{x}} \otimes J_{\hat{x}})$ and M is the fundamental multiplicative unitary of (W, Δ).*

Proof Since by the pentagonal equation for N we have

$$(b \otimes 1)N(1 \otimes (\omega(a\cdot) \otimes \iota)(N)) = (\iota \otimes \omega \otimes \iota)(((b \otimes a)N^* \otimes 1)N_{23}N_{12})$$

for $a, b \in B_0(V_x)$ and $\omega \in B(V_x)_*$, and since the closed span of the elements $(b \otimes a)N^*$ is $B_0(V_x) \otimes B_0 V_x)$, it suffices to check that elements of the form

$$(\iota \otimes \omega(\cdot b) \otimes \iota)((a \otimes 1 \otimes 1)N_{23}N_{12}) = (\iota \otimes \omega \otimes \iota)(N_{23}((a \otimes 1)N(1 \otimes b))_{12})$$

have $W_r \otimes B_0(V_x)$ as closed span, but this is clear from Proposition 16.3.8. □

For regular quantum groups we have a nice correspondence between coactions and continuous coactions.

Proposition 16.3.14 *If α is a coaction of a regular locally compact quantum group (W, Δ) on a von Neumann algebra U, and if for a C*-subalgebra A of U, we have $A_\alpha \equiv [\{(\omega \otimes \iota)\alpha(A) \mid \omega \in B(V_x)_*\}] \subset A$, then A_α is a C*-algebra and α restricts to a continuous coaction of (W, Δ) on it.*

Proof Now $AA_\alpha \subset A$, so $A_\alpha \supset [\{(\omega \otimes \iota)\alpha(A(\eta \otimes \iota)\alpha(A)) \mid \omega, \eta \in B(V_x)_*\}]$, and the latter space equals $[\{(\eta \otimes \omega \otimes \iota)(\alpha(A)_{23}N_{12}\alpha(A)_{13}) \mid \omega, \eta \in B(V_x)_*\}]$, which by Proposition 16.3.8 again equals the same closed span, but with N_{12} removed. Hence A_α is a C*-algebra. As $(A_\alpha)_\alpha = A_\alpha$, we may assume $A = A_\alpha$. Then

$$[\alpha(A)(W_r \otimes \mathbb{C})] = [\{\alpha((\omega \otimes \iota)\alpha(A))(W_r \otimes \mathbb{C}) \mid \omega \in B(V_x)_*\}],$$

which equals $[\{(\omega \otimes \iota \otimes \iota)(N_{12}\alpha(A)_{13}(\mathbb{C} \otimes W_r \otimes \mathbb{C})) \mid \omega \in B(V_x)_*\}]$, and by the previous proposition, this equals the same expression, but with N_{12} removed. Hence $[\alpha(A)(W_r \otimes \mathbb{C})] = W_r \otimes A$, which also proves $\alpha(A) \subset M(W_r \otimes A)$. □

Corollary 16.3.15 *Any coaction α of a regular locally compact quantum group (W, Δ) on a von Neumann algebra U is uniquely determined by the continuous coaction of (W, Δ) obtained by restricting α to the σ-strongly*-dense C*-subalgebra $[\{(\omega \otimes \iota)\alpha(U) \mid \omega \in B(V_x)_*\}]$ of U.*

Proof Let $A = U$ in the proposition above, and then use that every coaction has property A to get denseness and thus uniqueness. □

The following is an easy adaptation of a specialization of Theorem 16.1.12.

Theorem 16.3.16 *Let (W, Δ) be a regular locally compact quantum group, and let β be a continuous coaction of $(\hat{W}, \hat{\Delta}^{\mathrm{op}})$ on U. Suppose we have a unitary $Z \in M(W_r \otimes U)$ such that $(\iota \otimes \beta)(Z) = M_{12}Z_{13}$ and $(\Delta \otimes \iota)(Z) = Z_{13}Z_{23}$, and consider the $*$-homomorphism $f = Z^*\beta(\cdot)Z : U \to M(B_0(V_x) \otimes U)$. Then*

$$A \equiv [\{(\omega \otimes \iota)f(U) \mid \omega \in B(V_x)_*\}] \subset M(U)$$

is a C^*-algebra, and $\alpha = Z^*(1 \otimes (\cdot))Z$ is a continuous coaction of (W, Δ) on A, whereas f defines an isomorphism $U \to W_r \ltimes_\alpha A$ satisfying $\hat{\alpha} f = (\iota \otimes f)\beta$.

Exercises

For Sect. 16.1

1. Compare the Landstad theory for dual coactions with that of dual actions.
2. Suppose (W, Δ) is a finite dimensional locally compact quantum group, with a coaction α on a finite dimensional von Neumann algebra U of the opposite dual quantum group $(\hat{W}, \hat{\Delta}^{\mathrm{op}})$, and make the identification $W_* = \hat{W}$. Show that then there exists a cocycle coaction (β, X) of (W, Δ) on U^α with dual coaction isomorphic to α if and only if there exists a linear map $f : \hat{W} \to U$ such that $(\iota \otimes f)\hat{\Delta}^{\mathrm{op}} = \alpha f$ and $m(f^* \otimes f)\hat{\Delta}^{\mathrm{op}}(\cdot) = \hat{\varepsilon}(\cdot)1 = m(f \otimes f^*)\hat{\Delta}^{\mathrm{op}}(\cdot)$, where m is the multiplication map on U and $f^* = f\hat{S}((\cdot)^*)^*$. Can this be extended to the setting of general coactions of locally compact quantum groups?

For Sect. 16.2

1. Consider a cocycle bicrossed product (W, Δ). Prove that the left invariant faithful semifinite normal weight with semicyclic representation of $(\hat{W}, \hat{\Delta})$ considered as a cocycle bicrossed product is the same as the left invariant faithful semifinite normal weight with semicyclic representation obtained by considering $(\hat{W}, \hat{\Delta})$ as the quantum group dual to (W, Δ).
2. Let (W_i, Δ_i) be locally compact quantum groups with fundamental multiplicative unitaries M_i and left invariant faithful semifinite normal weights x_i. Let $\beta : W_1 \to \hat{W}_2$ be a normal $*$-homomorphism such that $\hat{\Delta}_2\beta = (\beta \otimes \beta)\Delta_1$, and consider the unitary $Z = (J_{x_1} \otimes J_{\hat{x}_2})(\iota \otimes \beta)(\hat{M}_2^*)(J_{x_1} \otimes J_{\hat{x}_2})$ on $V_{x_1} \otimes V_{x_2}$. Show that $\theta = Z(1 \otimes (\cdot))Z^*$ defines a coaction of $(\hat{W}_1, \hat{\Delta}_1^{\mathrm{op}})$ on W_2 satisfying $(\theta \otimes \iota)(M_2) = (\iota \otimes \beta)(\hat{M}_1)_{13}(M_2)_{23}$ and $(\iota \otimes \Delta_2^{\mathrm{op}})\theta = (\theta \otimes \iota)\Delta_2^{\mathrm{op}}$.
3. Let (W_i, Δ_i) be locally compact quantum groups. Call

$$(W_2, \Delta_2) \xrightarrow{\alpha} (W_3, \Delta_3) \xrightarrow{\beta} (\hat{W}_1, \hat{\Delta}_1)$$

a short exact sequence if $\alpha: W_2 \to W_3$ and $\beta: W_1 \to \hat{W}_3$ are faithful normal $*$-homomorphisms satisfying $\Delta_3\alpha = (\alpha \otimes \alpha)\Delta_2$ and $\hat{\Delta}_3\beta = (\beta \otimes \beta)\Delta_1$ and $\alpha(W_2) = W_3^\theta$, where θ is associated to β as in the previous exercise. How is this definition related to short exact sequences of abelian locally compact groups and their homomorphisms? In the case when all the algebras involved are finite dimensional, prove that

$$(W_1, \Delta_1) \xrightarrow{\beta} (\hat{W}_3, \hat{\Delta}_3) \xrightarrow{\alpha} (\hat{W}_2, \hat{\Delta}_2)$$

is also a short exact sequence. Can you prove it in the general case?

4. Let (W, Δ) be a cocycle bicrossed product of locally compact quantum groups (W_1, Δ_1) and (W_2, Δ_2) with associated α and β. Show that

$$(W_2, \Delta_2) \xrightarrow{\alpha} (W, \Delta) \xrightarrow{\beta} (\hat{W}_1, \hat{\Delta}_1)$$

is a short exact sequence in the sense of the previous exercise. When all the algebras involved are finite dimensional, can you prove a converse of this result, providing a cocycle matching from a short exact sequence satisfying some additional property? Say (W', Δ') is a cocycle bicrossed product of locally compact quantum groups (W_1, Δ_1) and (W_2, Δ_2) with associated α' and β' for some other cocycle matching. Show that the corresponding short exact sequences are isomorphic in a natural way by some isomorphism π of the bicrossed products that respects the α's and β's if and only if it is implemented by a unitary in $W_1 \bar{\otimes} W_2$ that also implements the canonical dual isomorphism $\hat{\pi}$, and that relates the cocycle matchings in a fairly natural way.

5. Let (W, Δ) be a cocycle bicrossed product of locally compact quantum groups (W_1, Δ_1) and (W_2, Δ_2). Show that $(\hat{W}, \hat{\Delta})$ is coamenable if and only if (W_1, Δ_1) and $(\hat{W}_2, \hat{\Delta}_2)$ are coamenable. Prove that (W, Δ) is amenable if and only if $(\hat{W}_1, \hat{\Delta}_1)$ and (W_2, Δ_2) are amenable.

For Sect. 16.3

1. We say a multiplicative unitary N on a Hilbert space V is semiregular if the closed span of $\{(\iota \otimes \omega)(FN) \mid \omega \in B(V)_*\}$ contains $B_0(V)$, and that a locally compact quantum group (W, Δ) is semiregular if its fundamental multiplicative unitary M is semiregular. Prove that (W, Δ) is semiregular if and only if the closed span of $\{(\omega\otimes\iota)(M^*(\mathbb{C}\otimes\hat{W}_r)M) \mid \omega \in B(V_x)_*\}$ intersects $\hat{W}_r$ non-trivially. Show that then $\hat{W}_r$ is contained in the closed span, and that it is the whole closed span if and only if (W, Δ) is regular.
2. Show that the naturally associated bicrossed product quantum group of locally compact groups G_1 and G_2 is regular if and only if the matching $G_1 \times G_2 \to G$ is onto, while the bicrossed product associated to any matched pair is always semiregular. Allowing for 'matchings' onto non-open sets, can you still produce a 'bicrossed product' as a locally compact quantum group, and then show that it

is semiregular exactly when the image of the 'matching' is open? Do you see a way to construct managable multiplicative unitaries that are not semiregular?

3. Prove that a locally compact quantum group (W, Δ) is semiregular if and only if the closed span of $\{(\omega \otimes \iota)\alpha(U) \mid \omega \in B(V_x)_*\}$ is a C*-algebra for any C*-algebra U and non-degenerate $*$-homomorphism $\alpha\colon U \to M(W_r \otimes U)$ satisfying $(\iota \otimes \alpha)\alpha = (\Delta_r \otimes \iota)\alpha$. Conclude that this yields two different notions of 'continuous coactions' when the quantum group is not semiregular.

Chapter 17
Basic Construction and Quantum Groups

This technical chapter deals with the beautiful relation between coactions of locally compact quantum groups and subfactor theory. We show in detail how a repeated crossed product formation, starting with a certain type of coaction, yields a tower of factors contained in each other, and conversely, how the quantum group and coaction can be recovered from a certain type of abstract tower of subfactors. This is accomplished using relative modular theory of weights and spatial derivatives.

Recommended literature for this chapter is [12, 14].

17.1 Basic Construction for Crossed Products of Quantum Groups

Given an inclusion $U_0 \subset U_1$ of von Neumann algebras and a faithful semifinite normal weight z on U_1, we get by representing U_1 on V_z, an inclusion $U_0 \subset U_1 = J_z U_1' J_z \subset U_2 \equiv J_z U_0' J_z$ of three von Neumann algebras called the *basic construction*. Continuing the same way, by choosing a faithful semifinite normal weight on U_2, we get an infinite chain $U_0 \subset U_1 \subset U_2 \subset U_3 \subset \cdots$ of von Neumann algebras called the *Jones tower*. In this section we consider the basic construction for the inclusion $U_0 \equiv U^\alpha \subset U_1 \equiv U$, given a coaction α of a locally compact quantum group (W, Δ) on U, and study to what extend $U_2 = W \ltimes_\alpha U$.

We will use spatial derivatives, and phrase the theory more in the spirit of operator valued weights. So say $U \subset B(V)$ is a von Neumann algebra with faithful semifinite normal weights x and y on U and U', respectively. For $v \in V$ let $R^y(v)q_y(u) = uv$ for all $u \in N_y$, so $R^y(v) = R^y_v$ whenever v is y-bounded. Define $T^y(v)$ in the extended positive part of $B(V)$ by letting $T^y(v)(\omega_{v'})$ be $\|R^y(v)^* v'\|^2$ if $v' \in D(R^y(v)^*)$ and otherwise infinite, so $T^y(v) = R^y(v)R^y(v)^* + \infty P$, where P is the orthogonal projection onto the orthogonal complement of $D(R^y(v)^*)$. Then

L. Tuset, *Analysis and Quantum Groups*,
https://doi.org/10.1007/978-3-031-07246-8_17

$T^y(v)$ belongs to $\bar{U}_+$, and $\omega_v(dx/dy) = xT^y(v)$, where we have extended x to $\bar{U}_+$ on the right hand side. We shall also be using Theorem 10.13.6 frequently.

Theorem 17.1.1 *Let α be a coaction of a locally compact quantum group (W, Δ) on a von Neumann algebra U in the semicyclic representation of a faithful semifinite normal weight z. Then $U_2 \equiv J_z(U^\alpha)' J_z = (U \cup \{(\omega \otimes \iota)(K) \,|\, \omega \in W_*\})''$, where $K = J_{\tilde{z}}(J_{\hat{x}} \otimes J_z)$ is the canonical unitary implementation of α with respect to z. More profoundly, the coaction α is integrable if and only if there exists a normal $*$-epimorphism $f : W \ltimes_\alpha U \to U_2$ such that $f\alpha = \iota$ on U and $f((\omega \otimes \iota)(M) \otimes 1) = (\omega \otimes \iota)(K^*)$ for $\omega \in W_*$.*

Proof Since M is managable and $(\Delta \otimes \iota)(K) = K_{23}K_{13}$, the norm closure of $\{(\omega \otimes \iota)(K) \,|\, \omega \in W_*\}$ is a C*-algebra, so the weak closure is a von Neumann algebra. Those elements in U that commute with the elements of this von Neumann algebra constitute U^α, and we have proved the first statement since $J_z U J_z = U'$ and $(J_{\hat{x}} \otimes J_z)K = K^*(J_{\hat{x}} \otimes J_z)$.

Suppose we have an f as described. Pick a central orthogonal projection P in $W \ltimes_\alpha U$ such that $\mathrm{Ker}(f) = (W \ltimes_\alpha U)(1 - P)$, and let f_P be the restriction of f to the orthogonal complement, so f_P is a $*$-isomorphism. Let e be a faithful semifinite normal weight on $W \ltimes_\alpha U$. As its modular group fixes the central P, it restricts to a faithful semifinite normal weight e_P on $(W \ltimes_\alpha U)P$ with modular group given by restriction.

Let $\tilde{a}$ be the dual weight of a faithful semifinite normal weight a on U, and consider the faithful semifinite normal weight $\tilde{a}_P f_P^{-1}$ on U_2. Then a trivial computation shows that its modular group coincides with that of a on U. Moreover, the cocycle derivative of a with respect to another faithful semifinite normal weight b on $(W \ltimes_\alpha U)P$ coincides with the cocycle derivative of $\tilde{a}_P f_P^{-1}$ with respect to to the corresponding weight for b. By Theorem 10.13.3 there is a unique faithful semifinite normal operator valued weight E_2 from U_2 to U such that $aE_2 f_P = \tilde{a}_P$. By Theorem 10.13.6 there is a unique faithful semifinite normal operator valued weight E_1 from U to U^α such that $d(aE_2)/d(b_r) = da/d((bE_1)_r)$ for faithful semifinite normal weights a and b on U and U^α, respectively.

Chose a faithful semifinite normal weight z_0 on U^α, and set $z_1 = z_0 E_1$ and $z_2 = z_1 E_2$. If we change the weight z in the proposition, the chain $U^\alpha \subset U \subset U_2$ will change into a unitarily equivalent chain implemented by the unitary intertwining the two semicyclic representations of U, and by Proposition 15.4.9, this unitary also intertwines the two implementations of α, so f is changed accordingly. Thus we may assume that $z = z_1$. Since $d(z_2)/d((z_0)_r) = d(z_1)/d((z_1)_r)$ we get $d((z_0)_r)/d(z_2) = d((z_1)_r)/d(z_1) = \Delta_z^{-1}$ by Theorem 10.10.1, where we in the last step used that U is in the semicyclic representation of z_1. To compute the left hand side we need a convenient semicyclic representation for z_2. Now $z_2 f_P = \tilde{z}_P$, so if $(q_{\tilde{z}}, \iota)$ is the usual semicyclic representation of $W \ltimes_\alpha U$ on $V_x \otimes V_z$ with respect to $\tilde{z}$, then with $q_{z_2} = q_{\tilde{z}} f_P^{-1}$ on N_{z_2}, we see that (q_{z_2}, f_P^{-1}) is a semicyclic representation of U_2 on $P(V_x \otimes V_z)$ with respect to z_2. For $a \in N_{\hat{x}}$ and $u \in N_z$, we then have $q_{z_2}(f(a \otimes 1)u) = P(q_{\hat{x}}(a) \otimes q_z(u))$. Let b be in the Tomita algebra of z.

Then

$$T^{z_2}(q_z\sigma^z_{-i/2}(b))((z_0)_r) = \omega_{q_z\sigma^z_{-i/2}(b)}(d((z_0)_r)/d(z_2)) = \omega_{q_z\sigma^z_{-i/2}(b)}(\Delta_z^{-1}) = z(b^*b)$$

is finite. Pick $v_i \in V_z$ such that $(z_0)_r = \sum \omega_{v_i}$ on the positive part of $J_zU^\alpha J_z$. Then the expression gotten by replacing $(z_0)_r$ with ω_{v_i} is finite for each i, and equals $\|R^{z_2}(q_z\sigma^z_{-i/2}(b))^*v_i\|^2$, where the element we take the norm of belongs to $P(V_x \otimes V_z)$. For $\omega \in K \subset W_*$ and $u \in N_z$, we get by definitions that

$$(R^{z_2}(q_z\sigma^z_{-i/2}(b))^*v_i|q_{\hat{x}}(\omega \otimes \iota)(M) \otimes q_z(u)) = \bar{\omega}(\iota \otimes \omega_{v_i,q_z(u)})((1 \otimes J_zbJ_z)K),$$

and by continuity, we may replace $q_z(u)$ by any vector $v \in V_z$.

Now if there to $w \in W$ is $v' \in V_x$ such that $\omega(w^*) = (p(\omega)|v')$ for all $\omega \in K$, then $q_x(w) = v'$. Indeed, one see that $p(\omega(c\cdot)) = J_x\sigma^x_{i/2}(c^*)J_xp(\omega)$ for c in the Tomita algebra of x, so for a net $\{c_j\}$ in the Tomita algebra of x with $\sigma^x_z(c_j) \to 1$ σ-strongly* for all $z \in \mathbb{C}$, we get $q_x(w) = \lim q_x(wc_j) = \lim J_x\sigma^x_{i/2}(c_j)^*J_xv' = v'$.

Combining this with the last formula in the previous paragraph, we get

$$q_x(\iota \otimes \omega_{v_i,v})((1 \otimes J_zbJ_z)K) = (1 \otimes f_v^*)R^{z_2}(q_z\sigma^z_{-i/2}(b))^*v_i.$$

For any orthonormal basis $\{e_j\}$ of V_z, we therefore get

$$\|R^{z_2}(q_z\sigma^z_{-i/2}(b))^*v_i\|^2 = \sum \|(1 \otimes f_{e_j}^*)R^{z_2}(q_z\sigma^z_{-i/2}(b))^*v_i\|^2,$$

which equals

$$x(\iota\otimes\omega_{v_i})(K^*(1\otimes J_zb^*bJ_z)K) = x(J_{\hat{x}}(\iota\otimes\omega_{J_zv_i})\alpha(b^*b)J_{\hat{x}}) = (y\otimes\iota)\alpha(b^*b)(\omega_{J_zv_i}).$$

Summing over i and comparing with the formula above involving T^{z_2}, we get $z(b^*b) = z_0(y \otimes \iota)\alpha(b^*b)$, so $z_0(y \otimes \iota)\alpha$ is semifinite, and thus, so is $(y \otimes \iota)\alpha$.

Asume now that α is integrable, and consider U in the semicyclic representation of the faithful semifinite normal weight $z = z_0(y \otimes \iota)\alpha$ for a faithful semifinite normal weight z_0 on U^α, and write $z_0 = \sum \omega_{v_i}$ for vectors $v_i \in V_z$. Let $q\colon N_y \to V_x$ be the canonical map in the semicyclic representation (q, ι) for y. For $c \in N_{y\otimes\iota}$ define $(q \otimes \iota)(c) \in B(V_z, V_x \otimes V_z)$ by $(q \otimes \iota)(c)q_z(u) = (q \otimes q_z)(c(1 \otimes u))$ for $u \in N_z$. Then $(q\otimes\iota)(c)^*(q\otimes\iota)(c) = (y\otimes\iota)(c^*c)$. Let $E = (y\otimes\iota)\alpha$. For $u \in N_E \cap N_z$ the map $q_z(u) \mapsto \oplus(q \otimes \iota)\alpha(u)v_i$ well-defines an isometry $L\colon V_z \to \oplus_i(V_x \otimes V_z)$ because $\sum \|(q \otimes \iota)\alpha(c)v_i\|^2 = z(c^*c)$ and $N_E \cap N_z$ is a σ-strong*-norm core for q_z. We claim that the range of L is invariant under $W \ltimes_\alpha U$ when represented on $\oplus_i(V_x \otimes V_z)$. Obviously $\alpha(u')Lq_z(u) = Lu'q_z(u)$ for $u' \in U$. As for invariance under $\hat{W} \otimes \mathbb{C}$, observe first that $q(\omega_{h^{1/2}v,v'} \otimes \iota)\Delta(w) = (\omega_{v,v'} \otimes \iota)(M^*)q(w)$ for $w \in N_y$ and $v \in D(h^{1/2})$ and $v' \in V_x$. Hence

$$(q \otimes \iota)(\omega_{h^{1/2}v,v'} \otimes \iota \otimes \iota)(\Delta \otimes \iota)(c) = ((\omega_{v,v'} \otimes \iota)(M^*) \otimes 1)(q \otimes \iota)(c).$$

Thus by definitions $((\omega_{v,v'} \otimes \iota)(M^*) \otimes 1)Lq_z(u) = L((\omega_{v,v'} \otimes \iota)(K)q_z(u)$, proving our claim. Then $f = L^*(\cdot)L$ will by the previous calculations, and the first paragraph of this proof, evidently have the correct properties. □

Corollary 17.1.2 *Suppose we have a coaction α of a locally compact quantum group (W, Δ) on a von Neumann algebra U, which is in the semicyclic representation of a faithful semifinite normal weight z. Then there is a $*$-epimorphism $f: W \ltimes_\alpha U \to U_2 \equiv J_z(U^\alpha)' J_z$ that restricts to isomorphisms $\alpha(U) \to U$ and $\mathbb{C} \otimes U^\alpha \to U^\alpha$ if and only if α is cocycle equivalent to an integrable coaction β with the same fixed point algebra.*

Proof For the forward direction pick a unitary $u \in B(V_z)$ such that $f\alpha = u(\cdot)u^*$ on U and $uJ_z = J_z u$, and define $f': W \ltimes_\alpha U \to B(V_z)$ by $f' = u^* f(\cdot)u$. Then $f'\alpha = \iota$ on U, and f' will otherwise have the same properties as f, so we may from the outset assume that $f\alpha = \iota$ on U. Consider the unitaries in $W\bar{\otimes}B(V_z)$ given by $X = (J_{\hat{x}} \otimes J_z)(\iota \otimes f)(M \otimes 1)(J_{\hat{x}} \otimes J_z)$ and $Y = XK^*$. Then we see that $(J_{\hat{x}} \otimes J_z)Y(J_{\hat{x}} \otimes J_z) = (\iota \otimes f)(M \otimes 1)K$, and we can easily check that the right hand side commutes with $1 \otimes u$ for $u \in U$, so it belongs to $W\bar{\otimes}U'$. Thus $Y \in W\bar{\otimes}U$, and a little standard calculation shows that Y is an α-cocycle. Then the coaction $\beta = Y\alpha(\cdot)Y^* = X(1 \otimes (\cdot))X^*$ of (W, Δ) on U has the same fixed point algebra as α because the σ-strong*-closure of $\{(\omega \otimes \iota)(X) \,|\, \omega \in W_*\}$ equals $J_z f(\hat{W} \otimes \mathbb{C})J_z$. We claim β is integrable. By Proposition 15.4.10 the unitary implementation $K(\beta)$ of β equals $Y(\iota \otimes f)(M^* \otimes 1)$, and from the proof of the same proposition, we know that Y implements an isomorphism between $W \ltimes_\alpha U$ and $W \ltimes_\beta U$. So we can define a $*$-epimorphism $g: W \ltimes_\beta U \to U_2$ by $g = f(Y^*(\cdot)Y)$, so $g\alpha = \iota$ on U. From the cocycle identity for Y one sees that $(\iota \otimes g)(M^* \otimes 1) = K(\beta)$, and we thus conclude from the previous theorem that β is integrable.

Conversely, if Y is an α-cocycle such that $\beta = Y\alpha(\cdot)Y^*$ is integrable, then from the proof of Proposition 15.4.10, we know that $h = Y(\cdot)Y^*: W \ltimes_\alpha U \to W \ltimes_\alpha U$ is an isomorphism such that $h\alpha = \beta$ on U. By the previous theorem we thus get $f': W \ltimes_\beta U \to J_z(U^\beta)' J_z$ such that $f'\beta = \iota$ on U. Then $f = f'h$ will do. □

A coaction α of a locally compact quantum group (W, Δ) on a von Neumann algebra U is *outer* when $(W \ltimes_\alpha U) \cap \alpha(U)' = \mathbb{C}$.

Corollary 17.1.3 *Suppose we have an outer coaction α of a locally compact quantum group (W, Δ) on a von Neumann algebra U, which is in the semicyclic representation of a faifthful semifinite normal weight z. Then the chain $\mathbb{C} \otimes U^\alpha \subset \alpha(U) \subset W \ltimes_\alpha U$ of inclusions is isomorphic to the basic construction $U^\alpha \subset U \subset U_2$ if and only if α is integrable.*

Proof When α is integrable the map f from the theorem above is injective since $W \ltimes_\alpha U$ is a factor and doesn't allow for a central orthogonal projection onto a possible non-trivial kernel. Conversely, there exists by the corollary above an integrable coaction $\beta = Y\alpha(\cdot)Y^*$ for some α-cocycle Y such that $U^\beta = U^\alpha$. For u in this set, we have $1 \otimes u = \beta(u) = Y\alpha(u)Y^* = Y(1 \otimes u)Y^*$, so $Y \in W \otimes (U \cap (U^\alpha)')$. But by assumption $U_2 \cap U' = \mathbb{C}$, so $U \cap (U^\alpha)' = J_z(U_2 \cap U')J_z = \mathbb{C}$, and $Y = w \otimes 1$ for some $w \in W$. The cocycle identity then means that $\Delta(w) = w \otimes w$. As $y(w^*(\cdot)w)$ is a right invariant faithful semifinite normal weight on W, we get by uniqueness of such weights, that there is a scalar $c > 0$ such that $y(w^*(\cdot)w) = cy$ on W_+. Hence $(y \otimes \iota)\alpha = c(y \otimes \iota)\beta$ is semifinite. □

Suppose $U_0 \subset U_1$ is an inclusion of von Neumann algebras with U_1 in the semicyclic representation of a faithful semifinite normal weight z, and let $U_0 \subset U_1 \subset U_2$ be the basic construction. Let E_1 be a faithful semifinite normal operator valued weight from U_1 to U_0, and let E_2 be the faithful semifinite normal operator valued weight from U_2 to U_1 uniquely determined by $d(aE_2)/d(b_r) = da/d((bE_1)_r)$ for faithful semifinite normal weights a and b on U_1 and U_0, respectively The existence of the unique E_2 is guaranteed by Theorem 10.13.6. Continuing this way one can construct faithful semifinite normal operator valued weights E_i from U_i to U_{i-1} anywhere in the Jones tower. We say E_1 is *regular* when the restrictions of E_2 to $U_2 \cap U_0'$ and E_3 to $U_3 \cap U_1'$ are both semifinite.

Theorem 17.1.4 *Say α is an integrable coaction of a locally compact quantum group (W, Δ) on U, and assume that the map f in the previous theorem is injective. Let $E_1 = (y \otimes \iota)\alpha$ be the faithful semifinite normal operator valued weight from U to U^α, and let E_2 be the faithful semifinite normal operator valued weight from U_2 to U associated as above to the basic construction $U^\alpha \subset U \subset U_2$. Then $f^{-1}E_2 f$ is the canonical faithful semifinite normal operator valued weight $E \equiv (\hat{x} \otimes \iota \otimes \iota)\hat{\alpha}$ from $W \ltimes_\alpha U$ to $\alpha(U)$.*

Proof Notice that the dual weight $\tilde{z}$ on $W \ltimes_\alpha U$ of a faithful semifinite normal weight z on U is by definition $z\alpha^{-1}E$. Pick a faithful semifinite normal weight z_0 on U^α, put $z = z_0 E_1$ and $z_2 = \tilde{z} f^{-1}$. We claim that $z_2 = zE_2$. As in the proof of the previous theorem we may assume that U is in the semicyclic representation of z. Let $(q_{\tilde{z}}, \iota)$ be the usual semicyclic representation of the crossed product with respect to the dual weight. Put $q_{z_2} = q_{\tilde{z}} f^{-1}$. For any u in the Tomita algebra of z we assert that $\omega_{q_z\sigma^x_{-i/2}(u)}(d((z_0)_r)/dz_2) = z_0(y \otimes \iota)\alpha(u^*u)$. Indeed, picking a family of vectors $v_i \in V_z$ such that $z_0 = \sum \omega_{J_z v_i}$, we have $(y \otimes \iota)\alpha(u^*u)(\omega_{J_z v_i}) \leq z(u^*u) < \infty$, so $x(\iota \otimes \omega_{v_i})(K^*(1 \otimes J_z u^* u J_z)K) < \infty$ and we can define $v \in V_x \otimes V_z$ for any orthonormal basis $\{e_j\}$ of V_z by $v = \sum_j q_x(\iota \otimes \omega_{v_i, e_j})((1 \otimes J_z u J_z)K) \otimes e_j$. Then one checks that $(v_i | R^{z_2}(q_z\sigma^z_{-i/2}(u))q_{z_2}((\omega \otimes \iota)(K^*)u')) = (v | q_{z_2}((\omega \otimes \iota)(K^*)u'))$ for $u' \in N_z$ and $\omega \in K \subset W_*$. By continuity and density the same identity holds with $(\omega \otimes \iota)(K^*)u'$ replaced by any element in N_z. Hence $T^{z_2}(q_z\sigma^z_{-i/2}(u))(\omega_{v_i}) = \|v\|^2 = (y \otimes \iota)\alpha(u^*u)(\omega_{J_z v_i})$, which upon summing over i proves our assertion. As $z(u^*u) = \omega_{q_z\sigma^x_{-i/2}(u)}(\Delta_z^{-1})$, the assertion shows that $d((z_0)_r)/dz_2$ equals Δ_z^{-1} provided they commute strongly. In that case $d(zE_2)/d((z_0)_r) = dz/d(z_r) = \Delta_z = dz_2/d((z_0)_r)$, and $zE_2 = z_2$ as claimed. But then $zfE = \tilde{z} = z_2 f = zE_2 f$, and $fE = E_2 f$. It remains to show that strong commutation holds. It clearly suffices to show that $(z_0)_r$ and z_2 are invariant under $\Delta_z^{it}(\cdot)\Delta_z^{-it}$. The case of $(z_0)_r$ is immediate from $\Delta_z^{it} J_z u J_z \Delta_z^{-it} = J_z \sigma_t^z(u) J_z = J_z \sigma_t^{z_0}(u) J_z \in J_z U^\alpha J_z$ for $u \in U^\alpha$. Clearly the automorphism at hand leaves U_2 invariant, and for $u \in U$, we have $\Delta_z^{it} f\alpha(u)\Delta_z^{-it} = \sigma_t^z(u) = f\alpha\sigma_t^z(u) = f\sigma_t^{\tilde{z}}\alpha(u)$. As in the proof of the previous theorem, we may use Proposition 15.4.11. We can then easily check that $\Delta_z^{it} f((\omega\otimes\iota)(M)\otimes 1)\Delta_z^{-it} = f\sigma_{it}^{\tilde z}((\omega\otimes\iota)(M)\otimes 1)$ for $\omega\in B(V_x)_*$, where we have used that $(h\Delta_{\hat x})^{it}\otimes(h\Delta_{\hat x})^{it}$ commutes with M and that $\Delta_{\tilde z}^{it} = (h\Delta_{\hat x})^{it}\otimes\Delta_z^{it}$ by the proof of the same proposition. In conclusion $\Delta_z^{it} f(\cdot)\Delta_z^{-it} = f\sigma_t^{\tilde z}$ and the required invariance for z_2 holds. □

Later we shall see that injectivity of f is redundant.

Proposition 17.1.5 *A coaction α of a locally compact quantum group (W,Δ) on U is* semidual *if there exists a unitary $Y\in B(V_x)\bar\otimes U$ such that $(\iota\otimes\alpha)(Y) = Y_{13}N_{12}^*$, where $N = (J_{\hat x}\otimes J_{\hat x})\hat M(J_{\hat x}\otimes J_{\hat x})$. Every dual coaction is semidual. Semidual coactions are integrable and the map f in Theorem 17.1.1 is injective for them.*

Proof For the first claim, note that $N = M$ for $\hat\alpha$, so we may use $Y = M^*\otimes 1$. For the second statement, say we have Y for α with described properties. We see that the automorphism $Y(\cdot)Y^*$ on $B(V_x)\bar\otimes U$ respects the two coactions β and γ in the biduality theorem for crossed products, and the latter is again isomorphic to the bidual coaction, so β, and hence α is integrable; of course we are here using that α satisfies property A. Next, we may assume that U is in the semicyclic representation of a faithful semifinite normal weight z. Let $X = (J_{\hat x}\otimes J_z)Y(J_{\hat x}\otimes J_z)$ and consider $g = KX^*(1\otimes(\cdot))XK^*\colon U_2\equiv J_z(U^\alpha)'J_z\to B(V_x\otimes V_z)$, so $g=\alpha$ on U. Multiplying the equation $K_{23}Y_{13}K_{23}^* = Y_{13}N_{12}^*$ from both sides by $J_{\hat x}\otimes J_{\hat x}\otimes J_z$ gives $K_{23}^*X_{13}K_{23} = X_{13}M_{21}$, which upon flipping the first two legs yields $X_{23}^*K_{13}^*X_{23} = M_{12}K_{13}^*$. Therefore, using that K satisfies $(\Delta\otimes\iota)(K) = K_{23}K_{13}$, and that M implements Δ, we get $K_{23}X_{23}^*K_{13}^*X_{23}K_{23}^* = M_{12}$, so $g(\omega\otimes\iota)(K^*) = (\omega\otimes\iota)(M)\otimes 1$ for $\omega\in W_*$. Hence $gf=\iota$ and f is injective. □

Corollary 17.1.6 *Say α is an integrable coaction of a locally compact quantum group (W,Δ) on U, and assume that the map f in Theorem 17.1.1 is injective. Then the operator valued weight $(y\otimes\iota)\alpha$ from U to U^α is regular.*

Proof Identify the basic construction $U^\alpha\subset U\subset U_2$ with $\mathbb{C}\otimes U^\alpha\subset\alpha(U)\subset W\ltimes_\alpha U$. As $\hat W\otimes\mathbb{C}\subset(W\ltimes_\alpha U)\cap(\mathbb{C}\otimes U^\alpha)'$ and $N_{\hat x}\otimes\mathbb{C}\subset N_{E_2}$, the restriction of $E_2 = (\hat x\otimes\iota\otimes\iota)\hat\alpha$ to $U_2\cap(U^\alpha)'$ is semifinite.

Apply now the first paragraph to the integrable dual coaction $\hat\alpha$, which by the theorem above has an injective f. Since the fixed point algebra of $\hat\alpha$ is $\alpha(U)$, we conclude that the restriction of E_3 to $U_3\cap U_1'$ is semifinite. □

An inclusion $U_0\subset U_1$ of von Neumann algebras is *irreducible* if $U_1\cap U_0' = \mathbb{C}$, and it is *of depth 2* when $U_1\cap U_0'\subset U_2\cap U_0'\subset U_3\cap U_0'$ is a basic construction. Note that irreducible inclusions can only consist of factors.

Proposition 17.1.7 *Suppose α is a coaction of a locally compact quantum group (W,Δ) on U such that $\mathbb{C}\otimes U^\alpha\subset\alpha(U)\subset W\ltimes_\alpha U$ is a basic construction. Then the inclusion $U^\alpha\subset U$ is of depth 2.*

Proof Say U is in the semicyclic representation of a faithful semifinite normal weight z, with the canonical semicyclic representation $(q_{\tilde z},\iota)$ for the dual weight $\tilde z$. The basic construction for $\alpha(U)\subset W\ltimes_\alpha U$ is then $J_{\tilde z}\alpha(U)'J_{\tilde z} =$

$J_{\bar{z}}K(B(V_x)\bar{\otimes}U')K^*J_{\bar{z}} = B(V_x)\bar{\otimes}U$. We must therefore show that

$$\alpha(U \cap (U^\alpha)') \subset (W \ltimes_\alpha U) \cap (\mathbb{C} \otimes U^\alpha)' \subset B(V_x)\bar{\otimes}(U \cap (U^\alpha)')$$

is a basic construction. Clearly α restricts to a coaction β of (W, Δ) on $U\cap(U^\alpha)'$, so by the first part of the proof, it suffices to show that the middle part of this inclusion chain concides with $W\ltimes_\beta(U\cap(U^\alpha)')$. But is true as this latter von Neumann algebra equals $\{a \in B(V_x)\bar{\otimes}(U\cap(U^\alpha)') \mid (\iota\otimes\beta)(a) = N_{12}a_{13}N_{12}^*\}$ by the biduality theorem for crossed products, and as properties A, B, C hold for any coaction. While $W \ltimes_\alpha U = \{a \in B(V_x)\bar{\otimes}U \mid (\iota \otimes \alpha)(a) = N_{12}a_{13}N_{12}^*\}$ for the same reason. □

Hence, if α is a coaction of a locally compact quantum group (W, Δ), then the inclusion $U^\alpha \subset U$ is of depth 2 whenever α is integrable and the map f in Theorem 17.1.1 is injective, which clearly happens when $W \ltimes_\alpha U$ is a factor. By the second last proposition above, the inclusion $\alpha(U) \subset W \ltimes_\alpha U$ is therefore of depth 2 for any coaction α.

Corollary 17.1.8 *Let α be an integrable outer coaction of a locally compact quantum group (W, Δ) on a von Neumann algebra U. Then the operator valued weight $(y \otimes \iota)\alpha$ from U to U^α is regular, and the inclusion $U^\alpha \subset U$ is irreducible and of depth 2.*

Proof The map f in Theorem 17.1.1 is injective by outerness of α, so the previous corollary tells us that $(y\otimes\iota)\alpha$ is regular, and the paragraph before this corollary says that $U^\alpha \subset U$ is of depth 2. But it is also irreducible as $U\cap(U^\alpha)' = J_z(U_2\cap U')J_z = \mathbb{C}$ by outerness of α. □

In the next section we will prove a spectacular converse of this result, namely, that to any irreducible depth 2 inclusion $U_0 \subset U_1$ of factors with a regular faithful semifinite normal operator valued weight E_1 from U_1 to U_0, there is a locally compact quantum group (W, Δ) with $W = U_3\cap U_1'$ and an integrable outer coaction α of $(W, \Delta)'$ on U_1 such that the basic construction $U_0 = U^\alpha \subset U_1 \subset U_2$ is isomorphic to the inclusion $\mathbb{C} \otimes U^\alpha \subset \alpha(U) \subset W \ltimes_\alpha U$. Note that integrability is automatic here by Corollary 17.1.3. We end this section with a couple of results on minimality.

A coaction α of a locally compact quantum group (W, Δ) on U is *minimal* if $U \cap (U^\alpha)' = \mathbb{C}$ and $\{(\iota \otimes \omega)\alpha(u) \mid u \in U, \omega \in U_*\}'' = W$.

Proposition 17.1.9 *Minimal coactions are outer, while outer integrable coactions are minimal.*

Proof Say α is a coaction of a locally compact quantum group (W, Δ) on U. Now $(W \ltimes_\alpha U) \cap \alpha(U)' \subset (B(V_x)\bar{\otimes}U) \cap (\mathbb{C} \otimes U^\alpha) \subset B(V_x) \otimes \mathbb{C}$ when α is minimal. In this case any element in the first of these sets is of the form $a \otimes 1$ for $a \in B(V_x)$. By the biduality theorem and properties A, B, C for α, we see that any $b \in W \ltimes_\alpha U$ satisfies $(\iota\otimes\alpha)(b) = N_{12}b_{13}N_{12}^*$, so $(a \otimes 1)N = N(a \otimes 1)$ and $a \in \hat{W}$. As $a \otimes 1 \in \alpha(U)'$, we also get $a \in W'$ by minimality, so $a \in \hat{W} \cap W' = \mathbb{C}$, and α is outer. For the second statement say U is in the semicyclic representation of a faithful semifinite normal weight z, and consider the basic construction $U^\alpha \subset U \subset U_2$. By outerness the map f in Theorem 17.1.1 is injective, so $U \cap (U^\alpha)' = J_z(U_2 \cap U')J_z = \mathbb{C}$, so we are halfway to minimality. The rest will evidently follow if we can prove that $(\alpha(U) \cup (\mathbb{C} \otimes U'))'' = W\bar{\otimes}B(V_z)$. By the biduality theorem we certainly have $B(V_z)\bar{\otimes}U = ((W \ltimes_\alpha U) \cup (W \otimes \mathbb{C}))''$. So $B(V_x)\bar{\otimes}(\alpha(U) \cup (\mathbb{C} \otimes U'))''$ equals

$$((\iota \otimes \alpha)(B(V_x)\bar{\otimes}U) \cup (\mathbb{C} \otimes \mathbb{C} \otimes U'))'' = ((\iota \otimes \alpha)(W \ltimes_\alpha U) \cup (W \otimes \mathbb{C} \otimes U'))'',$$

which again equals $N_{12}((W \ltimes_\alpha U)_{13} \cup (N^*(W \otimes \mathbb{C})N\bar{\otimes}U'))''N_{12}^*$. From the proof of the previous proposition we have $B(V_x)\bar{\otimes}U = J_{\tilde{z}}\alpha(U)'J_{\tilde{z}}$, so by outerness of α, we get $(B(V_x)\bar{\otimes}U)\cap(W \ltimes_\alpha U)' = \mathbb{C}$, or $((\mathbb{C}\otimes U')\cup(W \ltimes_\alpha U))'' = B(V_x)\bar{\otimes}B(V_z)$. Plugging this into the answer of the previous calculation, we see that

$$B(V_x)\bar{\otimes}(\alpha(U)\cup(\mathbb{C}\otimes U'))'' = N_{12}((B(V_x)\otimes\mathbb{C}\otimes B(V_z))\cup(N^*(W\otimes\mathbb{C})N\otimes\mathbb{C}))''N_{12}^*$$

equals

$$((N(B(V_x)\otimes\mathbb{C})N^*\otimes B(V_z))\cup(W\otimes\mathbb{C}\otimes\mathbb{C}))'' = ((\Delta(W)\bar{\otimes}B(V_z))\cup((\hat{W}\cup W)\otimes\mathbb{C}\otimes\mathbb{C}))'',$$

which again equals $B(V_x)\bar{\otimes}W\bar{\otimes}B(V_z)$, and $(\alpha(U)\cup(\mathbb{C}\otimes U'))'' = W\bar{\otimes}B(V_z)$. □

Working on separable Hilbert spaces we have the following result.

Proposition 17.1.10 *Suppose α is a minimal integrable coaction of a locally compact quantum group (W, Δ) on a von Neumann algebra U, with both U and W σ-finite, and with U^α properly infinite. Then α is a dual coaction.*

Proof Define a coaction α_1 of (W, Δ) on $U_1 \equiv B(V_x)\bar{\otimes}U\bar{\otimes}M_2(\mathbb{C})$ by

$$\alpha_1(\begin{pmatrix} a_{11} & a_{12} \\ a_{21} & a_{22} \end{pmatrix}) = \begin{pmatrix} \beta(a_{11}) & \beta(a_{12})N_{21} \\ N_{21}^*\beta(a_{21}) & \gamma(a_{22}) \end{pmatrix},$$

for $a_{ij} \in B(V_x)\bar{\otimes}U$, where β and γ are defined in the biduality theorem for crossed products. Clearly $a \in U_1^{\alpha_1}$ if and only if $a_{11} \in B(V_x)\bar{\otimes}U^\alpha$ and $a_{22} \in W \ltimes_\alpha U$ and $a_{12}, a_{21}^* \in L \equiv \{b \in B(V_x)\bar{\otimes}U \mid (\iota \otimes \alpha)(b) = b_{13}N_{12}^*\}$. As usual we take U to be in the semicyclic representation of a faithful semifinite normal weight z. For $u \in N_{(y\otimes\iota)\alpha}$ and $v \in V_x$, we claim that the element $b = (q \otimes \iota)\alpha(u)(f_v^* \otimes 1) \in B(V_x \otimes V_z)$ actually belongs to L^*, where $q \otimes \iota$ was defined in the proof of Theorem 17.1.1. Let $w \in N_y$ and $u' \in N_z$ and $v' \in V_x$. Then $(\omega_{v',q(w)} \otimes \iota)(b) = (v'|v)(y \otimes \iota)((w^* \otimes 1)\alpha(u))$ and $(v'|v)(q(\iota \otimes \omega)\alpha(u)|q(u')) = ((\iota \otimes \omega)(b)v'|q(u'))$ for $\omega \in U_*$. Hence

$$(\omega_{v',q(w)} \otimes \iota \otimes \omega)(\iota \otimes \alpha)(b) = (\iota \otimes \omega)\alpha((v'|v)(y \otimes \iota)((w^* \otimes 1)\alpha(u)))$$

equals $(v'|v)(y \otimes \iota)((w^* \otimes 1)\Delta(\iota \otimes \omega)\alpha(u)) = (v'|v)(\omega_{q(\iota\otimes\omega)\alpha(z),q(w)} \otimes \iota)(N)$, which again equals $(\omega_{(\iota\otimes\omega)(b)v',q(w)} \otimes \iota)(N) = (\omega_{v',q(w)} \otimes \iota \otimes \omega)(N_{12}b_{13})$, so

$b^* \in L$, as claimed. Now α is outer by the previous proposition, so both $W \ltimes_\alpha U$ and U^α are factors, and as we have just seen that $L \neq \{0\}$, we conclude that $U_1^{\alpha 1}$ is a factor. It is also σ-finite since the Hilbert space V_x is separabel by assumption on W. The orthogonal projections $1 \otimes 1 \otimes e_{ii}$ clearly belong to $U_1^{\alpha 1}$, and as $\mathbb{C} \otimes U^\alpha \subset W \ltimes_\alpha U$, they are both infinite, so there exists $c \in U_1^{\alpha 1}$ such that $c^*c = 1 \otimes 1 \otimes e_{22}$ and $cc^* = 1 \otimes 1 \otimes e_{11}$. Pick a unitary $Y \in L$ such that $c = Y \otimes e_{12}$. Then the automorphism $Y^*(\cdot)Y$ of $B(V_x)\bar{\otimes}U$ respects the two coactions β and γ. Hence β is a dual coaction. But as U^α is properly infinite and V_x is separabel, this coaction on $B(V_x)\bar{\otimes}U$ is isomorphic to α. □

17.2 From the Basic Construction to Quantum Groups

Recall that if $W \subset B(V)$ is a von Neumann algebra with a faithful semifinite normal weight x, there exists an *x-basis of V* of x-bounded elements $v_i \in V$ such that $\sum T^x(v_i, v_i) = 1$ in the strong* topology, where $T^x(v, v') \equiv R_v^x(R_{v'}^x)^*$ for $v, v' \in D_x$. Letting $(v, v')_x \equiv (R_v^x)^* R_{v'}^x \in \pi_x(W)'$, we then get $R_v^x = \sum R_{v_i}^x (v, v_i)_x$.

Proposition 17.2.1 *Let x be a faithful semifinite normal weight on a von Neumann algebra $W \subset B(V)$. Then there is an x-basis of V of elements v_i such that $T^x(v_i, v_i)$ are pairwise orthogonal projections, and $(v_i, v_j)_x = 0$ when $i \neq j$.*

Proof Pick a family $\{v_j\}$ in the dense subset D_x of V such that $V = \oplus\overline{Wv_j}$. Consider the spectral decomposition $T^x(v_j, v_j) = \int \lambda \, dP^j(\lambda)$. Put

$$p_n^j = \int_{\|R_{v_j}^x\|^2/(n+1)}^{\|R_{v_j}^x\|^2/n} dP^j(\lambda) \quad \text{and} \quad v_n^j = \int_{\|R_{v_j}^x\|^2/(n+1)}^{\|R_{v_j}^x\|^2/n} \lambda^{-1/2} dP^j(\lambda) v_j.$$

Then $\sum_n p_n^j$ is an orthogonal sum of orthogonal projections in W', and it projects onto $\overline{Wv_j}$. Moreover, we have $v_n^j \in D_x$ and $T^x(v_n^j, v_n^j) = p_n^j$ adds up to 1, so v_i being v_n^j with $i = (j, n)$ when $p_n^j \neq 0$, forms an x-basis of V satisfying the first property in the proposition, and $(v_j, v_i)_x^*(v_j, v_i)_x = 0$. □

Say we in addition have an antiunitary J on V, so JWJ is isomorphic to the von Neumann algebra W^o with the opposite product. The obvious opposite weight x^o of x has $N_{x^o} = JN_xJ$, and we identify V_{x^o} with V_x by sending $q_{x^o}(JwJ)$ to $J_xq_x(w)$, so $R^{x^o}(v)J_xq_x(w) = JwJv$ and $R^{x^o}(v)J_x\pi_x(w)J_x = JwJR^{x^o}(v)$. We get $R^{x^o}(uv) = uR^{x^o}(v)$ for $u \in U \equiv (JWJ)'$. The span of the elements $T^{x^o}(v, v')$ for $v, v' \in D_{x^o}$ is a dense ideal of U, and $(v, v')_{x^o} \in \pi_x(W)$ with $q_x((v, v')_{x^o}) = (R_{v'}^{x^o})^*v$. So for any x^o-basis $\{v_i\}$ of V, we have $v = \sum R_{v_i}^{x^o} q_x((v, v_i)_{x^o})$. In this case the von Neumann algebra generated by the elements $R_{v_i}^{x^o}\pi_x(w)(R_{v_j}^{x^o})^*$ for all $w \in W$ and all i and j, equals U. It is clearly contained in U, and it exhausts U since $T^{x^o}(v, v') = \sum R_{v_i}^{x^o}(v, v_i)_{x^o}(v', v_j)_{x^o}^*(R_{v_j}^{x^o})^*$ for $v, v' \in D_{x^o}$.

Lemma 17.2.2 *With the above setting, for any bounded linear map* $u\colon V_x \to V$ *such that* $uJ_x\pi_x(\cdot)J_x = J(\cdot)Ju$ *on* W, *we have* $q_x(u^*R_v^{x^o}) = u^*v$ *for* $v \in D_{x^o}$.

Proof Now $u^*R_v^{x^o} \in \pi_x(W)$ and $x((R_v^{x^o})^*uu^*R_v^{x^o}) \leq \|u\|^2\|v\|^2$, and clearly $(q_x(u^*R_v^{x^o})|q_x(w)) = x(\pi_x(w)^*u^*R_v^{x^o}) = x((R^{x^o}_{uq_x(w)})^*R_v^{x^o}) = (u^*v|q_x(w))$. □

Let y be an additional faithful semifinite normal weight on U, so we may form the spatial derivative dy/dx^o. Then there exists an x^o-basis of V consisting of elements $v_i \in D_{x^o} \cap_{t\in\mathbb{R}} D((dy/dx^o)^t)$ such that the orthogonal projections $T^{x^o}(v_i, v_i)$ are pairwise orthogonal. Indeed, it is easy to see, using appropriate integrals, that the intersection above is dense in V. Then tracing the proof of the proposition above with v_j picked from this intersection, we get $p_n^j \in D(\sigma^y_{-i/2})$, so v_n^j belongs to the given intersection, and $(dy/dx^o)^{1/2}v_n^j = \sigma^y_{-i/2}(p_n^j)(dy/dx^o)^{1/2}v^j$, all checked using by now standard modular theory arguments involving analytic extensions.

A W_1-W_2-*bimodule* of von Neumann algebras W_i is a representation space for a normal representation of W_1 and a normal representation of W_2^o that commutes. If V_i are W_1-W_2-bimodules, we write $\mathrm{Hom}_{W_1,W_2^o}(V_1, V_2)$ for the space of W_1-W_2-*bimodule maps*, that is, linear maps $f\colon V_1 \to V_2$ such that $f(w_1vw_2) = w_1f(v)w_2$ for $w_i \in W_i$, and where we have written w_1v for the action of $w_1 \in W_1$ on $v \in V_i$, and vw_2 for the action of $w_2 \in W_2^o$ on $v \in V_i$. The cases of (left) modules and right modules are gotten by setting $W_2 = \mathbb{C}$ and $W_1 = \mathbb{C}$, respectively.

Given our x on $W \subset B(V)$, suppose we are given an additional W-module V'. Mimiking the internal tensor product between bimodules, we define the *relative tensor product* $V \otimes_x V'$ of V and V' as the Hilbert space completion of $D_{x^o} \odot V'$ with respect to the preinner product given by $(v_1 \otimes v_1'|v_2 \otimes v_2') = ((v_1, v_2)_{x^o}v_1'|v_2')$ for $v_i \in D_{x^o}$ and $v_i' \in V'$. We denote the image of $v \otimes v'$ in $V \otimes_x V'$ by $v \otimes_x v'$. Here we are using that $(v_1, v_2)_{x^o} \in \pi_x(W)$, and we have identified $\pi_x(W)$ with W for simplicity. It is straightforward to check that the relative tensor product changes only up to unitary isomorphism when using another weight than x, and it is also fairly easy to see that the relative tensor product is associative.

Let now $U_0 \subset U_1 \subset U_2 \subset \cdots$ be the Jones tower of an inclusion of von Neumann algebras, and consider each U_i represented on the Hilbert space V_i of the chosen semifinite weights y_i. Note that $T^{y_0^o}(v', v) \in U_2 = J_1U_0'J_1$ for $v, v' \in D_{y_0^o}$. If in addition $v \in D((dy_2/dy_0^o)^{1/2})$, then

$$y_2(T^{y_0^o}(v', v)^*T^{y_0^o}(v', v)) = \|v' \otimes_{y_0} J_1(dy_2/dy_0^o)^{1/2}v\|^2$$

where we have written J_i for J_{y_i}. Indeed, we see that the argument of y_2 equals $R^{y_0^o}(v)(v', v')_{y_0^o}R^{y_0^o}(v)^*$, and by the definition of the inner product on the relative tensor product, it suffices to show that the latter expression with $(v', v')_{y_0^o}$ replaced by any $u \in U_0$, equals $(uJ_1(dy_2/dy_0^o)^{1/2}v|J_1(dy_2/dy_0^o)^{1/2}v)$. But when u is analytic with respect to y_0, this last expression equals

$$((dy_2/dy_0^o)^{1/2}J_1\sigma^{y_0}_{-i/2}(u^*)J_1v|(dy_2/dy_0^o)^{1/2}v) = y_2(R^{y_0^o}(J_1\sigma^{y_0}_{-i/2}(u^*)J_1v)R^{y_0^o}(v)^*),$$

which again equals $y_2(R^{y_0^o}(v)uR^{y_0^o}(v)^*)$. Using normality of y_2 we see that equality still holds for any $u \in U_0$. We have proved the non-trivial part of the result below.

Proposition 17.2.3 *The formula* $q_{y_2}(T^{y_0^o}(v', v)) \mapsto v' \otimes_{y_0} J_1(dy_2/dy_0^o)^{1/2}v$ *defines by denseness and continuity an isomorphism* $V_2 \to V_1 \otimes_{y_0} V_1$, *where* J_2 *and* $\Delta_{y_2}^{it}$ *and* π_{y_2} *when transferred to the relative tensor product act respectively as* $v_1 \otimes_{y_0} v_2 \mapsto J_1 v_2 \otimes_{y_0} J_2 v_1$ *and* $v_1 \otimes_{y_0} v_2 \mapsto (dy_2/dy_0^o)^{it} v_1 \otimes_{y_0} J_1(dy_2/dy_0^o)^{-it} J_1 v_2$ *and* $u \mapsto u \otimes_{y_0} 1$ *for* $v_i \in D_{y_0^o}$ *and* $u \in U_2$. *In particular, letting* $\{v_i\}$ *be a* y_0^o*-basis of* V_1 *as described just below the previous lemma, then* $\{v_i \otimes_{y_0} J_1(dy_2/dy_0^o)^{1/2} v_i\}$ *is a* y_2^o*-basis for* $V_1 \otimes_{y_0} V_1$, *which we shall identify with* V_2.

We also identify $V_0 \otimes_{y_0} V_1$ with V_1 by sending $q_{y_0}(u) \otimes_{y_0} v$ to uv for $u \in N_{y_0}$ and $v \in V_1$. Note that $R^{y_0^o}(v) \in \mathrm{Hom}_{U_0^o}(V_0, V_1)$ for $v \in D_{y_0^o}$, so $R^{y_0^o}(v) \otimes_{y_0} 1$ belongs to $\mathrm{Hom}_{U_1^o}(V_0 \otimes_{y_0} V_1, V_1 \otimes_{y_0} V_1) = \mathrm{Hom}_{U_1^o}(V_1, V_2)$, and for $v' \in V_1$ and u analytic with respect to y_0, we have $(R^{y_0^o}(v) \otimes_{y_0} 1)uv' = R^{y_0^o}(v)q_{y_0}(u) \otimes_{y_0} v' = R^{y_0^o}(v)J_0 q_{y_0}(\sigma_{i/2}^{y_0}(u)^*) \otimes_{y_0} v'$, which equals $J_1\sigma_{i/2}^{y_0}(u)^* J_1 v \otimes_{y_0} v' = v \otimes_{y_0} uv'$. So we see that $R^{y_0^o}(v) \otimes_{y_0} 1 = L_v$, where $L_v(v') = v \otimes_{y_0} v'$. Hence, if in addition $v' \in D_{y_0^o}$, then $L_v^* L_{v'} = (v', v)_{y_0^o}$ and $L_{v'} L_v^* = \pi_{y_2}(T^{y_0^o}(v', v))$ and $R^{y_0^o}(v \otimes_{y_0} v') = L_v R^{y_0^o}(v')$, while $R^{y_1^o}(v \otimes_{y_0} v') = L_v \pi_l(v')$ when $v' \in q_{y_1}(N_{y_1})$. This shows that if $\{v_i\}$ is a y_0^o-basis of V_1 and $\{v_j'\}$ is a y_1^o-basis of V_1, then $\{v_i \otimes_{y_0} v_j'\}$ is a y_1^o-basis of V_2 and $\{v_i \otimes_{y_0} v_j\}$ is a y_0^o-basis of V_1.

By the proposition above we have $V_3 = V_2 \otimes_{y_1} V_2 = (V_1 \otimes_{y_0} V_1) \otimes_{y_1} (V_1 \otimes_{y_0} V_1)$, which we identify with $V_1 \otimes_{y_0} V_1 \otimes_{y_0} V_1$ by sending $(v_1 \otimes_{y_0} v_1') \otimes_{y_1} (v_2 \otimes_{y_0} v_2')$ to $v_1 \otimes_{y_0} \pi_l(v_1')v_2 \otimes_{y_0} v_2'$ for $v_i \in D_{y_0^o}$ and $v_1' \in q_{y_1}(N_{y_1})$ and $v_2' \in V_1$. Under this identification $L_{v_1 \otimes_{y_0} v_1'} = (L_{v_1} \otimes_{y_0} 1)\pi_{y_2}(\pi_l(v_1'))$ and $J_3(v_1 \otimes_{y_0} v_2 \otimes_{y_0} v_3) = J_1 v_3 \otimes_{y_0} J_1 v_2 \otimes_{y_0} J_1 v_1$, while $\pi_{y_3}(u) = u \otimes_{y_0} 1$ for $u \in U_3$. In a similar fashion we identify V_k with the k-times relative tensor product of V_1 with respect to y_0, and J_k then acts on elementary tensors by first reversing their order and then applying J_1 on each factor, while $\pi_{y_k}(u) = u \otimes_{y_0} 1$ for $u \in U_k$.

Theorem 17.2.4 *If* $U_0 \subset U_1 \subset U_2 \subset \cdots$ *is the Jones tower, then the chain* $U_0 \subset U_2 \subset U_4$ *of inclusions is a basic construction of the inclusion* $U_0 \subset U_2$.

Proof We will show that if $U_0 \subset U_2 \subset U$ is a basic construction, then $U_4 = U \otimes_{y_0} \mathbb{C}$. Let $\{v_i\}$ be a y_0^o-basis of V_1 as described just below the previous lemma, and let $\{v_j'\}$ be a y_1^o-basis of V_1. Then by the considerations above we know that $v = v_i \otimes_{y_0} \pi_l(v_j')v_k \otimes_{y_0} J_1(dy_2/dy_0^o)^{1/2}v_k$ form a y_2^o-basis of V_2, and we have moreover that $R_v^{y_0^o} = L_{v_i}\pi_l(v_j')T^{y_0^o}(v_k, v_k) \otimes_{y_0} 1$. By the last assertion before the previous lemma we thus see that U_4 is contained in the von Neumann algebra generated by the elements $L_{v_i} u L_{v_n}^* \otimes_{y_0} 1$ for $u \in U_2$. But we have equality since $\sum \pi_l(v_j')T^{y_0^o}(v_k, v_k)\pi_l(v_j')^* u \pi_l(v_s')T^{y_0^o}(v_t, v_t)\pi_l(v_s')^* = u$. On the other hand U is by the considerations above and by the last assertion before the previous lemma, generated by the elements $L_{v_i} R^{y_0^o}(v_k) u R^{y_0^o}(v_s)^* L_{v_t}^*$ for $u \in U_0$, and the inner sandwich generates U_2, again by the last assertion before the previous lemma. □

Corollary 17.2.5 *If an inclusion $U_0 \subset U_1$ of von Neumann algebras is irreducible and of depth 2, then every inclusion $U_i \subset U_{i+1}$ in the Jones tower is irreducible, and $U_i' \cap U_{i+3}$ is a type I factor. Letting $j_i = J_i(\cdot)^* J_i$ on $B(V_i)$, we have $j_i(U_{i-1}' \cap U_{i+1}) = U_{i-1}' \cap U_{i+1}$ and $j_{i+1}(U_{i-1}' \cap U_{i+3}) = U_{i-1}' \cap U_{i+3}$ and $j_{i+1}(U_{i-1}' \cap U_{i+1}) = U_{i+1}' \cap U_{i+3}$ and $j_{i+1}(U_{i-1}' \cap U_{i+2}) = U_i' \cap U_{i+3}$.*

Proof Since $U_i' \cap U_{i+1} = J_i(U_i \cap U_{i-1}')J_i$, irreducibility is clear by induction. We have $U_0' \cap U_3 = J_z(U_0' \cap U_1)'J_z$ for some faithful semifinite normal weight z on $U_0' \cap U_2$, and $U_0' \cap U_1 = \mathbb{C}$ shows that $U_0' \cap U_3 = B(V_z)$. The theorem yields $U_{i+4} = J_{i+2}U_i'J_{i+2}$, so as $U_{i+3} = J_{i+2}U_{i+1}'J_{i+2}$, we get $U_{i+1}' \cap U_{i+4} = J_{i+2}(U_i' \cap U_{i+3})J_{i+2}$, and $U_i' \cap U_{i+3}$ is a type I factor by induction. The equalities with j_{i+1} are clear from the theorem and the definition of a basic construction, and the equality with j_i does not even use the theorem. □

Proposition 17.2.6 *Let $U_0 \subset U_1$ be an irreducible depth 2 inclusion of factors. Then $V^i \equiv \mathrm{Hom}_{U_{i-1},U_i^o}(V_i, V_{i+1})$ is a Hilbert space with inner product given by $(f|g) = g^* \circ f$. Then $\pi_i(u)f = u \circ f$ defines a $*$-isomorphism $\pi_i : U_{i-1}' \cap U_{i+2} \to B(V^i)$, while $G_i(f \otimes v) = f(v)$ defines a unitary $G_i : V^i \otimes V_i \to V_{i+1}$ such that $u = G_i(\pi_i(u) \otimes 1)G_i^*$. The formula $F_i(f) = J_{i+2}(1 \otimes_{y_{i-1}} f)J_{i+1}$ defines a conjugate linear bijection F_i from V^i to V^{i+1} such that $H_i = F_i(\cdot)^* F_i^* : B(V^i) \to B(V^{i+1})$ satisfies $H_i\pi_i = \pi_{i+1}j_{i+1}$ on $U_{i-1}' \cap U_{i+2}$.*

Proof Clearly $g^* \circ f \in U_{i-1}' \cap (J_iU_iJ_i)' = U_{i-1}' \cap U_i = \mathbb{C}$, so we have an inner product on V^i. If $\{f_n\}$ is a Cauchy sequence in V^i, then $\{f_n(v)\}$ is Cauchy in V_{i+1} for any $v \in V_i$ because $\|f_n(v) - f_m(v)\|_{i+1}^2 = \|v\|_i(f_n - f_m)^* \circ (f_n - f_m)$. So $f_n(v)$ converges to some $f(v)$ in the Hilbert space V_{i+1}, which defines $f \in V^i$ such that $f_n \to f$, as is seen by picking any $v \neq 0$ in the identity above with $f_m = f$. So V^i is a Hilbert space. As $U_{i+2} = \mathrm{Hom}_{U_i^o}(V_{i+1}, V_{i+1})$, we see that π_i makes sense and is evidently a homomorphism, which is also $*$-preserving since $(\pi_i(u)f|g) = (u \circ g)^* \circ f = (\pi_i(u^*)f|g)$. If $u' \in V^i$, we see that $u' \circ u^* \in \mathrm{Hom}_{U_{i-1},U_i^o}(V_{i+1}, V_{i+1}) = U_{i-1}' \cap U_{i+2}$ and $\pi_i(u' \circ u^*)f = u' \circ u^* \circ f = (f|u)u'$, so π_i hits all rank one operators in $B(V^i)$ and must therefore be surjective. Evidently G_i is an isometry. As $V^i \subset \mathrm{Hom}_{U_i^o}(V_i, V_{i+1})$, we get $J_{i+1}aJ_{i+1}G_i = G_i(1 \otimes J_iaJ_i)$ for $a \in U_i$, so $G_iG_i^* \in (J_{i+1}U_iJ_{i+1})' = U_{i+2}$. As $V^i \subset \mathrm{Hom}_{U_{i-1}}(V_i, V_{i+1})$, we get $bG_i = G_i(1 \otimes b)$ for $b \in U_{i-1}$, so $G_iG_i^* \in U_{i-1}'$. Thus $G_iG_i^* \in U_{i-1}' \cap U_{i+2}$, but it also belongs to the commutant of this as we evidently have $uG_i = G_i(\pi_i(u) \otimes 1)$. So G_i is a unitary by the previous corollary, and then π_i must be injective. Now $1 \otimes_{y_{i-1}} f \in \mathrm{Hom}_{U_{i+1},U_i^o}(V_i \otimes_{y_{i-1}} V_i, V_i \otimes_{y_{i-1}} V_{i+1})$, and $V_i \otimes_{y_{i-1}} V_{i+1} = V_i \otimes_{y_{i-1}} V_i \otimes_{y_0} V_1 = V_{i+1} \otimes_{y_0} V_1 = V_{i+2}$, so $F_i(f) \in V^{i+1}$. In fact, similar reasoning shows that the maps $f \mapsto 1 \otimes_{y_{i-1}} f$ and $h \mapsto 1 \otimes_{y_{i+1}} h$ for $h \in \mathrm{Hom}_{U_{i+1},U_i^o}(V_{i+1}, V_{i+2})$ are inverses of each other, so F_i is a norm preserving surjection. Finally, we clearly have $F_i\pi_i(u)f = (J_{i+1}uJ_{i+1} \otimes_{y_{i-1}} 1)J_{i+2}(1 \otimes_{y_{i-1}} f)J_{i+1} = \pi_{i+1}j_{i+1}(u^*)F_if$. □

Corollary 17.2.7 *As sets we have* $H_i\pi_i(U_i' \cap U_{i+2}) = \pi_{i+1}(U_i' \cap U_{i+2})$ *and* $H_i\pi_i(U_{i-1}' \cap U_{i+1}) = \pi_{i+1}(U_{i+1}' \cap U_{i+3})$ *and* $U_{i+2} = G_i(B(V^i)\bar{\otimes}U_i)G_i^*$ *and* $U_{i+3} = G_i(B(V^i)\bar{\otimes}U_{i+1})G_i^*$ *and* $U_{i-1}' \cap U_{i+2} = G_i(B(V^i) \otimes \mathbb{C})G_i^*$ *and* $U_{i-1}' \cap U_{i+3} = G_i(B(V^i)\bar{\otimes}(U_{i-1}' \cap U_{i+1}))G_i^*$ *and finally* $G_i^* j_{i+1}(U_{i-1}' \cap U_{i+1})G_i$ *is contained in* $\pi_i(U_{i-1}' \cap U_{i+1})'\bar{\otimes}(U_{i-1}' \cap U_{i+1})$.

Proof The first two equalities are immediate from the previous proposition and corollary. The next one is clear from taking the commutant of the set relation involving G_i and $a \in U_i$, and for the next one let $a \in U_{i-1}$ and then use the previous theorem. The next one is immediate from the previous proposition, and for the next one take the commutant of the set relation involving G_i and $b \in U_{i-1}'$ and combine this with the fourth equality. For the final inclusion, use $U_{i+1} \subset U_{i+3}$ in combination with the corollary to get the desired set in $B(V^i)\bar{\otimes}(U_{i-1}' \cap U_{i+1})$. On the other hand, using the relation between G_i and π_i in the previous proposition, and then taking commutants, we get $G_i^*(U_{i-1}' \cap U_{i+1})'G_i \subset \pi_i(U_{i-1}' \cap U_{i+1})'\bar{\otimes}B(V_i)$, and need now only observe that $j_{i+1}(U_{i-1}' \cap U_{i+1}) \subset (U_{i-1}' \cap U_{i+1})'$. □

Define a unitary linear operator $1 \otimes_{y_{i-1}} G_i : V^i \otimes V_{i+1} \to V_{i+2}$ by the formula $(1 \otimes_{y_{i-1}} G_i)(f \otimes (v_1 \otimes_{y_{i-1}} v_2)) = v_1 \otimes_{y_{i-1}} f(v_2)$ for $f \in V^i$ and $v_k \in D_{y_{i-1}} \subset V_i$. We have used that $V_{i+1} = V_i \otimes_{y_{i-1}} V_i = V_i \otimes_{y_0} V_1$. Then it is straightforward to check that $G_{i+1} = J_{i+2}(1 \otimes_{y_{i-1}} G_i)(F_i^* \otimes J_{i+1})$. Similarly we can define a unitary $G_i \otimes_{y_{i-1}} 1 : V^i \otimes V_{i+1} \to V_{i+2}$ in an obvious way.

Proposition 17.2.8 *We have unitary maps* X_i, Y_i *from* $V^i \otimes V^i \otimes V_i$ *to* V_{i+2} *given by* $X_i(f \otimes g \otimes v) = (1 \otimes_{y_{i-1}} f)g(v)$ *and* $Y_i(f \otimes g \otimes v) = (g \otimes_{y_{i-1}} 1)f(v)$. *There is a multiplicative unitary* $M_i \in B(V^i \otimes V^i)$ *given by* $M_i \otimes 1 = X_i^* Y_i$. *It is determined by* $M_i(f \otimes g) = \sum f_k \otimes g_k$ *with convergence i norm, where* $(g \otimes_{y_{i-1}} 1)f = \sum(1 \otimes_{y_{i-1}} f_k)g_k$ *as a strong limit. Moreover, we have* $M_{i+1} = (H_i \otimes H_i)((M_i)_{21})$.

Proof We have $X_i = (1 \otimes_{y_{i-1}} G_i)(1 \otimes G_i)$ and $Y_i = (G_i \otimes_{y_{i-1}} 1)(1 \otimes G_i)(F \otimes 1)$, where F is the flip on V^i, so X_i and Y_i are unitary. We have $X_i(1 \otimes 1 \otimes J_i a J_i) = J_{i+2} a J_{i+2} X_i$ and $X_i(1 \otimes 1 \otimes b) = bX_i$ for $a \in U_i$ and $b \in U_{i-1}$, and similarly for Y_i, so $X_i^* Y_i \in B(V^i \otimes V^i) \otimes \mathbb{C}$ as $U_i \cap U_{i-1}' = \mathbb{C}$. So M_i is a well-defined unitary on $V^i \otimes V^i$. That it acts as described on $f \otimes g$ is clear. Note that $M_i^*(f \otimes g) = \sum f_k \otimes g_k$ when $(1 \otimes_{y_{i-1}} f)g = \sum(g_k \otimes_{y_{i-1}} 1)f_k$. That M_i satisfies the pentagonal equation is clear as both sides of the equation act on $f \otimes g \otimes h$ as $\sum f_k \otimes g_k \otimes h_k$ whenever

$$(h \otimes_{y_{i-1}} 1 \otimes_{y_{i-1}} 1)(g \otimes_{y_{i-1}} 1)f = \sum(1 \otimes_{y_{i-1}} 1 \otimes_{y_{i-1}} f_k)(1 \otimes_{y_{i-1}} g_k)h_k.$$

The final statement is straightforward. □

The last statement means that M_i and M_{i+2} are unitarily equivalent.

Proposition 17.2.9 *Let* $W_i = \pi_i(U_i' \cap U_{i+2})'$ *and* $\hat{W}_i = \pi_i(U_{i-1}' \cap U_{i+1})'$ *and* $\Delta_i = M_i^*(1 \otimes (\cdot))M_i$ *and* $\hat{\Delta}_i = \hat{M}_i^*(1 \otimes (\cdot))\hat{M}_i$, *where* M_i *is defined in the previous proposition and* $\hat{M}_i \equiv (M_i^*)_{21}$. *Then* $M_i \in W_i\bar{\otimes}\hat{W}_i$ *and* Δ_i *is a coproduct on* W_i, *while $\hat{\Delta}_i$ is a coproduct on $\hat{W}_i$. Moreover, we have $(\iota \otimes \pi_i)(G_i^* j_{i+1} j_i(\cdot) G_i) = \hat{M}_i(1 \otimes \pi_i(\cdot))\hat{M}_i^*$ on $U_{i-1}' \cap U_{i+1}$.*

Proof Since $M_{i+1} = (H_i \otimes H_i)((M_i)_{21})$ by the previous proposition, and $H_i(W_i) = W_{i+1}$ from the previous corollary, in order to prove $M_i \in W_i \bar{\otimes} \hat{W}_i$, it suffices to show that $\pi_i(u)$ commutes with $(\omega_{f,g} \otimes \iota)(M_i)$ for $u \in U_{i-1}' \cap U_{i+1}$ and $f, g \in V^i$. But $1 \otimes_{y_{i-1}} g^* \in \mathrm{Hom}_{U_{i+1}, U_i^o}(V_{i+2}, V_{i+1})$, so $(1 \otimes_{y_{i-1}} g^*)(u \otimes_{y_{i-1}} 1) = u(1 \otimes_{y_{i-1}} g^*)$, and multiplying both sides by $(h \otimes_{y_{i-1}} 1)f$ from the right, where $h \in V^i$, gives $(\omega_{f,g} \otimes \iota)(M_i)\pi_i(u)h = \pi_i(u)(\omega_{f,g} \otimes \iota)(M_i)h$ by the determining property for M_i in the previous proposition. Coassociativity holds on $B(V^i)$ for Δ_i and $\hat{\Delta}_i$ as they are defined from multiplicative unitaries. Proceeding to the last statement, note that $(1 \otimes_{y_{i-1}} G_i)(\pi_i(u) \otimes 1)(1 \otimes_{y_{i-1}} G_i^*) = 1 \otimes_{y_{i-1}} u = j_{i+1} j_i(u)$, so $X_i(\pi_i(u) \otimes 1)X_i^* = j_{i+1} j_i(u)$ and $M_i(\pi_i(u) \otimes 1)M_i^* \otimes 1 = (\iota \otimes \pi_i)(G_i^* j_{i+1} j_i(u) G_i)_{21} \otimes 1$ as desired. From the previous corollary we thus get $M_i^*(\pi_i(U_{i-1}' \cap U_{i+1}) \otimes \mathbb{C})M_i \subset \pi_i(U_{i-1}' \cap U_{i+1}) \bar{\otimes} \hat{W}_i$. Sandwiching this with $M_i(\cdot)M_i^*$ and taking commutants shows that $\hat{\Delta}_i(\hat{W}_i) \subset \hat{W}_i \bar{\otimes} \hat{W}_i$ as $M_i \in W_i \bar{\otimes} \hat{W}_i$, and similarly $\Delta_i(W_i) \subset W_i \bar{\otimes} W_i$. □

Note that $H_{i+1} H_i$ identifies the coproduct Δ_i on W_i with the coproduct Δ_{i+2} on W_{i+2}, and also the coproduct $\hat{\Delta}_i$ on $\hat{W}_i$ with the coproduct $\hat{\Delta}_{i+2}$ on $\hat{W}_{i+2}$.

Proposition 17.2.10 *The map $\alpha_i = G_i^*(\cdot)G_i$ defines an injective normal $*$-homomorphism $U_i \to W_i \bar{\otimes} U_i$ such that $(\iota \otimes \alpha_i)\alpha_i = (\Delta_i^{\mathrm{op}} \otimes \iota)\alpha_i$ and $U_{i-1} \subset U_i^{\alpha_i} \equiv \{u \in U_i \mid \alpha_i(u) = 1 \otimes u\} = U_i \cap (U_{i-1}' \cap U_{i+2})'$. Also $\alpha_{i+1}(U_{i-1}' \cap U_{i+1}) \subset W_{i+1} \bar{\otimes} (U_{i-1}' \cap U_{i+1})$ and $(H_i^{-1} \otimes j_i)\alpha_{i+1} = G_i^* j_{i+1}(\cdot) G_i$ on $U_{i-1}' \cap U_{i+1}$.*

Proof By the previous corollary we have $\alpha_i(u) \in B(V^i) \bar{\otimes} U_i$ for $u \in U_{i+2}$, and by definition of G_i we have $G_i^*(\cdot)G_i = \pi_i(\cdot) \otimes 1$ on $U_i' \cap U_{i+2}$, so if $u \in U_i$, then $\alpha_i(u) \in W_i \bar{\otimes} U_i$. In this case $(\iota \otimes \alpha_i)\alpha_i(u) = (F \otimes 1)Y_i^* u Y_i(F \otimes 1)$, where F is the flip on $V^i \otimes V^i$, while

$$(\Delta_i \otimes \iota)\alpha_i(u) = Y_i^* J_{i+2} G_{i+1}(F_i \otimes J_{i+1})(1 \otimes u)(F_i^* \otimes J_{i+1})G_{i+1}^* J_{i+2} Y_i$$

by definition of Δ_i and α_i and M_i and the discussion just after the previous corollary. Using $J_{i+2}(\cdot)J_{i+2}G_{i+1} = G_{i+1}(1 \otimes J_{i+1}(\cdot)J_{i+1})$ on U_{i+1}, we thus get $(\iota \otimes \alpha_i)\alpha_i = (\Delta_i^{\mathrm{op}} \otimes \iota)\alpha_i$. Next, we see that $u \in U_i$ commutes with every element in $U_{i-1}' \cap U_{i+2}$ if and only if $\alpha(u)$ commutes with every element in $G_i^*(U_{i-1}' \cap U_{i+2})G_i = \pi_i(U_{i-1}' \cap U_{i+2}) \otimes \mathbb{C} = B(V^i) \otimes \mathbb{C}$, meaning that $\alpha(u) \in \mathbb{C} \otimes U_i$, and this happens if and only if $\alpha_i(u) = 1 \otimes u$, as is seen using $(\iota \otimes \alpha_i)\alpha_i = (\Delta_i^{\mathrm{op}} \otimes \iota)\alpha_i$ and injectivity of α_i. The inclusion $U_{i-1} \subset U_i^{\alpha_i}$ is now obvious. If $u \in U_{i-1}' \cap U_{i+1}$, then we thus see that $\alpha_{i+1}(u)$ commutes with $\alpha_{i+1}(u') = 1 \otimes u'$ for every $u' \in U_{i-1}$, so we get the final inclusion in the proposition. From the identity involving G_{i+1} and F_i used earlier in this proof, we readily get the final identity in the proposition. □

Given the Jones tower $U_0 \subset U_1 \subset U_2 \subset \cdots$ and the faithful semifinite normal operator valued weights E_i from U_i to U_{i-1} gotten from E_1 and the faithful semifinite normal weights $y_i = y_{i-1}E_i$ on U_i with semicyclic representation spaces

V_i. Then the faithful semifinite normal operator valued weights E'_i from U'_{i-1} to U'_i given by $d(rE'_i)/ds = dr/d(sE_i)$ for faithful semifinite normal weights r and s on U'_i and U_{i-1}, respectively, satisfy $E'_i(T^s(v,v)) = T^{sE_i}(v,v)$ for $v \in D_{sE_i}$. This holds since in the special case $\mathbb{C} \subset U_{i-1}$, we have $s'((\cdot|v)v) = T^s(v,v)$, as is easily checked, so both sides equal $E'_i(s'((\cdot|v)v)) = (sE_i)'((\cdot|v)v)$. Hence we get $E_i(T^{y^o_{i-2}}(v,v)) = T^{(y_{i-2}E_{i-1})^o}(v,v)$ for $v \in D_{y_{i-1}}$.

Lemma 17.2.11 *Let $U_0 \subset U_1$ be any inclusion of von Neumann algebras and let E be a faithful semifinite normal operator valued weight from U_1 to U_0. Let $U \supset U_1$ be a factor with faithful normal representations ρ_i, and let E'_{ρ_i} be the faithful semifinite normal operator valued weight from $\rho_i(U_0)'$ to $\rho_i(U_1)'$ associated to $\rho_i E \rho_i^{-1}$ as described prior to the lemma. Then $\rho_2\rho_1^{-1}(E'_{\rho_1}\rho_1(u)) = E'_{\rho_2}\rho_2(u)$ for $u \in U'_0 \cap U$ with $E'_{\rho_1}\rho_1(u) \in \rho_1(U)$.*

Proof Consider the isomorphism $\rho_2\rho_1^{-1}\colon \rho_1(U) \to \rho_2(U)$ as a composition of an induction, amplification and a spatial isomorphism, and check each case individually, using $E'_{\rho_i}(T^s(v,v)) = T^{s\rho_i E\rho_i^{-1}}(v,v)$ and an s-basis. □

Proposition 17.2.12 *We keep the notation from above the previous lemma. Then $q_{E_1}(u)$, which sends $q_{y_0}(a)$ to $q_{y_1}(ua)$ for $a \in N_{y_0}$, defines a U_1-U_0-bimodule map $q_{E_1}\colon N_{E_1} \to \mathrm{Hom}_{U_0^o}(V_0, V_1)$. The ideal $N_{E_1} \cap N_{y_1}$ is weakly dense in U_1 and its image under q_{y_1} is dense in V_i, while the image of $N_{E_1} \cap N_{y_1}$ intersected with its adjoint is a core for $\Delta_{y_1}^{1/2}$. If $f^* \circ q_{E_1}(N_{E_1}) = \{0\}$ for $f \in \mathrm{Hom}_{U_0^o}(V_0, V_1)$, then $f = 0$. We have $q_{E_1}(u)^* \circ q_{y_1}(u') = q_{y_0}E_1(u^*u')$ for $u' \in N_{E_1} \cap N_{y_1}$. When $u' \in N_{E_1}$, we have $q_{E_1}(u)^* \circ q_{E_1}(u') = E_1(u^*u')$, so $\|q_{E_1}(u)\|^2 = \|E_1(u^*u)\|$ and q_{E_1} is injective. We also have $q_{E_1}(u') = R^{y_0^o}(q_{y_1}(u'))$ for $u' \in N_{E_1} \cap N_{y_1}$. The image under E_2 of the elements $q_{E_1}(u') \circ q_{E_1}(u^*)$ is $u'u^*$, and they generate U_2 as u, u' range over N_{E_1}. For f, g with $f \circ f^*, g \circ g^* \in M_{E_2}$, we have $E_2(f \circ g^*)^* E_2(f \circ g^*) \leq \|E_2(f \circ f^*)\| E_2(g \circ g^*)$, and when g is replaced by $q_{E_1}(u)$, we get the same inequality with $E_2(g \circ g^*)$ replaced by uu^*, while if we shift the $*$ to the right outside E_2 on the left hand side, we must replace uu^* by $\|u\|^2$ on the right hand side. There exists a unique $A_1(f) \in U_1$ such that $A_1(f)u^* = E_2(f \circ q_{E_1}(u)^*)$ for all $u \in N_{E_1}$. This gives an injective U_1-U_0-bimodule map $A_1\colon \mathrm{Hom}_{U_0^o}(V_0, V_1) \to U_1$ such that $A_1(q_{E_1}(u)) = u$ and $A_1(f)A_1(f)^* \leq E_2(f \circ f^*)$. For $h \in N_{E_2}$ and $u' \in N_{E_1} \cap N_{y_1}$, we have $q_{y_1}A_1(h^* \circ q_{E_1}(u')) = h^* q_{y_1}(u')$, and we have $h^* \circ q_{E_1}(u) = q_{E_1}A_1(h^* \circ q_{E_1}(u))$ and $A_1(h^* \circ q_{E_1}(u))^* A_1(h^* \circ q_{E_1}(u)) \leq \|E_2(h^* \circ h)\| E_1(u^*u)$ for $u \in N_{E_1}$.*

Proof Since $y_1(a^*u^*ua) = y_0(a^*E_1(u^*u)a) \leq \|E_1(u^*u)\| y_0(a^*a)$, we have a well-defined linear map q_{E_1}, because

$$q_{E_1}(u)J_0bJ_0q_{y_0}(a) = q_{E_1}(u)q_{y_0}(a\sigma^{y_0}_{-i/2}(b)^*) = q_{y_1}(ua\sigma^{y_1}_{-i/2}(b)^*) = J_1bJ_1q_{E_1}(u)q_{y_0}(a)$$

when $a, b \in U_0$ are analytic. Clearly $q_{E_1}(cub) = \pi_{y_1}(c)q_{E_1}(u)\pi_{y_0}(b)$ when $c \in U_1$ and $b \in U_0$. Weak density is now clear. Writing $E_1(c^*c) = \int \lambda\, dP(\lambda) + \infty P$ for $c \in N_{y_1}$, we get $ny_0(P) \leq y_1(c^*c)$, so $y_0(P) = 0$ and $P = 0$. Now

$E_1((cP(n))^*cP(n)) = P(n)E_1(c^*c)P(n) = \int^n \lambda\, dP(\lambda)$, so $cP(n) \in N_{E_1}$. With $p_n \equiv \int_{1/n}^n dP(\lambda)$ we also get $E_1(c^*c) \geq \int_{1/n}^n \lambda\, dP(\lambda) \geq p_n/n$, so $y_0(p_n) \leq ny_1(c^*c)$ and $p_n \in N_{y_0}$. Hence $cp_n = cP(n)p_n \in N_{E_1} \cap N_{y_1}$ by what we have already proved, and

$$\|q_{y_1}(c-cp_n)\|^2 = y_0((1-p_n)E_1(c^*c)(1-p_n)) = y_0(E_1(c^*c)-\int_{1/n}^n \lambda\, dP(\lambda)) \to 0$$

as $n \to \infty$. Thus we have proved the next density statement. If c belongs to the intersection in the next statement, then writing $E_1(cc^*) = \int \lambda\, dP'(\lambda)$ and $p'_n = \int_{1/n}^n dP'(\lambda)$, and arguing as above, we get $q_{y_1}(p'_n cp_n) \to q_{y_1}(c)$, while $q_{y_1}(p_n c^* p'_n) \to q_{y_1}(c^*)$, and we have the required core. As for the next statement, note that $f^* q_{y_1}(ua) = f^* \circ q_{E_1}(u)q_{y_0}(a) = 0$ and $f = 0$ by the density result we have just proved. In the next statement the only non-trivial part is to show $E_1(u^*u') \in N_{y_0}$. This is clear from the inequality $E_1(u^*u')^*E_1(u^*u') \leq \|E_1(u^*u)\|E_1(u'^*u')$, which holds for any $u, u' \in N_{E_1}$. To see that this estimate holds, first observe that

$$\begin{pmatrix} E_1(u'^*u') & E_1(u'^*u) \\ E_1(u^*u') & \|E_1(u^*u)\| \end{pmatrix} \geq \begin{pmatrix} E_1(u'^*u') & E_1(u'^*u) \\ E_1(u^*u') & E_1(u^*u) \end{pmatrix} = (E_1 \otimes \iota)(\begin{pmatrix} u' & u \\ 0 & 0 \end{pmatrix}\begin{pmatrix} u'^* & 0 \\ u^* & 0 \end{pmatrix})$$

is a positive matrix of the form (a_{ij}) with a_{11} a positive operator, and $a_{12} = a_{21}^*$ a bounded operator, and with a_{22} a positive scalar. If A is the operator on $V_0 \oplus V_0$ defined by such a matrix, then $(A(v_1 \otimes rv_2)|v_1 \otimes rv_2) \geq 0$ for $v_i \in V_0$ and $r \in \mathbb{C}$, which for an optimal r gives $|(a_{12}v_2|v_1)|^2 \leq a_{22}(a_{11}v_1|v_1)\|v_2\|^2$, and the desired inequality is obtained for $v_2 = a_{21}v_1$. The statements including the one with $R^{y_0^o}$ are therefore true. For the next one, note that $q_{E_1}(u') \circ q_{E_1}(u^*) \in \mathrm{Hom}_{U_0^o}(V_1, V_1) = U_2$ and for $e \in N_{y_0}$ we have $E_2(q_{E_1}(u)ee^*q_{E_1}(u)^*) = E_2(R^{y_0^o}(q_{E_1}(u)q_{y_0}(e))R^{y_0^o}(q_{E_1}(u)q_{y_0}(e))^*)$, which equals $E_2(T^{y_0^o}(q_{y_1}(ue), q_{y_1}(ue))) = T^{y_1^o}(q_{y_1}(ue), q_{y_1}(ue)) = ue(ue)^* = uee^*u$. Letting e be members of a net of increasing orthogonal projections tending to 1, we get the correct image under E_2 by polarization. That U_2 is generated by the preimages, is clear from the previous statement involving f. Concerning the estimates, let U be a von Neumann algebra of matrices (f_{ij}) with $i, j \in \{0, 1\}$ and $f_{ij} \in \mathrm{Hom}_{U_0^o}(V_j, V_i)$, so $f_{ii} \geq 0$ if such a matrix is positive, and we define a faithful semifinite normal weight E on U by $E(f_{ij}) = (g_{ij})$ with $g_{00} = f_{00}$ and $g_{11} = E_2(f_{11})$ and $g_{ij} = 0$ when $i \neq j$. Applying the inequality above in this proof to E instead of E_1, and with u and u' replaced by the matrices with f^* and g^* in the upper right corned and zeros otherwise, establishes the first estimate. The next two inequalities follow from the inequality we have just proved together with what we proved just before that. For the existence of A_1, pick an increasing sequence of orthogonal projections $p_n \in N_{E_1}$ tending to 1. Then $\|E_2(f \circ q_{E_1}(p_n)^*)\|^2 \leq \|E_2(f \circ f^*)\|$ by one of our estimates, and passing to a subsequence we may assume by compactness that $E_2(f \circ q_{E_1}(p_n)^*) \to A_1(f)$ weakly for some element $A_1(f) \in$

U_1. Then $A_1(f)(up_m)^* = \lim_n E_2(f \circ q_{E_1}(up_n p_m)^*) = E_2(f \circ q_{E_1}(up_m)^*)$. Taking weak limits as $m \to \infty$, and using one of our estimates, we get $A_1(f)u^* = E_2(f \circ q_{E_1}(u)^*)$, which also shows that $A_1(f)$ does not depend on $\{p_n\}$. Using again one of our estimates, we see that the only non-trivial part of the next sentence in the proposition is the matter of injectivity. Say $A_1(f) = 0$, so $E_2(f \circ q_{E_1}(u)^*) = 0$ for all $u \in N_{E_1}$. Let $u', u'' \in N_{E_1} \cap N_{y_1} \cap N_{E_1}^* \cap N_{y_1}^*$ and let $e_i \in N_{y_1}$ tend strongly to 1. Then as $q_{E_1}(u') \circ q_{E_1}(u'')^* = T^{y_0^o}(q_{y_1}(u'), q_{y_1}(u''))$, we may identify $q_{y_2}(q_{E_1}(u') \circ q_{E_1}(u'')^*)$ with $q_{y_1}(u') \otimes_{y_0} J_1 \Delta_1^{1/2} q_{y_1}(u'')$ which form a dense subset of $V_1 \otimes_{y_0} V_1 = V_2$. As $(q_{E_1}(u) \circ f^*)^* \circ (q_{E_1}(u) \circ f^*) = f \circ E_1(u^* u) \circ f^*$, we see that $q_{E_1}(u) \circ f^* \circ e_i \in N_{y_2}$ by what we have already shown at the beginning, and

$$(q_{y_2}(q_{E_1}(u') \circ q_{E_1}(u'')^*)|q_{y_2}(q_{E_1}(u) \circ f^* \circ e_i)) = y_1(e_i^* \circ E_2(f \circ q_{E_1}(u'' E_1(u^* u')^*)^*)) = 0.$$

So $q_{y_2}(q_{E_1}(u) \circ f^* \circ e_i) = 0$ by the established denseness in V_2, and using faithfulness of y_2 and taking the limit in i, we get $q_{E_1}(u) \circ f^* = 0$. Hence $f = 0$ by the first statement in the proposition involving f. Thus A_1 is injective. For the first property of A_1 involving h, we know that $h^* q_{y_1}(u') \in V_1$ is y_0^o-bounded, and $R^{y_0^o}(h^* q_{y_1}(u')) = h^* \circ q_{E_1}(u')$ Hence $T^{y_1^o}(h^* q_{y_1}(u'), h^* q_{y_1}(u')) = E_2 T^{y_0^o}(h^* q_{y_1}(u'), h^* q_{y_1}(u'))$ equals $E_2(h^* \circ q_{E_1}(u') \circ q_{E_1}(u')^* \circ h)$, which is bounded by $\|q_{E_1}(u')\|^2 E_2(h^* \circ h)$, so $h^* q_{y_1}(u') \in q_{y_1}(N_{y_1})$ as it is y_1^o-bounded. If $u'' \in N_{E_1} \cap N_{y_1}$ is analytic with respect to y_1, from what we have proved, we know that $T^{y_0^o}(q_{y_1}(u'), q_{y_1}(u'')) = q_{E_1}(u') \circ q_{E_1}(u'')^*$, so by multiplying with its adjoint from the left, we get an expression that is majorized by $\|q_{E_1}(u')\|^2 q_{E_1}(u'') \circ q_{E_1}(u'')^*$, and $T^{y_0^o}(q_{y_1}(u'), q_{y_1}(u''))$ belongs to N_{E_2}. Then $B \equiv E_2(h^* T^{y_0^o}(q_{y_1}(u'), q_{y_1}(u'')))^* E_2(h^* T^{y_0^o}(q_{y_1}(u'), q_{y_1}(u'')))$ is bounded, by the first inequality in this proof, and $y_1(B) \leq \|E_2(h^* \circ h)\|^2 \|q_{y_1}(u') \otimes_{y_0} J_1 \Delta_{y_1}^{1/2} q_{y_1}(u'')\|^2$, so $C \equiv E_2(h^* T^{y_0^o}(q_{y_1}(u'), q_{y_1}(u''))) \in N_{y_1}$. For $u''' \in N_{y_1}$, we get $y_1(u'''^* C) = y_1 T^{y_1^o}(u'''^* h^* q_{y_1}(u'), q_{y_1}(u'')) = (u'''^* h^* q_{y_1}(u')|\Delta_{y_1} q_{y_1}(u''))$ which equals $(J_1 \pi_l(\Delta_{y_1}^{1/2} q_{y_1}(u'')) J_1 h^* q_{y_1}(u')|q_{y_1}(u'''))$, where π_l is the left representation from Hilbert module theory. So $q_{y_1}(C) = J_1 \pi_l(\Delta_{y_1}^{1/2} q_{y_1}(u'')) J_1 h^* q_{y_1}(u')$. Now $\|A_1(h^* \circ q_{E_1}(u')) J_1 \Delta_{y_1}^{1/2} q_{y_1}(u'')\|^2 = y_1(u'' A_1(h^* \circ q_{E_1}(u'))^* A_1(h^* \circ q_{E_1}(u')) u''^*)$, so $\|A_1(h^* \circ q_{E_1}(u')) J_1 \Delta_{y_1}^{1/2} q_{y_1}(u'')\| = \|J_1 \pi_l(\Delta_{y_1}^{1/2} q_{y_1}(u'')) J_1 h^* q_{y_1}(u')\|$ by definition of A_1 and C. Let $c \in M_{E_1} \cap M_{y_1}$ Then as usual $c_n \equiv (n/\pi)^{1/2} \int_{-\infty}^{\infty} e^{-nt^2} \sigma_t^{y_1}(c)\, dt \in M_{E_1} \cap M_{y_1}$ is analytic with respect to y_1 and $c_n \to c$ strongly, while $q_{y_1}(c_n) \to q_{y_1}(c)$, and $\sigma_z^{y_1}(c_n) \in M_{E_1} \cap M_{y_1}$ for complex z. Plugging in $u'' = \sigma_{i/2}^{y_1}(c_n)$ and taking the limit, and then using one of the density result just proved, we conclude that $A_1(h^* \circ q_{E_1}(u'))$ belongs to N_{y_1}, and equals $h^* q_{y_1}(u')$. Next, letting $a \in N_{y_0}$, then $(E_1(A_1(h^* \circ q_{E_1}(u))^* A_1(h^* \circ q_{E_1}(u))) q_{y_0}(a)|q_{y_0}(a)) = y_1(A_1(h^* \circ q_{E_1}(ua))^* A_1(h^* \circ q_{E_1}(ua))) = \|h^* \circ q_{E_1}(u) q_{y_0}(a)\|^2$, so $E_1(A_1(h^* \circ q_{E_1}(u))^* A_1(h^* \circ q_{E_1}(u)))$ equals $q_{E_1}(u)^* \circ h \circ h^* \circ q_{E_1}(u)$. Thus $q_{E_1} A_1(h^* \circ q_{E_1}(u))$ belongs to N_{E_1} and equals $h^* \circ q_{E_1}(u)$. Finally, by combining what we have already proved, we easily check that $u' A_1(h^* \circ q_{E_1}(u))^* A_1(h^* \circ q_{E_1}(u)) u'^* \leq \|E_2(h^* \circ h)\| u' E_1(u^* u) u'^*$ for $u' \in N_{E_1}$, which gives the required inequality. □

We will henceforth assume that $U_0 \subset U_1$ is irreducible and of depth 2. Then this holds for any inclusion in the Jones tower. Note that $q_{E_2}(N_{E_2} \cap U_0') \subset V^1$, so V^1 is non-trivial when $N_{E_2} \cap U_0' \neq \{0\}$. Since any element of U_{i-1} commutes with any element of $E_{i+1}(U_{i-1}' \cap U_{i+1})$, we conclude by irreducibility that the restriction of E_{i+1} to $U_{i-1}' \cap U_{i+1}$ is a faithful normal weight x_i. Recall that E_i is regular if x_i and x_{i+1} are semifinite. The following result shows that for this to hold for all i, it suffices that E_1 is regular.

Proposition 17.2.13 *If x_i is semifinite, then x_{i+2} is semifinite and proportional to $x_i j_i j_{i+1}$.*

Proof Denote the representation of U_{i+1} on V_{i+k} by ρ_k for $k = 0, 1$. Then $(E_i)'_{\rho_0} = j_i E_{i+1} j_i$ on $\rho_0(U_{i-1})' = J_i U_{i+1} J_i$, so $(E_i)'_{\rho_0} = x_i j_i(\cdot)1$ on $U_{i-1}' \cap U_{i+1}$. Since $\rho_1(U_{i-1})' = J_{i+1} U_{i+3} J_{i+1}$ and $\rho_1(U_i)' = J_{i+1} U_{i+2} J_{i+1}$, then $j_{i+1}(E_i)'_{\rho_1} j_{i+1}$ is a faithful semifinite normal operator valued weight from U_{i+3} to U_{i+2}, and is therefore proportional to E_{i+3} by Corollary 10.13.4. So there is a positive scalar c_i such that $(E_i)'_{\rho_1} = c_i j_{i+1} E_{i+3} j_{i+1}$ on $J_{i+1} U_{i+3} J_{i+1}$. But $(E_i)'_{\rho_1} \rho_1(u) = x_i j_i(u)1$ for $u \in U_{i-1}' \cap U_{i+1}$ by the previous lemma, so $E_{i+3} j_{i+1}(u) = c_i^{-1} x_i j_i(u)1$, and then it remains to note that $j_{i+1}(u) \in U_{i+1}' \cap U_{i+3}$. □

Proposition 17.2.14 *If E_1 is regular, we have a unitary $f : V_{x_1} \to V^1$ which sends $q_{x_1}(u)$ to $q_{E_2}(u)$ for $u \in N_{x_1}$, and which intertwines π_{x_1} and π_1 when considered as representations on $U_0' \cap U_2$.*

Proof Clearly we get an isometry. The orthogonal projection $p \equiv ff^*$ belongs to $\pi_1(U_0' \cap U_3) = B(V^1)$. For $b \in U_0' \cap N_{E_2}$ and $c \in U_0' \cap N_{E_3}$, we have $\pi_1(c^*) q_{E_2}(b) = q_{E_2}(A_2(c^* \circ q_{E_2}(b))$ by Proposition 17.2.12 applied to $U_1 \subset U_2$. So $p\pi_1(c^*)p = \pi_1(c^*)p$, and p belongs to the commutant of $\pi_1(U_0' \cap U_3)$. Thus f is a unitary, and $\pi_1(a) f q_{x_1}(u) = q_{E_2}(au) = f \pi_{x_1}(a) q_{x_1}(u)$ for $a \in U_0' \cap U_2$. □

Proposition 17.2.15 *If x_i is semifinite, then $U_{i+1} = (U_i \cup (U_{i-1}' \cap U_{i+1})''$, so $G_i^* U_{i+1} G_i = (\alpha_i(U_i) \cup (\hat{W}_i' \otimes \mathbb{C}))''$, and $U_i^{\alpha_i} = U_{i-1}$. The von Neumann algebra generated by the elements $(\omega \otimes \iota)(M_i)$ for $\omega \in B(V^i)_*$ is $\hat{W}_i$. If x_{i+1} is semifinite, then the von Neumann algebra generated by $(\iota \otimes \omega)(M_i)$ for $\omega \in B(V^i)_*$ is W_i.*

Proof Clearly $U \equiv (U_i \cap (U_{i-1}' \cap U_{i+1}))'' \subset U_{i+1}$. Now U is invariant under $\sigma_t^{y_{i+1}}$ since this automorphism equals $\sigma_t^{y_i}$ and $\sigma_t^{y_{i-1}}$ on U_i and U_{i-1}, respectively. As $N_{x_i} N_{y_i} \subset N_{y_{i+1}}$ by Theorem 17.2.12, we also see that the restriction of y_{i+1} to U is semifinite. By Theorem 10.11.1 and its proof we thus have a conditional expectation $E : U_{i+1} \to U$ given by the projection onto the closure of $q_{y_{i+1}}(U \cap N_{y_{i+1}})$. But $q_{y_{i+1}}(ab) = G_i(q_{E_{i+1}}(a) \otimes q_{y_i}(b))$ for $a \in N_{x_i}$ and $b \in N_{y_i}$, and the set of all $q_{E_{i+1}}(a)$ is dense in V^i by the previous proposition, so $E = 1$ and $U_{i+1} = (U_i \cup (U_{i-1}' \cap U_{i+1})''$. In this equality we can obviously replace U_{i-1}' by U_{i-2}', and taking commutants in V_i, we then get $U_{i+1}' = U_i' \cap (U_{i-2}' \cap U_{i+1})'$, and applying j_i on both sides gives $U_{i-1} = U_i \cap (U'_{i-1} \cap U_{i+2})' = U_i^{\alpha_i}$. By the proof of Proposition 17.2.9 we see that $\pi_i(u)$ commutes with $(\omega_{f,g} \otimes \iota)(M_i)$ for $u \in U'_{i-1} \cap U_{i+1}$ and $f, g \in V^i$ if and only if $(1 \otimes_{y_{i-1}} g^*)(u \otimes_{y_{i-1}} 1) = u(1 \otimes_{y_{i-1}} g^*)$, and this latter identity holds for all g exactly when $u \in U_{i+2}^{\alpha_{i+2}}$. Hence, by Proposition 17.2.13 and by what we have already proved, the elements $(\omega \otimes \iota)(M_i)$ for $\omega \in B(V^i)_*$ generate the von Neumann algebra $\hat{W}_i$. The last statement is now clear from Proposition 17.2.8. □

Using $\Delta_{i+1} H_i = (H_i \otimes H_i)\hat{\Delta}_i$ and Proposition 17.2.10, it is easy to see that $\beta_i \pi_i = (\iota \otimes \pi_i)(G_i^* j_{i+1} j_i(\cdot) G_i) = (H_i^{-1} \otimes \pi_i j_i)\alpha_{i+1} j_i$ defines an injective normal $*$-homomorphism $\beta_i : \hat{W}'_i \to \hat{W}_i \bar{\otimes} \hat{W}'_i$ such that $(\iota \otimes \beta_i)\beta_i = (\hat{\Delta}_i^{\mathrm{op}} \otimes \iota)\beta_i$.

Proposition 17.2.16 *If E_1 is regular, then $(\iota \otimes x_i)\beta_i = x_i(\cdot)1$.*

Proof By Proposition 17.2.14 we identify $U'_{i-1} \cap U_{i+1}$ with its image $\hat{W}'_i$ under π_i and π_{x_i}. Now $G_i(\iota \otimes E_{i+1})(G_i^*(\cdot)G_i)G_i^*$ is clearly a faithful semifinite normal operator valued weight from U_{i+3} to U_{i+2}, which is proportional to E_{i+3} by Corollary 10.13.4. Hence $E_{i+3} j_{i+1}$ and $G_i(\iota \otimes x_i)(G_i^* j_{i+1}(\cdot)G_i)G_i^*$ are proportional on $U'_{i-1} \cap U_{i+1}$. By Proposition 17.2.13 there is therefore a scalar $c > 0$ such that $(\iota \otimes x_i)\beta_i = c x_i(\cdot)1$. Standard arguments allow us to define an operator M'_i on $V^i \otimes V^i$ by $(\omega \otimes \iota)(M'_i) q_{x_i} = q_{x_i}(\omega \otimes \iota)\beta_i \pi_i$ for ω in the predual of $\hat{W}_i$, such that $(M'_i)^* M'_i = c^{-1}$. Using the 'coaction' property of β_i we readily check that $(\hat{\Delta}_i \otimes \iota)(M'_i) = (M'_i)_{13}(M'_i)_{23}$, so $c^{-1} = c^{-2}$ and $c = 1$. □

Given the definitions of α_i and π_i, and as $U_i^{\alpha_i} = U_{i-1}$, we see that α_i is a minimal coaction whenever E_1 is regular. Note that $W'_i \cap \hat{W}_i = \mathbb{C}$. Indeed, from Proposition 17.2.15 we have $U'_{i+1} = U'_i \cap (U'_{i-1} \cap U_{i+1})'$, which when intersected with U_{i+2} is $\mathbb{C}$ by irreducibility. Now $\hat{W}_i = \pi_i(U'_{i-1} \cap U_{i+2} \cap (U'_{i-1} \cap U_{i+1})')$ as $\pi_i(U'_{i-1} \cap U_{i+2}) = B(V^i)$, so $W'_i \cap \hat{W}_i = \mathbb{C}$ as we get π_i of the previous intersection. Let $\mathrm{Ad}k \equiv k(\cdot)k^{-1}$.

Proposition 17.2.17 *Say E_1 is regular. Then there exists a positive invertible operator k_1 in V^1 such that $E_1E_2E_3 = (\mathrm{Tr}_{k_1} \otimes E_1)(G_1^*(\cdot)G_1)$ and $y_3 = (\mathrm{Tr}_{k_1} \otimes y_1)(G_1^*(\cdot)G_1)$ on the positive part of U_3. On $U'_0 \cap U_3$ we have $\pi_1 \sigma_t^{y_3} = k_1^{it}\pi_1(\cdot)k_1^{-it} \equiv \rho_t \pi_1$. While on U_1 we have $\alpha_1 \sigma_t^{y_1} = (\rho_t \otimes \sigma_t^{y_1})\alpha_1$. We also have $x_1 = x_1 \rho_t$, and k_1^{it} is the standard implementation of $\sigma_t^{y_2} = \sigma_t^{y_3}$ on $\pi_1(U'_0 \cap U_2)$, so it commutes with J_{x_1} and Δ_{x_1}, and $\Delta_{y_2}^{it} G_1 = G_1(k_1^{it} \otimes \Delta_{y_1}^{it})$. On the positive part of $U'_0 \cap U_3$ we have $E_2E_3 = \mathrm{Tr}(\Delta_{x_1}^{1/2}\pi_1(\cdot)\Delta_{x_1}^{1/2})$. We have $(\iota \otimes y_1)\alpha_1 = y_1(\cdot)\overline{\Delta_{x_1} k_1^{-1}}$ on the positive part of M_{y_1}. The elements $\Delta_{x_1}^{it} k_1^{-it}$ are group-like for Δ_1, and we have $x'_1 \rho_t = x'_1$ and $\sigma_t^{x'_1} \rho_t = \rho_t \sigma_t^{x'_1}$ and $\hat{\Delta}_1 \sigma_t^{x'_1} \rho_t = (\sigma_t^{x'_1} \rho_t \otimes \iota)\hat{\Delta}_1$, where as usual $x'_1 = x_1(J_{x_1}(\cdot)J_{x_1})$ on the positive part of $\pi_1(U'_0 \cap U_2)'$.*

Proof Consider the faithful semifinite normal operator valued weight $E = G_1^* E_1E_2E_3(G_1(\cdot)G_1^*)G_1$ from $B(V^1)\bar{\otimes}U_1$ to $\mathbb{C} \otimes U_0 = U_0$. By Corollary 10.13.4 we have $(DE : D(\mathrm{Tr}\otimes E_1))_t = (Dy_0E : D(\mathrm{Tr}\otimes y_1))_t \in B(V^1)\otimes\mathbb{C}$ for any faithful semifinite normal weight y_0 on U_0. Hence this cocycle equals $k_1^{it} \otimes 1$ for a positive invertible operator k_1 in V^1, and $E = \mathrm{Tr}_{k_1} \otimes E_1$. On $U_3 = G_1(B(V^1)\bar{\otimes}U_1)G_1^*$ we have $(\rho_t \otimes \sigma_t^{y_1})(G_1^*(\cdot)G_1) = G_1^*\sigma_t^{y_3}(\cdot)G_1$, so $\pi_1\sigma_t^{y_3} = \rho_t\pi_1$ on $U_0' \cap U_3$. Since $\sigma_t^{y_3} = \sigma_t^{y_1}$ on U_1, we also get $\alpha_1\sigma_t^{y_1} = (\rho_t \otimes \sigma_t^{y_1})\alpha_1$ on U_1. Since $E_2\sigma_t^{y_2} = \sigma_t^{y_1}E_1$ on U_2, we get $x_1 = x_1\rho_t$ on $U_0' \cap U_2$. Define a positive invertible operator k in V^1 by $k^{it}q_{x_1} = q_{x_1}\sigma_t^{y_3}$, so $\rho_t = k^{it}(\cdot)k^{-it}$ on $\pi_1(U_0' \cap U_2)$ is the standard implementation. Now

$$\Delta_{y_2}^{it}G_1(q_{x_1}(a)\otimes q_{y_1}(b)) = \Delta_{y_2}^{it}q_{y_2}(ab) = q_{y_2}\sigma_t^{y_2}(ab) = G_1(k^{it}q_{x_1}(a)\otimes\Delta_{y_1}^{it}q_{y_1}(b))$$

for $a \in N_{x_1}$ and $b \in N_{y_1}$, so $\Delta_{y_2}^{it}G_1 = G_1(k^{it}\otimes\Delta_{y_1}^{it})$ and $\alpha_1\sigma_t^{y_1} = (\mathrm{Ad}k^{it}\otimes\sigma_t^{y_1})\alpha_1$. By minimality of α_1, we thus see that $\rho_t = \mathrm{Ad}k^{it}$ on W_1. But they also coincide on $\hat{W}_1'$, and hence on $B(V^1)$ as $W_1' \cap \hat{W}_1 = \mathbb{C}$. Hence $k = k_1$. Next, if $a, b \in N_{x_1} \cap N_{x_1}^*$, then $\pi_1(q_{E_2}(a)q_{E_2}(b)^*) = (\cdot|q_{E_2}(b))q_{E_2}(a)$ and $E_3(q_{E_2}(a)q_{E_2}(b)^*) = ab^*$. Using this and the definition of $J_{x_1}\Delta_{x_1}^{1/2}$, we get $E_2E_3 = \mathrm{Tr}(\Delta_{x_1}^{1/2}\pi_1(\cdot)\Delta_{x_1}^{1/2})$ on the positive part of $U_0' \cap U_3$ by density and normality. For the $\overline{\Delta_{x_1}k_1^{-1}}$-invariance of y_1, let $a \in N_{E_1} \cap N_{y_1}$ be analytic with respect to y_1, and let $b \in U_0' \cap N_{E_2E_3}$. Then, from what we have already shown, we get

$$(\mathrm{Tr}_{k_1} \otimes E_1)(G_1^*(ab)^*(ab)G_1) = y_1E_2E_3(b^*a^*ab) = y_1(b^*b)\mathrm{Tr}(\Delta_{x_1}^{1/2}\pi_1(a^*a)\Delta_{x_1}^{1/2}).$$

If a is analytic with respect to y_3, then using the definition of α_1 and π_1 in combination with what we have already shown, the left hand side equals

$$\mathrm{Tr}(\pi_1(a)k_1^{1/2}(\iota \otimes y_1)\alpha_1(\sigma_t^{y_1}(b)\sigma_t^{y_1}(b)^*)k_1^{1/2}\pi_1(a^*)).$$

For any unit vector v in the domain of $k_1^{1/2}$ and $\Delta_{x_1}^{1/2}$, pick $a \in U_0' \cap U_3$ such that $\pi_1(a)$ is the orthogonal projection onto $\mathbb{C}v$. Then the last expression equals $\|(\iota \otimes y_1)\alpha_1(\sigma_t^{y_1}(b)\sigma_t^{y_1}(b)^*)k_1^{1/2}v\|^2$, while the right hand side in the identity above equals $y_1(b^*b)\|\Delta_{x_1}^{1/2}v\|^2$, and we have the required invariance. This invariance also shows that $\overline{\Delta_{x_1}k_1^{-1}}$ is affiliated with W_1, and for any v, v' in its domain, and with $u \in M_{y_1}$, we get $y_1(u)\|\overline{\Delta_{x_1}k_1^{-1}}v\|^2\|\overline{\Delta_{x_1}k_1^{-1}}v'\|^2 = (\omega_v \otimes \omega_{v'} \otimes y_1)(\iota \otimes \alpha_1)\alpha_1(u)$, which equals $(\omega_v'\otimes\omega_v)\Delta_1(\iota\otimes y_1)\alpha_1(u) = y_1(u)(\omega_v'\otimes\omega_v)\Delta_1(\overline{\Delta_{x_1}k_1^{-1}})$, and $\Delta_{x_1}^{it}k_1^{-it}$ are group-like for Δ_1. This gives a commutation relation with M_1, and then the rest is straightforward from what we have already established. □

From the above result, we see that $\tau_t \equiv \Delta_{x_1}^{-it}(\cdot)\Delta_{x_1}^{it}$ is a $*$-automorphism on W_1 such that $\Delta_1\tau_t\rho_t = (\tau_t\rho_t \otimes \tau_t\rho_t)\Delta_1$.

Proposition 17.2.18 *Say E_1 is regular. For $h \in N_{x_2}$ analytic with respect to y_3, and u in $N_{x_1} \cap N_{x_1}^*$, we have $q_{x_1}A_2(h^*q_{x_1}(u^*)) = \pi_1(h^*)q_{x_1}(u^*)$ and $q_{x_1}(A_2(h^*q_{x_1}(u^*))^*) = J_{x_1}\pi_1\sigma_{-i/2}^{y_3}(h^*)J_{x_1}q_{x_1}(u)$. If $a \in (N_{E_2} \cap N_{y_2}) \cap (N_{E_2} \cap N_{y_2})^*$, then we have the identity $q_{y_2}(A_2(h^* \circ q_{E_2}(a^*))^*) = J_2\sigma_{-i/2}^{y_3}(h^*)J_2q_{y_2}(a)$. When $a \in N_{E_2} \cap N_{E_2}^*$, we get $q_{E_2}(A_2(h^* \circ q_{E_2}(a^*))^*) = J_2\sigma_{-i/2}^{y_3}(h^*)J_2q_{E_2}(a)$.*

Proof The first identity is immediate from Proposition 17.2.12 and Proposition 17.2.14. The previous proposition gives

$$\pi_1\sigma_{-i/2}^{y_3}(h^*)J_{x_1}q_{x_1}(u) = k_1^{1/2}\pi_1(h^*)k_1^{-1/2}\Delta_{x_1}^{1/2}q_{x_1}(u^*) = \pi_1(h^*)\Delta_{x_1}^{1/2}q_{x_1}(u^*),$$

and applying J_{x_1} on both sides, the second identity in the proposition follows now from the first one. The third identity follows readily from Proposition 17.2.12 and $\sigma_{-i/2}^{y_3}(h^*) = \Delta_{y_2}^{1/2}h^*\Delta_{y_2}^{1/2}$ and the definition of $J_2\Delta_{y_2}^{-1/2}$ used twice. For the final identity, pick $b, c \in N_{y_1}$, and apply the third identity in the proposition to b^*ac instead of a. This gives $q_{y_2}(b^*A_2(h^* \circ q_{E_2}(a^*))^*c) = b^*J_2\sigma_{-i/2}^{y_3}(h^*)J_2q_{E_2}(a)q_{y_1}(c)$. Using this and letting bb^* tend increasingly to 1, one gets $A_2(h^* \circ q_{E_2}(a^*)) \in N_{E_2}$ and $q_{y_2}A_2(h^* \circ q_{E_2}(a^*)c) = q_{E_2}A_2(h^* \circ q_{E_2}(a^*))q_{y_1}(c)$. Letting $b \to 1$ increasingly again in the above identity, we then get the final result. □

Corollary 17.2.19 *Let E_1 be regular. Then $\pi_1 j_2 = J_{x_1}\pi_1(\cdot)^*J_{x_1}$ on $U_1' \cap U_3$, and $J_{x_1}(\cdot)^*J_{x_1}$ is an antimultiplicative $*$-preserving involution on W_1 that commutes with ρ_t.*

Proof The first identity is clear from the previous proposition and Proposition 17.2.14, and this then shows that the involution in question leaves W_1 invariant, and commutes with ρ_t as J_{x_1} commutes with k_1^{it}. □

Proposition 17.2.20 *Assume E_1 is regular. Then $x_3 j_2 j_1 = x_1$, while we have $(\iota \otimes E_2)(G_1^*(\cdot)G_1) = G_1^*E_4(\cdot)G_1$ and $y_4 = (\mathrm{Tr}_{k_1} \otimes y_2)(G_1^*(\cdot)G_1)$ on the positive part of U_4, and $(\iota \otimes x_1)(G_1^*(\cdot)G_1) = \pi_1 E_4(\cdot) \otimes 1$ on the positive part of $U_0' \cap U_4$, and $\beta_1\rho_t = (\rho_t \otimes \rho_t)\beta_1$, and $k_1^{it} \otimes k_1^{it}$ commutes with the operator M_1' defined in the proof of Proposition 17.2.16.*

Proof The first two identities are gotten by mimicking the proof of Proposition 17.2.13, and fixing the proportionality constant for the second identity by using the left invariance of β_1 under x_1. The third and the fourth identity are then clear from Proposition 17.2.17 and the definition of π_1. Using $j_2 j_1 \sigma_t^{y_2} = \Delta_{y_2}^{it} j_2 j_1(\cdot)\Delta_{y_2}^{-it}$ on $U_0' \cap U_2$ together with the definition of β_1 and the identity $\Delta_{y_2}^{it}G_1 = G_1(k_1^{it} \otimes \Delta_{y_1}^{it})$ from Proposition 17.2.17, we get $\beta_1\rho_t = (\rho_t \otimes \rho_t)\beta_1$. The last statement is then clear from the definition of M_1' in terms of β_1. □

Proposition 17.2.21 *Let E_1 be regular. Let p_v be the orthogonal projection onto $\mathbb{C}v$ for a unit vector $v \in V^1$, let F be the flip on $V^1 \otimes V^1$, and $a \in N_{x_1}$. Then*

$$(\iota \otimes \omega_{q_{x_1}(a),v})(FM_1')^*(\iota \otimes \omega_{q_{x_1}(a),v})(FM_1') = (\iota \otimes x_1)(\beta_1\pi_1(a^*)(p_v \otimes 1)\beta_1\pi_1(a))$$

with M_1' from the proof of Proposition 17.2.16.

Proof This is straightforward from definitions and the identity

$$(\omega_{v',v} \otimes \iota)(B)^*(\omega_{v',v} \otimes \iota)(B) = (\omega_{v'} \otimes \iota)(B^*(p_v \otimes 1)B),$$

which evidently holds for any $B \in B(V^1 \otimes V^1)$ and $v' \in V^1$. □

Proposition 17.2.22 *Let E_1 be regular, and let k_2 be defined as k_1 in Proposition 17.2.17 but now with respect to the inclusion $U_1 \subset U_2$. Then $F_1^* k_2 F_1 = \overline{k_1^{-1}}$. Let $\tilde{\Delta}_1 \equiv F_1^* \Delta_{x_2} F_1$. Then $\tilde{\Delta}_1^{is}$ and k_1^{it} commute, and $(\iota \otimes x_1 j_1)\beta_1 \pi_1 = x_1 j_1(\cdot)\tilde{\Delta}_1 k_1$ on the positive part of $U_0' \cap U_2$. We have $M_1(\tilde{\Delta}_1^{it} k_1^{it} \otimes 1)M_1^* = \tilde{\Delta}_1^{it} k_1^{it} \otimes \tilde{\Delta}_1^{it} k_1^{it}$.*

Proof Using $\pi_j \sigma_t^{y_{j+2}} = k_j^{it} \pi_j(\cdot) k_j^{-it}$ on $U_{j-1}' \cap U_{j+2}$ for $j = 1, 2$, and $\pi_2 j_2 = F_1 \pi_1(\cdot) F_1^*$ on $U_0' \cap U_3$, we get $F_1 k_1^{it} \pi_1(\cdot) k_1^{-it} = k_2^{it} F_1 \pi_1(\cdot) F_1^* k_2^{-it} F_1$ on $U_0' \cap U_3$, so $F_1^* k_2^{it} F_1 = k_1^{it}$ by irreducibility of π_1, and thus $F_1^* k_2 F_1 = \overline{k_1^{-1}}$ by conjugate linearity of F_1. Commutation is clear as it holds for $\Delta_{x_2}^{is}$ and k_2^{it} by Proposition 17.2.17. The same proposition tells us that $(\iota \otimes E_2)\alpha_2 = \overline{\Delta_{x_2} k_2^{-1}} \otimes E_2(\cdot)$ on the positive part of M_{E_2}, and using the relation between α_2 and β_1, we now get the invariance property for $x_1 j_1$ with respect to β_1. The next statement is proved in the same way as we established group-likeness in Proposition 17.2.17. □

Note that $\rho_t = \mathrm{Ad}\tilde{\Delta}_1^{-it}$ on $\pi_1(U_0' \cap U_2)$. Let $\hat{\tau}_t$ be the $*$-automorphism $\mathrm{Ad}\tilde{\Delta}_1^{-it}$ on $\hat{W}$.

Corollary 17.2.23 *Let E_1 be regular, and let $\tilde{J}_1 = F_1^* J_{x_2} F_1$. Then $j_3 j_2 = j_2 j_1$ and $\tilde{J}_1 \pi_1(\cdot)^* \tilde{J}_1 = \pi_1 j_1$ on $U_0' \cap U_2$, and $\tilde{J}_1 k_1 \tilde{J}_1 = k_1^{-1}$. We also have $\hat{\tau}_t \rho_s = \rho_s \hat{\tau}_t$ and $\hat{\Delta}_1 \hat{\tau}_{-t} \rho_t = (\hat{\tau}_{-t} \rho_t \otimes \hat{\tau}_{-t} \rho_t)\hat{\Delta}_1$.*

Proof Fo any operator u on V_1 we have $j_2 j_1(u) = 1 \otimes_{y_0} u$ and $j_3 j_2(u) = 1 \otimes_{y_0} u \otimes_{y_0} 1$, so they are equal under our identifications. Using this and $\pi_2 j_2 = F_1 \pi_1(\cdot) F_1^*$ together with the previous corollary, we get the next result. The rest is obvious from the previous proposition and Proposition 17.2.17. □

Proposition 17.2.24 *Given that E_1 is regular, we have $M_1' = \hat{M}_1$ with M_1' from the proof of Proposition 17.2.16.*

Proof We claim that $L \equiv (\iota \otimes \omega_{v,q_{x_1}(a)})(FM_1)(\iota \otimes \omega_{v,q_{x_1}(a)})(FM_1)^*$ equals the expression in Proposition 17.2.21. Indeed, by definition of M_1 we have the identity $((\iota \otimes \omega_{f_1,f_2})(FM_1)g_1|g_2) = g_2^*(1 \otimes_{y_0} f_2^*)(f_1 \otimes_{y_0} 1)g_1$, so $(\iota \otimes \omega_{f_1,f_2})(FM_1) = \pi_1((1 \otimes_{y_0} f_2^*)(f_1 \otimes_{y_0} 1))$. Hence L is π_1 of $(1 \otimes_{y_0} q_{E_2}(a)^*)(vv^* \otimes_{y_0} 1)(1 \otimes_{y_0} q_{E_2}(a))$, and this element, when acting on V_3, equals $q_{E_4} j_2 j_1(a)^*(vv^* \otimes_{y_0} 1 \otimes_{y_0} 1) q_{E_4} j_2 j_1(a)$ by Proposition 17.2.20. Applying $\omega_{q_{y_3}(c)}$ to this element, for $c \in N_{y_3}$, we get $y_4(c^* j_2 j_1(a)^* vv^* j_2 j_1(a) c) = (E_4(j_2 j_1(a^*) vv^* j_2 j_1(a)) q_{y_3}(c) | q_{y_3}(c))$. Hence the claim holds by Proposition 17.2.20 and the definition of β_1 and $\pi_1(vv^*) = p_v$. By our proved claim, and the identity in the proof of Proposition 17.2.20, we therefore get $\hat{M}_1^*(p_v \otimes 1)\hat{M}_1 = M_1'^*(p_v \otimes 1)M_1'$. Then $(M_1'\hat{M}_1^*)^*(p_v \otimes 1)(M_1'\hat{M}_1^*) = p_v \otimes 1$ and thus $M_1' M_1'^*(p_v \otimes 1) M_1' M_1'^* = (M_1'\hat{M}_1^*)(p_v \otimes 1)(M_1'\hat{M}_1^*)^*$ is an orthogonal projection, so $M_1'M_1'^*$ and $p_v \otimes 1$ commute, and there is an orthogonal projection q such that $M_1'M_1'^* = 1 \otimes q$. Now $(p_v \otimes 1)M_1' = (p_v \otimes 1)(M_1'M_1'^*M_1') = (p_v \otimes q)M_1'$, which equals $M_1'\hat{M}_1^*(p_v \otimes 1)\hat{M}_1$, so $M_1'\hat{M}_1^*$ and $p_v \otimes 1$ commute, and there is an isometry s such that $M_1'\hat{M}_1^* = 1 \otimes s$. But $(\hat{\Delta}_1 \otimes \iota)(1 \otimes s) = (\hat{\Delta}_1 \otimes \iota)(M_1'\hat{M}_1^*) = 1 \otimes 1 \otimes s^2$ by invoking the corepresentation properties of M_1' and $\hat{M}_1$. So $s^2 = s$ and $s = 1$ as s is injective. □

In view of this proposition and our previous results, we then see that M_1 and $k_1^{it} \otimes k_1^{it}$ commute, so $\hat{\Delta}_1\rho_t = (\rho_t \otimes \rho_t)\hat{\Delta}_1$, and similarly for Δ_1, and also with ρ replaced by $\hat{\tau}_t$ and τ_t, respectively. We also get $\hat{\Delta}_1\sigma_t^{x_1'} = (\sigma_t^{x_1'} \otimes \rho_{-t})\hat{\Delta}_1$.

Proposition 17.2.25 *Let E_1 be regular, and consider $R \equiv J_{x_1}(\cdot)^*J_{x_1}$ and $\hat{R} \equiv \tilde{J}_1(\cdot)^*\tilde{J}_1$ on W_1 and $\hat{W}_1$, respectively. Then τ_t and R commute, and so does $\hat{\tau}_t$ and $\hat{R}$. We have $\tau_{-i/2}(\iota \otimes \omega)(M_1) = R(\iota \otimes \omega)(M_1^*)$ and $\hat{\tau}_{-i/2}(\iota \otimes \eta)(\hat{M}_1) = \hat{R}(\iota \otimes \eta)(\hat{M}_1^*)$ for $\omega \in (\hat{W}_1)_*$ and $\eta \in (W_1)_*$ such that the left hand sides to make sense, and $(J_{x_1} \otimes \tilde{J}_1)M_1 = M_1^*(J_{x_1} \otimes \tilde{J}_1)$ and $(R \otimes \hat{R})(M_1) = M_1 = (\tau_t \otimes \hat{\tau}_t)(M_1)$. We have $\hat{\Delta}_1\hat{R}' = (\hat{R} \otimes \hat{R}')\beta_1$, where $\hat{R}'$ is the linear map from $\hat{W}_1'$ to $\hat{W}_1$ given by $J_{x_1}(\cdot)^*J_{x_1}$. We have $(\iota \otimes x_1')(\hat{\Delta}_1(a^*)(1 \otimes b)) = (\iota \otimes \omega_{q_{x_1'}(b),q_{x_1'}(a)})(\hat{M}_1)$ for $a, b \in N_{x_1'}$.*

Proof The first statement is obvious. By the previous proposition and the definition of M_1', we have $(\iota \otimes \omega)(M_1^*)q_{x_1} = q_{x_1}(\omega \otimes \iota)\beta_1\pi_1$. Using this twice with the definition of $J_{x_1}\Delta_{x_1}^{1/2}$, we get $(\iota \otimes \omega)(M_1^*)\Delta_{x_1}^{-1/2}J_{x_1}q_{x_1} = \Delta_{x_1}^{-1/2}J_{x_1}(\iota \otimes \omega)(M_1)^*q_{x_1}$ on $N_{x_1} \cap N_{x_1}^*$, which proves the identity with M_1. The one with $\hat{M}_1$ follows from the first identity applied to the inclusion $U_1 \subset U_2$ together with the relation between M_2 and M_1 involving F_1. Using both these identities, a straightforward calculation then shows that $((J_{x_1} \otimes \tilde{J}_1)M_1(J_{x_1} \otimes \tilde{J}_1)(a_1 \otimes b_1)|a_2 \otimes b_2) = (M_1^*(\Delta_{x_1}^{1/2}a_1 \otimes \tilde{\Delta}_1^{1/2}b_1)|\Delta_{x_1}^{-1/2}a_2 \otimes \tilde{\Delta}_1^{-1/2}b_2)$ for a_i and b_i in appropriate domains. Hence $(\Delta_{x_1}^{-1/2} \otimes \tilde{\Delta}_1^{-1/2})M_1^*(\Delta_{x_1}^{1/2} \otimes \tilde{\Delta}_1^{1/2}) \subset (J_{x_1} \otimes \tilde{J}_1)M_1(J_{x_1} \otimes \tilde{J}_1)$. Taking the adjoint of this and composing, we see that M_1 commutes with $\Delta_{x_1} \otimes \tilde{\Delta}_1$, which then proves the remaining formulas involving M_1. The formula involving β_1 is evident, and by definition of M_1', we get the final formula involving $\hat{\Delta}_1$ and $\hat{M}_1$ by using the second identity in the proposition. □

Corollary 17.2.26 *Let E_1 be regular. Then we have $\hat{\Delta}_1^{\mathrm{op}}\hat{R} = (\hat{R} \otimes \hat{R})\hat{\Delta}_1$ and $(\iota \otimes x_1')\hat{\Delta}_1 = x_1'(\cdot)1$. Hence $(\hat{W}_1, \hat{\Delta}_1)$ is a locally compact quantum group with left invariant faithful semifinite normal weight x_1', unitary coinverse $\hat{R}$ and scaling group $\{\hat{\tau}_t\}$, and its fundamental multiplicative unitary is $\hat{M}_1$. It is the dual of the locally compact quantum group (W_1, Δ_1), which has $x_2'H_1$ as left invariant faithful semifinite normal weight, and R as unitary coinverse and $\{\tau_t\}$ as scaling group, and M_1 as fundamental multiplicative unitary.*

Proof The first identity can be check on the dense set of elements of the form $(\eta \otimes \iota)(M_1)$, and for them it follows readily by using the identity in the previous proposition involving $\hat{R}$ and $\hat{\tau}_{-i/2}$ together with the definition of $\hat{\tau}_t$ and its commutation relation with $\hat{\Delta}_1$. The left invariance property of x_1' with respect to $\hat{\Delta}_1$ is clear from the left invariant property of x_1 with respect to β_1, and the relation between $\hat{\Delta}_1$ and β_1 given in the previous proposition. The remaining statements concerning $(\hat{W}_1, \hat{\Delta}_1)$ are clear from the previous proposition; the last statement in that proposition shows that $\hat{M}_1$ is the fundamental multiplicative unitary. That $x_2'H_1$ is left invariant for Δ_1 follows from left invariance of x_2' for $\hat{\Delta}_2$, which holds from what we have already proved by considering the inclusion $U_1 \subset U_2$, and then combining this with the relation between M_2 and M_1 involving F_1. The remaining dual statements are now clear. □

It is time to cut the chase and sum up.

Theorem 17.2.27 *Suppose $U_0 \subset U_1$ is an irreducible depth 2 inclusion of factors with a regular faithful semifinite normal operator valued weight E_1 from U_1 to U_0. Consider the Jones tower $U_0 \subset U_1 \subset U_2 \subset \cdots$ associated to a chosen faithful semifinite normal weight y_0 on U_0. Then for each i the associated faithful semifinite normal operator valued weight E_{i+1} from U_{i+1} to U_i is regular, and its restriction to $U_{i-1}' \cap U_{i+1}$ is a faithful semifinite normal weight x_i with semicyclic representation unitarily equivalent to the restriction of the representation π_i defined in Proposition 17.2.6, being surjective as a map from its full domain $U_{i-1}' \cap U_{i+2}$. Consider W_i and $\hat{W}_i$ and Δ_i and $\hat{\Delta}_i$ and α_i from Proposition 17.2.8 and Proposition 17.2.9 and Proposition 17.2.10. Then (W_i, Δ_i) and $(\hat{W}_i, \hat{\Delta}_i)$ are locally compact quantum groups which are dual to each other, and the quantum group (W_{i+1}, Δ_{i+1}) is isomorphic to $(\hat{W}_i, \hat{\Delta}_i)'$, that is, to the commutant of the dual. Moreover, we have that α_i is a coaction of $(W_i, \Delta_i^{\mathrm{op}})$ on U_i such that $U_i^{\alpha_i} = U_{i-1}$, and such that the inclusion $\alpha_i(U_i) \subset W_i \ltimes_{\alpha_i} U_i$ is isomorphic to the inclusion $U_i \subset U_{i+1}$, and with dual coaction $\hat{\alpha}_i$ isomorphic to α_{i+1}. So the Jones tower appears as a successive dual crossed product construction starting from α_1.*

Proof We have already proved that each E_{i+1} is regular with required restriction a weight x_i. From the previous corollary we thus know that $(\hat{W}_i, \hat{\Delta}_i)$ is a locally compact quantum group with left invariant faithful semifinite normal weight x_i', and is the dual of the locally compact quantum group (W_i, Δ_i), which has $x_{i+1}'H_i$ as left invariant faithful semifinite normal weight. Using $J_{x_i'}$ and F_i we see that $(W_{i+1}, \Delta_{i+1}) \cong (\hat{W}_i, \hat{\Delta}_i)'$. We know α_i is a coaction, and concerning the fixed point algebras and inclusions, see Proposition 17.2.15. By definition $\hat{\alpha}_i$ is a coaction of $(W_i, \Delta_i^{\mathrm{op}})^{\wedge \mathrm{op}} \cong (\hat{W}_i, \hat{\Delta}_i)'^{\mathrm{op}} \cong (W_{i+1}, \Delta_{i+1}^{\mathrm{op}})$ on $W_i \ltimes_{\alpha_i} U_i \cong U_{i+1}$, and it is straightforward to see that this is just the coaction α_{i+1}. □

This theorem also settles the briefer claim towards the end of the last section.

Exercises

For Sect. 17.1

1. Consider a von Neumann dynamical system (U, G, α) with the unitary implementation $\{u_t\}$ of α in the standard representation of U, and let $U^\alpha \subset U \subset U_2$ be the basic construction. Show that $U_2 = (U \cup \{u_t \mid t \in G\})''$. Prove directly that when G is finite, then there exists a $*$-homomorphism $f\colon L^\infty(G) \ltimes_\alpha U \to U_2$ such that $f(\lambda_t) = u_t$ and $f\alpha = \iota$ on U. In this case show that f is injective if the action of G is free and U is a II_1-factor.
2. Let α be a coaction on a von Neumann algebra U of a locally compact quantum group (W, Δ) with $\dim(W) < \infty$, and consider the basic construction $U^\alpha \subset U \subset U_2$. Prove directly that U_2 is the quotient of $W \ltimes_\alpha U$ induced by a $*$-epimorphism that sends $\alpha(v)$ to v for $v \in U$. Conclude that when $W \ltimes_\alpha U$ is a factor, the inclusion $\mathbb{C} \otimes U^\alpha \subset \alpha(U) \subset W \ltimes_\alpha U$ is the basic construction.
3. What does outerness of a coaction mean when it stems from an action of a locally compact group on a locally compact Hausdorff space?

For Sect. 17.2

1. Let $W \subset B(V)$ be a type II_1 factor with a tracial state x and with the Hilbert space V assumed to be separable. Define $\dim_W(V)$ to be 1 if $V = V_x$, and to be $\sum_{i=1}^n x(p_{ii})$ when there exists $n \in \mathbb{N}$ and a W-module isometry $u\colon V \to V_x \otimes \mathbb{C}^n$ with $(p_{ij}) \in M_n(W)$ corresponding to $uu^* \in (W \otimes \mathbb{C}1_n)'$ under the isomorphism $J_x(\cdot)J_x\colon W \to W'$. If no such n exists, let $\dim_W(V) = \infty$. Show that $\dim_W$ is well-defined, that it coincides on V_1 and V_2 if and only if they are equivalent representation spaces for W, that it respects countable direct sums of representation spaces, and that $\dim_W(V)\dim_{W'}(V) = 1$.
2. Using the the previous exercise, define the Jones' index $[W : U]$ of a subfactor U in a type II_1 factor W with tracial state x to be $\dim_U(V_x)$. Show that if $[W : U]$ is finite, then it equals $[U' : W']$, and that if U_1 is a subfactor of U of finite index, then $[W : U_1] = [W : U][U : U_1]$. What can be said of the index of the fixed point algebra in W of a finite group action?
3. Consider a finite index subfactor U_0 of a type II_1 factor U_1 with tracial state x, and consider the corresponding Jones tower $U_0 \subset U_1 \subset U_2 \subset \cdots$ Using that the unique tracial state y on U_0 is the restriction of x, we may assume $V_y \subset V_x$. Let p_1 be the orthogonal projection onto V_y. Repeating this, show that the Jones tower can be written as

$$U_1 \cap \{p_1\}' \subset U_1 \subset (U_1 \cup \{p_1\})'' \subset ((U_1 \cup \{p_1\})'' \cup \{p_2\})'' \subset \cdots,$$

where the so called Jones projections p_i satisfy $p_i p_{i\pm 1} p_i = [U_1 : U_0]^{-1} p_i$ while $p_i p_j = p_j p_i$ if $|i - j| \geq 2$, and $z(p_n w) = [U_1 : U_0]^{-1} z(w)$ for $w \in (U_1 \cup \{p_1, \ldots, p_{n-1}\})''$, where z is the unique tracial state extending x. Set $c_n = 4\cos^2(\pi/n)$. Prove Jones' famous result saying that $[U_1 : U_0]$ takes values in $\{c_n \mid n = 3, 4, \ldots\} \cup [4, \infty\rangle$ by defining polynomials f_i recursively according to $f_0 = f_1 = 1$ and $f_{n+2}(a) = f_{n+1}(a) - af_n(a)$ to arrive at the contradiction $z(q_{n-1}) = f_n([U_1 : U_0]^{-1}) < 0$ when $[U_1 : U_0]^{-1} \in \langle c_{n+1}^{-1}, c_n^{-1}\rangle$, where q_n are the orthogonal projections defined recursively by $q_0 = 1$ and $q_k = q_{k-1} - (f_{k-1}([U_1 : U_0]^{-1})/f_k([U_1 : U_0]^{-1}))q_{k-1}p_kq_{k-1}$ for $k \in \{1, \dots, n-1\}$.

4. Consider now an arbitrary factor U_1 with a subfactor U_0, and assume that we have a faithful semifinite normal operator valued weight E_1 from U_1 to U_0, and let E_i be the operator valued weights satisfying $y_i = y_{i-1}E_i$ associated to chosen faithful semifinite normal weights y_i on U_i in the Jones tower $U_0 \subset U_1 \subset \cdots$ Set $[U_1 : U_0] = E_2(1)$ when E_2 is bounded. Prove $w \leq [U_1 : U_0]E_1(w)$ for positive $w \in U_1$, which is a generalization of the Pimsner-Popa inequality for type II_1 factors. Show also that we recover the Jones' index in this case. Prove the famous result of Jones on the restriction of the index values in the general setting, that is beyond the type II_1 case.
5. Show that a locally compact quantum group (W, Δ) with W a finite factor, is a compact locally compact quantum group with Haar state the unique tracial state on W.

Chapter 18
Galois Objects and Cocycle Deformations

Suppose (W, Δ) is a locally compact quantum group with a left invariant faithful semifinite normal weight x. Let α be a right coaction of (W, Δ) on a von Neumann algebra U that is ergodic, i.e. with fixed point algebra $\mathbb{C}$, and is integrable, meaning that $y_1 = (\iota \otimes x)\alpha$ is a semifinite weight on U. Consider the basic construction $\mathbb{C} \subset U \subset B(V_{y_1})$, and the normal $*$-epimorphism $f : U \rtimes_\alpha W \to B(V_{y_1})$ uniquely determined by $f\alpha = \iota$ and $f(1 \otimes (\omega \otimes \iota)(N)) = (\iota \otimes \omega)(K_r)$ for $\omega \in W_*$, where $N = (J_{\hat{x}} \otimes J_{\hat{x}})\hat{M}(J_{\hat{x}} \otimes J_{\hat{x}})$ and K_r is the unitary implementation of α. We say that α is a Galois object if f is injective. One can then so called reflect $(\hat{W}, \hat{\Delta})$ across α to obtain another locally compact quantum group.

By a unitary dual 2-cocycle of (W, Δ) we mean a unitary $X \in \hat{W}\bar{\otimes}\hat{W}$ such that $(1 \otimes X)(\iota \otimes \hat{\Delta})(X) = (X \otimes 1)(\hat{\Delta} \otimes \iota)(X)$. From what we know of cocycle crossed products we can then construct a Galois object α for (W, Δ) such that the reflection of $(\hat{W}, \hat{\Delta})$ across α is the locally compact quantum group with $\hat{W}$ as von Neumann algebra but with coproduct $X\hat{\Delta}(\cdot)X^*$; the so called twisted coproduct $\hat{\Delta}_X$ of $\hat{\Delta}$. How to construct invariant semifinite weights directly on $(\hat{W}, \hat{\Delta}_X)$ is not obvious.

Quantization schemes from geometry tend to deform the product on some algebra of bounded observables subject to symmetries. In such situations one often has from the outset a classical theory, say with a commutative locally compact quantum group (W, Δ) and a continuous coaction α on a C*-algebra U. The deformed product can in some cases be extracted as a dual unitary 2-cocycle X of (W, Δ), and while this quantum group does not act anymore on the algebra with deformed product, the dual of $(\hat{W}, \hat{\Delta}_X)$ does. So quantum groups really are needed as 'symmetries' in the quantum world, especially when the underlying configuration spaces are curved.

One may start at the opposite end and construct the deformed algebra from the coaction and the dual unitary 2-cocycle. This abstract approach works reasonably well when the fundamental multiplicative unitaries are regular. One might even start with a locally compact quantum group, and quantize further, or dequantize, and we identify natural maps associated to this. Repeated deformation yields a duality theorem that ties in nicely with those of (cocycle) crossed products. Throughout this

L. Tuset, *Analysis and Quantum Groups*,
https://doi.org/10.1007/978-3-031-07246-8_18

discussion we work in the continuous setting, restricting to unitary dual 2-cocycles X that are continuous, i.e. live in $M(\hat{W}_r \otimes \hat{W}_r)$, and tend to be regular, meaning for instance that $(B_0(V_x) \otimes \mathbb{C})\hat{M}X^*(\mathbb{C} \otimes B_0(V_x)) \subset B_0(V_x) \otimes B_0(V_x)$.

Recommended literature for this chapter is [11, 36].

18.1 Galois Objects

Let U be a von Neumann algebra, and V_i two right U-modules with antirepresentations π_i. Define $Q_{ij} \equiv \{a \in B(V_j, V_i) \mid a\pi_j(u) = \pi_i(u)a, \forall u \in U\}$. The Q_{11}-Q_{22} bimodule Q_{12} is the space of *intertwiners* between the right modules V_2 and V_1, while the commutant $Q = (Q_{ij})$ of $\pi_1 \oplus \pi_2$ is the *linking algebra* between them, which then carries the balanced weight $z_1 \oplus z_2$ of weights z_i on Q_i.

Lemma 18.1.1 *Consider the basic construction $U_0 \subset U = U_1 \subset U_2$ with weights y_i and notation as in Proposition 17.2.12. Let Q be the linking algebra of the right U_0-modules V_{y_2} and V_{y_0}. If $u \in U$ is analytic with $\sigma_z^{y_1}(u) \in N_E$ for complex z, then $q_E(u)$ is analytic for $\sigma_t^{y_2 \oplus y_0}$ and $\sigma_z^{y_2 \oplus y_0}(q_E(u)) = q_E\sigma_z^{y_1}(u)$.*

Proof By the proposition $q_E(u) \in Q_{12}$. Let $a \in N_{y_0}$ and $b \in N_{y_1}$ and c be in the Tomita algebra of y_1. Then f given by

$$f(z) = (q_E\sigma_z^{y_1}(u)q_{y_0}(a) | J_{y_1}\sigma_{i/2}^{y_1}(c)J_{y_1}q_{y_1}(b)) = (\sigma_z^{y_1}(u)q_{y_1}(ac)|q_{y_1}(b))$$

is analytic. Since the modular automorphisms of y_1 and y_0 coincide on U_0, for $z = r + is$ we get

$$|f(z)| = |(q_E\sigma_{is}^{y_1}(u)\Delta_{y_0}^{-ir}q_{y_0}(a)|\Delta_{y_1}^{-ir}J_{y_1}\sigma_{i/2}^{y_1}(c)J_{y_1}q_{y_1}(b))|,$$

which by the Phragmen-Lindelöf theorem is bounded on every horizontal strip by $M\|\omega_{q_{y_0}(a), J_{y_1}\sigma_{i/2}^{y_1}(c)J_{y_1}q_{y_1}(b)}\|$, where M depends only on u and the chosen strip. By denseness of the span of such ω's, we thus get that $z \mapsto q_E\sigma_z^{y_1}(u)$ is bounded on compacts, so this Banach space valued function has a power series by standard contour integral techniques. As $\sigma_z^{y_2 \oplus y_0}$ is implemented by $\Delta_{y_1} \oplus \Delta_{y_0}$, and $\Delta_{y_1}^{it}q_E(u)\Delta_{y_0}^{-it} = q_E\sigma_t^{y_1}(u)$, we are done. □

With notation as above, we define the *Tomita algebra of E* to be the σ-weakly dense subspace A_E of U consisting of those $u \in N_E \cap N_E^*$ in the Tomita algebra of y_1 such that $\sigma_z^{y_1}(u) \in N_E \cap N_E^*$ for complex z. Denote by A_2 the span of the elements $q_E(u)q_E(u')^*$ for $u, u' \in A_E$.

Proposition 18.1.2 *The space A_2 is contained in the domain of q_{y_2}, is a σ-weakly dense $*$-subalgebra of U_2, and $q_{y_2}(A_2)$ is a sub left Hilbert algebra of $q_{y_2}(N_{y_2} \cap N_{y_2}^*)$ of $\sigma_t^{y_2}$-analytic elements having y_2 as associated weight.*

Proof The domain issue is resolved by $E_2(q_E(u)q_E(u')^*) = uu'^*$ from Proposition 17.2.12, and A_2 is certainly $*$-invariant. If also $v, v' \in A_E$, then

$$q_E(u)q_E(u')^*q_E(v)q_E(v')^* = q_E(uE(u'^*v))q_E(v')^*,$$

so A_2 is an algebra if we can show that $uE(u'^*v) \in A_E$. This element certainly belongs to $N_{y_1}^* \cap N_E \cap N_E^*$. Invoking the previous lemma with u' and v for u, and using $q_E(u')^*q_E(v) = E(u'^*v)$ together with the fact that both $\sigma_t^{y_2 \oplus y_0}$ and $\sigma_t^{y_1}$ restrict to $\sigma_t^{y_0}$ on U_0, we see that $uE(u'^*v)$ is analytic for $\sigma_t^{y_1}$ and indeed belongs to A_E. It remains to show density, and that we get y_2. For density it suffices to show that $q_E(A_E)$ is strongly dense in Q_{12}. Now $q_E(A_E)$ is closed under right multiplication by elements from the Tomita algebra of y_0 which is σ-weakly dense in U_0. Arguing as in the proof of Proposition 17.2.12, we need only show that any $a \in Q_{12}$ with $a^*q_E(u) = 0$ for all $u \in A_E$, must be zero. Pick $b \in N_{y_0}$ analytic for $\sigma_t^{y_0}$. Then $\pi_0\sigma_{i/2}^{y_0}(b)a^*q_{y_1}(u) = a^*q_E(u)q_{y_0}(b) = 0$, where π_0 is the antirepresentation of U_0. Letting $\pi_0\sigma_{i/2}^{y_0}(b)$ tend to 1, we get $a^*q_{y_1}(u) = 0$. Using integrals as in the proof of Proposition 17.2.12, we conclude that a^* also vanishes on $q_{y_1}(N_{y_1} \cap N_E)$ which is norm dense in V_{y_1}, so $a^* = 0$. That our Tomita algebra gives y_2 is now clear from Proposition 10.7.9. □

Theorem 18.1.3 *Keep notation as above. The image V_2 of $q_{y_1}(A_E) \odot q_{y_1}(A_E)$ in $V_{y_1} \otimes_{y_0} V_{y_1}$ is dense, and the linear map $V_2 \to V_{y_2}$ which sends $q_{y_1}(u) \otimes_{y_0} q_{y_1}(u')$ to $q_{y_2}(q_E(u)q_E(u'^*)^*)$, extends to a unitary $V_{y_1} \otimes_{y_0} V_{y_1} \to V_{y_2}$ that respects the U_2-U_2 bimodule structures.*

Proof This is clear from Proposition 17.2.12 and denseness of A_E and A_2. □

If V is a U-W bimodule, its *conjugate* W-U bimodule is the conjugate Hilbert space $\bar{V}$ with module actions $w\bar{v}u \equiv \overline{u^*vw^*}$. We identify V_{y_1} with $\overline{V_{y_1}}$ as U_0-U_2 bimodules by the unitary given by $q_{y_1}(u) \mapsto \overline{q_{y_1}^{\mathrm{op}}(u^*)}$, where $q_{y_1}^{\mathrm{op}}(u^*) = J_{y_1}q_{y_1}(u)$.

Lemma 18.1.4 *Let $a, b \in A_E$ and $u \in N_{y_2}$. Then $(q_{y_1}(a) \otimes_{y_0} q_{y_1}(b)|q_{y_2}(u)) = (q_{y_1}(a)|uq_{y_1}\sigma_{-i}^{y_1}(b^*))$, and conversely, any $v \in V_{y_2}$ replacing $q_{y_2}(u)$ in this identity with $u \in N_{y_2}$, must satisfy $v = q_{y_2}(u)$.*

Proof For the first claim we may by density of A_2 assume $u = q_E(c)q_E(d^*)^*$ for $c, d \in A_E$. Then both sides equal $(q_{y_1}(a) \otimes_{y_0} q_{y_1}(b)|q_{y_1}(c) \otimes_{y_0} q_{y_1}(d))$ by the previous theorem. Similarly, for the converse, it suffers to check $uq_{y_1}^{\mathrm{op}}(w) = \pi_2(w)v$ for $w = q_E(a)q_E(b^*)^*$, where π_2 is the right action of U_2 on V_{y_2}. As $q_{y_1}^{\mathrm{op}}(w) = q_{y_1}\sigma_{-i/2}^{y_1}(a) \otimes_{y_0} q_{y_1}\sigma_{-i/2}^{y_1}(b)$ by the previous theorem, one easily verifies the required identity by checking it on the inner product with $q_{y_2}(d')$ for a typical $d' \in A_2$. □

We need one more result about weights.

Lemma 18.1.5 *Let $U_{00} \subset U_{10} \subset U_{11}$ and $U_{00} \subset U_{01} \subset U_{11}$ be inclusions of von Neumann algebras, and let Q_i be the linking algebra between the right U_{i0}-modules $V_{y_{i0}}$ and $V_{y_{i1}}$ for faithful semifinite normal weights y_{ij} on U_{ij}. Suppose*

F_1 is a faithful semifinite normal operator valued weight from U_{11} to U_{10} with a faithful semifinite normal restriction F_0 from U_{01} to U_{00}. Then $q_{F_0}(a) \mapsto q_{F_1}(a)$ for $a \in N_{F_0}$ defines a normal embedding $Q_0 \to Q_1$.

Proof Let A_i be the $*$-algebras generated by the elements $q_{F_i}(a)$, and let W be the σ-weak closure of A_1. We claim our map is an isomorphism $Q_0 \to W$. Indeed, it is an isomorphism on the level of A_i since $\sum q_{F_1}(a_k)q_{F_1}(b_k)^* = 0$ if and only if $\sum q_{F_0}(a_k)q_{F_0}(b_k)^* = 0$ as is seen by applying $F_0 = F_1$ on both expressions. Then one readily shows that this isomorphism and its inverse are σ-weakly continuous by considering the unit in U_{00}. □

Now let (W, Δ) be a locally compact quantum group with left invariant faithful semifinite normal weight x, fundamental multiplicative unitary M and modular element h. We will in this section use right coactions of quantum groups. Recall that $\alpha\colon U \to U\bar{\otimes}W$ is a *right coaction* of (W, Δ) on U if $\alpha(\cdot)_{21}$ is a coaction of (W, Δ^{op}) on U, so $(\iota \otimes \Delta)\alpha = (\alpha \otimes \iota)\alpha$. The *right* crossed product $U \rtimes_\alpha W$ is the von Neumann algebra generated by $\alpha(U)$ and $1 \otimes \hat{W}'$, which is then the flip of the ordinary crossed product of $\alpha(\cdot)_{21}$. We recall the right coaction version of various results and notions as we need them.

So let α be an integrable right coaction of (W, Δ) on a von Neumann algebra $U = U_1$, meaning that the faithful normal operator valued weight $E_1 = E = (\iota \otimes x)\alpha$ from U_1 to $U_0 = U^\alpha = \{u \in U \mid \alpha(u) = u \otimes 1\}$ is semifinite. Let $y_1 = y_0 E$ for a chosen faithful semifinite normal weight y_0 on U_0. We use $(q_{\tilde{y}_1}, \iota)$ with $q_{\tilde{y}_1}((1 \otimes c)\alpha(u)) = q_{y_1}(u) \otimes q_{\hat{x}}^{\text{op}}(c)$ for $u \in N_{y_1}$ and $c \in N_{\hat{x}^{\text{op}}}$ as the semicyclic representation of the right dual weight $\tilde{y}_1$ of y_1. Let $K = J_{\tilde{y}_1}(J_{y_1} \otimes J_{\hat{x}})$ be the unitary implementation of α with respect to the h-invariant weight y_1. Hence $(\iota \otimes \omega_{v,v'})(K)q_{y_1} = q_{y_1}(\iota \otimes \omega_{h^{-1/2}v,v'})\alpha$ on N_{y_1} for $v \in D(h^{-1/2})$ and $v' \in V_x$. Consider the basic construction $U_0 \subset U_1 \subset U_2$ and the normal $*$-epimorphism $f\colon U \rtimes_\alpha W \to U_2$ from the right coaction version of Theorem 17.1.1. So f is determined by $f\alpha = \iota$ and $f(1 \otimes (\iota \otimes \omega)(N)) = (\iota \otimes \omega)(K)$ for $\omega \in W_*$, where $N = (J_{\hat{x}} \otimes J_{\hat{x}})\hat{M}(J_{\hat{x}} \otimes J_{\hat{x}})$. Retaining notation from above, we define an isometry $G\colon V_{y_1} \otimes_{y_0} V_{y_1} \to V_{y_1} \otimes V_x$ extending the linear map $V_2 \to V_{y_1} \otimes V_x$ that sends $q_{y_1}(a) \otimes_{y_0} q_{y_1}(b)$ to $(q_{y_1} \otimes q_x)(\alpha(a)(b \otimes 1))$ for $a, b \in A_E$. As usual $y_i = y_{i-1}E_i$.

Lemma 18.1.6 *Retain notation. Then $G^*(q_{y_1}(u) \otimes q_{\hat{x}}^{\text{op}}(c)) = q_{y_2}f((1 \otimes c)\alpha(u))$ for $u \in N_{y_1}$ and $c \in N_{\hat{x}^{\text{op}}}$, and $Gf(d) = dG$ for $d \in U \rtimes_\alpha W$, and $\Delta_{\tilde{y}_1}^{it} G = G\Delta_{y_2}^{it}$ and $J_{\tilde{y}_1} G = GJ_{y_2}$.*

Proof Picking c of the form $(\iota \otimes \omega)(N)$ with $\overline{\omega(S(\cdot)^*)}$ and $\overline{\omega(S(\cdot h^{-1/2})^*)}$ that extend to bounded normal functionals ω^* and ω_h^* on W. As $(\iota \otimes \omega)(N)^* = (\iota \otimes \omega^*)(N)$, we get from the formulas in the previous paragraph that

$$(q_{y_1}(a)|f(1 \otimes c)uq_{y_1}\sigma_{-i}^{y_1}(b^*)) = y_1(u^*(\iota \otimes \omega_h^*)\alpha(a)b)$$

equals $(G(q_{y_1}(a) \otimes_{y_0} q_{y_1}(b))|q_{y_1}(u) \otimes q_{\hat{x}}^{\text{op}}(c))$ for $a, b \in A_E$ since $\omega_h^* = (q_x(\cdot)|q_{\hat{x}}^{\text{op}}(c))$ on N_x by definition of $q_{\hat{x}}$. The equality holds (by using smoothing integrals) for any $c \in N_{\hat{x}^{\rm op}}$, and the first statement follows Lemma 18.1.4. The second statement is now clear using f, the definition of $U \rtimes_\alpha W$, and identifying $V_{y_1} \otimes_{y_0} V_{y_1}$ with V_{y_2}. Next, we have $\Delta_{y_2}^{it}(q_{y_1}(a) \otimes_{y_0} q_{y_1}(b)) = q_{y_1}\sigma_t^{y_1}(a) \otimes_{y_0} q_{y_1}\sigma_t^{y_1}(b)$, so concerning the relation with the modular operators, we need only check the claim

$$\Delta_{\tilde{y}_1}^{it}(q_{y_1} \otimes q_x)(\alpha(a)(b \otimes 1)) = (q_{y_1} \otimes q_x)(\alpha\sigma_t^{y_1}(a)(\sigma_t^{y_1}(b) \otimes 1)).$$

To this end, consider the one-parameter group of automorphisms $k_t = h^{-it}\tau_{-t}(\cdot)h^{it}$ on W with generator k given by $k^{it}q_x = q_x k_t$ for real t. Arguing as in the proof of Proposition 15.4.11 we then have $\Delta_{\tilde{y}_1}^{it} = \Delta_{y_1}^{it} \otimes k^{it}$. Since we know that $\sigma_t^{\tilde{y}_1}\alpha = \alpha\sigma_t^{y_1}$, we thus get $\Delta_{\tilde{y}_1}^{it}(\alpha(a)(q_{y_1}(b) \otimes v)) = \alpha\sigma_t^{y_1}(a)(q_{y_1}\sigma_t^{y_1}(b) \otimes k^{it}v)$ for $v \in V_x$. Pick any $w \in N_x$ analytic for σ_t^x. Then $\sigma_z k_t(a) = k_t\sigma_z(a)$ for complex z, and

$$\Delta_{\tilde{y}_1}^{it} B(q_{y_1} \otimes q_x)(\alpha(a)(b \otimes 1)) = C(q_{y_1} \otimes q_x)(\alpha\sigma_t^{y_1}(a)(\sigma_t^{y_1}(b) \otimes 1)).$$

with $B = 1 \otimes J_x\sigma_{i/2}^x(a)^* J_x$ and $C = 1 \otimes J_x k_t \sigma_{i/2}^x(a)^* J_x$, and these tend to 1 by choosing a appropriately. Concerning the relation with modular conjugations, note that the span of the elements $\alpha(a)q_{\tilde{y}_1}((1 \otimes c)\alpha(b))$ with $a, b \in A_E$ and c in $N_{\hat{x}^{\rm op}} \cap N_{\hat{x}^{\rm op}}^*$ is $S_{\tilde{y}_1}$-invariant and dense in $V_{\tilde{y}_1}$, and using what we have already proved, it is easy to see that $S_{y_2}G^*$ and $G^*S_{\tilde{y}_1}$ coincide on this span. □

Theorem 18.1.7 *With notation from above, let p the central orthogonal projection in $U \rtimes_\alpha W$ such that* Ker $f = (1-p)(U \rtimes_\alpha W)$. *Then $GG^* = p$. So G is unitary if and only if f is injective.*

Proof The lemma shows that $GG^* \in (U \rtimes_\alpha W)'$, and in fact GG^* belongs to the center as it commutes with $J_{\tilde{y}_1}$. As $G^*pG = f(p) = 1$, we have $GG^* \leq p$, but $f(GG^*) = G^*GG^*G = 1$, so $p = GG^*$. □

Let f_p be the restriction of f to $p(U \rtimes_\alpha W)$.

Corollary 18.1.8 *We have $y_2 = \tilde{y}_1 f_p^{-1}$ as weights on U_2.*

Proof The span of the elements $(1 \otimes c)\alpha(u)$ for $u \in N_{y_1}$ and $c \in N_{\hat{x}^{\rm op}}$ is a σ-strong* core for $q_{\tilde{y}_1}$. Hence $qf((1 \otimes c)\alpha(u)) = p(q_{y_1}(u) \otimes q_{\hat{x}}^{\rm op}(c))$ defines a semicyclic representation (q, ι) for $\tilde{y}_1 f_p^{-1}$. By the previous theorem and proposition, we see that (G^*q, ι) is a semicyclic representation for y_2 with $G^*q \subset q_{y_2}$ and modular group $\sigma_t^{y_2}$. Thus the result is clear from Corollary 10.7.10. □

This shows that injectivity of f in Theorem 17.1.4 is redundant. It also shows that G^* coincides with the map $Z\colon V_{y_1} \otimes V_x = V_{\tilde{y}_1} \to V_{y_2} = V_{y_1} \otimes_{y_0} V_{y_1}$ that sends $q_{\tilde{y}_1}(d)$ to $q_{y_2}f(d)$, and both are $U \rtimes_\alpha W$-bimodule maps in a natural way. We say α is *Galois* if the *Galois homomorphism* $f\colon U \rtimes_\alpha W \to U_2$ is injective, and this happens if and only if the *Galois isometry* $\tilde{G} \equiv FG\colon V_{y_1} \otimes_{y_0} V_{y_1} \to V_x \otimes V_{y_1}$ for α is unitary, where F is the flip.

Corollary 18.1.9 *Given our integrable right coaction α. Let $U_{00} = U^\alpha \otimes \mathbb{C}$, $U_{01} = \alpha(U)$, $U_{10} = U \otimes \mathbb{C}$ and $U_{11} = U\bar{\otimes}W$ with operator valued weights $F_0 = E\alpha^{-1}$ and $F_1 = \iota \otimes x$, and consider the corresponding linking algebras Q_i from Lemma 18.1.5. Then α is Galois if and only if the inclusion $Q_0 \subset Q_1$ is unital if and only if $(Q_0)_{12}$ is the space of $U \rtimes_\alpha W$-intertwiners.*

Proof Let $g\colon Q_0 \subset Q_1$ be the inclusion, and let Q be the linking algebra between the right $U \rtimes_\alpha W$-modules V_{y_1} and $V_{\bar{y}_1}$. We claim that $g(Q_0) \subset Q$. Indeed, we have $q_{F_1}\alpha(a)q_{y_1}(b) = (q_{y_1} \otimes q_x)(\alpha(a)(b \otimes 1))$ for $a \in N_E$ and $b \in N_{y_1}$. Hence $q_{F_1}\alpha(a) = GL_a$, where $L_a\colon V_{y_1} \to V_{y_1} \otimes_{y_0} V_{y_1}$ is the right $U \rtimes_\alpha W$-module map that sends v to $q_{y_1}(a) \otimes_{y_0} v$. So $q_{F_1}\alpha(a) \in Q_{12}$, and the claim holds. If g_{11} denotes the restriction of g to $(Q_0)_{11}$ and $g'\colon U_2 \to (Q_0)_{11}$ is the isomorphism which sends $q_E(c)q_E(d)^*$ to $q_{F_0}\alpha(c)q_{F_0}\alpha(d)^*$ for $c, d \in N_E$, then we see that $fg_{11}g' = \iota$. The corollary is now immediate from the lemma prior to the theorem above. □

By the right coaction version of Proposition 17.1.5, every right coaction α of (W, Δ) on U such that $(\iota \otimes \alpha)(B) = \hat{M}_{13}B_{12}$ for a unitary $B \in B(V_{y_1})\bar{\otimes}U$, is Galois. And so are outer right coactions.

Example 18.1.10 Let (W, Δ) and (U, Δ_U) be locally compact quantum groups such that $(\hat{W}, \hat{\Delta})$ is a *closed quantum subgroup* of $(\hat{U}, \hat{\Delta}_U)$, meaning that $\hat{\Delta}_U$ restricts to $\hat{\Delta}$. Consider the normal embedding $g\colon \hat{W}' \to \hat{U}'$ that respects the comultiplications. Then $\alpha = (g \otimes \iota)(N)((\cdot) \otimes 1)(g \otimes \iota)(N)^*$ is a right coaction of (W, Δ) on U, and it is easy to see that $U \rtimes_\alpha W$ is isomorphic to the von Neumann algebra generated by U and $g(\hat{W}')$, and this is our Galois homomorphism. So α is Galois. ◇

In the sequel we will focus on *Galois objects* for (W, Δ), by which we mean Galois coactions α of (W, Δ) on U that are *ergodic*, meaning $U^\alpha = \mathbb{C}$. Then $E = y_1$ and $q_E = q_{y_1}$ and $U \rtimes_\alpha W \cong U_2 = B(V_{y_1})$ under f, and $y_2 = \mathrm{Tr}(\cdot\Delta_{y_1})$. Under our identification $V_{y_2} \cong V_{y_1} \otimes V_{y_1}$ we have $\Delta_{y_2}^{it} = \Delta_{y_1}^{it} \otimes \Delta_{y_1}^{it}$ and $J_{y_2} = F(J_{y_1} \otimes J_{y_1})$, where F is the flip. By the previous lemma we have $\tilde{G}(u \otimes 1) = \alpha(u)_{21}\tilde{G}$ and $\tilde{G}(1 \otimes J_{\hat{x}}u^*J_{\hat{x}}) = (1 \otimes J_{\hat{x}}u^*J_{\hat{x}})\tilde{G}$ for $u \in U$, while $\tilde{G}(\hat{\pi}_l(a) \otimes 1) = (a \otimes 1)\tilde{G}$ and $\tilde{G}(1 \otimes J_{y_1}\hat{\pi}_l(a)^*J_{y_1}) = (J_{\hat{x}} \otimes J_{y_1})(\iota \otimes \hat{\pi}_l)\hat{\Delta}'^{\mathrm{op}}(a^*)(J_{\hat{x}} \otimes J_{y_1})\tilde{G}$ for $a \in \hat{W}'$, where $\hat{\pi}_l$ is the representation of $\hat{W}'$ on V_{y_1} given by $\hat{\pi}_l(a) = f(1 \otimes a)$. For the latter identity one also uses $K = (\hat{\pi}_l \otimes \iota)(N)$ and $N(J_x \otimes J_{\hat{x}}) = (J_x \otimes J_{\hat{x}})N^*$. We will use the symbol $\hat{\pi}_r$ for the antirepresentation $\hat{\pi}_l(J_{\hat{x}}(\cdot)^*J_{\hat{x}})$ of $\hat{W}$ on V_{y_1}. Let Q be the linking algebra between the right $\hat{W}$-modules V_x and V_{y_1}. From the previous identities we then have $\tilde{G} \in Q_{21}\bar{\otimes}Q_{12}$ and $\tilde{G}_{12}K_{13} = N_{13}\tilde{G}_{12}$ and $\tilde{G}(J_{y_1} \otimes J_{y_1})F = FKF(J_{\hat{x}} \otimes J_{y_1})\tilde{G}$. We also have the pentagonal type equation $\hat{M}_{12}\tilde{G}_{13}\tilde{G}_{23} = \tilde{G}_{23}\tilde{G}_{12}$. Indeed, using $(\iota \otimes \omega)(\tilde{G})q_{y_1} = q_x(\omega \otimes \iota)\alpha$ for $\omega \in B(V_{y_1})_*$ together with the coaction property of α, the left hand side equals $(\iota \otimes \alpha(\cdot)_{21})(\tilde{G})$, and we are done by one of the previous identities.

Example 18.1.11 Whenever U and W have separable preduals, the element $L' \equiv \tilde{G}(L \otimes 1)$ for any unitary $L\colon V_x \to V_{y_1}$ will satisfy $(\iota \otimes \alpha)(L') = \hat{M}_{13}L'_{12}$, so we have then a one-to-one correspondence between Galois objects and semidual right coactions. The trivial Galois object Δ for (W, Δ) has $\tilde{G} = \hat{M}$ and $U = W$. $\diamond$

By the pentagonal type equation, the σ-weak closure of the set of elements $(\omega \otimes \iota)(\tilde{G})$ for vector functionals ω, is an algebra, and since $(\omega_{q_{y_1}(u),q_x(w)} \otimes \iota)(\tilde{G}) = (\iota \otimes x)((1 \otimes w^*)\alpha(u))$, it coincides with the σ-weak closed span of the elements $(\iota \otimes \eta)\alpha(u)$ for $u \in U$ and $\eta \in W_*$, which is U by the biduality theorem for right crossed products. Similarly, the σ-weak closure of the set of elements $(\iota \otimes \omega)(\tilde{G})$ for $\omega \in B(V_{y_1})_*$, is Q_{21}. Indeed, by the pentagonal type equation this closure B is closed under left multiplication by $\hat{W}$, so just as we proved that A_2 was a Tomita algebra for y_2, it suffices to show that $B^* V_x$ is dense in V_{y_1}, and this is evident from $(\iota \otimes \omega_{q_{y_1}(a),q_{y_1}(b)})(\tilde{G}^*)q_{\hat{x}}^{\mathrm{op}}(c) = \hat{\pi}_l(c)aq_{y_1}\sigma^x_{-i}(b^*)$ for $a, b \in A_E$ in the Tomita algebra of y_1, and $c \in N_{\hat{x}^{\mathrm{op}}}$. It is also worth noticing that $\tilde{G}^*(W' \otimes \mathbb{C})\tilde{G} \subset U' \bar{\otimes} U$.

Proposition 18.1.12 *The* scaling group *of α is the one-parameter group on U defined by $\tau_t^U = P_U^{it}(\cdot)P_U^{-it}$, where $P_U^{it} = \Delta_{y_1}^{it}\hat{\pi}_l(J_{\hat{x}}\hat{h}^{-it}J_{\hat{x}})$. It satisfies the identities $\alpha\tau_t^U = (\tau_t^U \otimes \tau_t)\alpha = (\sigma_t^{y_1} \otimes \sigma_{-t}^{xR})\alpha$ and $\alpha\sigma_t^{y_1} = (\tau_t^U \otimes \sigma_t^x)\alpha$ and $y_1\tau_t^U = \nu^t y_1$. We also have $P^{it}q_{y_1} = \nu^{t/2}q_{y_1}\tau_t^U$ on N_{y_1}, and $\tilde{G}(\Delta_{y_1}^{it} \otimes \Delta_{y_1}^{it}) = (h^{-it}\Delta_{\hat{x}}^{-it} \otimes \Delta_{y_1}^{it})\tilde{G}$ and $\tilde{G}(\Delta_{y_1}^{it} \otimes P_U^{it}) = (\Delta_x^{it} \otimes P_U^{it})\tilde{G}$ and $\tilde{G}(P_U^{it} \otimes P_U^{it}) = (P_W^{it} \otimes P_U^{it})\tilde{G}$, where P_W is the usual implementation of τ_t.*

Proof Since $J_{\hat{x}}\hat{h}^{is}J_{\hat{x}}$ is invariant under the one-parameter group $k^{it}(\cdot)k^{-it}$ on $\hat{W}'$, and as $\sigma_t^{y_2}\hat{\pi}_l = \hat{\pi}_l(k^{it}(\cdot)k^{-it})$ from above, the elements $\Delta_{y_1}^{it}$ and $\hat{\pi}_l(J_{\hat{x}}\hat{h}^{-it}J_{\hat{x}})$ commute. Hence, to show that τ_t^U is well-defined, we need only show that U is invariant under the implementation by the latter element. But for any group-like element $a \in \hat{W}'$, we have for $u \in U$, that $\hat{\pi}_l(a)u\hat{\pi}_l(a)^*$ is fixed under $(f \otimes \iota)\hat{\alpha}f^{-1}$, where $\hat{\alpha}$ is the dual action, and such elements must be in U by the biduality theory for right crossed products. As for the various identities, by the previous lemma we have $\alpha\sigma_t^{y_1} = (\sigma_t^{y_1} \otimes k_t)\alpha$ and $\alpha\mathrm{Ad}(\hat{\pi}_l(J_{\hat{x}}\hat{h}^{-it}J_{\hat{x}})) = (\iota \otimes \mathrm{Ad}(\hat{\pi}_l(J_{\hat{x}}\hat{h}^{-it}J_{\hat{x}})))\alpha$. Combining this with $P_W^{it} = \Delta_x^{it}J_{\hat{x}}\hat{h}^{-it}J_{\hat{x}}$ gives $\alpha\tau_t^U = (\sigma_t^{y_1} \otimes \sigma_{-t}^{xR})\alpha$. Combined with the coaction property of α, its injectivity, and $\Delta\sigma_t^{xR} = (\sigma_t^{xR} \otimes \tau_t)\Delta$, one sees that $\alpha\tau_t^U = (\tau_t^U \otimes \tau_t)\alpha$. Now the last of these types of identities is clear. The second of these, and $x\tau_t = \nu^t x$, gives the identity involving $y_1 = (\iota \otimes x)\alpha$. The rest is clear. $\square$

Proposition 18.1.13 *We may write $\tilde{G}^*(J_x h^{it} J_x \otimes 1)\tilde{G} = m^{it} \otimes h_U^{it}$ for (unique up to a positive scalar) positive non-singular elements m and h_U affiliated with U' and U, respectively. We have $\alpha(h_U^{it}) = h_U^{it} \otimes h^{it}$ and $m = J_{y_1}h_U^{-1}J_{y_1}$ and $\sigma_t^{y_1}(h_U^{is}) = \nu^{ist}h_U^{is}$ and $\tau_t^U(h_U^{is}) = h_U^{is}$ for real s, t.*

Proof Let $H^{it} = \tilde{G}^*(J_x h^{it} J_x \otimes 1)\tilde{G}$. We show that $H^{it}(B \otimes \mathbb{C})H^{-it} = B \otimes \mathbb{C}$ for $B = B(V_{y_1}) = f(U \rtimes_\alpha W)$, and it suffices to show it when B is U and $\hat{\pi}_l(\hat{W}')$, and in these cases it is clear as $\mathrm{Ad}(J_x h^{it} J_x)(\hat{W}') = \hat{W}'$. Let m be the positive non-singular element affiliated with U' such that $\mathrm{Ad}(H^{it})(a \otimes 1) = \mathrm{Ad}(m^{it})(a) \otimes 1$ for $a \in B(V_{y_1})$. Then $H^{it} = m^{it} \otimes h_U^{it}$ for a positive non-singular element h_U affiliated with U. Using the pentagonal type equation for $\tilde{G}$, its implementation of $\alpha(\cdot)_{21}$, and the identity $\hat{M}^*(J_x h^{it} J_x \otimes 1)\hat{M} = J_x h^{it} J_x \otimes h^{it}$, we get $(\iota \otimes \alpha(\cdot)_{21})(H^{it}) = m^{it} \otimes h^{it} \otimes h_U^{it}$, so $\alpha(h_U^{it}) = h_U^{it} \otimes h^{it}$. For the next identity, we have

$$F(J_{y_1} \otimes J_{y_1}) H^{it} (J_{y_1} \otimes J_{y_1}) F = \tilde{G}^*(J_{\hat{x}} \otimes J_{y_1}) F K^* F(J_x h^{it} J_x \otimes 1) F K F (J_{\hat{x}} \otimes J_{y_1}) \tilde{G},$$

which equals H^{it} as $K \in B(V_{y_1}) \bar{\otimes} W$ and $J_{\hat{x}}$ commutes with h^{is} and J_x up to scalars. Hence $J_{y_1} h_U^{it} J_{y_1} \otimes J_{y_1} m^{it} J_{y_1} = m^{it} \otimes h_U^{it}$ and $m = J_{y_1} h_U^{-1} J_{y_1}$ holds. By the first of the last three identities in the previous proposition, we get

$$(\Delta_{y_1}^{it} \otimes \Delta_{y_1}^{it})(J_{y_1} h_U^{is} J_{y_1} \otimes h_U^{is})(\Delta_{y_1}^{-it} \otimes \Delta_{y_1}^{-it}) = J_{y_1} h_U^{is} J_{y_1} \otimes h_U^{is},$$

so there is a positive scalar r such that $\sigma_t^{y_1}(h_U^{is}) = r^{ist} h_U^{is}$. Picking a state $\omega \in U_*$ for $u \in M_{y_1}$ such that $y_1(h_U^{is} u) = x(\omega \otimes \iota)\alpha(h_U^{is} u)$. Then using $\alpha(h_U^{it}) = h_U^{it} \otimes h^{it}$ twice in combination with $x(h^{is} \cdot) = \nu^s x(\cdot h^{is})$, we get $y_1(h_U^{is} u) = \nu^s y_1(u h_U^{is})$ and $r = \nu$. From the previous proposition we have $\alpha \tau_t^U \sigma_{-t}^{y_1} = (\iota \otimes \tau_t \sigma_{-t}^x)\alpha$, which at this stage gives the last identity. □

Theorem 18.1.14 *Let $z_1 = y_1(h_U^{1/2} \cdot h_U^{1/2})$ be the faithful semifinite normal weight supplied by Connes' cocycle theorem as the deformation of y_1 by the cocycle $\nu^{it^2/2} h_U^{it}$. Then z_1 is invariant with respect to α, while y_1 is h-invariant, and they are the only such ones up to positive scalars.*

Proof We already know that y_1 is h-invariant, and for the uniqueness statement of both weights, see for example Lemma 15.4.6. It remains to prove invariance of z_1. Just as we did when $\alpha = \Delta$, one sees that $q(u) \equiv q_{y_1}(\overline{u h_U^{1/2}})$ for $u \in U$ a left multiplier of $h_U^{1/2}$ with closure $\overline{u h_U^{1/2}} \in N_{y_1}$, defines a semicyclic representation (q, ι) for y_1. With $v \in D(h^{-1/2})$ and $v' \in V_x$, the closure of $(\iota \otimes \omega_{v,v'})\alpha(u) h_U^{1/2}$ equals $(\iota \otimes \omega_{h^{-1/2}v,v'})\alpha(u h_U^{1/2})$ with image $(\iota \otimes \omega_{v,v'})(K) q(u)$ under q_{y_1}. By closedness of q we thus get $q(\iota \otimes \omega)\alpha = (\iota \otimes \omega)(K) q$ for any $\omega \in W_*$. Let $v, v_i \in V_x$ with $\{v_i\}$ an orthonormal basis, and let $u \in U$ with $u^* u \in M_{z_1}$. Then by lower semicontinuity of z_1 we get

$$z_1(\iota \otimes \omega_v)\alpha(u^* u) = \sum \|q(\iota \otimes \omega_{v,v_i})\alpha(u)\|^2 = \sum \|((\iota \otimes \omega_{v,v_i})(K) q(u)\|^2,$$

which equals $\omega_v(1) z_1(u^* u)$ by unitarity of K. □

Proposition 18.1.15 *Consider the one-parameter group on $B(V_{y_1})$ given by $\hat{\Delta}_{y_1}^{it} = P_U^{it} J_{y_1} h_U^{it} J_{y_1}$. Then we have $\hat{\Delta}_{y_1}^{-it} \hat{\pi}_l(\cdot) \hat{\Delta}_{y_1}^{it} = \hat{\pi}_l \sigma_t^{\hat{x}^{\mathrm{op}}}$ and $(\Delta_x^{it} \otimes \hat{\Delta}_{y_1}^{it})\tilde{G} = \tilde{G}(\Delta_{y_1}^{it} \otimes \hat{\Delta}_{y_1}^{it})$ and $(\Delta_{\hat{x}}^{it} \otimes P_U^{it})\tilde{G} = \tilde{G}(\hat{\Delta}_{y_1}^{it} \otimes P_U^{it} h_U^{it})$.*

Proof The one-parameter group is well-defined as P_U^{it} commutes with J_{y_1} and h_U^{it}, and we see that $\hat{\Delta}_{y_1}^{it} q_{z_1} = q_{z_1}(\tau_t^U(\cdot) h_U^{-it})$. Applying $(\iota \otimes \omega)(K)$ to this with $\omega \in W_*$, then the second last proposition gives

$$(\iota \otimes \omega)(K)\hat{\Delta}_{y_1}^{it} q_{z_1}(u) = q_{z_1}(\tau_t^U(\iota \otimes \omega(\tau_t(\cdot)h^{-it}))\alpha(u)h_U^{-it})$$

for $u \in N_{z_1}$. So $\hat{\Delta}_{y_1}^{-it}\hat{\pi}_l(\iota \otimes \omega)(N)\hat{\Delta}_{y_1}^{it} = \hat{\pi}_l(\iota \otimes \omega(\tau_t(\cdot)h^{-it}))(N) = \hat{\pi}_l\sigma_t^{\hat{x}^{\mathrm{op}}}(\iota \otimes \omega)(N)$. The last two identities are clear from the last two propositions. □

Denote by $\hat{\pi}_{r0} = J_{\hat{x}}(\cdot)^* J_{\hat{x}}$ the antirepresentation of $\hat{W}$ on V_x. Let $Q_{ij}\bar{\otimes}Q_{ij}$ be the σ-weak closure of $Q_{ij} \odot Q_{ij} \subset Q\bar{\otimes}Q$, so $Q_{12}\bar{\otimes}Q_{12}$ is just the space of intertwiners for the right $\hat{W}\bar{\otimes}\hat{W}$-modules $V_x \otimes V_x$ and $V_{y_1} \otimes V_{y_1}$. Let $\Delta_{22} = \hat{M}^*(1\otimes(\cdot))\hat{M}$ and $\Delta_{12} = \tilde{G}^*(1\otimes(\cdot))\hat{M}$ and $\Delta_{21} = \hat{M}^*(1\otimes(\cdot))\tilde{G}$ and $\Delta_{11} = \tilde{G}^*(1\otimes(\cdot))\tilde{G}$.

Proposition 18.1.16 *We have $\Delta_{ij}(Q_{ij}) \subset Q_{ij}\bar{\otimes}Q_{ij}$, so $\Delta_Q(a_{ij}) \equiv \Delta_{ij}(a_{ij})$ for $a_{ij} \in Q_{ij}$ satisfies $\Delta_Q(Q) \subset Q\bar{\otimes}Q$. Moreover, the non-unital $*$-homomorphism Δ_Q is coassociative, and $\Delta_Q R_Q = (R_Q \otimes R_Q)\Delta_Q^{\mathrm{op}}$, where $R_Q : Q \to Q$ is the $*$-antiisomorphism gotten by extending the map that sends $a \in Q_{12}$ to $(J_{y_1} a J_x)^*$.*

Proof Let $a \in Q_{12}$ and $b \in \hat{W}$. Then $\tilde{G}^*(1\otimes a)\hat{M}$ clearly commutes with $\hat{\pi}_r(b)\otimes 1$, and thus has the first leg in Q_{12}. That it also has the second leg there is also clear as it commutes with $1 \otimes \hat{\pi}_r(b)$, which holds as $a\hat{\pi}_r(b) = \hat{\pi}_r(b)a$ and $\hat{M}(1 \otimes \hat{\pi}_r(b)) = (\hat{R}\otimes\hat{\pi}_r)\hat{\Delta}(b)\hat{M}$ and $\tilde{G}^*(\hat{R}\otimes\hat{\pi}_r)\hat{\Delta}(b) = (1\otimes\hat{\pi}_r(b))\tilde{G}^*$ which again holds as $\tilde{G}$ is a right $U \rtimes_\alpha W$-module map. Hence $\Delta_{12}(Q_{12}) \subset Q_{12}\bar{\otimes}Q_{12}$. That this holds with the indices interchanged, is clear from $Q_{21} = Q_{12}^*$ and $\Delta_{21} = \Delta_{12}((\cdot)^*)^*$. When both indices are 1 it holds because $Q_{12}Q_{21}$ is σ-weakly dense in Q_{11}, and when both indices are 2, it holds because $(Q_{22}, \Delta_{22}) = (\hat{W}, \hat{\Delta})$. Note that $\Delta_Q(1_Q) = 1_{11} \otimes 1_{11} + 1_{22} \otimes 1_{22}$, which is not the unit in $Q\bar{\otimes}Q$ as $1_Q = 1_{11} + 1_{22}$. Coassociativity is clear from the pentagonal type equation for $\tilde{G}$. The formula for R_Q follows easily by applying the identity $\tilde{G}(J_{y_1} \otimes J_{y_1})F = FKF(J_{\hat{x}} \otimes J_{y_1})\tilde{G}$, which again is a restatement of the last identity in Lemma 18.1.6. □

Now $(Q_{22}, \Delta_{22}) = (\hat{W}, \hat{\Delta})$ is a locally compact quantum group, but it is perhaps more surprising that (Q_{11}, Δ_{11}) is also a locally compact quantum group, known as the *reflection of* $(\hat{W}, \hat{\Delta})$ *across the Galois object* α. To show this we need a left invariant weight. Recall that $\hat{\Delta}_U^{-it}$ implements the modular group of $\hat{x}^{\mathrm{op}}$ on V_{y_1}. Consider $\hat{\Delta}_U$ as the spatial derivative of a faithful semifinite normal weight x_1 on Q_{11} with respect to $\hat{x}^{\mathrm{op}}$. The existence of x_1 is guaranteed by Corollary 10.10.2, and we will show that it is left invariant for Δ_{11}. To this end, let $x_Q = x_1 \oplus \hat{x}$ be the balanced weight on Q. We get a natural semicyclic representation from it by letting $q_{x_Q} = (q_{ij})$, where q_{11} and q_{22} are the usual GNS-maps for x_1 and $\hat{x}$, while $q_{12} : N_{x_Q} \cap Q_{12} \to V_{y_1}$ is determined by $q_{12}(L_v) = v$, where $v \in V_{y_1}$ is such that the closure L_v of the map $q_{\hat{x}}^{\mathrm{op}}(a) = J_{\hat{x}} q_{\hat{x}}(a^*) \mapsto \hat{\pi}_r(a)v$ for $a \in N_{\hat{x}}^*$ is bounded. Such v are called *left bounded*, see the analogous definition of x-bounded elements. Finally, let $q_{21} : N_{x_Q} \cap Q_{21} \to \overline{V_{y_1}}$ be determined by $q_{21}(L_v^*) = \overline{\hat{\Delta}_U^{1/2} v}$ with v left bounded in the domain of $\hat{\Delta}_U^{1/2}$. Then the restriction J_{21} of J_{x_Q} to $q_{x_Q}(N_Q \cap Q_{21})$ is the antiunitary $\overline{V_{y_1}} \to V_{y_1}$ which sends $\bar{v}$ to v. Let $J_{12} = J_{21}^*$.

Lemma 18.1.17 *For $a \in N_{\hat{x}}$ and $b \in Q_{12} \cap N_{x_Q}$, we have $(q_{12} \otimes q_{12})(\Delta_{12}(b)(a \otimes 1)) = \tilde{G}^*(q_{\hat{x}}(a) \otimes q_{12}(b))$, and for $u \in N_{y_1} \cap N_{y_1}^*$ and w in the Tomita algebra of x, we have $(\omega_{q_{y_1}(u^*), q_x(\sigma_i^x(w)^*)} \otimes \iota)(\tilde{G}) = (\omega_{q_x(w), q_{y_1}(u)} \otimes \iota)(\tilde{G}^*)$. For $u \in N_{y_1}$ and u' in the Tomita algebra of y_1, the element $q_{y_1}(u\sigma_{-i}^{y_1}(u'^*)$ is left bounded with $L_{q_{y_1}(u\sigma_{-i}^{y_1}(u'^*))} = (\iota \otimes \omega_{q_{y_1}(u), q_{y_1}(u')})(\tilde{G}^*)$.*

Proof The map $q_{\hat{x}}(a) \otimes q_{12}(b) \mapsto (q_{12} \otimes q_{12})(\Delta_{12}(b)(a \otimes 1))$ extends to a well-defined isometry since

$$(\iota \otimes x_Q)((a^* \otimes 1)\Delta_{12}(b)^*\Delta_{12}(b)(a \otimes 1)) = (\iota \otimes \hat{x})((a^* \otimes 1)\hat{\Delta}(b^*b)(a \otimes 1)) = x_Q(b^*b)a^*a.$$

To show that it coincides with $\tilde{G}^*$, we need only prove that $\Delta_{12}(b)(q_{\hat{x}}(a) \otimes q_{\hat{x}}^{\mathrm{op}}(c))$ equals $(1 \otimes \hat{\pi}_l(c))\tilde{G}^*(q_{\hat{x}}(a) \otimes q_{12}(b))$ for $c \in N_{\hat{x}^{\mathrm{op}}}$. This follows easily from the definition of Δ_{12} and $\hat{M}$, and by $\tilde{G}(1 \otimes \hat{\pi}_l(c))\tilde{G}^* = FK(1 \otimes 1 \otimes J_{\hat{x}}\hat{R}'(c)^*J_{\hat{x}})K^*F$ and $(\hat{\pi}_l \otimes \iota)(N) = K$. The next identity is direct from definitions. To show the last identity, combining Lemma 18.1.4 with the map Z introduced just before the definition of Galois right coactions, we have that

$$((\iota \otimes \omega_{q_{y_1}(u), q_{y_1}(u')})(\tilde{G}^*)q_{\hat{x}}^{\mathrm{op}}(c)|q_{y_1}(u'')) = (q_{y_2}(\hat{\pi}_l(c)u)|q_{y_1}(u'') \otimes q_{y_1}(u'))$$

equals $(\hat{\pi}_l(c)uq_{y_1}\sigma_{-i}^{y_1}(u'^*)|q_{y_1}(u''))$ for $u'' \in N_{y_1}$. □

Lemma 18.1.18 *For $b \in Q_{12} \cap N_{x_Q}$ and $c \in Q_{21} \cap N_{x_Q}$ we have*

$$(q_{\hat{x}} \otimes q_{21})(\Delta_{21}(c)(b \otimes 1)) = (J_x \otimes J_{12})\tilde{G}(J_{y_1} \otimes J_{21})(q_{12}(b) \otimes q_{21}(c)).$$

Proof We show that $J_{21}q_{21}(\omega \otimes \iota)\Delta_{21}(c) = (\bar{\omega}(J_x(\cdot)^*J_{y_1}) \otimes \iota)(\tilde{G})J_{21}q_{21}(c)$ for c in the Tomita algebra of x_Q, and ω of the form $\omega_{q_{y_1}(u), q_x(w)}$ with u and w in the Tomita algebras of y_1 and x, respectively. By Proposition 18.1.15 the image of $(\omega \otimes \iota)\Delta_{21}(c)$ under $\sigma_{-i/2}^{x_Q}$ is $(\omega_{\Delta_{y_1}^{1/2}q_{y_1}(u), \Delta_x^{-1/2}q_x(w)} \otimes \iota)\Delta_{12}\sigma_{-i/2}^{x_Q}(c)$, while the image of the adjoint element under q_{12} is $(\overline{\omega} \otimes \iota)(\tilde{G}^*)q_{12}(c^*)$ by the previous lemma. Using the lemma and Proposition 18.1.15, we get the result from $J_{21}q_{21}(\omega \otimes \iota)\Delta_{21}(c) = \Delta_{x_Q}^{1/2}q_{21}((\omega \otimes \iota)\Delta_{21}(c)^*)$ and the definitions of J_{y_1} and J_x. □

Theorem 18.1.19 *For any Galois object α, the faithful semifinite normal weight x_1 is left invariant for the locally compact quantum group (Q_{11}, Δ_{11}).*

Proof By the previous two lemmas we have $(\iota \otimes x_1)\Delta_{11}(L_vL_v^*) = x_1(L_vL_v^*)$ for left bounded $v \in D(\hat{\Delta}_U^{1/2})$. Just as shown for x-bounded elements, any positive element $b \in M_{x_1}$ can be approached from below by elements of the form $\sum L_{v_i}L_{v_i}^*$ for left bounded elements v_i. These must also belong to $D(\hat{\Delta}_U^{1/2})$ since $b \in M_{x_1}$, and remembering that x_1 is lower semicontinuous, left invariance follows. Then x_1R_Q will be right invariant, and we have a locally compact quantum group. □

Let us apply the theory to *twisting* of $(\hat{W}, \hat{\Delta})$ by a *unitary 2-cocycle*, that is, by a unitary $X \in \hat{W}\bar{\otimes}\hat{W}$ such that $(1 \otimes X)(\iota \otimes \hat{\Delta})(X) = (X \otimes 1)(\hat{\Delta} \otimes \iota)(X)$. Then the *twisted coproduct* $\hat{\Delta}_X \equiv X\hat{\Delta}(\cdot)X^*$ on $\hat{W}$ will be coassociative. Note that X^* will then be a unitary 2-cocycle of $(\hat{W}, \hat{\Delta}_X)$, and this twisted version will, as we shall see, be a locally compact quantum group. Now (α_0, X) is a cocycle coaction of $(\hat{W}, \hat{\Delta})$ with the trivial (left) coaction $\alpha_0 \colon \mathbb{C} \to \hat{W} \otimes \mathbb{C}$. Let $U = \hat{W} \ltimes_{\alpha_0, X} \mathbb{C}$, so U is the von Neumann subalgebra of $B(V_x)$ generated by $(\omega \otimes \iota)(\hat{M}X^*)$ for $\omega \in \hat{W}_*$. It will follow from the result just below and the discussion after Example 18.1.11, that U is the σ-weak closure of all such elements. By Proposition 16.1.3 we have a dual coaction of α_0, or a right coaction α of (W, Δ) on U determined by $(\iota \otimes \alpha)(\hat{M}X^*) = \hat{M}_{13}\hat{M}_{12}X^*_{12}$. To see this note that $\alpha(\cdot)_{21}$ is then a (left) coaction of $(W, \Delta^{\mathrm{op}})$. Also, by Proposition 16.1.7 the right coaction α is ergodic, and the same proposition confirms the statement about σ-weak closure made just above. The next two propositions after Proposition 16.1.7 tells us that α is integrable with semicyclic representation for $y_1 = (x \otimes \iota)\alpha$ given by $q_{y_1}(\omega \otimes \iota)(\hat{M}X^*) = q_x(\omega \otimes \iota)(\hat{M})$ for appropriate $\omega \in \hat{W}_*$. Since α is semidual, it is therefore a Galois object.

Corollary 18.1.20 *With notation as in the previous paragraph, the Galois map $\tilde{G}$ of α is $\hat{M}X^*$, and the reflection of $(\hat{W}, \hat{\Delta})$ across α is $(\hat{W}, \hat{\Delta}_X)$, so the latter is a locally compact quantum group. Its fundamental multiplicative unitary is $(J_{y_1} \otimes J_{\hat{x}})X\hat{M}^*(J_x \otimes J_{\hat{x}})X^*$.*

Proof Let $v, v', v'' \in V_x$, and let $\{v_i\}$ be an orthonormal basis for V_x. Let a be in the Tomita algebra of $\hat{x}$, and set $\omega = \omega_{v'', q_{\hat{x}}(a)}$. Then

$$(\iota \otimes \omega_{v,v'})(\tilde{G})q_x(\omega \otimes \iota)(\hat{M}) = q_x(\omega_{v,v'} \otimes \iota)\alpha(\omega \otimes \iota)(\hat{M}X^*),$$

which equals

$$\sum q_x(\omega \otimes \omega_{v,v_i} \otimes \omega_{v_i,v'} \otimes \iota)(\hat{M}_{14}\hat{M}_{13}X^*_{12}) = (\iota \otimes \omega_{v,v'})(\hat{M}X^*)q_x(\omega \otimes \iota)(\hat{M}),$$

so $\tilde{G} = \hat{M}X^*$. Now the canonical unitary implementation K of α is N, so the representation of $\hat{W}'$ on V_{y_1} is just the ordinary representation on V_x, and thus $Q_{11} = \hat{W}$ as von Neumann algebras. But $\Delta_{11} = \tilde{G}(1 \otimes (\cdot))\tilde{G}^* = X\hat{M}^*(1 \otimes (\cdot))\hat{M}X^* = \hat{\Delta}_X$, so $(\hat{W}, \hat{\Delta}_X)$ is the locally compact quantum group (Q_{11}, Δ_{11}). As for its fundamental multiplicative unitary $\hat{M}_X$, by what we have already proved we know that the Galois map G_X associated to the unitary 2-cocycle X^* of $(\hat{W}, \hat{\Delta}_X)$ is $\hat{M}_X X^*$. Under the identifications $V_{y_1} = V_x = \overline{V_x}$, using $J_{\hat{x}}J_{21}$ for the latter, we get $q_{12} = q_{\hat{x}}$ and $q_{21} = q_{x_1}$, and then, since $\Delta_{21} = \hat{\Delta}(\cdot)X^*$, we get by the previous lemma that $G^*_X = (J_{y_1} \otimes J_{\hat{x}})X\hat{M}^*(J_x \otimes J_{\hat{x}})$. □

For the next result, see Theorem 16.1.10, which is relevant since $K_0 \equiv J_{y_1}J_x$ is the canonical unitary implementation of $\alpha_0(\cdot)_{21}$. Note also that $u_t = \hat{\Delta}^{it}_U\Delta^{-it}_{\hat{x}} \in \hat{W}$ is the cocycle derivative of x_1 with respect to $\hat{x}$, so $u_{s+t} = u_s\sigma^{\hat{x}}_s(u_t)$.

Corollary 18.1.21 *Both K_0 and $v_t \equiv \Delta_{y_1}^{it}\Delta_x^{-it}$ belong to $\hat{W}$, and $\{v_t\}$ is a cocycle with respect to $\hat{\tau}_t$. We also have $X^*(v_t \otimes v_t) = \hat{\Delta}(v_t)(\hat{\tau}_t \otimes \hat{\tau}_t)(X^*)$ and $X^*(K_0 \otimes K_0) = \hat{\Delta}(K_0)(\hat{R} \otimes \hat{R})(X_{21}^*)$.*

Proof Since $\Delta_{y_1}^{it}$ and Δ_x^{it} implement the same automorphisms on $\hat{W}'$, and similarly for J_{y_1} and J_x, the first statement is clear. Combining the theorem above with Proposition 18.1.15, that $\Delta_x^{it} \otimes \Delta_{\hat{x}}^{it}$ commutes with $\hat{M}$, and that Δ_x^{it} implements $\hat{\tau}_t$, we get $\hat{\Delta}(u_t)(\hat{\tau}_t \otimes \sigma_t^{\hat{x}})(X^*) = X^*(v_t \otimes u_t)$. Combining this with the cocycle property of $\{u_t\}$ and $\hat{\Delta}\sigma_s^{\hat{x}} = (\hat{\tau}_s \otimes \sigma_s^{\hat{x}})\hat{\Delta}$, we get $v_{s+t} \otimes u_{s+t} = v_s\hat{\tau}_s(v_t) \otimes u_s\sigma_s^{\hat{x}}(u_t)$, so $v_{s+t} = v_s\hat{\tau}_s(v_t)$. Since $v_t = P_U^{it}P_W^{-it}$, the last identity in Proposition 18.1.12 gives the second last statement. The last statement comes from combining $\tilde{G}(J_{y_1} \otimes J_{y_1})F = FKF(J_{\hat{x}} \otimes J_{y_1})\tilde{G}$ with $K = N$ and the previous corollary. □

By slicing the last identity in the corollary with $\omega \otimes \iota$, we see that $K_0 \in M(\hat{W}_r)$ for *continuous* unitary 2-cocycles X, that is, when $X \in M(\hat{W}_r \otimes \hat{W}_r)$.

18.2 Deformation of C*-Algebras by Continuous Unitary 2-Cocycles

Let (W, Δ) be a regular locally compact quantum group with left invariant faithful semifinite normal weight x and fundamental multiplicative unitary M, and let X be a continuous unitary 2-cocycle of the dual quantum group $(\hat{W}, \hat{\Delta})$, so X is a unitary in $M(\hat{W}_r \otimes \hat{W}_r)$ that satisfies $(1 \otimes X)(\iota \otimes \hat{\Delta})(X) = (X \otimes 1)(\hat{\Delta} \otimes \iota)(X)$. Using $(\hat{\Delta}\otimes\iota)(\hat{M}) = \hat{M}_{13}\hat{M}_{23}$, and the fact that $\hat{M}$ implements $\hat{\Delta}$, we may write the cocycle identity as $(\hat{\Delta} \otimes \iota)(\hat{M}X^*)X_{12}^* = (\hat{M}X^*)_{13}(\hat{M}X^*)_{23}$. The *reduced X-twisted group C*-algebra* of $(\hat{W}, \hat{\Delta})$ is the C*-subalgebra W_{Xr} of $B(V_x)$ generated by the elements $(\omega \otimes \iota)(\hat{M}X^*)$ for $\omega \in B(V_x)_*$. Its bicommutant W_X is simply the von Neumann algebra $\hat{W} \ltimes_{\alpha_0, X} \mathbb{C}$ introduced towards the end of last section. Multiplying the previous form of the cocycle identity with $B_0(V_x)\otimes\mathbb{C}\otimes\mathbb{C}$ on the right, slicing it with $\iota\otimes\omega\otimes\iota$, and using regularity $[(\mathbb{C}\otimes B_0(V_x))\hat{M}^*(B_0(V_x)\otimes\mathbb{C})] = B_0(V_x)\otimes B_0(V_x)$, we get $\hat{M}X^*(B_0(V_x)\otimes W_{Xr}) = B_0(V_x)\otimes W_{Xr}$, and similarly, we get such an identity with $\hat{M}X^*$ to the right. Hence $\hat{M}X^* \in M(B_0(V_x) \otimes W_{Xr})$. Using this and slicing the cocycle identity in the form $(\hat{M}X^*)_{13} = \hat{M}_{12}(\hat{M}X^*)_{23}(\hat{M}X^*)_{12}(\hat{M}X^*)_{23}^*$ with $\iota \otimes \omega \otimes \iota$, we get $\hat{M}X^* \in M(\hat{W}_r \otimes W_{Xr})$. In the last section we saw that the dual coaction of α_0 gave us a right coaction of (W, Δ) on W_X, which we here denote by β. So $(\iota \otimes \beta)(\hat{M}X^*) = \hat{M}_{13}(\hat{M}X^*)_{12}$ and $\beta = N((\cdot) \otimes 1)N^*$, which also equals $(\hat{M}X^*)_{21}(1 \otimes (\cdot))(\hat{M}X^*)_{21}^*$. It is a continuous right coaction of (W, Δ) on W_{Xr} since slicing $(B_0(V_x) \otimes \mathbb{C} \otimes W_r)(\iota \otimes \beta)(\hat{M}X^*) = (B_0(V_x) \otimes \mathbb{C} \otimes W_r)(\hat{M}X^*)_{12}$ with $\omega \otimes \iota \otimes \iota$ gives $[(\mathbb{C} \otimes W_r)\beta(W_{Xr})] = W_{Xr} \otimes W_r$. From the previous section we know that $(\hat{W}, \hat{\Delta}_X)$ with $\hat{\Delta}_X = X\hat{\Delta}(\cdot)X^*$ is a locally compact quantum group having X^* as a continuous unitary 2-cocycle. Assume that this quantum group is regular, and let W_{X^*r} be its reduced X^*-twisted group C*-algebra with bicommutant

W_{X^*}. By the formula for the fundamental multiplicative unitary $\hat{M}_X$ of $(\hat{W}, \hat{\Delta}_X)$ from the previous section, we see that $W_{X^*r} = J_{\hat{x}} W_{Xr} J_{\hat{x}}$, so using the unitary coinverse on the dual quantum group, we can convert the right β-coaction to a continuous coaction β_{X^*} of $(\hat{W}, \hat{\Delta}_X)^\wedge$ on W_{Xr}. If M_{X^*} denotes the fundamental multiplicative unitary of this dual quantum group, we have $\beta_{X^*} = M_{X^*}^*(1 \otimes (\cdot)) M_{X^*}$. This is clear from $N_{X^*} = (J_{\hat{x}} \otimes J_{\hat{x}}) \hat{M}_X (J_{\hat{x}} \otimes J_{\hat{x}})$ and $(\hat{M}_X)_{21} = (M_{X^*})^*$. Using the formula for $\hat{M}_X$ and the characterizing property of β_{X^*}, we also see that $(\iota \otimes \beta_{X^*})(\hat{M}X^*) = (\hat{M}X^*)_{13}((J_{y_1} \otimes J_{\hat{x}}) \hat{M}_X^* (J_{y_1} \otimes J_{\hat{x}}))_{12}$, where $y_1 = (\iota \otimes x)\beta$. From this and the characterizing property of β, we get $(\beta_{X^*} \otimes \iota)\beta = (\iota \otimes \beta)\beta_{X^*}$. Now for each $\omega \in B_0(V_x)_*$ define a linear *quantization* map $T_\omega \colon W_r \to W_{Xr}$ by $T_\omega = (\iota \otimes \omega)(\hat{M}_X X((\cdot) \otimes 1)(\hat{M}_X X)^*)$. Actually $[\cup_\omega T_\omega(W_r)]$ equals the closure of $\{(\omega \otimes \iota)(\hat{M}X^*) \,|\, \omega \in B(V_x)_*\}$, and $\beta T_\omega = (T_\omega \otimes \iota)\Delta$. Indeed, we have the identity $(\hat{M}_X X)_{23} \hat{M}_{12} (\hat{M}_X X)_{23}^* = (\hat{M}X)_{12} (\hat{M}_X X)_{13}$, which follows from the cocycle identity for X^* and the formula for $\hat{M}_X$. Slicing this identity on the first and third leg we see that the stated properties of T_ω hold; the formula involving β and Δ is clear as $N_{12}(\hat{M}_X X)_{13} = (\hat{M}_X X)_{13} N_{12}$. Since the span of $\cup_\omega T_\omega(W_r)$ is self-adjoint, while $\{(\omega \otimes \iota)(\hat{M}X^*) \,|\, \omega \in B(V_x)_*\}$ is an algebra by the cocycle identity, its closure is already the C*-algebra W_{Xr}. Obviously T_ω depends only on the restriction of ω to $W_{X^*} = J_{\hat{x}} W_X J_{\hat{x}}$, and it extends to a normal map $W \to W_X$, which we still denote T_ω. Since $\hat{M}X^* \in M(B_0(V_x) \otimes W_{Xr})$ and $\hat{M}_X X \in M(B_0(V_x) \otimes W_{X^*r})$, the crucial identity above used to prove the properties of T_ω, also shows that $\hat{M}_X X(W_r \otimes \mathbb{C})(\hat{M}_X X)^*$ is a non-degenerate C*-subalgebra of $M(W_{Xr} \otimes W_{X^*r})$, so T_ω restricts to a map $M(W_r) \to M(W_{Xr})$ that is strictly continuous on bounded sets. This allows us to extend the quantization maps to the case of a general continuous coaction α of (W, Δ) on a C*-algebra U. We just extend the maps $T_\omega \otimes \iota \colon W_r \otimes U \to W_{Xr} \otimes U$ to the multiplier algebras, so $(T_\omega \otimes \iota)(b) = (\omega \otimes \iota \otimes \iota)((\hat{M}_X X)_{21}(1 \otimes b)(\hat{M}_X X)_{21}^*)$ for $b \in M(W_r \otimes U)$.

Definition 18.2.1 The *X-deformation of U* is the C*-subalgebra U_{Xr} of $M(W_{Xr} \otimes U)$ generated by $\cup_\omega (T_\omega \otimes \iota)\alpha(U)$ with *quantization maps* $(T_\omega \otimes \iota)\alpha$.

By $\beta T_\omega = (T_\omega \otimes \iota)\Delta$, we get $U_{Xr} \subset \{b \in M(W_{Xr} \otimes U) \,|\, (\beta \otimes \iota)(b) = (\iota \otimes \alpha)(b)\}$. When $U = W_r$ and $\alpha = \Delta_r$, the same identity gives $(W_r)_{Xr} = \beta(W_{Xr}) \cong W_{Xr}$, which is somehow consistent. In general, if γ is a continuous right coaction of any locally compact quantum group (W_1, Δ_1) on U such that $(\iota \otimes \gamma)\alpha = (\alpha \otimes \iota)\gamma$, the restriction of $\iota \otimes \gamma \colon M(W_{Xr} \otimes U) \to M(W_{Xr} \otimes U \otimes W_{1r})$ to U_{Xr} is a continuous right coaction of (W_1, Δ_1) on U_{Xr}. Indeed, for any $\omega \in B(V_x)_*$, the space

$$[(\mathbb{C} \otimes \mathbb{C} \otimes W_{1r})(\iota \otimes \gamma)(T_\omega \otimes \iota)\alpha(U)] = [(T_\omega \otimes \iota \otimes \iota)(\alpha \otimes \iota)((\mathbb{C} \otimes W_{1r})\gamma(U))]$$

equals $[(T_\omega \otimes \iota)\alpha(U)] \otimes W_{1r}$, so $[(\mathbb{C} \otimes \mathbb{C} \otimes W_{1r})(\iota \otimes \gamma)(U_{Xr})] = U_{Xr} \otimes W_{1r}$ and $(\iota \otimes \gamma)(U_{Xr}) \in M(U_{Xr} \otimes W_{1r})$.

We claim that $\alpha_X = (M_{X^*}^* \otimes 1)(1 \otimes (\cdot))(M_{X^*} \otimes 1)$ defines a continuous coaction of $(\hat{W}, \hat{\Delta}_X)^\wedge$ on U_{Xr}. Let B be the von Neumann algebra of the latter quantum group. Combining $(N_{X^*})_{23}(\hat{M}_X)_{12}(N_{X^*})_{23}^* = (\hat{M}_X)_{13}(\hat{M}_X)_{12}$ with the fact that

$(N_{X^*})_{12}$ and X_{13} commute, we get $(\iota \otimes \alpha_X)\eta_X\alpha = (N_{X^*})_{12}\eta_X\alpha(\cdot)_{134}(N_{X^*})^*_{12}$, where $\eta_X : \alpha(U) \to M(W_{X^*r} \otimes W_{Xr} \otimes U)$ is given by $\eta_X\alpha = (\hat{M}_X X)_{21}(1 \otimes \alpha(\cdot))(\hat{M}_X X)^*_{21}$. Multiplying the above identity with $\mathbb{C} \otimes B \otimes \mathbb{C} \otimes \mathbb{C}$ and slicing on the first leg, we get from $N_{X^*} \in M(B_0(V_x) \otimes B)$ and $[(\mathbb{C} \otimes B)N^*_{X^*}(B_0(V_x) \otimes \mathbb{C})] = B_0(V_x) \otimes B$ by regularity of $(\hat{W}, \hat{\Delta}_X)^\wedge$, that

$$[\{(B \otimes \mathbb{C} \otimes \mathbb{C})\alpha_X(T_\omega \otimes \iota)\alpha(U) \mid \omega \in B_0(V_x)\}] = B \otimes [\{(T_\omega \otimes \iota)\alpha(U) \mid \omega \in B_0(V_x)\}],$$

which settles the claim

Proposition 18.2.2 *If $u \in M(\hat{W}_r)$ is unitary, then $X_u \equiv (u \otimes u)X\hat{\Delta}(u)$ is a continuous unitary 2-cocycle of $(\hat{W}, \hat{\Delta})$ cohomologous to X. Its twisted quantum group is regular, and the map* $\mathrm{Ad}(u \otimes 1)$ *defines an isomorphism $U_{Xr} \to U_{X_u r}$.*

Proof We need only prove the last statement. Let u' be the unitary on V_x defined by $u' q_{y_1} = q_{y_{1u}}(u(\cdot)u^*)$, where y_{1u} is defined as y_1 but associated to X_u rather than X. By the standard semicyclic representations for these weights, and $\hat{M}X^*_u = (1 \otimes u)\hat{M}X^*(u^* \otimes u^*)$, we get $u' q_x(\omega(\cdot u^*) \otimes \iota)(\hat{M}) = q_x(\omega \otimes \iota)(\hat{M})$ for appropriate $\omega \in B(V_x)_*$. But $q_x(\omega(\cdot u^*) \otimes \iota)(\hat{M}) = u^* q_x(\omega \otimes \iota)(\hat{M})$ as we have $(q_x(\eta \otimes \iota)(\hat{M})|q_{\hat{x}}(\cdot)) = \eta((\cdot)^*)$, so $u' = u$ and $J_{y_{1u}} = uJ_{y_1}u^*$. Hence

$$\hat{M}_{X_u}X_u = (J_{y_{1u}} \otimes J_{\hat{x}})X_u(J_x \otimes J_{\hat{x}})\hat{M} = (u \otimes J_{\hat{x}}uJ_{\hat{x}})(\hat{M}_X X)(1 \otimes J_{\hat{x}}u^*J_{\hat{x}}),$$

and $\eta_{X_u} = \mathrm{Ad}(J_{\hat{x}}uJ_{\hat{x}} \otimes u \otimes 1)\eta_X$. □

We can deform in stages.

Proposition 18.2.3 *If X' is a continuous unitary 2-cocycle of $(\hat{W}, \hat{\Delta}_X)$, then $X'X$ is a continuous unitary 2-cocycle of $(\hat{W}, \hat{\Delta})$. Let us suppose that $U_{Xr} = [\{(T_\omega \otimes \iota)\alpha(U) \mid \omega \in B(V_x)_*\}]$. Then $h = (\hat{M}_X X'^*)_{21}(1 \otimes (\cdot))(\hat{M}_X X'^*)^*_{21}$ defines an isomorphism $U_{X'Xr} \to (U_{Xr})_{X'r}$ such that $(\alpha_X)_{X'}h = (\iota \otimes h)\alpha_{X'X}$.*

Proof The first statement is obvious. Next, combining the cocycle identity for X^* with $\hat{M}_{X'X} = (J_{y'_1} \otimes J_{\hat{x}})X'(J_z \otimes J_{\hat{x}})\hat{M}_X X^*$, where z is the left invariant faithful semifinite normal weight of $(\hat{W}, \hat{\Delta}_X)^\wedge$, and then using that $(\hat{M}_X X)_{23}$ and $((J_{y'_1} \otimes J_{\hat{x}})X'(J_z \otimes J_{\hat{x}}))^*_{12}$ commute, we get the identity

$$(\hat{M}_X X)_{23}(\hat{M}_{X'X}X'X)_{12}(\hat{M}_X X)^*_{23} = (\hat{M}_{X'X}X')_{12}(\hat{M}_X X)_{13}.$$

Combining this with the cocycle identity we get from X', just as we got the one for X^* when we defined T_ω, a short computation gives

$$(\iota \otimes \eta_{X'}\alpha_X)\eta_X\alpha = (\hat{M}_X X)_{21}(\hat{M}_X X'^*)_{43}\eta_{X'X}\alpha(\cdot)_{245}(\hat{M}_X X'^*)^*_{43}(\hat{M}_X X)^*_{21}.$$

Slicing the first two legs of this, shows that h is a well-defined isomorphism. Using the first identity in this proof once more in combination with the pentagonal equation for $\hat{M}_{X'X}$, we get $(\hat{M}_{X'X})_{23}(\hat{M}_X X'^*)_{12} = (\hat{M}_X X'^*)_{12}(\hat{M}_{X'X})_{13}(\hat{M}_{X'X})_{23}$, which gives the last identity in the proposition. □

The deformation U_{1r} with respect to the trivial unitary 2-cocycle is $\alpha(U)$, so if X satisfies the assumption of the previous proposition, then this result shows that $\eta_X\alpha$ is an isomorphism $U \cong (U_{Xr})_{X^*r}$.

Deformation can be seen from the point of view of cocycle crossed products. To this end consider a continuous coaction α of $(\hat{W}, \hat{\Delta}^{\mathrm{op}})$ on U, and let X be a continuous unitary 2-cocycle of $(\hat{W}, \hat{\Delta})$. Then (α, X_{21}^*) is a cocycle coaction of $(\hat{W}, X_{21}\hat{\Delta}^{\mathrm{op}}(\cdot)X_{21}^*)$, and we may form the C*-algebra version of the cocycle crossed product. Applying $\mathrm{Ad}(J_{\hat{x}}K_0^*J_{\hat{x}} \otimes 1)$ to this C*-algebra we get the *reduced twisted crossed product* of α, which we denote by $\hat{W}_r \ltimes_{\alpha,X} U$. The dual coaction of (α, X_{21}^*) becomes a continuous coaction $\hat{\alpha}$ of $(\hat{W}, \hat{\Delta}_X)^\wedge$ on $\hat{W}_r \ltimes_{\alpha,X} U$, and

$$\hat{\alpha} = \mathrm{Ad}((1 \otimes J_x J_{\hat{x}} \otimes 1)(M_{X^*}^* \otimes 1)(1 \otimes J_x J_{\hat{x}} \otimes 1))(1 \otimes (\cdot)).$$

Moreover, the C*-subalgebra $\hat{W}_r \ltimes_{\alpha,X} U$ of $M(B_0(V_x) \otimes U)$ is the closed span of $(J_x J_{\hat{x}} W_{Xr} J_{\hat{x}} J_x \otimes \mathbb{C})\alpha(U)$. To verify these facts, note from above that the reduced X_{21}^*-twisted group C*-algebra $W_{X_{21}^*r}$ of $(\hat{W}, X_{21}\hat{\Delta}^{\mathrm{op}}(\cdot)X_{21}^*)$ is $J_{\hat{x}} W_{X_{21}r} J_{\hat{x}}$, where $W_{X_{21}r}$ is the reduced X_{21}-twisted group C*-algebra of $(\hat{W}, \hat{\Delta}^{\mathrm{op}})$. By the last statement in Corollary 18.1.21, we have

$$\hat{M}X^*(K_0 \otimes K_0) = (1 \otimes K_0)(J_x \otimes J_x)\hat{M}^{\mathrm{op}} X_{21}^*(J_x \otimes J_x),$$

where $\hat{M}^{\mathrm{op}} = (J_x \otimes J_x)M(J_x \otimes J_x)$ is the fundamental multiplicative unitary of $(\hat{W}, \hat{\Delta}^{\mathrm{op}})$. Slicing this on the first leg shows that $W_{Xr} = K_0 J_x W_{X_{21}r} J_x K_0$. Hence $J_x J_{\hat{x}} W_{Xr} J_{\hat{x}} J_x = J_{\hat{x}} K_0^* J_{\hat{x}} W_{X_{21}^*r} J_{\hat{x}} K_0^* J_{\hat{x}}$. Since $\mathrm{Ad}(J_{\hat{x}} K_0^* J_{\hat{x}} \otimes 1)\alpha = \alpha$, we thus see that $\hat{W}_r \ltimes_{\alpha,X} U$ is the C*-algebra generated by $(J_x J_{\hat{x}} W_{Xr} J_{\hat{x}} J_x \otimes \mathbb{C})\alpha(U)$. That it suffices to consider the closed span, stems from the coaction identity for α and $\hat{\Delta}^{\mathrm{op}}(a) = \mathrm{Ad}((1 \otimes J_x J_{\hat{x}})\hat{M}X^*(1 \otimes J_{\hat{x}} J_x))(1 \otimes a)$, since then $(1 \otimes J_x J_{\hat{x}} \otimes 1)(\hat{M}X^* \otimes 1)(1 \otimes J_{\hat{x}} J_x \otimes 1)(\mathbb{C} \otimes \alpha(U))$ equals $(\iota \otimes \alpha)\alpha(U)(1 \otimes J_x J_{\hat{x}} \otimes 1)(\hat{M}X^* \otimes 1)(1 \otimes J_{\hat{x}} J_x \otimes 1)$, and slicing this identity on the first leg and using $(B_0(V_x) \otimes \mathbb{C})\alpha(U) \subset B_0(V_x) \otimes U$, gives $(J_x J_{\hat{x}} W_{Xr} J_{\hat{x}} J_x \otimes \mathbb{C})\alpha(U) \subset \alpha(U)(J_x J_{\hat{x}} W_{Xr} J_{\hat{x}} J_x \otimes \mathbb{C})$.

Proposition 18.2.4 *The map* $\mathrm{Ad}((1 \otimes J_x J_{\hat{x}} \otimes 1)(\hat{M}X^*)_{21}^*(1 \otimes J_{\hat{x}} J_x \otimes 1))$ *defines an isomorphism* $(\hat{W}_r \ltimes_\alpha U)_{Xr} \to \mathbb{C} \otimes (\hat{W}_r \ltimes_{\alpha,X} U)$.

Proof We only need to observe that $(\hat{W}_r \ltimes U)_{Xr}$ is the C*-algebra generated by $(1 \otimes J_x J_{\hat{x}} \otimes 1)(\beta(W_{Xr}) \otimes \mathbb{C})(1 \otimes J_{\hat{x}} J_x \otimes 1)(\mathbb{C} \otimes \alpha(U))$, while $\hat{W}_r \ltimes_{\alpha,X} U$ is generated by $(J_x J_{\hat{x}} W_{Xr} J_{\hat{x}} J_x \otimes 1)\alpha(U)$. □

For more general coactions, we need to consider a restricted class of continuous unitary 2-cocycles. Recall that we have a continuous coaction β of (W, Δ) on W_{Xr}, so we may consider the reduced right crossed product $W_{Xr} \rtimes_\beta \hat{W}_r = [\beta(W_{Xr})(\mathbb{C} \otimes J_{\hat{x}} \hat{W}_r J_{\hat{x}})]$. Since $\beta = (\hat{M}X^*)_{21}(1 \otimes (\cdot))(\hat{M}X^*)_{21}^*$, conjugation by $(\hat{M}X^*)_{21}^*$, which commutes with $\mathbb{C} \otimes J_{\hat{x}} \hat{W}_r J_{\hat{x}}$, maps this right crossed product to

$\mathbb{C} \otimes [W_{X_r} J_{\hat{x}} \hat{W}_r J_{\hat{x}}]$. Hence we say that a continuous unitary 2-cocycle X of $(\hat{W}, \hat{\Delta})$ is *regular* if $[W_{X_r} J_{\hat{x}} \hat{W}_r J_{\hat{x}}] = B_0(V_x)$. By the biduality theorem for right crossed products, we know that the weak closure of this normed closed span is $B(V_x)$. So regularity is equivalent to the condition $W_{X_r} J_{\hat{x}} \hat{W}_r J_{\hat{x}} \subset B_0(V_x)$. Alternatively, since the representation of $W_{X_r} \rtimes_\beta W_r$ is faithful and irreducible, regularity of X means that this reduced right crossed product is isomorphic to the C*-algebra of compact operators on some Hilbert space. Regularity of the trivial 2-cocycle just means that the quantum group is regular, which is not automatic. Regularity of cocycles holds, however, when $(\hat{W}, \hat{\Delta})$ comes from a locally compact group, and the proof goes just as the one showing that such groups are regular. Also, when $(\hat{W}, \hat{\Delta})$ is a discrete locally compact quantum group, its continuous unitary 2-cocycles are all regular, since then already $\hat{W}_r \subset B_0(V_x)$. In general we have the following result.

Proposition 18.2.5 *A continuous unitary 2-cocycle X on $(\hat{W}, \hat{\Delta})$ is regular if and only if $(B_0(V_x) \otimes \mathbb{C})\hat{M} X^*(\mathbb{C} \otimes B_0(V_x)) \subset B_0(V_x) \otimes B_0(V_x)$.*

Proof Consider the right continuous coaction β of (W, Δ) on W_{X_r} as a coaction $\beta' = \beta(\cdot)_{21}$ of $(W, \Delta^{\mathrm{op}})$ on W_X. By Proposition 16.1.5 and Proposition 16.1.6 the unitary $Y \equiv \hat{N}_{21}^* X_{21}^*$ with $\hat{N}_{21}^* = \hat{M}^{\mathrm{op}}$ defines a coaction $\gamma = Y(1 \otimes (\cdot))Y^*$ of $(\hat{W}, \hat{\Delta})$ on $B(V_x)$, and $Y(\cdot)Y^*$ is an isomorphism $W_X \bar{\otimes} B(V_x) \to ((W \otimes \mathbb{C})\gamma(B(V_x))''$ that intertwines $\beta' \otimes \iota$ and $\hat{\gamma} = N_{21}(1 \otimes (\cdot))N_{21}^*$. By the last statement in Corollary 18.1.21, we have $\hat{M} X^*(J_{y_1} \otimes J_{y_1}) = (J_x \otimes J_{y_1})\hat{M}^{\mathrm{op}} X_{21}^*$, so we may also write $Y = (J_x \otimes J_{y_1})\hat{M} X^*(J_{y_1} \otimes J_{y_1})$, showing that the second leg belongs to $(W_X)'$.

Suppose first that X is regular, so $B_0(V_x) = [W_{X_r} J_{\hat{x}} \hat{W}_r J_{\hat{x}}]$. We claim that γ restricts to a continuous coaction $B_0(V_x)$. Indeed, it suffices to show that it restricts to a continuous coaction on $J_{\hat{x}} \hat{W}_r J_{\hat{x}}$, but this is clear as $\gamma(J_{\hat{x}}(\cdot)J_{\hat{x}}) = (J_x \otimes J_{\hat{x}})\hat{\Delta}^{\mathrm{op}}(\cdot)(J_x \otimes J_{\hat{x}})$ on $\hat{W}_r$. Hence $(B_0(V_x) \otimes \mathbb{C})Y(\mathbb{C} \otimes B_0(V_x))Y^*(B_0(V_x) \otimes \mathbb{C})$ equals $(B_0(V_x) \otimes \mathbb{C})\gamma(B_0(V_x))(B_0(V_x) \otimes \mathbb{C}) \subset B_0(V_x) \otimes B_0(V_x)$. By polarization and square-rooting this means that $(B_0(V_x) \otimes \mathbb{C})Y(\mathbb{C} \otimes B_0(V_x)) \subset B_0(V_x) \otimes B_0(V_x)$, which proves the forward implication.

For the converse, we claim again that γ restricts to a continuous coaction on $B_0(V_x)$. Indeed, by Proposition 16.3.14 it suffices to show that $B_0(V_x)_\gamma$ equals $B_0(V_x)$. Inclusion holds as $(B_0(V_x) \otimes \mathbb{C})Y(B_0(V_x) \otimes \mathbb{C}) \subset B_0(V_x) \otimes B_0(V_x)$ by assumption. But $B_0(V_x)_\gamma$ is a C*-algebra and it is σ-strongly* dense in $B(V_x)$, so our claim holds. Next, the map $Y(\cdot)Y^*$ in the first paragraph restricts to an isomorphism $W_{X_r} \otimes B_0(V_x) \to \hat{W}_r \ltimes_\gamma B_0(V_x)$. Indeed, arguing as in the von Neumann algebra case, by slicing the cocycle identity $Y_{23}^* \hat{M}_{12} Y_{23} = (\hat{M} X^*)_{12} Y_{13}^*$ on the first leg, we see that $\mathrm{Ad}(Y^*)$ maps $\hat{W}_r \ltimes_\gamma B_0(V_x)$ onto

$$[\{(\omega \otimes \iota \otimes \iota)((\hat{M} X^* \otimes 1)Y_{13}^*(\mathbb{C} \otimes \mathbb{C} \otimes B_0(V_x))) \mid \omega \in B(V_x)_*\}] = W_{X_r} \otimes B_0(V_x).$$

Hence X is regular provided we can show that $[(J_{\hat{x}}\hat{W}_r J_{\hat{x}} \otimes \mathbb{C})(\hat{W}_r \ltimes_\gamma B_0(V_x))]$. But $[J_{\hat{x}}\hat{W}_r J_{\hat{x}} W_r] = B_0(V_x)$, again by regularity of (W, Δ), and $[(B_0(V_x) \otimes \mathbb{C})\gamma(B_0(V_x))]$ equals $[(B_0(V_x)\hat{W}_r \otimes \mathbb{C})\gamma(B_0(V_x))] = B_0(V_x) \otimes B_0(V_x)$. □

Theorem 18.2.6 *Let X be a regular continuous 2-cocycle of a regular locally compact quantum group $(\hat{W}, \hat{\Delta})$ with the twisted quantum group also regular. For any continuous coaction α of (W, Δ) on U, the map* $\mathrm{Ad}((J_{\hat{x}}J_x \otimes 1 \otimes 1)Y_{21}^*(J_xJ_{\hat{x}} \otimes 1 \otimes 1))$ *defines an isomorphism $\hat{W}_r \ltimes_{\hat{\alpha},X} (W_r \ltimes_\alpha U) \to B_0(V_x) \otimes U_{Xr}$ that is trivial on $J_xJ_{\hat{x}}W_{Xr}J_{\hat{x}}J_x \otimes \mathbb{C} \otimes \mathbb{C}$, and maps $\hat{\Delta}^{\mathrm{op}}(a) \otimes 1$ to $a \otimes 1 \otimes 1$ and $1 \otimes \alpha(u)$ to $\mathrm{Ad}(J_{\hat{x}}K_0^*J_{\hat{x}} \otimes 1 \otimes 1)\eta_X\alpha(u)$ for $a \in \hat{W}_r$ and $u \in U$. We also have $U_{Xr} = [\{(T_\omega \otimes \iota)\alpha(U) \mid \omega \in B_0(V_x)_*\}]$.*

Proof Since $[J_xJ_{\hat{x}}W_{Xr}J_{\hat{x}}J_x\hat{W}_r] = B_0(V_x)$, the map in question maps onto $[(B_0(V_x) \otimes \mathbb{C} \otimes \mathbb{C})\mathrm{Ad}(J_{\hat{x}}K_0^*J_{\hat{x}} \otimes 1 \otimes 1)\eta_X\alpha(U)]$, which, as it is a C*-algebra, equals $B_0(V_x) \otimes [\{\omega \otimes \iota \otimes \iota)\eta_X\alpha(U) \mid \omega \in B_0(V_x)_*\}]$. □

We have the following neat picture.

Theorem 18.2.7 *Let X be a continuous 2-cocycle of a regular locally compact quantum group $(\hat{W}, \hat{\Delta})$ with twisted quantum group also regular, and consider a continuous coaction α of (W, Δ) on U. If $\hat{\alpha}_X \equiv X_{21}\hat{\alpha}(\cdot)X_{21}^*$ defines a continuous coaction of $(\hat{W}, \hat{\Delta}_X)^{\mathrm{op}}$ on $W_r \ltimes_\alpha U \subset M(B_0(V_x) \otimes U)$, and if $M_{X^*} \otimes 1$ is in $M(B_r \otimes (W_r \ltimes_\alpha U)) \subset M(B_r \otimes B_0(V_x) \otimes U)$, where B_r is the reduced C*-algebra of $(\hat{W}, \hat{\Delta}_X)^\wedge$. Then $W_r \ltimes_\alpha U = [(\hat{W}_r \otimes \mathbb{C})U_{Xr}]$ and $(M_{X^*})_{12}^*\hat{\alpha}_X(\cdot)(M_{X^*})_{12}$ is an isomorphism $W_r \ltimes_\alpha U \to B_r \ltimes_{\alpha_X} U_{Xr}$ that turns $\hat{\alpha}_X$ into the dual coaction of α_X.*

Proof This is immediate from Theorem 16.3.16 and the observation that A in that theorem is U_{Xr} here. □

Example 18.2.8 Let G be a locally compact group. Then the left and right regular representations are given by $\lambda(t) = (\iota \otimes \delta_{t^{-1}})(\hat{M})$ and $\rho(t) = (\iota \otimes \delta_t)(N)$, respectively. We identify the predual of $\hat{W} = W^*(G)$ as a $*$-algebra with the Fourier algebra $A(G) \subset C_0(G)$, by sending ω to the function $\omega\lambda$, so $(f \otimes \iota)(\hat{M}) = fS$ for $f \in A(G)$, where $S(t) = t^{-1}$. Given a continuous unitary 2-cocycle X of $(W^*(G), \hat{\Delta})$, where $\hat{\Delta}\lambda(t) = \lambda(t) \otimes \lambda(t)$, we define a new product $\star$ on $A(G)$ by $f \star g = (f \otimes g)(\hat{\Delta}(\cdot)X^*)$. The cocycle identity means that the product is associative, and the same identity assures us that $\pi_X(f) = (f \otimes \iota)(\hat{M}X^*)$ defines a representation onto $L^2(G)$ of $A(G)$ with this new product. We know that W_{Xr} is the closed span of $\pi_X(A(G))$, although the latter image need not even be a $*$-algebra. Since $q_{y_1}(f \otimes \iota)(\hat{M}X^*) = q_x(f \otimes \iota)(\hat{M}) = fS$ for $f \in A(G)$, we have $\pi_X(f)gS = (f \star g)S$ when $g \in C_c(G) \cap A(G)$. So π_X is the representation on $A(G)$ by left multiplication with respect to the new product and with the latter space completed with respect to the right Haar measure. Also, left translations of G on itself define automorphisms of $A(G)$ with respect $\star$, which on the level of W_{Xr} is β from above, so $\beta_t\pi_X(f) = \pi_X(f(t^{-1}\cdot))$.

Conversely, assume we have a product $\star$ on $A(G)$ such that $f \otimes g \mapsto (f \star g)(e)$ extends to a bounded linear functional on $(W^*(G)\bar{\otimes}W^*(G))_*$. Then there exists

$X \in W^*(G)\bar{\otimes}W^*(G)$ such that $(f \star g)(e) = (f \otimes g)(X^*)$, so $f \star g = (f \otimes g)(\hat{\Delta}(\cdot)X^*)$ and X is a 2-cocycle. When G is finite and the algebra $A(G)$ with product $\star$ is semisimple, then X is invertible, and it is also easy to see that there exists $a \in W^*(G)$ such that $(a \otimes a)X\hat{\Delta}(a)^{-1}$ is unitary; it is obviously continuous. The question of regularity of continuous unitary 2-cocycles beyond finite and abelian locally compact groups is rather open. ◇

Assume that G is a locally compact group with a continuous action γ on a C*-algebra B. A positive element $b \in B$ is *γ-integrable* if there is $b_0 \in M(B)$ such that $\int z\gamma_t(b)\,dt = z(b_0)$ for all states z on B. The element b_0 is uniquely determined by b, and $b_0 \in M(B)^\gamma$. In fact, since $(aba^*)_0 = ab_0a^*$ for $a \in M(B)^\gamma$, the span P of γ-integrable elements is an $M(B)^\gamma$-bimodule, whence P_0 is a $*$-ideal in $M(B)^\gamma$ with closure suggested to be the *generalized fixed point algebra* of γ. The continuous action γ is *integrable* if P is dense in B. This is evidently a stronger condition than integrability in the von Neumann algebra sense.

Proposition 18.2.9 *Let G be a locally compact group, and suppose X is a continuous unitary 2-cocycle of $(W^*(G), \hat{\Delta})$. Then the coaction β from above, seen as a continuous action of G on W_{X^r}, is integrable.*

Proof If ω is a normal state on W_{X^*} and z is a state on W_{X^r}, then ωT_z is a state on $C_0(G)$. For positive $f \in C_0(G) \cap L^1(G)$, we have

$$\int z\beta_t T_\omega(f)\,dt = \int zT_\omega(f(\cdot t))\,dt = zT_\omega(1)x(f) = \int f(t)\,dt = y_1 T_\omega(f)$$

since integrability, in both senses, of positive elements in $C_0(G)$ amounts to requiring that these elements belong to $L^1(G)$. □

Since β is ergodic, we see that $b_0 = y_1(b)1$ whenever b is β-integrable.

Corollary 18.2.10 *Let G be a locally compact group, and suppose X is a continuous unitary 2-cocycle of $(W^*(G), \hat{\Delta})$ with regular twisted quantum group. If α is a continuous action of G on a C*-algebra U, the diagonal action $\beta \otimes \alpha$ of G on $W_{X^r} \otimes U$ is integrable and U_{X^r} is in the closure of P_0. When G is compact, then U_{X^r} is all of $M(W_{X^r} \otimes U)^{\beta\otimes\alpha}$.*

Proof Integrability is immediate from the proposition. If ρ is the action of G on $C_0(G)$ by right translations, and $f \in C_0(G) \cap L^1(G)$ and $u \in U$ are positive, then $f \otimes a$ is $\rho \otimes \alpha$-integrable and $(f \otimes b)_0 = \alpha(a_f)$ with $a_f = \int f(t)\alpha_t(a)\,dt$, so $\alpha(a_f)(t) = \alpha_{t^{-1}}(a_f)$. Then as in the previous proof, we see that $T_\omega(f) \otimes a$ is $\beta \otimes \alpha$-integrable with $(T_\omega(f) \otimes a)_0 = (T_\omega \otimes \iota)\alpha(a_f)$. So U_{X^r} is contained in the closure of P_0, which is a C*-subalgebra of $M(W_{X^r} \otimes U)^{\beta\otimes\alpha}$. When G is compact, then all the positive elements of $W_{X^r} \otimes U$ are $\beta \otimes \alpha$-integrable, and we are done by what we have already shown. □

Exercises

For Sect. 18.1

1. Let (W, Δ) be a locally compact quantum group with left invariant faithful semifinite normal weight x, and let $\alpha\colon U \to U\bar{\otimes}W$ be a Galois object for (W, Δ) with a faithful semifinite normal weight y_1 on U. Let Q be the linking algebra between the right $\hat{W}$-modules V_x and V_{y_1}. By a projective α-corepresentation for the dual quantum group $(\hat{W}, \hat{\Delta})$ on some Hilbert space V, we mean a unitary $X \in Q_{12}\bar{\otimes}B(V)$ such that $(\Delta_{12}\otimes\iota)(X) = X_{13}X_{23}$, where $\Delta_{12} = \tilde{G}^*(1\otimes(\cdot))\hat{M}$ and $\tilde{G}$ is the unitary Galois isometry for α, and where $\hat{M}$ is the fundamental multiplicative unitary of $(\hat{W}, \hat{\Delta})$. Show that $\beta_X \equiv X^*(1\otimes(\cdot))X$ is a coaction of $(\hat{W}, \hat{\Delta})$ on the type I factor $B(V)$. Prove that $X \mapsto \beta_X$ surjects onto the set of coactions β of $(\hat{W}, \hat{\Delta})$ on type I factors by considering as Galois object the dual coaction of β restricted to the relative commutant of $\mathrm{im}(\beta)$ in the domain of the dual coaction of β, and using this in turn to construct X with $\beta_X = \beta$.
2. Let (W, Δ) be a locally compact quantum group with dual quantum group $(\hat{W}, \hat{\Delta})$. Two right coactions α_i of $(\hat{W}, \hat{\Delta})$ on a von Neumann algebra U are said to be outer equivalent if there is a unitary $Y \in \hat{W}\bar{\otimes}U$ satisfying the identities $(\hat{\Delta}\otimes\iota)(Y) = Y_{23}(\iota\otimes\alpha_1)(Y)$ and $\alpha_2 = Y\alpha_1(\cdot)Y^*$. Suppose W_* is separable. Use the previous exercise to show that there is a one-to-one correspondence between outer equivalence classes of right coactions of $(\hat{W}, \hat{\Delta})$ on $B(V)$ for V separable and infinite dimensional, and isomorphism classes of Galois objects for (W, Δ) acting on von Neumann algebras with separable preduals.

For Sect. 18.2

1. Let G be a locally compact group, and let X be a unitary 2-cocycle of the locally compact quantum group $(W^*(G), \hat{\Delta})$, where $\hat{\Delta}(\lambda_t) = \lambda_t \otimes \lambda_t$ for $t \in G$. Show that X is continuous if both X and X^* map $C_c(G\times G)$ into $L^1(G\times G)\cap L^2(G\times G)$.
2. Let G be a locally compact group, and write

$$(X(u)|v) = \int Y(s,t)((\lambda_s\otimes\lambda_t)(u)|v)\,ds\,dt.$$

What does the kernel $Y\colon G\times G \to \mathbb{C}$, at least formally, need to satisfy for X to be a well-defined unitary 2-cocycle of $(W^*(G), \hat{\Delta})$, where u, v should belong to some appropriate dense subspace of $L^2(G\times G)$? You may want to simplify to the case when G is finite.
3. Let G be a locally compact group and let α be an integrable ergodic right coaction of the corresponding locally compact quantum group $(L^\infty(G), \Delta)$ on a von Neumann algebra U. Show that $U \rtimes_\alpha L^\infty(G)$ is a type I factor if and only if α is a Galois object. In this case, show that the corresponding action of G is of the form $t \mapsto \mathrm{Ad}(u_t)$ for a projective irreducible unitary representation u of G that is square integrable, meaning that the function $t \mapsto |(u_t(v)|v')|$ is

square integrable for some non-zero vectors v, v'. Projectiveness means that it is not really a unitary representation on the Hilbert space, but merely a group homomorphism from G to the quotient of the unitary group on the Hilbert space by the unit circle. In this case, show also that the reflection of $(L^\infty(G), \Delta)$ across α is again $(L^\infty(G), \Delta)$ when G is abelian, while the reflected quantum group is neither commutative nor cocommutative when G is non-abelian. When $W^*(G)$ is a type I factor, can G be compact or discrete?

Chapter 19
Doublecrossed Products of Quantum Groups

Given a coaction α of a locally compact quantum group (W, Δ), we establish a biduality theorem for faithful semifinite normal weights on U. Given such a weight y we talk about its Radon-Nikodym derivative $(Dy\alpha : Dy)_t$ under α. This is an element $u_t \in W\bar{\otimes}U$ that satisfies the following cocycle type of identity $(\Delta \otimes \iota)(u_t) = (\iota \otimes \alpha)(u_t)(1 \otimes u_t)$, and is by definition the cocycle derivative of the bidual weight of y with respect to $\mathrm{Tr}_{\Delta_{\hat{x}}} \otimes y$.

The Radon-Nikodym derivative of a weight under a coaction is used to prove that for a locally compact quantum group, any von Neumann subalgebra invariant under the unitary coinverse and the scaling group forms a locally compact quantum group with coproduct given by restriction whenever that makes sense.

Another application is towards the study of doublecrossed products. The doublecrossed product of two locally compact quantum groups (W_1, Δ_1) and (W_2, Δ_2) is the locally compact quantum group $(W_1\bar{\otimes}W_2, (\iota \otimes Fm \otimes \iota)(\Delta_1^{\mathrm{op}} \otimes \Delta_2))$, where $F\colon W_1\bar{\otimes}W_2 \to W_2\bar{\otimes}W_1$ is the flip and m is a matching, that is, a cocycle matching of (W_1, Δ_1) and (W_2, Δ_2) with trivial cocycles. That we indeed get a locally compact quantum group this way requires of course a proof. We supply formulas for the standard quantum group quantities in terms of the quantities of the constituent quantum groups, and we do this also for the dual and commutant of the doublecrossed product. On the way we express corepresentations of the bicrossed product in terms of corepresentations of the constituent quantum groups.

A special case of double crossed products occurs when the matching is given by conjugation with a so called bicharacter. We speak then of generalized quantum doubles since the special case of the fundamental multiplicative unitary as a bicharacter yields a matching of $(W, \Delta^{\mathrm{op}})$ and $(\hat{W}, \hat{\Delta})$ leading to what is know as the quantum double. Of course, these constructions are well known in the purely algebraic setting.

We devote a section to morphisms between quantum groups, establishing a correspondence between these and right coactions of the target quantum groups

L. Tuset, *Analysis and Quantum Groups*,
https://doi.org/10.1007/978-3-031-07246-8_19

on the von Neumann algebras of the domain quantum groups. We then characterize doublecrossed products by a universal property.

Finally, we look at continuous, or reduced versions, of bicrossed product and doublecrossed products, focusing on regularity.

Recommended literature for this chapter is [2].

19.1 Radon-Nikodym Derivatives of Weights Under Coactions

Let (W, Δ) be a locally compact quantum group with left invariant faithful semifinite normal weight x and fundamental multiplicative unitary M. Fix a coaction α of (W, Δ) on a von Neumann algebra U with a faithful semifinite normal weight y. Its bidual weight $\tilde{\tilde{y}}$ on $B(V_x)\bar{\otimes}U \cong \hat{W} \ltimes_{\hat{\alpha}} (W \ltimes_\alpha U)$ has GNS-map determined by $q_{\tilde{\tilde{y}}}((ab \otimes 1)\alpha(u)) = q'_x(a) \otimes q_{\hat{x}}(b) \otimes q_y(u)$, where $q'_x = J_x q_x(J_x(\cdot)J_x)$ is the canonical GNS-map on the commutant quantum group $(W, \Delta)'$ dual to $(W, \Delta)^{\wedge\mathrm{op}}$. One readily checks that $\tilde{\tilde{y}}(\alpha(u^*)(a^*b \otimes 1)\alpha(u')) = \mathrm{Tr}_{\Delta_{\hat{x}}}(a^*b)y(u^*u')$ for $a, b \in N_{\mathrm{Tr}_{\Delta_{\hat{x}}}}$ and $u, u' \in N_y$, and that $(b \otimes 1)\alpha(u')$ span a core for the GNS-map of $\tilde{\tilde{y}}$. We have the following *biduality theorem for weights*.

Proposition 19.1.1 *We have* $(D\tilde{\tilde{y}} : D(\mathrm{Tr}_{\Delta_{\hat{x}}} \otimes y))_t = \Delta_{\tilde{y}}^{it}(\Delta_{\hat{x}}^{-it} \otimes \Delta_y^{-it})$.

Proof Let $E_{\hat{\hat{\alpha}}}$ be the operator valued weight from $B(V_x)\bar{\otimes}U$ to $W \ltimes_\alpha U$ associated with $\hat{\hat{\alpha}}$. So $E_{\hat{\hat{\alpha}}} = J_{\tilde{y}}(E_{\hat{\alpha}})'(J_{\tilde{y}}(\cdot)J_{\tilde{y}})J_{\tilde{y}}$ by Theorem 17.1.4. Hence

$$(\frac{d\tilde{\tilde{y}}}{d(1 \otimes y(J_y(\cdot)J_y))})^{-it} = (\frac{d(\tilde{y}(J_{\tilde{y}}(\cdot)J_{\tilde{y}})(E_{\hat{\alpha}})')}{d(y\alpha^{-1})})^{it} = (\frac{d(\tilde{y}(J_{\tilde{y}}(\cdot)J_{\tilde{y}})}{d(y\alpha^{-1}E_{\hat{\alpha}})})^{it} = \Delta_{\tilde{y}}^{-it},$$

and multiplying this from the right with the cocycle derivative in question, yields $(d(\mathrm{Tr}_{\Delta_{\hat{x}}} \otimes y)/d(1 \otimes y(J_y(\cdot)J_y)))^{-it} = \Delta_{\hat{x}}^{-it} \otimes \Delta_y^{-it}$. □

We denote the quantity in the previous proposition by $(Dy\alpha : Dy)_t$ and call it the *Radon-Nikodym derivative of* y *under* α.

Proposition 19.1.2 *We have* $u_t \equiv (Dy\alpha : Dy)_t \in W\bar{\otimes}U$ *and* $(\Delta \otimes \iota)(u_t) = (\iota \otimes \alpha)(u_t)(1 \otimes u_t)$.

Proof Since $\tilde{y}$ is $\hat{h}$-invariant, we see from the proof of Proposition 15.4.11 that $\sigma_t^{\tilde{\tilde{y}}}(a \otimes 1) = \hat{h}^{-it}\Delta_x^{-it}a\Delta_x^{it}\hat{h}^{it} \otimes 1 = \Delta_{\hat{x}}^{it}a\Delta_{\hat{x}}^{-it} \otimes 1$ for $a \in W'$, as one readily checks that $\hat{h}^{-it}\Delta_x^{-it} = \Delta_{\hat{x}}^{it}h^{it}$. Hence $u_t \in W\bar{\otimes}U$. Let $y^\alpha = \tilde{\tilde{y}}$, and consider the coaction $\beta = (F \otimes \iota)(\iota \otimes \alpha)$ of (W, Δ) on $B(V_x)\bar{\otimes}U$, where F is the flip. When $u \in N_y$ and $b, c \in N_{\mathrm{Tr}_{\Delta_{\hat{x}}}}$, the GNS-map for the bidual weight $(y^\alpha)^\beta$ on $B(V_x \otimes V_x)\bar{\otimes}U$ sends $(b \otimes c \otimes 1)\beta\alpha(u)$ to $q_{\mathrm{Tr}_{\Delta_{\hat{x}}}}(b) \otimes q_{\mathrm{Tr}_{\Delta_{\hat{x}}}}(c) \otimes q_y(u)$, and the former elements span a core for this GNS-map. Working on this core, and observing that $\beta\alpha = (\Delta^{\rm op} \otimes \iota)\alpha$ and $N_{21}^*(\Delta_{\hat{x}} \otimes \Delta_{\hat{x}})N_{21} = Q \otimes \Delta_{\hat{x}}$, where Q is the closure of $h\Delta_{\hat{x}}$, we get $(y^\alpha)^\beta \text{Ad}(N_{21} \otimes 1) = \text{Tr}_Q \otimes y^\alpha$ and $(\text{Tr}_{\Delta_{\hat{x}}} \otimes \text{Tr}_{\Delta_{\hat{x}}} \otimes y)\text{Ad}(N_{21} \otimes 1) = \text{Tr}_Q \otimes \text{Tr}_{\Delta_{\hat{x}}} \otimes y$. Hence the cocycle derivative $(D((y^\alpha)^\beta) : D(\text{Tr}_{\Delta_{\hat{x}}} \otimes \text{Tr}_{\Delta_{\hat{x}}} \otimes y))_t$ equals

$$(N_{21} \otimes 1)(D(\text{Tr}_Q \otimes y^\alpha) : D(\text{Tr}_Q \otimes \text{Tr}_{\Delta_{\hat{x}}} \otimes y))_t(N_{21} \otimes 1)^* = (\Delta^{\rm op} \otimes \iota)(u_t)$$

on the one hand, and on the other hand, it equals $(D((y^\alpha)^\beta) : D((\text{Tr}_{\Delta_{\hat{x}}} \otimes y)^\beta))_t$ times $(D((\text{Tr}_{\Delta_{\hat{x}}} \otimes y)^\beta) : D(\text{Tr}_{\Delta_{\hat{x}}} \otimes \text{Tr}_{\Delta_{\hat{x}}} \otimes y))_t$, and the first of these factors equals $\beta((D(y^\alpha) : D(\text{Tr}_{\Delta_{\hat{x}}} \otimes y))_t)$, whereas in the second factor we can use that $(\text{Tr}_{\Delta_{\hat{x}}} \otimes y)^\beta$ is $(\text{Tr}_{\Delta_{\hat{x}}} \otimes y^\alpha)(F \otimes \iota)$, so the product equals $(F \otimes \iota)((\iota \otimes \alpha)(u_t)(1 \otimes u_t))$. □

Lemma 19.1.3 *Let α_i be coactions of locally compact quantum groups (W_i, Δ_i) on U with a faithful semifinite normal weight y. Suppose there is a $*$-isomorphism $m: W_1 \bar{\otimes} W_2 \to W_2 \bar{\otimes} W_1$ such that $(\iota \otimes \alpha_2)\alpha_1 = (mF \otimes \iota)(\iota \otimes \alpha_1)\alpha_2$ and $m(\tau_t^1 \otimes \tau_t^2) = (\tau_t^1 \otimes \tau_t^2)m$. Then the expression $(\iota \otimes \alpha_2)((Dy\alpha_1 : Dy)_t)(1 \otimes (Dy\alpha_2 : Dy)_t)$ equals $(mF \otimes \iota)((\iota \otimes \alpha_1)((Dy\alpha_2 : Dy)_t)(1 \otimes (Dy\alpha_1 : Dy)_t))$.*

Proof Let $\alpha_{ia} \equiv (F \otimes \iota)(\iota \otimes \alpha_i)$ be the amplified coaction of (W_i, Δ_i) on $B(V_{x_{i-1}})\bar{\otimes} U$, where $x_0 = x_2$, and x_i is the left invariant faithful semifinite normal weight of (W_i, Δ_i). Arguing as in the previous proof, we get

$$(D((y^{\alpha_i})^{\alpha_{ja}}) : D(\text{Tr}_{\Delta_{\hat{x}_j}} \otimes \text{Tr}_{\Delta_{\hat{x}_i}} \otimes y))_t$$

$$= (F \otimes \iota)((\iota \otimes \alpha_j)((Dy\alpha_i : Dy)_t)(1 \otimes (Dy\alpha_j : Dy)_t)).$$

Consider the operator $L \equiv J_{x_1}h_1J_{x_1} \otimes J_{x_2}h_2J_{x_2}$ affiliated with $W_1'\bar{\otimes} W_2'$. Then $L \otimes 1$ is invariant under the modular automorphism group of $\text{Tr}_{\Delta_{\hat{x}_1}} \otimes \text{Tr}_{\Delta_{\hat{x}_2}} \otimes y$ and commutes with $(D((y^{\alpha_2})^{\alpha_{1a}}) : D(\text{Tr}_{\Delta_{\hat{x}_1}} \otimes \text{Tr}_{\Delta_{\hat{x}_2}} \otimes y))_t \in W_1 \bar{\otimes} W_2 \bar{\otimes} U$, so it is invariant under the modular automorphism group of $(y^{\alpha_2})^{\alpha_{1a}}$, and we may define the faithful semifinite normal weight $y_{21} \equiv ((y^{\alpha_2})^{\alpha_{1a}})_{L \otimes 1}$ with GNS-map $(d \otimes 1)\alpha_{1a}\alpha_2(u) \mapsto q_{\text{Tr}_{P_1 \otimes P_2}}(d) \otimes q_y(u)$ for $u \in N_y$ and $d \in N_{\text{Tr}_{P_1 \otimes P_2}}$ since $P_1 \otimes P_2$ is the closure of $(\Delta_{\hat{x}_1} \otimes \Delta_{\hat{x}_2})L$. This determines the GNS-map since we have defined it on elements that span a core for it. Let $Z \in B(V_{x_1} \otimes V_{x_2})$ be the canonical unitary implementation of m. Now m commutes with $\tau_t^1 \otimes \tau_t^2$ and $P_1^{it} \otimes P_2^{it}$ implements the latter, so Z commutes with $P_1 \otimes P_2$. Working on the previous core and using the other identity for m in the lemma, we get $y_{21} = y_{12}\text{Ad}(F'Z \otimes 1)$, where F' is the flip on the Hilbert space and $y_{12} \equiv ((y^{\alpha_1})^{\alpha_{2a}})_{L_{21} \otimes 1}$. The first major identities in this proof can then be written

$$(Dy_{ji} : D(\text{Tr}_{P_i \otimes P_j} \otimes y))_t = (F \otimes \iota)((\iota \otimes \alpha_i)((Dy\alpha_j : Dy)_t)(1 \otimes (Dy\alpha_i : Dy)_t)),$$

which are related to each other by $\text{Ad}(F'Z \otimes 1)$. □

The converse of the following result on closed quantum subgroups is obvious.

Theorem 19.1.4 *Let (W, Δ) be a locally compact quantum group, and let U be a von Neumann subalgebra of W such that $\Delta(U) \subset U\bar{\otimes}U$. Then $(U, \Delta|U)$ is a locally compact quantum group if R and τ_t leave U invariant.*

Proof Consider the coaction $\alpha \equiv \Delta|_U$ of (W, Δ) on U, and the coaction $\beta \equiv \Delta^{\mathrm{op}}|U$ of $(W, \Delta^{\mathrm{op}})$ on U. Fix a faithful semifinite normal weight y on U, and let $u_t = (Dy\alpha : Dy)_t \in W\bar{\otimes}U$ and $w_t = (Dy\beta : Dy)_t \in W\bar{\otimes}U$. Arguing as in the proof of Proposition 15.4.8 we get $u_t, w_t \in U\bar{\otimes}U$ from our assumptions. By the previous proposition, we may write $(\Delta \otimes \iota)(\tilde{u}_t M) = (\tilde{u}_t M)_{13}(\tilde{u}_t M)_{23}$ with $\tilde{u}_t = (J_{\hat{x}} \otimes J_x)u_t^*(J_{\hat{x}} \otimes J_x) \in W\bar{\otimes}W'$. Hence there exists a faithful normal $*$-homomorphism $\pi_t\colon \hat{W} \to B(V_x)$ such that $(\iota \otimes \pi_t)(M) = \tilde{u}_t M$. Similarly, there exists a faithful normal $*$-homomorphism π_t‘: $\hat{W}' \to B(V_x)$ such that $(\iota \otimes \pi_t')(N_{21}^*) = \tilde{w}_t N_{21}^*$, where $\tilde{w}_t = (J_{\hat{x}} \otimes J_x)w_t^*(J_{\hat{x}} \otimes J_x)$. Using the previous lemma with m the identity, we find that $(\iota\otimes\Delta)(w_t)(1\otimes u_t) = (F\otimes\iota)((\iota\otimes\Delta^{\mathrm{op}}(u_t)(1\otimes w_t))$, which means that $(N_{21}^* w_t)_{13}$ and $(Mu_t)_{23}$ commute, or that the ranges of π_t and π_t' commute. One checks then that $\mathrm{Ad}((\pi_t'\otimes\iota)(N^*)M^*(\pi_t\otimes\iota)((J_x\otimes J_x)M_{21}(J_x\otimes J_x)))$ maps $1 \otimes a$ to $\pi_t(a) \otimes 1$ for $a \in \hat{W}$, while it maps $1 \otimes w$ to $w \otimes 1$ for $w \in W$. Since $W\hat{W}$ spans a strongly dense subset of $B(V_x)$, we thus get a faithful normal $*$-endomorphism f_t on $B(V_x)$ such that the previous Ad-map sends $1 \otimes b$ to $f_t(b) \otimes 1$ for $b \in B(V_x)$. Clearly $t \mapsto f_t(b)$ is strongly* continuous. Say b belongs to the commutant of the image of f_t. Since $f_t = \iota$ on W, we may write $c \equiv J_x b J_x \in W$. As $b \in \pi_t(\hat{W})'$, we get $\Delta(c) = u_t(1 \otimes c)u_t^*$, so $\Delta^{\mathrm{op}}(c) \in W\bar{\otimes}U$. But $(\Delta^{\mathrm{op}} \otimes \iota)\Delta^{\mathrm{op}}(c) = (\iota \otimes \beta)\Delta^{\mathrm{op}}(c)$, so $c \in U$, and $\alpha(c) = u_t(1 \otimes c)u_t^* = \Delta_{\tilde{y}}^{it}(1 \otimes \sigma_{-t}^y(c))\Delta_{\tilde{y}}^{-it}$, where the dual weight $\tilde{y}$ refers to α. This shows that $\Delta\sigma_{-t}^y(c) = 1 \otimes \sigma_{-t}^y(c)$, so $c, b \in \mathbb{C}$, and f_t is surjective. Pick a strongly continuous one parameter group of unitaries $\tilde{v}_t$ that implements f_t. Since $f_t = \iota$ on W, we have $v_t \equiv J_x \tilde{v}_t^* J_x \in W$, and $u_t = \Delta(v_t^*)(1\otimes v_t)$ by definition of π_t and $\tilde{v}_t$. Hence $\Delta(v_t) \in U\bar{\otimes}W$, and the same reasoning as above shows that $v_t \in U$, so $u_t = \alpha(v_t^*)(1 \otimes v_t)$. By definition of u_t we know that $u_{s+t} = u_t(\tau_t \otimes \sigma_t^y(u_s)$. Hence $\alpha(v_{t+s}^*)(1 \otimes v_{t+s})$ equals

$$\Delta_{\tilde{y}}^{it}(\Delta_{\hat{x}}^{-it} \otimes \Delta_y^{-it})(\tau_t \otimes \sigma_t^y)(\alpha(v_s^*)(1 \otimes v_s)) = \alpha(\sigma_t^y(v_s^*)v_t^*)(1 \otimes v_t\sigma_t^y(v_s)),$$

and $\Delta(v_t\sigma_t^y(v_s)v_{t+s}^*) = 1\otimes v_t\sigma_t^y(v_s)v_{t+s}^*$, which shows that $v_{t+s} = g(t, s)v_t\sigma_t^y(v_s)$ for a continuous function $g\colon \mathbb{R}^2 \to \mathbb{T}$ satisfying a cocycle identity. This equation can be easily solved yielding a continuous function $h\colon \mathbb{R} \to \mathbb{T}$ such that $g(t, s) = h(t)h(s)\overline{h(t+s)}$. Replacing v_t by $h(t)v_t$, we get a (σ^y)-cocycle $\{v_t\}$ such that $u_t = \alpha(v_t^*)(1 \otimes v_t)$. Connes' cocycle theorem then gives a faithful semifinite normal weight x_U on U such that $(Dx_U : Dy)_t = v_t$. But $(D\tilde{x}_U : D(\mathrm{Tr}_{\Delta_{\hat{x}}} \otimes x_U))_t$ equals

$$(D\tilde{x}_U : D\tilde{y})_t(D\tilde{y} : D(\mathrm{Tr}_{\Delta_{\hat{x}}}\otimes y))_t(D(\mathrm{Tr}_{\Delta_{\hat{x}}}\otimes y) : D(\mathrm{Tr}_{\Delta_{\hat{x}}}\otimes x_U))_t = \alpha(v_t)u_t(1\otimes v_t^*),$$

which means that $(\iota \otimes x_U)\alpha = x_U(\cdot)1$, so x_U is left invariant for $(U, \Delta|U)$. But $x_U R$ is then right invariant. □

19.2 Doublecrossed Products

Throughout this section (W_i, Δ_i) denotes locally compact quantum groups with left invariant faithful semifinite normal weights x_i and fundamental multiplicative unitaries M_i, and so on. We say m is a *matching* of (W_1, Δ_1) and (W_2, Δ_2) if $(m, 1, 1)$ is a cocycle matching of these quantum groups, which in this special case means that m is a normal $*$-endomorphism on $W_1 \bar{\otimes} W_2$ satisfying $(\Delta_1 \otimes \iota)m = m_{23}m_{13}(\Delta_1 \otimes \iota)$ and $(\iota \otimes \Delta_2)m = m_{13}m_{12}(\iota \otimes \Delta_2)$. Then $\alpha = m(1 \otimes (\cdot))$ is a coaction of (W_1, Δ_1) on W_2, while $\beta = m((\cdot) \otimes 1)$ is a right coaction of $(W_2, \Delta_2^{\mathrm{op}})$ on W_1. Recall that the *bicrossed product* of (W_1, Δ_1) and (W_2, Δ_2) associated to m is the locally compact quantum group (W, Δ) with $W = W_1 \ltimes_\alpha W_2 = (\alpha(W_2) \cup (\hat{W}_1 \otimes \mathbb{C}))''$ and with Δ determined by $\Delta\alpha = (\alpha \otimes \alpha)\Delta_2$ and the identity $(\iota \otimes \Delta)(M_1 \otimes 1) = M_{1,14}((\iota \otimes \alpha)\beta \otimes \iota)(M_1)_{1453}$. Its left invariant faithful semifinite normal weight x is the dual weight of x_2 with canonical GNS-map determined by $q_x((a \otimes 1)\alpha(w)) = q_{\hat{x}_1}(a) \otimes q_{x_2}(w)$ for $a \in N_{\hat{x}_1}$ and $w \in N_{x_2}$, whereas $M \equiv (\alpha \otimes \iota)(M_2)_{124}(\iota \otimes \beta)(\hat{M}_1)_{134}$ is its fundamental multiplicative unitary. Its dual quantum group $(\hat{W}, \hat{\Delta})$ has $\hat{W} = W_1 \rtimes_\beta W_2 = (\beta(W_1)(\mathbb{C} \otimes \hat{W}_2))''$ and coproduct determined by $\hat{\Delta}\beta = (\beta \otimes \beta)\Delta_1$ and $(\hat{\Delta} \otimes \iota)(1 \otimes \hat{M}_2) = (\iota \otimes (\beta \otimes \iota)\alpha)(\hat{M}_2)_{2345}\hat{M}_{2,25}$. With $K_\alpha \equiv J_x(J_{\hat{x}_1} \otimes J_{x_2}) \in W_1 \bar{\otimes} B(V_{x_2})$ and $K_\beta = J_{\hat{x}}(J_{x_1} \otimes J_{\hat{x}_2}) \in B(V_{x_1}) \bar{\otimes} W_2$ we have $\alpha = K_\alpha(1 \otimes (\cdot))K_\alpha^*$ and $\beta = K_\beta((\cdot) \otimes 1)K_\beta^*$. We reserve unindexed quantities to bicrossed products.

By the *doublecrossed product* of (W_1, Δ_1) and (W_2, Δ_2) associated to the matching m we mean that locally compact quantum group (W_m, Δ_m) with $W_m = W_1 \bar{\otimes} W_2$ and $\Delta_m \equiv (\iota \otimes Fm \otimes \iota)(\Delta_1^{\mathrm{op}} \otimes \Delta_2)\colon W_m \to W_m \bar{\otimes} W_m$, where $F\colon W_1 \bar{\otimes} W_2 \to W_2 \bar{\otimes} W_1$ is the flip. Although Δ_m is clearly coassociative, it remains to prove that we indeed get a locally compact quantum group this way, which is the main objective of this section. The index m will here relate to quantities of doublecrossed products.

Lemma 19.2.1 *The element $Z \equiv J_x J_{\hat{x}}(J_{\hat{x}_1} J_{x_1} \otimes J_{\hat{x}_2} J_{x_2})$ implements m.*

Proof For $w \in W_1$ we have

$$Z(w \otimes 1)Z^* = J_{\hat{x}} K_\alpha (J_{x_1} w J_{x_1} \otimes 1) K_\alpha^* J_{\hat{x}} = J_{\hat{x}}(J_{x_1} w J_{x_1} \otimes 1) J_{\hat{x}} = m(w \otimes 1),$$

and similarly $m(1 \otimes w') = Z(1 \otimes w')Z^*$ for $w' \in W_2$. □

Proposition 19.2.2 *The element $M_m \equiv N_{1,31}^* Z_{34}^* M_{2,24} Z_{34}$ is a multiplicative unitary for Δ_m, implements it, and $(m \otimes \iota \otimes \iota)(M_m) = Z_{34}^* M_{2,24} Z_{34} N_{1,31}^*$.*

Proof For $w \in W_m$ we have that

$$M_m^*(1 \otimes 1 \otimes w)M_m = Z_{34}^* M_{2,24}^* Z_{34}(\Delta_1^{\text{op}} \otimes \iota)(w)_{134} Z_{34}^* M_{2,24} Z_{34}$$

equals $Z_{34}^*((\iota \otimes \iota \otimes \Delta_2)(\iota \otimes m)(\Delta_1^{\text{op}} \otimes \iota)(w))_{1324} Z_{34} = \Delta_m(w)$. As for multiplicativity, by Lemma 16.3.9 we know that $(F(1 \otimes J_x J_{\hat{x}})M)^3 \in \mathbb{C}$, and it follows then that $(F(1 \otimes Z')M_m)^3 \in \mathbb{C}$, where $Z' \equiv (J_{x_1} J_{\hat{x}_1} \otimes J_{x_2} J_{\hat{x}_2})Z$, which is proportional to Z'^*. Let $U \equiv ((\hat{W}_1' \otimes \mathbb{C}) \cup Z^*(\mathbb{C} \otimes \hat{W}_2)Z)''$. We claim that $Z'UZ'^* \subset U'$. Clearly $Z'Z^*(\mathbb{C} \otimes \hat{W}_2)ZZ'^* = \mathbb{C} \otimes \hat{W}_2'$, which evidently commutes with $\hat{W}_1' \otimes \mathbb{C}$, but is also commutes with $Z^*(\mathbb{C} \otimes \hat{W}_2)Z$ since this amounts to $[\mathbb{C} \otimes \hat{W}_2, J_x J_{\hat{x}}(\mathbb{C} \otimes \hat{W}_2) J_{\hat{x}} J_x] = \{0\}$ and this holds since $\mathbb{C} \otimes \hat{W}_2 \subset \hat{W}$. Since Z' is proportional to $Z^*(J_{x_1} J_{\hat{x}_1} \otimes J_{x_2} J_{\hat{x}_2})$, one shows similarly that $Z'(\hat{W}_1' \otimes \mathbb{C})Z'^* \subset U'$, which settles the claim. Hence $(1 \otimes Z')M_m(Z' \otimes Z') \in B(V_{x_1} \otimes V_{x_2})\bar{\otimes}U'$, and we know from above that if we multiply this element from the left by $F\hat{N}_m^* \hat{M}_m^*$, where $\hat{N}_m = (Z'^* \otimes 1)M_m^*(Z' \otimes 1)$, we get a scalar, so $\hat{M}_m^*(1 \otimes 1 \otimes u)\hat{M}_m = \hat{N}_m(u \otimes 1 \otimes 1)\hat{N}_m^*$ for $u \in U$. Combining this with

$$\hat{N}_m(Z^*(1 \otimes a)Z \otimes 1 \otimes 1)\hat{N}_m^* = (Z'^* \otimes Z^*)M_{2,24}^*(1 \otimes J_{\hat{x}_2} J_{x_2} a J_{x_2} J_{\hat{x}_2} \otimes 1 \otimes 1)M_{2,24}(Z' \otimes Z)$$

for $a \in \hat{W}_2$, gives $M_m(Z^*(1 \otimes a)Z \otimes 1 \otimes 1)M_m^* = (Z^* \otimes Z^*)\hat{\Delta}_2^{\text{op}}(a)_{24}(Z \otimes Z)$. And one easily checks that $M_m(b \otimes 1 \otimes 1 \otimes 1)M_m^* = (\hat{\Delta}_1')^{\text{op}}(b)_{13}$ for $b \in \hat{W}_1'$, where $\hat{\Delta}_1'$ is the usual comultiplication on the commutant. Then both $M_{m,3456}M_{m,1234}M_{m,3456}$ and $M_{m,1234}M_{m,1256}$ equal $N_{1,31}^* N_{1,51}^* Z_{34}^* Z_{56}^* M_{2,24} M_{2,26} Z_{34} Z_{56}$, which in turn gives $(\Delta_m \otimes \iota \otimes \iota)(M_m) = M_{m,1256}M_{m,3456}$. The last identity in the proposition is straightforward. □

Lemma 19.2.3 *The matching m is an isomorphism satisfying* $m(\tau_{-t}^1 \otimes \tau_t^2) = (\tau_{-t}^1 \otimes \tau_t^2)m$ *and* $m(R_1 \otimes R_2) = (R_1 \otimes R_2)m^{-1}$ *and* $(R_1 \otimes R_2)mT_{-i/2}(\iota \otimes \iota \otimes \omega)(M_m) = (\iota \otimes \iota \otimes \omega)(M_m^*)$ *for* $\omega \in B(V_{x_1} \otimes V_{x_2})_*$, *where* $t \mapsto T_t = \tau_{-t}^1 \otimes \tau_t^2$ *for real t has an analytic extension with core consisting of the elements* $(\iota \otimes \iota \otimes \omega)(M_m)$.

Proof The elements in question belong to the domain of the closed map $T_{-i/2}$ since $(\iota \otimes \iota \otimes \omega)(M_m) = \sum(\iota \otimes \omega_{v_i,v})(N_{1,21}^*) \otimes (\iota \otimes \omega_{v',v_i})(Z_{23}^* M_{2,13} Z_{23})$ for $v, v' \in V_{x_1} \otimes V_{x_2}$ and an orthonormal basis $\{v_i\}$ for this space. This also shows that $(R_1 \otimes R_2)mT_{-i/2}(\iota \otimes \iota \otimes \omega)(M_m) = (\iota \otimes \iota \otimes \omega)(N_{1,31} Z_{34}^* M_{2,24}^* Z_{34})$. Let D be the domain of the closure of the restriction of $T_{-i/2}$ to the elements in question. We claim that $D(T_{-i/2}) = D$. Replacing ω by $\omega((1 \otimes w)(\cdot)(a \otimes 1)Z^*)$ with $w \in W_2$ and $a \in B(V_{x_1})$, we get $b \equiv (\iota \otimes \iota \otimes \omega)(N_{1,31}^* Z_{34}^*(\alpha(w)(a \otimes 1))_{34} M_{2,24}) \in D$ and $(R_1 \otimes R_2)mT_{-i/2}(b) = (\iota \otimes \iota \otimes \omega)(N_{1,31} Z_{34}^*(\alpha(w)(a \otimes 1))_{34} M_{2,24}^*)$. Now any element in $B(V_{x_1})\bar{\otimes}W_2$ can be approximated strongly* by a bounded net of elements in the span of the elements $\alpha(w)(a \otimes 1)$, and approximating $1 \otimes c$ with $c \in W_2$ this way, we may replace $\alpha(w)(a \otimes 1)$ by $1 \otimes c$ in the formula for b and its image under $(R_1 \otimes R_2)T_{-i/2}$. Then replacing ω once more by $\omega((\cdot)(1 \otimes d))$ with $d \in \hat{W}_2'$, and using that the span of the elements cd is dense in $B(V_{x_2})$, we see that

$(\iota\otimes\omega_1)(N^*_{1,21})\otimes(\iota\otimes\omega_2)(M_2)\in D$, so our claim holds. In particular, the elements $(\iota\otimes\iota\otimes\omega)(Z^*_{34}M_{2,24}Z_{34}N^*_{1,31})$ are dense in $W_1\bar{\otimes}W_2$ since their adjoints are dense in the image of $(R_1\otimes R_2)T_{-i/2}$, which in turn is dense in $W_1\bar{\otimes}W_2$. By the previous proposition m is surjective, and $(R_1\otimes R_2)mT_{-i/2}(\iota\otimes\iota\otimes\omega)(M_m)=(\iota\otimes\iota\otimes\omega)(M^*_m)$. Similarly, the elements $(\iota\otimes\iota\otimes\omega)(M^*_m)$ form a core for $T_{i/2}$ and $(R_1\otimes R_2)mT_{i/2}(\iota\otimes\iota\otimes\omega)(M^*_m)=(\iota\otimes\iota\otimes\omega)(M_m)$. Hence $m^{-1}(R_1\otimes R_2)T_{-i/2}=(m^{-1}(R_1\otimes R_2)T_{i/2})^{-1}=T_{-i/2}(R_1\otimes R_2)m$, so composing these in two different ways gives $(R_1\otimes R_2)mT_{-i}=T_{-i}(R_1\otimes R_2)m=(R_1\otimes R_2)T_{-i}m$ and $mT_{-i}=T_{-i}m$. Hence $mT_t=T_tm$ and $mT_{-i/2}=T_{-i/2}m$ and $m^{-1}(R_1\otimes R_2)=(R_1\otimes R_2)m$. □

Corepresentations of the bicrossed product split into corepresentations of the constituent quantum groups, which is useful in studying doublecrossed products.

Proposition 19.2.4 *Let V be a Hilbert space. For any unitary $A\in W\bar{\otimes}B(V)$ with $(\Delta\otimes\iota)(A)=A_{125}A_{345}$ there are unique unitaries $B\in W_2\bar{\otimes}B(V)$ and $C\in\hat{W}_1\bar{\otimes}B(V)$ with $(\Delta_2\otimes\iota)(B)=B_{13}B_{23}$ and $(\hat{\Delta}_1\otimes\iota)(C)=C_{13}C_{23}$ satisfying $(\alpha\otimes\iota)(B)C_{13}=A$. Conversely, when B and C satisfy the above conditions, then $A\equiv(\alpha\otimes\iota)(B)C_{13}$ satisfy the above conditions if and only if $C^*_{13}B_{23}Z^*_{12}C_{13}Z_{12}\in W'_1\bar{\otimes}B(V_{x_2}\otimes V)$.*

Proof The formula $(\gamma\otimes\iota\otimes\iota)(M)=M_{1245}(\iota\otimes\beta)(\hat{M}_1)_{345}$ defines a right coaction γ of $(\hat{W}_1,\hat{\Delta}_1)$ on W with $W^\gamma=\alpha(W_2)$, while $\delta\equiv F(R\otimes\hat{R}_1)\gamma R$ is a coaction of $(\hat{W}_1,\hat{\Delta}_1)$ on W with $W^\delta=\alpha(W_2)$. Since $(\Delta\otimes\iota)\gamma=(\iota\otimes\iota\otimes\gamma)\Delta$ we get $(\Delta\otimes\iota\otimes\iota)(A^*_{124}(\gamma\otimes\iota)(A))=1\otimes1\otimes A^*_{124}(\gamma\otimes\iota)(A)$, so there is a unique unitary $C\in\hat{W}_1\bar{\otimes}B(V)$ with $(\gamma\otimes\iota)(A)=A_{124}C_{34}$. But $\gamma\otimes\iota\otimes\iota$ and $\iota\otimes\iota\otimes\hat{\Delta}_1\otimes\iota$ applied to the left hand side of this latter equation give the same result, so $(\hat{\Delta}_1\otimes\iota)(C)=C_{13}C_{23}$. Let $f=(\cdot)\otimes1\colon\hat{W}_1\to W$, so $\gamma f=(f\otimes\iota)\hat{\Delta}_1$. If we apply $\iota\otimes\iota\otimes\gamma f\otimes\iota$ to $(\gamma\otimes\iota)(A)=A_{124}C_{34}$, we get $(\gamma\otimes\iota)(C_{13})=C_{14}C_{34}$, so $(\gamma\otimes\iota)(AC^*_{13})=(AC^*_{13})_{124}$. Since $W^\gamma=\alpha(W_2)$, there is therefore a unitary $B\in W_2\bar{\otimes}B(V)$ such that $AC^*_{13}=(\alpha\otimes\iota)(B)$. Using the corepresentation identity for A, and $(\gamma\otimes\iota)(A)=A_{124}C_{34}$ and $(\gamma\otimes\iota\otimes\iota)\Delta=(\iota\otimes\iota\otimes\delta)\Delta$, we get $(\delta\otimes\iota)(A)=C_{14}A_{234}$. Combining this with the splitting of A that we just got, and that $W^\delta=\alpha(W_2)$, we get $(\delta\otimes\iota)(C_{13})=(\alpha\otimes\iota)(B^*)_{234}C_{14}(\alpha\otimes\iota)(B)_{234}C_{24}$, which together with $\Delta f=(f\otimes\iota)\delta f$, gives $(\Delta\otimes\iota)(C_{13})=(\alpha\otimes\iota)(B^*)_{345}C_{15}(\alpha\otimes\iota)(B)_{345}C_{35}$. Hence, from the corepresentation identity for A and its splitting into B and C, we get $((\alpha\otimes\alpha)\Delta_2\otimes\iota)(B)=(\alpha\otimes\iota)(B)_{125}(\alpha\otimes\iota)(B)_{345}$ and $(\Delta_2\otimes\iota)(B)=B_{13}B_{23}$. Uniqueness of the splitting is clear. Now $C^*_{13}B_{23}Z^*_{12}C_{13}Z_{12}\in W'_1\bar{\otimes}B(V_{x_2}\otimes V)$ when C is a corepresentation, means that $\hat{M}^*_{1,12}Z^*_{23}C_{24}Z_{23}\hat{M}_{1,12}=B^*_{34}C_{14}B_{34}Z^*_{23}C_{24}Z_{23}$. Applying $\mathrm{Ad}Z_{23}$ and using $\Delta(a\otimes1)=(\iota\otimes\beta)(\hat{M}^*_1)_{134}(1\otimes1\otimes a\otimes1)(\iota\otimes\beta)(\hat{M}_1)_{134}$ and that B is a corepresentation and $\Delta\alpha=(\alpha\otimes\alpha)\Delta_2$, one sees that the above identity holds if and only if $A=(\alpha\otimes\iota)(B)C_{13}$ is a corepresentation. □

Lemma 19.2.5 *In this lemma (W,Δ) can in principle be any locally compact quantum group. Any group-like unitary u of W commutes with h^{it}.*

Proof By uniqueness of the left invariant faithful semifinite normal weight x, we find a positive scalar c such that $x(u(\cdot)u^*)=cx$, so $\sigma^x_t(u)=c^{it}u$ for real

t. But $S(u) = u^*$ by the formula for any unitary corepresentation, and as u^* is also group-like, we get $S^2(u) = u$, so $\tau_t(u) = u$ and $R(u) = u^*$. Hence $\sigma_t^{xR}(u) = R\sigma_{-t}R(u) = c^{it}u$, so u is fixed under $\sigma_t^{xR}\sigma_{-t}^x$, and the modular element h commutes with it. □

Proposition 19.2.6 *There are strictly positive self-adjoint group-like operators r_j affiliated with W_j such that $h^{it} = \alpha(r_2^{it})(\hat{h}_1^{it} \otimes 1) = (\hat{h}_1^{it} \otimes 1)\alpha(r_2^{it})$ and $\hat{h}^{it} = \alpha(r_1^{it})(1 \otimes \hat{h}_2^{it}) = (1 \otimes \hat{h}_2^{it})\alpha(r_1^{it})$. We also have $\nu = \nu_2/\nu_1$ and $J_x h^{it} J_x = J_{\hat{x}_1}\hat{h}_1^{it}J_{\hat{x}_1} \otimes r_2^{-it}h_2^{it}J_{x_2}h_2^{it}J_{x_2}$ and $J_{\hat{x}}\hat{h}^{it}J_{\hat{x}} = r_1^{-it}h_1^{it}J_{x_1}h_1^{it}J_{x_1} \otimes J_{\hat{x}_2}\hat{h}_2^{it}J_{\hat{x}_2}$. The Radon-Nikodym derivative $(Dx_2\alpha : Dx_2)_t = m(r_1^{-it/2} \otimes r_2^{-it/2})(r_1^{-it/2}h_1^{it} \otimes r_2^{it/2}h_2^{-it})$, and $r_1^{-it/2}h_1^{it} \otimes r_2^{it/2}h_2^{-it}$ is m-invariant, and there are positive scalars λ_j such that $\sigma_t^{x_j}(r_j^{is}) = \lambda_j^{ist}r_j^{is}$ and $\lambda_2/\lambda_1 = \nu$.*

Proof By the previous proposition there are unique unitary group-like elements $v_t \in \hat{W}_1$ and $u_t \in W_2$ such that $h^{it} = (v_t \otimes 1)\alpha(u_t)$. Using that α, β respect coproducts, one shows that $\Delta^{\rm op}(v_t \otimes 1) = v_t \otimes 1 \otimes \hat{h}_1^{it} \otimes 1$, so $\Delta^{\rm op}(h^{it}) = (v_t \otimes 1)\alpha(u_t) \otimes (\hat{h}_1^{it} \otimes 1)\alpha(u_t)$, and as h^{it} is group-like, we get $v_t = \hat{h}_1^{it}$. By the previous lemma we know that h^{it} and $\alpha(u_s)$ commute, so $\{(\hat{h}_1^{-it} \otimes 1)h^{it}\}_t$ is a one-parameter group of unitaries, and there is a positive self-adjoint operator r_2 affiliated with W_2 such that $h^{it} = (\hat{h}_1^{it} \otimes 1)\alpha(r_2^{it})$, and similarly we find r_1, proving the first part of the proposition. Let $a \in N_{\hat{x}_1}$ and $w \in N_{x_2}$. Then $(a \otimes 1)\alpha(w)h^{it} = (a\hat{h}_1^{it} \otimes 1)\alpha(r_2^{it}h_2^{-it}wh_2^{it})$ and $q_x((a \otimes 1)\alpha(w)h^{it}) = (\nu_2/\nu_1)^{-t/2}(J_{\hat{x}_1}\hat{h}_1^{-it}J_{\hat{x}_1} \otimes r_2^{it}h_2^{-it}J_{x_2}h_{x_2}^{-it}J_{x_2})q_x((a \otimes 1)\alpha(w))$, but the left hand side also equals $\nu^{-t/2}J_x h^{-it}J_x q_x((a \otimes 1)\alpha(w))$, which proves the next sentence in the proposition. According to the proof of the previous lemma there are positive numbers λ_j such that $\sigma_t^{x_j}(r_j^{is}) = \lambda_j^{ist}r_j^{is}$. But as $\mathrm{Ad}\hat{h}^{it} = \tau_{-t}\mathrm{Ad}(h^{-it})\sigma_{-t}^x$ in any locally compact quantum group, we get the formula for the Radon-Nikodym derivative from $\Delta^{it} = \hat{h}^{-it/2}h^{-it/2}J_x h^{-it/2}J_x J_{\hat{x}}\hat{h}^{it/2}J_{\hat{x}}$, which clearly also holds in any locally compact quantum group. We get $\hat{h}^{-is}\alpha(r_2^{it})\hat{h}^{is} = \lambda_2^{ist}\alpha(r_2^{it})$ and $h^{it}\beta(r_1^{is})h^{-it} = \lambda_1^{-ist}\beta(r_1^{is})$. Hence $h^{it}\hat{h}^{is} = (\lambda_2/\lambda_1)^{ist}\hat{h}^{is}h^{it}$ and $\lambda_2/\lambda_1 = \nu$. Now $(Dx_2\Delta_2^{\rm op} : Dx_2)_t = h_2^{-it} \otimes 1$. Since $(\iota \otimes \Delta_2^{\rm op})\alpha = (mF \otimes \iota)(\iota \otimes \alpha)\Delta_2^{\rm op}$ and $m(\tau_{-t}^1 \otimes \tau_t^2) = (\tau_{-t}^1 \otimes \tau_t^2)m$ by the second lemma above, we get from Lemma 19.1.3, that $(\iota \otimes \Delta_2^{\rm op})((Dx_2\alpha : Dx_2)_t) = (mF \otimes \iota)(h_2^{-it} \otimes (Dx_2\alpha : Dx_2)_t)(1 \otimes h_2^{it} \otimes 1)$, which combined with the previous expression for $(Dx_2\alpha : Dx_2)_t$ completes the proof. □

Lemma 19.2.7 *Thanks to the previous proposition and the lemma we have self-adjoint strictly positive operators k_j determined by $k_j^{it} = r_j^{it}h_j^{-it}$. Consider the right invariant faithful semifinite normal weights $y_j \equiv x_j R_j = (x_j)_{h_j}$. Then $q_{y_2}(\omega_{v,v'} \otimes \iota)(K_\beta^*\alpha(w)) = (\omega_{v,k_1^{1/2}v'} \otimes \iota)(K_\beta^*K_\alpha)q_{y_2}(w)$, and similarly for q_{y_1} by interchanging α and β, and slicing on the other side and using k_2 instead of k_1.*

Proof Consider the dual weight $\tilde{y}_2$ on W with GNS-map determined by $q_{\tilde{y}_2}((a \otimes 1)\alpha(w)) = q_{\hat{x}_1}(a) \otimes q_{y_2}(w)$. Since this weight equals $x_{\alpha(h_2)}$, we get from the proof of Theorem 10.8.1, that $\Delta_{\tilde{y}_2}^{it} = J_x\alpha(h_2^{it})J_x\alpha(h_2^{it})\Delta_x^{it}$ and $S_{\tilde{y}_2} = \nu_2^{i/4}J_x\Delta_{\tilde{y}_2}^{1/2}$. Using that K_α and K_β are the canonical implementations of α and β, in combination with the previous proposition, including the Radon-Nikodym formula and the properties of m, one then checks that $\Delta_{\tilde{y}_2}^{it} = \beta(k_1^{-it})(\Delta_{\hat{x}_1}^{it} \otimes \Delta_{y_2}^{it})$, which shows that $\beta(k_1)$ and $\Delta_{\hat{x}_1} \otimes \Delta_{y_2}$ commute strongly. Hence $\Delta_{\tilde{y}_2}^{1/2}\alpha(w^*)(v_1 \otimes q_{y_2}(w')) = \nu_2^{i/4}\beta(k_1^{-1/2})\alpha\sigma_{-i/2}^{y_2}(w^*)(\Delta_{\hat{x}_1}^{1/2}v_1 \otimes J_{x_2}q_{y_2}(w'^*))$, where $w' \in N_{y_2} \cap N_{y_2}^*$ and $w \in N_{y_2}$ is $\sigma_t^{y_2}$-analytic and $v_1 \in D(\Delta_{\hat{x}_1}^{1/2})$. Using instead the formula above with $S_{\tilde{y}_2}$, and its definition, we get another expression for the left hand side, which gives

$$\beta(k_1^{-1/2})\alpha\sigma_{-i/2}^{y_2}(w^*)(\Delta_{\hat{x}_1}^{1/2}v_1 \otimes J_{x_2}q_{y_2}(w')) = (1 \otimes J_{x_2}w'J_{x_2})K_\alpha(\Delta_{\hat{x}_1}^{1/2}v_1 \otimes J_{x_2}q_{y_2}(w)).$$

Applying K_β^* to this and using $J_{x_2}q_{y_2}(\cdot) = q_{y_2}\sigma_{-i/2}^{y_2}((\cdot)^*)$ together with density of analytic elements, one gets the stated identity in the lemma. The second one is obtained analogously. □

Theorem 19.2.8 *The doublecrossed product (W_m, Δ_m) of locally compact quantum groups (W_1, Δ_1) and (W_2, Δ_2) is again a locally compact quantum group (and using notation from above) with left invariant faithful semifinite normal weight $x_m \equiv y_1 \otimes (x_2)_{k_2}$ and fundamental multiplicative unitary M_m as defined before. Its right invariant faithful semifinite normal weight y_m is $(x_1)_{k_1} \otimes y_2$ and the multiplicative unitary N_m given by $N_m(q_{y_m}(w) \otimes q_{y_m}(w')) = (q_{y_m} \otimes q_{y_m})(\Delta_m(w)(1 \otimes w'))$ equals $Z_{12}^*\hat{M}_{1,13}Z_{12}N_{2,24} = (\iota \otimes \beta)(\hat{M}_1)_{132}K_{\alpha,32}N_{2,24}$. Moreover, we have that $R_m = m^{-1}(R_1 \otimes R_2) = (R_1 \otimes R_2)m$ and $\tau_t^m = \tau_{-t}^1 \otimes \tau_t^2$ and $\nu_m = \nu_2/\nu_1$ and $J_m = \lambda_1^{i/4}(J_{x_1} \otimes J_{x_2})$ and $h_m^{it} = r_1^{it}h_1^{-2it} \otimes r_2^{-it}h_2^{2it}$ and $Q_m = Q_1^{-1} \otimes Q_2$ and $J_{\hat{x}_m} = \lambda_1^{i/4}(J_{\hat{x}_1} \otimes J_{\hat{x}_2})J_xJ_{\hat{x}}(J_{\hat{x}_1}J_{x_1} \otimes J_{\hat{x}_2}J_{x_2})$ and $\hat{h}_m^{it} = (J_{\hat{x}_1}\hat{h}_1^{-it}J_{\hat{x}_1} \otimes 1)Z^*(1 \otimes \hat{h}_2^{it})Z$ and $J_{\hat{x}_m}\hat{h}_m^{it}J_{\hat{x}_m} = (1 \otimes J_{\hat{x}_2}\hat{h}_2^{it}J_{\hat{x}_2})Z^*(\hat{h}_1^{-it} \otimes 1)Z$. The dual quantum group has $\hat{W}_m = ((\hat{W}_1' \otimes \mathbb{C}) \cup Z^*(\mathbb{C} \otimes \hat{W}_2)Z)''$ with commutant quantum group having $\hat{W}_m' = (Z^*(\hat{W}_1 \otimes \mathbb{C})Z \cup (\mathbb{C} \otimes \hat{W}_2'))''$, where $(\hat{W}_1', \hat{\Delta}_1')$ and $(\hat{W}_2, \hat{\Delta}_2)$ are considered closed quantum subgroups of (W_m, Δ_m). We also have the relations $J_xJ_{\hat{x}}(J_{x_1} \otimes J_{x_2}) = (\nu_1/\nu_2)^{i/4}(J_{x_1} \otimes J_{x_2})J_xJ_{\hat{x}}$ and $(\iota \otimes m)(K_{\beta,13}N_{1,12}) = N_{1,12}K_{\beta,13}$ and $(m \otimes \iota)(K_{\alpha,13}N_{2,32}) = N_{2,32}K_{\alpha,13}$.*

Proof Let $s_n \equiv (\iota \otimes \omega_{v,v'} \otimes \iota)((K_\alpha \otimes 1)(\iota \otimes \Delta_2(w))(1 \otimes P_n \otimes 1)(K_\beta^*\beta(\iota \otimes \omega_{u,u'})\Delta_1(w') \otimes w''))$, where $P_n \equiv \chi_{[1/n,n\rangle}(k_2)$, and $w' \in N_{y_1}$ and $w, w'' \in N_{(x_2)_{k_2}}$ with w'' analytic with respect to $(x_2)_{k_2}$, and $v, v' \in V_{x_2}$ and $u, u' \in V_{x_1}$, and let s be s_n without the factor $1 \otimes P_n \otimes 1$. Then $s = (\omega_{u\otimes v,u'\otimes v'} \otimes \iota \otimes \iota)\Delta_m(w' \otimes w)(1 \otimes w'')$ and $s_n = \sum s_{nj}$, where $s_{nj} \equiv (\iota \otimes \omega_{v_j,v'} \otimes \iota)((K_\alpha \otimes 1)(1 \otimes \Delta_2(w)))((\iota \otimes \omega_{v,P_nv_j})(K_\alpha^*\beta(\iota \otimes \omega_{u,u'})\Delta_1(w')) \otimes w'')$ for an orthonormal basis $\{v_j\}$ of V_{x_2}. By the previous lemma we have $q_{x_m}(s_{nj}) = (\iota \otimes \omega_{v_j,v'} \otimes \iota)((K_\alpha \otimes 1)(1 \otimes \Delta_2(w)))((\iota \otimes \omega_{v,k_2^{1/2}P_nv_j})(K_\alpha^*K_\beta)v'' \otimes q_{(x_2)_{k_2}}(w''))$, where $v'' \equiv (\iota \otimes \omega_{u,u'})(N_1)q_{y_1}(w')$. Summing over j we get

$$q_{x_m}(s_n) = (\iota \otimes \omega_{v,v'} \otimes \iota)((\alpha \otimes \iota)(\Delta_2(w)(k_2^{1/2}P_n \otimes 1))(K_\beta \otimes 1))(v'' \otimes q_{(x_2)_{k_2}}(w'')).$$

Using that k_2 is group-like, for $u'' \in V_{x_1}$ we thus see that

$$(f^*_{u''} \otimes \iota)q_{x_m}(s_n) = q_{(x_2)_{k_2}}((\omega_{K_\beta(v''\otimes v),u''\otimes v'}\alpha \otimes \iota)(\Delta_2(w)(k_2^{1/2}P_n \otimes 1))w'')$$

equals $J_{x_2}\sigma^{(x_2)_{k_2}}_{i/2}(w'')^* J_{x_2}(\omega_{\alpha(P_n)K_\beta(v''\otimes v),u''\otimes v'}\alpha\otimes\iota)(M_2^*)q_{(x_2)_{k_2}}(w)$. So $q_{x_m}(s_n) = (1 \otimes J_{x_2}\sigma^{(x_2)_{k_2}}_{i/2}(w'')^* J_{x_2} \otimes 1)(\iota \otimes \omega_{v,v'} \otimes \iota)((\alpha \otimes \iota)(M_2^*)(\alpha(P_n)K_\beta \otimes 1))(v'' \otimes q_{(x_2)_{k_2}}(w))$. Letting $n \to \infty$ and $w'' \to 1$, we thus see that $q_{x_m}((\omega_{u\otimes v,u'\otimes v'} \otimes \iota \otimes \iota)\Delta_m(w' \otimes w)) = (\omega_{u\otimes v,u'\otimes v'} \otimes \iota \otimes \iota)((\alpha \otimes \iota)(M_2^*)_{324}K_{\beta,32}N_{1,31})q_{x_m}(w' \otimes w)$. This shows that x_m is left invariant, and since clearly $\Delta_m R_m = (R_m \otimes R_m)\Delta_m^{\mathrm{op}}$ with $R_m \equiv (R_1 \otimes R_2)m$, we conclude that (W_m, Δ_m) is a locally compact quantum group, and that the element $\tilde{M} \equiv N^*_{1,31}K^*_{\beta,32}(\alpha \otimes \iota)(M_2)_{324}$ is a multiplicative unitary for it. Analogously, one shows that $(x_1)_{k_1} \otimes y_2$ is right invariant with a similar formula involving the GNS-map for $N_m \equiv (\iota \otimes \beta)(\hat{M}_1)_{132}K_{\alpha,32}N_{2,24}$. Using that $(\Delta_m \otimes \iota \otimes \iota)(\tilde{M}) = \tilde{M}_{1256}\tilde{M}_{3456}$ and computing the left hand side by using that the various factors are corepresentations, together with the matching properties of m and the definitions of α and β, an easy computation shows that $(\iota \otimes m)(K_{\beta,13}N_{1,12}) = N_{1,12}K_{\beta,13}$. The identity in the proposition involving $m \otimes \iota$ is gotten by considering N_m. The equality $\tilde{M} = M_m$ follows from the claim $Z^*_{23}M_{2,13}Z_{23} = (\alpha \otimes \iota)(M_2^*)_{213}K_{\beta,21}$. But the left hand side equals $(J_{\hat{x}_2} \otimes J_{x_1} \otimes J_{\hat{x}_2})K^*_{\alpha,23}K^*_{\beta,21}N_{2,31}K_{\alpha,23}(J_{\hat{x}_2} \otimes J_{x_1} \otimes J_{\hat{x}_2})$ as is seen by using that N_2 implements Δ_2^{op} and that K_β is a corepresentation. Combining this with the fact that Z implements m, and the formula involving $m \otimes \iota$ that we just proved, together with the properties of the scaling constants in relation to the modular involutions, we see that our claim holds. A similar calculation proves the second formula for N_m in the proposition. Lemma 19.2.3 tells us that R_m is the unitary coinverse of (W_m, Δ_m) with $\{\tau^1_{-t} \otimes \tau^2_t\}$ as its scaling group. The it-th power of the Radon-Nikodym derivative of $(x_1)_{k_1} \otimes y_2$ with respect to x_m is $r_1^{it}h_1^{-2it} \otimes r_2^{-it}h_2^{2it}$ and is group-like by the previous proposition. The former weight is proportional to $x_m R_m$ whose Radon-Nikodym derivative with respect to x_m is also group-like, which is only possible if $(x_1)_{k_1} \otimes y_2 = x_m R_m$. This also gives the expressions for ν_m and h_m in the proposition. The formula for J_m is clear from the previous proposition and the proof of Theorem 10.8.1. By the formulas for M_m and N_m we have

$$(1 \otimes 1 \otimes L_m)(N^*_{1,31}Z^*_{34}M_{2,24}Z_{34}(1 \otimes 1 \otimes L_m^*) = Z^*_{34}M^*_{1,13}Z_{34}N_{2,42},$$

where $L_m \equiv J_{\hat{x}_m}J_{x_m}$. Taking slices of this yields $L_m(a \otimes 1)L_m^* = Z^*(u_1 a u_1^* \otimes 1)Z$ for $a \in \hat{W}_1'$, where $u_j = J_{\hat{x}_j}J_{x_j}$. Shifting instead the $L_m \sim L_m^*$ over onto the other side before taking slices gives $L_m(1 \otimes b)L_m^* = Z^*(1 \otimes u_2 b u_2^*)Z$ for $b \in \hat{W}_2'$. Hence $J_{\hat{x}_m}cJ_{\hat{x}_m} = Z^*(J_{\hat{x}_1} \otimes J_{\hat{x}_2})c(J_{\hat{x}_1} \otimes J_{\hat{x}_2})Z$ for $c \in \hat{W}_1' \otimes \hat{W}_2'$. But both sides equal $R_m(c^*)$ when $c \in W_1 \otimes W_2$, so $J_{\hat{x}_m} \sim Z^*(J_{\hat{x}_1} \otimes J_{\hat{x}_2})$. Hence the square of the latter operator is one, which means that $(J_{x_1} \otimes J_{x_2})J_x J_{\hat{x}} = J_{\hat{x}}J_x(J_{x_1} \otimes J_{x_2})$. Using this one finds by standard manipulations that $J_{\hat{x}_m} = \lambda^{i/4}Z^*(J_{\hat{x}_1} \otimes J_{\hat{x}_2})$, and one gets the formula for $J_{\hat{x}_m}$ in the proposition. Clearly $Q_m = Q_1^{-1} \otimes Q_2$. Since $\Delta_{x_m}^{it} = \Delta_{y_1}^{it} \otimes \Delta_{x_2}^{it} k_2^{it} J_{x_2} k_2^{it} J_{x_2}$ and $\Delta_{y_m}^{it} = \Delta_{x_1}^{it} k_1^{it} J_{x_1} k_1^{it} J_{x_1} \otimes \Delta_{y_2}^{it}$ one readily finds that $\hat{h}_m^{it} = J_{\hat{x}_1} \hat{h}_1^{-it} J_{\hat{x}_1} k_1^{-it} J_{x_1} k_1^{-it} J_{x_1} \otimes \hat{h}_2^{it}$. But the previous proposition tells us that $J_{\hat{x}}(1 \otimes \hat{h}_2^{it}) J_{\hat{x}} = k_1^{-it} J_{x_1} k_1^{-it} J_{x_1} \otimes J_{\hat{x}_2} \hat{h}_2^{it} J_{\hat{x}_2}$, and since $1 \otimes \hat{h}_2^{it}$ is group-like, the proof of the second lemma above shows that it is invariant under $J_x(\cdot)J_x$, and this then gives us the first formula for $\hat{h}_m^{it}$ in the proposition. The second one is proved similarly. □

A *bicharacter* of (W_1, Δ_1) and (W_2, Δ_2) is a unitary $X \in W_1 \bar{\otimes} W_2$ satisfying $(\Delta_1 \otimes \iota)(X) = X_{23} X_{13}$ and $(\iota \otimes \Delta_2)(X) = X_{13} X_{12}$. Then $m = X(\cdot)X^*$ is clearly a matching of (W_1, Δ_1) and (W_2, Δ_2) and the doublecrossed product (W_m, Δ_m) is called a *generalized quantum double*. The *quantum double* of a locally compact quantum group (U, Δ_U), is the generalized quantum double of $(U, \Delta_U^{\mathrm{op}})$ and $(\hat{U}, \hat{\Delta}_U)$ with $X = M_U$ the fundamental multiplicative unitary of (U, Δ_U).

Proposition 19.2.9 *Say we have a matching $m = X(\cdot)X^*$ for a bicharacter X of (W_1, Δ_1) and (W_2, Δ_2). Then $\mathrm{Ad}X^* \colon W \to \hat{W}_1 \bar{\otimes} W_2$ is an isomorphism that satisfies $(\mathrm{Ad}X^* \otimes \mathrm{Ad}X^*)\Delta = (\iota \otimes F \otimes \iota)(\hat{\Delta}_1 \otimes \Delta_2)\mathrm{Ad}X^*$. We also have $J_x = X(J_{\hat{x}_1} \otimes J_{x_2})X^*$ and $J_{\hat{x}} = X(J_{x_1} \otimes J_{\hat{x}_2})X^*$. The elements k_j^{it} are group-like for (W_j, Δ_j) and satisfy $X^*(\hat{h}_1^{it} \otimes 1)X = \hat{h}_1^{it} \otimes k_2^{-it}$ and $X^*(1 \otimes \hat{h}_2^{it})X = k_1^{-it} \otimes \hat{h}_2^{it}$, and $Z = \nu_1^{i/4} X(J_{x_1} \otimes J_{x_2})X(J_{x_1} \otimes J_{x_2})$. The generalized quantum double is unimodular if and only if $k_j = h_j$, which is the case for the quantum double.*

Proof One checks that $(X^* \otimes X^*)M(X \otimes X) = M_{2,24}\hat{M}_{1,13}$, so the first statement holds, while the formulas for J_x and $J_{\hat{x}}$ are clear from Proposition 15.4.10. Since X^* is a corepresentation for (W_1, Δ_1), we have $S_1(\iota \otimes \omega)(X^*) = (\iota \otimes \omega)(X)$, and similarly for S_2, so X is invariant under $\tau_{-t}^1 \otimes \tau_t^2$ and $R_1 \otimes R_2$, and the formula for Z follows. Now $f_t = \hat{h}_1^{it}(\cdot)\hat{h}_1^{-it}$ is a one-parameter groups of automorphisms of W_1 such that $\Delta_1 f_t = (f_t \otimes \iota)\Delta_1$. So $(f_t \otimes \iota)(X^*)X$ is invariant under $\Delta_1 \otimes \iota$, and there exists a unitary $k_2^{it} \in W_2$ such that $X^*(\hat{h}_1^{it} \otimes 1)X = \hat{h}_1^{it} \otimes k_2^{-it}$. Applying $\iota \otimes \Delta$ to this identity shows that k_2^{it} is group-like for (W_2, Δ_2). Since the modular element of $(\hat{W}_1 \bar{\otimes} W_2, (\iota \otimes F \otimes \iota)(\hat{\Delta}_1 \otimes \Delta_2))$ is $\hat{h}_1 \otimes h_2$, we get $h^{it} = X(\hat{h}_1^{it} \otimes h_2^{it})X^* = (\hat{h}_1^{it} \otimes 1)\alpha(k_2^{it} h_2^{it})$ and $r_2^{it} = k_2^{it} h_2^{it}$, and similarly for k_1. The previous theorem tells us that (W_m, Δ_m) is unimodular if and only if $k_j = h_j$. If (U, Δ_U) is a locally compact quantum group and $(W_1, \Delta_1) = (U, \Delta_U^{\mathrm{op}})$, then $\hat{h}_1^{it} = J_{\hat{x}_U} \hat{h}_U^{-it} J_{\hat{x}_U}$ and $X^*(\hat{h}_1^{it} \otimes 1)X = \hat{h}_1^{it} \otimes h_2^{-it}$, and similarly for $(W_2, \Delta_2) = (\hat{U}, \hat{\Delta}_U)$. So the quantum double of (U, Δ_U) is unimodular. □

19.3 Morphisms of Quantum Groups and Associated Right Coactions

Let (W_i, Δ_i) be locally compact quantum groups with fundamental multiplicative unitaries M_i, and let (W_{iu}, Δ_{iu}) be their associated universal quantum groups with surjective $*$-homomorphisms $\pi_i : W_{iu} \to W_{ir}$ determined by $\pi_i \hat{\lambda}_{iu} = \hat{\lambda}_i$, and with $\mathcal{M}_i \in M(W_{iu} \otimes \hat{W}_{ir})$ such that $(\pi_i \otimes \iota)(\mathcal{M}_i) = M_i$.

Definition 19.3.1 By a *morphism* $(W_1, \Delta_1) \to (W_2, \Delta_2)$ we mean a non-degenerate $*$-homomorphism $f: W_{1u} \to M(W_{2u})$ such that $(f \otimes f)\Delta_{1u} = \Delta_{2u} f$.

One checks that $f S_{1u} \subset S_{2u} f$, where on the right hand side we have taken the strict closure of the universal coinverse. One sees that f respects universal unitary coinverses, scaling groups, and counits as $(\varepsilon_{2u} f \otimes \iota)(\mathcal{M}_1) = 1 = (\varepsilon_{1u} \otimes \iota)(\mathcal{M}_1)$. Compositions of morphisms are clearly morphisms. Morphisms of quantum groups are in one-to-one correspondence with *associated* special right coactions. The associated coactions are then gotten by using unitary coinverses.

Proposition 19.3.2 *To any morphism $f: (W_1, \Delta_1) \to (W_2, \Delta_2)$ there is a unique right coaction α of (W_2, Δ_2) on W_1 determined by $\alpha\pi_1 = (\pi_1 \otimes \pi_2 f)\Delta_{1u}$. This right coaction satisfies $(\Delta_1 \otimes \iota)\alpha = (\iota \otimes \alpha)\Delta_1$. Conversely, every right coaction of (W_2, Δ_2) on W_1 satisfying this identity, comes from a unique morphism this way.*

Proof It is straightforward to check that

$$\alpha = (R_2\pi_2 f \otimes R_1\pi_1)(\mathcal{M}_1)_{21}((\cdot) \otimes 1)(R_2\pi_2 f \otimes R_1\pi_1)(\mathcal{M}_1)^*_{21}$$

has the required properties, where we are considering R_1 with domain $B(V_{x_1})$. For the converse, say we are given α as prescribed, then the special identity which it fulfills, shows that $y_1 \equiv x_1 R_1$ satisfies $y_1(\iota \otimes \omega)\alpha = y_1$ for any normal state ω on W_2. Hence there is an isometry $L \in B(V_{x_1} \otimes V_{x_2})$ such that $(\iota \otimes \omega)(L)q_{y_1} = q_{y_1}(\iota \otimes \omega)\alpha$. This implies that $\alpha(\cdot)L = L((\cdot) \otimes 1)$ and $(\iota \otimes \Delta_2)(L) = L_{12}L_{13}$, and one shows by now standard arguments that $L \in \hat{W}'_1 \bar{\otimes} W_2$, and consequently that the element $L' \equiv (J_{y_2} \otimes J_{y_1})(L^*_{21})(J_{y_2} \otimes J_{y_1})$ is a unitary in $W_2 \bar{\otimes} \hat{W}_1$ that satisfies $(\Delta_2 \otimes \iota)(L') = L'_{13}L'_{23}$ and $(\iota \otimes \hat{\Delta}_1)(L') = L'_{13}L'_{12}$. In fact, we have by standard arguments that $L' \in M(W_{2r} \otimes \hat{W}_{1r})$. Then we know that there is a unitary $L'' \in M(W_{2u} \otimes \hat{W}_{1r})$ such that $(\pi_2 \otimes \iota)(L'') = L'$ and $(\iota \otimes \hat{\Delta}_{1u})(L'') = L''_{13}L''_{12}$. Let $f: W_{1u} \to M(W_{2u})$ be the non-degenerate $*$-homomorphism such that $(f \otimes \iota)(\mathcal{M}_1) = L''$. Then $(f \otimes f)\Delta_{1u} = \Delta_{2u} f$, and it is easy to check that $(R_2\pi_2 f \otimes R_1\pi_1)(\mathcal{M}_1)_{21} = L$. Uniqueness of f is by now standard. □

19.4 More on Doublecrossed Products

Again we let (W_i, Δ_i) be locally compact quantum groups, with (W, Δ) their bicrossed product, while (W_m, Δ_m) denotes the doublecrossed product of them.

Remark 19.4.1 Assume for the moment that W_i are finite dimensional with counits ε_i. Applying $\iota \otimes \varepsilon_2 \otimes \iota$ to the matching identity $(\iota \otimes \Delta_2)m = m_{13}m_{12}(\iota \otimes \Delta_2)$, using injectivity of m, and applying $\iota \otimes \varepsilon_2$, one gets $(\iota \otimes \varepsilon_2)m = \iota \otimes \varepsilon_2$. Using this one sees that $\iota \otimes \varepsilon_2$ is a quantum group morphism $(W_m, \Delta_m) \to (W_1, \Delta_1^{\mathrm{op}})$ with associated coaction $((\iota \otimes \varepsilon_2) \otimes \iota \otimes \iota)\Delta_m$ of $(W_1, \Delta_1^{\mathrm{op}})$ on W_m, and since $(\iota \otimes \varepsilon_2)m = \iota \otimes \varepsilon_2$, one actually sees that this coaction equals $\Delta_1^{\mathrm{op}} \otimes \iota$. Similarly the morphism $(\varepsilon_1 \otimes \iota)m = \varepsilon_1 \otimes \iota$ has associated right coaction $(\iota \otimes \iota \otimes (\varepsilon_1 \otimes \iota))\Delta_m = \iota \otimes \Delta_2$ of (W_2, Δ_2) on W_m. If we also convert this right coaction to a coaction of (W_2, Δ_2) on W_m by using the unitary coinverses, it takes the form $F_{12}m_{12}(\iota \otimes \Delta_2)$, which also makes sense in general on the von Neumann level. As expected $(\iota \otimes \Delta_m)(\Delta_1^{\mathrm{op}} \otimes \iota) = (\Delta_1^{\mathrm{op}} \otimes \iota \otimes \iota \otimes \iota)\Delta_m$. $\diamond$

Suppose we are given a coaction γ of (W_m, Δ_m) on a von Neumann algebra U. Then we define coactions γ_1 of $(W_1, \Delta_1^{\mathrm{op}})$ on U and γ_2 of (W_2, Δ_2) on U by $(\iota \otimes \gamma)\gamma_1 = (\Delta_1^{\mathrm{op}} \otimes \iota \otimes \iota)\gamma$ and $(\iota \otimes \gamma)\gamma_2 = (F_{12}m_{12}(\iota \otimes \Delta_2) \otimes \iota)\gamma$. Consider the crossed product $L \equiv W_1 \ltimes_{\gamma_1} U$. Applying $\iota \otimes \gamma_2$ to L and using $(\iota \otimes \gamma_2)\gamma_1 = \gamma$, we get the representation $(\gamma(U) \cup (\hat{W}_1' \otimes \mathbb{C} \otimes \mathbb{C}))''$ of L, so we get the inclusions $U \subset L \subset W_m \ltimes_\gamma U$. In fact, we have the following result.

Proposition 19.4.2 *There is a unique coaction γ' of (W, Δ) on L such that $\gamma'\gamma_1 = (\alpha \otimes \gamma_1)\gamma_2$ and $\gamma'(J_{\hat{x}_1}(\cdot)J_{\hat{x}_1} \otimes 1) = (J_{\hat{x}} \otimes J_{\hat{x}_1})\alpha_1'(\cdot)(J_{\hat{x}} \otimes J_{\hat{x}_1}) \otimes 1$ on $\hat{W}_1$, where α_1' is the coaction of $(W, \Delta^{\mathrm{op}})$ on $\hat{W}_1$ defined by $\alpha_1'(\cdot) \otimes 1 = \Delta^{\mathrm{op}}((\cdot) \otimes 1)$. Furthermore, the coaction γ' is outer if and only if γ is outer. Finally, the inclusions $U \subset L \subset W_m \ltimes_\gamma U \subset W \ltimes_{\gamma'} L \subset \hat{W}_m \ltimes_{\hat{\gamma}} (W_m \ltimes_\gamma U)$ hold.*

Proof Working with the representation of L in $B(V_{x_1} \otimes V_{x_2})\bar{\otimes}U$ described above, introduce $T \equiv (J_{\hat{x}} \otimes J_{\hat{x}_1})(\beta \otimes \iota)(M_1)(J_{\hat{x}_1} \otimes J_{\hat{x}_2} \otimes J_{\hat{x}_1})$ and define $\gamma' = T_{123}(\iota \otimes \iota \otimes \gamma)(\cdot)T_{123}^*$. Then it is easy to see that we get a coaction as described in the proposition. Since $W \ltimes_{\gamma'} L = (\gamma'(L) \cup (\hat{W} \otimes \mathbb{C} \otimes \mathbb{C}))''$, we see that

$$\gamma(U) \subset L \subset (\gamma(U) \cup (\hat{W}_m \otimes \mathbb{C}))'' \subset W \ltimes_{\gamma'} L \subset B(V_{x_1} \otimes V_{x_2})\bar{\otimes}U.$$

with $W \ltimes_{\gamma'} L = (L \cup ((J_{\hat{x}_1} \otimes J_{\hat{x}_2})J_{\hat{x}}\hat{W}J_{\hat{x}}(J_{\hat{x}_1} \otimes J_{\hat{x}_2}) \otimes \mathbb{C}))''$. If γ is outer and $w \in (W \ltimes_{\gamma'} L) \cap L'$, then $w \in (B(V_{x_1} \otimes V_{x_2})\bar{\otimes}U) \cap \gamma(U)'$, which equals $W_1'\bar{\otimes}W_2'\bar{\otimes}\mathbb{C}$ by outerness of γ, so $w = a \otimes 1$ for $a \in W_1'\bar{\otimes}W_2'$. As w also commutes with $\hat{W}_1' \otimes \mathbb{C} \otimes \mathbb{C}$, we can write $a = 1 \otimes b$ with $b \in W_2'$. Since $(\iota \otimes \iota \otimes \gamma_2)\gamma = (\iota \otimes \Delta_2 \otimes \iota)\gamma$, we have $\iota \otimes \iota \otimes \gamma_2 = N_{2,23}(\cdot)_{124}N_{2,23}^*$ on $W \ltimes_{\gamma'} L$, so $b \in \hat{W}_1$ and $b \in \mathbb{C}$. Conversely, if γ' is outer and $w \in (B(V_{x_1} \otimes V_{x_2})\bar{\otimes}U) \cap (W_m \ltimes_\gamma U)'$, then $T_{123}(\iota \otimes \iota \otimes \gamma)(\cdot)T_{123}^*$ maps $B(V_{x_1} \otimes V_{x_2})\bar{\otimes}U = (\gamma(U) \cup (B(V_{x_1} \otimes V_{x_2}) \otimes \mathbb{C}))''$ into $B(V_{x_1} \otimes V_{x_2})\bar{\otimes}L$. So $T_{123}(\iota \otimes \iota \otimes \gamma)(w)T_{123}^* \in (B(V_{x_1} \otimes V_{x_2})\bar{\otimes}L) \cap \gamma'(L)'$, which equals $W' \otimes \mathbb{C} \otimes \mathbb{C} \otimes \mathbb{C}$ by outerness of γ'. So there is $a \in W'$ such that $(\iota\otimes\iota\otimes\gamma)(w) = T^*_{123}(a\otimes 1\otimes 1\otimes 1)T_{123}$. But $\gamma(U) \subset W_1\bar{\otimes}W_2\bar{\otimes}U$, while the last leg of T sits in $\hat{W}_1'$, so $w = w' \otimes 1$ for $w' \in B(V_{x_1} \otimes V_{x_2})$. Writing $w'' = (J_{\hat{x}_1} \otimes J_{\hat{x}_2})w'(J_{\hat{x}_1} \otimes J_{\hat{x}_2})$ and $b = J_{\hat{x}} a J_{\hat{x}}$, we see that $f(w'') = b \otimes 1$, where $f \equiv (\beta \otimes \iota)(M_1)((\cdot) \otimes 1)(\beta \otimes \iota)(M_1)^*$ is a right coaction of $(\hat{W}_1, \hat{\Delta}_1^{\mathrm{op}})$ on $B(V_{x_1} \otimes V_{x_2})$. Hence $f(b) \otimes 1 = b \otimes 1 \otimes 1$ and $w'' = b \in W' \cap \beta(W_1)' = (W_1'\bar{\otimes}W_2') \cap (\hat{W}_1 \otimes \mathbb{C})' = \mathbb{C} \otimes W_2'$. So $w = 1 \otimes d \otimes 1$ for $d \in W_2'$. Since $w \in (W_m \ltimes_\gamma U)'$ we know that $1 \otimes d$ commutes with $Z^*(\mathbb{C} \otimes \hat{W}_2)Z$. Up to a scalar Z^* equals $(J_{\hat{x}_1} J_{x_1} \otimes 1)K^*_\beta J_x(J_{x_1} \otimes J_{x_2})$, and the first part belongs to $B(V_{x_1})\bar{\otimes}W_2$, so $1 \otimes d$ commutes with $J_x(\mathbb{C} \otimes \hat{W}_2)J_x$. Hence $\alpha(J_{x_2} d J_{x_2})$ commutes with $\mathbb{C} \otimes \hat{W}_2$ and must belong to $W_1 \otimes \mathbb{C}$. Thus d and w are scalars. □

We have the following characterization of doublecrossed products.

Proposition 19.4.3 *Let (U, Δ_U) be a locally compact quantum group with morphisms from it to $(W_1, \Delta_1^{\mathrm{op}})$ and (W_2, Δ_2) with associated coaction α_1 and right coaction α_2, respectively. Suppose we have an isomorphism $f : U \to W_1\bar{\otimes}W_2$ that intertwines α_1 with $\Delta^{\mathrm{op}} \otimes \iota$ and α_2 with $\iota \otimes \Delta_2$, then there is a matching m of (W_1, Δ_1) and (W_2, Δ_2) and a group-like unitary u of $(\hat{U}, \hat{\Delta}_U)$ such that the isomorphism $f\mathrm{Ad}(u) : U \to W_1\bar{\otimes}W_2$ respects Δ_U and Δ_m.*

Proof Let M_U be the fundamental multiplicative unitary of (U, Δ_U). Since α_i come from morphisms we have a corepresentation $X_1 \in W_1\bar{\otimes}\hat{U}$ of $(W_1, \Delta_1^{\mathrm{op}})$ such that $(\alpha_1 \otimes \iota)(M_U) = X_{1,13}M_{U,23}$ and a corepresentation $X_2 \in W_2\bar{\otimes}\hat{U}$ of (W_2, Δ_2) such that $(\alpha_2 \otimes \iota)(M_U) = M_{U,13}X_{2,23}$. Let $M_0 = (\pi \otimes \iota)(M_U)$. Due to the intertwining properties of π we thus have $(\Delta_1^{\mathrm{op}} \otimes \iota \otimes \iota)(M_0) = X_{1,14}M_{0,234}$ and $(\iota \otimes \Delta_2 \otimes \iota)(M_0) = M_{0,124}X_{2,34}$. The first of these identities shows that $X^*_{1,13}M_0$ is invariant under $\Delta_1^{\mathrm{op}} \otimes \iota \otimes \iota$, so there is $Y \in W_2\bar{\otimes}\hat{U}$ such that $M_0 = X_{1,13}Y_{23}$. The second identity shows that $(\Delta_2 \otimes \iota)(Y) = Y_{13}X_{2,23}$, so YX_2^* is invariant under $\Delta_2 \otimes \iota$ and there is a unitary $u \in \hat{U}$ such that $M_0 = X_{1,13}(1 \otimes 1 \otimes u)X_{2,23}$. Since $(\iota \otimes \iota \otimes \hat{\Delta}_U)(M_0) = M_{0,124}M_{0,123}$ we thus get $(\iota \otimes \hat{\Delta}_U)(X_i) = X_{i,13}X_{i,12}$ and $\hat{\Delta}_U(u) = u \otimes u$, or $(\mathrm{Ad}u \otimes \iota)(M_0) = (1 \otimes u^*)M_0$. Hence $\mathrm{Ad}u$ leaves U invariant, and the formula for α_2 above shows that it commutes with $\mathrm{Ad}u$. Also, we see that $(\tilde{\alpha}_1 \otimes \iota)(M_U) = \tilde{X}_{1,13}M_{U,23}$ with $\tilde{\alpha}_1 = (\iota \otimes \mathrm{Ad}u^*)\alpha_1\mathrm{Ad}u$ and $\tilde{X}_1 = (1 \otimes u^*)X_1(1 \otimes u)$. The map $\tilde{\pi} = \pi\mathrm{Ad}u$ satisfies $(\tilde{\pi} \otimes \iota)(M_U) = \tilde{X}_{1,13}X_{2,23}$, and intertwines $\tilde{\alpha}_1$ with $\Delta_1^{\mathrm{op}} \otimes \iota$ and α_2 with $\iota \otimes \Delta_2$. So we may assume that $u = 1$. Then $M_0 = X_{1,13}X_{2,23}$. Since morphisms commute with coinverses we have $\alpha_1\tau_t^U = (\tau_{-t}^1 \otimes \tau_t^U)\alpha_1$ and $\alpha_2\tau_t^U = (\tau_t^U \otimes \tau_t^2)\alpha_2$, so $\pi\tau_t^U = (\tau_{-t}^1 \otimes \tau_t^2)\pi$. As X_i are corepresentations we have $(S_1^{-1} \otimes S_2)(\iota \otimes \iota \otimes \omega)(X_{1,13}X_{2,23}) = (\iota \otimes \iota \otimes \omega)(X^*_{1,13}X^*_{2,23})$ for $\omega \in \hat{U}_*$. Letting $m = (R_1 \otimes R_2)\pi R_U\pi^{-1}$ and using $S_U(\iota\otimes\omega)(M_U) = (\iota\otimes\omega)(M_U^*)$, we therefore get $(m \otimes \iota)(X_{1,13}X_{2,23}) = X_{2,23}X_{1,13}$. This yield $((\iota \otimes Fm \otimes \iota)(\Delta_1^{\mathrm{op}} \otimes \Delta_2) \otimes \iota)(M_0) = M_{0,125}M_{0,345}$ and $(\iota \otimes Fm \otimes \iota)(\Delta_1^{\mathrm{op}} \otimes \Delta_2)\pi = (\pi \otimes \pi)\Delta_U$. We also get

$$((\Delta_1\otimes\iota)m\otimes\iota)(X_{1,13}X_{2,23}) = X_{2,34}X_{1,24}X_{1,14} = m_{23}m_{13}(\Delta_1\otimes\iota\otimes\iota)(X_{1,13}X_{2,23}).$$

The other matching identity is proved similarly. □

We include some results on C*-algebraic aspects of bicrossed and doublecrossed products. Using $(\iota \otimes \Delta_2)(K_\beta) = K_{\beta,13}K_{\beta,12}$ and $K_\beta \in M(B_0(V_{x_1}) \otimes W_{2r})$, one readily sees that $W_r = [(\hat{W}_{1r} \otimes \mathbb{C})\alpha(W_{2r})]$, and similarly $\hat{W}_r = [(\mathbb{C} \otimes \hat{W}_{2r})\beta(W_{1r})]$. The previous theorem shows that $\hat{W}_{mr} = [(J_{\hat{x}_1}\hat{W}_{1r}J_{\hat{x}_1} \otimes \mathbb{C})Z^*(\mathbb{C} \otimes \hat{W}_{2r})Z]$.

Proposition 19.4.4 *We have that (W, Δ) is regular if and only if (W_i, Δ_i) are regular and either $W_{mr} = W_{1r} \otimes W_{2r}$ or $m(W_{1r} \otimes W_{2r}) = W_{1r} \otimes W_{2r}$. While (W_m, Δ_m) is regular if and only if (W_i, Δ_i) are regular.*

Proof If (W, Δ) is regular, then since $(\alpha \otimes \iota)(M_1)$ is a corepresentation of it, we have $[(B_0(V_{x_1}) \otimes B_0(V_{x_2}) \otimes \mathbb{C})(\alpha \otimes \iota)(M_1)(\mathbb{C} \otimes \mathbb{C} \otimes B_0(V_{x_1})] = B_0(V_{x_1}) \otimes B_0(V_{x_2}) \otimes B_0(V_{x_1})$, and as α is implemented by K_α, we see that (W_1, Δ_1) is regular, and the same is true for (W_2, Δ_2). With $C_M \equiv \{(\iota \otimes \omega)(FM) \mid \omega \in B(V_x)_*\}$, we get $[C_M] = [\{(\iota \otimes \iota \otimes \omega \otimes \eta)(F_{13}F_{24}M_{2,24}K_{\beta,34}K^*_{\alpha,12}\hat{M}_{1,13}) \mid \omega \in B(V_{x_1})_*,\ \eta \in B(V_{x_2})_*\}]$, and by using $(\Delta_1 \otimes \iota)(K_\alpha) = K_{\alpha,23}K_{\alpha,13}$ and $(\iota \otimes \Delta_2)(K_\beta) = K_{\beta,13}K_{\beta,12}$, we get $[C_M] = K_\beta[(\mathbb{C} \otimes C_{M_2})K^*_\beta K_\alpha(C_{\hat{M}_1} \otimes \mathbb{C})]K^*_\alpha$, so $B_0(V_x) = [(\mathbb{C} \otimes B_0(V_{x_2}))Z^*(B_0(V_{x_1}) \otimes \mathbb{C})]$. As $W_{mr} = [\{(\iota \otimes \iota \otimes \omega \otimes \eta)(N^*_{1,31}Z^*_{34}M_{2,24}) \mid \omega \in B(V_{x_1})_*,\ \eta \in B(V_{x_2})_*\}]$, we thus get $W_{mr} = W_{1r} \otimes W_{2r}$. Conversely, if this identity holds and (W_i, Δ_i) are regular, then since $W_{1r} \otimes W_{2r} = [W_{mr}(W_{1r} \otimes \mathbb{C})]$, we get $m(W_{1r} \otimes W_{2r}) = [(W_{1r} \otimes W_{2r})\beta(W_{1r})]$, and $[W_r\hat{W}_r] = [(\hat{W}_{1r} \otimes \mathbb{C})m(W_{1r} \otimes W_{2r})(\mathbb{C} \otimes \hat{W}_{2r})] = B_0(V_{x_1}) \otimes [W_{2r}\hat{W}_{2r}]$ as $K_\beta \in M(B_0(V_{x_1}) \otimes W_{2r})$ implements β. Therefore (W, Δ) is regular. If $m(W_{1r} \otimes W_{2r}) = W_{1r} \otimes W_{2r}$, we instead get $[W_r\hat{W}_r] = [\hat{W}_{1r}W_{1r} \otimes W_{2r}\hat{W}_{2r}]$, and (W, Δ) is regular. As $[C_{M_m}] = [\{(\iota \otimes \iota \otimes \omega)((FN_1)^*_{13}Z^*_{32}) \mid \omega \in B(V_{x_1})_*\}(\mathbb{C} \otimes C_{M_2})]$, we see that (W_m, Δ_m) is regular whenever (W_i, Δ_i) are, and the converse is now clear. □

Assuming that (W_i, Δ_i) are regular, then from Proposition 16.3.14 and its corollary, we know that α restricts to a continuous coaction of (W_{1r}, Δ_{1r}) on $T_2 \equiv [\{(\omega \otimes \iota)\alpha(W_{2r}) \mid \omega \in B(V_{x_1})_*\}]$ and that β restricts to a continuous right coaction of $(W_{2r}, \Delta^{\mathrm{op}}_{2r})$ on $T_1 \equiv [\{(\iota \otimes \omega)\beta(W_{1r}) \mid \omega \in B(V_{x_2})_*\}]$, and it easy to check that $W_r = W_{1r} \ltimes_{\alpha,r} T_2$, while $\hat{W}_r = T_1 \rtimes_{\beta,r} W_{2r}$.

Proposition 19.4.5 *Suppose that (W_i, Δ_i) are regular. Then*

$$W_{mr} = [\{(\iota \otimes \iota \otimes \omega)m_{13}(\iota \otimes \Delta^{\mathrm{op}}_2)(W_{1r} \otimes W_{2r}) \mid \omega \in B(V_{x_2})_*\}].$$

The map $\Delta^{\mathrm{op}}_1 \otimes \Delta_2 \colon W_{mr} \to M(W_{1r} \otimes W_{mr} \otimes W_{2r})$ defines a continuous coaction γ_1 of $(W_{1r}, \Delta^{\mathrm{op}}_{1r})$ on W_{mr} and a continuous right coaction γ_2 of (W_{2r}, Δ_{2r}) on W_{mr}. Also we have $W_r \rtimes_{\Delta,r} W_r \cong W_{1r} \ltimes_{\gamma_1,r} W_{mr} \rtimes_{\gamma_2,r} W_{2r}$.

Proof Since $W_{1r} \otimes W_{2r} = [W_{mr}(W_{1r} \otimes \mathbb{C})]$, the right hand side of the identity equals $[\{(\iota \otimes \iota \otimes \omega)m^{-1}_{12}((\mathbb{C} \otimes \mathbb{C} \otimes W_{2r})(\iota \otimes \Delta^{\mathrm{op}}_2)m(W_{mr}))m(W_{1r} \otimes \mathbb{C})_{13} \mid \omega \in B(V_{x_2})_*\}]$. By Proposition 19.2.2 the closed span of the argument of m^{-1}_{12} above equals $m(W_{mr}) \otimes W_{2r}$, so the right hand side of the desired identity equals $[W_{mr}(T_1 \otimes \mathbb{C})]$. This equals $[\{(\omega \otimes \iota \otimes \iota)(((T_1 \otimes B_0(V_{x_2}))\beta(W_{1r}))_{21}K_{\alpha,21}N_{2,13}) \mid \omega \in B(V_{x_2})_*\}]$ by the previous theorem. We claim $[(T_1 \otimes B_0(V_{x_2}))\beta(W_{1r})] = [(\mathbb{C} \otimes B_0(V_{x_2}))\beta(W_{1r})]$, and combining this claim with the previous identity and previous theorem, we get the desired identity. As for the claim, it suffices to show that $[T_1 W_{1r}] = W_{1r}$, but $T_1 = [\{(\omega \otimes \iota \otimes \eta)((\iota \otimes m)(K^*_{\beta,13})N_{1,12}) \mid \omega \in B(V_{x_1})_*, \ \eta \in B(V_{x_2})_*\}]$ by the previous theorem, and then $N_1 \in M(B_0(V_{x_1}) \otimes W_{1r})$ and $K_\alpha \in M(W_{1r} \otimes B_0(V_{x_2}))$ gives $[T_1 W_{1r}] = W_{1r}$. The properties of γ_i are clear, so it remains to prove the last identity. Using R_m in the first identity, gives

$$W_{mr} = [\{(\iota \otimes \iota \otimes \omega)(\iota \otimes \Delta_2)m^{-1}(W_{1r} \otimes W_{2r}) \mid \omega \in B(V_{x_2})_*\}],$$

which by invoking N_2, gives $W_{mr} = [m^{-1}(W_{1r} \otimes W_{2r})(\mathbb{C} \otimes J_{\hat{x}_2} \hat{W}_{2r} J_{\hat{x}_2})]$. Hence $W_{1r} \ltimes_{\gamma_1,r} W_{mr} \rtimes_{\gamma_2,r} W_{2r}$ by definition equals $(J_{\hat{x}_1} \otimes J_{\hat{x}_2})[W_r \hat{W}_r](J_{\hat{x}_1} \otimes J_{\hat{x}_2}) = [W_r J_{\hat{x}} \hat{W}_r J_{\hat{x}}] = W_r \rtimes_{\Delta,r} W_r$. □

Exercises

For Sect. 19.1

1. Suppose G is a locally compact group with closed subgroup H. Consider the quotient space G/H of G by H with coaction α of the usual locally compact group $(L^\infty(G), \Delta)$ on $L^\infty(G/H)$ associated to the natural action $G \times (G/H) \to G/H$. Let x be the Haar integral on G. Describe in group terms the Radon-Nikodym derivative of the weight x under α.
2. Suppose (W, Δ) is a locally compact quantum group with unitary coinverse R and scaling group $\{\tau_t\}$. Say U is a von Neumann subalgebra of W such that $\Delta(U) \subset U \bar{\otimes} U$. Show that if $(U, \Delta|U)$ is a locally compact quantum group in its own right, then $R(U) = U = \tau_t(U)$ for real t.
3. Produce an example of a locally compact quantum group (W, Δ) with a von Neumann subalgebra U of W such that $\Delta(U) \subset U \bar{\otimes} U$, but where $(U, \Delta|U)$ is not a locally compact quantum group.

For Sect. 19.2

1. Can you come up with two different matchings between two fixed locally compact quantum groups such that the associated doublecrossed products are non-isomorphic and non-trivial?
2. Describe the quantum double of the locally compact quantum group naturally associated to your favourite finite group.
3. Describe in terms of generators the quantum double of the compact locally compact quantum group $SU_q(2)$ by comparing with the group $SL(2, \mathbb{C})$ of complex 2×2-matrices of determinant one, thus viewing the quantum double as a q-deformation $SL_q(2, \mathbb{C})$ of $SL(2, \mathbb{C})$.

For Sect. 19.3

1. Show that a morphism between locally compact quantum groups is also a morphism between the corresponding opposite quantum groups. What is the situation between the corresponding commutant quantum groups, and between the corresponding dual quantum groups, and the corresponding quantum doubles?
2. Show that a continuous homomorphism between two locally compact groups induces a natural morphism between the corresponding locally compact quantum groups. What is the associated right coaction in this case?

For Sect. 19.4

1. Can you interpret the characterization result of the doublecrossed product given in this section in the classical case when the von Neumann algebras of the constituent locally compact quantum groups are commutative?
2. Show that the bicrossed product (W, Δ) of locally compact quantum groups (W_1, Δ_1) and (W_2, Δ_2) with a matching is semiregular if and only if both (W_i, Δ_i) are semiregular and $W_{1r} \otimes W_{2r} \subset W_r$.
3. Show that the doublecrossed product is semiregular if and only if both the constituent quantum groups are semiregular.
4. Can you establish a continuous, or C*-algebraic, version of the result in the previous section providing a correspondence between unitary corepresentations of the bicrossed product and unitary corepresentations of the constituent quantum groups.

Chapter 20
Induction

Let (W_2, Δ_2) be a locally compact quantum group with a locally compact quantum subgroup (W_1, Δ_1), and let L be a unitary corepresentation of the latter, meaning that $L \in W_1 \bar{\otimes} B(V)$ is unitary and $(\Delta_1 \otimes \iota)(L) = L_{13}L_{23}$, where V is a Hilbert space. Then we define the so called induced corepresentation $M(L)$ of L, and do indeed prove, using modular theory, that $M(L)$ is a unitary corepresentation of (W_2, Δ_2) on a larger Hilbert space; a construction that reduces to a well known one in the classical situation of a unitary representation of a subgroup of a group.

Recommended literature for this chapter is [28].

20.1 Inducing Corepresentations Using Modular Theory

Let (W_i, Δ_i) be locally compact quantum groups. We shall think of (W_1, Δ_1) as a quantum subgroup of (W_2, Δ_2) in that we have a right coaction α of (W_1, Δ_1) on W_2 satisfying $(\Delta_2 \otimes \iota)\alpha = (\iota \otimes \alpha)\Delta_2$, see Proposition 19.3.2. Fix a faithful semifinite normal weight z on W_2^α. Then $\beta\pi_z = (\iota \otimes \pi_z)\Delta_2$ evidently defines a coaction β of (W_2, Δ_2) on $\pi_z(W_2^\alpha) = W_2^\alpha$, and we let K be the canonical unitary implementation of it. Fix a unitary $L \in W_1 \bar{\otimes} B(V)$ such that $(\Delta_1 \otimes \iota)(L) = L_{13}L_{23}$. We shall see that this data induces a unitary corepresentation $M(L)$ of (W_2, Δ_2).

Let V' be a Hilbert space. Then one easily checks that the space

$$G_{V'} \equiv \{X \in B(V')\bar{\otimes} W_2 \bar{\otimes} B(V) \mid (\iota \otimes \alpha \otimes \iota)(X) = L_{34}^* X_{124}\}$$

for $X \in G_{V'}$ satisfies $(Y \otimes 1)X, XY, Y^*X \in G_{V'}$ whenever $Y \in B(V')\bar{\otimes} W_2^\alpha$ and $Y \in B(V')\bar{\otimes} W_2^\alpha \bar{\otimes} B(V)$ and $Y \in G_{V'}$, respectively. Let $H_{V'}$ be the Hilbert space completion of $G_{V'} \odot (V' \otimes V_z \otimes V)$ with respect to the positive sesquilinear form given by $(X \otimes v | X' \otimes v') = ((\iota \otimes \pi_z \otimes \iota)(X'^* X)v|v')$, and let $[X \otimes v] \in H_{V'}$ denote the

L. Tuset, *Analysis and Quantum Groups*,
https://doi.org/10.1007/978-3-031-07246-8_20

equivalence class of $X \otimes v$. Let $G \equiv G_{\mathbb{C}}$ and $H \equiv H_{\mathbb{C}}$. Clearly $(\omega \otimes \iota \otimes \iota)(X) \in G$ for $X \in G_{V'}$ and $\omega \in B(V')_*$, while $1 \otimes X \in G_{V'}$ when $X \in G$.

Lemma 20.1.1 *The formula* $U_{V'}(v \otimes [X \otimes u]) = [(1 \otimes X) \otimes (v \otimes u)]$ *for* $v \in V'$, $X \in G$ *and* $u \in V_z \otimes V$ *well-defines a unitary map* $U_{V'} \colon V' \otimes H \to H_{V'}$.

Proof It is straightforward to check that one gets a well-defined isometry, and surjectivity is clear from the easily established identity

$$U_{V'}(\sum v_i \otimes [(\omega_{v,v_i} \otimes \iota \otimes \iota)(Y) \otimes u]) = [Y \otimes (v \otimes u)],$$

which holds for $Y \in G_{V'}$ and any orthonormal basis $\{v_i\}$ of V'. □

This lemma allow us to define for $X \in G_{V'}$ an operator $X_* \colon V' \otimes V_z \otimes V \to V' \otimes H$ by $X_*(v) = U^*_{V'}([X \otimes v])$. Then the isometry $G_{V'} \to B(V' \otimes V_z \otimes V, V' \otimes H)$ which sends X to X_* is strongly-strongly (and strongly*-strongly*) continuous on bounded nets. We also have $(Y_*)^* X_* = (\iota \otimes \pi_z \otimes \iota)(Y^* X)$ for $X, Y \in G_{V'}$, and the closed span of the elements of the form $X_*(v)$ is $V' \otimes H$. We also see that $(XY)_* = X_*(\iota \otimes \pi_z \otimes \iota)(Y)$ for $Y \in B(V') \bar{\otimes} W_2^{\alpha} \bar{\otimes} B(V)$, while $((a \otimes 1 \otimes 1)X)_* = (a \otimes 1)X_*$ for $a \in B(V')$.

Proposition 20.1.2 *The* induced corepresentation *of* L *is* $M(L) \in W_2 \bar{\otimes} B(H)$ *given by* $M(L)^*(v \otimes X_*(u)) = (\Delta_2 \otimes \iota)(X)_* K_{12}(v \otimes u)$ *for* $v \in V_{x_2}$ *and* $X \in G$ *and* $u \in V_z \otimes V$. *It satisfies* $M(L)M(L)^* = 1$ *and* $(\Delta_2 \otimes \iota)(M(L)) = M(L)_{13} M(L)_{23}$.

Proof The formula for $M(L)^*$ clearly gives a well-defined isometry which commutes with $a \otimes 1$ for $a \in W_2'$. As for the final identity, first note that

$$(\iota \otimes \iota \otimes \alpha \otimes \iota)((\Delta_2 \otimes \iota)\Delta_2 \otimes \iota)(X) = L^*_{34}((\Delta_2 \otimes \iota)\Delta_2 \otimes \iota)(X)_{124}$$

so $((\Delta_2 \otimes \iota)\Delta_2 \otimes \iota)(X) \in G_{V_{x_2} \otimes V_{x_2}}$. Using the identity in the proof of the previous lemma two times, one checks that

$$(M_2)^*_{12}(v \otimes (\Delta_2 \otimes \iota)(X)_*(u)) = ((\Delta_2 \otimes \iota)\Delta_2 \otimes \iota)(X)_*(M_2)^*_{12}(v \otimes u)$$

for elementary tensors u. Replacing u by $K_{12}(v' \otimes u)$ for $v' \in V_{x_2}$, and then $v \otimes v'$ by $M_2(v_1 \otimes v_2)$ for $v_i \in V_{x_2}$, we thus get

$$(\Delta_2 \otimes \iota)(M(L)^*)(v_1 \otimes v_2 \otimes X_*(u)) = ((\Delta_2 \otimes \iota)\Delta_2 \otimes \iota)(X)_* K_{23} K_{13}(v_1 \otimes v_2 \otimes u),$$

where we used that the left hand side is $((M_2)^*_{12} M(L)^*_{23})(M_2(v_1 \otimes v_2) \otimes X_*(u))$ and that K is a corepresentation. One then shows that the right hand side of this identity is $(M(L)^*_{23} M(L)^*_{13})(v_1 \otimes v_2 \otimes X_*(u))$ by again using the identity in the proof of the previous lemma repeatedly. □

We claim that $M(L)$ is independent of K, or of z, up to unitary equivalence. Indeed, say z' is another such weight with induced corepresentation $M(L)'$ on H'. Then we know that there is a canonical unitary map $u \colon V_z \to V_{z'}$ such that $\pi_{z'} =$

$u\pi_z(\cdot)u^*$. Let $U\colon H \to H'$ be the unitary map uniquely determined by $UX_* = X_*(u \otimes 1)$ for all $X \in G$. Then $(1 \otimes U)Y_* = Y_*(1 \otimes u \otimes 1)$ for any $Y \in G_{V'}$. Letting K' be the canonical unitary implementation of the coaction β' of (W_2, Δ_2) on $\pi_{z'}(W_2^\alpha) = W_2^\alpha$ given by $\beta'\pi_{z'} = (\iota \otimes \pi_{z'})\Delta_2$, then $K' = (1 \otimes u)K(1 \otimes u^*)$ by Proposition 15.4.9, and one easily checks that $M(L)' = (1 \otimes U)M(L)(1 \otimes U^*)$, so the two induced corepresentations are canonically unitarily equivalent.

We aim to show that $M(L)$ is unitary under the following mild assumption, which we will make for the rest of this section.

Proposition 20.1.3 *The coaction α is integrable if and only if there is a non-zero positive element $w \in W_2$ such that $\alpha(w) \in M_{\iota\otimes x_1}$.*

Proof The σ-weak closure of the left ideal $N_{(\iota\otimes x_1)\alpha}$ of W_2 is of the form W_2P for an orthogonal projection $P \in W_2$. For $\omega \in (W_2)_*$ we have $(\omega \otimes \iota)\Delta_2(w') \in N_{(\iota\otimes x_1)\alpha}$ when $w' \in N_{(\iota\otimes x_1)\alpha}$ as $(\omega\otimes\iota)\Delta_2(w')^*(\omega\otimes\iota)\Delta_2(w') \leq \|\omega\|(|\omega|\otimes\iota)\Delta_2(w'^*w')$ and $(\iota \otimes \alpha)\Delta_2 = (\Delta_2 \otimes \iota)\alpha$. Hence $(\omega \otimes \iota)\Delta_2$ leaves W_2P invariant, and in particular $(\omega \otimes \iota)\Delta_2(P) = (\omega \otimes \iota)\Delta_2(P)P$, so $\Delta_2(P) \leq 1 \otimes P$, and $P = 1$ or $P = 0$. □

We need to produce a convenient dense subset of H. To get this we will use the following result with an easy, standard proof.

Lemma 20.1.4 *We have $(\iota \otimes x_1 \otimes \iota)(L_{23}(\alpha \otimes \iota)(X)) \in G$ for $X \in W_2\bar{\otimes}B(V)$ satisfying $L_{23}(\alpha \otimes \iota)(X) \in M_{\iota\otimes x_1\otimes\iota}$.*

Define a norm continuous one-parameter group σ on the predual of W_1 by $\sigma_t(\omega) = \omega\sigma_t^{y_1}$ for real t, where as usual $y_i = x_iR_i$. Then $\bar{\omega} \in D(\sigma_{\bar{z}})$ if and only if $\omega \in D(\sigma_z)$ if and only if there is $\eta \in (W_1)_*$ such that $\omega\sigma_z^{y_1} \subset \eta$. Then $\sigma_z(\omega) = \eta = \overline{\sigma_{\bar{z}}(\bar{\omega})}$. Let A be the analytic elements of σ. We also skip the easy proof of the following result.

Lemma 20.1.5 *For $\omega \in D(\sigma_{i/2})$ and $\eta \in B(V)_*$, we have the following identity $\sigma_{i/2}((\iota \otimes \eta)((1 \otimes (\omega \otimes \iota)(L^*))L^*)) = R_1((\iota \otimes \eta)((1 \otimes (\sigma_{i/2}(\omega)R_1 \otimes \iota)(L))L))$.*

Lemma 20.1.6 *Let $T = J_{x_1}(\cdot)^*J_{x_1}\colon W_1 \to W_1'$, $\omega \in D(\sigma_{i/2})$, $w \in N_{x_1}$. Then $(q_{x_1}\otimes\iota)((w\otimes(\omega\otimes\iota)(L^*))L^*) = (1\otimes(\sigma_{i/2}(\omega)R_1\otimes\iota)(L))(TR\otimes\iota)(L)(q_{x_1}(w)\otimes 1)$.*

Proof The proof is straightforward; use the previous lemma and the identity $(q_{x_1} \otimes \iota)(\cdot)v = \sum q_{x_1}(\iota\otimes\omega_{v,v_i})(\cdot)\otimes v_i$ on $N_{x_1\otimes\iota}$, where $v \in V$ and $\{v_i\}$ is an orthonormal basis for V. □

Lemma 20.1.7 *Let $\omega \in D(\sigma_{i/2})$ and $X \in N_{\iota\otimes x_1} \subset W_2\bar{\otimes}W_1$. Then $\iota \otimes q_{x_1} \otimes \iota$ applied to $(X \otimes (\omega \otimes \iota)(L^*))L_{23}^*$ equals*

$$(1 \otimes 1 \otimes (\sigma_{i/2}(\omega)R_1 \otimes \iota)(L))(TR_1 \otimes \iota)(L)_{23}((\iota \otimes q_{x_1})(X) \otimes 1).$$

Proof Pick a bounded net $\{X_i\}$ in $W_2 \odot N_{x_1}$ that converges strongly* to X and such that $(\iota\otimes q_{x_1})(X_i)$ form a bounded net that converges strongly* to $(\iota\otimes q_{x_1})(X)$. Using that $\iota \otimes q_{x_1} \otimes \iota$ is σ-strongly*-closed together with the previous lemma, then gives the result. □

Lemma 20.1.8 *There are nets $\{a_j\}$ in $N_{(\iota\otimes x_1)\alpha}$ and $\{\omega_j\}$ in A over the same directed set such that the net*

$$\{(\iota \otimes x_1 \otimes \iota)(L_{23}(\alpha(a_j^* a_j) \otimes (\omega_j \otimes \iota)(L)(\omega_j \otimes \iota)(L)^*)L_{23}^*)\}$$

converges strongly to 1 *from below.*

Proof Since α is integrable we have a positive net $\{b_l\}$ in $M_{(\iota\otimes x_1)\alpha}$ that converges strongly to 1 from below. Since slices of L on the right by appropriate normal functionals form a non-degenerate $*$-subalgebra of $B(V)$, we get by Kaplansky's density theorem a net $\{\eta_p\}$ in A such that $(\sigma_{i/2}(\bar{\eta}_p)R_1 \otimes \iota)(L)^*(\sigma_{i/2}(\bar{\eta}_p)R_1 \otimes \iota)(L) \to 1$ strongly from below. Let F be the directed set of finite subsets of $V_{x_2}\otimes V$, and consider $J \equiv F \times \mathbb{N}$ with product order. To $j \equiv (f, n) \in J$ pick l_j such that $((\iota \otimes x_1)\alpha(b_{l_j} b_{l_j}^*) \otimes 1)v - v$ is less that $1/2n$ in norm for all $v \in f$. By the previous lemma we may pick p_j such that

$$(\iota\otimes x_1\otimes\iota)(L_{23}(\alpha(b_{l_j}^* b_{l_j})\otimes(\eta_{p_j}\otimes\iota)(L)(\eta_{p_j}\otimes\iota)(L)^*)L_{23}^*)v-((\iota\otimes x_1)\alpha(b_{l_j} b_{l_j}^*)\otimes 1)v$$

is less than $1/2n$ in norm for all $v \in f$. Then $a_j = b_{l_j}$ and $\omega_j = \eta_{p_j}$ will work. □

Lemma 20.1.9 *The Hilbert space H is the closed span of elements of the form $(\iota \otimes x_1 \otimes \iota)(L_{23}(\alpha \otimes \iota)((a^* \otimes (\omega \otimes \iota)(L))X))_* v$, where $a \in N_{(\iota\otimes x_1)\alpha}$ and $\omega \in A$ and $v \in V_z \otimes V$ and $X \in W_2\bar{\otimes}B(V)$ with $(\alpha \otimes \iota)(X) \in N_{\iota\otimes x_1\otimes\iota}$.*

Proof This is immediate from the previous lemma since multiplying the elements in the resulting net with $Y \in G$ from the left one gets a bounded net that converges strongly to Y. □

Lemma 20.1.10 *For $X \in N_{\iota\otimes x_1\otimes\iota} \subset W_2\bar{\otimes}W_1\bar{\otimes}B(V)$ and $\omega \in B(V)_*$ we have $(\iota\otimes\iota\otimes\omega)(X) \in N_{\iota\otimes x_1}$, and $(\iota\otimes q_{x_1}\otimes\iota)(v'\otimes v) = \sum(\iota\otimes q_{x_1})(\iota\otimes\iota\otimes\omega_{v,v_i})v'\otimes v_i$, where $v' \in V_{x_2}$ and $v \in V$ and $\{v_i\}$ is an orthonormal basis of V.*

Proof The first statement basically uses an inequality of the kind we used in the proof of the previous proposition. We have that $\sum \|(\iota \otimes q_{x_1})(\iota \otimes \iota \otimes \omega_{v,v_i})v'\|$ equals $((\iota\otimes x_1)(\iota\otimes\iota\otimes\omega_{v,v})(X^*X)v'|v') < \infty$ by σ-weak lower semicontinuity of $\iota \otimes x_1$. The proof is then completed by repeated use of the identity in the proof of the fourth last lemma above. □

Lemma 20.1.11 *The Hilbert space H is the closed span of elements of the form $(\iota\otimes x_1\otimes\iota)(L_{23}(\alpha(w)\otimes(\omega\otimes\iota)(L)))_* v$, where $w \in M_{(\iota\otimes x_1)\alpha}$, $\omega \in A$ and $v \in V_z\otimes V$.*

Proof For the elements spanning H in the second last lemma above we may replace v by $u \otimes w$, where $u \in V_z$ and $w \in V$. Extend $(\pi_z(\cdot)u|u)$ to a positive normal functional on W_2, which, since W_2 is in standard form, may be written $\omega_{u'}$ for $u' \in V_{x_2}$. Then the previous lemma shows that a typical element spanning a dense subset of H is a limit of the net

$$\{\sum_{i\in J}(\iota \otimes x_1 \otimes \iota)(L_{23}(\alpha(a^*(\iota \otimes \omega_{w,v_i})(X)) \otimes (\omega \otimes \iota)(L)))_*(u \otimes v_i)\}_J,$$

where J is a finite subset of the index set of an orthonormal basis $\{v_i\}$ of V, and we consider the collection of such subsets as a directed set under inclusion. □

Let $N_0 = \{(\omega \otimes \iota)\Delta_2(w) \mid \omega \in (W_2)_*, w \in N_{(\iota\otimes x_1)\alpha}\}$. From the proof of the last proposition above we have $N_0 \subset N_{(\iota\otimes x_1)\alpha}$ and $\|(\iota \otimes q_1)\alpha(\omega \otimes \iota)\Delta_2(w)\| \leq \|\omega\| \cdot \|(\iota\otimes q_1)\alpha(w)\|$. Define $H_0 \subset H$ to be the closed span of elements of the form $(\iota \otimes x_1 \otimes \iota)(L_{23}(\alpha(w^*w') \otimes (\omega \otimes \iota)(L)))_* v$, where $w, w' \in N_0$ and $\omega \in A$ and $v \in V_z \otimes V$, see the previous lemma. We will ultimately show that $H_0 = H$.

Lemma 20.1.12 *We have $M(L)^*(V_{x_2} \otimes H) \subset V_{x_2} \otimes H_0$.*

Proof For $w, w' \in N_{(\iota\otimes x_1)\alpha}$ and $\omega \in A$ and $v \in V_{x_2}$ and $v' \in V_z \otimes H$, the element $M(L)^*(v\otimes(\iota\otimes x_1\otimes\iota)(L_{23}(\alpha(w^*w')\otimes(\omega\otimes\iota)(L)))_* v')$ equals $T'_* K_{12}(v\otimes v')$, where $T' \equiv (\Delta_2 \otimes \iota)(\iota \otimes x_1 \otimes \iota)(L_{23}(\alpha(w^*w') \otimes (\omega \otimes \iota)(L)))$. From the identity in the proof of the first lemma of this section we know that for an orthonormal basis $\{v_i\}$ of V_{x_2} and $u \in V_{x_2}$ and $u' \in V_z\otimes V$, the net of sums $\sum_{i\in J} v_i \otimes (\omega_{u,v_i} \otimes\iota\otimes\iota)(T')_* u'$ over finite J converges to $T'_*(u\otimes u')$ as the sets J get larger. By the previous lemma we are therefore done if we can show that $(\omega_{u,v_i} \otimes\iota\otimes\iota)(T')_* u' \in H_0$. But the latter element can be approached by the net of finite sums $\sum_{k\in J'}(\iota\otimes x_1 \otimes\iota)((\omega_{v_k,v_i} \otimes\iota\otimes \iota\otimes\iota)(L_{34}((\Delta_2\otimes\iota)\alpha(w^*)\otimes(\omega\otimes\iota)(L)))(\omega_{u,v_k} \otimes\iota\otimes\iota\otimes\iota)((\Delta_2\otimes\iota)\alpha(w')\otimes 1))_* u'$, and these summands belong to H_0. □

Lemma 20.1.13 *The closed span of the elements $(d \otimes 1)M(L)^*v$ with $d \in W_2$ and $v \in V_{x_2} \otimes H_0$ contains $V_{x_2} \otimes H_0$.*

Proof Observe first that for $w \in N_{(\iota\otimes x_1)\alpha}$ and $u, u' \in V_{x_2}$ and $a \in B(V_{x_2} \otimes V')$ with V' any Hilbert space, we have the inequality

$$\|(\iota\otimes\iota\otimes q_{x_1})(\iota\otimes\alpha)(\omega_{u,w^*u'} \otimes\iota\otimes\iota)(N^*_{13}(a\otimes 1))\| \leq \|u\|\|u'\|\|a\|\|(\iota\otimes q_{x_1})\alpha(w)\|$$

by invoking the standard properties of $N \equiv (J_{\hat{x}_2} \otimes J_{\hat{x}_2})(M^*_2)_{21}(J_{\hat{x}_2} \otimes J_{\hat{x}_2})$. One also checks that $\|(\iota \otimes q_{x_1})\alpha(\omega_{u,w^*u'} \otimes \iota)(N^*)\| \leq \|u\|\|u'\|\|(\iota \otimes q_{x_1})\alpha(w)\|$, and that $\|(\iota\otimes\iota\otimes q_{x_1})(\Delta_2\otimes\iota)\alpha(w)\| = \|(\iota\otimes q_{x_1})\alpha(w)\|$. Replacing u by $q_{y_2}(bc)$ and u' by $q_{y_2}(a)$ with $a, b \in N_{y_2}$ and c in the Tomita algebra of $y_2 = x_2R_2$, we thus see that the closed span of the elements $(\iota\otimes\iota\otimes q_{x_1})(\Delta_2\otimes\iota)\alpha(w')$ with $w' \in N_0$ contains the closed span of the elements $(\iota \otimes q_{x_1})\alpha(\omega_{u,w^*u'} \otimes \iota)(N^*)$. Hence the closed span S of the elements $(\iota\otimes\iota\otimes q_{x_1})(\Delta_2\otimes\iota)\alpha(w')(d\otimes 1)$ with $d \in W_{2r}$ contains the closed span of the elements $(\iota\otimes\iota\otimes q_{x_1})(\iota\otimes\alpha)(\omega_{u,w^*u'} \otimes\iota\otimes\iota)(N^*_{13}(N^*(w''\otimes d))_{12})$ with $w'' \in B_0(V_{x_2})$. Therefore standard manipulations show that S contains the closed span of the elements $d \otimes (\iota \otimes q_{x_1})\alpha(\omega_{w''u,w^*u'} \otimes \iota)(N^*)$. This statement still holds if we replace $(\omega_{w''u,w^*u'} \otimes \iota)(N^*)$ by $(\omega_{p,q} \otimes \iota)\Delta_2(b^*w)$ with $p, q \in V_{x_2}$.

On the other hand, the closed span S' of the elements $(d^* \otimes 1)M(L)^*v$ with $d \in W_{2r}$ equals the closed span of the elements

$$((d_2^*\otimes 1\otimes 1)(\iota\otimes\iota\otimes x_1\otimes\iota)(L_{34}((\Delta_2\otimes\iota)(\alpha(w_2^*w_1)\otimes(\eta\otimes\iota)(L)))(d_1\otimes 1\otimes 1))_*(u\otimes v')$$

with $d_i \in W_{2r}$ and $w_i \in N_0$ and $v' \in V_z \otimes V$ and $\eta \in A$. By Lemma 20.1.7 what is inside $(\cdot)_*$ of this long expression equals $(((\iota\otimes\iota\otimes q_{x_1})((\Delta_2\otimes\iota)\alpha(w_2))(d_2\otimes 1))^*\otimes 1)$ times $T''(((\iota \otimes \iota \otimes q_{x_1})((\Delta_2 \otimes \iota)\alpha(w_1))(d_1 \otimes 1)) \otimes 1)$, where

$$T'' \equiv (TR_1 \otimes \iota)(L^*)(1 \otimes (\sigma_{i/2}(\bar{\eta})R_1 \otimes \iota)(L)^*).$$

The last inclusion in the previous paragraph tells us therefore that S' contains the closed span of the elements $d_2^* d_1 u \otimes (\iota \otimes x_1 \otimes \iota)(L_{23}(\alpha((\omega_{p_2,q_2} \otimes \iota)\Delta_2(b_2^* w_2)^*(\omega_{p_1,q_1} \otimes \iota)\Delta_2(b_1^* w_1)) \otimes (\eta \otimes \iota)(L)))_* v'$, where $b_i \in N_{y_2}$ and $p_i, q_i \in V_{x_2}$. Replacing b_i by nets in N_{y_1} that converge strongly* to 1, and using that W_2 is in standard form, now gives the desired inclusion. □

Theorem 20.1.14 *The operator $M(L)$ is unitary, and $H = H_0$.*

Proof We claim that $M(L)^*(V_{x_2} \otimes H_0) = V_{x_2} \otimes H_0$. Let p be the orthogonal projection onto H_0, and define $r = M(L)^*(1 \otimes p)$, so $r^*r = 1 \otimes p$. Let $q = rr^* = M(L)^*(1\otimes p)M(L)^*$. Then $(\Delta_2\otimes\iota)(q) = M(L)_{23}^* q_{13} M(L)_{23}$. Since $M(L)^*$ leaves $V_{x_2} \otimes H_0$ invariant, we have $q(1 \otimes p) = q$, so $(\Delta_2 \otimes \iota)(q)q_{23} = (\Delta_2 \otimes \iota)(q)$ and $(\Delta_2 \otimes \iota)(q) = q_{23}$. Hence $q \in \mathbb{C} \otimes B(H)$, and there is a closed subspace H_1 of H such that q projects orthogonally onto H_1. Now $M(L)^*(V_{x_2} \otimes H_0) = r(V_{x_2} \otimes H_0) = V_{x_2} \otimes H_1$, and the previous lemma then shows that $V_{x_2} \otimes H_0 = V_{x_2} \otimes H_1$, so $H_1 = H_0$, proving the claim. The second last lemma now shows that $M(L)^*(V_{x_2} \otimes H) \subset M(L)^*(V_{x_2} \otimes H_0)$, so $H = H_0$ as $M(L)^*$ is an isometry. But now we know that it is also surjective. □

Exercises

For Sect. 20.1

1. Suppose we have a locally compact group G_2 with a closed subgroup G_1, and consider the naturally associated locally compact quantum groups (W_i, Δ_i). Describe the unitary representation of G_2 naturally associated to the induced unitary corepresentation $M(L)$ of a unitary corepresentation L of (W_1, Δ_1), which again comes from a unitary representation of G_1.
2. Consider the compact locally compact quantum group $SU_q(2)$ with the circle $\mathbb{T}$ group as a quantum subgroup. What are the unitary corepresentations induced up from unitary representations of $\mathbb{T}$?
3. Consider the bicrossed product of two locally compact quantum groups, and study the unitary corepresentations induced up from unitary corepresentations of the constituent subgroups. What is the situation for the locally compact quantum group q-deformation of the $ax + b$ group?
4. Consider the doublecrossed product of two locally compact quantum groups, and study the unitary corepresentations induced up from unitary corepresentations of the constituent subgroups.

5. Trace the proof of unitarity of the induced unitary corepresentation in the Kac algebra case.
6. Can you formulate a continuous, or C*-algebraic, version of an induction result for locally compact quantum groups? Describe the induction process from the point of view of representations of the corresponding dual quantum groups.

Appendix

In section one we introduce basic notions in set theory, and in section two we prove various results on cardinality of sets, and show that any vector space not only has a linear basis, but that any two bases have the same cardinality, so it makes sense to talk about the dimension of a vector space.

The next three sections are devoted to point set topology. We introduce and prove the basic concepts focusing on locally compact Hausdorff spaces, nets and induced topologies. We also state and prove the Stone-Weierstrass theorem, saying that any self-adjoint subalgebra of the algebra of complex valued continuous functions on a compact Hausdorff space is dense in the uniform norm provided the subalgebra separate points in the space.

We include six sections on measure theory, starting from scratch with measures on σ-algebras. We then move on to L^p-spaces, complex measures, duality, product measures, arriving at the existence and uniqueness of Haar measures on locally compact groups and the Banach $*$-algebra of regular complex Borel measures on such groups, where the self-adjoint norm closed ideal of L^1-functions with respect to the Haar measure are identified as the measures that are absolutely continuous with respect to the Haar measure.

Returning to the origin of measure theory, recall that a Riemann partition on an interval $[a, b]$ is a finite sequence $P = \{x_i\}$ such that $a = x_0 < x_1 < \cdots < x_{n-1} < x_n = b$. Given a bounded real valued function f on $[a, b]$. Let $S = \sum M_i(x_i - x_{i-1})$ and $s = \sum m_i(x_i - x_{i-1})$, where M_i and m_i is the supremum and infinum, respectively, of f on $[x_{i-1}, x_i]$. Then f is said to be Riemann integrable if $\inf_P S = \sup_P s$, and the Riemann integral $\int_a^b f(x)\,dx$ of f is this common value. We extend it by linearity to bounded complex valued functions.

The Riemann integral of f can thus be approximated by finite sums of the form $\sum f(y_i)m(A_i)$, where $\{A_i\}$ are disjoint intervals with union $[a, b]$, and $m(A_i)$ is the length of A_i, and $y_i \in A_i$. The basic idea of measure theory is to develop an integration theory better suited to handle limit processes by allowing more general sets A_i, so called measurable ones. We can then integrate measurable functions f,

L. Tuset, *Analysis and Quantum Groups*,
https://doi.org/10.1007/978-3-031-07246-8

that is, functions with inverse images of open sets being measurable. We replace the length function m by a general measure μ, which basically measures the amount of something A_i, say length, volume, mass, energy, probability etc., by systematically attaching a non-negative extended number to each measurable subset of any set. The resulting Lebesgue integration theory can be seen as a completion of the Riemannian one, much like the real (or complex) numbers completes the rational ones. The approach is axiomatic.

We produce measures m extending the length function on the real line. This goes via linear functionals on function spaces. In fact, a general theme in measure theory is the interplay between measures on locally compact Hausdorff spaces and linear functionals on vector spaces of continuous functions on the topological spaces. On the one hand, the integral with respect to a Radon measure on such a topological space is a positive linear functional on the vector space of continuous functions with compact support on the topological space. On the other hand, any abstract positive linear functional on such a function space is the integral with respect to a unique Radon measure. Similarly, there is a bijective correspondence between continuous functionals on spaces of continuous functions vanishing at infinity and complex Borel measures. In all cases, the integral offers a way to extend linear functionals to more general function spaces, namely the various L^p-spaces.

In the last part of the appendix we discuss holomorphic functional calculus. We introduce the needed background material from complex function theory in the more general context of Banach space valued holomorphic functions. As an example we look at simple applications to linear algebra and differential equations. Finally, we include some standard theorems of Carleson, Runge and Phragmen-Lindelöf needed in modular theory.

Recommended literature for this chapter is [18, 42, 53].

A.1 Set Theoretic Preliminaries

We recall here a few things from naive set theory.

A set X is given in terms of its members, and we write $x \in X$ to indicate that x is a *member* or an *element* of X. Two sets are equal if their members are the same.

Sets can be indicated by listing their members in brackets, like $\{x, y, z\}$. We write $\{x \mid P\}$ for the set of all elements x with property P. Attention should be made to avoid self-referring statements, like the set of all sets, which is meaningless, and the set that is not a member of itself; a notorious statement known as Russel's paradox.

We are allowed to form various sets from other ones. The *union* $\cup_i X_i$ and *intersection* $\cap_i X_i$ of any collection of sets consists of those elements that belong to at least one of the sets, respectively, to each one of them. A *subset* $Y \subset X$ of a set X is a set Y with members only from X, and its *complement* $X \backslash Y$ consists of those elements in X that do not belong to Y.

The useful *deMorgan's laws* are the easily proved statements that

$$\cap_i (X\backslash X_i) = X\backslash \cup_i X_i \quad \text{and} \quad \cup_i (X\backslash X_i) = X\backslash \cap_i X_i$$

for subsets X_i of a set X.

The *(Cartesian) product* $X \times Y$ of two sets X and Y consists of all *ordered pairs* (x, y) with $x \in X$ and $y \in Y$. By construction (x, y) is the subset $\{\{x\}, \{x, y\}\}$ of $X \cup Y$. Two ordered pairs coincide $(x, y) = (x', y')$ if and only if $x = x'$ and $y = y'$. To see this, suppose $\{\{x\}, \{x, y\}\} = \{\{x'\}, \{x', y'\}\}$. Either $\{x\} = \{x, y\} = \{x'\}$, and then all elements are equal, or $\{x\} = \{x'\}$ and $\{x, y\} = \{x', y'\}$, and then $x = x'$ and $y = y'$, or there are two other similar alternatives, and in these cases also $x = x'$ and $y = y'$.

By a *relation* on a set X, we mean any subset R of $X \times X$, and we write $x \sim y$ for $(x, y) \in R$.

A *function* or a *map* $f: X \to Y$ from X to Y is a relation $X \times Z$ on $X \cup Y$ with $Z \subset Y$ such that there is exactly one element $(x, y) \in X \times Z$ for each $x \in X$, and we then write $y = f(x)$. If there is only one such $x \in X$ to each $y \in Z$, then we say that f is *injective*, and f is *surjective* if $Z = Y$, and if it is both injective and surjective, then it is *bijective*. The relation $X \times Z$ is also called the *graph* of f with *domain* X and *image* Z.

The *composition* of $f: X \to Y$ and $g: Y \to Z$ is the function $g \circ f: X \to Z$ given by $g \circ f(x) = g(f(x))$. We sometimes write gf for $g \circ f$. Note that $h(gf) = (hg)f$ for $h: Z \to W$, so we often skip parentheses and write hgf for $(hg)f$.

If $Z = X$ and gf equals the identity map $\iota: X \to X$, then clearly g is surjective and f is injective, so if also $fg = \iota$, now with ι the identity map on Y, then both f and g are bijective. Also, one map is uniquely determined by the other because if also $g'f = \iota$ and $fg' = \iota$, then

$$g' = g'\iota = g'(fg) = (g'f)g = \iota g = g$$

and visa-versa. We say that g is the *inverse* map of f and write f^{-1} for g. Thus $(f^{-1})^{-1} = f$. In this uniqueness argument we only used $fg = \iota$ and $g'f = \iota$.

If f is bijective, then it has an inverse, namely the map g given by $g(f(x)) = x$. This definition makes sense firstly because any element of Y is of the form $f(x)$ for some $x \in X$ as f is surjective, and secondly because if $f(x) = f(y)$ then $x = y$ by injectivity of f, and thus $g(f(x)) = g(f(y))$. Hence $gf = \iota$, and g is obviously bijective, so by the same argument with the roles of f and g swapped, we also get $gf = \iota$. So f is invertible with inverse g.

A map on a finite set is injective if and only if it is surjective. Indeed, let f be a map on a finite set, and assume f is injective. To hit an element x, apply f to x repeatedly till repetitions $f^m(x) = f^n(x)$ occur. Then peal off f's till $x = f(f^k(x))$ for some k. Conversely, if f is not injective, then its image will contain too few elements for it to be surjective.

A *binary operation* on a set X is a map $X \times X \to X$, and one often writes xy or $x \cdot y$ or $x + y$, etc. for the image of (x, y) depending on context and what further properties the binary operation might have.

If a bijective map $f: X \to Y$ preserves binary operations on X and Y, say $f(x + y) = f(x) \cdot f(y)$, then so will its inverse because

$$f^{-1}(f(x) \cdot f(y)) = f^{-1}f(x + y) = x + y = f^{-1}(f(x)) + f^{-1}(f(y)),$$

and this is true with even more general (algebraic) operations on X. Any bijective map that preserves all the relevant operations on the collection of sets under consideration is called an *isomorphism*. Often one specifies with an adjective under what operations the map is an isomorphism. Maps that preserve the operations without being necessary bijective are often called *morphisms* or *homomorphisms*.

The *product* $\prod_{i \in I} X_i$ of sets $\{X_i\}$ over any (index) set I consists of all functions $f: I \to \cup X_i$ with $f(i) \in X_i$ for all $i \in I$. We write X^I for $\prod_{i \in I} X_i$ when all $X_i = X$. When $I = \{1, \ldots, n\}$ we write X^n for X^I, so X^n consists of all *n-tuples* $(x_1, \ldots, x_n)$ with $x_i \in X$.

We denote the set $\{1, 2, 3, \ldots\}$ of natural numbers by $\mathbb{N}$.

A *sequence* $\{x_n\}$ of elements $x_n \in X$ is a function $f: \mathbb{N} \to X$ with $x_n = f(n)$, or in other words, we have $\{x_n\} = f \in \prod_{n \in \mathbb{N}} X_n$ with $X_n = X$ for all n.

Say we have a function $f: X \to Y$ and subsets $A \subset X$ and $B \subset Y$. Then the *image* $f(A)$ of A and the *inverse image* $f^{-1}(B)$ of B are defined as .

$$f(A) = \{f(x) \mid x \in A\} \text{ and } f^{-1}(B) = \{x \in X \mid f(x) \in B\}.$$

The *power set* $P(X)$ of X is the set of all subsets of X. The *characteristic function* of $Y \subset X$ is the function $\chi_Y: X \to \{0, 1\}$ such that $\chi_Y(x) = 1$ if $x \in Y$ and $\chi_Y(x) = 0$ if $x \notin Y$.

Note that the map which sends a subset of X to its characteristic function is a bijection from $P(X)$ to $\{0, 1\}^X$. So the number of elements in $P(\{1, \ldots, n\})$ equals 2^n, hence the terminology *power set*.

A relation $\sim$ on a set X is called an *equivalence relation* if it is *reflexive*, $x \sim x$, *symmetric*, $x \sim y \Leftrightarrow y \sim x$, and *transitive*, $(x \sim y) \wedge (y \sim z) \Rightarrow x \sim z$. One can then form the *quotient set* $X/\sim$ of equivalence classes, and the *equivalence class of* $y \in X$ is the subset $\{x \in X \mid x \sim y\}$. Because $\sim$ is an equivalence relation, the quotient set is a *partition* of X, meaning that X is a disjoint (e.g. pairwise non-intersecting) union of equivalence classes.

To explain why every element of X belongs to exactly one equivalence class, first notice that due to reflexivity, any $x \in X$ belongs to its own equivalence class. And if x belongs to another equivalence class, say to that of an element y, then $x \sim y$, and any element z in the equivalence class of x, will because of transitivity, also belong to the equivalence class of y, so the equivalence class of x will be contained in the one of y. But by symmetry, we see that we also have inclusion the other way, so x belongs to only one class.

In fact, it is easy to see that to any partition of a set X, there is a unique equivalence relation $\sim$ having $X/\sim$ as the partition; just define $\sim$ by $x \sim y$ if x and y belong to the same block of the partition.

By an *order* $>$ on a set X we mean any relation that is transitive and such that for any $x, y \in X$ exactly one of the statements $x > y, x = y, y > x$ holds. Then $x \geq y$, meaning $x > y$ or $x = y$, defines a *partial order* $\geq$ on X, that is, a relation which is transitive, reflexive and *antisymmetric*, $(x \geq y) \wedge (y \geq x) \Rightarrow x = y$. Conversely, any partial order where all elements are pairwise comparable, i.e. either $x \geq y$ or $y \geq x$, defines an order $>$ with $x > y$ if $x \geq y$ and $x \neq y$.

A *chain* in a partially ordered set is any subset of pairwise comparable elements. This is a notion that plays an important role in *Zorn's lemma*:

AXIOM A.1.1 If every chain in a partially ordered non-empty set S has an upper bound, then S has a maximal element, i.e. with no elements superseding it.

This axiom is equivalent to the *axiom of choice*:

AXIOM A.1.2 To every collection of non-empty sets, there is a function that chooses exactly one element from each set.

Or equivalently, the product of any collection of non-empty sets is non-empty, containing at least one *choice function*.

The axiom of choice is again equivalent to Cantor's *well-ordering principle*:

AXIOM A.1.3 Every non-empty set can be endowed with a partial order for which it is well-ordered.

Clearly, a choice function would then be one that picks out the minimal element in each non-empty set. The converse direction is harder to prove, and normally goes via Zorn's lemma.

Despite the controversy around the axiom of choice, due to e.g. the non-intuitive requirement that the set $\mathbb{R}$ of real numbers can be well-ordered, we accept it as a set theoretic axiom along with the others, and these ones are commonly agreed upon to be those formulated by Zermelo-Frankel. The axiom of choice is independent of the ZF-axioms provided these are consistent, so neither the claim nor its negation can be proved from potentially consistent ZF-axioms.

A.2 Cardinality and Bases of Vector Spaces

Two sets have the same *cardinality* if there is a bijection between then, and a set is *countable* if it has the same cardinality as $\mathbb{N}$.

Proposition A.2.1 *If there is a surjection from one set to another, then there exists an injection in the other direction.*

Proof Say $f: X \to Y$ is a surjection. Then any choice function $g: Y \to X$ that picks an element $g(y) \in f^{-1}(\{y\})$ for every $y \in Y$ will obviously be injective. □

A host of injections are provided by the following result.

Proposition A.2.2 *Given any two non-empty sets, there is at least one injection between them.*

Proof Consider non-empty sets X and Y. Let $\mathcal{F}$ be the family of all injections from any subset of X to Y. Then $\mathcal{F}$ is non-empty because for any elements $x \in X$ and $y \in Y$, it contains the function $\{x\} \to Y$ which sends x to y. Now partially order $\mathcal{F}$ by saying that $f \leq g$ for injections $f: A \to Y$ and $g: B \to Y$ with A, B subsets of X, if $A \subset B$ and $g(x) = f(x)$ for $x \in A$, so g is an *extension* of f.

Every chain $\{f_i: A_i \to Y\}$ in $\mathcal{F}$ has an upper bound, namely the function $f: \cup A_i \to Y$ defined to be $f(x) = f_i(x)$ for $x \in A_i$. It is well-defined because if also $x \in A_j$ for some j, then either $A_i \subset A_j$ or $A_j \subset A_i$, and in both cases $f_i(x) = f_j(x)$. Also f is injective because if $x \neq x'$, with $x \in A_i$ and $x' \in A_j$, then again both $x, x' \in A_i$, in which case $f_i(x) \neq f_i(x')$ as f_i is injective, or both $x, x' \in A_j$, and then $f_j(x) \neq f_j(x')$. So $f \in \mathcal{F}$ and it is obviously an upper bound for the chain.

By Zorn's lemma there is a maximal element $g: B \to Y$ in $\mathcal{F}$. Now either $B = X$, and we are done, or g is surjective, and then by Proposition A.2.1, we are done. The last option is that g is not surjective and $B \neq X$. But then we can pick $x \in X \backslash B$ and $y \in Y \backslash g(B)$, and define a function $h: B \cup \{x\} \to Y$ by $h(x) = y$ and $h = g$ on B. Clearly $h \in \mathcal{F}$ and $g < h$, which contradicts maximality of g. □

The following result says that you can partition an infinite set in blocks that are countable, and not just at most countable.

Lemma A.2.3 *Any infinite set can be written as a disjoint union of countable subsets.*

Proof Say X is an infinite set. By Proposition A.2.2 either there is an injection from $\mathbb{N}$ to X, or there is an injection the other way, but then as X is infinite, it has to be countable. So in either case there is an injection $\mathbb{N} \to X$.

This shows that the family $\mathcal{F}$ of all disjoint countable subsets of X is non-empty, and we can partially order $\mathcal{F}$ by inclusion. Also every chain in $\mathcal{F}$ has an upper bound, namely the collection of all subsets of the chain, because any pair of subsets in this collection will belong to a common subcollection which requires them to be disjoint.

By Zorn's lemma $\mathcal{F}$ has a maximal element C of disjoint countable subsets of X. If the union Y of the members of the collection C is not the whole of X, there are two options. Either $X \backslash Y$ is finite, in which case these finitely many elements can be joined to a member of C, and we are done. Or $X \backslash Y$ is infinite. But then as in the first paragraph of this proof, we can get a copy Z of $\mathbb{N}$ inside $X \backslash Y$, and the collection $C \cup \{Z\}$ will be disjoint and strictly greater than C, which is impossible. □

Proposition A.2.4 *For any infinite set X we have $|X \times \mathbb{N}| = |X|$.*

Proof By Lemma A.2.3 we can write X as a disjoint union $\cup X_i$ of countable subsets X_i of X. Then $X \times \mathbb{N}$ equals the disjoint union $\cup(X_i \times \mathbb{N})$.

But clearly $|X_i \times \mathbb{N}| = |X_i|$, so by the axiom of choice there is a family $\{f_i: X_i \to X_i \times \mathbb{N}\}$ of bijections. Using this family we can define a function

$f: X \to X \times \mathbb{N}$ between the disjoint union of the X_i's and that of the $X_i \times \mathbb{N}$'s by $f(x) = f_i(x)$ for $x \in X_i$, and this f is obviously bijective. □

A *linear basis* of a vector space V over a field, is a set $S \subset V$ for which every vector can be written uniquely as a (finite) linear combination of elements in S. The subspace of all linear combinations of a subset of a vector space is the *span* of the subset.

Theorem A.2.5 *Every non-trivial vector space has a linear basis; any linear independent subset S of a vector space V can be enlarged to a basis S' for V, and by enlarged we mean $S \subset S'$.*

Proof The first assertion follows from the second because any subset consisting of a non-zero element v of V is linear independent; otherwise $av = 0$ for $a \neq 0$, and then $0 = a^{-1}0 = a^{-1}(av) = (a^{-1}a)v = 1v = v \neq 0$, which is absurd.

To prove the second claim, let $\mathcal{F}$ be the family of all linear independent subsets of V that contain S. This family is non-empty because S belongs to it. Order $\mathcal{F}$ by inclusion. Then every chain has an upper bound, namely the union of all members of the chain. By Zorn's lemma $\mathcal{F}$ has a maximal element S', which we claim is a basis for V.

We need only show that S' spans V. If not, pick any $v \in V$ outside the span. Then $S' \cup \{v\}$ is linear independent. To convince ourselves of this, say $S' = \{v_i\}$, and suppose that $\sum a_i v_i + av = 0$ for some scalars a_i and a. If $a \neq 0$, then

$$v = \sum(-a^{-1}a_i)v_i$$

is in the span of S', which is impossible. So $a = 0$, and then $\sum a_i v_i = 0$ forces all the a_i's to be 0 as the v_i's are linear independent.

So $S' \cup \{v\}$ is linear independent and obviously contains S, so it belongs to $\mathcal{F}$. But it is strictly larger than S' as $v \notin S'$, contradicting the maximality of S'. □

Theorem A.2.6 *Any two bases in a vector space have the same cardinality.*

Proof Say we have bases $\{v_i\}_{i\in I}$ and $\{w_j\}_{j\in J}$ of a vector space V. Aiming for a contradiction, we may assume by Proposition A.2.2 that we have an injection $J \to I$ which is not surjective. We have to distinguish two cases.

Suppose first that I is infinite. By assumption any y_j is a linear combination

$$w_j = \sum_{i\in F_j} a_{ij} v_i$$

of the v's, where F_j is a finite subset of I.

By the axiom of choice we have a family $\{f_j\}_{j\in J}$ of injections $f_j : F_j \to \mathbb{N}$ as each F_j is finite. Hence we have an injection

$$\bigcup_{j\in J} F_j \to J \times \mathbb{N},$$

which sends $i \in F_j$ to $(j, f_j(i))$. Together with Proposition A.2.4, this shows that the cardinality of $\cup F_j$ is less than that of I. So there is a $k \in I$ that does not belong to F_j for any $j \in J$.

But by assumption it must be possible to write v_k as a linear combination of the w_j's, and each of these w_j's can be written as a sum of the form

$$\sum_{i \in F_j} a_{ij} v_i$$

and k does not belong to F_j for any $j \in J$. So v_k is a linear combination of the v_i's, none of which can be v_k, and this shows that $\{v_i\}_{i \in I}$ is linear dependent; a contradiction.

Next suppose that I is finite, say with the same cardinality as $\{1, \ldots, n\}$, and say J has the same cardinality as $\{1, \ldots, m\}$, so $m < n$. Then

$$\{v_1, w_1, w_2, \ldots, w_m\}$$

spans V as $\{w_i\}$ already spans V. But $v_1 \neq 0$ because $\{v_i\}$ is linear independent. Writing v_1 as a linear combination of the w's, then at least one of the coefficients is non-zero, and the corresponding w can therefore be written as a linear combination of v_1 and the other w's. Removing this w we still get vectors that span V. Upon renumbering the indices of the w's, we therefore get a list

$$v_1, w_1, \cdots, w_{m-1}$$

of m vectors spanning V.

Next, the list with v_2 included will certainly also span V. Writing $v_2 \neq 0$ as a non-trivial linear combination of v_1 and the $m - 1$ new w's, then not all the coefficients of the w's can be 0 as otherwise v_2 would be a rescaling of v_1, and $\{v_i\}$ is linear independent. Removing any w with non-trivial coefficient, we get m vectors that still span V, say

$$v_1, v_2, w_1, \cdots, w_{m-2},$$

where again we have renumbered the indices of the w's.

We can continue this way and inductively remove w's, till we get a list

$$v_1, \ldots, v_n, w_1, \cdots, w_{m-n}$$

of m vectors that still span V. In particular, we see that $n \leq m$, which is a contradiction. □

We now know that a basis exists, and that any two of them have the same cardinality, so the following definition makes sense.

Definition A.2.7 The *dimension* dim V of a vector space V is the cardinality of any basis for it. If this cardinality is the same as that of $\{1, \ldots, n\}$, we talk about a *finite dimensional* vector space of dimension n. Otherwise we say that the vector space is *infinite dimensional.*

A.3 Topology

We include some results from topology. Recall that a topology on a set X is any collection of subsets, called open subsets, stable under formation of unions and finite intersections, and with ϕ, X declared open. All unions of all balls in a metric space is a typical example of a topology; the one *induced* by the metric.

Definition A.3.1 Let X be a topological space, that is, a set with a given topology. A *neighborhood* of a point in X is an open subset that contains the point. We say X is *Hausdorff* if every pair of distinct points have disjoint neighborhoods. A subset of X is *closed* if its complement is open. The *closure* $\overline{Y}$ of a subset Y of X is the intersection of all closed subsets that contain Y, so it is the smallest closed set that contains Y, and Y is *dense* in $\overline{Y}$. We say X is *compact* if every *open cover*, i.e. a collection of open sets with union X, has a finite *subcover*, i.e. any subcollection with union X. Finally, we say X is *locally compact* if every point in X has a neighborhood with compact closure. We say X is σ-compact if it is a countable union of compact subsets with respect to the *relative topology*, i.e. an *open set of a subset* Z of X is the intersection of Z with an open set in X.

The *discrete topology* on a set X is the collection of all subsets, so points are closed and compact. Thus X with the discrete topology is locally compact Hausdorff, and it is σ-compact if and only if X is countable.

Note that a point belongs to the closure of a set if and only if every neighborhood of the point intersects the set non-trivially. Any closed subset A of a compact set B is compact since we may include A^c in any open cover of A to get a cover of B, and then we can remove A_c in a any finite subcover of B to obtain a finite subcover of A. In a Hausdorff space B is closed since for $x \in B^c$ and $y \in B$, we may pick pairs of disjoint neighborhoods A_y and B_y of x and y, respectively, and obtain a finite subcover $\{B_{y_i}\}$ of B. Then $\cap A_{y_i} \subset B^c$ is a neighborhood of x, so B^c is open.

Definition A.3.2 A function $f : X \to Y$ between topological spaces is *continuous* if $f^{-1}(A)$ is open in X for every open $A \subset Y$, while it is *open* if $f(B)$ is open in Y for every open $B \subset X$. If f is a bijection that is both continuous and open, then it is a *homeomorphism*, and X and Y are *homeomorphic*, and we sometimes write $X \cong Y$, indicating that we are identifying X and Y as topological spaces. We say that f is *continuous at a point* $x \in X$ if for every neighborhood A of $f(x)$, we can find a neighborhood B of x such that $f(B) \subset A$.

Continuity at a point stems from the $\delta - \varepsilon$ definition of continuity of functions between metric spaces. It is easy to see that a function between topological spaces is continuous if and only if it is continuous at every point.

The *Heine-Borel theorem* says that the compact subsets of $\mathbb{R}^n$ with the topology given by the usual Euclidean metric are exactly those that are both closed and *bounded*, i.e. with M such that $\|x\| < M$ for all x. The forward direction is clear since we can cover the compact set with finitely many open balls of radius 1. For the converse, we may assume that the subset is a closed n-cube. If an open covering of this cube has no finite subcover, then by halving sides of cubes we get a sequence of smaller cubes contained in each other, each having no finite subcover. The centers of these cubes form a Cauchy sequence with limit x. Any neighborhood of x from the cover will obviously contain a small enough cube, which is a contradiction.

So $\mathbb{R}^n$ is locally compact Hausdorff and σ-compact. Another consequence of the Heine-Borel theorem is that a real continuous function on a compact set attains both a maximum and minimum. This is immediate from the fact that the continuous image of a compact set is compact, and this holds since for any open cover of the image, inverse images provide a cover of the domain.

A small variation on the proof of the Heine-Borel theorem shows that a metric space is compact if and only if it is complete and *totally bounded*, meaning that for any $r > 0$, the space can be covered by finitely many balls of radius r. So in a complete metric space the subsets with compact closures are precisely the totally bounded ones.

Any compact metric space is separable because for each $n \in \mathbb{N}$ we can cover it by finitely many balls $B_{1/n}(x_i)$, $i \in I_n$, so the sequence $\{x_i\}$ as $i \in \cup_n I_n$ is dense since some $x_i \in B_\varepsilon(x)$ when $1/n < \varepsilon/2$.

A topological space is *connected* if it is not the disjoint union of two non-empty open sets. A *connected component* of a topological space is the union of all subsets which are connected in the relative topology and contain a given point. The connected component is connected since any union of connected subsets with a point in common is clearly connected. The closure of a connected subset in a topological space is evidently connected, so connected components are closed in the space, and they obviously partition it into maximal connected subsets.

Continuous images of connected spaces are connected because inverse images of open subsets disconnecting an image would also disconnect the domain. The interval $[0, 1]$ is connected because if the open sets A and B disconnected it with say $1 \in B$, then every neighborhood of $a = \sup A$ would intersect both A and B. A *curve* between points x and y in a topological space is a continuous map f from $[0, 1]$ into the space such that $f(0) = x$ and $f(1) = y$. The space is *arcwise connected* if there is a curve between every pair of points. Arcwise connected spaces are connected because if the open sets A and B disconnected it and f was a curve from $x \in A$ to $y \in B$, then $f^{-1}(A)$ and $f^{-1}(B)$ would disconnect $[0, 1]$.

A function $f : X \to Y$ between metric spaces is *uniformly continuous* if there is $\delta > 0$ to $\varepsilon > 0$ such that $d(f(x), f(y)) < \varepsilon$ when $d(x, y) < \delta$. So the same δ works for all elements of X, which is not required when we merely talk about continuity at a point.

Proposition A.3.3 *Any continuous function from a compact metric space to any metric space is uniformly continuous.*

Proof Suppose $f : X \to Y$ is a continuous map from a compact metric space to a metric space, and let $\varepsilon > 0$. Since f is continuous at every $x \in X$, there is $\delta_x > 0$ such that $d(f(x), f(y)) < \varepsilon/2$ when $d(x, y) < \delta_x$. The balls

$$B_x \equiv \{y \in X \mid d(x, y) < \delta_x/2\}$$

form an open cover of X which has a finite subcover $\{B_{x_i}\}$. Let $\delta > 0$ be half the size of the smallest of these deltas δ_{x_i}. To x, y with $d(x, y) < \delta$ there is i such that $x \in B_{x_i}$, so $d(x, x_i) < \delta_{x_i}/2$. Hence

$$d(y, x_i) \leq d(y, x) + d(x, x_i) < \delta + \delta_{x_i}/2 \leq \delta_{x_i}$$

and

$$d(f(x), f(y)) \leq d(f(x), f(x_i)) + d(f(x_i), f(y)) < \varepsilon/2 + \varepsilon/2 = \varepsilon.$$

□

Definition A.3.4 The *support* supp(f) of a complex function f on a topological space is the closure of the subset of points where f is non-zero.

Let $C_c(X)$ denote the set of all continuous complex functions with compact support on a topological space X. It is an algebra under pointwise operations because if $f, g \in C_c(X)$, then the supports of $f + g$ and fg belong to the union and intersection, respectively, of the supports of f and g, and finite unions and intersections of compact closed sets are compact. For similar reasons the space $C_0(X)$ of continuous complex functions f on X *vanishing at infinity*, meaning that $|f(x)| < \varepsilon$ for all x in the complement of some compact closed set, is an algebra under pointwise operations, and clearly $C_c(X) \subset C_0(X)$.

Proposition A.3.5 *To every non-compact topological space X the disjoint union $X \cup \{\infty\}$ can be made a compact space containing X as a dense subset and with the topology on X as its relative topology. When X is a locally compact Hausdorff space, then $X \cup \{\infty\}$ is Hausdorff.*

Proof By a neighborhood of ∞ we mean a subset A of $Y \equiv X \cup \{\infty\}$ such that $\infty \in A$ and $Y \backslash A$ is compact and closed in X. Joining these to the open sets of X it is easily checked that one gets a topology on Y which induces the relative topology on X; closed subsets of compact sets are compact. Since X is not compact, every neighborhood of ∞ meets X, so X is dense in Y.

Any open cover of Y must contain a neighborhood A of ∞, and the remaining members cover the compact set $Y \backslash A$ from which we may pick a finite subcover, so Y is compact.

If X is locally compact Hausdorff, then pick a neighborhood A of $x \in X$ with compact closure. Then A and $Y \backslash \overline{A}$ are disjoint neighborhoods of x and ∞, so Y is Hausdorff.

□

We call $X \cup \{\infty\}$ the *one-point compactification* of the topological space X, and we always consider X as a subset with relative topology. Note that $f \in C_0(X)$ if and only if there is a continuous function g on the one-point compactification of X such that $g(\infty) = 0$ and its restriction to X is f.

Proposition A.3.6 *For any locally compact Hausdorff space X, the completion of $C_c(X)$ is $C_0(X)$ with respect to the* uniform norm $\|f\|_u = \sup_{x\in X} |f(x)|$.

Proof For $f \in C_0(X)$ and $\varepsilon > 0$ pick a compact set A such that $|f(x)| < \varepsilon$ outside A. By Urysohn's lemma below there is a function $g \in C_c(X)$ such that $0 \le g \le 1$ and $g(x) = 1$ on A. Then $fg \in C_c(X)$ and $\|f - fg\|_u < \varepsilon$, so $C_c(X)$ is dense in $C_0(X)$.

The pointwise limit f of a Cauchy sequence $\{f_n\}$ in $C_0(X)$ is continuous because to $\varepsilon > 0$ and $x \in X$, there is n and a neighborhood B of x such that $\|f - f_n\|_u < \varepsilon/3$ and $|f_n(y) - f_n(x)| < \varepsilon/3$ for all $y \in B$, and

$$|f(y)-f(x)| < |f(y)-f_n(y)|+|f_n(y)-f_n(x)|+|f_n(x)-f(x)| < \varepsilon/3+\varepsilon/3+\varepsilon/3.$$

But f_n converges to f in the uniform norm because we may pick N such that $\|f_m - f_n\|_u < \varepsilon/2$ for $n, m > N$, and if we to x also see to that $|f(x) - f_m(x)| < \varepsilon/2$, then

$$|f(x) - f_n(x)| \le |f(x) - f_m(x)| + |f_m(x) - f_n(x)| < \varepsilon.$$

As $|f_n|$ is less than ε outside some compact set E, we see that $|f|$ is less than 2ε outside E, so $f \in C_0(X)$ and $C_0(X)$ is a Banach space, even a Banach algebra. □

Lemma A.3.7 *Let E be a compact subset of an open subset A in a locally compact Hausdorff space. Then there is an open subset B with compact closure such that*

$$E \subset B \subset \bar{B} \subset A.$$

Proof We cover E by finitely many open sets $\{A_i\}$ with compact closures, so $C = \cup A_i$ has compact closure. If A is the whole space, let $B = C$.

For each $x \in A^c$ we claim there is an open set B_x that contains E with $x \notin \bar{B}_x$. Indeed, there are finitely many points in E with neighborhoods U_i disjoint from neighborhoods V_j of x such that $E \subset \cup U_i \equiv B_x$ and $\bar{B}_x \subset (\cap V_j)^c$.

Hence the collection $\{A^c \cap \bar{C} \cap \bar{B}_x\}$ as x ranges over A^c has empty intersection, and so has a finite subcollection with empty intersection, as is seen by going to complements in the definition of compactness. The intersection B of C and the corresponding finitely many B_x's has evidently the required properties. □

The following result, known as *Urysohn's lemma*, provides a rich supply of continuous functions.

Theorem A.3.8 *Let E be a compact subset of an open subset A in a locally compact Hausdorff space X. Then there exists a function $f \in C_c(X)$ with $0 \le f \le 1$ that is 1 on E and has support inside A.*

Proof By the lemma there is an open set A_1 with compact closure such that $E \subset A_1 \subset \bar{A}_1 \subset A$. Then there is an open set $A_{1/2}$ with compact closure such that $E \subset A_{1/2} \subset \bar{A}_{1/2} \subset A_1$. Again by the lemma there are open sets $A_{1/4}$ and $A_{3/4}$ such that

$$E \subset A_{1/4} \subset \bar{A}_{1/4} \subset A_{1/2} \subset \bar{A}_{1/2} \subset A_{3/4} \subset \bar{A}_{3/4} \subset A_1.$$

By induction we obtain for each $r = m2^{-n}$ with m any natural number in $[1, 2^n]$, an open set A_r containing E and such that $\bar{A}_r \subset A_s$ for $r < s$.

Define $g: X \to [0, 1]$ by letting $g(x) = 1$ for $x \notin \cup A_r = A_1$ and otherwise $g(x) = \inf\{r \mid x \in A_r\}$. Clearly g is 0 on E and 1 on A^c.

It remains to show that g is continuous. Since inverse images preserve unions, intersections and complements, and in view of the relative topology on $[0, 1]$, it suffices to show that $g^{-1}([0, t\rangle)$ is open and that $g^{-1}([0, s])$ is closed for every s and t with $0 \leq s < t \leq 1$. But $g(x) < t$ precisely when $x \in A_r$ for some $r < t$. Thus $g^{-1}([0, t\rangle) = \cup_{r<t} A_r$, which is indeed open. Next observe that $g(x) \leq s$ precisely when for each $r > s$ the element x belongs to A_p for some $p < r$, so

$$g^{-1}([0, s]) = \cap_{r>s} \cup_{p<r} A_p.$$

The latter expression is certainly contained in $\cap_{q>s} \bar{A}_q$, but we also have inclusion the other way because if $x \in \bar{A}_q$ for every $q > s$, then for every $r > s$ we can by construction find p and q with $r > p > q > s$, and then $x \in \bar{A}_q \subset A_p$. Thus

$$g^{-1}([0, s]) = \cap_{q>s} \bar{A}_q,$$

which is closed. Now set $f = 1 - g$. □

Proposition A.3.9 *If E is a compact set contained in a finite union of open sets V_i in a locally compact Hausdorff space, then there are continuous functions h_i with range in* $[0, 1]$ *and compact support inside V_i and such that $\sum h_i = 1$ on E.*

Proof By the lemma above and compactness of E, there are finitely many points $x_{ij} \in E$ having neighborhoods U_{ij} with compact $\overline{U}_{ij} \subset V_i$ that cover E. Let $B_i = \cup \overline{U}_{ij}$. By the theorem above pick continuous functions $0 \leq g_i \leq 1$ that are one on B_i and have support in V_i. Set $h_i = (1 - g_1)(1 - g_2) \cdots (1 - g_{i-1}) g_i$. □

Such a collection of functions h_i is called a *partition of unity on E associated to the open sets V_i*.

The following result by Arzela-Ascoli is useful.

Theorem A.3.10 *Let X be a compact space. Suppose we have $M \subset C(X)$ such that* $\sup_{f\in M} |f(x)| < \infty$ *for every $x \in X$ and that M is* equicontinuous, *i.e. that every x has a neighborhood A such that $|f(y) - f(x)| < \varepsilon$ for $y \in A$ and $f \in M$. Then M is totally bounded, and hence has compact closure, in the uniform norm.*

Proof Pick $x_1, \ldots, x_n \in X$ with neighborhoods V_i so that $|f(x) - f(x_i)| < \varepsilon$ for $x \in V_i$ and $f \in M$. Hence $\sup_{f \in M} \|f\|_u = c < \infty$. Let $D = \{a \in \mathbb{C} \mid |a| \leq c\}$ and define $p(f) = (f(x_1), \ldots, f(x_n)) \in D^n$ for $f \in M$. As D^n is a finite union of sets of diameter ε, there are $f_1, \ldots f_m$ in M such that every $p(f)$ lies within ε of some $p(f_k)$. Hence the balls $B_{3\varepsilon}(f_k)$ cover M. □

Definition A.3.11 A function $f\colon X \to \langle -\infty, \infty]$ on a topological space is *lower semicontinuous* if $f^{-1}(\langle a, \infty])$ is open for all $a \in \mathbb{R}$. It is *upper semicontinuous* if instead $f^{-1}(\langle -\infty, a\rangle)$ are open.

A real function is continuous if and only if it is both upper and lower semicontinuous. The characteristic function of an open (closed) set is lower (upper) semicontinuous. The sum and supremum (infinum) of lower (upper) semicontinuous functions are easily seen to be lower (upper) semicontinuous. If f is a non-negative lower semicontinuous function on a locally compact Hausdorff space X, then $f(x)$ is the supremum of $g(x)$ over all $g \in C_c(X)$ with $0 \leq g \leq f$. Indeed, for $a \in \langle 0, f(x)\rangle$ pick by Urysohn's lemma a continuous function g with range in $[0, a]$ that is a on x and has compact support in the open set $f^{-1}(\langle a, \infty])$.

A.4 Nets and Induced Topologies

We would like to talk about convergence in general topological spaces.

Definition A.4.1 An *upward filtered (ordered) set* is a partially ordered set I such that to each pair $i, j \in I$ there is $k \in I$ with $k \geq i$ and $k \geq j$. A *net* $\{x_i\}$ in a set X is a function from an upward filtered ordered set I to X which sends i to x_i. The net is *frequently* in a subset Y if for each i there is some $j \geq i$ such that $x_j \in Y$. It is *eventually* in Y if there is some i such that $x_j \in Y$ for all $j \geq i$. In a topological space a net *converges* to x if it eventually is in every neighborhood of x, and we then write $x_i \to x$ or $x = \lim x_i$. A point x in a topological space is an *accumulation point* of a net if the net is frequently in every neighborhood of x.

Nets are obvious generalizations of sequences and allow for index sets consisting of subsets. Indeed, the power set $P(X)$ is upward filtered $A \geq B$ under reverse inclusion $A \subset B$ because $A \cap B \geq A$ and $A \cap B \geq B$. For a topological space this order restricts to an upward filtered order on the neighborhoods of a fixed point.

The next result shows that topologies are determined by their families of convergent nets.

Proposition A.4.2 *A point in a topological space belongs to the closure of a subset if and only if there is a net in the subset that converges to the point.*

Proof If a net in the subset converges to a point, then it eventually is in every neighborhood of the point, so the point belongs to the closure of the subset.

Conversely, if $x \in \bar{X}$, then every neighborhood A of x intersects X. By the axiom of choice, pick a member x_A in each intersection $A \cap X$. Consider the neighborhoods of x as an upward filtered ordered set under reverse inclusion. Clearly the net $\{x_A\}$ converges to x. □

A net can converge to several points.

Proposition A.4.3 *A topological space is Hausdorff if and only if each net converges to at most one point.*

Proof If the space is Hausdorff, we can separate two distinct points by disjoint neighborhoods. A net cannot eventually be in both of these.

Conversely, say $x \neq y$ cannot be separated by disjoint neighborhoods. By the axiom of choice we can then for any pair of neighborhoods A and B of x and y, respectively, pick $x_{(A,B)} \in A \cap B$. Consider the index set of all such pairs with the upward filtered order:

$$(A, B) \geq (A', B') \text{ if } A \subset A' \text{ and } B \subset B'.$$

Then $\{x_{(A,B)}\}$ is a net that converges to both x and y, a contradiction. □

Proposition A.4.4 *A function $f: X \to Y$ between topological spaces is continuous at $x \in X$ if and only if $f(x_i) \to f(x)$ for any net such that $x_i \to x$.*

Proof Suppose $x_i \to x$ and that B is a neighborhood of $f(x)$. If f is continuous, then $f^{-1}(B)$ is a neighborhood of x, so $\{x_i\}$ will eventually be in $f^{-1}(B)$. But then the net $\{f(x_i)\}$ will eventually be in B.

Conversely, suppose that B is a neighborhood of $f(x)$ such that every neighborhood A of x intersects $X\backslash f^{-1}(B)$. Then x belongs to the closure of $X\backslash f^{-1}(B)$. By Proposition A.4.2 there is a net $\{x_i\}$ in $X\backslash f^{-1}(B)$ that converges to x. Since $f(x_i) \notin B$ for any i, we cannot have $f(x_i) \to f(x)$. □

A function $f: X \to Y$ between metric spaces is continuous if and only if $f(x_n) \to f(x)$ for any sequence in X such that $x_n \to x$. This is clear from the proof of this proposition and of Proposition A.4.2, where neighborhoods should be replaced by open balls with radii $1/n$.

Definition A.4.5 A *subnet* of a net $f: I \to X$ in X is a net $g: J \to X$ and a map $h: J \to I$ with $g = f \circ h$ such that for every $i \in I$, there is an element $j \in J$ with $h(j') \geq i$ for every $j' \geq j$.

Clearly a point is an accumulation point of a net if some subnet converges to it. The converse is also true; all accumulation points occur this way. This follows as a corollary of the following slightly more flexible result needed later.

Proposition A.4.6 *Suppose $\mathcal{F}$ is a family of subsets of a set X that is upward filtered under reverse inclusion. Then any net in X that is frequently in every member of $\mathcal{F}$ has a subnet that is eventually in every member of $\mathcal{F}$.*

Proof Say $f: I \to X$ is a net that is frequently in every member of $\mathcal{F}$. Then the set

$$J \equiv \{(i, A) \in I \times \mathcal{F} \mid x_i \in A\}$$

is an upward filtered set under the partial order:

$$(i, A) \geq (j, B) \text{ if } i \geq j \text{ and } A \subset B.$$

This is so because for every (i, A) and (j, B) in J, there is $k \in I$ with $k \geq i$ and $k \geq j$ and $C \in \mathcal{F}$ contained in both A and B such that $x_k \in C$, as the net is frequently in C. Hence $(k, C) \in J$ and $(k, C) \geq (i, A)$ and $(k, C) \geq (j, B)$.

Then $g: J \to X$ given by $(i, A) \mapsto x_i$ together with the map $h: J \to I$ given by $h(i, A) = i$ is a subnet of f. Indeed, clearly $g = f \circ h$, and if $i \in I$, we can pick any $A \in \mathcal{F}$ and some $j \geq i$ such that $x_j \in A$ because the net is frequently in every member of $\mathcal{F}$. Hence $(j, A) \in J$ and $h(j, A) \geq i$, and since h preserves partial order, we have a subnet.

This subnet is eventually in every $A \in \mathcal{F}$ because as the net is frequently in every member of $\mathcal{F}$, plainly there exists $x_i \in A$, so $(i, A) \in J$, and if $(j, B) \geq (i, A)$, we see that $g(j, B) = x_j \in B \subset A$. □

We now apply this proposition to the collection $\mathcal{F}$ of all neighborhoods of a fixed point in a topological space.

Corollary A.4.7 *To every accumulation point of a net in a topological space there is a subnet that converges to the point.*

Not all subnets of sequences are necessarily subsequences, so sequences may not have convergent subsequences even when they have accumulation points.

Proposition A.4.8 *Any infinite subset of a compact space has a* limit point, *that is, a point x such that all its neighborhoods will contain points from the subset other than x.*

Proof Suppose an infinite subset E of a compact space X has no limit points. Then for each point of X we can find a neighborhood of it that contains at most one element from E. We get then an open cover of X that has no finite subcover. □

Theorem A.4.9 *A space is compact if and only if every net has a convergent subnet.*

Proof Suppose we have a net $\{x_i\}$ in a compact space. Then the closures F_i of the tails $\{x_j \mid j \geq i\}$ will be a family with non-empty finite intersections, so there is an element x in the intersection of all the members of this family. Every neighborhood A of x will therefore intersect $\{x_j \mid j \geq i\}$, so there is $j \geq i$ with $x_j \in A$, and x is an accumulation point of the net. By Corollary A.4.7 the net therefore has a convergent subnet.

Conversely, suppose we have an open cover of a space X, and that every net in X has a convergent subnet. The collection $\mathcal{F}$ of finite subsets of this open cover is an upward filtered set when ordered by inclusion. If no $i \in \mathcal{F}$ covers X, then by the axiom of choice, we can pick

$$x_i \in X \setminus \bigcup_{A \in i} A = \bigcap_{A \in i} (X \setminus A)$$

for every i. By assumption the net $\{x_i\}$ has a convergent subnet and hence an accumulation point $x \in X$. For any given A in the original open cover and any neighborhood B of x, there is therefore an element i such that $i \geq \{A\} \in \mathcal{F}$ with $x_i \in B$. But $i \geq \{A\}$ means that $A \in i$, and thus $x_i \in X\backslash A$. So $(X\backslash A) \cap B \neq \phi$. Hence $x \in X\backslash A$ for every A in the open cover of X, which is impossible. So some member of $\mathcal{F}$ must cover X, which is then a finite subcover. □

Definition A.4.10 A net in a set is *universal* if it is eventually in any subset or its complement.

A universal net in a topological space will therefore converge to all its accumulation points.

Theorem A.4.11 *A topological space is compact if and only if every universal net in it converges.*

This theorem is immediate from Theorem A.4.9 and the following non-trivial lemma, which shows that there are many universal nets beyond those that are eventually constant.

Lemma A.4.12 *Every net has a universal subnet.*

Proof By a *filter of a net* $\{x_i\}$ in a set X we mean a family of non-empty subsets of X closed under finite intersections, containing all subsets containing any member of the family, and such that the net is frequently in every member of the family. Since $\{X\}$ is such a family, the collection of all filters of the net is non-empty. We partially order this collection by inclusion. Then every chain $\{\mathcal{F}_i\}$ of filters has an upper bound, namely their union, and thus by Zorn's lemma, there is a maximal filter $\mathcal{F}$ of the net.

Let Y be any subset of X. We claim that either $Y \in \mathcal{F}$ or $X\backslash Y \in \mathcal{F}$. If for some i and $A, B \in \mathcal{F}$, both $x_j \notin A \cap Y$ and $x_j \notin B\backslash Y$ for all $j \geq i$, then the same would obviously hold with A and B replaced by the smaller set $A \cap B$. But the set theoretic identity

$$A \cap B = (A \cap B \cap Y) \cup (A \cap B\backslash Y)$$

shows then that the net cannot be frequently in $A \cap B \in \mathcal{F}$, which should hold as $\mathcal{F}$ is a filter of the net. Hence the net is frequently in $A \cap Y$ for every $A \in \mathcal{F}$, or it is frequently in $B\backslash Y$ for every $B \in \mathcal{F}$. If the first case holds, the family

$$\{Z \supset A \cap Y \mid A \in \mathcal{F}\}$$

is a filter of the net that includes $\mathcal{F}$ as $A \supset A \cap Y$ for any $A \in \mathcal{F}$. Since $\mathcal{F}$ is maximal, we must therefore have equality, so $Y \in \mathcal{F}$ as $Y \supset A \cap Y$ for any $A \in \mathcal{F}$. If the second case holds, we get $X\backslash Y \in \mathcal{F}$ by a similar argument.

Apply Proposition A.4.6 to our net $\{x_i\}$ and to this maximal $\mathcal{F}$ to get a subnet in X that is universal. This must be so because with any $Y \subset X$, either $Y \in \mathcal{F}$ and then $\{x_i\}$ is frequently in Y, forcing the subnet to be eventually in Y, or $X\backslash Y \in \mathcal{F}$ and then the subnet is similarly eventually in $X\backslash Y$. □

Given any family of subsets of a set X, the *weakest topology* that contains this family is the intersection of all topologies on X (including the discrete one) that contains the family. This is a topology, which consists of ϕ and X and all unions of finite intersections of members from the family, because this collection is stable under unions and finite intersections. Also it is clear that no strictly weaker topology, i.e. with less open sets, can contain the family.

Definition A.4.13 The *initial topology* on X induced by a family of functions $f: X \to Y_f$ into topological spaces Y_f is the weakest topology on X making all these functions continuous.

The initial topology induced by such a family $\mathcal{F}$ has as open sets all unions of finite intersections of the sets $f^{-1}(A)$ with $f \in \mathcal{F}$ and A open in Y_f.

Proposition A.4.14 *Consider X with initial topology induced by a family $\mathcal{F}$ of functions. Then a net $\{x_i\}$ in X converges to x if and only if $f(x_i) \to f(x)$ for every $f \in \mathcal{F}$.*

Proof The forward implication is obvious.

For the opposite direction, let A be a neighborhood of x. By definition of the initial topology there are finitely many open sets $A_n \subset Y_{f_n}$ such that

$$x \in \bigcap f_n^{-1}(A_n) \subset A.$$

Hence A_n is a neighborhood of $f_n(x)$ for every n, and since $f_n \in \mathcal{F}$, by assumption $f_n(x_i) \to f_n(x)$. Thus $f_n(x_i) \in A_n$ when $i \geq i(n)$ for some $i(n)$. Pick j such that $j \geq i(n)$ for all the finitely many n's. Then $x_i \in \cap f_n^{-1}(A_n) \subset A$ for all $i \geq j$. □

Corollary A.4.15 *Consider X with initial topology induced by a family $\mathcal{F}$ of functions. Then a function $g: Z \to X$ on a topological space Z is continuous if and only if $f \circ g$ is continuous for all $f \in \mathcal{F}$.*

Proof The forward implication is obvious, and the opposite direction follows immediately from Propositions A.4.4 and A.4.14. □

The *product topology* on the direct product $\prod X_\lambda$ of topological spaces is the initial topology induced by the family of canonical projections π_λ. By Proposition A.4.14 a net $\{x_i\}$ in $\prod X_\lambda$ converges to x in the product topology if and only if $\pi_\lambda(x_i) \to \pi_\lambda(x)$ for all $\lambda \in I$. Note that $\cap_{\lambda \in J} \pi_\lambda^{-1}(A_\lambda)$ equals $\prod B_\lambda$ with $B_\lambda = A_\lambda$ for $\lambda \in J$ and $B_\lambda = X_\lambda$ for $\lambda \notin J$. The projections π_λ are both continuous and open as $\pi_\lambda(\prod B_\lambda)$ equals either A_λ or X_λ, and because maps preserve unions.

The technology of universal nets allows for a short proof of the following cornerstone in point set topology known as *Tychonoff's theorem.*

Theorem A.4.16 *Any product of compact spaces is compact in the product topology.*

Proof Suppose $\{x_i\}$ is a universal net in a product $\prod X_\lambda$ of compact spaces X_λ with projections π_λ. If $Y \subset X_\lambda$, then $\{x_i\}$ is eventually in $\pi_\lambda^{-1}(Y)$ or eventually in $X_\lambda \backslash \pi_\lambda^{-1}(Y)$. Thus $\{\pi_\lambda(x_i)\}$ is a universal net in X_λ. By Theorem A.4.11 it must converge to a point $x_\lambda \in X_\lambda$. Let $x\colon \lambda \mapsto x_\lambda$. Then $x \in \prod X_\lambda$ and $\pi_\lambda(x_i) \to \pi_\lambda(x)$ for every λ, so $x_i \to x$ in the product topology. But then again by Theorem A.4.11, we see that $\prod X_\lambda$ is compact. □

The axiom of choice is essential in this proof. Tychonoff's theorem is actually equivalent to it.

Proposition A.4.17 *Tychonoff's theorem implies the axiom of choice.*

Proof Given a family of sets X_i, consider the disjoint union $Y_i = X_i \cup \{i\}$ as a compact space with open sets all subsets having finite complement, together with the empty set and $\{i\}$. By Tychonoff's theorem the direct product $\prod Y_i$ is compact. Let $\pi_i\colon \prod Y_i \to Y_i$ be the canonical projections. For any finite subfamily $\{X_i\}_{i\in J}$ we have $f \in \cap_{i\in J}\pi_i^{-1}(X_i)$, where $f|J$ is any element of $\prod_{i\in J} X_i$ and $f(i) = i$ for $i \notin J$. The full intersection $\prod X_i = \cap\pi_i^{-1}(X_i)$ of closed sets is therefore non-empty, and any element of $\prod X_i$ is clearly a choice function for the family of sets X_i. □

Definition A.4.18 A family of functions on a set *separates points* $x \neq y$ in the set if $f(x) \neq f(y)$ for some member f of the family.

Proposition A.4.19 *Any space X with initial topology induced from a separating family of functions $f\colon X \to Y_f$ is Hausdorff whenever all Y_f are Hausdorff.*

Proof If $x \neq y$ in X, there exists a member f of the family that separates x and y. Since Y_f is Hausdorff, there are disjoint neighborhoods of $f(x)$ and $f(y)$, respectively. But then their inverse images under f are disjoint neighborhoods of x and y, respectively. □

Corollary A.4.20 *Any product of Hausdorff spaces is Hausdorff in the product topology.*

Proof The family of canonical projections separates points. □

Definition A.4.21 The *final topology* on a set Y induced by a family of functions $f\colon X_f \to Y$ on topological spaces X_f has as open sets all subsets A of Y such that $f^{-1}(A)$ is open in X_f for every member f of the family.

The final topology induced by a family of functions is indeed a topology, and it is the strongest possible for which all the members of the family are continuous.

Proposition A.4.22 *Consider Y with final topology induced by a family $\mathcal{F}$ of functions. Then a function $g\colon Y \to Z$ into a topological space Z is continuous if and only if $g \circ f$ is continuous for all $f \in \mathcal{F}$.*

Proof The forward implication is obvious.

If A is open in Z and $g \circ f$ is continuous for every $f \in \mathcal{F}$, then $f^{-1}(g^{-1}(A))$ is open in X_f for every $f \in \mathcal{F}$, which means that $g^{-1}(A)$ is open in Y. □

Example A.4.23 The *quotient topology* is the final topology on the quotient set $\tilde{X}$ induced by the quotient map $f : X \to \tilde{X}$. Although f is continuous, it is not always open. It is open precisely when

$$\tilde{A} \equiv \{x \in X \,|\, x \sim y \in A\}$$

is open in X for every open subset A of X, because $\tilde{A} = f^{-1}(f(A))$.

Also it can easily be verified that points in $\tilde{X}$ are closed if and only if each equivalence class is closed in X. ◇

A.5 The Stone-Weierstrass Theorem

We include here an important approximation result.

Lemma A.5.1 *Let A be a vector space of continuous real functions on a compact space X that contains the maximum and minimum of two vectors. Then every continuous real function on X that can be approximated from A in every pair of points can be approximated in the uniform norm.*

Proof Let f be a continuous real function that can be approximated in every pair $x, y \in X$. To each such pair and $\varepsilon > 0$ there is $f_{xy} \in A$ such that both x and y belong to the open sets U_{xy} and V_{xy} where $f < f_{xy} + \varepsilon$ and $f_{xy} < f + \varepsilon$, respectively. By compactness there are y_i such that $U_{xy_1}, \ldots, U_{xy_n}$ cover X. Then $f < f_x + \varepsilon$ for $f_x = \sup f_{xy_i} \in A$, and $f_x < f + \varepsilon$ on $W_x = \cap V_{xy_i}$. Now there are x_j such that $W_{x_1}, \ldots, W_{x_m}$ cover X, and $g = \inf f_{x_j} \in A$ satisfies $g - \varepsilon < f < g + \varepsilon$. □

Lemma A.5.2 *Any algebra A of continuous real functions on a topological space that is closed under the uniform norm is also closed under the maximum and minimum of any two vectors.*

Proof Considering power series we know there is a polynomial p on $[0, 1]$ such that $|(t + \varepsilon^2)^{1/2} - p(t)| < \varepsilon$ for every $t \in [0, 1]$. To $f \in A$ with $|f| \leq 1$ we have

$$\|p(f^2) - p(0) - |f|\|_u \leq \sup_t |p(t) - p(0) - t^{1/2}| \leq 3\varepsilon + \sup_t |(t + \varepsilon^2)^{1/2} - t^{1/2}| \leq 4\varepsilon,$$

so $|f| \in A$. Now we are done since

$$\max\{f, g\} = (f + g + |f - g|)/2 \quad \text{and} \quad \min\{f, g\} = (f + g - |f - g|)/2.$$

□

A space A of complex functions is *self-adjoint* if $\bar{f} \in A$ for $f \in A$. Since $f = (f + \bar{f})/2 + i(f - \bar{f})/2i$, this happens if and only if $A = A_s + i\,A_s$, where A_s is the set of real functions in A.

Theorem A.5.3 *Let A be a unital self-adjoint algebra of continuous complex functions on a compact space X such that to any $x \neq y$, there is $f \in A$ with $f(x) \neq f(y)$. Then the closure of A in the uniform norm consists of all continuous complex functions on X.*

Proof By the last lemma the self-adjoint part $\bar{A}_s$ of the closure of A in the uniform norm satisfies the assumptions in the first lemma. Given a continuous real function g on X and $x \neq y$. Pick $f \in A$ with $f(x) \neq f(y)$. Since $\mathrm{Re} f, \mathrm{Im} f \in \bar{A}_s$ there is $h \in \bar{A}_s$ such that $h(x) = g(x)$ and $h(y) = g(y)$, so $g \in \bar{A}_s$ by the first lemma. Finally observe that $\bar{A} = \bar{A}_s + i\bar{A}_s$. □

Corollary A.5.4 *Let X be a topological space, and let A be a self-adjoint subalgebra of $C_0(X)$ that separates points in X and has no point in X as a common zero. Then A is uniformly dense in $C_0(X)$.*

Proof Clearly $A+\mathbb{C}$ is a unital self-adjoint algebra of continuous complex functions on the one-point compactification $X \cup \{\infty\}$ of X, which by assumption separates points in $X \cup \{\infty\}$. By the theorem there is to each $\varepsilon > 0$ and $f \in C_0(X)$, considered now as a continuous function on $X \cup \{\infty\}$, a $g \in A$ and $a \in \mathbb{C}$ such that $\|f - (g + a)\|_u < \varepsilon$. As $f(\infty) = g(\infty) = 0$, we get $|a| < \varepsilon$ and $\|f - g\|_u < 2\varepsilon$. □

A.6 Measurability and L^p-Spaces

We introduce here some important Banach spaces from integration theory.

Definition A.6.1 A σ*-algebra* in a set X is a collection M of subsets, so called *measurable subsets*, of X which contains X and is closed under formation of complements and countable unions. A function from X into a topological space is *measurable* if the inverse image of any open set is measurable. A *(positive) measure on* X is a function $\mu\colon M \to [0, \infty]$ that takes at least one value less than ∞ and has the *countably additivity property*

$$\mu(\cup_{n=1}^{\infty} A_n) = \sum_{n=1}^{\infty} \mu(A_n)$$

for pairwise disjoint $A_n \in M$, and we then say X is a *measure space*.

Note that a σ-algebra is closed under countable intersections, and contains the empty set ϕ. The composition of a measurable function with a continuous one is again measurable. Any measure μ is zero on ϕ, and it will always be *increasing*, in that $\mu(A) \leq \mu(B)$ when $A \subset B$.

Example A.6.2 The power set $P(X)$ of a set X is trivially a σ-algebra with the *counting measure* μ on X being the function which assigns the cardinality to any finite subset of X, and otherwise takes the value ∞. ◇

Given any collection N of subsets of a set X, the intersection M of all σ-algebras (including $P(X)$) that contain N is easily seen to be a σ-algebra; it is the smallest σ-algebra that contains N, and is therefore said to be *generated by* N.

Definition A.6.3 The σ-algebra generated by the open sets in a topological space X is the σ-algebra B of *Borel sets of* X.

Topological spaces are tacitly considered with the σ-algebra of Borel sets. Obviously every continuous function from X is then Borel measurable, or a *Borel map*. For any topological space Y and $f: X \to Y$, the collection of subsets in Y with measurable inverse images is a σ-algebra (this holds for any σ-algebra on X). Hence if f is measurable, then inverse images of Borel sets are Borel. Also, compositions of f with Borel maps are Borel maps.

We consider the *extended real half line* $[0, \infty]$ as a topological space by declaring all half-intervals $[0, a\rangle$ and $\langle a, \infty]$ for any positive real number a to be open sets. So points, being closed, are Borel measurable.

For any set X with a σ-algebra M, it follows from what we have just said that a function $f: X \to [0, \infty]$ is measurable if and only if $f^{-1}(\langle a, \infty]) \in M$ for all $a > 0$. The pointwise supremum, infinum and limit of a sequence $\{f_n\}$ of measurable functions $X \to [0, \infty]$ are measurable because

$$(\sup f_n)^{-1}(\langle a, \infty]) = \cup f_n^{-1}(\langle a, \infty]).$$

Proposition A.6.4 *The set of measurable complex functions on a set with a σ-algebra is an algebra under pointwise operations.*

Proof Say f and g are complex measurable functions and $a, b \in \mathbb{C}$. As multiplication by a fixed scalar is a continuous function $\mathbb{C} \to \mathbb{C}$, we see that af is measurable. Since $\mathrm{Re}, \mathrm{Im}: \mathbb{C} \to \mathbb{R}$ are continuous, the real and imaginary parts of a measurable complex function are again measurable. So to show that fg and $f+g$ are measurable, assume first that f and g take only real values.

Introduce a function h by $h(x) = (f(x), g(x))$. Then fg and $f + g$ are the compositions with the continuous functions $\mathbb{R}^2 \to \mathbb{R}$ given by $(s, t) \mapsto st$ and $(s, t) \mapsto s + t$, respectively, so fg and $f + g$ are measurable if h is, and to check this, first observe that any open set in $\mathbb{R}^2$ is a countable union of rectangles $I \times J$ for segments $I, J \subset \mathbb{R}$, and that $h^{-1}(I \times J) = f^{-1}(I) \cap g^{-1}(J)$ is measurable, and so is the countable union of inverse images of such rectangles.

This also shows that our assumption was legitimate. □

A *simple function* is any function $s: X \to [0, \infty\rangle$ with finite range, i.e. of the form

$$s = \sum_{i=1}^{n} a_i \chi_{A_i}$$

for disjoint $A_i \subset X$ and distinct real numbers a_i, where χ_{A_i} is the characteristic function of A_i. If M is a σ-algebra in X, then clearly s is measurable if and only if all $A_i = s^{-1}(a_i) \in M$. If we in addition have a measure μ on X, we define

$$\int_A s\,d\mu = \sum a_i \mu(A_i \cap A)$$

for $A \in M$ with the convention $0 \cdot \infty = 0$.

Definition A.6.5 Let $f\colon X \to [0, \infty]$ be a measurable function on a set X with measure μ. The *Lebesgue integral* of f over a measurable set A in X is

$$\int_A f\,d\mu = \sup \int_A s\,d\mu \in [0, \infty],$$

where the supremum is taken over all simple measurable functions s such that $0 \le s \le f$. When $A = X$ we often skip the index A on the integral sign.

The Lebesgue integral is clearly increasing in both f and A, and it is zero if either $\mu(A) = 0$ or $f = 0$ on A.

One of the strengths of the Lebesgue integral is its flexibility in limit processes.

Lemma A.6.6 *For any measure μ and measurable sets $A_1 \subset A_2 \subset \cdots$ we have $\mu(A_n) \to \mu(\cup A_m)$.*

Proof This is clear as the measurable sets $B_n = A_n \backslash A_{n-1}$ with $A_0 = \phi$ are disjoint and $A_n = B_1 \cup \cdots \cup B_n$ and $\cup B_n = \cup A_n$. □

Lemma A.6.7 *If s is a measurable simple non-negative function on a set X with measure μ, then $A \mapsto \int_A s\,d\mu$ defines a measure on X.*

Proof With $A = \phi$ we get zero, and if we have a countable disjoint union $\cup B_n$ of measurable sets, then

$$\int_{\cup B_n} s\,d\mu = \sum_i a_i \mu(A_i \cap (\cup B_n)) = \sum_i a_i \sum_n \mu(A_i \cap B_n) = \sum_n \int_{B_n} s\,d\mu$$

where $s = \sum_i a_i \chi_{A_i}$ with distinct a_i and pairwise disjoint A_i. □

The following result is known as the *Lebesgue monotone convergence theorem.*

Theorem A.6.8 *If $\{f_n\}$ is a pointwise increasing sequence of measurable functions $X \to [0, \infty]$ on a set with a measure μ, then*

$$\int f_n\,d\mu \to \int \lim f_n\,d\mu.$$

Proof The pointwise limit $f = \lim f_n$ does evidently exist as a measurable function $X \to [0, \infty]$, so the right hand side above makes sense. Let $a \le \int f\,d\mu$ be the limit as $n \to \infty$ of the increasing numbers on the left hand side. If s is a measurable simple function such that $0 \leq s \leq f$, and $c \in \langle 0, 1 \rangle$, then the measurable increasing sets

$$A_n = \{x \in X \mid cs(x) \leq f_n(x)\}$$

have union X, because $f(x) = 0 \Rightarrow x \in A_1$, and if $f(x) > 0$, then $cs(x) < f(x)$, so $x \in A_n$ for some n. By the previous two lemmas

$$\int f_n \, d\mu \geq c \int_{A_n} s \, d\mu$$

for all n, and $a \geq c \int s \, d\mu$. Hence $a \geq \int s \, d\mu$ and $a \geq \int f \, d\mu$. □

For functions f_n from a set X to $[0, \infty]$, let $\liminf f_n = \sup_{k \geq 1}(\inf_{i \geq k} f_i)$. The following result is known as *Fatou's lemma*.

Corollary A.6.9 *For any sequence of measurable functions $f_n \colon X \to [0, \infty]$ on a set with measure μ, we have*

$$\int \liminf f_n \, d\mu \leq \liminf \int f_n \, d\mu.$$

Proof Just apply the theorem to the sequence of functions $\inf_{i \geq k} f_i \leq f_k$. □

We need the following approximation result.

Lemma A.6.10 *For any measurable $f \colon X \to [0, \infty]$ there is an increasing sequence $\{s_n\}$ of non-negative simple measurable functions that converges pointwise to f. Moreover, they converge uniformly to f on any set where it is bounded. If f had range in $\mathbb{C}$, the same would be true, then with $0 \leq |s_1| \leq |s_2| \leq \cdots \leq |f|$.*

Proof For every natural number n and real number t, there is a unique integer $m_n(t)$ such that $m_n(t)2^{-n} \leq t < (m_n(t) + 1)2^{-n}$. Define a function h_n on $[0, \infty]$ by $h_n(t) = m_n(t)2^{-n}$ for $t \in [0, n\rangle$ and $h_n(t) = n$ for $t \in [n, \infty]$. Then $\{h_n\}$ is an increasing sequence of non-negative Borel functions that converges to ι pointwise. Hence the functions $s_n = h_n f$ have the desired properties for the first statement. As for the second statement just note that $0 \leq f - s_n \leq 2^{-n}$ on the set where $f \leq 2^n$. For the last statement pick the obvious linear combination of the simple functions from the first part of the proof that converges to the positive and negative parts of the real and complex parts of f. □

Let $g^+ = \max\{g, 0\}$ and $g^- = -\min\{g, 0\}$ be the positive and negative parts, respectively, of a real function g. So they are the minimal non-negative functions such that $g = g^+ - g^-$.

Definition A.6.11 Given a measure μ, let $L^1(\mu)$ be the set of complex valued measurable function f with $\int |f| \, d\mu < \infty$. We define the integral of $f \in L^1(\mu)$ by

$$\int f \, d\mu = \int (\mathrm{Re} f)^+ \, d\mu - \int (\mathrm{Re} f)^- \, d\mu + i \int (\mathrm{Im} f)^+ \, d\mu - i \int (\mathrm{Im} f)^- \, d\mu.$$

This definition makes sense since the entries in the integrals are non-negative and measurable functions with absolute value less than $|f|$, so that each of the four integrals is finite. Later on $L^1(\mu)$ will actually mean something slightly different.

Proposition A.6.12 *For any measure μ the set $L^1(\mu)$ is a vector space under pointwise operations, and the integral $\int : L^1(\mu) \to \mathbb{C}$ is a linear map, and*

$$|\int f\,d\mu| \leq \int |f|\,d\mu.$$

Proof Let $f, g \in L^1(\mu)$ and $a \in \mathbb{C}$. Assume first that f and g are non-negative simple functions on the measure space X. Let a_i and b_j be the distinct values of f and g, respectively, with $A_i = \{x \in X \mid f(x) = a_i\}$ and $B_j = \{x \in X \mid g(x) = b_i\}$. Then

$$\int_{A_i \cap B_j} (f+g)\,d\mu = (a_i + b_j)\mu(A_i \cap B_j) = \int_{A_i \cap B_j} f\,d\mu + \int_{A_i \cap B_j} g\,d\mu$$

and since X is a disjoint union of the sets $A_i \cap B_j$, we get $\int (f+g)\,d\mu = \int f\,d\mu + \int g\,\mu$ by Lemma A.6.7. By the lemma above and Lebesgue's monotone convergence theorem this also holds for non-negative $f, g \in L^1(\mu)$.

Hence if f, g were complex valued, then $\int |af+g|\,d\mu \leq \int (|a|\cdot|f|+|g|)\,d\mu = |a|\int |f|\,d\mu + \int |g|\,d\mu < \infty$, so $L^1(\mu)$ is a complex vector space.

By the first paragraph the integral is additive on real f and g because with $h = f+g$, it splits according to

$$h^+ + f^- + g^- = f^+ + g^+ + h^-,$$

which when reassembling the finite terms gives $\int (f+g)\,d\mu = \int f\,d\mu + \int g\,d\mu$. The complex case is now obvious.

The equality $\int af\,d\mu = a\int f\,d\mu$ for complex f obviously holds when $a \geq 0$. For negative a one uses relations like $(-g)^+ = g^-$ for real g. It remains to check it for $a = i$, and this holds since

$$\int if\,d\mu = \int (i\mathrm{Re}f) - \mathrm{Im}f)\,d\mu = -\int \mathrm{Im}f\,d\mu + i\int \mathrm{Re}f\,d\mu = i\int f\,d\mu.$$

Finally, pick a complex number b such that

$$|\int f\,d\mu| = b\int f\,d\mu = \int bf\,d\mu = \int \mathrm{Re}(bf)\,d\mu \leq \int |f|\,d\mu.$$

□

The next limit-result is known as *Lebesgue's dominated convergence theorem.*

Theorem A.6.13 *If $\{f_n\}$ is a sequence of measurable functions with $|f_n| \leq g$ for some $g \in L^1(\mu)$, then*

$$\lim \int f_n \, d\mu = \int f \, d\mu,$$

where $f = \lim f_n$ pointwise.

Proof By Fatou's lemma

$$\int 2g \, d\mu \leq \liminf \int (2g - |f_n - f|) \, d\mu = \int 2g \, d\mu + \liminf(- \int |f_n - f| \, d\mu)$$
$$= \int 2g \, d\mu - \inf_{m \geq 1} \sup_{n \geq m} \int |f_n - f| \, d\mu$$

and subtracting the finite terms $\int 2g \, d\mu$ on both sides, gives $\lim \int |f_n - f| \, d\mu = 0$ and the result follows from the previous proposition. □

Definition A.6.14 A function $g \colon \langle a, b \rangle \to \mathbb{R}$ with a or b being possibly $\pm\infty$, is *convex* on $\langle a, b \rangle$ if

$$g((1 - c)x + cy) \leq (1 - c)g(x) + cg(y)$$

for any $c \in [0, 1]$ and $x, y \in \langle a, b \rangle$.

Geometrically this means the graph of g lies beneath the straight line between $(x, g(x))$ and $(y, g(y))$. If the second derivative of g is positive on the interval, then it is convex there. It is also clear that

$$\frac{g(t) - g(s)}{t - s} \leq \frac{g(u) - g(t)}{u - t} \tag{A.6.1}$$

for $a < s < t < u < b$ since the slopes of the line segments between adjacent points of the graph increase.

The function $g(x) = e^x$ is convex on $\mathbb{R}$.

Proposition A.6.15 *Any real convex function on an open interval is continuous on that interval.*

Proof Consider a convex function $g \colon \langle a, b \rangle \to \mathbb{R}$ with a, b being possibly $\pm\infty$. Let us also assume $a < s < x < y < t < b$ for real numbers with corresponding points S, X, Y, T on the graph of g. Then Y is above the line through S and X, and below the line through X and T. Hence $Y \to X$ as $y \to x$ from the right. A similar argument shows that $g(y) \to g(x)$ when $y \to x$ from the left. □

The following result is known as *Jensen's inequality*.

Theorem A.6.16 *Let μ be a measure on a set X with $\mu(X) = 1$. If g is a real convex function on $\langle a, b\rangle$, then*

$$g(\int f\,d\mu) \leq \int (gf)\,d\mu$$

for any measurable function $f\colon X \to \langle a, b\rangle$ with $\int |f|\,d\mu < \infty$.

Proof As g is continuous by the previous proposition, the function gf is measurable, and the composition does indeed make sense as the range of f lies in the domain of g. Here we allow the integral of gf to be $+\infty$. It is clear that $t \equiv \int f\,d\mu \in \langle a, b\rangle$, so the left hand side of the inequality in the theorem also makes sense. Let c be the supremum of the left hand side of Eq. A.6.1 as s varies. Then c cannot be greater than any of the quotients on the right hand side of Eq. A.6.1 for any $u \in \langle t, b\rangle$. Hence

$$g(s) \geq g(t) + c(s - t)$$

for all $s \in \langle a, b\rangle$. In particular, we have $g(f(x)) \geq g(t) + c(f(x) - t)$ for all $x \in X$, which upon integrating gives the desired inequality, again as $\mu(X) = 1$. □

Example A.6.17 Using the exponential function in Jensen's inequality we get

$$\exp(\int f\,d\mu) \leq \int e^f\,d\mu,$$

which in the case of a finite set $X = \{p_i\}$ with $f(p_i) = x_i$ and $\mu(p_i) = a_i > 0$ and $\sum a_i = 1$ gives

$$\exp(\sum a_i x_i) \leq \sum a_i e^{x_i}.$$

Putting $y_i = e^{x_i}$ in this formula, we get $\prod y_i^{a_i} \leq \sum a_i y_i$, which therefore holds for any positive numbers y_i. Thus we get the familiar inequality

$$(y_1 \cdots y_n)^{1/n} \leq \frac{1}{n}(y_1 + \cdots + y_n)$$

between the *geometric mean* and *arithmetic mean* for n positive numbers. ◇

Definition A.6.18 A pair of *conjugate exponents* is either 1 and ∞ or any pair p and q of positive numbers such that $1/p = 1 - 1/q$.

So 2 and 2 is a pair of conjugate exponents.

Let X be a set with a measure μ. For any positive real number p and measurable complex function f, define

$$\|f\|_p = (\int |f|^p\,d\mu)^{1/p},$$

which makes sense as $|\cdot|^p$ is continuous. Use the same definition for measurable non-negative functions with the convention $|\infty|^p = \infty$. If $\|f\|_p = 0$, then $f = 0$ *almost everywhere*, meaning that f is only non-zero on a set of measure zero, because the set of $x \in X$ such that $|f|^p(x) > 0$ is a countable union of sets $A_n = \{x \in X \mid |f|^p(x) > 1/n\}$ and

$$\mu(A_n)/n \leq \int |f|^p \, d\mu = 0.$$

Proposition A.6.19 *If $f, g\colon X \to [0, \infty]$ are measurable functions on a measure space, then $\|fg\|_1 \leq \|f\|_p \|g\|_q$ and $\|f+g\|_p \leq \|f\|_p + \|g\|_p$ for any pair $p, q \in \langle 1, \infty\rangle$ of conjugate exponents.*

The first of these inequalities is *Hölder's* and the second is *Minkowski's*.

Proof To prove Hölder's inequality, we may by the remark prior to the proposition assume that both $\|f\|_p$ and $\|g\|_q$ are positive real numbers. Using $\prod y_i^{a_i} \leq \sum a_i y_i$ with $a_1 = 1/p$ and $a_2 = 1/q$ and $y_1 = (|f(x)|/\|f\|_p)^p$ and $y_2 = (|g(x)|/\|g\|_q)^q$ for x such that these are positive real numbers, we get

$$\frac{|f(x)g(x)|}{\|f\|_p \|g\|_q} \leq \frac{|f(x)|^p}{p\|f\|_p^p} + \frac{|g(x)|^q}{q\|g\|_q^q}.$$

Integrating this gives the desired result.

To get Minkowski's inequality, note that

$$\|f(f+g)^{p-1}\|_1 \leq \|f\|_p \|(f+g)^{p-1}\|_q$$

by Hölder's inequality. Interchanging the roles of f and g gives a similar inequality, and adding the two gives

$$\|f+g\|_p^p \leq \|f+g\|_p^{p/q}(\|f\|_p + \|g\|_p).$$

We only need to consider the case when the left hand side is greater than zero and the right hand side is less than infinity. We are done if we can divide by the first factor in the above inequality, and this is allowed as long as $\|f+g\|_p < \infty$, which holds since

$$|(f+g)/2|^p \leq (|f|^p + |g|^p)/2$$

by convexity of $x \mapsto |x|^p$ on $\mathbb{R}$. □

Let $p \in [1, \infty\rangle$. By Minkowski's inequality the set of measurable complex functions f with $\|f\|_p < \infty$ on a set with a measure is a complex vector space V, and $\|\cdot\|_p$ is a seminorm on it since obviously $\|af\|_p = |a| \cdot \|f\|_p$ for $a \in \mathbb{C}$. We denote the normification of this seminormed space by $L^p(\mu)$, so it consists of the equivalence classes of functions in V that are equal almost everywhere. We write

$\|f\|_p$ for the *p-norm* of a representative f of an equivalence class. In practice we hardly distinguish between a representative and the class it belongs to.

Theorem A.6.20 *The normed space $L^p(\mu)$ is a Banach space for $p \in [1, \infty\rangle$.*

Proof Given a Cauchy sequence, pick a subsequence $\{f_n\}$ such that

$$\|f_{n+1} - f_n\|_p \leq 1/2^n.$$

It suffices to show that this sequence converges in p-norm to some $f \in L^p(\mu)$. Let $g_m = \sum^m |f_{n+1} - f_n|$ and $g = \sum |f_{n+1} - f_n|$. By Minkowski's inequality we have $\|g_k\|_p \leq 1$, so Fatou's lemma tells us that $\|g\|_p \leq 1$. Hence $g < \infty$ almost everywhere, and the series

$$f_1(x) + \sum (f_{n+1}(x) - f_n(x))$$

converges absolutely almost everywhere. Let $f(x)$ be the limit where it converges, and set $f(x) = 0$ for those x where the series diverges. Then f is measurable and $f = \lim f_n$ pointwise almost everywhere. To $\varepsilon > 0$ take large enough k so that $\|f_n - f_k\|_p < \varepsilon$ for large enough n. Bu Fatou's lemma, we get

$$\int |f - f_k|^p \, d\mu \leq \liminf \int |f_n - f_k|^p \, d\mu \leq \varepsilon^p,$$

so $f = f - f_k + f_k \in L^p(\mu)$ and $\|f - f_k\|_p \to 0$ as $k \to \infty$. □

In the proof we showed that any Cauchy sequence in $L^p(\mu)$ with limit f for $p \in [1, \infty\rangle$ has a subsequence that converges pointwise almost everywhere to f.

Definition A.6.21 We say that the *essential supremum* of a measurable function $f: X \to [0, \infty]$ on a set with a measure μ is the infinum of all $a \in [0, \infty\rangle$ with $\mu(f^{-1}((a, \infty])) = 0$. If there are no such a, the essential supremum is set to ∞. If g is a complex measurable function on X, we let $\|g\|_\infty$ be the essential supremum of $|g|$. We say g is *essentially bounded* if $\|g\|_\infty < \infty$.

If f and g are essentially bounded, then $\|f + g\|_\infty \leq \|f\|_\infty + \|g\|_\infty$ as $|f + g| \leq |f| + |g|$. Since also $\|cf\|_\infty = |c| \cdot \|f\|_\infty$ for any $c \in \mathbb{C}$, we see that the set of essentially bounded measurable functions on a set with a measure μ is a complex vector space V with seminorm $\|\cdot\|_\infty$. Let $L^\infty(\mu)$ be the normification of V. As above we often identify the classes in $L^\infty(\mu)$ with the essentially bounded functions, where those that belong to the same class are equal almost everywhere.

Since $|fg| \leq \|f\|_\infty |g|$ almost everywhere, we see that Hölder's inequality also holds for the pair $p = \infty$ and $q = 1$ of conjugate exponents. We have also the following result.

Proposition A.6.22 *The normed space $L^\infty(\mu)$ is a Banach space.*

Proof Let $\{f_n\}$ be a Cauchy sequence in $L^\infty(\mu)$. The countable union A over k and (m, n) of the sets of x's with $|f_k(x)| > \|f_k\|_\infty$ and $|f_n(x) - f_m(x)| > \|f_n - f_m\|_\infty$

has measure zero, and $\{f_n(x)\}$ converges to some complex number $f(x)$ for $x \in A^c$. By letting $f(x) = 0$ for $x \in A$, we get a function $f \in L^\infty(\mu)$ such that $\|f - f_n\|_\infty \to 0$ as $n \to \infty$. □

If μ is the counting measure on a countable set X, we write $l^p(X)$ for $L^p(\mu)$. An element f of this Banach space may be regarded as a complex sequence $\{a_n\}$ with p-norm

$$\|f\|_p = (\sum_{n=1}^{\infty} |a_n|^p)^{1/p}$$

when $p \in [1, \infty\rangle$, otherwise the norm is $\|f\|_\infty = \sup |a_n|$.

Proposition A.6.23 *The class of measurable complex simple functions s with $\mu(\{x \,|\, s(x) \neq 0\}) < \infty$ is dense in $L^p(\mu)$ for $p \in [1, \infty\rangle$.*

Proof Obviously the image of this class belongs to $L^p(\mu)$. To see that it is dense, take $f \in L^p(\mu)$. We may obviously assume $f \geq 0$. Pick a sequence $\{s_n\}$ as in Lemma A.6.10, which obviously belongs to this class, and $\|f - s_n\|_p \to 0$ by Lebesgue's dominated convergence theorem. □

A.7 Radon Measures

Definition A.7.1 A *Radon measure* μ on a locally compact Hausdorff space X is a Borel measure that is finite on every compact set, and satisfies

$$\mu(A) = \inf\{\mu(U) \,|\, A \subset U \text{ open}\} \quad \text{and} \quad \mu(B) = \sup\{\mu(K) \,|\, B \supset K \text{ compact}\}$$

for every Borel set A and open set B. The Radon measure is *regular (at B)* if this holds for every (one specific) Borel set B.

Proposition A.7.2 *Radon measures on σ-compact locally compact Hausdorff spaces are regular.*

Proof Say μ is a Radon measure on a countable union $\cup A_n$ of compact sets. Let E be a Borel set. Pick open sets $B_n \supset A_n \cap E$ such that $\mu(B_n \backslash (A_n \cap E)) < \varepsilon/2^{n+1}$ for $\varepsilon > 0$, so $\mu(B \backslash E) < \varepsilon/2$ with $B = \cup B_n$. Similarly, for E^c we get a closed set A such that $A \subset E \subset B$ and $\mu(B \backslash A) < \varepsilon$, so μ is regular as $A = \cup(A \cap A_n)$. □

The theorem below goes by the name of *Lusin's theorem*.

Theorem A.7.3 *Let X be a locally compact Hausdorff space with a Radon measure μ that is regular at every Borel set of finite measure. Suppose f is a complex measurable function function on X that vanishes outside a set A of finite measure. For $\varepsilon > 0$ there is $g \in C_c(X)$ that only differs from f on a set of measure less than ε, and we can arrange that $\|g\|_u \leq \sup_{x \in X}|f(x)|$. The (image of the) vector space $C_c(X)$ is dense in $L^p(\mu)$ for $p \in [1, \infty\rangle$.*

Proof Assume first that $0 \le f < 1$ and that A is compact. Pick a sequence $\{s_n\}$ as in Lemma A.6.10, and let $t_n = s_n - s_{n-1}$ with $s_0 = 0$, so $2^n t_n$ is the characteristic function on some set $T_n \subset A$ and $f = \sum t_n$ pointwise. Fix an open set $V \supset A$ with compact closure, and pick compact sets K_n and open sets V_n such that $K_n \subset T_n \subset V_n \subset V$ and $\mu(V_n \backslash K_n) < 2^{-n}\varepsilon$. By Urysohn's lemma pick $h_n \in C_c(X)$ with $0 \le h_n \le 1$ that is 1 on K_n and has support in V_n. Define $g = \sum 2^{-n} h_n$ pointwise, so $g \in C_0(X)$ by Proposition A.3.6, and clearly $\text{supp}(g) \subset \bar{V}$. As $2^{-n}h_n = t_n$ except on $V_n \backslash K_n$, we have $g = f$ except on $\cup(V_n \backslash K_n)$ which has measure less than ε.

This evidently also holds for bounded complex measurable f and for any Borel set A of finite measure. It also holds for unbounded f, because the sequence of sets $B_n = \{x \in X \mid |f(x)| > n\}$ have empty intersection, so $\mu(B_n) \to 0$ and f coincides with the bounded function $\chi_{B_n^c} f$ except on B_n.

To get the inequality in the proposition, compose g by $h \colon \mathbb{C} \to \mathbb{C}$ given by $h(z) = z$ if $|z| \le R \equiv \sup_{x \in X} |f(x)|$ and $h(z) = Rz/|z|$ if $|z| > R$.

As for the last statement, pick by Proposition A.6.23 a subclass S of measurable complex simple functions that is dense in $L^p(\mu)$. For $s \in S$, pick $g \in C_c(X)$ with $|g| \le \|s\|_\infty$ that only differs from s on a set of measure less than any preassigned $\varepsilon > 0$. We are now done since $\|g - s\|_p \le 2\varepsilon^{1/p}\|s\|_\infty$. □

For $p \in [1, \infty\rangle$ the two results above together with the completness of $L^p(\mu)$ says that the latter is the p-norm completion (of the normification) of $C_c(X)$ when X is σ-compact. However, the ∞-norm completion of $C_c(X)$ is not in general $L^\infty(\mu)$. If the measure μ is positive on non-empty open sets, then since two distinct continuous functions differ on a non-empty open set, we need not perform any normification, and the ∞-norm coincides with the uniform norm on $C_c(X)$, which has the completion $C_0(X)$.

A *positive linear functional on* $C_c(X)$ is a linear map $x : C_c(X) \to \mathbb{C}$ such that $x(f) \in [0, \infty\rangle$ for $f \in C_c(X)$ with $f \ge 0$. The following result is known as the *Riesz representation theorem.*

Theorem A.7.4 *Let X be a locally compact Hausdorff space. To any positive linear functional x on $C_c(X)$, there is a unique Radon measure μ such that*

$$x(f) = \int f \, d\mu$$

for all $f \in C_c(X)$. Moreover, the measure is regular at every Borel set of finite measure. In fact, the measure extends to a σ-algebra M for which it is complete, meaning that $A \in M$ whenever A is contained in some $B \in M$ with $\mu(B) = 0$, and the previous statement holds with Borel sets replaced by measurable sets in M.

Proof Say we have another such measures ν. To $\varepsilon > 0$ and compact set A pick open set $U \supset A$ such that $\nu(U) < \nu(A) + \varepsilon$. By Urysohn's lemma there is a continuous function f with range in $[0, 1]$ that is 1 on A and has support in U. Hence

$$\mu(A) = \int \chi_A \, d\mu \le x(f) \le \int \chi_U \, d\nu < \nu(A) + \varepsilon,$$

so $\mu(A) = \nu(A)$ by swopping the roles of μ and ν. This shows uniqueness, and that any μ relating to x this way must be finite on compact sets.

We turn now to existence. For an open set U let $\mu(U)$ be the supremum of $x(f)$ over all $f \in C_c(X)$ with $0 \leq f \leq 1$ that have support inside U. By Urysohn's lemma there is at least one such f, and clearly μ is a non-negative increasing function on the topology of X. For $B \subset X$ let $\mu(B)$ be the infinum of $\mu(U)$ over all open $U \supset B$. Let N be the collection of $B \subset X$ such that $\mu(B) = \sup_{A \subset B} \mu(A) < \infty$ with A compact. Let M be all $B \subset X$ with $B \cap A \in N$ for every compact set A.

It remains to prove that M and μ have the required properties. The function μ is clearly a non-negative increasing function on M, and if it is a measure, then it is certainly complete. We henceforth settle various less obvious claims.

Claim 1 Any compact set A belongs to N, and $\mu(A)$ is the infinum of $x(f)$ over all continuous functions f with range $[0, 1]$ that is one on A, so $\mu(A)$ is in particular finite.

For any such f and $a \in \langle 0, 1\rangle$, let U consist of those $x \in X$ with $f(x) > a$. Then $A \subset U$ and $ag \leq f$ for any continuous function g with range in $[0, 1]$ and support in U. Hence $\mu(A) \leq \mu(U) \leq a^{-1}x(f)$. Letting $a \to 1$, we conclude that $\mu(A) \leq x(f)$, so $A \in N$. Also, to $\varepsilon > 0$, there is open $B \supset A$ with $\mu(B) < \mu(A) + \varepsilon$. Pick by Urysohn's lemma a continuous function h with range in $[0, 1]$ that is one on A and has support in B. Then $x(f) \leq \mu(B) < \mu(A) + \varepsilon$ proves the claim.

Claim 2 Every open set U with $\mu(U) < \infty$ belongs to N.

Let a be a real number such that $a < \mu(U)$. Then there is a continuous function f with range in $[0, 1]$ and support in U such that $a < x(f)$. If A is any open set that contains the support B of f, then $x(f) \leq \mu(A)$, so $x(f) \leq \mu(B)$.

Claim 3 If $A = \cup A_n$ for a sequence of subsets A_n, then $\mu(A) \leq \sum \mu(A_n)$. We have equality when they are pairwise disjoint members of N, and if in addition $\mu(A) < \infty$, then $A \in N$.

If U_i are open, then $\mu(U_1 \cup U_2) \leq \mu(U_1) + \mu(U_2)$ because to any continuous function g with range in $[0, 1]$ and support in $U_1 \cup U_2$, pick by Proposition A.3.9 continuous functions h_i with range in $[0, 1]$ and support in U_i such that $h_1 + h_2 = 1$ on the support of g. Then $x(g) = x(h_1 g) + x(h_2 g) \leq \mu(U_1) + \mu(U_2)$.

We may obviously assume that all $\mu(A_n) < \infty$. To $\varepsilon > 0$ pick open sets $U_n \supset A_n$ with $\mu(U_n) < \mu(A_n) + \varepsilon 2^{-n}$. Let f be continuous with range in $[0, 1]$ and support in $\cup U_n$, hence in $\cup^m U_n$ for some m. Then $x(f) \leq \mu(\cup^m U_n) \leq \sum^m \mu(U_n) \leq \sum \mu(A_n) + \varepsilon$, so $\mu(A) \leq \mu(\cup U_n) \leq \sum \mu(A_n) + \varepsilon$, which proves the first part of our claim.

If B_1 and B_2 are disjoint compact sets, then $\mu(B_1 \cup B_2) = \mu(B_1) + \mu(B_2)$ because to $\varepsilon > 0$, there is by Urysohn's lemma a continuous function f with compact support and range in $[0, 1]$ that is one on B_1 and vanishes on B_2. By Claim 1 there is a continuous function g with compact support that is one on $B_1 \cup B_2$ such that $x(g) < \mu(B_1 \cup B_2) + \varepsilon$, whereas $\mu(B_1) + \mu(B_2) \leq x(fg) + x((1 - f)g) = x(g)$.

We may clearly assume that $\mu(A) < \infty$. Pick now compact sets $B_i \subset A_i$ with $\mu(B_i) > \mu(A_i) + 2^{-i}\varepsilon$. Then $\mu(A) \geq \lim_n \mu(\cup^n B_i) = \lim_n \sum^n \mu(B_i) > \sum \mu(A_i) + \varepsilon$, and this inequality also shows that $A \in N$.

Claim 4 For $A \in N$ and $\varepsilon > 0$ there is a compact set B and an open set U such that $B \subset A \subset U$ and $\mu(U \backslash B) < \varepsilon$.

Pick B and U such that $\mu(U) - \varepsilon/2 < \mu(A) < \mu(B) + \varepsilon/2$. Then $U \backslash B \in N$ by Claim 2, so $\mu(B) + \mu(U \backslash B) = \mu(U) < \mu(B) + \varepsilon$ by Claim 3.

Claim 5 The set N is closed under formations of finite unions, intersections and differences.

To $A_i \in N$ and $\varepsilon > 0$, there are by Claim 4 compact sets $B_i \subset A_i$ and open sets $U_i \supset A_i$ such that $\mu(U_i \backslash B_i) < \varepsilon$. As $A_1 \backslash A_2 \subset (U_1 \backslash B_1) \cup (B_1 \backslash U_2) \cup (U_2 \backslash B_2)$, we get $\mu(A_1 \backslash A_2) \leq \mu(B_1 \backslash U_2) + 2\varepsilon$ by Claim 3, so $A_1 \backslash A_2 \in N$. But then $A_1 \cup A_2 = (A_1 \backslash A_2) \cup A_2 \in N$ by Claim 3, and hence also $A_1 \cap A_2 = A_1 \backslash (A_1 \backslash A_2) \in N$.

Claim 6 The collection M is a σ-algebra that contains the Borel sets.

Let B be compact. If $A \in M$, then $A^c \cap B = B \backslash (A \cap B) \in N$ by Claim 5, so $A^c \in M$. If we have a sequence of elements $A_i \in M$, define $B_1 = A_1 \cap B$ and $B_n = (A_n \cap B) \backslash (B_1 \cup \cdots \cup B_{n-1})$. Then the pairwise disjoint sets B_i belong to N by Claim 5, and $(\cup A_i) \cap B = \cup B_i \in N$ by Claim 3, so $\cup A_i \in M$. Clearly M contains all closed sets, and in particular X.

Claim 7 The collection N consists of those $A \in M$ with $\mu(A) < \infty$.

If $A \in N$ and B is compact, then $A \cap B \in N$ by Claim 1 and Claim 5, so $A \in M$. Conversely, if $A \in M$ and $\mu(A) < \infty$, pick an open set $U \supset A$ with $\mu(U) < \infty$. By Claim 2 and Claim 4 there is to $\varepsilon > 0$ a compact set $B \subset U$ with $\mu(U \backslash B) < \varepsilon$. Since $A \cap B \in N$, there is a compact set $E \subset A \cap B$ with $\mu(A \cap B) < \mu(E) + \varepsilon$. As $A \subset (A \cap B) \cup (U \backslash B)$, we get $\mu(A) \leq \mu(A \cap B) + \mu(U \backslash B) < \mu(E) + 2\varepsilon$, so $A \in N$.

That μ is a measure on M is now clear. It remains to prove that $x(f) = \int f\, d\mu$ for $f \in C_c(X)$. To this end it suffices to show $x(f) \leq \int f\, d\mu$ for real f as then $-x(f) = x(-f) \leq \int (-f)\, d\mu = -\int f\, d\mu$.

Let B be the support of f, and pick a, b such that $\operatorname{im} f \subset [a, b]$. For $r > 0$ choose x_i with $x_i - x_{i-1} < r$ and $x_0 < a < x_1 < \cdots < x_n = b$. The disjoint Borel sets $A_i = B \cap f^{-1}(\langle x_{i-1}, x_i])$ have union B. Pick open sets $U_i \supset A_i$ with $\mu(U_i) < \mu(A_i) + r/n$ and $f(U_i) < x_i + r$. By Proposition A.3.9 there are continuous functions h_i with range in $[0, 1]$ and support in U_i such that $\sum h_i = 1$ on B, so $f = \sum h_i f$. Also $\mu(B) \leq x(\sum h_i) = \sum x(h_i)$ by Claim 1. Since $x_i - r < f$ on A_i and $h_i f \leq (x_i + r) h_i$, we therefore get

$$x(f) = \sum (x_i + r) x(h_i) \leq \sum (|a| + x_i + r)(\mu(A_i) + r/n) - |a|\mu(B),$$

which is not greater than $\int f\, d\mu + r(2\mu(B) + |a| + b + r)$. □

We say that x is *represented by the Radon measure* μ, and the integral is the unique extension of x from $C_c(X)$ to the Banach space completion $L^1(\mu)$. This theorem shows that under some technical assumptions, measures can be described in terms of functionals and vice versa; we have two equivalent pictures.

Definition A.7.5 The *Lebesgue measure* m is the Radon measure on $\mathbb{R}^n$ that represents the Riemann integral over $\mathbb{R}^n$.

The Lebesgue measure m on $\mathbb{R}^n$ is evidently regular, and it is positive on non-empty open sets because the measure of an open n-cube W is the volume of its closure $\overline{W}$. To see that this in turn holds, pick a sequence of open cubes V_k that shrink down to $\overline{W}$. By Urysohn's lemma we may further pick $f_k \in C_c(\mathbb{R}^n)$ with $0 \leq f_k \leq 1$ that is 1 on $\overline{W}$ and has support inside V_k. By Lebesgue's dominated convergence theorem, we get

$$m(\overline{W}) = \int \chi_{\overline{W}}\, dm = \lim \int f_k\, dm = \lim x(f_k) = \mathrm{vol}(\overline{W})$$

by virtue of the Riemann integral. By approaching W from the inside with closed cubes and using f_k's that are one there and have support inside W, we similarly get $m(W) = \mathrm{vol}(W)$. Hence the same result holds for *semiopen cubes* $\{(x_1, \ldots, x_n) \mid x_i \in [a_i, a_i + r\rangle\}$ with sides r and *corner at* $a = (a_1, \ldots, a_n)$, these then having both measure and volume r^n. Similar reasoning shows that m is zero on any hyperplane of $\mathbb{R}^n$

The measure is *translation invariant*, meaning that $m(x + A) = m(A)$ for every $x \in \mathbb{R}^n$ and Borel set A. To see this observe that $A \mapsto m(x + A)$ defines a Radon measure that coincides with m on all semiopen cubes, so the result follows from the following proposition.

Proposition A.7.6 *Every open set in $\mathbb{R}^n$ is a countable disjoint union of semiopen cubes.*

Proof For $k \in \mathbb{N}$ let A_k be set of semiopen cubes with sides 2^{-k} and corners at coordinates in $\mathbb{R}^n$ that are integral multiples of 2^{-k}. Then any open set B is a union of semiopen cubes which lie in B and belong to some A_k. From this collection select those that belong to A_1 and remove all smaller cubes in any of the selected cubes. Next select those in A_2 which lie in B and remove all smaller cubes in the selected ones. Proceeding this way we get the required collection because for fixed k every $x \in \mathbb{R}^n$ lies in exactly one cube in A_k, and two cubes belonging to distinct A_k and A_l are either disjoint, or one fully contained in the other. □

In fact, up to multiplication with a positive scalar, the Lebesgue measure is the unique non-trivial translation invariant Radon measure on $\mathbb{R}^n$. This is now clear since if we had another such measure $\tilde{m}$ having measure c on a semiopen cube W with sides one, then $\tilde{m} = cm$ since every semiopen cube is a partition of translates of appropriate divisions of W into smaller semiopen cubes.

Proposition A.7.7 *We have $mT = |\det T| m$ for $T \in \mathrm{End}(\mathbb{R}^n)$.*

Proof If $\det T = 0$, then $m(T(A)) = 0$ for any Borel set A as $T(A)$ belongs to some hyperplane, so the required identity holds in this case. When T is invertible, then T

provides a correspondence of open sets, so mT is a non-trivial translation invariant Radon measure on $\mathbb{R}^n$. By the arguments above the proposition we therefore only need to check the identity on any semiopen cube W with sides one. By uniqueness we also know that $mT = c(T)m$ for some positive scalar $c(T)$. For any other invertible $S \in \mathrm{End}(\mathbb{R}^n)$, we get $c(ST)m = mST = c(S)mT = c(S)c(T)m$. By multiplicativity of the determinant, and since every invertible matrix is the product of matrices implementing the elementary row operations, we therefore only need to check the identity for such matrices T. The effect of interchanging two rows leaves the volume of W unchanged, and $|\det T| = 1$ for the corresponding matrix T. Multiplying a row by a changes the volume by the factor $|a|$, and clearly $|\det T| = |a|$. It is also easy to see that the effect of adding a row to another leaves the volume unchanged, and here $\det T = 1$. □

In particular, the Lebesgue measure on $\mathbb{R}^n$ is rotation invariant.

A.8 Complex Measures

Definition A.8.1 A *complex (real)* measure is a countable additive function on a σ-algebra with values in the complex (real) numbers.

Note that a real measure is complex, while a measure need not be real. Sometimes we call measures *positive measures*. Every complex measure clearly vanishes on the empty set. We shall presently see that given a complex measure, there is a least measure that in some sense majorizes it.

Lemma A.8.2 *Among finitely many complex numbers z_n there is a subset C such that $|\sum_{c\in C} c| \geq \sum |z_n|/\pi$.*

Proof For $t \in [-\pi, \pi]$, let $C(t)$ be the set of z_n's such that $\mathrm{Re}(z_n e^{-it}) > 0$. Let t_0 maximize $t \mapsto \mathrm{Re}(\sum_{c\in C(t)} ce^{-it})$, and set $C = C(t_0)$. Then $|\sum_{c\in C} c| \geq \frac{1}{2\pi}\int_{-\pi}^{\pi} \mathrm{Re}(\sum_{c\in C(t)} ce^{-it})\, dt \geq \sum |z_n|/\pi$. □

Proposition A.8.3 *Given a complex measure μ, the supremum $|\mu|(A)$ of $\sum |\mu(A_i)|$ over all countable partitions $\{A_i\}$ of a measurable set A into measurable sets A_i, defines a finite measure $|\mu|$ called the* total variation *of μ.*

Proof Consider a countable partition of a measurable set A into measurable sets A_i. Let $t_i < |\mu|(A_i)$. Each A_i has a countable partition into measurable sets A_{ij} with $\sum |\mu(A_{ij})| > t_i$. Then $\sum t_i \leq \sum |\mu(A_{ij})| \leq |\mu|(A)$, so $\sum |\mu|(A_i) \leq |\mu|(A)$.

On the other hand, if we have a countable partition of A into measurable sets B_j, then $\{B_j \cap A_i\}$ are countable partitions of B_j and A_i. Hence

$$\sum |\mu(B_j)| \leq \sum |\mu(B_j \cap A_i)| \leq \sum |\mu|(A_i),$$

so $|\mu|(A) \leq \sum |\mu|(A_i)$, and clearly $|\mu|(\phi) = 0$.

Next we prove that $|\mu|(X) < \infty$ for the whole space X. If $|\mu|(C) = \infty$ for some measurable set C, there are finitely many pairwise disjoint measurable subsets C_i such that

$$\sum |\mu(C_i)| > t \equiv \pi(1 + |\mu(C)|).$$

By the lemma, a union A of some of the sets C_i satisfies $|\mu(A)| > t/\pi > 1$, while its complement B in C satisfies $|\mu(B)| \geq |\mu(A)| - |\mu(C)| > t/\pi - |\mu(C)| = 1$. Also, either $|\mu|(A) = \infty$ or $|\mu|(B) = \infty$.

Now, if $|\mu|(X) = \infty$, split X into A_1, B_1 with $|\mu(A_1)| > 1$ and $|\mu|(B_1) = \infty$. Split B_1 into A_2, B_2 with $|\mu(A_2)| > 1$ and $|\mu|(B_2) = \infty$ to obtain a sequence $\{A_n\}$ with $|\mu(A_n)| > 1$, which is absurd by countable additivity of μ. □

The complex measures on a space X is clearly a normed vector space under

$$(\mu + \nu)(A) = \mu(A) + \nu(A) \quad \text{and} \quad (c\mu)(A) = c\mu(A) \quad \text{and} \quad \|\mu\| = |\mu|(X)$$

for any measurable set $A \subset X$.

The *positive-* and *negative variations* of a real measure μ are the finite measures given by $\mu_\pm = (|\mu| \pm \mu)/2$. They give the *Jordan decomposition* $\mu = \mu_+ - \mu_-$ of μ, and we also note that $|\mu| = \mu_+ + \mu_-$. A measure or a complex measure ν is *absolutely continuous* with respect to a measure μ, and we write $\nu \ll \mu$, if $\nu(A) = 0$ whenever $\mu(A) = 0$. We say ν is *concentrated on A* if $\nu(B) = \nu(A \cap B)$ for all measurable sets B. Since $\nu(B) = \nu(B/(B \cap A)) + \nu(B \cap A)$ and $(B/(B \cap A)) \cap A = \phi$, we see that ν is concentrated on A if and only if $\nu(C) = 0$ when $C \cap A = \phi$. Given another measure or a complex measure λ, we say ν and λ are *mutually singular*, and write $\nu \perp \lambda$, if they are concentrated on disjoint measurable sets.

The following facts are easily verified, assuming still that μ is a measure: If ν is concentrated on a measurable set, then $|\nu|$ is concentrated on the same set. If $\nu \perp \lambda$, then $|\nu| \perp |\lambda|$. If ν and λ are absolutely continuous with respect to μ, then so are $\nu + \lambda$ and $|\nu|$. If $\nu \ll \mu$ and $\mu \perp \lambda$, then $\nu \perp \lambda$. If $\nu \ll \mu$ and $\nu \perp \mu$, then $\nu = 0$. Finally, if $\nu \perp \mu$ and $\lambda \perp \mu$, then $\nu + \lambda \perp \mu$; in fact, if ν and λ are concentrated on measurable sets A_1 and A_2, respectively, and μ is concentrated on measurable sets B_i with $A_i \cap B_i = \phi$, then $\nu + \lambda$ is concentrated on $A_1 \cup A_2$, while μ is concentrated on the disjoint set $B_1 \cap B_2$.

A measure is *σ-finite* if its underlying measure space is partitioned into measurable sets each with finite measure.

Lemma A.8.4 *Any σ-finite measure μ has $w \in L^1(\mu)$ with range in $\langle 0, 1\rangle$.*

Proof Pick a sequence of measurable sets A_n with finite measure having union the whole space, and let $w = \sum_{n=1}^{\infty} \chi_{A_n} 2^{-n}/(1 + \mu(A_n))$. □

Lemma A.8.5 *If $f \in L^1(\mu)$ for a finite measure μ, and $A_B(f) = \int_B f\, d\mu/\mu(B)$ lie in a closed subset S of the complex plane for any measurable set B with $\mu(B) > 0$, then $f(x) \in S$ almost everywhere.*

Proof It suffices to show that $B \equiv f^{-1}(B_r(a))$ has measure zero for $B_r(a)$ in S^c since S^c can be covered by countably many such discs. But if $\mu(B) > 0$, then $|A_B(f) - a| \leq \int_B |f - a| \, d\mu/\mu(B) < r$. □

The following central result is known as the *Lebesgue-Radon-Nikodym theorem.*

Theorem A.8.6 *Let μ be a σ-finite measure and let ν be a complex measure. Then there are unique complex measures ν_a and ν_s such that $\nu = \nu_a + \nu_s$ and $\nu_a \ll \mu$ and $\nu_s \perp \mu$. They are positive if ν is positive. There is a unique $h \in L^1(\mu)$ such that $\nu_a(A) = \int_A h \, d\mu$ for every measurable set A.*

Proof If λ_a and λ_s were other ones, then $\lambda_s - \nu_s = \nu_a - \lambda_a \ll \mu$ and $\lambda_s - \nu_s \perp \mu$, so $\lambda_s = \nu_s$ and $\lambda_a = \nu_a$. Uniqueness of $[h]$ is also clear.

Assume first that ν is positive. Pick w to μ as in the first lemma, so $\nu + w\mu$ is a finite measure. By the Cauchy-Schwarz inequality, the linear functional $f \mapsto \int f \, d\nu$ on the Hilbert space $L^2(\nu + w\mu)$ is bounded, so there exists g in it such that

$$\int f \, d\nu = \int fg \, d\nu + \int fgw \, d\mu$$

for all $f \in L^2(\nu + w\mu)$. Plugging in $f = \chi_A$ such that $\nu + w\mu$ has positive measure a on A, we get $0 \leq a^{-1}(\int_A g \, d\nu + \int_A gw \, d\mu) = \nu(A)/a \leq 1$, so we may assume that $\operatorname{im} g \subset [0, 1]$ by the second lemma.

Let $A = g^{-1}([0, 1\rangle)$ and $B = g^{-1}(\{1\})$ and define measures ν_a and ν_s by $\nu_a(C) = \nu(A \cap C)$ and $\nu_s(C) = \nu(B \cap C)$. Plugging in $f = \chi_B$ and using $w > 0$, we get $\mu(B) = 0$, so $\nu_s \perp \mu$. Plugging in $f = (1 + g + \cdots + g^n)\chi_C$, we get

$$\int_C (1 - g^{n+1}) \, d\nu = \int_C g(1 + g + \cdots + g^n) w \, d\mu.$$

Letting $n \to \infty$ and using Lebesgue's monotone convergence theorem, we get $\nu_a(C) = \nu(A \cap C) = \int_C h \, d\mu$ with $h = \lim g(1 + g + \cdots + g^n)w$ pointwise. Since ν is a finite measure, we see that $h \in L^1(\mu)$, and clearly $\nu_a \ll \mu$.

Starting with an arbitrary complex measure ν, write it as $\nu = \nu_1 + i\nu_2$ with ν_i real measures, and then apply the first part of the proof on the positive and negative variations of ν_i. □

The decomposition of ν above is called the *Lebesgue decomposition* of ν relative to μ. The function h is called the *Radon-Nikodym derivative* of ν_a, with the suggestive notation $h = d\nu_a/d\mu$.

The following result explains the terminology 'absolutely continuous'.

Proposition A.8.7 *Let μ be a measure and let ν be a complex measure. Then $\nu \ll \mu$ if and only if to $\varepsilon > 0$ there is $\delta > 0$ such that $|\nu(A)| < \varepsilon$ when $\mu(A) < \delta$.*

Proof The backward implication is trivial. Conversely, if there is $\varepsilon > 0$ and measurable sets A_n with $\mu(A_n) < 2^{-n}$ while $|\nu|(A_n) \geq |\nu(A_n)| \geq \varepsilon$, then $|\nu|$ cannot be absolutely continuous with respect to μ since for $B_n = \cup_{i=n}^{\infty} A_i$ and

$B = \cap_{n=1}^{\infty} B_n$, we get $\mu(B) = 0$ as $\mu(B_n) < 2^{-n+1}$ and $B_{n+1} \subset B_n$, whereas $|\nu|(B) \geq \varepsilon$. □

We can also polar decompose a complex measure.

Proposition A.8.8 *For any complex measure ν there is $h \in L^1(|\nu|)$ such that $|h| = 1$ and $d\nu = h\,d|\nu|$.*

Proof Let $h \in L^1(|\nu|)$ be the Radon-Nikodym derivative of ν with respect to $|\nu|$. Consider a countable partition of $A_r = |h|^{-1}([0, r\rangle)$ for $r > 0$ into measurable sets B_m. Then $\sum |\nu(B_m)| \leq \sum r|\nu|(B_m) = r|\nu|(A_r)$, so $|\nu|(A_r) \leq r|\nu|(A_r)$ and $|h| \geq 1$ almost everywhere. On the other hand, for any measurable set A, we have $|\int_A h\,d|\nu|| = |\nu(A)| \leq |\nu|(A)$, so $|h| \leq 1$ almost everywhere by the last lemma. □

The next *Hahn decomposition theorem* shows that the space of a real measure ν is a partition into two sets $A_\pm$ that carries the positive and negative masses of ν.

Corollary A.8.9 *The space any real measure ν is defined on is a disjoint union of two sets $A_\pm$ such that $\nu_\pm(B) = \pm\nu(B \cap A_\pm)$ for any measurable set B.*

Proof By the proposition we may assume there is $h \in L^1(|\nu|)$ with range in $\{\pm 1\}$ and such that $d\nu = h\,d|\nu|$. Put $A_\pm = h^{-1}(\pm\{1\})$. Then

$$\nu_+(B) = 2^{-1} \int_B (1 + h)\,d|\nu| = \int_{B \cap A_+} h\,d|\nu| = \nu(B \cap A_+)$$

and $\nu_-(B) = \nu_+(B) - \nu(B) = -\nu(B \cap A_-)$. □

The Jordan decomposition of a real measure has a minimum property.

Corollary A.8.10 *If for a real measure ν we can write $\nu = \nu_1 - \nu_2$ for measures ν_i, then $\nu_1 \geq \nu_+$ and $\nu_2 \geq \nu_-$.*

Proof In the terminology above $\nu_+(B) = \nu(B \cap A_+) \leq \nu_1(B \cap A_+) \leq \nu_1(B)$. □

We have the following important duality result.

Theorem A.8.11 *Let p and q be conjugate exponents with $p \in [1, \infty\rangle$, and let μ be a σ-finite measure. Let $S_g(f) = \int fg\,d\mu$. Then $g \mapsto S_g$ is an isometric isomorphism from $L^q(\mu)$ to the dual of the Banach space $L^p(\mu)$. When $p > 1$ the same conclusion holds without assuming σ-finiteness.*

Proof By Hölder's inequality $\|S_g\| \leq \|g\|_q$. It remains to show equality, and that the map is surjective. Assume first that μ is finite, so $\chi_A \in L^p(\mu)$ for every measurable set A. Suppose x is a bounded linear functional on $L^p(\mu)$. Let $\nu(A) = x(\chi_A)$ for every measurable set A. Then the additive function ν is a complex measure because if we have a countable partition of A into measurable sets A_i, then the characteristic functions of $\cup_{i=1}^n A_i$ tend to χ_A in p-norm as $n \to \infty$. Clearly $\nu \ll \mu$, so ν has a Radon-Nikodym derivative $g \in L^1(\mu)$ with respect to μ. Since any function in $L^p(\mu)$ can be approximated by a sequence of simple functions, it follows by linearity and continuity that $x(f) = \int fg\,d\mu$ for all $f \in L^p(\mu)$.

For $p = 1$ we therefore get $|\int_A g\,d\mu| \leq \|x\|\|\chi_A\|_1 = \|x\|\mu(A)$, so $\|g\|_\infty \leq \|x\|$ by the last lemma.

For $p \in \langle 1, \infty\rangle$ there is a measurable function h with $|h| = 1$ and $hg = |g|$. Let $f = |g|^{q-1} h\chi_{B_n} \in L^\infty(\mu)$, where $B_n = |g|^{-1}([0, n])$. Since $|f|^p = |g|^q$ on B_n, we get

$$\int_{B_n} |g|^q \, d\mu = \int fg\,d\mu = x(f) \leq \|x\|(\int_{B_n} |g|^q)^{1/p},$$

or $\int_{B_n} |g|^q \, d\mu \leq \|x\|^q$, so $\|g\|_q \leq \|x\|$ by Lebesgue's monotone convergence theorem.

In the σ-finiteness case we can always reduce ourselves to a finite measure λ given by $d\lambda = w\,d\mu$ with w from the first lemma because multiplication by $w^{1/p}$ is an isometric isomorphism from $L^p(\lambda)$ to $L^p(\mu)$.

Let us see how we can circumvent σ-finiteness when $p > 1$. By the first part of the proof we have for each A with $\mu(A) < \infty$, a unique $g_A \in L^q(\mu)$ such that $x(f) = \int \chi_A f g_A \, d\mu$ for all $f \in L^p(\mu)$. Then $g_B = g_A$ almost everywhere on $A \cap B$, so we can define a function g to be g_A on A if $\mu(A) < \infty$ and zero where there is no such A. Let a be the supremum of $\|g_A\|_q$ over all such A's. Then $a \leq \|x\|$. Since $A \subset B$ implies $\|g_A\| \leq \|g_B\|$, there is a sequence $A_1 \subset A_2 \subset \cdots$ such that $\|g_{A_n}\|_q \to a$. We claim that $g_A = 0$ whenever A is disjoint from $\cup A_n$. To see this, pick measurable functions h_n with $|h_n| = 1$ and $h_n g_{A\cup A_n} = |g_{A\cup A_n}|$. Since $q < \infty$ is a conjugate exponent to p, the function $f = |g_{A\cap A_n}|^{q-1} h_n$ belongs to $L^p(\mu)$. Plugging this f into $\int \chi_{A\cup A_n} f g_{A\cup A_n} \, d\mu = \int \chi_A f g_A \, d\mu + \int \chi_{A_n} f g_{A_n} \, d\mu$, we get $\|g_{A\cup A_n}\|_q^q = \|g_A\|_q^q + \|g_{A_n}\|_q^q$, which tends to $\|g_A\|_q^q + a^q$ as $n \to \infty$. Thus $g_A = 0$. So g is zero off $\cup A_n$. Hence g is measurable and $\|g\|_q = a$.

Now given $f \in L^p(\mu)$, let $B_n = |f|^{-1}([n, \infty\rangle)$. Then $\mu(B_n) < \infty$ and $f \neq 0$ exactly on $\cup B_n$. By Lebesque's dominated convergence theorem $\|\chi_{B_n} f - f\|_p \to 0$ and $x(f) = \lim x(\chi_{B_n} f) = \lim \int \chi_{B_n} f g_{B_n} \, d\mu = \int fg\,d\mu$, so $\|x\| \leq \|g\|_q \leq \|x\|$. □

A complex Borel measure ν on a locally compact Hausdorff space is *regular* if $|\nu|$ is a Radon measure, and hence a regular Radon measure. With $h \in L^1(|\nu|)$ associated to ν as in Proposition A.8.8, we define the *integral* of $f \in L^1(|\nu|)$ with respect to ν by $\int f\,d\nu = \int fh\,d|\nu|$. Denote the normed space of regular complex Borel measures on a locally compact Hausdorff space X by $M(X)$; the sum of two regular complex Borel measures ν and μ is indeed a measure of the same kind because $|\mu + \nu| \leq |\mu| + |\nu|$. We have the following *Riesz representation theorem* for bounded functionals on $C_0(X)$.

Theorem A.8.12 *Let ν be a regular complex Borel measure on a locally compact Hausdorff space X. Set $S_\nu(f) = \int f\,d\nu$. Then $\nu \mapsto S_\nu$ is an isometric isomorphism from the normed space $M(X)$ to the dual of the Banach space $C_0(X)$.*

Proof To prove injectivity, assume $S_\nu = 0$ and pick h as above. Choose a sequence of functions $f_n \in C_c(X)$ such that $f_n \to \bar{h}$ in $L^1(|\nu|)$. Then $|\nu|(X) = \int(\bar{h} - f_n)h\, d|\nu| \leq \int |\bar{h} - f_n|\, d|\nu| \to 0$ as $n \to \infty$, so $\nu = 0$.

To prove surjectivity, we construct from a bounded functional a positive one, and apply the Riesz representation theorem for such ones to get the measure. Thus, given a functional x on $C_0(X)$ with $\|x\| = 1$ and a non-negative $f_i \in C_c(X)$, let $y(f_i)$ be the supremum of $|x(h)|$ over all $h \in C_c(X)$ with $|h| \leq f_i$. Clearly y is a positive increasing function that respects multiplication with positive scalars. To $\varepsilon > 0$ there are $h_i \in C_c(X)$ such that $|h_i| \leq f_i$ and $y(f_i) \leq |x(h_i)| + \varepsilon$. Pick complex numbers a_i of modulus one such that $a_i x(h_i) = |x(h_i)|$. Then

$$y(f_1)+y(f_2) \leq |x(h_1)|+|x(h_2)|+2\varepsilon = x(a_1h_1+a_2h_2)+2\varepsilon \leq y(f_1+f_2)+2\varepsilon,$$

so $y(f_1) + y(f_2) \leq y(f_1 + f_2)$. To any $h \in C_c(X)$ with $|h| \leq f_1 + f_2$, define h_i to be $f_i h/(f_1 + f_2)$ when $f_1 + f_2 \neq 0$ and otherwise zero. Since $|h_i| \leq |h|$ and h is continuous, we see that $h_i \in C_c(X)$. Since $h = h_1 + h_2$ and $|h_i| \leq f_i$ we also have $|x(h)| \leq y(f_1) + y(f_2)$, so $y(f_1 + f_2) = y(f_1) + y(f_2)$. Decomposing functions in $C_c(X)$ into the usual sum of four positive functions, it is easy to see that we can extend y consistently to a linear functional on $C_c(X)$. The way we defined it on positive functions shows that $|x(f)| \leq y(|f|) \leq \|f\|_u$ for any $f \in C_c(X)$.

By the Riesz representation theorem for positive functionals on $C_c(X)$, there is a Radon measure μ on X such that $y(f) = \int f\, d\mu$. As $\mu(X)$ is the supremum of $\int f\, d\mu$ over all $f \in C_c(X)$ with $0 \leq f \leq 1$, we get $\mu(X) \leq 1$, so μ is regular. Since $|x(f)| \leq y(|f|) = \int |f|\, d\mu$, we know that x has norm not greater than one with respect to the 1-norm from μ. As the dual of the Banach space $L^1(\mu)$ is $L^\infty(\mu)$, there is a Borel function g with $|g| \leq 1$ such that $x(f) = \int fg\, d\mu$ for all $f \in C_0(X)$ by continuity and denseness of $C_c(X)$ in $L^1(\mu)$. But

$$\int |g|\, d\mu \geq \sup\{|x(f)| \mid f \in C_0(X),\ \|f\|_u \leq 1\} = \|x\| = 1,$$

so $\mu(X) = 1$ and $|g| = 1$ almost everywhere with respect to μ. Define $\nu \in M(X)$ by $d\nu = g\, d\mu$, and pick h to ν as in Proposition A.8.8. Then $S_\nu = x$ and $d|\nu| = \bar{h}\, d\nu = \bar{h}g\, d\mu = |g|\, d\mu = d\mu$ and $\|\nu\| \equiv |\nu|(X) = \mu(X) = 1 = \|x\|$, so S is also isometric. □

Example A.8.13 On any σ-algebra on a set X with an element x, define a positive measure k_x of norm one by declaring $k_x(A)$ for any measurable set A to be one if $x \in A$ and otherwise zero. Since $\int \chi_A\, dk_x = k_x(A) = \chi_A(x) = \delta_x(\chi_A)$, we see that $\int f\, dk_x = \delta_x(f)$ for $f \in L^1(k_x)$. When X is a locally compact Hausdorff space, then $k_x \in M(X)$ corresponds to $\delta_x \in C_0(X)^\star$ in the theorem above. ◇

The complex measures that are absolutely continuous with respect to a fixed measure on a space evidently form a subspace of all complex measures on the space.

Proposition A.8.14 *Let μ be a Radon measure on a locally compact Hausdorff space. Then $f \mapsto T(f)$ with $dT(f) = f\, d\mu$ is an isometric isomorphism* *from $L^1(\mu)$ onto the space of regular complex Borel measures that are absolutely continuous with respect to μ.*

Proof Let $f \in L^1(\mu)$. As $|f\,d\mu| = hf\,d\mu = |f|\,d\mu$ for a measurable function h with $|h| = 1$, the complex Borel measure given by $d\nu = f\,d\mu$ is regular by Proposition A.8.7, and it is certainly absolutely continuous with respect to μ. Hence T is an isometric linear map into the right space.

To see that T is surjective, consider any regular complex Borel measure ν that is absolutely continuous with respect to μ, which is, say, defined on X. Pick a σ-compact set A such that the finite Radon measure $|\nu|$ vanishes on A^c; just let $A = \cup A_n$, where A_n is compact and $|\nu|(X) - |\nu|(A_n) < 1/n$ for $n \in \mathbb{N}$. Then ν restricted to A is absolutely continuous with respect to μ restricted to A. By the Lebesgue-Radon-Nikodym theorem applied to this σ-finite case, there is $f \in L^1(\mu)$ that vanishes on A^c such that $T(f) = \nu$. □

A similar result clearly holds for any σ-finite measure μ. In this case the map T is an isometric isomorphism from $L^1(\mu)$ onto the space of complex measures that are absolutely continuous with respect to μ; the latter being a subspace of the normed space of all complex measures on the space.

In both cases the integral with respect to a complex measure ν with $d\nu = f\,d\mu$ for $f \in L^1(\mu)$ is $\int g\,d\nu = \int gf\,d\mu$ for $g \in L^1(|\nu|)$. This is trivially so because there is $h \in L^1(|\nu|)$ of modulus one such that $h\,d|\nu| = d\nu = f\,d\mu$.

Using the counting measure on $\mathbb{N}$, we see that the last theorem and proposition provide an isometric isomorphism $f\colon l^1(\mathbb{N}) \to C_0(\mathbb{N})^\star$ given by $f(x)(c) = \sum x_n c_n$ for $x = \{x_n\} \in l^1(\mathbb{N})$ and $c = \{c_n\} \in C_0(\mathbb{N})$.

Invoking the isometric isomorphism $L^\infty(\mu) = L^1(\mu)^\star$ for a Radon measure μ on a σ-compact locally compact Hausdorff space X, we claim that $C_c(X)$ is w^*-dense in $L^\infty(\mu)$. If not, there is by Proposition 2.4.19 a non-zero $g \in L^1(\mu)$ such that $\int fg\,d\mu = 0$ for all $f \in C_c(X)$, contradicting the last theorem.

A.9 Product Integrals

Recall that an *algebra of sets* in a non-empty set X is a collection of subsets of X closed under complements and finite unions, and contains X, whereas a *monotone class* in X is a collection of subsets of X closed under both countable increasing unions and countable decreasing intersections. Any σ-algebra on X is clearly an algebra and a monotone class in X. The intersection of monotone classes in X is again a monotone class in X, and the monotone class *generated* by any collection of subsets in X is the intersection of all monotone classes containing the given collection.

Lemma A.9.1 *The monotone class generated by an algebra of sets is the σ-algebra generated by the algebra of sets.*

Proof Since the σ-algebra M generated by the algebra A of sets is a monotone class, the monotone class C generated by A is contained in M. It remains to show that C is a σ-algebra. For $a \in C$ let $C(a)$ consist of all $b \in C$ such that $a \backslash b, b \backslash a, a \cap b \in C$. Note that $b \in C(a)$ if and only if $a \in C(b)$. Since A is an algebra, we have $A \subset C(a)$ when $a \in A$, and as $C(a)$ clearly is a monotone class, we get $C \subset C(a)$. Hence if $b \in C$ and $a \in A$, then $a \in C(b)$, so $C \subset C(b)$. Thus differences and finite intersections of elements in C belong to C, and as $X \in A \subset C$, we conclude that C is an algebra. Being a monotone class it must be a σ-algebra. □

The *product σ-algebra* of two measures μ and ν is the σ-algebra generated by the products $A \times B$ of measurable sets, whereas the *elementary sets* E of μ and ν are all finite disjoint unions of such products, or *rectangles*.

Corollary A.9.2 *The product σ-algebra of two measures μ and ν is the monotone class generated by the elementary sets E of μ and ν.*

Proof By the lemma we need only show that E is an algebra of sets. The identities $(a_1 \times b_1) \cap (a_2 \times b_2) = (a_1 \cap a_2) \times (b_1 \cap b_2)$ and

$$(a_1 \times b_1) \backslash (a_2 \times b_2) = ((a_1 \backslash a_2) \times b_1) \cup ((a_1 \cap a_2) \times (b_1 \backslash b_2))$$

show that $a \cap b, a \backslash b \in E$, and thus $a \cup b = (a \backslash b) \cup b \in E$, for $a, b \in E$. □

For $A \subset X \times Y$ and $x \in X$ and $y \in Y$, let A_x be the set of $z \in Y$ with $(x, z) \in A$, and let A^y be the set of $w \in X$ with $(w, y) \in A$.

Lemma A.9.3 *If A is measurable in the product σ-algebra of μ and ν, then all A_x and A^y are measurable, and all $f(x, \cdot)$, $f(\cdot, y)$ are measurable for f measurable.*

Proof Let D be the class of measurable A with all A_x measurable. Then every rectangle belongs to D, but as the operation $A \mapsto A_x$ preserves complements and countable unions, we see that D is a σ-algebra, so it contains the product σ-algebra. The proof for A^y is similar.

The set $f^{-1}(U)$ with U open is measurable, so $f(x, \cdot)^{-1}(U) = f^{-1}(U)_x$ is measurable by the first part of the proof. The proof for $f(\cdot, y)$ is similar. □

Lemma A.9.4 *If A belongs to the product σ-algebra of σ-finite measures μ and ν, then $x \mapsto \nu(A_x)$ and $y \mapsto \mu(A^y)$ are measurable and $\int \nu(A_x)\, d\mu(x) = \int \mu(A^y)\, d\nu(y)$.*

Proof Let M be the collection of measurable A for which the lemma holds. Clearly all rectangles belong to M. The union of any increasing sequence of members of M belongs to M by countable addititvity of ν and μ together with the Lebesgue monotone convergence theorem. The same is then true for the union of a sequence of disjoint members of M. Lebesgue's dominated convergence theorem shows that the intersection of a decreasing sequence of members of M belongs to M provided they are all contained in a rectangle $B \times C$ with $\mu(B) < \infty$ and $\nu(C) < \infty$.

Say the sequences $\{X_m\}$ and $\{Y_n\}$ partition the measure spaces for μ and ν, respectively, and that all members have finite measure. Let N be the class of any measurable A with $A_{mn} \equiv A \cap (X_m \times Y_n) \in M$ for all m and n. By what we have already said N is a monotone class that contains all the elementary sets, so N contains the product σ-algebra by the previous corollary. Hence all $A_{mn} \in M$ for A measurable, and as A is the disjoint union of them, we conclude that $A \in M$. □

The *product measure* $\mu \times \nu$ of σ-finite measures μ and ν is the measure on the product σ-algebra, which for a measurable set A is any of the two integrals in the last lemma. That it is indeed a measure is clear from the Lebesgue monotone convergence theorem, and it is then obviously σ-finite. It is clearly the unique measure that is $\mu(B)\nu(C)$ on rectangles $B \times C$.

The following *Fubini-Tonelli theorem* allows reversing the order of integration under fairly mild assumptions.

Theorem A.9.5 *Suppose h is a scalar valued measurable function with respect to the product σ-algebra of σ-finite measures μ and ν. If one of the integrals $\int |h|\, d(\mu \times \nu)$, $\int\int |h|\, d\mu d\nu$ and $\int\int |h|\, d\nu d\mu$ is finite, then with $|h|$ replaced by h, they all make sense, are finite, and do coincide.*

In the last two *iterated integrals* we have omitted brackets around the functions defined almost everywhere by the inner integrals.

Proof By Lemma A.9.3 the functions $x \mapsto \int h(x, \cdot)\, d\nu$ and $y \mapsto \int h(\cdot, y)\, d\mu$ make sense when h is non-negative. By the previous lemma, the integrals in the theorem coincide when h is the characteristic function of a measurable set. Thus they also coincide for h simple, and by invoking Lemma A.6.10 while using the Lebesgue monotone convergence theorem twice, we see that the integrals actually coincide for any non-negative h. The remaining part of the theorem is then seen to hold for real h by decomposing it into its positive and negative parts. For complex h one first decomposes it into its real and imaginary parts. □

We need the following result in the sequel.

Proposition A.9.6 *Let F be a family of non-negative lower semicontinuous functions on a locally compact Haudorff space X with the property that there is $f \in F$ to any $f_i \in F$ such that $f \geq f_1$ and $f \geq f_2$. For any Radon measure on μ on X we have $\int \sup_{f\in F} f\, d\mu = \sup_{f\in F} \int f\, d\mu$. In particular, the integral $\int f\, d\mu$ of a non-negative lower semicontinuous function is the supremum of the integrals $\int g\, d\mu$ over $g \in C_c(X)$ with $0 \leq g \leq f$.*

Proof Clearly $h = \sup_{f\in F} f$ is lower semicontinuous and Borel measurable. To prove $\int h\, d\mu \leq \sup_{f\in F} \int f\, d\mu$, let $s_n = 2^{-n} \sum_{i=1}^{2^{2n}} \chi_{U_{in}}$ with $U_{ni} = h^{-1}(\langle i2^{-n}, \infty\rangle)$, see also Lemma A.6.10. To $a < \int h\, d\mu$, there is by the Lebesgue monotone convergence theorem an n such that $\int s_n\, d\mu > a$. Pick compact sets $A_i \subset U_{ni}$ such that the integral of $t = 2^{-n} \sum \chi_{A_i}$ exceeds a. Pick $f_x \in F$ such that $f_x(x) > t(x)$. Now $f_x - t$ is lower semicontinuous, so $(f_x - t)^{-1}(\langle 0, \infty\rangle)$ is a neighborhood of x, and we can cover the compact set $\cup A_i$ with finitely many such sets as x varies. Picking $f \in F$ not less than any of the corresponding f_x, we get $\int f\, d\mu > a$. □

Proposition A.9.7 *If μ and ν are Radon measures on X and Y, respectively, then $C_c(X \times Y) \subset L^1(\mu \times \nu)$ and $\int f\, d(\mu \times \nu) = \int\int f\, d\mu d\nu = \int\int f\, d\nu d\mu$ for $f \in C_c(X \times Y)$.*

Proof Define $f \otimes g \in C_c(X \times Y)$ for $f \in C_c(X)$ and $g \in C_c(Y)$ by the formula $(f \otimes g)(x, y) = f(x)g(y)$. Note that $f \otimes g = (fp)(gq)$, where p and q are the canonical projections onto the first and second factor in $X \times Y$. So $f \otimes g$ is measurable with respect to the product σ-algebra. By the Stone-Weierstrass theorem the self-adjoint subalgebra A of $C_c(X \times Y)$ spanned by all such functions is dense in the Banach spaces $C_0(X \times Y)$. Since products, sums and pointwise limits of measurable functions remain measurable, we conclude that every $f \in C_c(X \times Y)$ is measurable. Moreover, since such f is supported and bounded on the rectangle $p(\operatorname{supp} f) \times q(\operatorname{supp} f)$ with finite measure, we get $C_c(X \times Y) \subset L^1(\mu \times \nu)$. We can then apply the Fubini-Tonelli theorem for the finite measures obtained by restricting μ to $p(\operatorname{supp} f)$ and ν to $q(\operatorname{supp} f)$. □

The *Radon product measure* of Radon measures μ and ν is the measure $\mu \otimes \nu$ associated to the positive linear functional $f \mapsto \int f\, d(\mu \times \nu)$ according to the Riesz representation theorem. Since the product σ-algebra is generated by all rectangles $A \times B$ with A and B open, we see that its measurable sets are Borel. Thus it makes sense to ask if $\mu \otimes \nu$ restricts to $\mu \times \nu$ on the product σ-algebra. In order to answer this we need a lemma.

Lemma A.9.8 *Let μ and ν be Radon measures on X and Y, respectively. If A is Borel measurable in $X \times Y$, then A_x and A^y are measurable. If f is a scalar valued Borel measurable function on $X \times Y$, then $f(x, \cdot)$ and $f(\cdot, y)$ are measurable, and if $f \in C_c(X \times Y)$, then the functions $x \mapsto \int f(x, \cdot)\, d\nu$ and $y \mapsto \int f(\cdot, y)\, d\mu$ are continuous.*

Proof The first two statements are proved like in Lemma A.9.3. The remaining part of the proof things on the fact that $x \mapsto f(x, \cdot)$ and $y \mapsto f(\cdot, y)$ are continuous with respect to the uniform norms, see also Lemma A.10.3. □

Proposition A.9.9 *Let μ and ν be σ-finite Radon measures on X and Y, respectively. If A is Borel measurable in $X \times Y$, then $x \mapsto \nu(A_x)$ and $y \mapsto \mu(A^y)$ are measurable, and $(\mu \otimes \nu)(A) = \int \nu(A_x)\, d\mu(x) = \int \mu(A^y)\, d\nu(y)$. Moreover, the restriction of $\mu \otimes \nu$ to the product σ-algebra is $\mu \times \nu$.*

Proof Fix open sets A and B with $\mu(A) < \infty$ and $\nu(B) < \infty$. Let M be the collection of Borel sets such that their intersection with $A \times B$ satisfy the theorem.

We claim that M contains every open set U. Indeed, let F be the family of all $f \in C_c(X \times Y)$ with $0 \leq f \leq \chi_U$. Then $\chi_U = \sup_{f \in F} f$, so we get $\chi_{U_x} = \sup_{f \in F} f(x, \cdot)$ and $\chi_{U^y} = \sup_{f \in F} f(\cdot, y)$. Hence $\nu(U_x) = \sup_{f \in F} \int f(x, \cdot)\, d\nu$ and $\mu(U^y) = \sup_{f \in F} \int f(\cdot, y)\, d\mu$ by Proposition A.9.6. From the last lemma it follows that $x \mapsto \nu(U_x)$ and $y \mapsto \mu(U^y)$ are lower semicontinuous, and thus measurable. Using Proposition A.9.6 twice together with the last proposition above, we get

$$(\mu \otimes \nu)(U) = \sup_{f \in F} \int f \, d(\mu \otimes \nu) = \sup_{f \in F} \int \int f(x, \cdot) \, d\nu d\mu(x) = \int \nu(U_x) \, d\mu(x)$$

and similarly $(\mu \otimes \nu)(U) = \int \mu(U^y) \, d\nu(y)$.

If $E_i \in M$ and $E_1 \subset E_2$, then $E_2 \backslash E_1 \in M$ because

$$(\mu \otimes \nu)(E_2 \cap (A \times B)) = (\mu \otimes \nu)(E_1 \cap (A \times B)) + (\mu \otimes \nu)((E_2 \backslash E_1) \cap (A \times B))$$

and likewise for $\nu((E_2 \cap (A \times B))_x)$ and $\mu((E_2 \cap (A \times B))^y)$.

We also see that M is closed under finite disjoint unions, under countable increasing unions by the Lebesgue monotone convergence theorem, and thus also under countable decreasing intersections.

Since $(a_1 \backslash b_1) \cap (a_2 \backslash b_2) = (a_1 \cap a_2) \backslash (b_1 \cup b_2)$ and $(a \backslash b)^c = ((X \times Y) \backslash a) \cup ((a \cap b) \backslash \phi)$, we see that the collection of finite disjoint unions of differences of open sets in $X \times Y$ is an algebra of sets. By Lemma A.9.1 the monotone class generated by this algebra coincides with the σ-algebra generated by the algebra, which is the σ-algebra of Borel sets on $X \times Y$. By what we have seen, and as $a \backslash b = a \backslash (a \cap b)$, the collection M certainly contains this monotone class, so M consists of all Borel sets.

Since μ and ν are σ-finite Radon measures, we may write $X = \cup_{m=1}^{\infty} A_m$ and $Y = \cup_{n=1}^{\infty} B_n$ for increasing sequences of open sets with finite measures. For any Borel set E in $X \times Y$, we have seen that the sets $E \cap (A_m \times B_n)$ satisfy the theorem, and so do their union E by the Lebesgue monotone convergence theorem.

If E is measurable in the product σ-algebra, we see then by invoking the Fubini-Tonelli theorem that $(\mu \otimes \nu)(E) = \int \nu(A_x) \, d\mu(x) = (\mu \times \nu)(A)$. □

Repeating the proof of the Fubini-Tonelli theorem, but using now instead the previous lemma and proposition, we get the following version of the Fubini-Tonelli theorem for Radon measures.

Theorem A.9.10 *Let μ and ν be Radon measures on X and Y, respectively. Assume $h \colon X \times Y \to \mathbb{C}$ is Borel measurable and that it vanishes almost everywhere outside a σ-compact set. If one of the integrals $\int |h| \, d(\mu \otimes \nu)$, $\int \int |h| \, d\mu d\nu$ and $\int \int |h| \, d\nu d\mu$ is finite, then with $|h|$ replaced by h, they all make sense, are finite, and do coincide.*

As for complex measures, consider locally compact Hausdorff spaces X_i and let $\nu_i \in M(X_i)$ with $h_i \in L^1(|\nu_i|)$ of moduli one such that $h_i d|\nu_i| = d\nu_i$. Then we define the product measure $\nu_1 \otimes \nu_2$ by $d(\nu_1 \otimes \nu_2) = (h_1 \otimes h_2) \, d(|\nu_1| \otimes |\nu_2|)$, where $|\nu_1| \otimes |\nu_2|$ is the Radon product measure of $|\nu_1|$ and $|\nu_2|$. By the Tonelli-Fubini theorem for Radon measures, we have $|\int h \, d(\nu_1 \otimes \nu_2)| \leq \|h\|_u \int d(|\nu_1| \otimes |\nu_2|) \leq \|h\|_u \|\nu_1\| \cdot \|\nu_2\|$ for $h \in C_0(X_1 \times X_2)$, so viewing $M(X_i)$ as the dual of the Banach space $C_0(X_i)$, we see that $\nu_1 \otimes \nu_2 \in M(X_1 \times X_2)$ with regularity guaranteed by Proposition A.8.14.

Note that $\int(f_1 \otimes f_2)\, d(\nu_1 \otimes \nu_2) = \int f_1\, d\nu_1 \int f_2\, d\nu_2$ for $f_i \in L^1(|\nu_i|)$ by the Tonelli-Fubini theorem for Radon measures. For $x_i \in X_i$ we have $\int h\, d(k_{x_1} \otimes \nu_2) = \int h(x_1, \cdot)\, d\nu_2$ since both sides are bounded linear functionals in $h \in C_0(X_1 \times X_2)$ and they evidently coincide for $h = f_1 \otimes f_2$ with $f_i \in C_0(X_i)$. Similarly, we have $\int h\, d(\nu_1 \otimes k_{x_2}) = \int h(\cdot, x_2)\, d\nu_1$.

A.10 The Haar-Measure

We consider here the basic property of translation invariance in a more general context.

Definition A.10.1 A *topological group* G is a group with a Hausdorff topology making $G \times G \to G;\ (s, t) \mapsto st^{-1}$ continuous, where $G \times G$ has the product topology.

So $\mathbb{R}^n$ is a locally compact (topological) group under addition since the topology on $\mathbb{R}^n$ is the product topology from $\mathbb{R}$. All matrix groups are evidently locally compact groups, and so is any group with discrete topology.

We have the following fundamental result.

Theorem A.10.2 *Every locally compact group G has a non-zero Radon measure μ that is* left invariant, *meaning that $\mu(sA) = \mu(A)$ for every Borel set A and $s \in G$, and which is unique up to multiplication with a positive scalar.*

Such a measure is called the *Haar measure* on the group, where an appropriate scaling is understood. Any Haar measure is positive on every non-empty open set A because the measure of every compact set B cannot be zero, and yet $B \subset \cup s_i A$ for finitely many s_i. We often write $\int f(t)\, d\mu(t)$, or simply $\int f(t) dt$, for the *Haar integral* of $f \in L^1(G) \equiv L^1(\mu)$. Left invariance of the measure then means that the integral is *left invariant*, meaning $\int f(st) dt = \int f(t) dt$ for every $s \in G$. For the forward implication, approximate f by measurable simple functions, and for the opposite direct, recall the construction of the measure from a linear functional in the Riesz representation theorem.

The unique (up to a positive scalar) Haar measure of the additive group $\mathbb{R}^n$ is the Lebesgue measure. For many groups concrete formulas for the Haar integrals are known. In these cases the uniqueness result is more important than existence. To prove Haar's result, we need the uniform norm part of the following result.

Lemma A.10.3 *Let G be a locally compact group and let $f \in C_c(G)$. There is to $\varepsilon > 0$ a neighborhood V of the unit $e \in G$ with $\| f(\cdot s) - f\|_u < \varepsilon$ for all $s \in V$. The same statement holds when u is replaced by $p \in [1, \infty\rangle$ with integration over any Radon measure μ on G and $f \in L^p(\mu)$.*

Proof Consider the uniform norm first. Since both the product in the group and the inverse operation are continuous maps, and since f is continuous, then for each $t \in \text{supp}(f) \equiv A$, we can find a neighborhood V_t of the unit e with $V_t^{-1} = V_t$ and such that $|f(ts) - f(t)| < \varepsilon/2$ for all $s \in V_t V_t$. Thus there are finitely many t_i in the compact set A so that the open sets $\{t_i V_{t_i}\}$ cover it. It is easy to check that for any $t \in G$, we get $|f(ts) - f(t)| < \varepsilon$ for all $s \in V \equiv \cap V_{t_i}$.

Next arrange so that V above has compact closure. Then for any $s \in V$, we have

$$\|f(\cdot s) - f\|_p^p = \int |f(\cdot s) - f|^p \, d\mu \leq \varepsilon^p \mu(A\overline{V})$$

and $\mu(A\overline{V}) < \infty$ as $A\overline{V}$ is compact, being the image under the continuous product map of $A \times \overline{V}$, which is again compact by Tychonoff's theorem. When $f \in L^p(\mu)$ invoke Lusin's theorem. □

Oviously, the statement with $f(\cdot s)$ replaced by $f(s\cdot)$ also holds, and both statements hold for any Borel measure on G that is finite on compact sets.

The construction of the Haar measure is done most conveniently by going via linear functionals. Let G be a locally compact group.

For positive $f, g \in C_c(G)$, finitely many $\{s \in G \,|\, g(s) > \|g\|_u/2\}$ cover $\mathrm{supp}(f)$, so

$$f \leq (2\|f\|_u/\|g\|_u) \sum g(s_i \cdot)$$

for finitely many $s_i \in G$. Let $(f : g)$ be the infinum over all finite sums $\sum c_i$ such that $f \leq \sum c_i g(s_i \cdot)$ for some $s_i \in G$. As $(f : g) \geq \|f\|_u/\|g\|_u > 0$, we may define positive numbers $x_g(f) = (f : g)/(f_0 : g)$ for fixed f_0. Clearly x_g is left invariant, and

$$(f_0 : f)^{-1} \leq x_g(f) \leq (f : f_0)$$

as $(\cdot : \cdot)$ is a transitive relation. Evidently x_g is subadditive on the positive continuous functions with compact support, and it respects multiplication by positive scalars. In fact, we have the following result.

Lemma A.10.4 *There is a neighborhood V of e to $\varepsilon > 0$ such that*

$$x_g(f_1) + x_g(f_2) \leq x_g(f_1 + f_2) + \varepsilon$$

whenever $\mathrm{supp}(g) \subset V$.

Proof Let $r > 0$. Pick positive $v \in C_c(G)$ that is one on $\mathrm{supp}(f_1 + f_2)$, and let $h = f_1 + f_2 + rv$. Define positive $h_i \in C_c(G)$ to be f_i/h on the support of f_i, and zero outside. By the lemma above there is neighborhood V of the unit e such that $|h_i(s) - h_i(t)| < r$ for $t^{-1}s \in V$. If $h \leq \sum c_j g(s_j \cdot)$, then for any t we have

$$f_i(t) = h(t)h_i(t) \leq \sum c_j g(s_j t)(h_i(s_j) + r),$$

so $(f_i : g) \leq \sum c_j(h_i(s_j) + r)$ and $(f_1 : g) + (f_2 : g) \leq \sum c_j(1 + 2r)$ as $h_1 + h_2 \leq 1$. Hence

$$x_g(f_1) + x_g(f_2) \leq (1 + 2r)x_g(h) \leq (1 + 2r)(x_g(f_1 + f_2) + r x_g(v)),$$

so it suffices to choose r such that

$$2r(f_1 + f_2 : f_0) + r(1 + 2r)(v : f_0) < \varepsilon.$$

□

Proof Of the existence of a Haar measure. Let X be the direct product of $[(f_0 : f)^{-1}, (f : f_0)]$ over all positive $f \in C_c(G)$, so X is compact in the product topology by Tychonoff's theorem and the Heine-Borel theorem, and $x_g \in X$ for every positive $g \in C_c(G)$. For each neighborhood V of the unit e, let $X(V)$ denote the closure in X of the set of all x_g with $\mathrm{supp}(g) \subset V$. There is an element x in the intersection of all such $X(V)$, which is non-empty since X is compact and $\cap X(V_i) \supset X(\cap V_i) \neq \phi$ for any finite collection V_i. Thus for any neighborhood V of e and finitely many positive $f_i \in C_c(G)$, there is a positive $g \in C_c(G)$ with $\mathrm{supp}(g) \subset V$ such that $|x(f_i) - x_g(f_i)| < \varepsilon$. Hence x is left invariant, and by the lemma above, we see that x is additive for positive continuous functions with compact support. We can then extend x to a positive functional on $C_c(G)$, just like we extended integrals. Applying the Riesz representation theorem to x we therefore get a Haar measure on G. □

We turn now to uniqueness.

Proof Of uniqueness of Haar measure. Given two such measures μ and ν, we must show that $\int f\,d\mu / \int f\,d\nu = \int g\,d\mu / \int g\,d\nu$ for positive $f, g \in C_c(G)$. Pick a neighborhood U of e with compact closure and such that $U^{-1} = U$. For $s \in U$, the functions $t \mapsto f(ts) - f(st)$ and $t \mapsto g(ts) - g(st)$ have support in the compact sets

$$A = \mathrm{supp}(f)\overline{U} \cup \overline{U}\mathrm{supp}(f) \quad \text{and} \quad B = \mathrm{supp}(g)\overline{U} \cup \overline{U}\mathrm{supp}(g).$$

To $\varepsilon > 0$ we may by Lemma A.10.3 pick a neighborhood $V \subset U$ of e such that these functions both have uniform norm less than ε when $s \in V$. Finally pick a positive $h \in C_c(G)$ with support in V and which satisfies $h(s) = h(s^{-1})$. Then

$$\int h\,d\nu \int f\,d\mu = \int\int h(s) f(t)\,d\mu(t) d\nu(s) = \int\int h(s) f(st)\,d\mu(t) d\nu(s)$$

while

$$\int h\,d\mu \int f\,d\nu = \int\int h(t) f(s)\,d\mu(t) d\nu(s) = \int\int h(s^{-1}t) f(s)\,d\mu(t) d\nu(s)$$

$$= \int\int h(t^{-1}s)f(s)\,d\nu(s)d\mu(t) = \int\int h(s)f(ts)\,d\nu(s)d\mu(t)$$
$$= \int\int h(s)f(ts)\,d\mu(t)d\nu(s)$$

by the Fubini-Tonelli theorem, so

$$|\int h\,d\nu \int f\,d\mu - \int h\,d\mu \int f\,d\nu| = |\int\int h(s)(f(ts) - f(st))\,d\mu(t)d\nu(s)|$$
$$\leq \varepsilon\mu(A)\int h\,d\nu$$

and similarly

$$|\int h\,d\nu \int g\,d\mu - \int h\,d\mu \int g\,d\nu| \leq \varepsilon\mu(B)\int h\,d\nu$$

Dividing the first of these inequalities by $\int h\,d\nu \int f\,d\nu$ and the second one by $\int h\,d\nu \int g\,d\nu$, and then adding them, gives

$$|\frac{\int f\,d\mu}{\int f\,d\nu} - \frac{\int g\,d\mu}{\int g\,d\nu}| \leq \varepsilon(\frac{\mu(A)}{\int f\,d\nu} + \frac{\mu(B)}{\int g\,d\nu})$$

and we are done. □

By symmetry there is also a unique up to a positive scalar right invariant non-zero Radon measure on any locally compact group G. This measure need not coincide with the Haar measure μ. However, for any $s \in G$, the formula $A \mapsto \mu(As)$ defines a left invariant Radon measure on G. By uniqueness there must be a positive number $\Delta(s)$ such that $\Delta(s)\int f(ts)dt = \int f(t)dt$ for any $f \in L^1(G)$. The function $\Delta\colon G \to \langle 0, \infty\rangle$ so obtained is called the *modular function*. It is a continuous map by Lemma A.10.3, and it is obviously multiplicative. We say G is *unimodular* if $\Delta = 1$. This happens if and only if μ is also right invariant. Abelian groups are obviously unimodular, and so are compact ones as we may use $f = 1$. Discrete groups are also unimodular with the counting measure μ as Haar measure, so $\int f(t)\,dt = \sum_{t\in G} f(t)$.

Here is a standard example of a non-unimodular group.

Example A.10.5 It is readily checked that the *$ax + b$-group*

$$G = \{\begin{pmatrix} a & b \\ 0 & 1 \end{pmatrix} \mid a, b \in \mathbb{R}, a > 0\}$$

of *affine transformations of* $\mathbb{R}$ has Haar measure $a^{-2}dadb$ and right invariant Radon measure $a^{-1}dadb$. ◇

Note that the Haar measure of a compact group is finite. The converse is also true, namely if the Haar measure is finite, then the group G must be compact. Suppose to the contrary that G is not compact. Consider a neighborhood U of the unit e with compact closure $\overline{U}$. Then no finite set of translates of $\overline{U}$ can cover G, so there is a sequence $\{t_n\}$ in G such that $t_n \notin \cup_{i<n} t_i \overline{U}$. Since $(s, t) \mapsto st^{-1}$ is continuous and every neighborhood of (e, e) in $G \times G$ contains some $V \times V$ with V open, there is an open set V with $VV^{-1} \subset \overline{U}$. Then $\{t_n V\}$ are pairwise disjoint and all have measure $\mu(V)$ by left invariance, so μ is not finite.

The Haar measure allows introducing an important Banach algebra.

Theorem A.10.6 *The space $L^1(G)$ is a Banach algebra with product*

$$f * g = \int f(t)g(t^{-1}\cdot)dt.$$

Proof We already know that $L^1(G)$ is a Banach space. By Lusin's theorem there are σ-compact sets A and B such that f and g are zero almost everywhere on A^c and B^c, respectively. Hence the measurable function on $G \times G$ given by $(s, t) \mapsto f(t)g(t^{-1}s)$ is zero almost everywhere outside the σ-compact set $AB \times A$. Now $\int\int |f(t)g(t^{-1}s)|\, ds dt = \|f\|_1 \|g\|_1 < \infty$ by left invariance, so by the Fubini-Tonelli theorem we conclude that $f * g \in L^1(G)$ and $\|f * g\|_1 \leq \|f\|_1 \|g\|_1$.

Using again the Fubini-Tonelli theorem in a similar fashion we see that the product is associative, and it is clearly distributive. □

Corollary A.10.7 *Let $p \in [1, \infty\rangle$, and let $f \in L^1(G)$ and $g \in L^p(G)$. Then $f * g \in L^p(G)$ and $\|f * g\|_p \leq \|f\|_1 \|g\|_p$.*

Proof The case $p = 1$ is the theorem above, so let $p \in \langle 1, \infty\rangle$ with conjugate exponent q. Pick sets A and B as in the proof of that theorem, and consider functions $h \in L^q(G)$ which are zero outside AB. From Hölder's inequality and left invariance, we have $\int\int |f(t)g(t^{-1}s)h(s)|\, ds\, dt \leq \|f\|_1 \|g\|_p \|h\|_q$, so by the Fubini-Tonelli theorem, the function $h \int f(t)g(t^{-1}\cdot)\, dt$ belongs to $L^1(G)$ with norm dominated by $\|f\|_1 \|g\|_p \|h\|_q$. Choosing h positive on AB, we see that $f * g$ is well-defined almost everywhere, and we are now done by Theorem A.8.11. □

One can use convolution to produce continuous functions from measurable ones. Suppose $f \in L^p(G)$ for $p \in [1, \infty\rangle$ and $\tilde{g} \in L^q(G)$ for the conjugate exponent q, where by tilde we mean $\tilde{g}(t) = g(t^{-1})$. Then $f * g$ is a bounded continuous function on G. Indeed, by Hölder's inequality and left invariance we have $|f * g(s)| \leq \int |f\tilde{g}(s^{-1}\cdot)|\, d\mu \leq \|f\|_p \|\tilde{g}\|_q$ and $|(f * g)(s) - (f * g)(e)| \leq \|f\|_p \|\tilde{g}(s^{-1}\cdot) - \tilde{g}\|_q$, which by the left version of Lemma A.10.3 can be made less than any $\varepsilon > 0$ for s in a corresponding neighborhood of the unit e. Moreover, when $p \in \langle 1, \infty\rangle$, then picking sequences $\{f_n\}$ and $\{g_n\}$ in $C_c(G)$ such that $\|f - f_n\|_p \to 0$ and $\|\tilde{g} - \tilde{g}_n\| \to 0$, we get

$$\|f_n * g_n - f * g\|_u \leq \|f_n\|_p \|\tilde{g}_n - \tilde{g}\|_q + \|f_n - f\|_p \|\tilde{g}\|_q \to 0,$$

so $f * g \in C_0(G)$.

We also see that $f * g \in C_c(G)$ for $f, g \in C_c(G)$ as $\operatorname{supp}(f * g) \subset \operatorname{supp}(f)\operatorname{supp}(g)$.

Recall that a $*$-operation on a complex algebra is a conjugate linear map from the algebra to itself that is antimultiplicative and which equals its own inverse.

Corollary A.10.8 *The map $f \mapsto f^\star$ on $C_c(G)$ with $f^\star(t) = \overline{f(t^{-1})}\Delta(t)^{-1}$ extends uniquely to an isometric $*$-operation $\star$ on the Banach algebra $L^1(G)$.*

Proof It is easy to see that $f \mapsto \int f(t^{-1})\Delta(t)^{-1}\,dt$ is a non-trivial left invariant positive functional on $C_c(G)$, so there is $c > 0$ such that

$$\int f(t^{-1})\Delta(t)^{-1}\,dt = c\int f(t)\,dt$$

for all $f \in C_c(G)$. Pick a neighborhood V of the unit with $V^{-1} = V$ and such that $|\Delta(t)^{-1} - 1| \leq |c - 1|/2$ for $t \in V$. Also pick a positive $f \in C_c(G)$ with support in V and such that $f(t^{-1}) = f(t)$ for all $t \in V$. Then

$$|c-1|\int f(t)\,dt = |\int f(t)(\Delta(t)^{-1} - 1)\,dt| \leq |c-1|\int f(t)\,dt/2$$

which is impossible unless $c = 1$. So $\|f^\star\|_1 = \|f\|_1$ for $f \in C_c(G)$. Hence by Lusin's theorem the isometric map $f \mapsto f^\star$ from $C_c(G)$ to $C_c(G)$ extends to an isometry $L^1(G) \to L^1(G)$, see the proof of Proposition 2.1.19. Involutiveness is now easily checked. □

Let $K(G)$ denote the dense $*$-subalgebra of $L^1(G)$ consisting of continuous functions with compact support.

Proposition A.10.9 *There is an approximate unit for the Banach algebra $L^1(G)$ consisting of positive functions $f_i \in K(G)$ of norm one such that $f_i^\star = f_i$.*

Proof To any neighborhood U of the unit e pick a continuous function g with support in U, that is one at e and has range in $[0, 1]$. Rescaling $g\tilde{g}$ to have integral one, and using as directed set the neighborhoods of e under reverse inclusion, we get a net $\{g_i\}$ of continuous positive functions that have integral one and have support in any neighborhood U of e for large enough i. Then $f_i = (g_i + g_i^\star)/2$ are in addition self-adjoint. By the left version of Lemma A.10.3 there is a neighborhood U of e to any $f \in L^1(G)$ and $\varepsilon > 0$ such that $\|f(t^{-1}\cdot) - f\|_1 < \varepsilon$ for $t \in U$. Pick i such that $\operatorname{supp}(f_i) \subset U$. By the Fubini-Tonelli theorem we then get

$$\|f_j * f - f\|_1 \leq \int\int f_j(t)|f(t^{-1}s) - f(s)|\,dt\,ds = \int f_j(t)\|f(t^{-1}\cdot) - f\|_1\,dt \leq \varepsilon$$

and $\|f^\star * f_j - f^\star\| = \|(f_j * f - f)^\star\| = \|f_j * f - f\| \leq \varepsilon$ for all $j \geq i$. □

We can also talk about convolution of measures.

Theorem A.10.10 *Suppose G is a locally compact group with unit e and product $p\colon G \times G \to G$. Then the Banach space $M(G)$ is a unital Banach $*$-algebra with* *identity element k_e and product and $*$-operation given by*

$$\int f\, d(\nu_1 * \nu_2) = \int fp\, d(\nu_1 \otimes \nu_2) \quad \text{and} \quad \int f\, d(\nu_1^\star) = \overline{\int f^a\, d\nu_1}$$

for $f \in C_0(G)$, where $f^a(t) = \overline{f(t^{-1})}$ and $M(G)$ is viewed as the dual of the Banach space $C_0(G)$. Moreover, the Banach $*$-algebra $L^1(G)$ sits as a self-adjoint closed ideal in $M(G)$, namely as those measures which are absolutely continuous with respect to the Haar measure.

Proof Now $|\int fp\, d(\nu_1 \otimes \nu_2) \leq \|f\|_u \|\nu_1\| \cdot \|\nu_2\|$, so $\nu_1 * \nu_2 \in M(G)$ with $\|\nu_1 * \nu_2\| \leq \|\nu_1\| \cdot \|\nu_2\|$. By the Fubini-Tonelli theorem, combined with associativity of the group product, we see that our multiplication in $M(G)$ is indeed associative. As $|\overline{\int f^a\, d\nu_1}| \leq \|f\|_u \|\nu_1\|$ we conclude that $\nu_1^\star \in M(G)$ with $\|\nu_1^\star\| \leq \|\nu_1\|$. Clearly $\star$ is involutive, so $\|\nu_1\| = \|\nu_1^{\star\star}\| \leq \|\nu_1^\star\|$ and $\star$ is isometric.

As for antimultiplicativity, define an isometric map $K : C_b(G\times G) \to C_b(G\times G)$ by $K(h)(s,t) = \overline{h(t^{-1}, s^{-1})}$, so the functional $\int K(\cdot)\, d(\nu_1 \otimes \nu_2)$ on $C_b(G \times G)$ is bounded by $\|\nu_1\| \cdot \|\nu_2\|$. Note that

$$\overline{\int (f \otimes g)\, d(\nu_2^\star \otimes \nu_1^\star)} = \int f^a\, d\nu_2 \int g^a\, d\nu_1 = \int K(f \otimes g)\, d(\nu_1 \otimes \nu_2)$$

for $f, g \in C_0(G)$, so $\int (\cdot)\, d(\nu_2^\star \otimes \nu_1^\star) = \overline{\int K(\cdot)\, d(\nu_1 \otimes \nu_2)}$ by denseness, linearity and continuity. Using $K(fp) = f^a p$, we therefore get

$$\int f\, d(\nu_2^\star * \nu_1^\star) = \int fp\, d(\nu_2^\star \otimes \nu_1^\star) = \overline{\int K(fp)\, d(\nu_1 \otimes \nu_2)} = \int f\, d((\nu_1 * \nu_2)^\star),$$

so $(\nu_1 * \nu_2)^\star = \nu_2^\star * \nu_1^\star$.

To see that k_e is the identity element of $M(G)$, we calculate

$$\int f\, d(k_e * \nu_1) = \int fp(e, \cdot)\, d\nu_1 = \int f\, d\nu_1 = \int f\, d(\nu_1 * k_e).$$

To say that $L^1(G)$ sits inside $M(G)$ as a Banach $*$-algebra means that the isometric linear map $T : L^1(G) \to M(G)$ from Proposition A.8.14 also preserves products and $*$-operations. We check $T(f_1 * f_2) = T(f_1) * T(f_2)$ for $f_i \in C_0(G)$ by calculating on the one hand

$$\int g\, dT(f_1 * f_2) = \int g(t)(f_1 * f_2)(t)\, dt = \int \int g(t) f_1(s) f_2(s^{-1}t)\, ds\, dt$$

for $f \in C_0(G)$, and on the other hand

$$\int g\, d(T(f_1) * T(f_2)) = \int gp\, d(T(f_1) \otimes T(f_2)) = \int g(st) f_1(s) f_2(t)\, ds\, dt$$

which equals the first expression by the Fubini-Tonelli theorem and left invariance. In the last step we also used that $\int (\cdot)\, d(T(f_1) \otimes T(f_2)) = \int (\cdot)(f_1 \otimes f_2)\, d(\mu \otimes \mu)$ for the Haar measure μ, and this is verified by checking for $g_1 \otimes g_2$ with $g_i \in C_0(G)$ and using denseness of their linear span together with linearity and continuity.

As for the $*$-operation it suffices to check $T(f^\star) = T(f)^\star$ for $f \in C_c(G)$ as T and the $*$-operations are all isometric maps. To this end, again by continuity, it is enough to check that $\int g\, dT(f^\star) = \int g\, d(T(f)^\star)$ for $g \in C_c(G)$. Now the left hand side equals $\int g(t) f^\star(t)\, dt = \int g(t) \overline{f(t^{-1})} \Delta(t)^{-1}\, dt$, whereas the right hand side equals $\overline{\int g^a\, dT(f)} = \overline{\int g^a(t) f(t)\, dt} = \int g(t^{-1}) \overline{f(t)}\, dt$, and they coincide by plugging in $h = \tilde{g}\overline{f} \in C_c(G)$ in the formula $\int h(t^{-1}) \Delta(t)^{-1}\, dt = \int h(t)\, dt$ from the proof of Corollary A.10.8.

So to see that $L^1(G)$ is an ideal in $M(G)$, we need only show $\nu * T(f) \in L^1(G)$ for $\nu \in M(G)$ and positive $f \in L^1(G)$. Just like in Proposition A.8.14, there is a σ-compact set A such that $|\nu|(A) = |\nu|(G)$. The function $h(s, t) = \chi_A(s) f(s^{-1}t)$ has σ-compact support and

$$\int \int h(\cdot, t)\, dt\, d\nu = \int \|f\|_1 \chi_A\, d\nu = \nu(G)\|f\|_1,$$

so by the Fubini-Tonelli theorem, the function $t \mapsto \int h(\cdot, t)\, d\nu = \int \tilde{f}(t^{-1} \cdot)\, d\nu$ belongs to $L^1(G)$ with $\int \int \tilde{f}(t^{-1} \cdot)\, d\nu\, dt = \nu(G)\|f\|_1$. Applying the Fubini-Tonelli theorem once more, we get

$$\int (g(t) \int \tilde{f}(t^{-1} \cdot)\, d\nu)\, dt = \int \int g(\cdot t) f(t)\, dt\, d\nu = \int g\, d(\nu * T(f))$$

for $g \in C_c(G)$, so $\nu * T(f)$ is the image under T of the function $t \mapsto \int \tilde{f}(t^{-1} \cdot)\, d\nu$. □

Corollary A.10.11 *A locally compact group G is discrete if and only if $L^1(G)$ is unital, and in this case $M(G) = L^1(G) = l^1(G)$ provided we normalize the Haar measure to be one at the unit e; it is then the counting measure.*

Proof If G is discrete, then $L^1(G) = l^1(G)$ is unital, so $L^1(G) = M(G)$ as $L^1(G)$ is an ideal in $M(G)$.

Conversely, if $L^1(G)$ is unital, then k_e is the identity element for $L^1(G) = M(G)$. If the Haar measure μ vanishes at e, then so does $|k_e|$ as $|k_e| \ll \mu$. By regularity there is a neighborhood U of e such that $|k_e|(U) < 1/2$. Pick $f \in C_c(G)$ with $f(e) = 1$ and range in $[0, 1]$. Then $1 = \delta_e(f) = |\int f\, dk_e| \leq \int |f|\, d|k_e| < 1/2$, so $0 < \mu(e) = \mu(t)$ for all $t \in G$ by left invariance. By regularity every compact set is therefore finite, and since G is locally compact, there are non-empty finite open sets. Using the Hausdorff property to one of these, there is an element which is open as a single element set. By left translations all single element sets are thus open, so G is discrete. □

Clearly $M(G)$ is commutative when G is an abelian locally compact group. The converse is also true because $t \mapsto k_t$ embeds any locally compact group G into $M(G)$, then with the w^*-topology gotten by regarding $M(G)$ as the dual of the Banach space $C_0(G)$, just like in Gelfand's theorem, and obviously $k_s * k_t = k_{st}$ holds for $s, t \in G$.

As for the convolution product between k_s and $\nu \in M(G)$ with $d\nu = f\,d\mu$ for $f \in L^1(G)$, we have

$$\int g\,d(k_s * \nu) = \int (gp)(s,t) f(t)\,dt = \int g(t) f(s^{-1}t)\,dt$$

for $g \in C_0(G)$, so $k_s * f = f(s^{-1}\cdot)$ and similarly $f * k_s = \Delta(s)^{-1} f(\cdot s^{-1})$.

In the case of a locally compact group we don't need to assume σ-compactness in Theorem A.8.11.

Proposition A.10.12 *As Banach spaces we have $L^1(G)^\star = L^\infty(G)$ for any locally compact group G.*

Proof Now $S_g(f) = \int f(t)g(t)\,dt$ for $g \in L^\infty(G)$ and $f \in L^1(G)$ clearly defines an isometric linear map $g \mapsto S_g$ from $L^\infty(G)$ to the dual of $L^1(G)$. As for surjectivity, pick a symmetric neighborhood of e with compact closure A, consider the σ-compact open subgroup $H = \cup_{n=1}^{\infty} A^n$ of G, and let $\{H_i\}$ be the partition of its left cosets. Now if $x \in L^1(G)^\star$, let $g = \sum \chi_{H_i} g_i$ where $g_i \in L^\infty(H_i)$ is associated via Theorem A.8.11 to x restricted to $L^1(H_i)$, now with respect to the restriction to H_i of the Haar measure on G. Since $\|g_i\|_\infty \leq \|x\|$, we get $\|g\|_\infty \leq \|x\|$. Let $f \in L^1(G)$ be non-negative. Since any uncountable sum of positive numbers is infinite, the set J of those i with H_i intersecting the support of f is countable. Then $x(f) = \sum_{i \in J} x(\chi_{H_i} f) = \sum_{i \in J} \int \chi_{H_i}(t) f(t) g_i(t)\,dt = \int f(t)g(t)\,dt$ by continuity of x and by Lebesgue's dominated convergence theorem. □

A.11 Holomorphic Functional Calculus

We can integrate functions taking values in more general spaces. Here is one approach which is more than general enough for our purposes.

Proposition A.11.1 *Let $f : X \to W$ be a continuous function from a compact Hausdorff space X with a Radon measure μ to a complex Banach space W. Then there exists a unique element $w \in W$ such that $x(w) = \int xf\,d\mu$ for all bounded linear functionals x on W.*

Proof Uniqueness is immediate from Theorem 2.3.4.

As for existence, first observe that $x \mapsto \int xf \, d\mu$ is a linear functional on $W^\star$ with norm not greater than $\mu(X)c$, where c is a bound on the continuous function $t \mapsto \|f(t)\|$ on the compact space X. So w belongs to the Banach space $W^{\star\star}$, which contains the Banach space W as a closed subspace.

To see that $w \in W$, to any $\varepsilon > 0$ we may by compactness of X pick finitely many $t_i \in X$ with neighborhoods U_i that cover X and such that $\|f(t) - f(t_i)\| < \varepsilon$ for $t \in U_i$. By Proposition A.3.9 we may pick a partition $\{h_i\}$ of unity on X associated to the open sets U_i. Then

$$\|f(t) - \sum_i f(t_i)h_i(t)\| \leq \sum h_i(t)\|f(t) - f(t_i)\| \leq \varepsilon$$

for all $t \in X$ as it holds for t in any U_i. By Theorem 2.3.4 there is $x \in W^\star$ with $\|x\| = 1$ and $\|w - \sum_i f(t_i) \int h_i \, d\mu\| = x(w - \sum_i f(t_i) \int h_i \, d\mu)$. But

$$x(w - \sum_i f(t_i) \int h_i \, d\mu) \leq \int x(f - \sum f(t_i)h_i) \, d\mu \leq \varepsilon\mu(X),$$

so $w \in W$ as it can be made arbitrarily close in norm to elements in W of the type $\sum_i f(t_i) \int h_i \, d\mu$. □

We denote w by $\int f \, d\mu$ determined by $x(\int f \, d\mu) = \int xf \, d\mu$ for all $x \in W^\star$. Clearly $A \int f \, d\mu = \int Af \, d\mu$ for $A \in B(W, V)$ with V any complex Banach space.

Definition A.11.2 A *curve in the plane* is a continuous map $[a, b] \to \mathbb{C}$. It is *closed* when a and b have the same image, and it is *simple* if it is without self-intersections. A *path* c is a piecewise continuously differentiable curve in the plane, so c is stitched together along some partition of the interval, and on each such stretch it has a continuous *derivative* $t \mapsto c'(t) = \lim_{s \to 0}(c(t + s) - c(t))/s$. A *chain* C *of paths* in $\mathbb{C}$ is a finite collection of paths.

Given a complex Banach space W, a chain C of paths $c_i : [a_i, b_i] \to \mathbb{C}$ and a continuous W-valued function f defined on the union of their images, then the *integral of* f *over* C is

$$\int_C f(z) \, dz \equiv \sum_i \int_{c_i} f(z) \, dz \equiv \sum_i \int_{a_i}^{b_i} f(c_i(t))c_i'(t) \, dt$$

with the proposition above applied to the continuous functions $t \mapsto f(c_i(t))c_i'(t)$ defined on the compact Hausdorff space $[a_i, b_i]$. We get the usual integral of complex valued functions over C when $W = \mathbb{C}$.

Reparametrization does not alter the integral. If we partition a path, the integral over it splits into a sum of integrals over the subpaths. The integral over a path c clearly changes sign when c is replaced by the *opposite path* $c^{op} : t \mapsto c(1 - t)$. We denote by C^{op} the chain with all paths in a chain C reversed.

We can extend the usual definition of complex valued holomorphic functions.

Definition A.11.3 A function $f\colon X \to W$ into a complex Banach space is *(weakly) holomorphic* on an open subset X of $\mathbb{C}$ if the *complex derivatives*

$$(xf)'(z) = \lim_{a \to 0} (xf(z+a) - xf(z))/a$$

exist for all $z \in X$ and $x \in W^\star$.

The set of all holomorphic functions from X to W is a vector space under pointwise operations. It is even an algebra when W is a Banach algebra. Holomorphic functions are obviously continuous. Compositions of complex valued holomorphic functions are holomorphic and obey a *chain rule* $(f \circ g)'(z) = f'(g(z))g'(z)$.

Definition A.11.4 An *admissible boundary* $C = \overline{X} \backslash X$ of a bounded open set X in $\mathbb{C}$ is a chain of simple closed paths each oriented such that when one travels along the path with increasing parameter, the set X is to the left.

The following result is a generalization of Cauchy's theorem from complex function theory.

Theorem A.11.5 *Let W be a complex Banach space, and let X be a bounded open set in $\mathbb{C}$ with an admissible boundary C. If $f : \overline{X} \to W$ is continuous, and holomorphic on X, then $\int_C f(z)\, dz = 0$. If $a \in X$, then*

$$f(a) = \frac{1}{2\pi i} \int_C \frac{f(z)}{z-a}\, dz.$$

Proof The usual Cauchy's theorem $\int_C xf(z)\, dz = 0$ holds for $z \mapsto xf(z)$ and $x \in W^\star$ due to Green's theorem (which is immediate from the fundamental theorem of calculus for fairly easy regions, but requires more geometry for more complicated regions) combined with the *Cauchy-Riemann equations* $u_x = v_y$ and $u_y = -v_x$, where the indices refer to partial derivatives of the real u and imaginary part v of $f(x+iy)$. These equations hold if and only if f is holomorphic; the relevant direction for us is verified by comparing the two expressions one gets for the limit $f'(z)$ by approaching z along the real- and imaginary axis. The first result now follows from the definition of the integral of f over C together with Hahn-Banach's theorem.

The last *Cauchy's integral formula* holds because if $c\colon t \mapsto a + re^{it}$ is a circle around a with r such that $\overline{B_r(a)} \subset X$, then $z \mapsto \frac{f(z)}{z-a}$ is holomorphic on $X \backslash \overline{B_r(a)}$; any doubt is swept away by the first resolvent formula written below this proof. Since $\int_c 1/(z-a)\, dz = 2\pi i$ by direct computation, then by the first part of the proof, we get

$$\begin{aligned} \left|\frac{1}{2\pi i}\int_C \frac{xf(z)}{z-a}\, dz - xf(a)\right| &= \left|\frac{1}{2\pi i}\int_{C \cup c^{op}} \frac{xf(z)}{z-a}\, dz + \frac{1}{2\pi i}\int_c \frac{xf(z)}{z-a}\, dz - xf(a)\right| \\ &= \left|\frac{1}{2\pi i}\int_c \frac{xf(z) - xf(a)}{z-a}\, dz\right| \end{aligned}$$

$$= |\frac{1}{2\pi i}\int_0^{2\pi}\frac{xfc(t)-xf(a)}{re^{it}}ire^{it}\,dt|$$
$$\leq \sup_{t\in[0,2\pi]}|xfc(t)-xf(a)|$$

which tends to zero as $r \to 0$ because xf is continuous. □

Every holomorphic function $f: X \to W$ on any open set X of $\mathbb{C}$ into a complex Banach space can be *represented by a power series in* X, meaning that for any open ball $B_r(c)$ in X, there are unique $w_n \in W$ such that

$$f(z) = \sum_{n=0}^{\infty} w_n(z-c)^n$$

for all $z \in B_r(c)$.

To see this, note that due to Hahn-Banach's theorem, instead of working with xf, we may proceed as if f is complex valued. Let C be an admissible boundary of $B_r(c)$. Observe that

$$\frac{z-c}{z-a} = (1-\frac{a-c}{z-c})^{-1} = \sum_{n=0}^{\infty}(\frac{a-c}{z-c})^n$$

for $a \in B_r(c)$ and z on the image of C. By Lebesgue's dominated convergence theorem, we thus get

$$f(a) = \frac{1}{2\pi i}\int_C \frac{f(z)}{z-a}\,dz = \frac{1}{2\pi i}\sum\int_C \frac{f(z)}{(z-c)^{n+1}}\,dz\,(a-c)^n,$$

which incidentally also shows that the n-th derivative of f at c is

$$f^{(n)}(c) = \frac{n!}{2\pi i}\sum\int_C \frac{f(z)}{(z-c)^{n+1}}\,dz.$$

Note that when $W = \mathbb{C}$, then $f(c+re^{it}) = \sum w_n r^n e^{int}$, so

$$|w_n|r^n = |\frac{1}{2\pi}\int_{-\pi}^{\pi} f(c+re^{it})e^{-int}\,dt| \leq \frac{1}{2\pi}\int_{-\pi}^{\pi}|f(c+re^{it})|\,dt.$$

We have arrived at *Liouville's theorem.*

Proposition A.11.6 *Every bounded holomorphic function $f: \mathbb{C} \to W$ into a complex Banach algebra is constant.*

Proof For any $x \in W^\star$, the right hand side in the last inequality with f replaced by xf, is bounded, and this is only possible if $w_n = 0$ for all $n \geq 1$, so xf is constant. To $a, b \in \mathbb{C}$ pick by Hahn-Banach's theorem an x such that $\|f(a) - f(b)\| = x(f(a) - f(b)) = xf(a) - xf(b) = 0$, so f is constant. □

In the notation just prior to this proposition Parseval's identity takes the form

$$2\pi \sum |w_n|^2 r^{2n} = \int_{-\pi}^{\pi} |f(c + re^{it})|^2 \, dt.$$

Hence, if $|f(c + re^{it})| \leq |f(c)|$ for all t, then $\sum |w_n|^2 r^{2n} \leq |f(c)|^2 = |w_0|^2$, so $w_1 = w_2 = \cdots = 0$ and f is constant on $B_r(c)$. We have arrived at the *Maximum modulus theorem.*

Proposition A.11.7 *If $f: X \to W$ is holomorphic on an open connected subset of $\mathbb{C}$, then $|f(c)| \leq \max_t |f(c + re^{it})|$ when the closure of $B_r(c)$ belongs to X. Equality occurs if and only if f is constant.*

Proof That f is constant throughout X is clear from Proposition A.13.1. □

Let W be a complex unital Banach algebra with $w \in W$, and let f be a complex valued holomorphic function on an open set in $\mathbb{C}$ that contains $\mathrm{sp}(w)$. We can then find a bounded smaller open set X with an admissible boundary C such that $\mathrm{sp}(w) \subset X$. This can be achieved by considering a grid over slightly more than $\mathrm{sp}(w)$ consisting of small enough squares; the paths can clearly even be made piecewise linear. Then we define an element of W by

$$f(w) \equiv \frac{1}{2\pi i} \int_C \frac{f(z)}{z - w} \, dz.$$

This element is in fact independent of what X one uses. Indeed, consider another such X_1 with admissible boundary C_1. To proceed we need a topological notion that allows us to deform and shrink paths without tearing them apart.

Definition A.11.8 Two closed curves c_0 and c_1 in a topological space Y having domain $I = [0, 1]$ are *homotopic in* Y if there is a continuous map $H: I^2 \to Y$ such that $H(s, i) = c_i(s)$ and $H(0, t) = H(1, t)$ for $s, t \in I$. If c_1 is the *constant curve*, meaning that it's image consists of one point, then c_0 is *null-homotopic*. We say Y is *simply connected* if every closed curve in Y is null-homotopic.

Using homotopy we can continuously deform the paths in C_1 to obtain an admissible boundary C_2 of an open set X_2 such that $\overline{X}_2 \subset X_1 \cap X$. Integration over the admissible boundaries $C_1 \cup C_2^{op}$ and $C \cup C_2^{op}$ of $X_1 \backslash \overline{X}_2$ and $X \backslash \overline{X}_2$, respectively, is zero by Cauchy's theorem as these sets do not intersect $\mathrm{sp}(w)$, so the integrands (omitted in the expression below) are holomorphic. Hence

$$\int_{C_1} = \int_{C_1 \cup C_2^{op}} + \int_{C_2} = \int_{C_2} = \int_{C \cup C_2^{op}} + \int_{C_2} = \int_C .$$

Thus if f_i are holomorphic W-valued functions on open sets containing $\mathrm{sp}(w)$, then $f_1(w) = f_2(w)$ if they are equal on a common smaller open set containing $\mathrm{sp}(w)$.

Holomorphic functional calculus is the map $f \mapsto f(w)$ in the following result.

Theorem A.11.9 *Let w belong to a complex unital Banach algebra W. Let $H(w)$ denote the unital algebra of holomorphic complex functions defined on some open set of $\mathbb{C}$ that contains* sp(w)*; the open set may vary with the function. Then the map $f \mapsto f(w)$ is a unital homomorphism from $H(w)$ to W. If f is represented by a power series $\sum a_n z^n$ on an open set containing* sp(w)*, then $f(w) = \sum a_n w^n$.*

Proof We have already checked that $f \mapsto f(w)$ is a well-defined linear map. To see that it is multiplicative, let $f, g \in H(w)$. Pick a bounded open set X with an admissible boundary C for g, and an open set X_1 with an admissible boundary C_1 for f such that $\overline{X}_1 \subset X$ and $\mathrm{sp}(w) \subset X_1$. By the Tonelli-Fubini theorem and Cauchy's theorem, we thus get

$$\begin{aligned} f(w)g(w) &= \frac{1}{2\pi i}\int_{C_1} \frac{f(z_1)}{z_1 - w}\left(\frac{1}{2\pi i}\int_C \frac{g(z)}{z - z_1}\, dz\right) dz_1 \\ &- \frac{1}{(2\pi i)^2}\int_C \frac{g(z)}{z - w}\left(\int_{C_1} \frac{f(z_1)}{z - z_1}\, dz_1\right) dz \\ &= \frac{1}{2\pi i}\int_{C_1} \frac{f(z_1)g(z_1)}{z_1 - w}\, dz_1 = (fg)(w). \end{aligned}$$

For $r > \|w\|$ and a non-negative integer k, consider the complex function $f(z) = z^k$ defined on $\overline{B_r(0)}$ with admissible boundary C. Then the resolvent map admits a power series representation and

$$f(w) = \frac{1}{2\pi i}\int_C \sum_{n=0}^{\infty} \frac{w^n}{z^{n+1-k}}\, dz = \frac{1}{2\pi i}\sum_{n=0}^{\infty} w^n \int_C \frac{1}{z^{n+1-k}}\, dz = w^k.$$

By linearity the map $f \mapsto f(w)$ thus sends polynomials in z to the same polynomials in w, and we are done by the dominated convergence theorem. □

Say w belongs to a unital complex Banach algebra W. We observe that if $\{f_n\}$ is a sequence of complex holomorphic functions on an open set X of sp(w) that converge in the uniform norm to a complex function f on X, then $f \in H(w)$ and $\|f(w) - f_n(w)\| \to 0$. The map $f \mapsto f(w)$ is the unique unital homomorphism from $H(w)$ to W with this convergence property and which sends the identity function to w. This is immediate from Runge's theorem.

Note that the image of the map above is a commutative unital complex subalgebra of W.

The following result is known as the *spectral mapping theorem.*

Corollary A.11.10 *If w is an element of a complex unital Banach algebra, and $f \in H(w)$, then* $\mathrm{sp}(f(w)) = f(\mathrm{sp}(w))$.

Proof If $a \notin f(\mathrm{sp}(w))$, then $g(z) = (f(z) - a)^{-1}$ defines $g \in H(w)$ with $g(w)(f(w) - a) = 1$, so $a \notin \mathrm{sp}(f(w))$.

Conversely, to $a \in \mathrm{sp}(w)$ there is $g \in H(w)$ such that $f(z) - f(a) = (z - a)g(z)$. Hence $f(w) - f(a) = (w - a)g(w)$ and $f(a) \in \mathrm{sp}(f(w))$. □

The next result says that here we can skip parentheses under compositions.

Proposition A.11.11 *If w is an element of a complex unital Banach algebra, and $g \in H(w)$ and $f \in H(g(w))$, then $(f \circ g)(w) = f(g(w))$.*

Proof By the spectral mapping theorem, the function fg is holomorphic on the open set containing $\mathrm{sp}(w)$ which consists of the points where g is holomorphic and have image in the open set containing $\mathrm{sp}(g(w))$ where f is holomorphic. Hence $fg \in H(w)$.

Pick a bounded open set X_1 with admissible boundary C_1 for f, and pick an open set X with admissible boundary C for $f \circ g$ such that $\overline{X} \subset g^{-1}(X_1)$. Since $h(z) = (z_1 - g(z))^{-1}$ is holomorphic on $g^{-1}(X_1)$ whenever z_i belongs to the image of C_1, we get by the Fubini-Tonelli theorem that

$$f(g(w)) = \frac{1}{2\pi i}\int_{C_1} f(z_1)h(w)\,dz_1 = \frac{1}{2\pi i}\int_{C_1} f(z_1)(\frac{1}{2\pi i}\int_C \frac{h(z)}{z-w}\,dz)\,dz_1$$
$$= \frac{1}{2\pi i}\int_C \frac{1}{z-w}(\frac{1}{2\pi i}\int_{C_1}\frac{f(z_1)}{z_1 - g(z)}\,dz_1)\,dz = \frac{1}{2\pi i}\int_C \frac{f \circ g(z)}{z-w}\,dz.$$

□

The logarithm has a power series representation

$$\ln(z+1) = z - z^2/2 + z^3/3 - \cdots$$

that converges only on $B_1(0)$, so this series fail to define $\ln(w+1)$ for invertible $w+1$ with $\|w\| \geq 1$ in a complex unital Banach algebra.

However, using holomorphic functional calculus, we have the following result.

Proposition A.11.12 *Suppose w belongs to a complex unital Banach algebra, and that there is a curve in $\mathrm{sp}(w)^c$ from the origin to some point outside any ball with center at the origin. Then $w = \exp(v)$ for some v, called a* logarithm *of w, that commutes with every operator commuting with w. To any $n \in \mathbb{N}$, there is u such that $w = u^n$, referred to as an n-th* root *of w. Moreover, when w has positive spectrum, then an n-th root can be chosen that also has positive spectrum.*

Proof By assumption we have a connected and simply connected open set Y that contains $\mathrm{sp}(w)$ but $0 \notin Y$. Let c be a path in Y from a fixed point to any point z in Y, which is possible to reach since Y is connected. As the reciprocal of the identity function is defined on Y, we can form the integral $f(z) \equiv \int_c z_1^{-1}\,dz_1$. If c_1 was another such path, then $c \cup c_1^{op}$ or the opposite path is an admissible path for a bounded open set in Y because Y is simply connected. So we get a well-defined function f by Cauchy's theorem, which clearly is holomorphic with $f'(z) = z^{-1}$. Hence $z \mapsto ze^{-f(z)}$ has zero derivative, and using again a similar argument with an integral over a path, we conclude that it must be constant on Y. Hence $z = e^{f(z)+a}$ for some scalar a as the exponential function surjects onto $\mathbb{C}\backslash\{0\}$. Hence $w = e^v$ with $v = f(w) + a$ by the previous proposition.

If the spectrum of w is positive, we can pick Y and c with image on the positive real axis, and also $a \in \mathbb{R}$. Then $f + a$ is real on $\mathrm{sp}(w)$, and v has real spectrum by the spectral mapping theorem.

Clearly $u = e^{v/n}$ is an n-th root of w, which by the spectral mapping theorem has positive spectrum whenever $\mathrm{sp}(w) \subset \langle 0, \infty\rangle$. □

Corollary A.11.13 *A complex quadratic matrix is invertible if and only if it is the exponential of some matrix, and in this case it has an n-th root. If an invertible complex matrix is positive, it has a unique positive square root, which moreover commutes with every matrix commuting with the original matrix. Any invertible complex matrix A admits a unique* polar decomposition, *that is, a unique positive $|A|$ and a unique unitary U such that $A = U|A|$.*

Proof The first statement is immediate from the proposition since the spectrum of an invertible complex $n \times n$-matrix in the Banach algebra $M_n(\mathbb{C})$ is finite and does not contain zero. For the next statement, since positive matrices are self-adjoint and thus diagonalizable, we may assume that the spectrum consists only of positive numbers. The square root $R \equiv e^{B/2}$ from the proposition has then positive spectrum, and it is self-adjoint, since $f + a$ in the proof of the proposition may be taken to be real, so B is self-adjoint, and so is R. Any other positive square root S of R^2 must commute with R. Hence $R - S$ vanishes on $\mathrm{im}(R + S)$. Then

$$\|R^{1/2}v\| = (Rv|v) \leq ((R + S)v|v) = 0$$

for $v \in \ker(R + S) = \mathrm{im}(R + S)^{\perp}$. Hence $Rv = R^{1/2}R^{1/2}v = 0$, and similarly $Sv = 0$, so $R = S$.

For the polar decomposition, let $|A|$ be the positive square root of A^*A, and consider the invertible matrix $U = A|A|^{-1}$. Then $U^*U = |A|^{-1}A^*A|A|^{-1} = 1 = U^{-1}U$, so $U^* = U^{-1}$ by multiplying with U^{-1} from the right.

As for uniqueness, say we had another decomposition $A = VS$ with V unitary and S positive. Then $A^*A = S^*S$. So $S = |A|$ by uniqueness of positive roots, and then also $V = U$. □

For the existence of the positive square root of any positive matrix A, we can use $A^{1/2} = PD^{1/2}P^{-1}$ for a diagonalization $A = PDP^{-1}$ of A with P unitary and $D^{1/2} = \mathrm{diag}(\lambda_1^{1/2}, \ldots, \lambda_n^{1/2})$, where $D = \mathrm{diag}(\lambda_1, \ldots, \lambda_n)$ with $\lambda_i \geq 0$. This is an example of continuous functional calculus for normal operators.

Proposition A.11.14 *Let $G_0(W)$ be the connected component of the unit element of the topological group $G(W)$ of invertible elements in a complex unital Banach algebra W. Then $G_0(W)$ is an open normal subgroup of $G(W)$ generated by* $\exp(W)$. *In particular, when W is commutative, then $G_0(W) =$* $\exp(W)$, *and only the unit element has finite order in the quotient group $G(W)/G_0(W)$.*

Proof Set $G = G(W)$ and $H = G_0(W)$. Since G is open, every point in H is the center of an open ball of W contained in G. This ball must be contain in H since such balls are obviously connected, so H is open. Since the continuous image of a connected set is connected, we see that $w^{-1}H$ is connected for $w \in H$, and as the unit 1 sits there, we get $w^{-1}H \subset H$, so H is a subgroup of G. It is also normal because $v^{-1}Hv \subset H$ for $v \in G$ since $w \mapsto v^{-1}wv$ is a continuous map from H to $v^{-1}Hv$, so $v^{-1}Hv$ is connected and obviously contains 1.

For $w \in W$, the map $[0, 1] \to G$ given by $t \mapsto e^{tw}$ is a path in G from 1 to e^w, and since the image of this path is connected, we see that $\exp(W)$ is a connected subset of H. The subgroup K of G generated by $\exp(W)$ is clearly the union of products of members of $\exp(W)$. Since the product map in G is continuous, induction shows that K is connected, so K is a subgroup of H.

By the first line in the proof of Proposition A.11.12, and since evidently the set $\{v \in W \mid \mathrm{sp}(v) \subset U\}$ is open in W for any open $U \subset \mathbb{C}$, we see that $\exp(W)$, and hence K, contains a non-empty open subset of G. But K is a group and left multiplication in G is a continuous map with a continuous inverse, so K is open in G and H. But K is also closed in H since all left cosets of K in H are open, and they partition H. Being both open and closed, we conclude that K equals H.

It remains to show that for W commutative, then $u \in H$ whenever $u \in G$ and $u^n \in H$ for some $n \in \mathbb{N}$. We already know that there is $v \in W$ with $u^n = e^v$. It suffices to show $w \equiv ue^{-v/n} \in H$. As $w^n = 1$, the spectral mapping theorem tells us that the spectrum of w consists of at most n points on the unit circle, so $w = e^r$ for some $r \in W$ by Proposition A.11.12. The map $t \mapsto e^{tr}$ from $[0, 1]$ to G is a path from 1 to w, so $w \in H$. □

The discrete group $G(W)/G_0(W)$ is referred to as the *index group* of the Banach algebra W.

A.12 Applications to Linear Algebra and Differential Equations

Say the spectrum of an element w in a complex unital Banach algebra is the finite disjoint union of closed subsets K_i. Pick disjoint open sets $U_i \supset K_i$, and let e_i be the holomorphic complex function on $\cup U_i$ that is 1 on U_i and otherwise zero, so $e_i e_j = \delta_{ij} e_i$ and $\sum e_i = 1$. By the holomorphic functional calculus the *spectral projections* $e_i(w)$ commute with w and satisfy $e_i(w)e_j(w) = \delta_{ij}e_i(w)$ and $\sum e_i(w) = 1$. If the Banach algebra is $B(V)$, we get the associated *spectral decomposition of* $A \in B(V)$ acting on the orthogonal invariant subspaces $\mathrm{im}\, e_i(A)$ of V.

When $V = \mathbb{C}^n$, the spectrum of A is finite, so the hypothesis on its spectrum automatically holds. Then we get the *Jordan decomposition* $V = \oplus \ker(A - \lambda_i)^{k_i}$ of the matrix A, where k_i is the *index of the eigenvalue* λ_i of A, meaning the least k_i with $\ker(A - \lambda_i)^{k_i} = \ker(A - \lambda_i)^l$ for $l \geq k_i$. Denote by P_i the projection $e_i(A)$ onto the *Jordan block* $\mathrm{im}\, e_i(A) = \ker(A - \lambda_i)^{k_i}$ corresponding to λ_i. Given

$f \in H(A)$, then expanding the holomorphic function fe_i around λ_i, we evidently get the following finite sum

$$f(A) = \sum_{\lambda_i \in \mathrm{sp}(A)} f(A)e_i(A) = \sum_{\lambda_i \in \mathrm{sp}(A)} \sum_{n=0}^{k_i-1} \frac{f^{(n)}(\lambda_i)}{n!}(A-\lambda_i)^n P_i.$$

In particular, for a real or complex square matrix A with eigenvalues λ_i of index k_i and with spectral projections P_i, we get

$$e^A = \sum P_i e^{\lambda_i + A - \lambda_i} = \sum e^{\lambda_i} P_i \sum_{k=0}^{k_i-1} \frac{(A-\lambda_i)^k}{k!},$$

as is also seen by using $e^{A+B} = e^A e^B$ for commuting matrices A and B. This has applications in the theory of differential equations.

Suppose V is a real or complex Banach space, and that $f\colon \mathbb{R} \times V \to V$ is *locally Lipschitz continuous*, meaning that to $x_0 \in V$ there are $a, b, c > 0$ such that $\|f(t, v) - f(t, w)\| \leq a\|v - w\|$ for $|t| < b$ and $v, w \in B_c(x_0)$. A function $x\colon \langle -b, b\rangle \to V$ that satisfies both the *system of first order differential equations* $x'(t) = f(t, x(t))$ for all t and the *initial condition* $x(0) = x_0$ is called a *local solution* of the system. Such solutions correspond to *fixed points* $F(x) = x$ of the integral operator $F(x) = x_0 + \int_0^t f(s, x(s))ds$ considered as a map $F\colon X \to X$ on the set X of all continuous functions $y\colon \langle -b, b\rangle \to V$ such that $\|y - x_0\|_\infty \leq c$, where we have shrunk b so that $bc < 1/2$ and c/b is greater than the maximum of $\|f(t, v)\|$ over $t \in \langle -b, b\rangle$ and $v \in B_c(x_0)$. Clearly X is complete under the metric d induced by $\|\cdot\|_\infty$, and $d(F(y_1), F(y_2)) < \frac{1}{2}d(y_1, y_2)$ for $y_i \in X$. Uniqueness and existence of a local solution is therefore guaranteed by the contraction principle. The resulting iteration process from x_0 towards a sufficiently accurate local solution $x_n(t)$ is known as *Picard's method*.

Example A.12.1 If we let $f(t, v) = A(v) + g(t)$ for $A \in B(V)$ and $v \in V$ and a V-valued continuous function g on $\langle -b, b\rangle$, then $x'(t) = f(t, x(t))$ is a system of first order *inhomogenous linear* differential equations. The unique local solution of this system is by the discussion above obviously

$$x(t) = e^{tA}(x_0) + e^{tA}\Big(\int_0^t e^{-sA}(g(s))ds\Big).$$

In the homogeneous case $g = 0$, we get $x_n(t) = (\sum_{k=0}^n (tA)^k/k!)(x_0)$, which converges to the exact solution $x(t) = e^{tA}(x_0)$. ◇

The system of *n-th order* differential equations

$$x^{(n)}(t) = h(t, x(t), x'(t), \ldots, x^{(n-1)}(t))$$

for $h\colon \mathbb{R}\times V^n \to V$ subject to the initial conditions $x^{(k)}(0) = x_k$ for $k = 0, \ldots, n-1$ can be translated into a first order system by letting $y(t) = (y_1(t), \ldots, y_n(t))^T$ with $y_k(t) = x^{(k-1)}(t)$ and getting

$$y'(t) = (y_2(t), y_3(t), \ldots, y_n(t), h(t, y_1(t), \ldots, y_n(t)))^T.$$

Uniqueness and existence results can then be translated accordingly.

Example A.12.2 Normally V is a higher dimensional real or complex Euclidean space. Let us consider a system of three inhomogeneous linear real first order differential equations, so $V = \mathbb{R}^3$. Let $g(t) = (60, 0, 0)^T$ and $x_0 = (29, 37, 43)^T$ and

$$A = \begin{pmatrix} 4 & -2 & 1 \\ 2 & 0 & 1 \\ 2 & -2 & 3 \end{pmatrix},$$

so we are seeking $x\colon \mathbb{R} \to \mathbb{R}^3$ such that $x'(t) = A(x(t)) + g(t)$ with $x(0) = x_0$. The characteristic polynomial of A is $f(u) = \det(u - A) = (u-3)(u-2)^2$, and

$$\frac{1}{f(u)} = \frac{1}{u-3} + \frac{1-u}{(u-2)^2}.$$

Thus the spectral projections associated to the eigenvalues $\lambda_1 = 3$ and $\lambda_2 = 2$ are

$$P_1 = (A-2)^2 = \begin{pmatrix} 2 & -2 & 1 \\ 2 & -2 & 1 \\ 2 & -2 & 1 \end{pmatrix} \text{ and } P_2 = (1-A)(A-3) = \begin{pmatrix} -1 & 2 & -1 \\ -2 & 3 & -1 \\ -2 & 2 & 0 \end{pmatrix}.$$

Hence

$$e^{tA} = e^{\lambda_1 t} P_1 + e^{\lambda_2 t} P_2 (1 + (A - \lambda_2)t) = \begin{pmatrix} 2e^{3t} - e^{2t} & -2e^{3t} + 2e^{2t} & e^{3t} - e^{2t} \\ 2e^{3t} - 2e^{2t} & -2e^{3t} + 3e^{2t} & e^{3t} - e^{2t} \\ 2e^{3t} - 2e^{2t} & -2e^{3t} + 2e^{2t} & e^{3t} \end{pmatrix},$$

which in turn gives the solution

$$x(t) = e^{tA}(x_0) + e^{tA}\left(\int_0^t e^{-sA}(g(s))ds\right) = \begin{pmatrix} -10 + 67e^{3t} - 28e^{2t} \\ 20 + 67e^{3t} - 50e^{2t} \\ 20 + 67e^{3t} - 44e^{2t} \end{pmatrix}$$

of the initial value problem.

Consider the third order differential equation

$$x'''(t) - 7x''(t) + 16x'(t) - 12x(t) = 0$$

with constant coefficients to be satisfied by complex valued functions x. The ansatz $x(t) = e^{\lambda t}$ gives the *characteristic equation* $\lambda^3-7\lambda^2+16\lambda-12 = (\lambda-3)(\lambda-2)^2 = 0$ of the differential equation. We translate this into a system $y'(t) = B(y(t))$ of first order linear equations for $y(t) = (y_1(t), y_2(t), y_3(t))^T$ with $y_k(t) = x^{(k-1)}(t)$ and

$$B = \begin{pmatrix} 0 & 1 & 0 \\ 0 & 0 & 1 \\ 12 & -16 & 7 \end{pmatrix}.$$

The characteristic equation of this matrix coincides with the characteristic equation of the differential equation; this will always be the case. Hence we have the formula

$$e^{tB} = e^{\lambda_1 t} Q_1 + e^{\lambda_2 t} Q_2(1 + (B - \lambda_2)t),$$

where Q_i are the spectral projections of B associated to the eigenvalues λ_i. In the solution $y(t) = e^{tB}(y_0)$ for some initial vector y_0, the t occurring due to the multiplicity of $\lambda_2 = 2$ corresponds to modifying the guessed solution $e^{\lambda_2 t}$ by multiplying it with t, so as to get another one which is linear independent to $e^{\lambda_2 t}$. This generalizes trivially to higher order multiplicities, and is caused by the presence of larger blocks in the decomposition of the analogue of B.

Of course, if a complex $n \times n$-matrix C has a basis of eigenvectors v_i with corresponding eigenvalues λ_i, then we can fix the scalars uniquely c_i in a *general solution* $z(t) = \sum c_i v_i e^{\lambda_i t}$ of $z'(t) = C(z(t))$ by any given initial condition.

Suppose we have an equation $w''(t) = D(w(t))$ for a function w with values in a Banach space V, and assume $D \in B(V)$ has a square root. Then the unique solution satisfying an initial condition of the form $w(0) = w_0$ and $w'(0) = w_1$ can be obtained by fixing the column vectors d_i in the general solution $w(t) = e^{t\sqrt{D}} d_1 + e^{-t\sqrt{D}} d_2$. ◇

A.13 The Theorems of Carleson, Runge and Phragmen-Lindelöf

Here we restrict to complex valued functions. A *limit point* of a subset in a topological space is a point of the subset such that every neighborhood of it contains other points of the subset. The collection of all limit points of the subset is clearly closed.

Proposition A.13.1 *A holomorphic function defined on an open connected subset of $\mathbb{C}$ is identically zero if it vanishes on a set that contains a limit point.*

Proof Let X be the set of all limit points of the set Y of all zeroes of the function, say f, defined on the open connected subset, say Z, of $\mathbb{C}$. So $X \subset Y$ is closed and non-empty, and we are done if we can show that it is also open. To $a \in X$ there is an open disc $B \subset Z$ with center a such that $f(z) = \sum_{n=0}^{\infty} c_n(z-a)^n$ for $z \in B$. Then $f(B) = \{0\}$ and X is open. Otherwise, let m be the smallest natural number with $c_m \neq 0$. Then $z \mapsto (z-a)^{-m} f(z)$ is holomorphic on Z and is non-zero on a, so by continuity, there is a neighborhood of a, where the function is non-zero at every point. Then f is also non-zero at every such point apart from at a, which is a contradiction. □

The following result is known as *Carleson's sampling theorem.*

Corollary A.13.2 *A holomorphic function f on the right half-plane that vanishes on $\mathbb{N} \subset \mathbb{C}$ and satisfies $|f(z)| \leq Ae^{B\mathrm{Re}(z)}$, must vanish identically.*

Proof The formula

$$g(w) = \int \frac{f(\frac{1}{2}+iy)}{\sin \pi(\frac{1}{2}+iy)} e^{iwy}\, dy$$

well-defines a function g on the strip $|\mathrm{Im}(w)| < \pi$ since f is bounded on the vertical line $\frac{1}{2}+iy$, and the remaining function within the integral tends to zero rapidly as $|y| \to \infty$. The Lebesgue dominated convergence theorem shows that g is holomorphic on the strip. Suppose in addition that $\mathrm{Re}(w) < -B$. Now $1/\sin(\pi \cdot)$ has only poles, and only simple ones, where f is zero, so substituting $\frac{1}{2}+iy$ by $x+iy$ in the whole function under the integral, produces a function of $z = x+iy$ that is holomorphic for $x > \frac{1}{2}$. The number $g(w)$ corresponding to $x = 2n+1/2$ will be unaltered by Cauchy's theorem, as the contour integral around a rectangle with vertical sides along $\frac{1}{2}+iy$ and $\frac{1}{2}+2n+iy$ has negligible contributions from the horizontal stretches when these move infinitely far away from the real axis. But as $n \to \infty$ the corresponding function under the integral tends to zero, again due to the growth condition on f, so $g(w) = 0$. By the proposition g must therefore vanish on the whole strip. But then f vanishes almost everywhere along $\frac{1}{2}+iy$, and by continuity vanishes everywhere along that line, and thus on the whole half-plane by the proposition. □

The *Riemann sphere* S^2 is the one-point compactification $\mathbb{C}\cup\{\infty\}$ of the complex plane, and the nature of a singularity of a function $f: S^2 \to \mathbb{C}$ at ∞ is by definition the same as that of the function $z \mapsto f(z^{-1})$ at zero. We would like to approximate holomorphic functions by rational functions, that is, functions which are fractions of polynomials.

Lemma A.13.3 *Any open set $A \subset \mathbb{C}$ is a union of a sequence of compact sets K_n, where $K_n \subset K_{n+1}^o$, and every compact subset of A lies in some K_m, and every component of $S^2 \backslash K_n$ contains a component of $S^2 \backslash A$.*

Proof The union of the compact sets

$$K_n = S^2 \backslash (\{\infty\} \cup \{z \in \mathbb{C} \,|\, n < |z|\} \cup \bigcup_{b \notin A} B_{n^{-1}}(b))$$

is A, and $K_n \subset K_{n+1}^o$ since if $z \in K_n$, then $B_{n^{-1}-(n+1)^{-1}}(z) \subset K_{n+1}$. Hence A is the union of the interiors of K_n, and any compact subset of A will be covered by finitely many such interiors. Clearly each component of $S^2\backslash K_n$ intersects $S^2\backslash A$, and no component of the latter can intersect two components of the former. □

Theorem A.13.4 *If $K \subset \mathbb{C}$ is compact with one point b_i in each component of $S^2\backslash K$, then for any holomorphic function f on an open set $A \supset K$ and $\varepsilon > 0$, there is a rational function g on S^2 with poles in $\{b_i\}$ such that $|f(z) - g(z)| < \varepsilon$ for $z \in K$. In particular, if $S^2\backslash K$ is connected, then g may be taken to be a polynomial.*

Proof We need to show that f is in the closure of the linear subspace of $C(K)$ consisting of the restrictions of those rational functions on S^2 that have all their poles in $\{b_i\}$. By the Hahn-Banach theorem and the Riesz representation theorem this happens if whenever there is a complex Borel measure μ on K with integral killing all such rational functions, then also $\int f\, d\mu = 0$. Consider the holomorphic function h on $S^2\backslash K$ given by $h(w) = \int (z' - w)^{-1}\, d\mu(z')$, and say A_i is a component of $S^2\backslash K$ such that $w \in B_r(b_i) \subset A_i$ and $b_i \neq \infty$. Then as $\sum_{n=0}^{N}(w - b_i)^n(z' - b_i)^{-n-1} \to (z' - w)^{-1}$ uniformly for $z' \in K$, by assumption h vanishes on A_i. For $b_i = \infty$ use $(z' - w)^{-1} = \lim \sum_{n=0}^{N}(z')^n w^{-n-1}$ for $|w| > r$. So h vanishes on $S^2\backslash K$. Pick an admissible curve $C \subset S^2\backslash K$ such that $f(z') = (2\pi i)^{-1} \int_C f(w)(w - z')^{-1}\, dw$. By the Fubini-Tonelli theorem we get $\int f(z')\, d\mu(z') = -(2\pi i)^{-1} \int_C f(w)h(w)\, dw = 0$. □

As a corollary we arrive at *Runge's theorem*.

Corollary A.13.5 *Let $A \subset \mathbb{C}$ be open, and let B consist of one point in each component of $S^2\backslash A$. For every holomorpic function f on A, there is a sequence of rational functions g_n on S^2 with poles in B such that $g_n \to f$ uniformly on compact subsets of A. When $S^2\backslash A$ is connected, we may take g_n to be polynomials.*

Proof Pick K_n as in the previous lemma. Then each component of $S^2\backslash K_n$ contains exactly one point of B, so by the theorem there is a rational function g_n on S^2 with its poles in B such that $|g_n(z) - f(z)| < 1/n$ for $z \in K_n$. If K is any compact subset of A, then it is contained in every K_n with n large enough, so for all such n we get $|g_n(z) - f(z)| < 1/n$ when $z \in K$. □

We finally include a version of the *Phragmen-Lindelöf theorem* relevant for us.

Theorem A.13.6 *Let f be continuous on the strip H and holomorphic on H^o, and say there are constants $a < \infty$ and $b < 1$ such that $|f(z)| < \exp(ae^{\pi b|\mathrm{Re}(z)|})$ on the strip. Then $|f|$ is not greater than one on the strip if it is not greater than one on its boundary.*

Proof Replacing z by $\pi^{-1}z + i/2$ we may identify our strip with the more symmetric one $S = \{x + iy \mid |y| \leq \pi/2\}$, so $|f(z)| < \exp(ae^{b|x|})$ and $|f(x \pm i\pi/2)| \leq 1$ for $z = x + iy \in S$. Pick $c > 0$ with $b < c < 1$, and define $g_\varepsilon(z) = \exp(-\varepsilon(e^{cz} + e^{-cz})$ for $\varepsilon > 0$. Let $d = \cos(c\pi/2)$. Then $\mathrm{Re}(e^{cz} + e^{-cz}) \geq d(e^{cx} + e^{-cx})$ on S, so $|g_\varepsilon f| \leq 1$ on the boundary of S and $|f(z)g_\varepsilon(z)| \leq \exp(ae^{b|x|} - \varepsilon d(e^{cx} + e^{-cx}))$ on S. As the latter exponent tends to $-\infty$ as $x \to \pm\infty$, we see that there is $x_0 > 0$

such that $|f(z)g_\varepsilon(z)| \leq 1$ for $|x| > x_0$. Then $|fg_\varepsilon| \leq 1$ by the maximum modulus principle applied to the rectangle with vertices $\pm x_0 \pm \pi i/2$, so $|fg_\varepsilon| \leq 1$ on S. Letting $\varepsilon \to 0$ we thus get $|f| \leq 1$ on S. □

Exercises

For Sect. A.1

1. Show that $(a, b) \sim (c, d)$ if $a + d = b + c$ defines an equivalence relation on $\mathbb{Z} \times \mathbb{Z}$, and identify the set of equivalence classes with a set you know.
2. Show that for the map $f : \mathbb{N} \to \mathbb{N}$ given by $f(x) = 2x + 1$, there are infinitely many maps $g : \mathbb{N} \to \mathbb{N}$ such $gf = \iota$ but none such that $fg = \iota$.
3. Given a positive irrational number a, then define $(x, y) \leq (z, w)$ on $\mathbb{Z} \times \mathbb{Z}$ if $a(z - x) \leq w - y$ on $\mathbb{R}$. Show that this defines an order, and sketch the set of all (x, y) with $(x, y) \geq (0, 0)$.
4. Show that any well-ordered set is ordered (by the same relation).
5. An order isomorphism between two partially ordered sets is a bijection that preserves the relation. Show that when two sets are well-ordered, then either they are order isomorphic, or one set is order isomorphic to a subset of the other set consisting of all elements less that some element.
6. Show that the equivalence class of well-ordered sets under the relation of order isomorphism is again a well-ordered set (or class).

For Sect. A.2

1. Show that if X, Y are non-empty sets with X infinite and Y finite, then $|X \times Y| = |X|$.
2. Show that if X, Y are non-empty sets with X infinite and $|Y| \leq |X|$, then $|X \cup Y| = |X|$.
3. Prove that $|X \times X| = |X|$ for an infinite set X.
4. Let X be an infinite set and let Y be the collection of finite subsets of X. Prove that $|Y| = |X|$.
5. Show that a finite dimensional vector space has twice the dimension considered as a real vector space.
6. Show that if S spans a vector space V and $|S| = \dim V$, then S is a linear basis for V. Is this true when V is infinite dimensional?

For Sect. A.3

1. Show that closed subsets of complete metric spaces are complete.
2. Prove that Hausdorffness, connectedness and compactness are topological invariants, in that these properties are preserved under homeomorphisms.
3. Show that $\tan : \langle -\pi/2, \pi/2 \rangle \to \mathbb{R}$ is a homeomorphism, so both spaces are non-compact and connected, whereas the interval is bounded and non-complete, while the image is unbounded and complete.

4. Show that the connected components of $\mathbb{Q} \subset \mathbb{R}$ in the relative topology are the points in $\mathbb{Q}$, so none of these are open.
5. Show that the continuous real image of a compact connected set is a closed interval.
6. Show that the graph of the function $f: \langle 0, \infty\rangle \to \mathbb{R}$ given by $f(x) = \sin(1/x)$ together with the origin is a connected subset of $\mathbb{R}^2$ in the relative topology, but that it is not archwise connected.
7. Show that $\langle 0, 1\rangle$ is open in $\mathbb{R}$, but not in $\mathbb{R}^2$.
8. Prove that the unit circle and $\mathbb{R}$ cannot be homeomorphic, and that none are homeomorphic to $\mathbb{R}^2$.
9. Produce a continuous function that is not uniformly continuous.
10. Prove that metric spaces are normal, in that disjoint closed subsets can be separated by (contained in) disjoint open subsets.

For Sect. A.4

1. Show that an infinite subset of a compact space has a limit point, that is, a point x such that all its neighborhoods will contain points from the subset other than x.
2. A topological space is second countable if it has a countable collection of open sets producing any open set as a union of this collection. Show that second countable topological spaces are separable in that they have a countable dense subset. Prove that separable metric spaces are second countable. Conclude that Euclidian spaces are second countable.
3. Show that compact Hausdorff spaces are rigid in that they have no stronger nor weaker topologies that are compact Hausdorff.
4. The Tychonoff cube $T = [0, 1]^{\mathbb{N}}$ is a compact Hausdorff space with product topology. Show that it is metrizable in that the topology is induced by a metric.
5. Consider the equivalence relation $\sim$ on $\mathbb{R}$ given by $a \sim b$ if $a - b \in \mathbb{Z}$. Show that $\mathbb{R}/\sim$ with the quotient topology is homeomorphic to the unit circle with relative topology from $\mathbb{R}^2$.
6. Show that a dense subset of a locally compact Hausdorff space is locally compact Hausdorff in the relative topology if and only if it is open in the original topology.
7. Show that a function $f: X \to Y$ between compact Hausdorff spaces is continuous if its graph $\{(x, f(x)) \mid x \in X\}$ is closed in $X \times Y$ with product topology.
8. Show that the product topology on $\prod X_i$ of non-empty locally compact spaces is locally compact if all but finitely many X_i's are compact.

For Sect. A.5

1. Show that the set of piecewise linear continuous complex functions on the interval $[0, 1]$ is uniformly dense in $C([0, 1])$.
2. Show that the span of functions on $\mathbb{R}$ of the form $x \mapsto (x - z)^n$ for $n \in \mathbb{N}$ and $z \in \mathbb{C}\backslash\mathbb{R}$ is dense in $C_0(\mathbb{R})$.
3. Can you give an example of a proper closed subalgebra of $C(D)$, where D is the closed unit disc of $\mathbb{C}$, that separates points and contains the constants?

For Sect. A.6

1. Let M be the collection of subsets A in an uncountable set such that either A or A^c is at most countable, and define accordingly $\mu(A) = 0$ in the first case, and $\mu(A) = 1$ in the second case. Prove that M is a σ-algebra with a measure μ.
2. Show that the points where a sequence of measurable functions converges is a measurable set.
3. Given a measure μ on X with a σ-algebra M. The completion of M consists of all $E \subset X$ having $A, B \in M$ with $A \subset E \subset B$ and $\mu(B \backslash A) = 0$. Show that this completion is a σ-algebra, and that $\mu(E) \equiv \mu(A)$ extends μ to a measure on it.
4. Show that if $f \in L^1(\mu)$ and $|\int f \, d\mu| = \int |f| \, d\mu$, then there is a scalar c such that $cf = |f|$ almost everywhere.
5. Assume $\sum_{n=1}^{\infty} \mu(A_n) < \infty$. Show that almost all elements in the space lie in at most finitely many A_n.
6. Prove that a real limit of a sequence of convex functions $\langle 0, 1 \rangle \to \mathbb{R}$ is convex.
7. Suppose $0 < p < q < \infty$ and consider a set X with measure μ. Show that $L^p(\mu) \not\subset L^q(\mu)$ if and only if X has subsets of arbitrarily small positive measure, and that $L^q(\mu) \not\subset L^p(\mu)$ if and only if X has subsets of arbitrarily large measure.
8. Say $p \in [1, \infty\rangle$ and $f, f_n \in L^p(\mu)$ with $f_n \to f$ almost everywhere. Show that $\|f - f_n\|_p \to 0$ if and only if $\|f_n\|_p \to \|f\|_p$.

For Sect. A.7

1. Can you produce an uncountable subset of $\mathbb{R}$ with Lebesgue measure zero?
2. Construct a compact subset with positive Lebesgue measure and with only single-point connected components.
3. Construct to $c \in [0, 1]$ an open dense subset A of $[0, 1]$ with $m(A) = c$.
4. Construct a Borel set $A \subset \mathbb{R}$ such that $0 < m(A \cap \langle a, b \rangle) < b - a$ for $b > a$. Can we then have $m(A) < \infty$?
5. Construct a sequence of continuous functions $f_n \colon [0, 1] \to [0, 1]$ that converges pointwise nowhere, but with $\lim \int_0^1 f_n(x) \, dx = 0$.
6. Given a regular Borel measure μ on a compact Hausdorff space with measure one. Prove that there is a smallest compact subset E such that $\mu(E) = 1$, called the support of μ. Prove that any compact subset of $\mathbb{R}$ is the support of some Borel measure.
7. Prove that any upper semicontinuous function on a compact space with range in $\langle -\infty, \infty \rangle$ attains its maximum.

For Sect. A.8

1. Show that the Lebesgue measure m on $\langle 0, 1 \rangle$ is absolutely continuous with respect to the counting measure μ on $\langle 0, 1 \rangle$, but that there is no $dm/d\mu \in L^1(m)$.
2. Let $p \in [1, \infty]$ with conjugate exponent q, let μ be a σ-finite measure, and let g be a measurable function with $fg \in L^1(\mu)$ for all $f \in L^p(\mu)$. Prove $g \in L^q(\mu)$.
3. Consider a measure μ on a 2-point set that is finite only on the empty set and one 1-point subset. Will $L^\infty(\mu)$ be the dual of $L^1(\mu)$?

4. Show that there exists a continuous linear functional on $L^\infty([0, 1])$ that vanishes on $C([0, 1])$, and that the dual of $L^\infty([0, 1])$ therefore cannot be $L^1([0, 1])$.
5. Consider the σ-algebra on $[0, 1]$ of subsets that are either countable or has a countable complement, and let μ be the counting measure. Show that the identity function g is not measurable, although $f \mapsto \int fg\,d\mu$ defines a continuous linear functional on $L^1(\mu)$. Conclude that the dual of $L^1(\mu)$ cannot be $L^\infty(\mu)$.

For Sect. A.9

1. Find a monotone class in $\mathbb{R}$ that is stable under formation of complements, but fails to be a σ-algebra.
2. Consider the unit interval with counting measure and the Lebesgue measure. Show that the conclusion of the Fubini-Tonelli theorem fails for the function on $[0, 1] \times [0, 1]$ that is one on the diagonal and vanishes otherwise.
3. Consider $\mathbb{N}$ with counting measure, and define a function f on $\mathbb{N} \times \mathbb{N}$ to be one on the diagonal, and $f(n, n+1) = -1$ and zero otherwise. Show that the two iterated integrals exist but are unequal.
4. Show that any measure μ on a set X can be completed, that is, extended to a measure on the completed σ-algebra, which by definition includes every subset of X between to measurable subsets with measure vanishing differences. Let m_n be the Lebesgue measure on $\mathbb{R}^n$. Show that the completion of the product measure $m_k \times m_n$ is m_{k+n}. Modify the Fubini-Tonelli theorem in this case.
5. Given a locally compact Hausdorff space X with a lower semicontinuous function f. Show that $f = \sup\{g \in C_c(X) \,|\, 0 \leq g \leq f\}$. Conclude that $\int f\,d\mu$ is the supremum of $\{\int g\,d\mu \,|\, g \in C_c(X),\ 0 \leq g \leq f\}$.
6. Suppose we have a family of Radon measure μ_i on compact Hausdorff spaces X_i with measure one. Prove that there is a unique Radon measure μ on the direct product $\prod X_i$ with product topology such that for any Borel set $A \subset \prod_{i \in F} X_i$ with F finite, we have $\mu(p_F^{-1}(A)) = (\prod_{i \in F} \mu_i)(A)$, where $p_F \colon \prod X_i \to \prod_{i \in F} X_i$ is the natural projection.

For Sect. A.10

1. Let G be a locally compact group and let $[G, G]$ denote the smallest subgroup of G containing all the commutators $sts^{-1}t^{-1}$ for $s, t \in G$. Show that $[G, G]$ is a normal subgroup of G, and that if the quotient group $G/[G, G]$ is finite, then G is unimodular.
2. Let G be a topological group. Show that the closure of $A \subset G$ is the intersection of AU over the neighborhoods U of the unit.
3. Suppose G is a locally compact group homeomorphic to an open subset of $\mathbb{R}^n$ for some n, and assume under this identification that $st = A_s(t) + b_s$ for all $s, t \in G$, where $b_s \in \mathbb{R}^n$ and $A_s \colon \mathbb{R}^n \to \mathbb{R}^n$ is a linear transformasjon. Show that the Haar measure on G is $|\det(A_s)|^{-1}ds$, where ds is the Lebesgue measure on $\mathbb{R}^n$. Conclude that the Haar measure for the multiplicative group in $\mathbb{C}$ of non-zero complex numbers $x + iy$ is $(x^2 + y^2)^{-1}dxdy$. Show also that the group of

real invertible $n \times n$-matrices A has Haar measure $|\det(A)|^{-n} dA$, where dA is the Lebesgue measure on $\mathbb{R}^{n^2}$.

4. Prove that $\mathbb{Q}$ with relative topology from $\mathbb{R}$ is not a locally compact group under addition, and that $\mathbb{Q}$ has no left invariant Borel measure that is finite on compact subsets.
5. Show that the product $\prod G_i$ of topological groups G_i is a topological group with obvious multiplication. Describe the Haar measure on $\prod G_i$ when all G_i are compact with Haar measure one.

For Sect. A.11

1. Prove that $\sum_{i=0}^{n} a_i z^i = 0$ for $a_i \in \mathbb{C}$ and $a_n = 1$ has n solutions $z \in \mathbb{C}$ when counted with multiplicity.
2. Suppose $\{f_n\}$ is a uniformly bounded convergent sequence of holomorphic complex valued functions defined on an open subset X of $\mathbb{C}$. Prove that the convergence is uniform on any compact subset of X.
3. Say W is a unital Banach algebra and that $X \subset \mathbb{C}$ is open. Show that the set $\{w \in W \mid \mathrm{sp}(w) \subset X\}$ is open in W.
4. Let W is a unital Banach algebra, and suppose f is a holomorphic complex valued function on a neighborhood of a compact subset E of $\mathbb{C}$. Show that the map $w \in \{w \in W \mid \mathrm{sp}(w) \subset E\} \mapsto f(w) \in W$ is continuous.
5. Say w is an element of a unital Banach algebra W such that $\mathrm{sp}(w)$ is not connected. Show that W contains an element $v \neq 0, 1$ such that $v^2 = 1$.
6. Let W be an element of a unital Banach algebra W, and assume that $f \in H(w)$ vanishes on $\mathrm{sp}(w)$. Show that $f(w) = 0$ if $\mathrm{sp}(w)$ is both infinite and connected, and that this can fail otherwise.
7. Does $\begin{pmatrix} 0 & 1 \\ 0 & 0 \end{pmatrix}$ have a positive square root in $M_2(\mathbb{C})$?

For Sect. A.12

1. Solve the initial value problem $x'(t) = Ax(t) + g(t)$ with $x(0) = (0, 0)^T$, where $A = \begin{pmatrix} i & 1 \\ 2 & i \end{pmatrix}$ and $g(t) = (3t, 5t^2)^T$ and $x(t) = (x_1(t), x_2(t))^T \in \mathbb{C}^2$. Calculate the first three steps in Picard's method.
2. Reduce $x''(t) + 2tx'(t) + e^t x(t) = 0$ to a system of first order differential equations.
3. Find the Taylor expansion of tan by using Picard's method to the differential equation $x'(t) = 1 + x(t)^2$.
4. Show that the differential equation $x'(t) = x(t)^{1/3}$ with initial condition $x(0) = 0$ has three local solutions.

For Sect. A.13

1. Prove that a holomorphic function that vanishes on $\{zn + w \mid n \in \mathbb{Z}\}$ for some $z \neq 0$, $w \in \mathbb{C}$ must be identically zero.

2. Suppose K is a compact set contained in an open subset A of $\mathbb{C}$, and that $S^2\backslash K$ is not connected. Produce a holomorphic function on A that is not a uniform limit on K of a sequence of polynomials.
3. Suppose f is a holomorphic function f on the open right half-plane that is continuous on the closure of it, and for which there are constants $a < \infty$ and $b < 1$ such that $|f(z)| < ae^{|z|^b}$ for z in the open right half-plane, and which satisfies $|f(iy)| \leq 1$ for real y. Show that $|f| \leq 1$ on the open right half-plane.

Bibliography

1. F. Albiac and N.J. Kalton; *Topics in Banach space theory*, Graduate Texts in Mathematics, vol. 233, Springer-Verlag, New York, 2006.
2. S.Baaj and S. Vaes; *Double crossed products of locally compact quantum groups* J. Inst. Math. of Jussieu **4**, 2005.
3. E. Bedos and R. Conti and L. Tuset; *On amenability and co-amenability of algebraic quantum groups and their corepresentations*, Canad. J. Math. **57**, 2005.
4. S. Baaj and G. Skandalis; *Unitaires multiplicatifs et dualité pour les produits croisés de C^*-algèbres*, Ann. scient. Ec. Norm. Sup. **26**, 1993.
5. E. Bedos and L. Tuset; *Amenability and co-amenability for locally compact quantum groups*, Internat. J. Math. **14**, 2003.
6. L. Bing-Ren; *Topics in Banach space theory*, World Scientific Publishing Co. Pte. Ltd., Singapore, 1992.
7. O. Bratteli and D.W. Robinson; *Operator algebras and quantum statistical mechanics 1*, Texts and Monographs in Physics, Springer-Verlag, Berlin, 1997.
8. A. Connes; *Noncommutative Geometry*, Academic Press, San Diego, 1994.
9. J.B. Conway; *A course in functional analysis*, Graduate Texts in Mathematics, vol. 96, Springer-Verlag, New York, 2007.
10. K.R. Davidson; *C^*-Algebras by Example*, Fields Institute Monographs, vol. 6, AMS, Providence, 1996.
11. K. De Commer: *Galois objects and cocycle twisting for locally compact quantum groups*, J. Operator Theory **66**, 2011.
12. M. Enock; *Inclusions irreductiblés de facteurs et unitaires multiplicatifs II*, J. Funct. Anal. **154**, 1998.
13. M. Enock; *Produit croisé d'une algèbre de von Neumann par une algèbre de Kac*, J. Funct. Anal. **26**, 1977.
14. M. Enock and R. Nest; *Irreducible inclusions of factors, multiplicative unitaries and Kac algebras* J. Funct. Anal. **137**, 1996.
15. M. Enock and J.M. Schwartz; *Kac algebras and duality of locally compact groups*, Springer-Verlag, Berlin, 1992.
16. M. Enock and J.M. Schwartz; *Produit croisé d'une algèbre de von Neumann par une algèbre de Kac II*, Publ. RIMS **16**, 1980.
17. P.A. Fillmore; *A users guide to operator algebras*, John Wiley & Sons, Inc., New York, 1996.
18. G.B. Folland; *Real Analysis: Modern Techniques and their Applications*, John Wiley and Sons Ltd., US, 1999.
19. U. Haagerup; *The standard form of von Neumann algebras*, Math. Scand. **37**, 1975.

L. Tuset, *Analysis and Quantum Groups*,
https://doi.org/10.1007/978-3-031-07246-8

20. U. Haagerup; *Operator valued weights in von Neumann algebras I*, J. Funct. Anal. **32**, 1979.
21. U. Haagerup; *Operator valued weights in von Neumann algebras II*, J. Funct. Anal. **33**, 1979.
22. R.B. Holmes; *Geometric functional analysis and its applications*, Graduate Texts in Mathematics, vol. 24, Springer-Verlag, New York, 1975.
23. R.V. Kadison and J.R. Ringrose; *Fundamentals of the Theory of Operator Algebras, vol. 1,2*, Academic Press, San Diego, 1983,1986.
24. J. Kustermans and L. Tuset; *A survey of C*-algebraic quantum groups. Part II*, Irish Math. Soc. Bull. **44**, 2000.
25. J. Kustermans and S. Vaes; *Locally compact quantum groups*, Ann. Sci. Ecole Norm. Sup. **33**, 2000.
26. J. Kustermans and S. Vaes; *Locally compact quantum groups in the von Neumann setting*, Math. Scan. **92**, 2003.
27. J. Kustermans and S. Vaes; *Weight theory for C*-algebraic quantum groups*, Technical report, University College Cork, 1999.
28. J. Kustermans; *Induced corepresentations of locally compact quantum groups*, J. Funct. Anal. **194**, 2002.
29. J. Kustermans; *Locally compact quantum groups in the universal setting*, Internat. J. Math. **12**, 2001.
30. J. Kustermans; *One-parameter representations on C*-algebras*, Technical report, Odense University, 1997.
31. E.C. Lance; *Hilbert C*-modules. A toolkit for operator algebraists*, London Math. Soc. Lecture Note Ser. 210, Cambridge University Press, Cambridge, 1995.
32. T. Masuda and Y. Nakagami and S.L. Woronowicz ; *A C*-algebraic framework for quantum groups*, Internat. J. Math. **14**, 2003.
33. R.E. Megginson; *An introduction to Banach space theory*, Graduate Texts in Mathematics, vol. 183, Springer-Verlag, New York, 1998.
34. G.J. Murphy; *Operator theory*, Academic Press, Inc., New York, 1990.
35. S. Neshveyev and L. Tuset; *Compact Quantum Groups and Their Representations Categories*, Courses Specialises 20, SMF, Paris, 2013.
36. S. Neshveyev and L. Tuset; *Deformation of C*-algebras by cocycles on locally compact quantum groups*, Adv. Math. **254**, 2014.
37. A.L.T. Paterson; *Amenability*, Mathematical Surveys and Monographs 29, AMS, Providence,1988
38. G.K. Pedersen; *Analysis now*, Graduate Texts in Mathematics, vol. 118, Springer-Verlag, New York, 1989.
39. G.K. Pedersen; *C*-Algebras and Their Automorphism Groups*, Academic Press Inc., San Diego, 1979.
40. M. Reed and B. Simon; *Methods of Modern Mathematical Physics I, Functional Analysis*, Academic Press, New York, 1980.
41. W. Rudin; *Functional analysis*, McGraw-Hill, New York, 1976.
42. W. Rudin; *Real and complex analysis*, McGraw-Hill, New York, 1966.
43. P.M. Soltan and S.L. Woronowicz; *A remark on manageable multiplicative unitaries*, Lett. Math. Phys. **57**, 2001.
44. P.M. Soltan and S.L. Woronowicz; *From multipicative unitaries to quantum groups II*, J. Funct. Anal. **252**, 2007.
45. S. Stratila; *Modular Theory in Operator Algebras*, Abacus Press, Turnbridge Wells, 1981.
46. V.S. Sunder; *Functional analysis, spectral theory*, Birkhäuser Advanced Texts, Basel, 1998.
47. M. Takesaki; *Theory of Operator Algebra I*, Encyclopedia of Mathematical Sciences, vol. 124, Springer-Verlag, New York, 2001.
48. M. Takesaki; *Theory of Operator Algebra II*, Encyclopedia of Mathematical Sciences, vol. 125, Springer-Verlag, New York, 2001.
49. T. Timmermann; *An Invitation to Quantum Groups and Duality*, EMS Textbooks in Mathematics, 2008.
50. S. Vaes; *A Radon-Nikodym theorem for von Neumann algebras*, J. Operator Theory **46**, 2001.

51. S. Vaes; *The unitary implementation of a locally compact quantum group action*, J. Funct. Anal. **180**, 2001.
52. S. Vaes and L. Vainerman; *Extensions of locally compact quantum groups and the bicrossed product construction*, Advances in Math. **175**, 2003.
53. S. Willard: *General Topology*, Addison-Wesley, London, 1968.
54. S.L. Woronowicz; *From multipicative unitaries to quantum groups*, Internat. J. Math. **7**, 1995.

Index

L. Tuset, *Analysis and Quantum Groups*,
https://doi.org/10.1007/978-3-031-07246-8

F

G

H

T

Printed in the United States
by Baker & Taylor Publisher Services

Printed in the United States
by Baker & Taylor Publisher Services